COMPREHENSIVE POLYMER SCIENCE

First Supplement

EDITORIAL BOARD

Sir Geoffrey Allen, FRS (Chairman)
Kobe Steel Ltd., London, UK

J. C. Bevington (Deputy Chairman)
University of Lancaster, UK

C. Booth
University of Manchester, UK

A. Ledwith
Pilkington plc Group Research, Ormskirk, UK

P. Sigwalt
Université Pierre et Marie Curie, Paris, France

G. C. Eastmond
University of Liverpool, UK

C. Price
University of Manchester, UK

S. Russo
Università di Sassari, Italy

S. L. Aggarwal
Global Polymer Technology Associates, Akron, OH, USA

INTERNATIONAL ADVISORY BOARD

H. C. Benoit
Université de Strasbourg, France

M. Hirooka
Sumitomo Chemical Co. Ltd., Tokyo, Japan

W. J. MacKnight
University of Massachusetts, Amherst, MA, USA

C. G. Overberger
University of Michigan, Ann Arbor, MI, USA

M. Tasumi
University of Tokyo, Japan

F. Danusso
Politecnico di Milano, Italy

Y. Imanishi
Kyoto University, Japan

J. H. O'Donnell
University of Queensland, St. Lucia, Australia

N. A. Platé
Russian Academy of Sciences, Moscow, Russia

G. Wegner
Max-Planck-Institut für Polymerforschung, Mainz, Germany

COMPREHENSIVE POLYMER SCIENCE

First Supplement

CHAIRMAN OF THE EDITORIAL BOARD
SIR GEOFFREY ALLEN, FRS
Kobe Steel Ltd., London, UK

VOLUME EDITORS
SUNDAR L. AGGARWAL
Global Polymer Technology Associates, Akron, OH, USA

&

SAVERIO RUSSO
Università di Sassari, Italy

PERGAMON PRESS
OXFORD · NEW YORK · SEOUL · TOKYO

U.K.	Pergamon Press Ltd., Headington Hill Hall, Oxford OX3 0BW, England
U.S.A.	Pergamon Press Inc., 660 White Plains Road, Tarrytown, New York 10591-5153, USA
KOREA	Pergamon Press Korea, KPO Box 315, Seoul 110-603, Korea
JAPAN	Pergamon Press Japan, Tsunashima Building Annex, 3-20-12 Yushima, Bunkyo-ku, Tokyo 113, Japan

Copyright © 1992 Pergamon Press Ltd.

All rights reserved. No part of this publication may be reproduced, stored in a retrieval system or transmitted in any form or by any means: electronic, electrostatic, magnetic tape, mechanical, photocopying, recording or otherwise, without permission in writing from the publishers.

First edition 1992

British Library Cataloguing-in-Publication Data

A catalogue record for this book is available from the British Library

0 08 037071 3

™The paper used in this publication meets the minimum requirements of the American National Standard for Information Sciences—Permanence of Paper for Printed Library Materials, ANSI Z39.48–1984.

Printed and bound in Great Britain by BPCC Wheatons Ltd, Exeter

Contents

	Preface	vii
	Contributors	ix

Polymerization Kinetics and Mechanisms

1	Inhibition and Retardation in Radical Polymerization C. H. BAMFORD, *University of Liverpool, UK*	1
2	Peculiarities of Radical Polymerization Conducted in the Presence of Nontraditional Initiators S. KUCHANOV, *Moscow State University, Russia*	23
3	Emulsion Copolymerization G. H. J. VAN DOREMAELE, H. A. S. SCHOONBROOD and A. L. GERMAN, *Eindhoven University of Technology, The Netherlands*	41
4	From Heterogeneous to Homogeneous Catalysis in Monoalkene Polymerization F. CIARDELLI, *University of Pisa, Italy* and C. CARLINI, *University of Bologna, Italy*	67

Functionalized Polymers

5	Anionic Synthesis of Polymers with Functional Groups R. P. QUIRK, *University of Akron, OH, USA*	83
6	Synthesis of Functional Polymers by Cationic Polymerization A. GANDINI, *Institut National Polytechnique de Grenoble, France*	107
7	Functionalized Polyalkenes T. SIMONAZZI, *Himont Italia, Ferrara, Italy*, A. DE NICOLA, *Himont USA, Elkton, MD, USA* and M. AGLIETTO and G. RUGGERI, *University of Pisa, Italy*	133

Fundamental Properties

8	Computer Modeling of Polymer Structure and Fundamental Properties W. L. MATTICE, *University of Akron, OH, USA*	159
9	Mutual Diffusion in Polymeric Systems B. D. FREEMAN, *North Carolina State University, Raleigh, NC, USA*	167
10	Fundamentals of the Formation, Structure and Properties of Polymer Networks R. F. T. STEPTO, *University of Manchester and UMIST, UK*	199

Degradation of Polymers

11	Thermal Degradation of Condensation Polymers G. MONTAUDO and C. PUGLISI, *University of Catania and Institute for the Chemistry and Technology of Polymeric Materials, Italy*	227
12	Photodegradation of Polymer Materials B. RÅNBY, *The Royal Institute of Technology, Stockholm, Sweden* and J. F. RABEK, *Karolinska Institute, Huddinge, Sweden*	253

13	Biodegradable Polymers A.-C. ALBERTSSON and S. KARLSSON, *The Royal Institute of Technology, Stockholm, Sweden*	285

Generic Polymer Systems and Applications

14	Molecular Engineering of Liquid Crystalline Polymers V. PERCEC and D. TOMAZOS, *Case Western Reserve University, Cleveland, OH, USA*	299
15	Rheological Behaviour of Liquid Crystalline Polymers V. G. KULICHIKHIN, V. S. VOLKOV and N. A. PLATÉ, *Russian Academy of Sciences, Moscow, Russia*	385
16	Polymers for Photonic Applications K.-S. LEE, M. SAMOC and P. N. PRASAD, *State University of New York at Buffalo, NY, USA*	407
17	Structure, Properties and Applications of Polymeric Langmuir–Blodgett Films K. MATHAUER, F. EMBS and G. WEGNER, *Max-Planck-Institut für Polymerforschung, Mainz, Germany*	449
18	Science and Technology of Polymer Composites J. M. KENNY and L. NICOLAIS, *University of Naples, Italy*	471
19	Polymers from Renewable Resources A. GANDINI, *Institut National Polytechnique de Grenoble, France*	527

Reactive Processing

20	Modeling and Computer Simulation of Reactive Processing L. J. LEE, *Ohio State University, Columbus, OH, USA*	575
21	Reactive Processing of Thermoplastic Polymers M. LAMBLA, *Université Louis Pasteur, Strasbourg, France*	619

Appendix

	Guide to the Polymer Literature R. MURRAY, *University of Akron, OH, USA*	643

Subject Index   659

Preface

Supplementary volumes to the seven-volume work *Comprehensive Polymer Science*, published in 1989, are designed to provide up-to-date and comprehensive reviews of polymer topics that were either not included in the original main work or for which recent developments justify authoritative critical reviews. Supplementary volumes are planned to be published every two years.

For this *First Supplement*, we have attempted to choose topics that provide a balance between the areas of polymer chemistry, physics and technology. Thus, in addition to several chapters on polymer chemistry, chapters on the following areas of current interest are included:

    Functionalized polymers, their synthesis and properties
    Computer modeling in polymer science
    Degradation and mutual diffusion of polymers
    Liquid crystalline polymers
    Structure and properties of polymers for photonic applications, for Langmuir–Blodgett films and for polymer composites
    Reactive processing

To assist readers of *Comprehensive Polymer Science* to explore specific areas of polymer science in greater depth, a guide to numerous books, handbooks and encyclopedias has also been included as an Appendix. This Appendix is intended for use with the whole series of volumes.

We would like to acknowledge the help of many of our colleagues in selecting topics for the *First Supplement* and to thank Dr Helen McPherson of Pergamon Press and her editorial team for their work in ensuring that the same high standards of production and publishing have been achieved as in the seven volumes of *Comprehensive Polymer Science*.

SUNDAR L. AGGARWAL
Akron, OH, USA
April 1992

SAVERIO RUSSO
Genova, Italy
April 1992

Contributors

Professor M. Aglietto
Dipartimento di Chimica e Chimica Industriale, Università di Pisa, Via Risorgimento 35, 56126 Pisa, Italy

Professor A.-C. Albertsson
Department of Polymer Technology, The Royal Institute of Technology, S-100 44 Stockholm, Sweden

Professor C. H. Bamford, FRS
Department of Clinical Engineering, Duncan Building, Royal Liverpool University Hospital, PO Box 147, Liverpool L69 3BX

Professor C. Carlini
Dipartimento di Chimica Industriale e dei Materiali, University of Bologna, Viale Risorgimento 4, 40136 Bologna, Italy

Professor F. Ciardelli
Dipartimento di Chimica e Chimica Industriale, University of Pisa, Via Risorgimento 35, 56126 Pisa, Italy

Dr A. De Nicola
Himont Research and Development Center, 800 Greenbank Road, Wilmington, DE 19808, USA

Dr F. Embs
Max-Planck-Institut für Polymerforschung, PO Box 3148, D-6500 Mainz, Germany

Professor B. D. Freeman
Department of Chemical Engineering, North Carolina State University, Box 7905, Raleigh, NC 27695-7905, USA

Professor A. Gandini
Institut National Polytechnique de Grenoble, Ecole Française de Papeterie et des Industries Graphiques, Domaine Universitaire, 461, Rue de la Papeterie, BP 65, 38402 Saint Martin d'Hères, France

Professor A. L. German
Laboratory of Polymer Chemistry & Technology, Eindhoven University of Technology, Den Dolech 2, PO Box 513, 5600 MB Eindhoven, The Netherlands

Dr S. Karlsson
Department of Polymer Technology, The Royal Institute of Technology, S-100 44 Stockholm, Sweden

Professor J. M. Kenny
Dipartimento di Ingegneria dei Materiali e della Produzione, Università di Napoli, P. le Tecchio, I-80131 Napoli, Italy

Professor S. I. Kuchanov
Department of Polymer Chemistry, Moscow State University, 119899 Moscow, Russia

Professor V. G. Kulichikhin
Institute of Petrochemical Synthesis, Russian Academy of Sciences, Leninsky Pr. 29,
117912 Moscow, Russia

Professor M. Lambla
Université Louis Pasteur, Strasbourg I, Ecole d'Application des Hauts Polymères,
4 rue Boussingault, F67000 Strasbourg, France

Dr K.-S. Lee
Photonics Research Laboratory, State University of New York at Buffalo, Buffalo,
NY 14214, USA

Professor L. J. Lee
121 Koffolt Laboratories, Ohio State University, 140 W 19th Avenue, Columbus,
OH 43210-1180, USA

Dr K. Mathauer
Max-Planck-Institut für Polymerforschung, PO Box 3148, D-6500 Mainz, Germany

Professor W. L. Mattice
Institute of Polymer Science, University of Akron, Akron, OH 44325-3909, USA

Professor G. Montaudo
Chemistry Department & CNR, Institute for the Chemistry and Technology of Polymeric
Materials, University of Catania, Viale Andrea Doria 6, I-95125 Catania, Italy

Mrs R. Murray
Rubber Division, American Chemical Society, University of Akron, Akron, OH 44325, USA

Professor L. Nicolais
Dipartimento di Ingegneria dei Materiali e della Produzione, Università di Napoli,
P. le Tecchio, I-80131 Napoli, Italy

Professor V. Percec
Department of Macromolecular Science, School of Engineering, Case Western Reserve
University, Cleveland, OH 44106, USA

Professor N. Platé
Institute of Petrochemical Synthesis, Russian Academy of Sciences, Leninsky Pr. 29,
117912 Moscow, Russia

Dr P. N. Prasad
Photonics Research Laboratory, State University of New York at Buffalo, Buffalo,
NY 14214, USA

Dr C. Puglisi
Chemistry Department & CNR, Institute for the Chemistry and Technology of Polymeric
Materials, University of Catania, Viale Andrea Doria 6, I-95125 Catania, Italy

Professor R. P. Quirk
Institute of Polymer Science, University of Akron, Akron, OH 44325-3909, USA

Professor J. F. Rabek
Polymer Research Group, Department of Dental Materials and Technology, Karolinska Institute,
Box 4064, S-141 04 Huddinge (Stockholm), Sweden

Professor B. Rånby
Department of Polymer Technology, The Royal Institute of Technology, S-100 44 Stockholm, Sweden

Dr G. Ruggeri
Dipartimento di Chimica e Chimica Industriale, Università di Pisa, Via Risorgimento 35, 56126 Pisa, Italy

Dr M. Samoc
Photonics Research Laboratory, State University of New York at Buffalo, Buffalo, NY 14214, USA

Professor H. A. S. Schoonbrood
Laboratory of Polymer Chemistry & Technology, Eindhoven University of Technology, Den Dolech 2, PO Box 513, 5600 MB Eindhoven, The Netherlands

Dr T. Simonazzi
Centro Ricerche G. Natta, Himont Italia, P. Le Privato G. Donegani 12, I-44100 Ferrara, Italy

Professor R. F. T. Stepto
Polymer Science and Technology Group, Manchester Materials Science Centre, University of Manchester and UMIST, Grosvenor Street, Manchester M1 7HS

Dr D. Tomazos
Department of Macromolecular Science, School of Engineering, Case Western Reserve University, Cleveland, OH 44106, USA

Dr G. H. J. Van Doremaele
Laboratory of Polymer Chemistry & Technology, Eindhoven University of Technology, Den Dolech 2, PO Box 513, 5600 MB Eindhoven, The Netherlands

Dr V. S. Volkov
Institute of Petrochemical Synthesis, Russian Academy of Sciences, Leninsky Pr. 29, 117912 Moscow, Russia

Professor G. Wegner
Max-Planck-Institut für Polymerforschung, PO Box 3148, D-6500 Mainz, Germany

1
Inhibition and Retardation in Radical Polymerization

CLEMENT H. BAMFORD
University of Liverpool, UK

1.1	INTRODUCTION	1
1.2	SOME ASPECTS OF THE CHEMISTRY OF INHIBITION AND RETARDATION	2
1.3	THE KINETICS OF IDEAL INHIBITION AND RETARDATION	3
	1.3.1 Ideal Inhibition	3
	1.3.1.1 Note on the use of the stationary state assumption	6
	1.3.2 Ideal Retardation	6
1.4	NUMBER AVERAGE DEGREES OF POLYMERIZATION AND MEAN KINETIC CHAIN LENGTHS IN INHIBITED AND RETARDED POLYMERIZATIONS	12
	1.4.1 Number Average Degrees of Polymerization, $\bar{P}_n$	12
	1.4.2 Mean Kinetic Chain Lengths, $\bar{v}$	12
	1.4.3 The Kinetic Parameter, k_p^2/k_t	13
1.5	POLYMERIZATIONS WITH SLOW REINITIATION	13
	1.5.1 Degradative Transfer versus Degradative Addition	17
1.6	POLYMERIZATIONS WITH SIZE DEPENDENT TERMINATION	18
1.7	REFERENCES	20

1.1 INTRODUCTION

It is a familiar observation that free-radical polymerizations are frequently very sensitive to the presence of small quantities of 'impurities'. Substances which produce very large decreases in the rate of polymerization when added in low concentrations are classed as inhibitors or retarders. These terms have been taken over from classical chain reaction theory, in which an inhibitor is defined as a substance which deactivates initiating centres and a retarder as one which interrupts propagation. In classical systems distinction between the two is often clear-cut since the initiating and propagating species may be quite different in type and reactivity. Although this is so in some polymerizations, more frequently the two types of species are similar, generally organic free radicals. A familiar (extreme) example is the polymerization of methacrylonitrile initiated by azobisisobutyronitrile; here the initiating radicals are 2-cyanoisopropyl ($Me_2\dot{C}CN$) and the chain carriers —$CH_2\dot{C}(Me)CN$. Generally, in practice, the distinction between inhibitors and retarders of polymerization is rather blurred and the same compound may function in either capacity, depending on its concentration and the nature of the system.

Inhibitors and retarders react with initiating or propagating radicals, respectively, to form species of lower reactivity which are either unable to reinitiate chains or reinitiate only slowly. If the products are inactive, the inhibitor or retarder is classed as ideal, and in such a case the kinetic behaviour of the system is relatively simple. On the other hand, slow reinitiation gives rise to more complex kinetics, which may be difficult to interpret unequivocally without additional evidence.

Inhibition and retardation may be either radical addition or radical abstraction processes. Basically these are copolymerization or chain transfer reactions, respectively, differing from 'normal' copolymerizations or chain transfers in that the radical products have low or zero reactivity.

Conversely, a chain transfer becomes a retardation if it yields an effectively inert product; it is then referred to as a degradative chain transfer.

It should be clear from the above that the extent to which a substance (additive) shows inhibiting or retarding properties depends on the nature of the system of which it is a part. The activities of the monomer and derived radicals play critical roles; the latter determines the rate of interaction with a given additive while the former strongly influences the rate of reinitiation. For example, the polymerization of vinyl acetate, unlike that of many vinyl monomers, is retarded by low concentrations of styrene. In this system highly active vinyl acetate radicals add rapidly to the reactive styrene monomer, producing styryl-type radicals which are unable to add to the relatively unreactive vinyl acetate monomer.

1.2 SOME ASPECTS OF THE CHEMISTRY OF INHIBITION AND RETARDATION

Inhibitors and retarders are of diverse chemical types and include stable free radicals, quinones, aromatic derivatives (especially nitro compounds), but also hydrocarbons, some elements (O_2, I_2, S) and many oxidizing cations such as Fe^{3+}, Cu^{2+}, Ce^{4+}, Hg^{2+}, Tl^{2+} and Ag^+ which enter into redox reactions. The literature reveals divergent views of the nature of the mechanisms involved. These have been reported in several texts[1-3] and we shall not elaborate them, but merely indicate some types of reaction involved.

Polymerizations of monomers which can undergo degradative chain transfer show 'self-retardation' (in the absence of added retarder). Allyl derivatives often behave in this way; for example, allyl acetate on hydrogen atom abstraction yields the acetoxyallyl radical $CH_2\text{---}\dot{C}H\text{---}CHOCOMe$, which is too stable to reinitiate polymerization readily.[4,5] Such inactive radicals normally interact with other radicals in the system. Isopropenyl acetate behaves similarly.[6] The participation of these reactions lowers the molecular weight of the product and introduces departures from classical kinetics.

Systems containing stable free radicals, Fe^{3+} and Cu^{2+} are discussed later (Sections 1.3.1 and 1.3.2 respectively).

Quinones are generally thought to add to propagating chains through an oxygen, giving a rather unreactive oxygen-centred radical (1) which may terminate a second chain (equation 1; $R_r^{\bullet}$, $R_s^{\bullet}$ represent radicals containing r, s, monomer units, respectively). Tracer studies have shown that with styrene several quinone residues may be incorporated in a single chain, suggesting that the adduct radical (1) in equation (1) is able to reinitiate polymerization of styrene.[7] Thus quinone can behave as a comonomer under appropriate conditions. However, with methyl methacrylate the reactions in equation (1) predominate, the polymeric product containing two initiator fragments and one quinone residue per molecule.[8]

$$R_r^{\bullet} + O{=}\!\!\bigcirc\!\!{=}O \longrightarrow R_r\text{--}O\text{--}\bigcirc\text{--}O\bullet \xrightarrow{R_s^{\bullet}} R_r O\text{--}\bigcirc\text{--}OR_s \quad (1)$$

(1)

Hydroquinone and its derivatives (*e.g.* hydroquinone methyl ether) are often used to stabilize commercial monomers. They are less powerful retarders than quinone; in the presence of oxygen retardation is enhanced by quinone formation from hydroquinone. These phenolic types are effective scavengers of oxygen-centred radicals, *e.g.* those which may arise from peroxide impurities.

Bagdasar'ian and Sinitsina[9] emphasized that active (polymer) radicals enter into addition reactions with aromatic hydrocarbons (even benzene) in preference to abstracting hydrogen atoms from the rings. The resulting radicals are relatively inactive, so that the aromatic derivatives function as retarders. Values of k_z/k_p were reported for many such additions (k_z, k_p are the rate coefficients for addition and propagation reactions, respectively), and it was shown that a linear relation exists between $\log k_z$ and the logarithm of the methyl affinity of the hydrocarbon.

The same workers[9] suggested that nitrobenzene retards through transfer of a β-hydrogen from the propagating radical (equation 2).

$$\sim\!\!\!\sim\!\!\overset{\bullet}{\diagup}\!\!R + \bigcirc\!\!-NO_2 \longrightarrow \sim\!\!\!\sim\!\!\diagup\!\!R + \bigcirc\!\!-N\!\!\overset{OH}{\underset{O\bullet}{\diagdown}} \quad (2)$$

Alternative routes proposed include those in equation (3), where P_s is a polymer molecule with s units,[10-12] and equation (4a).[13-15]

$$R_r^\bullet + \text{C}_6\text{H}_4\text{-NO}_2 \longrightarrow [\text{cyclohexadienyl intermediate}] \xrightarrow{R_s^\bullet} R_r\text{-C}_6\text{H}_4\text{-NO}_2 + P_s \quad (3)$$

$$R_r^\bullet + \text{C}_6\text{H}_5\text{-NO}_2 \longrightarrow \text{C}_6\text{H}_5\text{-N(O-R}_r\text{)(O}^\bullet\text{)} \quad (4a)$$
$$(2)$$

Some reactions of the radical (2) which have been suggested are shown in equations (4b) to (4e).[14,16-18]

$$(2) + R_s^\bullet \longrightarrow \text{C}_6\text{H}_5\text{-NO} + R_r\text{OR}_s \quad (4b)$$

$$(2) \longrightarrow \text{C}_6\text{H}_5\text{-NO} + R_r\text{O}^\bullet \quad (4c)$$

$$(2) + R_s \longrightarrow \text{C}_6\text{H}_5\text{-N(OR}_r\text{)(OH)} + P_s \quad (4d)$$

$$(2) + R_s \longrightarrow \text{C}_6\text{H}_5\text{-N(OR}_r\text{)(OR}_s\text{)} \quad (4e)$$

Aromatic nitrocompounds are generally less effective than quinones. They illustrate the influence of the monomer on retarding activity. The materials effectively inhibit the polymerization of vinyl acetate, behave as retarders with styrene and have little effect on methyl methacrylate or methyl acrylate polymerizations.

1.3 THE KINETICS OF IDEAL INHIBITION AND RETARDATION

1.3.1 Ideal Inhibition

The participating reactions in pure inhibition are represented in Scheme 1, in which I represents the initiator and Z the inhibitor. $R_0^\bullet$ and $R_r^\bullet$ are initial and propagating radicals, respectively, and M represents monomer; f is the efficiency of initiation.

Reaction (c) with rate coefficient k_{zi} represents the inhibition process. Assuming stationary conditions for the radical concentrations, we find

$$[R_0] = \frac{2fk_d[I]}{k_i[M] + k_{zi}[Z]} \quad (5a)$$

so that the rate of initiation $\mathscr{I}$ is

$$\mathscr{I} = k_i[M][R_0] = \frac{2fk_dk_i[I][M]}{k_i[M] + k_{zi}[Z]} \quad (5b)$$

Evidently the rate of initiation, and hence the overall rate of polymerization, is decreased in the presence of the inhibitor; if $\mathscr{I}_0$ is the rate of initiation in the absence of inhibitor, we have from

$$I \xrightarrow{2fk_d} 2R_0^\bullet \quad (a)$$

$$R_0^\bullet + M \xrightarrow{k_i} R_1^\bullet \quad \mathscr{I} \quad (b)$$

$$R_0^\bullet + Z \xrightarrow{k_{zi}} \text{inactive products} \quad (c)$$

$$R_1^\bullet + M \xrightarrow{k_p} R_2^\bullet \quad (d)$$

$$R_r^\bullet + M \xrightarrow{k_p} R_{r+1}^\bullet \quad (e)$$

$$R_r^\bullet + R_s^\bullet \xrightarrow{k_t} \text{polymer} \quad (f)$$

Scheme 1

equation (5)

$$\mathscr{I} = \mathscr{I}_0 \frac{k_i[M]}{k_i[M] + k_{zi}[Z]} \quad (6)$$

When $k_{zi}[Z] \gg k_i[M]$, $\mathscr{I}$ and the rate of polymerization become effectively zero. During this inhibition period the inhibitor is being consumed by reaction with the radicals generated by decomposition of the initiator (reaction a; Scheme 1) and the consequent decrease in [Z] is accompanied by a corresponding increase in $\mathscr{I}$, according to equation (6). Therefore polymerization gradually becomes perceptible, and steadily increases in rate; eventually when all the inhibitor has been consumed the rate of polymerization assumes its normal value, corresponding to the uninhibited reaction. A typical 'inhibited polymerization' thus shows three stages: an inhibition period with effectively zero rate of polymerization; a transition period with gradually increasing rate; and finally uninhibited polymerization at the normal rate. If the inhibitor is a powerful one (i.e. if k_{zi} is large) a low concentration suffices to give effectively complete inhibition and the end of the inhibition period will be relatively sharp; otherwise a reduced initial rate will be observed which slowly increases with time. In this simple discussion it has been assumed that the concentrations of I and M do not change significantly. A good early example of the phenomena we have been describing is provided by the work of Bevington, Ghanem and Melville[19] on the polymerization of styrene with quinone as inhibitor (Figure 1).

The length of the inhibition period t_{in} is usually measured by back extrapolation of the final, almost linear, portion of the conversion–time curve to the time axis. This period represents the time required for all the inhibitor to react with initial radicals (reaction c; Scheme 1) and since these are produced at a uniform rate $\mathscr{I}_0$ given by

$$\mathscr{I}_0 = 2fk_d[I] \quad (7)$$

we have

$$\mathscr{I}_0 t_{in} = [Z]_0; \quad \mathscr{I}_0 = [Z]_0/t_{in} \quad (8)$$

where $[Z]_0$ is the initial value of [Z], if each radical reacts with one inhibitor molecule. More generally, when m radicals react with each Z molecule

$$\mathscr{I}_0 = m[Z]_0/t_{in} \quad (9)$$

According to equations (8) and (9) a plot of t_{in} versus $[Z]_0$ (such as may be derived from the results in Figure 1, for example) should be linear with slope $1/\mathscr{I}_0$ or $m/\mathscr{I}_0$. The measurements therefore provide a method for the direct determination of the rate of initiation $\mathscr{I}_0$ in the absence of inhibitor. Further, if $\mathscr{I}_0$ is known, the efficiency of initiation may also be estimated from equation (7). If necessary, a correction for the decrease in [I] during the induction period may be applied.

Although, in principle, the use of inhibitors provides a direct and convenient method of measuring rates of initiation, it is subject in practice to some uncertainties. A very powerful inhibitor may

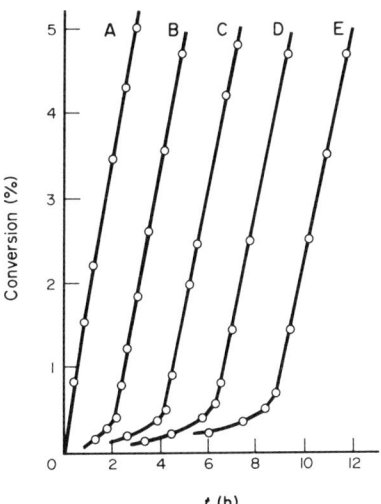

Figure 1 Conversion–time curves for polymerization of styrene at 60 °C initiated by azobisisobutyronitrile (0.5 g dm^{-3}) in the presence of quinone as inhibitor.[19] Concentrations of quinone (g dm^{-3}): A, zero; B, 0.050; C, 0.075; D, 0.123; E, 0.157 (reproduced by permission of the Royal Society of Chemistry from ref. 19)

interfere with (secondary) geminate or cage recombination;[20–23] this latter normally occurs between radicals formed in pairs from an initiator molecule and is responsible for the efficiency of initiation (f) being less than unity. Common vinyl monomers cannot compete significantly with this process, but if an inhibitor does so its use will lead to spuriously high values of $\mathscr{I}_0$.

In the second place, the value of m is often debatable. This may be illustrated by reference to the familiar stable free radical 2,2-diphenyl-1-picrylhydrazyl (DPPH), one of the most widely used inhibitors. The high extinction coefficient of DPPH at visible wavelengths of light greatly facilitates spectrophotometric estimation of low concentrations. According to Bengough,[24] in the polymerization of methyl methacrylate inhibited by DPPH the final disappearance of colour corresponds closely with the end of the induction period, in conformity with the mechanism discussed above. However, some other monomers behave in less simple fashion. Thus with styrene, Matheson et al.[25] found that the final rates of polymerization were lower than those measured in the absence of DPPH. With vinyl acetate little polymerization occurs for some time after the colour has disappeared. These results suggest that the products of reaction between DPPH and initiator radicals may not be inert, as required for an ideal inhibitor, but may themselves be inhibitors or retarders. Hammond, Sen and Boozer[26] expressed doubts about the value of DPPH as a reagent for counting radicals. They presented evidence indicating the occurrence of a 'double transfer' with radicals from azobisisobutyronitrile (equation 10). As a consequence of this chain process a single DPPH radical might react with several initiator radicals. A similar mechanism for reaction with polymer radicals was proposed by Burnett and Cowley.[27]

$$\begin{array}{c} \rangle\!\!-\!\!\overset{\bullet}{}\!\!-\!\!CN \;+\; DPPH \;\longrightarrow\; =\!\!\langle_{CN} \;+\; HDPPH \\[4pt] HDPPH \;+\; \rangle\!\!-\!\!\overset{\bullet}{}\!\!-\!\!CN \;\longrightarrow\; \rangle\!\!\overset{H}{\vphantom{x}}\!\!-\!\!CN \;+\; DPPH \end{array} \qquad (10)$$

Bevington[28] has shown that the products of reaction between radicals from azobisisobutyronitrile and DPPH do not contain major proportions of isobutyronitrile or the corresponding hydrazine, as required by the reactions in equation (10). Consequently, Bevington disputes the conclusions of Hammond et al., although he agrees with these authors that DPPH is probably not a good reagent for counting radicals in exact work.

The Koelsch radical (1,3-bis(diphenylene)-2-phenylallyl; **3**) has been used apparently successfully to determine rates of initiation.[29,30]

Triphenylmethyl is also an inhibitor, but is able to initiate polymerization in some systems.

The foregoing suggests that systems showing pure inhibition without retardation will be relatively uncommon, since an inhibitor for which $k_{zi}[Z]_0 \gg k_i[M]$ is also likely to conform to $k_z[Z]_0 \gg k_p[M]$,

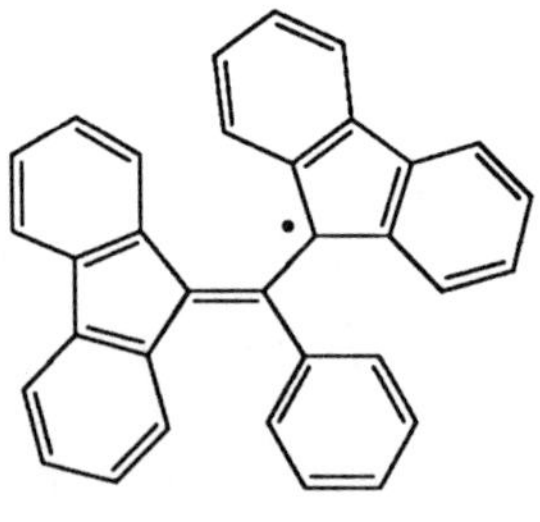

(3)

where k_z refers to reaction between Z and propagating chains. This will not invalidate determination of $\mathscr{I}_0$ by the technique outlined. For a pure inhibitor the sufficient condition for this measurement is that all initiating radicals should be trapped by the inhibitor during the inhibition period; for an inhibitor and retarder it is sufficient that bimolecular termination should be completely suppressed (*i.e.* all $R_0^{\cdot}$ and $R\cdot$ radicals trapped) during the inhibition period.

The kinetic equations for polymerization in the presence of a substance which is both an inhibitor and a retarder may be obtained by replacing $\mathscr{I}_0$ in the appropriate relations in Section 1.3.2 (equations 17, 18 and 20–22) by $\mathscr{I}$ from equation (6).

1.3.1.1 Note on the use of the stationary state assumption

It is appropriate to note here that stationary state equations should not be applied indiscriminately to systems in which radical concentrations are likely to change rapidly at some stage, *e.g.* near the end of an inhibition period. Thus in the system considered above the rates of change of $[R_0^{\cdot}]$ and $[R\cdot]\left(=\sum_{1}^{\infty}R_r^{\cdot}\right)$ are given by

$$\frac{d[R_0^{\cdot}]}{dt} = 2fk_d[I] - k_i[M][R_0^{\cdot}] - k_{zi}[Z][R_0^{\cdot}] \qquad (11)$$

$$\frac{d[R\cdot]}{dt} = k_i[M][R_0^{\cdot}] - k_t[R\cdot]^2 \qquad (12)$$

The stationary state assumption $d[R_0^{\cdot}]/dt = d[R\cdot]/dt = 0$ implies that

$$\frac{d[R\cdot]}{dt} \ll 2fk_d[I]$$

$$\ll (k_i[M] + k_{zi}[Z])[R_0^{\cdot}] \qquad (13)$$

and

$$\frac{d[R\cdot]}{dt} \ll k_i[M][R_0^{\cdot}]$$

$$\ll k_t[R\cdot]^2 \qquad (14)$$

In doubtful cases when the stationary state method has been used to derive numerical parameters, the validity of the above inequalities should be examined with the aid of their proposed values.

1.3.2 Ideal Retardation

The component reactions in a polymerization showing ideal retardation are set out in Scheme 2. Here Z represents the retarder which enters into reaction with propagating chains ($R_r^{\cdot}$) to form a dead polymer molecule containing r units, as shown in reaction (e). $\mathscr{I}_0$ is the rate of initiation, assumed unaffected by the presence of the retarder.

Some simple and interesting conclusions follow from this mechanism. First, we see immediately from Scheme 2 that

$$\frac{d[R\cdot]}{dt} = \mathscr{I}_0 - k_z[Z][R\cdot] - k_t[R\cdot]^2 \qquad (15)$$

$$\text{I} \xrightarrow{2f_d} 2\text{R}_0^\bullet \quad \text{(a)}$$

$$\text{R}_0^\bullet + \text{M} \xrightarrow{k_i} \text{R}_1^\bullet \qquad \mathscr{I}_0 \quad \text{(b)}$$

$$\text{R}_1^\bullet + \text{M} \xrightarrow{k_p} \text{R}_2^\bullet \quad \text{(c)}$$

$$\text{R}_r^\bullet + \text{M} \xrightarrow{k_p} \text{R}_{r+1}^\bullet \quad \text{(d)}$$

$$\text{R}_r^\bullet + \text{Z} \xrightarrow{k_z} \text{P}_r + \text{inert products} \quad \text{(e)}$$

$$\text{R}_r^\bullet + \text{R}_s^\bullet \xrightarrow{k_t} \text{polymer} \quad \text{(f)}$$

Scheme 2

and

$$-\frac{d[Z]}{dt} = k_z [Z][\text{R}\cdot] \tag{16}$$

where $[\text{R}\cdot] = \sum_r [\text{R}_r^\bullet]$ is the total concentration of propagating chains.

A retarder present in sufficiently high concentration terminates a large proportion of the growing chains and the rate of bimolecular termination $k_t[\text{R}\cdot]^2$ becomes infinitesimally small. In such conditions $d[\text{R}\cdot]/dt$ is effectively zero and we may write equation (15) in the following form

$$\frac{d[\text{R}\cdot]}{dt} = 0 = \mathscr{I}_0 - k_z[Z][\text{R}\cdot] \tag{17}$$

so that

$$[\text{R}\cdot] = \mathscr{I}_0 / k_z[Z] \tag{18}$$

In general, if the mean degree of polymerization $\bar{P}_n$ (or the mean kinetic chain length $\bar{v}$) is $\gg 1$, the rate of polymerization is conventionally taken as

$$\omega = -\frac{d[\text{M}]}{dt} = k_p[\text{M}][\text{R}\cdot] \tag{19}$$

Retarded polymerization may, however, give rise to short chains and in these circumstances consumption of monomer by reaction (b) in Scheme 2 is not necessarily negligible. This necessitates a correction to equation (19), which becomes

$$\omega = k_p[\text{M}][\text{R}\cdot] + k_i[\text{M}][\text{R}_0^\bullet]$$
$$= k_p[\text{M}][\text{R}\cdot] + \mathscr{I}_0 \tag{20}$$

In strongly retarded polymerizations expressions (19) and (20), together with equation (18), give for the retarded rate ω_r,

$$\omega_r = \frac{k_p[\text{M}]}{k_z[Z]} \mathscr{I}_0 \tag{21}$$

from equation (19), and

$$\omega_r = \left(\frac{k_p[\text{M}]}{k_z[Z]} + 1 \right) \mathscr{I}_0 \tag{22}$$

from equation (20).

Thus (for pure retardation) the rate is first order in the rate of initiation and inversely proportional (or linear in) $1/[Z]$. These are, of course, instantaneous relations. As the retarder is consumed (equation 16) the radical concentration and the rate of polymerization increase until finally both reach stationary values corresponding to the unretarded reaction (equations 23 and 24).

$$[R\cdot] = \left(\frac{\mathscr{I}_0}{k_t}\right)^{1/2} \qquad (23)$$

$$\omega = k_p[M]\left(\frac{\mathscr{I}_0}{k_t}\right)^{1/2} \qquad (24)$$

As the polymerization proceeds the order in $\mathscr{I}_0$ therefore decreases from 1.0 to 0.5.

Secondly, from equations (16) and (18) we see that under conditions of strong retardation

$$-\frac{d[Z]}{dt} = \mathscr{I}_0 \qquad (25)$$

that is, the rate of consumption of retarder is equal to the rate of initiation. This conclusion from the simple Scheme 2 is indeed obvious, since under the conditions specified the great majority of the chains started by the initiator terminate by reaction with the retarder. Note that the occurrence of chain transfer in the system does not invalidate equation (25). If the concentration of Z can be conveniently monitored this result provides a method for determining rates of initiation. Measurements of the retarded rates of polymerization and application of equation (21) then provide values of k_z/k_p.

For powerful retarders, only very low concentrations can be present if rates of polymerization are to be measurable. In such cases [Z] cannot be regarded as constant during a single experiment, and use of equations (21) and (25) as described above is not appropriate.

Thirdly, it follows from equations (16) and (21) that

$$-\frac{d[Z]}{dt}\bigg/\left(-\frac{d[M]}{dt}\right) = \frac{k_z[Z][R\cdot]}{k_p[M][R\cdot]} = \frac{k_z[Z]}{k_p[M]}$$

or

$$\frac{d\ln[Z]}{d\ln[M]} = \frac{k_z}{k_p} \qquad (26)$$

Hence, if a polymerization conforms to Scheme 2, measurements of the rates of consumption of retarder and monomer enable k_z/k_p to be evaluated. This method does not involve assumption of a stationary state.

Bamford, Jenkins and Johnston[31,32] studied the free-radical polymerizations of acrylonitrile (AN), methacrylonitrile (MAN), methyl acrylate (MA) and styrene in N,N-dimethylformamide (DMF) in the presence of iron(III) chloride and concluded that the behaviour of these systems was that expected for ideal retardation. Retardation results from the redox reaction between propagating radicals and the salt, so that the latter is reduced to the iron(II) state, which does not react further. (Redox processes of this type for some radicals in aqueous systems were studied by Collinson, Dainton and McNaughton.[33]) There are clearly two possibilities, shown in equations (27a) and (27b). Data obtained by Entwistle[34] for the monomers mentioned indicated the presence of chlorine in the polymer and so support the reaction in equation (27a); however, the earlier workers[31] preferred the second alternative.

$$\sim\!\!\!\sim\!\!\!\overset{\bullet}{\diagup}\!\!R + FeCl_3 \longrightarrow \sim\!\!\!\sim\!\!\!\overset{Cl}{\diagdown}\!\!R + FeCl_2 \qquad (27a)$$

$$\sim\!\!\!\sim\!\!\!\overset{\bullet}{\diagup}\!\!R + FeCl_3 \longrightarrow \sim\!\!\!\sim\!\!\!\diagup\!\!R + FeCl_2 + HCl \qquad (27b)$$

Some results for the acrylonitrile–DMF–FeCl$_3$–azobisisobutyronitrile systems are presented in Figures 2(a) and 2(b).[31] The linear plot in Figure 2(a) is consistent with equation (21), $Z = FeCl_3$ with [M] and $\mathscr{I}_0$ effectively constant, while Figure 2(b) shows that with [M] and [Z] constant the rate of polymerization is directly proportional to $\mathscr{I}_0$, as required by equation (21). Apparently FeCl$_3$ and FeCl$_3\cdot 6H_2O$ behave similarly. Equation (21), when applied to the slope of the line in Figure 2(a) leads to $k_p/k_z = 0.30$. This value was combined with a published value of k_p to give[31]

$$k_z = 6.5 \times 10^3 \,\text{mol}^{-1}\,\text{dm}^3\,\text{s}^{-1} \qquad (28)$$

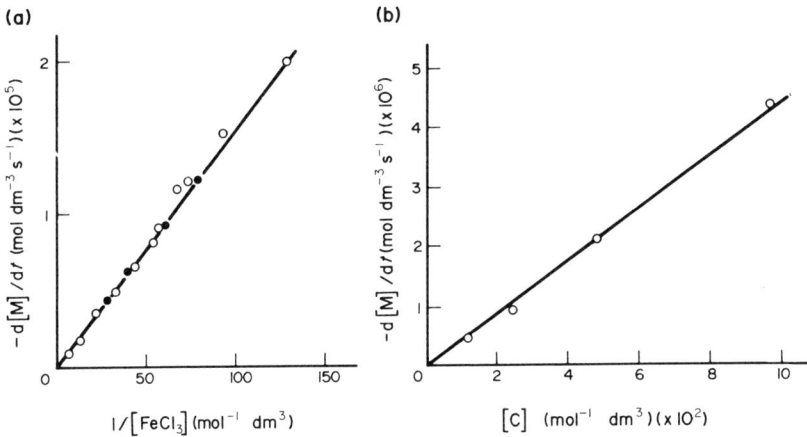

Figure 2 Retardation of acrylonitrile polymerization by iron(III) chloride in DMF solution at 60 °C. (a) Initial concentrations (mol dm^{-3}): acrylonitrile 2.51; azobisisobutyronitrile 1.14×10^{-2}. ○ represents FeCl$_3$·6H$_2$O; ● represents FeCl$_3$. Plot according to equation (21). (b) Dependence of rate of polymerization on initiator concentration for strong retardation (equation 21). Initial concentrations (mol dm^{-3}): acrylonitrile 3.63; FeCl$_3$·6H$_2$O 4.35×10^{-2} (reproduced by permission of the Royal Society of Chemistry from ref. 31)

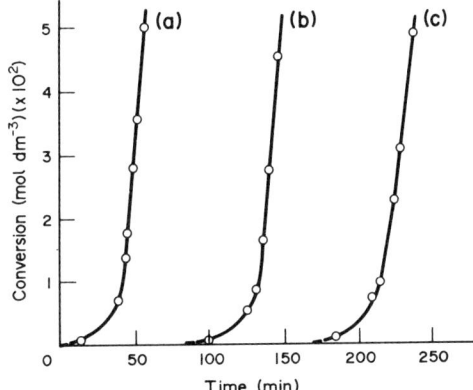

Figure 3 Apparent inhibition of styrene polymerization by iron(III) chloride. DMF solution at 60 °C. Initial concentrations (mol dm^{-3}): styrene 2.49; azobisisobutyronitrile 7.60×10^{-2}; FeCl$_3$·6H$_2$O (a) 2.62×10^{-3}, (b) 7.86×10^{-3}, (c) 13.1×10^{-3} (reproduced by permission of the Royal Society of Chemistry from ref. 31)

Iron(III) chloride reacts much more rapidly with polystyryl than with polyacrylonitrile radicals; the reaction in the former case is so fast that the salt behaves as an inhibitor of styrene polymerization in DMF solution, rather than a pure retarder.[31] This is illustrated by the data in Figure 3, which clearly show induction periods similar to those in Figure 1. Experiments established that the apparent inhibition does not arise from direct reaction of iron(III) chloride with initiator radicals.[31] Strong retardation such as this would be expected to lead to kinetic features of the type discussed for inhibition and in fact it was shown[31] that induction periods obtained from Figure 3 show the dependences expected from equation (9).

Bamford, Jenkins and Johnston[31] integrated the instantaneous stationary state equation over a period extending from the beginning of the strongly retarded reaction to the unretarded polymerization. A numerical justification of this procedure was provided (cf. Section 1.3.1.1). The following is an outline of the treatment.

The stationary state equation is obtained from equation (15) by setting $d[R\cdot]/dt = 0$. Differentiation then leads to equation (29) ($\mathscr{I}_0 = \mathscr{I}$)

$$\frac{d[Z]}{dt} = -\frac{\mathscr{I} + k_t[R\cdot]^2}{k_z[R\cdot]^2}\frac{d[R\cdot]}{dt} = -\mathscr{I} + k_t[R\cdot]^2 \tag{29}$$

By rearranging and integrating we find

$$-\frac{[R\cdot]_s}{[R\cdot]} + \ln\frac{[R\cdot]_s + [R\cdot]}{[R\cdot]_s - [R\cdot]} = k_z[R\cdot]_s t + A \tag{30}$$

where A is constant and $[R\cdot]$, $[R\cdot]_s$ are the radical concentration at time t and the final stationary concentration (corresponding to $[Z]=0$), respectively. It is convenient to express $[R\cdot]$ in equation (30) as a fraction ϕ_t of the final value when $[Z]=0$; if the monomer concentration is sensibly constant $\phi_t = (-d[M]/dt)_t/(-d[M]/dt)_s$, so that ϕ_t is simply the ratio of the prevailing rate of polymerization at time t to the final rate. ϕ_t has been called the 'reduced rate'. After making the appropriate substitution into equation (30) we have

$$\frac{1}{\phi_0} - \frac{1}{\phi_t} + \ln\left(\frac{1-\phi_0}{1+\phi_0}\cdot\frac{1+\phi_t}{1-\phi_t}\right) = k_z[R\cdot]_s t \quad (31)$$

where ϕ_0 is the initial value of ϕ_t and is given by

$$\phi_0 = \frac{(\mathscr{I}k_t)^{1/2}}{k_z[Z]_0} \quad (32)$$

If $\phi_0 \ll 1$ (i.e. strong initial retardation) equation (31) adopts the simpler form of equation (33).

$$\frac{1}{\phi_0} - \frac{1}{\phi_t} + \ln\left(\frac{1+\phi_t}{1-\phi_t}\right) = k_z[R\cdot]_s t \quad (33)$$

It is possible to choose a value of ϕ_t (i.e. a value of t) for which

$$-\frac{1}{\phi_t} + \ln\left(\frac{1+\phi_t}{1+\phi_t}\right) = 0 \quad (34)$$

at this value of $t(=\tau)$ equation (33) becomes

$$\frac{1}{\phi_0} = k_z[R\cdot]_s \tau \quad (35)$$

The solution of equation (34) is $\phi_t = 0.648$; thus τ is the time at which the rate of polymerization has reached 0.648 of its final stationary value. From equation (32) it follows that

$$\tau = \frac{[Z]_0}{\mathscr{I}} \quad (36)$$

so that τ is the exact 'equivalent induction period', i.e. the time required for the whole of the retarder to react with initial radicals if it had completely suppressed the polymerization reaction. Thus τ is a more reliable and useful quantity than the induction period obtained by back extrapolation from the final steady rate, which can be uncertain in some systems.[31]

The treatment described is general for retarded reactions provided: (a) the initial rate is sufficiently small; (b) the products of the retardation process do not react further (i.e. retardation is ideal); and (c) the monomer and initiator concentrations are effectively constant. When these conditions hold, measurement of τ enables the rate of initiation to be calculated from equation (36).

Equation (30) may be written in the form

$$-\frac{1}{\phi_t} + \ln\left(\frac{1+\phi_t}{1-\phi_t}\right) = k_z[R\cdot]_s t + A \quad (37)$$

which enables ϕ_t to be plotted against $k_z[R\cdot]_s t + A$. Such a plot is presented in Figure 4; no parameters specific to any particular reaction have been used in constructing this diagram which is therefore of general applicability. The curve indicates the calculated dependence of the reduced rate upon time and comparison with the experimental curve relating ϕ_t to time enables $k_z[R\cdot]_s$ to be determined. This is equal to $k_z(\mathscr{I}/k_t)^{1/2}$, and since $\mathscr{I}$ has been measured as described above $(k_z/k_t)^{1/2}$ may be calculated. From a knowledge of the unretarded rate of polymerization and $\mathscr{I}$, $(k_p/k_t)^{1/2}$ is known, hence k_z/k_p can be evaluated.

Bamford, Jenkins and Johnston[31] found that for styrene/FeCl$_3$ at 60 °C, $k_z/k_p = 536$ and, from literature values of k_p, concluded that

$$k_z = 5.4 \times 10^4 \text{ mol}^{-1}\text{dm}^3\text{s}^{-1} \quad (38)$$

This value is approximately nine times as great as that found for acrylonitrile/FeCl$_3$ at 60 °C (equation 28). The difference reflects the polar properties of the transition states in these reactions. Polystyrene radicals are nucleophilic and form a polar transition state with strongly electrophilic FeCl$_3$ (**4a**). On the other hand, polyacrylonitrile radicals are electrophilic and polarity is less well developed, since the two influences are opposed (**4b**). The polar contributions lead to a lowering

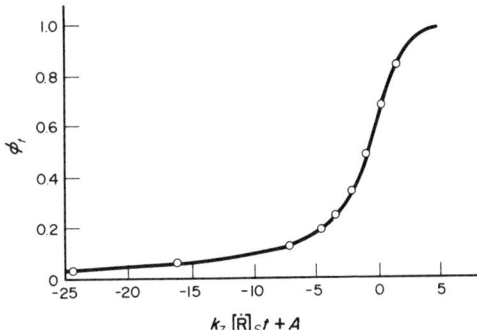

Figure 4 ϕ_t as a function of reaction time in an ideally retarded polymerization. Polymerization of styrene in DMF solution at 60 °C with azobisisobutyronitrile initiation and retardation by iron(III) chloride. Initial concentrations (mol dm^{-3}): styrene 2.61; azobisisobutyronitrile 7.34 × 10^{-2}; FeCl$_3$ 7.91 × 10^{-3}. ○ represents experimental points ($k_z[\text{R}\cdot]_s = 8.57 \times 10^{-3}\text{s}^{-1}$); curve calculated from equation (37) (reproduced by permission of the Royal Society of Chemistry from ref. 31)

of the energy of the transition state and a consequent increase in the rate coefficient. A more detailed discussion is given elsewhere.[35]

$$[\text{FeCl}_3\cdot\text{St}\sim] \qquad [\text{FeCl}_3\cdot\text{AN}\sim]$$

$$\text{(4a)} \qquad\qquad \text{(4b)}$$

Bengough and his colleagues[36-42] have reported related investigations of the retarding action of other transition metal salts on several free-radical polymerizations. Copper(II) chloride in DMF was found to be a more powerful retarder than iron(III) chloride. For methyl methacrylate polymerization at 60 °C Bengough and Fairservice[36,38] showed that copper(II) chloride is an ideal retarder and reported the values

$$\frac{k_z}{k_p} = 1050; \quad k_z = 7.7 \times 10^4 \text{ mol}^{-1}\text{dm}^3\text{s}^{-1} \tag{39}$$

As these workers point out, k_z is of the order of magnitude of the rate coefficients for some bimolecular radical terminations. It was also shown[37] that copper(II) chloride in DMF is not an ideal retarder of acrylonitrile polymerization; apparently the copper(I) derivative formed is also a strong retarder. Bengough and Fairservice's studies led to the following values for copper(II)

$$\frac{k_z}{k_p} = 100; \quad k_z = 1.93 \times 10^5 \text{ mol}^{-1}\text{dm}^3\text{s}^{-1} \tag{40a}$$

and for copper(I)

$$\frac{k_z}{k_p} = 5.7; \quad k_z = 1.1 \times 10^4 \text{ mol}^{-1}\text{dm}^3\text{s}^{-1} \tag{40b}$$

During retardation by copper(I) chloride the solution, initially colourless, becomes a deep yellow-brown. The course of this reaction has not been clarified.

Bengough and Fairservice[37] have developed a kinetic treatment for systems in which two retarders, e.g. Cu^{2+} and Cu$^+$, are present.

The same workers also discussed the differences between τ values and extrapolated induction periods.[36] For the CuCl$_2$/DMF/MMA system the two values differed by only about 1%, with τ being greater, as required by theory.[31]

Bamford, Jenkins and Johnston[31] considered the role of complexation between FeCl$_3$ and DMF in the retardation process, remarking that the complexes are less active than the free salt. This aspect has been investigated in detail by Dass and George.[43]

The retarding activities of several iron(III) salts in DMF on the polymerization of methyl methacrylate were examined by Entwistle[34] and found to vary widely: FeBr$_3$ appeared to be the most powerful retarder.

Other systems for which k_z/k_p has been measured include MMA/CuBr$_2$,[38] MAN/CuCl$_2$[41] and vinyl chloride/FeCl$_3$,[42] all with DMF as solvent.

1.4 NUMBER AVERAGE DEGREES OF POLYMERIZATION AND MEAN KINETIC CHAIN LENGTHS IN INHIBITED AND RETARDED POLYMERIZATIONS

These quantities may be calculated without difficulty from the kinetic Schemes 1 and 2 with the aid of the stationary state assumption. The resulting expressions are set out below, together with those of the uninhibited or unretarded ('uncomplicated') polymerization. A chain transfer component (nonretarding) has been added to the two schemes; S represents any transfer agent present and k_{fs} is the corresponding rate coefficient. The quantities k_{tc} and k_{td} are the termination coefficients for radical combination and disproportionation, respectively and x, the fraction of the total termination occurring by disproportionation, is

$$x = \frac{k_{td}}{k_{tc} + k_{td}}$$

As hitherto, $\mathscr{I}_0 = 2fk_d[I]$ is the rate of formation of primary radicals from the initiator and $\mathscr{I}$ the actual rate of chain initiation; thus for pure inhibition $\mathscr{I} < \mathscr{I}_0$ and for pure retardation $\mathscr{I} = \mathscr{I}_0$.

Obviously the following expressions for $\bar{P}_n$ and $\bar{v}$ may not be valid if application of the stationary state is questionable; see caveat in Section 1.3.1.1.

1.4.1 Number Average Degrees of Polymerization, $\bar{P}_n$

Uncomplicated

$$\frac{1}{\bar{P}_n} = \frac{(1+x)\mathscr{I}_0^{1/2}(k_{tc} + k_{td})^{1/2}}{2k_p[M]} + \frac{k_{fs}[S]}{k_p[M]} \tag{41}$$

Pure inhibition

$$\frac{1}{\bar{P}_n} = \frac{(1+x)\mathscr{I}^{1/2}(k_{tc} + k_{td})^{1/2}}{2k_p[M]} + \frac{k_{fs}[S]}{k_p[M]} \tag{42}$$

$\mathscr{I}$ is given by equation (6).
Pure retardation: (i) moderate retardation

$$\frac{1}{\bar{P}_n} = \frac{(1+x)(k_{tc} + k_{td})[R\cdot]}{2k_p[M]} + \frac{k_z[Z]}{k_p[M]} + \frac{k_{fs}[S]}{k_p[M]} \tag{43}$$

$[R\cdot]$ is given by equation (15) with $d[R\cdot]/dt = 0$.
(ii) strong retardation, $k_z[Z] \gg k_t[R\cdot]$

$$\frac{1}{\bar{P}_n} = \frac{k_z[Z] + k_{fs}[S]}{k_p[M] + k_z[Z]} \tag{44}$$

1.4.2 Mean Kinetic Chain Lengths, $\bar{v}$

Uncomplicated

$$\bar{v} = \frac{k_p[M]}{[\mathscr{I}_0(k_{tc} + k_{td})]^{1/2}} \tag{45}$$

Pure inhibition

$$\bar{v} = \frac{k_p[M]}{[\mathscr{I}(k_{tc} + k_{td})]^{1/2}} \tag{46}$$

$\mathscr{I}$ is given by equation (6).
Pure retardation: (i) moderate retardation

$$\bar{v} = \frac{k_p[M][R\cdot]}{\mathscr{I}_0} \tag{47}$$

$[R\cdot]$ is given by equation (15) with $d[R\cdot]/dt = 0$.
(ii) strong retardation, $k_z[Z] \gg k_t[R\cdot]$

$$\bar{v} = \frac{k_p[M]}{k_z[Z]} + 1 \tag{48}$$

1.4.3 The Kinetic Parameter k_p^2/k_t

From the above it follows that with pure inhibition degrees of polymerization are greater than those obtained in the uninhibited polymerization, unless chain transfer is completely dominating (equations 41 and 42). However, retardation reduces the values of $\bar{P}_n$ (equations 41 and 43). Similar conclusions follow for the mean kinetic chain length.

A corollary is that values of the kinetic parameter $k_p^2/k_t [\equiv k_p^2/(k_{tc}+k_{td})]$ deduced from measurements of ω and $\bar{v}$ with the aid of the conventional equations (49) for uncomplicated systems

$$\frac{\omega \bar{v}}{M^2} = k_p[M]\left(\frac{\mathscr{I}_0}{k_{tc}+k_{td}}\right)^{1/2} \cdot \frac{k_p[M]}{[\mathscr{I}_0(k_{tc}+k_{td})]^{1/2}}\frac{1}{[M]^2}$$

$$= \frac{k_p^2}{k_{tc}+k_{td}} \tag{49}$$

are unaffected by the occurrence of pure inhibition, since the terms in $\mathscr{I}_0$ cancel. Apparent values of k_p^2/k_t are, however, lowered by retardation. Evaluation of k_p^2/k_t in this way provides a simple method for distinguishing between pure inhibition and pure retardation. An example is furnished by the influence of carbon monoxide on the polymerization of methyl methacrylate initiated by molybdenum hexacarbonyl and carbon tetrachloride. The presence of CO reduces the rate of polymerization, but leaves k_p^2/k_t unchanged.[44] This is therefore a case of pure inhibition, CO interfering with chain initiation but not with chain growth.

1.5 POLYMERIZATIONS WITH SLOW REINITIATION

We now consider processes intermediate between ideal inhibition or retardation and chain transfer — those in which a component in the system (generally monomer or an additive) reacts with radicals giving products which reinitiate relatively slowly (compared to propagation). Retarders behaving in this fashion are very common and are of diverse chemical types. The kinetic treatment is generally more complex than those previously discussed and in efforts to obtain tractable forms various simplifications have been introduced. Inhibition is not considered specifically in the systems to be discussed and the symbol $\mathscr{I}$, representing as usual the rate of initiation, also denotes the rate of primary radical formation.

One of the earliest mechanistic proposals, advanced by Flory in 1953,[45] is presented in Scheme 3, reactions (a–g). The retarding reaction (e; Scheme 3) is written above as an addition, but for present purposes could be a transfer; reaction (f) is the reinitiation process. Z· radicals are less reactive than R· and could be present in relatively high concentrations; it is assumed that reaction (g) is the only significant termination, reactions $Z_r^\cdot + R_s^\cdot$ and $Z_r^\cdot + Z_s^\cdot$ occurring much less frequently. This assumption may not hold, as will be discussed later.

Stationary state treatment applied to [R·] and [Z·] ($=\sum_r[Z_r^\cdot]$) leads to the following relations

$$[R\cdot] = \frac{1}{k_z[Z]}\left\{\mathscr{I} + k_{pz}[M]\left(\frac{\mathscr{I}}{k_{tzz}}\right)^{1/2}\right\} \tag{50}$$

$$[Z\cdot] = \left(\frac{\mathscr{I}}{k_{tzz}}\right)^{1/2} \tag{51}$$

The rate of polymerization is given by

$$-\frac{d[M]}{dt} = k_p[M][R\cdot] + k_{pz}[M][Z\cdot] \tag{52}$$

and may be calculated from equations (50) and (51).
The rate of consumption of retarder is

$$-\frac{d[Z]}{dt} = k_z[R\cdot][Z\cdot] - qk_{tzz}[Z\cdot]^2 \tag{53}$$

in which q is unity or zero, depending on whether or not Z is regenerated in reaction (g). By substituting for [R·] and [Z·] we obtain from equation (53)

$$-\frac{d[Z]}{dt} = (1-q)\mathscr{I} + k_{pz}(\mathscr{I}/k_{tzz})^{1/2}[M]. \tag{54}$$

$$I \xrightarrow{2fk_d} 2R_0^\bullet \qquad (a)$$

$$R_0^\bullet + M \xrightarrow{k_i} R_1^\bullet \qquad \mathscr{I} \qquad (b)$$

$$R_1^\bullet + M \xrightarrow{k_p} R_2^\bullet \qquad (c)$$

$$R_r^\bullet + M \xrightarrow{k_p} R_{r+1}^\bullet \qquad (d)$$

$$R_r^\bullet + Z \xrightarrow{k_z} Z_r^\bullet \qquad (e)$$

$$Z_r^\bullet + M \xrightarrow{k_{pz}} R_{r+1}^\bullet \qquad (f)$$

$$Z_r^\bullet + Z_s^\bullet \xrightarrow{k_{tzz}} \text{inert products} \qquad (g)$$

$$R_r^\bullet + Z_s^\bullet \xrightarrow{k_{trz}} \text{inert products} \qquad (h)$$

$$R_r^\bullet + R_s^\bullet \xrightarrow{k_t} \text{polymer} \qquad (i)$$

Scheme 3

Hence, if [Z] can be monitored and $\mathscr{I}$ is varied, a plot of $(d[Z]/dt)/\mathscr{I}^{1/2}$ versus $\mathscr{I}^{1/2}$ in the early stages of reaction should serve to evaluate $(k_{pz}/k_{tzz})^{1/2}$ and q.

A more general mechanism for retardation is Scheme 3 with the two additional termination reactions (h) and (i). The most complete analysis of this scheme has been given by Kice,[46,47] who deduced the following relation (55) by use of stationary state equations

$$\frac{\mu^2[Z]}{1-\mu^2}\left\{1+\left[1+\frac{(1-\mu^2)}{\phi^2\mu^2}\right]^{1/2}\right\} = \frac{k_t\omega}{k_pk_z}\left(1+\frac{1-\mu^2}{\phi^2\mu^2}\right)^{1/2} + \frac{k_{pz}k_t[M]}{k_{trz}k_z} \qquad (55)$$

Here μ is the ratio of the rates of polymerization in the presence and absence of retarder, respectively, i.e. ω/ω_0, and $\phi = k_{trz}/(k_tk_{tzz})^{1/2}$. ϕ is a familiar parameter in copolymerization kinetics but its values in the systems under discussion are generally unknown. A plot of the left side of equation (55) against $\omega[1+(1-\mu^2)/\phi^2\mu^2]^{1/2}$ should be a straight line of slope k_t/k_pk_z and intercept $k_{pz}k_t[M]/k_{trz}k_z$ if [M] and [Z] remain effectively constant. Hence, if k_p and k_t are known k_z and k_{pz}/k_{trz} may be evaluated. Since a knowledge of ϕ is required to construct the plot, the procedure suggested by Kice is to cover a wide range of ϕ values and accept as correct that which gives the best straight line. The plot is insensitive to the value of ϕ if this is large; in these circumstances $(1-\mu^2)/\phi^2\mu^2 \ll 1$ and this term may be omitted from both sides of equation(55). The implication is that k_{tzz} is relatively very small, the termination $Z_r^\bullet + Z_s^\bullet$ being slow compared to $R_r^\bullet + Z_s^\bullet$ (equations g and h, Scheme 3, respectively). This method has been applied successfully by Kice to methyl acrylate[47] and methyl methacrylate[46] polymerizations in the presence of a variety of retarders. The derived values of k_z for the two monomers were discussed in terms of polar and steric effects. Kice's findings on the polymerization of methyl methacrylate retarded by quinone were consistent with those of Bevington, Ghanem and Melville.[8]

Allen, Merritt and Scanlan[48] also considered a similar scheme and applied their results to the polymerization of vinyl acetate in the presence of dihydromyrcene (which was found to retard the polymerization similarly to isopropylbenzene). These workers included chain transfer to monomer in their mechanism. They showed that, in general, the rate of polymerization is proportional to a power of $\mathscr{I}$ between 0.5 and 1.0, the value depending on the rate of the reinitiation reaction.

Jenkins[49,50] considered measurements made at different monomer concentrations, employing Kice's type of analysis. He considered three special cases, $\phi = 1$, $\phi \to \infty$ and $\phi \to 0$. These yielded simplified results; in each case the order of the rate of polymerization in $\mathscr{I}$ lay between 0.5 and 1.0, the precise value being dependent on the nature of the system.

Kice's treatment is a complete solution of the kinetics of retarded polymerization based on the scheme described; nevertheless it has some practical disadvantages, one of which is that the unretarded rate of polymerization ω_0 must be known. Clearly there are instances when this quantity cannot be measured, for example if the monomer itself is a retarder. If the monomer has highly active radicals (such as vinyl chloride) it is unlikely that a nonretarding solvent could be found. Even if this proved possible, the reaction may be subject to medium influences so that the values of ω_0 would not be applicable to the retarding solvent system of interest. Atkinson, Bamford and Eastmond[51] carried out an analysis in which the $Z_r^{\cdot} + Z_s^{\cdot}$ termination was omitted, arguing that this type of mutual radical destruction is unlikely to be significant except in special circumstances. This matter will be discussed later. The original scheme was conceived in terms of a retarding solvent rather than an additive, and for convenience it is reproduced below (Scheme 4). The retarding solvent is denoted by S; the S· + S· termination omitted in the treatment is included as reaction (i), Scheme 4. The nomenclature here follows that used in the earlier Scheme 3 with Z replaced by S throughout.

$$I \longrightarrow R_0^{\cdot} \qquad \text{(a)}$$

$$R_0^{\cdot} + M \xrightarrow{k_i} R_1^{\cdot} \qquad \text{(b)}$$

$$R_1^{\cdot} + M \xrightarrow{k_p} R_2^{\cdot} \qquad \text{(c)}$$

$$R_r^{\cdot} + M \xrightarrow{k_t} R_{r+1}^{\cdot} \qquad \text{(d)}$$

$$R_r^{\cdot} + S \xrightarrow{k_s} (P_r +)\,S\cdot \qquad \text{(e)}$$

$$S\cdot + M \xrightarrow{k_{ps}} R_1 \qquad \text{(f)}$$

$$R_r^{\cdot} + R_s^{\cdot} \xrightarrow{k_t} P_r + P_s \text{ or } P_{r+s} \qquad \text{(g)}$$

$$R_r^{\cdot} + S\cdot \xrightarrow{k_{trs}} P_r \qquad \text{(h)}$$

$$S\cdot + S\cdot \xrightarrow{k_{tss}} S_2 \qquad \text{(i)}$$

Scheme 4

In a stationary state we have (omitting reaction i)

$$(k_{ps}[M] + k_{trs}[R\cdot])(\mathscr{I} - k_t[R\cdot]^2) - 2k_{trs}k_s[S][R\cdot]^2 = 0 \qquad (56)$$

Unless $k_{ps} = 0$, it is always possible in principle to work at values of $\mathscr{I}$, and consequently of $[R\cdot]$, which are so low that

$$k_{trs}[R\cdot] \ll k_{ps}[M] \qquad (57)$$

Under these conditions the rate of destruction of S· by reaction (h), second order in radical concentration, is negligible compared to the rate of reaction (f), which is first order in radical concentration. We then obtain from equation (56) the following expression for the rate of polymerization provided the chains are not too short

$$\omega = k_p[M]\left(1 + 2abc\frac{[S]}{[M]}\right)^{-1/2}\left(\frac{\mathscr{I}}{k_t}\right)^{1/2} \tag{58}$$

where

$$\begin{aligned}a &= k_p[M]/k_t^{1/2}\\ b &= k_{trs}/(k_{ps}k_t^{1/2}[M])\\ c &= k_s/k_p\end{aligned} \tag{59}$$

Thus, for sufficiently low $\mathscr{I}$, ω is proportional to $\mathscr{I}^{1/2}$ as in a classical polymerization, but unlike the latter, the reaction shows an order in $[M] > 1$.

On the other hand, at sufficiently high $[R\cdot]$,

$$k_{trs}[R\cdot] \gg k_{ps}[M], \quad k_t[R\cdot] \gg k_s[S] \tag{60}$$

and the relation becomes the simple classical expression

$$\omega = k_p[M](\mathscr{I}/k_t)^{1/2} \tag{61}$$

The general expression for ω is the cubic

$$\frac{b}{a^3}\omega^3 + \left(1 + 2abc\frac{[S]}{[M]}\right)\frac{\omega^2}{a^2} - \frac{b\mathscr{I}}{a}\omega - \mathscr{I} = 0 \tag{62}$$

The plot of $\ln \omega$ versus $\ln \mathscr{I}$ is sigmoid; as already indicated, for extreme values of $\mathscr{I}$ the plot is linear of slope 0.5. Some properties of the curve are described in the original paper.

If no chain transfer to monomer occurs, the number average degree of polymerization is

$$\bar{P}_n = \omega\bigg/\frac{d[P]}{dt} = \left\{c\frac{[S]}{[M]}\left(1 + \frac{b\omega}{a+b}\right) + \frac{1}{2}\frac{\omega}{a^2}\right\}^{-1} \tag{63}$$

for termination by radical combination. If nonretarding chain transfer to monomer participates, the monomer transfer constant C_m should be included within the braces in equation (63). An alternative form, indicating the relation to the familiar Mayo equation is

$$\frac{1}{\bar{P}_n} = \frac{\mathscr{I}}{2\omega} + C_m + c\frac{[S]}{[M]} \tag{64}$$

Thus according to the mechanism under discussion the Mayo relation will not hold strictly unless $\mathscr{I}/\omega$ is kept constant while $[S]/[M]$ is varied. This corresponds to the classical condition that $1/P_n^0$ should be kept constant. If monomer is the only retarder present, the appropriate relations may be derived from the equation given above by substituting $S = M$.

Atkinson, Bamford and Eastmond[51] studied the polymerization of vinyl chloride in chlorobenzene, tetrahydrofuran and 1,2-dichloroethane solutions at 25 °C, using rates of initiation in the range 4×10^{-11} to 2×10^{-6} mol dm^{-3} s^{-1}. Photoinitiation by azobisisobutyronitrile, $(Mn_2(CO)_{10} + CCl_3CO_2Et)$, $(Mn_2(CO)_{10} + CCl_4)$ and thermal initiation by azobisisobutyronitrile were employed. Rates of polymerization were measured dilatometrically, using a 'dummy-dilatometer' technique.[52] Experimental data agreed well with the mechanism proposed; the $\log \omega$ versus $\log \mathscr{I}$ plot for the chlorobenzene system is presented in Figure 5. The sigmoid shape of the plot is clear, and the agreement between experimental data and calculation is excellent except for an anomaly occurring with azobisisobutyronitrile photoinitiation in the higher range of $\mathscr{I}$. We return to this point below. Calculated and observed values of $\bar{P}_n$ for polymerizations in chlorobenzene solution with photoinitiation by $Mn_2(CO)_{10} + CCl_3CO_2Et$ also showed good agrement with equation (63). However, an anomaly is present with photoinitiation by azobisisobutyronitrile, observed $\bar{P}_n$ values being smaller than those calculated. In the two cases mentioned azobisisobutyronitrile initiation gives simultaneously lower rates and lower degrees of polymerization than $Mn_2(CO)_{10}$ over the range of $\mathscr{I}$ we are considering. This behaviour is characteristic of primary radical termination, i.e. termination of a growing chain by a radical derived from the initiator.

The rates of polymerization in chlorobenzene (photoinitiation by $Mn_2(CO)_{10} + CCl_3CO_2Et$) over a range of monomer concentrations agreed well with those calculated from equation (62); the order of ω in $[M]$ was found to increase from 1.1 to 1.8 as $[M]$ increases from 1.44 to 7.20 mol dm^{-3}. A similar observation was made by Crosato-Arnaldi, Talamini and Vidotto.[53] Values of the kinetic

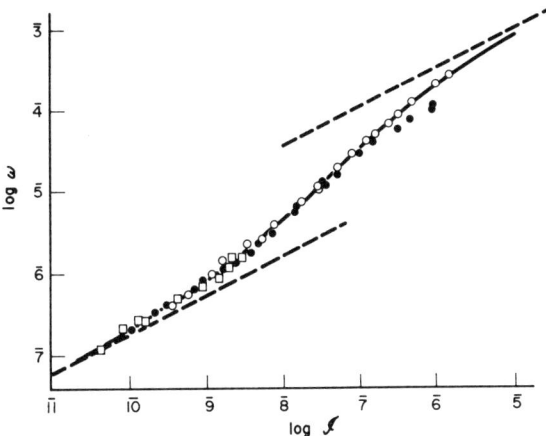

Figure 5 Polymerization of vinyl chloride in chlorobenzene at 25 °C; dependence of rate of polymerization on rate of initiation.[51] Initial concentrations (mol dm^{-3}): monomer 4.32; chlorobenzene 6.85. ○ represents $Mn_2(CO)_{10}$ photoinitiation at $\lambda = 436$ nm; ● azobisisobutyronitrile photoinitiation at $\lambda = 365$ nm; □ azobisisobutyronitrile thermal initiation. The curve was calculated from equations (59) and (62) with the parameters $k_p/k_t^{1/2} = 0.094$ mol$^{-1/2}$ dm$^{3/2}$ s$^{-1/2}$; $k_{trs}/k_{ps}k_t^{1/2} = 1.90 \times 10^6$ mol$^{1/2}$ dm$^{-3/2}$ s$^{1/2}$; and $k_s/k_p = 7.4 \times 10^{-4}$ (reproduced by permission of the Royal Society of Chemistry from ref. 51)

parameters $k_p/k_t^{1/2}$, $k_{trs}/k_{ps}k_t^{1/2}$ and k_s/k_p at 25 °C for the three solvents showed that the nonideality decreased in the order chlorobenzene, tetrahydrofuran, 1,2-dichloroethane. In the latter solvent at 25 °C kinetics were close to ideal, although the order in [M] appeared slightly greater than unity. Crosato-Arnaldi, Talamini and Vidotto[53] reported that in 1,2-dichloroethane at 50 °C with initiation by lauroyl peroxide the rate is proportional to $[M].\mathscr{I}^{1/2}$. The near-ideality of the system enabled Atkinson et al.[51] to determine the transfer constant of monomer; this was found to be low, 3.2×10^{-4}. Other values in the literature are 4.5×10^{-4} (50 °C,[54] 6.4–13.5 (50 °C)[55] and 3.2×10^{-4} (25 °C).[56]

The validity of the geometric mean approximation $\phi = 1$ was examined; expressions for ω and $\bar{P}_n$ based on this approximation were found to be incompatible with experimental data. By using ^{14}C-labelled chlorobenzene which had been rigorously purified from polymerizable impurities Atkinson et al.[51] were able to show that retardation with chlorobenzene results from an abstraction rather than an addition (copolymerization) process.

A study of the polymerization of vinyl chloride in chlorobenzene at temperatures between 30 and 45 °C has been reported by Park and Smith.[57,58] These authors concluded that the rate is proportional to $[M]^{1.23}\mathscr{I}^{0.56}$. The observed orders are ascribed to degradative reactions with the solvent, probably copolymerization rather than transfer. A high value, 1.9×10^{-3}, was obtained for transfer to monomer. These conclusions were contested in detail by Atkinson et al.[51]

1.5.1 Degradation Transfer *versus* Degradative Addition

Although, as we have seen, the formal kinetics of polymerization mechanisms which differ only in the nature of the retardation process (*i.e.* whether abstraction or addition) are identical, similar approximations may not be appropriate in the two cases. Bamford and Schofield[59] have discussed conditions under which different types of simplification are valid. *A posteriori* examination of the justification for omission of the S·/S· termination in Scheme 4 led to the conclusion that the ratio [S·]/[R·] is inversely proportional to $\lambda = k_{trs}/k_t$ (approximately). Two cases are of interest: (i) if $\lambda \sim 1$, [S·] is not necessarily negligible, and omission of the S·/S· termination is not justifiable unless S· is so unreactive that k_{tss} is extremely small; and (ii) on the other hand, if $\lambda \gg 1$, [S·] is relatively small and normally the S·/S· reaction may be omitted. These interradical reactions are likely to be diffusion controlled. The radical product of an abstraction is normally a small molecule (by polymer standards!) while that of an addition is a polymer in the systems we are considering. Abstraction would therefore be expected to correspond to $k_{trs} \gg k_t$ (case ii), since a small molecule is involved only in the R·/S· reaction. This is the situation with the vinyl chloride polymerization in chlorobenzene or tetrahydrofuran, so that neglect of the S·/S· termination is justified. Addition is more likely to lead to case (i) since both reactants in the R·/R· and R·/S· interactions are polymeric and of similar structure. For retardation by addition it is therefore appropriate to assume $k_t = k_{trs} = k_{tss}$.

Application of this assumption to Scheme 4 leads to equation (65) for the rate of polymerization[59]

$$\omega = k_p[M]\left(\frac{\mathscr{I}}{k_t}\right)^{1/2} \frac{k_{ps}[M] + (\mathscr{I}k_t)^{1/2}}{k_{ps}[M] + k_s[S] + (\mathscr{I}k_t)^{1/2}} \qquad (65)$$

As before, the retarding and reinitiation reactions are (e) and (f) of Scheme 4, respectively. This nomenclature is slightly different from that in the original paper, in which k_f, k_{pm}, are used instead of k_s, k_{ps}, respectively. Suitable rearrangements of equation (65) enable linear plots to be constructed from which the kinetic parameters $k_p/k_t^{1/2}$, $k_s/k_t^{1/2}$ and $k_{ps}/k_t^{1/2}$ may be evaluated. If monomer, rather than solvent, is the retarder, [S] in equation (65) must be replaced by [M] and, if desired, k_{ps} by k_{pm}, etc.

Bamford and Schofield[59] applied relation (65) to the polymerization of 1-vinylimidazole (VIM; **5**) initiated by azobisisobutyronitrile in ethanol at 70 °C. These workers presented evidence for the occurrence of degradative addition to monomer in polymerization. The order of ω in [M] decreases with increasing [M] from unity at low [M] and tending to zero at high [M], consistent with retardation by monomer. Retardation arises from attack of radicals on the 2-position of the monomer; if this is blocked by methyl substitution, retardation disappears.[60] Thus the rate of polymerization of 2-methyl-1-vinylimidazole in ethanol at 70 °C is directly proportional to the monomer concentration, and the rates are higher than with the unsubstituted monomer, especially at high [M]. Other observations support degradative addition, notably the existence of radical occlusion phenomena and the pH dependence of the reaction in aqueous solution.[60] The most convincing evidence is the absence of a primary isotope effect. Such an effect would be expected if the degradative step (reaction e; Scheme 4) were an abstraction and would result in $k_m(H) > k_m(D)$, where $k_m(H)$, $k_m(D)$ refer to reactions of VIM and the 2-deutero derivative (DVIM), respectively. The polymerization of DVIM would therefore be subject to less retardation and would proceed more rapidly. Experimentally[61] it was found that DVIM polymerizes rather more slowly. Estimates of the rates expected on the basis of C—H bond breaking (taking $k_m(D)/k_m(H) \sim 7$) show greatly enhanced rates for DVIM which are not compatible with experimental data. Possible reasons for the slightly lower rates found with DVIM have been discussed.[61] Thus the degradative process is apparently the addition shown in equation (66), which leads to a stabilized radical product.

$$R_r^\bullet + \underset{(5)}{\text{imidazole}} \xrightarrow{k_m} \underbrace{R_r\text{-adduct} \longleftrightarrow R_r\text{-adduct}^\bullet}_{S\bullet} \qquad (66)$$

Retarded polymerization of the types we have been discussing, incorporating either abstraction or addition steps, lead to kinetic schemes containing three adjustable parameters. Observations of conventional polymerization kinetics are therefore unlikely to provide a firm basis for distinguishing between abstraction and addition; these generally require supplementation by confirmatory evidence such as we have discussed in the cases of vinyl chloride and 1-vinylimidazole.

1.6 POLYMERIZATIONS WITH SIZE DEPENDENT TERMINATION

It is generally believed that the bimolecular termination coefficient in a free-radical polymerization is sensibly constant for large radicals, but shows a significant size dependence for smaller radicals. This is a manifestation of diffusion control, which is a feature of radical–radical interactions in general (unless the species are highly stabilized), including the termination reaction in polymerization.[3,62–66] Such behaviour introduces departures from classical kinetics, especially if small radicals are involved. Mean radical sizes are reduced by the presence of a retarder so that size dependent termination is likely to become significant; nevertheless it does not necessarily play a major role since termination by the retarder may predominate.

Bamford has discussed the kinetics of retarded polymerization using the group-termination coefficient procedure.[67–69] The familiar 'geometric mean' assumption[70,71] was made, that the termination coefficient is proportional to the product of the sizes (r, s) of the participating radicals raised to a small negative power

$$k_t \propto r^{-\beta} s^{-\beta} \qquad (67)$$

The treatment and general results are too lengthy to reproduce here, but may be found in the literature.[67-69] Explicit expressions were developed for $\omega, \bar{v}, \bar{r}$ (the mean radical size) and the molecular weight distributions and related quantities.

As an example of an ideal retarded polymerization (Scheme 2) the polymerization of methacrylonitrile at 60 °C in the presence of iron(III) chloride was considered.[68] Good agreement was obtained between the calculated and observed[31] plots of ω versus $1/[\text{FeCl}_3]$. β was taken equal to 0.1. The parameters evaluated were $k_p/k_{to}^{1/2} = 5.28 \times 10^{-3} \text{mol}^{-1/2} \text{dm}^{3/2} \text{s}^{-1/2}$ and $k_z/k_p = 2.77$ (k_{to} is the hypothetical termination coefficient for radicals of unit size); these may be compared with values from the classical treatment: $k_p/k_t^{1/2} = 6.97 \times 10^{-3} \text{mol}^{-1/2} \text{dm}^{3/2} \text{s}^{-1/2}$, $k_z/k_p = 3.08$.

A nonideal and more complex example is provided by acenaphthylene (**6**). The free-radical polymerization of this monomer in toluene solution was studied by Romani and Weale,[72] who recognized its nonclassical behaviour and proposed degradative transfer to monomer. Formation of high polymers takes place through addition at the 1,2-double bond as shown in equation (68a). This process is favoured by the relief of steric strain in the five-membered ring. We have already seen[9] that active radicals can add to aromatic derivatives which thus function as retarders. Acenaphthylene apparently behaves in this fashion, so that the polymerization is self-retarding.[73] The degradative addition of a propagating radical $R_r^{\bullet}$ to an acenaphthylene molecule shown in equation (68b) leads to the formation of a relatively unreactive S·-type radical. Molecular orbital calculations[73] reveal that radical addition is favoured over hydrogen abstraction and indicate several sites of comparable reactivity towards an attacking radical. They also confirm that S·-type radicals are relatively inactive in monomer addition.

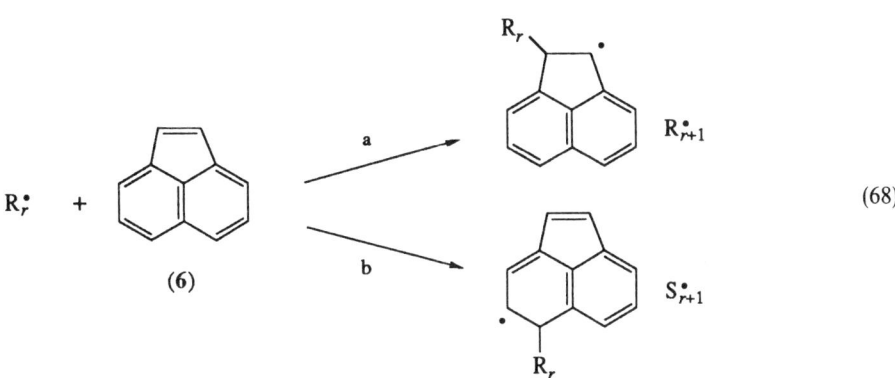

(68)

Bamford, Schofield and Eilers[73] introduced a simple modification into equation (65) to allow for the kinetic effects of size dependent termination. Rates of the three termination reactions (g, h, i) in Scheme 4 were taken as

$$k_{to}\left(\frac{\bar{r}}{r_0}\right)^{-2\beta}[R\cdot]^2, \quad k_{to}\left(\frac{\bar{r}}{r_0}\right)^{-\beta}\left(\frac{\bar{s}}{r_0}\right)^{-\beta}[R\cdot][S\cdot], \quad k_{to}\left(\frac{\bar{s}}{r_0}\right)^{-2\beta}[S\cdot]^2,$$

respectively. Here k_{to} is the rate coefficient for radicals of arbitrarily chosen size r_0 (taken as 50) and β was assumed to equal 0.16; $\bar{r}$ and $\bar{s}$ are mean sizes of the two types of radical and $[R\cdot] = \sum[R_r^{\bullet}]$, $[S\cdot] = \sum[S_r^{\bullet}]$. A further assumption was that $\bar{r} = \bar{s}$. These modifications were incorporated into the kinetic scheme with the aid of a function $F(r)$

$$F(r) = (\bar{r}/r_0)^{-\beta} \qquad (69)$$

which was evaluated from the following expression

$$\bar{r} = \frac{\omega}{\mathscr{I} + \omega C_X[X]/[M]}\left[(1-\beta)\frac{\Gamma 2/(1-\beta)}{\{\Gamma 1/(1-\beta)\}^2}\right] \qquad (70)$$

X is a nondegradative transfer agent with transfer constant C_X; in the present system it is the toluene solvent. Equation (70) is a simplified version of a more complex relation[67] of general validity. The basic kinetic equation (65) thus modified becomes (for S = M)

$$\omega = k_p F[M] \left(\frac{\mathscr{I}}{k_{to}}\right)^{1/2} \frac{k_{ps}F[M] + (\mathscr{I}k_{to})^{1/2}}{(k_{ps} + k_s)F[M] + (\mathscr{I}k_{to})^{1/2}} \tag{71}$$

which clearly reduces to the classical form (65) if $\beta = 0$, $F = 1$.

Equation (71) and the unmodified relation (65) appear to fit the experimental data equally well, probably within experimental error.[73] However, the evaluated kinetic parameters are different in the two cases, the main difference lying in the importance of reinitiation (reaction f; Scheme 4), which is virtually negligible in the classical case with k_t = constant.

A more general and detailed treatment, without the assumption $\bar{r} = \bar{s}$ (which does not hold exactly) has been given.[67-69] The general results have been presented in various forms, and allow estimation of [R·], [S·], $\bar{r}$, $\bar{s}$, ω, $\bar{v}$. The kinetic parameters deduced for the acenaphthylene system show the same pattern as those obtained by the approximate treatment.

1.7 REFERENCES

1. P. J. Flory, 'Principles of Polymer Chemistry', Cornell University Press, Ithaca, NY, 1953.
2. C. H. Bamford, W. G. Barb, A. D. Jenkins and P. F. Onyon, 'The Kinetics of Vinyl Polymerization by Radical Mechanisms', Butterworths, London, 1958.
3. G. C. Eastmond, in 'Comprehensive Chemical Kinetics', ed. C. H. Bamford and C. F. H. Tipper, Elsevier, Amsterdam, 1976, vol. 14A, chap. 2.
4. P. D. Bartlett and R. Altschul, *J. Am. Chem. Soc.*, 1945, **67**, 812, 816.
5. P. D. Bartlett and F. A. Tate, *J. Am. Chem. Soc.*, 1953, **75**, 91.
6. R. Hart and G. Smets, *J. Polym. Sci.*, 1950, **5**, 55.
7. J. C. Bevington, N. A. Ghanem and H. W. Melville, *J. Chem. Soc.*, 1957, **239**, 214.
8. J. C. Bevington, N. A. Ghanem and H. W. Melville, *Trans. Faraday Soc.*, 1955, **51**, 946.
9. Kh. S. Bagdasar'ian and Z. A. Sinitsina, *J. Polym. Sci.*, 1961, **52**, 31.
10. C. C. Price, *J. Am. Chem. Soc.*, 1943, **65**, 2380.
11. C. C. Price and D. A. Durham, *J. Am. Chem. Soc.*, 1943, **65**, 757.
12. C. C. Price and D. H. Read, *J. Polym. Sci.*, 1946, **1**, 44.
13. P. D. Bartlett and H. Kwart, *J. Am. Chem. Soc.*, 1950, **72**, 1051.
14. G. S. Hammond and P. D. Bartlett, *J. Polym. Sci.*, 1950, **6**, 617.
15. P. D. Bartlett and H. Kwart, *J. Am. Chem. Soc.*, 1952, **74**, 3969.
16. N. Inamoto and O. Simamura, *J. Org. Chem.*, 1958, **23**, 408.
17. F. Tüdös, *J. Polym. Sci.*, 1958, **30**, 343.
18. J. C. Bevington and N. A. Ghanem, *J. Chem. Soc.*, 1959, 2071.
19. J. C. Bevington, N. A. Ghanem and H. W. Melville, *J. Chem. Soc.*, 1955, 2822.
20. R. M. Noyes, *J. Am. Chem. Soc.*, 1955, **77**, 2042.
21. R. M. Noyes, *Z. Elektrochem.*, 1960, **64**, 153.
22. R. M. Noyes, in 'Progress in Reaction Kinetics', ed. G. Porter, Pergamon Press, Oxford, 1961, chap. 5.
23. R. M. Noyes, in 'Encyclopedia of Polymer Science and Technology', Interscience, New York, 1965, vol. 2, p. 796.
24. W. I. Bengough, *Chem. Ind. (London)*, 1955, 599.
25. M. S. Matheson, E. E. Auer, E. B. Bevilacqua and E. J. Hart, *J. Am. Chem. Soc.*, 1951, **73**, 1700.
26. G. S. Hammond, J. N. Sen and C. E. Boozer, *J. Am. Chem. Soc.*, 1955, **77**, 3244.
27. G. M. Burnett and P. R. E. J. Cowley, *Trans. Faraday Soc.*, 1953, **49**, 1490.
28. J. C. Bevington, *J. Chem. Soc.*, 1956, 1127.
29. A. L. Buchachenko, 'Stable Radicals', translated from the Russian by C. N. Turton and T. I. Turton, Consultants Bureau, New York, 1965.
30. M. Arnold and M. Raetzsch, *Acta Polym.*, 1980, **31**, 78.
31. C. H. Bamford, A. D. Jenkins and R. Johnston, *Proc. R. Soc. London*, 1957, **A239**, 214.
32. C. H. Bamford, A. D. Jenkins and R. Johnston, *Trans. Faraday Soc.*, 1962, **58**, 1212.
33. E. Collinson, F. S. Dainton and G. S. McNaughton, *J. Chim. Phys.*, 1955, **52**, 556.
34. E. R. Entwistle, *Trans. Faraday Soc.*, 1960, **56**, 284.
35. C. H. Bamford, in 'Encyclopedia of Polymer Science and Engineering', 2nd edn., ed. H. F. Mark, N. M. Bikales, C. G. Overberger, and G. Menges, Wiley, New York, 1988, p. 817.
36. W. I. Bengough and W. H. Fairservice, *Trans. Faraday Soc.*, 1965, **61**, 1206.
37. W. I. Bengough and W. H. Fairservice, *Trans. Faraday Soc.*, 1967, **63**, 382.
38. W. I. Bengough and W. H. Fairservice, *Trans. Faraday Soc.*, 1971, **67**, 414.
39. W. I. Bengough and I. C. Ross, *Trans. Faraday Soc.*, 1966, **62**, 2251.
40. W. I. Bengough, T. O'Neill, J. H. R. Clarke and L. A. Woodward, *Trans. Faraday Soc.*, 1968, **64**, 1014.
41. W. I. Bengough and T. O'Neill, *Trans. Faraday Soc.*, 1968, **64**, 2415.
42. W. I. Bengough and N. M. Chawdry, *J. Chem. Soc., Faraday Trans. 1*, 1972, **68**, 1807.
43. N. N. Dass and M. H. George, *J. Polym. Sci., Part A-1*, 1969, **7**, 269.
44. C. H. Bamford and C. A. Finch, *Trans. Faraday Soc.*, 1963, **59**, 118.
45. P. J. Flory, 'Principles of Polymer Chemistry', Cornell University Press, Ithaca, NY, 1953, p. 169.
46. J. L. Kice, *J. Am. Chem. Soc.*, 1954, **76**, 6274.
47. J. L. Kice, *J. Polym. Sci.*, 1956, **19**, 123.
48. P. W. Allen, F. M. Merrett and J. Scanlan, *Trans. Faraday Soc.*, 1955, **51**, 95.

49. A. D. Jenkins, *Trans. Faraday Soc.*, 1958, **54**, 1885.
50. A. D. Jenkins, *Trans. Faraday Soc.*, 1958, **54**, 1895.
51. W. H. Atkinson, C. H. Bamford and G. C. Eastmond, *Trans. Faraday Soc.*, 1970, **66**, 1446.
52. P. J. Flory and R. R. Garett, *J. Am. Chem. Soc.*, 1958, **80**, 4836.
53. A. Crosato-Arnaldi, G. Talamini and G. Vidotto, *Makromol. Chem.*, 1968, **111**, 123.
54. G. Vidotto, A. Crosato-Arnaldi and G. Talamini, *Makromol. Chem.*, 1968, **114**, 217.
55. J. Brandrup and E. H. Immergut (eds.), 'Polymer Handbook', Interscience, New York, 1966, vol. II.
56. M. Ryska, M. Kolinsky and D. Lim, *J. Polym. Sci., Part C*, 1967, **16**, 621.
57. G. S. Park and D. G. Smith, *Trans. Faraday Soc.*, 1969, **65**, 1854.
58. G. S. Park and D. G. Smith, *Makromol. Chem.*, 170, **131**, 1.
59. C. H. Bamford and E. Schofield, *Polymer*, 1983, **24**, 433.
60. C. H. Bamford and E. Schofield, *Polymer*, 1981, **22**, 1227.
61. C. H. Bamford and E. Schofield, *Polym. Commun.*, 1983, **24**, 4.
62. S. W. Benson and A. M. North, *J. Am. Chem. Soc.*, 1959, **81**, 1339.
63. S. W. Benson and A. M. North, *J. Am. Chem. Soc.*, 1962, **84**, 935.
64. A. M. North, 'Diffusion-Control of Homogeneous Free Radical Reactions', in 'Progress in High Polymers', ed. J. C. Robb and F. W. Peaker, Heywood, London, 1968, vol. 2.
65. A. M. North and G. A. Reed, *Trans. Faraday Soc.*, 1961, **57**, 859.
66. Ref. 35; p. 828 and refs. therein.
67. C. H. Bamford, *Eur. Polym. J.*, 1989, **25**, 683.
68. C. H. Bamford, *Polymer*, 1990, **31**, 1720.
69. C. H. Bamford, *Eur. Polym. J.*, 1990, **26**, 1245.
70. T. Yasukawa, T. Takahashi and K. Murakami, *J. Chem. Phys.*, 1973, **59**, 3937.
71. T. Yasukawa, T. Takahashi and K. Murakami, *Makromol. Chem.*, 1973, **174**, 235.
72. M. N. Romani and K. E. Weale, *Trans. Faraday Soc.*, 1966, **62**, 2264.
73. C. H. Bamford, E. Schofield and J. M. Eilers, *Polymer*, 1989, **30**, 540.

2
Peculiarities of Radical Polymerization Conducted in the Presence of Nontraditional Initiators

SEMION I. KUCHANOV
Moscow State University, Russia

2.1	GENERAL INTRODUCTION	23
2.2	POLYFUNCTIONAL INITIATORS	24
	2.2.1 Introduction	24
	2.2.2 Mechanism of Polymerization	24
	2.2.3 Superposition Principle	26
	2.2.4 Polyinitiators with One Type of Labile Group	28
	2.2.5 Polyinitiators with Several Types of Labile Group	30
	2.2.6 Synthesis of Block Copolymers	32
2.3	INIFERTERS	33
	2.3.1 Introduction	33
	2.3.2 Polymerization Mechanism and Kinetics	34
	2.3.3 Molecular Weight Distribution and its Characteristics	37
2.4	CONCLUSION	39
2.5	REFERENCES	39

2.1 GENERAL INTRODUCTION

The common feature of the majority of radical polymerization processes is the reduction of the molecular weight when the concentration of initiator increases. This property evidently imposes essential restrictions on the increase in the efficiency of production of the corresponding polymer materials. The explanation is that the length of a polymer chain in the absence of transfer agent equals either the length of a kinetic chain or its doubled value, according to whether the reaction is carried out according to the mechanism of combination or disproportionation, respectively. Therefore the increase in the rate of polymerization due to a rise in the rate of initiation will inevitably reduce the length of the kinetic chain in ordinary chain polymerization processes and will consequently decrease the molecular weight.

In order to avoid such an undesirable effect, it is advisable to apply nontraditional initiators of radical polymerization when the polymer chain comprises several elementary chains, the number of which continuously increases in the course of the process. As the molecular weight of polymer rises with the conversion, its molecular weight distribution simultaneously narrows. These peculiarities essentially differ from those which are usually observed for ordinary processes.

Some theoretical aspects of the quantitative description of such pseudo-living radical polymerization, initiated by two kinds of nontraditional compounds, are to be discussed in this chapter. The first kind includes polyfunctional initiators (*e.g.* polyperoxides), each molecule of which contains more than one labile (peroxide) group. The so-called 'iniferters' participating, in addition to the initiation, in the transfer and termination reactions belong to the second kind. Both kinds of nontraditional initiator are characterized by the formation during the polymerization process of macromolecules capable of subsequent reactivation and further growth of the polymer chain.

Some reference to these multifunctional initiators and iniferters can be found within *Comprehensive Polymer Science* (Volume 3, Chapters 8 and 10, respectively). However, in these chapters the authors focus mainly on the description of a variety of chemically different kinds of such compounds, so kinetics and mechanism of radical polymerization with the participation of nontraditional initiators are barely considered. An attempt to make up this deficiency on the basis of contemporary theoretical conceptions and experimental data is undertaken within this chapter.

2.2 POLYFUNCTIONAL INITIATORS

2.2.1 Introduction

The very appellation of such initiators implies the presence of more than one functional group which, after decomposition, forms free radicals. These groups can differ from each other by the chemical nature of the labile interatomic bond, for instance azo and peroxide groups. Nowadays, a considerable number of chemically distinct types of polyfunctional initiator are known, and the data concerning their structure and methods of synthesis are presented in many reviews.[1-6] The most investigated among such initiators, due to a great extent to the fundamental research performed by Ivanchev and coworkers,[7-16] are polyperoxides. For this reason, they have been chosen here to illustrate some general theoretical conclusions.

The molecules of polyfunctional initiators differ in their number of labile groups of various types. The simplest such initiator is diacyl diperoxide (**1**), which contains two labile groups. When the methylene bridge dividing them is long enough (k exceeds 6–8 units), both groups will decompose independently from each other.[12,15] All further consideration will concern only polyinitiators with such kinetically independent labile groups.

(**1**)

Peroxide groups in diperoxide (**1**), generally speaking, differ in thermostability since one of them contains an unsubstituted acyl bond, while the other contains a substituted one. The differences in the rate constants of decomposition of these groups can be noticeable, depending on the nature of substituent R. The kinetic parameters of thermolysis of a considerable number of polyperoxides have been determined by Ivanchev and coworkers.[9,10,12,15] They also found the dependence of these parameters on the length, k, of the bridge, the nature of substituent R and the number of peroxide groups in the molecule. Some data on the values of the mentioned parameters of decomposition of polyfunctional initiators are also given by other researchers.[17-20]

It has been ascertained[12,18] that polyperoxides with independent peroxide groups are more suitable for practical usage. The rate of polymerization initiated by such compounds will obviously coincide (where the Flory principle applies[21]) with the rate of polymerization influenced by traditional monofunctional initiators, which contain the same peroxide groups at the same overall concentration.[22,23] Thus, the aim of a quantitative theory of homopolymerization with polyfunctional initiator participation is to calculate molecular weight distribution (MWD) and its statistical moments; for copolymerization the above theory is also supposed to calculate composition distribution of obtained products.

2.2.2 Mechanism of Polymerization

The main peculiarity of polymerization in the presence of polyfunctional initiators is the formation in the course of the process of macromolecules containing labile groups in the main chain (inside and/or at its ends). The mechanism of formation of such macromolecules is schematically depicted in Figure 1 for the simplest case of synthesis of polymers, where termination occurs solely by combination (*e.g.* polystyrene) under the action of diperoxide with identical peroxide groups. On decomposition, each of them yields a pair of free radicals, which during the growth process add a

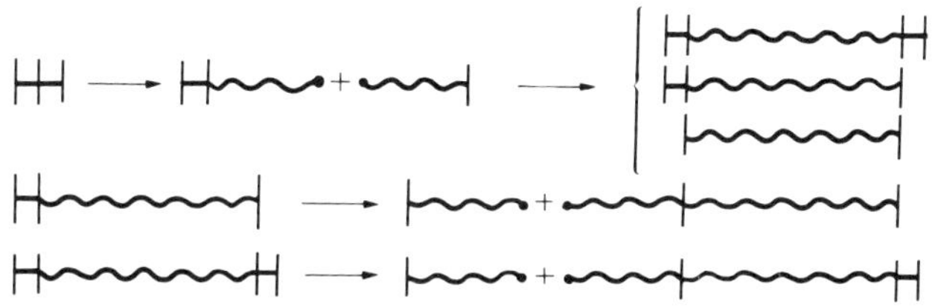

Figure 1 Scheme of polymer chain formation for polymerization initiated by a symmetric diperoxide

Figure 2 Schematic representation of a polymer chain formed in the course of the process depicted in Figure 1

number of monomers and then recombine with other analogous macroradicals, forming the 'dead' polymer macromolecules. These macromolecules contain one elementary polymer chain and can be distinguished from each other by the number of terminal peroxide groups, which on decomposition yield a pair of polymer radicals capable of further growth and recombination.

The set of such parallel consecutive reactions[1,23] exhaustively characterizes the chain polyrecombination mechanism of formation of macromolecules in the presence of the polyfunctional initiators under consideration. Naturally, if molecules of such initiators comprise more than two labile groups, molecules of macroinitiator formed during polymerization can contain these groups not only at the end of, but also inside the main chain. The structures of such macromolecules appearing in the course of successive decompositions of polyperoxide groups and recombination of macroradicals will be rather complicated in character (see Figure 2). An arbitrary polymer chain can consist of any number m of elementary chains divided by peroxide clusters which contain one, two, three, *etc.* peroxide groups. The term 'cluster' refers here to a sequence of such groups restricted either by two elementary chains or by one elementary chain and the end of a macromolecule. In the first case the cluster is termed 'closed', while in the second one it is 'semiopened'.

For polyinitiators with independent functional groups, the thermostability of each group does not depend on the cluster within which it is included or on its location inside this cluster. Thus the decomposition reaction constant of any such peroxide group will coincide with the analogous rate constant which charaterizes such a group in the initial molecule of polyinitiator. Cited examples of polyperoxides are well known.[1-3,5] The rate of polymerization, W, initiated by such compounds does not obviously depend on their functionality when the total concentration of peroxide groups, C, is the same. The quantity W can be described by chemical equations

$$W = k_p M (I/k_t)^{1/2}, \quad I = 2fk_d C \qquad (1)$$

usually applied in the framework of traditional theory of radical polymerization in the presence of monofunctional initiators.[21] Hereafter the following designations are used: I and f represent rate and efficiency of initiation; M represents monomer concentration; and k_d, k_p and k_t represent rate constants of elementary reactions of peroxide group decomposition, propagation and termination of the chain, respectively. In accordance with the Flory principle, these constants are the same for both mono- and poly-functional initiators. This statement can be convincingly confirmed in the example of peroxides (2)–(4) by experimental data presented in Figure 3.

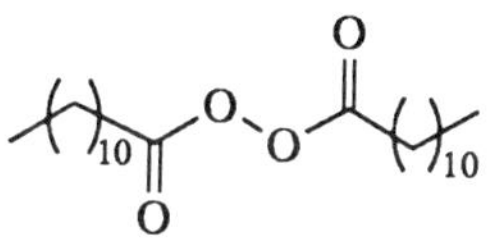

(2)

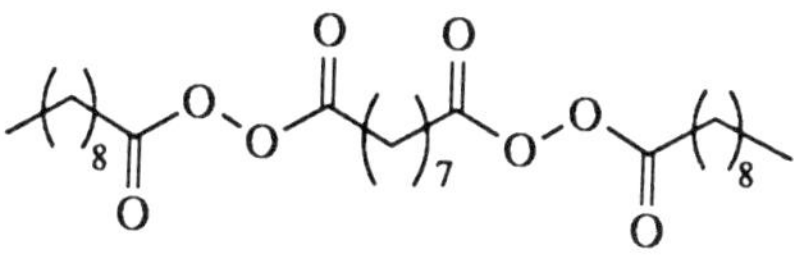

(3)

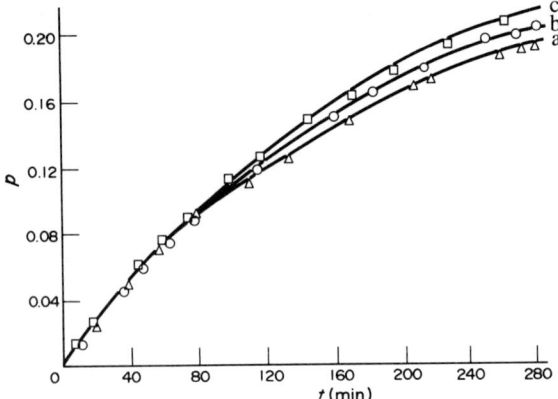

(4)

Figure 3 Kinetic curves of bulk polymerization of styrene initiated by (a) mono-, (b) di- and (c) poly-peroxide containing approximately 18 peroxide groups per molecule[23] ($T = 80\,°C$ and overall concentration of peroxide groups, $C^0 = 2.5 \times 10^{-3}\,mol\,l^{-1}$)

In contrast to kinetics, the calculation of molecular weight characteristics of a polymer cannot be performed within the framework of traditional theory of radical polymerization. So even the simplest formula of this theory[21]

$$P'_N = \frac{2}{1+\lambda}v, \quad v = \frac{W}{I}, \quad \lambda = \frac{k_{td}}{k_t}, \quad k_t = k_{td} + k_{tc} \tag{2}$$

connecting number average degree of polymerization P'_N of macromolecules with the kinetic chain length v as well as with kinetic constants of the elementary reactions of its termination by radical combination (k_{tc}) and disproportionation (k_{td}) turns out to be incorrect. Contrary to equation (2), the average molecular weight starts to rise right from the very beginning of polymerization if it is initiated by polyfunctional compounds.[1–3,5] This result naturally follows from the chain polyrecombination mechanism of macromolecules formation (see Figure 1), since each effectively cloven labile bond corresponds to the formation of one or two new elementary chains inside the polymer. Their number depends on the mechanism (combination or disproportionation) by which the termination reaction of the kinetic chain occurs.

2.2.3 Superposition Principle

To calculate the molecular weight distribution (MWD) of products of polymerization initiated by polyperoxides it is convenient to categorize each macromolecule by the number, m, of its elementary chains. The total number of their monomer units in a given macromolecule will characterize (when the contribution of peroxide clusters is neglected) its molecular weight. So, in order to define the number MWD $f_N(l,t)$ of polymer obtained up to time t, one can use the simple relation

$$f_N(l,t) = \sum_{m=1}^{\infty} P_m(t) W_m(l) \tag{3}$$

where $P_m(t)$ represents the fraction of macromolecules containing m elementary chains at the moment t, and $W_m(l)$ represents the conditional probability that each such macromolecule has the degree of polymerization l.

In accordance with general approaches of statistical macromolecular chemistry one may consider, instead of the MWD, its generating function

$$G_N(s,t) = \sum_{l=1}^{\infty} f_N(l,t)s^l \quad (4)$$

which is completely equivalent to the MWD but seems to be more suitable for the calculation of its statistical moments.[24] These characteristics determining number average, weight average and Z average molecular weight of polymer can be calculated as corresponding derivatives of generating function $G_N(s)$ at $s = 1$. The exact function of the MWD can be found in accordance with equation (4) via expansion of generating function $G_N(s)$ into the power series in s.

For the solution of the problem of MWD calculation one ought to find independently two generating functions

$$U(x,t) = \sum_{m=1}^{\infty} P_m(t)x^m, \quad g(s) = \sum_{l=1}^{\infty} \varphi_l s^l \quad (5)$$

The first of these functions corresponds to the distribution $P_m(t)$ of macromolecules for m elementary chains and the second is a generating function of the distribution of the chains for l monomer units. Then, the generating function of MWD can be obtained as a result of substitution in $U(x,t)$ of function $g(s)$ for argument x. This procedure yields the important relation[23]

$$G_N(s,t) = U(g(s),t) \quad (6)$$

which is the mathematical expression of the Superposition principle for distributions φ_l and P_m.

The former is the distribution of elementary chains for length l. Each such chain for the case $k_{td} = 0$ is a sum of two kinetic chain lengths. Their distribution for length l is described by the Most Probable Flory distribution with generating function

$$g_k(s) = \frac{(1-a)s}{1-as}, \quad a = 1 - \frac{1}{v} \quad (7)$$

determined only by the parameter a, depending merely on mean length v of the kinetic chain (equation 2). It is quite obvious that the distribution φ_l of elementary chains coincides with the MWD of the products of polymerization carried out in the presence of a traditional monofunctional initiator and it can be described by the generating function $g(s) = g_k^2(s)$. Its substitution in equation (6) allows one to calculate the MWD of polymer chains provided the generating function $U(x,t)$ of distribution $P_m(t)$ is known. The above procedure is true, however, only when the value of the kinetic parameter v does not change in the progress of polymerization due to the drift of monomer and initiator concentrations. If such a drift takes place, instead of $g(s)$ its averaged value

$$\langle g(s) \rangle = \int_0^t g(s,t')I(t')\,dt' \Big/ \int_0^t I(t')\,dt' \quad (8)$$

is to be inserted into equation (6). Throughout this chapter angle brackets represent the time-averaging operation of the term located between them.

If the polymerization is performed in the presence of a traditional monofunctional initiator (when each polymer chain consists of a single elementary chain) the equation (8) obviously describes the MWD of the polymerization products in the whole range of conversions and takes into account the consumption of monomer and/or initiator in the course of the process.

Such a consumption can give rise to undesirable conversional broadening of the MWD, quantitatively characterized by an appreciable increase with conversion in the polydispersity coefficient $K = P_W/P_N$. The latter, according to its definition, equals the ratio of the weight average to number average degree of polymerization. Equations (7) and (8) enable us to obtain for these statistical characteristics rather simple formulas

$$P_N' = 2\langle v \rangle, \quad K' = \frac{3\langle v^2 \rangle}{2\langle v \rangle^2} \quad (9)$$

The extent of the conversional broadening of the MWD can be considerably reduced if the polymerization is performed in the presence of polyperoxides comprising a great number of functional groups. This fundamental conclusion follows directly from the general equations which

$$P_N = P_N' P_N'', \quad K = K'' + \frac{K'-1}{P_N''} \quad (10)$$

can be easily deduced from the superposition principle (equation 6) with regard to equation (8). According to equation (10), the molecular weight and polydispersity coefficient of the polymer obtained depend on analogous characteristics of both the distribution for length of elementary chains (equation 9) and distribution P_m of macromolecules for m such chains. The mean value $P''_N = \bar{m}$ of this random magnitude, coinciding at the end of the polymerization with the average number of effectively cloven labile bonds in the molecule of polyperoxide, can be sufficiently large if its functionality is high. In this case the substitution of mono- for poly-functional initiator is accompanied by considerable (P''_N times) growth of average molecular weight, parallel with the narrowing of the MWD. If the value of P''_N is high enough, the second term in the expression for K in equation (10) can be neglected in comparison with the first. Therefore the application of polyperoxide as an initiator allows one to exclude the conversional contribution to inhomogeneity of the MWD. In this case, the latter is defined only by the polydispersity coefficient, K'', of distribution P_m, which, as will be seen from further considerations, is not too large.

The superposition principle can be easily extended to the polymerization of monomers like methyl methacrylate, where the fraction λ (2) of radicals, which lose their activity by disproportionation reaction, has a non-zero value. As a result of the elementary act of termination of the kinetic chain *via* such a mechanism, two interacting macroradicals lose their activity, turning into molecules of macroinitiator. This act is accompanied by the formation of an additional elementary chain at the end of each of them. The length of such a chain, however, in contrast to the case when the kinetic chain terminates *via* combination, will equal one, rather than two kinetic chain lengths. Thus, the formula

$$G_N(s,t) = U(\langle g_k^2(s) \rangle, \langle g_k(s) \rangle; t) \qquad (11)$$

will be an extension of equation (6). Here $U(x, y; t)$ is a generating function of the joint distribution of polymer chains for numbers m and i of elementary chains formed, respectively, as a result of combination and disproportionation of radicals. Proceeding from equation (11) it is easy to obtain a similar expression to equation (10) that contains the magnitudes

$$P'_N = \frac{2\langle v \rangle}{1+\lambda}, \quad K' = \frac{(\lambda+1)(3-\lambda)}{2}\left(\frac{\langle v^2 \rangle}{\langle v \rangle^2}\right) \qquad (12)$$

Naturally, advantages of polyinitiator application diminish with the increase in λ due to the decrease in the fraction of recombination reactions leading to the enlargement of polymer chains.

2.2.4 Polyinitiators with One Type of Labile Group

In accordance with the superposition principle for the calculation of the MWD, it is sufficient to find the generating function $U(x, t)$ of distribution $P_m(t)$ of macromolecules for m elementary chains. To calculate $U(x, t)$ Kuchanov *et al.*[23] suggested a corresponding set of kinetic equations for this distribution and managed to find its exact solution. Proceeding from this the distribution P_m has the generating function

$$U(x,t) = \frac{(1-\rho)x}{1-\rho x}, \quad P_m = (1-\rho)\rho^{m-1} \qquad (13)$$

at any initial distribution α_j of polyperoxide molecules for a number j of peroxide groups. The type of distribution α_j affects only the character of the dependence on time of the sole parameter ρ

$$\rho(t) = 1 - \sum_j \alpha_j [1-(1-\theta)^j] \Big/ \theta \sum_j j\alpha_j \qquad (14)$$

where $\theta = f(1 - e^{-\tau})$ and $\tau = k_d t$. The superposition principle (equations 10) in the case of function (13), with due regard for equation (7), leads to the simple expression for the MWD

$$f_N(l,t) = \frac{(1-\rho)}{2\sqrt{\rho}}(1-a)\{[a+(1-a)\sqrt{\rho}]^{l-1} - [a-(1-a)\sqrt{\rho}]^{l-1}\} \qquad (15)$$

which is obviously unimodal. The application of equations (10), together with equation (13), allows the equations

$$P_N = P'_N(1-\rho)^{-1}, \quad K = (1-\rho)K' + 2\rho \qquad (16)$$

to be obtained, which holds for the whole range of conversions if equations (9) are taken into account.

According to the first expression of equation (16), molecular weight increases in the course of the process from P'_N, obtained when monoperoxides are being employed, up to a certain maximum value P_N^{max} which depends on distribution α_j of peroxide groups in polyperoxide molecules as well as on initiator efficiency f. For polyinitiators of sufficiently high functionality the theory predicts linear dependence of polymer molecular weight on conversion in a wide range of its variations. At the same time a slope of this straight line, in accordance with equations (14) and (16), is proportional to polyinitiator functionality. These theoretical predictions are perfectly confirmed by experimental data given in Figure 4.

The extension of equation (13)

$$U(x, y; t) = \frac{V(x, y; t)}{V(1, 1; t)}, \quad V(x, y; t) = \frac{(1 + dy)^2}{1 - b^2 x} - 1 \qquad (17)$$

where

$$d = \lambda \rho (1 - \rho)^{-1} \quad \text{and} \quad b^2 = (1 - \lambda)\rho$$

taking into account the possibility of chain termination *via* disproportionation, has been obtained by means of a kinetic approach.[23] When kinetic chain length v is high and does not change with time the Superposition principle (equation 11) permits the derivation, using equation (17), of an expression for the MWD of polymer

$$f_N(l, t) = \frac{1 - b^2}{2b(d^2 + b^2 + 2d)} \{(d + b)^2 e^{-1/(1-b)v} - (d - b)^2 e^{-1/(1+b)v}\} \qquad (18)$$

which, as in equation (15), is unimodal.

The following equations are regarded as being extensions of equations (16)

$$P_N = \frac{\lambda + 1}{\lambda + 1 - \rho} P'_N, \quad K = \frac{\lambda + 1 - \rho}{\lambda + 1} K' + \frac{(2 - \lambda)^2 \rho (\lambda + 1 - \rho)}{2(1 - \rho + \lambda \rho)} \qquad (19)$$

These relations demonstrate to what extent the disproportionation reaction eliminates all the advantages of polyfunctional initiators. In fact, irrespective of the kind and values of polymerization kinetic parameters, according to equation (19) it turns out to be impossible to increase polymer molecular weight (in comparison with that resulting under the same conditions when ordinary monoperoxides are used) more than $(\lambda + 1)/\lambda$ times. For instance, in the polymerization of methyl methacrylate, where $\lambda \approx 0.5$, this factor is about 3. For such systems it follows from the second expression of equation (19), that the application of polyinitiators instead of monoinitiators is considerably less effective for the restriction of growth of conversional contribution to inhomogeneity of polymerization products. The above-mentioned factor can be regarded as the principal shortcoming in the practical application of polyperoxides.

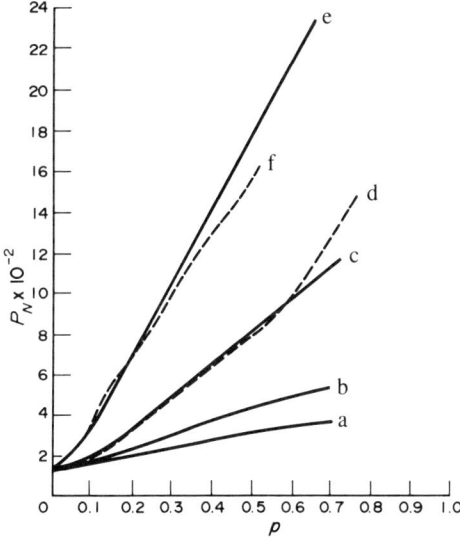

Figure 4 Alteration of number average molecular weight of polystyrene with conversion for bulk polymerization of styrene in the presence of oligoperoxides with different mean number of peroxide groups, which equals (a) 1, (b) 2, (c, d) 4.6, (e, f) 10.7. The solid lines are drawn in accordance with theory and dashed lines correspond to the experimental data[12]

2.2.5 Polyinitiators with Several Types of Labile Group

The use of polyperoxides containing peroxide groups with different thermostabilities has wide prospects for molecular design of polymers.[1-6] However, problems concerning the dependence of their statistical characteristics on functionality, composition and structure of initiating compounds usually arise. The quantitative theory,[31] some major results of which are given below, answers these questions.

General equations (3)–(10) for the calculation of molecular weight and MWD remain true for the substitution of mono- for poly-initiator with several types of peroxide group. However, the functions of distribution $P_m(t)$ of macromolecules for m elementary chains naturally differ. To find this distribution it is advisable to apply a statistical approach, following which any arbitrary macromolecule within the reaction system can be considered as the realization of some random process of conventional movement along it.[21,24] It turns out that such a process, describing products of homopolymerization initiated by arbitrary polyperoxides with independent groups, will be a Markov chain. This result is correct at any number r of types of labile groups, irrespective of how they are distributed among the molecules of the initial polyperoxide. The latter will influence exclusively the values of statistical parameters of the aforementioned Markov chain: matrix Q (with elements v_{ij}) of transition probabilities and vector v (with components v_i) of initial states. In the case under consideration, the extension of equation (13) yields

$$U(x,t) = \sum_{i=1}^{r} \sum_{j=1}^{r} v_i x [(E - \tilde{Q}x)^{-1}]_{ij} v_{j0} \tag{20}$$

where E represents the unit matrix (with elements δ_{ij}), $\tilde{Q}$ represents the transition matrix (with elements $v_{ij}, i \neq 0, j \neq 0$) between transient states and v_{j0} is the probability of transition from the jth state into zeroth one which is absorbent.

All these parameters of the Markov chain are expressed through kinetic and structural parameters of the initial polyinitiator. Both decomposition reaction constants k_{di} and initiation efficiencies f_i ($i = 1, \ldots r$) of all types of peroxide groups can be referred to the kinetic parameters, while fractions of chemically individual molecules of polyperoxide, with a fixed set of numbers of all types of groups with regard to the order of their alternation, belong to the structural ones. The structure of macromolecules for the case under consideration will resemble that depicted in Figure 2, the only exception being peroxide clusters which can include groups of different types. When applying the statistical approach, it seems to be convenient to differentiate (e.g. by coloring them conventionally) radicals, elementary chains and clusters, depending on the method of their formation. So the ith type labile group, on decomposition, forms a pair of macroradicals to which the same type i is attributed. Each act of recombination of pairs of such ith and jth type macroradicals is conducive to the formation of an elementary chain of (ij)th type. Following the same rule it is possible to differentiate peroxide clusters by their types, depending on the type of radicals which have been formed earlier at the ends of such a cluster. Hence, the elementary chain (ij) is connected either with the chain (kl) by a closed (jk)th type cluster or with the end of the macromolecule by semiopened cluster of $(j0)$th as well as $(0j)$th type, containing $0, 1, \ldots$ peroxide groups. It is convenient to refer to molecules of the initial polyperoxide as open clusters.

Concentrations N_{ij} of peroxide clusters of different types (ij) and rates I_i of generation of macroradicals of type $i = 1, \ldots r$

$$I_i = 2 f_i k_{di} C_i, \quad C_i = C_i^0 \exp(-k_{di} t), \quad I = \sum_{i=1}^{r} I_i \tag{21}$$

determine elements ρ_{ik} and σ_{kj} of matrices ρ and σ

$$\rho_{ik} = N_{ik} \bigg/ \sum_{k=0}^{r} N_{ik} \quad (i, k = 1, \ldots r), \quad \rho_{i0} = N_{i0} \bigg/ \sum_{k=0}^{r} N_{ik} \tag{22}$$

$$\sigma_{kj} = \int_0^t \frac{I_k I_j}{I} dt' \bigg/ \int_0^t I_k dt', \quad \rho_{0k} = N_{0k} \bigg/ \sum_{k=1}^{r} N_{0k} \tag{23}$$

through which in a simple way

$$v_{ij} = \sum_{k=1}^{r} \rho_{ik} \sigma_{kj} \quad (i, j = 1, \ldots r), \quad v_{i0} = \rho_{i0} \tag{24}$$

$$v_i = \sum_{k=1}^{r} \rho_{0k} \sigma_{ki} \tag{25}$$

all parameters of the Markov chain presented in equation (20) are expressed. Overall concentration of polymer molecules Π equals the difference between the initial I^0 and the current $I(t)$ values of the initiator concentration

$$\Pi = I^0 - I(t) = I^0 \left[1 - \sum_j \alpha(j) \prod_{i=1}^{r} (1 - \theta_i)^{j_i} \right] \tag{26}$$

Here $\alpha(j)$ is the fraction of initial polyperoxide molecules, which are characterized by functionality vector j with components $\{j_1, \ldots, j_i, \ldots, j_r\}$, equal to the numbers of functional groups of each of these types. Value

$$\theta_i = f_i[1 - \exp(-k_{di}t)] \quad (i = 1, \ldots, r) \tag{27}$$

has a probabilistic meaning for the arbitrarily chosen ith peroxide group to decompose effectively at the time t.

All equations in Section 2.2.3 are valid in the case under consideration. The mean number of elementary chains in the macromolecule

$$P_N'' = \sum_{i=1}^{r} \sum_{j=1}^{r} v_i [(E - \tilde{Q})^{-1}]_{ij} = \sum_{i=1}^{r} \frac{C_i^0 \theta_i}{\Pi} \tag{28}$$

and, consequently, in accordance with equation (10), the number average molecular weight of polymer do not depend [at a given distribution $\alpha(j)$] on the way in which different peroxide groups inside molecules of polyinitiator alternate. However, the polyperoxide microstructure does exert an influence on MWD and its degree of polydispersion, K (equation 10), in accordance with the following equation

$$K'' = 1 + \sum_{i=1}^{r} \sum_{j=1}^{r} v_i [(E - \tilde{Q})^{-2} \tilde{Q}]_{ij} \bigg/ \left\{ \sum_{i=1}^{r} \sum_{j=1}^{r} v_i [(E - \tilde{Q})^{-1}]_{ij} \right\}^2 \tag{29}$$

In the simplest and the most experimentally investigated case of initiators with two types of peroxide groups the general equation (20), can be reduced to the expression

$$U(x, t) = \frac{x(b_1 + b_2 x)}{1 - d_1 x + d_2 x^2} \tag{30}$$

where

$$b_1 = v_1 v_{10} + v_2 v_{20}, \quad b_2 = v_1(v_{12}v_{20} - v_{22}v_{10}) + v_2(v_{21}v_{10} - v_{11}v_{20})$$
$$d_1 = v_{11} + v_{22}, \quad d_2 = v_{11}v_{22} - v_{12}v_{21} \tag{31}$$

Provided that the conditions $b_2 < 0$ and $d_2 > 0$ hold, the analysis of equation (30) predicts the feasibility of the appearance at the initial stages of polymerization of bimodal MWD, which can later become unimodal. Just this character of MWD evolution has been observed in several experiments,[5,8,11,12] when diperoxides with two labile groups of different thermostability have been employed as initiators. The polydispersity coefficient K can be easily found according to equation (10), since equation (29), when $r = 2$, can be reduced to the following form

$$K'' = \frac{4(1 - d_2) - (b_1 + d_1)(3 - d_1 - d_2)}{(2 - b_1 - d_1)^2} \tag{32}$$

To calculate the MWD (equation 30) or its chracteristic parameters (equations 10 or 32), in accordance with equation (31) the values of the Markov chain parameters must first be found, which are completely characterized, following equations (24) and (25), by matrices ρ and σ. Their elements can be derived from equations (22) and (23), if the concentrations N_{ik} of clusters of different types (ik) are known. Several approaches to the calculation of concentrations N_{ij} for different kinds of initial polyperoxides have been suggested.[31] It is worth emphasizing that such a polyperoxide may not only be an individual chemical compound but a mixture of molecules differing from each other in the number of various types of peroxide groups (composition) as well as by the way in which they alternate (structure). The above mixtures are of great practical importance, since oligoinitiators are usually synthesized by means of polycondensation methods.[1-6] Products of such a process, due to the random character of its reactions, obviously show some distribution in composition and structure. The latter is described, as a rule, by some Markov chain[24] with r transient states corresponding to peroxide groups of r types and with a zeroth absorbing state, transition to which corresponds to overstepping the molecule border. If the matrix of transition probabilities between transient states of this Markov chain is designated as $\tilde{Q}^0$ and v^0 represents its initial vector, then the matrix ρ could be presented as follows

$$\rho = [E - \tilde{Q}^0(E - D)]^{-1}\tilde{Q}^0 D \tag{33}$$

where D represents the diagonal matrix with elements $\theta_i \delta_{ij}$. In the same way other parameters given in equations (22) and (23) can be defined

$$\rho_{i0} = \sum_{j=1}^{r} \{[E - \tilde{Q}^0(E - D)]^{-1}\}_{ij} v_{j0}^0 \tag{34}$$

$$\rho_{0i} = \frac{I^0}{\Pi} \sum_{j=1}^{r} v_j^0 [\theta_j \delta_{ji} + (1 - \theta_j)\rho_{ji}] \tag{35}$$

$$\Pi = I^0 - I(t) = I^0 \left[1 - \sum_{i=1}^{r} v_i^0 (1 - \theta_i)\rho_{i0} \right] \tag{36}$$

Equations (36) and (34), together with (28) and (10), present an exhaustive solution for the problem of calculating the number average molecular weight of the products of polymerization, initiated by polyperoxides with Markovian statistics of alternation of various labile groups.

The possibility of application of the statistical approach for the calculation of the MWD of the products of polymerization initiated by polyperoxides is connected with the independent decomposition of all peroxide groups. This is another illustration of Flory's idea,[21] which is of the greatest importance in polymer science. In the spirit of this idea, the MWD function depends on time exclusively through probability parameters, which for the systems under consideration are the elements σ_{kj} (equation 23) of the matrix σ and the components θ_i (equation 27) of vector $\boldsymbol{\theta}$.

As follows from rigorous kinetic consideration, all equations deduced by means of the statistical approach do not change in systems where temperature drifts during polymerization. However, the dependence of probability parameters on time in these formulas will be determined by more general expressions in comparison with those in the isothermic process. One such expression, by extension of equation (27), is as follows

$$\theta_i(t) = \int_0^t f_i(t') k_{di}(t') \exp\left[-\int_0^{t'} k_{di}(t'') dt'' \right] dt' \tag{37}$$

In this case, when calculating values of σ_{kj} (equation 23), the time dependence of kinetic parameters f_i and k_{di} in functions I_i (equation 21) is to be taken into account. Adduced elucidations allow the above theory to embrace polymerization processes with controlled rise of temperature which, compared with isothermic processes, exhibit a number of advantages.

2.2.6 Synthesis of Block Copolymers

Polymerization processes involving the participation of polyinitiators comprising two types of groups with a strongly pronounced distinction in thermostability seem to be of the greatest interest for practical purposes. Such polyperoxides are characterized by decomposition constants such that $k_{d1} \gg k_{d2}$ and thus it is possible to perform polymerization in two stages. During the first stage, carried out at a specific temperature, practically only the more labile first-type peroxide groups decompose. As a result, the formation of polymer molecules, containing in the main chain clusters of second-type peroxide groups which are capable of further transformations, occurs. Then, on increasing polymerization temperature the process enters its second stage where the decomposition of the more stable second-type groups is accompanied by both the growth and the subsequent recombination of macroradicals formed. When, after the completion of the first stage, one monomer is replaced by any other during the second stage, molecules containing blocks of different monomers are formed. Such a procedure for the synthesis of block copolymers seems to be rather simple and prospective.[13-15,25-30]

Changing the polyperoxide composition and structure can influence the architecture and statistical characteristics of the block copolymers obtained. In the simplest case it is possible to choose as an initiator an asymmetrical diperoxide such as compound (**1**) with the corresponding substituent R, which provides sufficient difference in thermostability of the two peroxide groups. The choice, in particular, of boron atom as R results in a distinction between decomposition rate constants k_{d1}

and k_{d2} of these groups by a factor of more than 20^{12}, which ensures the performance of the two-stage block copolymer synthesis.

The formation in both stages of the synthesis of homopolymer molecules as by-products can be regarded as a serious drawback of the above procedure. It is possible to obviate this drawback in the first stage by substituting di- for symmetric tri-peroxide which contains a more stable central second-type group and two identical first-type groups.[10,32] In order to reduce the formation of homopolymer in the second stage also, the process should be conducted in the presence of a polyperoxide with as high a functionality as possible and preferably with alternating groups of both types. The latter factor promotes the increase of composition homogeneity of copolymer products, since it ensures regular alternation in macromolecules of blocks, each of them consisting of a sole elementary chain.

In the general case of an arbitrary polyperoxide block of i-type, monomer units chosen at random in a block copolymer molecule can comprise any number, n, of elementary chains of corresponding monomer. The random value n is characterized by distribution function $(1-\rho_{ii})\rho_{ii}^{n-1}$, so the mean number of elementary chains in one block equals $(1-\rho_{ii})^{-1}$. The length distribution of i-type monomer blocks at an invariable length, v, of kinetic chain (equation 7) is described by equation (15) where, however, ρ will be substituted for ρ_{ii}.

To calculate block copolymer composition distribution one can also resort to the superposition principle, but apply it in a more general version. In contrast to homopolymers, binary block copolymer macromolecules are characterized, instead of by degree of polymerization l, by a composition vector $\boldsymbol{l}$ with components l_1 and l_2 equal to the numbers of first- and second-type monomer units. Each of these types corresponds to a type of homopolymer elementary chain, numbered m_1 and m_2, which in a macromolecule can be also considered as vector $\boldsymbol{m}$ components. With such designations the main relations (equations 3–6) of the superposition principle will remain valid if one replaces the scalars l and m by vectors $\boldsymbol{l}$ and $\boldsymbol{m}$. Thus, it is enough to find the generating function $U(\boldsymbol{x},t)$ of the distribution of polymer chains for the vector, $\boldsymbol{m}$. The solution of this problem gives the following expression for the function $U(x_1,x_2)$[31]

$$U(x_1,x_2) = \frac{\rho_{01}\rho_{10}x_1 + \rho_{02}\rho_{20}x_2 + (\rho_{01}\rho_{12}\rho_{20} + \rho_{02}\rho_{21}\rho_{10} - \rho_{01}\rho_{22}\rho_{10} - \rho_{02}\rho_{11}\rho_{20})x_1x_2}{1 - \rho_{11}x_1 - \rho_{22}x_2 + (\rho_{11}\rho_{22} - \rho_{12}\rho_{21})x_1x_2} \quad (38)$$

By subsequently substituting x_1 and x_2, respectively, for the generating functions $\langle g_1(s_1)\rangle$ and $\langle g_2(s_2)\rangle$ of distributions for the lengths of elementary chains of each type, one arrives at the expression for the generating function $G_N(s)$ of distribution $f_N(l,t)$.

Proceeding from equation (38) it is easy to calculate the mean numbers $\bar{m}_1$ and $\bar{m}_2$ of elementary chains of both types referring to a single polymer molecule

$$\frac{\bar{m}_1}{\Pi} = \frac{\rho_{12} + (\rho_{11}+\rho_{21})\rho_{10}}{\rho_{12}\rho_{20} + \rho_{21}\rho_{10} + \rho_{10}\rho_{20}}, \quad \frac{\bar{m}_2}{\Pi} = \frac{\rho_{21} + (\rho_{22}+\rho_{12})\rho_{20}}{\rho_{12}\rho_{20} + \rho_{21}\rho_{10} + \rho_{10}\rho_{20}} \quad (39)$$

which characterize the average composition of the block copolymer

$$\langle \zeta_i \rangle = \frac{\langle v_i \rangle \bar{m}_i}{\langle v_1 \rangle \bar{m}_1 + \langle v_2 \rangle \bar{m}_2} \quad (i=1,2) \quad (40)$$

where $2\langle v_i \rangle$ represents the average length of i-type elementary chain.

Important characteristics of the synthesis of block copolymers are parameters e_1 and e_2, which are equal, by definition, to the fractions of monomer units of each type in the homopolymer by-products. The theory suggests a simple expression for the calculation of these parameters

$$e_i = \frac{\Pi \rho_{0i}\rho_{i0}}{C_i^0 \theta_i (1-\rho_{ii})^2} \quad (i=1,2) \quad (41)$$

which is convenient in practice for evaluating the efficiency of application of different polyinitiators when one obtains block copolymers.

2.3 INIFERTERS

2.3.1 Introduction

The term 'iniferter', introduced for the first time by Otsu *et al.*,[33] is a result of the fusion of three words: *ini*tiator, trans*fer* agent and *ter*minator. The special feature of an iniferter is its participation in each of the three reactions mentioned. To date, some peculiarities of the polymerization of various monomers have been investigated in the presence of several iniferters, including a variety of phenylazomethanes, dithiocarbamates, disulfides, *etc.*[33-50] Of particular significance in this field is the contribution made by Japanese scientists, Otsu and coworkers especially,[33,34,36,41-52] who managed to reveal a number of important regularities of polymerization in the presence of iniferters. Wide potential for the application of such compounds for molecular design of synthetic polymers, has also been demonstrated.[42,44,45,50-52]

The common feature of all iniferters is that, despite possible essential differences in structural formulas, their molecules contain labile interatomic bonds, capable of cleavage in the course of the synthesis with subsequent formation of a pair of primary radicals. At least one of them does not turn into a polymer radical, although it does take part in the chain termination reaction. An example of such is S-benzyl-N,N-diethyldithiocarbamate (**5**; Scheme 1), which has been used[49,55-57] for initiating the photopolymerization of styrene. The decomposition of this initiating molecule yields two primary radicals A· (benzil) and B· (diethylidithiocarbamate) which are considerably different in their chemical activity. The primary radical A· after monomer addition becomes capable of further growth, whilst the radical B·, due to its low activity, scarcely reacts with the monomer during the polymerization process. Besides such asymmetrical iniferters, symmetrical ones also exist, for example tetraethylthiuram disulfide (**6**; Scheme 2), which by decomposition gives two similar inactive radicals B·. Concurrent to their mutual recombination such radicals, when their concentration is low enough, can add a monomer and initiate the polymerization.

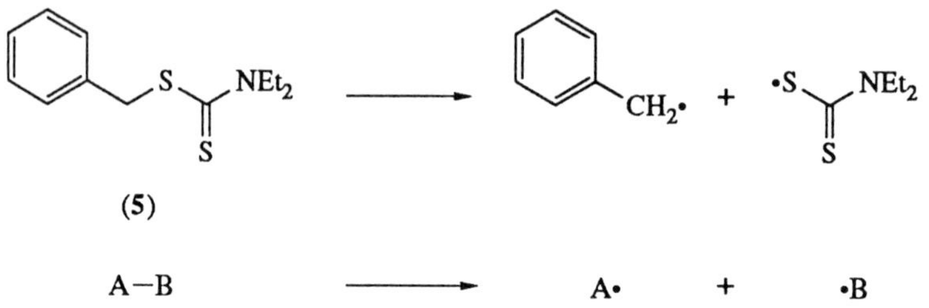

Scheme 1

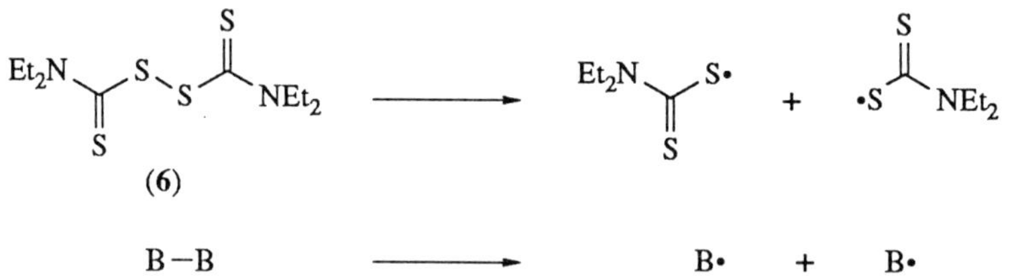

Scheme 2

Thermo- and photo-iniferters can be distinguished, depending on the factor responsible for the labile bond cleavage. The aptitude of various chemical compounds to act as iniferters has been discussed in a remarkable review by Moad *et al.*[59]

2.3.2 Polymerization Mechanism and Kinetics

During the course of the polymerization process with participation of the iniferter the formation of polymer chains with an active terminal bond is observed. Subsequently, each such chains can decompose under the action of heat or light into two compounds: the active macroradical and the small radical B·, which has to be stable enough to avoid the initiation of a new polymer chain. In its turn, after its propagation the macroradical recombines with the stable radical B· and eventually becomes a polymer molecule of macroiniferter. The latter decomposes again into a pair of radicals differing in their activity and so on. According to this scheme of stepwise polymerization, every polymer chain can successively be in two states: alive (as a macroradical) and dead (as a macromolecule). Since the polymer chain lives for so short a time compared with the time of its inactive state, macromolecule growth is believed to occur by separate steps. At every such step the molecular weight increases by just one length of the elementary chain.

Peculiarities of the mechanism of radical polymerization in the presence of iniferters are easy to understand on the basis of the rather general kinetic scheme of this process, which for an asymmetric iniferter I like (2) occurs as shown in Scheme 3. Besides the ordinary reactions of traditional radical polymerization this scheme includes the reinitiation reaction 10, which is undoubtedly the main feature of the process in question. Its second fundamental peculiarity is termination according to reaction 6 only, which occurs *via* combination of polymer and inactive primary radicals.

$$
\begin{array}{rlll}
1 & I \longrightarrow A\cdot + B\cdot & k'_d & \left.\right\} \text{Initiation} \\
2 & A\cdot + M \longrightarrow R^\bullet_1 & k'_{pA} & \\
3 & R^\bullet_l + M \longrightarrow R^\bullet_{l+1} & k_p & \text{Propagation} \\
4 & R^\bullet_l + R^\bullet_n \longrightarrow P_{l+n} & k_{tc} & \\
5 & R^\bullet_l + R^\bullet_n \longrightarrow P_l + P_n & k_{td} & \\
6 & R^\bullet_l + B\cdot \longrightarrow P_l B & k'_t & \left.\right\} \text{Termination} \\
7 & B\cdot + B\cdot \longrightarrow BB & k''_t & \\
8 & R^\bullet_l + I \longrightarrow P_l B + A\cdot & k_I & \\
9 & R^\bullet_l + P_n B \longrightarrow P_l B + R^\bullet_n & k_{PB} & \left.\right\} \text{Transfer} \\
10 & P_l B \longrightarrow R^\bullet_l + B\cdot & k_d & \text{Reinitiation}
\end{array}
$$

Scheme 3

In fact an accurate analysis[55] of the kinetic equations corresponding to Scheme 3 supports the possibility for termination reactions 4, 5 and 7 to be neglected as compared to 6. The indispensable condition of such an assumption is a considerably larger value of the dimensionless parameter $\Phi = k'_t/\sqrt{k_t k''_t}$, which depends on termination rate constants only. This condition is always true since such constants for diffusion controlled termination reactions are connected by the relations $k''_t \sim k'_t \sim (10^2 - 10^3) k_t$.

The theoretical expression for the rate of polymerization in the presence of iniferters[55]

$$W = k_p M \sqrt{I/k^e_t}, \quad k^e_t = k_t \Phi \qquad (42)$$

looks similar to that for traditional polymerization, with the only distinction being that the ordinary termination rate constant k_t is substituted for its effective value k^e_t. The ratio of the latter to the former equals Φ and thus, even at the same initiation rate I, the usage of iniferters instead of ordinary initiators inevitably leads to a decrease of polymerization rate of more than 10-fold. The simplest systems are those in which the cleavage rate constant k_d of the terminal labile bond in the macromolecule does not depend on its length and coincides with the decomposition rate constant k'_d of iniferter. For such ideal systems, where Flory's principle[21] of equal reactivity is valid for all elementary reactions, the reduced polymerization rate

$$\frac{W}{M} = k_p \sqrt{\frac{k_d C^0}{k^e_t}}, \quad I = k_d C^0 \qquad (43)$$

remains invariable and proportional to the square root of initial iniferter concentration C^0_I.

However, in addition to the ideal systems, where the above condition holds,[41,47] a considerable number of monomer–iniferter pairs, the polymerization rate of which changes from the very beginning of the process[41,42,44] have been experimentally discovered. A typical shape of the kinetic curve for such nonideal systems is represented in Figure 5, which clearly shows that, within the range of a few percent of conversion, the rate of its change decreases from a certain initial value W_0, gradually approaching the stationary value W_s.

To interpret such a phenomenon there has been advanced, and then experimentally proved, an assumption about possible violations of the Flory principle for polymerization initiation reactions proceeding in the presence of iniferter.[55] For this purpose several researchers[55-57] managed by means of the spin trapping technique to measure the decomposition rate of terminal labile groups in narrow fractions of macroiniferter obtained during styrene photopolymerization in the presence of iniferter (5). The results of this experiment, presented in Figure 6, demonstrate a rather dramatic decrease of the reinitiation rate constant k_d with increase in the oligoiniferter degree of polymerization l up to ca. 100 monomer units, and a further tendency to approach the limit value. Just after reaching the value of $P_N \sim 100$ in the course of the polymerization process its rate acquires the constant value in complete agreement with equation (42). The latter also predicts proportionality between initial W_0 as well as stationary W_s rates of polymerization and the square root of iniferter concentration. This has been experimentally confirmed[55-57] for styrene photopolymerization in the presence of iniferter (5) within a wide range of concentrations. Moreover, the experimental ratio $W_0/W_s \sim 3$ in this range turned out to be practically invariable. These results just coincide with theoretical predictions if the dependence of decomposition rate constant k_d of macroiniferters on their length as depicted in Figure 6 is taken into account.

Since such dependence in this system is important only for macroiniferters with less than 100 monomer units, for higher molecular homologs an unchanged value of the polymerization rate W from the very beginning of the process might be expected. Besides, this value W has to coincide with the stationary rate of polymerization W_s carried out under the same conditions but in the presence of low molecular iniferter (5). Indeed, the experimental data presented in Figure 5 completely support these theoretical conclusions.

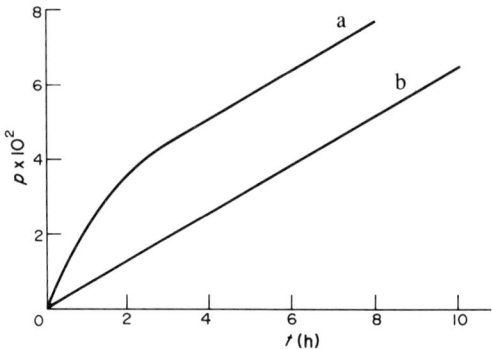

Figure 5 Kinetic curves for photopolymerization of styrene in benzene solution in the presence of (a) iniferter (5) and (b) macroiniferter containing about 170 styrene units[55-57]

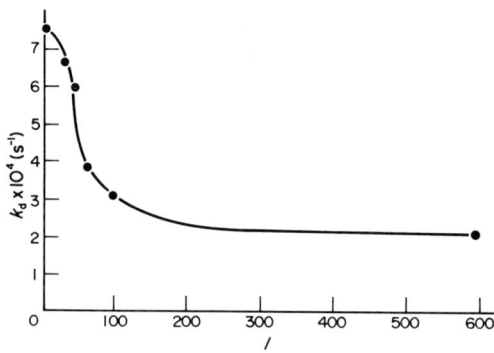

Figure 6 Dependence of the rate constant of decomposition of a macroiniferter molecule (oligostyrene with a thiocarbamate terminal group) on its degree of polymerization[55-57]

The very fact that, for a polymerization process in the presence of iniferter, termination reactions *via* interaction between polymer radicals are practically absent brings to light some rather interesting inferences concerning the kinetics of some highly important industrial processes. Of these, the bulk polymerization of some vinyl monomers, essential for the manufacturing of plastics and especially organic glasses, is a matter of particular concern. During such processes a drastic increase of the polymerization rate is normally observed within the range of intermediate conversions. Such a phenomenon, well known as the 'gel effect', is very undesirable when projecting industrial processes since it gives rise to local overheating of the reaction system and consequently deteriorates polymer service characteristics.

The cause of this 'gel effect' is a considerable decrease of the rate constants k_{tc} and k_{td} of reactions 4 and 5 (Scheme 3), which takes place due to a dramatic decrease of the mutual diffusion rates of the macroradicals when the polymer–monomer solution reaches high polymer concentrations in the course of the polymerization process.

Since during the polymerization in the presence of iniferter chain termination reactions are not of vital importance, it might be hoped that the 'gel effect' will not occur in the course of such a process. This theoretical conclusion has been convincingly proved experimentally[58] for the bulk polymerization of methyl methacrylate as well as styrene in the presence of N,N,N',N'-tetraethylthiuram disulfide as iniferter.

Concluding this section, we can safely formulate the quantitative conditions for the applicability of the aforementioned iniferter mechanism of radical polymerization. Firstly, it is important to answer the question of whether a specific initiator will or will not exhibit iniferter properties. Besides the effectiveness of the reinitiation reaction, the answer also depends on the choice of the reaction path by the less active of the two primary radicals formed from initiator decomposition.

There are two possibilities for primary radical B· to react: either with monomer M or with polymer radical R·. Which of these reactions will be predominant depends on the ratio of their respective probabilities, $k'_{pB}M$ and k'_0R. Apparently, if the activation reaction of primary radical B· prevails, the regime of traditional radical polymerization is being realized. In the opposite case, *i.e.* $k'_{pB}M \ll k'_0R$, when termination reaction prevails for B·, the polymerization will proceed by an iniferter regime. The larger the ratio of rate constants k'_{pA}/k'_{pB} of monomer addition to active A· and nonactive B· primary radicals the wider is the range of polymerization rates corresponding to this regime. Only if the difference in their activities is large enough can the following inequalities

$$1 \ll \frac{W}{W^*} \ll \frac{k'_{pA}}{k'_{pB}}, \quad W^* = \frac{k'_{pB}k_p}{k'_t} M^2 \tag{44}$$

ensuring an iniferter mechanism of polymerization hold. The first of them indicates that the polymerization rate should noticeably exceed some characteristic value W^* in order to prevent primary radicals B· from initiating polymer chains. At the same time the rate W, in accordance with the second inequality, must be considerably less than values $(k'_{pA}/k'_{pB})W^*$ in order to exclude the participation of primary radicals A· in the termination reaction leading to the formation of polymer molecules unable to reinitiate.

2.3.3 Molecular Weight Distribution and its Characteristics

In the case of the iniferter mechanism of polymerization, each macromolecule usually consists of several elementary chains. This fact allows the superposition principle to be applied for the calculation of polymer MWD in ideal systems, as formulated in Section 2.2.3. The generating function

$$U(x,t) = \frac{e^{(x-1)\tau} - e^{-\tau}}{1 - e^{-\tau}}, \quad \tau = k_d t \tag{45}$$

found when neglecting the chain transfer reaction, allows one to obtain the simple expressions

$$P_N = \frac{\tau \langle v \rangle}{1 - e^{-\tau}}, \quad K = \frac{1 - e^{-\tau}}{\tau} \left(\tau + 2 \frac{\langle v^2 \rangle}{\langle v \rangle^2} \right) \tag{46}$$

for the number average degree of polymerization and the polydispersity coefficient. The rise of the number of elementary chains in the macromolecules naturally stimulates, as suggested by equation (46), the increase of their molecular weight as well as the narrowing of the MWD. However, just the opposite effect is caused by the decrease of the kinetic chain length $v = v^0 M/M^0$ in the course of polymerization from its initial value $v = v^0$ to $v = 0$ as the monomer concentration is reduced from M^0 to 0. Thus, equations (46) become

$$P_N = \frac{v^0(1 - e^{-\beta\tau})}{\beta(1 - e^{-\tau})}, \quad \beta = \frac{C^0}{M^0}v^0, \quad v^0 = \frac{k_p M^0}{(k_t \Phi k_d C^0)^{1/2}} \tag{47}$$

$$K = (1 - e^{-\tau})[\beta_{\text{cth}}(\beta\tau/2) + 1] \tag{48}$$

The magnitude P_N grows with time from $P_N^0 = v^0$ up to $P_N^\infty = v^0\beta^{-1}$, where the values of the parameter β^{-1}, which refer to the average number, $\bar{m}$, of elementary chains in the macromolecule of the final polymer, are expected to be rather high for the iniferter mechanism of polymerization. In this case, the stationary value $K = 1 + \beta$ of the polydispersity coefficient (equation 48) only slightly differs from unity, thus testifying to the rather high homogeneity of MWD of the final polymerization products. Equations (47) and (48) predict monotonous change of P_N and K values with time.

As for the MWD function, its expression can be given for two specific and important cases. One of them corresponds to polymerization at low conversions, when the MWD looks as follows

$$f_N(l, t) = \frac{\tau}{v^0} \exp\left[-\tau\left(\frac{z}{\tau} - 1\right)^2\right] \phi(z)$$

$$\phi(z) = e^{-2z} I_1(2z)/z, \quad z = (l\tau/v^0)^{1/2} \tag{49}$$

where $I_1(2z)$ is the modified Bessel function of the first kind of order 1 and argument $2z$. In another case, corresponding to the completion of polymerization, its final products are described by the expression

$$f_N(l, \infty) = \left(v^0 \Gamma \frac{1}{\beta}\right)^{-1} \left(\frac{1}{v^0}\right)^{1/(\beta - 1)} \exp\left(\frac{-1}{v^0}\right) \tag{50}$$

where Γ is the Euler gamma function.

By taking into consideration reactions 8 and 9 (Scheme 3), i.e. by accounting for the chain transfer with iniferter, simple modifications to equations (45)–(48) are obtained. All of them, being functions of τ, remain invariable if one substitutes the average length v of the kinetic chain for the average length $v[1 + k_t C^0(k_t \Phi k_d C^0)^{-1/2}]^{-1}$ of the elementary chain.

In the framework of a more general model of ideal polymerization, labile bonds within the initial iniferter and polymer molecule are differentiated kinetically. At strongly pronounced differences ($k'_d \gg k_d$) between decomposition rate constants k'_d and k_d the polymerization will practically proceed in two stages. In the first of them, a comparatively short one, all molecules of the initial iniferter will decompose and macromolecules, each of them consisting of exactly one elementary chain, will be formed. These molecules of macroiniferter will then initiate subsequent polymerization for the second, longer stage. It is of particular importance that the MWD of the obtained polymer will exhibit a tendency at first to broadening and then to narrowing. Such nonmonotonous character of the polydispersity coefficient, K, can be observed in systems where the constant k_d drastically decreases when the length of the macroiniferter molecule increases. This theoretical result follows from the solution of the set of kinetic equations deduced[55] for the arbitrary dependence $k_d(l)$. Experimental curves $K(t)$ for polymerization of styrene in the presence of iniferter (5) actually show the maximum[56] confirming the theoretical conclusions.

One of the most specific features of polymerization in the presence of iniferters is the linear character of the growth with conversion, p, of number average molecular weight of obtained products.[38,41–43,48,53–57] Such dependence $P_N = (M^0/I^0)p$ follows directly from the iniferter mechanism of polymerization within the range of conversions $p \gg p^* = k_p(I^0/k_t \Phi k'_d)^{1/2}$, when the initial initiator has practically already vanished in the reaction system. Since each initiator molecule after its decomposition has initiated a polymer chain, their concentration coincides with I^0 when $p \gg p^*$. For conversions sufficiently low in comparison with p^*, the expression $P_N = (M^0/I^0)p$ is naturally invalid due to the increase with time of the number of polymer chains. By accomplishing the correct extrapolation of the dependence P_N to the point $p = 0$ one can obtain the expression $P_N = v^0 = k_p M(k_t \Phi k'_d C^0)^{-1/2}$, which means that macromolecules formed at the very beginning of polymerization are composed of only one elementary chain.

Summarizing Section 2.3, which is devoted to the peculiarities of polymerization initiated by iniferters, one could draw the conclusion that there is now an experimentally well-grounded understanding of the main specific features of its mechanism. Naturally, for particular processes of polymerization it is possible to take into consideration, while calculating the kinetics of polymerization and MWD of its products, some side reactions not presented in Scheme 3. Proper accounting of these side reactions is connected with specificity of the chosen monomer–iniferter pair, each component of which possesses its individual particular peculiarities. However, the most general features inherent to the iniferter mechanism of polymerization have been reflected in its simplest kinetic scheme (Scheme 3).

2.4 CONCLUSION

Radical polymerization in the presence of polyfunctional initiators or in the presence of iniferters, in spite of some fundamental distinctions in the mechanism of formation of polymer chains, have a number of common features. The most significant of these, which differentiate these processes from ordinary polymerization with participation of traditional initiators will be enumerated here.

The most important feature is the continuous growth with conversion of polymer molecular weight, which is usually described by a linear law if the mean number of elementary chains in the macromolecule is large enough. In this case the initiator vanishes at the very beginning of the process. After this the number of polymer chains remains invariable and only their length increases.

The second peculiarity of the processes in question is that the very mechanism of macromolecule formation promotes the decrease of conversional broadening of the MWD, originated by the consumption of monomer and initiator during the process. An appreciable rise of composition homogeneity of the copolymerization products obtained at high conversions is to be expected for this reason.

The above theoretical assumption, made for the first time by the author proceeding from general considerations, has been recently experimentally supported[60] for the bulk binary copolymerization of styrene with methyl acrylate and with acrylonitrile. In these systems, over a wide range of monomer feed compositions, the products of high conversion copolymerization, performed in the presence of traditional initiators, are turbid incompatible copolymers. On the other hand, copolymerization products of the same compositions and molecular weights, obtained at the same temperature but in the presence of iniferters, retain their optical transparency in the whole range of conversions.[60] The ability of iniferters to reduce drastically the composition inhomogeneity of the products of the high-conversion copolymerization they have initiated is especially promising for the synthesis of transparent homophase copolymers in the course of free radical processes.

The third common feature is the formation of macroinitiator molecules in the course of both processes with participation of the nontraditional initiators considered in this review. These molecules in contact with the monomer are able under suitable conditions to further growth of their molecular weight. This property allows one to apply nontraditional initiators of both types for the synthesis of block copolymers. If polyperoxides enable one to obtain macromolecules with a sufficiently large number of blocks, iniferters are normally used to synthesize diblock copolymers.

A basic shortcoming of polyfunctional initiators is their ineffectiveness for the polymerization of monomers (like methyl methacrylate) where the fraction of radicals losing their activity by a disproportionation mechanism is large enough. This shortcoming, however, is not peculiar to polymerization initiated by iniferters. Besides, their application permits the practically complete elimination of an undesirable 'gel effect' taking place for the bulk polymerization of many monomers.

ACKNOWLEDGEMENTS

I am very much obliged to Professor S. Ivanchev, who first attracted my attention to polyfunctional initiators and introduced me to the range of numerous problems of radical polymerization in which they participate. It is a matter of special pleasure for me to express my sincere gratitude to Academician V. Kabanov, heading the investigations of iniferters in the Department of Polymer Chemistry at Moscow State University as well as to my colleagues Professor V. Golubev and Drs. M. Zaremskiy, A. Olenin and E. Garina, with whom I have talked over some of the above-mentioned problems. I am also very grateful to Professor S. Russo and Professor D. Sherrington for fruitful discussions in Alma-Ata, Genoa and Glasgow which left me with the warmest memories.

2.5 REFERENCES

1. S. S. Ivanchev, *Vysokomol. Soedin.*, 1978, **A20**, 1923.
2. S. S. Ivanchev, 'Radical Polymerization' (in Russian), Khimia, Leningrad, 1985, p. 47.

3. C. I. Simionescu, E. Comanita, M. Pastravanu and S. Dumitriu, *Progr. Polym. Sci.*, 1986, **12**, 1.
4. O. Nuyken and R. Weidner, *Adv. Polym. Sci.*, 1986, **73–74**, 145.
5. V. V. Konovalenko and S. S. Ivanchev, in 'Reactions in Polymer Systems' (in Russian), ed. S. S. Ivanchev, Khimia, Leningrad, 1987, p. 5.
6. T. Mukundan, K. Kishore, *Progr. Polym. Sci.*, 1990, **15**, 475.
7. S. S. Ivanchev, V. I. Galibey and A. I. Yurzhenko, *Vysokomol. Soedin.*, 1965, **7**, 74.
8. S. S. Ivanchev, A. I. Prisyazhnuk and L. I. Efimova, *Vysokomol. Soedin.*, 1970, **B12**, 726.
9. A. I. Prisyazhnuk and S. S. Ivanchev, *Vysokomol. Soedin.*, 1970, **A12**, 450.
10. T. A. Tolpigina, V. I. Galibey and S. S. Ivanchev, *Vysokomol. Soedin.*, 1972 **A14**, 1027.
11. S. S. Ivanchev and Yu. L. Zherebin, *Dokl. Akad. Nauk SSSR*, 1973, **208**. 664.
12. S. S. Ivanchev and Yu. L. Zherebin, *Vysokomol. Soedin.*, 1974, **A16**, 829.
13. Yu. L. Zherebin, S. S. Ivanchev and N. M. Domareva, *Vysokomol. Soedin.*, 1974, **A16**, 893.
14. S. S. Ivanchev, N. G. Ivanova and Yu. L. Zherebin, *Plaste Kautsch.*, 1976, **23**, 5.
15. S. G. Erigova and S. S. Ivanchev, *Vysokomol. Soedin.*, 1969, **A11**, 2082.
16. N. G, Ivanova, S. S. Ivanchev and N. M. Domareva, *Vysokomol. Soedin.*, 1976, **A18**, 2788.
17. N. S. Tzvetkov and R. F. Markovskaya, *Izvestia Vuzov.*, *Chem. Ser.*, 1968, **11**, 936.
18. N. S. Tzvetkov and R. F. Markovskaya, *Kinet. Katal.*, 1969, **10**, 57.
19. N. S. Tzvetkov, R. F. Markovskaya, Yu. A. Saprykin and V. Ya. Zhukovski, *Vysokomol. Soedin.*, 1972, **A14**, 2072.
20. V. I. Galibey and E. G. Arkhipova-Kalenchenko, *Zh. Org. Khim.*, 1977, **13**, 227.
21. P. J. Flory, 'Principles of Polymer Chemistry', Cornell University Press, Ithaca, 1953.
22. N. S. Zvetkov, *Vysokomol. Soedin.*, 1961, **3**, 408.
23. S. I. Kuchanov, N. G. Ivanova and S. S. Ivanchev, *Vysokomol. Soedin.*, 1976, **A18**, 1870.
24. S. I. Kuchanov, 'Methods of Kinetic Calculations in Polymer Chemistry' (in Russian), Khimia, Moscow, 1978.
25. G. Smets and A. E. Woodward, *J. Polym. Sci.*, 1954, **14**, 126.
26. A. E. Woodward and G. Smets, *J. Polym. Sci.*, 1955, **17**, 51.
27. N. G. Ivanova, S. S. Ivanchev and N. M. Domareva, *Vysokomol. Soedin.*, 1976, **A18**, 2788.
28. I. Piirma and L.-P. H. Chou, *J. Appl. Polym. Sci.*, 1979, **24**, 2051.
29. B. Z. Gunesin and I. Piirma, *J. Appl. Polym. Sci.*, 1981, **26**, 3103.
30. B. Hazer, A. Ayas, N. Besirli and N. Saltek, *Makromol. Chem.*, 1989, **190**, 1987.
31. V. A. Ivanov, S. I. Kuchanov and S. S. Ivanchev, *Vysokomol. Soedin.*, 1977, **A19**, 1684.
32. S. S. Ivanchev, T. M. Artamonova. I. B. Belov, T. I. Nasonova and V. I. Kuznetzov, *Vysokomol. Soedin.*, 1979, **B21**, 426.
33. T. Otsu and M. Yoshida, *Makromol. Chem., Rapid Commun.*, 1982, **3**, 127.
34. T. Otsu and K. Nayatani, *Makromol. Chem.*, 1958, **27**, 149.
35. T. Ferrington and A. Tobolsky, *J. Am. Chem. Soc.*, 1958, **77**, 4510.
36. T. Otsu, Y. Kinoshita and M. Imoto, *Makromol. Chem.*, 1964, **73**, 225.
37. I. Beniska, *Vysokomol. Soedin.*, 1971, **A13**, 1790.
38. E. Borsig, M. Lazar and M. Capla, *Makromol. Chem.*, 1967, **105**, 212.
39. A. Bledzki and H. Balard, *Makromol. Chem.*, 1981, **182**, 3195.
40. G. S. Misra, A. Hafeez and K. S. Sharma, *Makromol. Chem.*, 1962, **51**, 123.
41. T. Otsu, M. Yoshida and T. Tazaki, *Makromol. Chem., Rapid Commun.*, 1982, **3**, 133.
42. T. Otsu, M. Yoshida and A. Kuriyama, *Polym. Bull. (Berlin)*, 1982, **7**, 45.
43. T. Otsu and M. Yoshida, *Polym. Bull. (Berlin)*, 1982, **7**, 197.
44. T. Otsu and A. Kuriyama, *Polym. Bull. (Berlin)*, 1984, **11**, 135.
45. T. Otsu and A. Kuriyama, *J. Macromol. Sci.*, 1984, **A21**, 961.
46. T. Otsu, K. Yamashita and K. Tsuda, *Macromolecules*, 1986, **19**, 287.
47. T. Otsu and T. Tazaki, *Polym. Bull. (Berlin)*, 1986, **16**, 277.
48. T. Otsu, A. Matsumoto and T. Tazaki, *Polym. Bull. (Berlin)*, 1987, **17**, 323.
49. T. Otsu, T. Matsunaga, A. Kuriyama and M. Yoshioka, *Eur. Polym. J.*, 1989, **25**, 643.
50. T. Otsu and A. Kuriyama, *Polym. J.*, 1985, **17**, 97.
51. A. Kuriyama and T. Otsu, *Polym. J.*, 1984, **16**, 511.
52. T. Otsu, T. Ogawa and T. Yamamoto, *Macromolecules*, 1986, **19**, 2087.
53. M. Niwa, T. Matsumoto and H. Izumi, *J. Macromol. Sci.*, 1987, **A24**, 567.
54. M. Niwa, Y. Sako and M. Shimizu, *J. Macromol. Sci.*, 1987, **A24**, 1315.
55. S. I. Kuchanov, M. Yu. Zaremskii, A. V. Olenin, E. S. Garina and V. B. Golubev, *Dokl. Akad. Nauk SSSR*, 1989, **309**, 371.
56. M. Yu. Zaremskii, E. S. Garina, A. V. Olenin, S. I. Kuchanov and V. B. Golubev, paper presented in Institute of Scientific Information of the USSR Moscow, 1989, VINITI, N 4690-B89.
57. M. Yu. Zaremskii, A. V. Olenin, E. S. Garina, S. I. Kuchanov, V. B. Golubev, V. A. Kabanov, *Vysokomol. Soedin.*, 1991, **A33**, 2167.
58. M. Yu. Zaremskii, S. M. Melnikov, A. V. Olenin, S. I. Kuchanov, E. S. Garina, M. B. Lachinov, V. B. Golubev and V. A. Kabanov, *Vysokomol. Soedin.*, 1990, **B32**, 404.
59. G. Moad, E. Rizzardo and D. H. Solomon, in 'Comprehensive Polymer Science', ed. G. Allen, Pergamon Press, Oxford, 1989, vol. 3, chap. 10, p. 143.
60. S. I. Kuchanov and A. V. Olenin, *Vysokomol. Soedin.*, 1991, **B33**, N 8, 563.

3
Emulsion Copolymerization

GERARD H. J. VAN DOREMAELE, HAROLD A. S. SCHOONBROOD
and ANTON L. GERMAN
Eindhoven University of Technology, The Netherlands

3.1	INTRODUCTION	41
3.2	MODELLING OF EMULSION COPOLYMERIZATION	42
	3.2.1 Introduction	42
	3.2.2 Emulsion (Co)polymerization Models	43
	3.2.3 Theoretical Aspects of Monomer Partitioning	43
	3.2.4 Particle Formation in the case of Emulsion Copolymerization	46
	3.2.5 Propagation Rate Models	47
	3.2.5.1 The ultimate and the penultimate model for copolymerization	47
	3.2.5.2 Other propagation rate models	48
	3.2.6 Kinetic Events Affecting the Average Number of Radicals per Particle	49
	3.2.6.1 Radical production	49
	3.2.6.2 Chain transfer	49
	3.2.6.3 Desorption of radicals	49
	3.2.6.4 Termination rate constant in the particles	50
	3.2.6.5 Radical entry and number of radicals per particle	50
	3.2.7 Copolymer Microstructure	51
3.3	MICROSTRUCTURE OF BATCH EMULSION COPOLYMERS	52
	3.3.1 Introduction	52
	3.3.2 (Molar Mass) Chemical Composition Distribution	53
	3.3.3 Copolymer Sequence Distribution	54
3.4	THE EFFECT OF COMPOSITION DRIFT ON COPOLYMERIZATION RATE	55
	3.4.1 Introduction	55
	3.4.2 Emulsion Copolymerizations of Vinyl Acetate–Butyl Acrylate and Styrene–Methyl Acrylate	55
3.5	EMULSION TERPOLYMERIZATION	57
3.6	INTRINSIC CHARACTERISTICS INFLUENCING COPOLYMER LATEX PROPERTIES	58
	3.6.1 Introduction	58
	3.6.2 Functionalized Latices	58
	3.6.2.1 Carboxylated latices	59
	3.6.2.2 Reactive latices	59
3.7	CONTROL OF COPOLYMER LATEX PROPERTIES BY SEMICONTINUOUS PROCESSES	59
	3.7.1 Introduction	59
	3.7.2 Process–Structure–Property Relationships	60
	3.7.3 Optimal Addition Profile	61
3.8	REFERENCES	63

3.1 INTRODUCTION

Conventional radical emulsion polymerization involves the dispersion of a monomer, an unsaturated organic molecule, in a continuous aqueous phase stabilized by an oil-in-water emulsifier, followed by free radical addition polymerization started usually with a water-soluble initiator. This results in a reaction medium consisting of submicron polymer particles swollen with the monomer and dispersed in an aqueous phase. The monomer droplets are usually 1–10 µm in diameter, whereas the polymer particles have a diameter ranging from about 50–500 nm. The final product is called a latex and consists of a colloidal dispersion of polymer particles in water. It is possible to isolate the polymer from the latex by controlled coagulation.

Copolymerization offers the possibility of modifying the properties of homopolymers into tailor-made products. Nowadays, emulsion copolymerization is a widely used process. Modern synthetic latices find a broad range of applications in the coating, ink, plastic and adhesive industries.

Despite many complexities, emulsion (co)polymerization displays the following advantageous properties, making it a very attractive process for the commercial production of polymeric materials: (i) the emulsion (co)polymerization process allows the synthesis of high molecular weight (molar mass) (co)polymers at high polymerization rates; (ii) heat can be excellently transferred as a result of the relatively low viscosity of the continuous aqueous phase, giving good temperature control; (iii) toxic and flammable organic solvents do not have to be used and the reaction can proceed to high conversion; (iv) the molecular weight can easily be controlled by the use of chain-transfer agents; and (v) the produced latex can be formulated directly into many final products. A disadvantage in certain applications must be mentioned, namely the relatively poor film formation properties as compared with solvent-based systems, partially due to the presence of surfactants.

Research on the modelling, optimization and control of the emulsion copolymerization process has been expanding rapidly since the demand for new and improved latex products has been increasing. Excellent reviews and textbooks on emulsion (co)polymerization and (co)polymer colloids have been published by Blackley,[1] Basset and Hamielec,[2] Piirma,[3] Penlidis et al.,[4] Guillot and Pichot,[5,6] and Candau and Ottewill.[7] Napper and Gilbert have given an excellent review on the modelling of homopolymerization,[8] whereas the present authors focus on emulsion copolymerization.

3.2 MODELLING OF EMULSION COPOLYMERIZATION

3.2.1 Introduction

A distinct aspect of (emulsion) copolymerization as compared with (emulsion) homopolymerization is the occurrence of composition drift. This phenomenon is mainly responsible for the chemical heterogeneity of the copolymers formed. Composition drift is a consequence of the difference between instantaneous copolymer composition and overall monomer feed composition. This difference is determined by: (i) the reactivity ratios of the monomers (kinetics); and (ii) the difference between the monomer ratio in the main loci of polymerization (viz. latex particles) and the overall monomer ratio of the feed (as added according to the recipe), which in turn is caused by differences in the partitioning of the two monomers over the phases that are present.

Batch processes are known to give two-peaked distributions of copolymer composition when a strong composition drift occurs during the course of the (emulsion) copolymerization. Moreover, in emulsion copolymerization the degree of bimodality appears to depend on the monomer/water ratio.[9,10] (Semi)continuous processes (i.e. addition of monomer during polymerization) can be used to prepare more homogeneous copolymers. Dynamic mechanical spectroscopy or differential scanning calorimetry and transmission electron microscopy combined with preferential staining techniques have been used to determine the possible occurrence of phase separation due to double-peaked chemical composition distributions (CCDs).

In industry many emulsion polymerization processes have been empirically developed, often resulting in process conditions far from optimal. Modelling of emulsion (co)polymerization is essential to gain a better understanding of the process and to enable a better control of the polymer synthesis and product quality. The most important variables that determine product quality include latex stability, particle size distribution (PSD), particle morphology,[11-13] gel content and copolymer microstructure.

The copolymer microstructure can be characterized in terms of sequence distribution, tacticity and molecular weight (molar mass) distribution (MMD) and chemical composition distribution (CCD). These can be combined in a three-dimensional distribution of molar mass and chemical composition (MMCCD). In addition to the intramolecular sequence distribution, molecular microstructure not only comprises the averages of molecular weight and chemical composition, but also their complete distributions as a whole (MMCCD). The MMCCD can be considered as a fingerprint of all the molecular events that contribute to polymer growth and it constitutes the linkage between the fundamental mechanistic chemical processes occurring in the reaction loci[14,15] and copolymer properties.[16-19] Therefore, it is generally recognized that detailed revelation of copolymer microstructure is a major factor contributing to a better understanding of both process and polymer properties, and that in detailed modelling of the emulsion copolymer microstructure, modelling of the process is a prerequisite.

The quintessence of modelling emulsion copolymerization is the proper description of the heterogeneity of the emulsion in combination with the characteristics of the monomers, *i.e.* their reactivity and water solubility, and their ability to swell the latex particles. Several researchers have worked in this field. Basic theories valid for emulsion homopolymerization were mathematically extended to emulsion copolymerization. In most of these descriptions, reaction rate, composition drift and copolymer composition were predicted and measured.

3.2.2 Emulsion (Co)polymerization Models

When an emulsion copolymerization starts, the monomers are usually present at an overall concentration higher than their saturation concentration in water, resulting in monomer droplets dispersed in water and stabilized by surfactant. Usually, chain-transfer agents are added to reduce the (co)polymer molecular weight. Electrolytes are used to control the particle size.

According to the Harkins–Smith–Ewart theory[20,21] the process of emulsion polymerization can be divided into three distinct intervals. Interval I comprises particle nucleation. A third phase is formed consisting of monomer-swollen polymer particles. In intervals II and III, ideally the particle number remains constant. During interval II the polymerization proceeds in the presence of monomer droplets, while micelles have then disappeared. The monomer droplets are totally consumed at the end of interval II, and the monomer remaining in the particles is polymerized during interval III. Therefore, intervals II and III are called the particle growth stages.

Stockmayer,[22] O'Toole[23] and Ugelstad[24] gave general solutions for the calculation of the average radical number per particle in the steady-state situation. Other major contributions to the model development of the emulsion polymerization process concerning particle nucleation, PSD and kinetics were made by Ray,[25–27] Hansen and Ugelstad,[28] Gilbert and Napper[8,29–35] and Chen.[36,37] Mathematical extensions to emulsion copolymerization kinetics were made by the research groups of Nomura,[38–41] Guillot,[42–44] Dougherty,[45,46] Poehlein,[47,48] Storti,[49] Gilbert,[50,51] Forcada[52] and Broadhead.[53] (Lack of space does not permit a review of each of these articles.)

Regarding the emulsion polymer microstructure, Lichti, Gilbert and Napper[54] and also Lin[55] have developed theories to describe the MMD (MWD) of emulsion homopolymers.

Ballard *et al.*[51] and also Guillot and coworkers[56,57] studied the sequence distribution of emulsion copolymers. Storti *et al.*[58,59] developed a model to describe the (MM)CCD of the (instantaneously) formed emulsion copolymer (Section 3.2.7), but did not compare the model predictions with experimental MMCCD data. However, detailed experimental elucidation of the copolymer microstructure is a prerequisite in this kind of study. Several experimental cross-fractionation techniques have now become available to determine not only the average copolymer composition, but also the complete MMCCD.[60,61] Together with experimental sequence distribution information obtained by means of NMR,[62] these cross-fractionation techniques have recently been used to provide the necessary experimental data to check the validity of the model calculations.

In the following, several aspects of modelling emulsion copolymerization will be briefly discussed: monomer partitioning, particle formation, and radical entry, exit, transfer, propagation and termination. Furthermore, models describing copolymerizations will be mentioned.

3.2.3 Theoretical Aspects of Monomer Partitioning

The monomers are distributed between the particles, the aqueous phase and, if present, the monomer droplets. It is generally recognized that in emulsion (co)polymerization the monomer partitioning is determined by thermodynamic equilibrium. Equilibrium requires the chemical potential (μ) of the monomers in all phases to be equal. Morton *et al.*[63] have developed a model that describes the monomer partitioning in the case of homopolymers, when monomer droplets are still present. This model is based on the classical Flory–Huggins lattice theory for monomer–(homo)polymer mixtures and includes an interfacial energy term. Morton assumed that even in those cases where the monomer is a good solvent for the polymer, only a limited amount of monomer is absorbed by the latex particles, because the increase in surface free energy on swelling counteracts the loss of free energy of mixing. Moreover, the particles are assumed to be homogeneous.

Morton's equation, describing the monomer partitioning during emulsion homopolymerization in the presence of monomer droplets, reads

$$\ln(1 - \Phi_p) + (1 - m_{ip})\Phi_p + \chi\Phi_p^2 + 2\gamma V_m/(rRT) = 0 \tag{1}$$

where

γ is the interfacial tension between the aqueous phase and the polymer particle swollen with monomer (N m^{-1});

Φ_p is the volume fraction of polymer in the polymer particle;

V_m is the molar volume of monomer (m^3 mol^{-1});

R is the gas constant (J K^{-1} mol^{-1});

T is the temperature (K);

r is the radius of the swollen polymer particle (m) ($r = r_{\text{nonswollen}}/\Phi_p^{1/3}$);

χ is the (temperature-dependent) Flory–Huggins interaction parameter; and

$m_{i,p}$ is the ratio of molar volumes of the monomer i and the polymer.

It directly follows from equation (1) that the monomer concentration inside the particles will increase with decreasing interfacial tension. An increase in temperature would result in an increase in monomer concentration in the particles. The complexity of this matter is demonstrated by experimentally determined data for styrene (ranging from 5.9 mol L^{-1} at 25 °C to 4.8 mol L^{-1} at 70 °C) reported by Van der Hoff,[64] exhibiting the opposite behaviour. This unexpected behaviour has to be explained by the temperature dependence of χ.

An excellent paper on this matter has been written by Gardon,[65] leading to the following conclusions.

(i) The interfacial tension (γ) mainly depends on the type and concentration of emulsifier, ionic strength, type of polymer and polar (sulfate) chain ends at the particle–water interface.

(ii) Latices prepared under different process conditions often have a similar swelling behaviour due to the self-compensating effects of some of these parameters. For example, a higher emulsifier concentration during emulsion polymerization reduces the interfacial tension between the particles and the aqueous phase which would lead to a higher extent of swelling, but it also results in the formation of more, smaller particles that have the tendency to swell less with monomer. In most cases these two opposite tendencies appear to compensate each other to a large extent.

(iii) Furthermore, Gardon noticed that (among the monomers that are good solvents for their polymers in bulk) in many cases the monomers exhibiting a higher water solubility swell their polymer particles to a larger extent than the poorly water-soluble monomers. This was attributed to the fact that monomer dissolved in the aqueous phase reduces the interfacial tension between the swollen particles and the aqueous phase.

(iv) For styrene and also for methyl methacrylate, below saturation (i.e. in the absence of monomer droplets), the monomer partitioning was found to adjust itself in such a way that the aqueous phase is closer to monomer saturation than the latex particles. For MMA this behaviour, together with the insensitivity of the monomer concentration in the particles to surfactant and ionic strength, was also observed by Ballard et al.[32]

Vanzo[66] extended Morton's equation to unsaturated systems (no monomer droplets present) by taking into consideration the chemical potential of the monomer in the water phase ($RT^* \ln a_i$), approximating the activity (a_i) of the monomer in the water phase by p/p_0, i.e. the ratio of the vapour pressure at a given volume fraction of polymer in the particles to the vapour pressure at equilibrium swelling.

Maxwell et al.[67] reconsidered the matter and compared the results of several studies on the partitioning of a number of monomers with predictions based on Vanzo's theory, setting a_i equal to $M_{\text{aq}}/M_{\text{aq,sat}}$, where M_{aq} is the monomer concentration in the aqueous phase and $M_{\text{aq,sat}}$ is the saturation value of M_{aq}. Furthermore, they predicted the monomer partitioning only using the values of the saturation concentrations in the aqueous phase and polymer particle.

Guillot[68] extended the thermodynamic monomer partitioning treatment of Morton in an effort to describe the monomer partitioning during emulsion copolymerization by introducing interaction terms for the monomers. Guillot gave equations for calculating the monomer and polymer volume fractions and (partial molar) free energies of both monomers in the three phases. The monomer chemical potential is defined as the difference between the chemical potential of a monomer in a given particular phase and the chemical potential of pure monomer under reference conditions ($\Delta\mu_{i,\text{phase}} = \mu_{i,\text{phase}} - \mu_i^*$). He assumed that the chemical potentials of the monomers will be equal in each phase (i.e. equilibrium). The polymerization within the particles tends to change the chemical potential of the monomers in the particles ($\mu_{i,p}$) continuously. The resulting infinitesimal differences in μ_p of the monomers lead to very rapid transfer of monomer towards the particles from the aqueous phase and, if present, from the monomer droplets in order to reestablish the equality of the chemical potential in all phases (phase equilibrium). The chemical potentials ($\Delta\mu_i$) of monomer i in the particles, droplets and aqueous phase are given by equations (2), (3) and (4), respectively. If necessary, additional terms accounting for crosslinks in the particles and surface electric charge can be implemented.[43]

$$\Delta\mu_{i,p} = RT[\ln(\Phi_{i,p}) + (1 - m_{ij})\Phi_{j,p} + (1 - m_{ip})\Phi_{P,p} + \chi_{ij}\Phi_{j,p}^2 + \chi_{ip}\Phi_{P,p}^2 \\ + \Phi_{j,p}\Phi_{P,p}(\chi_{ij} + \chi_{ip} - \chi_{jp}m_{ij})] + 2\gamma V_i/r_p \quad (2)$$

$$\Delta\mu_{i,d} = RT[\ln(\Phi_{i,d}) + (1 - m_{ij})\Phi_{j,d} + \chi_{ij}\Phi_{j,d}^2] + 2\gamma_d V_i/r_d \quad (3)$$

$$\Delta\mu_{i,a} = RT[\ln(\Phi_{i,a}) + (1 - m_{ij})\Phi_{j,a} + (1 - m_{iw})\Phi_{w,a} + \chi_{ij}\Phi_{j,a}^2 + \chi_{i,w}\Phi_{w,a}^2 + \Phi_{j,a}\Phi_{w,a}(\chi_{ij} + \chi_{iw} - \chi_{jw}m_{ij})] \quad (4)$$

where

V_i is the molar volume of monomer i (m³ mol⁻¹);
P,i,j are the polymer, monomer i and monomer j, respectively;
p   is the polymer particle;
$\Phi_{i,p}$ is the volume fraction of monomer i in the swollen polymer particle;
$\Phi_{P,p}$ is the volume fraction of polymer in the swollen polymer particle;
m_{ij} is the ratio of molar volumes (V_i/V_j);
r_p is the radius of swollen polymer particle (m);
d   is the monomer droplet;
γ_d is the interfacial tension between monomer droplets and the aqueous phase (N m⁻¹);
r_d is the radius of the monomer droplet;
a   is the aqueous phase; and
w   is water.

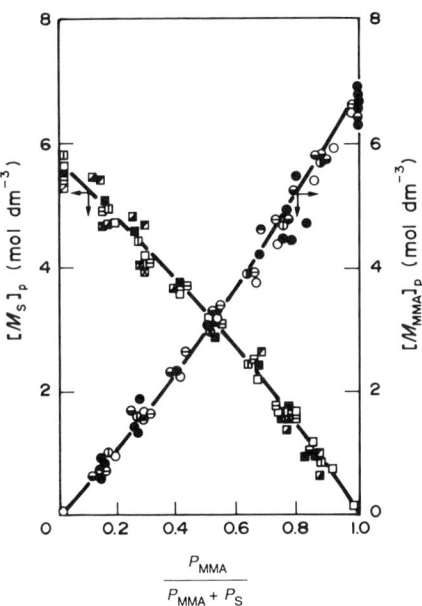

$[M_{MMA}]_p$	$[M_S]_p$	$P_{MMA}/(P_{MMA} + P_S)$	d_p (nm)
●	■	1.0	90
⊖	⊟	0.85	75
⊕	⊞	0.70	80
◔	☐	0.50	55
●	▨	0.10	90
⊖	▨	0	80
⊗	⊠	0.50	25
◉	▣	0.50	110

Figure 1 Effects of particle size and copolymer composition on the concentration of each monomer ($[M_S]_p$ and $[M_{MMA}]_p$) in monomer-swollen polymer particles *versus* weight fraction of MMA in the droplets (W_{dMMA}). Temperature = 50 °C; ionic strength $\mu = 0$; interfacial tension $\sigma = 57$ dyne cm⁻¹ (reproduced with permission from *J. Appl. Polym. Sci.*, 1982, **27**, 2483)

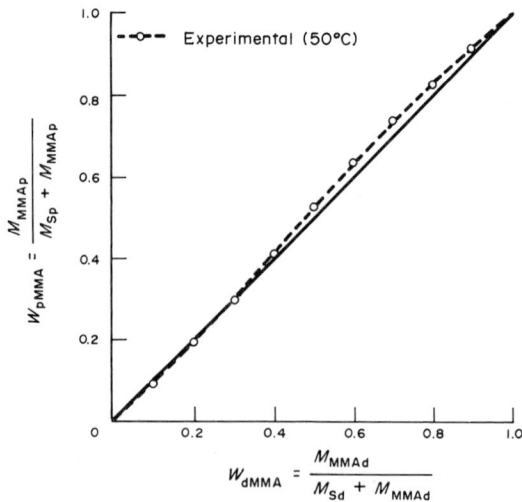

Figure 2 Experimental demonstration of the approximate validity of the assumption that the monomer weight fractions of methyl methacrylate in the swollen polymer particles (W_{pMMA}) and in the monomer droplets (W_{dMMA}) are equal (reproduced with permission from *Makromol. Chem. Suppl.*, 1985, **10/11**, 25)

Furthermore, Ugelstad[69,70] studied the effect of low molecular weight additives on the swelling of seed latex particles to be used for the preparation of large monodisperse polymer particles. Tseng et al.[71] described the effect of the presence of water in the latex particles and in the monomer droplets on the partitioning of relatively hydrophilic monomers. In the case of emulsion copolymerization, Guillot[72] and also de la Cal et al.[73] implemented the existence of monomer concentration gradients within the particles due to changing thermodynamic interactions between the monomers and copolymer molecules, varying in chemical composition as a function of the radius of the heterogeneous (co)polymer particles.

During the last few years the thermodynamic treatment has been applied by several investigators.[47,48,52,74] More frequently, however, experimentally determined partition coefficients have been used to describe the monomer partitioning, because the required interaction parameters and interfacial tension parameters are generally unknown.

For instance, Nomura and coworkers[75] measured the concentrations of styrene (S) and methyl methacrylate (MMA) in the latex particles as a function of the monomer phase composition (fraction MMA) in an S–MMA emulsion system by means of equilibrium experiments (Figure 1). They demonstrated that the monomer ratio (or the monomer weight fractions) in the latex particles is equal to the monomer ratio (or monomer weight fractions) in the monomer droplets (Figure 2).[40]

3.2.4 Particle Formation in the case of Emulsion Copolymerization

In the literature, many mechanisms have been proposed for particle nucleation. These theories can be divided into three main categories according to a different *locus* of nucleation, *i.e.*: (i) monomer-swollen micelles (micellar nucleation);[20] (ii) monomer droplets;[76,77] and (iii) aqueous phase (homogeneous or coagulative nucleation).[29,78]

(i) Small oligomeric radicals generated in the aqueous phase enter monomer-swollen micelles (usually about 5 nm in size) and initiate polymerization to form monomer-swollen polymer particles. Because the total surface area of the micelles is orders of magnitude larger than that of the monomer droplets, normally no nucleation occurs in the monomer droplets (mechanism ii), except under certain extreme circumstances, *e.g.* if the monomer droplet size is decreased significantly as in miniemulsions.[76] When the micelles disappear, the particle nucleation stage ends. According to the 'initiation in the aqueous phase' mechanism (iii)[28,29,78,79] that is believed to be dominant when the emulsifier concentration is below the critical micelle concentration (cmc) or in the case of polar monomers, radicals generated in the aqueous phase add monomer until the oligomeric radicals so formed reach a critical degree of polymerization. This critical length is suggested to be the point at which they reach their solubility limit or the length at which the oligomers become surface active and (co)precipitate. The precipitated oligomeric chains will absorb monomer and adsorb emulsifier to form primary particles. These primary particles are colloidally unstable and can coagulate with each other or with already existing latex particles (heterocoagulation).

Based on the micellar nucleation theory Nomura[41] developed a mathematical, kinetic model for unseeded emulsion copolymerization systems, where he discriminated between charged and uncharged radicals in view of their different roles in the particle nucleation process. The negatively charged radicals consist of the initiator radicals and the oligomeric radicals with initiator end groups. The other (oligomeric) radicals are uncharged and are formed by transfer to monomer and successive desorption from the already existing particles. Batch unseeded emulsion copolymerization experiments with styrene and methyl methacrylate demonstrated that almost all polymer particles are generated by charged radicals stemming from the initiator. The effect of emulsifier concentration, initiator concentration and monomer composition on the number of particles and therefore on the characteristic kinetic features, could be successfully described by this model.

Based on both homogeneous and heterogeneous (micellar) nucleation Forcada and Asua[80] developed an alternative mathematical model for unseeded emulsion copolymerization of styrene (S) and methyl methacrylate (MMA). The effect of monomer feed ratio on the number of particles was analyzed with this model and could be explained on the basis of two combined effects: (i) an increase in the particle surface area covered by a single surfactant molecule with an increase in MMA content; and (ii) an increase in the radical desorption rate coefficient with an increase in the MMA content of the latex particles. Furthermore, induction periods in the conversion–time plots could be explained by enhanced radical desorption during the nucleation stage.

Nomura and coworkers[81] tried to determine the locus of particle formation by investigating the difference in chemical composition of (styrene–methyl methacrylate and styrene–methyl acrylate) copolymers initially formed in both systems in the presence and absence of emulsifier, respectively. They found that the initially formed copolymer was enriched in MMA (the most water-soluble monomer) in the absence of emulsifier, in comparison with the experiments in the presence of emulsifier. This indicates the occurrence of homogeneous nucleation (aqueous phase polymerization). On the other hand, the experiments may not be conclusive, because the conversion (0.1%) was still 10 times too high[82] for the aqueous phase to be the main locus of polymerization.

3.2.5 Propagation Rate Models

Applying the pseudo homopolymerization rate approach, the copolymerization rate can be calculated by equation (5) where $[M_t]_p$ is the total monomer concentration in the particles (mol L^{-1}), $\bar{n}_t$ is the average number of radicals per particle and $\bar{k}_p$ is the average propagation rate constant. In Sections 3.2.5.1 and 3.2.5.2 a model for calculating $\bar{k}_p$ is described.

$$R_p = \bar{k}_p \bar{n}_t [M_t] N / N_{av} \tag{5}$$

3.2.5.1 The ultimate and the penultimate model for copolymerization

The model most frequently used to describe copolymerization kinetics, sequence distribution and chemical composition of copolymers (for homogeneous copolymerization processes, *viz.* bulk and solution copolymerization), is the ultimate model, also known as the terminal model. In 1944 this model was independently introduced by Alfrey and Goldfinger,[83] and Mayo and Lewis.[84] In this model the monomer addition rate only depends on the nature of the terminal group (see Table 1), and therefore obeys first-order Markov statistics.[85]

The reactivity ratios of monomers A and B are defined by $r_a = k_{aa}/k_{ab}$ and $r_b = k_{bb}/k_{ba}$, respectively, where k_{ij} is the propagation rate constant of the propagation reaction between radical i and monomer j.

Table 1 Copolymerization Scheme According to the Ultimate Model

Terminal group	Added monomer	Rate	Result
~A*	[A]	$k_{aa}[A^*][A]$	~AA*
~A*	[B]	$k_{ab}[A^*][B]$	~AB*
~B*	[A]	$k_{ba}[B^*][A]$	~BA*
~B*	[B]	$k_{bb}[B^*][B]$	~BB*

*Denotes free radical activity.

Table 2 Copolymerization Scheme According to the Penultimate Model

Terminal group	Added monomer	Rate	Result
~AA*	[A]	$k_{aaa}[AA^*][A]$	~AAA*
~AA*	[B]	$k_{aab}[AA^*][B]$	~AAB*
~BA*	[A]	$k_{baa}[BA^*][A]$	~BAA*
~BA*	[B]	$k_{bab}[BA^*][B]$	~BAB*
~AB*	[A]	$k_{aba}[AB^*][A]$	~ABA*
~AB*	[B]	$k_{abb}[AB^*][B]$	~ABB*
~BB*	[A]	$k_{bba}[BB^*][A]$	~BBA*
~BB*	[B]	$k_{bbb}[BB^*][B]$	~BBB*

*Denotes free radical activity.

Combining the equations for monomer consumption and the steady-state assumption for the radicals, the instantaneous copolymerization equation giving the mol fraction monomer *a* units in the instantaneously formed copolymer (F_a) as a function of reactivity ratios and monomer feed composition ($f_a = 1 - f_b$), can easily be derived, applying the long chain approximation

$$F_a = \frac{r_a f_a^2 + f_a f_b}{r_a f_a^2 + 2 f_a f_b + r_b f_b^2} \tag{6}$$

The average propagation rate constant ($\bar{k}_p$) can then be calculated using

$$\bar{k}_p = \frac{r_a f_a^2 + 2 f_a f_b + r_b f_b^2}{(r_a f_a / k_{aa}) + (r_b f_b / k_{bb})} \tag{7}$$

In systems where the nature of the penultimate unit has a significant effect on the absolute rate constants in copolymerization, the penultimate model has to be used, where in principle eight different reactions have to be considered (see Table 2).[86]

The six reactivity ratios of monomer i and j are defined by $r_i = k_{iii}/k_{iij}$, $r_i' = k_{jii}/k_{jij}$, and $s_i = k_{jii}/k_{iii}$ (with $i = a, b$ and $j = b, a$, successively).

The average propagation rate constant and copolymer composition can now be calculated using equations (8) to (11).

$$\bar{k}_p = \frac{\bar{r}_a f_a^2 + 2 f_a f_b + \bar{r}_b f_b^2}{(\bar{r}_a f_a / \bar{k}_{aa}) + (\bar{r}_b f_b / \bar{k}_{bb})} \tag{8}$$

and

$$F_a = \frac{\bar{r}_a f_a^2 + f_a f_b}{\bar{r}_a f_a^2 + 2 f_a f_b + \bar{r}_b f_b^2} \tag{9}$$

where

$$\bar{k}_{ii} = \frac{k_{iii}(r_i f_i + f_j)}{r_i f_i + (f_j / s_i)} \tag{10}$$

and

$$\bar{r}_i = \frac{r_i'(r_i f_i + f_j)}{r_i' f_i + f_j} \tag{11}$$

As recently shown for several styrene–(meth)acrylate monomeric systems, in some cases the (overall) rate behaviour may exhibit significant deviations from the ultimate model, whereas composition drift and sequence distribution can be described by the ultimate model. Numerical values of average propagation rate constants of those systems appeared to follow the penultimate model.[87,88] The apparently contradictory situation that the polymerization rate can be described by the penultimate model, whereas the composition drift and sequence distribution can still be described by the ultimate model, can be easily verified from equations (8) to (11) by substituting $r_i = r_i'$.

3.2.5.2 *Other propagation rate models*

Other mechanisms for copolymerization that have to be mentioned are the complex participation model[89] and the complex dissociation model.[90] Under certain conditions even depropagation has

3.2.6 Kinetic Events Affecting the Average Number of Radicals per Particle

3.2.6.1 Radical production

The radical production rate (R_i) per litre of water (mol L^{-1} s^{-1}) can be calculated using equation (12)

$$R_i = 2k_d f [I]_w \tag{12}$$

where k_d is the initiator decomposition rate constant (s^{-1}), f is the initiator decomposition efficiency and $[I]_w$ is the initiator concentration in the aqueous phase (mol L^{-1}).

3.2.6.2 Chain transfer

In the emulsion copolymerization models the chain-transfer rate can be calculated by taking into account all possible chain-transfer reactions (viz. transfer to monomer, polymer and chain-transfer agent), occurring both in the aqueous and the organic phase. For instance, equation (13) gives the rate of transfer to monomer (r_{mf}) in the particles[40]

$$r_{mf} = (k_{maa}[M_a]_p + k_{mab}[M_b]_p)\bar{n}_a N_T + (k_{mbb}[M_b]_p + k_{mba}[M_a]_p)\bar{n}_b N_T \tag{13}$$

where k_{mij} is the chain-transfer constant of a polymer radical of type i to monomer j, $[M_i]_p$ is the concentration of monomer i in the particles, $\bar{n}_i$ is the average number of radicals of type i in the particles and N_T is the total number of particles.

3.2.6.3 Desorption of radicals

It is generally recognized that free radicals can desorb from the polymer particles. Exit phenomena may play a crucial role, even for relatively water-insoluble monomers such as styrene.

According to the transfer/diffusion theory of Nomura and coworkers[38,75,92] radical desorption can be regarded as a four-step process: (i) a propagating polymer chain transfers its free radical activity to a monomer molecule or to a (low molecular weight) chain-transfer agent; (ii) this small molecule (radical) then diffuses towards the surface of the particle; (iii) it subsequently desorbs from the particle provided this radical has not undergone significant propagation; and (iv) the exit process is completed by diffusion away from the particle into the bulk of the aqueous phase. More detailed considerations concerning the theory of monomer radical transport in the particles were given by Mead and Poehlein.[93] Asua et al.[94] also included aqueous phase reactions (propagation, termination).

According to the theory of Nomura, in copolymerization the two monomeric radical species may desorb at different rates, also depending on the monomer concentration ratios within the particles and the rates of chain transfer to both monomers. The desorption rate constant for monomer a radicals (K_{o_a}) (s^{-1}) can be calculated according to equation (14)

$$K_{o_a} = \frac{12 D_{w_a} \delta_a z}{m_{p,a} d_p^2} \tag{14}$$

with

$$\delta_a = [1 + 6D_{w_a}/(m_{p,a} D_{p_a})]^{-1} \tag{15}$$

and

$$m_{p,a} = [M_a^*]_p / [M_a^*]_w \tag{16}$$

where $m_{p,a}$ is the partition coefficient for the a radicals between the water phase and polymer particle, respectively, and δ_a is the ratio of film mass transfer resistance to overall mass transfer resistance of monomer a. D_{p_a} and D_{w_a} are the diffusion coefficients for the exiting species (free a radicals) in the particle and in the water phase, respectively, and z is the average degree of polymerization of the exiting free radicals.

Normally, it is assumed that only monomeric radicals can desorb ($z = 1$). When 1-dodecanethiol (DT) is used as the transfer agent, the desorption of free radicals stemming from chain transfer to thiol can be neglected due to the extremely low water solubility of DT.[95] So under these conditions, exit is only possible after transfer to monomer and if the monomer radical neither initiates (propagates), nor transfers or terminates its radical activity. Equation (17) then gives an expression for the net desorption rate for a radicals

$$k_{fa} = K_{oa} \cdot \left[\frac{r_a C m_{aa}[M_a]_p + C m_{ba}[M_b]_p}{r_a([M_a]_p + K_{oa}\bar{n}_t/k_{paa}) + [M_b]_p} \right] \quad (17)$$

A similar equation can be deduced for k_{fb}. The average rate coefficient for radical desorption from a polymer particle (s^{-1}) can then be calculated by means of equation (18)

$$\bar{k}_f = (n_a k_{fa} + n_b k_{fb})/\bar{n}_t \quad (18)$$

Several other researchers have used this approach, e.g. Mead and Poehlein[47] and Richards et al.[50] The latter extended Nomura's equations by taking into account transfer to the chain-transfer agent.

3.2.6.4 Termination rate constant in the particles

In general, termination processes in emulsion polymerization will be different from those operative in low viscosity noncompartmentalized systems. North and coworkers[96,97] have pointed out that termination is strongly diffusion controlled.

It is to be expected that the 'effective' k_{ta} and k_{tb} will dramatically decrease at higher weight fractions of polymer (w_p) in the latex particles.[32,98] Termination rate coefficients may also be a function of the molecular weight of the (co)polymer matrix and the degree of polymerization of the two mutually terminating chains.[99] Since at present the dependence of k_t on w_p is not known, this parameter has to be estimated within certain realistic limits. Usually, in copolymerization, one calculates an average termination rate constant $\bar{k}_t$ (equation 19)[52,47]

$$\bar{k}_t = k_{taa} P_a^2 + 2k_{tab} P_a P_b + k_{tbb} P_b^2 \quad (19)$$

where k_{tij} is the termination rate constant between two radical species i and j and P_i is the time-averaged probability of finding a free radical of type i.

3.2.6.5 Radical entry and number of radicals per particle

Several models have been proposed in order to describe the radical entry rate (i.e. the efficiency of the radical production rate). The rate-determining steps for radical entry were proposed to be: (i) diffusion of the radical in the aqueous phase to the particle;[100] (ii) surfactant displacement;[101] (iii) colloidal entry;[102] and (iv) recently, Maxwell et al.[103,104] have provided theoretical as well as experimental evidence strongly suggesting that in most cases aqueous phase propagation to a critical degree of polymerization (whereupon capture of the oligomeric radical is essentially instantaneous) will be the controlling factor. Until that critical length is reached a radical may undergo several desorption and re-entry events.

The average number of a and b radicals in the latex particles can be calculated according to the method of Nomura,[40] based on a proper averaging. The steady-state assumption (propagation is faster than changes in the numbers of radicals) can be used to calculate the ratio (A) of a and b radicals. Assuming that propagation is much faster than chain transfer, it can then easily be shown that[40]

$$A = \frac{\bar{n}_b}{\bar{n}_a} = \frac{k_{p,aa} r_b [M_b]_p}{k_{p,bb} r_a [M_a]_p} \quad (20)$$

In the well-known Smith–Ewart equations that provide the fraction of latex particles having i radicals in the steady-state situation, the parameters have their usual meanings

$$\frac{dN_n}{dt} = (\rho/N_T)N_{n-1} + k_f(n+1)N_{n+1} + c(n+2)(n+1)N_{n+2} - (\rho/N_T)N_n - k_f n N_n - c n(n-1)N_n = 0 \quad (21)$$

where the termination rate constant c (s^{-1}) can be calculated from $k_{tp} v_p^{-1} N_{av}^{-1}$, k_{tp} is the termination rate constant (L mol^{-1} s^{-1}), v_p is the volume of a single swollen latex particle and N_{av} is the Avogadro's constant.

Equation (21) is valid for homopolymerization. However, by applying the pseudo homopolymerization approach by using average rate constants, as introduced by Nomura[40] and further developed by Storti et al.,[49] this equation can also be used for copolymerization.

When employing pseudo homopolymerization constants, the radical entry rate (ρ_e) is expressed in the same form as in emulsion homopolymerization by solving the mass balance equation of the radicals in the aqueous phase

$$\rho_e = \bar{k}_a [R^*]_w N_t = R_i + \sum_{n=1}^{\infty} \bar{k}_e n N_n - 2\bar{k}_{tw}[R^*]_w^2 \qquad (22)$$

where n is $n_a + n_b$, $[R^*]_w$ is the concentration of all free radicals in the aqueous phase (mol L^{-1}), $\bar{k}_{tw}$ is the average free radical termination rate constant in the aqueous phase (L mol^{-1} s^{-1}), $\bar{k}_e$ is the average rate constant for radical desorption (mol's^{-1}) and $\bar{k}_a$ is the average radical absorption constant for latex particles (L s^{-1}).

Equations (21) and (22) can also be written in dimensionless parameters

$$\alpha N_{n-1} + m(n+1)N_{n+1} + (n+2)(n+1)N_{n+2} = \alpha N_n + mnN_n + n(n-1)N_n \qquad (23)$$

and

$$\alpha = \alpha' + m \cdot \bar{n} - Y \cdot \alpha^2 \qquad (24)$$

with

$$\alpha = \frac{\rho_e v_p N_{av}^2}{\bar{k}_{tp} N_T}; \quad \alpha' = \frac{R_i v_p N_{av}^2}{\bar{k}_{tp} N_T}; \quad m = \frac{\bar{k}_e v_p N_{av}^2}{\bar{k}_{tp}}; \quad Y = \frac{2\bar{k}_{tw}\bar{k}_{tp}}{\bar{k}_a^2 N_T v_p N_{av}^2}$$

If there is no water phase termination (thus 100% re-entry or 100% radical capture efficiency is assumed), Y is equal to zero.

The general solution of equation (23) gives the average number of radicals per particle

$$\bar{n}_t = \frac{\sum n N_n}{N_t} = \frac{a}{4} \cdot \frac{I_m(a)}{I_{m-1}(a)}; \quad a = (8\alpha)^{0.5} \qquad (25)$$

where I_m and I_{m-1} represent the modified Bessel functions of the first kind, defined by the following recursive equations

$$I_{m-1}(a) - I_{m+1}(a) = (2m/a)I_m(a)$$

Because ρ_e, α and therefore also a are unknown, α has to be eliminated. Ugelstad et al.[24] solved equations (24) and (25) simultaneously, thereby eliminating the parameter α that contains ρ_e. Assuming a fixed value for Y (e.g. $Y = 0$, i.e. negligible water phase termination compared with the entry rate), $\bar{n}_t$ can be expressed as a function of the predictable parameters α' and m only.[24]

3.2.7 Copolymer Microstructure

The instantaneous sequence distribution in terms of triad fractions can be calculated assuming first order Markovian kinetics.[15] The a-centred triad fractions are given by

$$F_{aaa} = [1 - P(B/A)]^2 \qquad (26)$$

$$F_{aab} = F_{baa} = 2P(B/A)[1 - P(B/A)] \qquad (27)$$

$$F_{bab} = [P(B/A)]^2 \qquad (28)$$

$$P(B/A) = 1/(1 + r_a[A]/[B]) \qquad (29)$$

$P(B/A)$ is the conditional probability of an a radical propagating with a b monomer. For the b-centred triads similar equations can be deduced. Ballard[105] derived more general and more complex copolymer sequence distribution functions in terms of the rate coefficients for free radical entry, exit, homo- and cross-propagation, transfer and termination. Forcada et al.[106] have shown that the simpler Markovian kinetics approach predicts the copolymer microstructure as accurately.

A limited number of models for calculating the (instantaneous) (MM)CCD and sequence distribution of emulsion copolymers have been developed in the last few years using either kinetic (i.e. stochastic)[107-109] or probabilistic approaches.[59] The general aspects will be discussed rather than details of the models. In nearly all cases the macromolecular chains are presumed to be linear.

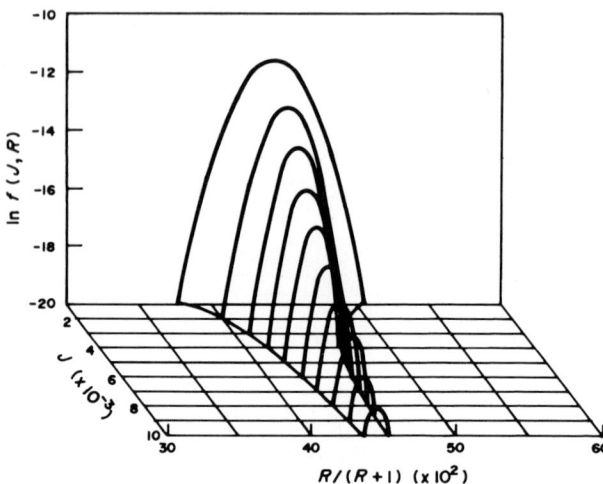

Figure 3 Example of a calculated three-dimensional distribution of degree of polymerization (J) and chemical composition ($R/(R+1)$) of an emulsion copolymer (reproduced with permission from *J. Polym. Sci., Polym. Chem. Ed.*, 1988, **26**, 2307)

On the basis of the mathematics of Markov chains, Storti *et al.*[109] recently proposed a general and rigorous model that determines the required two-dimensional distribution function (according to molecular weight and chemical composition) that allows the evaluation of the MMCCD for any number of radicals per particle. In earlier papers Storti *et al.* developed model calculations of the MMD and CCD, limited to the 'zero–one' case[59] and the 'zero–one–two' case[107] using the probabilistic approach. In Figure 3 an example is given of a three-dimensional distribution according to the degree of polymerization and chemical composition of an emulsion copolymer, calculated by Storti *et al.*[59]

In the most general case (*i.e.* any number of active chains per particle), an approximate solution can always be obtained through the pseudo homopolymerization approaches as reported by Broadhead *et al.*[53] and German and coworkers.[10] The pseudo homopolymerization approach was also applied by Nomura (polymerization rate, average chemical composition, and average degree of polymerization) and Ballard (sequence distribution). German and coworkers applied the equations valid for the instantaneous MWD of homopolymers given by Lichti, Gilbert and Napper[54] to copolymers after the introduction of pseudo homokinetic rate constants. In combination with the MMD equations they used a modified Stockmayer equation (valid for homogeneous solution and bulk systems) in order to account for the instantaneous CCD.

The widely applied pseudo homopolymerization approach reduces the computational effort for MMD calculations to that of an equivalent one-component system. The use of the pseudo homopolymerization approach necessarily results in the loss of part of the information.[49] Storti *et al.*[107] demonstrated that these approximated models are reliable owing to the high molecular weights usually encountered in emulsion polymerization.

3.3 MICROSTRUCTURE OF BATCH EMULSION COPOLYMERS

3.3.1 Introduction

The intermolecular microstructure of emulsion copolymers in terms of the molar mass chemical composition distribution (MMCCD) was mathematically studied by Giannetti *et al.*[58,59] The authors did not compare, however, the MMCCD model calculations with experimental distribution measurements. German and coworkers developed methods of experimentally measuring (MM)CCDs of emulsion copolymers of styrene–ethyl methacrylate, but in the first instance did not make the comparison with applicable model calculations.[60,110] Later they also developed the MMCCD determination of emulsion styrene–methyl acrylate copolymers[10] and compared the results with appropriate model calculations. Guillot and coworkers[111,120] investigated several styrene–acrylate (S–MA) emulsion copolymers using differential scanning calorimetry (DSC) in an attempt to determine the copolymer CCD. Unfortunately, this technique does not give the full MMCCD, but only average compositions.

Regarding the intramolecular microstructure Guillot and coworkers compared model calculations of sequence distribution at the triad level with experimental data, determined by means of ^{13}C NMR.[57,120] Later Van Doremaele et al.[112] measured and performed model calculations on the sequence distribution (in terms of triads) of S–MA batch emulsion copolymers. Furthermore, a detailed model study of the sequence distribution of emulsion copolymers was performed by Ballard.[105]

In the next sections some examples are given of the determination of the intermolecular microstructure (i.e. MMCCD) and the intramolecular microstructure (sequence distribution) of batch emulsion copolymers.

3.3.2 (Molar Mass) Chemical Composition Distribution

In the emulsion copolymerization of S and MA, which was studied by Van Doremaele,[10] the difference between the water solubilities of MA and S leads to a situation where the occurrence of azeotropy and the azeotropic composition depend on the overall monomer to water ratio. When using an overall initial monomer feed ratio of $(S/MA)_0 = 3$ (mol mol^{-1}) (this is the azeotropic composition of S–MA solution copolymerization), at an initial monomer to water ratio of $(M/W)_0 = 0.2$ (g g^{-1}), composition drift during emulsion copolymerization is negligible. Hence it might be expected that the copolymer formed is homogeneous. The MMCCD of this particular emulsion S–MA copolymer was determined experimentally (by means of SEC–TLC/FID).[4] The copolymers were separated according to molecular weight by means of size exclusion chromatography (SEC) and each SEC fraction was subsequently analyzed according to chemical composition by means of either gradient elution quantitative thin layer chromatography (TLC/FID) or gradient high performance liquid chromatography (HPLC). From the MMCCD it could be concluded that under these conditions, at least up to 90 mol% conversion, the copolymer formed is homogeneous with respect to the chemical composition. Because the composition distribution was narrow, it was concluded that the polymer particles are the main site of polymerization. However, under different conditions (e.g. higher temperature and higher initiator concentrations) it was found that non-negligible polymerization in the aqueous phase and polymerization inside the very small precursor particles during the early stages of emulsion polymerization (interval I) may lead to anomalous CCDs, because the monomer ratio is different at the various sites of (co)polymerization.[113]

Under the same reaction conditions, but by applying a different recipe (nonazeotropic conditions), asymmetrically shaped and even bimodal MMCCDs can be obtained (Figure 4).

Guillot and coworkers reported aqueous phase polymerization in the case of S–MA,[114] vinyl acetate–butyl acrylate (VAc–BA)[115] and styrene–ethyl acrylate (S–EA)[116] batch emulsion copolymerization (Figure 5). Capek et al.[117] reported a similar behaviour for acrylonitrile–butyl acrylate. Polymerization in the aqueous phase results in the formation of copolymer that is enriched in the water-soluble monomer. This effect is enhanced by low monomer/water ratios, and is expected

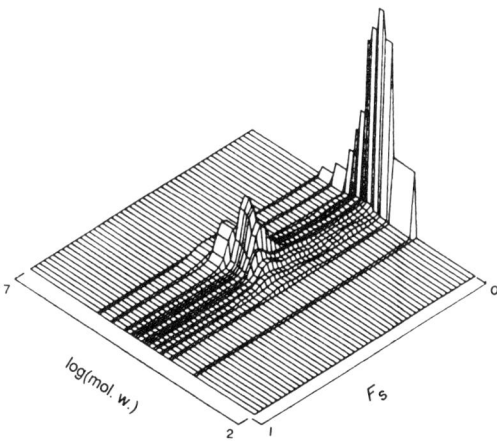

Figure 4 Experimental MMCCD of a high conversion (95 mol%) S–MA emulsion copolymer. $(S/MA)_0 = 0.33$ (mol mol^{-1}), $(M/W)_0 = 0.5$ (g g^{-1}), 1 wt% DT, 50 °C, and $M_w = 110\,000$ (g mol^{-1}) (reproduced with permission from G. H. J. Van Doremaele, Ph.D. Thesis, Eindhoven University of Technology, The Netherlands, 1990)

to show up most explicitly during interval I (nucleation stage). During intervals II and III the (co)oligomers formed in the aqueous phase were shown to be scavenged by the latex particles, already at relatively short length.[103,104]

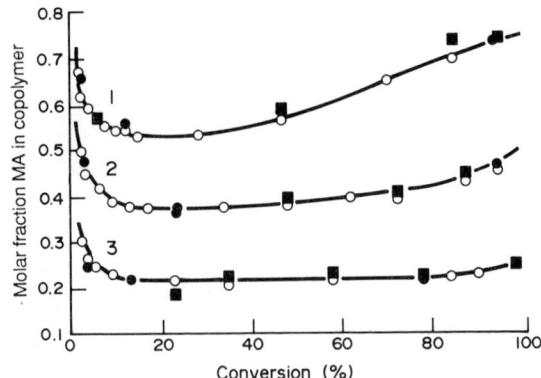

Figure 5 Fraction of methyl acrylate in copolymer in batch emulsion copolymerization of styrene and methyl acrylate as a function of conversion for various initial monomer compositions f_{MA}/f_S: (1) = 75/25, (2) = 50/50, (3) = 25/75. Monomer to water ratio is 0.2. (■) ^{1}H NMR; (○) element analysis; (●) gas chromatography (reproduced with permission from *Makromol. Chem.*, 1988, **189**, 361)

Neglecting aqueous phase polymerization, which generally comprises less than 1% of the total amount of polymer formed,[48,81,118] the monomer ratio inside the latex particles, together with the reactivity ratios, governs the instantaneous copolymer composition. The local monomer ratio inside the latex particles is generally equal to the monomer ratio in the droplets but differs from the overall monomer feed ratio in the system. In the present case, the latex particles will contain more styrene as compared with the overall monomer ratio. Therefore, the copolymer initially formed will also be richer in styrene as compared with expectations based on homogeneous systems, thus neglecting monomer partitioning. This results in a strong composition drift during polymerization towards compositions richer in the less reactive and more water-soluble monomer, *i.e.* MA, eventually leading to a considerable amount of pure PMA formation at high conversion. This bimodality in copolymer CCD is reflected in the occurrence of two glass transition temperatures due to phase separation: one at *ca.* 15 °C characteristic of the PMA rich domains, and one at a temperature between 15 and 100 °C depending on the chemical composition of the (mixed) copolymer peak domain. Concerning the effect of composition drift on the occurrence of glass transition temperatures, Guillot and coworkers (see *e.g.* refs. 119 and 120), investigated the DSC curves of several copolymer systems and were able to model these curves (Figure 6).

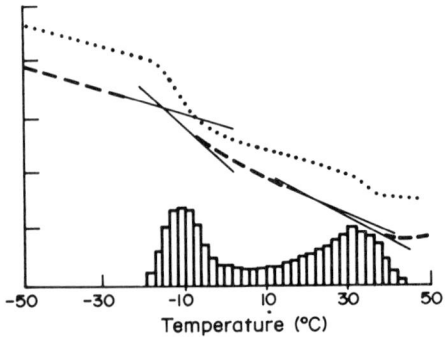

Figure 6 Styrene/ethyl acrylate batch emulsion copolymer DSC curves and histogram. Initial monomer feed: 80 mol% EA; $(M/W)_0 = 0.2 (g\,g^{-1})$. Overall conversion 98%. Simulation (···); experimental data (---) (reproduced with permission from *Eur. Polym. J.*, 1990, **26**, 1017)

3.3.3 Copolymer Sequence Distribution

The intramolecular microstructure of S–MA batch emulsion copolymers in terms of triad fractions was studied by German and coworkers.[112] In Figure 7 the theoretically predicted cumulative triad

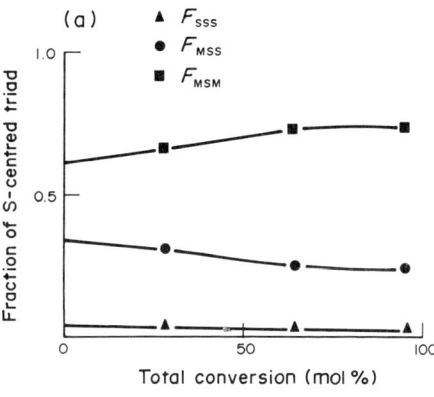 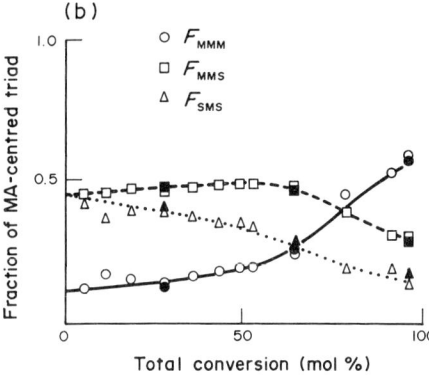

Figure 7 Cumulative emulsion copolymer triad fractions as a function of conversion. $(S/MA)_0 = 0.33$ (mol mol^{-1}), $(M/W)_0 = 0.5$ (g g^{-1}). Open symbols = ^{1}H NMR; closed symbols = ^{13}C NMR; lines = model calculations (reproduced with permission from *Macromolecules*, 1990, **23**, 4206)

fractions (lines) together with the measured triads have been plotted *versus* conversion for an emulsion copolymerization with $(S/MA)_0 = 0.33$ (mol mol^{-1}) and $(M/W)_0 = 0.5$ (g g^{-1}).

In this particular experiment, at *ca.* 60 mol% conversion all styrene has already been depleted and after this moment almost pure PMA is formed, thus resulting in a strong increase of the cumulative MMM (M = MA here) triad fractions at the expense of the cumulative MMS and SMS triad fractions. Because almost no free styrene was present after 60 mol% conversion, the cumulative styrene-centred triad fractions remain constant.

3.4 THE EFFECT OF COMPOSITION DRIFT ON COPOLYMERIZATION RATE

3.4.1 Introduction

In the following, some examples of batch emulsion copolymerizations will be treated that have typical conversion–time curves. As stated before, composition drift is a distinct aspect of emulsion copolymerization as compared with conventional emulsion homopolymerization. It is characterized by the preferential consumption of one of the monomers. Depending on whether or not the more reactive comonomer is also the least water-soluble one, this may result in either a broader or a narrower chemical composition distribution, as the water content increases.

Apart from the effect of composition drift on the chemical heterogeneity of the copolymers there can also be an effect on the copolymerization rate. It can be easily understood that the kinetic characteristics of the polymerizing system change as the composition of the monomer feed in the main locus of polymerization changes. In the literature several examples are given of kinetic investigations concerning this aspect of composition drift (see *e.g.* ref. 121). Several models have been developed to describe the conversion–time curves of batch emulsion copolymerizations (see *e.g.* refs. 41, 50 and 52).

3.4.2 Emulsion Copolymerizations of Vinyl Acetate–Butyl Acrylate and Styrene–Methyl Acrylate

A well-known example is the emulsion copolymerization of vinyl acetate (VAc) with butyl acrylate (BA). This monomer pair has been extensively described in the literature (see *e.g.* refs. 115 and 122). In this case BA is more reactive than VAc and less water-soluble than VAc ($r_{ba} \approx 5.5$, $r_{vac} \approx 0.04$).[123] This results in a strong composition drift. Figure 8 shows the conversion–time plot of this copolymerization. It can be clearly seen that in the first part of the reaction BA is preferentially built-in and when there is virtually no BA left, VAc practically homopolymerizes.

Similar results are obtained in the batch emulsion copolymerization of styrene (S) and methyl acrylate (MA).[10,114] Also in this case the more reactive monomer (S) is also the less water-soluble one. However, not only the large difference between the reactivity ratios but also the large difference in water solubility is a major cause of the strong composition drift. The conversion–time plot is

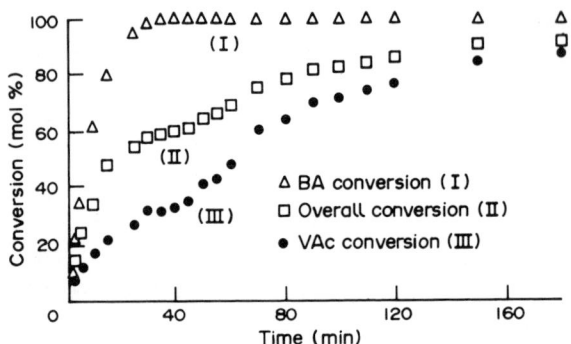

Figure 8 Conversion–time plot of batch emulsion vinyl acetate/butyl acrylate copolymerization at 60 °C with $(VAc/BA)_0 = 0.5\,(g\,g^{-1})$, $(M/W)_0 = 0.5\,(g\,g^{-1})$ (reproduced with permission from *Eur. Polym. J.*, 1988, **24**, 485)

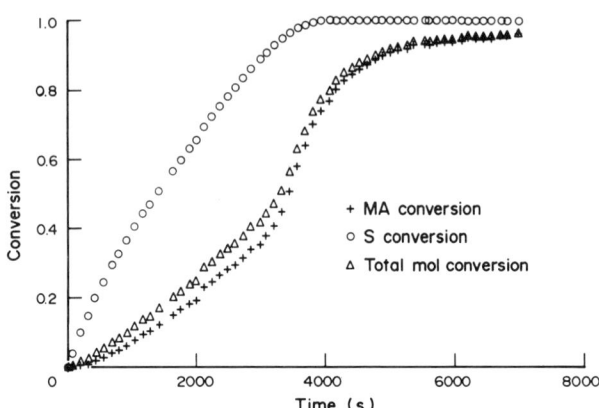

Figure 9 Conversion–time plot of batch emulsion styrene/methyl acrylate copolymerization at 50 °C with $(S/MA)_0 = 0.14\,\mathrm{mol\,mol^{-1}}$, $(M/W)_0 = 0.2\,(g\,g^{-1})$, an SDS (sodium dodecyl sulfate) concentration of 0.0116 $(\mathrm{mol\,L^{-1}})$, a $K_2S_2O_8$ concentration of 1.233 $(\mathrm{mmol\,L^{-1}})$, an $NaHCO_3$ concentration of 1.233 $(\mathrm{mmol\,L^{-1}})$ and 1 wt % DT (reproduced with permission from G. H. J. Van Doremaele, Ph.D. Thesis, Eindhoven University of Technology, The Netherlands, 1990)

shown in Figure 9. Again it is clear that in the first part copolymerization occurs favouring S consumption and after depletion of S copolymer very rich in MA is made.

Both examples show a remarkable phenomenon. From these plots it is obvious that the copolymerizations pass through several stages. Polymerization starts with the particle nucleation stage. After *ca.* 10% conversion an almost constant total copolymerization rate is observed until *ca.* 40% conversion. Copolymerization rate then decreases in the case of VAc–BuA and this was also found for S–MA with $(S/MA)_0 = 0.33\,(\mathrm{mol\,mol^{-1}})$.[10] This decrease in copolymerization rate is attributed to a decrease in monomer concentrations in the latex particles. At the point at which BA or S is (almost) totally depleted, the polymerization rate is at its lowest. From this point on the reaction rate suddenly shows a significant increase as VAc or MA homopolymerizes. Depletion of VAc or MA, the preferential presence of VAc or MA in the aqueous phase and diffusion controlled propagation finally result in a decrease in polymerization rate at high conversions. Thus composition drift strongly affects (both the VAc–BA and the S–MA) copolymerization rates. The increase in rate occurring when the more reactive monomers are depleted can in principle be attributed to several possible mechanisms, since according to equation (5) polymerization rate is proportional to: (i) the monomer concentrations inside the particles; (ii) the number of latex particles; (iii) the number of radicals per particle; and (iv) the average propagation rate constant.

Mechanism (a): during interval III of the emulsion polymerization an increase in monomer concentration in the particles is of course impossible. However, a (relative) increase in local monomer concentration in the shells of the polymer particles cannot be completely ruled out, although this phenomenon, if occurring at all, is not believed to be a major cause of the strong rate increase observed.

Mechanism (b): the rate acceleration may be caused by an increase in particle number (secondary nucleation) at the moment at which the homopolymerization of the more water-soluble monomer starts. More water-soluble (hydrophilic) monomers have the propensity to form more latex particles in emulsion polymerization as compared with less water-soluble monomers (at equal surfactant concentrations).

Mechanism (c): the Trommsdorff–Norrish gel effect, causing an increase of $\bar{n}$, might cause the observed acceleration. Also, radical entry, exit and bimolecular termination are likely to depend on monomer feed and radical chain composition.

Mechanism (d): the cause of the sudden increase in rate may also be found in the average propagation rate constant ($\bar{k}_p$), since this constant strongly depends on the composition of the monomer feed.

The increase in rate of the batch VAc/BA emulsion copolymerization was attributed by Kong et al.[115] to an increase of $\bar{k}_p$, as the monomer feed in the particles is enriched in VAc as a result of composition drift. The $\bar{k}_p$ value increases rapidly with the mol fraction VAc in the VAc rich region, as appears from the ultimate propagation model, since k_p of VAc is much higher than that of BA.

Delgado et al.[122] explained the increase in rate by an increase of $\bar{k}_p$, as did Kong et al., but also by an increase in the average number of radicals per particle ($\bar{n}$) due to the gel effect, although $\bar{n}$ is lower than 0.5 at the moment at which the rate increase is observed.

While studying emulsifier-free S–BA emulsion copolymerization Guillaume et al.[124] noticed a similar rate increase and also attributed it to an increase in the average number of radicals per particle.

Nomura and coworkers[40] explained the increase in rate observed during the copolymerization of styrene with methyl methacrylate (MMA) (at high initial fractions of MMA in the feed) by an increase of $\bar{n}$ due to the gel effect, because styrene is consumed faster than MMA, which would favour the gel effect.

Van Doremaele et al.[10] showed that in the case of S–MA emulsion copolymerization the rate increase must be mainly attributed to an increase in $\bar{k}_p$ when going from a monomer feed with ca. 90% to 100% MA. This is in complete agreement with the finding of Davis et al.[87] that the penultimate rather than the ultimate model should be applied to describe the kinetic behaviour, leading to a sharp increase in $\bar{k}_p$ between ca. 90% and 100% MA in the feed.

3.5 EMULSION TERPOLYMERIZATION

In the fundamental investigations described in the literature dealing with emulsion copolymerization, most attention has been given to binary copolymerization, i.e. polymerization of two monomers. Far less attention has been paid to ternary emulsion copolymerization (three monomers), hereafter referred to as terpolymerization. Emulsion terpolymerization investigations, mostly dealing with properties and applications, have been published mainly as patents.

It is obvious that the typical aspects that distinguish emulsion copolymerization from homopolymerization, e.g. monomer partitioning, dependence of kinetics on the local monomer concentration ratio etc., become more complex when three monomers are involved, not to mention the complications in terpolymer analysis.

However, since it can be easily understood that using three monomers gives the possibility of obtaining an even larger variety and refinement of copolymer properties, the time may have come to put more effort into fundamental research on emulsion terpolymerization, although it is to be expected that there will be little or no mechanistic differences between binary and ternary emulsion copolymerization systems.

Storti and Carrà[49] generalized the Smith–Ewart theory by developing a system of particle balances for describing the kinetics of emulsion polymerization processes involving any number of monomer species. By neglecting all other processes but chain propagation processes (the propagation reactions are at quasi-steady-state conditions with respect to the other reactions: termination, initiation, transfer etc.), they could solve the system easily.

This pseudo homopolymerization approach, which can be applied without any significant loss of accuracy, results in the classical Smith–Ewart population balance for homopolymerization where the kinetic parameters such as entry (ρ), exit (k), termination (c) etc. are replaced by the pseudo homopolymerization parameters ρ^*, k^*, c^* etc., defined as follows for emulsion terpolymerization

$$\rho^* = \rho_1 + \rho_2 + \rho_3 \tag{30}$$

$$k^* = k_1 P_1 + k_2 P_2 + k_3 P_3 \tag{31}$$

$$c^* = c_{11} P_1^2 + c_{22} P_2^2 + c_{33} P_3^2 + 2c_{12} P_1 P_2 + 2c_{13} P_1 P_3 + 2c_{23} P_2 P_3 \tag{32}$$

where ρ_1, k_1 etc. are the values for the individual monomeric species.

More generally, the following equations can be applied for n monomers

$$\rho^* = \sum \rho_j \tag{33}$$

$$k^* = \sum k_j P_j \tag{34}$$

$$c^* = \sum \sum c_{ij} P_i P_j \tag{35}$$

with $i, j = 1, \ldots, n$ and P_i is the probability associated with a single radical of type i.

So far, however, the majority of papers in the literature about terpolymerization only relate average terpolymer composition to terpolymer properties (see e.g. refs. 125–127).

Recently, the microstructure of emulsion terpolymers of vinyl chloride, vinylidene chloride and hydroxyethyl acrylate, prepared in batch and semicontinuous reactions, were studied by means of DSC and ^{13}C NMR.[128]

In many cases one uses two relatively water-insoluble comonomers (e.g. styrene, butyl acrylate, methyl methacrylate) and small amounts of a third, highly water-soluble comonomer[129–131] (e.g. (meth)acrylic acid, 2-hydroxyethyl methacrylate) or even a surface active comonomer.[132] These water-soluble comonomers are generally introduced to obtain functionalized latices (Section 3.6.2), for example, to improve adhesive product properties.

In some cases low conversion emulsion terpolymerization was used to verify the applicability of binary reactivity ratios to terpolymerization.[126] The present authors believe, however, that this approach should be considered with some scepticism, since in fact 'apparent' reactivity ratios are determined in this manner. Much work has been devoted to the applicability of binary reactivity ratios in low conversion solution terpolymerization.

Only few authors looked at the emulsion terpolymerization over the whole range of conversion and monitored the partial conversions of the monomers during batch[133] or semicontinuous[134,135] emulsion terpolymerization.

3.6 INTRINSIC CHARACTERISTICS INFLUENCING COPOLYMER LATEX PROPERTIES

3.6.1 Introduction

Copolymer molecular structure and particle morphology are directly reflected in the properties of the emulsion copolymers. It is therefore important to study which parameters influence the structure and in what way. In general, the molecular structure and morphology of the polymer particles are governed by intrinsic characteristics of the system like the reactivity ratios, the monomer partitioning, the specific chemical properties of the monomers and by the (in)compatibility (the mutual miscibility or phase separation) of the different polymeric materials formed during the entire process.

However, the emulsion polymerization strategy, i.e. the kind of process, can also have a considerable effect on the molecular structure and particle morphology. The intrinsic factors and the process conditions both determine colloidal aspects of the copolymer latex (particle diameter, surface charge density, colloidal stability, etc.), as well as the characteristics of the polymeric material in the particles (MMCCD) and the structure of the particles (copolymer composition as a function of particle radius, etc.). In turn, these factors determine the properties of the latex or the copolymer. The influence of the process conditions will be treated in Section 3.7.2.

Barton[136] gave a review on the influence of intrinsic characteristics, in which three cases were distinguished: (i) both monomers have almost the same water solubility (e.g. methyl methacrylate and ethyl acrylate); (ii) one monomer is much more water soluble than the other (e.g. acrylonitrile and butyl acrylate); and (iii) one monomer is very water insoluble (e.g. styrene) and the other is extremely water soluble (e.g. acrylic acid, acrylamide).

For example, Zosel et al.[137] investigated the influence of the chemical heterogeneity (distribution of monomer units in the latex particles), caused by intrinsic characteristics of the system, on the structure and properties of emulsion copolymers (film formation). These authors looked at copolymers containing acrylic acid, acrylamide and acrylamide ethers, examples of monomers frequently used in functionalized latices.

3.6.2 Functionalized Latices

One can change the properties of a latex by introducing comonomers with specific chemical properties, leading to functional(ized) latices. These are prepared by emulsion copolymerization of relatively water-insoluble monomers (*e.g.* styrene) and small amounts of very water-soluble comonomers (*e.g.* acrylic acid, acrylamides, hydroxylic (meth)acrylates) or specific, reactive comonomers. The use of these functional monomers offers the possibility of producing latices with an enormous variety of (specialized) properties, with great interest in *e.g.* print/coating, pharmaceutical/medical applications[138] and in catalysis.[139–141]

The use of highly water-soluble comonomers is widespread, a well-known example is the use of (meth)acrylic acid in carboxylated latices.

3.6.2.1 Carboxylated latices

In general polymers are called 'carboxylated' if they contain up to 10% monomer units that bear a carboxylic acid group like acrylic acid (AA).[142] Nowadays these carboxylated polymers are mainly used in the form of latices. Carboxylated styrene–butadiene latices are the most common examples of this class.

Carboxylated latices are used: (i) because of their improved chemical and mechanical stability; (ii) because of their capability of reacting with specific reagents; and (iii) because of their better adhesion to various substrates.

The latices are produced using emulsion copolymerization, generally under acidic conditions (pH < 4) in order to reduce aqueous phase (homo)polymerization of AA (the equilibrium distribution of AA, obviously, is shifted to the aqueous phase as pH increases) and to favour real copolymerization within the particles.

An important feature of these latices is the distribution of the acidic monomer units in the latex. In principle, there are five possible sites where these units can reside: (i) in the aqueous phase in monomeric form; (ii) in the aqueous phase in the form of water-soluble polymer chains; (iii) in surface-active polymer chains adsorbed onto the surface of the particles; (iv) copolymerized in the particles at or near the surface; and (v) copolymerized within the interior of the particles (buried). If the latex is used as such, site (iv) is normally favoured.

This distribution is influenced by the intrinsic characteristics of the system as stated in Section 3.6.1 (reactivity ratios, water solubility of the monomers and glass transition of the copolymer)[143] and the way the process is carried out (pH of the aqueous phase, batch or semicontinuous process;[143,144] Section 3.7.2). Generally applicable, predictive models are still lacking here, and empirical approaches are mostly resorted to.

3.6.2.2 Reactive latices

This specific category of functional latices will not be discussed here, since it would necessitate the treatment of much 'chemistry' not directly related to emulsion copolymerization. Upson[145] has given a short overview of the various possibilities.

In principle any kind of functional group can be attached to the polymeric material in the latex particles. For instance, by using a comonomer bearing the desired group the direct incorporation of hydroxylic comonomers[146] (of which also the distribution over the latex particles was studied[147]) or the incorporation of acrylamides[148] or acrylamide ethers[131] (which cause crosslinking when the polymer is heated[149]) can be effected. Also, by using specific comonomers that can react (after the copolymerization has been completed) with reagents bearing the desired group, the reactive group is thus attached to the polymer. For example, if a nucleophilic comonomer like vinylbenzyl chloride is incorporated,[145] it can react with an electrophilic reagent. In the same way nucleophilic reagents can attack electrophilic comonomers like (vinylic) tertiary amines.

3.7 CONTROL OF COPOLYMER LATEX PROPERTIES BY SEMICONTINUOUS PROCESSES

3.7.1 Introduction

The ultimate goal of most of the investigations on emulsion copolymerization is to be able to control the process in such a way as to produce a copolymer latex with the desired properties. For this purpose the semicontinuous (sometimes called semibatch) emulsion copolymerization process is widely used in industry. The main advantages of this process as compared with conventional emulsion batch processes include a convenient control of emulsion polymerization rate in relation with heat removal, and control of the chemical composition of the copolymer and particle morphology. These are important features in the preparation of specialty or high performance polymer latices. Semicontinuous emulsion copolymerization processes can be performed by applying various monomer addition strategies. The most widely investigated and described procedure is the addition of a given mixture of the monomers (sometimes preemulsified monomers) at a constant rate.[134,150–157] For instance, this procedure is followed in many papers dealing with the semicontinuous emulsion copolymerization of vinyl acetate and butyl acrylate.[74,123,158–160] With respect to the monomer addition rate two main situations can be distinguished: (i) flooded conditions—the addition rate is higher than the polymerization rate; and (ii) starved conditions—the monomers are added at a rate lower than the maximum attainable polymerization rate (if more monomer were present). The latter process (starved conditions) is often applied in the preparation of homogeneous copolymers/latex particles. In this case at some time during the reaction, because of the low addition rates, a steady state is attained in which the polymerization rate of each monomer is equal to its addition rate and a copolymer is made with a chemical composition identical to that of the monomer feed. Sometimes semicontinuous processes with a variable feed rate (power feed) are used to obtain latex particles with a core–shell morphology.[161]

3.7.2 Process–Structure–Property Relationships

A huge number of papers on process–structure–property relationships have appeared. It is therefore not possible to treat this subject extensively here. Only some characteristic examples will be given to illustrate the effect of the polymerization process on the properties of the copolymers.

As stated in Section 3.4.1, composition drift may result in chemically heterogeneous copolymer and may even lead to particle structure, i.e. a copolymer composition not evenly distributed over the particle radius. In the case of vinyl acetate (VAc) and butyl acrylate (BA), it was shown by specific staining techniques that the latices produced by batch reactions had a BA rich core and a VAc rich shell,[162] whereas latices produced semicontinuously under starved conditions were more homogeneous, although these also contained a small BA rich core. Pichot et al.[13] also showed, with the help of ^{13}C NMR spectra, that semicontinuous copolymers were homogeneous, whereas batch copolymers contained a considerable fraction of vinyl acetate homopolymer.

El-Aasser and coworkers[159] studied the effect of copolymer composition and polymerization process on the colloidal aspects and the molecular weight distribution (MWD) of VAc–BA latices. These investigators found that the molecular weights of the semicontinuous latices were much lower as compared with the batch latices. Also the MWD of the semicontinuous latices exhibited bimodality. They attributed the lower molecular weight to the lower monomer concentrations in the particles under starved conditions and the bimodality was explained by long chain branching, enhanced by the low monomer concentrations (chain transfer to polymer). Also, smaller particle diameters were found for the semicontinuous latices, and micellar nucleation was assumed to be more dominant in the batch process resulting in a short nucleation stage. By contrast, during the semicontinuous process nucleation was supposed to take place in the water phase (homogeneous nucleation) extending over the whole addition period, thus leading to a larger particle number. The authors further assumed that this phenomenon is enhanced by the following. Because the semicontinuous particles are more or less homogeneous and have a surface that contains a significant amount of butyl acrylate, whereas the batch particles have a VAc shell, the semicontinuous particles can adsorb more soap molecules and therefore the small particles or oligomeric species are better stabilized during the nucleation stage in the semicontinuous process. It was also found that semicontinuous latices had higher surface concentrations of acid end groups, allegedly due to the lower molecular weight. These acid end groups are introduced by hydrolysis of BA units. But hydrolysis also causes the formation of (poly)vinyl alcohol segments (providing steric stability) from vinyl acetate units. It was found that the presence of BA at the surface impeded the formation of poly(vinyl alcohol) but enhanced the concentration of acid end groups. Upon hydrolysis the batch latices had a higher stability against electrolyte and it was therefore concluded that the surface of the batch particles contained almost no BA, because the high electrolyte stability (only high electrolyte concentrations caused destabilization) was supposed to originate from the steric stability brought about by poly(vinyl alcohol).

Kong et al.[163] also studied the colloidal characteristics and particle morphology of VAc–BA latices by means of surface end group (sulfate or carboxylic acid) titration, and soap titration with sodium dodecyl sulfate (SDS) and sodium hexadecyl sulfate (HDS). It was shown that in the case of batch copolymerization the surface was enriched in VAc as might be expected from the copolymerization behaviour, *viz.* the occurrence of VAc homopolymerization during the last part of the reaction. The particles of the batch reaction showed the same characteristics as the particles of latices produced by means of core-shell polymerization in which one monomer is polymerized onto a (seed) latex of the other polymer. It even made no difference whether VAc was polymerized on a BA seed latex or BA on a VAc seed latex, indicating the possibility of migration of poly(vinyl acetate) to the particle surface and a probable phase rearrangement. On the other hand, latices produced semicontinuously showed the characteristics of a more homogeneous VAc-BA copolymer.

Cruz-Rivera et al.[166] examined the influence of the process on the structure and the properties of butyl acrylate-styrene (BA-S) latices prepared in composition-controlled batch (on-line controlled addition of S), core-shell and multistage (copolymerization of a BA-S mixture on an S seed followed by a homopolymerization of BA) polymerizations. The authors applied a simulation computer program to predict the copolymer microstructure. They predicted and experimentally found that homogeneously made copolymers showed only one glass transition, whereas core-shell copolymers had two and the multistage copolymers even had three. In a second paper in this series[164] models were presented for predicting the viscoelastic properties of the different copolymers. By investigating these properties it could be demonstrated that the copolymer materials that were expected to be homogeneous indeed consisted of one phase and that the core-shell and multistage copolymers consisted of more than one phase.

One of the main problems of semicontinuous polymerization performed under starved conditions is the extremely long reaction time required for the preparation of homogeneous copolymers. A more advanced method is the semicontinuous process performed in a controlled composition reactor.[68,74,135,165-167] The overall monomer concentrations then have to be monitored by means of *e.g.* on-line GLC. The monomer concentrations are kept at the constant level which is required to obtain a desired copolymer composition by controlled feeding of the separate monomers into the reactor. Due to scatter of the GLC data it is usually difficult to maintain constant levels of the monomer concentrations in the reactor. Furthermore, the required optimal overall monomer concentrations may shift away from the initial values, because the volume ratio of organic to aqueous phase continuously changes and subsequently the monomer partitioning changes. A method described in the literature that overcomes these problems is treated in the next section.

3.7.3 Optimal Addition Profile

Arzamendi et al.[118,168] developed the so-called optimal monomer addition strategy. By using this method Arzamendi *et al.* demonstrated that within a relatively short period of time homogeneous vinyl acetate (VAc)-methyl acrylate (MA) emulsion copolymers can be prepared in spite of the large difference between the pertaining reactivity ratios. The reactor was initially charged with all of the less reactive monomer (*viz.* VAc) plus the amount of the more reactive monomer (*viz.* MA) needed to initially form a copolymer of the desired composition. Subsequently, the more reactive monomer (MA) was added at a computed (time variable) flow rate (optimal addition profile) in such a way as to ensure the formation of a homogeneous copolymer.

The key problem in this method is the calculation of the amount of methyl acrylate to be initially charged in the reactor and the optimal addition rate profile of the remaining amount of methyl acrylate. The calculations are based on the following assumptions: (i) copolymerization is carried out starting from a monodisperse seed latex of the desired composition; (ii) the number of particles remains constant during the reaction; (iii) aqueous phase polymerization is negligible; and (iv) thermodynamic equilibrium determines the various monomer concentrations.

By applying the instantaneous copolymer composition equation, the desired monomer concentration ratio inside the latex particles is calculated. In combination with the thermodynamic equilibria equations, this ratio allows the calculation of the amount of methyl acrylate to be initially charged in the reactor.

Similar calculations can be used to determine the optimal addition profile of methyl acrylate. These equations then must be coupled to the differential equations accounting for initiator decomposition, consumption of vinyl acetate (and MA) and the change of phase volumes.

Since the differential equation accounting for vinyl acetate consumption assumes $\bar{n}$ to be known, it is necessary to apply equations describing the balance of radicals in the water phase and inside the latex particles. However, it is generally impossible to reliably predict $\bar{n}$ due to the lack of detailed knowledge of the mechanisms that control $\bar{n}$, not to mention the lack of values for the various parameters.

Therefore, Arzamendi et al. applied a semiempirical method to calculate the time dependent evolution of $\bar{n}$. This evolution is calculated from a semicontinuous experiment carried out under similar conditions as the final optimal process, but applying an estimated, constant addition rate of methyl acrylate. The evolution of $\bar{n}$ is correlated with the volume fraction (Φ_p) of polymer in the particles. This correlation is then used to calculate an addition profile. Another semicontinuous experiment is then carried out using this addition profile. If copolymer composition deviates too much from the desired value, another correlation of $\bar{n}$ with Φ_p is then calculated from the last experiment. This procedure can be repeated until the addition profile is optimal.

Alternatively, Van Doremaele[10] applied an even more pragmatic approach. This method can be applied without actually calculating $\bar{n}(t)$ or $\bar{n}(\Phi_p)$ and may therefore be more generally applicable. This method was applied to the emulsion copolymerization of styrene (S) and methyl acrylate (MA). The batch emulsion copolymerization of S and MA is known often to produce highly heterogeneous copolymers (styrene being the more reactive and less water-soluble monomer).

Rather than a large difference between the reactivity ratios (VAc–MA), here the large difference between the water solubilities of S and MA is the main problem. As stated, the time-evolution of $\bar{n}$ was not actually calculated but it was set equal to 0.5 as a first estimation. It would be highly fortuitous if the first estimated addition profile, based on $\bar{n} = 0.5$, were optimal, because the average number of radicals will generally deviate from this first estimation (i.e. $\bar{n} \ne 0.5$). Nevertheless, a first addition profile was calculated, presuming $\bar{n} = 0.5$. Separately, the correlation between the amount of styrene to be added and the conversion that would lead to the desired copolymer composition F_S was calculated from thermodynamic equilibrium data. Combining the results, i.e. the conversion–time curve from the experiment carried out with this addition profile and the correlation between the amount of styrene to be added and the conversion, a new addition profile could be calculated. In the case of the S–MA system the iteration converges rapidly, only four iteration steps appeared to be required in S–MA emulsion copolymerization to arrive at indistinguishable monomer addition rate profiles.

In order to evaluate the results, Van Doremaele[10] analyzed the copolymers formed by means of high performance liquid chromatography (HPLC), providing detailed microstructural information (viz. chemical composition distribution, CCD) of the copolymers.[61]

In Figure 10 the CCDs are depicted of three high conversion S/MA copolymers having the same average chemical composition but prepared by different processes. The one prepared by the conventional batch process exhibits bimodality, has two glass transition temperatures and has a minimum film formation temperature of 17 °C. Both the one prepared in a semibatch process under starved conditions (32 h) and the one obtained in a semicontinuous process while applying the optimal monomer addition strategy (5 h), are homogeneous with respect to chemical composition and have a minimum film formation temperature of 27 °C.

Thus the strategy developed and first applied by Arzamendi et al. resulted in a copolymer of constant composition. When applying constant addition rates, however, the copolymer composition varied considerably during the entire course of the reaction: only very low addition rates resulted

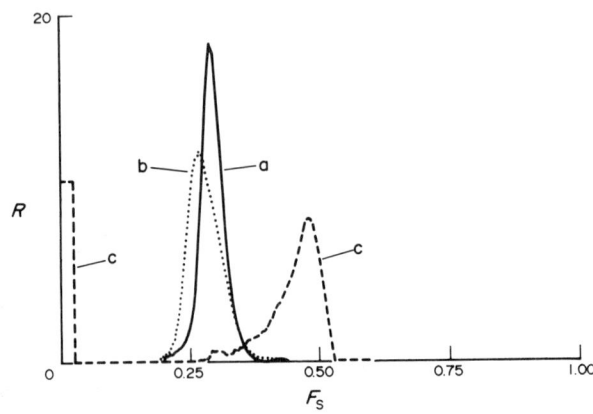

Figure 10 CCDs, experimentally determined with HPLC, of three styrene/methyl acrylate emulsion copolymers, all with $F_S = 0.25$ and $(M/W)_0 = 0.2 (g\,g^{-1})$ (R = relative weight). (a) Semicontinuous, starved conditions (32 h); (b) semicontinuous, optimal addition profile (5 h); and (c) conventional batch process (3 h) (reproduced with permission from G. H. J. Van Doremaele, Ph.D. Thesis, Eindhoven University of Technology, The Netherlands, 1990)

in nearly constant copolymer compositions. What is more, the reactions with the optimal addition rate profile proceed more rapidly than the reactions with constant addition rates.

3.8 REFERENCES

1. D. C. Blackley, 'Emulsion Polymerisation', Applied Science, London, 1975.
2. D. R. Basset and A. E. Hamielec, *ACS Symp. Ser.*, 1981, **165**.
3. I. Piirma, 'Emulsion Polymerization', Academic Press, New York, 1982.
4. A. Penlidis, J. F. MacGregor and A. E. Hamielec, *AIChE J.*, 1985, **31**, 881.
5. J. Guillot and C. Pichot, 'Emulsion Copolymerization', Hüthig & Wepf Verlag, Basel, 1985.
6. J. Guillot and C. Pichot, *Makromol. Chem., Macromol. Symp.*, 1990, **35/36**, 269.
7. F. Candau and R. H. Ottewill, 'Scientific Methods for the Study of Polymer Colloids and Their Applications', Kluwer, Dordrecht, 1990.
8. D. H. Napper and R. G. Gilbert, in 'Comprehensive Polymer Science', ed. G. Allen and J. C. Bevington, Pergamon Press, Oxford, 1989, vol. 4, chap. 11.
9. J. Guillot, *New J. Chem.*, 1987, **11**, 787.
10. G. H. J. Van Doremaele, Ph.D. Thesis, Eindhoven University of Technology, The Netherlands, 1990.
11. M. Lambla, B. Schlund, E. Lazarus and T. Pith, *Makromol. Chem., Suppl.*, 1985, **10/11**, 463.
12. M. Hirooka and T. Kato, *J. Polym. Sci., Polym. Lett. Ed.*, 1974, **12**, 31.
13. C. H. Pichot, M. Llauro and Q. Pham, *J. Polym. Sci., Polym. Chem. Ed.*, 1981, **19**, 2619.
14. W. H. Stockmayer, *J. Chem. Phys.*, 1945, **13**, 199.
15. J. L. Koenig, 'Chemical Microstructure of Polymer Chains', Wiley, New York, 1980.
16. J. M. Barton, *J. Polym. Sci., Part C*, 1970, **30**, 573.
17. N. W. Johnston, *Appl. Polym. Symp.*, 1974, **25**, 19.
18. B. J. Schmitt, *Angew. Chem.*, 1979, **91**, 286.
19. F. R. Rodriguez, 'Principles of Polymer Systems', McGraw-Hill, Japan, 1983.
20. W. D. Harkins, *J. Am. Chem. Soc.*, 1947, **69**, 1428.
21. W. V. Smith and R. H. Ewart, *J. Chem. Phys.*, 1948, **16**, 592.
22. W. H. Stockmayer, *J. Polym. Sci.*, 1957, **24**, 314.
23. J. T. O'Toole, *J. Appl. Polym. Sci.*, 1965, **9**, 1291.
24. J. Ugelstad, P. C. Mørk and J. O. Aasen, *J. Polym. Sci., Polym. Chem. Ed.*, 1967, **5**, 2281.
25. J. B. Rawlings and W. H. Ray, *Polym. Eng. Sci.*, 1988, **28**, 237.
26. J. B. Rawlings and W. H. Ray, *Polym. Eng. Sci.*, 1988, **28**, 257.
27. K. W. Min and W. H. Ray, *J. Macromol. Sci., Rev. Macromol. Chem.*, 1974, **C11**, 177.
28. F. K. Hansen and J. Ugelstad, *J. Polym. Sci., Polym. Chem. Ed.*, 1978, **16**, 1953.
29. G. Lichti, R. G. Gilbert and D. H. Napper, *J. Polym. Sci., Polym. Chem. Ed.*, 1983, **21**, 269.
30. G. Lichti, B. S. Hawkett, R. G. Gilbert and D. H. Napper, *J. Polym. Sci., Polym. Chem. Ed.*, 1981, **19**, 925.
31. R. G. Gilbert and D. H. Napper, *J. Macromol. Sci., Rev. Macromol. Chem.*, 1983, **C23**, 127.
32. M. J. Ballard, D. H. Napper and R. G. Gilbert, *J. Polym. Sci., Polym. Chem. Ed.*, 1984, **22**, 3225.
33. D. H. Napper and R. G. Gilbert, *Makromol. Chem., Macromol. Symp.*, 1987, **10/11**, 503.
34. P. J. Feeney, D. H. Napper and R. G. Gilbert, *Macromolecules*, 1984, **17**, 2520.
35. P. J. Feeney, D. H. Napper and R. G. Gilbert, *Macromolecules*, 1987, **20**, 2922.
36. S. Chen and K. Wu, *Polymer*, 1988, **29**, 545.
37. S. Chen and K. Wu, *J. Polym. Sci., Polym. Chem. Ed.*, 1988, **26**, 1487.
38. M. Nomura and M. Harada, *J. Appl. Polym. Sci.*, 1981, **26**, 17.
39. M. Nomura, M. Kubo and K. Fujita, *J. Appl. Polym. Sci.*, 1983, **28**, 2767.
40. M. Nomura and K. Fujita, *Makromol. Chem., Suppl.*, 1985, **10/11**, 25.
41. M. Nomura, I. Horie, M. Kubo and K. Fujita, *J. Appl. Polym. Sci.*, 1989, **37**, 1029.
42. J. Guillot, in 'Polymer Reaction Engineering', ed. K. M. Reichert and W. Geiseler, 1986, p. 147.
43. J. Guillot, in 'Future Directions in Polymer Colloids', ed. M. S. El-Aasser and R. M. Fitch, Nijhoff, Amsterdam, 1987, p. 65.
44. J. Guillot, *Makromol. Chem., Suppl.*, 1985, **10/11**, 235.
45. E. P. Dougherty, *J. Appl. Polym. Sci.*, 1986, **32**, 3051.
46. E. P. Dougherty, *J. Appl. Polym. Sci.*, 1986, **32**, 3079.
47. R. N. Mead and G. W. Poehlein, *Ind. Eng. Chem. Prod. Res. Dev.*, 1988, **27**, 2283.
48. R. N. Mead and G. W. Poehlein, *Ind. Eng. Chem. Prod. Res. Dev.*, 1989, **28**, 51.
49. G. Storti, S. Carrà, M. Morbidelli and G. Vita, *J. Appl. Polym. Sci.*, 1989, **37**, 2443.
50. J. R. Richards, J. P. Congalidis and R. G. Gilbert, *J. Appl. Polym. Sci.*, 1989, **37**, 2727.
51. M. J. Ballard, D. H. Napper and R. G. Gilbert, *J. Polym. Sci., Polym. Chem. Ed.*, 1981, **19**, 939.
52. J. Forcada and J. M. Asua, *J. Polym. Sci., Polym. Chem. Ed.*, 1990, **28**, 987.
53. T. O. Broadhead, A. E. Hamielec and J. F. MacGregor, *Makromol. Chem., Macromol. Symp.*, 1985, **10/11**, 105.
54. G. Lichti, R. G. Gilbert and D. H. Napper, *J. Polym. Sci., Polym. Chem. Ed.*, 1980, **18**, 1297.
55. C. C. Lin and W. Y. Chiu, *J. Appl. Polym. Sci.*, 1979, **23**, 2049.
56. M. F. Llauro, C. Pichot, J. Guillot, L. Rios G., M. A. Cruz E. and C. Guzman C., *Polymer*, 1986, **27**, 889.
57. M. F. Llauro, C. Pichot, J. Guillot, L. Rios G., M. A. Cruz E. and C. Guzman C., *Polym. Mater. Sci. Eng.*, 1986, **54**, 613.
58. E. Giannetti, G. Storti and M. Morbidelli, *J. Polym. Sci., Polym. Chem. Ed.*, 1988, **26**, 1835.
59. E. Giannetti, G. Storti and M. Morbidelli, *J. Polym. Sci., Polym. Chem. Ed.*, 1988, **26**, 2307.
60. J. C. J. F. Tacx and A. L. German, *Polymer*, 1989, **30**, 918.
61. G. Glöckner, in 'Comprehensive Polymer Science', ed. G. Allen and J. C. Bevington, Pergamon Press, Oxford, 1989, vol. 1, chap. 16.
62. H. J. Harwood, in 'Natural and Synthetic High Polymers', ed. P. Diehl, Springer-Verlag, Berlin, 1970, vol. 4, 71.

63. M. Morton, S. Kaizerman and M. W. Altier, *J. Colloid Sci.*, 1954, **9**, 300.
64. B. M. E. Van Der Hoff, *J. Polym. Sci.*, 1960, **44**, 241.
65. J. L. Gardon, *J. Polym. Sci., Polym. Chem. Ed.*, 1968, **6**, 2859.
66. E. Vanzo, R. H. Marchessault and V. Stannet, *J. Colloid Sci.*, 1965, **20**, 62.
67. I. A. Maxwell, J. Kurja, G. H. J. van Doremaele and A. L. German, *Makromol. Chem.*, submitted.
68. J. Guillot, *Acta Polym.*, 1981, **32**, 593.
69. J. Ugelstad, H. R. Mfutakamba, P. C. Mørk, T. Ellingsen, A. Berge, R. Schmid, L. Holm, A. Jørgedal, F. K. Hansen and K. Nustad, *J. Polym. Sci., Polym. Symp.*, 1985, **72**, 225.
70. J. Ugelstad, P. C. Mørk, I. Nordhuus, H. R. Mfutakamba and E. Soleimany, *Makromol. Chem., Suppl.*, 1985, **10/11**, 215.
71. C. M. Tseng, M. S. El-Aasser and J. W. Vanderhoff, *Am. Chem. Soc., Div. Org. Coat. Plast. Chem.*, 1981, 317.
72. J. Guillot, in 'Future Directions in Polymer Colloids', ed. M. S. El-Aasser and R. M. Fitch, Nijhoff, Amsterdam, 1987, p. 72.
73. J. C. de la Cal, R. Urzay, A. Zamora, J. Forcada and J. M. Asua, *J. Polym. Sci., Polym. Chem. Ed.*, 1990, **28**, 1011.
74. J. Dimitratos, C. Georgakis, M. S. El-Aasser and A. Klein, *Comput. Chem. Eng.*, 1989, **13**, 21.
75. M. Nomura, K. Yamamoto, I. Horie, K. Fujita and M. Harada, *J. Appl. Polym. Sci.*, 1982, **27**, 2483.
76. J. Ugelstad, M. S. El-Aasser and J. W. Vanderhoff, *J. Polym. Sci., Polym. Lett. Ed.*, 1973, **11**, 503.
77. J. Ugelstad, F. K. Hansen and S. Lange, *Makromol. Chem.*, 1974, **175**, 507.
78. W. J. Priest, *J. Phys. Chem.*, 1952, **56**, 1077.
79. R. M. Fitch, *Br. Polym. J.*, 1973, **5**, 467.
80. J. Forcada and J. M. Asua, *J. Polym. Sci., Polym. Chem. Ed.*, 1990, **28**, 990.
81. M. Nomura, U. S. Satpathy, Y. Kouno and K. Fujita, *J. Polym. Sci., Polym. Lett. Ed.*, 1988, **26**, 385.
82. R. M. Fitch, *Makromol. Chem., Macromol. Symp.*, 1990, **35/36**, 549.
83. T. Alfrey and G. Goldfinger, *J. Chem. Phys.*, 1944, **12**, 205.
84. F. R. Mayo and F. M. Lewis, *J. Am. Chem. Soc.*, 1944, **66**, 1594.
85. J. L. Koenig, 'Chemical Microstructure of Polymer Chains', Wiley, New York, 1980.
86. E. Merz, T. Alfrey and G. Goldfinger, *J. Polym. Sci.*, 1946, **1**, 75.
87. T. P. Davis, K. F. O'Driscoll, M. C. Piton and M. A. Winnik, *Polym. Intern.*, 1991, **24**, 65.
88. T. Fukuda, Y. Ma and H. Inagaki, *Macromolecules*, 1985, **18**, 17.
89. R. E. Cais, R. G. Farmer, D. J. T. Hill and J. H. O'Donnell, *Macromolecules*, 1979, **12**, 835.
90. D. J. T. Hill, J. H. O'Donnell and P. W. O'Sullivan, *Macromolecules*, 1983, **16**, 1295.
91. J. A. Howell, M. Izu and K. F. O'Driscoll, *J. Polym. Sci., Polym. Chem. Ed.*, 1970, **8**, 699.
92. M. Nomura, in 'Emulsion Polymerization', ed. I. Piirma, Academic Press, New York, 1982, p. 191.
93. R. N. Mead and G. W. Poehlein, *J. Appl. Polym. Sci.*, 1989, **38**, 105.
94. J. M. Asua, E. D. Sudol and M. S. El-Aasser, *J. Polym. Sci., Polym. Chem. Ed.*, 1989, **27**, 3903.
95. M. Nomura, Y. Minamino, K. Fujita and M. Harada, *J. Polym. Sci., Polym. Chem. Ed.*, 1982, **20**, 1261.
96. A. M. North, *Polymer*, 1963, **4**, 134.
97. J. N. Atherton and A. M. North, *Trans. Faraday Soc.*, 1962, **58**, 2049.
98. M. E. Adams, B. S. Casey, M. F. Mills, G. T. Russell, D. H. Napper and R. G. Gilbert, *Makromol. Chem., Macromol. Symp.*, 1990, **35/36**, 1.
99. M. E. Adams, G. T. Russell, B. S. Casey, R. G. Gilbert and D. H. Napper, *Macromolecules*, 1990, **23**, 4624.
100. J. Ugelstad and F. K. Hansen, in 'Emulsion Polymerization', ed. I. Piirma, Academic Press, New York, 1982, p. 51.
101. V. I. Yeliseeva, in 'Emulsion Polymerization', ed. I. Piirma, Academic Press, New York, 1982, p. 247.
102. R. H. Ottewill, in 'Emulsion Polymerization', ed. I. Piirma, Academic Press, New York, 1982, p. 7.
103. I. A. Maxwell, Ph.D. Thesis, Sydney University, Australia, 1990.
104. I. A. Maxwell, B. R. Morrison, D. H. Napper and R. G. Gilbert, *Macromolecules*, 1991, **24**, 1629.
105. M. J. Ballard, D. H. Napper and R. G. Gilbert, *J. Polym. Sci., Polym. Chem. Ed.*, 1981, **19**, 939.
106. J. Forcada and J. M. Asua, *J. Polym. Sci., Polym. Chem. Ed.*, 1985, **23**, 1955.
107. G. Storti, G. Polotti, P. Canu, S. Carrà and M. Morbidelli, *Makromol. Chem., Macromol. Symp.*, 1990, **35/36**, 213.
108. G. H. J. Van Doremaele, A. M. Van Herk and A. L. German, *Makromol. Chem., Macromol. Symp.*, 1990, **35/36**, 231.
109. G. Storti, G. Polotti, S. Carrà and M. Morbidelli, *J. Polym. Sci., Polym. Chem. Ed.*, submitted.
110. J. C. J. F. Tacx and A. L. German, *J. Polym. Sci., Polym. Chem. Ed.*, 1989, **27**, 817.
111. W. Ramirez-Marquez, Ph.D. Thesis, University Claude Bernard, Lyon, France, 1987.
112. G. H. J. Van Doremaele, A. L. German, N. K. de Vries and G. P. M. Van der Velden, *Macromolecules*, 1990, **23**, 4206.
113. G. H. J. Van Doremaele, A. M. Van Herk, J. L. Ammerdorffer and A. L. German, *Polym. Commun.*, 1988, **29**, 299.
114. W. Ramirez-Marquez and J. Guillot, *Makromol. Chem.*, 1988, **189**, 361.
115. X. Z. Kong, C. Pichot and J. Guillot, *Eur. Polym. J.*, 1988, **24**, 485.
116. S. Djekhaba, C. Graillat and J. Guillot, *Eur. Polym. J.*, 1988, **24**, 109.
117. I. Capek, J. Barton and E. Orolinova, *Acta Polym.*, 1985, **36**, 187.
118. G. Arzamendi and J. M. Asua, *Makromol. Chem., Macromol. Symp.*, 1990, **35/36**, 249.
119. W. Ramirez-Marquez and J. Guillot, in 'Proceedings of the IUPAC Symposium, Santa Margherita Ligure, Italy, 1987', Parodi off. Grafica, Genova, 1987, p. 170.
120. S. Djekhaba and J. Guillot, *Eur. Polym. J.*, 1990, **26**, 1017.
121. I. Capek, J. Barton and L. Quang Tuan, *Makromol. Chem.*, 1987, **188**, 1723.
122. J. Delgado, M. S. El Aasser, C. A. Silebi and J. W. Vanderhoff, *J. Polym. Sci., Polym. Chem. Ed.*, 1990, **28**, 777.
123. T. Makgawinata, M. S. El-Aasser, A. Klein and J. W. Vanderhoff, *J. Dispersion Sci. Technol.*, 1984, **5**, 301.
124. J. L. Guillaume, C. Pichot and J. Guillot, *J. Polym. Sci., Polym. Chem. Ed.*, 1990, **28**, 119.
125. J. N. Coker, *J. Polym. Sci., Polym. Chem. Ed.*, 1975, **13**, 2473.
126. K. Saric, Z. Janovic and O. Vogl, *J. Polym. Sci., Polym. Chem. Ed.*, 1983, **21**, 1913.
127. E. Wallace, Jr., C. Y. Chen, C. U. Pittman Jr. H. Kwiatkowski, C. F. Cook, Jr. and J. N. Helbert, *Polym. Eng. Sci.*, 1985, **25**, 83.
128. N. Pourahmady and P. I. Bak, *Polym. Prepr., Am. Chem. Soc., Div. Polym. Chem.*, 1990, **31**(1), 603.
129. S. H. Ronel and D. H. Kohn, *J. Appl. Polym. Sci.*, 1975, **19**, 2379.
130. B. P. Huo, A. E. Hamielec and J. F. McGregor, *J. Appl. Polym. Sci.*, 1988, **35**, 1409.

131. C. Bonardi, Ph. Christou, M. F. Llauro-Darricades, J. Guillot, A. Guyot and C. Pichot, *New Polym. Mater.*, 1991, **2**, 295.
132. J. L. Guillaume, C. Pichot and J. Guillot, *J. Polym. Sci., Polym. Chem. Ed.*, 1990, **28**, 137.
133. L. Rios, C. Pichot and J. Guillot, *Makromol. Chem.*, 1980, **181**, 677.
134. J. Šnupárek and F. Krška, *J. Appl. Polym. Sci.*, 1977, **21**, 2253.
135. L. Rios and J. Guillot, *Makromol. Chem.*, 1982, **183**, 531.
136. J. Barton, *Makromol. Chem., Macromol. Symp.*, 1990, **35/36**, 41.
137. A. Zosel, W. Heckmann, G. Ley and W. Mächtle, *Makromol. Chem., Macromol. Symp.*, 1990, **35/36**, 423.
138. M. Okubo, Y. Yamamoto, M. Uno, S. Kamei and T. Matsumoto, *Colloid Polym. Sci.*, 1987, **265**, 1061.
139. W. T. Ford, R. Chandran and H. Turk, *Pure Appl. Chem.*, **60**, 395.
140. A. M. Van Herk, K. H. Van Streun, J. Van Welzen and A. L. German, *Br. Polym. J.*, 1989, **21**, 125.
141. A. Bahadur and P. Bahadur, *Indian J. Biochem. Biophys.*, 1985, **22**, 107.
142. D. C. Blackley, in 'Science and Technology of Polymer Colloids', ed. G. W. Poehlein, R. H. Ottewill and J. W. Goodwin, Nijhoff, The Hague, 1983, vol. 1, p. 203.
143. K. L. Hoy, *J. Coat. Technol.*, 1979, **51**, 27.
144. L. W. Morgan, O. P. Jensen and S. C. Johnson, *Makromol. Chem., Macromol. Symp.*, 1985, **10/11**, 59.
145. D. A. Upson, *J. Polym. Sci., Polym. Symp.*, 1985, **72**, 45.
146. S. A. Chen and H. S. Chang, *J. Polym. Sci., Polym. Chem. Ed.*, 1990, **28**, 2547.
147. M. Okubo, Y. Yamamoto and S. Kamei, *Colloid Polym. Sci.*, 1989, **267**, 861.
148. Y. Ohtsuka, H. Kawaguchi and Y. Sugi, *J. Appl. Polym. Sci.*, 1981, **26**, 1637.
149. E. Penzel and A. Zosel, *Angew. Makromol. Chem.*, 1981, **99**, 23.
150. J. Šnupárek, *Angew. Makromol. Chem.*, 1972, **25**, 105.
151. R. A. Wessling and D. S. Gibbs, *J. Macromol. Sci., Chem.*, 1973, **A7**, 647.
152. J. Šnupárek and F. Krška, *J. Appl. Polym. Sci.*, 1976, **20**, 1753.
153. J. Šnupárek and K. Kašpar, *J. Appl. Polym. Sci.*, 1981, **26**, 4081.
154. A. Garcia-Rejon, C. Guzman, J. C. Mendez and L. Rios, *Chem. Eng. Commun.*, 1983, **24**, 71.
155. J. Šnupárek, *Makromol. Chem., Suppl.*, 1985, **10/11**, 129.
156. L. Rios, M. A. Cruz, J. Palacios, L. M. Ruiz and A. Garcia-Rejon, *Makromol. Chem., Suppl.*, 1985, **10/11**, 477.
157. S. Omi, M. Negishi, K. Kushibiki and M. Iso, *Makromol. Chem., Suppl.*, 1985, **10/11**, 149.
158. K. Chujo, Y. Harada, S. Tokuhara and K. Tanaka, *J. Polym. Sci., Part C*, 1969, **27**, 321.
159. M. S. El-Aasser, T. Makgawinata, J. W. Vanderhoff and C. Pichot, *J. Polym. Sci., Polym. Chem. Ed.*, 1983, **21**, 2363.
160. S. C. Misra, C. Pichot, M. S. El-Aasser and J. W. Vanderhoff, *J. Polym. Sci., Polym. Chem. Ed.*, 1983, **21**, 2383.
161. D. R. Basset, in 'Science and Technology of Polymer Colloids', ed. G. W. Poehlein, R. H. Ottewil and J. W. Goodwin, Nijhoff, The Hague, 1983, vol. 1, p. 220.
162. S. C. Misra, C. Pichot, M. S. El-Aasser and J. W. Vanderhoff, *J. Polym. Sci., Polym. Lett. Ed.*, 1979, **17**, 567.
163. X. Z. Kong, C. Pichot and J. Guillot, *Colloid. Polym. Sci.*, 1987, **265**, 791.
164. B. Schlund, J. Guillot, C. Pichot and A. Cruz, *Polymer*, 1989, **30**, 1883.
165. J. Guillot and C. Rios-Guerrero, *Makromol. Chem.*, 1982, **183**, 1979.
166. A. Cruz-Rivera, L. Rios-Guerrero, C. Monnet, B. Schlund, J. Guillot and C. Pichot, *Polymer*, 1989, **30**, 1872.
167. A. Guyot, J. Guillot, C. Graillat and M. F. Llauro, *J. Macromol. Sci., Chem.*, 1984, **A21**, 683.
168. G. Arzamendi and J. M. Asua, *J. Appl. Polym. Sci.*, 1989, **38**, 2019.

4

From Heterogeneous to Homogeneous Catalysis in Monoalkene Polymerization

FRANCESCO CIARDELLI
University of Pisa, Italy

and

CARLO CARLINI
University of Bologna, Italy

4.1	INTRODUCTION	67
4.2	POLYMERIZATION OF MONOALKENES WITH HOMOGENEOUS (SOLUBLE) CATALYSTS	68
	4.2.1 Polymerization of Ethylene	68
	4.2.2 1-Alkenes	70
	4.2.3 Copolymers of Ethylene with 1-Alkenes	76
	4.2.4 Cycloalkenes	77
4.3	A COMPARISON BETWEEN HETEROGENEOUS AND HOMOGENEOUS CATALYSTS	78
4.4	REFERENCES	80

4.1 INTRODUCTION

The award of the Nobel Prize for chemistry to Ziegler and Natta in 1963 was intended to acknowledge their revolutionary studies which had allowed them to obtain new polymeric materials with exceptionally controlled molecular structure and excellent properties. The polymerization catalyst they discovered and developed was indeed heterogeneous, and derived from the reaction of a transition metal compound, usually titanium tetra- or tri-chloride, and an alkylaluminum derivative.[1,2] The heterogeneous structure, with crystalline titanium trichloride supporting the active species (coordinatively unsaturated alkyltitanium derivatives) on its surface, was from the beginning[3] considered to be fundamental to ensuring good efficiency in the reaction. Moreover, it appeared necessary for the high linearity of ethylene polymers and the high isotactic stereoregularity of polypropylene and other 1-alkene polymers.[4,5] Similarly, another catalytic system which acquired relevant industrial importance for producing polyethylene with high linearity was also heterogeneous, being based on chromium oxide supported over silica and/or alumina.[6] Further generations of catalysts were also heterogeneous, up to the high activity systems supported on magnesium chloride presently used in industrial processes.[7,8]

In the meantime the fundamental studies aimed at understanding the active site structure and the polymerization mechanism lead to the generally accepted hypothesis[9] that the propagation step consists of the insertion of an activated monoalkene molecule into the transition metal–carbon bond. On the other hand, the isotactic stereocontrol was associated with the chiral structure of the transition metal complex,[10,11] the growing chain end playing a complementary role.

These indications suggested that monoalkene polymers with high molecular weight and structural regularity could in principle be obtained with homogeneous transition metal organometallic derivatives, provided the structure of the active complex could be reproduced. In such a way several attempts have appeared in the literature dealing with soluble transition metal complexes which were very useful for mechanistic studies, but not competitive in their performances against the heterogeneous analogs.

The real breakthrough occurred about ten years ago with the discovery that metallocene complexes of titanium, zirconium and hafnium, when activated with alkylalumoxanes, gave extremely active, soluble catalysts for the polymerization of ethylene[13] to highly linear macromolecules, and of 1-alkenes to highly isotactic chains.[14,15]

The polymerization of monoalkenes with heterogeneous transition metal catalysts was extensively described in the main work of *Comprehensive Polymer Science* by Tait[16] and by Corradini,[17] while only brief mention was made by Tait[18] concerning the homogeneous systems. Therefore it was considered worthwhile to dedicate a short chapter of this volume to soluble, metallocene-based catalysts, which have shown an elegant confirmation of the hypotheses previously formulated for their heterogeneous precursors, and offered new exciting projections for future applications of transition metals in polymer synthesis. While according to the title this chapter should be concerned with the comparison of the two catalyst types, it will be necessary to first describe the main features and performances of the metallocene soluble catalysts which were not described in detail in the main volumes. Therefore this introduction is followed by a descriptive section devoted to the polymerization of monoalkenes in the presence of soluble metallocene transition metal complexes activated with alkylalumoxanes. Subsections concerning respectively, ethylene, 1-alkenes, copolymers and cycloalkenes are reported. Finally, the last section contains a comparison between heterogeneous and homogeneous catalysts, where reference will be made to the concepts anticipated by Pino in the abstract[19] of a lecture he was expected to present at the *International Symposium on Homogeneous and Heterogeneous Catalysis, 6th, Pisa, 1989*. Due to his untimely death in July of the same year, Pino could not present his Plenary Lecture; however, the contents of the abovementioned abstract already covers all of the main features of the relationship between the two systems, and is reported here with the mere addition of recently reported data which substantially confirm his statements.

4.2 POLYMERIZATION OF MONOALKENES WITH HOMOGENEOUS (SOLUBLE) CATALYSTS

In the main volumes of this series Tait[18] stated correctly that while some soluble catalysts can initiate monoalkene polymerization, the polymer is very often, if not always, precipitated as it is formed. Accordingly, the polymerization cannot be considered as a real homogeneous process. This holds not only for the scattered examples of soluble systems tested in several laboratories since the discovery of Ziegler–Natta polymerization,[12] but also for the more recently developed metallocene–alkylalumoxane catalysts that are the main subject of this chapter. The considerable amount of data that have appeared in the last few years lead to the presentation in detail of background knowledge concerning these latter systems based on soluble catalysts, before the comparison of homogeneous *versus* heterogeneous catalysis. Indeed, while several soluble catalysts were reported in the relevant literature, in particular vanadium-based systems for syndiotactic polypropylene[20] and for ethylene–propylene rubber,[21] the metallocene catalysts seem to provide a real alternative to the classical heterogeneous systems even if several technical problems need to be solved before a real industrial competition starts.

The following sections are organized according to the type of monomer, as this plays a considerable role in determining polymerization features. Thus the first section will be devoted to ethylene, which has also been polymerized with other soluble catalysts activated by alkylaluminums or chloroalkyls instead of alkylalumoxanes, but with much lower activity.[22,23,24] The second section is concerned with 1-alkenes, in particular propylene, which could not practically be polymerized to isotactic, high molecular weight polymers by homogeneous systems before the use of alkylalumoxane. Successively, two minor sections are devoted to the copolymerization of monoalkenes and to the homopolymerization of cycloalkenes.

4.2.1 Polymerization of Ethylene

In the early 1980s Sinn, Kaminsky and coworkers[13] demonstrated that dichlorobis(cyclopentadienyl)- or bis(cyclopentadienyl)dimethyl-zirconium, when activated with a large excess (Al:Zr =

Table 1 Polymerization of Ethylene with Bis(cyclopentadienyl)zirconium Dichloride/Methylalumoxane Catalyst (95 °C, 8 bar, 330 mL Toluene)[26]

Characteristic	
Productivity [(kg PE)(g Zr)$^{-1}$ h^{-1}]	39.8×10^3
Zirconocene concentration (mol L^{-1})	6.2×10^{-8}
Methylalumoxane concentration (MW 1200) (mol L^{-1})	7.1×10^{-4}
PE molecular weight	78 000
Number of macromolecules [(Zr atom)$^{-1}$ h^{-1}]	46 000
Time of formation of one macromolecule (s)	0.087
Turnover time (s)	3.1×10^{-5}

1000:1) of methylalumoxane (MAO), could polymerize ethylene in toluene solution with an activity 10 000 times greater than in the presence of alkylaluminums.[25] Titanium and hafnium metallocenes are also active under the same conditions, even if to a lesser extent than zirconium, particularly at temperatures above 50 °C. Typical characteristics of ethylene polymerization are reported in Table 1.[26]

The polyethylene (PE) obtained in the presence of the zirconocene/MAO catalyst shows relatively narrow molecular weight dispersity (MWD) with respect to heterogeneous systems, with $\bar{M}_w/\bar{M}_n =$ 1.6–2.4. The control of molecular weight (MW) can be realized in a number of ways. The increase of zirconocene concentration leads to a linear decrease of molecular weight. The effect of temperature is remarkable, with MW falling from 1 500 000 at 10 °C to only 90 000 at 90 °C, whereas oligomeric 1-alkenes with an even number of carbon atoms are formed at temperatures above 100 °C. Finally, the system is more sensitive to hydrogen than most other heterogeneous catalysts, only traces being needed to reduce the chain length considerably.[27,28]

Several zirconocene derivatives can be used in combination with MAO to polymerize ethylene with very high activity, but this depends appreciably on the type of transition metal complex employed. In particular, the influence of methyl substitution on the cyclopentadienyl ring is shown by the data reported in Table 2.[28]

The increase of molecular weight observed when using methyl-substituted cyclopentadienyl derivatives is accompanied by the broadening of molecular weight distribution with $\bar{M}_w/\bar{M}_n$ in the range of 6 to 15. This result was attributed[28] to the formation of at least two types of active center due to prereactions at different rates between the transition metal complex and MAO. The occurrence of prereactions involving the catalyst components and leading to the creation of the active species is demonstrated[28] by the reduction of the reaction time necessary to reach the maximum polymerization rate when increasing the aging time of the zirconocene in the presence of MAO and in the absence of ethylene.[28]

On this basis it was possible to reduce the amount of cocatalyst by one order of magnitude. Such a technically important result was obtained by using a more concentrated solution of zirconocene (10^{-4} M) in an external prereactor with a concentrated solution of MAO. The resulting catalyst was very active and stable for long periods; otherwise, the lowering of the Al:Zr ratio is accompanied by a reduction in activity.[29]

The characteristics of the PE prepared by metallocene-based soluble catalysts can be modified to meet the needs of industrial applications. Thus an even broader molecular weight distribution can be obtained by using metallocenes of different metals in the same reactor. This concept in the case of heterogeneous catalysts was used industrially by supporting two different transition metals on magnesium chloride.[30,31] In this way it was possible to obtain broader molecular weight distributions in the case of different polyethylenes, thus satisfying the needs for different applications. This approach was extended to soluble metallocene catalysts quite successfully. Indeed, Ewen obtained polyethylenes with bimodal distribution in the presence of a mixture of zirconocene and titanocene derivatives activated with methylalumoxane.[32] As the catalyst systems containing titanocenes are

Table 2 Polymerization of Ethylene with Different Zirconocenes in the Presence of MAO at 70 °C[28]

Catalyst	Polymerization time (min)	Productivity [(kg PE)(g Zr)$^{-1}$ h^{-1} bar^{-1}]	$\bar{M}_v \times 10^{-3}$
[ZrCp$_2$Cl$_2$]	15	323	120
[Zr(Me$_5$Cp)CpCl$_2$]	45	62	770
[Zr(Me$_5$Cp)$_2$Cl$_2$]	45	42	270

not stable in the case of long polymerization times and at temperatures above 50 °C, bimetallic catalytic systems based on [ZrCp$_2$Cl$_2$]/[HfCp$_2$Cl$_2$]/MAO (Cp = cyclopentadienyl) and [HfCp$_2$Cl$_2$]/racemic ethylenebis(indenyl)zirconium dichloride/MAO (in this chapter, ethylenebis(indenyl)zirconium dichloride will be represented by [Zr{(Ind)$_2$C$_2$H$_4$}Cl$_2$], where Ind = indenyl) were investigated with the same aim of broadening the molecular weight distribution of ethylene polymers.[33] While the polymer obtained in the presence of a sole transition metal, either Hf or Zr, had relatively narrow MWD with $\bar{M}_w/\bar{M}_n$ in the range 1.9 to 2.3, the system [HfCp$_2$Cl$_2$]/[Zr{(Ind)$_2$C$_2$H$_4$}Cl$_2$] with MAO gave a bimodal MWD with a polydispersity ratio up to 10. In any case, the degree of polydispersion and the occurrence of monomodal or bimodal distribution appear to be largely dependent on polymerization conditions, type of catalytic complexes and the molar ratio between them.

A broad molecular weight distribution is in general associated with the presence of different catalytic centers. Thus the use of two different metallocene complexes in the same polymerization process appears to be able to provide this feature to soluble catalysts, whereas with a single metallocene complex largely only one type of active center seems to be present.[33]

PE with density from 0.95 to 0.89 g mL^{-1} can also be obtained by copolymerizing different amounts of 1-alkenes (see Section 2.3).[34]

4.2.2 1-Alkenes

The Group IVB metallocene/MAO systems have been reported to be the first homogeneous catalysts active in the polymerization of 1-alkenes.

Previous efforts by employing bis(cyclopentadienyl)M^{IV} (M = Ti, Zr) derivatives and MAO as cocatalyst yielded largely atactic polypropylene.[14,25,35] However, when polymerization experiments were carried out at low temperature (−45 to −65 °C) by using the [TiCp$_2$Ph$_2$]/MAO catalytic system,[15] isotactic polypropylene with —mmmrmmmmrmmm— stereoblock sequences, as represented by (1), was obtained. This type of microstructure is consistent with Bernoullian statistics, the stereochemical control of the polymerization process being due to the configuration of the last inserted monomeric unit (chain end control mechanism). It is worth noting that with the above system high molecular weights (in the 70 000–300 000 range) were obtained in narrower distribution ($\bar{M}_w/\bar{M}_n$ = 1.6–1.9), as compared with those of the corresponding polypropylenes prepared by heterogeneous Ziegler–Natta catalysts.

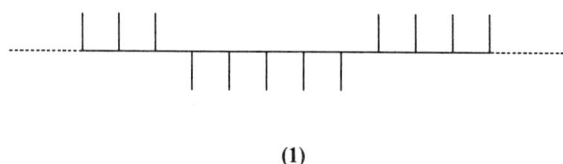

(1)

Zambelli,[36] by adding ^{13}C-enriched Al(^{13}CH$_3$)$_3$ or Al(^{13}CH$_2$Me)$_3$ to the [TiCp$_2$Ph$_2$]/MAO system at −40 °C succeeded in partially initiating the macromolecules on the resulting ^{13}C-enriched titanium–alkyl bonds formed by alkyl/aryl exchange reactions between [TiCp$_2$Ph$_2$] and various alkylaluminums. The ^{13}C NMR analysis of the enriched methyls or methylene carbons in the end groups of the polypropylene and poly(1-butene) obtained, confirmed that the steric control in this case was essentially due to the last chiral unit of the growing chain end.

^{1}H NMR analysis of the polymer produced from cis-1,2,3,3,3-pentadeuteropropylene by using [TiCp$_2$Ph$_2$]/MAO catalyst at −60 °C has proven[15] the presence of an erythro diisotactic structure consistent with alkene insertion occurring through cis addition to the double bond and with retention of configuration at the coordination carbon (Scheme 1).

On the other hand, ^{13}C NMR analysis of the end groups of the polymer chains obtained in the presence of [TiCp$_2$Ph$_2$]/MAO at low temperatures shows primary insertion of propylene, i.e. the methylene carbon atom of the last monomeric unit is bound to the transition metal.[15] Analogous results were obtained with the [Zr(Me$_5$Cp)CpCl$_2$]/MAO system at −30 °C.[15] For both cis opening of the double bond and primary insertion, the aforementioned soluble catalysts behave in the same way as the heterogeneous systems.[9]

Chiral homogeneous catalysts obtained from stereorigid, racemic, ethylene-bridged indenyl transition metal derivatives such as ethylenebis(indenyl)titanium dichloride and MAO as cocatalyst, were

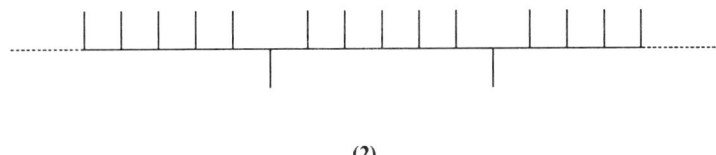

Scheme 1 Stereochemistry of double-bond opening by [TiCp$_2$Ph$_2$]/MAO catalyst[15]

the first metallocene systems able to form isotactic polypropylene with —mmmmrrmmmmrrmmm— stereosequences, represented by (**2**). Such steric sequences are analogous to those found in the presence of conventional and high activity MgCl$_2$-supported heterogeneous Ziegler–Natta catalysts, where an enantiomorphic site stereochemical control mechanism was demonstrated.[37] More recently,[38] the polymerization of propylene at low temperatures by the chiral racemic bis-[(1-phenylethyl)cyclopentadienyl]zirconium dichloride gave, in combination with MAO as cocatalyst, isotactic polypropylene consisting of a mixture of (**1**) and (**2**), whereas a 55:45 mixture of *meso* and *rac* forms of the above zirconocene complex afforded isotactic polypropylene with sequences having essentially the structure (**1**).

(**2**)

Noteworthy stereochemical differences were detected by ^{13}C NMR analysis[36] of samples of polypropylene and poly(1-butene), obtained in the presence of *rac*-[Ti{(Ind)$_2$C$_2$H$_4$}Cl$_2$]/MAO catalyst with added Al(^{13}CH$_3$)$_3$ or Al(^{13}CH$_2$Me)$_3$. Indeed, the end groups containing the enriched methylene of polypropylene and of poly(1-butene) prepared in the presence of enriched AlEt$_3$ resulted almost exclusively in an isotactic arrangement. On the contrary, the end groups enriched on the methyl carbon of polypropylene and of poly(1-butene), obtained in the presence of enriched AlMe$_3$, were partially isotactic and partially syndiotactic, it being irrelevant whether the insertion of the first monomeric unit leads to a chiral center or not. Therefore the placement of the enriched carbon is stereoregular when initiation occurs on the metal–ethyl bonds, while it is stereoirregular when it occurs on metal–methyl bonds. A similar trend was previously observed[39] for the polymer prepared in the presence of heterogeneous, isotactic specific catalysts. This result was explained by attributing the isotactic steric control to the presence of a chiral, stereorigid counterion rather than to the chiral carbon of the last unit inserted into the growing chain. These results clearly indicate that chiral active sites are responsible for the isotactic specific polymerization of propylene in the presence of a metallocene/MAO catalyst, thus demonstrating that highly isotactic poly(1-alkenes) can be prepared even with homogeneous catalytic systems, provided stable chiral transition metal compounds can be used. As an additional confirmation the symmetric *meso*-[Ti{(Ind)$_2$C$_2$H$_4$}Cl$_2$]/MAO system yields atactic polypropylene only.[15]

The most peculiar features of the above-reported catalysts are summarized in Table 3, in terms of stereospecificity and stereochemical control.

Following these observations several stereorigid chiral metallocenes of different transition metals with various ligands were synthesized and characterized in racemic or enantiomorphic forms.[16,40] These metallocenes have been used by several research groups in combination with MAO, in order to investigate the influence of the type of metal and/or ligand on the catalyst productivity in the polymerization, as well as on the average molecular weight, its distribution, and the microstructure of the obtained polypropylenes.

By use of a racemic mixture of ethylenebis(indenyl)zirconium dichloride or ethylenebis(4,5,6,7-tetrahydro-1-indenyl)zirconium dichloride, *rac*-[Zr{(Ind-H$_4$)$_2$C$_2$H$_4$}Cl$_2$], where Ind-H$_4$ = 4,5,6,7-

Table 3 Stereochemistry of Propylene Polymerization by Different Metallocene Derivatives

Metallocene	Polymerization temperature (°C)	Stereosequence	Stereocontrol mechanism	Ref.
[ZrCp$_2$Cl$_2$]	20	Atactic	—	14, 25
[Zr(Me$_5$Cp)CpCl$_2$]	−30	Atactic	—	15
[TiCp$_2$Ph$_2$]	25	Atactic	—	15
[TiCp$_2$Ph$_2$]	−45	Isotactic stereoblocks (1)	Chain end	15
rac-[Ti{(Ind)$_2$C$_2$H$_4$}Cl$_2$]	−60	Isotactic (2)	Enantiomorphic site	15
meso-[Ti{(Ind)$_2$C$_2$H$_4$}Cl$_2$]	−60	Atactic	—	15

Table 4 Polymerization of Propylene with rac-[Zr{(Ind)$_2$C$_2$H$_4$}Cl$_2$] or rac-[Zr{(Ind-H$_4$)$_2$C$_2$H$_4$}Cl$_2$] and MAO at Different Temperatures in Toluene Solution

Catalyst	T (°C)	Productivity [(kg PP)(g Zr)$^{-1}$ h^{-1}]	$\bar{M}_w \times 10^{-3}$	$\bar{M}_w/\bar{M}_n$	Ref.
[Zr{(Ind)$_2$C$_2$H$_4$}Cl$_2$]	15	16	n.d.[b]	n.d.	26
[Zr{(Ind)$_2$C$_2$H$_4$}Cl$_2$]	21	175	n.d.	n.d	26
[Zr{(Ind)$_2$C$_2$H$_4$}Cl$_2$]	35	471	n.d.	n.d.	26
[Zr{(Ind)$_2$C$_2$H$_4$}Cl$_2$]	50	232	28	2.1	46
[Zr{(Ind-H$_4$)$_2$C$_2$H$_4$}Cl$_2$]	−20	0.9	300[a]	n.d.	41
[Zr{(Ind-H$_4$)$_2$C$_2$H$_4$}Cl$_2$]	−10	3.3	305	2.6	41
[Zr{(Ind-H$_4$)$_2$C$_2$H$_4$}Cl$_2$]	0	9.6	144	2.4	41
[Zr{(Ind-H$_4$)$_2$C$_2$H$_4$}Cl$_2$]	15	32	62	2.0	41
[Zr{(Ind-H$_4$)$_2$C$_2$H$_4$}Cl$_2$]	20	52	45	1.9	41
[Zr{(Ind-H$_4$)$_2$C$_2$H$_4$}Cl$_2$]	60	84	0.8[a]	n.d.	41
[Zr{(Ind-H$_4$)$_2$C$_2$H$_4$}Cl$_2$]	80	279	9	2.1	46

[a]$\bar{M}_v$ values. [b]n.d. = not determined.

tetrahydro-1-indenyl, with MAO as cocatalyst, highly isotactic polypropylene can be produced.[41–49] The production of polypropylene increases with increasing polymerization temperature for both catalysts, even if temperature dependence is more pronounced for the former system, which also reaches much higher activities (Table 4).

The average molecular weight of the polypropylenes obtained is strongly affected by polymerization temperature, becoming rather low above room temperature (Table 4).

A much narrower distribution ($\bar{M}_w/\bar{M}_n$ = 1.9–2.6) of the molecular weights is also obtained with both catalysts, as compared with that of the polypropylenes produced by heterogeneous Ziegler–Natta catalysts.

The influence of the type of ligand and of the substituents on the C_5 ring as well as of the bridge linking the C_5 rings has been studied by Ewen[47] in the case of propylene polymerization in the presence of chiral, stereorigid zirconium metallocene/MAO catalyst systems. As the results reported in Table 5 show, the tetrahydroindenyl derivative is more active than the corresponding indenyl complex. In addition, the presence of a methyl group in the 3-position of the cyclopentadienyl ring of the indenyl ligand drastically reduces the yield of polypropylene (PP).

The molecular weight of polypropylene obtained with rac-[Zr{(3-MeInd)$_2$C$_2$H$_4$}Cl$_2$] (3-MeInd = 3-methylindenyl) at 30 °C is surprisingly high for a zirconium complex. The more pronounced dependence of the molecular weights on polymerization temperature for this latter complex, as compared with the other complexes reported in Table 5, has been explained[47] according to an unusually large difference between both the enthalpies and entropies of activation for insertion *versus* termination.

The microstructure of the polymers obtained[47] is a consequence of the enantiomorphic site control mechanism, the *meso* placements being connected by single units of the opposite handedness (2). In all cases the fraction of atactic polymer increases with increasing polymerization temperature due to the thermal instability of the catalysts, the more stable complex being rac-[Zr{(Ind-H$_4$)$_2$C$_2$H$_4$}Cl$_2$].[47] However, in the isotactic fractions some regioirregularities have been observed. It was concluded[47] that most of the head–head and tail–tail enchainments result from 2,1-insertions with the chiral species. Regioirregularities are unimportant for heterogeneous catalysts

Table 5 Propylene Polymerization by Different Chiral Stereorigid Zirconocene/MAO Catalyst Systems at Different Temperatures[47]

Chiral stereorigid zirconocene	Polymerization temperature (°C)	PP yield (g)	$\bar{M}_v \times 10^{-3}$	m.p.[a] (°C)	Inversions[b] (mol%)
rac-[Zr{(3-MeInd)$_2$C$_2$H$_4$}Cl$_2$]	30	—	85	—	—
	50	33	31	none	28
	80	—	7	—	—
rac-[Zr{(Ind)$_2$SiMe$_2$}Cl$_2$]	30	—	53	—	—
	50	164	52	144	1.5
	80	—	24	134	3.1
rac-[Zr{(Ind)$_2$C$_2$H$_4$}Cl$_2$]	30	—	41	—	—
	50	—	30	135	2.5
	80	355	14	126	4.7
rac-[Zr{(Ind-H$_4$)$_2$C$_2$H$_4$}Cl$_2$]	30	—	18	147[c]	1.2[c]
	50	—	12	137	—
	80	469	7	—	—

[a] Determined by DSC measurements. [b] Evaluated by ^{13}C NMR analysis. [c] Sample prepared at 20 °C.

and this explains the rather low melting points of the obtained polypropylene as compared with the corresponding samples prepared by heterogeneous catalysts. Melting points and percentage of stereochemical inversions (Table 5), clearly indicate that the stereorigid, dimethylsilyl-bridged species with only one atom bridging the two indene ligands is more stereospecific than the more flexible ethylene-bridged indenyl analogs. The almost complete absence of stereospecificity with rac-[Zr{(3-MeInd)$_2$C$_2$H$_4$}Cl$_2$] (Table 5) is in accordance with opposing steric effects from the methyl substituent on the C$_5$ ring and from the C$_6$ ring attached to the other cyclopentadienyl ligand.[47]

The influence of the transition metal in chiral, stereorigid metallocene/MAO isospecific propylene polymerization catalysts has also been studied (Table 6).[46,47]

rac-[Ti{(Ind)$_2$C$_2$H$_4$}Cl$_2$] is the only member of this series with a high activity at subambient conditions.[15] The higher fundamental activity for titanium is consistent with both the order of metal–carbon σ-bond strengths (Ti < Zr < Hf) and the increased stability of titanium(IV) at low temperature.[46] On the other hand, the zirconium and hafnium polymerization catalysts both have rather high activities at 50–80 °C because of their relatively more stable M^{IV} oxidation states. The important finding is that the polypropylene molecular weights obtained by chiral stereorigid hafnocenes are an order of magnitude higher than those produced by the zirconium analogs (Table 6). The chiral zirconocenes produce brittle polypropylene waxes in conditions where appreciable yields are obtained, whereas the soluble chiral hafnium catalysts produce plastics with high molecular weights.[46] Several patents[49–51] and reports[52] have also appeared, dealing with the synthesis and properties of polypropylenes for different applications, by using such soluble catalysts. The polymer DSC melting points indicate that the hafnium-produced polypropylenes are slightly more stereoregular than those obtained with zirconium. ^{13}C NMR analysis of the isotactic polymers[46] shows that the racemic zirconocenes are 95–99% stereospecific at 20–80 °C. The soluble catalysts are therefore similar to heterogeneous δ-TiCl$_3$ systems in this respect. However, rac-[Zr{(Ind-H$_4$)$_2$C$_2$H$_4$}Cl$_2$] and rac-[Zr{(Ind)$_2$C$_2$H$_4$}Cl$_2$] isomerization and/or decomposition products produce 6–15% by weight of perfectly atactic polymer at 20–80 °C and their 'as polymerized' samples contain about 5% regioirregularities.[46]

Table 6 Propylene Polymerization by Chiral Stereorigid Metallocene/MAO Catalyst Systems of Different Transition Metals

Chiral stereorigid metallocene	Polymerization temperature (°C)	$10^{-6} R_p$ [(g PP)(mol M)$^{-1}$ h^{-1}]	$\bar{M}_w \times 10^{-3}$	$\bar{M}_w/\bar{M}_n$	m.p. (°C)[a]	Ref.
meso:rac-[Ti{(Ind)$_2$C$_2$H$_4$}Cl$_2$] (44:56)	−60	0.02	154	1.6	94[b]	15
rac-[Zr{(Ind)$_2$C$_2$H$_4$}Cl$_2$]	50	21	28	2.1	134	46
rac-[Hf{(Ind)$_2$C$_2$H$_4$}Cl$_2$]	50	44	304	2.3	136	46
rac-[Zr{(Ind-H$_4$)$_2$C$_2$H$_4$}Cl$_2$]	80	26	9	2.1	117	46
rac-[Hf{(Ind-H$_4$)$_2$C$_2$H$_4$}Cl$_2$]	80	35	42	2.4	127	46

[a] Determined by DSC measurements. [b] n-Pentane insoluble fraction at 25 °C.

Table 7 Characterization of Fractions of Polypropylene Obtained by the rac-[Zr{(Ind-H$_4$)$_2$C$_2$H$_4$}Cl$_2$]/MAO Catalytic System at Polymerization Temperature Below Zero

Sample	Polymerization temperature (°C)	$\bar{M}_n^a \times 10^{-3}$	$\bar{M}_w/\bar{M}_n^a$	m.p. (°C)^b	I_{mm}^c	I_{mmmm}^c	Ref.
Heptane soluble	−10	17.7	2.8	149	96.14	—	43
Heptane insoluble		89	2.6	160	98.57	—	43
Heptane soluble	−15	56.8	2.74	151	—	97.2	45
Heptane insoluble		75.1	1.95	153	—	98.2	45

^{a}Determined by GPC measurements. ^{b}Evaluated by DSC analysis. ^{c}Isotacticity index calculated by ^{13}C NMR spectroscopy.

A more detailed investigation on the structure of polypropylene obtained by the rac-[Zr{(Ind-H$_4$)$_2$C$_2$H$_4$}Cl$_2$]/MAO system has been performed[43,45] by solvent fractionation, ^{13}C NMR, GPC and DSC analyses (Table 7). In particular, the isotactic polypropylene was found to be partially soluble in boiling toluene. Such a profound difference in solubility, as compared with polypropylenes produced with the usual heterogeneous Ziegler–Natta catalysts, must be attributed to a structural difference in the two cases. Soga[43] found that the ratio $\bar{M}_w/\bar{M}_n$ of the total polymer is broad, whereas that of the fractionated polymers is narrow, suggesting that the catalyst incorporates at least two kinds of active species.

As reported in Table 7 the heptane insoluble polymer fraction shows a melting point of 160 °C, which is comparable to the usual isotactic polypropylene,[2] whereas the melting point of the heptane soluble fraction is found to be lower (149 °C). According to ^{13}C NMR data and to the isotacticity index, in terms of triads the occurrence of (3) was suggested.

$$\begin{array}{c} \quad\ \ C \quad\ \ C \quad\quad\quad\quad\ C \quad\ \ C \\ \quad\ \ | \quad\ \ | \quad\quad\quad\quad\ | \quad\ \ | \\ -C-C-C-C-C-C-C-C-C-C-C- \end{array}$$

(3)

Therefore it was concluded[43] that hydrogen transfer polymerization takes place to some extent, in addition to the normal polymerization, especially with the catalyst species that give a lower molecular weight polymer, soluble in boiling heptane. However, Tsutsui et al.[45] reported that the spectra of both polypropylene fractions obtained with the same catalyst show a number of small peaks arising from head–head and tail–tail enchainments which are arranged with *meso* placements, suggesting that 1,2-insertion necessarily takes place just after 2,1-insertion, as schematically represented by (4).

$$\begin{array}{c} \quad\ \ C \quad\ \ C\ \ C \quad\quad\ C \quad\ \ C \\ \quad\ \ | \quad\ \ |\ \ | \quad\quad\ | \quad\ \ | \\ -C-C-C-C-C-C-C-C-C-C- \end{array}$$

(4)

The data reported in Table 7 and in particular the lower melting points of both fractions (151 °C and 153 °C) with respect to isotactic polypropylenes obtained by the usual heterogeneous Ziegler–Natta catalysts can be explained[45] by the presence of some regioirregularities (2,1-insertions).

A similar conclusion was reported in the case of polypropylene obtained by rac-[Ti{(Ind)$_2$C$_2$H$_4$}Cl$_2$]/MAO catalyst.[44] The higher solubility and the lower melting point of the polymer obtained with respect to the usual isotactic polypropylene was related to the characteristic microstructure of the polymer, i.e. the presence of head–head or tail–tail enchainments and the —(CH$_2$)$_4$— units. These structures derive from 2,1- and 1,3-insertions of propylene monomer, respectively.

The above catalyst was confirmed to display,[53] in the polymerization of propylene, lower stereochemical and regiochemical controls than the heterogeneous Ziegler–Natta catalysts. In particular, the degree of control decreases greatly with increase of temperature and by lowering the Al:Zr ratio. These results have been interpreted[53] by assuming that the ansa-bridged indenyl ligand is relatively stereorigid at low temperature but is fluxional at high temperature. Polypropylenes soluble in n-heptane and lower boiling solvents exhibit[53] a distinct propensity towards crystallization in a thermally stable γ-modification contrary to what is observed for polypropylenes obtained by heterogeneous Ziegler–Natta catalysts which crystallize in the α-modification.

Table 8 Polymerization of 1-Butene with the $rac\text{-}[Zr\{(Ind\text{-}H_4)_2C_2H_4\}Cl_2]/$ MAO System in Toluene[41]

Polymerization temperature (°C)	Time (h)	Productivity [kg(g Zr)$^{-1}$h^{-1}]	$\bar{M}_v \times 10^{-3}$
−10	5.5	5.5	150
20	3.5	29	50

Table 9 Optical Rotation of the Mixture of Hydrogenated Oligomers Obtained with $(-)\text{-}(R)\text{-}[Zr\{(Ind\text{-}H_4)_2C_2H_4\}Me_2]$ (Z1) or $(+)\text{-}(S)\text{-}[Zr\{(Ind\text{-}H_4)_2C_2H_4\}Cl_2]$ (Z2) Catalytic Systems[55,56]

$\bar{M}_n$	Catalytic system	Experimental value $[\Phi]_D^{25}$	Calculated value $[\Phi]_D^{20}$	Content of (5) (%)	Prevailing configuration[a]
310	Z1	+15.90	+16.67	72	(S)
380	Z1	+14.63	+18.00	71	(S)
490	Z1	+14.52	+19.20	71	(S)
670	Z1	+13.64	+19.74	71	(S)
445	Z2	−10.07	−12.37	38	(R)
585	Z2	−9.16	−12.96	38	(R)
900	Z2	−7.24	−13.68	38	(R)

[a]Of the first asymmetric carbon atom following the n-propyl end group.

Few examples are reported in the literature dealing with the homopolymerization of 1-alkenes higher than propylene in the presence of soluble catalysts based on metallocene derivatives.

Poly(1-butene) prepared at −40 °C in the presence of the $[TiCp_2Ph_2]$/MAO catalyst is much less stereoregular than polypropylene obtained with the same system, whereas poly(4-methyl-1-pentene) is almost completely atactic.[36]

No polymerization of vinylcyclopropane occurs with the same catalytic system.[36] However, the use of $rac\text{-}[Ti\{(Ind)_2(C_2H_4)Cl_2\}]$[36] or $rac\text{-}[Zr\{(Ind\text{-}H_4)_2C_2H_4\}Cl_2]$,[41] in combination with MAO as cocatalyst, gives rise to highly isotactic poly(1-butene). In particular, with the latter system rather high productivity and molecular weight are also obtained (Table 8).

The discovery of the $rac\text{-}[Zr\{(Ind\text{-}H_4)_2C_2H_4\}X_2]$ (X = Me, Cl)/MAO catalysts that polymerize propylene to isotactic polymer, and the possibiliy of separating a single antipode of the above zirconium derivatives, has offered for the first time the possibility of preparing well-defined chiral catalysts of a single configuration for the stereospecific polymerization of 1-alkenes. Only a negligibly small optical activity can be expected for high molecular weight polypropylene in solution.[54] However, for propylene oligomers, if the end groups are, as usual, different, a measurable optical activity is expected. Therefore Pino et al.[55] polymerized propylene in the presence of hydrogen by using $(-)\text{-}(R)\text{-}[Zr\{(Ind\text{-}H_4)_2C_2H_4\}Me_2]$ or $(+)\text{-}(S)\text{-}[Zr\{(Ind\text{-}H_4)_2C_2H_4\}Cl_2]$ and MAO as cocatalyst. The polymers were fractionated by solvent extraction and oligomeric fractions were isolated and characterized (Table 9).[56]

All the fractions were optically active, the optical rotation being positive using the former complex and negative using the latter one as catalyst precursor. ^{1}H NMR spectra indicate that all the oligomer fractions have a highly isotactic structure.

A prevailing (S) absolute configuration has been assigned to the trimers prepared by $(-)\text{-}(R)\text{-}[Zr\{(Ind\text{-}H_4)_2C_2H_4\}Me_2]$ and having prevailing structures (5; n = 1) and (6; n = 1).

Optical purity of at least 70% could be estimated for (5; n = 1). On the contrary, the configuration of the asymmetric carbon atom following the n-propyl group in (5) and (6) is (R) when $(+)\text{-}(S)\text{-}[Zr\{(Ind\text{-}H_4)_2C_2H_4\}Cl_2]$ was used as catalyst.

(5)

(6)

In view of the very high regio- and stereo-specificity of the process and of the high optical purity of the catalyst precursors a higher enantiomeric excess would be expected. Possible origins of the relatively low optical purity could be a racemization of the zirconocene complex during the interaction with MAO and/or the presence of two conformers in the zirconium catalyst precursors. Pino et al.[55] suggested that regioselectivity and stereospecificity are controlled by very peculiar steric features of ion couple species (7) rather than Zr moieties (8). The enantioselective hydrooligomerization (proto- or deutero-) of propylene, 1-pentene and 4-methyl-1-pentene in the presence of the diastereomerically pure $(-)$-(R)-$[Zr\{(Ind-H_4)_2C_2H_4\}][(S)$-1,1'-bis(2-naphtholate)] or enantiomerically pure $(-)$-(R)-$[Zr\{(Ind-H_4)_2C_2H_4\}Me_2]$ catalyst precursors and MAO gave low and opposite enantioselectivity for the two reactions.[57] This result emphasizes the role of the growing chain for enantioface selection. On the other hand, asymmetric oligomerization to 1-alkenes in the absence of hydrogen gave in some cases high stereoselectivity ($\sim 80\%$), at least in the formation of the second chiral center. These reactions were carried out at high catalyst and low alkene concentration to control the molecular weight, and with (S)-$[Zr\{(Ind-H_4)_2C_2H_4\}][O$-acetyl-$(R)$-mandelate$]_2$ as the optically active catalyst precursor.[58]

$$[Zr\{(Ind-H_4)_2C_2H_4\}Me]^+ \ [Al_nMe_{n-1}O_nX_2]^-$$

(7)

$$[Zr\{(Ind-H_4)_2C_2H_4\}Me(X)] \cdot [Al_nMe_{n-1}O_nX]$$

(8)

Syndiotactic polypropylene was also obtained[59] by using metallocene derivatives, such as isopropyl(cyclopentadienyl-1-fluorenyl)-hafnium(IV) or -zirconium(IV) dichlorides. Their crystal structure reveals a bent, sandwich complex with trihapto bonding of the fluorenyl ligand to the metal. The complex is prochiral, C-2 and C-5 in the C_5 fluorene ring are chiral and of opposite handedness, and the hydrocarbon ligand system is stereorigid.

The polymerizations were carried out above room temperature in the presence of MAO as cocatalyst. The zirconocene-produced polymers show an —rrrrmmrrrrrmmrr— microstructure. Such a stereosequence indicates site stereochemical control with chain migratory insertions resulting in site isomerizations and occasional reversals of diasteroface selectivity.

The syndiospecific complexes are stereospecific over a wide range of temperatures, and high molecular weight and polymer yield are obtained (Table 10).

The morphology of the polymer particles is typical of heterogeneous catalysts, whereas the narrow polydispersities indicate chemically homogeneous active species.

4.2.3 Copolymers of Ethylene with 1-Alkenes

In one of the early papers on soluble metallocene catalysts, Kaminsky and coworkers[14] reported the copolymerization of ethylene with 1-hexene in the presence of bis(cyclopentadienyl)zirconium dimethyl/MAO catalyst. The formation of copolymer is supported by the lower melting point and

Table 10 Syndiotactic Polymerization of Propylene by $[M(Pr^iCp$-1-fluorenyl$)Cl_2]$/MAO (M = Hf, Zr) Catalyst Systems[59]

M	Polymerization temperature (°C)	Yield (g)	$\bar{M}_w \times 10^{-3}$	$\bar{M}_w/\bar{M}_n$	I_{rrrr}[a]
Zr	25	26	133	1.9	0.86
Zr	50	162	69	1.8	0.81
Zr	60	185	52	1.8	—
Zr	70	158	55	2.4	0.76
Hf	50	27	777	2.3	0.74
Hf	70	96	474	2.6	—

[a] Syndiotacticity index based on pentads, as revealed by ^{13}C NMR analysis.

Table 11 Properties of Ethylene/1-Hexene Copolymers Obtained by $[ZrCp_2Me_2]$/MAO Catalyst as Compared to Polyethylene[14]

1-Hexene (mol%)	$\bar{M}_w$ (GPC) × 10^{-3}	$\bar{M}_w/\bar{M}_n$	Density (g cm^{-3})	m.p. (°C)
0	112–143	1.93–2.19	0.97–0.98	139–140
4.6	88	2.64	0.92	115
7.7	41	3.15	0.90	105

Table 12 Copolymerization of Ethylene with Propylene in the Presence of Different Catalysts

Catalyst	T (°C)	$r_{C_2H_4}{}^a$	$r_{C_3H_6}{}^a$	$r_{C_2H_4} \times r_{C_3H_6}{}^a$	Ref.
$[ZrCp_2Cl_2]$/MAO	25	6.61	0.06	0.40	34
	50	6.26	0.11	0.69	34
$[ZrCp_2Me_2]$/MAO	—	20.1	0.015	0.30	34
$[TiCp_2Me_2]$/AlMe$_2$·H$_2$O	26	18.6	0.032	0.60	66
$[VO(acac)Cl]$/AlEt$_2$Cl	26	16.4	0.012	0.30	67
TiCl$_3$/Al(C$_6$H$_{13}$)$_3$	75	15.7	0.11	1.73	68

$^a r$ = reactivity ratio.

density of the polymeric product with respect to the polymer obtained under the same conditions by homopolymerization of ethylene (Table 11). In this way it was shown that linear low density polyethylene (LLDPE) could also be obtained starting with this soluble catalyst. Reactivity ratios for ethylene/propylene copolymerization were also determined in the presence of various soluble catalysts and at different temperatures (Table 12). Experiments were carried out in a laboratory scale bubble column reactor. The dependence of the kinetic behavior and of the reactivity ratios on polymerization temperature suggests the possibility of controlling copolymer properties as a function of the experimental conditions.[34] Terpolymers of ethylene/propylene and ethylidene/norbornene can also be prepared in the presence of $[ZrCp_2Me_2]$/MAO and $[ZrCp_2Cl_2]$/MAO catalysts,[60,61] but this last system was not able to give terpolymers when conjugated or longer chain dienes were used.[28] On the other hand, the chiral catalyst $[Zr\{(Ind)_2C_2H_4\}Cl_2]$ was able to give terpolymers also with 1,5-hexadiene[62] allowing the incorporation of up to 6 mol% of this diene, which is higher than the usual amount (2–3 mol%) necessary for the subsequent vulcanization. The content of hexadiene in the terpolymers was calculated from the reactivity ratios, the result always being higher than that measured by ^{1}H NMR, the deviation increasing with the diene content. This discrepancy, as well as the increase in molecular weight with increase in content of 1,5-hexadiene, suggests that crosslinking reactions probably take place involving the polymerization at the level of both double bonds of the diene. These terpolymers are transparent amorphous materials and their T_g is higher at higher diene contents also suggesting the formation of long branches.

4.2.4 Cycloalkenes

Chiral stereorigid metallocene/MAO catalysts facilitate the homopolymerization of cycloalkenes such as cyclopentene and cycloheptene to give isotactic polycycloalkenes.[26] Contrary to what is observed by using other Ziegler–Natta-type catalysts, with the above homogeneous systems no ring opening occurs and the resulting polymers are highly crystalline and are obtained in high yields. As an example, the rac-$[Zr\{(Ind)_2C_2H_4\}Cl_2]$/MAO catalytic system at 30 °C gives polycyclopentene with an activity of 2.2 (kg polymer)(g Zr)$^{-1}$ h^{-1}.[26]

The polymer produced is insoluble in all solvents and its X-ray diffraction pattern shows three sharp peaks, indicating a very high crystallinity, with a melting point probably exceeding 300 °C. It is of interest to remember that wholly amorphous polypentenamer was obtained from the same cyclopentene monomer using heterogeneous Ziegler–Natta catalysts. IR and solid state ^{13}C NMR analyses suggest high stereoregularity of isotactic type in the macromolecules with a prevailing *trans*-1,2 structure (**9**).

(9)

The copolymerization of cyclopentene with ethylene in the presence of a zirconocene catalyst is very rapid and comparable to ethylene homopolymerization.[63] Poly(cyclopentene-co-ethylene) was obtained with a cycloalkene content of 0.3 to 37 mol % with a mainly random distribution, with short blocks present at higher content. The incorporation of cyclopentene decreases with higher temperature and lower catalyst concentration. The process can be carried out at a temperature of −10 to 20 °C with the reactivity ratio for ethylene going from 80 to 300 and that of the cycloalkene from 0.05 to 0.007, respectively.

With cyclopentene as well as with cycloheptene and cyclooctene no ring opening occurs during the copolymerization with ethylene in the presence of the chiral rac-[Zr{(Ind)$_2$C$_2$H$_4$}Cl$_2$]/MAO catalyst. Also it is remarkable that the composition of feed has little effect on the reaction rate and molecular weight, and that relatively small amounts of cycloalkene units can substantially reduce the melting point, thus leading to a new type of linear low density polyethylene.[63] Finally, it is of interest to remember that with a heterogeneous catalyst alternating copolymers of ethylene with cycloalkenes are obtained.[64]

The activity of the [Zr{(Ind)$_2$C$_2$H$_4$}Cl$_2$]/MAO catalyst in cycloalkene polymerization increases from cyclopentene to cyclobutene, but decreases markedly for norbornene. In spite of this sensitivity to steric hindrance it was possible to incorporate in ethylene copolymer up to 12 mol % of dimethanooctahydronaphthalene.[65]

4.3 A COMPARISON BETWEEN HETEROGENEOUS AND HOMOGENEOUS CATALYSTS

As briefly mentioned in the 'Introduction' the subject comparison was made by Pino in mid 1989.[19] What he said at that time is reported here with a moderate updating which takes into account recent results which have appeared in the last year and a half (Table 13).

Table 13 Comparison between Soluble and Heterogeneous (Supported) Ziegler–Natta Catalysts for 1-Alkene Polymerization (after Pino)[19]

Polymerization process variables and polymer properties	Catalytic systems	
	rac-[Zr{(Ind-H$_4$)$_2$C$_2$H$_4$}X$_2$]	TiCl$_4$/MgCl$_2$/AlR$_3$/ethylbenzoate
Monomers	Ethylene, linear 1-alkenes, β-branched 1-alkenes, cycloalkenes	Ethylene, linear 1-alkenes, α- and β-branched 1-alkenes
Active sites	Zr^{IV}; one species	Ti^{III}/Ti^{IV}; several species
Polymerization rate	Proportional to [M]	Proportional to [M]
Main polymer chain termination process	β-Hydrogen elimination	Hydrogen transfer to the monomer
Productivity	200–500 (kg PP)(g Zr)$^{-1}$ h^{-1}	Up to 1000 (kg PP)(g Ti)$^{-1}$ h^{-1}
$\bar{M}_n$	Linearly increasing with [M], large decrease with increasing temperature from 0 to 80 °C	No influence by [M]; modest influence by temperature from 10 to 60 °C
$\bar{M}_w/\bar{M}_n$	1.5–2	>5 (in general)
Macromolecule microstructure	Macromolecules having the same degree of stereoregularity	Mixture of stereoirregular, partially stereoregular and highly isotactic macromolecules
Regioselectivity (1,2-insertion)	High ($k_{1,2}/k_{2,1} \sim 80$)	Very high particularly in the isotactic fractions ($k_{1,2}/k_{2,1} > 98$)
Stereospecificity	Very high (>98%); strongly dependent on polymerization temperature between 0 and 80 °C	High (>90%); little dependence on temperature between 0 and 80 °C

His presentation started with the statement that with both heterogeneous ($MgCl_2$-supported $TiCl_4$, activated with triisobutylaluminum in the presence of a Lewis base) and soluble (racemic or optically active $[Zr\{(Ind\text{-}H_4)_2C_2H_4\}X_2]$ activated with MAO) catalysts the polymerization process can be divided into three main steps: formation of catalytic centers, growth, and termination of the macromolecular chain.

The formation of the catalytic centers through reaction between the transition metal compound and the alkylaluminum compound is still not well understood. In the case of soluble catalysts the alkylation of the transition metal compound has been proven and, on the basis of the investigation of model compounds, the formation of a tight ion couple bearing a positive charge on the transition metal atom seems very likely. Only indirect indications exist about the Lewis acid nature of the active sites of the $MgCl_2/TiCl_4/AlR_3$ heterogeneous catalytic systems. In both cases, the catalytic centers polymerizing propylene form very rapidly at temperatures between 40 and 60 °C. However, a rapid decrease of the polymerization rate with time occurs which is attributed to the deactivation of the catalytic centers, probably due to the decrease of the oxidation state of the transition metal. This behavior is largely dependent on the type of monomer, on the temperature at which the centers are formed and, in case of the $MgCl_2$-supported catalysts, on the type of Lewis base added to improve stereospecificity.

The growth of the macromolecular chain is similar in homogeneous and heterogeneous catalytic systems, consisting of an insertion of the monoalkene (1,2-insertion for 1-alkene) into a transition metal–carbon bond, as shown by the structure of the terminal groups. However, recent research has shown that with heterogeneous catalysts small amounts of syndiotactic polypropylene are formed, as a consequence of 2,1-insertion, which is responsible for syndiotactic specific polymerization with traditional Ziegler–Natta catalysts.[69]

Measurements made after stopping the polymerization with radioactive carbon monoxide indicate that the fraction of metal atoms active in catalysis is near to 100% in the soluble zirconium catalyst, but is much lower (up to 10%) in the $MgCl_2/TiCl_4/AlEt_3$ catalytic systems. The productivity in propylene polymerization with the soluble and the heterogeneous catalysts considered is of the same order of magnitude and increases with temperature at least between 30 and 60 °C. The main difference between the homogeneous and the heterogeneous catalysts concerns the substrate selectivity. In fact the homogeneous zirconium catalysts polymerize cycloalkenes and 1-alkenes that are linear or have no branching in the 3-position, but do not homopolymerize conjugated dialkenes. On the contrary, the heterogeneous catalysts also polymerize 1-alkenes branched in the 3-position, but do not homopolymerize cyclic alkenes.

The catalytic centers producing highly isotactic polymers have higher regioselectivity in the heterogeneous than in the homogeneous catalysts, $k_{1,2}/k_{2,1}$ being about 80 for the latter catalysts but higher than 95 for the former. Stereoregularity in the most insoluble polypropylene fractions obtained with both types of catalytic system is about the same. Differences in melting point between isotactic polypropylenes obtained with a soluble or with a heterogeneous catalyst can be attributed to different distributions of the steric irregularities in the main chains.

The predominant macromolecular chain termination process in propylene polymerization is different in the two types of catalytic system. In the homogeneous case the main termination process is β-hydrogen elimination, as shown by the linear increase of $\bar{M}_n$ by increasing monomer concentration. This result indicates that the monomer participates in the propagation process but not in transfer reactions. On the contrary, with the heterogeneous catalysts, no change of $\bar{M}_n$ is observed over a wide range of monomer concentration, indicating that the main termination reaction involves the monomer and probably consists of the migration of a hydrogen atom (or hydride ion) from the β-position of the last monomeric unit of the growing chain to the 2-position of the incoming monomer molecule. During propylene polymerization, temperature does not markedly influence the molecular weight in the presence of heterogeneous catalysts, whereas a dramatic effect is observed in the presence of homogeneous catalysts as expected, taking into account the higher energy of activation of the β-hydrogen abstraction. Both catalytic systems react with H_2 giving increased productivity and an evident decrease of molecular weight.

The above features (summarized in Table 13) suggest that the active sites in the homogeneous catalysts are sterically much better defined than those in the heterogeneous catalysts. Indeed, they produce polymers with a very narrow distribution of steric irregularities which can vary from 1–2% to 10% and depend more on temperature.[70]

The peculiar substrate selectivity of the homogeneous catalytic system and the negligible significance of chain termination *via* hydrogen transfer to the monomer are other aspects of the steric requirements which characterize this type of catalyst.

The narrower distribution of molecular weights for the soluble catalyst as compared to the heterogeneous catalyst may indicate a partially living character of the polymerization in the former case.

The above comparison also confirms the present views on the nature of the heterogeneous catalytic systems with active sites with different characteristics existing on the surface. This situation leads to reduced substrate selectivity and simultaneous production of stereoirregular, partially stereoregular and highly isotactic polymers. It is remarkable that even after the fractionation of the polymer into largely homogeneous fractions with respect to stereoregularity, the molecular weight distribution is still very broad. This also indicates that the highly stereospecific centers can be described as a large family of catalytic centers with some similarities in the steric structure but large differences in reactivity which affect the ratio of chain growth to chain termination rate.

The catalytic centers of the heterogeneous catalysts are extremely reactive and derive from the interaction between at least three components. Thus the detailed investigation of their structure is still not possible with the physical methods available. Therefore, it is necessary to continue to rely mainly on indirect methods and particularly on the detailed investigation of the structure of the macromolecules, which fortunately gives interesting information on the nature of the catalytic centers. These indications, on the other hand, have allowed a better understanding and rationalization of the behavior concerning the soluble catalysts based on transition metal metallocenes.

4.4 REFERENCES

1. K. Ziegler, *Angew Chem.*, 1952, **64**, 323.
2. G. Natta, P. Pino, P. Corradini, F. Danusso, E. Mantica, G. Mazzanti and G. Moraglio, *J. Am. Chem. Soc.*, 1955, **77**, 1708.
3. G. Natta, P. Pino and G. Mazzanti (Montecatini S.p.A.), *Ital. Pat.* 526 101 (1954).
4. G. Natta, *J. Inorg. Nucl. Chem.*, 1958, **8**, 589.
5. E. J. Arlman and P. Cossee, *J. Catal.*, 1964, **3**, 99.
6. J. P. Hogan and R. L. Banks (Phillips Petroleum Co.), *US Pat.* 2 825 721 (1958) (*Chem. Abstr.*, 1958, **52**, 8621h).
7. L. L. Böhm, *Polymer*, 1978, **19**, 553.
8. A. Mayr, P. Galli, E. Susa, G. Di Drusco and E. Giachetti (Montedison S.p.A.), *Ger. Pat.* 1 958 488 (1970) (*Chem. Abstr.*, 1970, **73**, 35 929u).
9. P. Pino, U. Giannini and L. Porri, in 'Encyclopedia of Polymer Science and Engineering', ed. H. F. Mark, N. M. Bikales, C. G. Overberger, G. Menges and S. I. Kroschwitz, Wiley, New York, 1985, vol. 8, p. 147.
10. P. Pino, A Oschwald, F. Ciardelli, C. Carlini and E. Chiellini, in 'Coordination Polymerization of α-Olefins', ed. J. C. W. Chien, Elsevier, New York, 1975, p. 25.
11. A. Zambelli, G. Bajo and E. Rigamonti, *Makromol. Chem.*, 1978, **179**, 1249.
12. See ref. 9, p. 163.
13. H. Sinn, W. Kaminsky, H. J. Vollmer and R. Woldt, *Angew Chem., Int. Ed. Engl.*, 1980, **19**, 390.
14. W. Kaminsky, M. Miri, H. Sinn and R. Woldt, *Makromol. Chem., Rapid Commun.*, 1983, **4**, 417.
15. J. A. Ewen, *J. Am. Chem. Soc.*, 1984, **106**, 6355.
16. P. J. T. Tait, in 'Comprehensive Polymer Science', ed. G. Allen and J. C. Bevington, Pergamon Press, Oxford, 1989, vol. 4, p. 1; P. J. T. Tait and N. D. Watkins, in 'Comprehensive Polymer Science', ed. G. Allen and J. C. Bevington, Pergamon Press, Oxford, 1989, vol. 4, p. 533.
17. P. Corradini, V. Busico and G. Guerra, in 'Comprehensive Polymer Science', ed. G. Allen and J. C. Bevington, Pergamon Press, Oxford, 1989, vol. 4, p. 29.
18. P. J. T. Tait, in 'Comprehensive Polymer Science', ed. G. Allen and J. C. Bevington, Pergamon Press, Oxford, 1989, vol. 4, p. 15.
19. P. Pino, in 'Proceedings of the 6th International Symposium on Relationships between Homogeneous and Heterogeneous Catalysis II, Pisa, 1989', Pacini, Pisa, 1989, p. 7.
20. A. Zambelli, G. Natta and I. Pasquon, *J. Polym. Sci., Part C*, 1963, **4**, 411.
21. A. Valvassori and G. Sartori, *Adv. Polym. Sci.*, 1967, **5**, 24.
22. U. Giannini, U. Zucchini and E. Albizzati, *J. Polym. Sci., Part B*, 1970, **8**, 405.
23. G. Henrici-Olivè and S. Olivè, *J. Polym. Sci., Part B*, 1970, **8**, 271.
24. G. Natta, G. Mazzanti, A. Valvassori, G. Sartori and A. Barbagallo, *J. Polym. Sci.*, 1961, **51**, 429.
25. H. Sinn and W. Kaminsky, *Adv. Organomet. Chem.*, 1980, **18**, 99.
26. W. Kaminsky, A. Bark, R. Spiehl, N. Möller-Lindenhof and S. Niedoba, in 'Transition Metals and Organometallics as Catalysts for Olefin Polymerizations', ed. W. Kaminsky and H. Sinn, Springer-Verlag, Berlin, 1988, p. 291.
27. W. Kaminsky, in 'Transition Metals Catalyzed Polymerizations', ed. R. P. Quirk, Cambridge University Press, Cambridge, 1988, p. 37.
28. W. Kaminsky and M. Schlobohm, *Makromol. Chem., Macromol. Symp.*, 1986, **4**, 103.
29. W. Kaminsky and R. Steiger, *Polyhedron*, 1988, **7**, 2375.
30. R. Invernizzi and F. Marcato (EniChimica Secondaria S.p.A.), *Jpn. Pat.* 60 101 104 (1985) (*Chem. Abstr.*, 1985, **103**, 142 527w).
31. F. Masi, S. Malquori, L. Barazzoni, C. Ferrero, A. Moalli, F. Menconi, R. Invernizzi, N. Zandoné, A. Altomare and F. Ciardelli, *Makromol. Chem. Suppl.*, 1989, **15**, 147.
32. J. A. Ewen, in 'Studies in Surface Science and Catalysis', 'Catalytic Polymerization of Olefins', ed. T. Keii and K. Soga, Kodansha and Elsevier, Tokyo and New York, vol. 25, 1986.
33. A. Ahlers and W. Kaminsky, *Makromol. Chem., Rapid Commun.*, 1988, **9**, 457.

34. H. Drögemüller, K. Heiland and W. Kaminsky, in ref. 26, p. 303.
35. P. Pino and R. Mulhaupt, *Angew. Chem., Int. Ed. Engl.*, 1980, **19**, 857.
36. A. Zambelli and P. Ammendola, in ref. 26, p. 329.
37. P. Pino, B. Rotzinger and E. von Achenbach, in 'Catalytic Polymerization of Olefins', ed. T. Keii and K. Soga, Kodansha and Elsevier, Tokyo and New York, 1986, p. 461.
38. G. Erker, R. Nolte, Y.-H. Tsay and C. Krüger, *Angew. Chem., Int. Ed. Engl.*, 1989, **28**, 628.
39. A. Zambelli and C. Tosi, *Adv. Polym. Sci.*, 1974, **15**, 38.
40. H. H. Britzinger, in ref. 26, p. 249 and refs. therein.
41. W. Kaminsky, K. Külper, H. H. Britzinger and F. R. W. P. Wild, *Angew. Chem.*, 1985, **97**, 507.
42. W. Kaminsky in 'Catalytic Polymerization of Olefins', ed. T. Keii and K. Soga, Kodansha and Elsevier, Tokyo and New York, 1986, p. 293.
43. K. Soga, T. Shono, S. Takemura and W. Kaminsky, *Makromol. Chem., Rapid Commun.*, 1987, **8**, 305.
44. T. Tsutsui, N. Ishimaru, A. Mizuno, A. Toyota and N. Kashiwa, *Polymer*, 1989, **30**, 1350.
45. T. Tsutsui, A. Mizuno and N. Kashiwa, *Makromol. Chem.*, 1989, **190**, 1177.
46. J. A. Ewen, L. Haspeslagh, J. L. Atwood and H. Zhang, *J. Am. Chem. Soc.*, 1987, **109**, 6544.
47. J. A. Ewen, L. Haspeslagh, M. J. Elder, J. L. Atwood, H. Zhang and H. N. Cheng, in ref. 26, p. 281.
48. A. Toyota, T. Tsutsui and N. Kashiwa, *J. Mol. Catal.*, 1989, **56**, 237.
49. J. A. Ewen (Cosden Technology, Inc.) Eur. Pat. Appl. 284 707 (1988) (*Chem. Abstr.*, 1989, **110**, 95 999j).
50. J. A. Ewen (Cosden Technology, Inc.) Eur. Pat. Appl. 284 708 (1988) (*Chem. Abstr.*, 1989, **110**, 155 019b).
51. J. A. Ewen (Fina Technology, Inc.) Eur. Pat. Appl. 310 734 (1988) (*Chem. Abstr.*, 1989, **111**, 154 553e).
52. A. Toyota, T. Tsutsui, M. Kioka and N. Kashiwa, *Prepr. Am. Chem. Soc., Div. Pet. Chem.*, 1989, **34**, 609.
53. B. Rieger, X. Mu, D. T. Mallin, M. D. Rausch and J. C. W. Chien, *Macromolecules*, 1990, **23**, 3559.
54. P. Pino, F. Ciardelli and M. Zandomeneghi, *Annu. Rev. Phys. Chem.*, 1970, **21**, 561.
55. P. Pino, P. Cioni and J. Wei, *J. Am. Chem. Soc.*, 1987, **109**, 6189.
56. P. Pino, P. Cioni, M. Galimberti, J. Wei and N. Piccolrovazzi, in ref. 26, p. 269.
57. P. Pino, M. Galimberti, P. Prada and G. Consiglio, *Makromol. Chem.*, 1990, **191**, 1677.
58. W. Kaminsky, A. Ahlers and N. Möller-Lindenhof, *Angew. Chem., Int. Ed. Engl.*, 1989, **28**, 1216.
59. J. A. Ewen, R. L. Jones, A Razavi and J. D. Ferrara, *J. Am. Chem. Soc.*, 1988, **110**, 6255.
60. W. Kaminsky and M. Miri, *J. Polym. Sci., Polym. Chem. Ed.*, 1985, **23**, 2151.
61. W. Kaminsky, 'History of Polyolefins', ed. R. B. Seymour and T. Cheng, Reidel, Dordrecht, 1986, p. 257.
62. W. Kaminsky and H. Drögemüller, *Makromol. Chem., Rapid Commun.*, 1990, **11**, 89.
63. W. Kaminsky and R. Spiehl, *Makromol. Chem.*, 1989, **190**, 515.
64. G. Dall'Asta and G. Mazzanti, *Makromol. Chem.*, 1963, **61**, 178.
65. W. Kaminsky, A. Bark and I. Däke, in 'Catalytic Polymerization of Olefins', ed. T. Keii and K. Soga, Kodansha and Elsevier, Tokyo and New York, 1990, vol. 56, p. 425.
66. V. Busico, L. Mevo, G. Palumbo, A. Zambelli and T. Tancredi, *Makromol. Chem.*, 1983, **184**, 2193.
67. C. Cozewith and G. Ver Strate, *Macromolecules*, 1971, **4**, 482.
68. G. Natta, G. Mazzanti, A. Valvassori and G. Sartori, *Chim. Ind. (Milan)*, 1958, **40**, 896.
69. A. Zambelli, G. Bajo and E. Rigamonti, *Makromol. Chem.*, 1978, **179**, 1249.
70. F. Ciardelli, A. Altomare and C. Carlini, *Prog. Polym. Sci.*, 1991, **16**, 259.

5
Anionic Synthesis of Polymers with Functional Groups

RODERIC P. QUIRK
University of Akron, OH, USA

5.1	INTRODUCTION	84
	5.1.1 Anionic Polymerization	84
	5.1.2 Characterization Methodology for Functionalized Polymers	85
	5.1.2.1 Molecular weight	85
	5.1.2.2 Functional group titration	85
	5.1.2.3 NMR spectroscopy	86
	5.1.2.4 FTIR spectroscopy	86
	5.1.2.5 UV–visible spectroscopy	86
	5.1.2.6 TLC and column chromatography	86
	5.1.2.7 Chemical reactions	87
5.2	SPECIFIC FUNCTIONALIZATION REACTIONS	87
	5.2.1 Carbonation	87
	5.2.1.1 Carbon dioxide	87
	5.2.1.2 Phthalic anhydride	89
	5.2.1.3 Protected oxazoline initiator	90
	5.2.2 Hydroxylation	90
	5.2.2.1 Ethylene oxide	90
	5.2.2.2 Protected initiators	90
	5.2.2.3 Carbonyl compounds	91
	5.2.3 Amination	92
	5.2.3.1 Protected imines	92
	5.2.3.2 Methoxyamine/methyllithium	93
	5.2.3.3 Functionalized and protected initiators	93
	5.2.3.4 ω-Halo-α-aminoalkanes	94
	5.2.4 Sulfonation	95
	5.2.4.1 Direct reaction	95
	5.2.4.2 Ethylene oxide end capping	95
	5.2.4.3 1,1-Diphenylethylene end capping	96
	5.2.5 Oxidation	97
	5.2.6 Summary of Specific Functionalization Reactions	97
5.3	1,1-DIPHENYLETHYLENE FUNCTIONALIZATION REACTIONS	97
	5.3.1 Background	97
	5.3.2 Phenol Functionality	99
	5.3.3 Amine Functionality	99
	5.3.4 Carboxy Functionality	100
	5.3.5 Copolymerization	101
	5.3.6 Fluorescent Group Labeling	102
	5.3.7 Conclusions	102
5.4	OTHER FUNCTIONALIZATION METHODOLOGIES	102
5.5	REFERENCES	103

5.1 INTRODUCTION

There has been growing interest and research on new synthetic methods for the preparation of well-defined polymers with in-chain and chain-end functional groups.[1-14] These functional groups in polymers can participate in: (a) reversible ionic association; (b) chain extension, branching or cross-linking reactions with polyfunctional reagents; (c) coupling and linking with reactive groups on other oligomer or polymer chains; and (d) initiation of polymerization of other monomers. In order to exploit this unique potential of functionalized polymers, it is important to consider the scope and limitations of current functionalization methodology using anionic polymerization.

The primary and review literature abounds with tabular and text descriptions of functionalization reactions for carbanions of living polymers with a variety of electrophilic species.[15-22] Unfortunately, many of these functionalization reactions have not been well characterized. Thus, it is not obvious to the uninitiated which specific procedures would be suitable for the introduction of many functional groups. This review will provide insight into the state of the art with regard to anionic functionalization chemistry for polymers. A variety of specific functionalization reactions will be reviewed and critically evaluated. In addition, the scope and limitations of a general, functionalization reaction methodology will be described.

5.1.1 Anionic Polymerization

The methodology of living anionic polymerization is particularly suitable for syntheses of functionalized polymers with well-defined structures in certain systems which proceed in the absence of chain termination and chain transfer reactions.[15,21-28] A variety of linear, branched and cyclic molecular architectures can be prepared with control of structural variables such as molecular weight, molecular weight distribution, copolymer composition and microstructure, tacticity and diene microstructure, and chain end functionality. For example, living block copolymers with block segments of controlled, predictable molecular weight can be prepared by sequential addition of monomers.[29,30] Star-branched polymers can be prepared either by using multifunctional initiators or by addition of postpolymerization, electrophilic linking agents.[31-33] Cyclic macromolecules can be synthesized by first preparing α,ω-polymeric dianions using difunctional initiators, followed by coupling reactions with difunctional electrophiles under conditions of high dilution to minimize intermolecular reactions.[34,35] Since these living polymerizations generate stable, anionic polymer chain ends when all of the monomer has been consumed, postpolymerization reactions with a variety of electrophilic species can be used to generate a diverse array of functional groups.[15-22,36-41] With this functional-

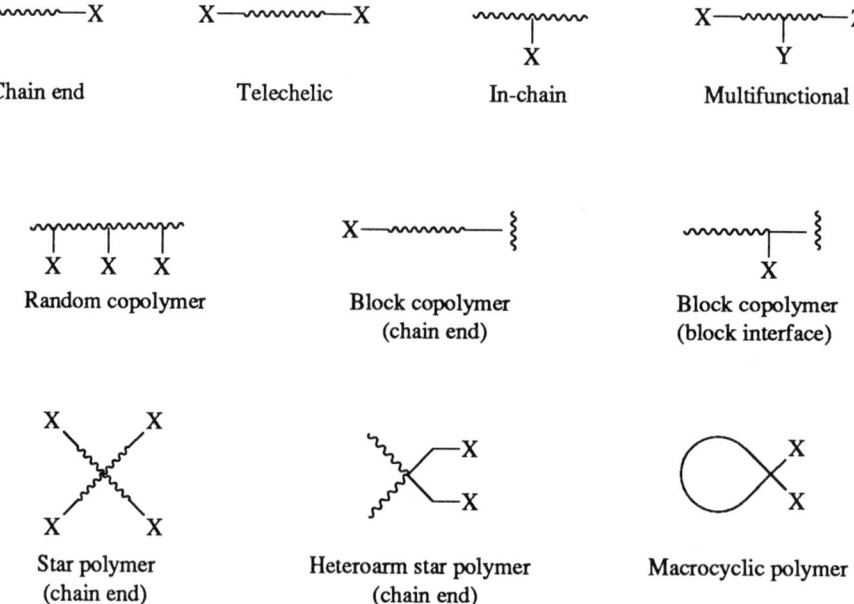

Figure 1 Molecular architecture for functionalized polymers prepared by anionic polymerization

ization chemistry and the control of polymer structure outlined above, functionalized polymers with diverse molecular structures can, in principle, be prepared, as illustrated in Figure 1. The ability to realize this tremendous potential will be discussed further in this review. In general, the scope of this review will be limited to the styrene, diene and alkyl methacrylate monomers with lithium as the counterion, since the use of organolithium compounds provides the maximum control of the major variables which affect polymer properties relative to other counterions.

5.1.2 Characterization Methodology for Functionalized Polymers

In order to evaluate the literature on anionic functionalization reactions for polymers, the following guidelines are recommended. The first premise is that it is the obligation of the authors to prove as unambiguously as possible that a given functionalization reaction has occurred as described. Precedent from model reactions with small molecule analogs is insufficient. There are numerous examples which show deviations from expected behavior for the polymeric systems compared to the analogous small molecules.[42-44] The second important consideration is the necessity for the utilization of a number of probes to determine the course of the reaction and the purity of the functionalized polymer. In general, no one method can be relied upon to provide unambiguous, quantitative information on the course of a given functionalization reaction. This problem follows directly from the experimental errors inherent in most analytical methods and the fact that the concentration of functional groups is generally quite low even for relatively low molecular weight polymers. A problem which is obvious, but important, is the purity of the polymer. The presence of small amounts of low molecular weight impurities can complicate quantitative determinations of functionality. More specific details of the types of analytical methods which can and should be used to determine polymer functionality are as follows.

5.1.2.1 Molecular weight

In order to reliably count the functional group content of a polymer, it is essential to determine the number average molecular weight as accurately as possible. For many living anionic polymerization systems, the molecular weight calculated based on the stoichiometry of the living polymerization (equation 1) can be used as a first approximation[22]

$$\bar{M}_n = \frac{\text{mass of monomer (g)}}{\text{moles of initiator}} \quad (1)$$

Further delineation of $\bar{M}_n$ can be obtained by size exclusion chromatography (SEC) for polystyrenes and other polymers for which suitable standard, well-characterized samples are available and can be used to generate a calibration curve of $\bar{M}_n$ versus elution volume.[45-48] If such information is not available, it may be necessary to determine the intrinsic viscosity of the polymer and then to utilize the universal calibration approach to determine $\bar{M}_n$.[49] Even if suitable SEC calibration information is available and especially if it is not, the $\bar{M}_n$ values can be determined using various counting methods such as membrane osmometry or vapor phase osmometry.[46,50] It should be recognized, however, that all of these methods have inherent limits of error which generate a certain amount of ambiguity in the interpretation of the degree of functionalization of polymers. Obviously, if the error limits of a given analytical method are $\pm 5\%$ or $\pm 10\%$, the corresponding $\bar{M}_n$ values and functionalization results are subject to the same types of error limits. This is not a problem if it is recognized and taken into account. This is one of the factors which dictates the use of a number of analytical probes for functionality determinations.

5.1.2.2 Functional group titration

A variety of titration methods are available for functional groups which are acidic or basic, or which form complexes with specific reagents.[51,52] When such an analytical method is available for a given functional group, it should be utilized as one criterion for the degree of functionalization. However, it should be recognized that titrations have inherent error limits associated with them, and that these are often magnified when applied to functional groups in polymers because of their concentration. For example, the determination of the number of functional groups per polymer molecule requires that the resultant value of $\bar{M}_n$ be compared with the $\bar{M}_n$ value which has been determined independently; thus, in addition to titration errors, this method is limited by the errors inherent in the independent determination of $\bar{M}_n$. Another possible source of problems in titration analysis for functional groups is insolubility of the polymer either initially or as titrant is added. Even if the polymer is soluble, there can be collapse of the polymer chain and inaccessibility of the functional groups to the titrant when a nonsolvent for the polymer, *e.g.* a theta solvent, is involved. These cautionary notes are only mentioned herein to further emphasize the point that no single criterion should be used for determination of polymer functionality.

5.1.2.3 NMR spectroscopy

The availability of high resolution NMR spectrometers has provided a powerful probe for determining the structures of polymers using primarily ^{1}H and ^{13}C NMR spectroscopy.[53–61] Since the ^{1}H and especially the ^{13}C chemical shifts are sensitive to the local chemical environment, NMR can often be utilized to identify functional groups or atoms adjacent to these functional groups.[62–65] In principle, electronic or manual integration can be used to determine the concentration of a given type of atom or group of atoms in a given environment within the inherent error limits of the method. This is less ambiguous for ^{1}H NMR than for ^{13}C NMR determinations because of the complications caused by relaxation effects and nuclear Overhauser effects which affect the signal areas in the latter.[60] Despite this, NMR spectroscopy provides a unique opportunity to directly observe signals pertaining to a given functional group, whether or not quantitative data can be extracted.

5.1.2.4 FTIR spectroscopy

Infrared spectroscopy provides a sensitive probe for specific functional groups in polymers.[58,59,61,66–70] The enhanced sensitivity of modern FTIR spectrometers and the ability to perform spectral subtractions using unfunctionalized polymer models has provided a powerful, often under-utilized probe for functional groups and analysis.[71] The potential exists for establishing quantitative Beer's Law correlations of absorption intensities *versus* the concentration of functional groups, but the limitations of these methods should be noted.[58,59,71] The simple observation of infrared absorptions characteristic of a given functional group is a necessary but not sufficient condition for the presence of that functional group in the polymer.

5.1.2.5 UV–visible spectroscopy

UV–visible spectroscopy provides another powerful method for evaluating the functionality of polymers when the functional group is an appropriate chromophore which is separated in wavelength from those of the polymer.[72,73] With respect to quantitative determinations, it is often more sensitive than other spectroscopic methods. When this method is applicable, there are instruments available which provide for its use as a detection method for SEC. The detector can be set at a specific wavelength for the functional group and has the added advantage that it provides a means of determining the uniformity of the functionalization as a function of molecular weight.

5.1.2.6 TLC and column chromatography

TLC and column chromatographic methods have been extensively used for many years as a means of separating, identifying and quantitatively evaluating low molecular weight organic compounds. Recently, the value of these procedures for identifying, characterizing and purifying functionalized polymers has been recognized.[74–81] In general, the presence of a polar functional group in a polymer is sufficient to alter significantly the elution behavior of the polymer and permit separation and identification relative to the unfunctionalized polymer. TLC can often be used to determine the purity of a functionalized polymer, since the presence of the unfunctionalized polymer in concentrations as low as 1% can be detected. Such techniques have been used to characterize the purity of polymers functionalized with hydroxy,[62] phenol,[63] amine[64,65,80,81] and carboxy[44,80] groups. Thus, this criterion of purity is often more sensitive and reliable than other quantitative measures of functionality which have much larger error limits on their determinations. Many syntheses of functionalized polymers produce either side reaction products or unfunctionalized polymers. Not only can this situation be detected by TLC, but preparative column chromatography can be used to separate and purify the functionalized polymer from other reaction products. It should be noted that there is a sensitivity of these chromatographic separations to the nature of the functional group, the stationary phase and the molecular weight of the functionalized polymer.[78] For example, chain end sulfonated polystyrenes do not elute from normal silica gel or alumina columns even with methanol as eluent; the use of annealed silica gel[47,82] was effective, however.

5.1.2.7 *Chemical reactions*

Functional groups often undergo chain extension reactions[79,81] or further reactions to generate other functional groups which can then be quantitatively evaluated.[44,83] These transformations and their quantitative evaluation are subject to the same limitations as discussed above. For example, if a given functional group cannot be readily identified, chemical transformations can provide a useful analytical procedure. Condensation-type, chain extension reactions which result in an increase in molecular weight have traditionally been utilized to evaluate functionality, especially for difunctional, telechelic polymers.[1,79,81] Provided the limitations generated by the sensitivity of this chemistry to stoichiometry are recognized, chain extension can be used as another probe for functionality.

Although there are serious problems associated with the determination of the functionality and purity of functionalized polymers, careful investigators will endeavor to provide as much experimental confirmation as possible for the efficiency of functionalization procedures and the purity of the functionalized polymers. Otherwise, the reader is encouraged to regard claims of quantitative functionalization with the healthy skepticism which they deserve.

5.2 SPECIFIC FUNCTIONALIZATION REACTIONS

The traditional approach for the anionic synthesis of chain-end functionalized polymers utilizes the postpolymerization reactions of the living anionic polymers with specific electrophiles for each different functional group.[15-23] Thus, it is necessary to develop, analyze and optimize new procedures for each different functional group. Optimization procedures often utilize variables such as chain-end structure, solvent, temperature, concentration, stoichiometry, mode of addition of reagents and polar additives since these factors can have dramatic effects on yield and product distributions. This review section will provide a critical overview of some recent developments in the use of specific functionalization reactions to prepare polymers labeled with carboxy, hydroxy, amino and sulfonate end groups primarily *via* alkyllithium-initiated polymerization methods.

5.2.1 Carbonation

5.2.1.1 *Carbon dioxide*

The carbonation of polymeric carbanions using carbon dioxide is one of the most useful and widely used functionalization reactions. However, there are special problems associated with the simple carbonation of polymeric organolithium compounds.[44,80] For example, when carbonations with high purity, gaseous carbon dioxide are carried out in benzene solution at room temperature using standard high vacuum techniques, the carboxylated polymer is obtained in only 46–66% yields for (polystyryl)lithium, (polyisoprenyl)lithium and (polystyrene-*block*-isoprenyl)lithium. The functionalized polymer is contaminated with dimeric ketone (25–30%) and trimeric alcohol (7–24%), as shown in equation (2), where P represents a polymer chain. It was proposed that the formation of these side products is favored relative to the desired carboxylated polymer by aggregation of the chain ends in hydrocarbon solution.[80] For example, it is known that (polystyryl)lithium is primarily associated into dimers in benzene solution while the degree of association of (polydienyl)lithiums in hydrocarbon solvents is at least dimeric.[23] It has been reported that sufficient quantities of Lewis bases such as tetrahydrofuran (THF)[23,84] and N,N,N',N'-tetramethylethylenediamine (TMEDA)[23,85] can reduce or even eliminate the association of the polymeric organolithium chain ends. In accord with these considerations, it was found that addition of either large amounts of THF (25 vol %)[80] or TMEDA (10–12 molar excess relative to [Li])[44] was effective in favoring the carbonation reaction to the extent

$$\text{PLi} \xrightarrow{\text{CO}_2} \xrightarrow{\text{H}_3\text{O}^+} \text{PCO}_2\text{H} + \text{P}_2\text{CO} + \text{P}_3\text{COH} \qquad (2)$$

that the carboxylated polymer was obtained in yields >99% for (polystyryl)lithium and (polyisoprenyl)lithium. A 40 molar excess of TMEDA was necessary to eliminate dimer and trimer formation for (polybutadienyl)lithium. The actual mechanism of this reaction is complex, since the product distribution is dependent on chain end concentration, solvent, temperature, the rate of stirring and the pressure of carbon dioxide.[44] For example, the yield of carboxylated polystyrene is reduced from >90% to 60% when the solution is stirred during the addition of carbon dioxide even in the presence of 25 vol % THF. In general, the yield of carboxylated polymer is increased by lower chain end concentration and higher carbon dioxide pressure.

In general, it is observed that the amount of dimer and trimer contaminants is higher for (polydienyl)lithium *versus* (polystyryl)lithium. Thus, under conditions where the yields of carboxylated polymer, dimer and trimer are 47%, 27% and 26%, respectively, for (polystyryl)lithium the corresponding yields are 27%, 23% and 50% for the analogous (polystyrene-*block*-butadienyl)lithium.[86] These results are consistent with evidence which suggests that (polydienyl)lithiums are either more highly associated or more strongly associated compared to (polystyryl)lithium.[87-89]

The effect of chain end structure (stability and steric requirements) has also been investigated. The steric and electronic nature of the anionic chain end can be modified by reaction with 1,1-diphenylethylene as shown in equation (3).[90-97] When the direct carbonation is effected in benzene at room temperature for the diphenylakyllithium species (**1**), formed by addition of (polystyryl)lithium ($M_n = 2.0 \times 10^3$ g mol^{-1}) to 1,1-diphenylethylene, the carboxylated polymer can be isolated in 98% yield compared to only a 47% yield for the analogous (polystyryl)lithium without end capping under the same conditions (equation 3).[86] These 1,1-diphenylalkyllithium species are reported to be associated into dimers in hydrocarbon solution, although it would be anticipated that the strength of the dimeric association (*e.g.* K_{assoc}) would be decreased by the increased steric requirements of the chain end.[98] It is tentatively concluded that the competing reaction to form dimeric (and trimeric) side products is quite sensitive to the steric requirements of the chain end.

$$\text{PSLi} + \underset{\text{Ph}}{\overset{\text{Ph}}{=}} \longrightarrow \text{PS}\underset{\text{Li}}{\overset{\text{Ph Ph}}{\diagdown}} \xrightarrow[\text{ii, H}_3\text{O}^+]{\text{i, CO}_2} \text{PS}\underset{\text{CO}_2\text{H}}{\overset{\text{Ph Ph}}{\diagdown}} \quad (3)$$
$$(\mathbf{1})$$

The important conclusion is that the carbonation reaction of polymeric organolithium compounds with gaseous carbon dioxide can be carried out in essentially quantitative yield by adding sufficient quantities of Lewis bases such as tetrahydrofuran or TMEDA prior to the functionalization reaction. It is particularly important to note that this procedure ensures that functionalized polydienes with high 1,4-enchainment can be prepared since the Lewis base is not present during the diene polymerization.[15,23,25,99,100]

A rather specialized solid state carbonation procedure can be used to carbonate (polystyryl)lithium and other living polymers with backbones which have glass transition temperatures significantly above room temperature. Thus, freeze drying of benzene solutions of (polystyryl)lithium generates a porous solid which can be carbonated in the solid state to give minimal amounts of dimeric ketone products (1–4%).[44] In addition, essentially quantitative yields of carboxylated polystyrene were obtained from freeze-dried solutions of (polystyryl)lithium which was complexed with 1–2 equiv. of TMEDA.[44] No dimer was detected by SEC or TLC analyses. In contrast, a freeze-dried sample of (polystyrene-*block*-butadienyl)lithium formed the corresponding carboxylated polymer in only 58% yield.[86] When this (polydienyl)lithium chain end was complexed with 3 equiv. of TMEDA, a 93% yield of the carboxylated polymer was obtained using the freeze-dried polymer.[86]

From a practical, synthetic point of view, it should be noted that it has been reported that >90% yields of carboxylated polymers can be obtained simply by pouring a hydrocarbon/THF solution of the (polystyryl)lithium onto solid carbon dioxide.[101] In the absence of THF, a 78% yield of carboxylated polymer was reported.[102] Conversion to the corresponding Grignard reagent prior to gaseous CO_2 termination has also been reported to produce >90% yields of the carboxylated polymer.[101] It is also noteworthy that essentially quantitative carboxylation has been reported when potassium is the counterion in THF.[103]

The carbonation of α,ω-dilithiopolymers is complicated by the occurrence of gelation phenomena which produce severe mixing problems.[104] In general, heteroatom derivatives of lithium such as lithium carboxylate salts are highly associated in solution;[105,106] therefore, the polymeric α,ω-dicarboxylate salts will form an insoluble, three-dimensional network during the functionalization reaction. A variety of procedures have been described to minimize these effects, including the use of solvents with low solubility parameters (<7.2),[107] reaction in a T-tube mixer[108] and the use of a two-substance jet with a high flow rate and high CO_2:PLi ratio.[104]

The carbonation reaction is somewhat ideal since it is possible to analyze the reaction products using a variety of probes including osmometry, SEC, end group titration, ^{13}C NMR, FTIR and TLC. In addition, the pure functionalized polymer can be separated from unfunctionalized polymer and dimeric ketone products by SiO_2 chromatography using toluene as eluent. For example, column chromatography has been used to separate about 1% unfunctionalized polybutadiene with $\bar{M}_n = 98 \times 10^3$ g mol^{-1} from the corresponding carboxy-functionalized polymer using this technique.[86] Furthermore, it was possible to detect <1% of the unfunctionalized polybutadiene by SiO_2 TLC using toluene as eluent.[86]

5.2.1.2 Phthalic anhydride

It has been reported that α,ω-disodiopolystyrene can be functionalized using phthalic anhydride to form the corresponding aromatic carboxy-functionalized polymer.[109] It should be noted, however, that the reactions of simple organolithium compounds with anhydrides are reported to produce the corresponding tertiary alcohols[36] and only moderate yields of the desired ketone addition products.[110-112] When (polystyryl)lithium ($\bar{M}_n = 2 \times 10^3$ g mol^{-1}) was added to a two-fold excess of phthalic anhydride at 30 °C in THF/C_6H_6 (v/v = 1/3), the desired carboxy-functionalized polymer was obtained in only 16% yield (v[IR] = 1697 and 1717 cm^{-1}).[113] It should be noted that this functionalized polymer could be present either as the ring-opened form or in the ring-closed form (see equation 4). The major product was the dimeric lactone (72% yield; v[IR] = 1774 cm^{-1}) accompanied by the trimeric ketone (12% yield; v[IR] = 1676 and 3550 cm^{-1}). When (polystyryl)lithium ($\bar{M}_n = 1.1 \times 10^3$ g mol^{-1}) was added to 2 equiv. of phthalic anhydride at -78 °C in THF, a 50% yield of the desired aromatic carboxylated polymer was obtained; this was accompanied by a 50% yield of the dimeric lactone. In order to minimize dimer and trimer formation, (polystyryl)lithium was reacted with 1,1-diphenylethylene prior to the phthalic anhydride functionalization reaction as shown in equation (5). Addition of the resulting diphenylalkyllithium to 2 equiv. of phthalic anhydride in THF at -78 °C produced the corresponding keto acid in 80% yield (v[IR] = 1717 and 1697 cm^{-1}). The only other polymeric product was the unfunctionalized polystyrene homopolymer which could be easily separated by silica gel column chromatography.[113]

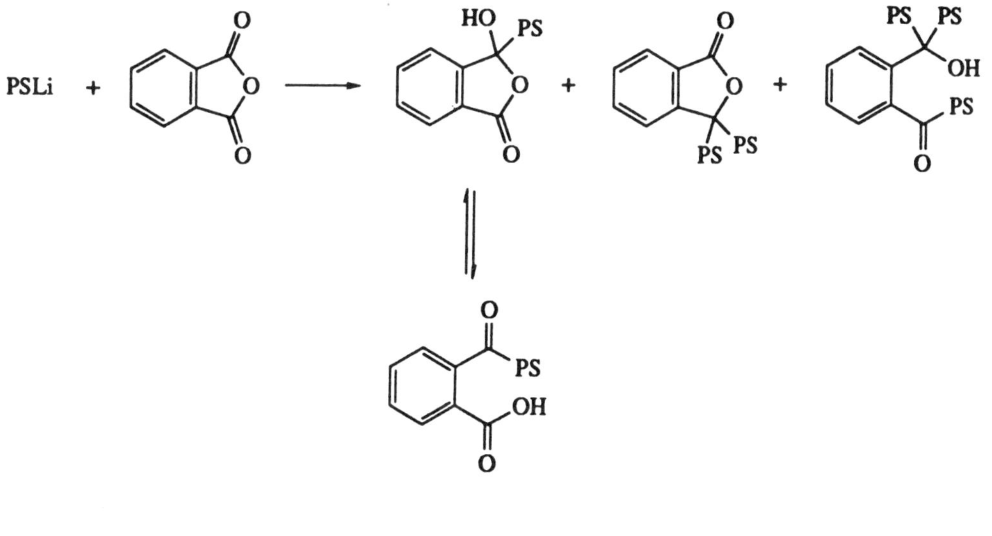

(4)

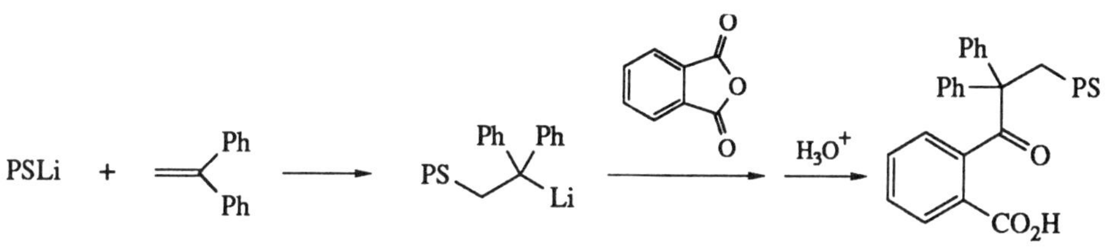

(5)

5.2.1.3 *Protected oxazoline initiator*

The use of an organolithium initiator with an oxazoline protecting group has been examined for the preparation of α-carboxy-functionalized polymers.[114] The initiator was prepared by lithiation of 2,4,4-trimethyl-2-oxazoline, as shown in equation (6). This initiator solution was used without further analysis or characterization to initiate the polymerization of styrene, 2-vinylpyridine and methyl methacrylate at $-78\,°C$ in THF. Low initiator efficiencies (4% for styrene) and broad molecular weight distributions ($\bar{M}_w/\bar{M}_n = 2$ for polystyrene) were observed for methyl methacrylate and styrene polymerizations. The polymerization of 2-vinylpyridine with this initiator provided the stoichiometric molecular weight, but a very broad molecular weight distribution ($\bar{M}_w/\bar{M}_n = 3$). These results were interpreted as an indication of the weakness of the initiator.

$$\text{BuLi} + \underset{\text{(oxazoline)}}{\bigg|} \xrightarrow[\text{THF}]{-78\,°C} \underset{\text{Li}}{\bigg|} \tag{6}$$

The carboxyl functionalization results which have been discussed in Sections 5.2.1.1–5.2.1.3 are typical of the types of problems which are encountered in developing methods for quantitative chain end functionalization. Many of the procedures reported in the literature for a variety of functionalization reactions cannot be reproduced to provide high yields of the desired functionalized polymers.

5.2.2 Hydroxylation

5.2.2.1 *Ethylene oxide*

The preparation of hydroxy-terminated polymers from polymeric organolithium compounds by reaction with ethylene oxide is one of the few simple, efficient functionalization reactions. It also serves to emphasize the uniqueness of organolithium-initiated polymerizations. The direct reaction of (polystyryl)lithium with excess ethylene oxide in benzene solution produces the corresponding hydroxyethylated polymer in quantitative yield without formation of detectable amounts of oligomeric ethylene oxide blocks (equation 7).[62] For example, ^{13}C NMR analysis of a hydroxy-ethylated polystyrene ($\bar{M}_n = 1.3 \times 10^3$ g mol^{-1}; $\bar{M}_w/\bar{M}_n = 1.08$) showed no evidence for the formation of any ether linkages expected for oligomerization of ethylene oxide. This result is surprising in view of the steric strain[115] and intrinsic reactivity of ethylene oxide toward nucleophiles.[116,117] Although organolithium compounds react the most rapidly with ethylene oxide itself compared to other organoalkali compounds,[118] the lithium alkoxides are the least reactive among the alkali metal alkoxides for anionic polymerization of ethylene oxide.[119–122] Apparently the high degree of aggregation of lithium alkoxides and the strength of this association even in polar solvents renders them unreactive.[123] However, it was recently reported that poly(ethylene oxide) is the major product from the reaction of ethylene oxide with the metalation products from the interaction of ferrocene with *n*-butyllithium in hexane in the presence of an equivalent amount of *N,N,N',N'*-tetramethyl-ethylenediamine (TMEDA); poly(ethylene oxide) was also reportedly formed from the direct reaction of ethylene oxide with Bu^{n}Li/TMEDA in hexane.[124] Lithium alkoxides will also polymerize ethylene oxide in a dipolar aprotic solvent such as dimethylsulfoxide at elevated temperature.[125–127]

$$\text{PSLi} + \underset{\triangle}{O} \xrightarrow{H_3O^+} \text{PS}\frown\text{OH} \tag{7}$$

Telechelic dihydroxypolymers can be prepared from the corresponding α,ω-dilithium polymers by treatment with ethylene oxide. For example, functionalities of 1.9–2.0 have been reported for the ethylene oxide termination reaction for α,ω-dilithiopolyisoprene.[128] It should be noted, however, that termination of α,ω-dilithio polymers with ethylene oxide in hydrocarbon solvents results in gel formation which required a period of 1–4 days to achieve complete termination.

5.2.2.2 *Protected initiators*

Hydroxy-terminated polymers have also been prepared using organolithium initiators with protected hydroxy functionality.[2,129] Thus, using initiators such as 2-(6-lithio-*n*-hexoxy)tetrahydropyran and ethyl 6-lithiohexyl acetaldehyde acetal, it was possible to prepare narrow molecular weight distribution polybutadiene polymers ($\bar{M}_w/\bar{M}_n = 1.05-1.08$) with hydroxy functionalities of 0.87–1.02 per chain after mild acid hydrolysis of the acetal groups. However, since these initiators are prepared in diethyl ether, the polybutadienes had relatively high 1,2-microstructures (36–54%). Difunctional α,ω-dihydroxypolybutadienes (functionality = 1.77–2.04) were formed by either terminating with ethylene oxide or coupling with dichlorodimethylsilane followed by mild acid hydrolysis. Either of these approaches avoids the gel formation which occurs when α,ω-dilithio polymers are reacted with ethylene oxide.[128,130] These initiators are reported to be insoluble in hexane, but are soluble in benzene or diethyl ether.

5.2.2.3 *Carbonyl compounds*

The review literature generally describes the reaction of polymeric organolithium compounds with carbonyl compounds in terms of the simple addition reaction shown in equation (8).[15,16,19,20] However, our investigations suggest that these reactions do not proceed simply to provide quantitative yields of the desired hydroxy-terminated polymers. For example, the addition reaction of polymeric organolithium compounds with *p*-(dimethylamino)benzaldehyde was described as a simple reaction, relatively free of side reactions because of the absence of enolizable α-protons.[131] Careful reexamination of this reaction showed that the yellow polymeric reaction products included carbonyl chain end functionality and significant dimer formation (10–33%) for polymers with $\bar{M}_n$ values in the range of $0.5-6.4 \times 10^3 \text{ g mol}^{-1}$.[42] A Cannizzaro reaction of the initially formed carbonyl addition product to form the corresponding acetophenone-type end functionalized polymer (**2**) was suggested based on the isolation and identification of the expected reduction product, *p*-(dimethylamino)benzyl alcohol (see Scheme 1). The formation of dimer product (**3**) results from the addition of a second mole of polymeric organolithium compound to the acetophenone-functionalized polymers.

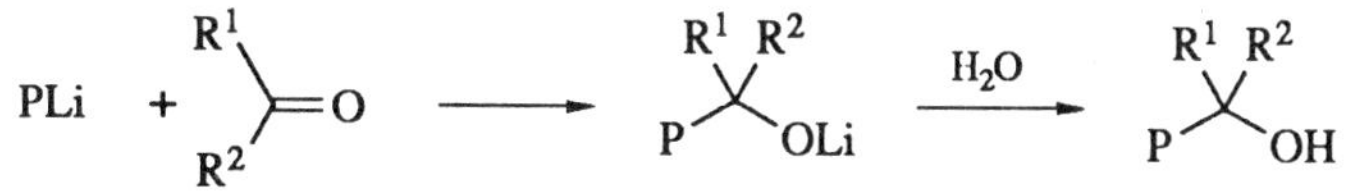

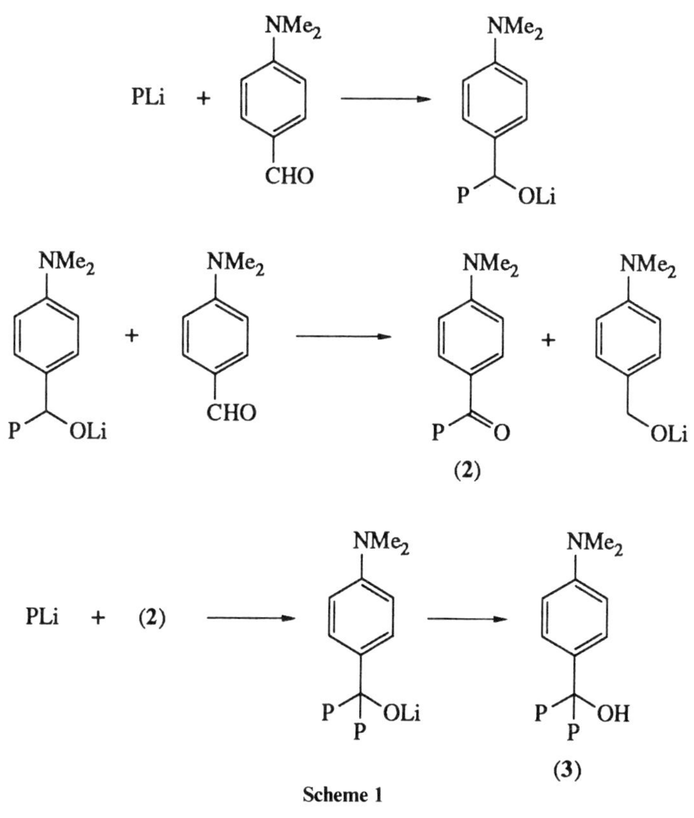

Scheme 1

Another presumably simple reaction is the addition of polymeric organolithium compounds to benzophenone and its derivatives as shown in equation (9). Preliminary investigation of the reaction of (polystyryl)lithium with benzophenone (1.2–1.5 molar excess) in benzene indicated that although the desired hydroxy-functionalized polymer was formed in 88% yield, a dimeric product was formed in approximately 8% yield, accompanied by 4% unfunctionalized polymer.[132] The formation of significant amounts of dimer and unfunctionalized polymer is consistent with the intervention of an electron transfer pathway[133,134] for this functionalization reaction as shown in Scheme 2, where SH represents a solvent molecule. Work is in progress to examine the effects of reaction variables on this functionalization reaction and to study the effects of substituents in the aromatic ketone on the yields of the functionalized polymers. This is especially important in view of the observations that functionalization of (polybutadienyl)lithium with 4,4′-bis(diethylamino)benzophenone effects significant improvement in the properties of tires utilizing the functionalized polymer in their formulation compared to the corresponding unfunctionalized polybutadiene.[135] Preliminary results with this ketone suggest that the functionalized product is formed in only 50% yield in benzene at room temperature; the remainder of the product consists of the unfunctionalized polymer (6%) and dimer (1%).[132]

$$\text{PSLi} + \text{Ph}_2\text{C=O} \xrightarrow{\text{H}_2\text{O}} \text{PS-C(Ph)}_2\text{OH} \quad (9)$$

$$\text{PLi} + \text{Ph}_2\text{C=O} \longrightarrow [\text{P}^\bullet \; \text{Li}^+ \; \text{Ph}_2\text{C-O}^-] \longrightarrow \text{P-C(Ph)}_2\text{OLi}$$

$$2\,\text{P}^\bullet \longrightarrow \text{P}\cdots\text{P}$$

$$\text{P}^\bullet + \text{SH} \longrightarrow \text{PH} + \text{S}^\bullet$$

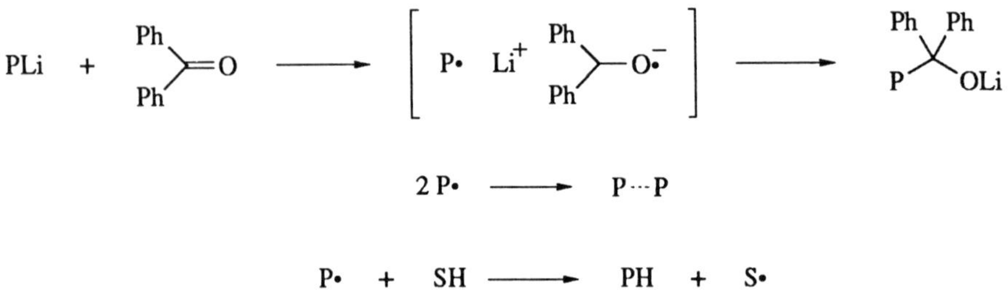

Scheme 2

5.2.3 Amination

5.2.3.1 Protected imines

The primary amination of polymeric organolithium compounds is a challenge because of the acidity of nitrogen–hydrogen bonds.[136] Thus, chain end amination reactions require indirect methods such as the use of protecting groups.[137] Nakahama and coworkers[138,139] have reported that high yields (96–100%) of primary amine functionalized polymers could be obtained by reaction of (polystyryl)lithium with 1.5–2 equiv. of the protected imine, N-(benzylidene)trimethylsilylamine, in benzene at room temperature (equation 10). The analogous functionalization reaction of (polyisoprenyl)lithium in cyclohexane proceeded in 90% yield.

$$\text{PLi} + \text{PhCH=N-SiMe}_3 \xrightarrow{\text{H}_3\text{O}^+} \text{PhCH(P)NH}_2 \quad (10)$$

Attempts to reproduce these results in our laboratories with (polystyryl)lithium were not successful.[65] The aminated polystyrenes were contaminated with dimeric products (15–19% yields). In fact, the aminated polystyrene was obtained in only 69% yield for (polystyryl)lithium with $\bar{M}_n = 3 \times 10^3$ g mol^{-1} and an acetophenone-type functionalized polymer (12% yield) was formed in addition to the dimeric product (19% yield). It is proposed that the acetophenone-type functionality and the dimeric side reaction product result from a Cannizzaro-type reaction (Scheme 1), as previously described for p-(dimethylamino)benzaldehyde under the carbonyl functionalization reactions (see Scheme 3).

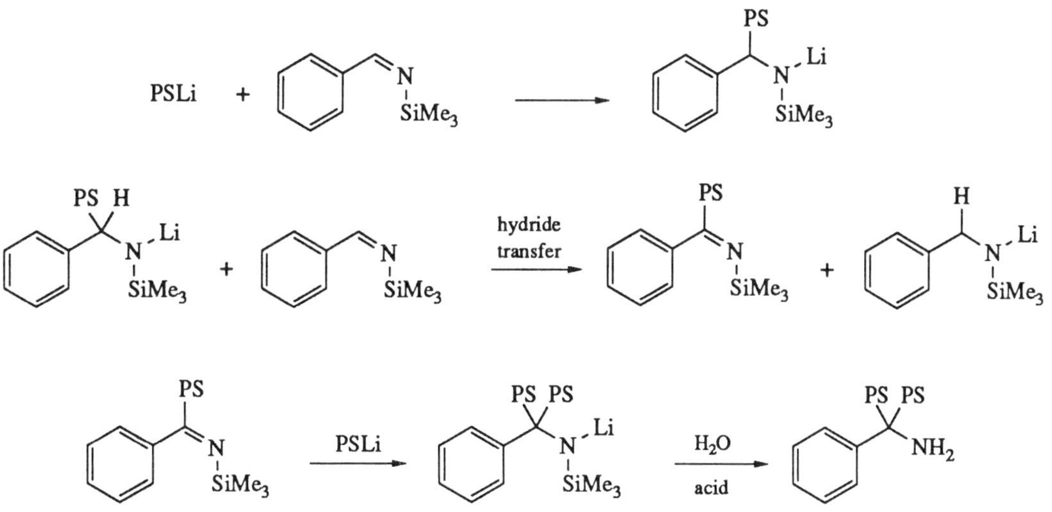

Scheme 3

5.2.3.2 Methoxyamine/methyllithium

Primary amine functionalized polystyrenes have been prepared by the reaction of (polystyryl)-lithium with the product of the reaction of methoxyamine and methyllithium at low temperatures, as shown in equation (11).[79,140,141] Using a two-fold excess of the methoxyamine/methyllithium reagent at $-78\,°C$ in a mixture of THF/benzene/hexane, (polystyryl)lithium ($\bar{M}_n = 2 \times 10^3\,\mathrm{g\,mol^{-1}}$) was aminated in 92% yield after methanol work-up.[79] The telechelic diamine, α,ω-diaminopolystyrene, was prepared in 80% yield by amination of α,ω-dilithiopolystyrene ($\bar{M}_n = 10 \times 10^3\,\mathrm{g\,mol^{-1}}$) using similar procedures.[140] It was possible to isolate pure, amine-functionalized polymers using silica gel column chromatography.

$$\mathrm{PSLi} \xrightarrow[-78\,°C]{\mathrm{MeONH_2/MeLi}} \xrightarrow{\mathrm{ROH}} \mathrm{PSNH_2} \qquad (11)$$

5.2.3.3 Functionalized and protected initiators

Eisenbach et al.[142] first reported the use of the dimethylaminopropyllithium initiator for the preparation of tertiary amine functionalized polymers. Although the initiator produced polystyrenes with high amine functionalities, their procedure was limited because the initiator was prepared in a polar ethereal solvent. Recently Stewart and coworkers[143] described a procedure for the preparation of a benzene solution of 3-dimethylaminopropyllithium in >90% yields by lithiation of the corresponding chloride in hexane at $20\,°C$, followed by replacement of the solvent by benzene. Use of this initiator solution for butadiene polymerization in hexane produced the corresponding α-(dimethylamino)polybutadienes with high (76–86%) 1,4-polybutadiene microstructure but unspecified molecular weight distributions. These workers noted that the amount of vinyl content in the resulting polybutadiene increased with the amount of diamine formed by Wurtz coupling during the synthesis of the initiator.

Primary amine terminated polymers have also been prepared using organolithium initiators with protected amine functionality. The p-lithio-N,N-bis(trimethylsilyl)aniline initiator was prepared from the corresponding aryl bromide by reaction with lithium metal in diethyl ether.[137] A major limitation of this initiator is the fact that it is insoluble in hydrocarbon solvents and therefore diene polymerizations must be carried out in mixtures of hexane and diethyl ether. Using this initiator, relatively narrow molecular weight distribution ($\bar{M}_w/\bar{M}_n = 1.06$–$1.25$) polybutadienes were prepared with functionalities of 0.69–1.0 as determined by titration after acid hydrolysis of the amine protecting group. However, the use of diethyl ether for the initiator solution produced polybutadienes and polyisoprenes with 39–43% and 45–50% vinyl microstructures, respectively. Termination of the corresponding isoprene polymerizations with dichlorodimethylsilane followed by hydrolysis of the protecting group generated polymeric α,ω-diaminopolyisoprenes with functionalities of 1.7–1.9 and relatively broad molecular weight distributions ($\bar{M}_w/\bar{M}_n = 1.49$–$2.22$).

It has recently been reported that a useful protected primary amine initiator can be generated by the reaction of s-butyllithium with p-bis(trimethylsilyl)aminostyrene in benzene or cyclohexane solution at 25 °C by careful control of the stoichiometry of the reaction, as shown in equation (12).[144]

$$Bu^sLi + \text{CH}_2=\text{CH}-C_6H_4-N(SiMe_3)_2 \longrightarrow Bu^s-CH_2-CH(Li)-C_6H_4-N(SiMe_3)_2 \quad (12)$$

Although oligomerization was observed when THF was added or when n-butyllithium was used as the initiator, it was reported that no oligomerization was observed for s-butyllithium in hydrocarbon, as deduced by a combination of gas chromatography and ^{1}H NMR analysis of the acetic acid quenched reaction. This initiator was used to prepare primary amine functionalized poly(dimethylsiloxane) using the cyclic trimer D_3 as monomer in the presence of the promoter hexamethylphosphoramide, followed by acid hydrolysis with dilute aqueous hydrochloric acid. Reported amine functionalities were 0.94–0.97 for the α-aminopolydimethylsiloxanes and 1.9 for the corresponding telechelic polymers obtained by coupling with dichlorodimethylsilane. It remains to be demonstrated that a useful protected primary amine functionalized initiator for styrene and diene polymerization can be generated using these procedures.

5.2.3.4 ω-Halo-α-aminoalkanes

Richards et al.[145] have reported that (polybutadienyl)lithium can be functionalized efficiently by reaction with N-(3-chloropropyl)dialkylamines, as shown in equation (13). The reported yields were >95%.

$$PBDLi + Cl-CH_2CH_2CH_2-NR_2 \longrightarrow PBD-CH_2CH_2CH_2-NR_2 \quad (13)$$

Recently Teyssie and coworkers[146] have applied this procedure to the preparation of α,ω-bis-(dimethylamino)polyisoprene. Sodium naphthalenide was used as the difunctional initiator to prepare α,ω-disodiopolyisoprene at −78 °C. Termination with excess 1-chloro-3-(dimethylamino)-propane produced α,ω-bis(dimethylamino)polyisoprene with $\bar{M}_w/\bar{M}_n = 1.2$, presumably high vinyl microstructure, and a functionality of 1.85 as determined by perchloric acid titration. All of these results are surprising in view of the fact that normally the coupling reactions of alkyl halides with organolithium compounds are complicated by side reactions such as lithium–halogen exchange and dehydrohalogenation (elimination) reactions.[36]

Nakahama and coworkers[147] have recently extended this approach to include a study of the use of α-halo-ω-aminoalkanes with a protected amino functionality, as shown in equation (14). It was reported that the yields of >94% were readily obtained for the functionalization reactions of (polystyryl)lithium, (polyisoprenyl)lithium, and α,ω-dipotassiopolyisoprene with the α-bromo- and α-chloro-ω-silyl-protected aminopropanes and the α-bromo-ω-silyl-protected aminoethane to produce the corresponding monoamine and telechelic amine functionalized polymers. Precipitation into methanol was sufficient to remove the silyl protecting groups. Evidence for Wurtz coupling reactions leading to the formation of dimer was observed for the α-iodo derivative. The typical procedure involved reaction with excess reagent at −78 °C in the presence of THF, followed by standing for 1 h at 25 °C. The products were characterized by SEC analysis of the benzoyl derivatives, TLC, TLC with flame ionization detection, VPO, and end group titration with $HClO_4$. In addition, chain extension reactions of the telechelic α,ω-diaminopolystyrene ($\bar{M}_n = 5.35 \times 10^3$ g mol^{-1}) with 2,4-tolyl diisocyanate produced a segmented polyurea with $\bar{M}_n = 230 \times 10^3$ g mol^{-1} and no base polymer was detected by SEC. This result is consistent with high functionality for the diamine.

$$PLi + X-(CH_2)_n-N(SiMe_2-)_2 \longrightarrow P-(CH_2)_n-N(SiMe_2-)_2 \quad (14)$$

5.2.4 Sulfonation

5.2.4.1 Direct reaction

The simplest procedure for the sulfonation of polymeric organolithium compounds is the direct reaction with sultones. Thus, the functionalization of (polystyryl)lithium ($\bar{M}_n = 2.7$–4.7×10^3 g mol^{-1}), generally with a two-fold molar excess of 1,3-propanesultone (equation 15), was investigated as a function of solvent and temperature.[148] At temperatures from 3–80 °C in benzene solution, only 25–30% of the terminally sulfonated polymer was formed, the remainder of the product being the unfunctionalized polymer. At -78 °C in THF using a molar excess of sultone, a yield of 53% of the terminally sulfonated polystyrene was obtained compared to the yields of 67–72% reported by Omeis et al.[149] using a five-fold excess of sultone. Presumably the relatively low yields of sulfonated polymer are a result of the competing metallation reaction of (polystyryl)lithium with the acidic α-hydrogens of the sultone, as shown in equation (16). Analogous reactions of butyllithium with 1,3-propanesultone produce the metallated sultone in 65–85% yields.[150,151]

$$\text{PSLi} + \text{sultone} \longrightarrow \text{PS}\frown\frown\text{SO}_3\text{Li} \qquad (15)$$

$$\text{PSLi} + \text{sultone} \longrightarrow \text{PSH} + \text{metallated sultone} \qquad (16)$$

These results with (polystyryl)lithium are quite different from the results of Eisenbach et al.,[142] who reported sulfonation yields of >90% for the reaction of [poly(α-methylstyryl)]lithium with 1,3-propanesultone in THF at -78 °C. In order to examine this effect of chain end structure, (polystyryl)lithium ($\bar{M}_n = 4.7 \times 10^3$ g mol^{-1}; $\bar{M}_w/\bar{M}_n = 1.03$) was reacted with α-methylstyrene at -78 °C in THF to form [poly(styrene-block-α-methylstyryl)]lithium ($\bar{M}_n$ (SEC, apparent) = 6.0×10^3 g mol^{-1}; $\bar{M}_w/\bar{M}_n = 1.16$). When this block copolymeric organolithium compound was reacted with a molar excess of 1,3-propanesultone, the corresponding sulfonated diblock copolymer was obtained in 94% yield.[148] This is in excellent agreement with the reported results of Eisenbach et al.[142] However, when the sulfonation reaction of an analogous diblock copolymeric organolithium compound was effected in toluene under analogous conditions with an almost three-fold excess of sultone, the corresponding sulfonated product was obtained in only 28% yield. It appears that the competition between ring opening and metallation is quite sensitive to solvent and chain end structure. These results indicate that although the direct sulfonation of [poly(α-methylstyryl)]lithium with 1,3-propanesultone is an efficient reaction in THF at -78 °C,[142] this is not a generally useful procedure for (polystyryl)lithium since only 53–72% yields can be obtained.[148,149] This dramatic effect of chain end structure suggested that other methods of attenuating the reactivity of the polymeric organolithium chain end might also be effective. A sulfonation procedure which could be effected at room temperature was also desirable.

5.2.4.2 Ethylene oxide end capping

1,3-Propanesultone and 1,4-butanesultone are very reactive sulfoalkylating agents. It has been reported that a variety of nucleophiles react with these sultones to open the ring and produce the corresponding sulfoalkylation product.[152] For example, even tertiary amines react efficiently to produce the corresponding sulfonated zwitterionic product, as shown in equation (17).[153,154] In addition, since it has been reported that alkoxides are sufficiently reactive to open the sultone ring (equation 18),[152,155,156] it was envisioned that an effective method for attenuating the reactivity of the anionic chain end would be to carry out the hydroxyethylation reaction with ethylene oxide prior to reaction with the sultone, as shown in Scheme 4. This approach has been effectively used to generate macromonomers by reaction of (polystyryl)lithium with ethylene oxide prior to reaction with methacryloyl chloride.[157,158]

A series of hydroxyethylated polystyrenes were prepared by reaction of (polystyryl)lithium with excess ethylene oxide at room temperature in benzene followed by methanol quenching of aliquots

$$R_3N \ + \ \underset{}{\overset{O\ O}{\underset{O-S}{\bigtriangleup}}} \longrightarrow R_3\overset{+}{N}\frown SO_3^- \quad (17)$$

$$RO^- \ + \ \underset{}{\overset{O\ O}{\underset{O-S}{\bigtriangleup}}} \longrightarrow RO\frown SO_3^- \quad (18)$$

$$PSLi \ + \ \triangle^O \longrightarrow PS\frown OLi \xrightarrow{\text{sultone}} PS\frown O\frown SO_3Li$$
$$\qquad\qquad\qquad\qquad\qquad (4) \qquad\qquad\qquad\qquad (5)$$

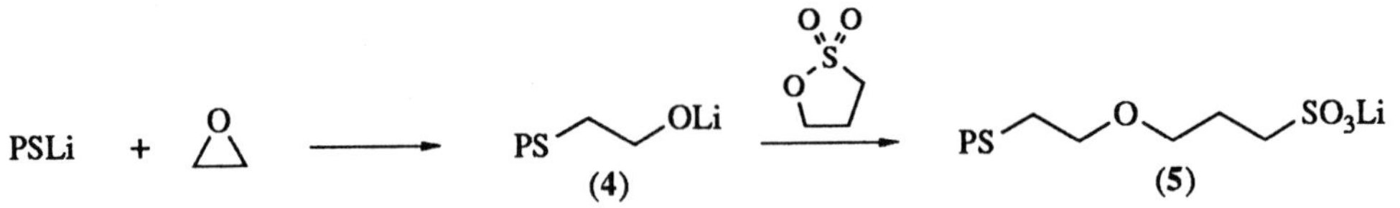

Scheme 4

for analyses. The corresponding lithium alkoxide derivatives (**4**; Scheme 4) were reacted with either 1,3-propanesultone or 1,4-butanesultone under a variety of conditions. The reaction of the polymeric lithium alkoxide derivative with 1,3-propanesultone produced the corresponding sulfonated polymer in quantitative yield in benzene solution using a four-fold excess of sultone at 40 °C after approximately 16 h, as determined by colorimetric titration.[148] A 96% yield was obtained in a 1:10 (v/v) mixture of THF/benzene with a three-fold excess of sultone at 40 °C. Lower yields were obtained using a smaller excess of sultone, lower temperatures or 1,4-butanesultone. These results are consistent with previous work which has shown that 1,4-butanesultone is less reactive toward nucleophiles compared to 1,3-propanesultone.[152] However, it was possible to obtain a 93% yield when an eight-fold excess of 1,4-butanesultone was used in benzene solution.

Attempted silica gel column chromatographic purification of the sulfonated polymers synthesized as shown in Scheme 4 indicated that these polymers were not stable to the chromatographic conditions. TLC analyses of the products eluting from the silica gel column indicated the formation of the hydroxyethylated polymer ($R_f = 0.7$) and a spot corresponding to an unfunctionalized product ($R_f = 1.0$) in addition to the sulfonation product ($R_f = 0$). Analogous results, i.e. desulfonation, were observed when the sulfonation products were treated with an alcoholic hydrochloric acid solution (1 M) at room temperature for 24 h. Pure sulfonated polymer obtained via the lithium alkoxide route could be purified and separated from the excess sultone without desulfonation by using a modified silica gel which had been annealed at ca. 600 °C and stored in a desiccator;[47,82] the eluents were toluene followed by THF.

5.2.4.3 1,1-Diphenylethylene end capping

Because of the apparent hydrolytic instability of the sulfonation product from the lithium alkoxide reaction (Scheme 4), an alternative method of attenuating the reactivity of the carbanionic chain end was sought. End capping of simple alkyl or polymeric organolithium compounds with 1,1-diphenylethylene produces the corresponding 1,1-diphenylalkyllithium compound quantitatively without homopolymerization, as shown in equation (19).[24,25,36,93-97]

The decreased basicity and increased steric requirements of this carbanion have been effectively used to generate an efficient initiator for the anionic polymerization of alkyl methacrylates.[90-92] Thus, end capping of (polystyryl)lithium with 1,1-diphenylethylene prior to reaction with a sultone (see Scheme 5) was investigated as a method for increasing the efficiency of the sulfonation reaction as well as generating a hydrolytically stable sulfonation product. The sulfonation reaction of the polymeric diphenylalkyllithium compounds (**6**) with 1,3-propanesultone and 1,4-butanesultone were

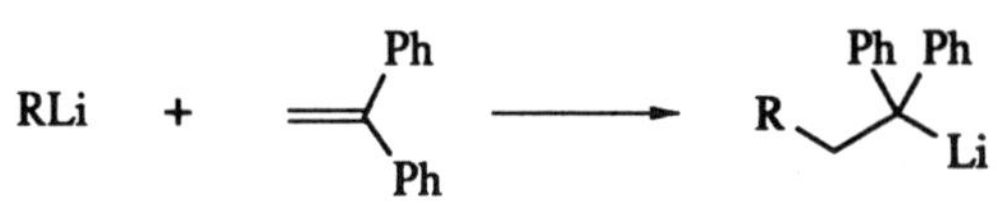

$$RLi \ + \ \underset{Ph}{\overset{Ph}{=\!\!\!<}} \longrightarrow R\underset{Li}{\overset{Ph\ Ph}{\diagup\!\!\!\diagdown}} \quad (19)$$

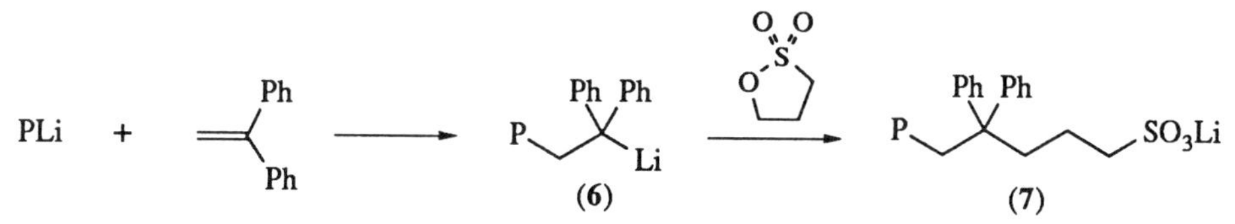

$$\text{PLi} + \underset{\text{Ph}}{\overset{\text{Ph}}{=}} \longrightarrow \underset{\text{(6)}}{\text{P}\diagdown\overset{\text{Ph Ph}}{\diagup}\text{Li}} \xrightarrow{\overset{\text{O}\diagdown\overset{\text{O}}{\text{S}}\diagup\text{O}}{\square}} \underset{\text{(7)}}{\text{P}\diagdown\overset{\text{Ph Ph}}{\diagup}\diagdown\text{SO}_3\text{Li}}$$

Scheme 5

investigated.[148] Using a 1:6 (v/v) ratio of THF/benzene and at least a 1.5-fold excess of 1,3-propanesultone at 25–30 °C, a 93% yield of the sulfonated polymer (7) could be obtained *via* the 1,1-diphenylethylene end-capping route for (polystyryl)lithium. Lower sulfonation yields (54–76%) were obtained in benzene solution and the yield decreased when a large excess of sultone (nine-fold) was used. Sulfonation yields with 1,4-butanesultone were generally lower than the corresponding sulfonations with 1,3-propanesultone under analogous conditions; however, the effects of solvent and excess sultone were similar to the effects observed with 1,3-propanesultone.

5.2.5 Oxidation

The reaction of (polystyryl)lithium with molecular oxygen produces a complex mixture of products.[159–162] The products after work-up and the range of yields obtained by allowing molecular oxygen to diffuse into an unstirred solution at room temperature under various conditions are shown in equation (20).[162] Within the yield ranges indicated in equation (20), the product distribution was relatively insensitive to the presence of polar additives such as THF or TMEDA in benzene solution.[162] It is noteworthy that the expected formation of the dimeric peroxide by a free radical chain oxidation mechanism[163] has been generally overlooked.[17] In contrast to the solution oxidation results, the oxidation of (polystyryl)lithium complexed with 1 equiv. of TMEDA in the solid state, obtained by freeze-drying the corresponding benzene solution, produced ω-hydroperoxypolystyrene in 95% yield.[159,162] This functionalization reaction generates a potentially useful polymeric free radical initiator, which could in turn be used to generate new block copolymers.

$$\text{PSLi} + \text{O}_2 \longrightarrow \underset{18-22\%}{\text{P--P}} + \underset{3-18\%}{\text{P--O--O--P}} + \underset{2-9\%}{\text{P--O--OH}} + \underset{58-69\%}{\text{P--OH}} \quad (20)$$

5.2.6 Summary of Specific Functionalization Reactions

This brief overview of chain end functionalizations using specific reactions for each functional group illustrates the complexity of effecting the quantitative preparation of functionalized polymers with a variety of electrophilic species. The ability to prepare chain end functionalized or labeled polymers is not limited by the range of potential and reported functionalization and labeling reactions for living carbanionic chain ends. It is limited by the fact that many functionalization reactions which have been reported in the literature have not been adequately characterized or optimized for general utility.[14] Another limitation is the necessity of developing, optimizing and characterizing new reactions for each different functional group as illustrated by the previous discussion. Recently, we embarked on a search for general, quantitative functionalization reactions. We have explored the use of the addition reactions of polymeric carbanions to 1,1-diphenylethylene derivatives as a general functionalization method, independent of the nature of the functional group.

5.3 1,1-DIPHENYLETHYLENE FUNCTIONALIZATION REACTIONS

5.3.1 Background

The reaction of polymeric organolithium compounds with substituted 1,1-diphenylethylene derivatives is an excellent system for the development of a general functionalization reaction (equation 21) because: (a) these addition reactions are simple and quantitative; (b) only monoaddition, *i.e.* no oligomerization, has been reported; (c) the rate and efficiency of the crossover reaction can be monitored by UV–visible spectroscopy; (d) copolymerization of substituted 1,1-diphenylethylene derivatives with other monomers will result in polymers with multiple functional groups along the polymer chain; and (e) a variety of substituted 1,1-diphenylethylenes with functional groups on the aromatic ring can be readily prepared.[24,25,36,93-97] Unlike most electrophilic functionalization reactions, this reaction is not in itself a termination reaction. The product of the addition reaction of a simple or polymeric organolithium compound to a substituted 1,1-diphenylethylene is a carbanionic species (1,1-diphenylalkyllithium) which could initiate anionic polymerization of an additional monomer such as isoprene[164] or methyl methacrylate[90-92] to extend the chain or form a new block. Thus, this procedure can be described as a 'living functionalization reaction'. This method can be used to prepare polymers with functional groups at the initiating end of the polymer chain (see equation 22) or within the polymer chain (see equation 23), in addition to the terminating end (equation 21). Thus, it offers the potential of providing a versatile, general anionic functionalization procedure with which one can rationally design and place functional groups at essentially any position in a polymer molecule. Furthermore, this methodology can be applied to the synthesis of polymers with two or more functional groups at the initiating end, at the interface between two blocks or at the terminating end, as shown in equation (24), where X and Y can either be the same or different functional groups. The α,α- or ω,ω-difunctional products of these functionalization reactions could behave as macromolecular monomers for condensation-type copolymerization reactions. These applications of 1,1-diphenylethylene functionalization chemistry are currently under active investigation.[165,166]

The following sections describe the applications of 1,1-diphenylethylene functionalization reactions for the preparation of phenol-, amino- and carboxyl-functionalized polymers as well as for the preparation of polymers with aromatic functional groups which provide fluorescent labels. In general, the methodology described in equations (21)–(23) has been realized. However, the ability to either prepare polymers with labels at the initiating end or at the interface between two blocks is limited by the reactivity of the diphenylalkyllithium species as an initiator. Since diphenylmethane, toluene and propene have estimated pK_a values of 32, 42 and 43,[167,168] respectively, it is surprising that diphenylalkyllithium species can initiate the polymerization of styrene and diene monomers.[164] However, we have observed that diphenylalkyllithiums, either unsubstituted or with electron-donating substituents on the aromatic ring, function as effective initiators for the polymerization of styrene, diene and methacrylate monomers, i.e. polymers with predictable molecular weights and relatively narrow molecular weight distributions ($\bar{M}_w/\bar{M}_n \leq 1.1$[169]) are obtained.[63,64] However, when electron-withdrawing functional groups such as the oxazoline and tertiary N,N-dialkylamide protecting groups are substituted on the aromatic ring, broad molecular weight distribution polymers are obtained ($\bar{M}_w/\bar{M}_n > 1.1$).[170] These results have been ascribed to the effects of the conjugated electron-withdrawing groups in stabilizing the intermediate 1,1-diphenylalkyllithium species, which reduces their reactivity as initiators for styrene and diene monomers. Nakahama and coworkers[12,170–173] have observed similar limitations on the ability to utilize living polymers formed from styrene monomers with these electron-withdrawing groups for block copolymer formation with styrene and α-methylstyrene. Aside from these limitations, the anionic functionalization methodology utilizing substituted 1,1-diphenylethylenes as outlined herein provides the most versatile, general functionalization currently available.

5.3.2 Phenol Functionality

The first application of this methodology was directed to the preparation of a phenol-terminated polymer.[63] The *t*-butyldimethylsilyl protecting group was chosen for the aromatic hydroxy group based on the results reported by Nakahama and coworkers.[174,175] (Polystyryl)lithiums with $\bar{M}_n = 1.5–13.1 \times 10^3$ g mol^{-1} were functionalized with 1-(4-*t*-butyldimethylsiloxyphenyl)-1-phenylethylene (**8**) to form the corresponding phenol end functionalized polystyrenes in >99% yield after hydrolysis with 1% HCl in THF. The efficiency of these functionalization reactions was evaluated by end group titration, elemental analyses, ^{1}H NMR and ^{13}C NMR analyses, as well as thin layer chromatography.[63] All of the available evidence suggests that this is an essentially quantitative functionalization reaction. It is noteworthy that Heitz and Hocker[165] have carried out similar conversions using 1,1-(4,4'-dimethoxyphenyl)ethylene.

Ph (**8**)

5.3.3 Amine Functionality

We have also investigated the use of 1-(4-dimethylaminophenyl)-1-phenylethylene (**9**) to prepare amine-terminated polymers.[64] This substituted diphenylethylene was readily prepared from the corresponding benzophenone derivative *via* the Wittig reaction. The reaction of (polystyryl)lithiums ($\bar{M}_n = 3.1–10.2 \times 10^3$ g mol^{-1}) in benzene solution with (**9**) could be monitored by UV–visible spectroscopy at 406 nm. Analyses of these functionalization reactions indicated that the amine-terminated polymers are obtained in >99% yields.

Ph (**9**)

Telechelic α,ω-bis(dimethylamino)-terminated polystyrene was prepared by using the adduct of *s*-butyllithium with (**9**) as the initiator for styrene polymerization to form the α-dimethylamino-(polystyryl)lithium (**10**) as shown in equation (25). An aliquot of (**10**) was removed from the reactor and terminated with methanol to form the corresponding monofunctional polymer. This polymer had $\bar{M}_n = 2.05 \times 10^4$ g mol^{-1} (SEC), $\bar{M}_w/\bar{M}_n = 1.04$ (SEC) and the functionality of this polymer was 1.2 as determined by amine end group titration. The remainder of the living polymer (**10**) was then functionalized by addition of (**9**) to form the corresponding α,ω-dimethylaminopolystyrene (**11**) after methanol termination, as shown in equation (26). The α,ω-bis(dimethylamino)polystyrene exhibited $\bar{M}_n$ (SEC) $= 2.05 \times 10^4$ g mol^{-1}, $\bar{M}_n$ (VPO) $= 2.19 \times 10^4$ g mol^{-1} and $\bar{M}_w/\bar{M}_n = 1.04$. The amine group functionality of this telechelic polymer was 2.1 as determined by end group titration.

The ability to place functional groups within polymer chains was demonstrated by first functionalizing (polystyryl)lithium with (**9**).[63] A portion of this functionalized, living polymer (**12**) was removed from the reactor and terminated with methanol to give a monofunctional ω-dimethylamino-functionalized polystyrene. The functionality of this polymer as calculated from the end group titration was 0.96. This amine-functionalized polystyrene exhibited only one spot by TLC analysis and it was estimated that less than 1% unfunctionalized polystyrene was formed. The living, functionalized polymer (**12**) was chain extended by addition of butadiene monomer to form a diblock polymer, poly(styrene-*block*-butadiene) (**13**) with the dimethylamino functional group at the interface between the blocks, as shown in equation (27).

The block copolymer (**13**) exhibited $\bar{M}_n$ (membrane osmom) = 2.13×10^4 g mol^{-1}, $\bar{M}_n$ (titration) = 2.19×10^4 g mol^{-1} with $\bar{M}_w/\bar{M}_n$ (SEC) = 1.01. The functionality calculated from the titration results was 0.97 and the polymer exhibited only one spot by TLC analysis with no unfunctionalized polystyrene base polymer detected. These results show that functionalized diphenylalkyllithiums with electrondonating substituents are effective initiators for anionic polymerization of styrene and diene monomers, *i.e.* polymers with narrow molecular weight distributions are obtained.

Recent work suggests that the N,N-bis(trimethylsilyl)-protected 1-(4-aminophenyl)-1-phenylethylene can be used to prepare the corresponding primary amine functionalized polymers, as shown in equation (28).[176] The polymer functionality was 0.97 as determined by end group titration for amination of (polystyryl)lithium ($\bar{M}_n = 3.8 \times 10^3$ g mol^{-1}) in benzene at room temperature and no unfunctionalized base polymer was detected by SiO$_2$ TLC analysis.

5.3.4 Carboxy Functionality

The use of substituted 1,1-diphenylethylenes to prepare end-functionalized polymers has also been utilized to prepare carboxy-functionalized polymers. The carboxy functionality has been protected using the oxazoline group. The oxazoline-substituted 1,1-diphenylethylene was not stable to the anionic chain end at room temperature, however. The functionalization reaction was effected in toluene/THF mixtures (4:1, v/v) at $-78\,°C$ to produce the carboxy-functionalized polystyrene ($\bar{M}_n = 2.4$–$14.6 \times 10^3\,g\,mol^{-1}$) in quantitative yield after acid hydrolysis as shown in equation (29).[170]

The carboxy group has also been protected using the diisopropylamide derivative.[170] The corresponding diisopropylamide-functionalized 1,1-diphenylethylene was synthesized and characterized by elemental analysis and a variety of spectroscopic methods. The diisopropylamide-functionalized 1,1-diphenylethylene was not stable to the anionic chain end at room temperature. The functionalization reaction was effected in toluene/THF mixtures, (4:1, v/v) at $-78\,°C$ to produce the amide-functionalized polystyrene in 92–100% yields as shown in equation (30) ($\bar{M}_n = 2.3$–$12.6 \times 10^3\,g\,mol^{-1}$). Although it was somewhat difficult to hydrolyze the amide, heating under reflux in toluene with toluenesulfonic acid was found to be effective in generating the carboxy-functionalized polymers.

$$\text{PSLi} + \text{[1,1-diphenylethylene with para-C(O)NPr}^i_2\text{]} \xrightarrow{\text{ROH}} \text{PS-CH}_2\text{-C(Ph)-[aryl-C(O)NPr}^i_2\text{]} \quad (30)$$

5.3.5 Copolymerization

The anionic copolymerization of substituted 1,1-diphenylethylene derivatives with copolymerizable monomers (M) will result in polymers with multiple functional groups along the polymer chain, as illustrated in equation (31). The number of functional groups per polymer molecule can be controlled by the monomer feed ratio and the molecular weight.

$$n\left(\text{[1,1-diphenylethylene-X]}\right) + m\text{M} \xrightarrow{\text{initiator}} \text{polymer with X groups} \quad (31)$$

The anionic copolymerization of styrene and 1-(4-dimethylaminophenyl)-1-phenylethylene in benzene has been investigated.[177] Yuki and coworkers[178–180] have developed the formalism for analyzing the kinetics of copolymerization of 1,1-diphenylethylene (M_2) with styrene and diene monomers (M_1). It is assumed that the 1,1-diphenylethylene derivative M_2 does not add to itself due to steric effects, i.e. $k_{22} = 0$. Thus, the monomer reactivity ratio for M_2 is zero, i.e. $r_2 = k_{22}/k_{21} = 0$. It is also assumed that the styrene monomer is completely consumed at the end of the polymerization ([M_1] = 0) and that M_2 is in excess, i.e. there is still unreacted 1-(4-dimethylaminophenyl)-1-phenylethylene after the copolymerization. The resulting copolymerization equation is given below (equation 32)

$$\ln\frac{[M_2]}{[M_2]_0} + \frac{1}{r_1 - 1}\ln\left[\frac{[M_1]_0}{[M_2]_0}(r_1 - 1) + 1\right] = 0 \quad (32)$$

where r_1 is not unity and $[M_1]_0$, $[M_2]_0$ and $[M_1]$, $[M_2]$ are the initial and final monomer concentrations, respectively.

The copolymerization of a 1.5 molar excess of styrene with 1-(4-dimethylaminophenyl)-1-phenylethylene produced a copolymer ($\bar{M}_n = 1.6 \times 10^4\,g\,mol^{-1}$) with 24 amine groups per chain. Analogously, the copolymerization of a 3.6 molar excess of styrene with 1-(4-dimethylaminophenyl)-1-phenylethylene produced a copolymer ($\bar{M}_n = 3.8 \times 10^4\,g\,mol^{-1}$) with 37 amine groups per chain. These copolymers did not move on TLC plates in toluene, i.e. $R_f = 0$, due to the large number of amine groups in the polymer molecules. The average value for r_1 was determined to be 5.6, which means that styrene is 5.6 times as reactive as the amine-substituted 1,1-diphenylethylene towards the (polystyryl)lithium anion. The corresponding value of r_1 is 0.71 for the copolymerization of styrene and 1,1-diphenylethylene in benzene at $30\,°C$.[178] These results are consistent, since the addition reaction of 1,1-diphenylethylene derivatives to (polystyryl)lithium in benzene at room temperature has a Hammett ρ value of $+1.8$.[94] The crossover reaction rate of (polystyryl)lithium with 1-(4-dimethylaminophenyl)-1-phenylethylene (k_{12}) is expected to be much less than that of (polystyryl)lithium to 1,1-diphenylethylene, due to the strong electron-donating effect of the dimethylamino group ($\sigma = -0.83$).[181] Thus, the monomer reactivity ratio, r_1 ($r_1 = k_{11}/k_{12}$), is expected to be larger for the copolymerization of styrene with 1-(4-dimethylaminophenyl)-1-phenylethylene than in the copolymerization of styrene and 1,1-diphenylethylene as observed.

It is anticipated that the copolymerization of substituted 1,1-diphenylethylenes with dienes such as butadiene and isoprene will be complicated by the very unfavorable monomer reactivity ratio for the addition of (polydienyl)lithium compounds to 1,1-diphenylethylene.[179,180] Yuki and coworkers[179,180] calculated values of $r_1 = 54$ and $r_1 = 29$ in hydrocarbon solutions for the copolymerization of 1,1-diphenylethylene (M_2) with butadiene (M_1) and isoprene (M_1), respectively. Although the corresponding values in THF are r_1 (butadiene) $= 0.13$ and r_1 (isoprene) $= 0.12$, this would not be an acceptable solution since THF is known to form polymers with high 1,2-microstructures.[15,23,25,99,100] Work is underway to evaluate the use of tertiary potassium alkoxides as copolymerization promoters based on their effectiveness in promoting the random copolymerization of styrene and dienes without the formation of high vinyl diene microstructure.[182,183]

5.3.6 Fluorescent Group Labeling

Anionic functionalization methodology based on addition reactions to 1,1-diphenylethylenes and their analogs can also be utilized for the preparation of polymers labeled with fluorescent groups. We have reported that (polystyryl)lithium can be quantitatively labeled with a fluorescent naphthyl end group *via* the reaction with 1-(2-naphthyl)-1-phenylethylene.[73] The adduct of 1-(2-naphthyl)-1-phenylethylene with butyllithium has been used to initiate the anionic polymerization of methyl methacrylate at low temperature in THF; the degree of labeling was reported to be 93%.[184] This method has also been utilized for the preparation of pyrene end-labeled polymers,[126,185] as shown in equation (33). Spectroscopic analyses (λ_{max} (dioxane) = 330, 346 nm; ^{1}H and ^{13}C NMR) indicate that this reaction proceeds quantitatively to produce the polymers with fluorescent labels.[126,185] This methodology is also being investigated as a means to introduce the anthracene group.[186]

$$\text{PSLi} + \text{[1-phenyl-1-pyrenyl-ethylene]} \xrightarrow{\text{ROH}} \text{PS-CH(Ph)-pyrenyl} \tag{33}$$

5.3.7 Conclusions

These results show that the addition reactions of simple and polymeric organolithium compounds with substituted 1,1-diarylethylene derivatives provide a general method for the synthesis of functionalized and labeled polymers. We are continuing to investigate the scope and limitations of this functionalization methodology. It is anticipated that with appropriate protecting groups and reaction conditions, a wide variety of well-characterized, quantitatively functionalized and labeled polymers and copolymers can now be prepared.

5.4 OTHER FUNCTIONALIZATION METHODOLOGIES

It should be noted that Nakahama and coworkers[12] have reviewed their work on the anionic polymerization of monomers with protected functional groups. In general, the anionic homopolymerization or copolymerization of monomers with stable or protected functional groups offers a methodology for the introduction of functional groups into polymers which complements the specific functionalization reactions described herein. Bergbreiter and coworkers[18,187] have investigated the use of anionic polymerization to prepare end-functionalized polyethylenes. In general, *n*-butyllithium complexed with *N,N,N',N'*-tetramethylethylenediamine was used as the initiator for ethylene polymerization at 30 psi (1 psi = 6895 Pa) and the resulting insoluble (polyethylenyl)lithium oligomer was functionalized with various electrophiles. The functionalized polyethylenes were characterized primarily by spectroscopic methods. Although the preparation of a variety of functionalized polyethylenes has been described, it would be desirable to have more thorough characterization and purification data available. The organolithium-initiated, anionic polymerization of alkyl methacrylates at low temperatures produces living polymers with narrow molecular weight distributions which can undergo functionalization reactions with electrophiles such as allyl and benzyl halides.[188,189] However, the scope and limitations of electrophilic functionalization reactions for living poly(alkyl methacrylate) enolate anions has not been delineated. The use of functional monomers offers a reliable methodology for the preparation of α-functionalized poly(alkyl methacrylates) and, after electrophilic coupling reactions, α,ω-telechelic polymers.[64,189]

ACKNOWLEDGEMENTS

The author is indebted to many capable coworkers who have contributed their ideas and laboratory skills to the research described herein. I am especially grateful to recent work by Dr Jian Yin (carboxylation), Dr Jing-Jing Ma (hydroxylation), Dr Jungahn Kim (sulfonation), Dr Linfang Zhu (phenol and aromatic amine), Dr Gabriel Summers (aromatic carboxy), Mr Thomas Lynch (aromatic amine) and Dr Laurel E. Schock (pyrene label). The research work described herein and carried out at the University of Akron was supported by The Exxon Education Foundation, Mobil Chemical Company, Dow Chemical Company, Gencorp, The Ferro Corporation and the National Science Foundation (Grant DMR-8 706 166).

5.5 REFERENCES

1. 'Telechelic Polymers: Synthesis and Applications', E. J. Goethals (ed.), CRC Press, Boca Raton, FL, 1989.
2. D. N. Schulz, J. C. Sanda and B. G. Willoughby, *ACS Symp. Ser.*, 1981, **166**, 427.
3. F. W. Harris and H. J. Spinelli (eds.), *ACS Symp. Ser.*, 1985, **282**.
4. P. F. Rempp and E. Franta, *Adv. Polym. Sci.*, 1984, **58**, 1.
5. J.-C. Brosse, D. Derouet, F. Epaillard, J.-C. Soutif, G. Legeay and K. Dusek, *Adv. Polym. Sci.*, 1986, **81**, 167.
6. A. Akelah, *J. Mater. Sci.*, 1986, **21**, 2977.
7. T. Hjertberg and J.-E. Lakso, *J. Appl. Polym. Sci.*, 1989, **37**, 1287.
8. T. C. Chung, M. Raate, E. Berluche and D. N. Schulz, *Macromolecules*, 1988, **21**, 1903.
9. D. N. Schulz, S. R. Turner and M. A. Golub, *Rubber Chem. Technol.*, 1982, **55**, 809.
10. G. Paulus, R. Jerome and P. Teyssie, *Br. Polym. J.*, 1987, **19**, 361.
11. K.-W. Lee and T. J. McCarthy, *Macromolecules*, 1988, **21**, 3353.
12. S. Nakahama and A. Hirao, *Prog. Polym. Sci.*, 1990, **15**, 299.
13. Y. Gnanou and P. Lutz, *Makromol. Chem.*, 1989, **190**, 577.
14. D. E. Bergbreiter and C. R. Martin (eds.), 'Functional Polymers', Plenum Press, New York, 1989.
15. M. Morton, 'Anionic Polymerization: Principles and Practice', Academic Press, New York, 1983, chap. 11.
16. D. H. Richards, G. C. Eastmond and M. J. Stewart, in 'Telechelic Polymers: Synthesis and Applications', ed. E. J. Goethals, CRC Press, Boca Raton, FL, 1989.
17. M. Fontanille, in 'Comprehensive Polymer Science', ed. G. Allen, Pergamon Press, Oxford, 1989, vol. 3, chap. 27, p. 425.
18. D. E. Bergbreiter, J. R. Blanton, R. Chandran, M. D. Hein, K.-J. Huang, D. R. Treadwell and S. A. Walker, *J. Polym. Sci., Polym. Chem. Ed.*, 1989, **27**, 4205.
19. L. J. Fetters, *J. Polym. Sci., Part C*, 1969, **26**, 1.
20. M. Morton and L. J. Fetters, *J. Polym. Sci. Macromol. Rev.*, 1967, **2**, 71.
21. S. Bywater, *Prog. Polym. Sci.*, 1975, **4**, 27.
22. P. Rempp, E. Franta and J.-E. Herz, *Adv. Polym. Sci.*, 1988, **86**, 145.
23. R. N. Young, R. P. Quirk and L. J. Fetters, *Adv. Polym. Sci.*, 1984, **56**, 1.
24. M. Szwarc, *Adv. Polym. Sci.*, 1983, **49**, 1.
25. S. Bywater, in 'Encyclopedia of Polymer Science and Engineering', 2nd edn., ed. J. I. Kroschwitz, Wiley-Interscience, New York, 1985, vol. 2, p. 1.
26. M. van Beylen, S. Bywater, G. Smets, M. Szwarc and D. J. Worsfold, *Adv. Polym. Sci.*, 1988, **86**, 87.
27. T. E. Hogen-Esch and J. Smid (eds.), 'Recent Advances in Anionic Polymerization', Elsevier, New York, 1987.
28. J. E. McGrath (ed.), *ACS Symp. Ser.*, 1981, **166**, 427.
29. R. P. Quirk, D. J. Kinning and L. J. Fetters, in 'Comprehensive Polymer Science', ed. G. Allen, Pergamon Press, Oxford, 1989, vol. 7, chap. 1, p. 1.
30. G. Riess and G. Hurtrez, in 'Encyclopedia of Polymer Science and Engineering', 2nd. edn., ed. J. I. Kroschwitz, Wiley-Interscience, New York, 1985, vol. 2, p. 324.
31. S. Bywater, *Adv. Polym. Sci.*, 1979, **30**, 89.
32. B. J. Bauer and L. J. Fetters, *Rubber Chem. Technol.*, 1978, **51**, 406.
33. H. L. Hsieh, *Rubber Chem. Technol.*, 1976, **49**, 1305.
34. P. Lutz, C. Strazielle and P. Rempp, in 'Recent Advances in Anionic Polymerization', ed. T. E. Hogen-Esch and J. Smid, Elsevier, New York, 1987, p. 403.
35. J. Roovers, *Rubber Chem. Technol.*, 1989, **62**, 33.
36. B. J. Wakefield, 'The Chemistry of Organolithium Compounds', Pergamon Press, Elmsford, NY, 1974.
37. B. J. Wakefield, 'Organolithium Methods', Academic Press, San Diego, CA, 1988.
38. L. Brandsma and H. D. Verkruijsse, 'Preparative Polar Organometallic Chemistry', Springer-Verlag, Berlin, 1987, vol. 1.

39. R. B. Bates and C. A. Ogle, 'Carbanion Chemistry', Springer-Verlag, Berlin, 1983.
40. J. L. Wardell, in 'The Chemistry of the Metal–Carbon Bond', ed. F. R. Hartley, Wiley, New York, 1987, vol. 4, p. 1.
41. B. J. Wakefield, in 'Comprehensive Organometallic Chemistry', ed. G. Wilkinson, Pergamon Press, Oxford, 1982, vol. 7, chap. 44.
42. R. P. Quirk and M. Alsamarraie, *Ind. Eng. Chem. Prod. Res. Dev.*, 1986, **25**, 381.
43. C. D. Eisenbach, H. Schnecko and W. Kern, *Makromol. Chem.*, 1975, **176**, 1587.
44. R. P. Quirk, J. Yin and L. J. Fetters, *Macromolecules*, 1989, **22**, 85.
45. J. Cazes, *J. Chem. Educ.*, 1970, **47**, A461, A625.
46. E. A. Collins, J. Bares and F. W. Billmeyer, Jr., 'Experiments in Polymer Science', Wiley-Interscience, New York, 1973, p. 154.
47. G. Glöckner, 'Polymer Characterization by Liquid Chromatography', Elsevier, Berlin, 1986.
48. L. H. Tung and J. C. Moore, in 'Fractionation of Synthetic Polymers. Principles and Practices', ed. L. H. Tung, Dekker, New York, 1977, p. 545.
49. Z. Grubisic, P. Rempp and H. Benoit, *J. Polym. Sci., Part B*, 1967, **5**, 753.
50. D. McIntyre (ed), 'Characterization of Macromolecular Structure', National Academy of Sciences, Washington, DC, 1968.
51. G. H. Jeffery, J. Bassett, J. Mendham and R. C. Denney (eds.), 'Vogel's Textbook of Quantitative Chemical Analysis', 5th edn., Longman, Essex, 1989.
52. S. Siggia and J. G. Hanna, 'Quantitative Organic Analysis via Functional Groups', Wiley-Interscience, New York, 1979.
53. F. A. Bovey, 'High Resolution NMR of Macromolecules', Academic Press, New York, 1972.
54. Q.-T. Pham, R. Petiaud and H. Waton, 'Proton and Carbon NMR Spectra of Polymers', Wiley-Interscience, 1980, vol. 1; 1983, vol. 2; 1984, vol. 3.
55. J. Schaefer, in 'Topics in C-13 NMR Spectroscopy', ed. G. C. Levy, Wiley, New York, 1974, vol. 1, p. 149.
56. 'Sadtler Carbon-13 NMR of Monomers and Polymers' Spectra', Sadtler Research Laboratories, Philadelphia, PA, 1979, vol. 1.
57. E. Breitmaier and G. Bauer, '^{13}C-NMR Spectroscopy. A Working Manual with Exercises', Harwood Academic, MMI Press, New York, 1984.
58. J. L. Koenig, 'Chemical Microstructure of Polymer Chains', Wiley-Interscience, New York, 1980.
59. W. Klopffer, 'Introduction to Polymer Spectroscopy', Springer-Verlag, Berlin, 1984.
60. F. W. Wehrli, A. P. Marchand and S. Wehrli, 'Interpretation of Carbon-13 NMR Spectra', 2nd edn., Wiley, New York, 1988.
61. R. M. Silverstein, G. C. Bassler and T. C. Morrill, 'Spectrometric Identification of Organic Compounds', 4th edn., Wiley, New York, 1981.
62. R. P. Quirk and J.-J. Ma, *J. Polym. Sci., Polym. Chem. Ed.*, 1988, **26**, 2031.
63. R. P. Quirk and L. Zhu, *Makromol. Chem.*, 1989, **190**, 487.
64. R. P. Quirk and L. Zhu, *Br. Polym. J.*, 1990, **23**, 47.
65. R. P. Quirk and G. J. Summers, *Br. Polym. J.*, 1990, **22**, 249.
66. 'The Infrared Spectra Atlas of Monomers and Polymers', Sadtler Research Laboratories, Philadelphia, PA, 1980.
67. A. D. Cross, 'An Introduction to Practical Infra-red Spectroscopy', 2nd edn., Butterworths, London, 1964.
68. L. J. Bellamy, 'The Infrared Spectra of Complex Molecules', Methuen, London, 1958.
69. L. J. Bellamy, 'Advances in Infrared Group Frequencies', Methuen, London, 1968.
70. K. Nakanishi, 'Infrared Absorption Spectroscopy. Practical', Holden-Day, San Francisco, CA, 1962.
71. H. E. Diem and S. Krimm, *Appl. Spectrosc.*, 1981, **35**, 421.
72. A. I. Scott, 'Interpretation of the Ultraviolet Spectra of Natural Products', Pergamon Press, Frankfurt, 1964.
73. R. P. Quirk, S. Perry, F. Mendicuti and W. L. Mattice, *Macromolecules*, 1988, **21**, 2294.
74. H. Inagaki, in 'Fractionation of Synthetic Polymers. Principles and Practices', ed. L. H. Tung, Dekker, New York, 1977.
75. H. Inagaki and T. Tanaka, in 'Developments in Polymer Characterization 3', ed. J. V. Dawkins, Applied Science, London, 1982, p. 1.
76. Y. Ikada, *Adv. Polym. Sci.*, 1978, **29**, 47.
77. T. I. Min, T. Miyamoto and H. Inagaki, *Rubber Chem. Technol.*, 1977, **50**, 63.
78. D. R. Iyengar and T. J. McCarthy, *Polym. Prepr., Am. Chem. Soc., Div. Polym. Chem.*, 1989, **30**(2), 154.
79. R. P. Quirk and P. L. Cheng, *Macromolecules*, 1986, **19**, 1291.
80. R. P. Quirk and W.-C. Chen, *Makromol. Chem.*, 1982, **183**, 2071.
81. K. Ueda, A. Hirao and S. Nakahama, *Macromolecules*, 1990, **23**, 939.
82. J. M. Bather and R. A. C. Gray, *J. Chromatogr.*, 1976, **122**, 159.
83. R. C. Larock, 'Comprehensive Organic Transformations. A Guide to Functional Group Preparations', VCH, New York, 1989.
84. M. Morton, L. J. Fetters, R. A. Pett and J. F. Meier, *Macromolecules*, 1970, **3**, 327.
85. R. Milner, R. N. Young and A. R. Luxton, *Polymer*, 1983, **24**, 543.
86. R. P. Quirk and J. Yin, *J. Polym. Sci., Polym. Chem. Ed.*, in press.
87. H. L. Hsieh and A. G. Kitchen, *ACS Symp. Ser.*, 1983, **212**, 291.
88. D. J. Worsfold and S. Bywater, *Macromolecules*, 1972, **5**, 393.
89. A. Hernandez, J. Semel, H.-C. Broecker, H. G. Zachmann and H. Sinn, *Makromol. Chem., Rapid Commun.*, 1980, **1**, 75.
90. D. Freyss, P. Rempp and H. Benoit, *J. Polym. Sci., Part B*, 1964, **2**, 217.
91. D. M. Wiles and S. Bywater, *J. Polym. Sci., Part B*, 1964, **2**, 1175.
92. D. M. Wiles and S. Bywater, *Trans. Faraday Soc.*, 1965, **61**, 150.
93. Z. Laita and M. Szwarc, *Macromolecules*, 1969, **2**, 412.
94. R. Busson and M. van Beylen, *Macromolecules*, 1977, **10**, 1320.
95. K. Ziegler and H. G. Gellert, *Justus Liebigs Ann. Chem.*, 1950, **567**, 179.
96. G. Köbrich and I. Stöber, *Chem. Ber.*, 1970, **103**, 2744.
97. G. Wittig and U. Schollkopf, *Chem. Ber.*, 1954, **87**, 1318.
98. L. J. Fetters and R. N. Young, *ACS Symp. Ser.*, 1981, **166**, 95.
99. S. Bywater, in 'Comprehensive Polymer Science', ed. G. Allen, Pergamon Press, Oxford, 1989, vol. 3, chap. 28, p. 433.

100. T. A. Antkowiak, A. E. Oberster, A. F. Halasa and D. P. Tate, *J. Polym. Sci., Part A-1*, 1972, **10**, 1319.
101. P. Mansson, *J. Polym. Sci., Polym. Chem. Ed.*, 1980, **18**, 1945.
102. D. P. Wyman, V. R. Allen and T. Altares, Jr., *J. Polym. Sci., Part A*, 1964, **2**, 4545.
103. J. Pannell, *Polymer*, 1971, **12**, 547.
104. L. Weber, *Makromol. Chem., Macromol. Symp.*, 1986, **3**, 317.
105. W. N. Setzer and P. Von Rague Schleyer, *Adv. Organomet. Chem.*, 1985, **24**, 353.
106. L. J. Fetters, W. W. Graessley, N. Hadjichristidis, A. D. Kiss, D. S. Pearson and L. B. Younghouse, *Macromolecules*, 1988, **21**, 1644.
107. K. Komatsu, A. Nishioka, N. Ohshima, M. Takahashi and H. Hara, *US Pat.*, 4 083 834 (1978); *Jpn. Pat.*, 73-17 231 (1973); *Ger. Pat.*, 2 406 092 (1974) (*Chem. Abstr.*, 1974, **81**, 153 157d).
108. C. A. Wentz and E. E. Hopper, *Ind. Eng. Chem., Prod. Res. Dev.*, 1967, **6**, 209.
109. P. Rempp and M. H. Loucheux, *Bull. Soc. Chim. Fr.*, 1958, 1497.
110. R. Finding and U. Schmidt, *Angew. Chem., Int. Ed. Engl.*, 1970, **9**, 456.
111. A. G. Brook, J. M. Duff and D. G. Anderson, *Can. J. Chem.*, 1970, **48**, 561.
112. W. E. Parham and R. M. Piccirilli, *J. Org. Chem.*, 1976, **41**, 1268.
113. G. Summers, S.-H. Guo and X.-W. Hu, unpublished results.
114. G. Broze, R. Jerome and P. Teyssie, *Makromol. Chem.*, 1978, **179**, 1383.
115. H. Sawada, *J. Macromol. Sci., Rev. Macromol. Chem.*, 1970, **C5**(1), 151.
116. F. E. Bailey and J. V. Koleske, 'Poly(ethylene Oxide)', Academic Press, New York, 1976.
117. S. Patai (ed.), 'The Chemistry of the Ether Linkage', Interscience, New York, 1967.
118. C. J. Chang, R. F. Kiesel and T. E. Hogen-Esch, *J. Am. Chem. Soc.*, 1973, **95**, 8446.
119. L. E. St. Pierre and C. C. Price, *J. Am. Chem. Soc.*, 1956, **78**, 3432.
120. N. N. Lebedev and Y. I. Baranov, *Vysokomol. Soedin., Ser. A*, 1966, **8**, 198 (*Chem. Abstr.*, 1966, **65**, 808g).
121. N. P. Doroshenko and Y. L. Spirin, *Vysokomol. Soedin., Ser. A*, 1970, **12**, 2481 (*Chem. Abstr.*, 1971, **74**, 54 245e).
122. I. Cabasso and A. Zilkha, *J. Macromol. Sci., Chem.*, 1974, **A8**(8), 1313.
123. V. Halaska, L. Lochmann and D. Lim, *Collect. Czech. Chem. Commun.*, 1968, **33**, 3245.
124. K. Gonsalves and M. D. Rausch, *J. Polym. Sci., Polym. Chem. Ed.*, 1986, **24**, 1419.
125. R. P. Quirk and N. S. Seung, *ACS Symp. Ser.*, 1985, **286**, 37.
126. R. P. Quirk, I. Kim, K. Rodrigues and W. L. Mattice, *Polym. Prepr., Am. Chem. Soc., Div. Polym. Chem.*, 1990, **31**(1), 87.
127. S. Kobayashi, M. Kaku, T. Mizutani and T. Saegusa, *Polym. Bull.*, 1983, **9**, 169.
128. M. Morton, L. J. Fetters, J. Inomata, D. C. Rubio and R. N. Young, *Rubber Chem. Technol.*, 1976, **49**, 303.
129. D. N. Schulz, A. F. Halasa and A. E. Oberster, *J. Polym. Sci., Polym. Chem. Ed.*, 1974, **12**, 153.
130. S. F. Reed, *J. Polym. Sci., Part A-1*, 1972, **10**, 649.
131. W. J. Trepka, *Macromolecules*, 1984, **17**, 497.
132. R. P. Quirk, T. Takizawa, G. Lizarraga and L.-F. Zhu, *J. Appl. Polym. Sci., Symp. Ser.*, 1991, **41**, 23.
133. L. Eberson, 'Electron Transfer Reactions in Organic Chemistry', Springer-Verlag, Heidelberg, 1987.
134. H. Yamataka, N. Fujimura, Y. Kawafuji and T. Hanafusa, *J. Am. Chem. Soc.*, 1987, **109**, 4305.
135. N. Nagata, T. Kobatake, H. Watanabe, A. Ueda and A. Yoshioka, *Rubber Chem. Technol.*, 1987, **60**, 837.
136. W. C. E. Higginson and N. S. Wooding, *J. Chem. Soc.*, 1952, 760.
137. D. N. Schulz and A. F. Halasa, *J. Polym. Sci., Polym. Chem. Ed.*, 1977, **15**, 2401.
138. A. Hirao, I. Hattori, T. Sasagawa, K. Yamaguchi, S. Nakahama and N. Yamazaki, *Makromol. Chem., Rapid Commun.*, 1982, **3**, 59.
139. I. Hattori, A. Hirao, K. Yamaguchi, S. Nakahama and N. Yamazaki, *Makromol. Chem.*, 1983, **184**, 1355.
140. R. P. Quirk, W.-C. Chen and P.-L. Cheng, *ACS Symp. Ser.*, 1985, **282**, 139.
141. P. Beak and B. J. Kokko, *J. Org. Chem.*, 1982, **47**, 2822.
142. C. D. Eisenbach, H. Schnecko and W. Kern, *Makromol. Chem.*, 1975, **176**, 1587.
143. M. Stewart, N. Shepherd and D. M. Service, *Br. Polym. J.*, 1990, **22**, 319.
144. W. H. Dickstein and C. P. Lillya, *Macromolecules*, 1989, **22**, 3882.
145. D. H. Richards, D. M. Service and M. J. Stewart, *Br. Polym. J.*, 1984, **16**, 117.
146. P. Charlier, R. Jerome and P. Teyssie, *Macromolecules*, 1990, **23**, 1831.
147. K. Ueda, A. Hirao and S. Nakahama, *Macromolecules*, 1990, **23**, 939.
148. R. P. Quirk and J. Kim, *Macromolecules*, 1991, **24**, 4515.
149. J. Omeis, E. Muhleisen and M. Moller, *Polym. Prepr., Am. Chem. Soc., Div. Polym. Chem.*, 1986, **27**(1), 213.
150. T. Durst and J. D. Manoir, *Can. J. Chem.*, 1969, **47**, 1230.
151. W. E. Truce and D. J. Vencur, *Can. J. Chem.*, 1969, **47**, 860.
152. D. S. Breslow and H. Skolnik, 'Compounds of Heterocyclic Compounds. Multi-Sulfur and Sulfur and Oxygen Five- and Six-Membered Heterocycles', Interscience, New York, 1966, chap. 4.
153. W. E. Truce and F. D. Hoerger, *J. Am. Chem. Soc.*, 1955, **77**, 2496.
154. N. S. Davidson, L. J. Fetters, W. G. Funk, W. W. Graessley and N. Hadjichristidis, *Macromolecules*, 1988, **21**, 112.
155. J. H. Helberger and W. Grublewski, *Ger. Pat.*, 901 288 (1954) (*Chem. Abstr.*, 1955, **49**, 3249a).
156. P. M. van der Velden, B. Rijpkema, C. A. Smolders and A. Bantjes, *Eur. Polym. J.*, 1977, **13**, 37.
157. G. O. Schulz and R. Milkovich, *J. Appl. Polym. Sci.*, 1982, **27**, 4773.
158. G. O. Schulz and R. Milkovich, *J. Polym. Sci., Polym. Chem. Ed.*, 1984, **22**, 1633.
159. J.-M. Catala, J. F. Boscato, E. Franta and J. Brossas, *ACS Symp. Ser.*, 1981, **166**, 483.
160. J.-M. Catala, G. Reiss and J. Brossas, *Makromol. Chem.*, 1977, **178**, 1249.
161. L. J. Fetters and E. R. Firer, *Polymer*, 1977, **18**, 306.
162. R. P. Quirk and W.-C. Chen, *J. Polym. Sci. Polym. Chem. Ed.*, 1984, **22**, 2993.
163. E. J. Panek and G. M. Whitesides, *J. Am. Chem. Soc.*, 1972, **94**, 8768.
164. L. J. Fetters and M. Morton, *Macromolecules*, 1969, **2**, 453.
165. T. Heitz and H. Hocker, *Makromol. Chem.*, 1988, **189**, 777.
166. Y. Wang, unpublished work.
167. A. Streitwieser, Jr., E. Juaristi and L. L. Nebenzahl, in 'Comprehensive Carbanion Chemistry. Part A — Structure and Reactivity', ed. E. Buncel and T. Durst, Elsevier, Amsterdam, 1980, p. 323.

168. R. B. Bates and C. A. Ogle, 'Carbanion Chemistry', Springer-Verlag, Berlin, 1983.
169. L. J. Fetters, in 'Encyclopedia of Polymer Science and Engineering', 2nd edn., ed. J. I. Kroschwitz, Wiley-Interscience, New York, 1987, vol. 10, p. 19.
170. G. Summers, unpublished results.
171. Y. Ishino, A. Hirao and S. Nakahama, *Macromolecules*, 1986, **19**, 2307.
172. A. Hirao, Y. Ishino and S. Nakahama, *Macromolecules*, 1988, **21**, 561.
173. A. Hirao and S. Nakahama, *Polymer*, 1986, **27**, 309.
174. A. Hirao, K. Yamaguchi, K. Takenaka, K. Suzuki, S. Nakahama and N. Yamazaki, *Makromol. Chem., Rapid Commun.*, 1982, **3**, 941.
175. A. Hirao, K. Takenaka, S. Packirisamy, K. Yamaguchi and S. Nakahama, *Makromol. Chem.*, 1985, **186**, 1157.
176. T. Lynch, unpublished work.
177. R. P. Quirk and L.-F. Zhu, *Polym. Int.*, 1992, **27**, 1.
178. H. Yuki, J. Hotta, Y. Okamoto and S. Murahashi, *Bull. Chem. Soc. Jpn.*, 1967, **40**, 2659.
179. H. Yuki and Y. Okamoto, *Bull. Chem. Soc. Jpn.*, 1969, **42**, 1644.
180. H. Yuki and Y. Okamoto, *Bull. Chem. Soc. Jpn.*, 1970, **43**, 148.
181. J. A. Dean, 'Lang's Handbook of Chemistry', 13th edn., McGraw-Hill, New York, 1985, **3**, 136.
182. H. L. Hsieh and C. F. Wofford, *J. Polym. Sci., Part A-1*, 1969, **7**, 449.
183. C. F. Wofford and H. L. Hsieh, *J. Polym. Sci., Part A-1*, 1969, **7**, 461.
184. L. Chen, M. A. Winnik, E. T. B. Al-Takrity, A. D. Jenkins and D. R. M. Walton, *Makromol. Chem.*, 1987, **188**, 2621.
185. R. P. Quirk and L. E. Schock, *Macromolecules*, 1991, **24**, 1237.
186. L. G. Hong, unpublished work.
187. D. E. Bergbreiter, in 'Functional Polymers', ed. D. E. Bergbreiter and C. R. Martin, Plenum Press, New York, 1989, p. 143.
188. B. C. Anderson, G. D. Andrews, P. Arthur, Jr., H. W. Jacobson, L. R. Melby, A. J. Playtis and W. H. Sharkey, *Macromolecules*, 1981, **14**, 1599.
189. G. D. Andrews and L. R. Melby, in 'New Monomers and Polymers', ed. B. M. Culbertson and C. U. Pittman, Jr., Plenum Press, New York, 1984, p. 357.

6
Synthesis of Functional Polymers by Cationic Polymerization

ALESSANDRO GANDINI
Institut National Polytechnique de Grenoble, France

6.1	INTRODUCTION	107
6.2	MONOFUNCTIONAL STRUCTURES	110
	6.2.1 Alkenyl Monomers	110
	6.2.1.1 Predominant transfer	111
	6.2.1.2 The inifer technique	112
	6.2.1.3 Living systems	113
	6.2.1.4 Macromonomers	117
	6.2.2 Heterocyclic Monomers	117
	6.2.2.1 Cyclic amines	118
	6.2.2.2 Cyclic ethers	118
	6.2.2.3 Cyclic acetals	119
	6.2.2.4 Lactones	119
	6.2.2.5 2-Oxazolines	120
	6.2.2.6 Cyclic siloxanes	120
	6.2.3 Conclusion	121
6.3	DIFUNCTIONAL STRUCTURES	121
	6.3.1 Alkenyl Monomers	121
	6.3.1.1 Vinyl ethers	122
	6.3.1.2 Isobutene	122
	6.3.1.3 1,4-Divinylbenzene	123
	6.3.2 Heterocyclic Monomers	124
	6.3.2.1 Cyclic ethers	124
	6.3.2.2 Cyclic acetals	125
	6.3.2.3 N-t-Butylaziridine	126
	6.3.2.4 Lactones	126
	6.3.2.5 2-Oxazolines	126
	6.3.2.6 Cyclic siloxanes	127
	6.3.3 Conclusion	127
6.4	POLYFUNCTIONAL STRUCTURES	127
	6.4.1 Alkenyl Monomers	127
	6.4.2 Heterocyclic Monomers	129
6.5	CONCLUSION	129
6.6	REFERENCES	129

6.1 INTRODUCTION

The cationic polymerization of alkenyl monomers has recently undergone a major revival following the discovery of several systems which display living features. For several decades, indeed from the very inception of that discipline with the pioneering investigations of the 1940s,[1] much effort had been devoted to the search of sound ways to reduce or minimize chain-breaking reactions and attain conditions which could approach the ideal living features; but up to a few years ago, only some heterocyclic monomers had been found to respond more or less adequately to this quest. This situation is amply reflected in the major treatises, reviews and conference proceedings published up to the mid 1980s.[2-7]

The most serious obstacles to the goal of obtaining living behaviour, which in the meantime had become a readily accessible reality for anionic polymerization following Szwarc's brilliant contribution, had to do with: (i) the obstinacy of transfer reactions (mostly with alkenyl monomers); and (ii) the difficulty of controlling the relative abundance and reactivity of the different propagating species (ester molecules, ion pairs, unpaired ions). Other disturbing interactions included: (iii) termination or retardation mechanisms; (iv) formation of cyclic oligomers (mostly with n-donor monomers); and (v) slow initiation with respect to propagation.

This state of affairs did not, however, hinder a number of successful attempts at synthesizing functionalized oligomers and polymers by cationic techniques. In fact, some specific systems could be brought to behave according to near-living conditions, while others exhibited specific features which allowed a good degree of structural control, hence possible functionalizations.

The purpose of the present review is to discuss and illustrate the various ways recently made available to obtain macromolecules possessing functional groups through the use of cationic polymerization. The further modifications and/or applications of these materials will not be discussed in detail, since these aspects fall outside the scope of the review, but occasional mentions as to their interest and possible uses seem justified in order to place the issues at stake in the broader context of material science. The plan is to summarize the achievements in the field of functional macromolecules preceding the very recent advances related to living cationic polymerization and to deal in a more-detailed fashion with the results and potentials related to these advances. An appraisal of the more fundamental issues concerning the nature of the active species in these novel living systems will also be attempted, given its relevance to the topic under study.

For a comprehensive analysis of the state of the art of cationic polymerization up to about 1989, the reader should refer to the chapters devoted to it in Volume 3 of *Comprehensive Polymer Science*.[8] Only a brief reminder of the basic features of this type of polyaddition is given here.

Initiation of cationic polymerization requires the activation of some monomer molecules (usually a very small proportion of all those present) by an electrophilic reaction capable of generating chain carriers. These active species are either cationic entities possessing various degrees of association with the corresponding anions, ranging from tight ion pairs to essentially unpaired ('free') cations, or covalent structures bearing a polarized bond, capable of inducing monomer propagation with variable reactivities. A large variety of acidic initiators can be used to accomplish this activation,[4-6] mostly Lewis and Brönsted acids of very different strengths (from acetic acid to 'superacids') added to the polymerization systems, but also created *in situ* by external energy furnished to them in the form of heat, photons or electrons. Only monomers with a well-defined nucleophilic character can be effectively turned into active species, namely: (i) π-donors like certain aliphatic and aromatic alkenes and polyenes: (ii) n,π-donors like vinyl ethers and N-vinylcarbazole (one of the most nucleophilic alkenyl monomers); and (iii) n-donors like certain cyclic ethers and acetals and other saturated heterocycles containing sulfur, nitrogen, phosphorus and silicon. Scheme 1 summarizes very succinctly the initiation patterns with alkenyl and heterocyclic monomers.[4-6,8] Here E^+ denotes any electrophilic initiator, including of course a neutral species, such as an undissociated protonic acid (in which case the meaning of the positive charge is the H^+ arising from it in the reaction with the nucleophilic monomer).

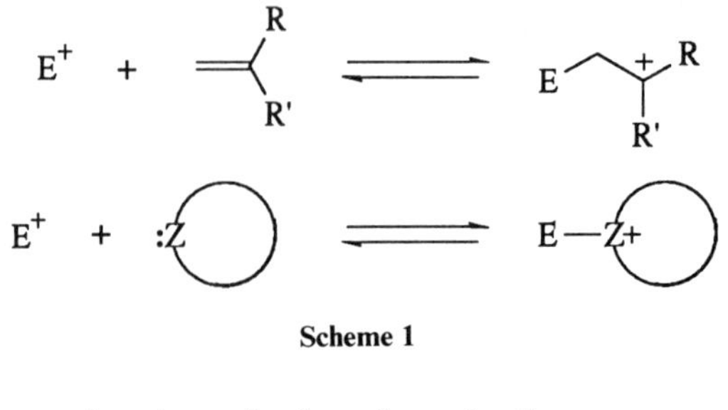

Scheme 1

The very fact that chain carriers in cationic polymerization possess an electrophilic character, demands that the most appropriate solvents (if one is used at all) should be neutral or (slightly) acidic, *viz.* either aprotic inert molecules such as heptane and dichloromethane or aprotic dipolar protophobic structures such as nitromethane.

The propagation step of this chain reaction can be a complex affair involving different types of often coexisting mechanisms depending on the nature of the active species (ionic or covalent) and of the monomer.[5,6,8] Thus, the classical electrophilic addition of a cation to a π-system, e.g. a C=C or a C=O bond or its insertion counterpart when the active species is an ester, differ substantially from the S_N2 mechanism which characterizes the propagation step with n-donor monomers, as shown in Scheme 2 which shows the two distinct ionic pathways. To these mechanistic alternatives one should add the kinetic differences connected to the variable reactivity of chain carriers as a function of their extent of dissociation from tight ion pairs to unpaired cations, bearing in mind that in most systems the whole spectrum of species is represented. This multiplicity of possible chain carriers and modes of propagation has been a source of scientific debate over the last decades[5,6,8] and the recent contributions to it are briefly discussed below.

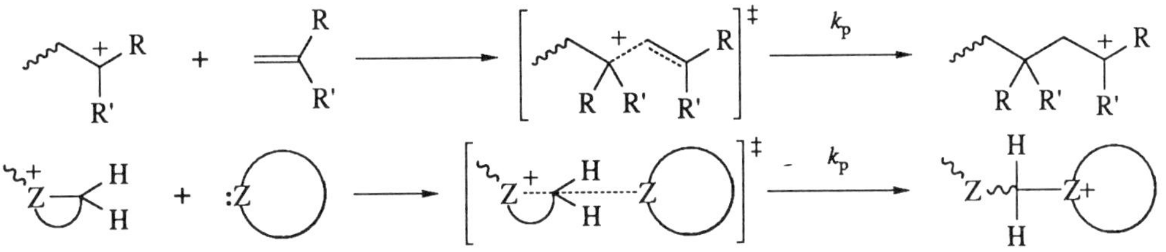

Scheme 2

Chain-transfer reactions are important in cationic polymerization, particularly with alkenyl monomers.[5,6,8] These include: (i) 'spontaneous' mechanisms, e.g. protonic acid formation from an active species; (ii) reorganizations of the ionic moieties; and (iii) bimolecular interactions involving the monomer or the polymer chain. Scheme 3 illustrates one of the most common mechanisms for either type of monomer.

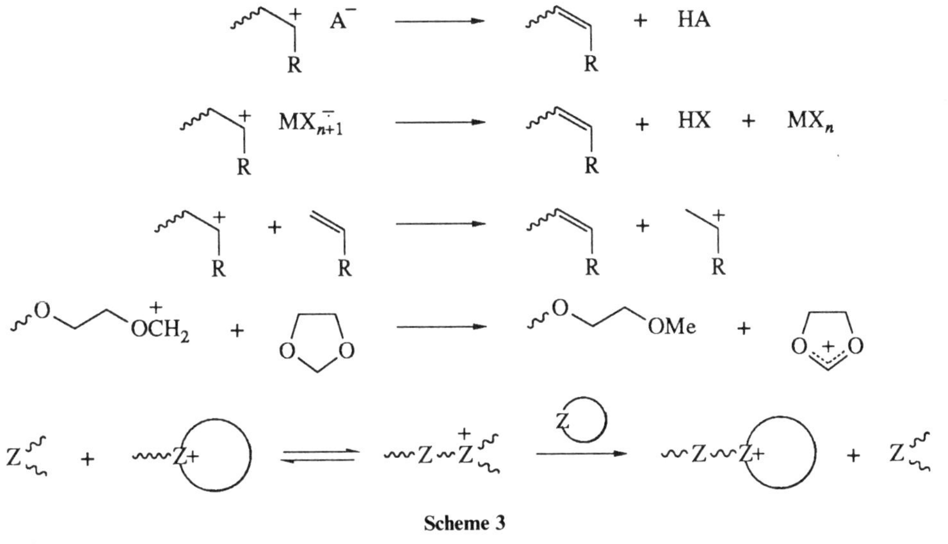

Scheme 3

Termination is not a very general feature of cationic polymerizations, except of course when nucleophilic impurities are present. The most frequent mechanisms provoking the permanent disappearance of chain carriers[5,6,8] in an uncontaminated system are: (i) their conversion into inactive esters or halides; (ii) the formation of allylic or onium ions which are too stabilized to ensure propagation; and (iii) the consumption of the cocatalyst, e.g. water, by its chemical incorporation into the polymer structure. Scheme 4 gives examples of these reactions.

Ideally, if one wishes to prepare oligomeric or polymeric structures bearing a specific type and number of functionalities in a controlled and reliable way using cationic systems, living conditions must be ensured, namely: (i) the initiation reaction should be rapid with respect to propagation; (ii) transfer reactions should be absent or negligible; and (iii) no termination should occur.

This set of three basic restrictions provides the only kinetic solution to the problem of attaining a chain polymerization in which all the active species are generated 'instantaneously' relative to

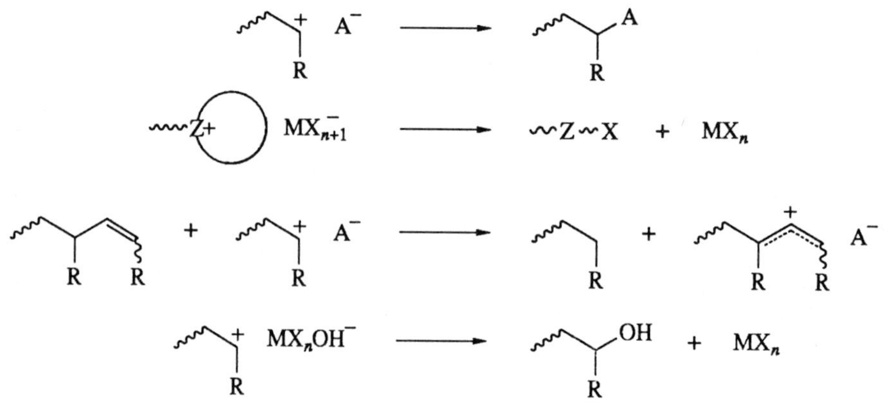

Scheme 4

propagation and thereafter the only reactions taking place are chain growth events. Under these conditions the ensuing (first-order) monomer consumption is accompanied by a corresponding increase in the number-average degree of polymerization of the growing macromolecules. The polymer thus obtained has a narrow Poisson molecular weight distribution.

The privileges resulting from such exclusive situations are related to the rich possibilities of structural games one can play and of fundamental studies one can conduct to unravel kinetic and mechanistic problems. As already mentioned, this is precisely what has been achieved in the field of anionic polymerization, which readily affords living systems if thorough experimental procedures are respected. Cationic polymerization has shown a strong resistance to yield to numerous research efforts[9,10] mostly because, as outlined above, the chain carriers of this type of polyaddition are prone to take side pathways with respect to the main road of propagation much more readily than their anionic counterparts. Indeed, physical organic chemistry has repeatedly shown that carbenium ions such as those arising from alkenyl monomers, and to a lesser extent the various onium ions derived from n-donor monomers, are fragile intermediates whose reactivity cannot be easily channelled into a single, exclusive reaction mode.

The understanding of these important events, alternative to propagation, has provided first a way to make good use of them towards the preparation of functionalized structures and later some solutions towards their minimization, *i.e.* the development of living or near-living systems, which of course opens the way to a much wider scope in terms of readily accessible functionalized structures. Both approaches are treated in this chapter, but of course the latter receives much more attention than the former.

The following sections are organized according to the number of functionalities *per* macromolecule and the type of monomer(s) polymerized.

6.2 MONOFUNCTIONAL STRUCTURES

Monofunctional macromolecules bearing the relevant group at one end of a linear chain find widespread applications related: (i) either to the different polar (*e.g.* hydrophilic) character of the end group with respect to that (*e.g.* hydrophobic) of the polymer backbone; or (ii) to the possibility of synthesizing diblock copolymers through the specific reactivity of the functionality appended, *viz.* the possibility of coupling with another polymer bearing a complementary function or of initiating the polymerization of another monomer from that site. Although some of the basic approaches to the preparation of these structures by cationic polymerization are common to all types of monomers, some features are appropriate only to one given family and hence the following subdivision.

6.2.1 Alkenyl Monomers

Within the realm of the cationic polymerization of alkenyl monomers, there is still a major difference in behaviour between relatively weak nucleophiles like isobutene and strong ones like vinyl ethers. The expedients and conditions adopted to introduce functionalities into the polymers can therefore vary according to this criterion, although some techniques seem to be more generally applicable than others.

6.2.1.1 Predominant transfer

The idea of introducing one functional terminal group per polymer molecule by cationic polymerizations, characterized by predominant transfer reactions onto an added agent, has been applied to phenols and furans. Sterically hindered phenols like 2,6-di-*t*-butylphenol have been shown to act as very powerful chain-transfer agents in the cationic polymerization of isobutene.[11] This facile electrophilic substitution tends to obscure all other transfer reactions intrinsic to the system and provides an interesting possibility of synthesizing, for example, oligo(isobutenes) (**1**) bearing one phenolic moiety as the end group, as shown in equation (1).

$$\sim\!\!\!\!+ \text{SnCl}_4\text{ArO}^- + \text{2,6-di-}t\text{-Bu-C}_6\text{H}_3\text{OH} \longrightarrow \sim\!\!\!\!-\text{C}_6\text{H}_2(\text{Bu}^t)_2\text{OH} + \text{SnCl}_4\!\cdot\!\text{ArOH} \quad (1)$$

(**1**)

The C-5 position of the furan ring provides an even more attractive site for such functionalization procedures. This aptitude was discovered while studying the cationic polymerization of 2-vinylfuran and its homologues.[12] The extent of branching was very pronounced with monomers having an unsubstituted C-5 position, whereas linear chains were obtained when a methyl group was appended at that position. Model compounds simulating the branching structures were prepared by simply adding a substantial amount of 2-methylfuran to these systems, which resulted in such an important transfer by electrophilic substitution at C-5 that the monomeric active species could be trapped predominantly to give the corresponding difuranic compound, *e.g.* (**2**) from 2-isopropenylfuran, the other products being oligomers (**3**), all bearing the terminal 2-methylfuryl group.

(**2**) (**3**)

The extreme propensity and regiospecificity of this electrophilic substitution has been exploited to synthesize monofunctional oligomers from a variety of alkenyl monomers susceptible to cationic polymerization.[13] Typically, these reactions were carried out under conventional conditions, adapted to the specific monomer, but in the presence of small amounts of 2-methylfuran (the methyl group at C-2 is a good activator of the C-5 position for this reaction). Thus, for example, polystyrene, polyisobutene and poly(vinyl ether)s, macromolecules with average degrees of polymerization ranging from unity to about 100 and possessing a terminal furanic functionality close to one, could be readily prepared. The average length of the functionalized chain could be modulated by the relative amount of furan derivative added. A representative transfer reaction applied to the formation of monofuranic polyisobutene (**4**) is shown in equation (2).

$$\sim\!\!\!\!+ \text{A}^- + \text{2-methylfuran} \longrightarrow \sim\!\!\!\!-\text{(2-methylfuryl)} + \text{HA} \quad (2)$$

(**4**)

The presence of the furanic moieties at one end of the macromolecules was detected by various spectroscopic techniques and assessed quantitatively by ^{1}H NMR spectroscopy.

The possible exploitation of these monofunctionalized polymers stems from the reactivity of the furan ring which allows such chemical modifications as those arising from the Diels–Alder reaction, *e.g.* with maleic anhydride to give two terminal carboxylic acid groups placed at the same chain end. The fields of interest of such macromolecular structures bearing a nonpolar chain terminated by a polar (and reactive) extremity include surface-active materials, compatibilizers and the preparation of block copolymers.

The limitations of the synthetic technique based on predominant transfer are the relatively low DPs obtained. Indeed, if the transfer to the functionalizing agent is to be maintained at a preponderant level relative to other chain-breaking reactions, its frequency must be kept high and thus the average molecular weight of the resulting polymer molecules will tend to be affected by this limitation. It was found that even by working at low temperatures, the maximum DPs compatible with proper monofunctionality were around 100. However, the range of possible applications of these materials remains large and the simplicity of their synthesis should constitute a significant advantage over alternative approaches.

6.2.1.2 *The inifer technique*

The term inifer was coined more than a decade ago[14] to denote an agent capable of inducing *ini*tiation and trans*fer* in a cationic polymerization system. In the specific context of monofunctional products, the typical structure is 1-chloro-1-methylethylbenzene, which, when used in conjunction with BCl_3, can operate as inifer for the polymerization of isobutene to give polymers bearing a chlorine atom at one end of the chain.[15] The reaction (Scheme 5) illustrates the complex situation involved with these systems,[15] which can, however, be simplified considerably by adding a hindered pyridine[16] to inhibit all steps involving HCl without interfering with the reactions of the carbenium ions and thus giving better controlled structures.[15]

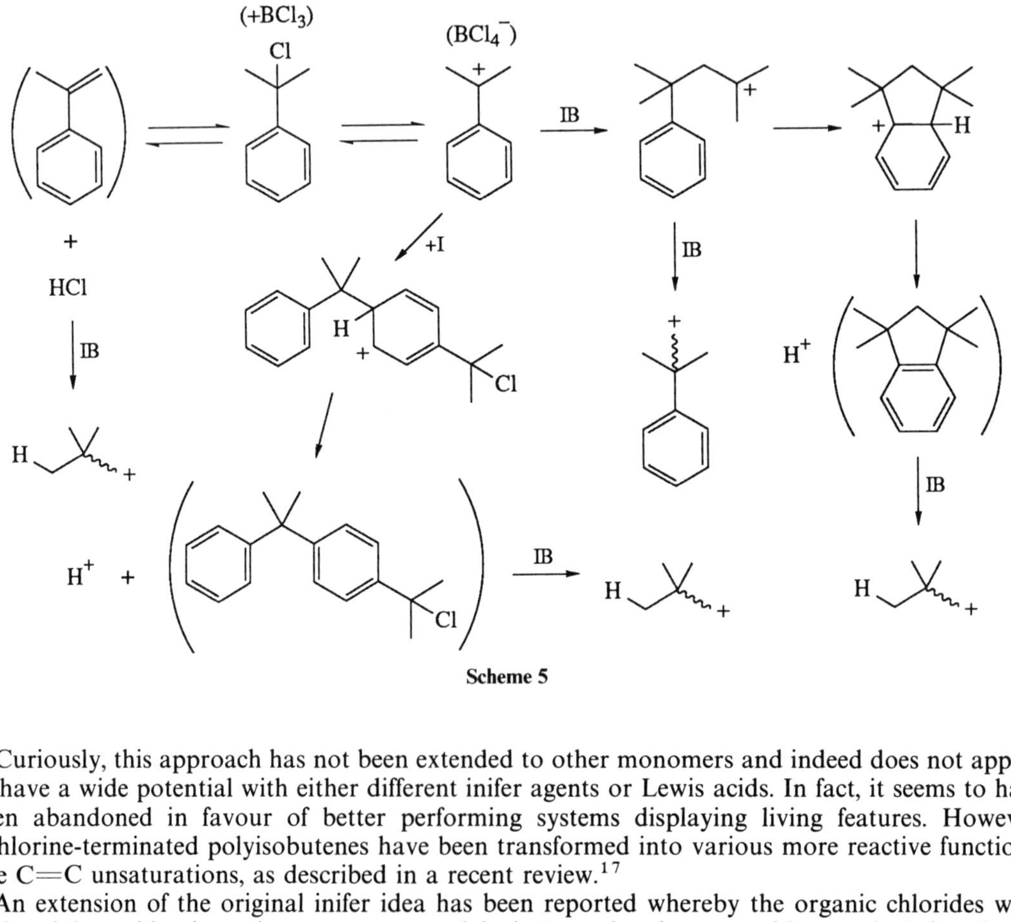

Scheme 5

Curiously, this approach has not been extended to other monomers and indeed does not appear to have a wide potential with either different inifer agents or Lewis acids. In fact, it seems to have been abandoned in favour of better performing systems displaying living features. However, *t*-chlorine-terminated polyisobutenes have been transformed into various more reactive functions, like C=C unsaturations, as described in a recent review.[17]

An extension of the original inifer idea has been reported whereby the organic chlorides were replaced by azides in order to prepare polyisobutenes bearing an azido terminal function.[18] 2-Azido-2-phenylpropane used in conjunction with $TiCl_4$, BF_3 and Et_2AlCl gave the expected structures, whereas with BCl_3, SbF_5 and $EtAlCl_2$ no incorporation of azido groups in the polyisobutene occurred. 2-Azido-2,4,4-trimethylpentane appeared to be a less efficient agent even with the most suitable Lewis acids and the products were mixtures of functionalized and conventional structures. This work was a follow-up on a previous study in which hydroazoic acid was shown to operate as an incorporating cocatalyst in the polymerization of isobutene with $TiCl_4$, giving rise to monoazido polymers. As with the classical inifer systems, it would be interesting to know whether other monomers can be polymerized to give N_3-functionalized oligomers or polymers.

6.2.1.3 Living systems

Vinyl ethers were the first group of alkenyl monomers to be polymerized by a living cationic system.[19] Following the original catalytic couple, HI/I_2, a number of other combinations have been successfully tested in the last several years.[20] They include a rather large variety of combinations which can, however, be reduced to two major families:[20] (i) HI (which adds onto the monomer to give the corresponding iodide) or an iodide bearing a structure closely resembling that of the iodide derived from the monomer, in conjunction with a weak Lewis acid (e.g. iodine, zinc or tin(IV) halides) or an ammonium salt; and (ii) a strong Lewis (e.g. $EtAlCl_2$) or Brönsted (e.g. CF_3SO_3H) acid in conjunction with a Lewis base like an ester, an ether, a sulfide or even an amine.

With the first group of catalytic combinations, the iodide alone is inactive and the role of the added Lewis acid or ammonium salt is to promote these passive molecules to a chain carrier status, achieving moreover a mechanistic situation in which living conditions are attained. More recently, other halides[21] and carboxylates[22] derived from the addition of HCl or HBr and RCO_2H (R = Me, CH_2Cl, $CHCl_2$, CCl_3, CF_3), respectively, to isobutyl vinyl ether have been shown to behave in a similar manner.

With the second type of catalytic combinations, the activity without the Lewis base is high, but the polymerization has no living character. The moderating effect of the added base induces a considerable reduction of the rate of monomer consumption, but also bestows onto the systems a progressively more pronounced living connotation.

The reaction temperature and the polarity of the polymerization medium are also important variables for the establishment of experimental conditions which are best suited to the ideal situation.

N-vinylcarbazole is more nucleophilic than the vinyl ethers. HI alone is sufficiently acidic to promote its polymerization: in other words, the iodide arising from the addition of HI on this monomer is a chain carrier in its own right, and gives living polymerizations under certain conditions.[23] However, comparatively little attention has been devoted to this monomer in the search of optimal living features.

The nucleophilic character of styrenes can be varied considerably by changing the nature of the *para* substituents. *p*-Methoxystyrene, a highly nucleophilic alkene, behaves like vinyl ethers in the context of the most appropriate catalytic systems capable of inducing a living polymerization.[24] The *p*-methyl homologue gives living systems with $HI/ZnCl_2$[25] and BCl_3 plus esters.[26] Styrene itself was not easy to master in terms of living behaviour, despite numerous efforts.[20] The conditions which bring about a satisfactory response[27,28] are more sensitive to small changes in the nature of the catalytic species, in the reaction temperature and in the polarity of the medium than with other monomers.[21]

Isobutene has been studied very actively in this context. The first reports of its living cationic polymerization[29] called upon boron chloride/carboxylic esters systems. Since then, several catalytic combinations have been shown to fulfil the same task.[20] One finds again here two basic combinations among these initiators: (i) BCl_3 plus tertiary esters, ethers or alcohols, which seem to operate according to the same basic premises as the HI/I_2-type systems, and (ii) Lewis acids (BCl_3, $TiCl_4$) with added Lewis bases, which are likely to act as the corresponding combinations discussed above for the vinyl ethers. The use of ammonium salts of the same halide as that of the Lewis acid to convert a 'normal' cationic polymerization of isobutene into a living one has also been reported lately.[30]

This sudden explosion of publications, which tends to show that virtually all nucleophilic alkenyl monomers can be placed in a specific context suitable for their living cationic polymerization, represents a dramatic qualitative change from the general scepticism of a few years ago concerning such an 'extravagant' possibility. The state of the art in cationic polymerization has therefore undergone a major upheaval, although for the moment the feeling is that the positive results obtained are due more to an empirical testing procedure intended to cover as wide a ground as possible in terms of the variables explored (catalyst combinations, solvent, temperature with each monomer), rather than to the consequence of logical reasoning based on scientific arguments and predictions. Rationalizations based on solid organic physicochemical considerations are just beginning to appear, *a posteriori*.

However, it can be stated safely that the cationic polymerization of alkenyl monomers is witnessing a revolution in which two distinct aspects must be reassessed. On the one hand, a vast domain of exploitation has been opened in terms of new materials, namely macromonomers, functionalized polymers, block copolymers, *etc.* (some of which have already been described as discussed below), although not yet high polymers, since for the moment the maximum DPs obtained with most of these systems tends to be around 100. On the other hand, these polymerizations are a stimulating challenge for the fundamental polymer chemist: what makes them so sensitive to apparently minor changes in the experimental conditions which can readily destroy their living character?; what is the nature of the active species?; is there only one kind or several types of them? In other words, what are the basic kinetic, mechanistic and structural factors which determine their living and nonliving behaviour?

Answering these questions obviously means being able to move from an empirical and rather superficial approach based on trial and error to an argumented planning built on sound principles which would provide more rewarding results and therefore materials. The debate on these fundamental issues is presently in full swing, as testified by some of the lectures and communications presented at the latest symposium on cationic polymerization.[31] The issue of whether the active species are ionic or covalent in living systems has surfaced again, after the long discussion which had followed the discovery of pseudocationic polymerization in 1964.[4,6] Indeed, the conditions favouring a living character are also those which tend to suppress unpaired ions and give low value monomodal DP distributions of the resulting polymers. It can be argued that both Lewis bases and nucleophilic anions from added salts are efficient suppressors of unpaired carbocations.

There is at this point a more or less general consensus as to the fact that for most of the living systems described above, the vast majority of chain ends are in the form of covalent species. The disagreement is as to whether these covalent, but polarized, moieties can be the actual sites of propagation through an insertion mechanism or whether a very modest proportion of ion pairs in equilibrium with them are responsible for monomer chain addition, whereas the ester sites are dormant and therefore inactive.

The most convincing arguments in favour of the latter interpretation are based on the lack of any direct evidence pointing to the possibility of the electrophilic addition of an alkene across a covalent bond without its (transient) ionization.[32] However, the absence of major detrimental effects of rather high concentrations of added moisture on the rate of pseudocationic polymerizations[33] remains a puzzling feature which is hard to reconcile with the survival of carbocationic ion pairs present in extremely low concentrations. Moreover, results from a recent study[34] of leaving group activity from acetates derived from isobutyl vinyl ether by dynamic proton NMR spectroscopy have been interpreted as suggesting that activated esters can indeed be the chain carriers in the living (pseudo)cationic polymerization of such monomers.

The initially proposed propagation mechanisms based on chain carriers in the form of activated esters bear a striking resemblance to those originally proposed for pseudocationic polymerization.[33] Lately, a wide variety of structures and a correspondingly rich terminology have been put forward, in practice oscillating periodically between polarized covalent species and various types of ion pairs. Scheme 6 gives a selection of these together with the insertion mechanism published in 1965 for

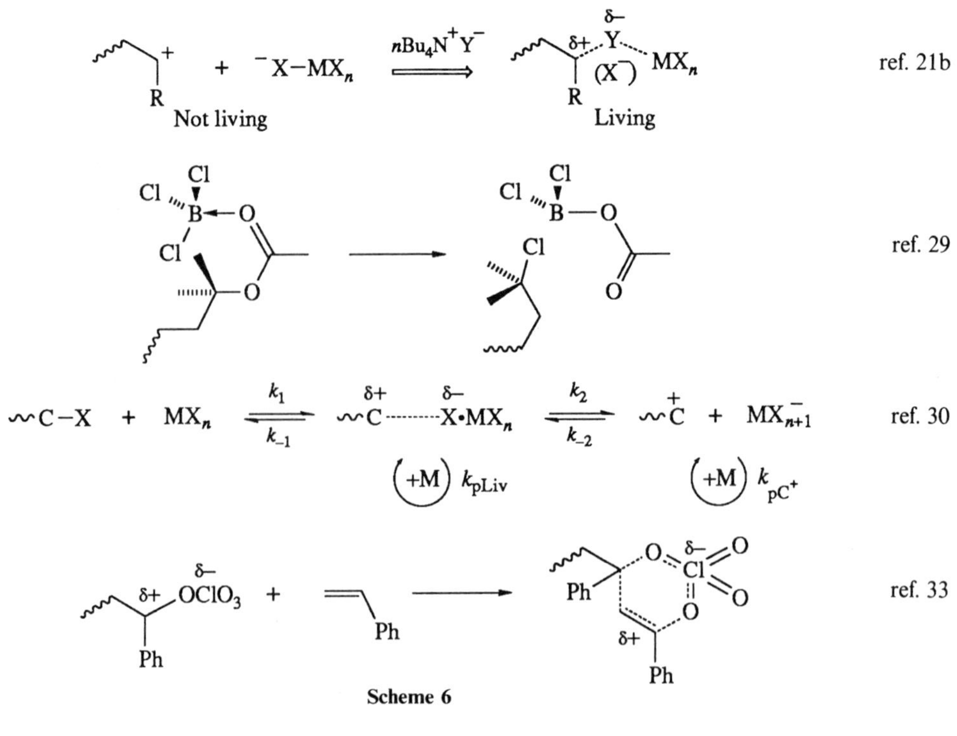

Scheme 6

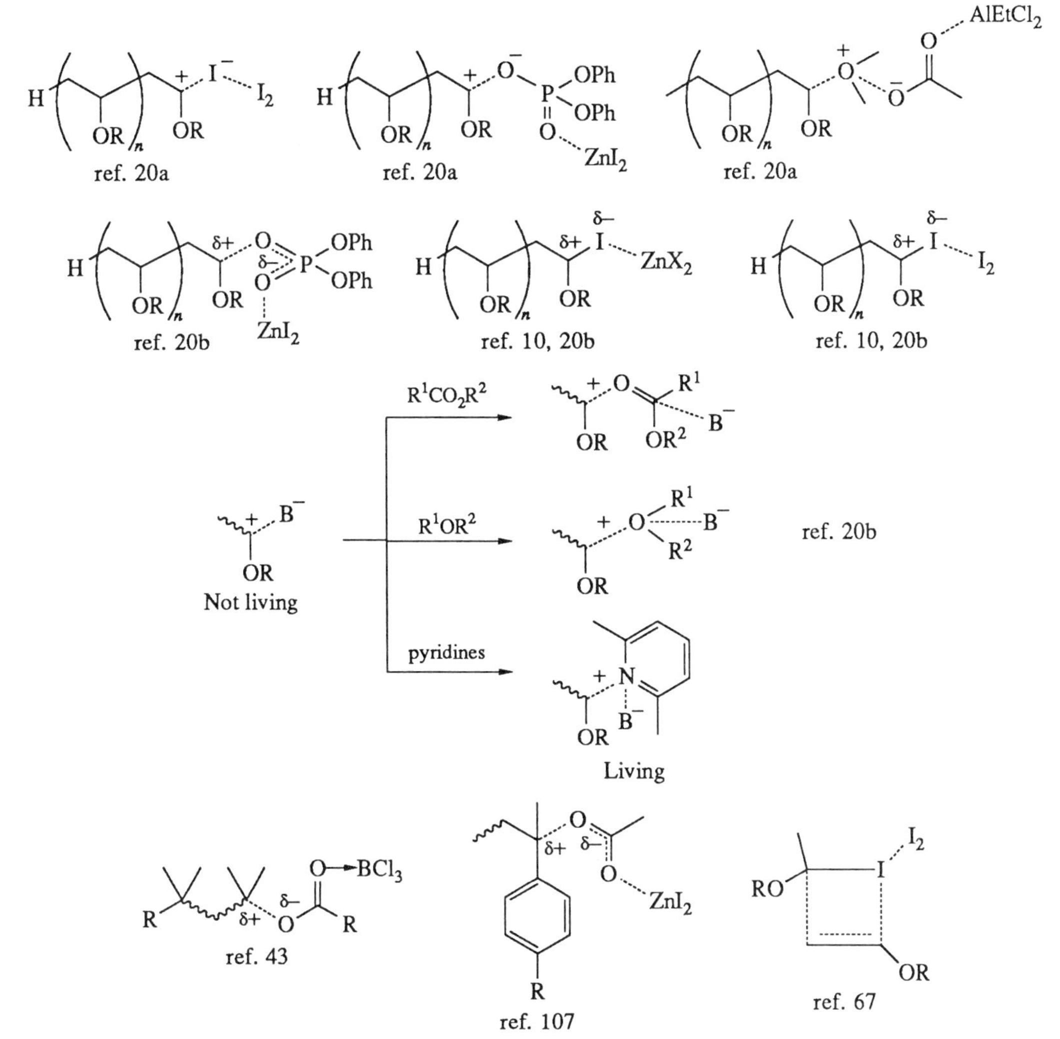

Scheme 6 (continued)

the styrene/perchloric acid system. Clearly, the configurations with incomplete charge separation seem to be the preferred way to depict the chain ends responsible for the living behaviour, whereas the ionic moieties are often held responsible for the faster, nonliving polymerization, even in the latest publications. However, the consistency in these matters is still rather poor.

Notwithstanding the fact that these important fundamental considerations obviously need further investigation, the progressive generalization of systems possessing living features in the field of the cationic polymerization of alkenyl monomers has already produced an impressive number of papers dealing with the possibility of preparing functionalized macromolecules. The structures relevant to this section, namely monofunctional oligomers and polymers, cover a broad range of (mostly) poly(vinyl ether)s and polyisobutenes, many of which have already been reviewed.[8,10] Two alternative modes of inserting one terminal function in virtually all chains produced by a living polymerization have been successfully tested and made available: (i) the use of a functionalized coinitiator; or (ii) the deactivation of the chain carrier with a functionalized quencher, as shown in the general Schemes 7 and 8, respectively.[20]

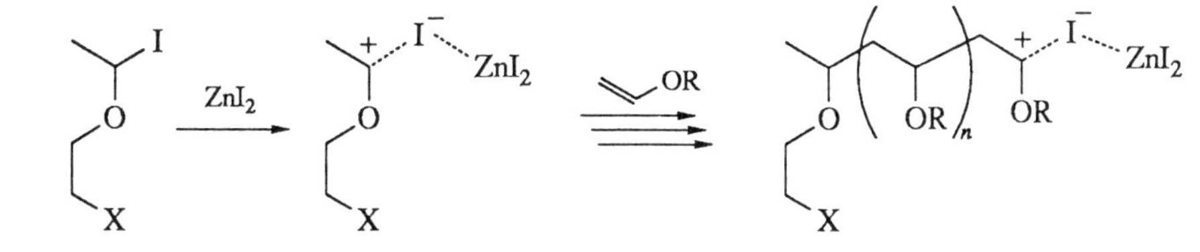

Scheme 7

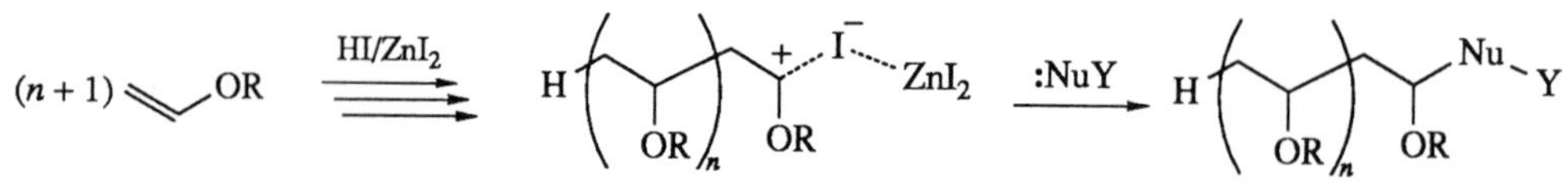

Scheme 8

Within the former approach based on the mechanism sketched in Scheme 7, the fact that systems of the type HI/mild Lewis acid can initiate the living polymerization of vinyl ethers through the formation of the corresponding iodide (addition of HI across the C=C double bond) and its activation by the Lewis acid, provides a useful means of synthesizing monofunctional poly(vinyl ether)s. The idea is to generate active species from equimolar amounts of HI and a functional vinyl ether in the presence of a mild Lewis acid, e.g. zinc chloride, and to add thereafter the 'normal' monomer, as shown in Scheme 9.[36]

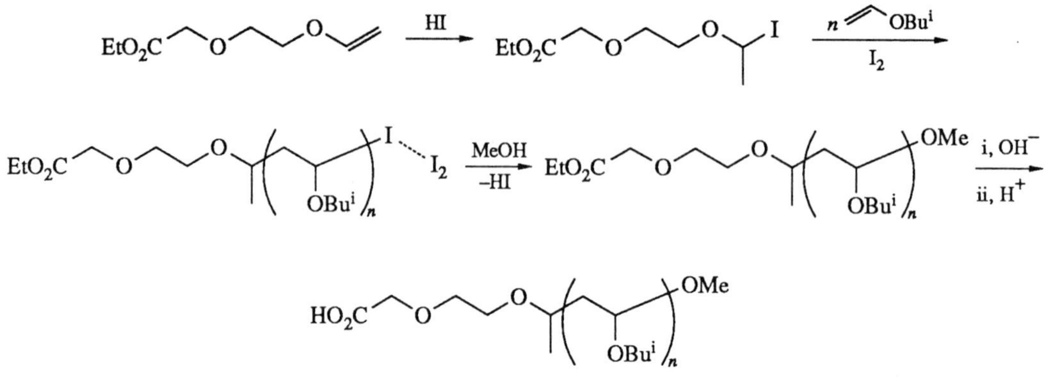

Scheme 9

Initiation by the activated functional iodide ensures that the polymer chains thus produced will all bear the chosen functionality as one end group. Among the moieties introduced in this manner, or through a similar approach with trifluoroacetates and $EtAlCl_2$, one finds CO_2H,[35-37] OH[37] and NH_2,[36,37] mostly with poly(isobutyl vinyl ether). A styryl end group was introduced into poly(chloroethyl vinyl ether) by the same type of procedure.[38] In all these investigations polymerizations were quenched with an alcohol and therefore the other terminal moiety was the corresponding acetal group.

A similar approach was applied to systems comprising an initiator resulting from the addition of trimethylsilyl iodide onto dioxolane.[39] This primary alkoxy iodide bearing a silyloxy tail induced the living polymerization of vinyl ethers under certain specific conditions. Killing these reactions with methanol and hydrolyzing the siloxane function *in situ* afforded poly(vinyl ether)s with one hydroxy group per chain.

In the numerous reports discussing the living polymerizations of isobutene,[20,40,41] most of the emphasis has been devoted to the elaboration of novel initiating systems and to the mechanisms which seem appropriate to justify the living character and the nature of the active species. The idea of using functionalized initiator systems has not been taken up nearly as frequently as with vinyl ethers. One rare example describes the living polymerization of this monomer with lactone/BCl_3 complexes,[42] and shows that the cyclic ester moiety is incorporated as a terminal group upon initiation. Quenching the living macromolecules with methanol converts these functions into methyl esters, whereas quenching with pyridine gives the corresponding CO_2H group. No report has been found on the functionalization of polystyrenes by the method of Scheme 7.

The killing of an active species with a functional terminating agent in the living polymerization of vinyl ethers, according to Scheme 8, has been applied in several instances. This end-capping procedure works well with such nucleophiles as sodiomalonates,[35] diamines[43] and various substituted anilines.[44] More recently, the use of a silyl ketene acetal[45] and of azo alcohols[46] as terminating agents, giving easy ways of appending a CO_2H and an N=N group, respectively, at the end of poly(vinyl ether) chains, has also been reported. Whether as a result of the actual quenching reaction according to Scheme 8 or after a facile *in situ* transformation of the product of that reaction, acidic, e.g. CO_2H, or basic, e.g., NH_2, groups can thus be appended at one end of each poly(vinyl ether) chain. The reactivity of such moieties provides many interesting structural consequences, like chain extension, macromolecular coupling or further, more specific functionalizations.

Practically all initiators used in the living polymerizations of isobutene[20,39-41] provide a terminal tertiary chloride group at the end of each polyisobutene molecule. This is an intrinsic feature related to the actual systems used and to the mechanism of propagation, i.e., when the reactions are quenched with standard nucleophiles the active species collapses into an inactive (but potentially reactivatable) —CMe_2Cl moiety. Thus, one can consider that all these polymerizations are 'self-functionalized' in the sense that, as mentioned above in the context of the inifer systems, this moiety can be converted into various more specific functions through more-or-less complex modification procedures.

The same considerations apply to the living polymerization of styrenes which mostly generate polymers bearing a terminal secondary chloride or bromide group.[21,47] However, a recent paper[48] describes the synthesis of monofunctional polyisobutenes with living systems in which the quenching strategy illustrated by Scheme 8 was successfully applied. The living polymerizations of isobutene initiated by $TiCl_4$/ester combinations were terminated by allyltrimethylsilane, thereby generating polymer chains bearing an allylic group resulting from this end-capping operation. The authors also describe the transformation of the terminal unsaturation into a hydroxy or an oxirane function.

6.2.1.4 Macromonomers

The synthesis of an oligomer with one terminal functionality which is a polymerizable moiety leads to a macromonomer. The principles discussed above for preparing monofunctional polymers apply to the present context, but only products with modest DP are usually sought for this specific application. Only the direct syntheses of macromonomers will be reviewed here. The chemical modification of monofunctional oligomers, such as those discussed in the preceding section, to arrive at macromonomers falls outside the scope of this chapter. A careful listing of these modifications, mostly of polyisobutenes obtained by the inifer technique has already been drawn up elsewhere.[20]

Acrylic-type oligo(vinyl ether) macromonomers have been prepared by calling upon the principle of a functional initiator in living polymerization, as shown in Schemes 7 and 9. Here the iodide $CH_2=CMeC(O)OCH_2CH_2OCHMeI$ provided the polymerizable moiety.[49] The end-capping method of Scheme 8 was also explored: quenching the poly(vinyl ether) active species with a carbanion bearing a vinyl ether function[50] provided macromonomer (5) in which both the oligomeric structure and the (cationically) polymerizable moiety are of the same nature.

(5)

Another termination reaction was used in the living polymerization of vinyl ethers by triflic acid in the presence of thiolane.[51] There, quenching with a tertiary amine carrying a methacrylic group gave a methacrylate macromonomer and killing with allyl alcohol gave an allyloxy-terminated poly(ethyl vinyl ether).

The inifer technique has been used to prepare a macromonomer of isobutene bearing an oxirane function.[52]

6.2.2 Heterocyclic Monomers

It is with n-donor monomers that the first cationic systems without appreciable transfer and termination reactions were achieved,[5,8,9] and thereby the first functionalized polymers.[53] A recent review[9] examines the living character of cationic ring-opening polymerizations with a detailed analysis of the kinetic and mechanistic features which can contribute to spoil the ideal behaviour. Of course, as with any near-living polymerization system, slow initiation relative to propagation is detrimental to the sharpness of the molecular weight distribution of the polymers.

It was pointed out in the previous section that transfer reactions, particularly to monomer, are undoubtedly the most pernicious events marring the living character of the polymerization of alkenyl monomers, hence the search for systems capable of 'stabilizing' the active species against these and other interactions. With heterocyclic monomers, only transfer to polymer must be considered as a possible major drawback. Termination reactions in the absence of quenching impurities are possible in ring-opening polymerization and consist mainly of recombination of the ionic chain carriers to give an inert molecule.

Finally, the pronounced reversible character of certain ring-opening polymerizations, associated with the low ceiling temperatures of heterocyclic monomers bearing five to seven members, is a specific feature which must be taken into account when considering the quality of polymers obtained from a living system; indeed, when conversions close to the equilibrium monomer concentration are reached, the DP distribution of the polymer broadens from a Poisson towards a Gaussian form.

In spite of these difficulties, cationic ring-opening polymerizations have frequently been mastered to yield living behaviour which is not far from ideal. The major difference with the corresponding systems based on alkenyl monomers stems from the fact that onium (oxonium, sulfonium, ammonium, phosphonium) ions are usually less likely than carbenium ions to react in steps other than propagation steps.

6.2.2.1 Cyclic amines

Among the cyclic amines,[8] N-t-butylaziridine[54] is the best example of a well-behaved monomer capable of excellent living performances. The use of initiators like methyl triflate generates polymer chains bearing a methyl head group and an active cyclic ammonium triflate end which can be quenched with carboxylic acids to generate an ester function, as shown in equation (3). Naturally, the triflic acid 'liberated' in this reaction goes on to protonate a nitrogen atom on the polymer.

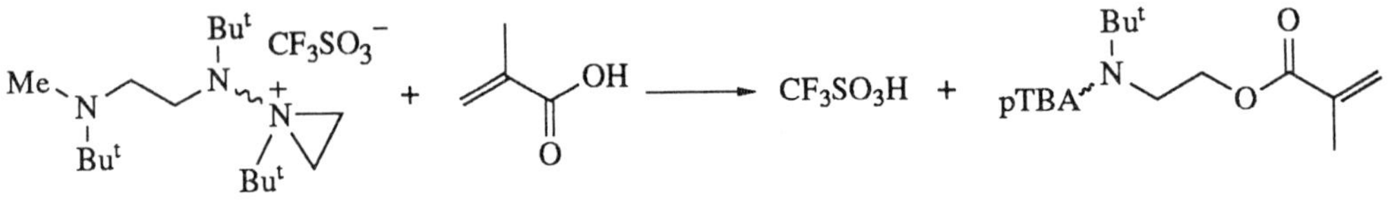

$$\text{Me-N(Bu}^t\text{)-...-N}^+(\text{Bu}^t)\text{-N(Bu}^t) \quad \text{CF}_3\text{SO}_3^- \; + \; \text{CH}_2\text{=C(CH}_3\text{)COOH} \longrightarrow \text{CF}_3\text{SO}_3\text{H} + \text{pTBA-N(Bu}^t\text{)-...-O-C(=O)C(CH}_3\text{)=CH}_2 \quad (3)$$

Other functionalities can be introduced depending on the nucleophilic terminating agent used, e.g. OH groups from hydroxides and NHR groups from the corresponding primary amines.

1,3,3-trimethylazetidine[55] and 1-azabicyclo[4.2.0]octane[56] also give proper living polymerizations, but functionalizations do not seem to have been attempted with these systems.

6.2.2.2 Cyclic ethers

Cyclic ethers[8] and, more specifically, oxiranes and tetrahydrofuran have attracted much attention in the context of the preparation of functional oligomers and polymers. The cationic polymerizations of oxiranes tend to be marred by back-biting reactions which give macrocyclic products. In order to circumvent this intramolecular transfer reaction a new mode of propagation has been devised whereby a protonated or otherwise electrophilically activated monomer molecule adds onto the neutral hydroxy-terminated polymer molecule, viz. the reverse situation with respect to traditional propagation. These systems, termed activated monomer (AM) polymerizations,[57] require the use of hydroxylic compounds to induce the first addition of the protonated monomer, as shown in Scheme 10.

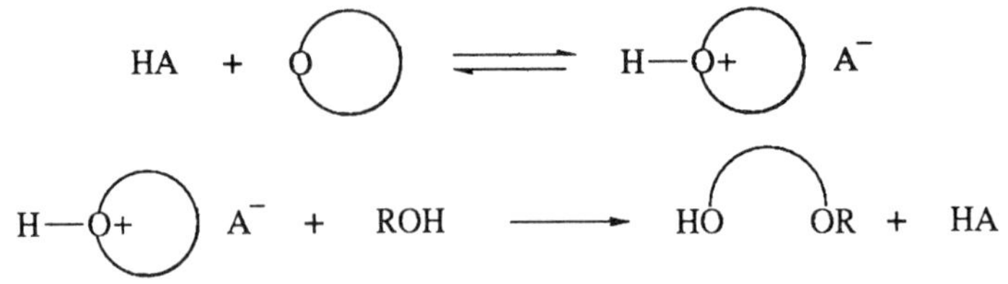

$$\text{HA} + \text{(cyclic ether)} \rightleftharpoons \text{H-O}^+\text{(cyclic)} \; \text{A}^-$$

$$\text{H-O}^+\text{(cyclic)} \; \text{A}^- + \text{ROH} \longrightarrow \text{HO-...-OR} + \text{HA}$$

Scheme 10

Ethylene oxide can thus be polymerized without major interference from back-biting reactions. However, the protonated monomer can be 'lost' in reactions with ether oxygen atoms in the existing polymer chains to give oxonium ions which undergo scrambling and some dioxane formation. The result is that the AM systems involving ethylene oxide are not quite living polymerizations.[57] When substituted oxiranes such as methyloxirane and chloromethyloxirane are used, this side reaction does not take place to any relevant extent and the living character of the polymerization is enhanced.[57,58] Clearly, then, when the promoting reagent is an alcohol, the DP_n of the polymer obtained is the ratio of the concentrations of consumed monomer and alcohol added. Thus, the AM mechanism applied in this way provides a direct means of preparing monohydroxylated polyethers with the desired molecular weight and a narrow DP distribution.

The living polymerization of tetrahydrofuran is the first and most extensively studied system of this family of systems.[5,8,53] The monofunctional active species generated with initiators like methyl triflate, oxonium and monooxocarbenium salts can be quenched with a variety of nucleophiles to yield monofunctional oligomers and polymers bearing ether, ester and amino groups, among others. If thiolane is the terminating agent, the resulting thiolanium end group is stable towards water, but readily reacts with charged nucleophiles to give a wide array of terminal functionalities,[59] as in the example in equation (4).

$$(4)$$

Initiation of THF polymerization by protonic acids gives monohydroxylic structures intrinsically.

6.2.2.3 Cyclic acetals

1,3-Dioxolane is by far the most important monomer in this section in terms of the number and quality of investigations which have been devoted to it,[5,8] although 1,3-dioxepane and trioxane have also received some attention. Cyclic acetals have a tendency to give chain transfer to polymer and to monomer, the latter by hydride ion abstraction. However, under carefully chosen conditions these reactions can be minimized and near-living features attained, particularly with 1,3-dioxolane and 1,3-dioxepane.[60] When initiators giving monofunctional active species were used, *e.g.* triethyloxonium hexafluoroantimonate, it was shown that killing them with tertiary phosphines or amines leads to the corresponding polymers bearing one phosphonium or one ammonium end group. Little more has been published on monofunctional polyacetals, compared with the important recent contributions dealing with the bifunctional structures described below.

6.2.2.4 Lactones

A recent revival of interest concerning the cationic polymerization of lactones has widened the scope of this field both in terms of the mechanistic understanding and of the products obtained. The monomers studied include the classical four-, six- and seven-membered cyclic esters[61] (**6**), L,L-dilactide[62] (**7**) and cyclic carbonates[63] (**8**).

(6)   (7)   (8)

The polymerization of (**6**) and (**7**) by methyl triflate was shown to proceed, as shown in Scheme 11, *via* the open chain ester which is in equilibrium with the corresponding cyclic dioxocarbenium ions, which are, surprisingly, less active entities in this context.

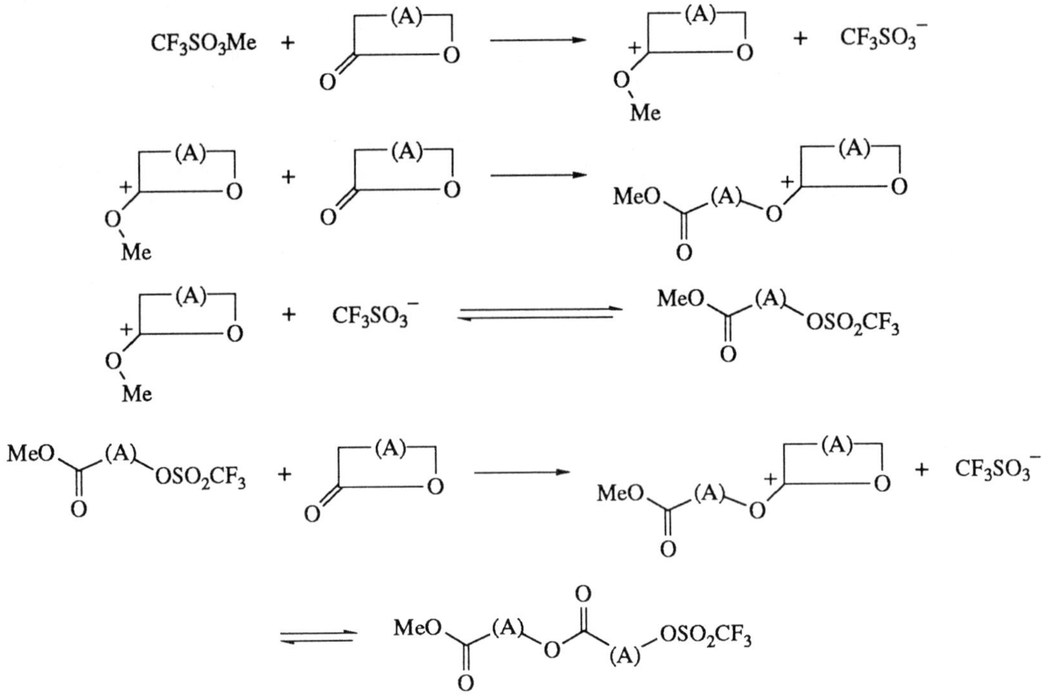

Scheme 11

These systems therefore provide a source of polyesters bearing an alkyl ester and a triflate end group, the latter being a potential source of ready functionalization, although it does not appear to have been exploited for this purpose.

The cationic polymerization of cyclic carbonates follows the same general mechanism, but a side reaction giving ether groups was found to occur. Polycarbonates free of ether linkages were, however, prepared with weak Lewis acids, but here the propagation took place through an insertion mechanism, as with lactones and metal bromides.

6.2.2.5 *2-Oxazolines*

Following the discovery that 2-oxazolines (**9**) readily undergo living cationic polymerization, a vast amount of work was done with these monomers.[64] Initiation by esters of strong acids or halides gives macromolecules with the corresponding predominantly ionic or predominantly covalent active species, respectively, as shown in Scheme 12.

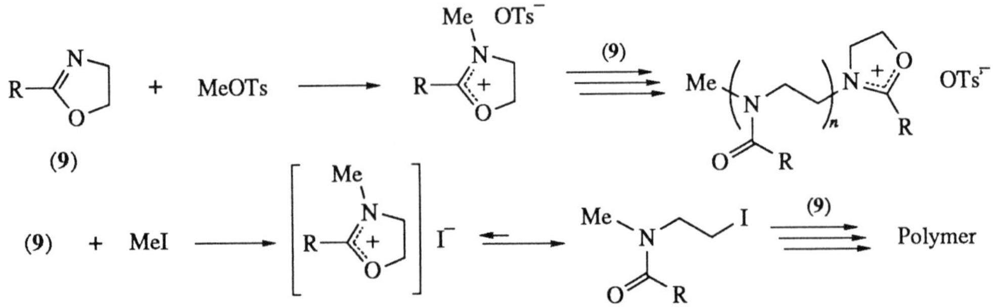

Scheme 12

Specific reactions with added nucleophiles have been conducted to transform the latter (active) end group into convenient functionalities which include hydroxy, amino, and carboxy groups and polymerizable functions like vinylic and acrylic moieties. Thus, monofunctional poly(*N*-acylaziridine)s, *i.e.* a peculiar kind of polyamide, as well as the corresponding macromonomers, can be readily prepared from these monomers.

6.2.2.6 Cyclic siloxanes

The cationic polymerization of cyclic siloxanes proceeds through a complex mechanism involving the formation of linear and macrocyclic products.[65] Under 'conventional' conditions it is therefore impossible to envisage proper functionalization of the ensuing polymers. However, the fact that the siloxane sites on monomers, cyclic oligomers and linear polymers display similar reactivities has been exploited to prepare functionalized poly(dimethylsiloxane)s through the addition of specific 'end-blockers', *i.e* compounds bearing a siloxane group and a given function R, as shown in Scheme 13.

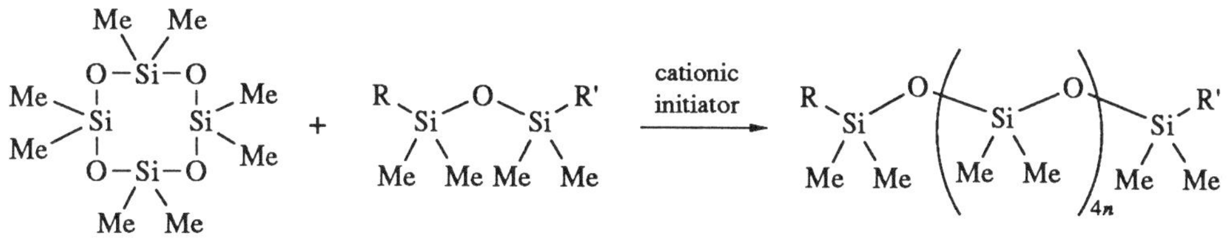

Scheme 13

The applications and consequences of this original approach will be discussed in the section on difunctional polymers (Section 6.3.2.6).

6.2.3 Conclusion

It is mostly thanks to the possibility of achieving living or near-living conditions that the cationic polymerization of both alkenyl and heterocyclic monomers can be used to synthesize macromolecules with one terminal functionality. The procedures usually call upon the use of initiators capable of generating chain carriers bearing a single active end. At the end of the polymerization, quenching with an appropriate nucleophile transforms that active moiety into the desired function. An alternative approach consists in using a functionalized reagent which can add onto the monomer upon initiation, thus introducing the desired group at the very onset of the polymerization. In this latter context, killing is carried out more conventionally, since it is not required to provide functionalization.

With alkenyl monomers, for which living conditions are more difficult to attain, some other simple procedures have been studied, namely the imposition of predominant transfer onto an agent which constitutes the source of functionalization: this can be an added compound or the anion formed in the initiation reaction, as in the inifer technique.

6.3 DIFUNCTIONAL STRUCTURES

Difunctional macromolecules bearing the two relevant groups at both ends of a linear chain owe their importance mostly as precursors to tri- or multi-block structures. These are obtained essentially by coupling reactions with other mono-, di- or multi-functional polymers possessing complementary functionalities or by inducing the polymerization of another monomer from the two terminal sites, the latter procedure giving only triblock structures. The fields of application of such materials grow incessantly as a function of the variety of original architectures obtained in the laboratory. Again, it was deemed convenient to analyze the recent achievement of cationic polymerization in this realm according to the nature of the monomers involved.

6.3.1 Alkenyl Monomers

Only two types of alkenyl monomer have been extensively studied for the synthesis of bifunctional telechelic linear polymers, *i.e.* those which have been found to respond more readily to the search for living polymerization conditions: the vinyl ethers and isobutene. The latter is also conducive to the alternative approach based on the inifer technique. The following examples are therefore limited to these monomers, although the very recent extension of living features to styrenes opens the way to the application of the concepts described here to their polymers.

6.3.1.1 *Vinyl ethers*

The living polymerization of these n,π-donors described in Section 6.2.1.3 is certainly the best-controlled process in cationic polymerization. Two techniques were discussed in the context of the preparation of monofunctional polymers, *viz*: (i) that which calls upon the use of a functionalized initiator (see Scheme 7); and (ii) that whose principle is to quench the active ends with a functionalized nucleophile (see Scheme 8). Clearly, the combined use of both stratagems represents a convenient method of preparing difunctional oligomers and polymers.

Indeed, homotelechelic structures have been prepared in this way with linear poly(isobutyl vinyl ether)s and poly(methyl vinyl ether)s bearing malonate functions at each end.[35] The malonate moieties were readily hydrolyzed and decarboxylated to give the corresponding CH_2CO_2H groups. More recently, this approach has been reapplied[66] to the preparation of homo- and hetero-telechelic poly(isobutyl vinyl ether)s bearing: (i) two malonate end groups, transformed after the synthesis into two CO_2H functions; (ii) a malonate and an acetate group, transformed subsequently into a carboxylic and a hydroxylic function, respectively; and (iii) a malonate and an imido group, later transformed into a carboxylic and an NH_2 function, respectively.

A different way of preparing difunctional poly(vinyl ether)s was recently suggested and put into practice. The general idea is that since the use of difunctional initiators, such as dihalides derived from divinyl ethers,[21,43,67] *e.g.* the diiodide $ICH(Me)O(CH_2)_4OCH(Me)I$,[43] provides living chains with two terminal active sites, quenching them with the appropriate nucleophiles should yield the corresponding telechelic difunctional macromolecules. The experiments[68] were carried out with chloroethyl vinyl ether and two diiodides: living conditions were attained, but not all the chains seemed to grow at both ends, as shown by the GPC tracings of the polymers. This was attributed to the formation of some inactive vicinal diiodo moieties. However, the *in situ* transformation of the active ends into aldehyde functions and their subsequent reduction to primary hydroxy groups, followed by the analytical determination of an average functionality close to two, showed the basic validity of the original proposal. Surprisingly, however, this route to bisfunctionalization has not been exploited more widely.

6.3.1.2 *Isobutene*

The inifer concept discussed in Section 6.2.1.2 has been extensively applied to the preparation of polyisobutenes bearing two terminal tertiary chloride moieties. This relatively 'old' work has been thoroughly reviewed[16,18,69] in the late 1980s, including the topic of the further chemical modification of these telechelic polymers, and only a brief reminder is given here. Scheme 5 gave the essential mechanistic elements of the inifer principle applied to the synthesis of monofunctional polyisobutenes. The use of difunctional reagents **(10)** and **(11)**, the latter being more suitable at the higher polymerization temperatures because it hinders cyclization (see Scheme 5), which gives indane dead ends, in conjunction with boron chloride, under carefully chosen conditions, has made it possible to prepare structure **(12)** with a near-ideal functionality over a fairly wide range of DPs and a relatively narrow DP distribution.

The only other monomer successfully tested with the above difunctional inifer is β-pinene,[70] which yielded the telechelic structure (13).

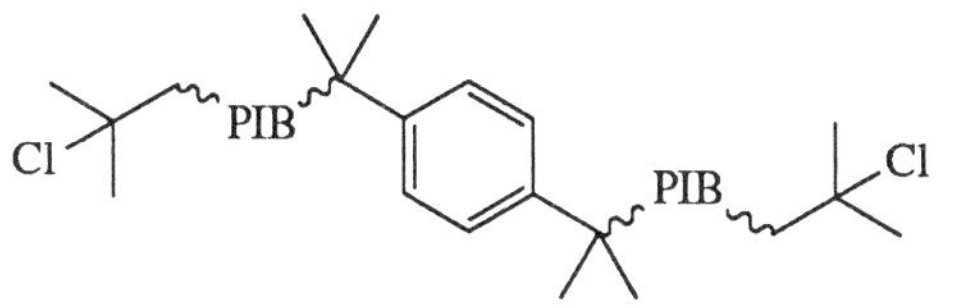

(12)

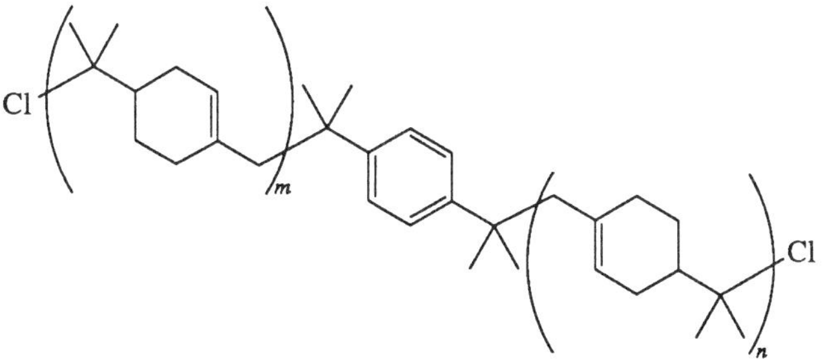

(13)

The discovery of living conditions adapted to isobutene polymerization,[20,40,41,43] has introduced the possibility of synthesizing structure (12) with a chlorine functionality close to two using BCl_3 in conjunction with various diethers[71] and diesters.[72] However, their polydispersity was no better than with the inifer technique, viz. 1.5 to 2.5, except when dimethyl sulfoxide was used as electron donor[73] in order to reduce the activity of the chain carriers, but also, more importantly, the extent of side reactions.

As already mentioned in the context of monofunctional polymers, the use of living conditions based on $TiCl_4$ in combination with diethers or diesters can give access to diallylic polyisobutenes by quenching the difunctional active sites with allyltrimethylsilane.[47]

A more recent exploitation of the living polymerization of isobutene[42] makes use of the concept of a functional initiator to prepare heterotelechelic polymers bearing an ester group at one end and a tertiary chloride moiety at the other. Similar structures have been prepared in a very recent investigation in which one of the ester groups appended was a methacrylate function.[74]

6.3.1.3 1,4-Divinylbenzene

The discovery that the cationic polymerization of 1,4-divinylbenzene by acetyl perchlorate[75] proceeds by a succession of addition/transfer cycles yielding the linear polyunsaturated structure (14) with two terminal vinyl groups was the first report of a telechelic polymer from that monomer. A subsequent study[76] involved the copolymerization of divinylbenzene and various substituted styrenes in which the addition of the monovinylic monomer onto an active chain end interrupted the possibility of its further growth. Thus the styrenes acted as predominant chain-transfer agents giving difunctional oligomers (15), with DPs ranging from a few units to a few tens depending on the initial ratio of divinylbenzene to substituted styrene.

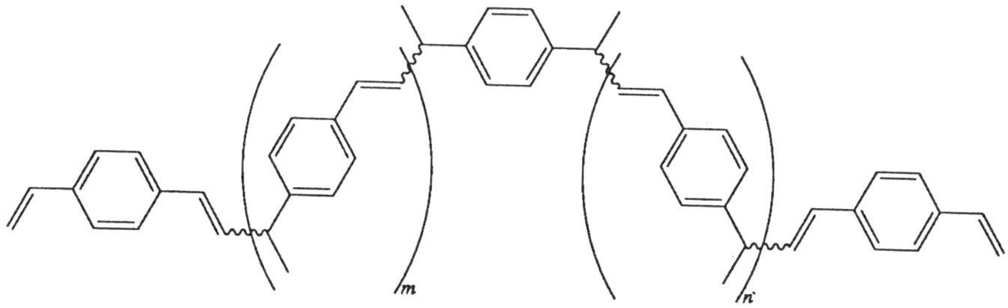

(14)

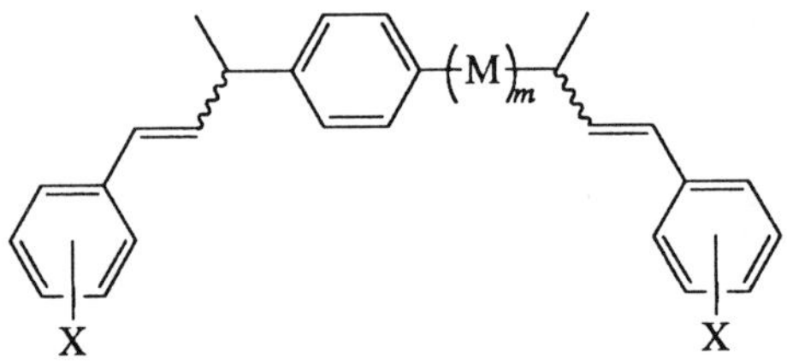

X = p-Me, p-OCOMe, p-OH, p-CH$_2$Cl or m-CH$_2$Cl

(15)

Curiously, when 1,4-diisopropenylbenzene is submitted to cationic activation with typical Brönsted and Lewis acids,[77] the major product is a poly(1,1,3-trimethylindane) (16) arising from successive addition/intramolecular alkylation/transfer cycles. Attempts to functionalize these polymers by copolymerization with substituted styrenes,[78] as in the work with 1,4-divinylbenzene summarized above, failed, except when p-methylstyrene was used, but the structure obtained had to be functionalized by a subsequent severe oxidation reaction which converted the methyl groups into carboxylic moieties.

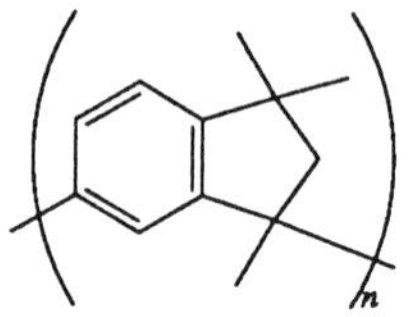

(16)

6.3.2 Heterocyclic Monomers

The possibility of generating growing cationic chains bearing two active ends was conceived and put into practice for the first time in the ring-opening polymerization of tetrahydrofuran nearly 20 years ago,[79] and led to the preparation of poly(trimethylene oxide) diols following the simple mechanism shown in Scheme 14.

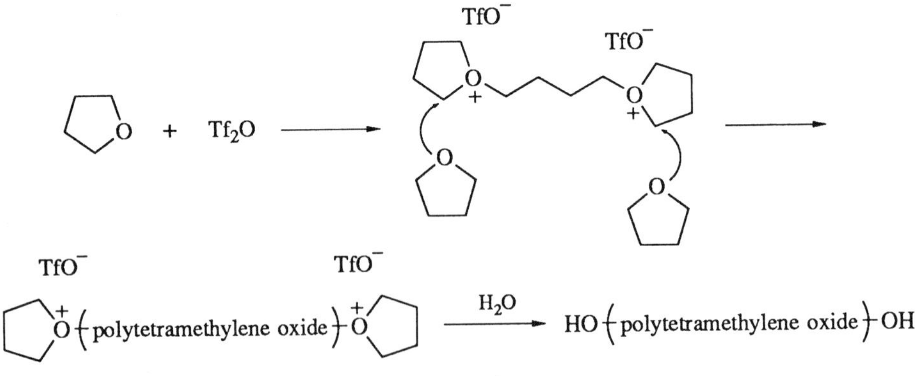

Scheme 14

Since then much progress has been made in the extension of this concept to many n-donor monomers and to the synthesis of difunctional polymeric structure.[5,9,53,80] This section is devoted to bringing this topic up to date.

6.3.2.1 Cyclic ethers

The activated monomer (AM) polymerization of oxiranes[57,58] seems ideally suited to the elaboration of polyethers bearing two terminal functionalities if instead of an alcohol (as shown in Scheme 10) a diol is used as the initial reagent for the protonated monomer, thus giving rise to chain growth at both ends of the linear macromolecule. Indeed, at any stage of the polymerization one already has *de facto* the presence of a macrodiol in the reaction medium.

Ethylene oxide,[57,58] methyloxirane[57,58] and chloromethyloxirane[57,58,81] have indeed been converted into the corresponding macromolecular diols by this technique. As already pointed out, whereas the polymerization of ethylene oxide is disturbed by some spurious chain growth from the ether groups in the polymer, substituted oxiranes give a clean behaviour: thus, the polymerization of chloromethyloxirane by $BF_3 \cdot THF$ in the presence of ethylene glycol[81] produces macrodiols with narrow DP distribution and an excellent agreement between the expected and the calculated DP, *i.e.* proof of a living process, up to about 30. The percentage of cyclic oligomers was less than unity in these systems.

Phosphorous and phosphoric acid have been known to induce the polymerization of oxiranes, but only very recently has the mechanism of these reactions been scrutinized.[82] The addition reaction of ethylene oxide onto a P—OH moiety in the absence of added acidic catalyst proceeds *via* the formation of an AM mechanism involving the hydrogen-bonded species (17). Chain growth stops when all P—OH functions have been consumed because of the lack of monomer activator. In the presence of HBF_4 as added catalyst, the monomer is activated by it to give the corresponding oxonium species (see Scheme 10) in conjunction with complexes (17). Whereas the concentration of the latter decreases as the monomer insertion reaction proceeds, the former survives and allows chain growth to continue. Phosphorous acid thus gives macrodiols (18) and phosphoric acid the corresponding triols (19).[83]

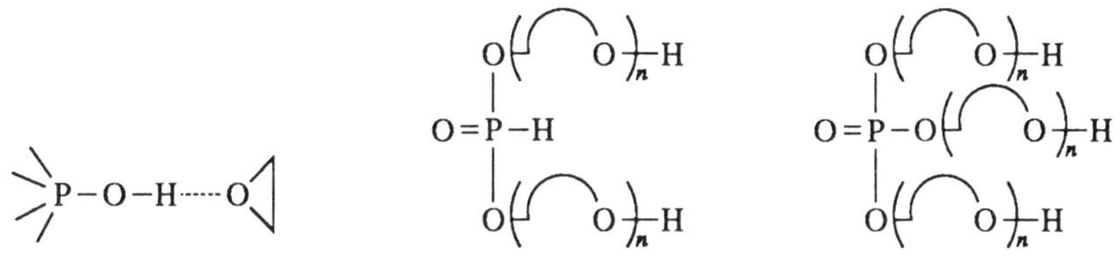

The variety of tetrahydrofuran difunctional oligomers and polymers includes an impressive choice of end groups prepared as a direct result of the living polymerizations initiated by difunctional acids or by suitable subsequent chemical modifications.[53] Among recent additions to this rich catalogue is the extension of equation (3) to difunctional active chains[59] to give the corresponding bisthiolanium derivatives and telechelics therefrom and the preparation of dicarboxylic oligomers[84] by quenching with the anion of diethyl malonate, hydrolyzing and decarboxylating the product.

6.3.2.2 *Cyclic acetals*

The concept of AM polymerization has been successfully applied to 1,3-dioxolane using diols and macrodiols as initiating hydroxy reagents.[85] The mechanism of this process is shown in Scheme 15. The resulting product is thus a polymer diol: if the original diol is a small molecule, *e.g.* ethylene

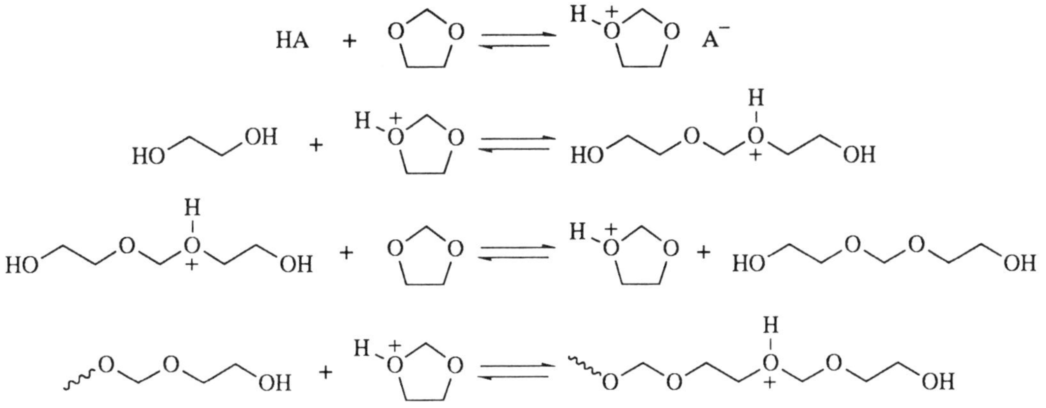

Scheme 15

glycol, the product is essentially the telechelic poly(1,3-dioxolane), but if macromolecular diols are used, like poly(ethylene oxide) glycols or poly(tetramethylene oxide) glycols, then triblock copolymers are obtained with two external poly(1,3-dioxolane)—OH chains. All of these polymers can of course be used to prepare speciality materials.[86]

The AM mechanism in the polymerization of 1,3-dioxolane has recently been revised[87] and a new heterotelechelic structure, bearing an OH group at one end and an isopropenyl moiety at the other, has been reported.

A novel living system[88] for the polymerization of 1,3-dioxolane to give difunctional active species involved the use of terephthaloyl ditriflate. Treatment of the chain carriers at the end of the polymerization with triethylamine gave the corresponding quaternized amino ether end groups (20), which could be converted into various functions, e.g. OH and acrylate moieties, by reaction with appropriate carboxylic anions.

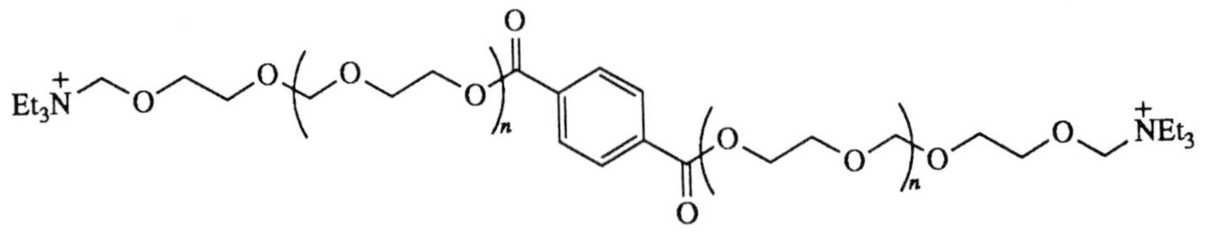

(20)

Coupling (20) with a polymeric dicarboxylate gave multiblock structures.[89] Treatment of the active dication with formals, used as end-blockers, provided a means of preparing other difunctional telechelic poly(1,3-dioxolane)s bearing alkenyl and chloride terminal groups,[90] following the reaction pathway shown in Scheme 16 (R = allyl, acrylate and CH_2CH_2Cl).

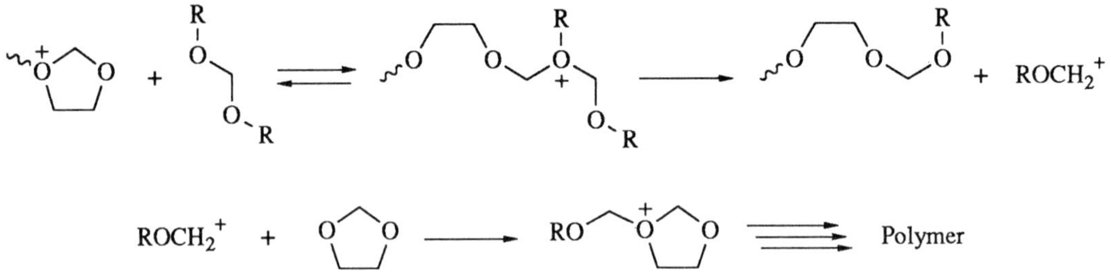

Scheme 16

6.3.2.3 N-t-Butylaziridine

The only report on the preparation of a living system involving the difunctional poly(N-t-butylaziridine) (21) called upon the addition of the monomer to the corresponding poly(tetrahydrofuran) dication, prepared according to Scheme 14.[91] The respective lengths of the outer and inner segments could be varied widely and in particular the central polyether chain could be reduced to a minimum so that (21) would be close to a homopolymeric chain carrier. The neutralization of (21) to give various functionalities was conducted following the same principles as those discussed above for the diammonium-terminated poly(1,3-dioxolane) (20).

6.3.2.4 Lactones

The cationic polymerization of ε-caprolactone by ammonium or phosphonium salts in the presence of diols leads to telechelic macromolecules with terminal hydroxy groups.[92] The mechanism postulated for these systems was based essentially on monomer activation, although other reactions were thought to intervene.

6.3.2.5 2-Oxazolines

The difunctional initiator (22) has recently been used to induce the living polymerization of 2-oxazolines to give chain carriers with two active ends.[64] Killing with strong nucleophiles such as NaOH, ammonia or carboxylate anions gave the corresponding dihydroxy, diamino and diester telechelics.

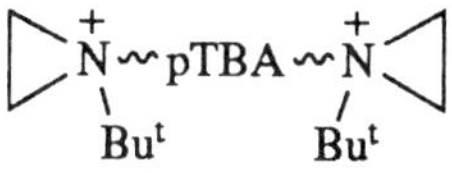

(21) 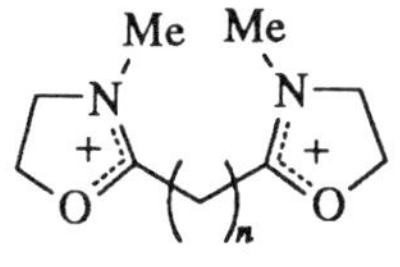   (22)

6.3.2.6 Cyclic siloxanes

Difunctional poly(dimethylsiloxane)s have been prepared by the cationic polymerization of suitable cyclic monomers in the presence of end-blockers with two reactive groups[93] (R=R') according to the idea illustrated in Scheme 13. Thus telechelic structures were prepared bearing two terminal amino, CO_2H or OH groups and with varying molecular weights.

6.3.3 Conclusion

The access to linear macromolecules with two reactive ends by cationic polymerization has been widened substantially with the recent discovery of living systems for alkenyl monomers. However, the potential of these techniques has just begun to be exploited and structures containing units from styrenic or aliphatic alkenes should be among the next targets.

With heterocyclic monomers the progress was more modest in the last few years, but much more had been done previously on account of the existing knowledge on how to obtain living or near-living conditions. Thus, research in this area in now focussing primarily on new materials and their properties, *i.e.* one step further than the synthesis of precursors.

6.4 POLYFUNCTIONAL STRUCTURES

This section includes branched macromolecules possessing functional chain ends and linear polymers with numerous functional side groups.

6.4.1 Alkenyl Monomers

The general principle of inifer (Scheme 5) obviously also applies to trifunctional initiating structures and indeed three-armed star-shaped polyisobutenes have been prepared by that technique.[16,18,69] Likewise, the living polymerization of isobutene initiated with triesters and quenched with allyltrimethylsilane[47] produces the three-branch structure (23), as already discussed for the mono- and di-functional homologues. In the absence of a specific nucleophilic quencher, this type of process yields a macromolecule like (23), but with tertiary chloride end groups.[94]

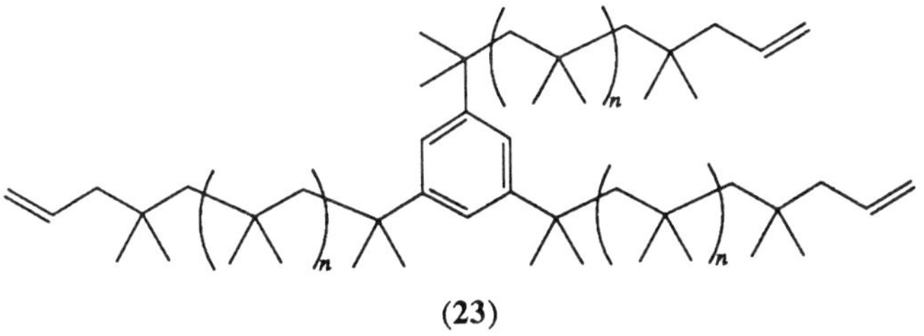

(23)

Even a four-armed star-shaped polyisobutene has been reported.[95] It was prepared using a tetraester in conjunction with boron chloride in typical living conditions and had again tertiary chloride terminal functions.

With vinyl ethers, a similar approach is just being published. Tris(trifluoroacetate)s bearing a resemblance to the addition product of trifluoroacetic acid onto vinyl ethers were used together with EtAlCl$_2$ and 1,4-dioxane to induce the three-armed living cationic polymerization of isobutyl vinyl ether alone[96] or in sequential blocking with 2-acetoxyethyl vinyl ether.[97]

The living cationic polymerization or copolymerization of vinyl ethers possessing a functional group yields narrow molecular-weight distribution products with a large number of functionalities pendant to the linear chains. A whole variety of these macromolecules has been synthesized,[20] starting from a simple homopolymer of 2-acetoxyethyl vinyl ether, subsequently hydrolyzed to the corresponding poly(2-hydroxyethyl vinyl ether;[98] equation 5), and going on to progressively more elaborate structures with two[99] and three[100] ester groups, or a silyloxy moiety[101] per monomer unit.

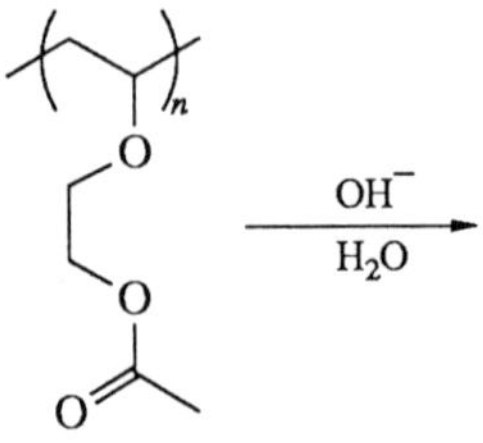

$$\text{(5)}$$

Block copolymers of two vinyl ethers have also been constructed in order to obtain amphiphilic macromolecular materials, both linear[102,103] and star shaped.[97,104] The representative structure (24) of an amphiphilic block copolymer illustrates the type of pendant group introduced.

A similar approach has been sketched for styrenic monomers, but only one report has appeared thus far, namely the living polymerization of 4-t-butoxystyrene with the aim of preparing narrow DP distribution poly(vinylphenol)s.[105] This led to the only synthesis of a block copolymer made from a vinyl ether (isobutyl vinyl ether) and a styrenic monomer (4-t-butoxystyrene) and its subsequent transformation into the amphiphilic macromolecular material (25).[106]

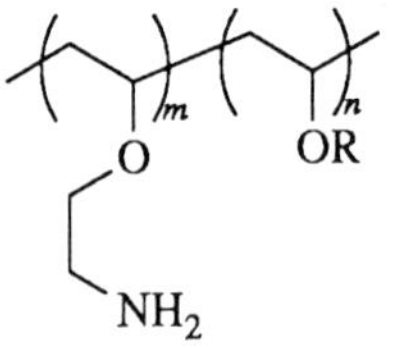

(24) R = Bu^n, i-C$_8$H$_{17}$, n-C$_{16}$H$_{33}$ 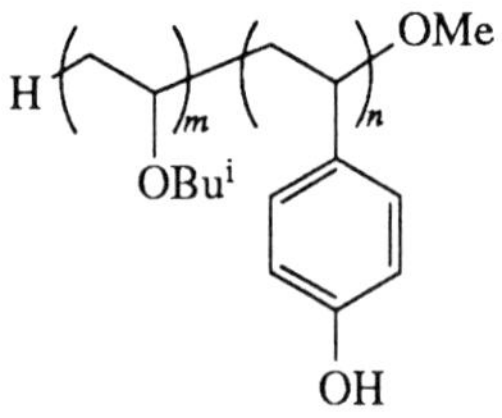   (25)

An interesting development due to the mild conditions associated with the living cationic polymerization of vinyl monomers is that structures bearing two potentially sensitive sites can be selectively activated. Thus, 1-isopropenyl-4-glycidyloxybenzene was made to polymerize exclusively *via* the alkenyl moiety[107] to give the linear polymer (26) with oxirane groups as pendant substituents on each monomer unit. The interest of such structures obviously stems from the possibility of subsequent crosslinking, *e.g.* with photochemical activation, as recently reported for similar structures involving a poly(vinyl ether) main chain and pendant acrylic or cinnamic groups (27).[108]

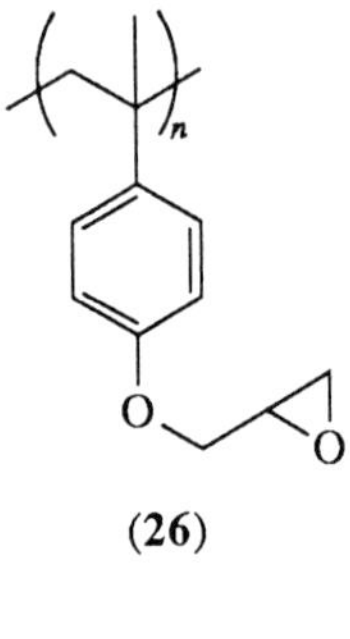

(26)

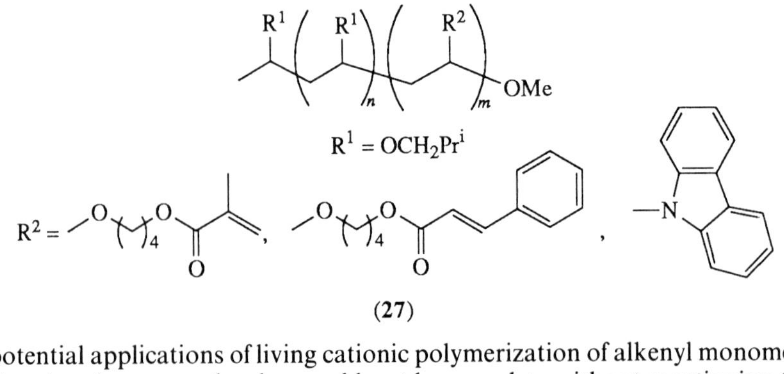

$R^1 = OCH_2Pr^i$

$R^2 = $

(27)

The panorama of the potential applications of living cationic polymerization of alkenyl monomers in the direction of multifunctional macromolecules would not be complete without mentioning the possibility of regulating the sequences of monomer units in oligomers made up of different vinyl ethers.[109] Trimers and tetramers, with as many monomers, were carefully synthesized using HI/ZnI$_2$

catalysis. Some of the monomers carried a functional side group which could be exploited for further use of these oligomers.

Poly(vinyl ether)-bearing pendant mesogenic[110] or perfluoroalkyl[111] groups have also been prepared recently.

6.4.2 Heterocyclic Monomers

Three-branch star-shaped poly(tetrahydrofuran)s were among the first nonlinear functionalized polymers produced by a cationic mechanism.[112] The initiator used was the tris(oxocarbenium) salt (**28**), and obviously the type of nucleophile used to kill the reaction determined the nature of the groups introduced as chain ends.

(**28**)

The search for multifunctional structures by ring-opening polymerization has not been given high priority and the number of reports to this effect in the literature is quite limited. The most recent examples include the already mentioned three-armed polyols arising from the use of phosphoric acid as initiator of the polymerization of oxiranes[83] and the use of chloroformates of triols to induce the polymerization of 2-oxazolines,[113] each ester group being the site of monomer activation.

In a different vein, polyhydroxylic macromolecules were obtained upon the polymerization of 1,3-dioxolane in the presence of glycidol,[90] which acted both as a comonomer (through ring opening) and as a transfer agent (through its hydroxy group). Structures could be modulated as a function of the relative amount of glycidol used and of the mode of its addition: if it was added all at the beginning of the reaction, star-like macromolecules were obtained, whereas its slow introduction during the polymerization resulted in a branched configuration. Obviously, however, each chain end consisted of a hydroxy function, irrespective of the polymer topology.

6.5 CONCLUSION

There is no question that a qualitative change has occurred to cationic polymerization in the last few years with the inception of living systems applied to alkenyl monomers. The fact that somehow the field managed to grow out of a rather prolonged infancy has suddenly opened new perspectives in terms of both scientific knowledge and useful polymer structures. This review has attempted to show how the various leading laboratories specializing in cationic polymerization are responding to those challenges, although more emphasis was placed on discussing the latter, more practical, topic, rather than the former, given the scope of the survey. In view of the very sustained research effort being conducted in many of these laboratories, as testified by the avalanche of publications, the reader should consider the present chapter as a progress report instead of a conclusive assessment.

6.6 REFERENCES

1. P. H. Plesch (ed.), 'Cationic Polymerisation and Related Complexes', Heffer, Cambridge, 1953.
2. P. H. Plesch (ed.), 'The Chemistry of Cationic Polymerisation', Pergamon Press, Oxford, 1963.
3. J. P. Kennedy and E. Maréchal, 'Carbocationic Polymerization', Wiley Interscience, New York, 1982.
4. A. Gandini and H. Cheradame, *Adv. Polym. Sci.*, 1980, **34/35**, 1.
5. S. Penczek, P. Kubisa and K. Matyjaszewski, *Adv. Polym. Sci.*, 1980, **37**, 1; 1985, **68/69**, 1.
6. A. Gandini and H. Cheradame, in 'Encyclopedia of Polymer Science and Engineering', ed. J. I. Kroschwitz, Wiley Interscience, New York, 1985, vol. 2, p. 729.
7. E. J. Goethals (ed.), 'Cationic Polymerization and Related Processes', Academic Press, New York, 1984.
8. G. C. Eastmond, A. Ledwith, S. Russo and P. Sigwalt, (eds.), 'Comprehensive Polymer Science', Pergamon Press, Oxford, 1989, vol. 3, chap. 39–53.
9. S. Penczeck and P. Kubisa, in 'Encyclopedia of Polymer Science and Engineering', ed. J. I. Kroschwitz, Wiley Interscience, New York, 1989, suppl. vol., p. 380.

10. M. Sawamoto and T. Higashimura, in 'Encyclopedia of Polymer Science and Engineering', ed. J. I. Kroschwitz, Wiley Interscience, New York, 1989, suppl. vol., p. 399.
11. K. E. Russell and L. G. M. C. Vail, *Can. J. Chem.*, 1979, **57**, 2355; B. K. Hunter, E. Redler, K. E. Russell, W. G. Schnarr and S. L. Thomson, *J. Polym. Sci., Polym. Chem. Ed.*, 1983, **21**, 435.
12. A. Gandini and R. Martinez, *Makromol. Chem.*, 1983, **184**, 1189.
13. H. Razzouk, K. Bouridah, A. Gandini and H. Cheradame, in ref. 7, p. 355.
14. J. P. Kennedy and R. A. Smith, *J. Polym. Sci., Polym. Chem. Ed.*, 1980, **18**, 1523 and 1539.
15. O. Nuyken, S. D. Pask, A. Vischer and M. Walter, *Makromol. Chem., Macromol. Symp.*, 1986, **3**, 129.
16. A. Gandini and A. Martinez, *Makromol. Chem., Macromol. Symp.*, 1988, **13/14**, 211.
17. O. Nuyken and S. D. Pask, in 'Telechelic Polymers: Synthesis and Applications', ed. E. J. Goethals, CRC Press, Boca Raton, FL, 1988, p. 95.
18. J. Habimana, Ph.D. Thesis, National Polytechnic Institute, Grenoble, France, 1989; H. Cheradame, J. Habimana, E. Rousset and F. J. Chen, *Makromol. Chem.*, 1991, **192**, 2777; H. Cheradame, J. Habimana and F. J. Chen, Abstracts of Communications C01 and C02 Presented at the 10th International Symposium on Cationic Polymerization, Balatonfüred, Hungary, 1991, ed. T. Kelen, L. Kossuth University, Debrecen, 1991.
19. M. Miyamoto, M. Sawamoto and T. Higashimura, *Macromolecules*, 1984, **17**, 265.
20. (a) M. Sawamoto, *Prog. Polym. Sci.*, 1991, **16**, 111. (b) M. Sawamoto and T. Higashimura, *Makromol. Chem., Macromol. Symp.*, 1990, **32**, 131.
21. M. Schappacher and A. Deffieux, *Macromolecules*, 1991, **24**, 2140 and 4221; M. Sawamoto and T. Higashimura, in ref. 31.
22. Y. H. Kim and T. Heitz, *Makromol. Chem., Rapid Commun*, 1990, **11**, 525; M. Kamigaito, M. Sawamoto and T. Higashimura, *Macromolecules*, 1991, **24**, 3988.
23. M. Sawamoto, J. Fujimori and T. Higashimura, *Macromolecules*, 1987, **20**, 916.
24. K. Kojima, M. Sawamoto and T. Higashimura, *Macromolecules*, 1990, **23**, 948.
25. K. Kojima, M. Sawamoto and T. Higashimura, *J. Polym, Sci., Polym. Chem. Ed.*, 1990, **28**, 3007.
26. R. Faust and J. P. Kennedy, *Polym. Bull.*, 1988, **19**, 29 and 35.
27. R. Faust and J. P. Kennedy, *Polym. Bull.*, 1988, **19**, 21.
28. Y. Ishihama, M. Sawamoto and T. Higashimura, *Polym. Bull.*, 1990, **23**, 366 and **24**, 201.
29. R. Faust and J. P. Kennedy, *J. Polym. Sci., Polym. Chem. Ed.*, 1987, **25**, 1847.
30. T. Pernecker and J. P. Kennedy, *Polym. Bull.*, 1991, **26**, 305.
31. Proceedings of the 10th International Symposium on Cationic Polymerization, *Makromol, Chem., Macromol. Symp.*, in press.
32. K. Matyjaszewski, in ref. 31.
33. A. Gandini and P. H. Plesch, *J. Polym. Sci., Part B.*, 1965, **3**, 127; *J. Chem. Soc.*, 1965, 4826; *SCI Monogr.*, 1966, **20**, 107; P. H. Plesch, *Makromol. Chem., Macromol. Symp.*, 1988, **13/14**, 375.
34. Y. H. Kim, *Macromolecules*, 1991, **24**, 2122.
35. M. Sawamoto, T. Enoki and T. Higashimura, *Macromolecules*, 1987, **20**, 1.
36. T. Hashimoto, E. Takeuchi, M. Sawamoto and T. Higashimura, *J. Polym. Sci., Polym. Chem. Ed.*, 1990, **28**, 1137.
37. H. Shohi, M. Sawamoto and T. Higashimura, *Polym. Bull.*, 1989, **21**, 357.
38. M. Schappacher and A. Deffieux, *Makromol. Chem., Rapid Commun.*, 1991, **12**, 447.
39. D. Van Meirvenne, N. Haucourt and E. J. Goethals, *Polym. Bull.*, 1990, **23**, 185.
40. B. Ivan and J. P. Kennedy, *Macromolecules*, 1990, **23**, 2880; R. Faust, B. Ivan and J. P. Kennedy, *J. Macromol. Sci., Chem.*, 1991, **A28**, 1.
41. J. E. Puskas, G. Kaszas and M. Litt, *Macromolecules*, in press.
42. A. F. Fehervari, R. Faust and J. P. Kennedy, *J. Macromol. Sci., Chem.*, 1990, **A27**, 1571.
43. M. Miyamoto, M. Sawamoto and T. Higashimura, *Macromolecules*, 1985, **18**, 123.
44. M. Sawamoto, T. Enoki and T. Higashimura, *Polym. Bull.*, 1987, **18**, 117.
45. A. Verma and J. S. Riffle, *Polym. Prepr., Am. Chem. Soc., Div. Polym. Chem.*, 1990, **31**(1), 590.
46. O. Nuyken, H. Kröner and S. Aechtner, *Polym. Bull.*, 1990, **24**, 513.
47. J. P. Kennedy, *Makromol. Chem., Macromol. Symp.*, 1991, **47**, 55.
48. B. Ivan and J. P. Kennedy, *J. Polym. Sci., Polym. Chem. Ed.*, 1990, **28**, 89.
49. T. Higashimura, K. Ebara and S. Aoshima, *J. Polym. Sci., Polym. Chem. Ed.*, 1989, **27**, 2937.
50. M. Sawamoto, T. Enoki and T. Higashimura, *Polym. Bull.*, 1986, **16**, 117.
51. E. J. Goethals, N. H. Haucourt, A. M. Verheyen and J. Habimana, *Makromol. Chem., Rapid Commun.*, 1990, **11**, 623.
52. J. P. Kennedy and J. D. Carter, *Macromolecules*, 1990, **23**, 1238.
53. E. J. Goethals (ed.), 'Telechelic Polymers: Synthesis and Applications', CRC Press, Boca Raton, FL, 1988, p. 115.
54. A. Munir and E. J. Goethals, *J. Polym. Sci., Polym. Chem. Ed.*, 1981, **19**, 1985.
55. E. J. Goethals, E. H. Schacht, E. H. Bogaert, S. I. Ali and Y. Tezuka, *Polym. J.*, 1980, **12**, 571.
56. K. Matyjaszewski, *Makromol. Chem.*, 1984, **185**, 37 and 51.
57. P. Kubisa, *Makromol. Chem., Macromol. Symp.*, 1988, **13/14**, 203; S. Penczek and P. Kubisa in 'Frontiers of Macromolecular Science', ed. T. Saegusa, T. Higashimura and A. Abe, Blackwell, Oxford, 1989, p. 107.
58. M. Bednarek, P. Kubisa and S. Penczek, *Makromol. Chem., Suppl.*, 1989, **15**, 49; M. Bednarek, T. Biedron, P. Kubisa and S. Penczek, *Makromol. Chem., Macromol. Symp.*, 1991, **42/43**, 475.
59. F. D'Haese and E. J. Goethals, *Br. Polym. J.*, 1988, **20**, 103.
60. W. Chwialkowska, P. Kubisa and S. Penczek, *Makromol. Chem.*, 1982, **183**, 753; R. Szymanski, P. Kubisa and S. Penczek, *Macromolecules*, 1983, **16**, 1000.
61. H. R. Kricheldorf, J. M. Jonté and R. Dunsing, *Makromol. Chem.*, 1986, **187**, 771; H. R. Kricheldorf, R. Dunsing and A. Serra, *Macromolecules*, 1987, **20**, 2050; H. R. Kricheldorf and M. Sumbél, *Makromol. Chem.*, 1988, **189**, 317.
62. H. R. Kricheldorf and R. Dunsing, *Makromol. Chem.*, 1986, **187**, 1611; H. R. Kricheldorf and M. Sumbél, *Eur. Polym. J.*, 1989, **25**, 585.
63. H. R. Kricheldorf, R. Dunsing and A. Serra, *Makromol. Chem.*, 1987, **188**, 2453; H. R. Kricheldorf and J. Jenssen, *J. Macromol. Sci., Chem.*, 1989, **A26**, 631; H. R. Kricheldorf and J. Jenssen, in ref. 31.
64. T. Saegusa, *Makromol. Chem., Macromol. Symp.*, 1988, **13/14**, 111; T. Saegusa and Y. Chujo, *Makromol. Chem., Macromol. Symp.*, 1990, **33**, 31; S. Kobayashi, *Prog. Polym. Sci.*, 1990, **15**, 751.

65. T. C. Kendrick, B. M. Parbhoo and J. W. White, ref. 8, vol. 4, chap. 25; P. Sigwalt, *Makromol. Chem., Macromol. Symp.*, 1990, **32**, 217; P. Sigwalt, C. Gobin, P. Nicol, M. Moreau and M. Masure, *Makromol. Chem., Macromol. Symp.*, 1991, **42/43**, 229.
66. H. Shohi, M. Sawamoto and T. Higashimura, *Macromolecules*, in press.
67. O. Nuyken, H. Kröner and S. Aechtner, *Makromol. Chem., Rapid Commun.*, 1988, **9**, 671.
68. V. Héroguez, A. Duffieux and M. Fontanille, *Makromol. Chem., Macromol. Symp.*, 1990, **32**, 199.
69. R. Faust, A. Fehervari and J. P. Kennedy, *ACS Symp. Ser.*, 1985, **282**, 125.
70. J. P. Kennedy, T. P. Liao, S. Guhaniyogi and V. S. C. Chang, *J. Polym. Sci., Polym. Chem. Ed.*, 1982, **20**, 3219.
71. B. Wang, M. K. Mishra and J. P. Kennedy, *Polym. Bull.*, 1987, **17**, 205 and 213.
72. R. Faust, A. Nagy and J. P. Kennedy, *J. Macromol. Sci., Chem.*, 1987, **A24**, 595.
73. M. Zsuga and J. P. Kennedy, *Polym. Bull.*, 1989, **21**, 5.
74. L. Balogh, A. Takacs and R. Faust, *Polym. Prepr., Am. Chem. Soc., Div. Polym. Chem.*, 1992, in press.
75. H. Hasegawa and T. Higashimura, *Macromolecules*, 1980, **13**, 1350.
76. T. Higashimura, S. Aoshima and H. Hasegawa, *Macromolecules*, 1982, **15**, 1221.
77. T. Dittmer, F. Gruber and O. Nuyken, *Macromol. Chem.*, 1989, **190**, 1755 and 1771.
78. O. Nuyken, G. Maier, D. Yang and M. B. Leitner, in ref. 31.
79. S. Smith and A. J. Hubin, *J. Macromol. Sci., Chem.*, 1973, **A7**, 1399.
80. E. J. Goethals, *Makromol. Chem., Macromol. Symp.*, 1991, **42/43**, 51.
81. T. Biedron, P. Kubisa and S. Penczek, *J. Polym. Sci., Polym. Chem. Ed.*, 1991, **29**, 619.
82. T. Biela and P. Kubisa, *Makromol. Chem.*, 1991, **192**, 473.
83. P. Kubisa and T. Biela, in ref. 31.
84. S. Kobayashi, H. Sato, S. Nishihara and S. Shoda, *Macromolecules*, 1990, **23**, 2861.
85. L. Reibel, H. Zouine and E. Franta, *Makromol. Chem., Macromol. Symp.*, 1986, **3**, 221; E. Franta, P. Kubisa, J. Refai, S. Ould Kada and L. Reibel, *Makromol. Chem., Macromol. Symp.*, 1988, **13/14**, 127; E. Franta, J. Refai, C. Durand and L. Reibel, *Makromol. Chem., Macromol Symp.*, 1990, **32**, 169; E. Franta and L. Reibel, *Makromol. Chem., Macromol. Symp.*, 1991, **47**, 141.
86. E. Franta, E. Gérard, Y. Gnanou, L. Reibel and P. Rempp, *Makromol. Chem.*, 1990, **191**, 1689.
87. E. Franta, P. Kubisa, S. Ould Kada and L. Reibel, in ref. 31.
88. D. Van Meirvenne and E. J. Goethals, *Makromol. Chem. Suppl.*, 1989, **15**, 61.
89. D. Van Meirvenne and E. J. Goethals, *New Polym. Mater.*, 1990, **1**, 281.
90. E. J. Goethals, R. D. De Clercq, H. C. De Clercq and P. J. Hartmann, *Makromol. Chem., Macromol. Symp.*, 1991, **47**, 151.
91. M. Van de Velde and E. J. Goethals, *Makromol. Chem., Macromol. Symp.*, 1986, **6**, 271.
92. B. A. Rozenberg, in ref. 31.
93. I. Yilgör, J. S. Riffle and J. E. McGrath, *ACS Symp. Ser.*, 1985, **282**, 161; P. M. Sormani, R. J. Minton and J. E. McGrath, *ACS Symp. Ser.*, 1985, **286**, 147.
94. M. Zsuga, L. Balogh, T. Kelen and J. Borbély, *Polym. Bull.*, 1990, **23**, 335.
95. K. J. Huang, M. Zsuga and J. P. Kennedy, *Polym. Bull.*, 1988, **19**, 43.
96. H. Shohi, M. Sawamoto and T. Higashimura, *Macromolecules*, 1991, **24**, 4926.
97. H. Shohi, M. Sawamoto and T. Higashimura, *Polym. Bull.*, 1991, **25**, 529.
98. S. Aoshima, T. Nakamura, N. Uesugi, M. Sawamoto and T. Higashimura, *Macromolecules*, 1985, **18**, 2097.
99. T. Higashimura, T. Enoki and M. Sawamoto, *Polym. J.*, 1987, **19**, 515.
100. M. Minoda, M. Sawamoto and T. Higashimura, *Polym. Bull.*, 1987, **17**, 107.
101. T. Higashimura, K. Ebara and S. Aoshima, *J. Polym. Sci., Polym. Chem. Ed.*, 1989, **27**, 2937.
102. M. Minoda, M. Sawamoto and T. Higashimura, *Macromolecules*, 1987, **20**, 2045.
103. S. Kanaoka, M. Minoda, M. Sawamoto and T. Higashimura, *J. Polym. Sci., Polym. Chem. Ed.*, 1990, **28**, 1127; M. Minoda, M. Sawamoto and T. Higashimura, *Macromolecules*, 1990, **23**, 1897.
104. S. Kanaoka, M. Minoda, M. Sawamoto and T. Higashimura, *Macromolecules*, 1991, **24**, 2309.
105. T. Higashimura, K. Kojima and M. Sawamoto, *Makromol. Chem., Suppl.*, 1989, **15**, 127.
106. K. Kojima, M. Sawamoto and T. Higashimura, *Macromolecules*, 1991, **24**, 2658.
107. T. Hashimoto, Y. Sano, M. Sawamoto, T. Higashimura, N. Saito and S. Kanagawa, *J. Polym. Sci., Polym. Chem. Ed.*, 1991, **29**, 339.
108. O. Nuyken and S. Aechtner, Communication at the one day Meeting on New Developments in the Cationic Polymerization of Vinyl Ethers, University of Ghent, Oct. 22, 1991, ed. E. J. Goethals, University of Ghent, Belgium, 1991.
109. M. Minoda, M. Sawamoto and T. Higashimura, *Polym. Bull.*, 1990, **23**, 133; *Macromolecules*, 1990, **23**, 4889.
110. V. Héroguez, M. Schappacher, E. Papon and A. Deffieux, *Polym. Bull.*, 1991, **25**, 307.
111. J. Höpken, M. Möller, M. Lee and V. Percec, *Makromol. Chem.*, in press.
112. E. Franta, L. Reibel, J. Lehmann and S. Penczek, *J. Polym. Sci., Polym. Symp.*, 1976, **56**, 139.
113. A. Dworak and R. C. Schulz, *Bull. Soc. Chim. Belg.*, 1990, **99**, 881; *Makromol. Chem.*, 1991, **192**, 437.

7
Functionalized Polyalkenes

TONINO SIMONAZZI
Himont Italia, Ferrara, Italy

ANTHONY DE NICOLA
Himont USA, Elkton, MD, USA

MAURO AGLIETTO and GIACOMO RUGGERI
University of Pisa, Italy

7.1	INTRODUCTION	133
7.2	PARTIAL CHEMICAL MODIFICATION OF POLYALKENES	134
	7.2.1 Functionalization of Polyalkenes	135
	7.2.1.1 Surface functionalization	135
	7.2.1.2 Bulk functionalization	137
	7.2.1.3 Attachment of low molecular weight compounds to a polyalkene backbone	140
7.3	FREE RADICAL GRAFTING	144
	7.3.1 Radiation Grafting Techniques	144
	7.3.1.1 Radiation peroxidation	144
	7.3.1.2 Simultaneous (direct) radiation grafting	145
	7.3.1.3 Trapped radical techniques	147
	7.3.2 Chemically Initiated Graft Polymerization	147
	7.3.2.1 Hydroperoxidation	147
	7.3.2.2 Chain transfer polymerization	148
7.4	FUNCTIONALIZATION OF POLYPROPYLENE (PP) AND RELATED AREAS OF APPLICATION	150
	7.4.1 Main Applications of Functionalized PP	151
	7.4.1.1 Joint reinforcements	151
	7.4.1.2 Anticorrosive coatings	152
	7.4.1.3 Film and multilayer sheets	153
	7.4.1.4 Polymer alloys	153
	7.4.1.5 Painting and gluing	154
7.5	REFERENCES	155

7.1 INTRODUCTION

Functionalized polyalkenes are a new class of polymers which can be obtained by: (i) copolymerization of alkenes with suitable polar comonomers; (ii) partial chemical modification of preformed polyalkenes; and (iii) free radical graft polymerization.

These functionalized polymers have new properties due to the presence of functional polar groups attached to the backbone but also maintain, to a large extent, the original properties of the polyalkenes.

Copolymerization cannot be applied as a general method, as Ziegler–Natta catalysts, which are the only convenient catalysts for polyalkenes, can be sensitive to polar monomers.

However, ethylene copolymers with polar monomers have been obtained by a free radical mechanism[1,2] or by group VIII transition metal catalysts[3] (preferably nickel derivatives).[4] CO/ethylene copolymers obtained using Pd catalysts are also known.[5]

The first attempt at obtaining functionalized polyalkenes using Ziegler–Natta catalysts was made by Giannini et al.[6] with the polymerization of nitrogen-containing and oxygen-containing monomers.

Recently Ciardelli et al.,[7] using $MgCl_2$-supported Ziegler–Natta catalysts, obtained a copolymer with a low content of polar comonomer in the backbone by copolymerization of 1-heptene or 4-methyl-1-pentene with 4-vinylbenzoic acid methyl ester.

Purgett and Vogl[8] effected the copolymerization of the 2,6-dimethyl phenyl ester of 10-undecenoic acid with α-alkenes using an Al-activated $TiCl_3/R_2AlCl$ coordination initiating system.

Chung[9–11] described a new approach to preparing polyalkenes containing functional groups, using stable borane monomers in the presence of transition metal catalysts.

For more detailed information about the possibility of obtaining functionalized polyalkenes by copolymerization, the recent review by Padwa should be consulted.[12]

The first and the second parts of the present chapter are devoted to describing briefly all possible ways of modifying polyalkenes, as well as providing detailed information about the methods and reactions which permit new polymer structures to be produced and the structure of introduced functional groups to be determined. The structural analysis of new functional polymers[13] is extremely useful for predicting the availability of the functional groups for assessing the compatibility of the interactions of modified polyalkenes with other polymers as well as their interfacial interactions with other substrates.

The third and final part of this review is devoted to functionalization of polypropylene and its main areas of application.

7.2 PARTIAL CHEMICAL MODIFICATION OF POLYALKENES

There has been a growing interest in recent years in extending functionalization reactions to polyalkenes, due to the convenient mechanical, chemical and environmental properties of such materials.

Such a goal can be approached by introducing controlled amounts of functional groups into the saturated hydrocarbon backbone of the unmodified alkane structure, *i.e.* by chemical modification of preformed polymers.

This method offers several practical advantages such as: (i) the use of commercial polyalkenes (such as polyethylene or polypropylene) as starting materials; (ii) a large selection of functional groups available by using different functionalizing agents and methods; (iii) the retention, at least to a large extent, of the original properties of the polyalkene; (iv) the possibility of surface functionalization; and (v) the possibility of performing functionalization during processing (*e.g.* extrusion).

Free radical functionalization of polyalkenes has been the most investigated approach. Many scientists have devoted their efforts to achieving such a goal and the literature contains a wide variety of articles (scientific papers, reviews, short communications and patents) related to this topic, treating either different polymers or different modification routes (initiators, functional monomers, reaction conditions).

Ethylene–propylene rubber (EPR), polyethylene (PE) and polypropylene (PP) have been the most investigated as starting polymer materials; experiments performed on these polymers give the most meaningful results and constitute the core of this review. Different substances have been used as functional, low molecular weight reagents, from polar organic acid derivatives (maleates and fumarates or anhydrides, with suitable free radical initiators) to more reactive molecules (from chlorine up to oxidation mixtures of chromic acid) to introduce hydroxy groups into polyethylene films.[14–17]

Where a free radical initiator was needed, it was compulsory to find a compromise between the relatively high temperature which is necessary to melt the polyalkene (in order to allow good diffusion of the functionalizing mixture in the polymer), and the kinetics of the decomposition of the initiator. At the proposed temperature, 2,2'-azobisisobutyronitrile (AIBN) had a too rapid decomposition kinetics to provide good functionalization efficiency. Peroxides, however, in particular dicumyl and dibenzoyl peroxide, with their longer decomposition times, showed better possibilities for initiating and propagating the functionalization reaction, and their use has found practical applications in different ways.[18]

Among free radical initiation methods, UV radiation is also worth mentioning. Its use either alone or coupled with suitable sensitizers, such as aromatic ketones, enabled the functionalization of polymeric films (by grafting of polar monomers) in order to improve their adhesion to different substrate surfaces or to vary the hydrophilicity of the substrate.[19-23]

Radical functionalization is usually performed in bulk as the reactions in solution show that the radical initiator might also react with the solvent, with a drastic decrease in the reaction efficiency.

The polymer is heated up to incipient melting and then a mixture of functionalizing monomer and radical initiator is added. The whole mixture is allowed to react for a reasonable length of time and then the polymer is collected and worked up.[14,18] In industrial production systems (extruders, Brabender or Banbury mixers) these two steps may be combined and solid functional monomers (*e.g.* maleic anhydride) are added together with the polymer and the initiator in the charge hopper.

Functionalization in bulk has been widely studied in Italy by research groups in Pisa and Arco Felice (Napoli) and the possibility of performing a clean modification of an alkane structure, while substantially avoiding undesired side reactions (crosslinking and degradation), has now been fully demonstrated;[24] indeed, it is possible to freeze the macroradical initially formed by hydrogen abstraction from the macromolecular backbone by the primary radical, if the functionalizing monomer is an unsaturated molecule bearing electron-withdrawing groups. As a result of its reactivity, it can immediately react with the unpaired electron and can avoid coupling, disproportionation and degradation of macroradicals (see Scheme 1).[13]

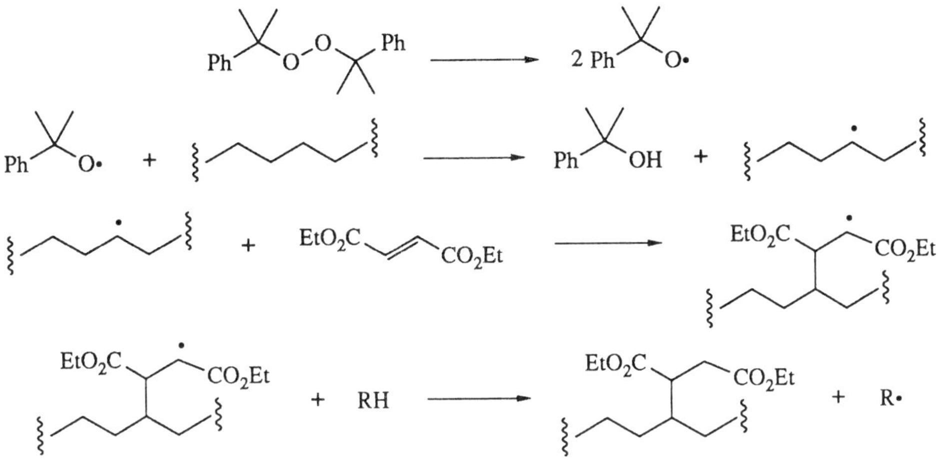

Scheme 1 Free radical promoted functionalization of polyethylene with diethyl maleate and dicumyl peroxide. Side reactions such as coupling and crosslinking, disproportionation and β-scission are not reported

Under some reaction conditions ($T = 200\,°C$ and polymer–diethyl maleate ratio equal to 1 by weight), side reactions such as disproportionation and crosslinking of macroradicals seem to be completely avoided. In the case of polyethylene, a clear demonstration of the insertion of single and isolated units of functional monomer has been obtained using ^{13}C NMR.[24] This insertion enables broadening of the molecular weight distribution of the polymer, as revealed by gel permeation chromatography,[13] to be avoided completely.

7.2.1 Functionalization of polyalkenes

7.2.1.1 Surface functionalization

(i) Polyethylene

In 1969, Olsen and Osteraas[25] described sulfur insertion onto polyethylene surfaces. Atomic sulfur is isoelectronic with carbenes and nitrenes, which react with PE and PE surfaces. Thus atomic sulfur, generated by the pyrolysis of carbonyl sulfide or by photolysis of carbonyl sulfide and carbon disulfide, has been shown to modify irreversibly the surface of polyethylene (an insertion of atomic sulfur into the hydrogen–carbon bond) to form a surface thiolic group. A functionalization rating can be performed by wettability measurements and frustrated multiple internal reflection spectroscopy (FMIR).[26]

Surfaces may be modified by various methods, e.g. by flame oxidation, corona discharge, or with a strong acid.[27] Other methods include plasma treatment,[28] use of surfactants, and surface grafting.[29]

Allmer, Hult and Rånby have published some work on the surface modification of polymers, especially polyethylene.[19-23] Light in the near UV region is advantageous as it is selectively absorbed by the initiator without affecting the PE. The initiator most commonly used for grafting is benzophenone or its derivatives which, when UV irradiated, remove hydrogen from the polymer substrate, thus creating reactive grafting sites.[29]

The surface of LLDPE and HDPE has been modified by grafting with acrylic acid,[19] using benzophenone and acrylic acid in the vapour phase with UV irradiation (Scheme 2).

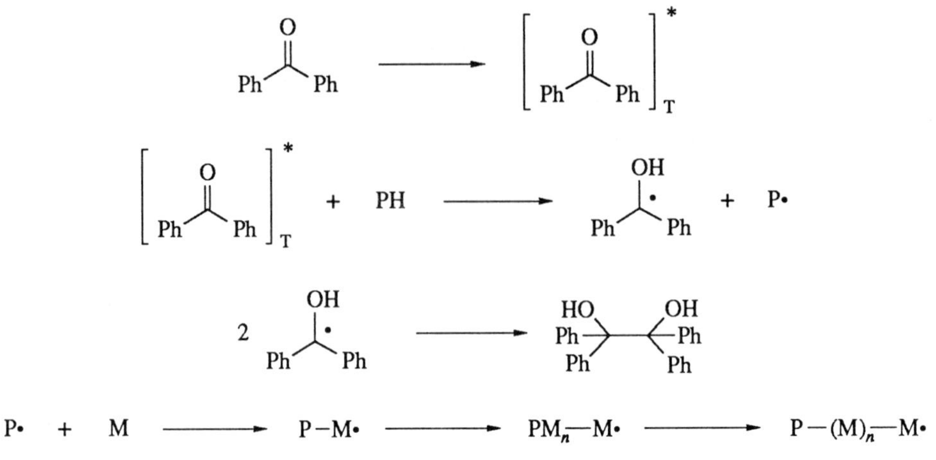

$$Ph-CO-Ph \longrightarrow [Ph-CO-Ph]^*_T$$

$$[Ph-CO-Ph]^*_T + PH \longrightarrow Ph-\overset{OH}{\underset{\bullet}{C}}-Ph + P\bullet$$

$$2\ Ph-\overset{OH}{\underset{\bullet}{C}}-Ph \longrightarrow Ph-\overset{HO}{\underset{Ph}{C}}-\overset{OH}{\underset{Ph}{C}}$$

$$P\bullet + M \longrightarrow P-M\bullet \longrightarrow PM_n-M\bullet \longrightarrow P-(M)_n-M\bullet$$

PH = polymer backbone
M = monomer

Scheme 2 Mechanism of surface modification by irradiation in the presence of benzophenone and functional monomer

In a heterogeneous system it is possible to ensure that the UV-sensitized grafting is located on the surface. ESCA, FTIR-ATR and contact angle measurements have been used to characterize the surfaces and the stability of the grafted layer in different surroundings and temperatures.

The solvent is very important: (i) it must be a solvent for and carrier of both components in the vapour phase (benzophenone and acrylic acid); and (ii) it must have a sensitizing effect on the grafting reaction. Acetone proved to be the best solvent.

The grafted PE surface is stable at room temperature, and the wettability decreases very slowly as the temperature at the PE surface increases.

The same process has been applied to LLDPE grafting with glycidyl acrylate (GA) and glycidyl methacrylate (GMA) by photoinitiation,[20] with or without 2,4-dihydroxybenzophenone (DHBP) or 4-aminosalicylic acid as stabilizer.[20]

After 10 minutes of grafting with UV irradiation, ESCA measurements on the grafted surface indicated that a 72% yield for GA and a 52% yield for GMA had been obtained. The solvent used was acetone or EtOH: acetone gave slightly higher grafting on the surface.

The grafted surfaces react with amines (i.e. aniline and propylamine in EtOH) by opening of epoxy bonds.

A stabilizer was attached to the oxirane group grafted onto the surface; the amount varied depending on the polymer substrate.[22]

PE films grafted with GMA have been reacted with poly(ethylene glycol) (PEG) and heparin.[23] ESCA measurements have shown that heparin is grafted to the surface and the heparinized PE surface was shown to give reduced thrombus formation by performing in vivo blood clotting tests.

The size of the alkyl group R (PE–CONHR or PE–CO$_2$R) is the main factor in determining the wettability by water of surface-functionalized PE.[30]

(ii) Polypropylene

Recently, surface modification of polypropylene (PP) induced by cold plasma has been reported. The formation of peroxide radicals, after treatment with CO_2, O_2 or other cold plasmas, allows the subsequent grafting of acrylate monomers on the polypropylene surface.[31,32] This method may represent a potential new route in the field of surface functionalization of polyalkenes.

Polypropylene films were oxidized by treatment with ozone and ozone–UV light.[33] The oxygen species, which are the main products of ozone photolysis (ozone, atomic oxygen and singlet oxygen), react with the polymer surface: the PP surface is oxidized more rapidly in the presence of atomic oxygen than with ozone. The surface oxidation products are mainly carbonyl groups with a lower level of hydroperoxide groups.

The introduction of hydroxy groups, alkenes, ketones and esters into a thin layer at the surface of the polypropylene films was performed by oxidation at room temperature of the PP film with chromium(VI) in acetic acid[17] or by plasma, and by radiation-induced polymerization of acrylic acid and hydroxyethyl methacrylate.[34]

The effects of exposing the PP substrates to fluorocarbon plasmas[35] can be summarized as follows: (i) CF_3Cl and CF_3Br plasmas cause chlorination and bromination, respectively, of the surface, and fluorination to a minor extent; the net tendency to graft onto the PP surface appears to follow the order $Cl > Br > CF_3$; there is no evidence of plasma polymerization; (ii) CF_3H undergoes polymerization in a plasma, forming a film which contains fluorinated functionalities; and (iii) CF_4 fluorinates the PP surface directly without deposition of a thin, plasma-polymerized film. CF_4 plasma reactions are dominated by F atoms.

ESCA and contact angle measurement were used for the analysis of the reaction products, and indicated that chemical modification of the PP does not extend beyond a depth of ca. 70 Å.

7.2.1.2 Bulk functionalization

(i) Polyethylene

Polyethylene functionalization can be achieved by insertion of halogen atoms in the polymer backbone and also by chlorosulfonation and chlorocarboxylation.

Chlorination in solution tends to be uniform[36] and in this situation crystallization of the polymer may be prevented by a chlorine content of less than 30%. In the chlorination of PE[36], a chlorine atom takes the place of a hydrogen atom on the polymer chain (equation 1).

$$\{\!\sim\!\} + Cl_2 \longrightarrow \{\!\sim\!\}_{Cl} + HCl \text{ (gas)} \qquad (1)$$

Usually a free radical mechanism is operating, either initiated by UV light or various initiators.

The insertion of chlorine atoms destroys the polyethylene crystallinity because chlorination starts at the amorphous areas and proceeds by melting of the crystalline surface and subsequent exposure of new areas for chlorination.

In a series of papers published in 1986, Joshi and Natu[37-39] presented a complete study on the PE–chlorocarboxylation reaction. With chlorine and maleic anhydride (MA) a simultaneous two-step process produced chlorocarboxylated PE (CCPE) (Schemes 3 and 4).

The chlorine radicals abstract hydrogen from PE, generating the macroradical (**1**): this can react with chlorine or MA. The reaction of (**1**) with chlorine produces chlorinated PE and a chlorine radical is generated.

The reaction with MA has two different paths: (i) generation of an MA radical; and (ii) generation of a new macroradical (**2**), which does not react with further MA molecules. The formation of MA radicals can be minimized by appropriate reaction conditions.

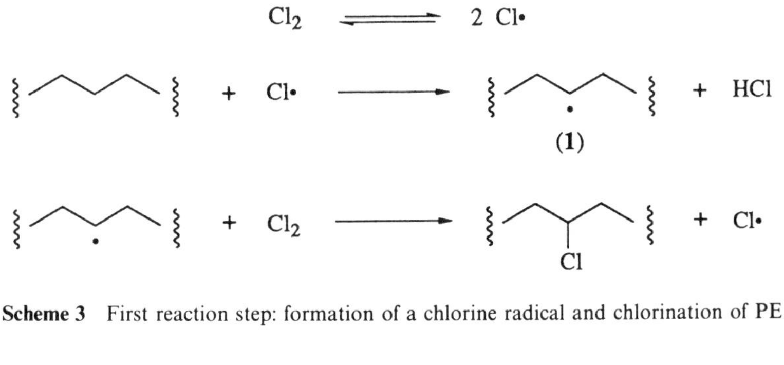

$$Cl_2 \rightleftharpoons 2 Cl\cdot$$

$$\{\!\sim\!\} + Cl\cdot \longrightarrow \{\!\sim\!\}_{\cdot} + HCl$$
$$(\mathbf{1})$$

$$\{\!\sim\!\}_{\cdot} + Cl_2 \longrightarrow \{\!\sim\!\}_{Cl} + Cl\cdot$$

Scheme 3 First reaction step: formation of a chlorine radical and chlorination of PE

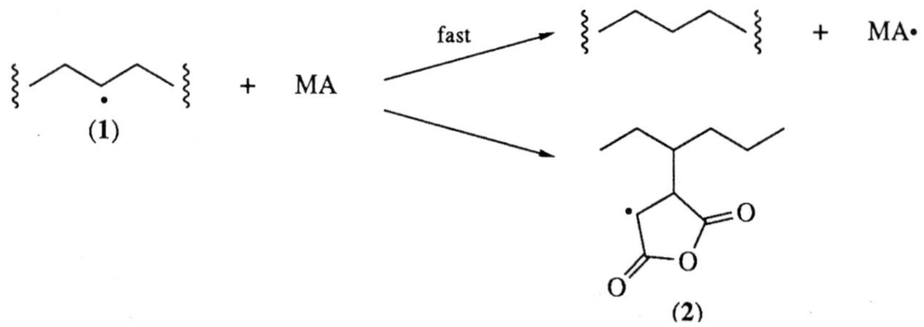

Scheme 4 Second reaction step: reaction of MA with PE

The macroradical (2) can follow different reaction paths (Schemes 5–8).

The carboxylated CCPE (7) was obtained by hydrolyzing CCPE in benzene with mild alkali.

The fluorination of PE has been widely studied. Using various mixtures of F_2 and O_2, Adcock, Inore and Lagow[40] obtained poly(tetrafluoroethylene) with varying degrees of functionalization (5–60%). This 'oxyfluorination' process can be used with different hydrocarbon polymers. Acid fluoride groups are produced primarily by oxidation of pendant Me groups or other alkyl groups in LDPE or in PP rather than from cleavage of C—C bonds (Scheme 9).

HDPE and LLDPE were functionalized by using elementary fluorine.[41] The depth of fluorination (6 μm) was measured by scanning electron microscopy.

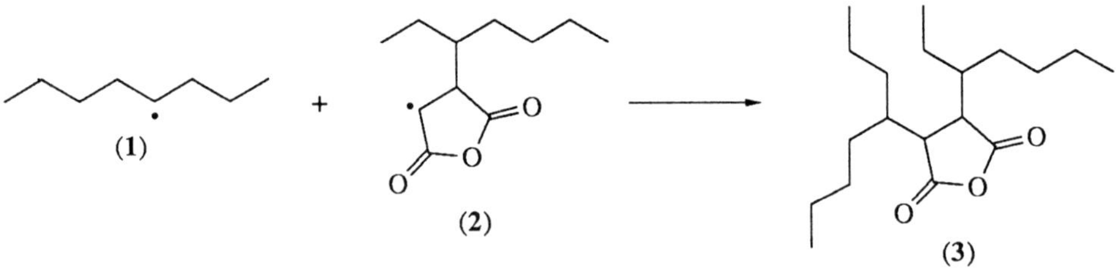

Scheme 5 Coupling with PE macroradical (1) to give crosslinked CCPE (3) or coupling by itself to give expanded crosslinked CCPE (4)

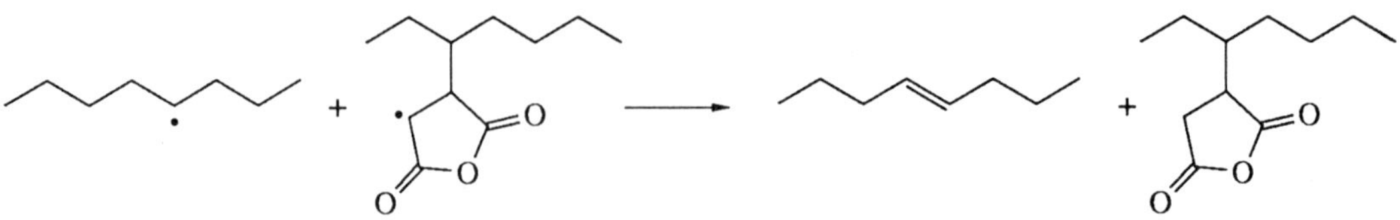

Scheme 6 Disproportionation by reaction with macroradical (1) giving an unsaturated bond over PE chain and MA-functionalized PE

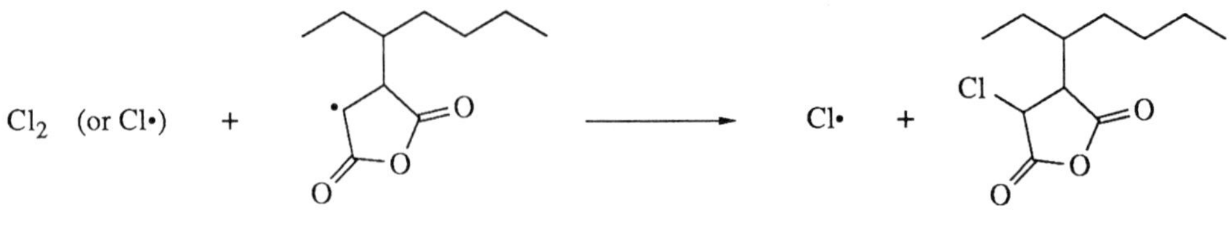

Scheme 7 Reaction with chlorine radical or chlorine molecules, reducing the probability of forming graft copolymer on the PE chains

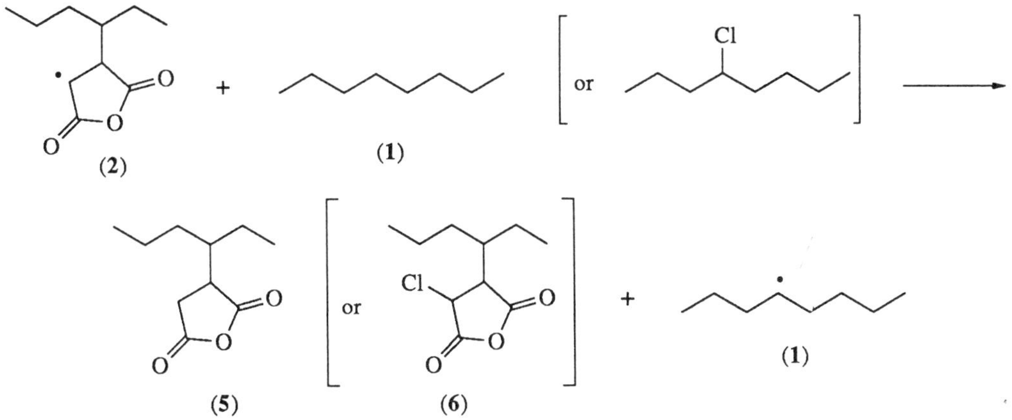

Scheme 8 Reaction with PE or chlorinated PE segments by transfer of hydrogen or Cl atom to CCPE, generating MA-functionalized PE (**5**) or chlorinated MA-functionalized PE and the PE macroradical

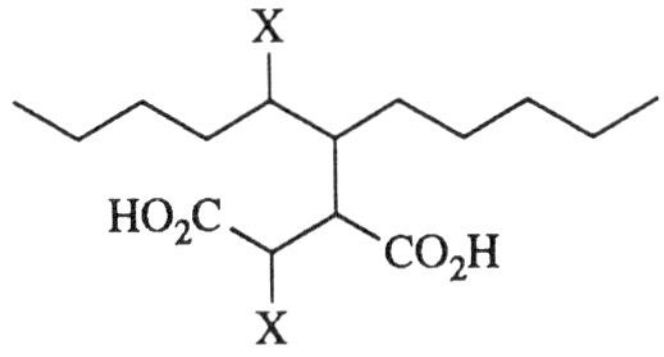

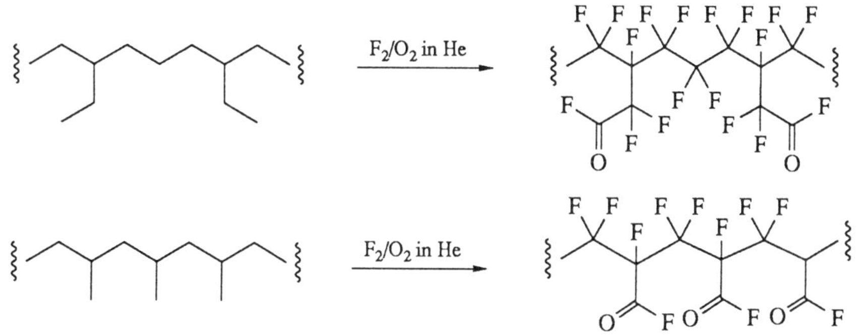

Scheme 9 Oxyfluorination process over LDPE and PP

Chlorosulfonated PE (CSPE) forms the basis of a large family of elastomers: these can be prepared by simultaneous insertion of a chlorine atom and a sulfonic group with a free radical initiator or light. The polymer can be prepared only in solution and the solvents must dissolve both PE and CSPE and be inert towards the chlorinating agent.[42]

(ii) Polypropylene

Chlorinated polypropylene (chlorine content 30% by weight)[43] with $M_w \approx 15\,000$ can be grafted with radical-reactive unsaturated alkoxysilanes to give polymers with good adhesion and cured film properties.

The effect of gas phase photochlorination on the wettability of PP was studied by Hartig and Hüttinger.[44] Chlorination with Cl_2 and UV causes an increase only in the surface tension dispersion. Different treatment in aqueous solution causes an increase in the polar contribution. Heat treatment diminishes the wettability by both liquids to a small extent: all these phenomena are explained by conformational changes of the polymer chains at the surface.

7.2.1.3 *Attachment of low molecular weight compounds to a polyalkene backbone*

(i) Polyethylene

By using a range of different electrophiles (E^+), various end groups can be introduced at the terminus of the living oligomer (**8**) produced by anionic oligomerization of ethylene using *n*-butyllithium as initiator (Scheme 10).[45,46]

For example, ClPPh$_2$ or CO$_2$, followed by acid hydrolysis, can be used to prepare polyethylene diphenylphosphine or a carboxylated ethylene oligomer. These functionalized polyethylenes, with M_w in the range 1000–3000, can be easily transformed into alternative, soluble, recyclable polymer-bound phase transfer catalysts.[45]

Functionalized polyethylene (0.4–5.6 functional groups for 100 methylenes) was obtained by Schlecht and Pearce[47,48] by radical-promoted addition of hexafluoroacetone to LLDPE in the presence of Bu$_2^t$O$_2$ in benzene. ^{19}F NMR showed that 80% of the pendant groups introduced were fluoroalkanol groups and 20% were fluoroalkyl ether groups.

Recently[49] it has been shown that moderate functionalization (3–5 mol %) of different polyethylenes (HDPE and LLDPE) with polar carboxylate groups can be simply performed in bulk, at 200 °C, by using diethyl maleate and dicumyl peroxide as initiator. In these conditions, as shown in Scheme 1, it is possible to reduce markedly or eliminate side reactions, in particular degradation and crosslinking.

The unequivocal determination of the structure of the functional groups introduced into polyethylenes[24] has been performed by synthesizing two low molecular weight model compounds of the functionalized polymer, ethyl 3-ethoxycarbonyl-4-*n*-propylheptanoate and ethyl 3-ethoxycarbonyl-4-*n*-hexyldecanoate, and comparing their ^{13}C NMR spectra with analogous spectra of functionalized PE.

A different method for the functionalization of polyalkenes, based on carbene insertion into C–H bonds has also been described.[50,51]

Functionalization of molten LLDPE with 2-(dimethylamino)ethyl methacrylate (DMAEMA) was studied in an intermeshing, corotating twin-screw extruder using a peroxide initiator.[52] The same authors have reported[53,54] the preparation of a tertiary amine functionalized polymer by grafting DMAEMA onto the molten linear LLDPE in a batch-type reactor (Scheme 11).

The functionalization of LLDPE with DMAEMA involves three competing reactions: (i) the desired grafting of DMAEMA onto LLDPE backbones; (ii) homopolymerization of DMAEMA; and (iii) crosslinking of ungrafted and/or grafted LLDPE.

In this process a significant predominance of homopolymerization over the desired grafting takes place in the twin-screw compared to the batch mixer. One possible reason could be the insolubility of the monomer/initiator mixture in the melt, thus preventing intimate contact with the polymer.

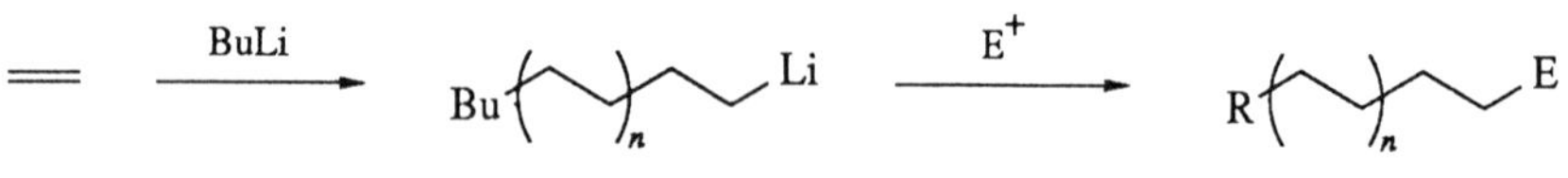

Scheme 10 Anionic oligomerization of ethylene

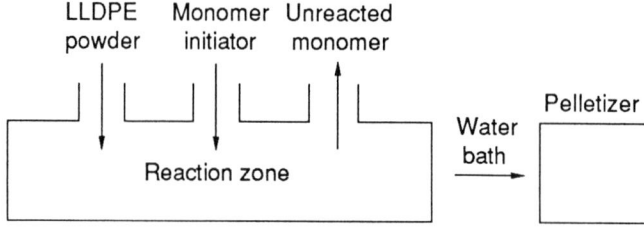

Scheme 11 Design of processing apparatus

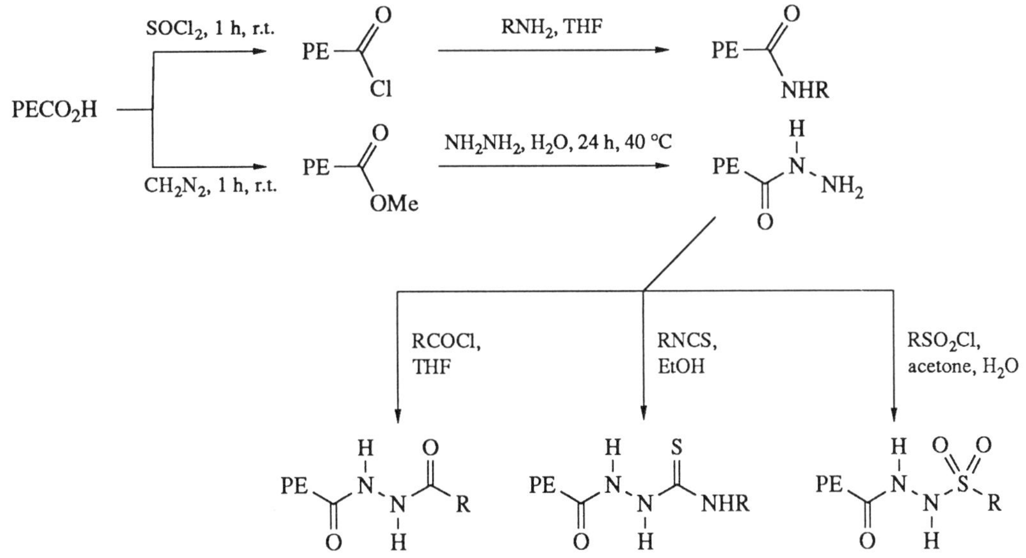

Scheme 12 Chemical modification of PECO$_2$H

Samples of LLDPE film, modified by introducing carbonyl hydrazide groups (PECONHNH$_2$) on the surface, can react with 4-thiocyanato-2,2,6,6-tetramethylpiperidin-1-oxyl and 5-dimethylamino-1-naphthalenesulfonyl chloride to yield spin-labelled (PE–CONHNHCSNH–TMPO) and fluorescence-labelled (PE–CONHNH–DANS) films, respectively (Scheme 12).

Thus PECO$_2$H and its derivatives appear to be suitable materials for use as the basis for detailed studies of the relationships between polyethylene surface composition and structure, and polyethylene surface properties.[55]

Maleic anhydride (MA) moieties are grafted onto molten LLDPE at 140–180 °C in the presence of radical precursors,[15] and onto LDPE at 160 °C in 1,2-dichlorobenzene solution in the presence of 2,5-di(t-butylperoxy)-2,5-dimethyl-3-hexyne.[56] The crosslinking of LLDPE resulting from reaction with DCP at 180 °C increased in the presence of MA.

The presence of electron-donating, nitrogen-containing, phosphorus-containing and sulfur-containing compounds which inhibit the homopolymerization of MA, but not that of methyl methacrylate, prevents crosslinking and yields soluble PE containing MA units.[15]

The reaction of different polyalkenes such as PP (atactic and isotactic), EPR and also HDPE with diazo esters at 210 °C in bulk has been shown to give a product containing the carboxylate functionality attached to the polymer backbone (Scheme 13).[50] The reaction temperature allowed the polymer to remain in the molten state and enabled complete decomposition of the diazo compounds, with formation of carbenic species, :CH—CO$_2$Et and :CH—CO$_2$C$_2$H$_4$Cl, respectively. By this method 0.4–1 mol% functionalized polyethylenes were obtained and their molecular structure was determined by ^{13}C NMR.[51]

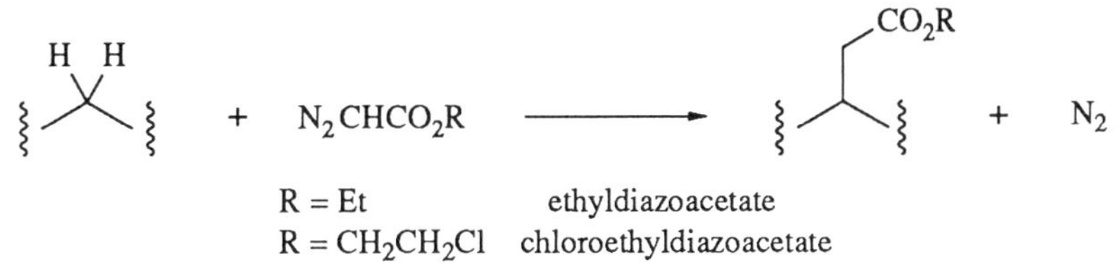

R = Et   ethyldiazoacetate
R = CH$_2$CH$_2$Cl   chloroethyldiazoacetate

Scheme 13 Insertion of carbene into a C—H bond

(ii) Polypropylene

Terminally functionalized polypropylene having vinyl, phenyl or hydroxy groups at the chain end was prepared by adding common monomers such as butadiene, styrene and 1,2-epoxypropane during the living coordination polymerization of propylene with a soluble V(acac)$_3$AlEt$_2$Cl catalyst

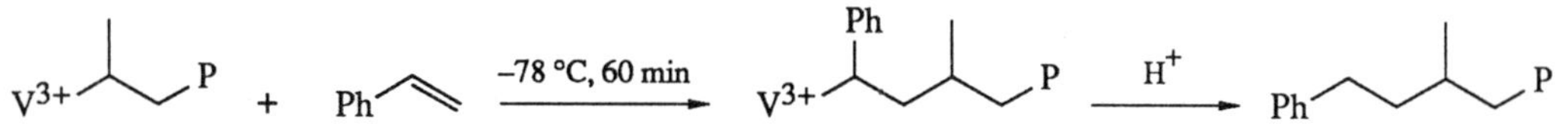

Scheme 14 Preparation of PP with phenyl end groups[57]

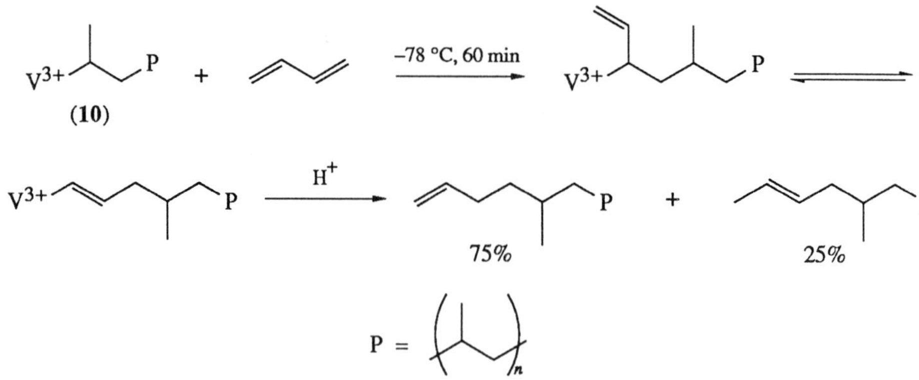

Scheme 15 Preparation of PP with vinyl end groups[57]

(Schemes 14 and 15).[57] These new types of terminally functionalized polypropylenes were well characterized by ^{1}H NMR analysis.

PP with hydroxy functions was prepared by an alternative method based on the hydrogenation of PP containing aldehyde functions with LiAlH$_4$.[57]

Propylene polymers have been carboxylated by treatment with azidosulfonylbenzoic acid at $T \geqslant 180\,°C$ in the absence of a radical initiator: the resulting polymer presents improved viscosity and is useful in polymer blends.[58]

Functionalization experiments on isotactic polypropylene (IPP) and on atactic polypropylene (APP) with α-methylstyrene, diethyl fumarate and diethyl maleate have been carried out in solution or in bulk in the presence of dicumyl peroxide as initiator.[18] The extent of functionalization, determined by IR spectroscopy, depends on the microstructure of the macromolecules being 0.5–1% by weight for IPP and 3–8% by weight for APP.

Atactic polypropylene was functionalized by reaction with dicumyl peroxide and eugenol methyl ether, allyl phenyl ether, allylurea, N-allylurea, diethyl allylmalonate and N-allylamine at 170 °C.[59]

The reaction rate in functionalization[60] of polypropylene melts with maleic anhydride had a maximum value during the first few minutes and did not change significantly afterwards. Functionalization proceeded best at low concentrations of MA and peroxide since side reactions took place when the reaction time was greater than 20 min.

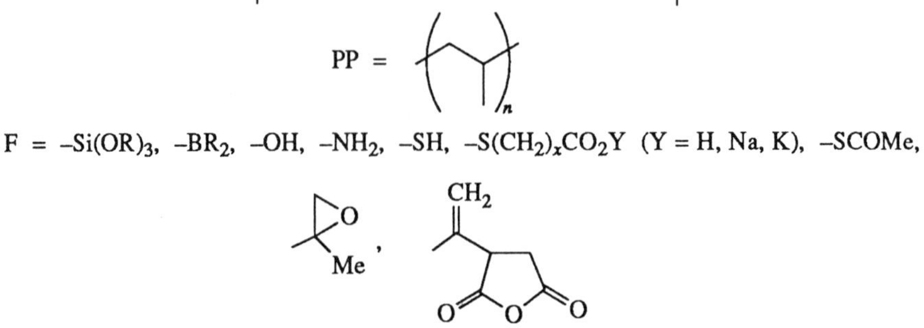

$F = -Si(OR)_3, -BR_2, -OH, -NH_2, -SH, -S(CH_2)_xCO_2Y$ (Y = H, Na, K), $-SCOMe$,

Figure 1 Polypropylenes with polar end groups

Monofunctional polypropylenes containing a variety of end groups (*e.g.* anhydrides, esters, amines, carboxylic acids, silanes, boranes, alcohols and thiols) were obtained from monoalkene-terminated polypropylenes. These alkenic polypropylene intermediates have been prepared with soluble metallocene catalysts, which give excellent control of polypropylene molecular weight, molecular weight distribution, microstructure and alkenic end groups (Figure 1).[61]

On the basis of the alkene-terminated polypropylene feedstocks, highly diversified polymeric systems can be obtained, including block, branched and star-shaped propylene polymers. When containing thiol end groups, polypropylene can be used as a chain transfer agent in free radical polymerizations of monomers such as styrene, acrylic acid or acrylonitrile, which grow onto the polypropylene chain end to form propylene block copolymers.[61]

(iii) Ethylene–propylene copolymers

Functionalization of an amorphous ethylene–propylene copolymer (EPR) has been performed by grafting reactions of dibutyl maleate, diethyl fumarate and maleic anhydride initiated by free radicals.[62] Different reactivities of the unsaturated molecules have been observed.

The functionalization of saturated ethylene–propylene rubber has been performed by radical grafting of 2-(diethylamino)ethyl methacrylate onto EPR or by reaction of a succinic anhydride grafted EPR with N,N'-dimethylethylenediamine.[63] For the insertion of 2-(dimethylamino)ethyl methacrylate (DAEM) onto the EPR backbone the reaction mechanism in Scheme 16 was proposed for the DCP free radical promoted grafting.

The polymeric product recovered from the reaction mixture contains DAEM molecules along the backbone; in fact, in the IR spectrum new absorptions are present, at $1730\,cm^{-1}$ due to the carbonyl stretching band of the ester group and at $1170\,cm^{-1}$ due to the C—N stretching band of the dimethylamino group.

The authors suggest that short poly(DAEM) side chains may arise by successive additions of unsaturated monomer units M to PM· macroradicals and by coupling of growing poly(DAEM) oligoradicals with P· and PM· macroradicals.

The increase of viscosity observed for EPR—*g*–DAEM products can be explained by considering that chain propagation and termination reactions may compete with chain scission reactions on EPR, such as β-scission reactions of P· and PM· macroradicals.

Evaluation of thermal and mechanical properties[64] indicated that the insertion of succinic anhydride, 2-(diethylamino)ethylsuccinimide and 2-(diethylamino)ethyl methacrylate deeply influenced the elastic response of the rubber.

A drastic improvement in mechanical properties has been found in the presence of a small number of crosslinks, which eliminate flow processes. Preliminary data show that the presence of strong ionic linkages between EPR chains exert an effect on the mechanical response of the materials which follows patterns different from those found in the case of weak, polar interactions. A semiquantitative correlation between the decrease of the mechanical properties at a high grafting degree and degradation of the polyalkene backbone is given.[65]

Amorphous EPR[14] has been functionalized by utilizing a bulk mixing process at 160 °C. EPR was first mixed at 90 °C with a solution of DBM and DCP, then the temperature of the mixing chamber was allowed to rise rapidly up to 160 °C. The primary radical R· derived from DCP thermal decomposition extracts a hydrogen atom from the polymer backbone and these highly reactive radical polymer species P· may add to activated unsaturated molecules, such as DBM, forming PM· macroradicals.

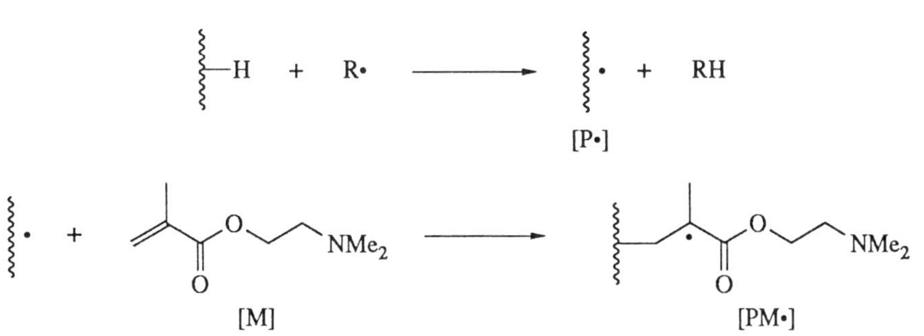

Scheme 16 DAEM functionalization of EPR

The amount of grafted groups can be controlled by process parameters such as reaction time, temperature and composition of the reaction mixture, but it is impossible to avoid a partial chain degradation.

The premixing step at lower temperatures is useful for controlling the subsequent reaction in terms of the degree of grafting and chain degradation.[66]

Another parameter[67] which plays a determining role in the reaction kinetics is the initial reaction mixture composition. The initial reaction rate and the grafting process efficiency were increased by enhancing the DCP initial concentration in the reaction mixture. Different degrees of grafting have been obtained by changing the initial composition and it was possible to determine the influence of the degree of grafting on the structural and superreticular order. It was found[68] that functionalization leads to a decrease in the residual crystallinity present in the parent copolymer. Linear relationships between the grafting degree and the crystallinity degree X_c, evaluated by both DSC and WAXS, have been obtained. The results suggest that the grafting occurs preferentially onto the longer or less-substituted methylene sequence.

The extent of crosslinking and/or degradation which accompanies the EPR–MA reaction can be studied by viscosity measurements and MA content determination.[69]

The crosslinking of EPR, in the presence of radicals from peroxide decomposition, is attributable to the attack on the secondary CH_2 moieties and the generation of macroradicals which couple either with similar macroradicals or macroradicals generated by attack on the tertiary CH moieties in the polypropylene units of the chain. Increasing the DCP concentration results in an increase in the amount of soluble polymer and a decrease in the molecular weight of the polymer due to increased tertiary CH attack, followed by disproportionation.[69]

7.3 FREE RADICAL GRAFTING

Among the synthetic methods available for modifying the properties of polyalkenes, in general, and polypropylene specifically, free radical graft polymerization is by far the most widely known and most frequently studied approach. The ability to create radical sites along the polymer backbone is an essential condition for free radical grafting chemistry. Polyethylene and ethylene–propylene rubber have also been successfully modified *via* free radical techniques.

Numerous articles have been published on the subject of free radical grafting to polyalkene substrates, going back to the 1950s. The large number of patents issued related to grafting technology attests to the fact that there is significant industrial interest in this field of technology, primarily because of the versatility and potentially attractive economics of the technology. Ionizing radiation, both gamma and electron beam, and peroxide-initiated chemistry are the most common approaches, although photoinitiated graft polymerizations have been described.

7.3.1 Radiation Grafting Techniques

7.3.1.1 Radiation peroxidation

Chapiro was one of the first to describe the effects of high energy radiation exposure of polypropylene and polyethylene in the presence of air.[70,71] The sequence of reactions can be written as shown in Scheme 17.[72]

$$\text{Initiation:} \quad PH \longrightarrow P\cdot + H\cdot \quad (1)$$

$$\text{Propagation:} \quad P\cdot + O_2 \longrightarrow PO_2^\cdot \quad (2)$$

$$PO_2^\cdot + PH \longrightarrow PO_2H + P\cdot \quad (3)$$

$$\text{Termination:} \quad PO_2^\cdot + P\cdot \longrightarrow PO_2P \quad (4)$$

$$PO_2^\cdot + PO_2^\cdot \longrightarrow PO_2P + O_2 \quad (5)$$

Scheme 17 Hydroperoxide and diperoxide formation

$$\text{POOP} \xrightarrow{} 2\,\text{PO}\cdot \xrightarrow{\text{monomer}} \text{POM} \qquad (6)$$

$$\text{POOH} \xrightarrow{} \text{PO}\cdot + \text{HO}\cdot \xrightarrow{\text{monomer}} \text{POM}_n + \text{HOM}_n \qquad (7)$$

Scheme 18 Initiation sites for graft polymerization

$$\text{POOH} + \text{Fe}^{2+} \longrightarrow \text{PO}\cdot + \text{OH}^- + \text{Fe}^{3+} \qquad (8)$$

$$\text{PO}\cdot + n\text{M} \longrightarrow \text{POM}_n^\cdot \qquad (9)$$

$$\text{PO}\cdot + \text{Fe}^{2+} \longrightarrow \text{PO}^- + \text{Fe}^{3+} \qquad (10)$$

Scheme 19 Peroxide species decomposition by reducing agents

Hydroperoxide formation results from the propagation reactions of steps (2) and (3), while diperoxide structures result from termination reactions (4) and (5). The labile tertiary hydrogen in polypropylene makes it particularly susceptible to chain-propagated oxidation, leading to a predominance of hydroperoxide functionalities. Polyethylene irradiation leads predominantly to diperoxide crosslinks.[71,73] For both polymers, the relative balance of hydroperoxide and diperoxide is temperature dependent, a higher temperature favouring hydroperoxide formation.[72]

The breakdown of the hydroperoxide and diperoxide links upon heating (in the absence of oxygen) produces alkoxy free radical initiation sites for subsequent graft polymerization, as shown in steps (6) and (7), Scheme 18.

Hydroperoxide decomposition will lead to both graft copolymer and homopolymer, whereas diperoxide decomposition yields only graft chains. Elevated temperatures are generally required to achieve any appreciable rate of thermal decomposition of the peroxide and hydroperoxide intermediates. The temperature range reported is typically from 60 to 120 °C. Hence the polyalkene is generally contacted at an elevated temperature below its melting point.

Homopolymer production is generally considered to be an undesirable side effect, formed as a by-product of reaction sequence (7). A reducing agent, such as iron(II) sulfate ($FeSO_4 \cdot 7H_2O$) or iron(II) acetylacetonate ($Fe(acac)_2$), is often used to decompose the peroxy species and inactivate the HO· radical,[74-76] thereby minimizing the production of homopolymer chains, as described by steps (8) and (9) in Scheme 19.

An optimal level of reducing agent exists beyond which a retarding effect is shown,[74] due to a competing radical scavenging reaction step (10 in Scheme 19), which effectively reduces the level of active PO· initiation sites available for graft polymerization.

The presence of a redox agent also serves to reduce the temperature required for appreciable decomposition of the peroxide and hydroperoxide intermediates. Room temperature graft polymerization is therefore possible.

Because the graft polymerization is generally conducted in the solid state, the accessibility of the active sites and the ability of the monomer to diffuse into the polymer matrix play an important role. Misra et al.[77] have observed enhancements in the extent of graft polymerization in the presence of certain polar solvents. The magnitude of this enhancement is dependent upon solvent polarity and concentration. They attribute this effect to swelling of the solid polymer which facilitates accessibility and diffusion of monomer to the active sites. Very high solvent concentrations, however, reduce the extent of grafting, presumably due to chain transfer to solvent causing termination of the growing chains.

The effect of dose rate and total exposure dose on the rate of grafting of acrylic acid to radiation-peroxidized polypropylene films has been described.[74] Initial polymerization rates exhibit a square root dependence on the radiation dose but is independent of the dose rate over a range of 1.6–8.0 Mrad s^{-1}.

7.3.1.2 Simultaneous (direct) radiation grafting

An alternative technique for radiation-initiated grafting to polyalkene substrates is the so-called simultaneous or direct method.[78-80] In contrast to the peroxidation method in which high energy radiation serves as an indirect initiation source *via* the *in situ* formation of peroxide and hydroperoxide intermediates, the direct method involves simultaneous exposure of both the polymer

Initiation: $\quad PH \longrightarrow P\cdot + H\cdot \quad$ (11)

Propagation: $\quad P\cdot + nM \longrightarrow PM_n^\cdot \quad$ (12)

$\quad PM_n^\cdot + PH \longrightarrow PM_nH + P\cdot \quad$ (13)

Homopolymerization: $\quad M \longrightarrow M\cdot \quad$ (14)

$\quad M\cdot + nM \longrightarrow M_{n+1}^\cdot \quad$ (15)

Scheme 20 Radiation-initiated graft polymerization

substrate and the monomer to be grafted in an atmosphere substantially free of oxygen. Direct irradiation is a nonselective process in that all components are exposed and respond independently to the radiation. Hence the chemistry is more complex.

Initiation of polymerization occurs directly through carbon-centred radical sites (Scheme 20) and homopolymer formation is a more significant factor in this case. Levels of free homopolymer can represent 50% or more of the total amount of polymer produced.[81] Homopolymer can result from direct interaction of radiation with monomer.

The presence of solvents and other additives can greatly affect the grafting yield and the amount of homopolymer formed.

Ang et al.[82] observed increases in the grafting efficiency of styrene to polyethylene and polypropylene film substrates with the addition of mineral acids and polyfunctional comonomers. Mineral acid increases both the grafting yield and the yield of homopolymer, the former preferentially, so that graft efficiency is effectively increased. The effects are generally attributed to greater radiolytic yields of radical intermediates, which create more initiation sites for grafting as well as homopolymerization of the monomer. Alternatively, the presence of inhibitors, such as p-t-butylcatechol, in low concentrations can effectively suppress homopolymer production without a corresponding reduction in graft yield, thereby effectively increasing graft efficiency.[83]

Odian et al.[84] have observed enhancements in the rate of graft polymerization to polypropylene and polyethylene in the presence of methanol. They attribute this enhancement of rate to a Trommsdorff-type effect due to the insolubility of the growing chains (in this case polystyrene) in methanol. An increase in the viscosity of the polymer matrix causes a reduction of termination rates, thereby producing an autoacceleration of the polymerization rate.[85] Methanol also effectively suppresses homopolymer production in the solution without a corresponding suppression of polymerization within the polymer matrix.[81]

The use of monomer in vapour form instead of liquid or solution can also help to minimize the generation of homopolymer by effectively reducing the concentration of free monomer not present in the polymer matrix.[86,87] Furuhashi et al.[87] investigated simultaneous radiation grafting of butadiene to polyethylene and polypropylene films. The studies involved exposure to both liquid and gaseous butadiene. The amount of occluded polybutadiene homopolymer formed was significantly higher for liquid butadiene graft polymerization. At a reaction temperature of 50 °C, grafting rates to high density polyethylene were in the range of 0.5–2.0% h^{-1} over a dose rate range of 0.2–2.0 Mrad h^{-1}, the rate increasing with dose rate. Grafting rates in the core of low density polyethylene and polypropylene were consistently lower.

Oxygen is a strong inhibitor of vinyl polymerization and is in most studies excluded from the reaction mixture. In certain circumstances, however, it is possible to produce graft polymerization in the presence of air. Pinkerton et al.[88,89] were able to polymerize methyl methacrylate onto polypropylene films via simultaneous grafting techniques with a methanol solution of the monomer open to air, provided the solution contained water in excess of 10 wt%. At concentrations below 10%, polymerization was effectively prevented. This effect was attributed to a reduction of oxygen solubility in the grafting solution upon the addition of water. The solubility of oxygen in water is approximately an order of magnitude lower than in methanol or methyl methacrylate.

The rate of monomer grafting and the extent of grafting are functions of the dose rate and total dose. Increasing dose rate increases the rate of grafting, the dependence being proportional to dose rate to the power 0.3 to 0.9 for various systems. The extent of grafting and graft efficiency, however, decline with increasing dose rates. For a fixed dose rate, increasing the total dose (and hence polymerization time) results in an approximately linear increase in percent graft. Graft efficiency generally declines with increasing dose.[90]

The duration of exposure of the polymer substrate to the monomer or solution following irradiation also influences the extent of grafting. In general, the time dependence is linear.[80] Temperature effects tend to be more complex due to the competing effects of enhanced polymerization rates and enhanced radical termination rates. While initial rates are enhanced due to higher temperature, the total amount polymerized is lower, with a higher temperature during the postirradiation stage.[79] This is due to an increase in termination rates. Since the total radical population is fixed by the radiation dose, radicals terminated during the postirradiation step are not replaced and the number of sites available for initiation decreases.

The effect of irradiation temperature on grafting yields also shows a complex dependence. Furuhashi et al.[86] observed an increase in the grafting rate of butadiene (in liquid and gaseous form) to polyethylene films with temperature. Maximum rates were observed at 50–60 °C, further increases in temperature causing a reduction in polymerization rates. This behaviour was attributed to the competing effects of accelerated propagation reactions, which increase the rate, and enhanced chain termination reactions, which serve to decrease the rate.

7.3.1.3 Trapped radical techniques

An alternative technique for radiation grafting to semicrystalline polyalkene substrates is the so-called trapped radical technique. Exposure of polyalkene substrates to ionizing radiation in the absence of oxygen produces polymer-bound carbon-centred free radicals. When the irradiation is conducted at room temperature or below, a significant portion of the radical population is prevented from recombining due to the high viscosity of the polymer matrix. The trapped radicals are capable of initiating polymerization when exposed to monomer vapour or liquid.

Graft efficiencies are typically very high because the monomer is not exposed to the ionizing radiation and the initiation sites are part of the polymer backbone structure. Higher dose exposures are typically required and the extent of grafting is dependent upon the backbone polymer's ability to trap radical intermediates. The degree of crystallinity and polymer morphology can strongly influence the polymer's radical-trapping ability and hence the extent of monomer grafting.[91]

Sundardi[92] investigated the graft polymerization of vinyl pyrrolidone and acrylic acid to polypropylene fibres *via* a trapped radical technique. The extent of grafting increased in direct proportion to the log of irradiation dose and reaction time, consistent with kinetic predictions. An optimum reaction temperature of approximately 70 °C was observed, resulting from the competing effects of temperature on chain propagation and radical termination reactions.

Trapped radical grafting technology, applied to both polyethylene and polypropylene, has been the subject of a number of patents.[93–96]

7.3.2 Chemically Initiated Graft Polymerization

Chemically initiated graft polymerization has been the subject of extensive investigation and is the topic of many technical papers and patents. Several approaches to chemical initiation have been explored, including hydroperoxidation, and chain transfer graft polymerization with polyalkenes in solid, melt and solution forms.

7.3.2.1 Hydroperoxidation

Chemically initiated hydroperoxidation grafting is similar in many ways to radiation-initiated peroxidation. Both involve the formation of hydroperoxide and diperoxide groups along the backbone chain with subsequent decomposition of these groups to produce initiation sites for grafting. They differ in the method by which the backbone groups are created and also the distribution of these groups within the polymer substrate.

Several methods for introducing hydroperoxide functionality into polyalkene precursor polymers have been described in the technical literature and patents. Decomposition of organic or inorganic peroxide in the presence of the polyalkene substrate, in a finely divided form, with oxygen sparging has been described. This approach is in many ways analogous to radiation hydroperoxidation in that backbone polymer radicals are generated, which react with oxygen to form peroxy radical intermediates. The intermediates then undergo chain-propagating reactions as described in reaction sequences (2) to (5) of Scheme 18.

Ozonolysis of the precursor polymer has been shown to be a particularly effective means of introducing hydroperoxide structures into the backbone. Addition of a peroxide agent is not required in this case since ozone is sufficiently reactive to undergo abstraction reactions with the polymer backbone.

Jabloner and Mumma[97] found potassium persulfate in combination with oxygen sparging to be particularly effective in creating polypropylene hydroperoxide in an aqueous, surfactant-stabilized slurry of powdered polymer. Reaction times of approximately one hour at 100 °C were sufficient for hydroperoxide formation. Subsequent addition of a metal redox system in the presence of vinyl monomers provided a good yield of grafted polypropylene. This indicates that oxidation of the polypropylene preferentially occurs at the surface of the suspended particles, leading to a heterogeneous graft product in which much of the backbone polymer chains are not grafted.

The use of oxygen or an oxygen/ozone mixture to peroxidize polyalkene powders has been described by a number of researchers.[98,99] Citovicky et al.[100] in their investigation of grafting reactions on polypropylene powder observed a linear dependence of both hydroperoxide and diperoxide concentration with exposure time to an oxygen/ozone mixture. The reaction was conducted at room temperature. The concentration of hydroperoxide was approximately twice that of diperoxide. Redox graft polymerization of styrene in an aqueous suspension produced good yields of grafted copolymer.

Oxygen or oxygen/ozone hydroperoxidation and subsequent graft polymerization with redox agents has been the subject of a number of patents.[101,102] Powder slurries in both organic[103] and aqueous[104] dispersants as well as fluidized bed[105] hydroperoxidation conditions have been described.

7.3.2.2 Chain transfer polymerization

Decomposition of organic peroxides in the presence of polyalkene polymers produces polymer backbone radicals *via* a chain transfer reaction as schematically described in Scheme 21.

The polymeric radicals are capable of initiating graft polymerization of vinyl monomers and the termination can occur *via* recombination of radical end groups or chain transfer to polymer backbone, monomer, solvent, *etc.* Homopolymer is produced as well by direct addition of the alkoxy radical to the monomer.

$$\text{Initiation:} \quad ROOR \longrightarrow 2\,RO\bullet \quad (16)$$

$$RO\bullet + PH \longrightarrow P\bullet + ROH \quad (17)$$

$$\text{Propagation:} \quad P\bullet + nM \longrightarrow PM_n^\bullet \quad (18)$$

$$RO\bullet + M \longrightarrow ROM\bullet \quad (19)$$

$$ROM\bullet + nM \longrightarrow ROM_{n+1}^\bullet \quad (20)$$

Scheme 21 Chain transfer polymerization mechanism

(i) Solid state precursors

Peroxide-initiated graft polymerization *via* chain transfer to polyalkene solids below their melting point has been the subject of numerous patents.[106-109] Canterino[110] describes the graft polymerization of various acrylate monomers to finely divided polypropylene powder in the absence of solvent with benzoyl peroxide. A high surface area powder is recommended due to the surface selective nature of the polymerization. The peroxide and monomers were combined and sprayed onto the polypropylene powder and allowed to polymerize at elevated temperatures. While adequate monomer conversions are achieved, graft efficiencies are generally low.

Grafting in an aqueous suspension is described in several patents. First impregnating the dispersed polyalkene with monomer at an elevated temperature below the decomposition temperature of the peroxide activator and subsequently raising the temperature to initiate polymerization provided a more even distribution of the grafting monomer and hence a more uniform product.[111,112]

The presence of an organic solvent capable of swelling the precursor polymer is advantageous for polar monomers which do not readily diffuse into the polymer matrix.[111] The effects of solvents on the graft polymerization of maleic anhydride onto polypropylene powders below the melting point of the polymer has been discussed by Lee et al.[113] The presence of a solvent, such as toluene or decalin, in quantities sufficient to swell the polymer but insufficient to cause any appreciable dissolution, yields a significant increase in grafting yields at 120 °C. Increases of the order of 50% were observed with the addition of approximately 15% solvent. In contrast, tetralin reduced the grafting efficiency under similar conditions. This was attributed to surface dissolution without swelling, which lowered the effective surface area of the polymer inhibiting the grafting process.

Homopolymer formation is a significant side reaction, generally considered to be undesirable. The extent of homopolymerization is dependent upon a number of factors, most notably initiator selection. Initiators which exhibit a strong tendency towards hydrogen abstraction are generally preferred. Peresters and peroxy carbonates have been found to be particularly effective in producing high graft efficiency in polyalkene graft polymerizations.[114,115]

(ii) Grafting in the melt

Peroxide-initiated grafting to polyalkenes in the molten state has several attractive features. The heterogeneous nature of solid phase reactions with semicrystalline polyalkenes and the tendency towards preferential grafting at the particle surface can be avoided in the melt. The use of extruders to effect melt mixing of the polymer and reactive monomer/initiator is also of obvious commercial and economic advantage.

Melt reactive extrusion grafting is, however, more limited in terms of the level of monomer which can be incorporated and the temperature range possible. The melt extrusion approach is therefore generally limited to polymer functionalization with the incorporation of up to approximately 15% monomer. Crosslinking in the case of polyethylene and ethylene–propylene rubber and excessive degradation in the case of polypropylene are also potential pitfalls of the technique. Despite these limitations, the technique has been widely used industrially and is the subject of a number of technical articles and patents.

Maleic anhydride and acrylic acid, alone or in combination with other copolymerizable monomers, have been the most extensively studied.[116–118] Polypropylene, polyethylene, and ethylene–propylene rubber functionalization has been described.[119–122]

The issues of crosslinking of polyethylene and ethylene–propylene rubber, and degradation of polypropylene during melt extrusion grafting have been investigated by Gaylord et al.[123–125] Both polyethylene and ethylene–propylene rubber will crosslink upon exposure to decomposing peroxide in the melt state. The extent of gelation is dependent upon the particular peroxide selected and the reaction temperature. The presence of maleic anhydride further increases the gel level.[123,124] This enhancement in the extent of crosslinking in the presence of maleic anhydride has been attributed to homopolymerization and coupling of grafted poly(maleic anhydride) chains. The presence of an electron donor, such as stearamide or dimethylformamide (DMF), inhibits homopolymerization of maleic anhydride and reduces or eliminates gelation of the polyethylene or EPR-modified products.

In contrast to the gelation effects of peroxide on polyethylene and EPR, polypropylene is severely degraded in the molten state in the presence of peroxide. Tertiary carbons in the chain are particularly susceptible to attack.

At elevated temperatures, disproportionation is favoured over radical coupling reactions. The extent of degradation is enhanced by the presence of both peroxide and maleic anhydride.[125]

Electron donors, such as DMF or dimethylacetamide (DMAC), effectively suppress this degradation tendency. Once again the ability of electron donors to suppress homopolymerization of the anhydride has been implicated. Certain amine coagents reportedly enhance graft efficiency of maleic anhydride, with or without styrene as a comonomer, to polypropylene in the molten state.[126]

Greco et al.[66–68] investigated melt reactions of DBM with ethylene–propylene rubber in the presence of dicumyl peroxide. A reduction in intrinsic viscosity, indicative of degradation of the EPR backbone, occurred over a temperature range of 140–200 °C. This is in contrast to the findings of Gaylord for maleic anhydride functionalization of EPR. Severe gelation was observed in the absence of dibutyl maleate. Initial reaction rates exhibited a square root dependence on peroxide concentration and the extent of grafting increased with peroxide level. An optimum grafting temperature was also observed which shifted to lower temperature with increasing reaction times.

Styrene is known to undergo thermally induced homopolymerization and copolymerization in the absence of a free radical catalyst. In the presence of polyalkene 'trunk' polymers which contain labile or active hydrogens, grafting to the trunk polymer has been reported[127,128] Hence, styrene/maleic anhydride graft copolymers have been produced *via* melt extrusion of the monomer mixture with polypropylene and polyethylene in the absence of added free radical catalysts.

The synthesis of graft copolymers of PE is described by Boutevin and Robin.[129] Grafting is achieved, using PE activation by ozone, in bulk with various acrylic monomers. Also their applications as adhesives for composites and as hot melt adhesives are described.

(iii) Grafting in solution

Peroxide-initiated grafting of vinyl monomers to polyalkene precursors in homogeneous solution has been the subject of a number of technical articles and patents. A homogeneous medium has the attractive feature of providing a more uniform distribution of grafted chains randomly attached to the precursor backbone. However, semicrystalline polymers, such as polyethylene and polypropylene, are relatively difficult to dissolve and require elevated temperatures for long periods of time. Alkene rubbers such as EPR and EPDM are significantly easier to bring into solution. Polymer concentrations are generally limited to less than 5% by weight due to the very high viscosity exhibited by the solutions. This is of obvious commercial and economic disadvantage.

Maleation of polyethylene in a homogeneous solution has been described by Porejko *et al.*[130,131] Their studies were restricted to grafting conditions in which very large excesses of maleic anhydride were present. Graft levels greater than 50% were achieved. An optimum grafting temperature was observed, attributed to the temperature sensitivities of competing chain propagation and termination reactions. Multiple maxima were observed in the grafting level as a function of polymer concentration, the first attributed to an enhancement of termination events with increasing polymer concentration. The second maximum was associated with the occurrence of a gel or Trommsdorf effect due to the high viscosity of the concentrated solutions.

Maleic anhydride grafting to EPR in xylene solution is strongly influenced by the particular initiator selected.[59] Dicumyl peroxide was found to be significantly more efficient than dibenzoyl peroxide. Under similar reaction conditions, maleic anhydride grafting consistently occurred more readily than for dibutyl maleate or diethyl fumarate. Solvent selection also influences grafting efficiency. Not surprisingly, solvents with high chain transfer constants will tend to reduce graft efficiency; hence, xylene is a poorer solvent than chlorobenzene.

Recently, grafting of butyl acrylate (BA) onto LDPE has been successfully demonstrated.[132] It is observed that it is difficult to get high percentages of BA photografting (from 5.6 to 17.6 wt%) under the given reaction conditions: 90 °C in toluene using BP as initiator. No further change in grafting percentage and grafting efficiency was observed with increase in reaction time.

Solvent grafting techniques are described in a number of patents, the primary focus being related to the technology for grafting onto EPR or EPDM.[133-137]

7.4 FUNCTIONALIZATION OF POLYPROPYLENE (PP) AND RELATED AREAS OF APPLICATION

One of the main objectives of scientific research is to find new synthetic materials which have the same performance levels as traditional materials but are cheaper to produce. Over the last 10 years this goal has been reached due to the availability of a wide range of plastics, which have been rapidly appearing on the market, and their usage is on the increase in a number of interesting ways in various sectors. Side by side with this development, however, plastics have clearly recognizable limits, particularly in the building sector, where they have proved to be unsuitable due to the fact that the end products usually undergo particularly strained conditions. To overcome this obstacle, compound systems made up of various plastics only or polymers and traditional materials have recently been studied and produced.

Up until now polypropylene (PP) has had a marginal role, the reason being that although it does present singularly interesting characteristics (such as low cost, low specific weight, low energetic content, high chemical inactivity, *etc.*), it cannot be used in compound multilayer structures due to its apolar nature; it is also impossible to use paint, glue or ink on this material. Leading firms in this field have for some time now been trying to overcome these drawbacks and have been investing heavily in vast research programmes in order to obtain functionalized polyalkenes, and in particular a functionalized PP.

According to publications and patents,[138-143] the best results have been achieved by grafting onto the polyalkene chain monomers containing polar groups (carbon monoxides, anhydrides, esters, *etc.*) through radicals. Using this method the properties of the alkenic macromolecules are modified, so that it is possible to water-wet, to dye or to paint the material; it adheres to metals and it is compatible with polymers of a different chemical nature and physical mechanical characteristics.

As already mentioned in the introduction, the functionalization of PP can be done in two ways: (i) by copolymerizing the propylene with polar vinyl monomers; and (ii) by grafting the above monomers on the preformed PP.

Using the former method block copolymers as well as graft copolymers can be obtained, whereas with the second technique only grafted copolymers are obtained.

Functionalized PP can be used in: (i) coupling agents for reinforced PP with mineral fillers; (ii) anticorrosive coatings for metal pipes for use in underground or submarine pipelines for the transport of fluids, large containers that can be used in severe environmental conditions, small and medium sized parts for household electrical equipment, garden furniture, *etc.*; (iii) adhesives for metal–plastic laminates for structural use in the car industry, household electrical goods and furnishings; (iv) tie-resin for the production of film and multilayer sheets of paper for chemical and food packaging; (v) compatibilizing agents for polymeric alloys with a PP base which have similar characteristics to technopolymers but with lower costs for the same performance level; and (vi) PP which is thermoformable, expandable and resistant to hot creep.

7.4.1 Main Applications of Functionalized PP

7.4.1.1 *Joint reinforcements*

It is a well-known fact that a compound made up of fibrous or laminate reinforcements can break; this is due either to the breakage of the reinforcement itself or to the unthreading of the matrix. Whether the former or the latter takes place depends on the form of reinforcement adopted, and this in turn depends on the relationship between the length and diameter and thickness of the laminate according to whether the reinforcement is fibrous or laminate.

A critical aspect ratio is defined as the value above or below which one or the other type of breakage occurs.

$$RF_{cr} = \frac{\sigma_r}{2Tm/r}$$

where RF_{cr} is the critical aspect ratio, σ_r is the tensile strength of the reinforcing agent and Tm/r is the shear strength of the matrix–reinforcing agent interface. The shear strength Tm/r, and consequently the critical aspect ratio, depend on the degree of adhesion between the reinforcing agent and the matrix. It is therefore obvious that the relationship of critical form depends on what kind of adhesion is used between the two phases. During the stage where the compound is prepared and transformed the fibres and lamellae undergo a mechanical action, which leads to shattering more or less under pressure. This in turn brings about a reduction in their form ratio. Bearing this in

Table 1 Mechanical Characteristics of PP Reinforced with 30% Fibreglass Containing Functionalized PP

Modified PP	Weight %	Force of breakage traction (MPA) ASTM D638	Hot 1820 KPA (°C) ASTM D648	Creep during flexion at 120°C, for 8h with charge of 10 MPA (% of deformation)
—	—	60.3	105	2.00
With 3% of maleic anhydride	0.5	79.9	143	0.83
With 3% of IPM[a]	0.5	81.6	143	0.94
With 2% of acrylic acid	4.0	77.3	143	0.97

[a] IPM = *N,N'*-isoforon–bismaleamic acid.

mind, it is clear that the adhesion between the two stages is of the utmost importance.[144] Since PP does not adhere to inorganic reinforcements, due to the fact that it is apolar, it must be grafted into the monomer's hydrocarbon chain with polar groups. The use of functionalized PP is well known with maleic anhydride, bismaleamic acids and acrylic acids as combining agents for the preparation of reinforced PP fibreglass.[145]

Table 1 shows the improvement of the mechanical properties of PP when reinforced with 30% of fibreglass brought about by the presence of small quantities of functionalized PP.

7.4.1.2 Anticorrosive coatings

Functionalized polyalkenes are also used in anticorrosive coatings. For metallic pipes and the forging of large containers coatings are produced by extrusion, whereas fluid bed technology is adopted for the coatings of small and medium-sized spare parts for household electrical goods, garden furniture, *etc.*

Over the last few years temperatures in oil ducts have reached rather high levels near pumping stations and in the heating areas which have been designed to reduce the viscosity of the oil. For this reason the use of PP for coatings has been suggested as compared with those made of polyethylene; these are able to withstand higher temperatures and have a greater resistance to penetration and abrasion.[146]

Table 2 shows the main comparative characteristics of bitumen, epoxide, polyethylene and polypropylene coatings. Polypropylene coatings have a wider temperature lag, compared to other types, better mechanical properties, which allow a reduction in thickness of about 20%, a low rate of water absorption, an excellent adhesion rate and an almost nonexistent cathode separation.

The assets of functionalized PP, which are characterized by good adhesive properties to various metal underlayers, have opened up whole new possibilities in extremely important sectors such as large structural components and thin materials, which are at the moment the monopoly of metallic sheets.

Advantages in terms of performance per unit weight and cost can be achieved by using to advantage the reinforcement qualities of the metallic sheet. So, for example, a sandwich structure made up of two thin external steel sheets with an intermediary layer of PP can weigh up to 70% less and cost as much as 25% less compared to a steel sheet of the same rigidity.[147] Bearing this in mind, it is clear that plastic laminates used in the above way are ideal materials when performance under flexure stress is the most important criterion of any project.

Table 2 A Comparison of Characteristics Between Coating in Bitumen, Epoxide, Polyethylene and Polypropylene

	Bitumen	*Epoxide resin*	*PE*	*PP*
Temperature (°C) lag	$-20/+50$	$-40/+90$	$-50/+70$	$-50/+100$
Resistance to penetration at 23 °C, DIN 30670 (mm)	2.3	—	0.1	0.04
Resistance to impact at 23 °C, DIN 30670 (Nm mm^{-1})	Almost nonexistent	Almost nonexistent	10	20
Water absorption at 23 °C, ASTM D 570 (% 24h)	—	—	0.01	0.005
Peeling value at 23 °C (N mm^{-1})	—	—	20	Cannot be removed
Cathode detachment DIN 30671 (mm^2)	—	<500	<500	10

7.4.1.3 Film and multilayer sheets

The packing sector is substituting more and more metal materials by plastics, so that the demand has never been so urgent for materials with a low permeability to gases (oxygen, water steam), in order to guarantee the long term preservation of foods. Also, in the field of chemical packaging (paints, pesticides, solvents), plastic containers for solvents need to be less permeable.

Polymer materials such as poly(vinylidene chloride) (PVDC) and poly(ethylene vinyl alcohol) (EVOH) are excellent barriers to oxygen, but with their poor resistance to water and their inferior mechanical characteristics they can only be inserted in multilayer structures together with other polymers.[148]

Of the existing polymers PP is the best choice, as can be seen by the performance demands in this sector shown in Table 3.

Film and multilayer sheets which are impermeable both to water and oxygen can only be produced if adhesives are available to guarantee excellent adhesion between different kinds of polymers.

The adhesives which satisfy all these demands are maleic anhydride functionalizing PP for multilayer PP/EVOH structures and functionalized PP with butyl methacrylate for multilayer PP/PVDC structures.[149,150]

Multilayer coextruded films now offer what was once the prerogative of special metallized laminates. Their most important characteristic is low permeability to oxygen (lower than 1 cm^3 m^{-2} (24 h)$^{-1}$ at 23 °C and 50% relative humidity; see Table 4).

Table 3 Performance Demands in the Food-packing Sector

Rigidity for liquid packing	HIPS–ABS–PP–HDPE–PAN[a]
	Hard PVC–PMMA–CPET[b]
Flexibility for the packing of solid products	LLDPE–PP–ionomers
	EVA–PA
Barrier to gas	PVCD–EVOH–PAN–PA
Barrier to water	LLDPE–HDPE–PP–EVA
Baking in oven (+130 °C)	CPET
Possibility of sterilizing (+130 °C)	PP–PA–CPET
Weldability	LLDPE–HDPE–PP–EVA

[a]PAN = polyacrylonitrile. [b]CPET = crystallized PET.

Table 4 Permeability to Oxygen and Water Vapour of PP, PVDC, EVOH and their Multilayer Structures

	Distribution of layers (μm)	Permeability to oxygen (mL (mq 24h)$^{-1}$ 23 °C, 50% relative humidity)	Permeability to water vapour (g (mq 24h)$^{-1}$ 38 °C, 90% relative humidity)
PP	25	3600	8
PVDC	25	1.6	1.6
EVOH	25	0.4	132
PP/PVDC/PP	300/75/625	0.3	0.15
PP/EVOH/PP	300/40/660	0.5–1.0[a]	0.25

[a]The exact value depends on the humidity content of the sample.

7.4.1.4 Polymer alloys

Over the last few years there has been an evident growth in the use of graft and block copolymers in the polymer alloy sector, the aim being to obtain certain morphological structures with better mechanical characteristics. The properties of a polymeric alloy are a function both of the morphology and adhesion between the phases. It is possible to act upon these two aspects by adding a compatibilizing agent to the alloy dispersion of the discrete phase and, at the same time, it allows for better adhesion between the two phases.

To be efficient the compatibilizer must preferably be placed at the interface of the two phases and for this block copolymers are usually better than graft copolymers, which have multiple ramifications and may therefore limit the penetration of the compatibilizer at the homopolymer phase.[151]

In the case of the PP alloys it has been demonstrated that functionalized PP with styrene is quite a good compatibilizing agent both for the PP/polystyrene polymer system (PS) and the PP/poly(phenylene ether) polymer system (PPE).[152]

Many scientific papers have been published about PP/polyamide alloys and the use of functionalized PP with maleic anhydride as the compatibilizing agent. A few of these products are already on the market and are gaining commercial ground in the car industry.[153]

For some time now it has been known that functionalized PP with polar groups, e.g. carboxy and anhydride, can react with terminal PA groups. The resulting copolymers act as surface active agents capable of lowering the interfacial tension between the two polymers. This is of great importance to the mechanical characteristics of the alloy. Figure 2 shows the pattern of the IZOD resilience with an incision of blends of PP/PA 6 compatibilizer.[154]

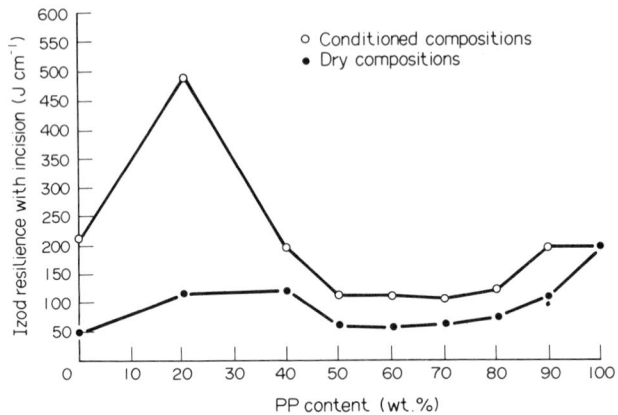

Figure 2 IZOD resilience with incision at 23 °C of blends of PP/PA 6 compatibilizer

7.4.1.5 Painting and gluing

PP is a difficult material to paint and glue, because it is unable to interact strongly with adhesives. The reason for this is that it is quite impossible to wet, due to its low surface tension.

At present, various treatments are used to paint and glue PP (corona discharge, cold plasma, flame hardening, photochemical, chemical), which have one thing in common: the ability to modify the structure substantially at a superficial level without altering the physical–mechanical characteristics inside the mass.

Recently, there has been an explosion of interest in plasma techniques for polymer surface modification. A great body of literature on this subject has been produced.[155] Inert gases (helium and argon), reactive gases (O_2, N_2, F_2, NO, NO_2) and polymerizable gases (tetrafluoroethylene, methane, acetylene) can be used to generate plasma.[156-161]

Cold gas plasma involves an energetic process, where gas is 'energy excited' into a plasma state in a vacuum chamber. This chemically active plasma modifies the molecular surface structure of the plastic, through three molecular reactions which alter the polymer simultaneously: ablation or etching, that is the removal by evaporation of surface materials; crosslinking, the connection of two or more polymer chains; and activation, the substitution of atoms in the polymer chain with chemical groups from the plasma.

Hydroxy, carbonyl and carboxy groups are but a few of the functional groups that can be generated with an oxygen plasma, as determined by XPS.[162,163]

In the polymer film packaging industry, the surface modification of hydrocarbon polymers that is crucial to many end uses (printing, coating and lamination) is most often achieved by corona discharge treatment (air plasma at atmospheric pressure).[164-166]

Flame treatments have often been used industrially for achieving reliable paint adhesion to thermoplastic alkene polymer body surfaces, particularly in the automotive industry. The radical mechanism of the surface oxidation of polypropylene by flame treatment has recently been confirmed by XPS analysis.[167]

The first step of the proposed mechanism involves the formation of hydroperoxide species, reacting in time to form a number of different products (hydroxyl, carbonyl, carboxy, ether, peroxide, *etc.*).

The formation of carboxylic products is particularly evident in flame treatments. This would suggest an ulterior oxidation of already existing oxidized functions (C—OH or C=O), whose amount decreases with the third and fourth flame treatments, as the major source of carboxy groups.

The amount of polar groups shows a constant increase with the number of flame treatments, the depth of oxidation is then a function of the number of treatments.[168]

From the practical point of view, a single flame treatment is sufficient to obtain strong adhesion to paint coatings (about 30 times stronger adhesion in comparison to the untreated sample). Further treatments do not significantly increase the strength of adhesion.

Photochemical treatment has proved to be the most interesting treatment on account of the excellent results obtained and the low investment costs. The success of the photochemical treatment is due to the use of suitable photoinitiators, which are activated through UV irradiation; then, after eventual internal conversions and intersystem crossing, they are deactivated by removing the tertiary H atoms of PP.[169]

The photoinitiator can be added in the mass, in a molten state or spread over the surface with solvents. The latter is without doubt the most efficient technique, particularly if an aromatic or chloride solvent is adopted to swell the amorphous fraction of PP.[166-170] By using this technique, deep modifications of over 1000 nm are made and are therefore able, unlike flame hardening, to penetrate below the 'weak boundary layer'.[168] The theory of the 'weak boundary layer' states that to obtain good adhesion it is necessary to modify the polymer to a depth beyond the thin layer (to a thickness of about 50 nm), which contains impurities, low molecular weights, additives, *etc.*[171]

Figure 3 shows the values of surface tension and adhesion to a bicomponent polyurethane resin paint, during irradiation of PP pretreated by heat with 1,1,1-trichloroethane and trichloroethylene vapours (1:1) for 1 min.[154]

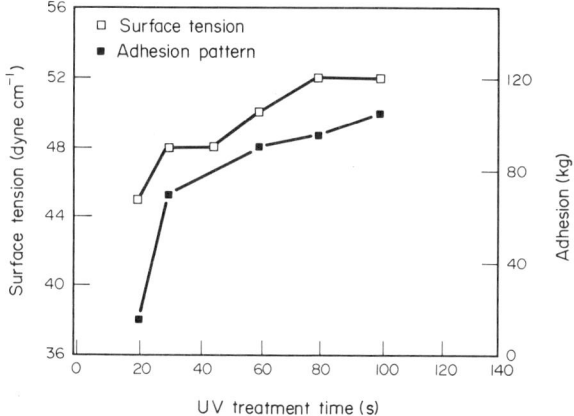

Figure 3 Values of surface tension and adhesion during irradiation time of PP pretreated by heat with 1,1,1-trichloroethane and trichloroethylene vapours (1:1) for 1 min

Apart from having a retarding action, the 1,1,1-trichloroethane facilitates the permeation of the trichloroethylene into the PP. Increases in the surface tension and adhesion are mainly due to the photooxidative processes which are triggered off in the PP following the photodecomposition of the C—Cl bonds in trichloroethylene.

7.5 REFERENCES

1. H. F. Mark, N. M. Bikales, C. G. Overberger and G. Mendes (eds.), 'Encyclopedia of Polymer Science and Engineering', 2nd edn., Wiley, New York, 1985.
2. V. F. Maghan and J. W. Crary, ref. 1, vol. 1, p. 325.
3. U. Klabunde, J. C. Calabrese, W. C. Fultz, T. Herskovitz, A. H. Janowicz, R. Mülhaupt, D. C. Roe, T. H. Tulip and S. D. Ittel, *J. Polym. Sci., Polym. Chem. Ed.*, 1987, **25**, 1989.

4. A. Ostoja Starzewski, J. Witte, K. M. Reichert and G. Vasiliou, in 'Transition Metals and Organometallics as Catalysts for Olefin Polymerization', ed. W. Kaminsky and H. Sinn, Springer-Verlag, Berlin, 1988.
5. H. Starweather, ref. 1, vol. 10, p. 369.
6. U. Giannini, G. Bruckner, E. Pellino and A. Cassata, *J. Polym. Sci., Part C*, 1968, **22**, 157.
7. F. Ciardelli, F. Menconi, A. Altomare and C. Carlini, 'XVIth Congresso Nazionale di Chimica', BononiaChem '88', Bologna, 1988, Abstracts, p. 278.
8. M. D. Purgett and O. Vogl, *J. Polym. Sci., Polym. Chem. Ed.*, 1989, **27**, 2051.
9. T. C. Chung, *ACS Polym. Prepr.*, 1988, **29**, 461.
10. E. Berluche, T. C. Chung and S. Ramakrishnan, *Macromolecules*, 1990, **23**, 378.
11. T. C. Chung and D. Rhubright, *Macromolecules*, 1991, **24**, 970.
12. A. R. Padwa, *Prog. Polym. Sci.*, 1988, **14**, 811.
13. R. Bertani, Tesi di Dottorato di Ricerca in Scienze Chimiche, University of Pisa, 1990.
14. R. Greco, G. Maglio, E. Martuscelli, P. Musto and R. Palumbo, *Polym. Proc. Eng.*, 1986, **4**, 253.
15. N. G. Gaylord and R. Metha, *J. Polym. Sci., Polym. Chem. Ed.*, 1988, **26**, 1189.
16. W. L. Young and R. R. Blauchard, ref. 1, vol. 6, p. 495.
17. K. W. Lee and T. J. McCarthy, *Macromolecules*, 1988, **21**, 309.
18. G. Ruggeri, M. Aglietto, A. Petragnani and F. Ciardelli, *Eur. Polym. J.*, 1983, **19**, 863.
19. K. Allmer, A. Hult and B. Rånby, *J. Polym. Sci., Polym. Chem. Ed.*, 1988, **26**, 2099.
20. K. Allmer, A. Hult and B. Rånby, *J. Polym. Sci., Polym. Chem. Ed.*, 1989, **27**, 1641.
21. K. Allmer, A. Hult and B. Rånby, *J. Polym. Sci., Polym. Chem. Ed.*, 1989, **27**, 3405.
22. K. Allmer, A. Hult and B. Rånby, *J. Polym. Sci., Polym. Chem. Ed.*, 1989, **27**, 3419.
23. K. Allmer, J. Hilborn, P. H. Larsson, A. Hult and B. Rånby, *J. Polym. Sci., Polym. Chem. Ed.*, 1990, **28**, 173.
24. M. Aglietto, R. Bertani, G. Ruggeri and A. L. Segre, *Macromolecules*, 1990, **23**, 1928.
25. D. A. Olsen and A. J. Osteraas, *J. Polym. Sci., Part A-1*, 1969, **7**, 1913.
26. D. A. Olsen and A. J. Osteraas, *J. Polym. Sci., Part A-1*, 1969, **7**, 1927.
27. D. M. Brewis, 'Surface Analysis and Pretreatment of Plastic and Metals' Applied Science Publishers, London, 1982, p. 199.
28. H. K. Yasuda, *ACS Symp. Ser.*, 1985, **287**, 89.
29. M. Palsson, B. Rånby and A. Hult, *Macromolecules Preprints*, paper presented at 'Macromolecules' 86', Oxford, England, p. 123.
30. M. D. Wilson, G. S. Ferguson and G. M. Whitesides, *J. Am. Chem. Soc.*, 1990, **112**, 1244.
31. F. Poncin-Epaillard J. C. Brosse and G. Legeay, *Makromol. Chem.*, 1988, **189**, 2293.
32. F. Poncin-Epaillard, B. Chevet and J. C. Brosse, *Eur. Polym. J.*, 1990, **26**, 333.
33. J. F. Rabek, J. Lucki, B. Rånby, Y. Watanabe and B. J. Qu, *ACS Symp. Ser.*, 1988, 364.
34. P. Gatenholm, *Adv. Org. Coat. Sci. Technol. Ser.*, 1991, **13**, 335 (*Chem. Abstr.*, 1991, **115**, 93 546p).
35. M. Strobel, S. Corn, C. S. Lyons and G. A. Korba, *J. Polym. Sci., Polym. Chem. Ed.*, 1985, **23**, 1125.
36. O. Bayer and W. Becker, *US Pat.* 2 748 105 (1956) (*Chem. Abstr.*, 1956, **50**, 11 055f).
37. S. G. Joshi and A. A. Natu, *Angew Makromol. Chem.*, 1986, **140**, 99.
38. S. G. Joshi and A. A. Natu, *Angew. Makromol. Chem.*, 1986, **143**, 115.
39. S. G. Joshi and A. A. Natu, *Angew. Makromol. Chem.*, 1986, **144**, 113.
40. J. L. Adcock, S. Inore and R. J. Lagow, *J. Am. Chem. Soc.*, 1978, **100**, 1948.
41. T. Volkmann and H. Widdecke, *Makromol. Chem., Macromol. Symp.*, 1989, **25**, 243.
42. G. D. Andrews and R. L. Dawson, ref. 1, vol. 6, p. 513.
43. M. Mori, M. So and K. Yamaji, *Jpn. Pat.* 61 101 509 (1986) (*Chem. Abstr.*, 1986, **105**, 209 609).
44. H. Harttig and K. J. Hüttinger, *J. Colloid Interface Sci.*, 1983, **93**, 63.
45. D. E. Bergbreiter and J. R. Blanton, *J. Org. Chem.*, 1985, **50**, 5828.
46. D. E. Bergbreiter, J. R. Blanton, R. Chandran, M. D. Hein, K. J. Huang, D. R. Treadwell and S. A. Walker, *J. Polym. Sci., Polym. Chem. Ed.*, 1989, **27**, 4205.
47. M. F. Schlecht, E. M. Pearce, T. K. Kwey and W. Cheung, *Polym. Prepr.*, 1986, **27**, 63.
48. M. F. Schlecht, E. M. Pearce, T. K. Kwey and W. Cheung, *ACS Symp. Ser.*, 1988, **364**, 300.
49. M. Aglietto, R. Bertani, G. Ruggeri and F. Ciardelli, *Makromol. Chem.*, 1992, **193**, 173.
50. M. Aglietto, R. Bertani, G. Ruggeri, R. Alterio and F. Galleschi, *Polymer*, 1989, **30**, 1133.
51. M. Aglietto, R. Bertani, G. Ruggeri, P. Fiordiponti and A. L. Segre, *Macromolecules*, 1989, **22**, 1492.
52. Z. Song and W. E. Baker, *J. Appl. Polym. Sci.*, 1990, **41**, 1299.
53. A. Simmons and W. E. Baker, *Polym. Eng. Sci.*, 1989, **29**, 1117.
54. Z. Song and W. E. Baker, *Angew. Makromol. Chem.*, 1990, **181**, 1.
55. J. R. Rasmussen, D. E. Bergbreiter and G. M. Whitesides, *J. Am. Chem. Soc.*, 1977, **99**, 4746.
56. N. C. Liu, W. E. Baker and K. E. Russel, *J. Appl. Polym. Sci.*, 1990, **41**, 2285.
57. Y. Doi, G. Hizal and K. Soga, *Makromol. Chem.*, 1987, **188**, 1273.
58. A. C. Udding, *Eur. Pat.* 318 115 (1989) (*Chem. Abstr.*, 1989, **111**, 135 014p).
59. E. Borsig and D. Braun, *Angew. Makromol. Chem.*, 1987, **150**, 1.
60. O. Laguna, J. Vigo, J. Taranco, J. L. Oteo and E. P. Collar, *Rev. Plast. Mod.*, 1989, **58**, 221, 228 (*Chem. Abstr.*, 1990, **112**, 78 110m).
61. R. Mülhaupt, T. Duschek and B. Rieger, *Makromol. Chem., Macromol. Symp.*, 1991, **48/49**, 317.
62. G. De Vito, N. Lanzetta, G. Maglio, M. Malinconico, P. Musto and R. Palumbo, *J. Polym. Sci., Polym. Chem. Ed.*, 1984, **22**, 1335.
63. B. Immirzi, N. Lanzetta, P. Laurenzio, G. Maglio, M. Malinconico, E. Martuscelli and R. Palumbo, *Makromol. Chem.*, 1987, **188**, 951.
64. P. Greco, P. Laurenzio, G. Maglio, M. Malinconico and E. Martuscelli, *Makromol. Chem.*, 1987, **188**, 961.
65. A. D'Amore, P. Laurenzio, B. Immirzi, M. Malinconico, E. Martuscelli and R. Greco, *Makromol. Chem.*, 1989, **190**, 1457.
66. R. Greco, G. Maglio, P. Musto, *J. Appl. Polym. Sci.*, 1987, **33**, 2513.
67. R. Greco, G. Maglio, P. Musto and G. Scarinzi, *J. Appl. Polym. Sci.*, 1989, **37**, 777.
68. R. Greco, P. Musto, F. Riva and G. Maglio, *J. Appl. Polym. Sci.*, 1989, **37**, 789.

69. N. G. Gaylord, M. Metha and R. Metha, *ACS Polym. Prepr.*, 1986, **27**, 105.
70. A. Chapiro, *J. Polym. Sci.*, 1958, **34**, 439.
71. A. Chapiro, *J. Polym. Sci.*, 1960, **48**, 109.
72. A. Chapiro, *J. Polym. Sci., Polym. Symp.*, 1975, **50**, 181.
73. A. Chapiro, 'Radiation Chemistry of Polymeric Systems', Wiley-Interscience, New York, 1962.
74. T. O'Neill, *J. Polym. Sci., Part A-1*, 1972, **10**, 569.
75. L. Minnema, J. F. A. Hazenberg, L. Callaghan and S. Pinner, *J. Appl. Polym. Sci.*, 1960, **4**, 246.
76. A. Dessouki, E. Hegazy and M. Shaker, *Radiat. Phys. Chem.*, 1987, **29**, 111.
77. B. Misra, D. Sood and I. Mehta, *J. Polym. Sci., Polym. Chem. Ed.*, 1985, **23**, 1749.
78. L. Sidorova, A. Aliev, V. Zlobin, R. Aliev, A. Chalykh, and V. Kabanov, *Radiat. Phys. Chem.*, 1986, **28**, 407.
79. R. Stamm, E. Hosterman, C. Felton and C. Chen, *J. Appl. Polym. Sci.*, 1963, **7**, 753.
80. H. Matsuo, K. Iino and M. Kondo, *J. Appl. Polym. Sci.*, 1963, **7**, 1833.
81. A. Mukherjee and B. Gupta, *J. Appl. Polym. Sci.*, 1985, **30**, 2655.
82. C. Ang, J. Garnett, M. Long and R. Levot, *Radiat. Phys. Chem.*, 1983, **22**, 831.
83. F. Hartley, D. McCaffrey, S. Murray and P. Nicholson, *J. Organomet. Chem.*, 1981, **206**, 347.
84. G. Odian, T. Acker and M. Sobel, *J. Appl. Polym. Sci.*, 1963, **7**, 245.
85. J. Wilson, *J. Macromol. Sci., Chem.*, 1976, **A10**, 1441.
86. A. Furuhashi and M. Kadonaga, *J. Appl. Polym. Sci.*, 1966, **10**, 127.
87. A. Armstrong and H. Rutherford, *Text. Res. J.*, 1963, **33**, 264.
88. P. Burchill, D. Pinkerton and R. Stacewicz, *J. Polym. Sci., Polym. Symp.*, 1976, **55**, 303.
89. D. Pinkerton and R. Stacewicz, *J. Polym. Sci., Polym. Lett. Ed.*, 1976, **14**, 287.
90. A. Mukherjee and B. Gupta, *J. Appl. Polym. Sci.*, 1985, **30**, 2643.
91. N. Mironov and V. Nikolshii, *Vysokomol. Soedin.*, 1975, **A17**, 2540.
92. F. Sundardi, *J. Appl. Polym. Sci.*, 1978, **22**, 3163.
93. H. Fujiwara, A. Sugushita and K. Asano, *US Pat.* 3 970 534 (1976) (*Chem. Abstr.*, 1975, **83**, 180 255g).
94. N. S. Marans and W. D. Addy, *US Pat.* 3 137 674 (1964) (*Chem. Abstr.*, 1964, **61**, 5873g).
95. N. S. Marans and F. A. Wessells, *US Pat.* 3 310 605 (1967) (*Chem. Abstr.*, 1967, **66**, 105 459x).
96. J. Burkus, *US Pat.* 3 314 904 (1967) (*Chem. Abstr.*, 1965, **62**, 12 000b).
97. H. Jabloner and R. Mumma, *J. Polym. Sci., Part A-1*, 1972, **10**, 763.
98. K. Minsker, I. Shapiro and G. Razuvayev, *Polym. Sci. USSR*, 1963, **4**, 112.
99. Y. Yamauchi, K. Ikemoto and A. Yamaoka, *Makromol. Chem.*, 1977, **178**, 2483.
100. P. Citovicky, D. Mikulasova, V. Chrastova, J. Reznicek and G. Benc, *Eur. Polym. J.*, 1977, **13**, 661.
101. G. W. Stanton and T. G. Traylor, *US Pat.* 3 049 508 (1962) (*Chem. Abstr.*, 1962, **57**, 15 388).
102. F. Engelbart, *US Pat.* 3 485 660 (1969) (*Chem. Abstr.*, 1969, **71**, 116 390x).
103. E. Vandenberg, *US Pat.* 2 837 496 (1958) (*Chem. Abstr.*, 1959, **53**, 1851).
104. T. Patton, *US Pat.* 3 870 692 (1975) (*Chem. Abstr.*, 1975, **83**, 061 458).
105. P. Bernstein, J. Caffey and A. Varker, *US Pat.* 4 131 637 (1978) (*Chem. Abstr.*, 1979, **90**, 122 460).
106. D. W. Eastman and L. E. Walker, *US Pat.* 4 605 704 (1986).
107. J. L. Olener and L. E. Walker, *US Pat.* 4 536 545 (1985).
108. C. Favie, W. Dellsperger and P. Meline, *US Pat.* 3 646 165 (1972) (*Chem. Abstr.*, 1970, **72**, 67 779r).
109. H. Yui, T. Kakizaki, H. Sano, M. Arai and H. Matsui, *US Pat.* 4 097 554 (1978) (*Chem. Abstr.*, 1977, **87**, 136 801r).
110. P. J. Canterino, *US Pat.* 3 162 697 (1964) (*Chem. Abstr.*, 1965, **62**, 7953).
111. U. Grigo, J. Merten and R. Binsack, *US Pat.* 4 370 450 (1983) (*Chem. Abstr.*, 1982, **96**, 105 203d).
112. L. E. Walker, *US Pat.* 4 806 581 (1989).
113. S. Lee, R. Rengarajan and V. Parameswaran, *J. Appl. Polym. Sci.*, 1990, **41**, 1891.
114. Y. Moriya, N. Suzuki and H. Goto, *US Pat.* 4 879 347 (1989) (*Chem. Abstr.*, 1989, **110**, 58 368g).
115. D. F. Knaack, *US Pat.* 3 644 581 (1972) (*Chem. Abstr.*, 1970, **73**, 57 093u).
116. R. A. Steinkamp and T. J. Grail, *US Pat.* 3 862 265 (1975) (*Chem. Abstr.*, 1973, **79**, 006 161).
117. R. A. Steinkamp and T. J. Grail, *US Pat.* 4 001 172 (1977) (*Chem. Abstr.*, 1977, **86**, 091 227).
118. R. A. Steinkamp and T. J. Grail, *US Pat.* 3 953 655 (1976) (*Chem. Abstr.*, 1976, **85**, 047 537).
119. R. T. Swiger and P. C. Juliano, *US Pat.* 4 147 740 (1979) (*Chem. Abstr.*, 1979, **91**, 005 675).
120. R. R. Gallucci and R. C. Going, *J. Appl. Polym. Sci.*, 1982, **27**, 425.
121. R. L. McConnell, R. B. Taylor and P. M. Grant, *US Pat.* 3 862 266 (1975) (*Chem. Abstr.*, 1974, **80**, 097 513).
122. N. G. Gaylord, R. Metha, D. Mohan and W. Kumar, *Polym. Mater. Sci. Eng.*, 1991, **64**, 151.
123. N. Gaylord and M. Mehta, *J. Polym. Sci., Polym. Lett. Ed.*, 1982, **20**, 481.
124. N. Gaylord, M. Mehta and R. Mehta, *J. Appl. Polym. Sci.*, 1987, **33**, 2549.
125. N. Gaylord and M. Mishra, *J. Polym. Sci., Polym. Lett. Ed.*, 1983, **21**, 23.
126. W. Shyu and D. Woodhead, *US Pat.* 4 753 997 (1988) (*Chem. Abstr.*, 1987, **107**, 155 001c).
127. C. Wong, *US Pat.* 4 857 254 (1989) (*Chem. Abstr.*, 1989, **110**, 39 572c).
128. N. Gaylord, *US Pat.* 3 708 555 (1973) (*Chem. Abstr.*, 1972, **76**, 15 572f).
129. B. Boutevin and J. J. Robin, *Eur. Polym. J.*, 1990, **26**, 559.
130. S. Porejko, W. Gabara and J. Kulesza, *J. Polym. Sci., Part A-1*, 1967, **5**, 1563.
131. S. Porejko, W. Gabara, T. Blazejewicz and M. Lecka, *J. Polym. Sci., Part A-1*, 1969, **7**, 1647.
132. H. Raval, Y. P. Singh, M. H. Metha and S. Devi, *Polym. Int.*, 1991, **24**, 99.
133. F. O'Shea, *US Pat.* 3 642 950 (1972) (*Chem. Abstr.*, 1970, **73**, 67 222z).
134. H. S. Witt, *US Pat.* 3 489 822 (1970) (*Chem. Abstr.*, 1964, **61**, 16 276c).
135. C. Meredith and G. von Bodungen, *US Pat.* 3 657 395 (1972) (*Chem. Abstr.*, 1972, **77**, 035 551).
136. C. Meredith, R. Barrett and W. Bishop, *US Pat.* 3 538 190 (1970) (*Chem. Abstr.*, 1969, **71**, 113 610h).
137. A. Fournier and C. Paddock, *US Pat.* 4 166 081 (1979) (*Chem. Abstr.*, 1982, **96**, 140 963s).
138. R. Marzola, E. Garagnani and A. Moro (Montedison SpA), *US Pat.* 4 350 797 (1982) (*Chem. Abstr.*, 1982, **96**, 21 435t).
139. P. Tacke, H. Korber, J. Merten and D. Neuray (Bayer Aktjengesellschaft), *US Pat.* 4 499 237 (1985) (*Chem. Abstr.*, 1981, **95**, 62 962k).
140. C. S. Liu (Hercules Inc.), *US Pat.* 4 510 286 (1985) (*Chem. Abstr.*, 1984, **101**, 132 092v).

141. D. W. Klosiewicz (Hercules Inc.), *US Pat.* 4 595 726 (1986) (*Chem. Abstr.*, 1986, **105**, 192 449j).
142. D. A. Krueger and T. W. Odorzynski (American Can Company), *US Pat.* 4 617 240 (1986) (*Chem. Abstr.*, 1983, **98**, 35 752x).
143. A. B. Clayton and B. D. Kramer (Hercules Inc.), *US Pat.* 4 675 210 (1987) (*Chem. Abstr.*, 1986, **105**, 153 770n).
144. E. Garagnani, R. Marzola and A. Moro, *Mater. Plast. Elastomeri*, 1982, **5**, 298.
145. R. Marzola and E. Garagnani (Montedison SpA), *US Pat.* 4 278 586 (1981) (*Chem. Abstr.*, 1980, **93**, 47 854s).
146. G. P. Guidetti, R. Locatelli, R. Marzola and G. L. Rigosi, 'Proceedings of the 7th International Conference on the Internal and External Protection of Pipes', London, 1987.
147. A. F. Johnson and G. D. Simms, *Composites*, 1986, **17**, 321.
148. M. Boysen, *Kunststoffe*, 1987, **77**, 522.
149. C. Tremblay and R. E. Prud'homme, *J. Polym. Sci., Polym. Phys. Ed.*, 1984, **22**, 1857.
150. R. J. Ashley, *Adhesion*, 1988, **12**, 239.
151. A. Rudin, *J. Macromol. Sci., Rev. Macromol. Chem.*, 1980, **C19**, 267.
152. L. Del Giudice (Montedison SpA), *US Pat.* 4 713 416 (1987), (*Chem. Abstr.*, 1985, **103**, 105 783f).
153. W. Witt, *Kunststoffe*, 1987, **77**, 1009.
154. G. Guidetti and E. Garagnani, 'IX Convegno Italiano di Scienza e Tecnologia delle Macromolecole', Bologna, 1989, Abstracts, p. 53.
155. D. T. Clark and M. M. Abu Shabak, *J. Polym. Sci., Polym. Chem. Ed.*, 1984, **22**, 17.
156. H. Schonhorn and R. H. Hansen, *J. Appl. Polym. Sci.*, 1967, **11**, 1461.
157. C. A. L. Westerdahl, J. R. Hall, E. C. Schramm and D. W. Levi, *J. Colloid Interface Sci.*, 1974, **47**, 610.
158. R. H. Hansen and H. Schonhorn, *J. Polym. Sci., Part B*, 1966, **4**, 203.
159. N. Inagaki and H. Yatsuda, *J. Appl. Polym. Sci.*, 1981, **26**, 3333.
160. A. Moshonov and Y. Avny, *J. Appl. Polym. Sci.*, 1980, **25**, 771.
161. N. Inagaki, S. Tasaka and J. Ohkubo, *J. Appl. Polym. Sci., Appl. Polym. Symp.*, 1990, **46**, 399.
162. R. G. Nuzzo and G. Smolinsky, *Macromolecules*, 1984, **17**, 1013.
163. D. A. Franzia *et al.*, 'SPE ANTEC 1991', Montreal, 1991.
164. D. Briggs, 'Surface Analysis and Pretreatment of Plastic and Metals', Applied Science, London, 1982, p. 199.
165. D. Briggs, C. R. Kendall, R. Blithe and A. B. Wootton, *Polymer*, 1983, **24**, 47.
166. D. Briggs and C. R. Kendall, *Polymer*, 1979, **20**, 1053.
167. F. Garbassi, E. Occhiello and F. Polato, *J. Mater. Sci.*, 1987, **22**, 207.
168. F. Garbassi, E. Occhiello, F. Polato and A. Brown, *J. Mater. Sci.*, 1987, **22**, 1450.
169. J. F. Rabek, 'Mechanism of Photophysical Processes and Photochemical reactions in Polymers', Wiley, New York, 1987, p. 273.
170. G. Cecchin and F. Polato, *Ital. Pat.* 20812A/88 (1988).
171. S. Wu, 'Polymer Blends', ed. D. R. Paul and S. Newman, Academic Press, New York, 1978, vol. 1, p. 243.

8

Computer Modeling of Polymer Structure and Fundamental Properties

WAYNE L. MATTICE
University of Akron, OH, USA

8.1	INTRODUCTION	159
	8.1.1 Motivation for Computer Modeling	159
	8.1.2 Software and Hardware	160
8.2	MOLECULAR DYNAMICS IN THE DEVELOPMENT OF STATIC MODELS	160
	8.2.1 The Multiple Minimum Problem for Ordered Polymers	160
	8.2.2 Rotational Isomeric State (RIS) Theory	161
8.3	INTRACHAIN DYNAMICS ON LONG TIMESCALES	162
	8.3.1 The Dynamic Rotational Isomeric State (DRIS) Model	162
	8.3.2 DRIS for Calculation of Dynamics at Long Times	163
8.4	GLASSY POLYMERS AT BULK DENSITY	164
	8.4.1 The Theodorou–Suter Model for Glassy Polypropylene	164
	8.4.2 Extensions to Other Polymers	165
8.5	REFERENCES	165

8.1 INTRODUCTION

8.1.1 Motivation for Computer Modeling

Computer modeling of polymers can be pursued with a variety of motivations. An important intellectual motivation is the development of a fundamental understanding, at the atomic level, of the origin of the physical properties of macromolecules. The emphasis on an atom-based model for the modeling is stressed here, because then the modeling is performed with structures that make immediate and unambiguous contact with the language used by the chemists who synthesize the polymers. The atom-based models also permit incorporation in the software of information obtained from firmly based experimental and theoretical studies of appropriately chosen molecules of low molecular weight.

Successful interpretation of the known physical properties of existing polymers leads quite naturally to the prediction of the physical properties that would be observed for related systems that have not yet been prepared by the chemists. Here we encounter a second motivation, which is the channeling of product-oriented research into paths that are most likely to lead to the goal, and away from alternative paths that would lead to dead ends. An obvious use is the successful prediction, *via* molecular modeling, of a new polymer system that, after the investment of the resources required for development, finds important uses by society. A less glamorous use is the prevention of the expenditure of resources on a line of research that would ultimately be abandoned, because the modeling showed that the system proposed for development would not really have the desired properties, even if it could be prepared by the chemists.

The main body of this article will discuss areas of computer modeling using fully atomistic descriptions of flexible macromolecules, giving a few representative examples of current work. The focus on fully atomistic descriptions means that lattice models are not considered, nor are models that use semiatomistic descriptions, *i.e.* descriptions that do not explicitly incorporate all hydrogen atoms as distinct, interacting centers. The first area is the application of molecular dynamics trajectories to the evaluation of equilibrium properties of synthetic polymers. Next we will look at methods that have been used to extend, by orders of magnitude, the computational efficiency for the evaluation of intrachain dynamics. This article will conclude with methods used to evaluate static properties and the elastic deformation of glassy polymers packed at bulk density.

The examples cited in this article are meant to be illustrative of the points made. No attempt is made to provide a comprehensive review of all of the recent work in these areas. Those readers who want to study a more comprehensive review of recent activity in this area are referred to the recent book edited by Roe.[1] This book contains articles based on a five day symposium on the molecular modeling of polymers, held in Miami Beach, September 1989.

8.1.2 Software and Hardware

Recent developments in computers and in commercially available software have contributed greatly to the dramatic increase in the application of computer modeling to polymers. Two types of hardware play an especially important role. High performance workstations with high resolution graphics are readily available at prices that are attractive to individual scientists and to small groups. Large supercomputers have also become readily accessible at national or regional centers, and can also be found in the computer centers of large organizations. The computational power supplied by these larger machines is indispensible for many current applications of computer modeling to polymers. However, for those problems which do not lend themselves to vectorization, the computational advantage of the supercomputer over the workstation is much smaller than the ratio of their prices.[2]

Also of great importance is the increasing availability of sophisticated, user friendly software for molecular modeling of polymers. Much of this software traces its origins to applications in the biomedical and pharmaceutical fields. The developers of the software have quite properly noticed that their potential market is greatly expanded if their software evolves so that it can be applied to the problems presented by synthetic polymers.

The body of this article does not attempt to evaluate by model the current software packages or the hardware platforms for those packages, but the magnitude of computer power needed for a successful attack on the problems discussed is indicated.

8.2 MOLECULAR DYNAMICS IN THE DEVELOPMENT OF STATIC MODELS

8.2.1 The Multiple Minimum Problem for Ordered Polymers

The multiple minimum problem has a long history with biophysical chemists who seek the prediction of the single conformation of lowest energy for complicated biopolymers, from knowledge of the covalent structure and fundamental physics. The conformational energy surface for these macromolecules contains a multitude of local minima. The local minimum located by the application of a routine for minimization of the conformational energy will depend on the conformation selected for the initiation of the optimization. Even after several iterations, starting each time from a different conformation, it is difficult to know whether the lowest minimum detected is really the global minimum.

Recent studies suggest that molecular dynamics may greatly assist efforts to locate the global minimum. The approach utilizes a molecular dynamics trajectory calculated at a very high temperature, perhaps in excess of 1000 K. Thermal energies are sufficient to permit rapid crossings of moderate conformational energy barriers that separate local minima. Therefore a trajectory at elevated temperature will permit the macromolecule to sample several conformational energy minima. For small macromolecules, it may be possible to calculate a trajectory that would sample all important regions of the conformational energy surface. Refinement of the important conformations can then be obtained either with a series of energy minimizations starting in each distinct region of low conformational energy, or by additional molecular dynamics simulations performed at lower temperatures. A recent example of the successful application of this approach is supplied by the study of the antigen-combining site of McPC 603.[3]

8.2.2 Rotational Isomeric State (RIS) Theory

Equilibrium rotational isomeric state theory provides the link between the local conformational properties of small fragments of macromolecules and the conformational properties of the entire chain under theta conditions.[4] Recent developments, which will be addressed in subsequent sections of this article, have adapted conventional rotational isomeric state theory to the evaluation of the bulk properties of glassy polymers[5–7] and to the intramolecular dynamics, on long timescales, of individual chains.[8–10] In this section we describe the recent utilization of molecular dynamics for the development of more reliable rotational isomeric state models of chain molecules.

Central to all applications of rotational isomeric state theory is a conformation partition function, Z, that is assembled as a sequential product of n statistical weight matrices[11]

$$Z = U_1 U_2 \cdots U_n \tag{1}$$

where U_1 is a row, U_n is a column, and the internal U_i are rectangular matrices. In most applications, in which rotations about bonds i and $(i-1)$ are interdependent, U_i has a number of rows and columns identical with the number of rotational isomers accessible to bonds $(i-1)$ and i, respectively. For the familiar case of a chain with a three-fold symmetric rotation potential, the internal U_i are of the form

$$U_i = \begin{bmatrix} \tau & \sigma & \sigma \\ 1 & \sigma\psi & \sigma\omega \\ 1 & \sigma\omega & \sigma\psi \end{bmatrix} \tag{2}$$

where the order of indexing for rows and columns is t, g^+, g^-. A common interpretation of the four statistical weights in this matrix is

$$\sigma = \exp[-(E_{g^+} - E_t)/kT] \tag{3}$$

$$\tau = \exp[-(E_{g^+} + E_{tt} - E_{g^+t} - E_t)/kT] \tag{4}$$

$$\psi = \exp[-(E_{g^+g^+} + E_t - E_{g^+t} - E_{g^+})/kT] \tag{5}$$

$$\omega = \exp[-(E_{g^+g^-} + E_t - E_{g^+t} - E_{g^-})/kT] \tag{6}$$

where E_η and $E_{\xi\eta}$ might be identified with conformational energies at the appropriate minima in conformational energy versus ϕ_i or conformational energy versus ϕ_i and ϕ_{i-1}, respectively. The role of σ, τ, ψ, and ω is to provide the correct weighting for the different regions of conformational space, taking into account not only the conformations at the minima, but also neighboring areas of the conformational energy surface that have an energy within a few factors of kT of the energy at the minima. They should be assigned so that they correctly reproduce the population of the different regions of conformational space by bonds i and $(i-1)$.

A molecular dynamics trajectory, of sufficient length so that it samples all of conformational space for ϕ_i and ϕ_{i-1}, will give directly the probability for occupation of any small element in that space. Interpreting the symbols t, g^+ and g^- now to represent the ranges of ϕ_i of $180° \pm 60°$, $300° \pm 60°$ and $60° \pm 60°$, respectively, the probabilities for occupancy of t and g^+ states are given by

$$z = \int P \, d\phi_i \tag{7}$$

$$P_t = z^{-1} \int P \, d\phi_i \tag{8}$$

$$P_{g^+} = z^{-1} \int P \, d\phi_i \tag{9}$$

where the range of the integration is over $360°$ in equation (7), and over the appropriate $120°$ segment in equations (8) and (9). In a formulation of Z that employs first-order interactions only, i.e.

$$Z = (1 + 2\sigma)^{n-2} \tag{10}$$

the best assignment for the first-order statistical weight σ is

$$\sigma = P_{g^+}/P_t \tag{11}$$

which permits calculation of σ from equations (7)–(9).

An extension of this approach gives the statistical weights for the chain subject to both first- and second-order interactions. For example, in the case of τ the data from the molecular dynamics trajectory is plotted as P_{ϕ_i} versus ϕ_i for just those conformations where ϕ_{i-1} is in the angular range defined as *trans*. Integration of this probability distribution in a manner perfectly analogous to equations (7)–(9) yields P_{tt} and P_{tg^+}. A similar approach yields $P_{g^+g^+}$ and $P_{g^+g^-}$. The best set of statistical weights in the matrix presented in equation (2) is found from the P_η and $P_{\xi\eta}$ evaluated from the molecular dynamics trajectory and

$$P_{g^+} = \frac{1}{2}\frac{\partial \ln \lambda}{\partial \ln \sigma} \tag{12}$$

$$P_{tt} = \frac{\partial \ln \lambda}{\partial \ln \tau} \tag{13}$$

$$P_{g^+g^+} = \frac{1}{2}\frac{\partial \ln \lambda}{\partial \ln \psi} \tag{14}$$

$$P_{g^+g^-} = \frac{1}{2}\frac{\partial \ln \lambda}{\partial \ln \omega} \tag{15}$$

$$\lambda = (1/2)(\tau + \sigma(\psi + \omega) + \{[\tau - \sigma(\psi + \omega)]^2 + 8\sigma\}^{1/2}) \tag{16}$$

If the second-order interactions are weak, a good approximation is provided by equations of the type

$$\tau = P_{tt}P_{g^+}/P_{tg^+}P_t \tag{17}$$

The advantage to obtaining the statistical weights *via* a molecular dynamics trajectory is that this trajectory permits the molecule to sample all conformational space that is accessible by variation in dihedral angles, bending of bond angles and stretching of bonds. All degrees of freedom are sampled, the influence of these degrees of freedom on P_{ϕ_i} is assessed, and the statistical weights are calculated directly from P_{ϕ_i}. By simultaneous examination of ϕ_{i-2}, ϕ_{i-1} and ϕ_i, the molecular dynamics trajectory can be used to investigate whether third-order interactions are important.

Recently this method was used for the evaluation of realistic statistical weights for poly-(dimethylsiloxane).[12] The trajectory was calculated in a few days using a fast workstation. The molecular dynamics approach was absolutely essential in this case in order to correctly capture the influence of the high degree of flexibility in the molecule, including the unusually low resistance to distortion at the Si—O—Si bond angle.[13,14] This new rotational isomeric state model for poly(dimethylsiloxane) provides a much better quantitative description of the measured macrocyclization equilibrium constants[15] than does an earlier model[16] that used a more traditional approach for the evaluation of the statistical weights. An extension of this type of analysis can be applied to chains in which the second-order interactions are not weak.[17]

Molecular dynamics trajectories will also greatly simplify the treatment of macromolecules with a high density of articulated side chains attached to atoms in the main chain, as occurs in the higher poly(1-alkenes) and poly(dialkylsiloxanes). Here the molecular dynamics method can be used to efficiently evaluate the influence of the conformations accessible to the articulated side chains on the probability distribution for the dihedral angles at the bonds in the main chain.

8.3 INTRACHAIN DYNAMICS ON LONG TIMESCALES

8.3.1 The Dynamic Rotational Isomeric State (DRIS) Model

Conventional rotational isomeric state theory[4,11] treats the conformation dependent physical properties of a chain at equilibrium. The conformation partition function Z must account for the location and relative weighting of all locations in conformational space that have high probabilities for occupancy. Regions with energies more than several factors of kT above the minimum energy can be ignored in the construction of Z because they have very low probabilities of occupancy, and therefore make a negligible contribution to the conformation dependent physical properties of chains at equilibrium.

If a chain has discrete rotational isomers, the energy barriers that separate two stable regions must be higher in energy than the local minima by at least several factors of kT. For such chains, the heights of the barriers do not enter into the formulation of Z because the barriers represent regions of space that have very small probabilities of occurrence. A much different situation is encountered in the treatment of the dynamics of rotational isomerism in the chain. Now, the heights of the saddle points separating two stable rotational isomers are of enormous importance, because these heights determine the size of the rate constant for crossing the saddle point from one rotational isomer to the other. Expressing the rate using Kramers' result for the high friction medium,[18] we have

$$r = (\gamma\gamma^*)^{1/2}(2\pi\zeta)^{-1}\exp(-E_a/kT) \qquad (18)$$

where E_a denotes the activation energy, ζ is the friction coefficient, and γ and γ^* characterize the curvature of the reaction coordinate in the initial well and at the saddle point, respectively.

Dynamic rotational isomeric state theory incorporates such rates into a matrix formalism for a time dependent conformation partition function, Z_τ, as was shown initially by Jernigan.[19] Bahar and coworkers[8-10] have recently extended Jernigan's pioneering work in order to completely develop the analogy between Z_τ and Z. The expression for Z_τ is of the form[9]

$$Z_\tau = \boldsymbol{J}^* \boldsymbol{V}_2(\tau) \boldsymbol{V}_3(\tau) \cdots \boldsymbol{V}_{n-1}(\tau) \boldsymbol{J} \qquad (19)$$

where $\boldsymbol{J}^*$ and $\boldsymbol{J}$ are a row and column, respectively, that play roles equivalent to U_1 and U_2 in equation (1). The $V_i(\tau)$ of equation (14) are stochastic weights that depend on the conditional probability of finding conformation $\xi'\eta'$ as bonds $(i-1)$ and i, given conformation $\xi\eta$ at time τ earlier. The $V_i(\tau)$ are of larger dimensions than the U_i of equation (1), because they must couple the conformations at times t and $t+\tau$. The matrix expression for Z_τ can then be manipulated in a manner quite analogous to the classical manipulations of Z, but with the result being a characterization of the intrachain dynamics, rather than the evaluation of static properties.[10] The matrix formalism common to Z_τ and Z means that both types of calculations are extremely efficient. For this reason, the treatment of the intrachain dynamics *via* Z_τ can easily be extended to timescales that are orders of magnitude longer than can be achieved by the more traditional ways of calculating intrachain dynamics.[20]

Before turning in the next section to an example that well illustrates the power of the dynamic rotational isomeric state method, we briefly define conditions under which other methods should take precedence. DRIS, in its present stage of development, is a single-chain theory. The environment affects the dynamics only through the friction coefficient ζ in the expression for the rate, equation (13). For this reason, other methods should be adopted for the study of the intrachain dynamics of chains in highly constrained environments, as illustrated by the successful application of traditional molecular dynamics to a poly(1,4-*trans*-butadiene) chain confined to a channel in crystalline perhydrotriphenylene.[21] The calculation *via* Z_τ yields the intrachain dynamics, but has nothing to say about motions of the center of mass. The suitability of DRIS for a particular chain is intimately entwined with the details of the conformational energy surface. The strength of DRIS is in dealing with the internal dynamics of chains or fragments of chains, such as polyethylene,[8,10,22-24] poly(oxyethylene)[23,25] and poly(oxymethylene),[20] in which stable rotational isomers are well defined. Application to chains such as poly(dimethylsiloxane), which has a rather flat conformational energy surface, with ill-defined rotational isomers separated by very low energy barriers,[12] is more problematic.

8.3.2 DRIS for Calculation of Dynamics at Long Times

The power of the dynamic rotational isomeric state method is well illustrated by a recent application to poly(oxymethylene).[20] The internal bonds in this chain have a strong preference for *gauche* states, and a strong penalty is imposed for occupancy by consecutive bonds of *gauche* states of opposite sign. The preferred local conformation is a *gauche* helix of either chirality, $g^\pm g^\pm \cdots g^\pm$. At sufficiently low temperatures, the two *gauche* helices of opposite chirality completely dominate the ensemble. Other conformations become competitive as the temperature increases. The dependence on T and n of the transition from the helix to other conformations, at equilibrium, has been characterized using Z.[26,27] A single, sharp transition is found when the transition is monitored as $P(g^\pm g^\pm \cdots g^\pm)$ versus T, *i.e.* as the probability that all internal bonds will occupy *gauche* states of the same sign as a function of T. Although the equilibrium calculation shows a single thermal transition, a recent study of the dynamics shows that this transition can have two distinct relaxations in the time domain.[20]

Dynamic rotational isomeric state theory was used to calculate $P_\tau(g^\pm g^\pm \cdots g^\pm / g^\pm g^\pm \cdots g^\pm)$, which denotes the conditional probability that a chain forming a perfect helix at time t can be found in the same conformation at time $t+\tau$. A more concise representation of this term is $P_\tau(h/h)$, where h denotes the conformation in which all internal bonds adopt *gauche* placements of the same sign. The time dependence of $P_\tau(h/h)$ for a poly(oxymethylene) chain with $n=40$ is depicted in Figure 1. At low temperatures, the decay of $P_\tau(h/h)$ shows two distinct relaxations that are separated by

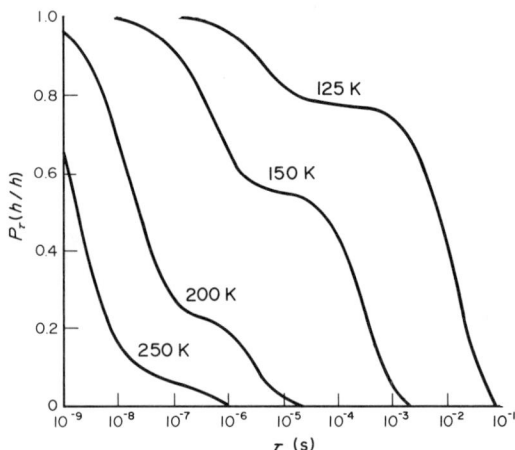

Figure 1 Time dependence of $P_r(h/h)$ for a chain with $n = 40$ at the four temperatures identified on the figure. Reprinted with permission from I. Bahar and W. L. Mattice, *Macromolecules*, 1991, **24**, 877. Copyright (1991) American Chemical Society.

several orders of magnitude in time. The faster of the two processes becomes dominant as the temperature increases. It completely dominates the dynamics at sufficiently high T.

A detailed analysis shows convincingly that the two relaxations at the lower temperatures arise from the mechanism

$$(g^{\pm})^n \rightleftarrows (g^{\pm})^{n-1} t \rightleftarrows \text{other} \tag{20}$$

The first step in equation (20) is the equilibration between the perfect helix and the family of conformations in which one of the $g^{\pm}$ states has become a t state. At 125 K helical chains that switch one bond to a *trans* state are more likely to reverse this transition than to produce any other type of rotational isomerization. Thus there is a metastable condition in which the only conformations present are $(g^{\pm})^n$ and $(g^{\pm})^{n-1}t$. This equilibration accounts quantitatively for the fast relaxation seen at low T in Figure 1.[20] The slower relaxation at low T is the further disordering of $(g^{\pm})^{n-1}t$ to other conformations that contain fewer than $n-1$ bonds in a *gauche* state of the same sign. The timescales of the two relaxations are 10^{-6} to 10^{-5} s and 10^{-2} s at 125 K. Two distinct relaxations are also seen in the orientation autocorrelation function calculated at 125 K by dynamic rotational isomeric state theory.[20]

The two relaxations observed in Figure 1 would not have been detected if the intrachain dynamics had been studied using conventional molecular dynamics simulations or Brownian dynamics, because of the enormous amount of computational time required to extend these methods into the millisecond regime. The extraordinary computational efficiency of the dynamic rotational isomeric state formalism makes this problem very easy indeed. (All computations required for the preparation of Figure 1 were performed rapidly using an Apple Macintosh IIci.) The computational efficiency of the formulation of intrachain dynamics based on equation (19) makes it feasible to explore processes that occur on the timescale of seconds or longer.

8.4 GLASSY POLYMERS AT BULK DENSITY

8.4.1 The Theodorou–Suter Model for Glassy Polypropylene

Theodorou and Suter recently described a novel application of rotational isomeric state theory that permits an atomistic modeling of the static properties of a vinyl polymer glass.[5-7] The initial application of their method was to atactic polypropylene. This molecule offered three important advantages that made it an excellent choice for this purpose. Firstly, the force field is relatively simple because all of the atoms are either carbon or hydrogen, thereby avoiding strong dipolar interactions, and unsaturation is absent. The methyl side chain was treated as a united atom, but in subsequent applications to other polymers a fully atomic description was used for the repeat unit. Secondly, the distance scale of the correlations is quite small, thereby permitting the use of periodic boxes of relatively small size, with an accompanying saving in the computational resources required for the project. Finally, previous applications of rotational isomeric state theory to polypropylene had produced an accurate expression for the conformational partition function Z.[28] Their method starts with Z, and hence this quantity must be known with confidence.

A classical use of Z has been the extraction from it of *a priori* and conditional probabilities that can be used in efficient Monte Carlo calculations for ensembles of unperturbed chains.[4,11] One type of application might be the generation of a representative sample of atactic polypropylene chains with x monomer units, as they might exist in dilute solution in a theta solvent. One of the modifications of the classical scheme by Theodorou and Suter[5] was the growth of the chain in a periodic cubic box, with the dimensions of the box selected so that, when filled with x monomer units, the density would be 0.892 g cm^{-3}, corresponding to a well-relaxed glass at $-40\,°C$. The chain generation scheme used by Theodorou and Suter also differs from the classical approach in that, at each step, the conditional probabilities are recalculated so that the influence of all interactions of long range is taken into account. After the chain is generated, the system is relaxed *via* a three-step optimization in which the dihedral angles at the internal bonds are the variables. Bond lengths and bond angles are held constant. The final step is computationally intensive. It requires access to a very fast workstation or a supercomputer.

Fifteen different cubes, each containing 76 monomer units in a box 18.15 Å on a side, were generated in the initial study.[5] Averages over these 15 replicas gave a calculated value of the Hildebrand solubility parameter δ of $14.2(\pm 0.8)(\text{J cm}^{-3})^{1/2}$, which is quite close to the experimental estimate of δ. Subsequent elaboration of the model allowed the estimate of elastic constants that were, in general, within $\sim 10\%$ of those obtained from experiment.[6] Consequently this clever modification of a rotational isomeric state model is capable of providing an atom-based description of properties of a polymer glass.

8.4.2 Extensions to Other Polymers

Recently the method introduced by Theodorou and Suter has been applied to the atom-based modeling of the properties of more-complicated polymers. In atactic poly(vinyl chloride) the methyl group of polypropylene is replaced by a chlorine atom.[2,29] This simple replacement greatly complicates the modeling because electrostatic interactions must be added to the force field. The high extension of chain segments rich in racemic dyads suggests poly(vinyl chloride) might require a much larger cube than did polypropylene for accurate modeling. Nevertheless, the estimate of δ from 10 cubes with 76 monomer units each is within $\sim 20\%$ of the experimental value (which itself has an uncertainty of $\sim 10\%$).[29] The agreement between calculation and experiment would probably improve if the modeling were to use larger cubes. The deterrent to the use of large cubes is the amount of computer time necessary for relaxation of the initial guess of the conformation of the chain.

Aromatic rings are introduced in the modeling of amorphous bisphenol A polycarbonate.[30] Here the initial objective of the modeling was an estimate of the barrier height for the ring flip. To estimate this quantity, a ring in a well-relaxed cube is driven through a ring flip by changing its orientation by a small angle, then relaxing the remainder of the structure, and continuation of this procedure until the flip has been completed. The barrier to the ring flip obtained in this manner agrees quite well with experimental estimates.

8.5 REFERENCES

1. R.-J. Roe (ed.), 'Computer Simulation of Polymers', Prentice–Hall, Englewood Cliffs, NJ, 1991.
2. P. J. Ludovice, M. G. Davidson and U. W. Suter, *ACS Symp. Ser.*, 1987, **353**, 162.
3. R. E. Bruccoleri and M. Karplus, *Biopolymers*, 1990, **29**, 1847.
4. P. J. Flory, 'Statistical Mechanics of Chain Molecules', Wiley, New York, 1969.
5. D. N. Theodorou and U. W. Suter, *Macromolecules*, 1985, **18**, 1467.
6. D. N. Theodorou and U. W. Suter, *Macromolecules*, 1986, **19**, 139.
7. D. N. Theodorou and U. W. Suter, *Macromolecules*, 1986, **19**, 379.
8. I. Bahar and B. Erman, *Macromolecules*, 1987, **20**, 1368.
9. I. Bahar, *J. Chem. Phys.*, 1989, **91**, 6525.
10. I. Bahar and W. L. Mattice, *Macromolecules*, 1990, **23**, 2719.
11. P. J. Flory, *Macromolecules*, 1974, **7**, 381.
12. I. Bahar, I. Zuniga, R. Dodge and W. L. Mattice, *Macromolecules*, 1991, **24**, 2986, 2993.
13. S. Grigoras and T. H. Lane, *J. Comput. Chem.*, 1988, **9**, 25.
14. S. Grigoras and T. H. Lane, *Adv. Chem. Ser.*, 1990, **24**, 7.
15. J. F. Brown and G. M. Slusarczuk, *J. Am. Chem. Soc.*, 1965, **87**, 931.
16. P. J. Flory, V. Crescenzi and J. E. Mark, *J. Am. Chem. Soc.*, 1964, **86**, 146.

17. W. L. Mattice, I. Zuniga, R. Dodge and I. Bahar, *Comput. Polym. Sci.*, 1991, **1**, 35.
18. H. A. Kramers, *Physica*, 1940, **7**, 284.
19. R. L. Jernigan, in 'Dielectric Properties of Polymers', ed. F. E. Karasz, Plenum Press, New York, 1972, p. 99.
20. I. Bahar and W. L. Mattice, *Macromolecules*, 1991, **24**, 877.
21. R. Dodge and W. L. Mattice, *Macromolecules*, 1991, **24**, 2709; Y. Zhan and W. L. Mattice, *J. Chem. Phys.*, in press.
22. I. Bahar, B. Erman and L. Monnerie, *Macromolecules*, 1989, **22**, 431.
23. I. Bahar and W. L. Mattice, *J. Chem. Phys.*, 1989, **90**, 6783.
24. I. Bahar, B. Erman and L. Monnerie, *Macromolecules*, 1990, **23**, 1174.
25. I. Bahar, B. Erman and L. Monnerie, *Macromolecules*, 1989, **22**, 2396.
26. J. G. Curro and J. E. Mark, *J. Chem. Phys.*, 1985, **82**, 3820.
27. J. G. Curro, K. S. Schweizer, D. Adolf and J. E. Mark, *Macromolecules*, 1986, **19**, 1739.
28. U. W. Suter and P. J. Flory, *Macromolecules*, 1975, **8**, 765.
29. P. J. Ludovice and U. W. Suter, in 'Computer Simulation of Polymers', ed. J. Bicerano, Dekker, New York, in press.
30. M. Hutnik, A. S. Argon and U. W. Suter, *Macromolecules*, 1991, **24**, 5956.

9
Mutual Diffusion in Polymeric Systems

BENNY D. FREEMAN
North Carolina State University, Raleigh, NC, USA

9.1	INTRODUCTION, SCOPE AND OVERVIEW	167
9.2	DIFFUSION OF POLYMERS IN SOLVENTS AT INFINITE DILUTION	168
	9.2.1 Influence of Thermodynamic Interaction and Molecular Weight on the Infinite Dilution Mutual Diffusion Coefficient	169
9.3	INFLUENCE OF CONCENTRATION, TEMPERATURE AND PRESSURE ON THE MUTUAL DIFFUSION COEFFICIENT IN DILUTE AND SEMIDILUTE SOLUTIONS	170
	9.3.1 The Effect of Polymer Concentration on the Mutual Diffusion Coefficient	170
	9.3.2 The Effect of Temperature on the Mutual Diffusion Coefficient	174
	9.3.3 The Effect of Pressure on the Mutual Diffusion Coefficient	175
9.4	APPLICATION OF FREE VOLUME THEORY TO DESCRIBE MUTUAL DIFFUSION IN MORE-CONCENTRATED POLYMERIC SYSTEMS	176
9.5	DIFFUSION OF SMALL PENETRANT MOLECULES IN POLYMERS	179
	9.5.1 Summary of Sorption and Transport Phenomena in Amorphous and Semicrystalline Polymers	179
	9.5.1.1 Sorption and transport of gases and vapors in amorphous polymers	179
	9.5.1.2 Sorption and transport of gases in semicrystalline polymers	181
	9.5.1.3 Sorption, transport and permeation of gases and vapors in liquid crystalline polymers	182
	9.5.1.4 Sorption, transport and permeation of gases and vapors in substituted polyacetylenes	183
	9.5.2 Review of Theoretical Approaches	184
	9.5.2.1 Sorption models	184
	9.5.2.2 Diffusion flux models	188
9.6	NOMENCLATURE	191
9.7	APPENDIX	193
	9.7.1 Unifying Analytical Fundamentals	193
	9.7.1.1 Reference frames and flux definitions	193
	9.7.1.2 Fick's first law	194
	9.7.1.3 Solution of diffusion problems	194
	9.7.1.4 Tracer diffusion coefficients	195
9.8	REFERENCES	196

9.1 INTRODUCTION, SCOPE AND OVERVIEW

A comprehensive and unifying discussion of binary mutual diffusion in polymeric systems, ranging from the behavior of a single chain in solution to the behavior of trace amounts of penetrants in solids, is presented. Mutual diffusion is important in a broad spectrum of practical applications, from the drying and curing of lacquers and enamels at low polymer concentrations to the removal of trace residuals of solvents and monomers at low penetrant concentrations. Applications at intermediate concentration include solvent spinning, pervaporation, barrier packaging and gas membrane separation.

The review begins with a discussion of mutual diffusion in polymer–solvent systems in which solvent is the major component and the polymer chains undergo Brownian motion independently of one another. Afterwards, mutual diffusion is considered in polymer–penetrant systems in which the concentration of polymer is high enough to cause interpenetration of the polymer chains, an effect which has a strong influence on the dynamic properties of the polymer chains. The chapter ends with a discussion of gas and vapor diffusion into solid polymers. In all sections of the chapter, an attempt is made to explain current notions of the physical basis for observed behavior and to discuss models used to describe these phenomena.

The discussion is limited to binary systems consisting of a linear polymer and a solvent or penetrant. Thus, ternary systems (i.e. two solvents and a polymer or one solvent in a polymer blend) are not included. An excellent review of transport of small penetrant molecules in polymer blends is provided by Hopfenberg and Paul.[1] A more recent review of the transport of small molecules in binary polymer blends has been written by Petropoulos.[2] The flow and diffusion of ions in porous membranes are not discussed, but a good recent review of the thermodynamic basis for the theory may be found in the work of Baranowski.[3] The theory of ion transport in charged membranes is discussed lucidly by Cwirko and Carbonell.[4,5] The discussion will be restricted to diffusion in systems where the solvent molecular weight is much lower than that of the polymer. Excellent reviews of polymer-polymer interdiffusion are available elsewhere.[6-8] A description of the unifying analytical base for the discussion which follows may be found in the Appendix (Section 9.7).

9.2 DIFFUSION OF POLYMERS IN SOLVENTS AT INFINITE DILUTION

This discussion begins with a description of the diffusion of a single flexible polymer chain in an infinite expanse of solvent. Many of the principal concepts used to understand diffusion in polymer solutions are observed most clearly in this limit. An excellent review of the modern theory of hydrodynamic properties of polymer chains in solution is provided in the article by Joanny and Candau in *Comprehensive Polymer Science*.[9] Several of the key results from their work are reproduced here to provide continuity to the discussion. As indicated in the Appendix, in solutions which are very dilute in polymer, the binary mutual diffusion coefficient characterizing the mixture is determined by the tracer diffusion coefficient of the entire polymer chain. In the limit of very low concentrations of polymer molecules, the polymer molecules move independently of one another, and each isolated polymer chain undergoes random Brownian motion in a sea of solvent. In this limit, the polymer molecules exist as separate coils in the solution and the frictional resistance between the surrounding solvent and a polymer molecule is related to the volume occupied by the chain (and interstitial solvent) in space.

Classic models of diffusion of a single linear, flexible polymer chain in solution have been formulated by Rouse,[10] Kirkwood and Riseman,[11] and Zimm.[12] Lucid discussions of these models are available in the literature.[6,8,13-15] A critical review of the physical foundations of such models is provided by deGennes.[14] In the following discussion, the Kirkwood and Riseman model is briefly reviewed since this model seems to capture the essential physical processes important in single chain diffusion and is the basis for modern renormalization group theories of polymer diffusion in solution. The polymer chain is treated as a covalently connected set of monomers, approximated as spheres, floating in solvent, which is treated as a continuum. The monomer units are distributed in space according to their equilibrium pair distribution function, g(r). The probability of finding a monomer at a position $r = r'$ in a small region dr', given that a monomer is located at $r = 0$, is $g(r')dr'$. Hydrodynamic interaction between monomers is included by solving the linearized Navier-Stokes equations to determine the solvent velocity at a position $r = r'$ due to the motion of a monomer sphere at $r = 0$. In this way, the motion of a particular monomer affects the solvent flow field which, in turn, modifies the motion of other monomers in the chain. From this calculation, the friction coefficient of the polymer chain is determined. As described in the Appendix, the friction coefficient is the proportionality constant between a force applied to the polymer chain and the resultant velocity. The result for the friction coefficient of the polymer is given by[11]

$$f_2 = 6\pi N_A \eta_s R_h \tag{1}$$

where R_h, the hydrodynamic radius of the polymer chain, is defined as[14]

$$R_h^{-1} = \frac{\int_V \frac{g(\mathbf{r})}{|\mathbf{r}|} d\mathbf{r}}{\int_V g(\mathbf{r}) d\mathbf{r}} \tag{2}$$

where the equilibrium monomer radial distribution function is $g(r)$. The integral is over all space occupied by the chain molecule. The solvent viscosity is η_s. As explained in the Appendix, in the limit of infinite dilution the mutual diffusion coefficient is given by

$$D^0_{12}(\rho_2 \to 0) = D^*_2 = \frac{kT}{f_2} \tag{3}$$

With the result of the Kirkwood analysis for the friction coefficient, the mutual diffusion coefficient of a single polymer chain in solution is given by

$$D^0_{12}(\rho_2 = 0) = \frac{kT}{6\pi\eta_s R_H} = \frac{kT}{6\pi\eta_s} \frac{\int_V \frac{g(\mathbf{r})}{|\mathbf{r}|} d\mathbf{r}}{\int_V g(\mathbf{r}) d\mathbf{r}} \tag{4}$$

This result is equivalent to the result for the diffusion coefficient of a single solid sphere of radius R_h in solution, as predicted by the Stokes–Einstein theory.[16] A key feature of this theory, an example of the so-called 'mode–mode coupling' theories used extensively to describe critical phenomena,[14] is the direct connection of a dynamic transport property, the mutual diffusion coefficient, to a static or equilibrium property, the monomer radial distribution function.

In flexible, linear chains, the dependence of the infinite dilution mutual diffusion coefficient on polymer molecular weight and the nature of thermodynamic interactions with the surrounding solvent environment are contained in the hydrodynamic radius. The temperature and pressure dependence of the infinite dilution value of the mutual diffusion coefficient depends upon the temperature and pressure dependence of both the solvent viscosity and the hydrodynamic radius. An excellent summary of available experimental data for the dependence of the infinite dilution diffusion coefficient on molecular weight and thermodynamic interactions is provided in *Comprehensive Polymer Science*.[9] The key results from this work are cited below and, afterwards, a description of the temperature, pressure and concentration dependence of the mutual diffusion coefficients is presented.

9.2.1 Influence of Thermodynamic Interaction and Molecular Weight on the Infinite Dilution Mutual Diffusion Coefficient

The nature of the thermodynamic interactions between the monomer units and solvent molecules influences the monomer radial distribution function and, in turn, the hydrodynamic radius and mutual diffusion coefficient. Some recent theories of polymer solution thermodynamics are based upon renormalization group methods.[6,17,18] This approach has been used quite successfully to describe critical phenomena. A well-written introduction to the subject of renormalization group theory is provided by Wilson.[19] In renormalization group theories of polymer solution thermodynamics, polymer chains are approximated as continuous curves or strings in space, and the solvent is treated as a continuum. In this approximation, the details of the primary chemical structure of the polymer chain do not enter directly into the model. The model treats interactions between the polymer and solvent as a parameter which is determined from experimental data. Such models cannot predict polymer properties without externally supplied data, such as virial coefficients, which characterize the interactions between the polymer chain and its surrounding environment of solvent. The model imposes a balance between two competing effects: (i) an elastic (or entropic) force which characterizes the entropy of mixing the polymer chain and the solvent molecules; and (ii) an energetic interaction force which prohibits more than one part of the chain from occupying the same position in space. The first effect tends to cause the polymer chain to be compact in space, while the second effect leads to a swelling of the polymer chain in space. A parameter, called the excluded volume, characterizes the importance of the second effect relative to the first.

The excluded volume parameter is directly proportional to χ, the so-called Flory–Huggins interaction parameter[14]

$$v = a_{ev}(1 - 2\chi) \tag{5}$$

where a_{ae} is a scale factor for the excluded volume parameter. The parameter χ depends upon the potential energy of pairwise interaction between monomer units and solvent[20]

$$\chi = [\Gamma_{MS} - (\Gamma_{MM} + \Gamma_{SS})/2]/kT \tag{6}$$

where Γ_{MM}, Γ_{MS} and Γ_{SS} are the potential energies of interaction between two monomer units, one monomer unit and one solvent molecule, and two solvent molecules, respectively. A good solvent is one in which χ is zero, indicating that the monomer unit and solvent molecules have very similar interaction energies. A poor solvent is one in which χ is greater than zero. In the classic theory due to Flory and Huggins, $\chi = \frac{1}{2}$ defines the limit, in high molecular weight polymers, for complete miscibility.[20] If χ is greater than $\frac{1}{2}$, the polymer solution will exist, in certain concentration ranges, as two separate phases. When $\chi = \frac{1}{2}$ in a polymer solvent system the solvent is called a 'theta solvent' for the polymer. A rough estimate of the Flory–Huggins interaction parameter is provided by the following relation, valid for nonpolar systems[20]

$$\chi = \frac{\hat{V}_1}{RT}(\delta_1 - \delta_2)^2 \quad (7)$$

where δ_1 and δ_2 are the solvent and polymer solubility parameters, respectively, and $\hat{V}_1$ is the solvent molar volume. Tables of solubility parameters are available in the literature.[21,22] When $\chi = \frac{1}{2}$, the excluded volume is zero and the polymer chain is distributed in space according to a Gaussian probability function.[14] As χ is decreased below $\frac{1}{2}$ the excluded volume parameter is increased, and the polymer chain expands in solution.

The renormalization group approach may be used to calculate the relationship between the infinite dilution diffusion coefficient and the polymer molecular weight. The results for good solvents (equation 8) and theta solvents (equation 9), in the limit of infinite length monodisperse chains, are[9]

$$D_{12}^0(\rho_2 = 0) \propto M^{-3/5} \quad (8)$$

and

$$D_{12}^0(\rho_2 = 0) \propto M^{-1/2} \quad (9)$$

Equations (8) and (9) illustrate a characteristic of the theoretical treatment. By focusing attention on the large scale behavior of the polymer chains in solution, the theory predicts the overall dependence of the diffusion coefficient on molecular weight but does not predict prefactors for equations (8) and (9).

Mixtures of polystyrene dissolved in toluene have been widely studied. The composite results for the infinite dilution value of the mutual diffusion coefficient at 20 °C are[9]

$$D_{12}^0(\rho_2 = 0) = 3.64 \times 10^{-8} M_w^{-0.577} \quad (10)$$

for weight average molecular weights, M_w, in the range of 130 000 to 40×10^6 (kg kg·mol^{-1}). The difference between the observed (0.577) and predicted (0.6) scaling exponent is attributed to the finite length of the chains and to the fact that the excluded volume interaction used in the derivation of equation (8) may be too strong (*i.e.* toluene may not be as good a solvent for the polymer as assumed by the theory).[9]

In theta systems, solutions of polystyrene dissolved in cyclohexane are among the most widely studied. The theta temperature for this system is 35 °C. The experimental result for the infinite dilution mutual diffusion coefficient of polystyrene in cyclohexane at 35 °C is[9]

$$D_{12}^0(\rho_2 = 0) = 1.3 \times 10^{-8} M_w^{-0.497} \quad (11)$$

which is considered to be in excellent agreement with the theoretical prediction (equation 9).

The behavior of systems between theta and good solvents can be seen clearly in the work by Pritchard and Caroline[23] and, in *Comprehensive Polymer Science*, by King.[24] These researchers determined the dependence of the mutual diffusion coefficient on molecular weight, concentration and temperature. Their work is summarized in the following section.

9.3 INFLUENCE OF CONCENTRATION, TEMPERATURE AND PRESSURE ON THE MUTUAL DIFFUSION COEFFICIENT IN DILUTE AND SEMIDILUTE SOLUTIONS

9.3.1 The Effect of Polymer Concentration on the Mutual Diffusion Coefficient

As the polymer concentration is increased, the motion of an individual polymer coil begins to be coupled to and, therefore, to depend upon the motion of the coils surrounding it. Eventually the coils overlap and interpenetrate, forming entanglements between sections of the chains. The mutual diffusion coefficient becomes less dependent upon the motion of the entire chain and more dependent upon the cooperative motions of sections of chains between entanglement points. Recent work has focused upon the use of renormalization group methods to extend the expression for the mutual diffusion coefficient to finite concentrations.[25-27] The model by Oono predicts the distribution of the polymer segments in space. Using this model for the distribution of the chain segments in solution [i.e. g(**r**)] and the approach of Kirkwood and Riseman[11] to calculate the frictional resistance to the motion of segments in the solvent, Oono finds the following for the mutual diffusion coefficient[26]

$$\frac{D^0_{12}(\rho_2)}{D^0_{12}(\rho_2=0)} = (1+X)^{-3/8\{Z[1-(1+Z)^{-3/4}]/(Z+1)\}} \left(1 + \left\{\left(1 + \frac{1}{8}\frac{Z}{Z+1}\right)X + \frac{1}{4}\frac{Z}{Z+1}\left[\frac{\ln(1+X)}{X}-1\right]\right\}\right.$$
$$\left. \times \exp\left\{\frac{1}{4}\frac{Z}{Z+1}\left[\frac{1}{X}+\left(1-\frac{1}{X^2}\right)\ln(1+X)\right]\right\}\right) \quad (12)$$

where Z is a parameter which characterizes the strength of the excluded volume interaction. The case of $Z=0$ corresponds to a theta solvent; $Z=\infty$ corresponds to a good solvent.[26] X is related to the second osmotic virial coefficient, B_2, and the polymer concentration as follows[26]

$$X = 16\frac{1+Z}{8+9Z}B_2\rho_2 \quad (13)$$

where ρ_2 is the polymer concentration. The theory has been derived for the case of a good solvent, and the authors indicate that it is not intended to be used in solutions near the theta point (i.e at Z values less than approximately 1).[26] The model predictions are not very sensitive to the value of Z for $Z > 10$.[26] Most importantly, this model predicts that the mutual diffusion coefficient depends only upon $D^0_{12}(\rho_2=0)$, the infinite dilution mutual diffusion coefficient, X, which is proportional to the product of the second osmotic virial coefficient and the polymer concentration, and Z, which depends upon the quality of the solvent.

Figure 1 presents the concentration dependence of the mutual diffusion coefficient of several molecular weight samples of polystyrene in toluene at 25 °C. The highest polymer concentration shown in this figure corresponds roughly to a polymer weight fraction of 0.27 (assuming volume additivity). This figure shows that, at a high enough concentration of polymer, all of the data points lie on the same curve. Thus, above a critical concentration, the diffusion coefficient is no longer sensitive to the polymer molecular weight. As the polymer concentration increases from infinite dilution and the chains become entangled at higher concentrations, the mutual diffusion coefficient ceases to depend upon the motion of the polymer chain as a whole but depends, rather, on the mobility of segments of the polymer chain between entanglement points.[14] The mutual diffusion coefficient, therefore, becomes independent of the polymer molecular weight when the chains begin to interpenetrate strongly. A rough estimate of the concentration at which overlap begins is given by[6]

$$\tfrac{4}{3}\pi R_g^3 = \frac{M_2}{\rho_2^* N_a} \quad (14)$$

where R_g is the radius of gyration of the polymer chain, M_2 is the polymer molecular weight, and ρ_2^* is the critical polymer concentration above which the chains should interpenetrate and begin to be entangled. The concentration ρ_2^* marks the transition from the dilute to the semidilute

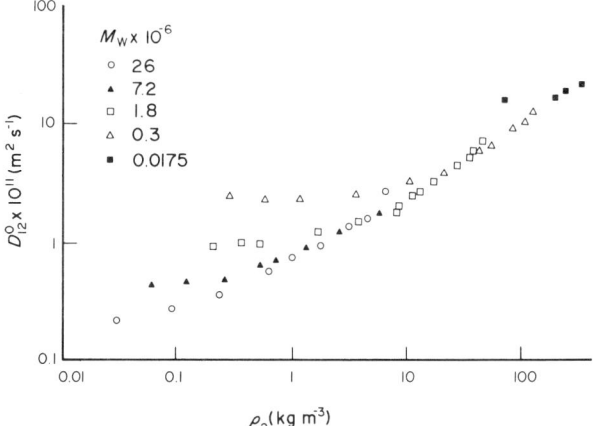

Figure 1 Influence of polymer concentration on the mutual diffusion coefficient of polystyrene in toluene at 25 °C (reprinted with permission from Wiltzius et al., Phys. Rev. Lett., 1984, **53** (8), 834)

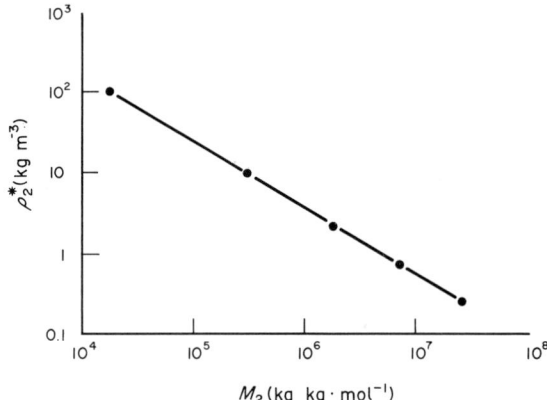

Figure 2 Effect of polymer molecular weight on the critical crossover concentration from dilute to semidilute solution for polystyrene in toluene at 25 °C

concentration range. The left hand side of equation (14) is a rough estimate of the volume occupied by a single isolated polymer chain. The right hand side is a measure of the volume available for a single polymer molecule in solution. For polystyrene in toluene at 20 °C, the radius of gyration[9] is given by $R_g = 1.107 \times 10^{-11} M_2^{0.605}$. The radius of gyration of polystyrene in toluene (a good solvent) is independent of temperature in the range 10–110 °C.[28] Therefore, equation (14) may be used to construct a plot of the critical concentration for entanglement as a function of the molecular weight of polystyrene in toluene for the data in Figure 1. This relation is presented in Figure 2. The data points presented in this figure correspond to the molecular weights studied by Wiltzius et al.[29] At concentrations above the critical concentrations shown in Figure 2, the mutual diffusion coefficient (see Figure 1) becomes independent of molecular weight, indicating that equation (14) provides a conservative estimate of the polymer concentration at the transition from dilute to semidilute solution for this polymer in a good solvent. Figure 1 also indicates that the binary mutual diffusion coefficient increases with increasing polymer concentration. From the Appendix, the mutual diffusion coefficient depends upon the product of a thermodynamic factor and a friction factor. The friction factor tends to decrease the mutual diffusion coefficient as concentration increases, while the thermodynamic factor acts to increase the mutual diffusion coefficient as concentration increases.[30] In good solvents, the thermodynamic factor is a stronger function of composition than the friction factor and, consequently, the mutual diffusion coefficient increases with increasing polymer concentration.[30]

Figure 3 presents a comparison of experimental determinations of the binary mutual diffusion coefficient in polystyrene–toluene mixtures in the manner suggested by the theory of Oono (equation 13). The experimental data presented in this figure are from Wiltzius et al.[29] The theory predicts that the abscissa of Figure 3 should be a unique function of the ordinate. The experimental data, however, do not fall upon a single master curve. As presented in Figure 4, Wiltzius et al. discovered that a nearly universal curve could be obtained if the relative diffusion coefficients were plotted

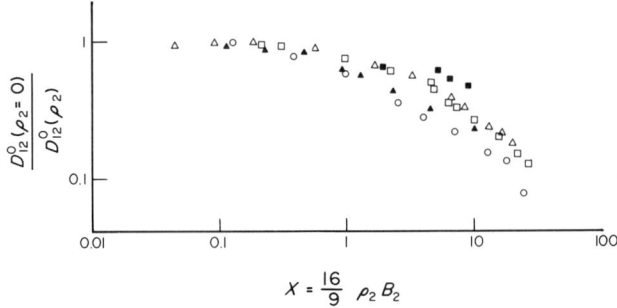

Figure 3 Experimentally determined mutual diffusion coefficients of polystyrene in toluene at 25 °C plotted in the manner suggested by the renormalization group theory of Oono. (reprinted with permission from Oono and Baldwin, *Phys. Rev. A*, 1986, **33** (5), 3391)

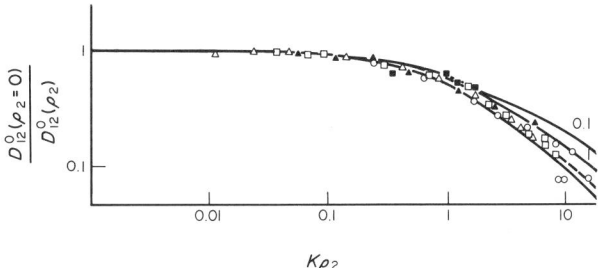

Figure 4 Dependence of the mutual diffusion coefficients of polystyrene in toluene at 25 °C on the product of the polymer concentration and K. The numbers beside the curves are values of the excluded volume parameter Z from the theory (reprinted with permission from Oono and Baldwin, *Phys. Rev. A*, 1986, **33** (5), 3391)

against the product of the polymer concentration and K, where K is defined as[29]

$$K \equiv \frac{d[D^0_{12}/D^0_{12}(\rho_2 = 0)]}{d\rho_2}\bigg|_{\rho_2 = 0} \tag{15}$$

The theory does not predict this result *a priori*.

The preceding section discussed the concentration dependence of the mutual diffusion coefficient in a good solvent. Renormalization group theories are only available for good solvents, and there are no equivalently well-developed theories for poor solvents. An example of the mutual concentration dependence of the mutual diffusion coefficient of polystyrene in a poor solvent, *trans*-decalin, is presented in Figure 5.[30] *trans*-Decalin is a theta solvent for polystyrene at 20 °C. At and just above the theta temperature, the diffusion coefficient initially decreases with increasing polymer concentration, indicating that the concentration dependence of the friction factor (see equation 17) is stronger than that of the thermodynamic factor. Osmotic pressure determinations,[31] light scattering studies[28] and available theoretical considerations[14] indicate that in poor solvents the concentration dependence of the thermodynamic factor (see equation 17) is much weaker in poor solvents than in good solvents and can be a strong function of temperature. As temperature increases above the theta temperature, the concentration dependence of the thermodynamic factor increases and, consequently, the mutual diffusion coefficient becomes less dependent upon concentration. At 40 °C in the polystyrene–*trans*-decalin system presented in Figure 5, there is a balance between the concentration and friction effects, and the diffusion coefficient is independent of concentration over the range of concentrations studied. At higher temperatures, the solvent quality should be better (*i.e.* the Flory–Huggins χ parameter should be smaller), and the mutual diffusion coefficient should begin to increase with increasing concentration. Such behavior has been observed by Pritchard and Caroline, who determined the temperature and concentration dependence of the mutual diffusion coefficient of polystyrene in the poor solvent cyclohexane.[23] The results of their study are shown in Figure 6.

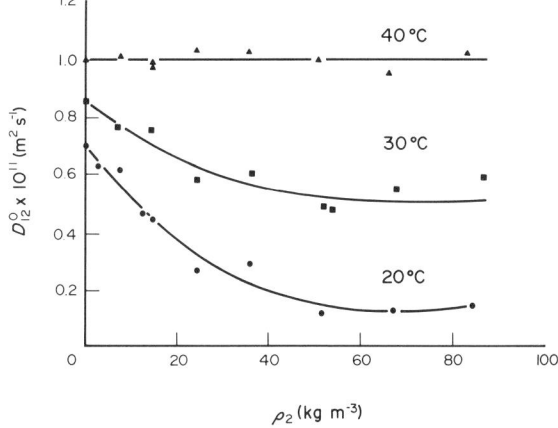

Figure 5 Dependence of binary mutual diffusion coefficient on concentration and temperature in the system polystyrene–*trans*-decalin. The polymer molecular weight is approximately 400 000 kg kg·mol^{-1}. The highest concentration studied corresponds to a mass fraction of polymer of approximately 0.11

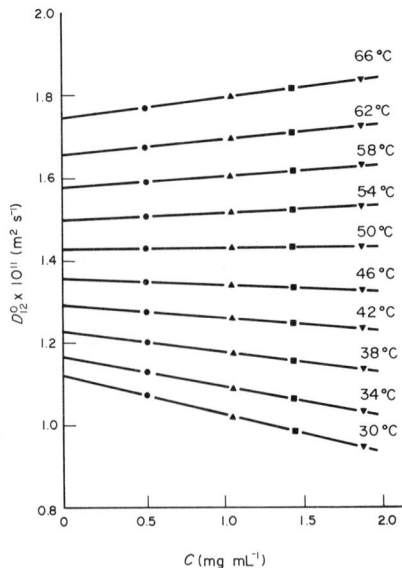

Figure 6 Effect of temperature and concentration on the mutual diffusion coefficient in the system polystyrene–cyclohexane. The polymer molecular weight is $1.26 \times 10^6 \, \text{kg kg} \cdot \text{mol}^{-1}$ (reprinted with permission from Pritchard and Caroline, *Macromolecules*, 1981, **14**, 424)

The results presented here represent examples of typically observed behavior. Extensive tabulations of values for infinite dilution mutual diffusion coefficients and for the concentration dependence of the mutual diffusion coefficient for a wide variety of polymers are available in the literature.[32]

9.3.2 The Effect of Temperature on the Mutual Diffusion Coefficient

Temperature can be used to change the thermodynamic interactions between monomer and solvent and, in turn, influence the mutual diffusion coefficient. The influence of temperature on the value of the mutual diffusion coefficient depends upon thermodynamic interactions between the polymer and solvent, concentration of polymer in the mixture and on temperature dependence of solvent viscosity.[33] The effect of temperature on the mutual diffusion coefficient is demonstrated in Figure 7 for polystyrene dissolved in the poor solvent *trans*-decalin. If the diffusion coefficient is interpreted according to an Arrhenius relation

$$D^0_{12}(T) = D_0 \exp\left(-\frac{\Delta E_d}{RT}\right) \tag{16}$$

where D_0 is a temperature independent prefactor, then an activation energy for diffusion, ΔE_d, may be calculated. Roots *et al.*[30] and Rehage *et al.*,[34] who studied the temperature dependence of the diffusion coefficient of polystyrene dissolved in cyclohexane, have found similar results. The activation energy is independent of temperature and increases with increasing concentration in the dilute region. From Figure 7, the activation energy at infinite dilution is $13.7 \, \text{kJ mol}^{-1}$ and $19 \, \text{kJ mol}^{-1}$ at a polymer concentration of $7 \, \text{kg m}^{-3}$. The data of Pritchard and Caroline (Figure 6) also show an increase in activation energy with polymer concentration. In their work, the activation energy of diffusion increases from approximately $12 \, \text{kJ mol}^{-1}$ to $15 \, \text{kJ mol}^{-1}$ as the polymer concentration increases from $0.5 \, \text{kg m}^{-3}$ to $1.8 \, \text{kg m}^{-3}$.

The data presented in Figure 7 show that at higher concentrations, the activation energy depends less upon concentration and, at the highest concentrations studied, begins to show a marked temperature dependence. The shape of the curve at the highest concentration in Figure 7 is not consistent with equation (16) (in which ΔE_d is assumed to be independent of temperature). The pronounced curvature occurs at temperatures near the theta temperature ($20.8 \, °C$). Roots *et al.* ascribe this behavior to changes in the solution thermodynamic interactions near the theta temperature. In good solvents, the activation energy for diffusion becomes sensitive to temperature near the glass transition temperature of the mixture, and this sensitivity is ascribed to the influence of free volume on the frictional resistance to segmental mobility (see Section 9.4).[35]

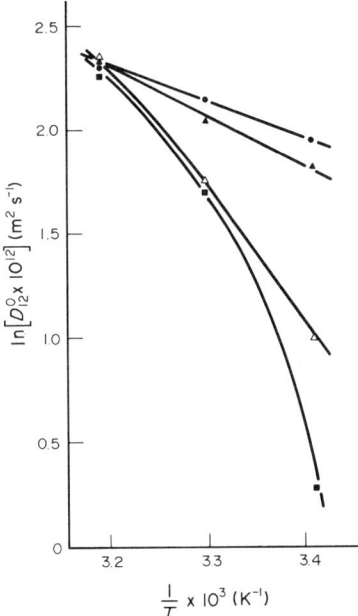

Figure 7 Effect of temperature and concentration on the mutual diffusion coefficient in the system polystyrene–*trans*-decalin. The polymer molecular weight is 3.9×10^5 kg kg·mol^{-1}. Polymer concentrations (kg m^{-3}): (●) 0; (▲) 7; (△) 25; (■) 67 (reprinted with permission from Nyström and Roots, *J. Macromol. Sci., Rev. Macromol. Chem.*, 1980, **C19** (1), 35)

9.3.3 The Effect of Pressure on the Mutual Diffusion Coefficient

The theory for predicting mutual diffusion coefficients as a function of molecular weight, solvent quality and concentration can be used to postulate the pressure dependence of the diffusion coefficient. From equations (12) and (13), the factors which could be pressure sensitive and have an impact on the mutual diffusion coefficient are the solvent viscosity, η_s; the parameter X; and the chain hydrodynamic radius, R_h. The solvent viscosity will change with pressure because of compression induced volume reduction. The variable X may change with pressure if either the second virial coefficient or the polymer concentration changes with pressure. The polymer concentration (in kg polymer m^{-3} solution) will increase with applied pressure since, generally, the solvent is more compressible than the polymer.[36] The hydrodynamic radius will change with pressure if there is a change in the polymer–solvent thermodynamic interactions with pressure which, in turn, affects the polymer segment pair correlation function.

Studies of the pressure dependence of the mutual diffusion coefficient in dilute and semidilute solutions are relatively rare. The pressure dependence of the mutual diffusion coefficient of polystyrene in toluene (a good solvent) has been determined by Roots and Nyström[37] and by Freeman *et al.*[38] An example of the pressure dependence of the mutual diffusion coefficient is presented in Figure 8 for the case of polystyrene in toluene. In the work of Freeman *et al.* and Roots and Nyström, the pressure dependence of the diffusion coefficient could be accounted for by considering the pressure-induced changes in the solvent viscosity and polymer concentration. The diffusion coefficients at higher concentration are more sensitive to pressure since the variation in polymer concentration with pressure is more pronounced at higher concentrations. Figure 9 shows the influence of hydrostatic pressure on the infinite dilution values of the mutual diffusion coefficient in a mixture of polystyrene and toluene. The solid line through the data is the relative fluidity (*i.e.* $\eta_s(P)/\eta_s(P_a)$) of toluene. From this figure, the principal effect of hydrostatic pressure on the infinite dilution mutual diffusion coefficient in this system is to change the viscosity of the solvent in which the polymer chains are floating. This result is in good agreement with other data. Schulz and Lechner[39] used static light scattering to determine the pressure dependence of the second virial coefficient of polystyrene in toluene at pressures up to 80 MPa. No change in the second virial coefficient with pressure was found at temperatures from 25–45 °C. McDonald and Claesson[40] also used static light scattering to determine the second virial coefficient, B_2, in the polystyrene–toluene system at pressures up to 400 MPa. Over the first 100 MPa, their data show no significant change in the second virial coefficient with pressure. The lack of change of B_2 with

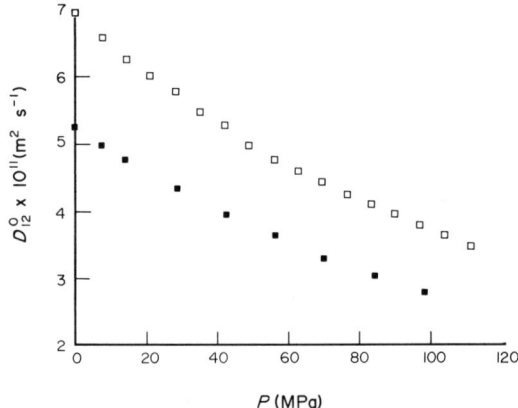

Figure 8 Effect of pressure on the mutual diffusion coefficient in polystyrene–toluene mixtures at two different concentrations. The polymer molecular weight is 1.84×10^5 kg kg·mol^{-1} and the measurements were performed at 25 °C. The filled and open symbols represent data at polymer concentrations of 12.0 and 25.2 kg m^{-3}, respectively (reprinted with permission from Freeman et al., Macromolecules, 1990, **23** (1), 245)

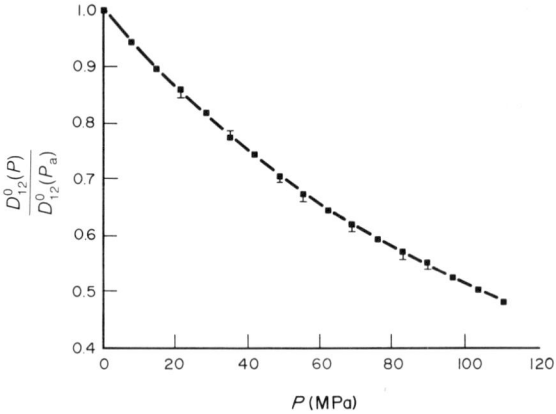

Figure 9 Comparison of the ratio of the infinite dilution mutual diffusion coefficient at pressure P to that at atmospheric pressure P_a to the relative fluidity of toluene [i.e. $\eta_s(P)/\eta_s(P_a)$]. The polymer molecular weight is 1.84×10^5 kg kg·mol^{-1} and the measurements were performed at 25 °C. The filled symbols represent data at a polymer concentration of 25.2 kg m^{-3} (reprinted with permission from Freeman et al., Macromolecules, 1990, **23** (1), 245)

pressure suggests that the pair correlation function and, in turn, the hydrodynamic radius of polystyrene in toluene may be insensitive to pressure as well, since these quantities are measures of the strength of polymer–solvent interactions.

Studies of the pressure dependence of the mutual diffusion coefficient in dilute and semidilute polymer solutions have only been carried out for good solvents. The composite results suggest that, in good solvents, the thermodynamic interactions between polymer and solvent are not strongly affected by pressure. In poor solvents, pressure-induced changes in the polymer–penetrant interactions could introduce additional effects which would alter the trends observed in good solvents.

9.4 APPLICATION OF FREE VOLUME THEORY TO DESCRIBE MUTUAL DIFFUSION IN MORE-CONCENTRATED POLYMERIC SYSTEMS

The renormalization group theoretical approach to calculating the mutual diffusion coefficient characterizing a binary mixture of a polymer and a thermodynamically good solvent can satisfactorily predict the concentration dependence of the diffusion coefficient at polymer concentrations in the dilute and semidilute ranges but not at higher concentrations. The expression provided by

Oono et al.[41] (equation 13) predicts that the mutual diffusion coefficient should increase over the entire range of concentration in a thermodynamically good solution, such as polystyrene–toluene. The experimental data, however, for this system exhibit a maximum in the mutual diffusion coefficient at higher polymer concentration.[42] In a similar system, polystyrene dissolved in ethylbenzene, the diffusion coefficient also exhibits a maximum.[34] Oono suggests that his theory will not be applicable at very high concentrations because: (i) screening of the hydrodynamic interaction of a given chain due to the presence of other chains is not included; and (ii) the theory only accounts for two-body interactions (i.e. interactions between one monomer unit and another monomer unit, between one monomer unit and a solvent molecule, and between two solvent molecules).[26] Another principal reason for the inability of dilute solution theories to predict the mutual diffusion coefficient over the entire composition range is that these theories assume that the frictional resistance to flow to relax a concentration gradient is dominated by the polymer contribution. As discussed in the Appendix, the mutual diffusion coefficient is often written as

$$D_{12}^0 = \left(\frac{\partial \ln a}{\partial \ln x_1}\right)_{T,P} \times (D_1^* x_2 + D_2^* x_1) \tag{17}$$

At high concentrations of solvent, the binary mutual diffusion coefficient is sensitive to the tracer diffusion coefficient of the polymer, D_2^*, which is inversely proportional to the friction coefficient of the polymer. Thus the mutual diffusion coefficient at high concentrations of solvent will depend upon the thermodynamic interactions between the polymer and solvent and the frictional resistance experienced by the polymer segments in the solvent. Conversely, at high polymer concentrations, the frictional resistance term in equation (17) is dominated by the solvent tracer diffusion coefficient.[43]

The crossover between the regime (at low polymer concentration) in which the polymer tracer diffusion coefficient dominates the friction factor and the regime (at high polymer concentration) in which the solvent or penetrant tracer diffusion coefficient dominates the friction factor is illustrated by the experimental results of von Meerwall et al.[44] These authors used pulsed field gradient NMR spectroscopy to determine the polymer concentration dependence of the tracer diffusion coefficient of polystyrene (PS), tetrahydrofuran (THF) and hexafluorobenzene (HFB) in a solution of these three components. HFB was added to the solution as a marker molecule and, in the following analysis, its presence is neglected. Figure 10 presents the tracer diffusion coefficient of polystyrene and THF as a function of polymer concentration. The tracer diffusion coefficients for both the polymer, D_{PS}^*, and the solvent, D_{THF}^*, decrease as a function of increasing polymer concentration, but the polymer tracer diffusion coefficient decreases more than two orders of magnitude while the solvent tracer diffusion coefficient decreases by less than one order of magnitude over the concentration range. Using these data, the components of the friction factor in equation (17) were calculated and are presented in Figure 11. This figure shows the orders of magnitude of the solvent ($x_{PS}D_{THF}$) and

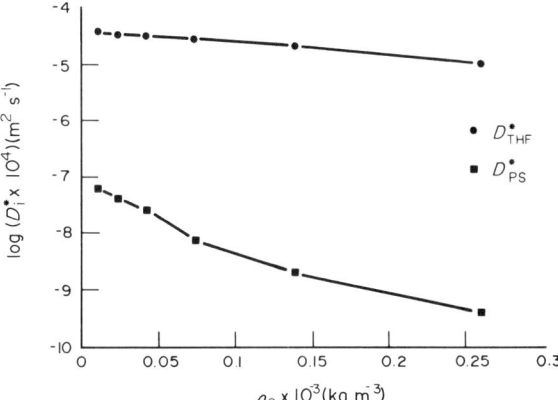

Figure 10 Dependence of the polystyrene and tetrahydrofuran tracer diffusion coefficients on polystyrene concentration. The polymer molecular weight is 498 000 kg kg·mol^{-1} and $T = 30.0\,°C$. The experimental data are from von Meerwall et al., Macromolecules, 1985, **18**, 260

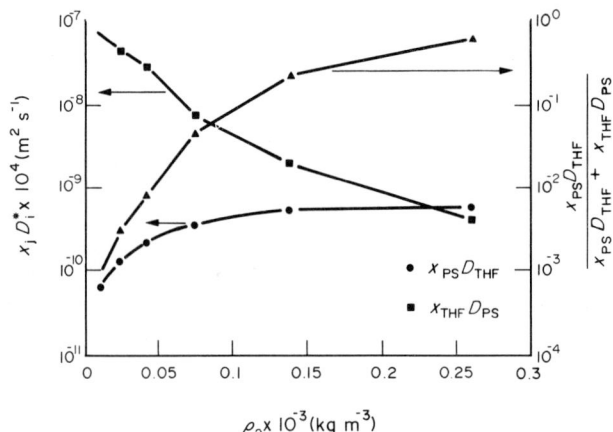

Figure 11 The effect of polystyrene concentration on the relative contributions of polymer and solvent to the mutual diffusion coefficient friction factor. The polymer molecular weight is 498 000 kg kg-mol^{-1}. The temperature is 30.0 °C. The experimental data are from von Meerwall et al., Macromolecules, 1985, **18**, 260

polymer ($x_{THF}D_{PS}$) contributions to the friction factor and the relative importance of the solvent contribution to the friction factor. At low polymer concentrations, the friction factor is dominated by the polymer contribution while, at the highest concentration of this study (260 kg m^{-3}, which corresponds to a polymer weight fraction of 0.28), the solvent contribution to the friction factor is more important than the polymer contribution.

Thus, models of mutual diffusion in dilute polymer–solvent mixtures, which consider only the polymer tracer diffusion coefficient contribution to the frictional resistance term, are not applicable when the polymer tracer diffusion coefficient becomes very low, and the frictional resistance term of the mutual diffusion coefficient is dominated by the solvent tracer diffusion coefficient. In this limit, other approaches, which focus attention on a determination of the solvent or penetrant tracer diffusion coefficient, are more useful. Among the most successful approaches to date are those based upon the concept of free volume.

In polymer–solvent mixtures at conditions between the glass transition temperature T_g and $T_g + 100$ K, free volume concepts may be used to explore the influence of variables such as temperature, pressure and concentration on dynamic properties of polymers and of polymer–solvent mixtures.[35,45] The fractional free volume at a given temperature and pressure is defined as follows

$$f_P = \frac{V - V_{occupied}}{V} = \frac{V_f}{V} \tag{18}$$

where V is the total volume of the sample, V_f is the free volume and $V_{occupied}$ is the so-called 'occupied volume', which is not available to participate in molecular motion. In a series of articles, Vrentas and Duda[43,46,47] have extended free volume theories (which have been used historically to describe the temperature, concentration and pressure dependence of rheological properties of rubbery polymers and other viscous liquids which are near their glass transition point[45]) to describe mutual diffusion in polymer–penetrant mixtures. The model is based upon equation (74) in the Appendix; that is, the mutual diffusion coefficient is taken to be proportional to the product of a thermodynamic factor and the solvent self-diffusion coefficient. The solvent self-diffusion coefficient is assumed to depend primarily upon the amount of free volume in the polymer–solvent mixture. Vrentas and. Duda have developed expressions to describe the influence of concentration, temperature, solvent and polymer type, and polymer molecular weight on the free volume of the polymer–solvent mixture. The details of their theoretical model, along with examples of the application of this theory to describe mutual diffusion in rubbery polymer–penetrant systems, are available in the literature.[42] An example of the application of this model to describe the mutual diffusion in the polystyrene–toluene system is presented in Figure 12. This model is most widely used to describe the concentration and temperature dependence of the mutual diffusion coefficient in rubbery polymers. It is particularly useful for solvent concentrations, such as those important in industrial applications like solvent devolatilization, which lie between dilute solution and the solid state.

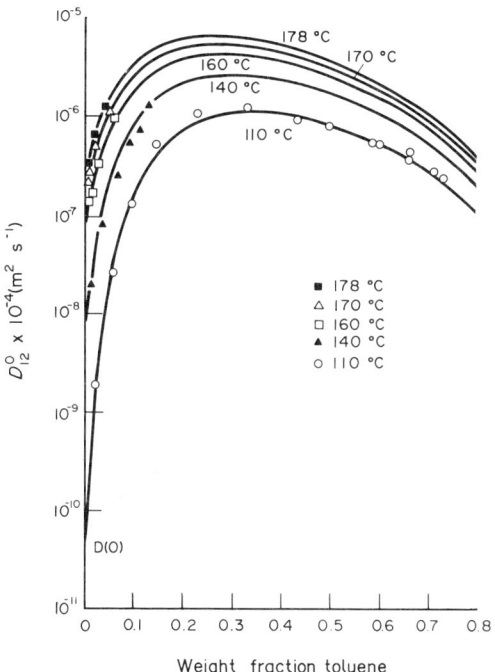

Figure 12 The use of free volume theory to describe the temperature and concentration dependence of the mutual diffusion coefficient of polystyrene in toluene (reprinted with permission from Duda *et al.*, *AIChE J.*, 1982, **28** (2), 279)

9.5 DIFFUSION OF SMALL PENETRANT MOLECULES IN POLYMERS

9.5.1 Summary of Sorption and Transport Phenomena in Amorphous and Semicrystalline Polymers

9.5.1.1 *Sorption and transport of gases and vapors in amorphous polymers*

Well-written introductions to the subject of sorption and transport of small molecules through solid polymers are provided by Naylor,[48] in the classic work by Stannett[49] and in a recent review article by Koros and Hellums.[50] In this work, therefore, the salient features governing sorption and transport of gases and vapors in organic polymers will only be summarized briefly.

In typical amorphous flexible polymers the transport of small molecules proceeds by a solution and diffusion mechanism.[49,51] Rubbery polymers behave essentially as high molecular weight liquids, and penetrants dissolve into the matrix of the rubber and diffuse by a cooperative process through the polymer.[50] In glassy polymers, penetrant molecules dissolve and diffuse into both densified regions of the polymer matrix and into preexisting gaps between the chain segments, which arise from the excess volume present due to the nonequilibrium nature of the glassy state.[52] The second mechanism of sorption and transport is not available in rubbery polymers, since rubbers are at equilibrium and, therefore, do not contain fixed regions of nonequilibrium excess volume.

The solubility, S, of a penetrant in a polymer is commonly defined as C/p, where C is the concentration of penetrant in the polymer, and p is the partial pressure of penetrant in equilibrium with the penetrant dissolved in the polymer. The temperature dependence of the solubility in a rubbery or glassy polymer can typically be expressed as a Van't Hoff relationship over temperature ranges in which no significant thermal transitions (such as glass transitions or, for semicrystalline polymers, melting transitions) occur[53]

$$S = S_0 \exp(-\Delta H_S/RT) \qquad (19)$$

where S_0 is a temperature independent prefactor and ΔH_S is the overall heat of sorption, which is often written as a sum of two terms

$$\Delta H_S = \Delta H_{cond} + \Delta H_{mix} \qquad (20)$$

where ΔH_{cond} is the heat of condensation, the enthalpy difference between the pure penetrant in the gas phase and in the condensed phase. If the temperature of interest is above the critical temperature of the penetrant, the condensed phase is taken to represent a hypothetical condensed phase. ΔH_{mix} is the energy of mixing pure, condensed penetrant and polymer. Penetrant condensation is typically exothermic while polymer–penetrant mixing is always endothermic.[49] The overall heat of sorption derived from the Van't Hoff relationship is the algebraic sum of these two effects. For permanent gases such as N_2, O_2 and H_2, which are well above their critical temperature at ambient conditions, solubility typically increases with increasing temperature since the heat of condensation is small, and the heat of sorption is determined, therefore, by the endothermic heat of mixing. However, for more condensible gases such as CO_2, SO_2 and hydrocarbon vapors, solubility frequently decreases with increasing temperature since the heat of condensation may be more important than the heat of mixing.[48] For liquid penetrants and solvents, the overall heat of sorption is determined by the heat of mixing since the penetrant is already condensed.

Penetrant diffusion in rubbery polymers is believed to occur by motion of penetrant molecules into fluctuating gaps which arise due to thermally induced, random segmental motion of the polymer chains.[50] Since the rates of translational and rotational motions of small penetrant molecules are quite rapid compared with those of the polymer, the rate-controlling step in penetrant diffusion is the creation of a gap, by the segmental dynamics of the polymer chains, of sufficient size to accommodate the penetrant.[50] Penetrant diffusion coefficients in rubbery polymers are typically observed to decrease with increasing penetrant size, approaching a plateau for large penetrant molecules, as shown in Figure 13. In natural rubber, for example, this plateau region is attained by penetrants such as *n*-butane, benzene or *n*-pentane. This plateau is believed to be dictated by the rate of polymer local segmental mobility and presumably corresponds to penetrant mobilities which are of the same order of magnitude as the polymer segmental mobility.[48]

In glassy polymers, penetrant diffusivity is lower than in rubbery materials, with the difference increasing markedly with penetrant size, as shown in Figure 13. For example, the infinite dilution diffusion coefficient of *n*-pentane in natural rubber at 25 °C is approximately 10^{-11} m^2 s^{-1}, while that of much smaller helium is of the order of 10^{-9} m^2 s^{-1}.[54] In glassy poly(vinyl chloride) at 30 °C, the diffusivity of *n*-pentane is approximately 10^{-18} m^2 s^{-1}, while that of helium is of the order of 10^{-10} m^2 s^{-1}.[55] In glassy poly(methyl methacrylate) the diffusivities of *n*-pentane and helium are approximately 10^{-19} and 10^{-11} m^2 s^{-1}, respectively.

The diffusion of a penetrant into a rubbery or glassy polymer is an activated process and may be modeled by an Arrhenius temperature dependence with a positive activation energy. The activation energy for the diffusion of a penetrant in a polymer generally increases with increasing

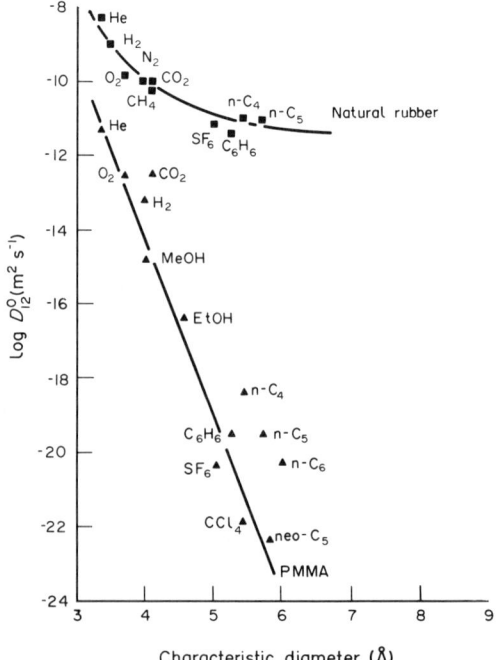

Figure 13 The effect of penetrant size on the infinite dilution diffusion coefficient of small molecules in rubbery and glassy polymers at 35 °C (reprinted with permission from Witchey–Lakshmanan *et al.*, *J. Membr. Sci.*, 1990, **48**, 321)

polymer cohesive energy, density, polymer chain rigidity and penetrant size.[50] For many rubbery and glassy polymers, the logarithm of the infinite dilution mutual diffusion coefficient is directly proportional to the energy of activation for diffusion.[48,69] In rubbery polymers, this rule of thumb is given by

$$\log[D^0_{12}(\rho_1 = 0)] = 0.001 \frac{\Delta E_d}{R} - 8 \qquad (21)$$

and for glassy polymers, the relationship is given by[48]

$$\log[D^0_{12}(\rho_1 = 0)] = 0.001 \frac{\Delta E_d}{R} - 9 \qquad (22)$$

In a series of chemically similar glassy polyimides, the diffusion coefficient of penetrant was observed to depend upon two factors: (i) the degree of packing of the polymer segments; and (ii) the local motion of segments of the polymer chain.[56] The effects of chain packing and torsional mobility on gas transport properties in glassy polymers are lucidly described in a series of articles by Chern and coworkers.[57-59] Penetrant diffusion coefficients typically increase as chain packing is disrupted and decrease as local rotational motions about flexible linkages on the polymer backbone are inhibited.[56,60]

9.5.1.2 Sorption and transport of gases in semicrystalline polymers

Early studies of the transport of gases in semicrystalline polymers such as polyethylene[61-64] indicated that small molecules are soluble in and diffuse through the amorphous regions of the polymer and are excluded from the crystalline region of the material. The crystalline regions in these polymers seem to act as nonsorbing, impermeable barriers dispersed in the amorphous polymeric medium. These experimental findings led to the development of a two-phase model to describe penetrant sorption and transport in semicrystalline polymers. In polyethylene, for example, the concentration of gases is well-described by

$$C = k_d \phi_a p \qquad (23)$$

where ϕ_a is the volume fraction of the polymer which is amorphous and k_d is the Henry's law constant for the purely amorphous polymer.[62] Penetrant diffusivity was observed to decrease with increasing crystallinity. This decrease was attributed to two allied effects: (i) the presence of impervious crystallites offers a more tortuous path to the penetrant; and (ii) the polymer segmental mobility seems to be restricted in the amorphous phase due to the presence of the crystallites.[62] The restriction of polymer segmental dynamics was inferred from activation energies for diffusion which were higher in samples with higher crystalline content. However, decreases in the observed activation energy of diffusion for helium in semicrystalline polymers have been observed, and this phenomenon is ascribed to diffusion along 'grain boundaries' between amorphous and crystalline regions of the polymer.[51] The arrangement of crystalline regions in polymers is believed to influence the transport of small molecules through the polymer; a review article by Petropoulos summarizes the predictions of models designed to describe the influence of morphology on gas transport.[2] Strong changes in the diffusion coefficients of gases in semicrystalline polymers are observed upon orientation of the crystalline regions by polymer processing operations. Good descriptions of these effects (in both amorphous and glassy semicrystalline) are available elsewhere.[50,68]

Recent studies of sorption and transport of small molecules through semicrystalline poly(4-methyl-1-pentene) (PMP) have led to the hypothesis that, when the crystalline density is sufficiently low, gases can dissolve into and diffuse through the crystalline phase of semicrystalline polymers.[65] The chemical structure of PMP is represented by (**1**). PMP is unusual because the crystalline phase density of this polymer is *lower* than the amorphous phase density. In contrast, the density of crystalline polyethylene is approximately 15% higher than the density of amorphous polyethylene.[61]

(**1**)

Puleo and Paul have observed that the low density crystallites found in PMP appear to permit solubility and mobility of methane and carbon dioxide in the crystalline regions.[65] The solubility of these gases in the crystalline regions appears, however, to be approximately 25–30% of that observed in the amorphous regions of this polymer. The diffusion coefficients of carbon dioxide and methane were approximately 60% and 300% higher, respectively, in the amorphous phase.[65] The authors caution, however, that the method for determining the diffusion coefficients in the crystalline phase is approximate.

9.5.1.3 Sorption, transport and permeation of gases and vapors in liquid crystalline polymers

There is a growing industrial and scientific interest in the properties of liquid crystalline polymers (LCPs). Thermotropic LCPs, which exhibit liquid crystalline order as a result of changes in temperature, are of particular interest for industrial melt processing applications, and several thermotropic copolyesters have been produced on a commercial scale.[66] These materials have unusually high stiffness and strength (comparable to metals), outstanding solvent resistance, coefficients of thermal expansion which are low relative to those of conventional amorphous polymers, low density (relative to metals), and very good thermal stability, allowing them to be used at high temperatures. Moreover, they exhibit low melt viscosity and, therefore, can be shaped and molded by conventional melt processing. Determinations of the transport properties of gases and vapors through LCPs are, as yet, rare.[51,67,68] The limited information available, however, suggests that LCPs may be among the most impermeable polymers currently available. In food and beverage packaging applications, where water resistance at high retorting temperatures is desirable, these materials, many of which are hydrophobic, may provide an alternative to the current technology which involves complex coextrusion techniques to protect a barrier layer of expensive and water sensitive poly(ethylene vinyl alcohol) from humid environments.[69]

Table 1 shows a comparison of the solubility, diffusivity and permeability (equal to the product of diffusivity and solubility in this case) of several gases in three liquid crystalline polymers and two organic glassy polymers. Permeability is the flux of gas, per unit pressure gradient driving force, across a flat membrane of uniform thickness.[50] Transient permeation experiments were used to determine the permeability of gases through the LCP and to extract solubility and diffusion coefficients.[67,68] The glassy polymers poly(acrylonitrile) (PAN) and poly(ethyleneterephthalate) (PET) selected by Paul and coworkers for comparison are particularly appropriate since PAN is a noted barrier plastic, and PET is similar chemically to two of the liquid crystalline polymers shown in the table. The PET was wholly amorphous. The LCP labelled HBA/HNA is a copolyester of p-hydroxybenzoic acid (HBA) and 6-hydroxy-2-naphthoic acid (HNA) similar to Vectra®, a

Table 1 Comparison of Sorption and Transport Behavior of Small Molecules in Liquid Crystalline Polymers and Common Barrier Polymers at 35 °C

Gas	Polymer	Solubility ($\times 10^3$ m^3(STP)/m^3 atm)	Diffusivity ($\times 10^{14}$ m^2 s^{-1})	Permeability ($\times 10^{15}$ cm^3(STP)cm/cm^2 s cmHg)[d]
O_2	HBA/HNA[a]	5.0	7.1	47
	PAN[a]	290	0.14	54
	PET/PHB (60)[b]	10	83	1100
	PET/PHB (80)[b]	8.8	41	500
	PET[b]	88	95	11 000
N_2	HBA/HNA[a]	1.6	1.4	3.0
	PAN[a]	52	0.042	2.9
	PET/PHB (60)[b]	2.9	32	120
	PET/PHB (80)[b]	3.9	11	50
	PET[b]	42	32	1700
CO_2	HBA/HNA[a]	51	0.96	70
	PAN[a]	9200	0.023	280
	PET/PHB (60)[b]	89	28	3300
	PET/PHB (80)[b]	76	12	120
	PET[b]	2000	16	43 000

[a] D. H. Weinkauf and D. R. Paul, *ACS Symp. Ser.*, 1990, **423**, 60. Data were taken at 35 °C; Celanese LCP is similar in structure to Vectra.
[b] D. H. Weinkauf and D. R. Paul, *Polym. Prepr.*, 1990, **30**, 3. Data were taken at 35 °C; Kodak LCP; 60 and 80 are the mol% rigid component (PHB). This material is a copolymer of PET and PHB. [c] 1 atm = 1.013×10^5 Pa. [d] 1 cm Hg ≈ 1330 Pa.

commercial polymer manufactured by the Hoechst–Celanese Corporation.[67] The LCPs labeled PET/PHB 60 and PET/PHB 80 are copolymers of PET and p-oxybenzoate (PHB) containing 60 and 80 mol % PHB, respectively. These composite data seem to indicate that the solubility of gases in the liquid crystalline polymers is markedly reduced relative to gas solubilities in PAN and PET, while the value of the diffusion coefficients in LCPs lies between values observed in PAN and PET. In contrast, in semicrystalline polymers, an increase in crystallinity is accompanied by a monotonic decrease in both the overall solubility and mobility of the penetrant. The activation energy for diffusion in the LCP was found to be 30–50% larger, for a given penetrant, than activation energies in either PAN or PET.[67] This result would seem to be consistent with the stiff nature of the polymer backbone but inconsistent with the large diffusion coefficients.

9.5.1.4 Sorption, transport and permeation of gases and vapors in substituted polyacetylenes

Polymers used as impermeable barriers are typically glassy polymers, while rubbery polymers are typically regarded as being highly permeable. However, Masuda et al.[70,71] have reported that the substituted polyacetylene poly(1-(trimethylsilyl)-1-propyne) (PMSP) exhibits oxygen permeability at ambient conditions an order of magnitude greater than rubbery poly(dimethylsiloxane), which had previously been recognized as the most permeable polymer to oxygen.[72] A two-dimensional representation of the chemical structure of PMSP is given by (2). Unlike highly permeable, flexible chain rubbery polymers, such as poly(dimethylsiloxane), PMSP exhibits a glass transition temperature in excess of 300 °C.[73] This polymer exists at room temperature as a stiff, high modulus, brittle, glassy, amorphous polymer.[73] Rotational freedom along the chain backbone is severely limited by the steric interference of the double substitution of the methyl group on C-2 and the trimethylsilyl group on the contiguous monomer unit. The steric hindrance and the double bonds in the polymer backbone are believed to lead to a rigid, twisted helix spatial arrangement of the polymer chain.[73] This chain shape presumably causes extremely poor interchain packing, resulting in a density in the range of 0.7 g cm^{-3}, well below that of conventional glassy or rubbery polymers.[72] Spectroscopic evidence suggests that the alternating carbon–carbon double bonds along the chain backbone are not conjugated, probably due to the twisted chain structure.[73] These substituted polyacetylenes are, therefore, much more resistant to air oxidation than conventional, unsubstituted polyacetylene.

(2)

Polyacetylenes containing a wide variety of substituents, similar in chemical structure to PMSP, have been synthesized. The oxygen permeability of many of these polymers has been determined.[74] Based on this information, Masuda et al.[74] recognized that 'The fact that poly(t-butylacetylene) and poly(o-(trifluoromethyl)phenylacetylene) are also very permeable indicates that the presence of a silyl group is not a requirement for high permeability'. Data obtained by Masuda et al., reproduced in Table 2, illustrate that the permeability of the silicone-containing members of this family of polymers may, in fact, be quite low relative to that of PMSP. There is no ready paradigm which is apparent from Table 2 which would provide a rational basis for relating the two-dimensional structural representation of these polymers to the observed permeability. The primary structure of this polymer series (the chemical formula) does not provide adequate insight into the subtle higher order structural features which, therefore, might be responsible for the observed permeation properties.

Disubstituted PMSP and liquid crystalline polymers such as Vectra® (a copolymer manufactured by Hoechst–Celanese containing HBA and HNA), which are very high and low permeability polymers, respectively, share important structural features. Both polymers apparently exist as very stiff, rigid glassy materials at ambient conditions. Chain-packing considerations seem to be very important in determining the permeation properties of these materials. While the double bonds in the chain backbone combine with the bulky sidegroup in PMSP to produce rigid, twisted polymer chain conformations which pack together very inefficiently, the rigid moieties in the backbone of main chain liquid crystalline polymers promote efficient chain packing to achieve liquid crystalline order and bulk densities of the order of 1.4 g cm^{-3}. Very permeable and very impermeable polymers appear, therefore, to be very rigid glassy materials which differ substantially in the ability of the

Table 2 A Comparison of the Oxygen Permeability of Several Substituted Polyacetylenes and Poly(dimethylsiloxane)

—(CR=CR')$_n$—		
R	R'	Oxygen permeability at 25 °C[a] ($\times 10^{-10}$ cm^3(STP)cm/cm^{2}cmHg s)
Me	SiMe$_3$	6100
Me	SiEt$_3$	860
H	Bu^t	130
Me	SiMe$_2$CH$_2$SiMe$_2$	81
Me	SiMe$_2$(n-C$_6$H$_{13}$)	17
Poly(dimethylsiloxane)		600[b]

[a] T. Masuda and T. Higashimura, in 'Advances in Polymer Science', Springer-Verlag, Berlin, 1987, vol. 81, p. 121. [b] K. Takada, H. Matsuya, T. Masuda and T. Higashimura, *J. Appl. Polym. Sci.*, 1985, **30**, 1605.

macromolecules to organize in space. This observation would suggest that higher order structural features, which control chain packing, appear to confer very high and very low permeability properties upon these materials.

9.5.2 Review of Theoretical Approaches

In order to determine completely the transport behavior of a penetrant in a polymer, two pieces of information are needed: (i) a description of the dependence of the penetrant chemical potential or activity in the polymer on penetrant concentration, temperature, pressure and the nature and state of the polymer; and (ii) a constitutive relation between the flux of the penetrant in the polymer and the driving force for mass transfer. In multicomponent systems in which there are no pressure or thermal diffusion effects and in which there are no ionic effects, the driving force for mass transfer is taken as the gradient in chemical potential.[75] In the section which follows, a review of chemical potential relationships or, equivalently, solubility expressions, and of flux constitutive relationships is given.

9.5.2.1 Sorption models

(i) Henry's law

The Henry's law model of the solubility of a penetrant in a polymer postulates a linear relationship between the concentration of a penetrant in the polymer and the partial pressure of the penetrant in a contiguous phase. For gas and vapor sorption, this relationship takes the form

$$C = k_d p \tag{24}$$

where C is the concentration of penetrant in the polymer, k_d is the Henry's law constant, and p is the partial pressure of the penetrant in the pure penetrant phase surrounding the polymer sample. This relationship may be written more generally in terms of the penetrant activity, a

$$C = k'_d a \tag{25}$$

where the penetrant activity is given by

$$a = \gamma p/p^0 \tag{26}$$

where γ is the activity coefficient of the penetrant in the phase contiguous to the polymer, p is the partial pressure of the penetrant in the penetrant phase, and p^0 is the saturation vapor pressure of the penetrant. Henry's law represents ideal dilute solution behavior. At a fixed temperature, the Henry's law constant is independent of penetrant concentration in the polymer. Henry's law typically describes well the sorption of fixed gases into rubbery polymers. The logarithm of the Henry's law constant has been found to be a linear increasing function of the Lennard–Jones force constant.[76,77] As discussed earlier, gas solubility and, therefore, k_d typically follow an Arrhenius relationship with temperature, with a heat of sorption which is positive or negative depending upon the relative magnitude of the polymer–penetrant interaction energy and penetrant heat of condensation. Compendia of Henry's law constants for various gaseous penetrants in polymers are available.[78]

(ii) Flory–Huggins

The sorption of organic vapors into uncrosslinked, amorphous, rubbery polymers and the sorption of water into hydrophobic polymers are often well described by the Flory–Huggins model

$$\ln a = \ln \Phi_1 + 1 - \Phi_1 + \chi(1 - \Phi_1)^2 \qquad (27)$$

where Φ_1 is the volume fraction of penetrant in the polymer, and χ is an energetic interaction term. The volume fraction is related to the concentration of penetrant in the polymer by

$$\Phi_1 = \frac{(C/0.0224)\bar{V}_1}{1 + (C/0.0224)\bar{V}_1} \qquad (28)$$

where $\bar{V}_1$ is the partial molar volume of the penetrant in the polymer. This expression may be inverted to give another useful expression

$$C = \frac{\Phi_1}{1 - \Phi_1} \frac{0.0224}{\bar{V}_1} \qquad (29)$$

The factor 0.0224 is the number of cubic meters of gas in one mole at standard temperature and pressure.

The Flory–Huggins model, like several models described in this section, is based upon a lattice theory approach to polymer thermodynamics. Polymer segments and solvent molecules are understood to occupy sites on a lattice, and the techniques of statistical thermodynamics are applied to determine the behavior of the polymer–penetrant mixture on the lattice. The first two terms on the right hand side of equation (27) are associated with the entropy change upon mixing polymer segments and penetrant molecules. The last term characterizes the energetics of interactions between pairs of polymer segments, pairs of penetrant molecules, and between pairs consisting of one polymer molecule and one solvent molecule.[20] The energetic interaction parameter χ may be estimated for nonpolar systems from the solubility parameters of the solvent and polymer, as shown in equation (7).

The Flory–Huggins model has also been extended to crosslinked polymeric systems. In this case, the penetrant concentration in the polymer at a given activity is

$$\ln a = \ln \Phi_1 + 1 - \Phi_1 + \chi(1 - \Phi_1)^2 + \frac{\bar{V}_1 N_e}{V_0}\left[(1 - \Phi_1)^{1/3} - \frac{1 - \Phi_1}{2}\right] \qquad (30)$$

where N_e is the effective number of crosslinks and V_0 is the volume occupied by the penetrant-free polymer.[79]

(iii) Dual Mode

The sorption of gases and vapors into glassy polymers is characterized by sorption levels which can be one order of magnitude larger than in rubbery polymers (at the same penetrant activity) and by sorption isotherms which are concave to the pressure or activity axis. Additionally, due to the nonequilibrium nature of the glassy state, the prior processing history of the polymer can strongly affect its gas sorption properties. The dual mode sorption concept has been used very successfully to model gas sorption in glassy polymers. The model is given by

$$C = k_d p + C'_H \frac{bp}{1 + bp} \qquad (31)$$

where k_d is the Henry's law constant, C'_H is the Langmuir capacity, and b is the Langmuir affinity. This model assumes that the penetrant may dissolve into the bulk of the densified polymer matrix, with a solubility characterized by the Henry's law constant. This mode of transport is the only one available in rubbery materials. In glassy polymers the model also allows penetrant sorption, *via* the Langmuir isotherm in equation (31), into the nonequilibrium excess volume associated with the glassy state.[50] The amount of excess volume available in the glass is characterized by C'_H, which has been shown to vary systematically with the excess volume in the glass, as determined from dilatometric studies.[77,80,81] Upon increasing temperature C'_H decreases and becomes zero at or near the polymer glass transition temperature, leaving only the Henry's law mode of sorption in the rubbery state. The Henry's law constant and the Langmuir affinity constant show an Arrhenian dependence upon temperature.[81]

The three models just described have provided the cornerstone for the understanding of the solubility of gases and vapors in both rubbery and glassy polymers. More recent models, based upon different physical assumptions, have been proposed to describe sorption equilibrium data in both glassy and rubbery polymers. The following sections describe two of the models which characterize current efforts in the field.

(iv) Lipscomb Model

The model proposed by Lipscomb attempts to provide a uniform framework for interpreting gas sorption data in glassy and rubbery polymers.[82] In this model, the Flory–Huggins theory is extended to allow for the influence of polymer deformation, due to the addition of solvent molecules to the polymer, and residual polymer stress, due to prior processing history, on the free energy of mixing and, therefore, on the penetrant chemical potential. The concentration of the penetrant in the polymer is given by

$$\ln a = \frac{\text{tr}(\sigma_R)}{3RT}\bar{V}_1 + \frac{B_0 \bar{V}_1^2}{RT}\frac{C}{0.0224} + \ln \Phi_1 + 1 - \Phi_1 + \chi(1-\Phi_1)^2 \qquad (32)$$

The residual stress in the solid polymer due to prior processing history is given by σ_R. B_0 is the bulk modulus of the undeformed polymer. The first term on the right hand side of equation (32) describes, through residual stress in the polymer, the influence of the processing history of the polymer prior to contact with the penetrant on the activity of the penetrant in the polymer. This term attempts to account for the well-known processing history effects which have been observed in glassy polymers. It is proposed that the residual stress term may be made negligibly small by annealing the polymer sample.[82] Additionally, the residual stress of a rubbery sample should be negligibly small. The second term on the right hand side of equation (32) describes the change in the penetrant activity due to internal stresses generated in the polymer when the penetrant enters and swells the polymer chains. This effect is proportional to the polymer bulk modulus and should, therefore, be substantially more important in glassy rather than rubbery polymers. Lipscomb proposes that this effect is negligible in rubbery polymers. Thus the Lipscomb model reduces to the Flory–Huggins model in rubbery polymers and contains two additional terms whose purpose is to capture, in glassy polymers, the influence of residual stresses due to prior processing history and penetrant-induced mechanical stress.

The Flory–Huggins model predicts sorption isotherms which are always concave upwards when plotted against penetrant activity and, therefore, cannot describe gas sorption isotherms observed in glassy polymers, which are typically concave downward at low penetrant activities. Due to the presence of the mechanical terms, the model proposed by Lipscomb can predict isotherms which are concave downwards. This model, like the Flory–Huggins model, cannot predict volume changes upon mixing the polymer and penetrant *a priori*.

(v) Sanchez

Pope *et al.* have used the Sanchez–Lacombe lattice fluid model to estimate the equilibrium solubility of several gases in silicone rubber.[83] In the Sanchez–Lacombe model, polymer chains are modeled as a connected set of interacting beads on a lattice. The so-called 'van der Waals' interaction potential (which assumes that the potential energy of interaction scales as the inverse of the volume) is used to describe interactions between components on neighboring lattice sites.[84] Dee and Walsh have asserted that equation of state models which use the van der Waals potential do not describe the behavior of pure polymeric fluids as well as those based on the Lennard–Jones interaction potential.[84] Like the Flory–Huggins model, the Sanchez–Lacombe model assumes that polymer chains and penetrant molecules mix together randomly on a lattice (the so-called mean field approximation). Unlike the Flory–Huggins model, the Sanchez–Lacombe model permits some lattice sites to be empty, which allows for holes or free volume in the fluid. The addition of free volume to the lattice permits, for example, the calculation of volume changes upon mixing the polymer and penetrant.[85] This model provides equations of state and chemical potential expressions for the pure components and the mixture. Only data on the pure components are required to predict mixture properties, which allows prediction of, for example, sorption isotherms and penetrant partial molar volume with no adjustable constants. The model has been used to predict the properties of low molecular weight pure fluids,[86] mixtures of low molecular weight fluids,[85] pure polymers in the liquid state,[87] polymer solutions[88] and gas–polymer mixtures.[89]

This model has been used to describe the sorption of gases into both glassy and rubbery polymers by Kiszka *et al.*[90] The model was found to be capable of describing gas sorption into rubbery polymers substantially better than gas sorption into glassy polymers. The model does not provide a description of the nonequilibrium properties of glassy polymers and would, therefore, be most suitable for modeling the sorption properties of penetrants into rubbery polymers.

Pope *et al.* used the Sanchez–Lacombe equation of state to model the equation of state properties of the pure penetrant phase, the polymer–penetrant mixture, and the chemical potential of the penetrant in both the penetrant and polymer phases. The amount of gas sorbed into a polymer at a given temperature and pressure is, therefore, determined from the solution of the following set of three coupled, nonlinear equations for the density of penetrant in the pure penetrant phase, the density of the polymer–penetrant mixture, and the volume fraction of penetrant in the polymer. The equations to be solved are the equation of state for pure penetrant in penetrant phase

$$\tilde{\rho}_1 = 1 - \exp\left[-\frac{\tilde{\rho}_1^2}{\tilde{T}_1} - \frac{\tilde{P}_1}{\tilde{T}_1} - \left(1 - \frac{1}{r_1}\right)\tilde{\rho}_1\right] \quad (33)$$

the equation of state for penetrant in penetrant–polymer mixture

$$\tilde{\rho} = 1 - \exp\left[-\frac{\tilde{\rho}^2}{\tilde{T}} - \frac{\tilde{P}}{\tilde{T}} - \left(1 - \frac{\Phi_1}{r_1}\right)\tilde{\rho}\right] \quad (34)$$

and the equation for equilibration of the penetrant chemical potential between the pure penetrant and penetrant–polymer mixture phases

$$\left[-\frac{\tilde{\rho}_1}{\tilde{T}_1} + \frac{\tilde{P}_1}{\tilde{T}_1\tilde{\rho}_1} + \frac{(1-\tilde{\rho}_1)\ln(1-\tilde{\rho}_1)}{\tilde{\rho}_1} + \frac{\ln\tilde{\rho}_1}{r_1}\right]r_1 = \ln\Phi_1 + 1 - \Phi_1 + \tilde{\rho}\frac{M_1}{\rho_1^*}\chi(1-\Phi_1)^2$$

$$+ \left[-\frac{\tilde{\rho}}{\tilde{T}_1} + \frac{\tilde{P}_1}{\tilde{T}_1\tilde{\rho}} + \frac{(1-\tilde{\rho})\ln(1-\tilde{\rho})}{\tilde{\rho}} + \frac{\ln\tilde{\rho}}{r_1}\right]r_1 \quad (35)$$

The reduced variables are defined by

$$\tilde{\rho}_1 \equiv \rho_1/\rho_1^* \quad (36)$$

$$\tilde{\rho} \equiv \rho/\rho^* \quad (37)$$

$$\tilde{P}_1 \equiv P/P_1^* \quad (38)$$

$$\tilde{P} \equiv P/P^* \quad (39)$$

$$\tilde{T}_1 \equiv T/T_1^* \quad (40)$$

$$\tilde{T} \equiv T/T^* \quad (41)$$

where ρ_1 is the density of the penetrant, and ρ is the density of the polymer–penetrant mixture. The bracketed term on the right hand side of equation (35) was not included in the analysis by Pope *et al.*, as it was expected to be negligibly small in the condensed phase.[91] Each component of the mixture is characterized by three parameters:[89] (i) P_i^*, the reduced pressure, which is the hypothetical cohesive energy density of component i at absolute zero; (ii) T_i^*, the reduced temperature, which does not have a clear physical significance; and (iii), ρ_i^*, the reduced density, which is the hypothetical mass density of component i at absolute zero. Here, $i = 1$ corresponds to penetrant and $i = 2$ corresponds to polymer. These parameters may be determined from pure component experimental PVT data.[83] The number of lattice sites occupied by a penetrant molecule, r_1, may be determined from the penetrant parameters as follows

$$r_1 = \frac{P_1^* M_1}{RT_1^* \rho_1^*} \quad (42)$$

For polymers, r_2 is set to infinity and does not appear in the chemical potential of equation of state expressions. The following mixing rules are used to obtain characteristic mixture parameters

$$P^* = \Phi_1 P_1^* + \Phi_2 P_2^* - \Phi_1\Phi_2[(P_1^*)^{1/2} - (P_2^*)^{1/2}]^2 \quad (43)$$

$$T^* = \frac{P^*}{\frac{\Phi_1 P_1^*}{T_1^*} + \frac{\Phi_2 P_2^*}{T_2^*}} \quad (44)$$

$$\frac{1}{\rho^*} = \frac{\omega_1}{\rho_1^*} + \frac{\omega_2}{\rho_2^*} \quad (45)$$

The volume fraction of penetrant in the polymer is related to the weight fraction of penetrant in the polymer as follows

$$\Phi_1 = \frac{(\omega_1/\rho_1^*)}{(\omega_1/\rho_1^*) + (\omega_2/\rho_2^*)} \qquad (46)$$

and the concentration of penetrant in the polymer, C, may be calculated from equation (28). The partial molar volume of the penetrant is required in equation (28), and it may be calculated directly from the theory.

A principal advantage of this approach is the ability to provide predictions of mixture properties using only pure component parameters. According to Pope et al.[83] the model provides estimates of gas solubility in silicone rubber over a wide pressure range which are within 15% of the experimentally determined sorption data. Model predictions of the partial molar volume of penetrants in silicone rubber also compare favorably with experimental results.

9.5.2.2 Diffusion flux models

For many polymer–penetrant diffusion problems, Fick's first law, with a composition independent diffusion coefficient, is combined with the mass conservation equations and solved to yield penetrant concentrations as a function of spatial position and time. In these problems, the average velocity which appears in the mass conservation relations is often set to zero.[92] Under these conditions, analytical solutions are available for a wide range of geometries and boundary and initial conditions.[93]

Penetrant diffusion through rubbery polymers is typically well described by the Fickian transport model.[92] Fickian behavior is also commonly observed in glassy polymers when the penetrant is very dilute. The signature of Fickian diffusion in a thin slab of polymer contacted with a penetrant is a weight increase in the slab, due to penetrant absorption by the polymer, which increases initially with the square root of contact time between the polymer and penetrant before monotonically approaching a fixed equilibrium value.[93]

In amorphous glassy polymers deviations from the ideal Fickian model are often observed, particularly when the penetrant is substantially below its critical temperature.[92] The time dependence of the weight uptake of penetrant by the polymer in the sorption experiment described in the previous paragraph may exhibit a rich spectrum of behavior. For example, the penetrant may sorb into the polymer in two stages, an initial Fickian-like stage followed by a protracted drift towards a final equilibrium uptake.[94] The penetrant weight uptake may be linear in contact time until reaching equilibrium, and penetrant concentration profiles inside the polymer may be very sharp. This behavior is known as 'case II' sorption to distinguish it from Fickian, or case I, sorption.[95,96] The initial rate of penetrant uptake may also lie between the poles of Fickian and case II diffusion.[97] Additionally, oscillations in the weight uptake have been observed as the weight uptake approaches its equilibrium value.[98] These examples provide a brief overview of the breadth of responses which may be observed. More detailed descriptions of non-Fickian phenomena are available elsewhere.[50,92,99] The common feature in these examples is a time dependence of penetrant uptake which cannot be described by a diffusion coefficient and penetrant activity which depends only upon local penetrant concentration (at fixed temperature and pressure).

The molecular processes leading to this behavior are commonly understood to be intimately related to the relative rates of solvent ingress and polymer chain deformation and relaxation accompanying the invasion of penetrant molecules into the polymer. If the polymer chains are not deformed significantly by the penetrant or if the chains reorganize very quickly relative to the rate of solvent penetration, Fickian diffusion is observed.[46] Between the bounds of no response and an infinitely fast response by the polymer chains to the presence of the penetrant molecules, there lies a regime where the rates of polymer deformation and swelling to accommodate the penetrant molecules and the rate of diffusion of the penetrant occur over similar timescales. Non-Fickian or anomalous diffusion, accompanied by significant swelling of the polymer sample, is commonly understood to occur in this regime. As the penetrant invades and swells the polymer, local stresses are built up when the polymer chains disentangle from one another.[50] These local stresses may be quite high, and can cause mechanical failure (such as crazing) in the polymer.[92] Recent efforts to model non-Fickian diffusion have focused on properly accounting for the influence of stresses and stress histories (since polymers and concentrated polymer solutions typically behave as viscoelastic materials) on the penetrant chemical potential. Several recently proposed models are described below which characterize the philosophy of the effort to model non-Fickian or anomalous diffusion of penetrants into amorphous glassy polymers.

(i) Carbonell and Sarti

A model for diffusion of a low molecular weight penetrant into a linear viscoelastic solid has been presented by Carbonell and Sarti.[92] In this model, the Helmholtz free energy (at fixed temperature and pressure) is postulated to be a function of penetrant concentration, penetrant concentration history in the polymer, polymer strain and polymer strain history. The stress in the polymer–penetrant mixture is calculated from a Maxwell identity which relates the stress to the derivative of the Helmholtz free energy with respect to strain. The expression for the Helmholtz free energy of the polymer–penetrant mixture is used to calculate the difference in chemical potential between the penetrant and polymer. The solvent mass flux is related to the gradient in solvent chemical potential as follows[75]

$$j_1 = -\frac{(\rho_2 M_1 + \rho_1 M_2)}{RT} D_{12} \omega_1 \nabla \mu_1^m \tag{47}$$

where μ_1^m is the mass-based chemical potential of the penetrant, which is related to the molar chemical potential of the penetrant as follows

$$\mu_1 = M_1 \mu_1^m \tag{48}$$

The expression obtained for the chemical potential difference between the penetrant and polymer for an isotropic solid is

$$\mu_1^m - \mu_2^m = \mu^{*m} - \frac{\beta}{\rho} \operatorname{tr} \sigma - \int_{-\infty}^{t} \omega_1(t') \frac{\partial}{\partial t'} \left[m(t-t') - \frac{3\beta^2}{\rho} G_2(t-t') \right] dt' \tag{49}$$

where μ^{*m} is the mass-based chemical potential difference between the penetrant and polymer due to a strain-free mixing of the components, which is taken to be a linear function of penetrant weight fraction in the polymer–penetrant mixture. Here, β is the normalized partial mass volume of the mixture, defined as the partial mass volume times the total mass of polymer and penetrant divided by the initial penetrant-free volume of the polymer. The stress in the polymer–penetrant mixture is represented by σ. The bulk modulus of the polymer–penetrant mixture is $G_2(t)$. The function $m(t)$ is a memory function which describes the second order effects of concentration history on the mixture Helmholtz free energy. The stress of the mixture is given by

$$\sigma = \int_{-\infty}^{t} G_1(t-t') \frac{\partial \hat{E}}{\partial t'} dt' + \left\{ \frac{1}{3} \int_{-\infty}^{t} [G_2(t-t') - G_1(t-t')] \frac{\partial \operatorname{tr} \hat{E}}{\partial t'} dt' \right\} \mathbf{I} \tag{50}$$

where $G_1(t)$ is the shear modulus of the polymer–penetrant mixture. The difference between the total strain tensor of the polymer–penetrant mixture and the strain due only to isotropic swelling is given by $\hat{E}$, and 'tr' signifies the trace of the tensor. Using expressions for the chemical potential of the polymer and penetrant in the absence of mechanical deformation, appropriate viscoelastic functions to describe the stress–strain relationship in the polymer mixture, and a suitable memory function for the concentration history of the polymer, this model may be used to estimate the influence of strain, and the attendant stresses generated by the strains, in the polymer on the flux of solvent. Moreover, the model provides a thermodynamically consistent approach for including the effect of stress on the free energy of the polymer–penetrant mixture and, therefore, on the penetrant diffusive flux.

(ii) Durning

A model has been proposed by Durning[100,101] to describe non-Fickian transport processes in polymer–penetrant systems. The model proposes an expression for the penetrant chemical potential which accounts for the influence of the deformation and flow of polymer on the penetrant chemical potential. The model treats the polymer as a network of intertwining chains which slowly disentangle and swell to accommodate penetrant molecules that enter the polymer. As the penetrant molecules enter the polymer, they induce deformation and flow of the polymer segments. Two molecular theories (transient network and reptation) are shown to give similar results for the change in the Gibbs free energy of the polymer–penetrant mixture due to deformation of the polymer chains. This change of free energy is added to the free energy of mixing the pure, undeformed polymer and penetrant to determine the overall Gibbs free energy change upon mixing the two components. The solvent chemical potential is then evaluated from the Gibbs free energy change upon mixing.

The solvent flux is written in terms of j_1^2, the mass flux of penetrant relative to the polymer velocity

$$j_1^2 = \rho_1(v_1 - v_2) \tag{51}$$

which is directly related to the gradient in penetrant chemical potential. In terms of the mass flux relative to fixed coordinates, n_1, defined in the Appendix, j_1^2 is given by

$$j_1^2 = n_1 - \frac{\rho_1}{\rho_2} n_2 \tag{52}$$

In the case of diffusion in one direction (the z direction), the following constitutive relation is proposed for the mass flux of penetrant

$$j_1^2 = -D\frac{\partial c_1'}{\partial \xi} - D'\frac{\partial}{\partial \xi}\int_{-\infty}^{t} \phi(t-t')\frac{\partial c_1'}{\partial t'}dt' \tag{53}$$

where c_1' is the concentration of penetrant (g penetrant per cm^3 polymer), $\phi(t-t')$ is the viscoelastic relaxation function, and D is a diffusion coefficient which can be written as

$$D \equiv D_{12}^0(\rho_2 \hat{V}_2)^2 \tag{54}$$

where D_{12}^0 is the binary mutual diffusion coefficient. The coefficient D' is given by

$$D' \equiv \frac{D\bar{V}_1}{RT}\frac{\rho \hat{V}_1}{\omega_2(\partial \ln a/\partial \omega_1)_{T,P}}G_0 \tag{55}$$

where G_0 is the instantaneous shear modulus of the mixture evaluated at the average composition of the mixture, and $G_0\phi(t)$ is the time dependent shear relaxation modulus of the mixture. In polymer systems which swell substantially upon the absorption of a penetrant, the dilation length scale, $d\xi$, is often used to model the diffusion phenomena. This length scale is related to the length scale in the laboratory frame, dz by

$$d\xi = \rho_2 \hat{V}_2 dz \tag{56}$$

Durning offers solutions of the model and comparison to experimental data for one-dimensional diffusion when the coefficients D and D' are constant and the polymer–penetrant mixture behaves as a Maxwell viscoelastic fluid. For a Maxwell fluid, the relaxation function and shear modulus are given by

$$\phi(t-t') = \begin{cases} \exp[-(t-t')/\tau] & t-t' \geq 0 \\ 0 & t-t' < 0 \end{cases} \tag{57}$$

$$G_0 = \frac{\eta}{\tau} \tag{58}$$

where η is the characteristic shear viscosity and τ is the characteristic relaxation time of the mixture. The relaxation time is treated as a constant. The natural logarithm of the penetrant activity is taken to be a linear function of the penetrant concentration. The results of model calculations are compared with experimental sorption measurements of the uptake of benzene by atactic polystyrene.[100] When all model parameters were determined without recourse to the experimental sorption data, the model seems to capture qualitatively the behavior of the system. The authors speculate that better quantitative agreement with the experimental data might be obtained by replacing the Maxwell fluid model with one which more reasonably approximates the viscoelastic behavior of the polymer–penetrant mixture.

(iii) Others

Neogi[102,103] and Camera–Roda and Sarti[104] have proposed similar expressions for the penetrant flux in a polymer–penetrant mixture. The flux includes two terms, one linear in the penetrant concentration gradient (*i.e.* a Fickian contribution) and another linear in the penetrant concentration gradient history. By allowing the flux contribution to the penetrant concentration gradient history to decay exponentially with a single relaxation time and by choosing an appropriate set of parameter values, the model was able to mimic qualitatively the range of anomalous diffusion phenomena which has been observed experimentally. Thus it would appear that a spectrum of relaxation times is not a required feature of a model for describing non-Fickian sorption and diffusion in polymers. The results of these calculations indicate that the type of diffusion behavior observed depends principally upon the Deborah number for diffusion, De, which is the ratio of the relaxation time of the polymer chains to the characteristic timescale of penetrant diffusion[46]

$$\mathrm{De} \equiv \frac{\tau}{L^2/D_{12}} \tag{59}$$

At very low or very high values of the Deborah number, Fickian diffusion is observed. When the Deborah number is near unity, case II transport, oscillations and overshoots are observed in the penetrant weight uptake curves.

9.6 NOMENCLATURE

a	activity of solvent
a_{ev}	excluded volume scale factor (m^3)
b	Langmuir affinity (Pa^{-1})
B_0	bulk modulus of undeformed polymer (Pa)
B_2	second osmotic virial coefficient of a polymer solution (m^3 kg^{-1})
C	concentration of penetrant [m^3(STP) m^{-3}(polymer)]
c_i	molar concentration of component i [mol m^{-3}(mixture)]
c	molar concentration of mixture [mol m^{-3}(mixture)]
c'_1	concentration of penetrant [kg(penetrant) m^{-3}(polymer)]
C'_H	Langmuir capacity [m^3(STP) m^{-3}(polymer)]
C_M	molar concentration of penetrant [mol(penetrant) m^{-3}(mixture)]
D	diffusion coefficient (m^2 s^{-1})
D'	parameter characterizing importance of memory effect on solvent flux in flux model by Durning (m^2 s^{-1})
D_i^*	tracer diffusion coefficient of component i (m^2 s^{-1})
D_{12}	binary mutual diffusion coefficient (m^2 s^{-1})
D_{12}^0	activity-corrected binary mutual diffusion coefficient (m^2 s^{-1})
$\hat{E}$	difference between total polymer–penetrant mixture strain tensor and the strain due only to isotropic swelling
f	friction coefficient per mol of polymer (kg mol^{-1} s^{-1})
f_i	friction coefficient per molecule of component i (kg mol^{-1} s^{-1} molecule^{-1})
f_p	fractional free volume of the polymer–solvent mixture
F_i	frictional force per molecule (N molecule^{-1})
G_0	shear modulus of polymer–penetrant mixture (Pa)
G_1	shear modulus of the polymer–penetrant mixture (Pa)
G_2	bulk modulus of the polymer–penetrant mixture (Pa)
$\mathbf{I}$	unit tensor
j_i	mass flux of component i relative to mass average velocity (kg m^{-2} s^{-1})
j_i^*	molar flux of component i relative to molar average velocity (mol m^{-2} s^{-1})
$j_i^{\square}$	mass flux of component i relative to volume average velocity (kg m^{-2} s^{-1})
j_1^2	mass flux of penetrant relative to polymer velocity (kg m^{-2} s^{-1})
k	Boltzmann's constant (1.38066 × 10^{-23} J mol^{-1})
k_d	Henry's law constant (m^3(STP) m^{-3}(polymer) Pa^{-1})
k'_d	Henry's law constant based on activity (m^3(STP) m^{-3}(polymer) unit activity^{-1})
K	low polymer concentration limit of the derivative of the relative mutual diffusion coefficient with respect to polymer concentration (m^3 kg^{-1})
m	memory function describing second order effects of concentration history on polymer–penetrant mixture Helmholtz free energy (J kg^{-1})
M_i	molecular weight of component i (kg mol^{-1})
M_w	weight average molecular weight of polymer (kg mol^{-1})
N_e	effective number of crosslinks in polymer (mol)
N_A	Avogadro's number
n_i	mass flux of component i relative to fixed coordinates (kg m^{-2} s^{-1})

N_i	molar flux of component i relative to fixed coordinates (mol m^{-2} s^{-1})
p	penetrant partial pressure (Pa)
p^0	penetrant vapor pressure (Pa)
P	pressure (Pa)
P_a	atmospheric pressure (Pa)
P^*	characteristic pressure of polymer–penetrant mixture (Pa)
$\tilde{P}_1$	reduced pressure of penetrant
$\tilde{P}$	reduced pressure of polymer–penetrant mixture
P_i^*	characteristic pressure of component i (Pa)
R	gas constant (J mol^{-1} K^{-1})
R_h	hydrodynamic radius of polymer chain (m)
R_g	radius of gyration of polymer chain (m)
r_i	number of lattice sites occupied by a molecule of component i
r_i^*	spatial position of a labeled molecule of species i (m)
S	solubility of penetrant (m^3(STP) m^{-3}(polymer) Pa^{-1})
S_0	temperature independent prefactor of penetrant solubility (m^3(STP) m^{-3}(polymer) Pa^{-1})
t, t'	time (h)
T	temperature (K)
T_g	glass transition temperature (K)
T^*	characteristic temperature of polymer–penetrant mixture (K)
T_i^*	characteristic temperature of component i (K)
$\tilde{T}$	reduced temperature of polymer–penetrant mixture
$\tilde{T}_1$	reduced temperature of penetrant
v	mass average velocity of mixture (m s^{-1})
v_i	velocity of component i (m s^{-1})
v_i^*	velocity of a labeled molecule of component i (m s^{-1})
v^*	molar average velocity of mixture (m s^{-1})
$\dot{v}$	volume average velocity of mixture (m s^{-1})
$\bar{V}_i$	partial molar volume of component i (m^3 mol^{-1})
$\hat{V}_i$	molar volume of component i (m^3 mol^{-1})
V_{occupied}	occupied volume (m^3)
V_f	free volume (m^3)
V	volume (m^3)
x_i	mole fraction of component i
X	normalized polymer concentration
z	length scale fixed relative to laboratory coordinates (m)
Z	excluded volume parameter

Greek Symbols

β	normalized partial mass volume of the mixture
γ	penetrant activity coefficient
ΔE_d	activation energy for diffusion (J mol^{-1})
Γ	memory function describing the penetrant concentration history of the polymer–penetrant mixture (J kg^{-1})
Γ_{MS}	monomer–solvent potential energy of interaction (J)
Γ_{SS}	solvent–solvent potential energy of interaction (J)
Γ_{MM}	monomer–monomer potential energy of interaction (J)
δ_i	solubility parameter of component i (Pa$^{-1/2}$)
ΔH_{mix}	energy of mixing the pure, condensed penetrant and the polymer (J mol^{-1})
ΔH_{cond}	heat of condensation of pure penetrant (J mol^{-1})
ΔH_S	overall heat of sorption (J mol^{-1})
η	pure solvent viscosity (kg m^{-1} s^{-1})
ρ^*	characteristic mass density of polymer-penetrant mixture (kg m^{-3})
ρ_i^*	characteristic mass density of component i (kg m^{-3})
$\tilde{\rho}$	reduced mass density of polymer–penetrant mixture
$\tilde{\rho}_1$	reduced mass density of penetrant
ρ	mass density of polymer–penetrant mixture [kg(mixture) m^{-3}(mixture)]
ρ_i	mass density of component i [kg m^{-3}(mixture)]

ρ_2^*	mass density of polymer at which entanglement effects begin to be observed [kg m^{-3}(mixture)]
μ_i^m	mass-based chemical potential of component i (J kg^{-1})
μ_i	molar-based chemical potential of component i (J mol^{-1})
μ^{*m}	mass-based chemical potential difference between the penetrant and polymer due to strain-free mixing of the components (J kg^{-1})
v	excluded volume parameter (m^3)
Φ_i	volume fraction of component i
Φ_a	volume fraction of amorphous polymer in a semicrystalline polymer
σ	stress of the polymer–penetrant mixture (Pa)
χ	Flory–Huggins interaction parameter
σ_R	residual stress tensor (Pa)
ω_i	mass fraction of component i
η	characteristic viscosity of polymer–penetrant mixture (kg m^{-1} s^{-1})
τ	characteristic relaxation time of polymer–penetrant mixture(s)
ξ	dilation length scale (m)

9.7 APPENDIX

9.7.1 Unifying Analytical Fundamentals

9.7.1.1 Reference frames and flux definitions

The complete description of mass transport phenomena in a binary mixture requires expressions for the component fluxes. A flux relation is always accompanied by a statement specifying the frame of reference relative to which the flux is measured. In practice, diffusion problems, typically employ fluxes defined relative to a fixed frame of reference or relative to frames of reference based upon the mass, molar or volume average velocity of the mixture. As discussed in more detail below, the choice of an appropriate frame of reference often simplifies the analytical solution of a problem. The mass average velocity of a binary mixture, v, is defined by

$$v \equiv \omega_1 v_1 + \omega_2 v_2 \tag{60}$$

whereas the molar average velocity is given by

$$v^* \equiv x_1 v_1 + x_2 v_2 \tag{61}$$

The volume average velocity is

$$\dot{v} \equiv \bar{V}_1 c_1 v_1 + \bar{V}_2 c_2 v_2 = \Phi_1 v_1 + \Phi_2 v_2 \tag{62}$$

where v_i is the velocity of component i, ω_i is the mass fraction of component i, $\bar{V}_i$ is the partial molar volume of component i, and Φ_i is the volume fraction of component i. In this chapter, $i = 1$ corresponds to solvent or penetrant and $i = 2$ corresponds to polymer. The molar concentrations of penetrant and polymer are c_1 and c_2, respectively; the mass densities of the penetrant and polymer are ρ_1 and ρ_2, respectively. The mass density is related to the molar concentration by $c_i = \rho_i/M_i$.

In Table 3, several fluxes are defined. In this table, the total mass density ρ is defined as the sum of the component mass densities (i.e. $\rho = \rho_1 + \rho_2$), and the total molar concentration c is defined as the sum of the component molar concentrations (i.e. $c = c_1 + c_2$). The component mole and mass fractions are defined as $x_i = c_i/c$ and $\omega_i = \rho_i/\rho$, respectively. Mass fluxes (n_i, j_i, $\bar{j}_i$) are particularly useful in the description of mass transport phenomena in liquid systems and in systems where trace amounts of gases are diffusing through liquids or solids, since the total mixture mass density ρ is often constant. Molar fluxes (N_j, J_i, $\bar{J}_i$) are often used to describe diffusion in gas mixtures since, at constant temperature and pressure, the total molar density of an ideal gas mixture, c, is constant. Interrelation of the various fluxes defined in Table 3 and definitions of other common fluxes may be found in standard mass transfer texts.[16,75] The flux relationships are used in expressions of continuity (i.e. conservation of mass) (see Table 3) to calculate the spatial and temporal dependence of the component concentration. These mass conservation relations apply when there is no homogeneous chemical reaction in the mixture. To close the mass conservation equations, a constitutive relationship is required to relate component fluxes to concentration gradients in the mixture.

Table 3 Flux Definitions[a]

Flux of component i	Definition	Fick's first law (binary mixtures)	Continuity relation
Mass flux relative to fixed coordinates	$n_i \equiv \rho_i v_i$	$n_i = -\rho D_{12}^0 \nabla \omega_i + \omega_i(n_i + n_j)$	$\dfrac{\partial \rho_i}{\partial t} = -\nabla \cdot n_i$
Molar flux relative to fixed coordinates	$N_i \equiv c_i v_i$	$N_i = -cD_{12}^0 \nabla x_i + x_i(N_i + N_j)$	$\dfrac{\partial c_i}{\partial t} = -\nabla \cdot N_i$
Mass flux relative to mass average velocity	$j_i \equiv \rho_i(v_i - v)$	$j_i = -\rho D_{12}^0 \nabla \omega_i$	$\rho \dfrac{\partial \omega_i}{\partial t} = -\nabla \cdot j_i - v \cdot \nabla \omega_i$
Molar flux relative to molar average velocity	$J_i^* \equiv c_i(v_i - v^*)$	$J_i^* = -cD_{12}^0 \nabla x_i$	$c \dfrac{\partial x_i}{\partial t} = -\nabla \cdot J_i^* - v^* \cdot \nabla x_i$
Mass flux relative to volume average velocity	$\dot{j}_i \equiv \rho_i(v_i - \dot{v})$	$\dot{j}_i = -D_{12}^0 \nabla \rho_i$	$\dfrac{\partial \rho_i}{\partial t} = -\nabla \cdot \dot{j}_i - \nabla \cdot (\rho_i \dot{v})$
Molar flux relative to volume average velocity	$\dot{J}_i \equiv c_i(v_i - \dot{v})$	$\dot{J}_i = -D_{12}^0 \nabla c_i$	$\dfrac{\partial c_i}{\partial t} = -\nabla \cdot \dot{J}_i - \nabla \cdot (c_i \dot{v})$

[a] $i, j = 1, 2; i \neq j$; note $c_i = \rho_i/M_i$.

9.7.1.2 Fick's first law

At fixed temperature and pressure, and in the absence of species specific body forces (such as electric fields used to cause mass transfer in ionic systems), the flux in a binary mixture is typically given by Fick's first law, shown in Table 3 for various reference frames.[75] In this table, D_{12}^0 is the binary mutual diffusion coefficient. It is related to the so-called 'activity corrected mutual diffusion coefficient', D_{12}, as follows[105]

$$D_{12}^0 \equiv \left(\frac{\partial \ln a}{\partial \ln x_1}\right)_{T,P} D_{12} \qquad (63)$$

where a is the solvent or penetrant activity. In Fick's first law, the diffusion coefficient is taken to be a function of, at most, composition, temperature and pressure. More complicated behavior, in which the mass flux depends upon composition history or the diffusion coefficient depends upon time, is called 'non-Fickian' diffusion and is discussed in the chapter.

9.7.1.3 Solution of diffusion problems

Fick's first law, when combined with a statement of conservation of mass, yields a partial differential equation describing the spatial and temporal distribution of components in a mixture. To solve this differential equation, the average velocity of the mixture must be specified. In systems in which the total mixture mass density ρ is constant, the mixture boundaries do not move, and the flow of matter is due only to diffusion (i.e. no forced convective flow of matter), the mass average velocity is zero. In mixtures which are in a container of fixed total volume and exhibit no volume changes upon mixing, the volume average velocity is zero.[106,107] In systems where one component, present in trace amounts, is diffusing through a liquid or solid (such as a gas diffusing in a polymer) the average mixture velocity is commonly set to zero. In these cases, the equation of mass conservation is given by

$$\frac{\partial \rho_i}{\partial t} = \nabla \cdot (D_{12}^0 \nabla \rho_i) \qquad (64)$$

or by

$$\frac{\partial c_i}{\partial t} = \nabla \cdot (D_{12}^0 \nabla c_i) \qquad (65)$$

Many solutions to equations (64) and (65), with constant and variable diffusion coefficients, are available in the excellent reference works by Crank[93] and Carslaw and Jaeger.[108] When the diffusion coefficient is constant, the resulting form of equations (5) and (6) is called Fick's second law.

When a gas or liquid diffuses through an immobile polymer (i.e. $n_2 = 0$), the mass flux expression is simplified as shown below

$$n_1 = -\frac{\rho D^0_{12}}{(1-\omega_1)} \nabla \omega_1 \tag{66}$$

The so-called 'frame of reference' correction factor $(1 - \omega_1)^{-1}$ is usually neglected when the penetrant concentration is low. For example, the diffusion of a sparingly soluble gas into a polymer is typically modeled using the approximation $(1 - \omega_1) \approx 1$.[50] However, at a penetrant or solvent mass fraction which is not small relative to unity, the factor can be important. This factor is included in the description of the steady state diffusion of liquids through highly swollen polymer systems.[79]

When the motion of a liquid mixture is due to convective flow driven by, for example, external pressure gradients, the average velocity of the mixture is often obtained from a solution of the equations of motion for the mixture. The mixture average velocity may also be determined by boundary conditions if the boundary is moving or if chemical reaction is occurring at a boundary. The solution of this class of problems is discussed in detail in mass transfer texts.[16,75]

9.7.1.4 Tracer diffusion coefficients

A tracer diffusion coefficient describes the Brownian motion of a single, labeled molecule in a mixture of unlabeled components,[105] and may be determined experimentally by methods such as spin-echo NMR.[44] It is a measure of the frictional resistance experienced by the molecule as it undergoes Brownian motion in the mixture. The tracer diffusion coefficient of component i is given by[109]

$$D^*_i \equiv \frac{kT}{f_i} \tag{67}$$

where f_i is the friction coefficient of component i. In a binary mixture, the friction coefficient is the proportionality constant between the average viscous frictional force experienced by a component molecule and the velocity of that molecule.[110] For a dilute suspension of spherical particles undergoing Brownian motion in a fluid $f_i = 6\pi \eta_s R_s$, where η_s is the fluid viscosity and R_s is the particle radius. Similarly, for polymer chains diffusing in solution or for penetrant molecules diffusing in a solid or liquid, the friction coefficient depends upon the size of the diffusing species and the effective viscosity of the mixture. The tracer diffusion coefficient of a mixture component is equal to the mutual diffusion coefficient in the limit of vanishing concentration of the component

$$D^*_i = \lim_{x_i \to 0} D^0_{12} \tag{68}$$

The tracer diffusion coefficient is related to the component velocity autocorrelation function through the Kubo formula[109]

$$D^*_i = \frac{1}{3} \int_0^\infty \langle v^*_i(0) \cdot v^*_i(t) \rangle \, dt \tag{69}$$

where $v^*_i(t)$ is the velocity of a labeled molecule of component i at time t. The angular brackets denote an ensemble average. The tracer diffusion coefficient is also related to the mean square displacement of a component molecule at time t, $r^*(t)$, relative to its initial position $r^*(0)$[111]

$$\langle |r^*(t) - r^*(0)|^2 \rangle = 6 D^*_i t \tag{70}$$

Equation (70) is used to determine diffusion coefficients of penetrant molecules in polymers by computer simulation.[112] The time dependence of the position of the penetrant molecule is determined by integrating the equations of motion of the penetrant and of segments of the polymer chains. This procedure, called molecular dynamics, is computationally intensive.[113] It has been used to track the motion of polymer chain segments and penetrant molecules in polymers over very short timescales, of the order of several hundred picoseconds.[112]

When the molecules of the two species in the mixture have similar size, shape and interaction potentials (or, more rigorously, when the radial distribution function is independent of composition) and the components mix ideally, the tracer diffusion coefficients are related by the inverse of their molar volumes[110]

$$\frac{D_1^*}{D_2^*} = \frac{\hat{V}_2}{\hat{V}_1} \qquad (71)$$

With these approximations, the binary mutual diffusion coefficient is given by

$$D_{12}^0 = \left(\frac{\partial \ln a}{\partial \ln x_1}\right)_{T,P} (D_1^* x_2 + D_2^* x_1)$$

or
(17)

$$D_{12}^0 = \text{(thermodynamic factor)(friction factor)}$$

The binary mutual diffusion coefficient depends, therefore, upon the product of a term which characterizes the thermodynamic interactions between the polymer and solvent and a term which describes the frictional resistance to motion in the mixture. A critical discussion of the applicability of equation (17) (often called the Hartley–Crank or Darken relation) to mixtures of small molecules is offered by McCall and Douglass.[114] These authors conclude that equation (17) provides a reasonable approximation for D_{12}^0 when the mixture components do not strongly associate. For example, equation (17) describes quantitatively diffusion coefficients in the system benzene–cyclohexane but can only qualitatively describe the concentration dependence of D_{12}^0 is strongly associating mixtures of acetone and H_2O. Systematic studies of the validity of equation (17) in polymer–solvent systems are not widely available, though equation (17) is often used as the starting point in a discussion of the concentration dependence of the mutual diffusion coefficient in mixtures of polymer and solvent.[50]

In the limit of a single polymer molecule diffusing in an infinite expanse of solvent, the mutual diffusion coefficient approaches the self diffusion coefficient of the polymer

$$D_{12}^0 = D_2^* \qquad (72)$$

In this limit, D_{12}^0 depends only upon the mobility of the polymer molecules.

In solutions containing a low concentration of solvent, the mobility of polymer molecules, as characterized by the polymer tracer diffusion coefficient, is typically many orders of magnitude less than that of low molecular weight penetrants or solvents.[50] In this limit, D_{12} can often be approximated as[50]

$$D_{12} = x_2 D_1^* \qquad (73)$$

According to estimations by Vrentas and Duda, equation (73) is valid to within 1% at solvent mass fractions up to 0.5 in the system polystyrene–ethylbenzene.[43,47] Thus the mutual diffusion coefficient may be written as follows for mixtures which are polymer rich

$$D_{12}^0 = D_1^*(1 - x_1)\left(\frac{\partial \ln a}{\partial \ln x_1}\right)_{T,P} \qquad (74)$$

9.8 REFERENCES

1. H. B. Hopfenberg and D. R. Paul, in 'Polymer Blends', ed. D. R. Paul and S. Newman, Academic Press, New York, 1978, vol. 1, p. 445.
2. J. H. Petropoulos, *J. Polym. Sci., Polym. Phys. Ed.*, 1985, **23**, 1309.
3. B. Baranowski, *J. Membr. Sci.*, 1991, **57**, 119.
4. E. H. Cwirko and R. G. Carbonell, *J. Colloid Interface Sci.*, 1989, **129**, 513.
5. E. H. Cwirko and R. G. Carbonell, *J. Membr. Sci.*, 1990, **48**, 155.
6. M. Doi and S. F. Edwards, 'The Theory of Polymer Dynamics', Clarendon Press, Oxford, 1986, p. 188.
7. M. Tirrell, *Rubber Chem. Technol.*, 1984, **57**, 523.
8. M. Muthukumar and S. F. Edwards, in 'Comprehensive Polymer Science', ed. G. Allen, Pergamon Press, Oxford, 1989, vol. 2, p. 1.
9. J. F. Joanny and S. J. Candau, in 'Comprehensive Polymer Science', ed. G. Allen, Pergamon Press, Oxford, 1989, vol. 2, p. 199.
10. P. E. Rouse, *J. Chem. Phys.*, 1953, **21**, 1273.
11. J. G. Kirkwood and J. Riseman, *J. Chem. Phys.*, 1948, **16**, 565.
12. B. Zimm, *J. Chem. Phys.*, 1956, **24**, 269.
13. H. Yamakawa, 'Modern Theory of Polymer Solutions', Harper and Row, New York, 1971.
14. P. G. deGennes, 'Scaling Concepts in Polymer Physics', Cornell University Press, New York, 1976.
15. R. B. Bird, O. Hassager, R. C. Armstrong and C. F. Curtiss, 'Dynamics of Polymeric Liquids', Wiley, New York, 1977.

16. R. B. Bird, W. E. Stewart and E. N. Lightfoot, 'Transport Phenomena', Wiley, New York, 1960, p. 499.
17. K. F. Freed, 'Renormalization Group Theory of Macromolecules', Wiley, New York, 1987, p. 1.
18. Y. Oono, in 'Advances in Chemical Physics', ed. I. Prigogine and S. Rice, Wiley, New York, 1985, vol. 61, p. 301.
19. K. G. Wilson, *Sci. Am.*, 1979, **241**, 158.
20. J. M. Prausnitz, 'Molecular Thermodynamics of Fluid-Phase Equilibria', Prentice Hall, Englewood Cliffs, NJ 1969, p. 295.
21. E. A. Grulke, in 'Polymer Handbook', ed. J. Brandrup and E. H. Immergut, Wiley, 1989, p. VII-519.
22. A. F. M. Barton, 'CRC Handbook of Polymer–Liquid Interaction Parameters and Solubility Parameters', CRC Press, Boca Raton, FL, 1990, p. 1194.
23. M. J. Pritchard and D. Caroline, *Macromolecules*, 1981, **14**, 424.
24. T. King, in 'Comprehensive Polymer Science', ed. G. Allen, Pergamon Press, Oxford, 1989, vol. 1, chap. 7, p. 911.
25. Y. Shiwa, *Phys. Rev. Lett.*, 1987, **58**, 2102.
26. Y. Oono and P. Baldwin, *Phys. Rev. A*, 1986, **33**, 3391.
27. B. Nyström and J. Roots, *J. Polym. Sci., Polym. Phys. Ed.*, 1990, **28**, 521.
28. G. C. Berry, *J. Chem. Phys.*, 1966, **44**, 4550.
29. P. Wiltzius, H. R. Haller, D. S. Cannell and D. W. Schaefer, *Phys. Rev. Lett.*, 1984, **53**, 834.
30. J. Roots, B. Nyström, L.-O. Sundelöf and B. Porsch, *Polymer*, 1979, **20**, 339.
31. J. Roots and B. Nyström, *Eur. Polym. J.*, 1978, **14**, 773.
32. M. D. Lechner and D. G. Steinmeier, in 'Polymer Handbook', ed. J. Brandrup and E. H. Immergut, Wiley, New York, 1989, p. VII-61.
33. B. Nyström and J. Roots, *J. Macromol. Sci., Rev. Macromol. Chem.*, 1980, **C19**, 35.
34. G. Rehage, O. Ernst and J. Fuhrmann, *Discuss. Faraday Soc.*, 1970, **49**, 208.
35. C. Cang, L. Bokobza, L. Monnerie, S. J. Clarson, J. A. Semlyen, J. Vandendriessche and F. C. DeSchryver, *Polymer*, 1987, **28**, 1561.
36. G. R. Andersson, *Ark. Kemi*, 1963, **20**, 513.
37. J. Roots and B. Nyström, *Macromolecules*, 1982, **15**, 553.
38. B. D. Freeman, D. S. Soane and M. M. Denn, *Macromolecules*, 1990, **23**, 245.
39. G. V. Schulz and M. Lechner, in 'Light Scattering From Polymer Solutions', ed. M. B. Huglin, Academic Press, Orlando, FL, 1972, p. 503.
40. C. J. McDonald and S. Claesson, *Chem. Scr.*, 1976, **9**, 36.
41. Y. Oono, P. R. Baldwin and T. Ohta, *Phys. Rev. Lett.*, 1984, **53**, 2149.
42. J. L. Duda, J. S. Vrentas, S. T. Ju and H. T. Liu, *AIChE J.*, 1982, **28**, 279.
43. J. S. Vrentas and J. L. Duda, *J. Polym. Sci., Polym. Phys. Ed.*, 1977, **15**, 403.
44. E. D. von Meerwall, E. J. Amis and J. D. Ferry, *Macromolecules*, 1985, **18**, 260.
45. J. D. Ferry, 'Viscoelastic Properties of Polymers', Wiley, New York, 1980, p. 264.
46. J. S. Vrentas and J. L. Duda, *J. Polym. Sci., Polym. Phys. Ed.*, 1977, **15**, 441.
47. J. S. Vrentas and J. L. Duda, *J. Polym. Sci., Polym. Phys. Ed.*, 1977, **15**, 417.
48. T. Naylor, in 'Comprehensive Polymer Science', ed. G. Allen, Pergamon Press, Oxford, 1989, vol. 2, chap. 20, p. 643.
49. V. T. Stennett, in 'Diffusion in Polymers', ed. J. Crank and G. S. Park, Academic Press, Orlando, FL, 1968, p. 41.
50. W. J. Koros and M. W. Hellums, in 'Encyclopedia of Polymer Science and Engineering', ed. J. I. Kroschwitz, Wiley, 1990, suppl. vol., p. 724.
51. D. H. Weinkauf and D. R. Paul, in 'Barrier Polymers and Barrier Structures', ed. W. J. Koros, American Chemical Society, Washington, DC, 1990, p. 60.
52. R. T. Chern, W. J. Koros, E. S. Sanders, S. H. Chen and H. B. Hopfenberg, in 'Proceedings of the American Chemical Society, Washington, D.C., 1983, ed. J. T. E. Whyte, C. M. Yon and E. H. Wagener, p. 47.
53. T. H. Kim, W. J. Koros and G. R. Husk, *J. Membr. Sci.*, 1989, **46**, 43.
54. R. T. Chern, W. J. Koros, H. B. Hopfenberg and V. T. Stannett, in 'Materials Science of Synthetic Membranes', ed. D. R. Lloyd, American Chemical Society, Washington, DC, 1985, p. 25.
55. A. R. Berens and H. B. Hopfenberg, *J. Membr. Sci.*, 1982, **10**, 283.
56. M. R. Coleman and W. J. Koros, *J. Membr. Sci.*, 1990, **50**, 285.
57. R. T. Chern, L. Jia, S. Shimoda and H. B. Hopfenberg, *J. Membr. Sci.*, 1990, **48**, 333.
58. R. T. Chern, *Sep. Sci. Technol.*, 1990, **25**, 1325.
59. F. R. Sheu and R. T. Chern, *J. Polym. Sci., Polym. Phys. Ed.*, 1989, **27**, 1121.
60. M. W. Hellums, W. J. Koros, G. R. Husk and D. R. Paul, *J. Membr. Sci.*, 1989, **46**, 93.
61. A. S. Michaels and R. B. Parker Jr., *J. Polym. Sci.*, 1959, **41**, 53.
62. A. S. Michaels and H. J. Bixler, *J. Polym. Sci.*, 1961, **50**, 413.
63. A. S. Michaels and H. J. Bixler, *J. Polym. Sci.*, 1961, **50**, 393.
64. A. S. Michaels, W. R. Vieth and J. A. Barrie, *J. Appl. Phys.*, 1963, **34**, 1.
65. A. C. Puleo, D. R. Paul and P. K. Wong, *Polymer*, 1989, **30**, 1357.
66. C. Noël, *Makromol. Chem., Macromol. Symp.*, 1988, **22**, 95.
67. J. S. Chiou and D. R. Paul, *J. Polym. Sci., Polym. Phys. Ed.*, 1987, **25**, 1699.
68. D. H. Weinkauf and D. R. Paul, in 'Proceedings of the ACS Symposium on Barrier Polymers, Washington, D.C., 1989, American Chemical Society, p. 3.
69. W. J. Koros, in 'Barrier Polymers and Barrier Structures', ed. W. J. Koros, American Chemical Society, Washington, DC, 1990, p. 1.
70. T. Masuda, E. Isobe and T. Higashimura, *Macromolecules*, 1985, **18**, 841.
71. K. Takada, H. Matsuya, T. Masuda and T. Higashimura, *J. Appl. Polym. Sci.*, 1985, **30**, 1605.
72. L. C. Witchey-Lakshmanan, H. B. Hopfenberg and R. T. Chern, *J. Membr. Sci.*, 1990, **48**, 321.
73. T. Masuda and T. Higashimura, in 'Advances in Polymer Science', Springer-Verlag, Berlin, 1987, vol. 81, p. 121.
74. T. Masuda, Y. Iguchi, B.-Z. Tang and T. Higashimura, *Polymer*, 1988, **29**, 2041.
75. J. C. Slattery, 'Momentum, Energy, and Mass Transfer in Continua', Krieger, Huntington, 1981.
76. D. R. Paul, *Ber. Bunsenges. Phys. Chem.*, 1979, **83**, 294.
77. K. Toi, G. Morel and D. R. Paul, *J. Appl. Polym. Sci.*, 1982, **27**, 2997.

78. S. Pauly, in 'Polymer Handbook', ed. J. Brandrup and E. H. Immergut, Wiley, New York, 1989, p. VI-435.
79. D. R. Paul and O. Ebra-Lima, *J. Appl. Polym. Sci.*, 1970, **14**, 2201.
80. W. J. Koros and D. R. Paul, *J. Polym. Sci., Polym. Phys. Ed.*, 1978, **16**, 1947.
81. W. J. Koros, G. N. Smith and V. T. Stannett, *J. Appl. Polym. Sci.*, 1981, **26**, 159.
82. G. G. Lipscomb, *AIChE J.*, 1990, **36**, 1505.
83. D. S. Pope, I. C. Sanchez, W. J. Koros and G. K. Fleming, *Macromolecules*, 1991, **24**, 1779.
84. G. T. Dee and D. J. Walsh, *Macromolecules*, 1988, **21**, 811.
85. R. H. Lacombe and I. C. Sanchez, *J. Phys. Chem.*, 1976, **80**, 2568.
86. I. C. Sanchez and R. H. Lacombe, *J. Phys. Chem.*, 1976, **80**, 2352.
87. I. C. Sanchez and R. H. Lacombe, *J. Polym. Sci., Polym. Lett. Ed.*, 1977, **15**, 71.
88. I. C. Sanchez and R. H. Lacombe, *Macromolecules*, 1978, **11**, 1145.
89. I. C. Sanchez, *Polymer*, 1989, **30**, 471.
90. M. B. Kiszka, M. A. Meilchen and M. A. McHugh, *J. Appl. Polym. Sci.*, 1988, **36**, 583.
91. G. K. Fleming, personal communication, 1991.
92. R. G. Carbonell and G. C. Sarti, *Ind. Eng. Chem. Res.*, 1990, **29**, 1194.
93. J. Crank, 'The Mathematics of Diffusion', Clarendon Press, Oxford, 1975, p. 105.
94. A. R. Berens and H. B. Hopfenberg, *Polymer*, 1978, **19**, 489.
95. H. B. Hopfenberg, in 'Membrane Science and Technology', ed. J. E. Flinn, Plenum Press, New York, 1970, p. 16.
96. N. L. Thomas and A. H. Windle, *Polymer*, 1981, **22**, 627.
97. A. R. Berens, *Research Tech.*, 1985, **57**, Nov.
98. N. M. Franson and N. A. Peppas, *J. Appl. Polym. Sci.*, 1983, **28**, 1299.
99. A. H. Windle, in 'Polymer Permeability', ed. J. Comyn, Elsevier, New York, 1985, p. 75.
100. S. Mehdizadeh and C. J. Durning, *AIChE J.*, 1990, **36**, 877.
101. C. J. Durning and M. Tabor, *Macromolecules*, 1986, **19**, 2220.
102. P. Neogi, M. Kim and Y. Yang, *AIChE J.*, 1986, **32**, 1146.
103. F. Adib and P. Neogi, *AIChE J.*, 1987, **33**, 164.
104. G. Camera-Roda and G. C. Sarti, *AIChE J.*, 1990, **36**, 851.
105. R. C. Reid, J. M. Prausnitz and T. K. Sherwood, 'The Properties of Gases and Liquids', McGraw–Hill, New York, 1977, p. 544.
106. J. S. Vrentas and J. L. Duda, *J. Appl. Polym. Sci.*, 1976, **20**, 2569.
107. G. D. Hartley and J. Crank, *Trans. Faraday Soc.*, 1949, **45**, 801.
108. H. S. Carslaw and J. C. Jaeger, 'Conduction of Heat in Solids', Clarendon Press, Oxford, 1959.
109. A. R. Altenberger and M. V. Tirrell, *J. Polym. Sci., Polym. Phys. Ed.*, 1984, **22**, 909.
110. R. J. Bearman, *J. Phys. Chem.*, 1961, **65**, 1961.
111. D. A. McQuarrie, 'Statistical Mechanics', Harper & Row, New York, 1976, p. 455.
112. S. Trohalaki, A. Kloczkowski, J. E. Mark, D. Rigby and R. J. Roe, in 'Computer Simulation of Polymers', ed. R. J. Roe, Prentice Hall, Englewood Cliffs, NJ, 1991, p. 220.
113. K. Kremer and G. S. Grest, in 'Computer Simulation of Polymers', ed. R. J. Roe, Prentice Hall, Englewood Cliffs, NJ, 1991, p. 167.
114. D. W. McCall and D. C. Douglass, *J. Phys. Chem.*, 1976, **71**, 987.

10
Fundamentals of the Formation, Structure and Properties of Polymer Networks

ROBERT F. T. STEPTO
University of Manchester and UMIST, UK

10.1	INTRODUCTION	199
	10.1.1 *Networks and Network Materials*	199
	10.1.2 *Gelation and Network Formation*	201
	10.1.3 *Scope of Chapter*	202
10.2	IDEAL GELATION AND NETWORK FORMATION	202
	10.2.1 *Flory–Stockmayer (F–S) Gel Point*	202
	10.2.2 *Perfect Network Formation*	205
10.3	INTRAMOLECULAR REACTION AND GELATION	207
	10.3.1 *Characterization of Intramolecular Reaction*	207
	10.3.2 *Theories and Models of Polymerization*	211
	10.3.3 *Gelation*	212
	10.3.4 *Percolation*	215
10.4	NETWORK FORMATION, STRUCTURE AND PROPERTIES	216
	10.4.1 *Introduction*	216
	10.4.2 *Experimental Results on Formation, Structure and Modulus*	216
	10.4.3 *Applications of Theories and Models of Polymerizations*	220
	10.4.3.1 Cascade theory	221
	10.4.3.2 Monte Carlo simulations	222
	10.4.3.3 Rate theory	223
	10.4.4 *Stress–Strain Behaviour of Networks*	223
10.5	REFERENCES	225

10.1 INTRODUCTION

10.1.1 Networks and Network Materials

A polymer network material consists of a network macromolecule or an assembly of interacting macromolecules of an extent limited only by the macroscopic dimensions of the sample of material. A network may be formed by the reaction of monomer or prepolymer molecules to give a covalent network[1] or by the intermolecular association of linear or branched polymers, commonly through hydrogen bonding, to give a physical network.[2,3] In addition, some network structures, such as thermoplastic elastomers, are formed from a mixture of covalent bonds and intermolecular associations.[4]

The basis of the network is the junction point, from which, ideally, at least three polymer chains emanate and continue, passing through other junction points, to define independent paths which extend macroscopically throughout the material sample.[5] In covalent networks, the junction points are defined by the structure of the reactant monomers or prepolymers. At least one of the reactant species must have a chemical functionality (number of reactive groups) greater than two. Network materials, such as polysiloxanes, polyurethanes and polyester resins, are often formed from reactants which carry their reactive groups at the ends of chains, and the chemical functionalities of the reactants define the functionalities of the junction points in the network. Such networks are networks formed by endlinking polymerizations. Covalent networks are also formed by the crosslinking of existing polymer chains using chemical functionalities (often double bonds) along the chain, as in the vulcanization of natural rubber. In these networks, only a small proportion of the much larger number of reactive groups on the reactant species, the polymer chains, need be reacted. Each crosslink chemically produces two trifunctional junction points, which, because of the short length of the crosslinking molecule compared with chemical chain lengths between pairs of reacted groups along the polymer chains, are considered as one tetrafunctional junction. The endlinking and crosslinking processes are illustrated in Figure 1.

The unique properties of polymer network materials are their elastomeric properties. They are determined principally (see Queslel and Mark, Volume 2, Chapter 9 and Vilgis, Volume 6, Chapter 8, in *Comprehensive Polymer Science*) by the changes of conformational entropy on deformation of the chains between junction points, to which must be added changes of conformational energy and the effects of topological chain entanglements[6-9] and chain interactions[1,2,4] (*e.g.* hydrogen bonding, crystallization). To a good approximation, the shear modulus (G) at low strains of a bulk elastomeric network is proportional to the concentration of elastically active chains between network junctions ($n_{e,c}$), according to the elassical equation

$$G = n_{e,c}kT = (\rho/M_c)kT \tag{1}$$

(The interpretation and limitations of this equation will be discussed later.) ρ is the density of the network and M_c the effective molar mass of elastic chains between junction points. For polydisperse chain lengths, M_c is the number-average value (see equation 29).

Equation (1) emphasizes a fundamental difference between the properties and structure of network polymers in comparison with linear polymers. The latter, once a certain threshold of molar mass is exceeded, have properties which are determined principally by the structure of the repeating unit of the chains. With networks, superimposed on such a dependence, are the effects of the value of $n_{e,c}$. Thus, the properties of network materials depend not only on chain structure and the average molar masses of chains (extending between and from junction points) but also on $n_{e,c}$, a characteristic of the molecular structure which relates directly to physical properties. For perfect networks, consisting only of elastically active junction points joined by elastically active chains, with no loose ends or topological entanglements, M_c is calculable from the reactant or primary polymer chain structures and the amount of reaction and chain interaction.

In addition to changing modulus, an effect of increasing junction point concentration is to increase T_g of the material due to an increase in restrictions of chain movements by the junction points.[10-12] Thus, for example, during the formation of networks of low M_c (thermosets), vitrification may

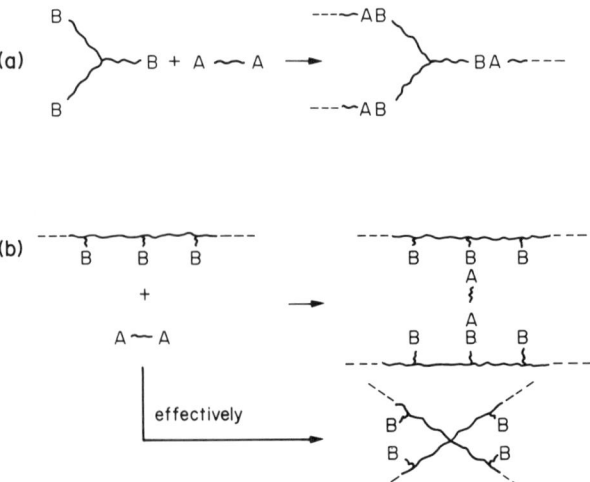

Figure 1 (a) Endlinking *via* an $RA_2 + RB_3$ polymerization; (b) crosslinking. One crosslink usually gives effectively one tetrafunctional junction point

occur if T_g increases to become equal to the temperature of reaction. The characterization of this phenomenon is important in the preparation of thermosets and has led to the concept of a time–temperature transformation (TTT) diagram introduced by Gillham and coworkers.[11,13]

Control of $n_{e,c}$ is to a large extent, but not completely, under the control of the chemist preparing a covalent network. Extension of the diagrams in Figure 1 to completed networks shows that

$$n_{e,c} = (f/2)n_{e,J} \qquad (2)$$

where f is the functionality of the junction points and $n_{e,J}$ is the concentration of elastically active junctions.[1,14] In forming covalent networks by endlinking, $n_{e,J}$ is determined by the molar masses and functionalities of the reactants and, in crosslinking, by the degree of crosslinking (fraction of groups crosslinked). In practice, deviations occur and the values of $n_{e,J}$ deduced from elastomeric properties are rarely those expected from the amount of chemical reaction which has occurred. Such deviations may be due to topological entanglements and chain interactions,[6–9,15] to side reactions, incomplete reaction in endlinking polymerizations (giving loose ends),[16,17] nonrandom reaction in crosslinking,[18–20] and, more fundamentally and generally, inelastic chain or loop formation, due to the intramolecular reaction of pairs of groups.[10,14,15,20–50]

It is obviously a key matter in predicting and interpreting the elastomeric properties of networks to be able to calculate the value of $n_{e,J}$ from the reactants and reaction conditions used. This topic is discussed in detail in later sections in relation to covalent networks. Control over $n_{e,J}$ for physical networks is more difficult as the actual structure of the junction points varies with conditions of preparation.[2,3]

10.1.2 Gelation and Network Formation

During the formation of covalent networks *via* endlinking or crosslinking, there are innumerable molecular structures which form and eventually join together to give the completed network. The structures become innumerable because as highly branched molecules grow they eventually acquire 'infinite' numbers of reactive groups[1] which can react further. The term 'infinite' implies limited only by the macroscopic amounts of reactants used in the polymerization.

The growth with degree of polymerization of the number of reactive groups on a molecule is illustrated in Figure 2 with respect to an $RA_2 + RB_4$ polymerization. Generalizing the figure, it can be seen that for an $RA_2 + RB_f$ polymerization, a molecule with n branch points has $n(f - 2) + 2$ reactive groups. Thus, for $f > 2$ the number of reactive groups increases with the number of branch points. The increase, linked with the assumption of equal reactivity of like groups, means that more complex molecules react in preference to simpler molecules leading to infinite or network species before complete reaction. The point in a nonlinear polymerization where infinite species first occur is the gel point. In the crosslinking of polymer chains the gel point occurs very early in the polymerization as the functionality of the original chains (f) is large.

The gel point, or point of incipient network formation,[1] is an important point in a polymerization from both a fundamental and a technological point of view. The fundamental aspects are discussed in detail later. Technologically, after the appearance of infinite species the viscosity of a polymerizing mixture becomes infinite; it starts to develop an elastic modulus which increases as $n_{e,J}$ increases. From the gel point onwards, a liquid to solid transformation starts to occur with the solid (network) fraction increasing as the reaction proceeds. Thus, in processing, liquid flow is no longer possible and, for example, in reaction injection moulding[51] and in the reactive processing of thermosets generally,[52] the transport into a mould and simultaneous reaction of the reactants has to occur before the gel point.

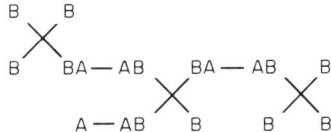

Figure 2 A molecule in an $RA_2 + RB_4$ polymerization with three RB_4 units. It has eight reactive groups

10.1.3 Scope of Chapter

Links between the formation, structure and properties of covalent networks are considered in detail. The idealized polymerization statistics of network formation, that is, how reactant molecules join to form a network are presented in Section 10.2. In reality, the statistics are more complex and not yet completely understood. Important aspects of studies in this area are experimental and theoretical investigations of pregel intramolecular reaction, the gel point, and changes of sol fraction (fraction of finite species) and modulus as a polymerization proceeds. Studies of pregel intramolecular reaction and the gel point are discussed in Section 10.3.

The polymerization systems used to prepare network materials are often complicated (see: Morrison and Porter, Volume 6, Chapter 4; Pappas, Chapter 5; Paul, Chapter 6; and Sperling, Chapter 13 in *Comprehensive Polymer Science*) and, in order to relate quantitatively theory and experiment, it is necessary to consider relatively simple polymerizations with well characterized reactants. In this respect, endlinking polymerizations are preferred to crosslinking polymerizations. The functionalities and molar masses of the reactants can be characterized with more certainty and reactants in which like groups have equal reactivities can be used. However, even with such model systems the molecular structures of the networks formed are not unique but are related, for example, to conditions of preparation, ratios of reactants, dilution of reactive groups and temperature. It should not be assumed that model networks are perfect networks.

If the structure of a network can be predicted through polymerization statistics, then its elastomeric properties can be interpreted in terms of that structure. As indicated, the key parameter is $n_{e,J}$, the concentration of elastically active junction points. Once $n_{e,J}$ and their functionalities are known, then absolute values of modulus can be predicted using rubber elasticity theories and compared with measured values. Such an approach allows a test of how such theories apply to actual networks, regarding, particularly, affine and phantom chain behaviour and the occurrence or not of additional physical junction points through chain interactions and topological entanglements. Progress in the prediction of $n_{e,J}$ is reviewed in Sections 10.4.1–10.4.3.

Finally, the stress–strain behaviour of model networks has been the subject of many investigations. Perfect network structures have normally been assumed in interpretations of data in terms of theories. Network defects will influence the values of the various theoretical parameters used in fitting data. In addition, recent work has shown that none of the existing theories can describe satisfactorily all types of deformation. These points are discussed in Section 10.4.4.

10.2 IDEAL GELATION AND NETWORK FORMATION

10.2.1 Flory–Stockmayer (F–S) Gel Point

The molecular changes underlying the phenomenon of gelation in polymerization were first elucidated by Flory[1] and by Stockmayer.[53,54] The gel point is that point in a polymerization where continuing structures first occur with unit probability. The statistical considerations are illustrated in Figure 3 for an $RA_4 + RB_3$ polymerization. Imagine A^1 is a randomly chosen reacted A group. A^2 is a group which is statistically equivalent to A^1 and could have been chosen instead. The chain from A^1 to A^2 defines the basic repeat structure of continuing chains. The probability that a chain continues from A^1 to A^2 becomes equal to unity at the gel point. That is, from that point onwards in the polymerization some infinite structures must exist.

For an $RA_{f_a} + RB_{f_b}$ polymerization the F–S gel point is simply[1,53]

$$(f_a - 1)p_a (f_b - 1)p_b = 1 \tag{3a}$$

where p_a and p_b are the extents of reaction of A and B groups, namely, the fractions of the two types of groups which have reacted. The assumptions leading to equation (3a) are that intramolecular reaction does not occur and that all A and all B groups have the same probabilities of being reacted.

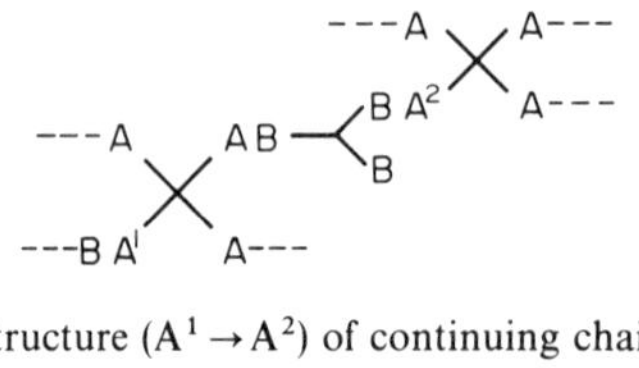

Figure 3 Depiction of the repeat structure ($A^1 \rightarrow A^2$) of continuing chains in an $RA_4 + RB_3$ polymerization

Subject to the same assumptions, Stockmayer[54] treated the general $\Sigma_i RA_{fai} + \Sigma_j RB_{fbj}$ polymerization of mixtures of reactants of different functionalities, with some reactants bearing A groups and some reactants bearing B groups and A groups reacting only with B groups. The reactant mixture is N_{a1} moles of $RA_{fa1}, \ldots, N_{ai}$ moles of $RA_{fai}, \ldots$ reacting with N_{b1} moles of $RB_{fb1}, \ldots, N_{bj}$ moles of $RB_{fbj}, \ldots$ Following a generalization of the statistical reasoning leading to equation (3a), the gel point condition becomes

$$(f_{aw} - 1)p_a \cdot (f_{bw} - 1)p_b = 1 \qquad (3b)$$

p_a and p_b are overall extents of A and B groups and f_{aw} and f_{bw} the so-called mass-average functionalities of the A- and B-bearing reactants. That is

$$f_{aw} = \frac{\sum_i N_{ai} f_{ai}^2}{\sum_i N_{ai} f_{ai}} = \frac{\sum_i N_{ai} f_{ai} \cdot f_{ai}}{\sum_i N_{ai} f_{ai}} \qquad (4)$$

and

$$f_{bw} = \frac{\sum_j N_{bj} f_{bj}^2}{\sum_j N_{bj} f_{bj}} = \frac{\sum_i N_{bj} f_{bj} \cdot f_{bj}}{\sum_j N_{bj} f_{bj}} \qquad (5)$$

In fact, the terminology 'mass-average functionality' is misleading. f_{aw} and f_{bw} are in fact functionalities averaged according to numbers of functional groups, and are so expressed in the second equalities in equations (4) and (5). In the derivation of equation (3b), when randomly choosing an A group, for example, the reactant to which it belongs contains f_{ai} groups with probability[55] $N_{ai} f_{ai} / \sum_i N_{ai} f_{ai}$.

Equation (3b), of course, includes equation (3a) as a special case. It also encompasses the gel point expressions discussed in Volume 6, Chapter 8 of *Comprehensive Polymer Science* and derived separately by Flory[1] for $RA_2 + RA_3 + RB_3$ polymerizations and the crosslinking of polymer chains. In the former case, $f_{bw} = f_b = 3$ and f_{aw} is evaluated from the relative amounts of RA_2 and RA_3 reactants used. In the latter case, the crosslinking molecules may be considered difunctional (equivalent to RA_2 reactants) and the chains to be crosslinked normally bear one reactive group or site (*e.g.* double bond) per structural unit. Thus, $f_w (= f_{bw}) = x_w$, where x_w is the mass-average degree of polymerization (DP). Hence, directly from equation (3b)

$$(x_w - 1)\rho = 1 \qquad (6)$$

at gel, where ρ is the fraction of crosslinked units. ρ is statistically equivalent to $p_a p_b$, although the chemistry of the crosslinking process often means that crosslinker molecules always react at both ends.

Often the reactants used for preparing network materials have like functional groups of intrinsic or induced unequal reactivities. Intrinsic unequal reactivity can occur, for example, when both primary and secondary hydroxy groups are present on reactants. Induced unequal reactivities (or the substitution effect) can occur when the similar groups on one molecule depend on how many groups on the molecule have already reacted. This can be, for example, for reasons of increased steric hindrance. The evaluation of gel points in simple polymerizations involving like groups having intrinsic unequal reactivities has been treated by Case[56] and others.[55] The evaluation of gel points for a number of polymerizations having induced unequal reactivities of like functional groups has been carried out by Dusek and Ilavsky and coworkers and others[20,32-42,57] using cascade theory, as developed for polymerizations by Gordon.[58]

As may be expected, the occurrence of infinite species at the gel point is accompanied by a divergence of the molar mass distribution, or, more fundamentally, the DP distribution, in that the mass average and higher averages become infinite. The expression for number-average DP for the general $RA_{fai} + RB_{fbj}$ polymerization is simply derived, independent of the distribution.

$$x_n = \frac{\text{number of units}}{\text{number of molecules}} = \frac{\sum x N_x}{\sum N_x} \qquad (7)$$

where N_x is the number of molecules of degree of polymerization x. In the absence of intramolecular reaction, the number of molecules is the initial number (*i.e.* the number of units) less the number lost (*i.e.* the number of A groups or B groups reacted $= n_{a0} p_a = n_{b0} p_b$). Hence

$$x_n = \frac{N_a + N_b}{N_a + N_b - n_{a0} \cdot p_a} \qquad (8)$$

Using the relationships

$$r = n_{a0}/n_{b0} = p_b/p_a \tag{9}$$

where r is the initial ratio of reactive groups, equation (8) reduces to

$$x_n = \frac{1}{1 - p_a\left(\frac{1}{f_{an}} + \frac{1}{f_{bn}r}\right)} \tag{10}$$

where f_{an} and f_{bn} are the 'number-average' functionalities of the A- and B-bearing reactants.

$$f_{an} = \frac{\sum N_{ai} f_{ai}}{\sum N_{ai}} = \frac{n_{a0}}{N_a} \tag{11}$$

and

$$f_{bn} = \frac{\sum N_{bj} f_{bi}}{\sum N_{bj}} = \frac{n_{b0}}{N_b} \tag{12}$$

The expression for x_w, which requires an expression for the distribution of species, is due to Stockmayer,[53] namely,

$$x_w = \frac{\sum x^2 N_x}{\sum x N_x} = 1 + \frac{p_a p_b f_{an} f_{bn}[p_a(f_{aw} - 1) + p_b(f_{bw} - 1) + 2]}{(p_a f_{an} + p_b f_{bn})[1 - p_a p_b(f_{aw} - 1)(f_{bw} - 1)]} \tag{13}$$

Comparison with equation (3b) shows that $x_w \to \infty$ yet x_n is finite at the gel point.

Another basic type of network-forming polymerization is the $\sum_i RA_{fai}$ self-polymerization, where A groups react with each other, as, for example, in the preparation of polysiloxanes. The condition for gelation and the expressions for x_n and x_w are in this case

$$(f_w - 1)p = 1 \tag{14}$$

$$x_n = \frac{1}{1 - f_n p/2} \tag{15}$$

and

$$x_w = \frac{1 + p}{1 - p(f_w - 1)} \tag{16}$$

Here, p is the extent of reaction (of A groups) and $f_n (= f_{an})$ and $f_w (= f_{aw})$ are defined by equations (11) and (4). In this case, the number of molecules lost is one-half of the number of A groups reacted.

An early criterion for gelation proposed by Carothers[59] that $x_n \to \infty$ at gel is unfortunately still often quoted in the literature (for example, see Volume 6, Chapter 8 in *Comprehensive Polymer Science*). It is based on the erroneous precept that all units are connected to form an infinite tree at the gel point. Thus, according to equation (15), for an RA_f polymerization gelation occurs when $p = 2/f$. This corresponds to on average just enough groups on each unit reacting for all units to be linked together to give only one ($\cong 0$) molecule(s). Similarly, in a stoichiometric $RA_{fa} + RB_{fb}$ polymerization mixture ($r = 1$), equation (10) would give $p_a = p = (1/f_a + 1/f_b)$, again the condition for the polymerizing mixture to connect to a single, infinite tree. These conditions for gelation are inconsistent with the random reaction of groups as they imply that groups so organize themselves that in RA_f polymerizations, for example, only two per original reactant molecule react.

The distinction between the condition $x_n \to \infty$ and $x_w \to \infty$ in fact illustrates well the phenomenon of gelation. $x_n \to \infty$ means that the average DP per molecule is infinite, because the number of molecules reduces to zero. On the other hand

$$x_w = \frac{\sum x \cdot x N_x}{\sum x N_x} = \sum x \cdot u_x \tag{17}$$

where

$$u_x = xN_x / \sum xN_x \tag{18}$$

is the unit fraction of x-mer. Hence, $x_w \to \infty$ means that the average DP per unit is infinite. That is, some units belong to infinite species.

10.2.2 Perfect Network Formation

The classical Flory–Stockmayer treatment of the gel point and the accompanying changes in distributions of species give a basic explanation of the phenomena to which the behaviour and changes in actual polymerizations may be related. The equations in the preceding section relate to changes from the beginning of a polymerization up to the gel point. As discussed in detail by Flory,[1] the infinite species which occur from the gel point to complete reaction cannot be enumerated as individual molecules. The sum of reactant units which have joined together to form infinite species defines the gel fraction, namely, the unit fraction (or approximately mass fraction) of gel (u_g). The gel fraction increases from zero at the gel point until complete reaction. For stoichiometric formulations, $u_g = 1$ at complete reaction. The complementary quantity, the unit fraction of sol (u_s), can be evaluated directly by summing over the units in finite species knowing the DP distribution. For example, for an RA_f self-polymerization

$$u_s = \sum_{x=1}^{\infty} u_x = 1 - u_g \tag{19}$$

where u_x is the fraction of reactant molecules (or units) in the species of degree of polymerization x. (For an RA_f polymerization, $u_s = w_s$, the so-called mass or weight fraction of sol, provided all units have the same mass. The equality neglects any loss of small molecules resulting from the reaction of functional groups. Generally, for any type of polymerization it is u_s, not w_s, which is the fundamental quantity.) Although x_w overall tends to infinity at the gel point and beyond, the sum in the numerator of equation (17) converges because N_x decreases with x more rapidly than x, provided the sum is restricted to species with enumerable numbers of unreacted ends, namely $x(f-2)+2$ for degree of polymerization x.

The detailed expression for u_x is

$$u_x = \frac{x(1-p)^2}{p} \cdot \frac{f[x(f-1)]!}{[x(f-2)+2]!x!} \cdot \beta^x \tag{20}$$

where

$$\beta = p(1-p)^{f-2} \tag{21}$$

The combinatorial factor relates to the number of distinguishable isomers of molecules of x units with $x(f-2)+2$ unreacted ends. The expression for u_s is

$$u_s = \frac{(1-p)^2 p^*}{(1-p^*)^2 p} \tag{22}$$

where p ($0 \leq p \leq 1$) is the extent of reaction and p^* is the lowest value of p which satisfies equation (21) for a given value of β.

The key to the difference between species in pregel and postgel is β; $d\beta/dp = 0$ at $p = 1/(f-1)$ defines a maximum value. Thus, $p^* \leq 1/(f-1)$, and as $p \to 1$, $p^* \to 0$. The reversion of the values of β from the gel point onwards means that the distribution of finite species also reverts and conjugate points (p and p^*) exist, pregel and postgel, which have identical distributions. The behaviour can be illustrated by combining equations (20) and (22) to give the unit fraction (u'_x) of species within the sol with

$$u'_x = \frac{u_x}{u_s} = \frac{x(1-p^*)^2}{p^*} \cdot \frac{f \cdot [x(f-1)]!}{[x(f-2)+2]!x!} \cdot \beta^x \tag{23}$$

Thus, at extent of reaction p, postgel, $\beta = p(1-p)^{f-2} = p^*(1-p^*)^{f-2}$ and the distribution is the same as that which occurred at p^*, pregel. Concurrently, equation (22) shows that $u_s = 1$ pregel ($p^* = p$), and decreases from the gel point onwards to zero at $p = 1$ ($p^* = 0$).

The behaviour is illustrated in Figure 4 for an RA_3 polymerization. In this case $p^* = (1-p)$ for $p > 1(f-1) = 1/2$, and the reversion is symmetrical about p at gel. For higher values of f, reversion is unsymmetrical about p at gel but the distributions at $p > 1/(f-1)$ map onto those at p^*, with p and p^* evaluated numerically through equation (21).

Reversion is a direct result of the unequal probability of growth of molecular species arising from the equal reactivity, or random reaction, of like functional groups, with only intermolecular reaction occurring between finite species. As discussed previously, the larger species have more unreacted groups and, hence, react with higher probabilities. As a polymerization proceeds past the gel point, more and more units are 'consumed' by the gel species and the finite species remaining

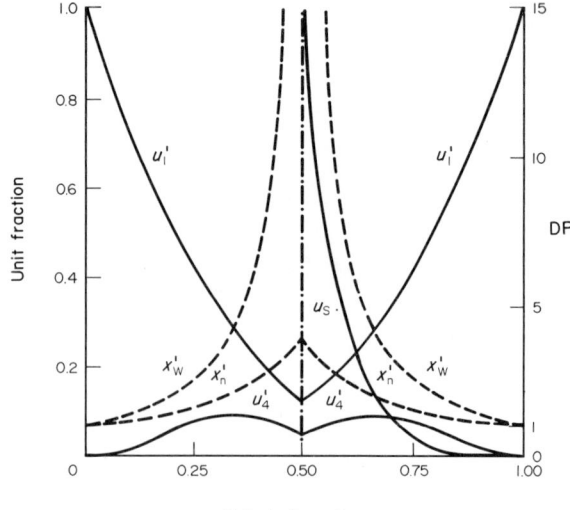

Figure 4 Illustrating the reversion of the distribution of sol species after the gel point in an RA_3 polymerization. u'_1 and u'_4 follow equation (20) and x'_n and x'_w equations (15) and (16) with p^* for p. u_s is given by equation (22). (unit fractions —, DP - - -)

become less and less and contain increasing fractions of smaller molecules. For stoichiometric $\sum_i RA_{fai} + \sum_j RB_{fbj}$ polymerization mixtures and for self-polymerizations, the vanishing sol fraction eventually contains only monomeric reactants and the gel fraction becomes the network which is formed by a polymerization.

Flory–Stockmayer theory says nothing concerning the detailed topology of the network, which grows and defines its structure through the random reaction of its reactive groups with other groups on the gel and with groups on sol species. To obtain a perfect network all reactions (sol–sol, sol–gel and gel–gel) are assumed to yield elastically active chains between junction points in the final network. This assumption is discussed in more detail in Section 10.4.

If a perfect network structure is assumed to be formed then the value of M_c can be calculated directly from the reactant structures, taking account of unreacted groups for nonstoichiometric reaction mixtures. Formulae for such calculations and for the calculations of sol and gel fractions have been given by Miller and Macosko.[60,61] It is also possible to calculate the same quantities directly from the distribution of molecular species and the fractions of unreacted groups on sol and gel molecules.[62]

Formulae for M_c from stoichiometric endlinking polymerizations can in fact be derived directly from considerations of the perfect network structure and its relationship to the structures of the reactants from which it was formed. The quantities required are derived from those in equation (1), namely

$$n_{e,c} = \frac{N_{e,c}}{W_{net}} \cdot \frac{W_{net}}{V_{net}} = \frac{1}{M_c} \cdot \rho \tag{24}$$

where $N_{e,c}$ is the number of elastic chains, and W_{net} and V_{net} are the mass and the volume of the network, respectively. It is assumed that any sol fraction has been removed. $1/M_c$ for networks formed from several reactants is an average value, denoted $\langle 1/M_c \rangle$, and expressions for $\langle 1/M_c \rangle$ may be obtained after further factorization to give

$$\left\langle \frac{1}{M_c} \right\rangle = \frac{N_{e,c}}{N_{e,J}} \cdot \frac{N_{e,J}}{W_{net}} \tag{25}$$

where $N_{e,J}$ is the number of elastic junctions.

For a $\sum_i RA_{fai}$ polymerization

$$\frac{N_{e,c}}{N_{e,J}} = \frac{\frac{1}{2}\sum' N_{ai} f_{ai}}{\sum' N_{ai}} = \frac{f'_{n,J}}{2} \tag{26}$$

where $'$ denotes summation over chains and junctions excluding RA_2 units ($f_{ai} = 2$). $f'_{n,J}$ is the number-average functionality of the junction points or the branched reactants (compare equation 11). Secondly

$$\frac{N_{e,J}}{W_{net}} = \frac{\sum' N_{ai}}{\sum' N_{ai} M_{ai}} \cdot \frac{\sum' N_{ai} M_{ai}}{\sum N_{ai} M_{ai}} \tag{27}$$

that is,

$$\frac{N_{e,J}}{W_{net}} = \frac{1}{M_{n,br}} \cdot w_{br} \tag{28}$$

where $M_{n,br}$ is the number-average molar mass and w_{br} is the mass fraction of branched reactants. The M_{ai} in equation (27) are the molar masses of the reactant molecules as they occur in the network (i.e. excluding any small molecule lost in the reaction $A + A$). Combination of equations (25), (26) and (28) shows that

$$\left\langle \frac{1}{M_c} \right\rangle = \frac{f'_{n,J}}{2} \cdot \frac{w_{br}}{M_{n,br}} \tag{29}$$

If no linear reactants are used and the junction functionalities are all the same ($=f$) then the usual expression

$$\left\langle \frac{1}{M_c} \right\rangle = \frac{1}{\langle M_c \rangle} = \frac{f}{2} \cdot \frac{1}{M_n} \tag{30}$$

is obtained. For $\sum_i RA_{fai} + \sum_j RB_{fbi}$ networks, $\langle 1/M_c \rangle$ is still given by equation (29), with $f'_{n,J}$, $M_{n,br}$ and w_{br} evaluated over both A- and B-bearing species.

10.3 INTRAMOLECULAR REACTION AND GELATION

10.3.1 Characterization of Intramolecular Reaction

There are two fundamental assumptions of Flory–Stockmayer statistics as applied to linear and to gelling and network-forming polymerizations — the random reaction of pairs of functional groups and the exclusion of intramolecular reaction between finite (sol) species.

Random reaction requires that like groups must be intrinsically equally reactive, a condition which can be met by the choice of suitable reactants, and, secondly, that all groups are uniformly accessible to all other groups in the polymerization. Thus, diffusion control cannot occur, nor intramolecular reaction. For finite species, on which the number of reactive groups is negligible compared with the total number of groups in a polymerization, random reaction means that intermolecular reaction will dominate, unless there are close spatial correlations between pairs of reactive groups on individual molecules. However, in a nonlinear polymerization the number of groups per molecule increases as the polymerization proceeds and, if gelation occurs, then species form which have unlimited numbers of groups. In Flory–Stockmayer statistics the groups on such species still undergo random reaction but because there are so many groups on the gel molecule some of that reaction is intramolecular. The author[62] has evaluated the amount of postgel intramolecular reaction in stoichiometric RA_f and $RA_2 + RB_f$ polymerizations according to Flory–Stockmayer statistics. For example, for RA_4 polymerizations 50% is intramolecular gel–gel reaction and for $RA_2 + RB_4$ polymerizations the amount is 25%. It is such a reaction which gives the network its final structure. If the reaction all leads to elastic chains then the perfect network results for stoichiometric formulations.

The neglect of intramolecular reaction between finite species in polymerizations is a fundamental assumption of Flory–Stockmayer statistics. It has been shown experimentally[63] that for linear polymerizations in bulk, in which molecules only have two reactive groups, the assumption is usually justified. However, for nonlinear polymerizations the increasing number of reactive groups per molecule together with the spatial correlations between groups on the same molecule mean that intramolecular reaction cannot generally be neglected. This point is illustrated by comparison of number fractions of rings species as functions of extent of reaction in Figure 5 for linear ($RA_2 + RB_2$) and nonlinear ($RA_2 + RB_3$) bulk polymerizations.[23]

The ring structures formed by intramolecular reaction cause gelation to be delayed because they reduce the probability of chains continuing to infinity. Thus, the extent of reaction at the gel point for the nonlinear polymerization in Figure 5 was[64] 0.765 rather than 0.707, as predicted by

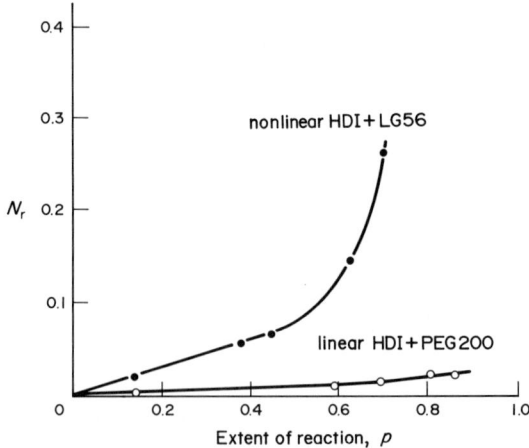

Figure 5 Number of ring structures per molecule (N_r) as a function of p for bulk, linear and nonlinear stoichiometric polyurethane-forming reactants ($r = [NCO]_0/[OH]_0 \cong 1$). ○ — linear, hexamethylene diisocyanate (HDI) + polyoxyethylene diol (PEG200) at 70 °C, $[NCO]_0 = 5.111$ mol kg^{-1}, $[OH]_0 = 5.188$ mol kg^{-1}, number of bonds in the chain forming the smallest ring (v) = 25.2. ● — nonlinear, HDI + polyoxypropylene (POP) triol (LG56) at 70 °C, $[NCO]_0 = 0.9073$ mol kg^{-1}, $[OH]_0 = 0.9173$ mol kg^{-1}, $v = 115$ (reproduced from ref. 23 by permission of the American Chemical Society)

equation (3a) for an $RA_2 + RB_3$ polymerization. In addition, some of the pregel and postgel intramolecular reaction will lead to inelastic loops in the final network. It is the evaluation and prediction of inelastic loops which is the key to the characterization and prediction of network structures and, hence, their physical properties.

The first treatment of the intramolecular reaction occurring during random polymerizations was due to Jacobson and Stockmayer[65] who treated linear ring–chain equilibria. The treatment has been applied extensively by Semlyen and coworkers for the analysis of ring-chain equilibriates and the preparation of macrocylics.[66] However, as with the data in Figure 5, most polymerizations yielding high molar mass polymers are carried out under essentially irreversible conditions. The balance between intermolecular and intramolecular reaction is then altered and changes as a polymerization proceeds.

The basic parameter characterizing intramolecular reaction introduced by Jacobson and Stockmayer[65] and by Haward[67] for radical polymerizations, is $P(0)$, the value of the probability density function of the end to end vector of a chain $P(r)$, for $r = 0$. That is, the probability density, or concentration of coincident reactive groups. Assuming a Gaussian form for $P(0)$ one may write

$$P(0) = \left(\frac{3}{2\pi \langle r_0^2 \rangle}\right)^{3/2} / N_{AV} \quad (\text{mol vol}^{-1}) \tag{31}$$

where N_{AV} is the Avogadro constant and $\langle r_0^2 \rangle$ is the mean-square end to end distance of the chain. In the context of a polymerization, only certain sizes of ring can form, as defined by the functionalities and molar masses of the reactants. This is illustrated in Figure 6 for $RA_2 + RB_2$ and $RA_2 + RB_f$ polymerizations. If the smallest possible chains forming ring structures contain v skeletal bonds then (to within about one bond) only rings formed from chains with $v, 2v, \ldots, iv$ skeletal bonds can occur. Thus, assuming independent Gaussian subchain statistics, the single parameter for the smallest ring structure

$$P_1 = \left(\frac{3}{2\pi v b^2}\right)^{3/2} / N_{AV} \tag{32}$$

is sufficient to characterize intramolecular reaction for all sizes of ring structure. In equation (32)

$$\langle r_0^2 \rangle = vb^2 \tag{33}$$

where b is the effective bond length of the chain of v bonds. To account for excluded volume, $\langle r_0^2 \rangle$ in equation (31) may be replaced by $\langle r^2 \rangle$. However, this is not normally necessary unless diluted reaction systems are of interest and v is greater than about 100. Chains of iv skeletal bonds have a concentration of coincident ends equal to

$$P_1 = \left(\frac{3}{2\pi ivb^2}\right)^{3/2} / N_{AV} = P_1/i^{3/2} \tag{34}$$

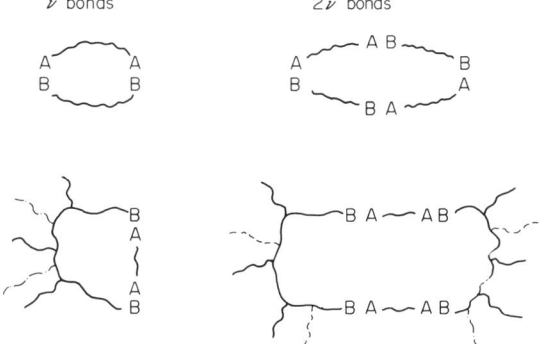

Figure 6 Illustrating the two smallest ring structures of v and $2v$ bonds in $RA_2 + RB_2$ and $RA_2 + RB_f$ polymerizations

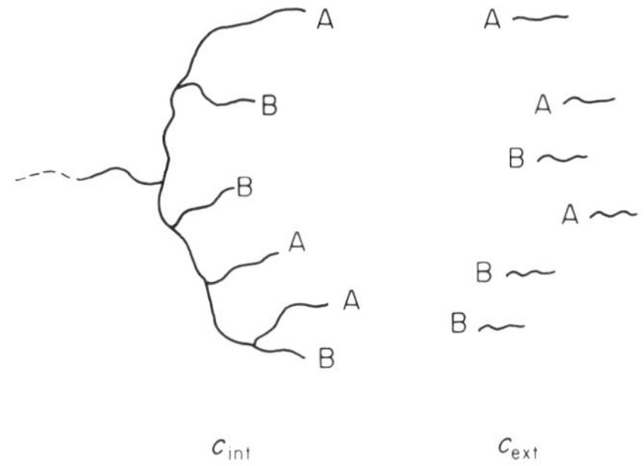

Figure 7 Illustrating the competition between intramolecular and intermolecular reaction in an A + B polymerization. c_{int}: internal concentration; total mutual concentration due to all pairs of A and B groups on the given molecule. c_{ext}: external concentration; due to A and B groups on other molecules

The competition between the intermolecular and intramolecular reaction of a molecule which occurs during a polymerization is characterized by P_1, by the number and positions of pairs of groups on the molecule which can react intramolecularly and by the concentrations of groups from other molecules. The situation is illustrated in Figure 7. The internal concentration, namely the total mutual concentration due to all pairs of groups on the given molecule, is c_{int}. It may be expressed by the equations

$$c_{int} = \sum c_{int,i} = \sum_i n_i P_i \tag{35}$$

where $c_{int,i}$ is the internal concentration due to all pairs of groups which could form ring structures of size i, and n_i is the number of such pairs. Depending upon how a polymerization theory or model is constructed, c_{int} can also refer to a given group rather than a given molecule. The quantity n_i is then the number of groups which can react intramolecularly with the given group. In either case, the general evaluation of n_i for highly branched and gel species is not feasible and theories of nonlinear polymerization, gelation and network formation which include intramolecular reaction essentially make different approximations for the n_i.

Intramolecular reaction is governed by c_{int} and intermolecular reaction by c_{ext}, which is the total concentration of A and B groups from the molecules which can react with the B and A groups on the given molecule. To a good approximation c_{ext} may be expressed simply as

$$c_{ext} = c_a + c_b = c_{a0}(1 - p_a) + c_{b0}(1 - p_b) \tag{36}$$

where c_{a0} and c_{b0} are the initial concentrations, and c_a and c_b the instantaneous concentrations. Equation (36) assumes that intermolecular reaction occurs by the random reactions of pairs of groups (as in Flory–Stockmayer statistics). That is, diffusion effects which would restrict the movement of molecules do not dominate. It also assumes that the number of groups on an individual molecule is not a significant proportion of $c_a + c_b$. This assumption, which has yet to be examined critically, will become less valid as complete reaction is approached.

Combination of equations (35) and (36) defines the ring-forming parameter

$$\lambda = \frac{c_{int}}{c_{int} + c_{ext}} \tag{37}$$

which is the probability of intramolecular *versus* intermolecular reaction.

An alternative parameter is

$$\lambda' = \frac{c_{int}}{c_{ext}} = \frac{\lambda}{1-\lambda} \tag{38}$$

λ and λ' will vary with molecular structure, through c_{int}, and with the overall concentration of reactive groups, through c_{ext}. The variation with molecular structure is reflected through the n_i, defined by the graph topology of the molecule, and through P_1. The latter quantity depends, through $\langle r_0^2 \rangle$, on molar mass (v) and chain stiffness (b). The dependence on c_{ext} means that λ' increases as a polymerization proceeds and c_{ext} decreases, and that λ' is larger for diluted systems.

Parameters equivalent to λ and λ' are used in all those theories and models of polymerization which express intramolecular reaction by taking the random reaction of groups (Flory–Stockmayer statistics) as a basis and modify probabilities of reaction due to the spatial correlations which must exist between reactive groups on the same molecule. For example, equivalent parameters are used in approaches based on differential equations[55]—kinetics theories, cascade theory, and rate theory—and on Monte Carlo simulations.[49,50] In addition, although the factors influencing intramolecular reactions have been discussed here in the context of endlinking polymerizations, similar considerations apply to the crosslinking of polymer chains.[20]

The interplay of factors which determine λ or λ' may be illustrated by reference to analysis of the data in Figure 5. Because of the lower equivalent masses of the reactants and their more flexible chains, v and b^2 for the linear polymerization are less than those for the nonlinear polymerization. Hence, c_{int} per opportunity for intramolecular reaction is larger for the linear polymerization. Again, because of lower equivalent masses, $c_{a0} + c_{b0}$ is larger for the linear polymerization. The ratio of the initial values of λ' is

$$\frac{\lambda'_{o,l}}{\lambda'_{o,nl}} = \frac{P_{1,l}}{(c_{a0} + c_{b0})_l} \cdot \frac{(c_{a0} + c_{b0})_{nl}}{P_{1,nl}} \cong 1.7 \tag{39}$$

where l denotes 'linear' and nl 'nonlinear'. Thus, if the same number of opportunities per molecule for intramolecular reaction existed in both polymerizations, larger ring fractions would occur in the linear polymerization. That the opposite effect occurs is directly attributable to the much larger average number of opportunities per molecule for intramolecular reaction (that is, larger values of n_i in equation 35) in the gelling system. This number is the dominating factor and emphasizes the importance of trying to take account of the complex structures which occur in nonlinear polymerizations.

Figure 5 shows results from polymerizations in bulk. In both the linear and nonlinear cases, dilution of the same reactants increases intramolecular reaction,[63,64] as expected from the decrease in c_{ext}. The resulting ring fraction data on the linear polymerizations at various dilutions have been interpreted satisfactorily in terms of a variety of theoretical approaches, kinetics theory, cascade theory, rate theory and modified Jacobson–Stockmayer theory.[68]

The corresponding nonlinear ring fraction data and associated gel point data[64] have not been fitted completely satisfactorily using cascade theory,[69] rate theory[69] or Monte Carlo simulations,[70] indicating that approximations in the theories, as applied to nonlinear polymerizations, can be significant (see Section 10.3.2). In this respect, it should be noticed that the use of total ring fraction data together with gel point data provides an extreme test of a theory.

It is evident from Figure 5 that a decrease in c_{ext} with dilution does not mean, conversely, that intramolecular reaction is absent in polymerizations in bulk. Reactive groups are still diluted by the inert units which connect them. Hence, the maximum values of c_{a0} and c_{b0} are always limited by the molar masses per reactive group (the equivalent masses). The limited values result in compensating effects on the amounts of intramolecular reaction which occur in related polymerizations in bulk, using reactants which differ only in molar masses. Decreases in molar masses mean increases in $c_{a0} + c_{b0}$ and decreases in v. Hence, both c_{int} and c_{ext} increase and the values of λ' at the start of a reaction can remain approximately constant. Thus, intermolecular reaction will not necessarily be reduced if reactant molar masses are increased.

10.3.2 Theories and Models of Polymerization

Theories and models of polymerization mentioned previously, cascade theory,[32-42] rate theory[21,26-31] and Monte Carlo simulation,[43-50,70] have been developed to account for intramolecular reaction occurring alongside the random intermolecular reaction of pairs of groups, particularly in nonlinear polymerizations. They can also account for the unequal reactivities of like functional groups. All have been described adequately elsewhere and only a brief description of their bases will be given here.

Cascade and rate theory consider in detail the reactions between subsets of significant structures in a polymerization and treat only approximately intramolecular reaction in larger species. Cascade theory was developed originally by Gordon[58] and applied particularly by Dusek and Ilvasky and coworkers[32-42] to a variety of polymerizations. It is based on the direct application of branching theory to define the probability of chains continuing from randomly chosen reactant units. Only units up to one unit (generation) away from a chosen unit are considered in detail and the relative probabilities of intramolecular reaction to give ring structures of different sizes are essentially approximated by P_i (equation 34), neglecting any variation in n_i (equation 35) with i. This so-called spanning-tree approximation means that only two probability generating functions (pgfs) are required for each reactant species. The zeroth-order pgf defines probabilities of different numbers of groups on a chosen unit being intermolecularly reacted, intramolecularly reacted and unreacted, and the first-order pgf defines the corresponding probabilities on units known to be attached to the chosen unit. Unreacted groups turn into intermolecularly and intramolecularly reacted groups as a polymerization proceeds, and this process is described by numerical solutions to a set of simultaneous differential equations describing the rates of change of the pgfs, starting from the initial condition that all groups are unreacted. The number of ring structures is equal to the number of pairs of groups reacted intramolecularly and the gel point is related to the first-order generating functions. Beyond the gel point the first-order pgf is used to define the fractions of intermolecularly reacted groups which give chains continuing indefinitely and those which do not in terms of an extinction coefficient. Only those junctions (branched units) which have three or more chains continuing indefinitely are elastically active in the final network. Dusek et al.[71] have recently improved the spanning-tree approximation by using kinetic rate equations to trace the detailed interconversion of local structures and then cascade theory to combine these structures. A similar approach has also been used by Samoria et al.[72]

Rate theory was developed by the author and coworkers.[21,26-31] It uses Flory–Stockmayer statistics to express the probabilities of existence of subsets of structures in a polymerization based on the reactant units. The rates of interconversion of the structures with respect to extent of reaction (rather than time) allows independent analytical expressions for the rates of interconversion to be derived. The set of independent (rather than simultaneous) differential equations is extended to account for all the pairs of groups which could react intramolecularly. Fractions of the rates corresponding to λ (equation 37) are used to give rates of intramolecular reaction with rates of intermolecular reaction described by $1 - \lambda$. Within the structures in a subset all the n_i (equation 35) are counted correctly. In comparison with cascade theory, the connectivity of a larger number of reactant units is accounted for locally, but outside the subset all reactions are assumed to be intermolecular. Although the rate equations can in principle be solved analytically, numerical integration is normally used to avoid the derivation of cumbersome expressions. The subset of structures contains unreacted, intermolecularly reacted and intramolecularly reacted groups, and sequences of reacted groups which give continuing chains. The number of pairs of intramolecularly reacted groups gives the number of ring structures and the probability of continuing chains defines the gel point. The gel fraction and the number of elastically active junctions in the gel fraction and in the final network are also counted in terms of numbers continuing chains from units. Rate theory has so far been applied to model endlinking polymerizations.

A method of Monte Carlo simulation for nonlinear polymerizations has been developed by Eichinger and coworkers and has been applied to endlinking polymerizations and cross-linking.[43-50,70] One reactant is designated a polymer and rotational isomeric state (RIS) statistics can be used to define probabilities of separations of the reactive groups on each molecule and, hence, the effective, mutual concentrations of pairs of groups which could eventually become parts of ring structures. The other reactant is designated a crosslinker and its molecules are treated as sticks of defined length. Several thousands ($\geqslant 10\,000$) of reactant molecules are placed at random in a box and distances between all pairs of A and B groups recorded. A capture sphere is defined, and all A and B pairs whose distances of separation fall within a sphere can react. The pairs are allowed to react sequentially in increasing order of their mutual separations and, as crosslinkers react, they are moved to join the groups on the polymer molecule with which they have reacted. A record is kept of the connectivity of all molecules during the linkage process *via* a spanning-tree algorithm (SPANFO), which can delineate gel and sol components. Structures, such as smallest loops in the gel, double-loop species and various types of dangling ends are also enumerated. A polymerization is followed by increasing the size of the capture sphere in steps. This allows more pairs of groups to react and the extent of reaction to increase. The molecular connectivities and structures are analyzed after each step.

The gel point is located as the mean of: (i) the extent of reaction at which the reduced mass-average degree of polymerization, *i.e.* the average over all species except the largest one, displays a maximum (lower bound); and (ii) the extent of reaction at which the rate of increase of the overall mass-average degree of polymerization is a maximum (upper bound).

As a reaction proceeds and capture spheres increase beyond a certain volume, the occurrence of gel–gel reaction is suppressed in the simulations, but gel–sol and sol–sol reactions are allowed to continue. The value of the capture-sphere radius chosen for the suppression of gel–gel reaction is $2\langle r_0^2\rangle^{1/2}$, where $\langle r_0^2\rangle$ is, for example the average mean-square end to end distance between two reactant groups on the polymer molecule. The value is chosen arbitrarily but is related to the magnitudes of junction point fluctuations in networks. It was introduced to reflect the possible reduction, imposed by molecular connectivity, in the mutual reactivity of pairs of groups in the gel, compared with the reactivity of pairs of groups of which at least one group is on a sol molecule. Significantly, it also leads to more efficient computations.

In addition to the approaches discussed, it is, of course, possible to write systems of simultaneous kinetics equations to describe the interconversion of molecular species including intramolecular reaction, rather than subsets based on the growth from randomly chosen reactant units. Temple[73] has shown that for a nonlinear polymerization, the number of simultaneous equations needed is excessive and intramolecular reaction is underestimated compared with cascade theory (and rate theory), for the same initial value of the ring-forming parameter λ (equation 37).

10.3.3 Gelation

Gelation has been studied mostly using endlinking polymerizations. As mentioned previously, in the crosslinking of polymer chains, the gel point is hardly significant (see Section 10.1.2 and equation 6). In endlinking polymerizations, reactants typically have functionalities from two to six and reaction mixtures are usually of low viscosity, initially. Thus, the gel point occurs well after the start of polymerization and is manifest by a dramatic increase in viscosity and the appearance of a nonzero elastic shear modulus.

The increase in viscosity and the appearance of an elastic modulus are not only important parameters in the processing of gelling polymerizations, they are also the phenomena used to determine the gel time.[74,75] For fundamental studies, the conversion *versus* time curve for one type of group is determined and extent(s) of reaction at the gel point can be found by extrapolation or interpolation, together with the use of equation (9) for —A + B— reactions. Other studies assume the chemical kinetics are known and use them to evaluate extents of reaction corresponding to the gel time.

Gelation rarely occurs at the extents of reaction predicted by equations (3). Delayed gelation always occurs because of intramolecular reaction and delayed or early gelation can occur because of the unequal reactivities of like functional groups. There have been a number of applications of cascade theory to interpret gel points in polymerizations, such as polyepoxide- and polyurethane-forming polymerizations.[34,35,39,41,42,71] Such studies can give quantitative interpretations of gel points but, if both unequal reactivities and intramolecular reaction are significant, it is possible to obtain the same gel point with different sets of parameters.

Intramolecular reaction is the more fundamental and universal phenomenon and, to elucidate unambiguously its effects on the gel point, it is necessary to choose reactants which have like groups with equal reactivities. Polymerizations with such reactants forming polyesters and polyurethanes have been used particularly by the author and coworkers[10,14,21–25,27–29,68,74,75] and by Dusek and coworkers.[32–38,40]

Rate theory[21,30] and Monte Carlo simulations[49,50] have been used successfully to interpret gel points in polyurethane-forming polymerizations carried out at various initial dilutions using hexamethylene diisocyanate (HDI) and polyoxypropylene (POP) triols and tetrols. Excess reaction at gelation has been shown to be consistent with intramolecular reaction superimposed upon random intermolecular reaction. The amounts of intramolecular reaction found experimentally

Table 1 Comparison[15] of Experimental and Monte Carlo (MC) Simulated Values of Extents of Reaction at Gelation in Stoichiometric HDI + POP Tetrol Polymerizations using two Different POP Tetrols

w_{solv} (%)[a]	$(c_{a0} + c_{b0})^{-1}$ (kg mol^{-1})[b]	$\alpha_c(expt)$[c]	$\alpha_c(MC)$
$v = 44$[d]			
0	0.198	0.445	0.456
14.9	0.233	0.462	0.462
30.0	0.282	0.492	0.480
48.9	0.393	0.515	0.522
69.9	0.652	0.545	0.577
$v = 66$			
0	0.302	0.445	0.432
10.2	0.337	0.462	—
20.4	0.378	0.471	0.446
30.0	0.433	0.491	—
39.7	0.503	0.502	0.498
50.0	0.604	0.512	0.507
59.9	0.747	0.535	0.528

[a] w_{solv} = mass fraction of solvent; [b] $c_{a0} + c_{b0}$ = initial dilution of reactive groups; [c] $\alpha_c = p_a p_b$ at gel; [d] v = number of bonds in chain forming the smallest ring structure (equations 32 and 33)

were shown to be entirely consistent with the spatial correlations between reactive groups being governed by chain conformational statistics. For example,[15] Table 1 gives experimental values of $p_a p_b (= \alpha_c)$ at gel[27–29] and values calculated by Monte Carlo simulation by Lee and Eichinger[49] for stoichiometric HDI + POP tetrol polymerizations in bulk and in nitrobenzene solvent. The F–S value of α_c is 1/3, hence, pregel intramolecular reaction is significant. Importantly, the calculated values of α_c were based on realistic RIS chain statistics and no adjustable parameters.

In order to predict and interpret gel points, it is not necessary to use theories and models of polymerizations. Theories which may be termed approximate gelation theories have been developed. They focus on the gel point and cannot be used to evaluate ring fractions (Figure 5) nor network defects (see later). They are from the outset approximate because they essentially assume a constant value of λ, which means that some average value c_{ext} is used. They have the advantage that they can give analytical expressions for some polymerizations. Computation is not required as for the polymerization theories.

The most useful of the theories is that due to Ahmad and Stepto[76] based on a correction and development of Kilb's theory.[77] The growing chain structure used for $RA_2 + RB_f$ polymerizations is shown in Figure 8. An essentially linear sequence of units is considered with pendant chains of limited complexity. (Structures are also simplified in polymerization theories.) The gel point is when there is unit probability of growth from B^1 to B^2, accounting for the intramolecular reaction of B^1 and A^1 with A or B groups on all of the pendant chains. The gelation condition is

$$\alpha_c(f-1)(1-\lambda_{ab})^2 = 1 \qquad (40)$$

where

$$\lambda_{ab} = \lambda = \frac{c_{int}}{c_{int} + c_{ext}} \qquad (41)$$

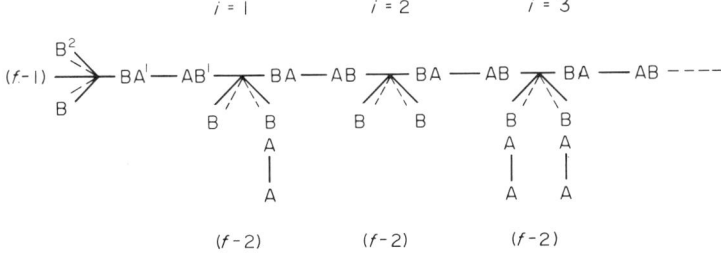

Figure 8 Ahmad and Stepto's development[76] of Kilb's linear sequence[77] to define the condition for gelation in an $RA_2 + RB_f$ polymerization (reproduced from ref. 76 by permission of Steinkopff)

and
$$c_{int} = (f-2)P_{ab} \cdot \phi(1,3/2)/N_{Av} \quad (42)$$

$$c_{ext} = (c'_a + c'_b) \quad (43)$$

P_{ab} is equivalent to P_1 of equation (32) and $ca' + cb'$ is some average value of the total concentration of A and B groups between the start of the polymerization and the gel point. $\phi(1,3/2) = \sum_i^\infty i^{-3/2} = 2.612$ arises from the occurrence of ring structures of all sizes (see equation 34).

Equation (40) has been used to interpret gel points from stoichiometric polyester-[76,78] and polyurethane-forming[22,24,25,27] polymerizations at different dilutions. For application to data it is best transformed to

$$\alpha_c^{1/2}(f-1)^{1/2} - 1 = \frac{\lambda_{ab}}{1-\lambda_{ab}} = \lambda'_{ab} \quad (44)$$

where

$$\lambda'_{ab} = \frac{c_{int}}{c_{ext}} = (f-2)\left(\frac{3}{2\pi v b^2}\right)^{3/2} \phi(1,3/2)/(c_{ext}N_{Av}) \quad (45)$$

allowing λ'_{ab}, evaluated from experimental values of α_c through equation (44), to be plotted *versus* a chosen value of dilution ($1/c_{ext}$). Examples of such plots are shown in Figure 9; and Table 2 gives values of b evaluated from the slopes of the plots.[76] Also shown in Table 2 are the values of b from the initial slopes of corresponding plots using $c_{ext} = c_{a,c} + c_{b,c}$, where $c_{a,c} = c_{a0}(1 - p_{a,c})$ and $c_{b,c} = c_{b0}(1 - p_{b,c})$, with $p_{a,c}$ and $p_{b,c}$ the extents of reaction at gel and $c_{a,c} + c_{b,c}$ the corresponding external concentrations.

The plots in Figure 9 and the data in Table 2 are worthy of discussion as they illustrate some of the general points discussed in Section 10.3.1. For stoichiometric reaction mixtures, λ'_{ab} is in fact the fractional increase in extent of reaction at gelation due to intramolecular reaction. That is, $\alpha_c = p_{a,c} \cdot p_{b,c} = p_c^2$ (as $p_a = p_b$), and without intramolecular reaction $p_c^0 = (f-1)^{-1/2}$. Thus, from

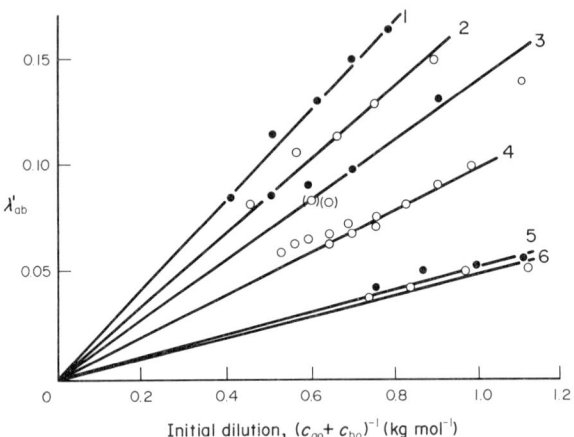

Figure 9 λ'_{ab} versus initial dilution, $(c_{a0} + c_{b0})^{-1}$, for $RA_2 + RB_3$ polymerizations using diacid chlorides and POP triols. Details of polymerization systems are given in Table 2 (reproduced from ref. 76 by permission of Steinkopff)

Table 2 Values of b Derived[76] from the Slopes of Plots of λ'_{ab} versus c_{ext}^{-1} according to Equation (45)

System[a]		v	v_2/v[b]	$b(i)$ (nm)	$b(ii)$ (nm)
2	SC + LHT240	41	0.27	0.318	0.508
1	AC + LHT240	37	0.19	0.313	0.480
4	SC + LHT112	70	0.16	0.293	0.433
3	AC + LHT112	66	0.11	0.270	0.399
6	SC + LG56	136	0.08	0.267	0.390
5	AC + LG56	132	0.05	0.260	0.371

[a]SC = sebacoyl chloride; AC = adipoyl chloride: POP triols = LHT240, LHT112, LG56. [b]v_2/v = fraction of bonds from the difunctional reactant in the chain forming the smallest ring structure; $b(i)$ — $c_{ext} = c_{a0} + c_{b0}$; $b(ii)$ — $c_{ext} = c_{ac} + c_{bc}$.

equation (44)

$$\lambda'_{ab} = \frac{p_c - p_c^0}{p_c^0} \qquad (46)$$

Figure 9 shows fractional increases of 4 to 16% depending on reactant molar mass (v) and dilution (c_{ext}^{-1}). The lowest values of λ'_{ab} occur in bulk for the two reactants with the highest molar masses. In comparison, the data for $RA_2 + RB_4$ polymerizations in Table 1 have λ'_{ab} varying from 16% (bulk) to 28%. The increase with functionality in the proportion of intramolecular reaction is expected because of the increase in average numbers of groups per molecule with functionality.

The slopes of the lines in Figure 9 increase as the molar masses of the reactants decrease. The increase results principally from a decrease in v, which gives a higher probability of ring formation. However, equation (45) shows that the slopes should be proportional to $(vb^2)^{-3/2}$, and the different values of b in Table 2 indicate that there is indeed a change in chain stiffness as the proportion of bonds from the two reactants in the chain structure changes.

The values of b found using $c_{ext} = c_{a,c} + c_{b,c}$ are in reasonable agreement with those expected from conformational statistics. The values found using $c_{ext} = c_{a0} + c_{b0}$ are too small. Thus, most of the ring structures are apparently formed near the gel point, as expected from the larger number of groups per molecule at that point (cf. Figure 5).

The general conclusions of interpretations of gel points of model polymerizations show clearly the amount of intramolecular reaction at gelation is dependent on functionalities, reactant molar masses (v), dilution of reactive groups and chain structure (b). In addition, because, even in bulk, reactive groups are diluted, gel points are always delayed beyond the F–S value. The range of fractional increases in extent reaction at gel shown by the polymerizations here, namely 4–16%, are typical for bulk trifunctional polymerizations.

10.3.4 Percolation

Percolation models for gelation and the gel state have been highlighted in other chapters (Volume 5, Chapter 8; Volume 6, Chapter 8 in *Comprehensive Polymer Science*) and it is important to put these in perspective regarding the formation and structure of polymer networks and the gel point.

Molecular percolation implies connectivity between the physical boundaries of an assembly of molecules. In the context of gelling polymerizations, it means that groups have reacted so that continuing, covalently bonded chains extend across a macroscopic sample. The onset of percolation is equated with the gel point. The sol–gel transition is treated in an analogous way, for example, to liquid–gas, magnetic and electrical phase transitions. Percolation phenomena are described by considering probabilities of connectivities of molecules which are placed on lattices. Such models automatically allow for intramolecular reaction. However, they do not allow for molecular movement and the resulting random intermolecular reaction between pairs of groups or reactive sites in the assembly. Reaction occurs between units on neighbouring sites, with units remaining stationary on sites throughout a polymerization. Hence, if they are used to describe a polymerization from start to finish, their very basis is at variance with F–S statistics and the wealth of experimental evidence supporting the assumptions of those statistics. For example, if applied to linear irreversible polymerizations they would not predict the most probable distribution from polycondensation and polyaddition polymerizations in bulk nor the Poisson distribution in living monomer-addition polymerizations. Further, the results are at variance with the experimental evidence discussed in the Sections 10.3.1 and 10.3.3 regarding total ring fractions (N_r) and delays in gel points. The former, as illustrated in Figure 5, shows almost no intramolecular reaction in bulk for a linear polymerization, which means no deviations from x_n as predicted by F–S statistics. In nonlinear polymerizations, the values of N_r and delays in gel point measured are entirely consistent with spatial correlations dictated by normal chain conformation statistics superimposed on the random reaction of pairs of groups.

Percolation concepts, as presently developed, apply to assemblies where the diffusional (centre of mass) movements of molecules can be neglected. Thus, they can apply to polymerization in the glassy state, where the reaction between permanent spatial neighbours predominates. They may also have some relevance to reactions between groups on the forming gel molecule in the neighbourhood of the gel point (typically $|(p - p_c)/p_c| < 0.01$), provided the positions of its reactive groups are considered fixed. Such a condition is of course at variance with the subchains between pairs of reactive groups following normal chain conformation statistics. In addition, in the usual homogeneous polymerization, gel–gel reaction in the sol–gel transition regime is a negligible fraction of the total amount of reaction occurring. Thus, percolation processes, whilst they may have some relevance to the detailed connectivities forming in the gel very near the gel point, and perhaps to gel–gel reaction towards the end of a polymerization, appear to have little relevance to the overall statistics of molecular growth throughout a polymerization.

Similar appraisals of the application of percolation theory to polymerizations have appeared elsewhere in the literature.[41,79]

10.4 NETWORK FORMATION, STRUCTURE AND PROPERTIES

10.4.1 Introduction

The close connection between network and structure and modulus is summarized in equations (1) and (2). In endlinking polymerizations, if each pair of groups reacting leads only to linear chains between junction points then the perfect network structure is formed. In the absence of physical interactions or topological entanglements between chains, concentrations of elastic chains and junctions ($n_{e,c}$ and $n_{e,J}$) can be calculated from the molar masses and functionalities of the reactants (equation 29). However, if intramolecular reaction forms a significant number of inelastic chains and junctions then $n_{e,c}$ and $n_{e,J}$ will be reduced below the values expected and materials of lower moduli will result. Thus, the first problem to be addressed is the investigation of $n_{e,c}$ and $n_{e,J}$ for given reactants and polymerization conditions. Having established the structure of a network in this way, measured moduli can be used to elucidate any effects (additional modulus) due to interactions or entanglements.

The usefulness of this approach relies on the correctness of equation (1). However, G varies with frequency of measurement and the decreases with magnitude of the strain (Volume 2, Chapter 9 in *Comprehensive Polymer Science*). Hence, G in equation (1) should refer to moduli measured at zero frequency, in the limit of zero strain (limiting static shear moduli).

The present section considers first experimental evidence for the close link between amount of intramolecular reaction and the absolute value of modulus. Also, effects due to chain interactions are illustrated. Second, results using theories of polymerization to predict network defects due to intramolecular reaction are discussed. Finally, present interpretations of the stress–strain behaviour of networks are discussed.

10.4.2 Experimental Results on Formation, Structure and Modulus

In order to fully characterize the amount of intramolecular reaction which occurs during the formation of a network one should aim to measure: (i) the variation of ring formation with extent of reaction; (ii) the gel point; and/or (iii) the variation of sol fraction (w_s) with extent of reaction.

The measurement of (i) is lengthy, requiring the determination of the change of number-average molar mass with conversion and has only been attempted[64] when linked with gel point determinations (Figure 5), not experiments on network preparation. The measurement of (ii) linked with the modulus of the network formed at complete reaction has been used by the author and coworkers.[10,14,22–25,27–29] Measurements of w_s at complete reaction, linked with modulus, have been used by Dusek and Ilavsky and coworkers.[33,35–40]

The use of sol fraction–modulus data at complete reaction has disadvantages. Calculations using polymerization theories have to be employed for interpretation, and stoichiometric reaction mixtures often yield negligible sol fractions, except at high dilutions. Hence, the data to be discussed in this section are values of gel points and limiting moduli at complete reaction of stoichiometric polymerizations.[14,23–25,27–29] The so-called, smallest loop analysis will be used,[14,27–29] enabling general relationships between formation, structure and properties to be established without recourse to detailed polymerization theories. (Interpretations using polymerization theories are discussed in the following section.)

Figure 10 shows results from polyurethane-forming polymerizations at various initial dilutions using HDI and POP triols and tetrols of various molar masses. The gel points for Systems 5 and 6 are in Table 1. The results are plotted as M_c/AM_c^0 versus $p_{r,c}$. M_c^0 is the value of M_c expected for the perfect network structure (Section 10.2.2) and M_c/A is the quantity evaluated from measured limiting shear moduli. M_c^0 is the molar mass of the chain v bonds which can form the smallest loop. A is equal to 1 and $(1 - 2/f)$, respectively, for affine and phantom chain behaviours (Volume 2,

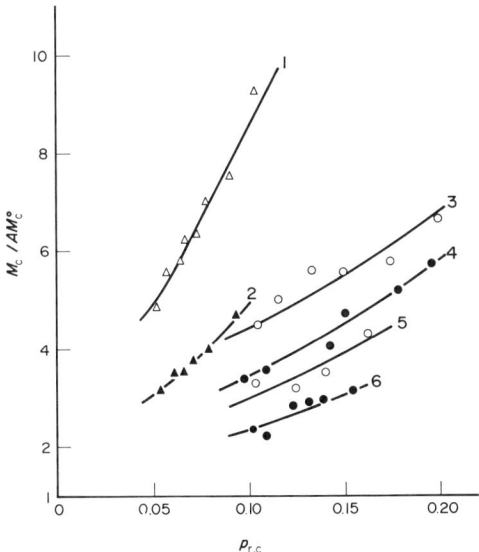

Figure 10 M_c/A from modulus measurements relative to M_c^0 versus $p_{r,c}$. Polyurethane networks formed at complete reaction using HDI and POP triols and tetrols. Systems: 1 and 2—HDI/POP triols; 1 $M_c^0 = 635 \text{ g mol}^{-1}$ ($v = 33$), 2 $M_c^0 = 1168 \text{ g mol}^{-1}$ ($v = 61$); 3 to 6—HDI/POP tetrols; 3 $M_c^0 = 500 \text{ g mol}^{-1}$ ($v = 29$), 4 $M_c^0 = 568 \text{ g mol}^{-1}$ ($v = 33$), 5 $M_c^0 = 789 \text{ g mol}^{-1}$ ($v = 44$), 6 $M_c^0 = 1220 \text{ g mol}^{-1}$ ($v = 66$)

Chapter 9 in *Comprehensive Polymer Science*). (In equation 1, $A = 1$.) The evaluation of M_c/A from shear modulus has been discussed elsewhere.[22] The values shown are averages from the moduli of the various networks in the dry and equilibrium swollen states.

$p_{r,c}$ is the extent of intramolecular reaction at the gel point, namely (see equation 46)

$$p_{r,c} = p_c - p_c^0 = \lambda'_{ab} \cdot p_c^0 \qquad (47)$$

The points for each reaction system at different values of $p_{r,c}$ come from reactions at different dilutions, with that at the lowest value of $p_{r,c}$ being for the reaction in bulk.

The trends in Figure 10 show how the various factors characterizing pregel intramolecular reaction to a large extent carry through to determine network structure and modulus. All values of M_c/AM_c^0 are greater than 1 and also greater than $f/(f-2)$. Hence, the majority of the networks have significantly lower moduli than those of the corresponding perfect networks, even on a phantom basis. In no case is a perfect network formed. If affine chain behaviour is assumed (see later), M_c/M_c^0 is equal to the reduction in modulus below that expected for the perfect network ($M_c/M_c^0 = G^0/G$). The total range of M_c/M_c^0 values is about 2.1 to 9, and for polymerizations in bulk 2.1 to 5.

There are direct correlations between $p_{r,c}$ and M_c/M_c^0. The former quantity may be taken as a measure of the propensity of a polymerization for intramolecular reaction. Thus, this propensity continues past the gel point and, for a given polymerization system, more inelastic chains are formed in the final network for larger values of $p_{r,c}$. Some of the inelastic chains will form pregel and some postgel (see later). In addition, relatively small values of $p_{r,c}$ can be symptomatic of large reductions in modulus.

More defects occur for trifunctional (curves 1 and 2) compared with tetrafunctional networks (curves 3–6). However, as expected, the delay in gel points is higher for the tetrafunctional polymerizations (see Section 10.3.3). The major differences between the two functionalities regarding network structure relates to the defects introduced per smallest loop in the two cases. Figure 11 shows that two elastic junction points (three elastic chains) are lost per smallest loop for $f = 3$, and only one elastic junction point (two elastic chains) for $f = 4$. Smaller loops must give inelastic junctions and chains. Larger loop structures can form parts of elastic chains. In addition, due to their smaller members of bonds, smallest loops are the ones most likely to form, per opportunity for intramolecular reaction. The actual distribution of ring sizes is a problem still to be solved, but the larger values of M_c/M_c^0 for $f = 3$ compared with $f = 4$ shows that the smallest loops are in fact important for determining network defects.

For a given value of $p_{r,c}$ and functionality, M_c/M_c^0 is larger for smaller values of M_c^0 (reactants of lower molar mass and lower values of v). Thus, the larger the proportion of the loops that give elastically inactive junctions, the smaller is the value of v. In contrast to the assumption made in

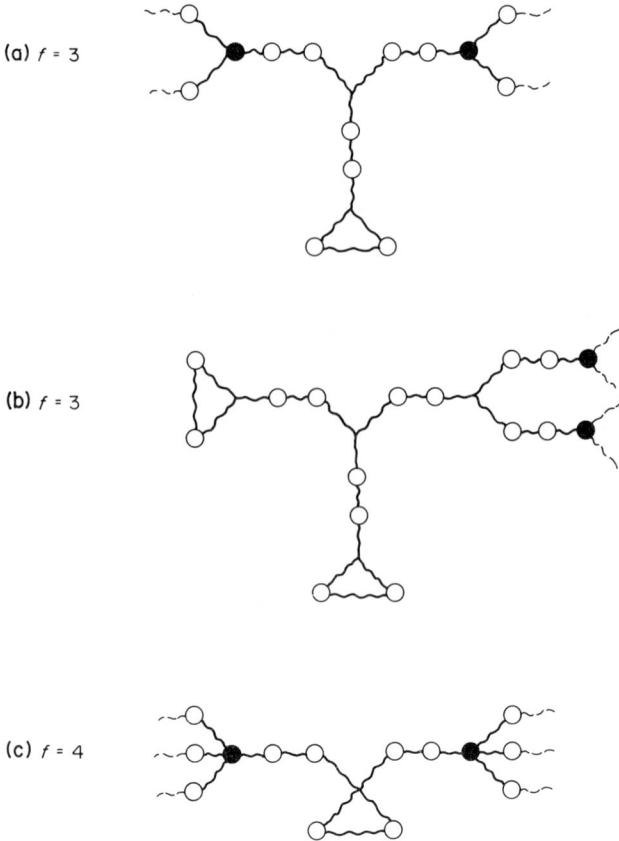

Figure 11 Smallest loop structures in: (a) and (b) $RA_2 + RB_3$ networks; and (c) $RA_2 + RA_4$ networks (○ — reacted pairs of groups ● — elastically active junctions)

all existing theories and models of polymerizations, $p_{r,c}$ does not uniquely define M_c/M_c^0. The statistics of postgel intramolecular reaction apparently have detailed correlations which are significant in determining the topology of a network. In this respect, $p_{r,c}$ is insensitive to loop size, whereas the value of M_c is expected to be more sensitive to the number of smallest loops than the numbers of larger loops.

The reductions in moduli due to intramolecular reaction below these expected for the perfect network structures can be analyzed further by converting the values of M_c/AM_c^0 from modulus measurements to values of extents of intramolecular which have formed inelastic loops by the end of the reaction ($p_{r,e}$). Assuming each inelastic loop or intramolecular reaction of a pair of groups in an $f = 3$ network leads to the loss of two junction points and in an $f = 4$ network to the loss of one junction point (conditions which may be satisfied on average even if loops larger than the smallest are also significant), the following expressions are obtained[14,27–29]

$$f = 3; \quad p_{r,e} = \frac{1}{6}\left[1 - \frac{1}{A}\cdot(AM^0/M_c)\right] \tag{48}$$

$$f = 4; \quad p_{r,e} = \frac{1}{4}\left[1 - \frac{1}{A}(AM_c^0/M_c)\right] \tag{49}$$

The factors 6 and 4, respectively, are equivalent to $2f$ and f, for $f = 3$ and $f = 4$.

Plots of $p_{r,e}$ versus $p_{r,c}$ are shown in Figure 12(a)–(c) for $A = 1$ and $(1 - 2/f)$. If all of the reaction which contributes to $p_{r,c}$ also contributes to $p_{r,e}$, again an assumption made in theories of polymerization, then $p_{r,e} \geqslant p_{r,c}$. It is apparent that this condition is satisfied only if $A = 1$, consistent with the affine chain behaviour expected from the small strain measurements used. On the same basis, the distances of the points above the lines $p_{r,e} = p_{r,c}$ in Figures 12(a) and 12(b) are the postgel contributions to $p_{r,e}$. For a given polymerization system, they are larger for the more concentrated reactions (lower values of $p_{r,c}$) indicating that spatial correlations between groups on the gel are

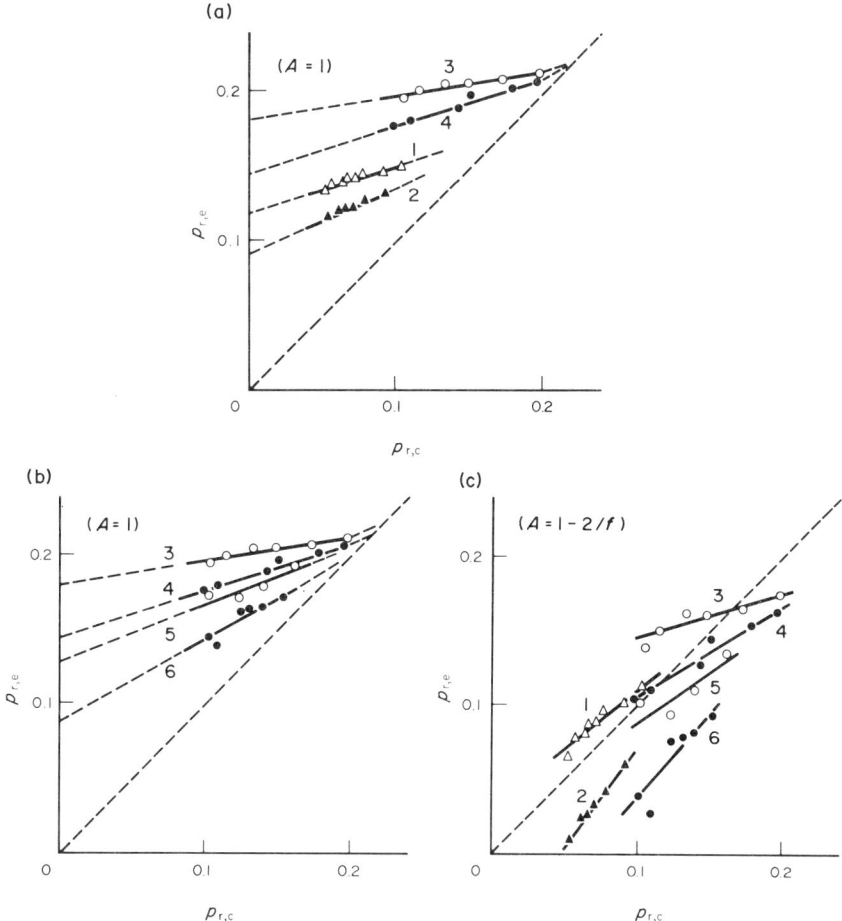

Figure 12 Extent of intramolecular reaction at complete reaction derived from modulus measurements ($p_{r,e}$) versus $p_{r,c}$. Systems as in Figure 10. (a) and (b) $A = 1$; (c) $A = 1-2/f$ (reproduced from ref. 27 by permission of Elsevier)

more important. Conversely, as dilution is increased, pregel intramolecular reaction yields larger proportions of the inelastic chains. Simple considerations show that, for pregel intramolecular reaction in a given polymerization system, the smaller the value of c_{ext} the larger is the proportion of loops formed early in a polymerization. Hence, the average loop size is smaller and the loops are more likely to be elastically inactive.

Other trends in Figures 12(a) and 12(b), with M_c^0 and f at constant values of $p_{r,c}$ correspond precisely to the trends in Figure 10. However, extrapolation of the points in Figure 12 to $p_{r,c} = 0$ shows that $p_{r,e} > 0$. In other words, even with no pregel intramolecular reaction, inelastic chains are formed postgel. Nonzero intercepts signify the universal occurrence of network imperfections. Some of the gel–gel reaction which must occur to give the final network leads to elastically inactive chains.

The intercepts in $p_{r,e}$ at $p_{r,c} = 0$ can be reconverted to values of M_c/M_c^0 for $A = 1$ in the limit $p_{r,c} = 0$ through equations (48) and (49). The reductions in moduli are significant. Thus, the curves in Figure 10 do not extrapolate to $M_c/M_c^0 = 1$ at $p_{r,c} = 0$ (see, for example, Figures 13 and 15). The resulting reductions in modulus at $p_{r,c} = 0$ can be analyzed further to show that, even in the limit of reactants of infinite molar mass (i.e. no spatial correlations between reactive groups on the same molecule), some of the random intramolecular reaction between groups on the gel still leads to inelastic chains.[14,27-29] Random gel–gel reaction is part of F–S statistics,[62] irrespective of the molar masses of the reactants, and does not lead to a perfect network. Obviously, detailed considerations of the spatial correlations governing gel–gel reaction and the structures which result are needed for the quantitive prediction of network structures and properties.

Due to intramolecular reaction, the polyurethane-forming polymerizations just discussed, based on HDI and POP polyols, form networks with moduli lower than those expected for perfect network structures. Related polymerizations using MDI instead of HDI give networks showing

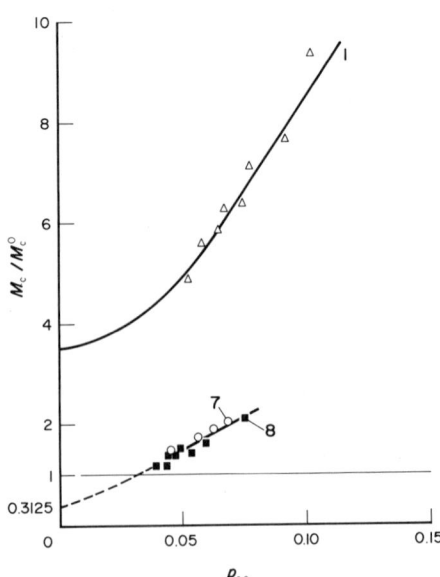

Figure 13 Comparison[15] of HDI and MDI networks. M_c from modulus measurements ($A = 1$) relative to M_c^0 versus $p_{r,c}$ 1 HDI + POP triol (LHT240)—system 1 from Figure 10; 7 MDI + POP triol (LHT240)/80 °C/in bulk and in nitrobenzene solution/$M_c^0 = 705$ g mol^{-1} (ref. 23–25); 8 MDI + POP triol (LHT240)/70 °C/in bulk and in xylene solution/$M_c^0 = 811$ g mol^{-1} (ref. 38). Curves: System 1—as in Figure 10 with extrapolated value of M_c/M_c^0 at $p_{r,c} = 0$ consistent with extrapolated value of $p_{r,c}$ at $p_{r,c} = 0$ in Figure 12. Systems 7, 8—see text and ref. 15 (reproduced from ref. 15 by permission of Prentice–Hall)

much smaller reductions in modulus.[20,23–25,35–41] Results are shown in Figure 13, in comparison[15] with the results from system 1 of Figure 10. The MDI- and HDI-based systems have similar values of M_c^0. The smaller values of $p_{r,c}$ for the polymerizations with MDI are explicable[20,23–25] in terms of a stiffer chain structure (larger value of b). The increase in M_c/M_c^0 with $p_{r,c}$ means that the effects of intramolecular reaction are still significant. However, the moduli are much larger than would be expected. Calculation[15] shows that the enhanced moduli are consistent with interchain interactions (incipient segmentation) increasing the effective numbers of junction points. The extrapolated value of $M_c/M_c^0 = 0.3125$ at $p_{r,c} = 0$ in fact corresponds to one extra (tetrafunctional) junction per elastic chain. For the bulk MDI-based polymerization the compensating effects of intramolecular reaction and interations combine to give an apparently perfect network.

The plots in Figure 13 illustrate the difficulty of separating the compensating effects of network imperfections and chain interactions. Topological entanglements in the MDI-based networks are unlikely because the system has a low value of M_c^0 and such entanglements are not evident in the HDI-based networks. However, there is much evidence for topological entanglements in networks of high M_c contributing to modulus[6–9] in amounts relating to the plateau moduli of corresponding (uncrosslinked) linear chains. Obviously, interpretations of absolute values of moduli have to take account of network imperfections before affine or phantom chain behaviour and the effects of topological entanglements and other interactions can be evaluated. To this end, the preparation of networks at different dilutions and the concomitant measurement of gel points should be of general applicability. A limitation is that methods of determining extents of reaction in the particular reaction used have to be devised. If this is not feasible, then the decrease of the sol (soluble) fraction as a function of time up to complete reaction could be used. However, as mentioned previously, the latter approach relies on the application of theories of polymerization to deduce extents of intramolecular and intermolecular reaction from the differences between experimental and F–S sol fractions. Thus, the interpretation of results rests heavily on the exactness of the theory being used and a critical test of theories is difficult to achieve.

10.4.3 Applications of Theories and Models of Polymerizations

Cascade theory,[32–42,58,71] rate theory[21,26–31] and Monte Carlo simulations[43–50,70] (Section 10.3.2) have been used to predict amounts of intramolecular reaction in network-forming polymerizations and, hence, pregel ring fractions, gel points, sol fractions and network defects. In addition, the earliest approach, cascade theory, has been applied with some succes to interpret gel point, gel fraction and limiting modulus data often from systems in which like groups have unequal reactivities.[34,35,39,41,42,71] In order to give some comparison of the predictions made, the present discussion is focused on interpretations of common data using different theoretical approaches.

10.4.3.1 Cascade theory

An example of the interpretation of results using cascade theory by Ilavsky and Dusek[38] is shown in Figure 14. It also illustrates the interplay of factors which have to be considered when interpreting results. The polymerizations used a POP triol reacting with MDI at various reactant ratios $r_H = [OH]_0/[NCO]_0$ and dilutions in xylene. The results plotted in Figure 13, in comparison with results from polymerization using HDI, are in fact derived from those at $r_H = 1$.

Figure 14 shows $\log G_r$ versus ϕ_2, where the reduced modulus G_r is equal to $A \cdot n_{e,c} = A \cdot \rho/M_c$. The solid and dashed curves give the theoretical predictions from cascade theory for affine and phantom chain behaviour. From the measurements of sol fraction at complete reaction, the effects of cyclization were predicted to be small. With no cyclization, G_r would be independent of ϕ_2. The experimental values of modulus lie between the predicted values for affine and phantom behaviour.

Dusek and Ilavsky have interpreted the results by assuming phantom chain behaviour and attributing the increases in modulus to pair entanglements between chains. The numbers of pairs of entanglement chains were assumed to be proportional to ϕ_2^2 and were also calculated using cascade theory. The interpretation appears to have two shortcomings. First, phantom chain behaviour is not expected as limiting moduli were measured. Second, concentrations of entanglements should relate to segmentation effects in the final polyurethane networks rather than topological entanglements during formation. An alternative explanation which avoids these shortcomings and is consistent with results on HDI-based networks is that given in the discussion of Figure 13, namely, affine chain behaviour coupled with intramolecular reaction and segmentation effects in the final networks.

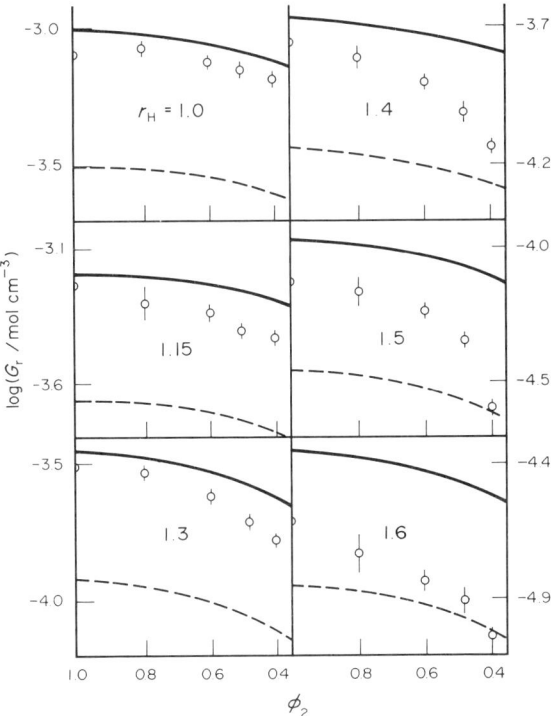

Figure 14 Comparison of experimental and calculated dependence of reduced moduli, G_r, using cascade theory *versus* volume fraction of reactants, ϕ_2, for MDI + POP triol polymerizations and networks (from ref. 38). $G_r = A \cdot \rho/M_c$; $r_H = [OH]_0/[NCO]_0$. Curves calculated using cascade theory accounting for intramolecular reaction: ——— $A = 1$; --- $A = 1/3$

10.4.3.2 Monte Carlo simulations

The Monte Carlo simulations developed by Eichinger and coworkers[43-50,70] have applied to several model network-forming, endlinking polymerizations and to the crosslinking of polymer chains. In the former applications, data on pregel number fractions of rings,[70] gel points[49,50,70] and modulus at complete reaction[49,50] have been interpreted.

Figures 15 and 16 show calculated and experimental values of M_c/M_c^0 on the basis of affine chain behaviour. Figure 15 refers to the HDI–POP tetrol systems 5 and 6 of Figure 10 and Figure 16 to the MDI–POP triol system of Figures 13 and 14 for stoichiometric polymerization mixtures. It can be seen that the calculated values of M_c/M_c^0 are somewhat lower (moduli somewhat higher) than the experimental ones for the HDI-based networks. The reverse is true for the MDI-based networks. As discussed in Section 10.3.3 (Table 1), the Monte Carlo results[15,49] give good predictions of measured gel points of the polymerizations in Figure 15 using realistic chain conformational statistics. However, for modulus predictions the simulations were stopped before complete reaction. In addition, gel–gel reaction was suppressed in the later stages of the polymerizations. The dashed curves $p_g = 1$ in Figures 15 and 16 give approximate extensions of the simulations to complete reaction neglecting the formation of any further inelastic chain formation. On the basis of the curves in Figure 15, the simulations appear to underestimate network defects, whereas in Figure 16 they appear to overestimate them. The apparent overestimation of defects in the MDI-based networks is attributable to the effects of chain interactions increasing the actual measured moduli (see Figure 13).[15] The underestimation of defects in the HDI-based networks would seem to relate

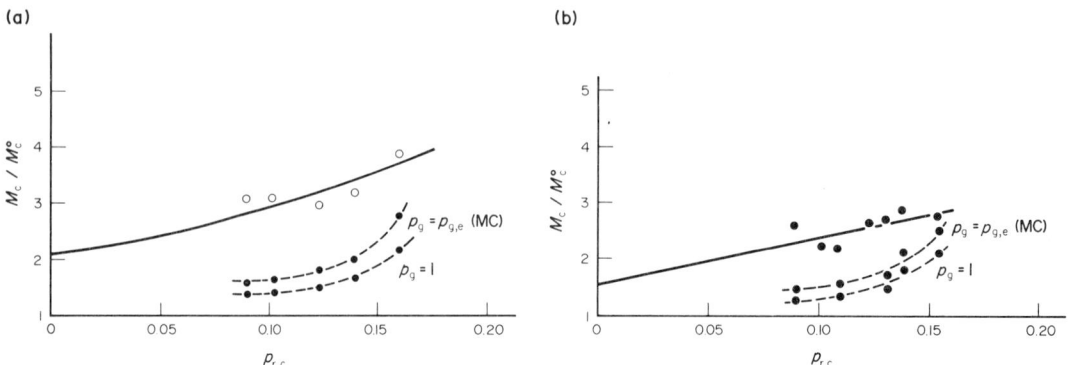

Figure 15 Comparison[15] of experimental and Monte Carlo simulated values of M_c/M_c^0 versus $p_{r,c}$ for the HDI + POP tetrol polymerizations and networks of systems: (a) 5; and (b) 6 of Figure 10. Full curves: experimental results and smallest-loop analysis with extrapolated values of M_c/M_c^0 at $p_{r,c} = 0$ consistent with extrapolated values of $p_{r,c}$ at $p_{r,c} = 0$ in Figure 12. Dashed curves: Monte Carlo (MC) results; $p_g = p_{g,c}$ (MC) gives results at the end of the MC simulations; $p_g = 1$ gives MC values of M_c/M_c^0 extrapolated to complete reaction in the gel (reproduced from ref. 15 by permission of Prentice–Hall)

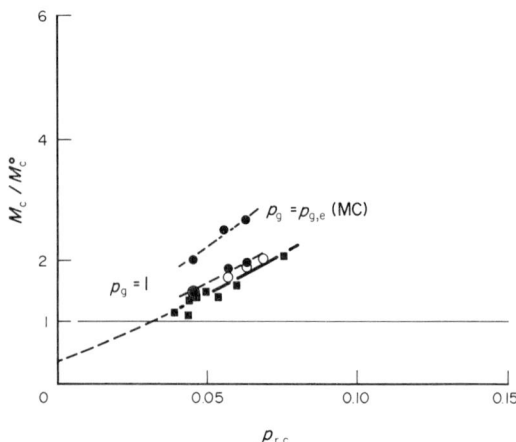

Figure 16 Comparison[15] of experimental and Monte Carlo simulated values of M_c/M_c^0 versus $p_{r,c}$ for the MDI + POP triol polymerizations and networks of systems 7 and 8 of Figure 13. Symbols and curves as for Figures 13 and 15 (reproduced from ref. 15 by permission of Prentice–Hall)

to the computationally expedient cessation of the simulations before complete reaction and the suppression of gel–gel reaction.

Basically, the Monte Carlo results are in agreement with those of the smallest-loop analysis of the same data (Section 10.4.2). They also confirm the predominance of smallest loops which were shown[15] to increase from 60–65% of the total loops for bulk reactions to 72–83% at 60–70% solvent. In principle, Monte Carlo simulations, although requiring longer computations than those using cascade theory or rate theory, are able to account in more detail for the structure of a formed network.

10.4.3.3 Rate theory

First-order rate theory, accounting for smallest loops, has so far been developed for RA_3, RA_4, $RA_2 + RB_3$ and $RA_2 + RB_4$ polymerizations.[21,26–31,80] Extensions to other polymerizations is straightforward.

Figures 17(a) and 17(b) show the predictions of modulus obtained for all the data in Figure 10. The values of the ring-forming parameter (P_{ab}/c_{a0}) used were those required to give agreement with the experimental gel points ($p_{r,c}$). Generally, the calculated moduli agree with the experimental values to better than a factor of two. Overall, the prediction of moduli is satisfactory, given that reductions in moduli range from a factor of two to a factor of 10. Even though only the smallest loops were counted, the theory seems to be able to predict moduli as well as the Monte Carlo simulations. The latter have been tested only against systems 5 and 6 with smallest numbers of defects (cf. Figure 15).

The values calculated using rate theory show approximately the correct relative differences in M_c/M_c^0 between trifunctional and tetrafunctional networks. However, they do not show sufficient variation with reactant molar masses (M_c^0 or v). This behaviour again indicates that the variation of distribution of loop sizes with M_c^0 and the details of gel–gel reaction are important (see discussions of Figures 10–12). The smaller variation of predicted values of M_c/M_c^0 with M_c^0 compared with that found experimentally is also a feature of cascade theory and Monte Carlo simulations. All give M_c/M_c^0 versus $p_{r,c}$ as universal curves. In other words, a given ring-forming parameter, which results in a certain delay in gel point, always gives the same reduction in modulus, independent of M_c^0.

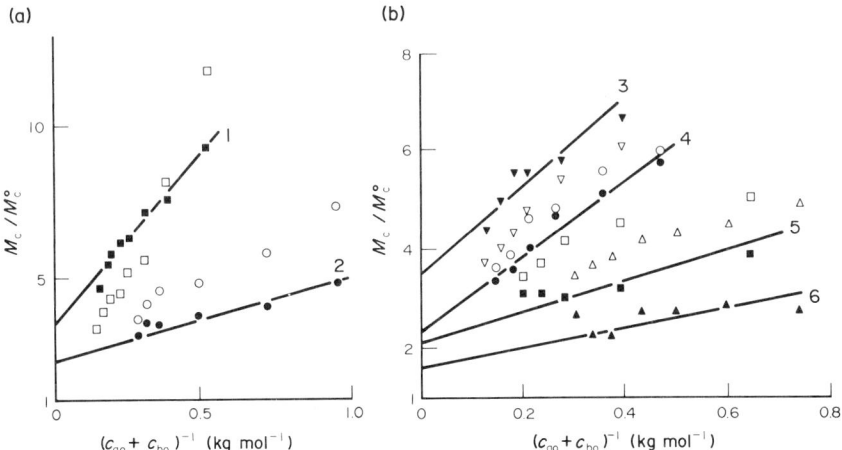

Figure 17 Comparison[80] of experimental and calculated values of M_c/M_c^0 using rate theory versus $(c_{a0} + c_{b0})^{-1}$ for the HDI-based polymerizations and networks of Figure 10. (a) HDI + POP triols; (b) HDI + POP tetrols. Full symbols: experimental results; open symbols: calculated values of M_c/M_c^0 consistent with the experimental values of $p_{r,c}$

10.4.4 Stress–Strain Behaviour of Networks

Theories of rubber-like elasticity have been reviewed in other chapters (Volume 2, Chapter 9; Volume 6, Chapter 8 in *Comprehensive Polymer Science*), and discussion here is limited to correlations with data illustrating some fundamental questions which still need to be addressed.

(i) Most of the model networks prepared to test theories of rubber-like elasticity have been assumed to have perfect structures (apart from loose ends). In view of the established marked dependences of network structures on the functionalities, molar masses and chain flexibilities of the reactants and on the reaction conditions (dilution of reactive groups and reactant ratios), the assumption of perfect structures must at least be viewed with caution. Even at their present stage of development, theories and models of gelation should be used to predict approximate levels of intramolecular reaction and network defects. Such predictions are particularly important if absolute values of moduli are discussed in relation to affine and phantom chain behaviour and whether topological entanglements or chain interactions are significant. For example, some of the model poly(dimethyl siloxane) (PDMS) networks discussed by Mark[81] have values of M_c^0 similar to those of the polyurethane networks in Figure 10.

(ii) Gottlieb and Gaylord[82,83] have recently tested several of the existing elasticity theories against stress–strain data on model networks and shown that none of the theories can predict correctly the forms of stress–strain behaviour for different types of mechanical deformation and levels of strain. Similar conclusions have been drawn regarding swelling behaviour.[84-87] The correlations focus on changes in the effective modulus or strain energy density function with deformation. Hence, they are insensitive to uncertainties in the absolute values of $n_{e,c}$ due to network defects.

(iii) Stress–strain behaviour is commonly studied using uniaxial deformation. Several models can successfully explain experimental data.[82,83] What is found experimentally and also predicted by various theories is that under uniaxial deformation (extension and compression) the modulus (or reduced stress) decreases as deformation increases. It is the magnitude and functional forms of the decreases which still need to be explained. For example, the most commonly used theory is the constrained junction theory discussed in detail in Volume 2, Chapter 9 of *Comprehensive Polymer Science*. At small strains, the affine behaviour of chains is predicted. Junction fluctuations are presumed to increase with strain and, hence, reduce the entropic force counteracting the applied force. Concentrating on limiting behaviour

$$G = An_{e,c}RT \qquad (50)$$

where $A = 1$ at zero strain and tends to $(1 - 2/f)$ at large strains. Thus, the ratio of small strain to large strain modulus should be given by

$$\frac{G_0}{G_\infty} = \frac{f}{f-2} \qquad (51)$$

The prediction of equation (51) is examined in Figure 18 for an interesting series of PDMS networks of different functionalities prepared by Mark and Llorente.[81,88,89] The small-strain moduli for the networks were approximately the same. It can be seen that the theoretical values of G_0/G_∞ are much larger than the experimental ones at low functionality but tend to unity as expected at higher functionalities. Apparently, the decreases in modulus with extension are less than those predicted by the transformation from affine to phantom chain behaviour.

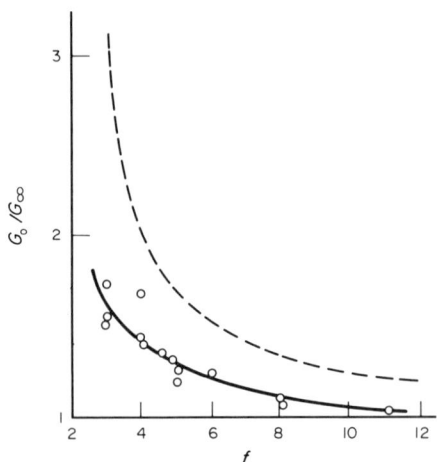

Figure 18 Ratio of small strain to large strain modulus, G_0/G_∞, versus junction point functionality, f, for poly(dimethylsiloxane) networks (derived from ref. 81). —— experimental, ---- G_0 (affine)/G_∞ (phantom) = $f/(f-2)$

The discrepancies between theories and experiment highlighted in points (ii) and (iii) probably relate to the theoretical *ab initio* factorizations of the partition function of a network into the product of the partition functions of individual chains, and, with regard to swelling, to the separation of elastic and osmotic free energies. Affine behaviour assumes ends of individual chains, the junctions, deform affinely with the macroscopic deformation. In addition, Gaussian chain statistics are usually assumed for the calculation of the mainly entropic force exerted by each chain. These assumptions may be expected to hold in the limit of zero deformation. As deformation increases, the very connectivity of the elastic chains in the network will reduce the average entropy available per chain and the measured modulus, in a manner which still needs to be defined.

ACKNOWLEDGEMENTS

My thanks are due to Professor Saverio Russo for the invitation to write this chapter and for his patience and encouragement as deadlines passed. My thanks also to Joan Stepto for typing the manuscript and her support during its preparation.

10.5 REFERENCES

1. P. J. Flory, 'Principles of Polymer Chemistry', Cornell University Press, Ithaca, New York, 1953, chap. 9, p. 347.
2. A. H. Clark and S. B. Ross-Murphy, *Adv. Polym. Sci.*, 1987, **83**, 57.
3. W. Burchard and S. B. Ross-Murphy (eds.), 'Physical Networks, Polymers and Gels', Elsevier, Barking, 1990.
4. N. R. Legge, V. G. Holden and H. E. Schroeder, 'Thermoplastic Elastomers: a Comprehensive Review', Hanser, Munich, 1987.
5. K. Dusek and W. Prins, *Adv. Polym. Sci.*, 1969, **6**, 102.
6. W. W Graessley, *Adv. Polym. Sci.*, 1974, **16**.
7. M. Doi and S. F. Edwards, *J. Chem. Soc., Faraday Trans. 2*, 1978, **74**, 1789; 1802; 1818.
8. G. Heinrich, E. Straube and G. Helmis, *Adv. Polym. Sci.*, 1988, **85**, 34.
9. O. Kramer, in 'Elastomeric Polymer Networks. A Memorial to Eugene Guth', ed. J. E. Mark and B. Erman, Prentice–Hall, Englewood Cliffs, NJ, 1992, chap. 17, p. 243.
10. R. F. T. Stepto, *Polymer*, 1979, **20**, 1324.
11. J. B. Enns and J. K. Gillham, *Coat. Plast. Prepr. Pap. Meet. (Am. Chem. Soc., Div. Org. Coat. Plast. Chem.)*, 1982, **47**, 575.
12. I. J. Goldfarb, C. Y.-C. Lee and C. C. Kuo, in 'Chemorheology of Thermosetting Polymers', ed. C, A. May, Am. Chem. Soc., Washington, D.C., 1983, chap. 3, p. 49.
13. M. J. Aronhime and J. K. Gillham, *Adv. Polym. Sci.*, 1986, **78**, 83.
14. R. F. T. Stepto, in 'Advances in Elastomers and Rubber Elasticity', ed. J. Lal and J. E. Mark, Plenum Press, New York, 1987, p. 329.
15. R. F. T. Stepto and B. E. Eichinger, in 'Elastomeric Polymer Networks. A Memorial to Eugene Guth', ed. J. E. Mark and B. Erman, Prentice–Hall, Englewood Cliffs, NJ, 1992, chap. 18, p. 256.
16. M. Gottlieb, C. W. Macosko, G. S. Benjamin, K. O. Meyers, and E. W. Merrill, *Macromolecules*, 1981, **14**, 1039.
17. C. W. Macosko and J. C. Saam, *Polymer Prepr., Am. Chem. Soc., Div. Polym. Chem.* 1985, **26**, 48.
18. K. Dusek, in 'Developments in Polymerisation—3', ed. R. N. Haward, Elsevier, Barking, 1982, chap. 4.
19. K. Dusek, H. Galina and J. Mikes, *Polym. Bull. (Berlin)*, 1980, **3**, 19.
20. S. B. Ross-Murphy and R. F. T. Stepto, in 'Cyclic Polymers', ed. J. A. Semlyen, Elsevier, Barking, 1986, chap. 10.
21. J. L. Cawse, J. L. Stanford and R. F. T. Stepto, in Proceedings of the 26th IUPAC International Symposium on Macromolecules, Mainz, 1979', IUPAC, Oxford, 1979, p. 393.
22. A. B. Fasina and R. F. T. Stepto, *Makromol. Chem.*, 1981, **182**, 2479.
23. J. L. Stanford and R. F. T. Stepto, in 'Elastomers and Rubber Elasticity', ed. J. E. Mark and J. Lal, Am. Chem. Soc., Washington, D.C., 1982, chap. 20, p. 379.
24. J. L. Stanford, R. F. T. Stepto and R. H. Still, in 'Reaction Injection Molding and Fast Polymerization Reactions', ed. J. E. Kresta, Plenum Press, New York, 1982, p. 31.
25. J. L. Stanford, R. F. T. Stepto and R. H. Still, in 'Characterization of Highly Cross-Linked Polymers', ed. S. S. Labana and R. A. Dickie, Am. Chem. Soc., Washington, D.C., 1984, chap. 1, p. 1.
26. A. C. Lloyd and R. F. T. Stepto, *Br. Polym. J.*, 1985, **17**, 190.
27. R. F. T. Stepto, in 'Biological and Synthetic Polymer Networks', ed. O. Kramer, Elsevier, Barking, 1988, chap. 10, p. 153.
28. R. F. T. Stepto, in 'Cross-Linked Polymers, Chemistry, Properties and Applications of Crosslinking Systems', ed. R. A. Dickie, S. S. Labana and R. Bauer, Am. Chem. Soc., Washington, D.C., 1988, chap. 2, p. 28.
29. R. F. T. Stepto, *Acta Polym.* 1988, **39**, 61.
30. H. Rolfes and R. F. T. Stepto, *Makromol. Chem., Macromol. Symp.*, 1990, **40**, 61.
31. H. Rolfes and R. F. T. Stepto, *Computat. Polym. Sci.*, 1991, **1**, 100.
32. K. Dusek and V. Vojta, *Br. Polym. J.*, 1977, **9**, 164.
33. K. Dusek, M. Hadhoud and M. Ilavsky, *Br. Polym. J.*, 1977, **9**, 172.
34. L. Matejka and K. Dusek, *Polym. Bull. (Berlin)*, 1980, **3**, 489.
35. K. Dusek, *Rubber Chem. Technol.*, 1982, **55**, 1.
36. K. Dusek and M. Ilavsky, in 'Elastomers and Rubber Elasticity', ed. J. E. Mark and J. Lal, Am. Chem. Soc., Washington, D.C., 1982, chap. 21, p. 403.
37. M. Ilavsky and K. Dusek, *Polymer*, 1983, **24**, 981.
38. M. Ilavsky and K. Dusek, *Macromolecules*, 1986, **19**, 2139.

39. K. Dusek, *Adv. Polym. Sci.*, 1986, **78**, 1.
40. K. Dusek and M. Ilavsky, in 'Biological and Synthetic Polymer Networks', ed. O. Kramer, Elsevier, Barking, 1988, chap. 14, p. 233.
41. K. Dusek and W. J. MacKnight, in 'Cross-Linked Polymers, Chemistry, Properties and Applications', ed. R. A. Dickie, S. S. Labana and R. S. Bauer, Am. Chem. Soc., Washington, D.C., 1988, chap. 1, p. 2.
42. J. Zelenka, M. Ilavsky, I. Dobas and K. Dusek, *Polym. Networks Blends*, 1991, **1**, 93.
43. Y.-K. Leung and B. E. Eichinger, in 'Characterization of Highly Cross-Linked Polymers', ed. S. S. Labana and R. A. Dickie, Am. Chem. Soc., Washington, D.C., 1984, chap. 2, p. 28.
44. Y.-K. Leung and B. E. Eichinger, *J. Chem. Phys.* 1984, **80**, 3877, 3885.
45. L. Y. Shy, Y.-K. Leung and B. E. Eichinger, *Macromolecules*, 1985, **18**, 983.
46. N. A. Neuburger and B. E. Eichinger, *J. Chem. Phys.*, 1985, **83**, 804.
47. L. Y. Shy and B. E. Eichinger, *Macromolecules*, 1986, **19**, 2787.
48. V. Galiastsatos and B. E. Eichinger, *J. Polym. Sci., Polym. Phys. Ed.*, 1988, **26**, 595.
49. K.-J. Lee and B. E. Eichinger, *Polymer*, 1990, **31**, 406.
50. K.-J. Lee and B. E. Eichinger, *Polymer*, 1990, **31**, 414.
51. C. W. Macosko, 'RIM: Fundamentals of Reaction Injection Molding', Hanser, Munich, 1989.
52. C. A. May (ed.), 'Chemorheology of Thermosetting Polymers', Am. Chem. Soc., Washington, D.C., 1983.
53. W. H. Stockmayer, *J. Chem. Phys.*, 1943, **11**, 45; 1944, **12**, 125.
54. W. H. Stockmayer, *J. Polym. Sci.*, 1952, **9**, 69; 1953, **11**, 424.
55. R. F. T. Stepto, in 'Developments in Polymerisation—3', ed. R. N. Haward, Elsevier, Barking, 1982, chap. 3, p. 81.
56. L. C. Case, *J. Polym. Sci.*, 1957, **26**, 333.
57. M. Ilavsky and K. Dusek, *Polym. Bull. (Berlin)*, 1982, **8**, 359.
58. M. Gordon, *Proc. R. Soc. London, Ser. A.*, 1962, **268**, 240.
59. W. H. Carothers, *Trans. Faraday Soc.*, 1936, **32**, 39.
60. C. W. Macosko and D. R. Miller, *Macromolecules*, 1976, **9**, 199.
61. D. R. Miller and C. W. Macosko, *Macromolecules*, 1976, **9**, 206.
62. R. F. T. Stepto, *Polym. Bull. (Berlin)*, 1990, **24**, 53.
63. R. F. T. Stepto and D. R. Waywell, *Makromol. Chem.*, 1972, **152**, 263.
64. J. L. Stanford and R. F. T. Stepto, *Br. Polym. J.*, 1977, **9**, 124.
65. H. Jacobson and W. H. Stockmayer, *J. Chem. Phys.*, 1950, **18**, 1600.
66. J. A. Semlyen (ed.), 'Cyclic Polymers', Elsevier, Barking, 1986.
67. R. N. Haward, *Trans. Faraday Soc.*, 1950, **46**, 204.
68. J. L. Stanford, R. F. T. Stepto and D. R. Waywell, *J. Chem. Soc., Faraday Trans. 1*, 1975, **71**, 1308.
69. V. Askitopoulos, M.Sc. Thesis, University of Manchester, 1981.
70. L. Y. Shy and B. E. Eichinger, *Br. Polym. J.*, 1985, **17**, 200.
71. K. Dusek, J. Somvarsky, M. Ilavsky and L. Matejka, *Computat. Polym. Sci.*, 1991, **1**, 90.
72. C. Samoria, E. Valles, D. R. Miller, *Makromol. Chem., Macromol. Symp.*, 1986, **2**, 69.
73. W. B. Temple, *Makromol. Chem.*, 1972, **160**, 277.
74. R. H. Peters and R. F. T. Stepto, in 'The Chemistry of Polymerization Processes', Monograph No. 20. Soc. Chem. Ind., London, 1966, p. 157.
75. W. Hopkins, R. H. Peters and R. F. T. Stepto, *Polymer*, 1974, **15**, 315.
76. Z. Ahmad and R. F. T. Stepto, *Colloid Polym. Sci.*, 1980, **258**, 663.
77. R. W. Kilb, *J. Phys. Chem.*, 1958, **62**, 969.
78. Z. Ahmad, R. F. T. Stepto and R. H. Still, *Br. Polym. J.*, 1985, **17**, 205.
79. S. I. Kuchanov, S. V. Korolev and S. V. Panyakov, *Adv. Chem. Phys.*, 1988, **72**, 115.
80. H. Rolfes and R. F. T. Stepto, *Polym. Prepr., Am. Chem. Soc., Div. Polym. Chem.*, 1992, **33** (1), in press.
81. J. E. Mark, *Adv. Polym. Sci.*, 1982, **44**, 1.
82. M. Gottlieb and R. J. Gaylord, *Polymer*, 1983, **24**, 1644.
83. M. Gottlieb and R. J. Gaylord, *Macromolecules*, 1987, **20**, 130.
84. L. Y. Yen and B. E. Eichinger, *J. Polym. Sci., Polym. Phys. Ed.*, 1978, **16**, 121.
85. R. W. Brotzman and B. E. Eichinger, *Macromolecules*, 1981, **14**, 1445; 1982, **15**, 531.
86. M. Gottlieb and R. J. Gaylord, *Macromolecules*, 1984, **17**, 2024.
87. M. Gottlieb, in 'Biological and Synthetic Polymer Networks', ed. O. Kramer, Elsevier, Barking, 1988, chap. 27, p. 403.
88. M. A. Llorente and J. E. Mark, *Macromolecules*, 1980, **13**, 681.
89. J. E. Mark, *Pure Appl. Chem.*, 1981, **53**, 1495.

11
Thermal Degradation of Condensation Polymers

GIORGIO MONTAUDO and CONCETTO PUGLISI
University of Catania and Institute for the Chemistry and Technology of Polymeric Materials, Italy

11.1	THE SEARCH FOR RULES	227
11.2	THERMAL REACTIVITY OF FUNCTIONAL GROUPS IMMOBILIZED IN POLYMERIC STRUCTURES	228
11.3	THERMAL STABILITY *VERSUS* STRUCTURE	229
11.4	DETECTION OF PRIMARY PYROLYSIS PRODUCTS	230
11.5	IONIC PROCESSES	231
	11.5.1 N—H Hydrogen Transfer	232
	11.5.2 O—H Hydrogen Transfer	234
	11.5.3 C—H Hydrogen Transfer	234
	11.5.3.1 β-C—H Hydrogen transfer	234
	11.5.3.2 α-C—H Hydrogen transfer	237
	11.5.4 Exchange Reactions	239
	11.5.4.1 Intermolecular	239
	11.5.4.2 Intramolecular	240
	11.5.4.3 Macrocyclization	242
	11.5.4.4 Alternating copolymers	242
11.6	FREE RADICAL PROCESSES	244
11.7	MOLECULAR REARRANGEMENTS	248
11.8	REFERENCES	249

11.1 THE SEARCH FOR RULES

Recently, it has become increasingly apparent that a great difference exists between the mechanisms of thermal decomposition of addition polymers, like polyalkenes, polydienes and polyvinyls, and those corresponding to condensation polymers. Addition polymers carry large numbers of aliphatic hydrogen atoms along the backbone, which are easily transferred by homolytic cleavage in the temperature range of 300–500 °C, thus initiating thermal degradation. The thermal degradations of the majority of addition polymers may appear rather monotonous from a mechanistic point of view. In fact, the initiation reaction invariably occurs through the formation of macroradicals, which are very reactive species. Their decay may take place *via* several parallel routes, involving bond cleavages, recombinations, hydrogen eliminations or abstractions, *etc.*[1-5]

The large number of pyrolysis products generated by these unselective radical reactions can usually be described by a mechanistic scheme that does not retain a great discriminatory character. In fact, the chemical structure of pyrolysis products can often be reconciled with one or more free radical reaction pathways, and therefore the exact mechanism of thermal degradation of the addition polymers becomes relatively unimportant, the phenomenological (kinetic) aspects of the overall process being more significant. Kinetic treatments of the thermal decomposition of polyhydrocarbons and polyvinyls are quite established. They are based on the general concepts developed in the framework of free radical chemistry. The free radical theory starts with carbon–carbon or carbon–hydrogen bond cleavage to yield macroradicals, then sets a few definite rules about the relative stability of macroradicals, and finally identifies a score of decay pathways for these macroradicals to yield the final pyrolysis products. Although the decomposition of macroradicals is often an unselective process dominated by multiple hydrogen transfer reactions occurring at very high rates, and the mixture of thermal fragmentation products finally detected may as a result be very complex, the remarkable informative content of the free radical theory and its interpretative power have allowed the kinetics of thermal degradation of addition polymers to be dealt with.[3,5]

The picture changes noticeably when one considers the thermal decomposition of condensation polymers, where no general rules had emerged until recently.[6] This lack of rules is due to a failure to realize that the inclusion of highly polar functional groups along the polymer backbone lowers the importance of decomposition reactions involving free radical processes, which predominate in addition polymers.[6] As a consequence, several classes of condensation polymers decompose by ionic cleavage processes. This fact has not yet been clearly recognized in the current literature, and most studies still insist in trying to interpret the pyrolysis data exclusively on the basis of radical processes.[4,5] Despite the fact that there are thermal decomposition pathways in condensation polymers that involve radical processes, ionic reactions do have an important role. The specific thermal decomposition pathways occurring in these polymers depend largely upon the nature of the functional groups present and also upon the chemical structure of the monomer units, which may vary considerably. An essential problem in polymer pyrolysis studies is therefore to derive a set of rules capable of rationalizing the nature of the thermal decomposition processes which occur in the wide range of condensation polymers existing.

Current studies in this field have particularly benefited from recent analytical improvements (see Section 11.4),[6,7] and therefore this chapter is intended to illustrate the new concepts which are emerging from the investigation of condensation polymers. In the following sections, evidence existing in the literature will be discussed that points towards the existence of a 'continuum' in reaction pathways which involve free radical intermediates in the case of functional groups of low polarity, and ionic intermediates in the case of highly polar groups.

11.2 THERMAL REACTIVITY OF FUNCTIONAL GROUPS IMMOBILIZED IN POLYMERIC STRUCTURES

Condensation polymers can be regarded as functional groups immobilized in a polymeric structure. This implies that when several structural units containing functional groups are linked together to form a polymer, they cease to be volatile and, when heated, their decomposition mechanisms will be dominated by the highly reactive functional groups included in the polymer structure.

Due to the highly polar character of some functional groups, the corresponding thermal decomposition reactions will be ionic and selective, rather than radical and unselective. For processes occurring at up to 450–500 °C, polyesters,[8-25] polyamides,[26-43] polycarbonates,[44-54] polyurethanes,[55-63] polyureas,[64-66] polysulfides,[67-70] polysiloxanes[71-79] and several others[80-87] undergo thermal degradation by ionic cleavage. If polymers belonging to the above classes decompose at higher temperatures, radical reactions are likely to prevail (see Section 11.6). Furthermore, polymers which contain less polar functional groups, like ethers[88-93] and azomethynes,[94] tend to undergo radical cleavage even when they decompose at relatively low temperatures (see Section 11.6).

The pyrolytic processes occurring in a condensation polymer are expected to originate either from the inner functional groups or from those at the end of the polymer chain. The most visible event in the pyrolysis of these polymers is the ultimate fate of the functional groups present, and it is therefore important to verify whether a given functional group was cleaved or not in the pyrolysis. This recognition affords an immediate characterization of the pyrolytic processes. In fact, when the pyrolysis products contain uncleaved functional groups, this implies that the reaction mechanism involves an exchange process (ionic). On the contrary, if the cleavage or the transformation of the functional groups occurs in the pyrolytic process, this indicates that the thermal decomposition pathway involves hydrogen transfer reactions, or molecular rearrangements. In some cases the latter processes are ionic and involve polar transition states, while in others they are free radical reactions, as illustrated below. A complicating factor in these studies arises from the dramatic influence of structural factors, which cause drastic changes in the pyrolysis mechanisms even within the same class of condensation polymers.

A systematic approach to the overall problem has been recently devised,[6] which consists of introducing systematic variations of structure into a series of condensation polymers, and then revealing the changes of significant thermal decomposition products by using direct methods of pyrolysis. This scheme has been applied to several classes of polymers, and a consistent set of data has been obtained on the decomposition pathways occurring.[6]

It must be remarked, however, that there is still a widespread tendency in the literature to deal with the thermal decomposition processes of condensation polymers as if they were occurring by free radical reactions, postulating the unselective rupture of a number of different bonds and the decay of macroradicals, as in the case of polyalkenes or polyvinyls.[4,5] For instance, the pyrolysis of nylon 6,6 has been described as occurring by random chain scissions that originate radicals, as indicated in Schemes 1 and 2. The process in Scheme 1 was believed to occur in the low temperature pyrolysis and that in Scheme 2 was believed to account for the pyrolysis at higher temperature, respectively.[5] Recent work[42] has shown that cyclohexane is not produced at all in the pyrolysis of nylon 6,6 and has afforded a more reasonable ionic mechanism for that reaction (see Scheme 26).

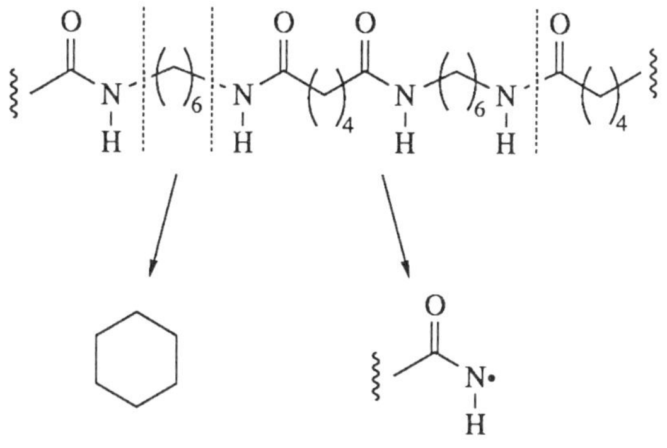

Scheme 1

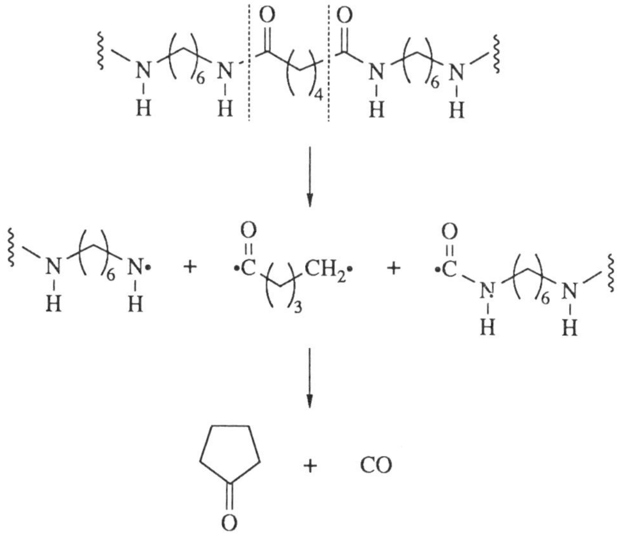

Scheme 2

11.3 THERMAL STABILITY *VERSUS* STRUCTURE

In general, the thermal stability of polymers is subject to the ease of the hydrogen transfer reactions compatible with their structure. For instance, polymers containing aliphatic units undergo thermal degradation at lower temperatures than aromatic polymers, because aliphatic hydrogens are more easily transferred. In the case of the well-known 'high temperature polymers', the thermal stability results from the elimination of hydrogen atoms from the polymer structure (only hydrogens attached to condensed aromatic rings are left), thus achieving a 'quasi-chair' stage already in the initial polymer structure. When these polymers are heated, they yield char and a few gaseous products, most likely by radical processes.

Another parameter that influences the thermal stability and needs to be taken into account, especially when dealing with condensation polymers which decompose by an exchange process, is the conformational rigidity of the polymer backbone. In these polymers, the structure of the repeating units may vary considerably even within a single class, and therefore the thermal stability may change noticeably.[6]

A convenient method of evaluating the thermal stability of polymers is to measure the temperature of the maximum rate of polymer degradation (the polymer decomposition temperature, PDT) under an inert gas atmosphere. Usually, thermogravimetric (TG) runs with a heating rate of $10\,^{\circ}C\,min^{-1}$ are performed to this purpose. Due to its simplicity and wide availability, the PDT criterion is widely adopted in the literature.[6,95] The molecular weights of the pyrolysis products also vary with the thermal stability of the starting materials, and polymers decomposing at relatively low temperatures (300–500 °C) usually yield higher molecular weight thermal fragments (originating from ionic processes), whereas polymers of higher thermal stability tend to generate fragments of lower molecular weight (possibly originating both from ionic and radical reactions).

11.4 DETECTION OF PRIMARY PYROLYSIS PRODUCTS

The thermal degradation of polymers was an active field of investigation long before modern instrumental techniques were available. Earlier studies favored the pyrolysis of large samples and indirect pyrolysis, since the analytical methods then available required sizeable quantities of compounds. This caused problems of product identification, which have been tolerated too often even in recent times, when the detection and identification of pyrolysis products has experienced remarkable progress.

In many studies, the structural assignments are only tentative and limited to low molecular weight compounds (pyrolysis products with molecular weights above 150 have seldom been identified). The identification of the primary thermal decomposition products is a difficult task, and in the current literature there is a noticeable shortage of data on the identification of pyrolysis products carrying significant structural information (*i.e.* polymer segments large enough to contain the repeating units or the sequential arrangement of subunits). Furthermore, primary pyrolysis products from polymers are often highly labile and short-lived at the temperatures at which they are generated, and therefore the residence time of the pyrolysis products in the hot zone and the time of transport to the detector must be very short in order to reveal them.

The choice of the pyrolysis technique is crucial for the quality of the data obtained. Flash pyrolysis, followed by gas chromatography mass spectrometry (GCMS) analysis, is extremely versatile and reproducible, and has gained wide popularity. However, only nonreactive and thermally stable, low molecular weight compounds usually emerge from GC columns and are detected.[6,7] Also thermal volatilization analysis (TVA) shows limitations,[4] since the high molecular weight pyrolysis products (which are most likely primary products, and therefore highly diagnostic) are not analyzed on-line, but only as cold ring fraction, and therefore the short-lived and reactive species may be lost.

Instead, the characterization of polymers by direct pyrolysis mass spectrometry (DPMS) yields important structural information. A general advantage of this technique is that pyrolysis is accomplished under high vacuum, and therefore the thermal products formed are volatilized and readily removed from the hot zone. This, together with the low probability of molecular collision and fast detection, reduces the occurrence of secondary reactions, so that primary products are detected. The information thus obtained is of particular importance in order to assess the primary thermal degradation mechanism of pyrolysis of a polymer. Furthermore, since pyrolysis is achieved very close to the ion source and no problem of transport exists, ions of high mass, which are often essential for the structural characterization of the polymer, can be detected, whereas they are often lost using other techniques. The main problem connected with this technique is, however, the identification of the products in the spectrum of the multicomponent mixture produced by thermal degradation. In fact, in the overall end spectrum of a polymer, the molecular ions of the oligomers will appear mixed with the fragment ions formed in the ionizing step.

Ionization methods that minimize the fragmentation processes of the molecular ions are very valuable in the mass spectral analysis of the mixture, since the resulting spectra are likely to contain intense molecular ions and therefore the identification of the components (primary pyrolysis products) is easier. For instance, it has been observed in several instances that electron impact ionization (EI) mass spectra do not allow conclusions to be drawn about the primary thermal

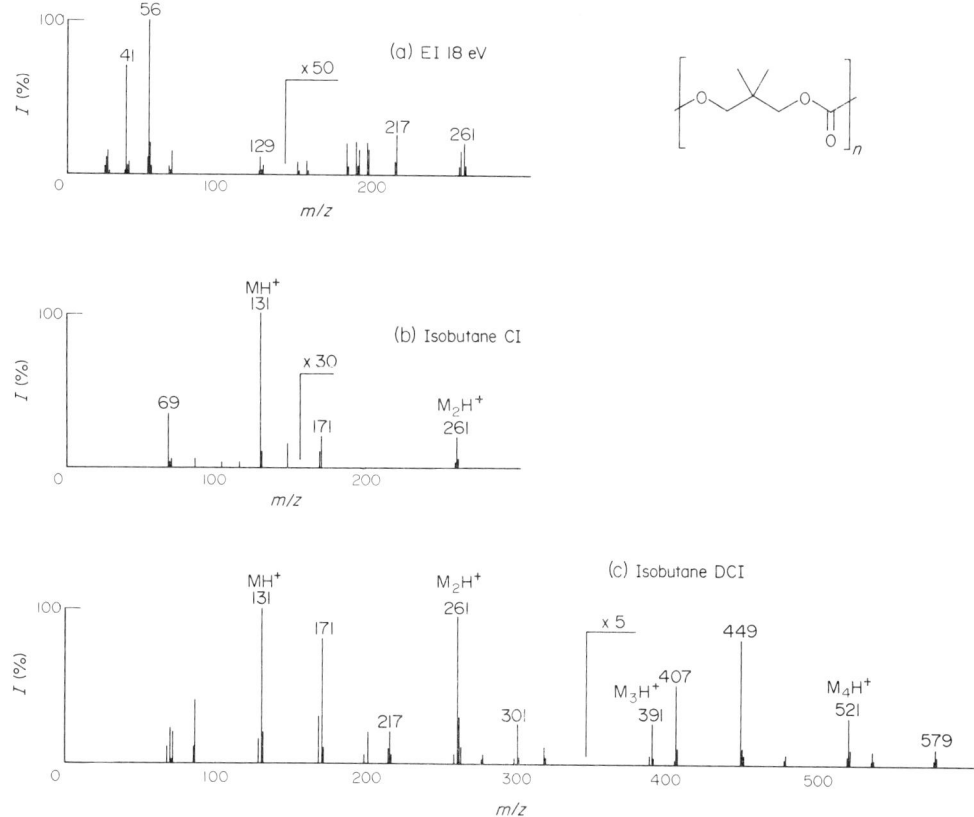

Figure 1 Direct pyrolysis mass spectra of poly(neopentylene carbonate) obtained by: (a) EI 18 eV; (b) isobutane CI; (c) isobutane DCI, ionization modes

decomposition mechanisms that operate in aliphatic polyesters,[21,23,25] polyamides[38,42] and polycarbonates,[51] since the corresponding cyclic aliphatic oligomers generated in the pyrolysis of these polymers do not exhibit significant molecular ions. EIMS studies of the thermal decomposition mechanism of aliphatic polymers are therefore hampered by the lability of aliphatic molecules towards EI.

In contrast, it has been established that the chemical ionization (CI) and collision-induced dissociation (DCI) mass spectra show intense molecular ions corresponding to cyclic oligomers originating from the aliphatic polymers mentioned above. A comparative EI/CI/DCI study, as shown in Figure 1, has been found appropriate in order to detect correctly the primary pyrolysis products originating from poly(neopentylene carbonate).[51]

In fact, no molecular ions are seen in the EI spectrum (Figure 1a). However, protonated molecular ions (MH$^+$) appear in the isobutane CI mass spectrum, corresponding to the cyclic carbonate monomer and dimer (Figure 1b). The low intensity of the peak corresponding to the cyclic dimer in Figure 1b is due to its lability, whereas the DCI spectrum (Figure 1c) allows the detection of peaks corresponding to the cyclic carbonates M_1–M_4, respectively.

Mild thermal gradients can be used in the pyrolysis of polymers in the hope of observing the early stages of their thermal decomposition processes. This allows the pyrolysis to be monitored continuously and higher molecular weight fragments from the thermally decomposing polymers to be detected. DPMS is nearly identical to a TG run (only the detector changes) and actually the differential thermogravimetric and DPMS curves are usually superimposable.[6,7]

11.5 IONIC PROCESSES

Some examples of the ionic processes which occur in the pyrolysis of the most common condensation polymers, and the polar transition states involved, are discussed in the following sections.

11.5.1 N—H Hydrogen Transfer

The intramolecular transfer of an N—H hydrogen to an oxygen atom is a remarkably low energy process. Therefore, the thermal stability of polyurethanes is rather low, independent of the presence of aliphatic or aromatic units along the polymer chain, and the PDT range is about 280–350 °C.[55-63] The transfer reaction may be described by a four-centered polar transition state (Scheme 3), which produces random cleavage of the functional groups along the chain and initiates a stepwise depolymerization process.

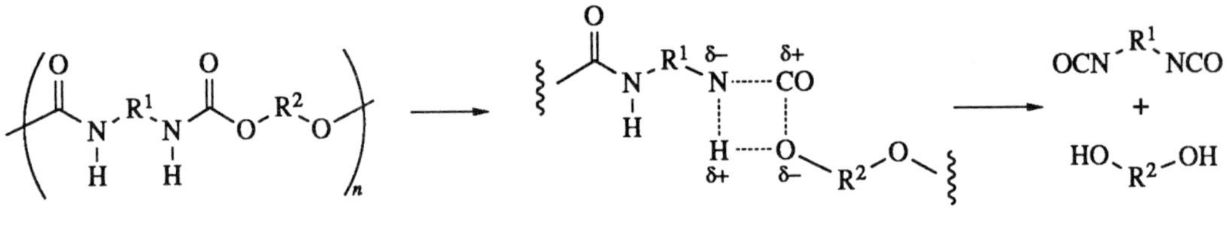

Scheme 3

A similar mechanism applies for polyureas,[64-66] since the hydrogen needs only to be scrambled from one nitrogen atom to the other (Scheme 4). These polymers, therefore, have a thermal stability comparable to that of polyurethanes. Alternatively, these reactions may be thought of as occurring through a tautomerization equilibrium, which would lead to a slightly different transition state (Scheme 5).

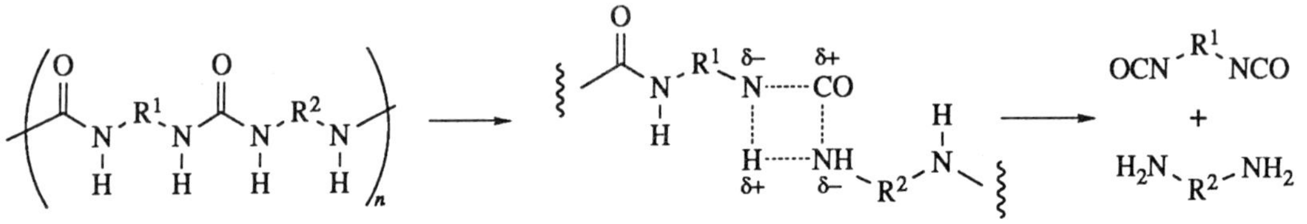

Scheme 4

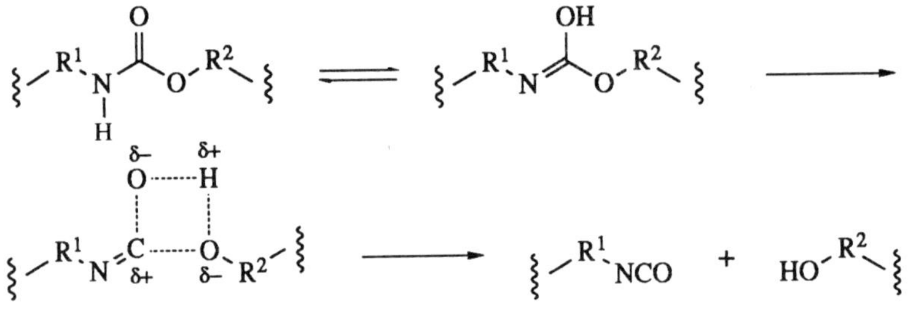

Scheme 5

Although of little practical relevance in the above cases, there are some N—H transfer processes which occur through an iminolization equilibrium (see Schemes 8 and 9).

Another case where a clear N—H transfer to the oxygen has been observed is in poly-(γ-methylglutamate) (PDT of 335 °C).[27] The primary pyrolytic process is the loss of a methanol molecule, with consequent formation of pyroglutamic units along the polymer chain. Also, in this case the reaction may be thought to occur through a polar, four-centered transition state (Scheme 6). In this reaction the amide linkage is not cleaved, but is transformed into an imide group.

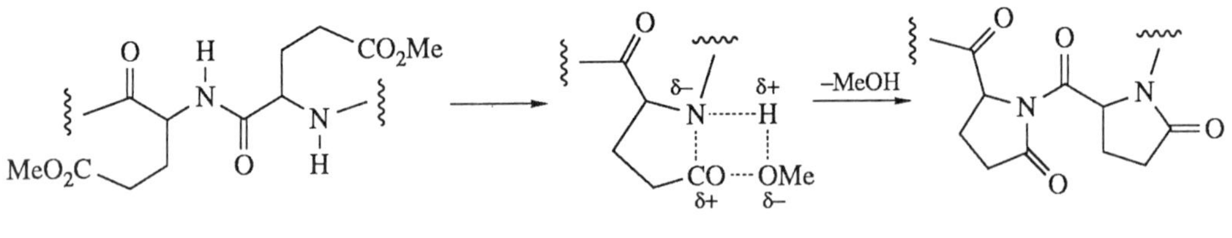

Scheme 6

The process is a typical example of a thermal reaction initiated at the side chain, and does not lead to the rupture of the main polymer chain.[27]

The unusually low thermal stability of aromatic polysuccinamides,[37] which show a PDT of 275–295 °C, as compared to a PDT of 370–470 °C for the polyamides of comparable structure,[37] is explained by the occurrence of a low energy N—H transfer process. Recent work[37] has shown that in polysuccinamides the polymer backbone is broken through the thermal cleavage of the amide bonds, and that a succinimide unit is formed together with an amine end group, as illustrated in Scheme 7.

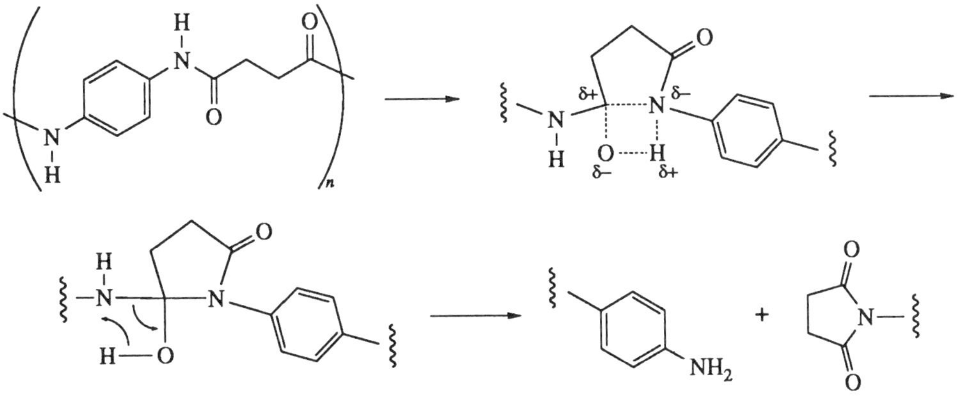

Scheme 7

Other N—H transfer processes, observed for polyoxamides and polyhydrazides,[41] are likely to occur by iminolization equilibria, as indicated in Schemes 8 and 9, respectively. Iminolization is also responsible for some of the pyrolytic processes occurring in totally aromatic polyamides.[34–36] The formation of a benzoxazole structure has been observed in the case of *ortho*-chlorinated polyamides,[34–36] which is explained through an iminolization–cyclization process with HCl elimination (Scheme 10).

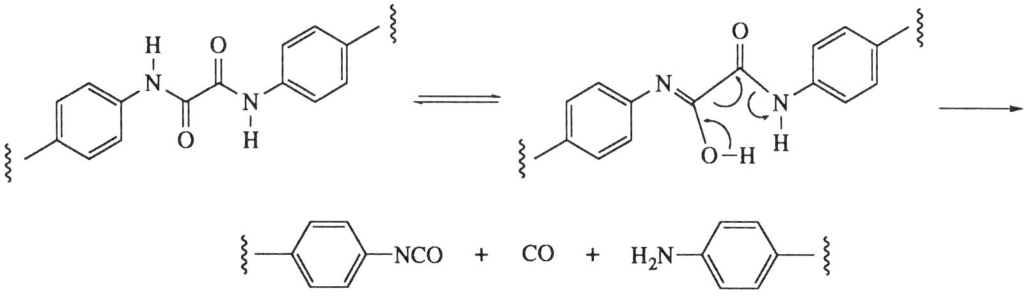

Scheme 8

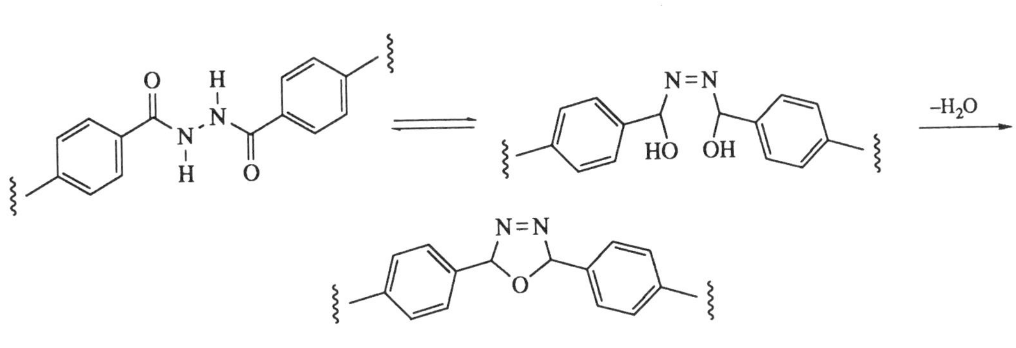

Scheme 9

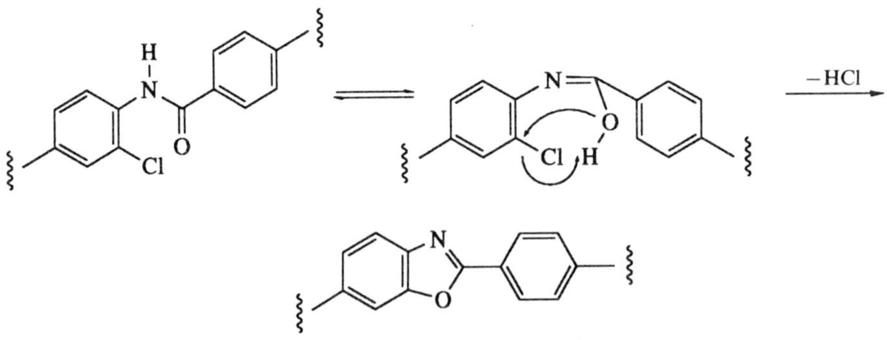

Scheme 10

11.5.2 O—H Hydrogen Transfer

Hydrogen transfer reactions from hydroxylic groups are not frequent. The best example is the side chain initiated cleavage of cellullose (Scheme 11). This reaction is very selective and leads to the formation of levoglucosane as the primary pyrolysis product, which may undergo further decomposition and polymerization.[82-87]

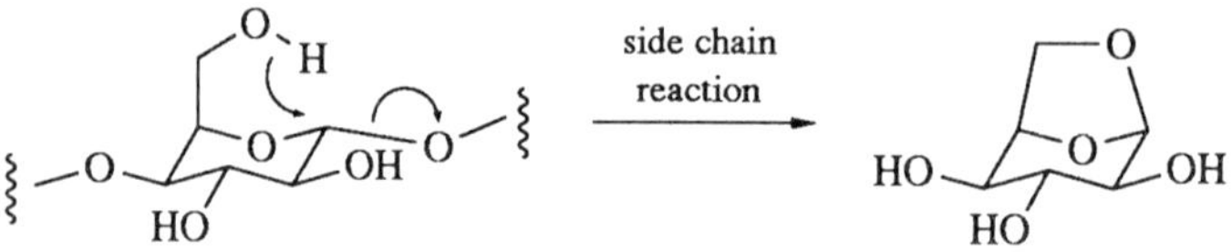

Scheme 11

11.5.3 C—H Hydrogen Transfer

C—H hydrogen transfer processes require higher activation energies than the N—H hydrogen transfer reactions mentioned above. Therefore, polymers which undergo thermal degradation through these processes usually have higher polymer decomposition temperatures (PDT values in the range of 350–450 °C). Two types of C—H hydrogen transfer processes have been observed: (i) β-C—H hydrogen transfer, which involves a hydrogen atom linked to a carbon in a β-position with respect to the heteroatom (**1**); and (ii) α-C—H hydrogen transfer, which occurs from a carbon in an α-position with respect to the heteroatom, as in (**2**). The latter process occurs when hydrogens in a β-position are absent in the structure[6] and therefore requires a higher activation energy than the β-C—H transfer.

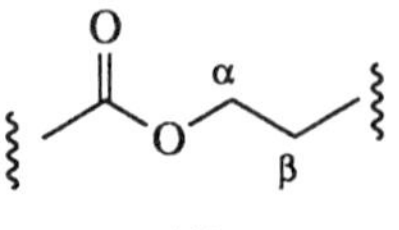

(**1**) 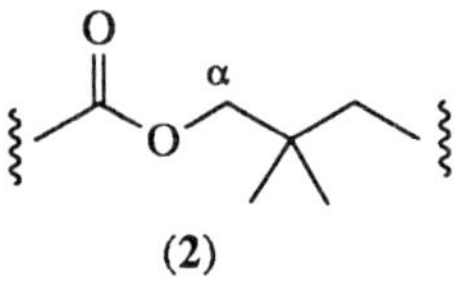   (**2**)

11.5.3.1 β-C—H Hydrogen transfer

β-C—H hydrogen transfer processes usually occur through concerted reactions which involve six-membered transition states, but some four-membered transition states have also been reported.

The thermal decomposition of N-methyl-substituted polyurethanes containing an aliphatic chain linked to the oxygen atom[56,58] occurs selectively via a β-C—H hydrogen transfer process involving a six-membered cyclic polar transition state, producing secondary amines, alkenes and CO_2 (Scheme 12).

A β-C—H hydrogen transfer process has also been found[16,17] to occur in the pyrolysis of polyesters originating from aromatic dicarboxylic acids and aliphatic diols (such as poly(ethylene terephthalate), PET, and poly(butylene terephthalate), PBT). This reaction (Scheme 13) produces compounds with alkene and carboxylic acid end groups and is concomitant with an ester exchange process leading to cyclic oligomers[15,18] (see Section 11.5.4).

Polyamides originating from aromatic dicarboxylic acids and aliphatic diamines[39] undergo thermal degradation through a selective β-C—H hydrogen transfer reaction involving a six-membered cyclic polar transition state, which produces compounds with alkene and amide end groups as primary pyrolysis products (Scheme 14). Similar behavior has been observed for polyoxamides containing aliphatic diamines (Scheme 15).[41]

The presence of amide or ester groups adjacent to the β-methylene hydrogen atoms in poly-(β-propiolactone) and in nylon 3, respectively, lowers the activation energy of the β-C—H hydrogen transfer process below that of ester exchange,[9,21,43] so that pyrolysis compounds with alkene and amide or carboxylic acid end groups have been almost exclusively detected as primary thermal degradation products (Schemes 16 and 17).[9,21,43]

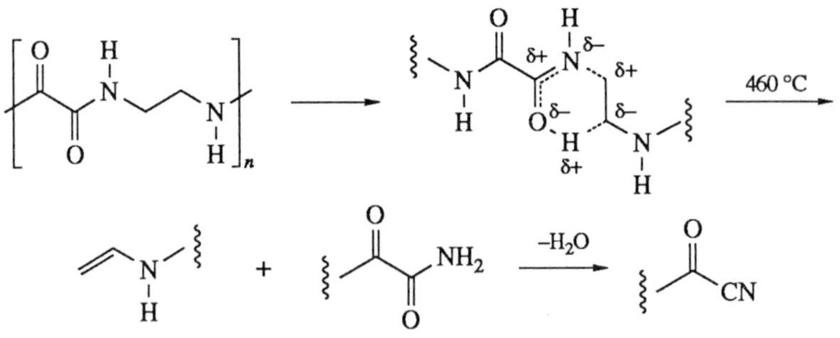

Scheme 15

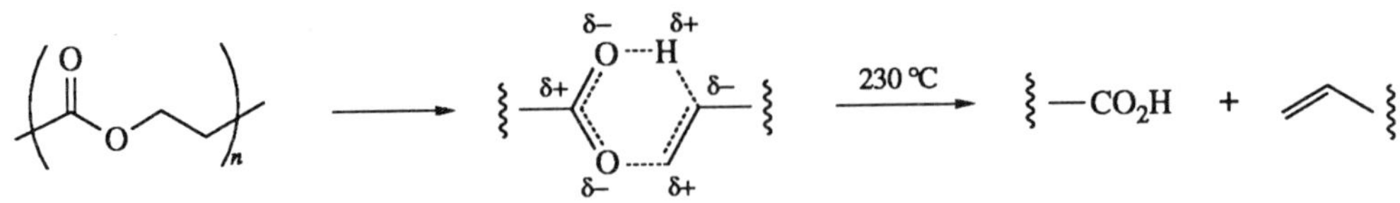

Scheme 16

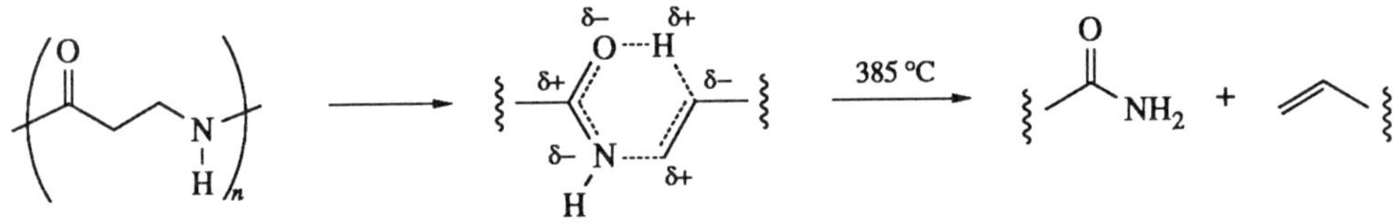

Scheme 17

It should be remarked that polylactams[43] and polylactones[21] generally decompose by intramolecular exchange processes (see Section 11.5.4) producing cyclic oligomers at temperatures between 400–450 °C. In fact, although the electron-withdrawing effect of oxygen and nitrogen induces a fractional charge on the methylene in the β-position, this is not sufficient to promote β-C—H hydrogen transfer over cyclization and the latter occurs only in the case of poly(β-propiolactone) and nylon 3. Accordingly, the PDT values for these polymers are substantially lower than the corresponding homolog polymers (230 °C for poly(β-propiolactone) and 385 °C for nylon 3).

The six-membered cyclic transition state mechanism has also been found[27] to occur selectively in the thermal degradation of poly(α-amino acids). This process involves the transfer of a hydrogen atom of the methyl group in the β-position, and produces alkenes and amide end groups as primary pyrolysis products (Scheme 18).

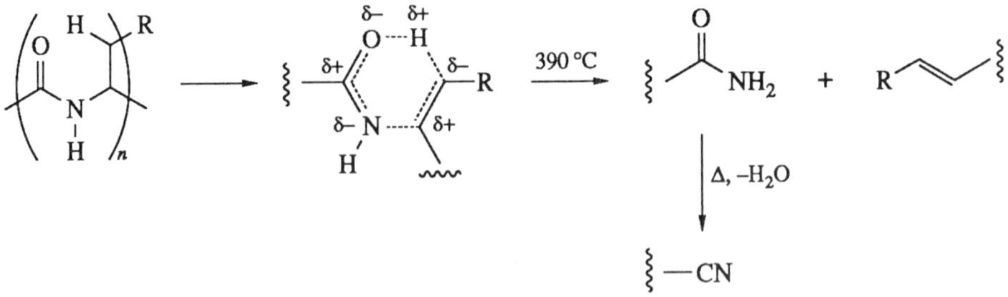

Scheme 18

Similarly, poly(α-esters) also decompose by a β-C—H hydrogen transfer process (Scheme 19),[8] which competes with the cyclization reaction.

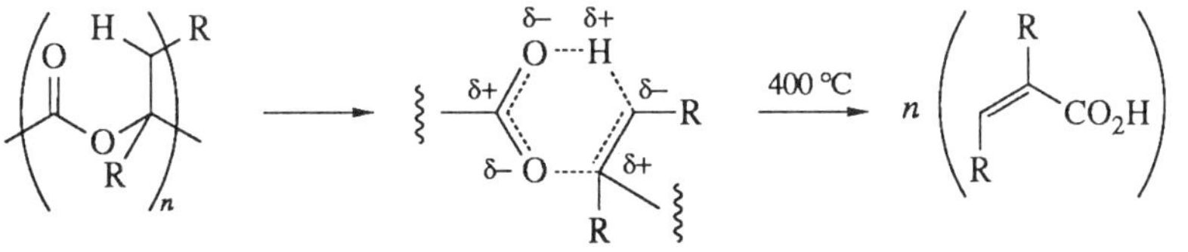

Scheme 19

The thermal decomposition of polysulfides containing an aliphatic chain[68,69] occurs selectively via a β-C—H hydrogen transfer process involving a four-membered cyclic polar transition state, which produces primary pyrolysis compounds with alkene and thiol end groups (Scheme 20).

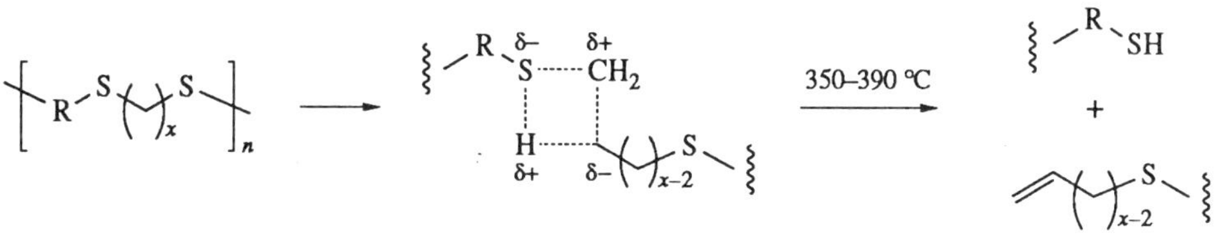

Scheme 20

Also a copoly(imide amine), obtained from bis(N-allyl)pyromellitimide and piperazine by a Michael addition reaction,[81] undergoes a selective depolymerization process via a four-membered β-C—H hydrogen transfer that produces the starting materials (Scheme 21). In this case the activation energy of the reaction is quite low (the two methylenes are highly activated, since they lie in β-positions with respect to two heteroatoms), and therefore the degradation reaction occurs at low temperature (280 °C).

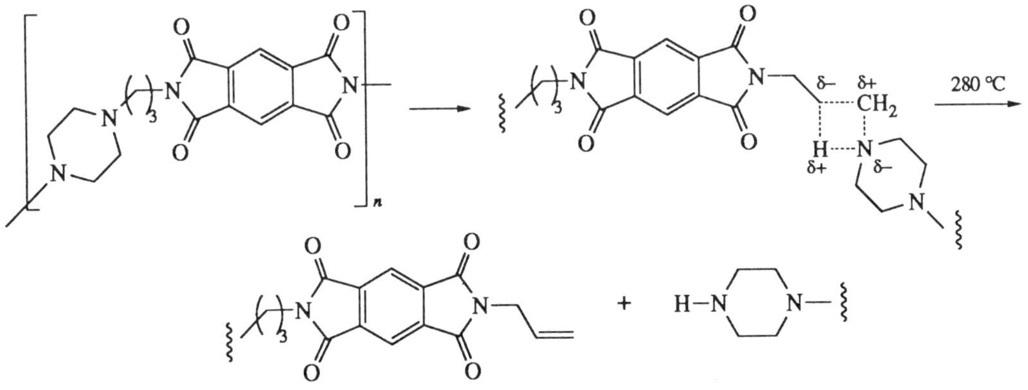

Scheme 21

11.5.3.2 α-C—H Hydrogen transfer

As mentioned above, the thermal degradation of polymers lacking hydrogens in the β-position occurs at higher temperatures through an α-C—H hydrogen transfer process, generally involving a five-membered polar transition state. Thus, N-methyl-substituted polyurethanes containing the neopentylglycol moiety along the chain decompose at 460 °C[56] by an α-C—H hydrogen transfer process involving a five-membered transition state which produces secondary amines, alkenes and CO_2, as shown in Scheme 22.

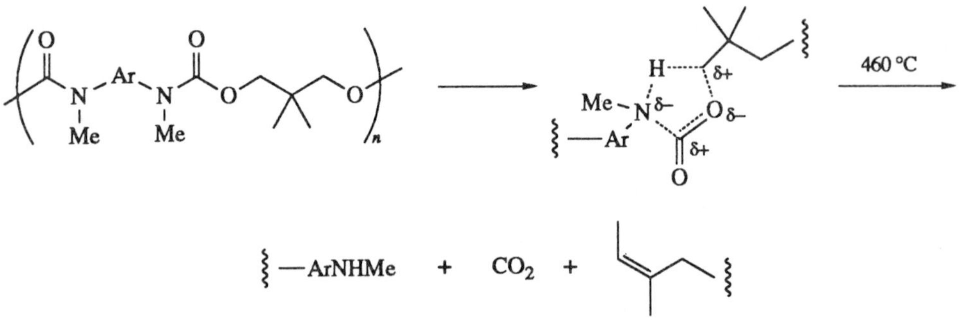

Scheme 22

N-Methyl-substituted aromatic polyhydrazides[41] decompose at 430 °C through an α-C—H hydrogen transfer reaction involving the hydrogen atom of the *N*-methyl group, as shown in Scheme 23, and produce primary pyrolysis compounds containing amide and/or imine end groups.[41]

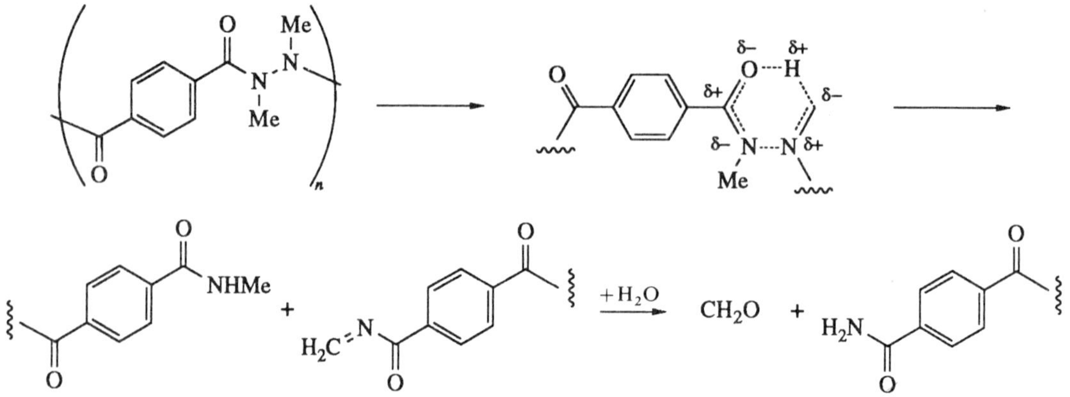

Scheme 23

Piperazine-containing polyurethanes[56,57] decompose by an α-C—H hydrogen transfer process occurring at high temperature (475 °C), involving a five-membered polar transition state, shown in Scheme 24. In this process, a hydrogen atom is transferred from the methylene in the α-position to the piperazine nitrogen, to the urethane oxygen. This effects the cleavage of the functional group, with consequent evolution of CO and formation of primary pyrolysis compounds with amine and phenol end groups.[56,57] A similar decomposition pathway has been observed [64,65] for poly(carboxypiperazine) (PDT of 480 °C), as described in Scheme 25.

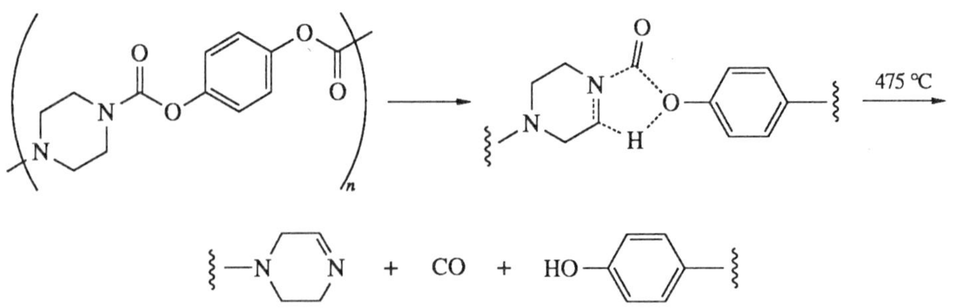

Scheme 24

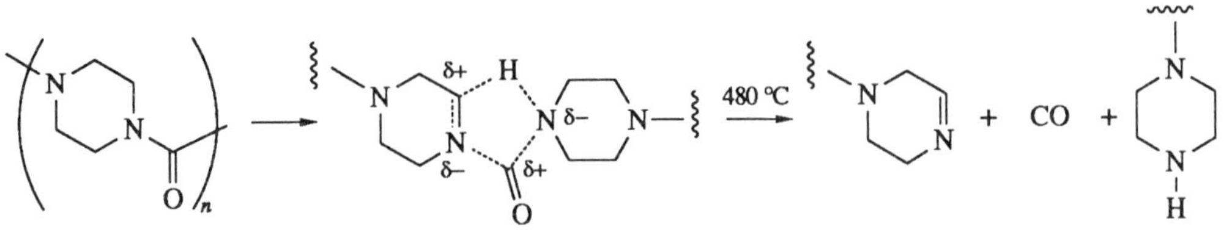

Scheme 25

Nylon 6,6,[42] and some polyamides[37] and polyhydrazides[41] which contain the adipic acid moiety along the polymer chain, have been found to decompose by an α-C—H hydrogen transfer process (Scheme 26) with formation of primary pyrolysis products with amine and cyclopentanone end groups, which decompose further and produce cyclopentanone and compounds with isocyanate end groups. The behavior of these polyamides differs from that of polyamides containing succinic units in the main chain,[37] which have been shown (Scheme 7) to decompose at much lower temperatures through an N–H transfer process, producing succinimide end groups.[37] Evidently, the thermal decomposition reaction evolves in each case towards the most stable cyclic products, which are succinimide and cyclopentanone, respectively.[37,42]

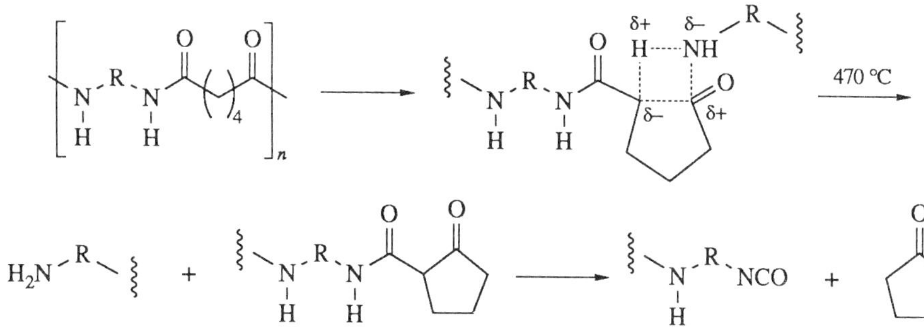

Scheme 26

11.5.4 Exchange Reactions

Exchange reactions between functional groups belonging to different chains (intermolecular exchange), or to the same polymer chain (intramolecular exchange), usually occur in the temperature range of 250–450 °C, and are among the most frequent processes responsible for the thermal cleavage of condensation polymers.

11.5.4.1 Intermolecular

Intermolecular exchange has no effect on the structure of homopolymers. However, the effect of intermolecular exchange is immediately apparent in the thermal reactions of polymer blends and of ordered copolymers, where it produces mixing of the chains and randomization of the structural units.

A typical example, shown in Scheme 27, is the ester interchange reaction occurring at 250–300 °C in the case of binary blends of bisphenol A polycarbonate (PC) with poly(ethylene terephthalate) (PET) and poly(butylene terephthalate) (PBT).[96–98]

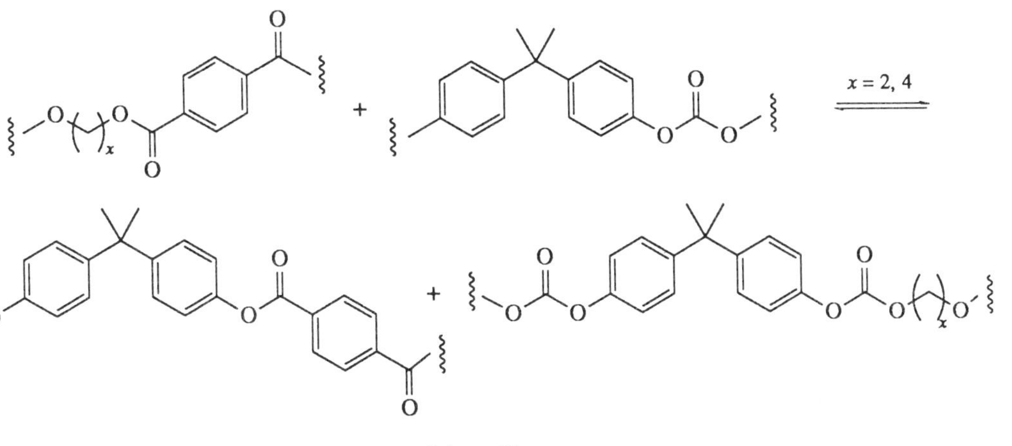

Scheme 27

This exchange might take place by one of two mechanisms, illustrated in Schemes 28 and 29. The inner functional groups of the two polymer chains can be exchanged through the four-centered polar transition state in Scheme 28. Alternatively, in Scheme 29 the exchange reaction occurs through the nucleophilic attack of the reactive end groups of one chain on the other.

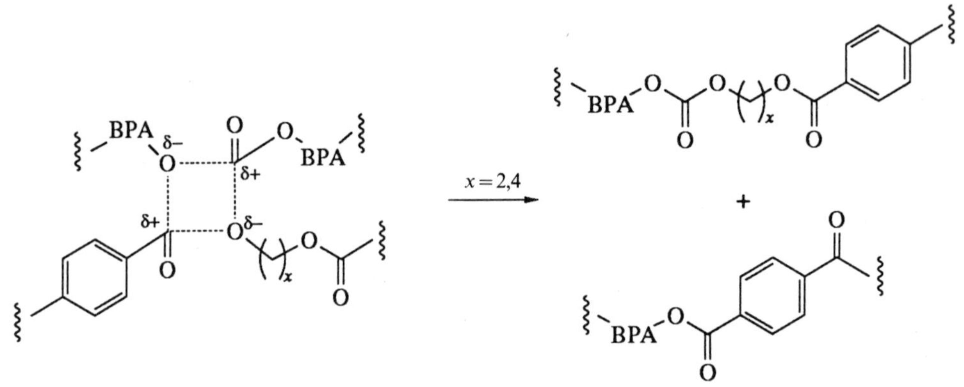

Scheme 28

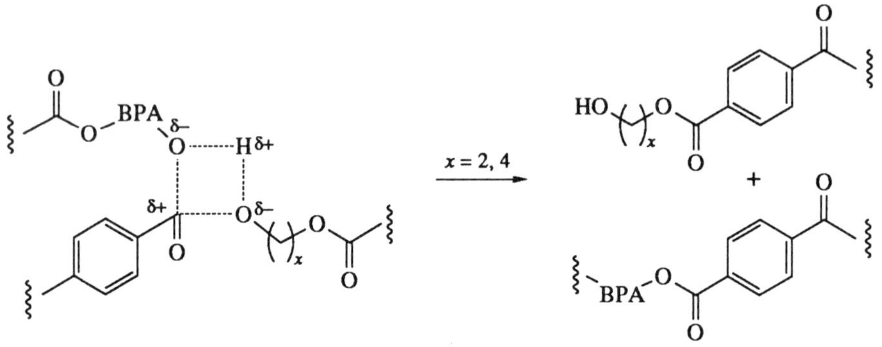

Scheme 29

Exchange by a mechanism of the type formulated in Scheme 28, although frequently postulated to occur in polyesters, polyamides, *etc.*, has seldom been proved.[99] However, a method that has been successfully used in order to distinguish between the two processes in the case of PC/PBT blends is that of adding to the blend opportune low molecular weight compounds containing reactive end groups, and measuring eventual changes induced in the reaction kinetics.[98] The results of that study indicate that the intermolecular exchange process is indeed occurring by the mechanism described in Scheme 28.

11.5.4.2 Intramolecular

At sufficiently high temperatures, intramolecular exchange processes become quite efficient in determining the cleavage of the polymer chains, through a reaction that produces cyclic oligomers. In fact, the formation of rings, as opposed to linear chains, is thermodynamically favored at high temperature (equation 1).[100,101]

$$\sim\!\!-R\!-\!X\!-\!R\!-\!X\!\sim \quad \rightleftharpoons \quad [-(R\!-\!X)_n-] \tag{1}$$

Depending on the nature of the functional groups involved, the exchange reactions leading to cyclization may occur by two alternative mechanisms, similar to those illustrated in Schemes 28 and 29. When the exchange occurs through the interaction of the inner functional groups, the process is called 'back-biting'. When it occurs by an attack of reactive end groups on the inner groups of the chain, then it is called 'end-biting'.[100]

The occurrence of end-biting processes has been recently ascertained in the case of some aliphatic polysulfides[68] that were shown to form cyclic sulfides by a nucleophilic attack of the —SH end group on the inner sulfide groups along the chain (Scheme 30). Both capped and uncapped polysulfides were available for the direct pyrolysis mass spectrometry (DPMS) experiments. In one case (Figure 2a), the total ion current curve exhibited two maxima, suggesting that the thermal degradation of the polysulfide occurs in two stages. A series of cyclic oligomers were produced in the first decomposition stage, and it was hypothesized that the processes had originated from macromolecular chains containing —SH end groups, as indicated in Scheme 30. This hypothesis was then tested by capping the —SH end groups of the polysulfide sample. The thermal degradation of the capped polymer was found to occur in a single stage, and only the decomposition stage at higher temperature was still present (Figure 2b). No formation of cyclic oligomers was observed among the pyrolysis products of the capped polysulfide, and it was therefore concluded that the reaction occurs through an end-biting process.[68]

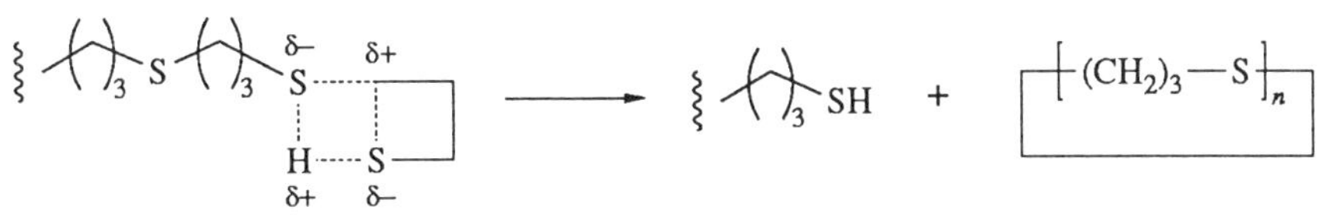

Scheme 30

However, end capping does not always provide an unequivocal answer. In the case of some polycarbonates,[51,53] the comparison between the pyrolysis of capped and uncapped polycarbonates showed that both polymers formed cyclic oligomers, although the uncapped polymer decomposed in two stages and the capped sample decomposed only in one stage.[51,53] The results might be interpreted as indicating that the first decomposition stage was due to an end-biting process and that the second decomposition stage was due to a back-biting process, both producing rings. Alternatively, one might think that the original polymer contained a sizeable portion of chains lacking reactive end groups (namely hydroxy groups), which were responsible for initiating the first thermal degradation step. Once the chains containing hydroxy groups were consumed, one might except that the latter were regenerated in the second stage of thermal decomposition by some other pyrolytic process (for instance, a C—H hydrogen transfer activated by the higher temperature). The reactive end groups would then start another end-biting process, so that the thermal decomposition of the polycarbonate would proceed by end biting in both stages. The end-capping experiment is not diagnostic in this case, because the reactive end groups might be generated by the thermal cleavage of the capping group.[51,53]

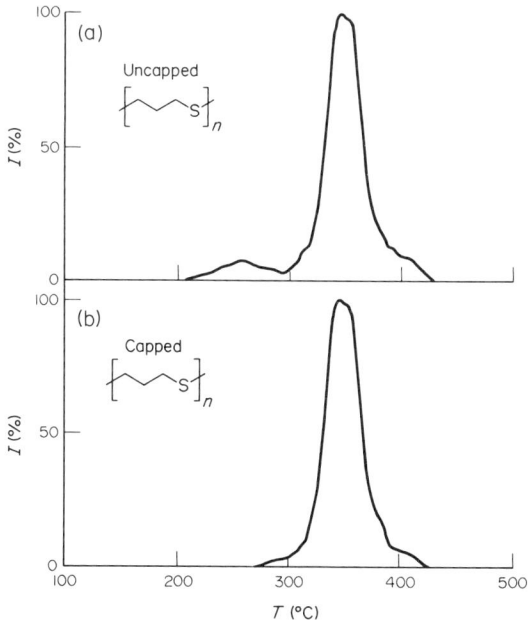

Figure 2 Total ion current curves of poly(trimethylene sulfide) obtained by DPMS: (a) uncapped; and (b) capped polymer

11.5.4.3 Macrocyclization

Intramolecular exchange is an ionic reaction, and one might therefore expect the process to be selective, *i.e.* that only a specific cyclic oligomer is formed in the cyclization reaction. The experimental evidence shows instead that a distribution of cyclic oligomers is produced in each case. This behavior is due to the fact that the cyclization reactions are dominated by thermodynamic and kinetic factors.[100,101] Macrocyclization equilibria[100,101] describe the behavior of polymer chains when they are allowed to reach thermodynamic equilibrium and generate cyclic oligomers. According to the theory, the cyclization probability is related to the mean separation of the reacting sites, and the equilibrium concentration of each oligomer is predicted to decrease proportionally to $n^{-2.5}$, where n is the ring size expressed as the number of repeating units present in the ring.

Systems that entirely follow the ring–chain equilibria have been found, but the theory has been verified in detail only for systems that fulfill the assumption of equal reactivity of reaction sites. This condition is usually satisfied by functional groups attached to the inner portions of a flexible chain molecule, and the cyclization reaction occurs through a back-biting process.[100,101] In most cases, however, the distribution of the cyclic oligomers generated does not reflect the thermodynamic equilibrium, mainly due to conformational preference restrictions present along the polymer chain, and therefore the relative abundance of the oligomers formed may as a result be very different from that predicted by theory.

A number of condensation polymers are thermally degraded to yield a distribution of cyclic oligomers. The formation of a distribution of cyclic oligomers is highly diagnostic, suggesting that they are primary thermal degradation products generated by an intramolecular exchange process.[100,111,112]

When highly reactive groups are present at the end of the polymer chain, and they are capable of reacting with the inner functional groups along the chain at a higher rate than that of the inner groups with themselves, an end-biting process takes place, which violates the equal reactivity principle of the functional groups. In these instances, a kinetically controlled distribution of cyclic oligomers results, and the concentration of each oligomer is predicted to decrease with $n^{-1.5}$, where n is the ring size expressed as the number of repeating units present in the ring.[100,101] Another situation where the thermodynamic distribution cannot be achieved is in the case of small rings, since the theory requires strainless rings.

The size and relative abundance of the cyclic oligomers produced in the pyrolytic intramolecular cyclization processes are dominated by the relative conformational stability of the compounds formed. A marked structural effect is observed on the distribution of the cyclic oligomers generated in the pyrolyses of *o*-, *m*- and *p*-poly(phenylene sulfide)s (PPS).[67] In the case of *o*-PPS, only the cyclic dimer is produced, whereas the trimer and tetramer predominate in the case of the *meta* isomer. The distribution of the cycles generated in the case of the *p*-PPS is still different, and the pentamer is then the most abundant product.[67] In each case, the most abundant cyclic oligomer produced corresponds to the energetically most stable compound.

In a further example, the pyrolysis of the two isomeric polyesters (**3**) and (**4**) yields compounds of different structure.[15] The cyclic dimer and trimer are produced in the pyrolysis of polyester (**3**) (where the *meta* substitution favors the formation of small rings). However, no cyclic oligomers are observed among the pyrolysis products of polyester (**4**), due to the conformational instability of the *para*-substituted cyclic dimer and trimer. Higher oligomers, which might be the primary pyrolysis products of the *p*-polyester (**4**), have a molecular weight too high to volatilize undecomposed and, therefore, are not revealed experimentally.

(3)   (4)

11.5.4.4 Alternating copolymers

Intramolecular exchange processes may affect the structure of ordered copolymers. Some interesting examples of this kind have been reported in the case of alternating copolymers, where the pyrolytic products have different structures from those of the original copolymer repeating units.[32,33,51,53,78]

A series of copolyamides from *p*-aminobenzoic acid and several lactams have been found[32,33] to decompose by an intramolecular exchange reaction, occurring through a four-centered polar transition state, which produces demixing of the alternating copolymer structure to the corresponding cyclic and aromatic polyamides (Scheme 31).

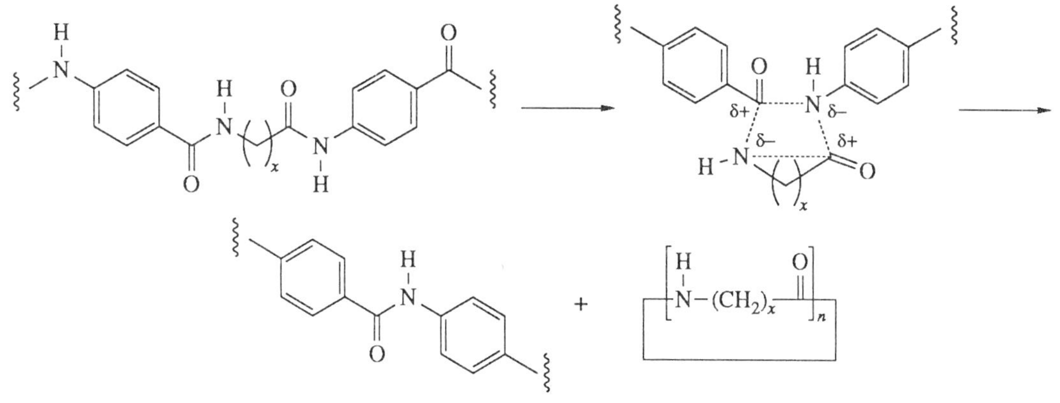

Scheme 31

A similar reaction has been reported in the thermal cleavage of ethylene–BPA and butylene–BPA copolycarbonates,[51] where ethylene carbonate and butylene carbonate were evolved in the first stage of the pyrolysis, and sequences of BPA polycarbonate were formed (Scheme 32). Furthermore, thermal products with structures that did not correspond to the original repeating units of the copolymer, but had instead random copolymer sequences, were detected.[51] Again in the case of propylene–BPA copolythiocarbonate (Scheme 33), the intramolecular exchange has been found to produce demixing of the copolymer repeating units, with formation of cyclic trimethylene thiocarbonate and of BPA polycarbonate.[53]

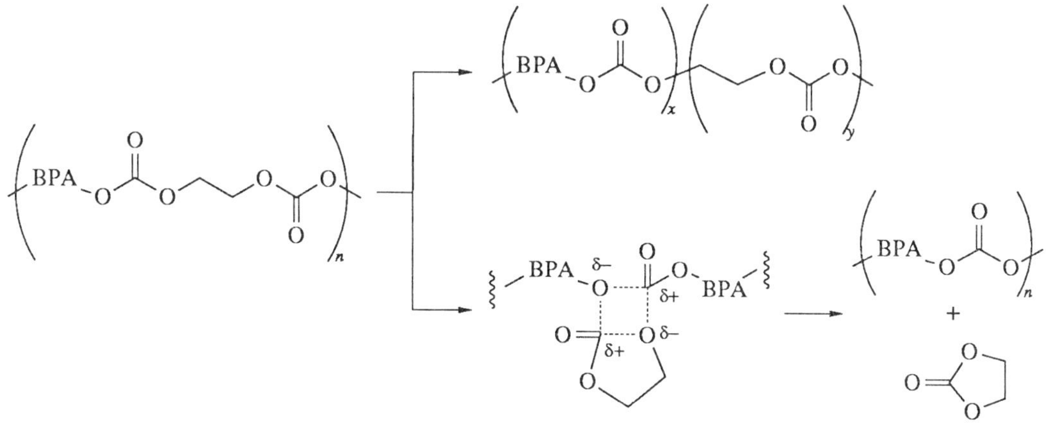

Scheme 32

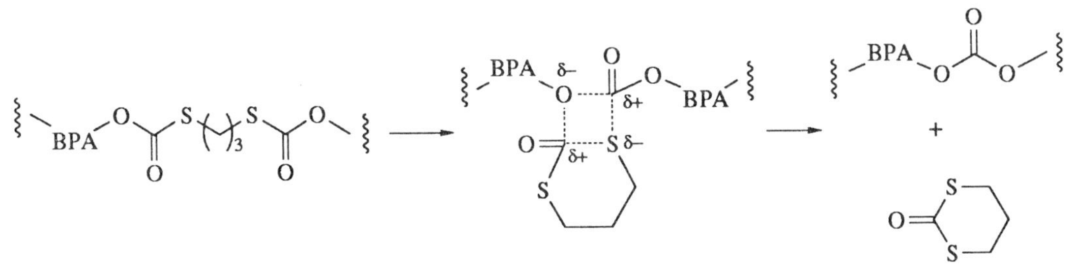

Scheme 33

A thermal rearrangement of the silarylene–siloxane units has also been observed in the case of some alternating silarylene–siloxane copolymers.[78] As in the cases discussed above, only part of the thermal decomposition compounds had structures corresponding to the original copolymer repeating unit, and a major portion of the pyrolysis products showed complete demixing, indicating that an Si—O/Si—O exchange process had occurred (Scheme 34).[78]

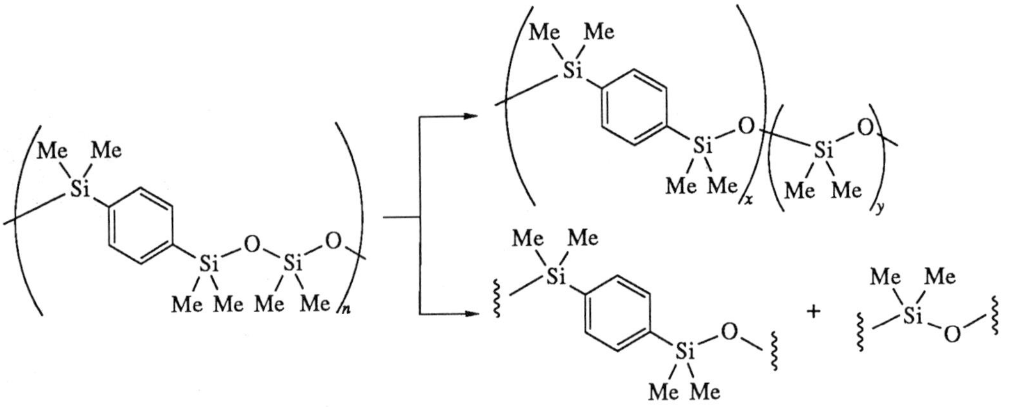

Scheme 34

All these examples show that when the thermal decomposition of alternating copolymers proceeds through exchange reactions, there is a strong tendency towards demixing and formation of homopolymer sequences, which are thermodynamically more stable.

11.6 FREE RADICAL PROCESSES

In the foregoing sections we have discussed a number of ionic thermal processes, induced by the presence of highly polar groups in the polymer chains. On the other side, free radical thermal reactions are predominating in the well-known and well-studied pyrolyses of polyalkenes, polydienes, polyvinyls and in general in cases where the polymer backbone has an apolar structure.[1-5] In between highly polar and completely apolar structures, there are polymers containing relatively nonpolar functional groups. The latter, when heated, may decompose by free radical processes. Furthermore, when the structure of some polymers containing highly polar groups is such that the thermal degradation occurs at relatively high temperatures, then free radical processes may take place.

For instance, the thermal degradation of aromatic polyamides, which decompose at temperatures higher than 500 °C, involves radical processes. In fact, small pyrolysis fragments are preferentially formed in this case, and water, CO_2, CO, nitriles, amines and carboxylic acids are found to be the major pyrolysis products.[102-104] The pyrolysis is not selective in this case, and all those compounds may arise by thermal degradation processes involving undefined multiple hydrogen transfer. When multiple hydrogen transfer must be invoked in order to explain a thermal degradation process, this simply means that the process is occurring *via* free radicals.

A well-known example of a radical process occurring in the thermal degradation of a condensation polymer is the unzipping of capped poly(methylene oxide) (PMO) to the monomer,[88] as depicted in Scheme 35. Evidently, the presence of the oxygen atom within the chain is not sufficient to induce the highly polar character which is necessary in order to activate an ionic process. However, if the pyrolysis of PMO is carried out in the presence of a proton-donating agent,[105] an acid-catalyzed

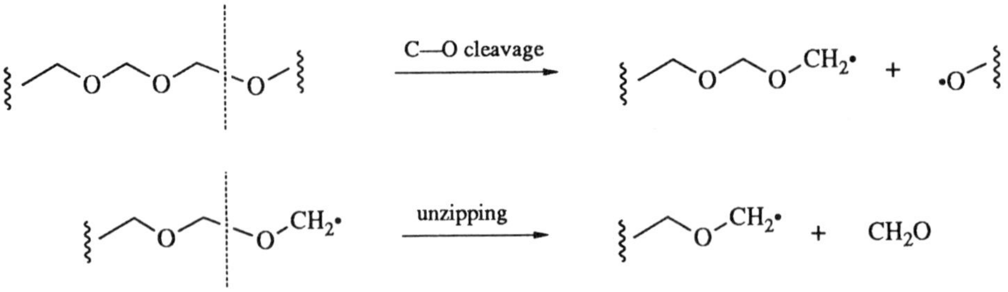

Scheme 35

process takes place, and a distribution of cyclic oligomers is generated by an electrophilic attack of the charged methylene group on the newly formed hydroxy group, as indicated in Scheme 36.

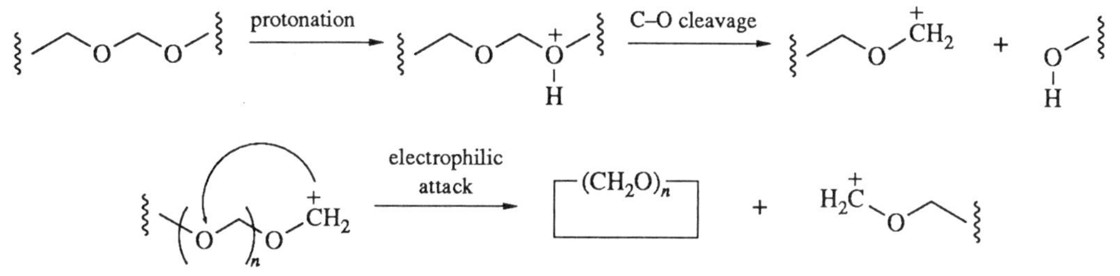

Scheme 36

A free radical reaction is also responsible for the thermal degradation of poly(ethylene oxide) (PEO).[89] This process involves a hydrogen elimination, followed by two competitive C—O cleavage steps (an α-cleavage and a β-cleavage, respectively), and by hydrogen abstraction, as illustrated in Scheme 37. The complex mixture of compounds formed[89] in this unselective cleavage is typical of processes involving free radical intermediates.

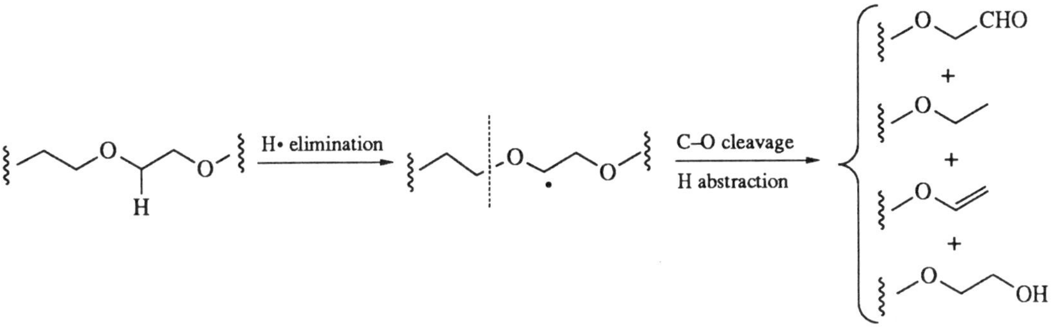

Scheme 37

Aromatic polyethers also undergo thermal degradation through radical processes.[90-93] An interesting example is given by poly(phenyleneoxyxylyl ether) (PPOX),[90] which decomposes as shown in shown in Scheme 38, to yield a mixture of compounds close to those detected in the case of PEO.[89]

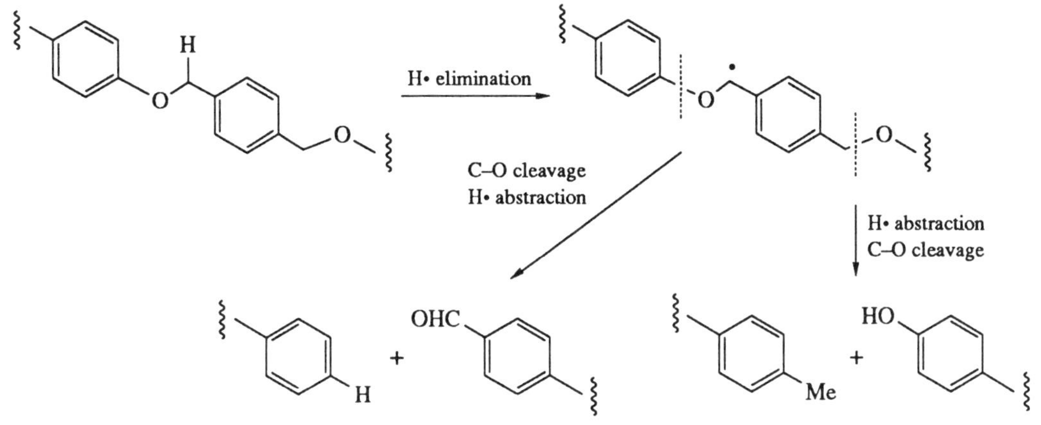

Scheme 38

Another class of condensation polymers where the low polarity of the functional groups present in the main chain allows the occurrence of radical processes, is given by the poly(azomethine)s.[94] Totally aromatic poly(azomethine)s have been found to pyrolyze by means of multiple hydrogen transfer reactions, which produce a mixture of compounds in a typically unselective radical cleavage process (Scheme 39). The inclusion of aliphatic units along the polymer chain[94] does not change the free radical nature of the pyrolysis process in these poly(azomethine)s, and the volatile reaction products appear to originate from extensive hydrogenation processes (Scheme 40).[94]

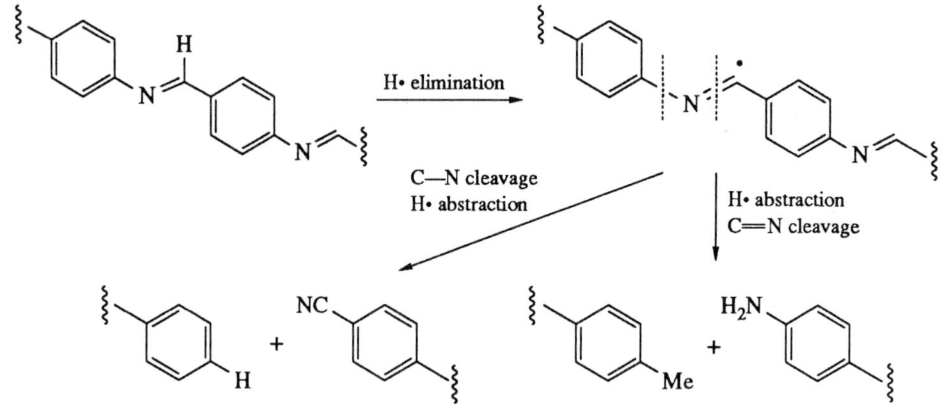

Scheme 39

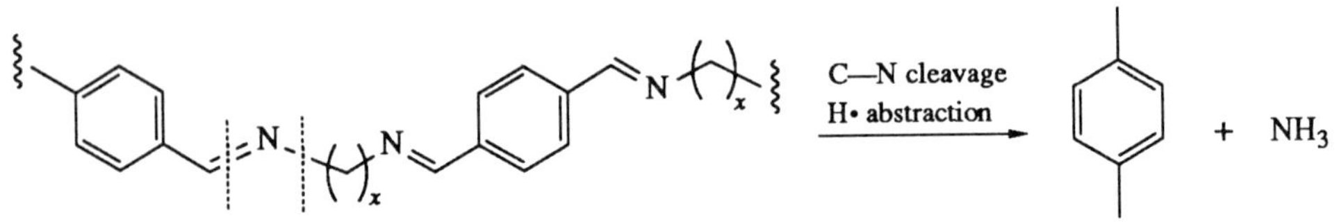

Scheme 40

Similarly, aliphatic polysulfones show a radical thermal decomposition mechanism,[106] as illustrated in Scheme 41.

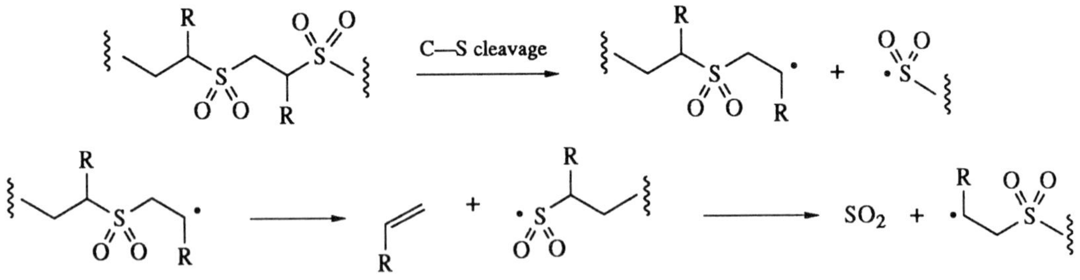

Scheme 41

Therefore, the existing evidence points towards the prevalence of radical processes in polyethers, poly(azomethine)s and in polysulfones, which contain functional groups of relatively low polarity and low thermal reactivity. Some other examples will now be considered, where more polar and reactive functional groups are involved.

The first example is the thermal degradation of poly(methylene sulfide) (PMS),[68] which decomposes by a typical radical mechanism with formation of CS_2 (Scheme 42). This represents a remarkable deviation from the behavior of aliphatic and aromatic polysulfides, which have been found to react thermally by ionic processes (see above).[67-69] However, when the pyrolysis of PMS is performed in the presence of a hydrogen-donating agent,[68] the mechanism of thermal decomposition changes drastically and becomes ionic, with consequent formation of cyclic sulfides (Scheme 43). This behavior is analogous to that of PMO (Schemes 35 and 36), and represents an interesting example of acid catalysis in the pyrolytic processes, which can be used to obtain a dramatic change in the pyrolysis products.[68]

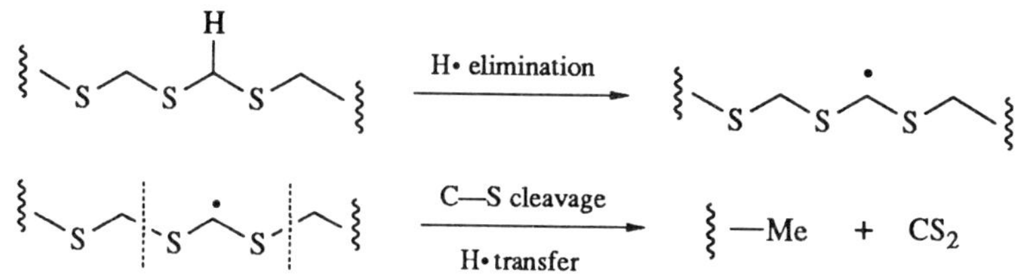

Scheme 42

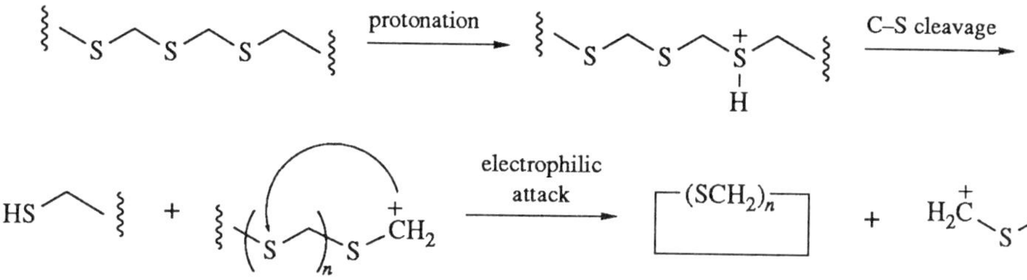

Scheme 43

In the cases where this catalysis occurs, it can also be exploited as a criterion to establish the nature of the reaction mechanisms occurring in the pyrolysis processes. For instance, the thermal decomposition of the N,N-disubstituted polyureas occurs at about 500 °C, with formation of methyl isocyanate and aromatic compounds,[66] and the mechanism is likely to involve free radical intermediates, as illustrated in Scheme 44. However, the urea group is highly polar and usually polyureas decompose thermally by ionic mechanisms, as discussed above.[64-66]

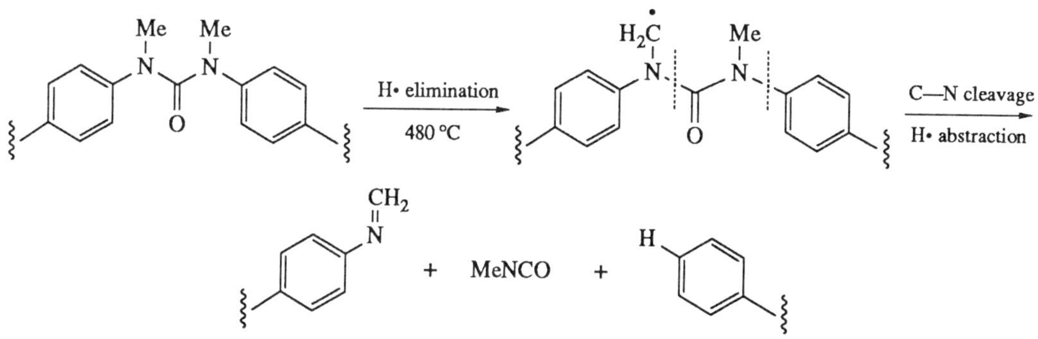

Scheme 44

When the pyrolysis is run in the presence of ammonium polyphosphate (APP), an acid precursor, the reaction products change[66] and the evolution of CO and methylarylamines is observed (Scheme 45). This suggests that these products should be generated if an ionic process were occurring in the case of the purely thermal degradation, and therefore supports the free radical mechanism proposed in Scheme 44 for the thermal process.

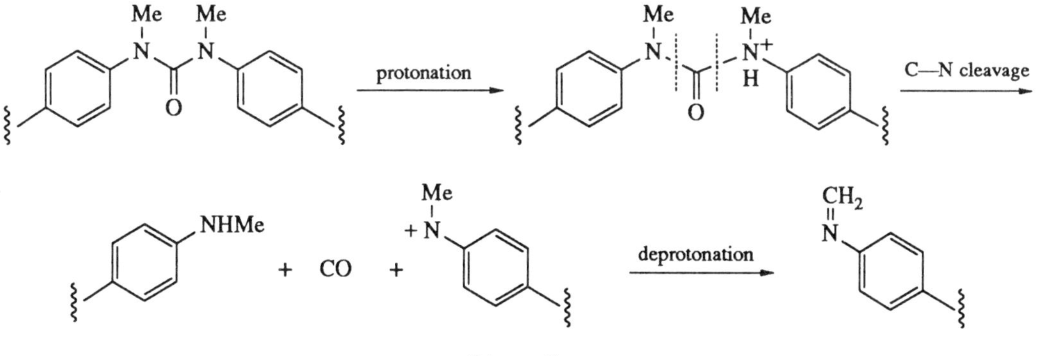

Scheme 45

11.7 MOLECULAR REARRANGEMENTS

A few molecular rearrangements that occur during the pyrolysis of condensation polymers, and that are capable of inducing the transformation of the functional groups originally present in the polymer will now be discussed. These rearrangements essentially occur by two mechanisms: (i) the extrusion of neutral molecules from the polymer backbone; and (ii) the isomerization of the polymer backbone.

An example of the first type is given by the pyrolysis of aliphatic polycarbonates, which yields aliphatic ether sequences and CO_2.[51] This may be interpreted by the process in Scheme 46, where the extrusion of CO_2 occurs through a polar, four-membered transition state. Although the initial decomposition stage in aliphatic polycarbonates is an ionic process (Scheme 46), the polyether sequences thus generated may undergo thermal decomposition by a radical mechanism (see Section 11.6), and this creates a case of concomitant occurrence of ionic and radical processes in the thermal degradation of a polymer.

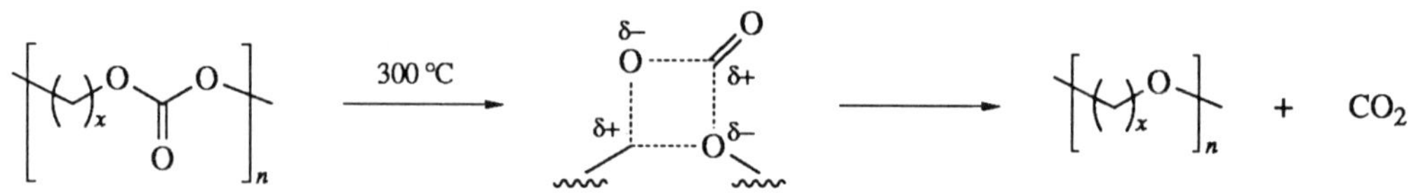

Scheme 46

In aromatic polycarbonates, the extrusion of CO_2 occurs by another molecular rearrangement (a Kolbe reaction)[44-48,52,113] that leads to polyether sequences and condensed aromatics, as illustrated in Scheme 47. In any case, polycarbonates rearrange to polyethers during pyrolysis.

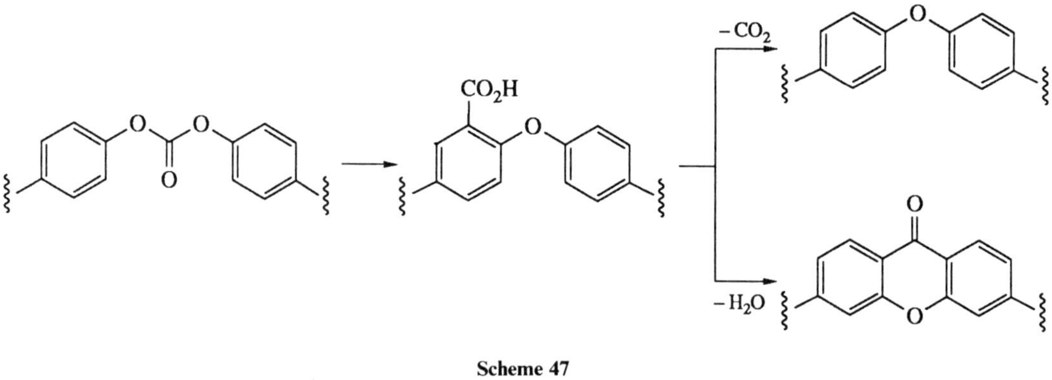

Scheme 47

Further examples of polymers undergoing rearrangements that result in the transformation of the functional groups through extrusion of sulfur and SO_2 include certain aromatic polydisulfides[107] and polysulfones (equations 2 and 3).[108,109]

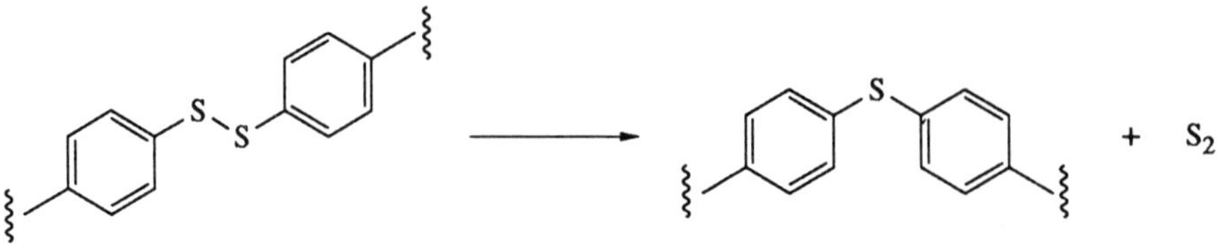

$$+ S_2 \quad (2)$$

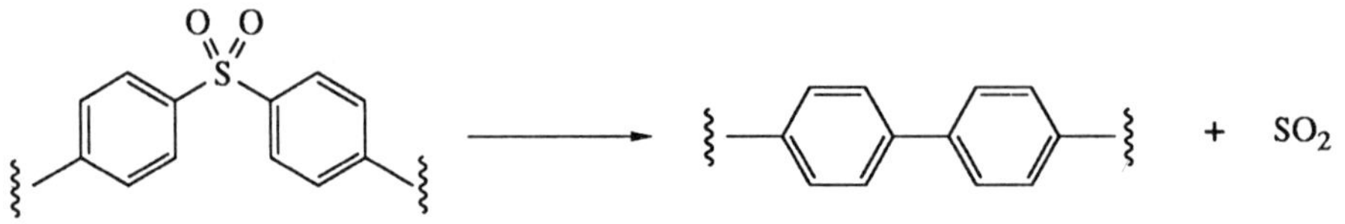

$$+ SO_2 \quad (3)$$

Finally, it is worth mentioning two cases where the transformation of the functional groups occurs by isomerization. In the pyrolysis of poly(2,6-dimethylphenylene oxide) (PPO), the formation of methylene bridges along the chain has been observed to occur through the rearrangement shown in Scheme 48.[92] A similar rearrangement might also be hypothesized in the case of the polysulfide in Scheme 49, which would fit the experimental results well.[110]

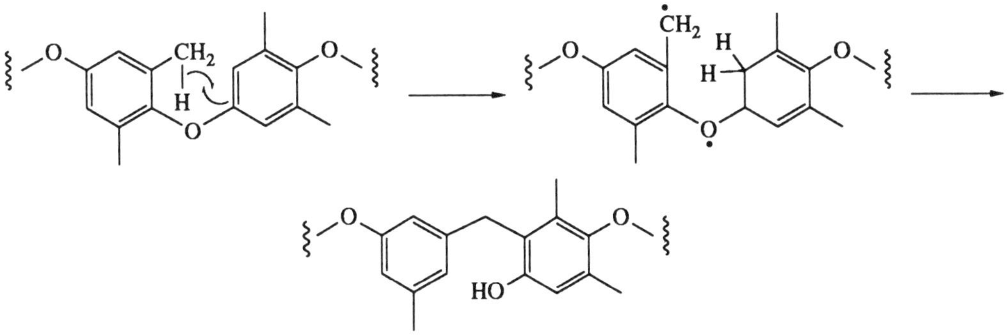

Scheme 48

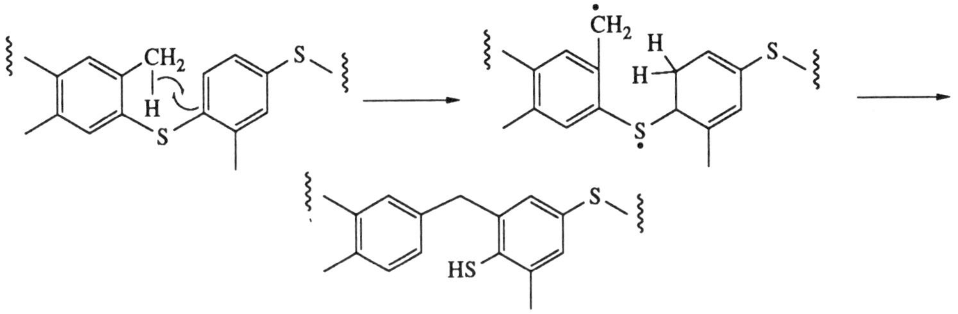

Scheme 49

ACKNOWLEDGEMENT

Partial financial support from the Italian Ministry for University and for Scientific and Technological Research (MURST), and from the National Council of Research (CNR) (Rome), Finalized Project of Fine and Secondary Chemistry, is gratefully acknowledged.

11.8 REFERENCES

1. F. O. Rice and K. K. Rice, 'The Aliphatic Free Radicals', Hopkins Press, Baltimore, 1935.
2. N. Grassie, 'Chemistry of High Polymer Degradation Processes', Butterworths, London, 1956.
3. C. David, in 'Degradation of Polymers', ed. C. H. Bamford and C. F. H. Tipper, Elsevier, Amsterdam, 1975.
4. McNeill, I. C. in 'Comprehensive Polymer Science', ed. G. C. Eastmond, A. Ledwith, S. Russo and P. Sigwalt, Pergamon Press, Oxford, 1989, vol. 6, p. 492.
5. H. H. G. Jellinek (ed.), 'Degradation and Stabilization of Polymers', Elsevier, Amsterdam, 1983, vol. 1.
6. G. Montaudo and C. Puglisi, in 'Developments in Polymer Degradation', ed. N. Grassie, Applied Science, London, 1987, vol. 7, p. 35.
7. G. Montaudo, *Br. Polym. J.*, 1986, **18**, 231.
8. B. J. Tighe, in 'Development in Polymer Degradation', ed. N. Grassie, Applied Science, London, vol. 5, 1984.
9. H. R. Kricheldorf and I. Luderwald, *Makromol. Chem.*, 1978, **179**, 421.
10. R. H. Wiley, *J. Macromol. Sci., Chem.*, 1970, **A4**, 1797.
11. N. Grassie and E. J. Murray, *Polym. Degradation Stab.*, 1984, **6**, 47, 95, 127.
12. R. J. Helleur, *Polym. Prepr., Am. Chem. Soc., Polym. Chem. Div.*, 1988, **29** (1), 609.
13. H. Morikawa and R. H. Marchessault, *Can. J. Chem.*, 1981, **59**, 2306.
14. S. Iwabuchi, W. Jaacks, F. Galil and W. Kern, *Markromol. Chem.*, 1973, **165**, 59; 1976, **177**, 2675.
15. S. Foti, M. Giuffrida, P. Maravigna and G. Montaudo, *J. Polym. Sci., Polym. Chem. Ed.*, 1984, **22**, 1201, 1217.
16. L. H. Buxbaum, *Angew. Chem., Int. Ed. Engl.*, 1968, **7**, 182.
17. R. M. Lum, *J. Polym. Sci., Polym. Chem. Ed.*, 1979, **17**, 203.
18. R. E. Adams, *J. Polym. Sci., Polym. Chem. Ed.*, 1982, **20**, 119.

19. H. G. Ramjit, *J. Macromol. Sci., Chem.*, 1983, **A19**, 41; 1983, **A20**, 659.
20. H. G. Ramjit and R. D. Sedgwick, *J. Macromol. Sci., Chem.*, 1976, **A10**, 815.
21. D. Garozzo, M. Giuffrida and G. Montaudo, *Macromolecules*, 1986, **19**, 1643.
22. M. Giuffrida, P. Maravigna, G. Montaudo and E. Chiellini, *J. Polym. Sci., Polym. Chem. Ed.*, 1986, **24**, 1643.
23. D. Garozzo, M. Giuffrida and G. Montaudo, *Polym. Degradation Stab.*, 1986, **15**, 143.
24. D. Garozzo, M. Giuffrida, G. Montaudo and R. W. Lenz, *J. Polym. Sci., Polym. Chem. Ed.*, 1987, **25**, 271.
25. A. Ballistreri, D. Garozzo, M. Giuffrida, G. Impallomeni and G. Montaudo, *Polym. Degradation Stab.*, 1988, **21**, 311.
26. B. Plage and H.-R. Schulten, *J. Anal. Appl. Pyrol.*, 1989, **15**, 197.
27. A. Ballistreri, M. Giuffrida, P. Maravigna and G. Montaudo, *J. Polym. Sci., Polym. Chem. Ed.*, 1985, **23**, 1145, 1731.
28. H. Ohtani, T. Nagaya, Y. Sugimura and S. Tsuge, *J. Anal. Appl. Pyrol.*, 1982, **4**, 117.
29. I. Luderwald, *Proc. Eur. Symp. Polym. Spectrosc., 5th, 1978*, Verlag Chemie, Weinheim, 1979, 217.
30. F. Wiloth, *Makromol. Chem.*, 1971, **144**, 283.
31. R. E. Adams, *Anal. Chem.*, 1983, **55**, 414.
32. H. R. Kricheldorf and E. Leppert, *Makromol. Chem.*, 1974, **175**, 1731.
33. I. Luderwald and H. R. Kricheldorf, *Angew. Makromol. Chem.*, 1976, **56**, 173.
34. Y. P. Khanna, E. M. Pearce, S. D. T. Burkitt, H. Njuguna, D. M. Hindenlang and B. D. Forman, *J. Polym. Sci., Polym. Chem. Ed.*, 1981, **19**, 2817.
35. Y. P. Khanna and E. M. Pearce, *J. Polym. Sci., Polym. Chem. Ed.*, 1981, **19**, 2835.
36. A. K. Chaudhuri, B. Y. Min and E. M. Pearce, *J. Polym. Sci., Polym. Chem. Ed.*, 1980, **18**, 2949.
37. A. Ballistreri, D. Garozzo, M. Giuffrida, P. Maravigna and G. Montaudo, *Macromolecules*, 1986, **19**, 2693.
38. A. Ballistreri, D. Garozzo, M. Giuffrida and G. Montaudo, *Polym. Degradation Stab.*, 1986, **16**, 337.
39. A. Ballistreri, D. Garozzo, M. Giuffrida and G. Montaudo, *J. Polym. Sci., Polym. Chem. Ed.*, 1987, **25**, 1049, 2351.
40. A. Ballistreri, D. Garozzo, M. Giuffrida and G. Montaudo, *J. Anal. Appl. Pyrol.*, 1987, **12**, 3.
41. A. Ballistreri, D. Garozzo, M. Giuffrida, G. Montaudo and A. Pollicino, *Polymer*, 1987, **28**, 139.
42. A. Ballistreri, D. Garozzo, M. Giuffrida and G. Montaudo, *Macromolecules*, 1987, **20**, 2991.
43. A. Ballistreri, D. Garozzo, M. Giuffrida, G. Impallomeni and G. Montaudo, *Polym. Degradation Stab.*, 1988, **23**, 25.
44. A. Davis and J. H. Golden, *Makromol. Chem.*, 1967, **110**, 180.
45. S. Tsuge, T. Okumoto, Y. Sujimura and T. Tacheuchi, *J. Chromatogr. Sci.*, 1969, **7**, 253.
46. R. H. Wiley, *Macromolecules*, 1971, **4**, 254.
47. S. Foti, M. Giuffrida, P. Maravigna and G. Montaudo, *J. Polym. Sci., Polym. Chem. Ed.*, 1983, **21**, 1567.
48. M. S. Lin, B. J. Bulkin and E. M. Pearce, *J. Polym. Sci., Polym. Chem. Ed.*, 1981, **19**, 2773
49. S. Inoue, T. Tsuruta, T. Takada, N. Miyazaki, M. Kambe and T. Takaoka, *Appl. Polym. Symp.*, 1975, **26**, 257.
50. D. D. Dixon, M. E. Ford and G. J. Mantell, *J. Polym. Sci., Polym. Lett. Ed.*, 1980, **18**, 131.
51. G. Montaudo, C. Puglisi and F. Samperi, *Polym. Degradation Stab.*, 1989, **26**, 285; 1991, **31**, 229.
52. A. Ballistreri, G. Montaudo, C. Puglisi, E. Scamporrino, D. Vitalini and S. Cucinella, *J. Polym. Sci., Polym. Chem. Ed.*, 1988, **26**, 2113.
53. G. Montaudo, C. Puglisi, C. Berti, E. Marianucci and F. Pilati, *J. Polym. Sci., Polym. Chem. Ed.*, 1989, **27**, 2277, 2657.
54. I. C. McNeill and A. Rinchon, *Polym. Degradation Stab.*, 1989, **24**, 59, 171; 1990, **27**, 35.
55. S. Foti, P. Maravigna, G. Montaudo, *Polym. Degradation Stab.*, 1982, **4**, 287.
56. G. Montaudo, C. Puglisi, E. Scamporrino and D. Vitalini, *Macromolecules*, 1984, **17**, 1605.
57. A. Ballistreri, S. Foti, P. Maravigna, G. Montaudo and E. Scamporrino, *Makromol. Chem.*, 1980, **181**, 2161.
58. A. Ballistreri, S. Foti, P. Maravigna, G. Montaudo and E. Scamporrino, *J. Polym. Sci., Polym. Chem. Ed.*, 1980, **18**, 1923.
59. S. Foti, P. Maravigna and G. Montaudo, *J. Polym. Sci., Polym. Chem. Ed.*, 1981, **19**, 1679.
60. S. Foti, M. Giuffrida, P. Maravigna and G. Montaudo, *J. Polym. Sci., Polym. Chem. Ed.*, 1983, **21**, 1583.
61. K. J. Voorhees and R. P. Lattimer, *J. Polym. Sci., Polym. Chem. Ed.*, 1982, **20**, 1457.
62. R. P. Lattimer, H. Muenster and H. Budzikiewicz, *J. Anal. Appl. Pyrol.*, 1990, **17**, 237.
63. B. Durairaj, A. W. Dimock, E. T. Samulski and M. T. Shaw, *J. Polym. Sci., Polym. Chem. Ed.*, 1989, **27**, 3211.
64. S. Caruso, S. Foti, P. Maravigna and G. Montaudo, *J. Polym. Sci., Polym. Chem. Ed.*, 1982, **20**, 1685.
65. S. Foti, A. Liguori, P. Maravigna and G. Montaudo, *Anal. Chem.*, 1982, **54**, 674.
66. G. Montaudo, E. Scamporrino and D. Vitalini, *J. Polym. Sci., Polym. Chem. Ed.*, 1983, **21**, 3321.
67. G. Montaudo, C. Puglisi, E. Scamporrino and D. Vitalini, *Macromolecules*, 1986, **19**, 2157.
68. G. Montaudo, C. Puglisi, E. Scamporrino and D. Vitalini, *J. Polym. Sci., Polym. Chem. Ed.*, 1987, **25**, 475.
69. G. Montaudo, C. Puglisi, E. Scamporrino and D. Vitalini, *Polymer*, 1987, **28**, 477.
70. G. Montaudo, C. Puglisi, E. Scamporrino and D. Vitalini, *J. Anal. Appl. Pyrol.*, 1987, **10**, 283.
71. H. R. Allcock, *J. Macromol. Sci., Rev. Macromol. Chem.*, 1970, **C4**, 149.
72. J. A. Semlyen, *Adv. Polym. Sci.*, 1976, **21**, 41.
73. B. Zelei, M. Blazsò and S. Dobos, *Eur. Polym. J.*, 1981, **17**, 503.
74. N. Grassie and I. G. Macfarlane, *Eur. Polym. J.*, 1978, **14**, 875.
75. N. Grassie, K. F. Francey, *Polym. Degradation Stab.*, 1980, **2**, 53.
76. N. Grassie, K. F. Francey and I. G. Macfarlane, *Polym. Degradation Stab.*, 1980, **2**, 67.
77. N. Grassie, S. R. Beattie, *Polym. Degradation Stab.*, 1984, **7**, 109; 1984, **7**, 231; 1984, **8**, 177; 1984, **9**, 23.
78. A. Ballistreri, G. Montaudo and R. W. Lenz, *Macromolecules*, 1984, **17**, 1848.
79. A. Ballistreri, D. Garozzo, and G. Montaudo, *Macromolecules*, 1984, **17**, 1312.
80. H. D. Stenzenberger, K. U. Heinen and D. O. Hummel, *J. Polym. Sci., Polym. Chem. Ed.*, 1976, **14**, 2911.
81. G. Montaudo, C. Puglisi, N. Biçak and A. Orzeszko, *Polymer*, 1989, **30**, 2237.
82. H. R. Schulten and W. Gortz, *Anal. Chem.*, 1978, **50**, 428.
83. W. E. Franklin, *Anal. Chem.*, 1979, **51**, 992.
84. F. Shafizadeh, R. H. Furneaux, T. G. Cochran, J. P. Scholl and Y. Sakai, *J. Appl. Polym. Sci.*, 1979, **23**, 3525.
85. H. R. Schulten, U. Bahr, W. Gortz, *J. Anal. Appl. Pyrol.*, 1981/1982, **3**, 229.
86. A. Van Der Kaaden, J. Haverkamp, J. J. Boon and J. W. DeLeeuw, *J. Anal. Appl. Pyrol.*, 1983, **5**, 199.
87. R. J. Evans, T. A. Milne, M. N. Soltys and H. R. Schulten, *J. Anal. Appl. Pyrol.*, 1984, **6**, 273.
88. A. Ballistreri, G. Montaudo and C. Puglisi, *J. Therm. Anal.*, 1984, **29**, 237.
89. E. Bortel and R. Lamot, *Makromol. Chem.*, 1977, **178**, 2617.

90. G. Montaudo, C. Puglisi, E. Scamporrino and D. Vitalini, *Macromolecules*, 1986, **19**, 870, 882.
91. G. Montaudo, M. Przybylski and H. Ringsdorf, *Makromol. Chem.*, 1975, **176**, 1763.
92. M. Kryszewski and J. Jachowiz, in 'Developments in Polymer Degradation', ed. N. Grassie, Applied Science, London, 1982, vol. 4, p. 1.
93. D. O. Hummel, H. J. Dussel, H. Rosen and K. Rubenacker, *Makromol. Chem., Suppl.*, 1975, **1**, 471.
94. A. Ballistreri, D. Garozzo, M. Giuffrida, P. Maravigna and G. Montaudo, *J. Polym. Sci., Polym. Chem. Ed.*, 1986, **24**, 331.
95. C. Arnolds, *J. Polym. Sci., Macromol. Rev.*, 1975, **14**, 265.
96. P. Godard, J. M. Dekoninck, V. Devlesaver and J. Devaux, *J. Polym. Sci., Polym. Chem. Ed.*, 1986, **24**, 3301, 3315.
97. G. Montaudo, C. Puglisi and F. Samperi, *Polym. Degradation Stab.*, 1991, **31**, 291; *J. Polym. Sci., Polym. Chem. Ed.*, 1992, **30**.
98. J. Devaux, P. Godard and J. P. Mercier, *J. Polym. Sci., Polym. Phys. Ed.*, 1982, **20**, 1875, 1881, 1895, 1901.
99. A. M. Kotliar, *J. Polym. Sci., Macromol. Rev.*, 1981, **16**, 367.
100. G. Montaudo and P. Maravigna, in 'Comprehensive Polymer Science', ed. G. C. Eastmond, A. Ledwith, S. Russo and P. Sigwalt, Pergamon Press, Oxford, 1989, vol. 5, p. 63.
101. J. A. Semlyen (ed.), 'Cyclic Polymers', Elsevier, New York, 1986.
102. G. F. L. Ehlers, K. R. Fish, W. R. Powell, *J. Polym. Sci., Part A-1*, 1970, **8**, 3511.
103. D. A. Chatfield, I. N. Einhorn, R. W. Mickelson and J. H. Futrel, *J. Polym. Sci., Polym. Chem. Ed.*, 1979, **17**, 1353, 1367.
104. J. R. Brown and A. J. Power, *Polym. Degradation Stab.* 1982, **4**, 379, 479.
105. V. K. H. Burg, H. D. Herman and H. Rehling, *Makromol. Chem.*, 1968, **111**, 181.
106. M. J. Bowden, L. F. Thompson, W. Robinson and M. Biolsi, *Macromolecules*, 1982, **15**, 1417.
107. S. Foti and G. Montaudo, in 'Analysis of Polymer Systems' ed. L. S. Bark and N. S. Allen, Applied Science, London, 1982.
108. A. Davis, *Makromol. Chem.*, 1969, **128**, 242.
109. G. F. L. Ehlers, K. R. Fish and W. R. Powell, *J. Polym. Sci., Part A-1*, 1969, **7**, 2955.
110. G. Bruno, S. Foti, P. Maravigna, G. Montaudo and M. Przybylski, *Polymer*, 1977, **18**, 1149.
111. G. Montaudo, *Macromolecules*, 1991, **24**, 5289.
112. D. Garotto and G. Montaudo, *Macromolecules*, 1991, **24**, 1416.
113. G. Montaudo and C. Puglisi, *Polym. Degradation Stab.*, 1992, **36**, 201.

12
Photodegradation of Polymer Materials

BENGT RÅNBY
The Royal Institute of Technology, Stockholm, Sweden

and

JAN F. RABEK
Karolinska Institute, Huddinge, Sweden

12.1	INTRODUCTION	253
	12.1.1 Early Work on Fibers, Rubber and Plastics	253
	12.1.2 Polymers of Different Photostability	254
	12.1.3 Definition of Photostability	254
	12.1.4 Bond Strengths and Energies of Quanta	255
12.2	GENERAL MECHANISMS OF POLYMER PHOTODEGRADATION	256
	12.2.1 Organic Compounds and Oxygen	256
	12.2.2 Radical Formation and Initiation	257
	12.2.3 Chain Propagation	258
	12.2.4 Chain Termination	259
12.3	PHOTODEGRADATION OF IMPORTANT POLYMER MATERIALS	259
	12.3.1 Molecular Mechanisms	259
	12.3.2 Sensitized Photodegradation	263
	12.3.3 Polyalkenes and Diene Copolymers	263
	12.3.4 Poly(vinyl chloride)	267
	12.3.5 Polydienes	269
	12.3.6 Polymers Containing Carbonyl Groups	269
	12.3.7 Polyacrylates	271
	12.3.8 Polystyrene	272
	12.3.9 Polyamides, Polyesters and Polycarbonates	272
	12.3.10 Polyurethanes, Polysulfones, Polysilanes and Polysiloxanes	275
	12.3.11 Polyphosphazenes	278
	12.3.12 Cellulose	279
	12.3.13 Degradation by Laser Radiation	279
12.4	REFERENCES	280

12.1 INTRODUCTION

12.1.1 Early Work on Fibers, Rubber and Plastics

It has been known for hundreds of years that polymer materials like native fibers are degraded by exposure to sunlight. This is called 'tendering' and applies to textile fabrics made of cotton, linen and other cellulosic fibers. Bleaching in sunlight of textiles from the slightly yellow virgin cellulose fibers has been common practice both by housewives and textile producers for a very long time and it is still used. By the 1880s it was already established by experiment that the brightening of off-white fabrics was accompanied by a gradual weakening of the fibers.[1] After 1900, when dyed textiles from cellulose fibers became common, it was observed that the presence of certain dyes further tendered the fibers during bleaching. When exposed to sunlight, the dyes accelerated the degradation.[2] It was also observed that the degradation is caused not only by the ultraviolet part of sunlight but also ultraviolet radiation from lamps. For dyed fibers, visible light of all wavelengths could also cause degradation, but the nature of the sensitizing mechanism of the reaction was not known at this time. The photodegradation of cellulosic materials is further treated in Section 12.3.12.

Photodegradation of rubber was also observed in the 1800s and was a problem until 1904, when reinforcement by addition of carbon black to rubber compounds before vulcanization was invented and commercially introduced. The carbon particles absorb light of all wavelengths, including ultraviolet radiation, and act as photostabilizers by protecting the polymer. Light colored rubber compounds could later be photostabilized by addition of stabilizers, e.g. phenols and aromatic amines.[3,4] More effective stabilizers were developed later.

The first synthetic plastic material in common usage since about 1910 was phenol–formaldehyde resin, which has cellulose fibers or sawdust as reinforcement, marketed as Bakelite®. It is inherently rather stable to photodegradation due to its dark color from the reaction by-products formed during the thermosetting of the resin. With the introduction of transparent thermoplastics during the 1920s and 1930s, photodegradation was observed during outdoor exposure. Polystyrene and poly(vinyl chloride) (PVC) exposed to sunlight turned yellow to brown and became gradually brittle. These observations prompted the development of stabilizing additives, both pigments that screened or reflected the light and aromatic organic compounds that absorbed the light and prevented its action on the polymer itself. Another early method was the addition of blue or fluorescent compounds which compensate the yellowing rather than prevent degradation.[5]

Systematic studies of photodegradation and photostabilization started in the 1940s when it was observed that photoinitiated degradation in air also involved a simultaneous oxidation. The well-established mechanism of photooxidation of polymers with initial free radical formation by chain scission, or abstraction of hydrogen, chlorine, etc. and subsequent addition of molecular oxygen which gives peroxy radicals, was published in the 1940s by Bolland and Gee.[6] This fundamental concept is the basis for the extensive studies and interpretations of the mechanisms of photooxidative degradation of polymer materials in air during the last 40 years.

12.1.2 Polymers of Different Photostability

Photodegradation of a polymer is in most cases initiated by chromophores which are extraneous groups on the chains, e.g. formed by oxidation, or impurities which can be metal ions, organometallic compounds, light-absorbing organic compounds or solid particles which absorb light. These sensitizing phenomena due to modified groups or impurities will be further treated in Section 12.3.2.

Polymers have different inherent photostability in air and can be divided into three groups:[7] (i) highly photostable polymers, commonly used without photostabilizer added, e.g. poly(tetrafluoroethylene) and poly(methyl methacrylate) with an outdoor lifetime of many years; (ii) moderately photostable polymers that can be used without photostabilizer, e.g. poly(ethylene terephthalate), aliphatic polyamides (nylon-6, -6,6, -6,10 and -11) and aliphatic polyurethanes with an outdoor lifetime without additives of a few years; and (iii) poorly photostable polymers which need extensive stabilization for outdoor use, e.g. polyalkenes, poly(vinyl chloride), polystyrene, aromatic polyamides and polyurethanes, diene elastomers, and many coatings with an outdoor lifetime of less than a year if compounded without stabilizers.

Most packaging materials and disposable items like cups and plates are used without additives to prevent contamination by migration. In such applications long-term photostability means a useful lifetime indoors of a few years. For application of polymers in building and machine construction and general outdoor use a lifetime of 10–50 years is usually required.

Photodegradation causes chain scission in the polymer material which results in the loss of important physical properties such as impact strength, tensile strength, elongation at break, etc. and can be observed as cracking, chalking and color changes. Photodegradation has been treated in *Comprehensive Polymer Science*, vol. 6, chap. 18, where general references are given.[8]

12.1.3 Definition of Photostability

The photostability of a polymeric material can be defined as its ability to withstand exposure to visible light and/or ultraviolet radiation under defined environmental conditions, e.g. atmosphere (air, reactive or inert gas or vacuum), moisture, reactive chemicals, temperature, mechanical stress etc. For stability outdoors in sunlight, the inherent absorption of ultraviolet radiation is of great

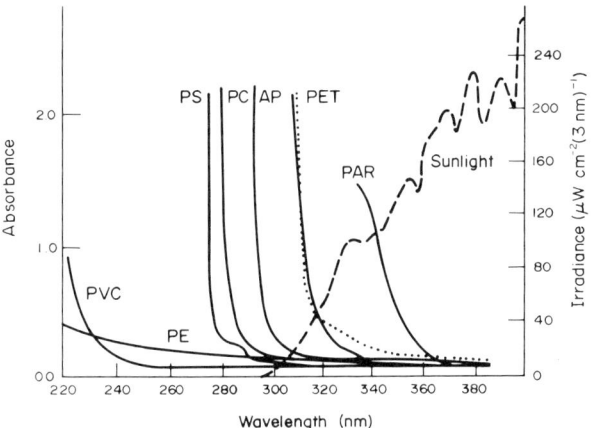

Figure 1 Ultraviolet absorption spectra of 0.05 mm polymer films and the spectral irradiance of July noon sunlight (direct beam) at 41° north latitude; AP, aromatic polyester; PAR, polyarylate; PC, polycarbonate; PE, polyethylene; PET, poly(ethylene terephthalate); PS, polystyrene; PSF, polysulfone; PVC, poly(vinyl chloride). Sunlight measured in irradiance, all other curves in absorbance (reproduced from ref. 10 by permission of Technomic)

importance. For a thin film (10 μm) of pure polymer[9] the 'cutoff' is defined as the wavelenth at which the absorbance reaches 10. The cutoff limit for polyalkenes is about 180 nm, for poly(methyl methacrylate) and aliphatic polyamides 240 nm, for polystyrene 270 nm, for polycarbonate, polyurethanes (based on bisphenol A) and poly(phenylene oxide) about 280 nm, for poly(ethylene terephthalate) about 310 nm and for aromatic polyamides 340–350 nm. This means that all but the last two of the above polymer types in pure form do not absorb sunlight at sea level, which contains UV irradiation of about 300 nm (Figure 1).[10]

The tail of light absorption for all polymers stretching towards the visible regions (∼ 400 nm) is due to modified groups and impurities which act as chromophores, as discussed in Section 12.1.2, and will be further treated in Section 12.3.2.

12.1.4 Bond Strengths and Energies of Quanta

The bond strengths or bond dissociation energies for the chemical bonds in the most common polymer materials are given in Table 1 and expressed as kcal mol^{-1}, kJ mol^{-1} and eV bond^{-1}.[11] These different units are recalculated using the relations 1 eV bond^{-1} is 23.1 kcal mol^{-1} or 96.5 kJ mol^{-1}. Chemical bonds of normal strength, *i.e.* main valence bonds or σ-bonds, have a bond strength from about 3 to 6 eV and weak bonds a strength <3 eV bond^{-1}.

The energy quanta of ultraviolet and visible light of different wavelengths are given in Table 2 in the same units as in Table 1. By comparing the data in these tables, it is concluded that the energy quanta of ultraviolet light are in the same range as the bond dissociation energies of the normal valence bonds. However, all of these bonds do not directly absorb UV radiation or visible light in the range 200–800 nm, and therefore cannot cause bond dissociation into free radicals. This

Table 1 Bond Dissociation Energies for Common Chemical Bonds[11]

Chemical bond	(kcal mol^{-1})	(kJ mol^{-1})	(eV bond^{-1})
H$_3$C—F	110	460	4.76
H$_3$C—H	102	426	4.40
H$_3$C—OH	89	372	3.85
H$_3$C—CH$_3$	84	351	3.63
H$_3$C—Cl	82	343	3.55
H$_3$C—NH$_2$	79	330	3.42
H$_3$C—Br	69	280	2.90
H$_3$C—I	53	222	2.29
HO—OH	51	213	2.20

Table 2 The Energies of Quanta of Ultraviolet and Visible Light

Wavelength λ (nm)		Energy of Quanta		
		(kcal mol^{-1})	(kJ mol^{-1})	(eV Quanta^{-1})
200		143	597	6.19
300	Ultraviolet	95	395	4.11
400		71	299	3.10
400		71	299	3.10
600	Visible light	48	199	2.06
800		36	149	1.55

is in agreement with the Grothus–Draper law (the first law of photochemistry) which states: "Only the light which is absorbed by a molecule can be effective in producing photochemical change in that molecule." The absorption of UV/visible radiation occurs only by chromophores containing π- and/or n-electrons (*e.g.* C=C, phenyl, C=O) in their structures.

12.2 GENERAL MECHANISMS OF POLYMER PHOTODEGRADATION

Polymer materials as plastics are used in an air atmosphere at moderate temperatures, *i.e.* below the glass temperature (T_g) and the melting point (T_m) where the materials are in the solid state. Elastomers are used above T_g and T_m. Only in exceptional cases is the material irradiated in an atmosphere of a reactive gas or vapor or in a vacuum. Photoreactions taking place under such unusual conditions will be treated in this chapter only as reference. This means that the mechanisms discussed here are primarily photooxidative degradations of plastics and rubber in air.

12.2.1 Organic Compounds and Oxygen

Common organic compounds like conventional polymer materials are held together by single and/or double bonds which physically contain electron pairs in σ- or π-orbitals, one σ-pair in a single bond and one σ- and one π-pair in a double bond. In addition, certain groups like ketone and carboxy groups contain a nonbonding electron orbital (an n-pair) on the CO oxygen atom. Aromatic groups like phenyl groups contain three π-electron pairs.

The ground state of molecular oxygen is a triplet state, which means that it contains two unpaired electrons and is a biradical. As a result, molecular oxygen cannot react directly with normal organic compounds containing electron pairs in all chemical bonds but only with free radicals, which have unpaired electrons.

The main reaction of molecular oxygen with polymers is its addition to polymer radicals formed by bond scission, as established by Bolland and Gee in the 1940s.[6] Innumerable later studies have verified this basic mechanism for polymer photooxidation and thermal oxidation, *i.e.* the primary polymer radicals are formed by photoinitiation and thermal initiation, respectively.[12]

Molecular oxygen can also exist in two excited states as singlet oxygen $^1O_2(^1\Delta_g)$ (22.6 kcal mol^{-1}) and singlet oxygen $^1O_2(^1\Sigma_g^+)$ (37.6 kcal mol^{-1}). The $^1O_2(^1\Delta_g)$ state is much more long-lived in the gas phase (*ca.* 45 min) in comparison to the $^1O_2(^1\Sigma_g^+)$ state which has a lifetime *ca.* 7 s. The $^1O_2(^1\Delta_g)$ state exists also in solvents with a lifetime from 2 to 700 s depending on the type of solvent, whereas $^1O_2(^1\Sigma_g^+)$ is deactivated in solvents. Singlet oxygen cannot be formed by direct absorption of light, because it is a forbidden transfer. It is formed by physical methods (*e.g.* absorption of microwave energy by oxygen molecules) or by chemical or photosensitizing methods.[13]

The main reactions of singlet oxygen with polymers are the 'ene' reactions with alkenes (equation 1) and cycloadditions to conjugated double bonds.

$$O_2 + \underset{R}{\underset{|}{R}}\overset{R}{\underset{|}{C}}=\overset{R}{\underset{|}{C}}R \longrightarrow R\overset{R}{\underset{|}{C}}=\overset{R}{\underset{|}{C}}\overset{O-OH}{\underset{R}{C}}R \tag{1}$$

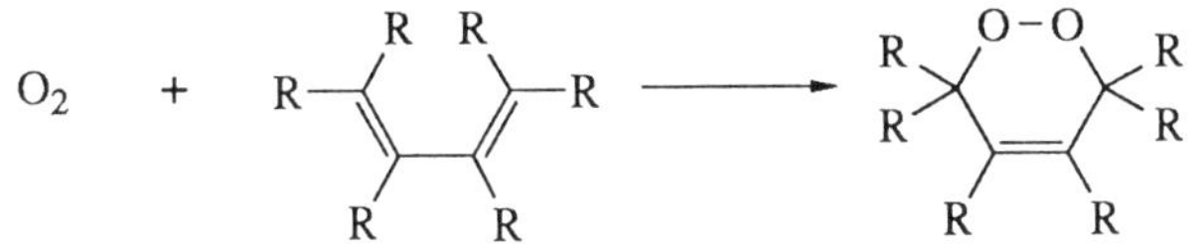

$$O_2 + \text{R}\underset{\underset{R}{|}}{\overset{\overset{R}{|}}{C}}=\underset{\underset{R}{|}}{\overset{\overset{R}{|}}{C}}\text{-R} \longrightarrow \text{(endoperoxide)} \quad (2)$$

The hydroperoxides (equation 1) and endoperoxides (equation 2) formed are further decomposed by heating or UV irradiation which involves bond scission and oxidation of the polymer. Singlet oxygen reactions mainly initiate photooxidation and will not be further described here. A recent review has been published by Clough et al.[14]

12.2.2 Radical Formation and Initiation

Polymer photodegradation in air (photooxidative degradation) is largely a free radical process which can lead to cleavage of the polymer backbone (chain scission), crosslinking, photorearrangement, unsaturation and/or formation of products of low molecular weight. All of these processes are responsible for the loss of mechanical and other physical properties of a polymer material such as color, gloss, impact strength, tensile strength, elongation at break, wettability, adhesion, etc. The polymer becomes brittle, cracks and holes are formed on the surface and gradually in the bulk. Oxygen and impurities, e.g. in water, are transported through the cracks and have much easier access to the bulk. Once started, an aging process on the surface spreads through the sample.

Despite the large number of studies dealing with the photodegradation of polymers, and careful reviews in a number of books,[15-22] the detailed mechanism of photooxidation is still not established. There are a number of disagreements concerning the proposed mechanisms, as the experimental conditions and purity of polymers investigated were different in the various works.

It is generally accepted that polymer photodegradation includes the following important steps: initiation, chain propagation and termination.

It is obvious that in the initiation step free radicals must participate. They can be formed from: (i) external low molecular weight impurities (RH; equation 3); (ii) internal absorbing chromophoric groups present in the polymer, i.e. carbonyl (C=O) and hydroperoxide (OOH) groups or double bonds (C=C); and (iii) charge-transfer (CT) complexes, formed between polymer (PH) and oxygen (Scheme 1)

$$\text{RH} \longrightarrow \text{R}\cdot + \text{H}\cdot \quad (3)$$

$$\text{PH} + O_2 \longrightarrow (\text{PH-}O_2) \xrightarrow{h\nu} (\text{PH-}O_2)^* \longrightarrow \text{P}\cdot + \text{HO}_2^\bullet$$

$$2\text{HO}_2^\bullet \longrightarrow H_2O_2 + O_2$$

$$H_2O_2 \longrightarrow 2\text{HO}\cdot$$

$$\text{HO}\cdot + \text{PH} \longrightarrow \text{P}\cdot + H_2O$$

Scheme 1

These three radicals: polymer alkyl radical (P·), hydroxyl radical (HO·) and hydroperoxy radical (HO$_2^\bullet$) participate in the following propagation reactions. If these radicals are formed in a solid polymer matrix, it is rather improbable that a polymer alkyl radical (P·) can migrate from the site at which it was formed, whereas other radicals like HO· and HO$_2^\bullet$ can escape from a cage almost immediately.

Oxygen charge-transfer (CT) complexes are one of the primary initiating species in the photooxidation of polyethylene[23-25] and polystyrene.[26-28] The photolysis of the CT complexes can result in the formation of two reactive forms of molecular oxygen: singlet oxygen $^1O_2(^1\Delta_g)$ and oxygen radical ion O_2^-.[29-31] The singlet oxygen ($^1\Delta_g$) may contribute to the oxidative degradation of many polymers, especially polydienes,[32-43] through the 'ene' reaction during which $^1O_2$ reacts,[31-44] shifting the location of a double bond and forming a hydroperoxide (equations 4 and 5; cf. Section 12.2.1).

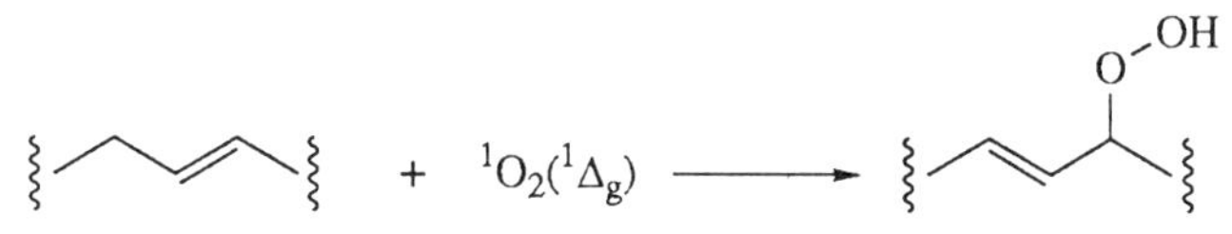

$$\{\sim\!\!\diagup\!\!\diagdown\!\!\sim\} + {}^1O_2(^1\Delta_g) \longrightarrow \{\sim\!\!\diagup\!\!\diagdown\!\!\overset{\overset{\displaystyle O\text{-OH}}{|}}{\sim}\} \quad (4)$$

$$\{\!\!-\!\!\diagup\!\!\diagdown + {}^1O_2({}^1\Delta_g) \longrightarrow \{\!\!-\!\!\diagup\!\!\diagdown\!\!\diagup\!\!-\!O\!-\!OH \quad (5)$$

12.2.3 Chain Propagation

In the first step of a propagation reaction, a polymer alkyl radical (P·) is formed and reacts immediately with an oxygen molecule to give a polymer peroxy radical ($PO_2^{\cdot}$, equation 6).

$$P\cdot + O_2 \longrightarrow PO_2^{\cdot} \quad (6)$$

The peroxy radical can easily abstract a hydrogen atom from the same or a neighboring macromolecule, yielding a polymer hydroperoxy group (POOH) and a polymer alkyl radical (P·, equation 7).

$$PO_2^{\cdot} + PH \longrightarrow POOH + P\cdot \quad (7)$$

A similar reaction may occur also with a hydroperoxy radical ($HO_2^{\cdot}$, equation 8).

$$HO_2^{\cdot} + PH \longrightarrow HOOH + P\cdot \quad (8)$$

However, both radicals $PO_2^{\cdot}$ and $HO_2^{\cdot}$ may carry an excess energy after absorption of UV radiation and participate in a number of reactions giving P·, H· and $HO_2^{\cdot}$ radicals and chain scission by disproportionation (equation 9).

$$PO_2^{\cdot} \xrightarrow{h\nu} \{\!\!-\!\!C(\!=\!O)\!-\!H + \diagup\!\!\diagdown\!\!-\}\ + HO\cdot \quad (9)$$

The secondary $(POOH)_s$ and tertiary $(POOH)_t$ polymer hydroperoxy groups are easily photolyzed to form carbonyl and vinyl groups and *trans*-vinylene bonds (equations 10–12).[25,45–47]

$$(POOH)_s + PH \xrightarrow{h\nu} P=O + H_2O + PH \quad (10)$$

$$(POOH)_t \xrightarrow{h\nu} P^1\!-\!C(\!=\!O)\!-\!R + =\!\!\diagdown_{P^2} + H_2O \quad (11)$$

$$(POOH)_t + P^1\!\!-\!\!\diagup\!\!\diagdown_{P^2} \xrightarrow{h\nu} POH + H_2O + P^1\!\!\diagdown\!\!=\!\!\diagup_{P^2} \quad (12)$$

However, the majority of the free hydroperoxides yield ketones and not *trans*-vinylene groups.

The polymer carbonyl groups (P=O) can also be further photolyzed according to two reactions (chain scissions): the Norrish type I reaction (equation 13) and the Norrish type II reaction (equation 14).

$$P=O \xrightarrow{h\nu} P^1\!-\!\overset{\cdot}{C}\!=\!O + \cdot CH_2P^2 \longrightarrow P^1\!\cdot + CO + \cdot CH_2P^2 \quad (13)$$

$$P=O \xrightarrow{h\nu} P^1\!-\!C(\!=\!O)\!-\!\text{H} + =\!\!\diagdown_{P^2} \quad \text{(nonradical rearrangement)} \quad (14)$$

The step regulating the rate of photooxidation of polyalkenes involves polymer peroxy radicals ($PO_2^{\cdot}$) competing between hydrogen abstraction and disproportionation of two polymer peroxy radicals (equations 7 and 15).[45–50] The former reaction accelerates degradation and the latter retards degradation.

$$\{\!\!-\!\!C(\!-\!O\!-\!O\cdot)\!\!-\!\! + PO_2^{\cdot} \longrightarrow \{\!\!-\!\!C(\!=\!O)\!\!-\!\! + POH + O_2 \quad (15)$$

12.2.4 Chain Termination

The termination reaction of the radical chain is where free radicals react with each other to give an inactive product (Scheme 2).

$$P\cdot + P\cdot \longrightarrow P-P$$
$$PO_2^\cdot + P\cdot \longrightarrow POOP$$
$$PO_2^\cdot + PO_2^\cdot \longrightarrow POOP + O_2$$
$$PO_2^\cdot + PO_2^\cdot \longrightarrow P=O + POH + O_2$$

Scheme 2

In the presence of sufficient amounts of oxygen, only the last two reactions in Scheme 2 are significant. However, in the presence of hydroperoxy radicals ($HO_2^\cdot$) formed in the photoreaction of polymer peroxy radicals ($PO_2^\cdot$), two additional termination reactions may occur. The first involves the interaction of two hydroperoxy radicals, according to equation (16).

$$HO_2^\cdot + HO_2^\cdot \longrightarrow HOOH + O_2 \tag{16}$$

The second termination reaction is between a hydroperoxy radical ($HO_2^\cdot$) and a polymer peroxy radical ($PO_2^\cdot$, equations 17 and 18).

$$PO_2^\cdot + HO_2^\cdot \longrightarrow POOH + O_2 \tag{17}$$
$$PO_2^\cdot + HO_2^\cdot \longrightarrow P=O + H_2O + O_2 \tag{18}$$

Secondary peroxy radicals ($PO_2^\cdot$) such as those encountered mainly in polyethylene react with each other in a bimolecular reaction to give a carbonyl group and a secondary alcoholic group. The interaction of tertiary peroxy radicals in polypropylene is nonterminating. In this case the termination involves at least one secondary or primary peroxy radical. However, in solid polymers, the probability of encounter of two polymer radicals is very low if they are not already close from the beginning (cage reaction). In a termination reaction the migration of free valency and/or low molecular weight radicals plays an important role.

12.3 PHOTODEGRADATION OF IMPORTANT POLYMER MATERIALS

12.3.1 Molecular Mechanisms

The detailed mechanism and kinetics of the photooxidative degradation of polymers depend on the following factors: (i) the chemical (presence of strongly light-absorbing chromophoric groups) and morphological (amorphous or semicrystalline states) structures of the polymer sample;[51,52] (ii) processing and storage conditions;[53,54] (iii) if the reaction is carried out in solution (type of solvent used)[55,56] or in a polymer solid state; (iv) thickness of the polymer sample (film, sheet or solid) — most photooxidative degradation reactions occur on the surface or close to the surface; (v) spectrum (monochromatic or polychromatic) and intensity (low or high) of UV/visible radiation;[47,57,58] (vi) concentration and pressure of air (oxygen) at the surface and in the polymer matrix;[26,28,59,60] (vii) temperature (photothermal degradation);[26,28,61-64] (viii) presence of impurities (photoinitiating species and/or inhibiting (terminating) species);[65,66] (ix) presence of crosslinked structures;[67-69] and (x) elongation and mechanical stress.[70,71]

Commercial polymers may contain a number of internal and/or external impurities which can be formed during the following reaction steps:[39,40] (i) polymer manufacture (unsaturation, catalyst residues, hydroperoxides and carbonyl groups by adventitious oxidation); (ii) processing and fabrication (hydroperoxides and carbonyl groups formed under high temperature oxidation, transition metal ions from machinery or compounding ingredients); and (iii) environmental exposure (polycyclic hydrocarbons from atmospheric pollution, carbonyl groups by photolysis of hydroperoxides, unsaturation by photolysis of ketones, transition metal ions).

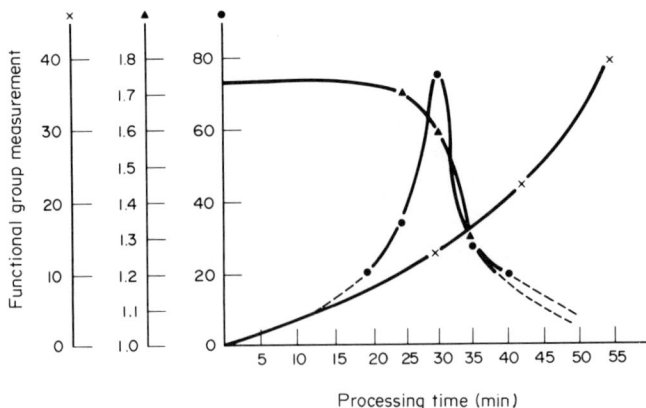

Figure 2 Change in functional group concentrations during processing of low density polyethylene: (▲) vinylidiene content, (x) carbonyl content and (●) hydroperoxide content (g mol^{-1} × 10^6) (reproduced from ref. 53 by permission of the American Chemical Society)

The processing time has an important role in the formation and disappearance of vinylidene, hydroperoxide and carbonyl groups in low density polyethylene (LDPE; Figure 2).

Ketonic groups increase in the autoaccelerating mode of oxidation and vinylidene groups decay rapidly after an induction period. The vinylidene decrease is associated with the rapid formation and rapid decay of hydroperoxide groups.

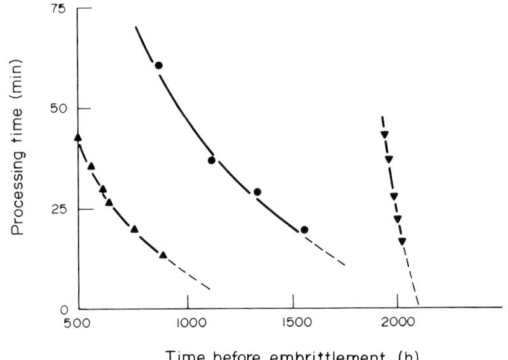

Figure 3 Effect of melt processing in air upon the time before embrittlement of: (▲) low density polyethylene, (▼) high density polyethylene, and (●) low density polyethylene containing Fe(acac)$_3$ (1 × 10^{-4} mol 100 g^{-1}) (reproduced from ref. 53 by permission of the American Chemical Society)

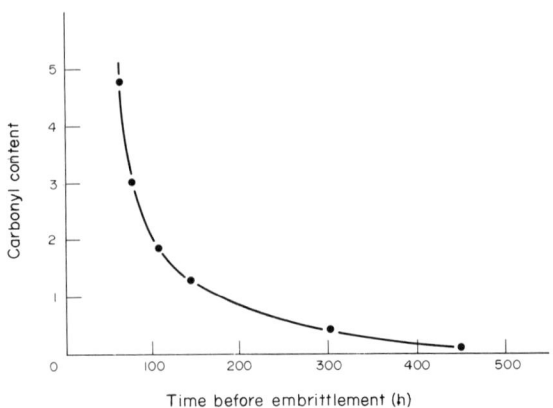

Figure 4 Relationship between carbonyl content and time before embrittlement for high density polyethylene (reproduced from ref. 53 by permission of the American Chemical Society)

Processing time has an effect on the time to embrittlement (Figure 3), which for polyalkenes during photooxidation is not directly related to the concentration of carbonyl groups formed (Figure 4).

This finding is based on the assumption that carbonyl groups are not responsible for the initiation of the photodegradation of polyalkenes. Allylic hydroperoxides, formed from thermal oxidation, and vinylidene groups are the most important photoinitiating groups initially present in thermally treated polyalkenes. The thermal and/or photochemical decomposition of hydroperoxy groups leads to: (i) chain scission reactions during which end aldehyde groups are formed (equation 19); (ii) a crosslinking reaction, which occurs only during UV irradiation, but not in thermal oxidation (equation 20), closely associated with changes in mechanical properties during the early stages of photooxidation; and (iii) carbonyl group formation (equation 21).

The rates of photochemical (and/or thermal) oxidation of polymer films are usually measured by monitoring the hydroxy (OH/OOH) group formation (I_{OOH}) at 3400–3450 cm^{-1} and/or carbonyl (CO) group formation (I_{CO}) at 1710–1720 cm^{-1} using the following expressions

$$I_{OOH} = \frac{\text{Absorbance at } 3400\text{–}3450 \text{ cm}^{-1}}{\text{Absorbance of reference peak (cm}^{-1})} \qquad (22)$$

$$I_{CO} = \frac{\text{Absorbance at } 1710\text{–}1720 \text{ cm}^{-1}}{\text{Absorbance of reference peak (cm}^{-1})} \qquad (23)$$

The absorbance of a reference peak, *e.g.* CH$_2$ compensates for changes in film thickness and should not change during oxidation time.

The carbonyl index (measured at 1710–1720 cm^{-1}) is commonly used for the determination of embrittlement time (mainly in polyalkenes), *i.e.* time (in hours) to reach 0.06 carbonyl units, from the plot of absorbance of the carbonyl group *versus* irradiation time[22]

$$\text{Carbonyl index} = [(\log I_0/I_t)/d] \cdot 100 \qquad (24)$$

where I_0 and I_t are intensities of incident and transmitted light and d is the film thickness. By extrapolation of the steep portion of the curve to the time axis at zero carbonyl, the induction period is determined.

As a result of photooxidative degradation of polymers, two main processes are observed: (i) chain scission (decreases in molecular weight and shifts in molecular weight distribution curves; Figure 5); and (ii) crosslinking (formation of nonsoluble macrogel and brittle structures).

The number of main chain scissions (S) can be calculated from measurements of the number-

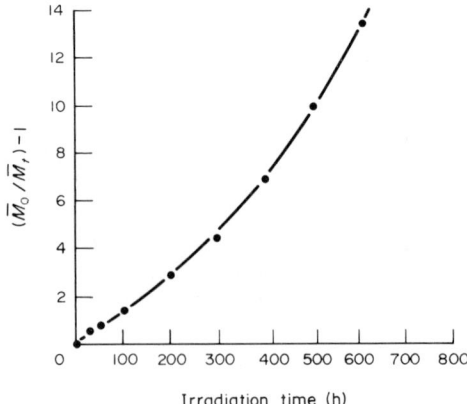

Figure 5 Rate of chain scission per polypropylene molecular chain as a function of radiation time (reproduced from J. H. Adam, *J. Polym. Sci., Polym. Chem. Ed.*, 1970, **8**, 1279 by permission of Wiley)

average molecular weight of a polymer before irradiation $(\bar{M})_0$ and after irradiation (t hours) $(\bar{M})_t$

$$S = \frac{(\bar{M})_0}{(\bar{M})_t} - 1 \tag{25}$$

The crosslinking number (X) can be determined from the amount of gel formed.

The quantitative relationship between the number of polymer molecules which are scissioned and the number of photons absorbed per unit time is given by the quantum yield of chain scission (ϕ_S)

$$\phi_S = N \frac{w_t}{w_0} \frac{[(\bar{M})_0/(\bar{M})_t - 1]}{I_a t} \tag{26}$$

where N is Avogadro's number; w_0 and w_t are the polymer weight before and after irradiation time (t time), respectively; and I_a is the intensity of light absorbed by the polymer given by

$$I_a = I_0(1 - 10^{-A}) \tag{27}$$

where I_0 is the incident irradiation intensity and A the polymer sample absorbance.

Table 3 Quantum Yield of Chain Scission (See Refs. 5 and 15).

Polymer	Wavelength (nm)	Quantum yield ϕ_{cs}
Poly(methyl acrylate)	254	1.3×10^{-2}
Poly(methyl methacrylate)	254	$1.7–3 \times 10^{-2}$
	254	3.2×10^{-2}
	254	2.3×10^{-3}
Poly(methyl vinyl ketone)	254	2.5×10^{-2}
	254	2.0×10^{-2}
Poly(methyl isopropyl ketone)	254	2.2×10^{-1}
Polystyrene	254	9×10^{-5}
Poly(α-methylstyrene)	254	1×10^{-3}
Poly(acrylonitrile)	254	$2–7.7 \times 10^{-4}$
Poly(vinylpyrrolidone)	254	4.3×10^{-4}
cis-1,4-Polyisoprene (natural rubber)	254	4×10^{-4}
Poly(ethylene terephthalate)	280–360	5.0×10^{-4}
Poly(tetramethylene sebacte-co-γ-ketopimelate)	250–300	4.0×10^{-3}
Cellulose	254	1.0×10^{-3}
	254	7×10^{-4}
Cellulose acetate	254	2.0×10^{-4}
	254–435	2.5×10^{-3}

Examples of the quantum yield of chain scission (ϕ_S) for various polymers at various wavelengths are given in Table 3.

12.3.2 Sensitized Photodegradation

In the initiation of photodegradation internal and/or external impurities (RH) play a special role. Under UV irradiation these compounds are photolyzed to low molecular weight free radicals R· which can initiate formation of polymer alkyl radicals (P·) by hydrogen extraction (Scheme 3).

$$RH \xrightarrow{h\nu} R\cdot + H\cdot$$
$$R\cdot + PH \longrightarrow P\cdot + RH$$

Scheme 3

Some of these compounds, called 'photoinitiators', are specially added to polymeric materials in order to cause their accelerated degradation *via* chain scission or crosslinking. Typical examples of such compounds are benzophenones,[64,72-87] quinones[88-90] and phenylacetophenone.[91]

Iron(III) chloride (FeCl$_3$) is also a well-known photoinitiator for the photodegradation of polymers.[92-94] For example, it accelerates photodehydrochlorination of poly(vinyl chloride) by the mechanism shown in Scheme 4.[84]

$$FeCl_3 \xrightarrow{h\nu} FeCl_2 + Cl\cdot$$

Scheme 4

Many inorganic and/or organic external impurities are introduced into polymers during processing (*e.g.* catalysts, inhibitors, terminators, solvents *etc.*), during storage (metal salts, metal oxides, environmental pollutants) and during processing. The extrusion, milling, chopping and compounding steps involved in polymer processing can all introduce traces of metal ions or even particles of metals, *e.g.* Fe, Ni and Cr.

Industrially produced polymers contain a number of light-absorbing internal impurities produced by side reactions during polymerization, processing and storage, such as hydroperoxides, ketones and unsaturated groups.

12.3.3 Polyalkenes and Diene Copolymers

Photooxidative degradation of polyalkenes is a very complex reaction. A photoreaction started in the presence of air may yield several polymer peroxy radicals (PO$_2^-$), which further generate a number of new secondary polymer alkyl radicals (P·) according to the reaction shown in equation (28).[47,95]

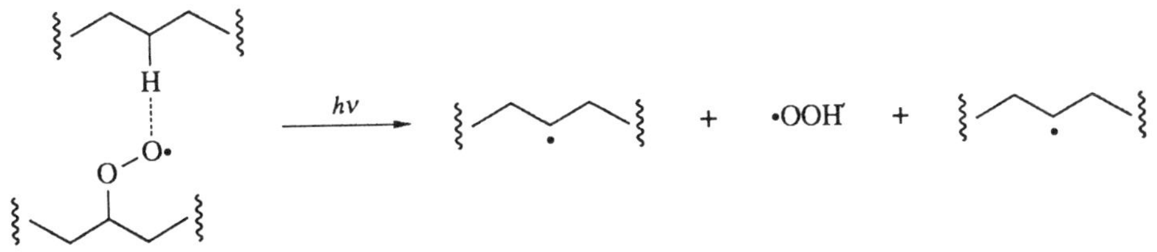

$$\tag{28}$$

The P· radicals may further participate in a number of complicated reactions, *e.g.* alkyl radical combination, peroxide group formation, and disproportionation of two alkoxyl radicals to an alcohol and a ketone group.

The main products of polyethylene photooxidation are carbonyl, alcohol and vinyl groups.[25,45-47,96-99] *Trans*-vinylene groups are also formed but in small amounts only. Photolysis of hydroperoxides is usually so fast that they do not accumulate. The carbonyl groups formed in polyethylene photooxidation are mainly of two types, ketones and carboxylic acids. Ketones are formed from:[98] (i) photolysis of secondary hydroperoxides (equation 29); (ii) interaction of secondary peroxy radicals (equation 30); and (iii) free radical oxidation of secondary alcohols (Scheme 5).

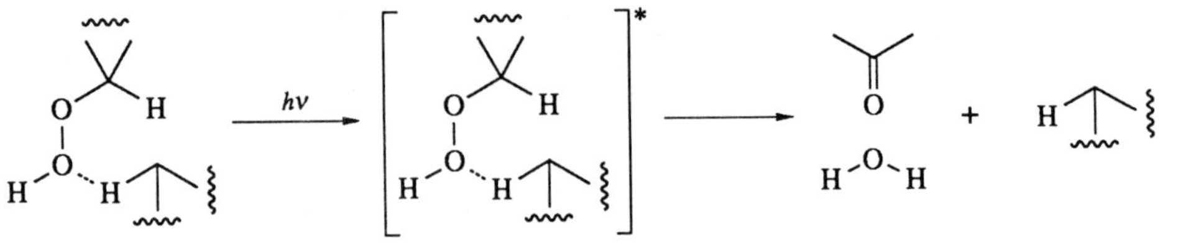

$$\text{(29)}$$

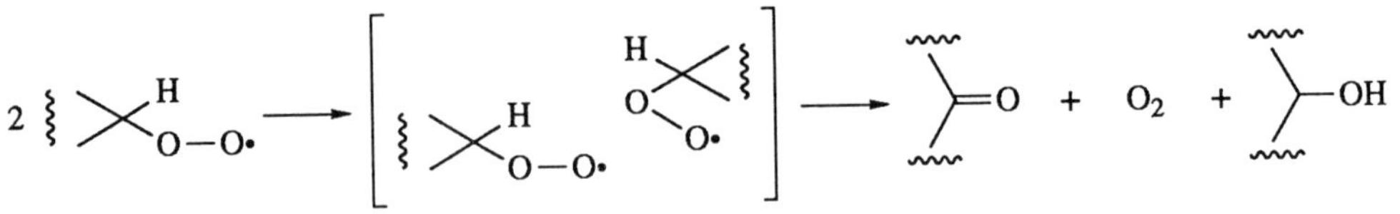

$$\text{(30)}$$

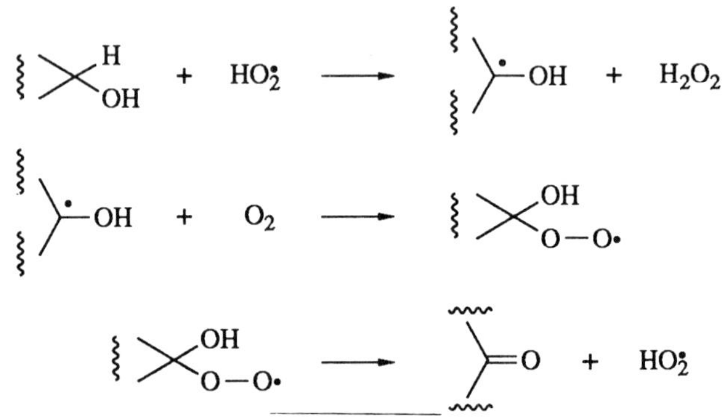

Scheme 5

The main source of carboxylic acid in polyethylene photooxidation is ketone photolysis according to the Norrish type I reaction (Scheme 6).[98]

However, carboxylic acid can also be formed in the reaction between ketones and hydroperoxides (equation 31).[48,49]

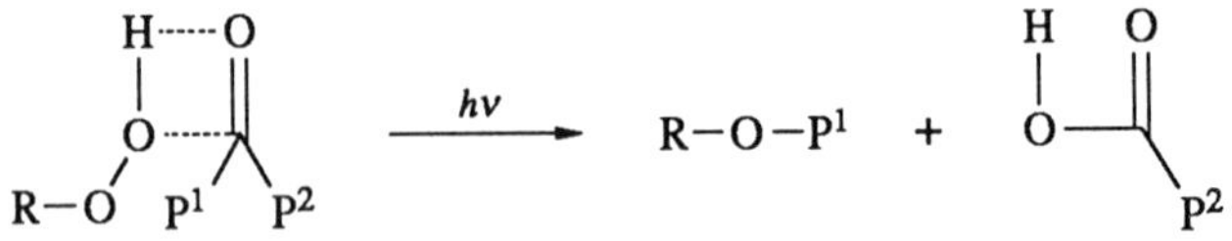

$$\text{(31)}$$

The vinyl group can be formed by:[98] (i) the Norrish type II reaction (equation 32); and (ii) intramolecular decomposition of tertiary hydroperoxides (equation 33).

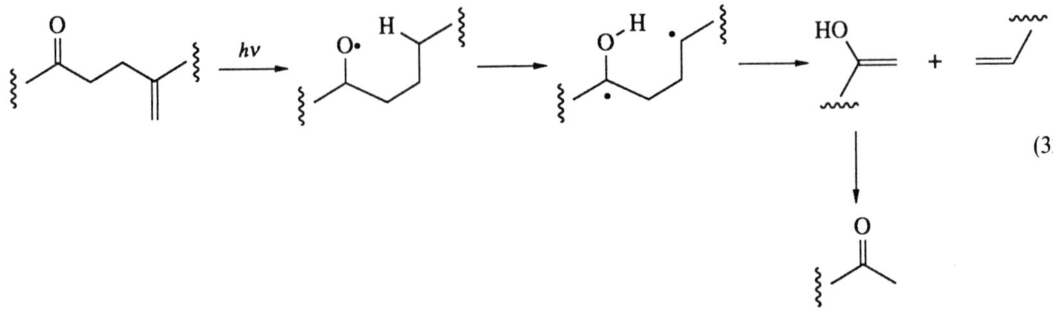

$$\text{(32)}$$

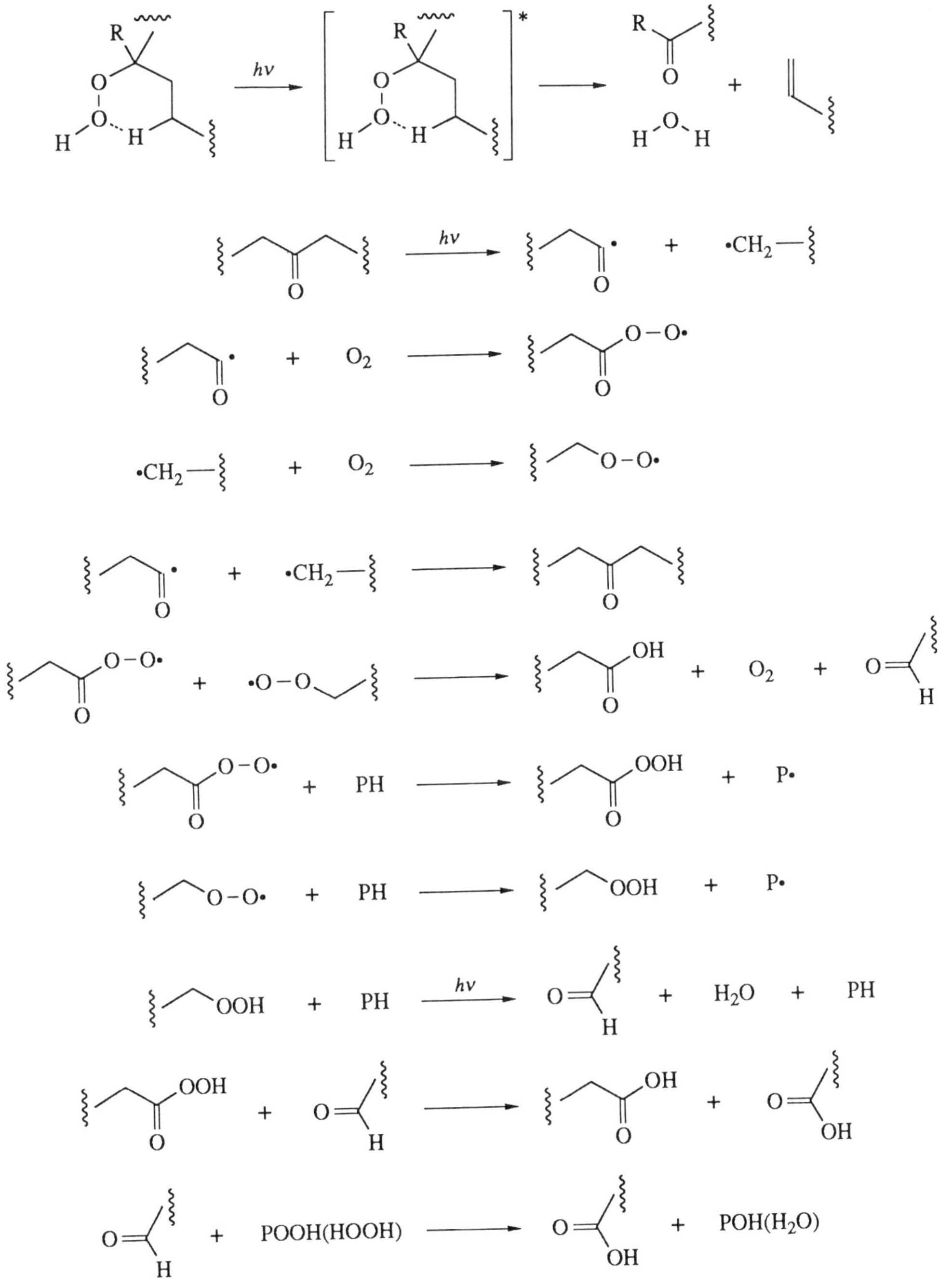

Scheme 6

Low density polyethylene chains (LDPE) are more easily oxidized than those of high density polyethylene (HDPE). In spite of this, the photostability of LDPE in bulk is higher than that of HDPE. This can be explained by the number of tie molecules between crystalline lamellae. Oxidation damage in the tie molecule region is important for determining the embrittlement time. In HDPE, due to its highly crystalline nature with few tie chains, a low amount of oxidation product formation causes great damage to the tie molecules.[52] The photostability of quenched LDPE is superior to annealed LDPE. Quenched polyethylene has crystallites smaller than those of annealed polyethylene. However, the degree of crystallinity does not depend on the method of film preparation. Consequently, quenched LDPE has larger amounts of tie molecules than annealed polymer and is able to attain higher degrees of oxidation without great damage to the tie molecules.

Photodegradation of polypropylene is more complicated than that of polyethylene because the initial peroxy radicals will be mainly tertiary. However, the second step radicals formed in the photoreaction may be tertiary, secondary or primary. The following reactions of tertiary with primary or secondary peroxy radicals can be expected.[42] Photoreaction with a primary hydrogen atom, e.g. equation (34) may be followed by: (i) interaction between two tertiary peroxy radicals (equation 35); (ii) interaction of a tertiary with a secondary peroxy radical (equation 36); and (iii) interaction of a tertiary with a primary peroxy radical (equation 37).

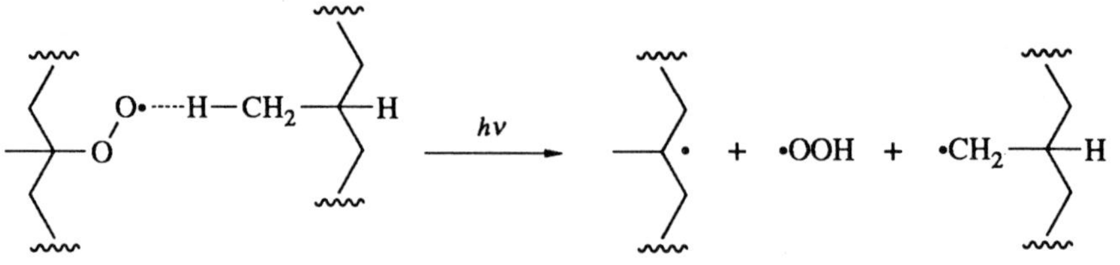

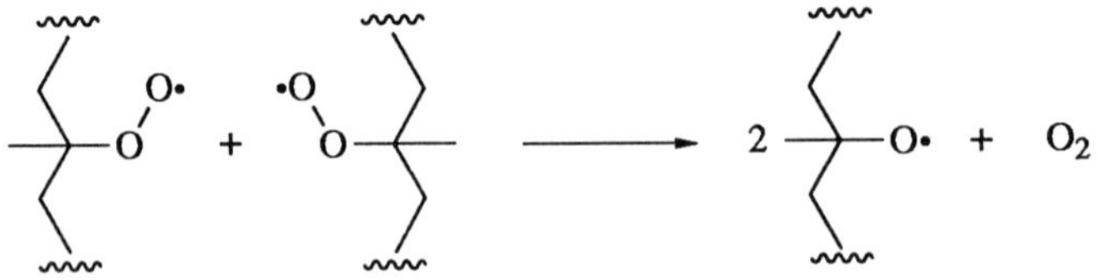

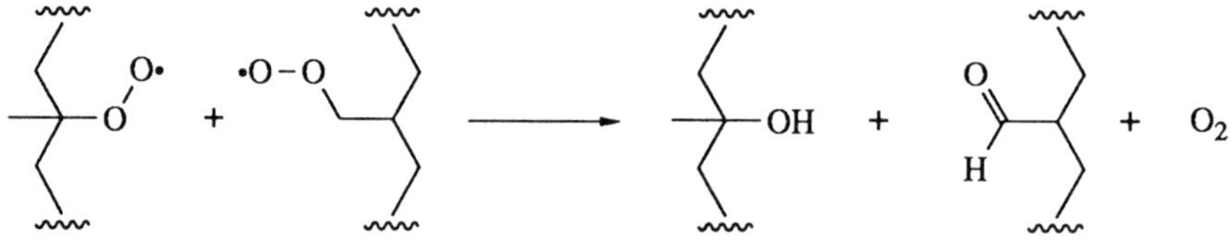

In isotactic polypropylene (at 25 °C) the half-life for photolysis of polypropylene peroxy radicals is less than 2400 s for a film exposed in the noon summer sunlight.[47]

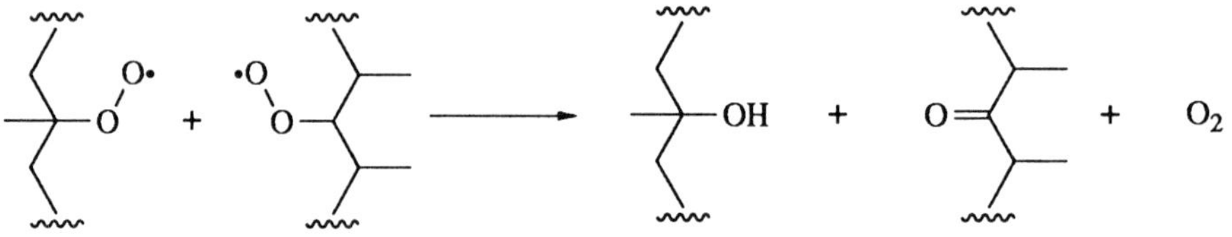

Photooxidative processes in polyalkenes occur mainly in the surface layer, which reaches a thickness of a few hundred μm up to 1 mm. This reaction, however, has dramatic effects on the mechanical properties of the polymer material and often causes transition from ductile failure to brittle failure. This decreases the residual strength considerably. Fracture occurs from surface cracks, which form easily in the degraded layer. The residual strength of a cracked polyalkenic material will depend on the progress of photooxidative degradation.

Ethylene–propylene (EP) copolymers are amorphous. During UV irradiation photodegradation processes are not complicated by semicrystalline structures as in the case of polyethylene and polypropylene.[48-50,100] The photooxidation of ethylene–propylene copolymers occurs by mechanisms in which: (i) Norrish type II cleavage of keto groups, which exist to some extent in EP, is the initial chain scission process; (ii) the photolysis of hydroperoxides does not contribute to main chain scission as a primary reaction; and (iii) as the oxidation proceeds to yield a high local concentration of hydroperoxides and keto groups, the main source of chain scission is the formation of carboxylic acids by photolysis of a ketone–hydroperoxide complex (equation 38).

$$\text{POOH} + \text{P}\overset{\text{O}}{\overset{\|}{\text{C}}}\text{P} \longrightarrow \text{P}\overset{\text{O}}{\overset{\|}{\text{C}}}\text{OH} + \text{POP} \tag{38}$$

Important studies on the photodegradation of elastomeric materials such as EPDM rubber terpolymers with the following structures have been made: (i) ethylene–propylene–1,4-phenylhexadiene (1);[83,100–104] main chain scission reactions occur with participation of hydroperoxides formed during photooxidation; and (ii) ethylene–propylene–ethylidenenorbornene (2).[83,105,106]

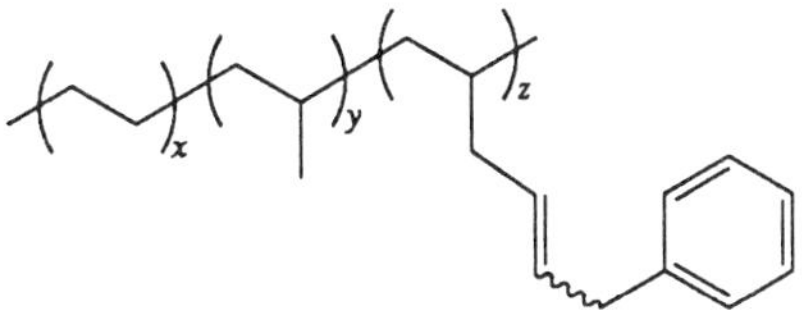

(1)

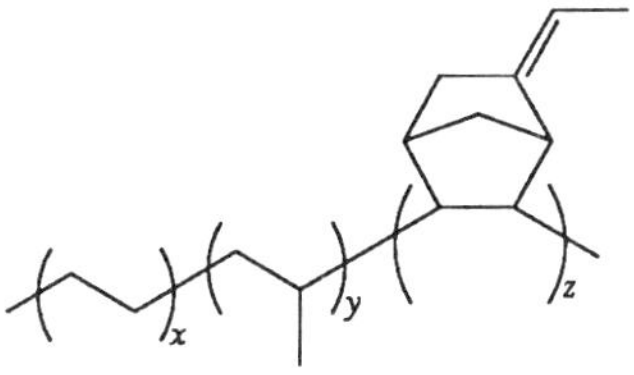

(2)

The specific photooxidation chemistry of the ethylidenenorbornene (ENB) group in EPDM is still not well understood. EPDM containing ENB is significantly more resistant to oxidation than EPDM containing dicyclopentadiene or methylendomethylenehexahydronaphthalene.[107] In the photooxidation of ENB–EPDM two photooxidation phases can proceed:[106] (i) the concentration of ethylidene remains constant, while both free and hydrogen-bonded hydroperoxides accumulate (equation 39); and (ii) the free hydroperoxide concentration reaches a stationary level and the hydrogen-bonded hydroperoxides continue to increase and the ethylidene sites disappear. Ethylidene disappearance is attributed to an addition reaction with radicals from the decomposition of hydrogen-bonded hydroperoxides.

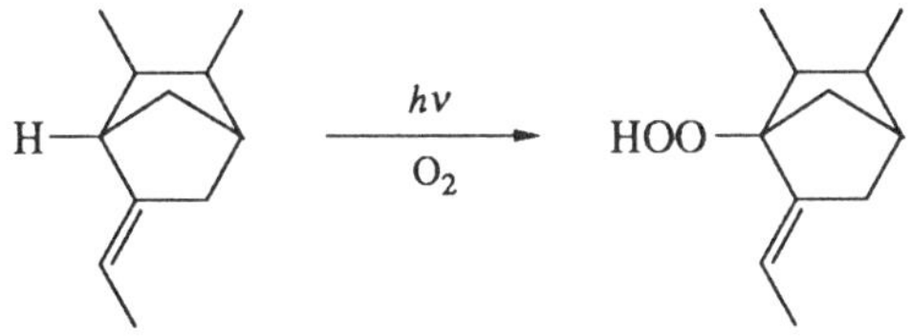

$$\text{H–} \xrightarrow[O_2]{h\nu} \text{HOO–} \tag{39}$$

12.3.4 Poly(vinyl chloride)

Poly(vinyl chloride) (PVC) is especially susceptible to photooxidative degradation.[54,62,67,108–121] Commercial PVC samples contain some light-absorbing chromophores like α-chlorinated short conjugated polyenes which are responsible for the initiation step (equation 40).

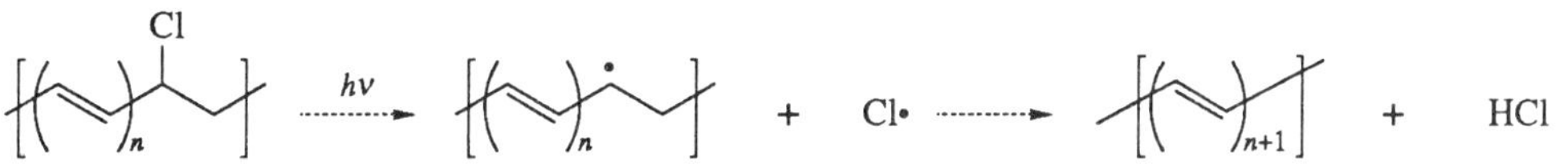

$$\cdots \xrightarrow{h\nu} \cdots + \text{Cl}\cdot \longrightarrow \cdots + \text{HCl} \tag{40}$$

The photodehydrochlorination of PVC involves a multistep excitation of polyene sequences. Each excitation leads to an increase of the length of conjugation ($n = 2\text{--}14$) with gradually increased absorption of visible light. Hydrogen abstraction by chlorine radical (Cl·) leads to the formation of two different types of polymer alkyl radicals, which react with molecular oxygen to give polymer peroxy radicals (equations 41 and 42).

Hydroperoxides are then formed by abstraction of hydrogen from the same or a neighboring macromolecule (PH) (equations 43 and 44).

$$\text{~CHCl-CH•-CHCl~} + O_2 \longrightarrow \text{~CHCl-CH(OO•)-CHCl~} \quad (41)$$

$$\text{~CH}_2\text{-CCl•-CH}_2\text{~} + O_2 \longrightarrow \text{~CH}_2\text{-CCl(OO•)-CH}_2\text{~} \quad (42)$$

$$\text{~CHCl-CH(OO•)-CHCl~} + PH \longrightarrow \text{~CHCl-CH(OOH)-CHCl~} + P• \quad (43)$$

$$\text{~CH}_2\text{-CCl(OO•)-CH}_2\text{~} + PH \longrightarrow \text{~CH}_2\text{-CCl(OOH)-CH}_2\text{~} + P• \quad (44)$$

Homolysis of the O—O bond of these hydroperoxides occurs under photolytic or thermolytic conditions (equations 45 and 46).

$$\text{~CHCl-CH(OOH)-CHCl~} \longrightarrow \text{~CHCl-CH(O•)-CHCl~} + HO• \quad (45)$$

$$\text{~CH}_2\text{-CCl(OOH)-CH}_2\text{~} \longrightarrow \text{~CH}_2\text{-CCl(O•)-CH}_2\text{~} + HO• \quad (46)$$

Finally, polymer oxy radicals (PO•) yield different ketonic groups (equations 47 and 48), which can easily be detected by IR spectroscopy.[116] An increase in temperature accelerates photodehydrochlorination of PVC.[115,122] After prolonged UV exposure the photodegraded PVC samples become yellow-brown or even black, are very brittle and lose their mechanical properties.

$$\text{~CHCl-CH(O•)-CHCl~} + HO• \longrightarrow \text{~CHCl-CO-CHCl~} + H_2O \quad (47)$$

$$\text{~CH}_2\text{-CCl(O•)-CH}_2\text{~} \longrightarrow \text{~CH}_2\text{-C(=O)Cl} + •CH_2\text{~} \quad (48)$$

The mechanism of photooxidative degradation of PVC is much more complicated in the presence of additives such as thermostabilizers,[123] plasticizers[54,124,125] and pigments.[126]

UV laser irradiation of chlorinated PVC causes rapid photodehydrochlorination to a pure carbon polymer with graphite-like structure (Scheme 7).[127,128]

Photooxidative degradation of vinyl chloride–vinylidene chloride copolymer[129] and poly(vinyl chloride) polymer blends[130] has also been investigated and shows similar dehydrochlorination reactions.

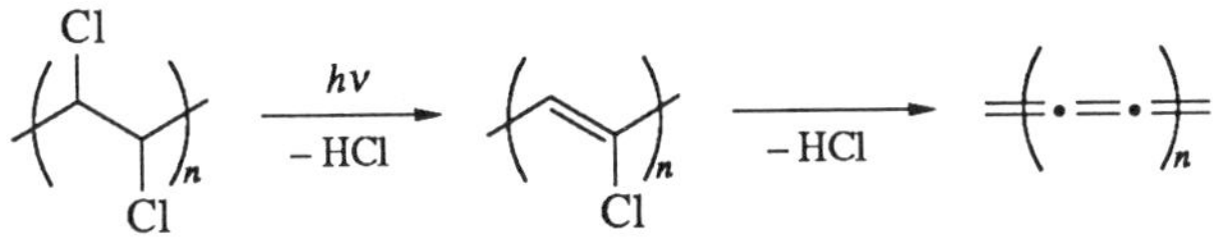

Degradation of Polymers 269

Scheme 7: PVC → chlorinated polyenes → graphite-like structure, with loss of HCl

Scheme 7

12.3.5 Polydienes

Photooxidative degradation of polydienes depends very much on their structure. The unsaturation present in these polymers can be combinations of three structural units (3), (4) and (5).

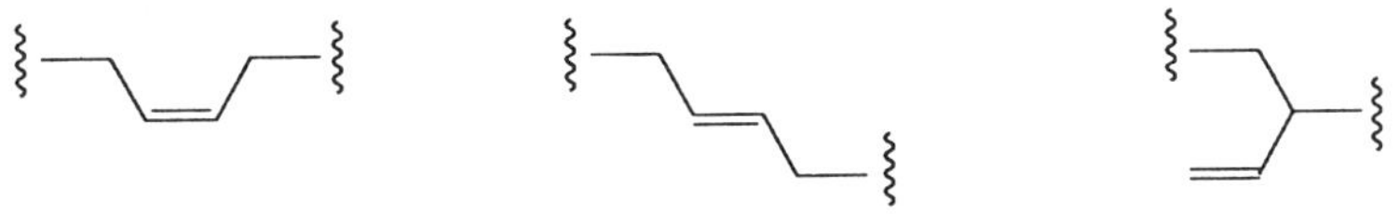

Structures of (3) cis-1,4-unit, (4) trans-1,4-unit, (5) 1,2-vinyl unit

(3) *cis*-1,4-unit   (4) *trans*-1,4-unit   (5) 1,2-vinyl unit

Photooxidation of polydienes having structure 1,4 may occur with participation of molecular oxygen $^3O_2(^3\Sigma_g^+)$ and/or with singlet oxygen $^1O_2(^1\Delta_g)$.[13,15,31,53] Polydienes are also susceptible to thermal oxidation during which hydroperoxy groups are formed. Present data available indicate that the mechanisms for initial photochemical and thermal oxidation of polydienes are almost identical.

The *cis*- and *trans*-1,4 units can be directly attacked by molecular oxygen in the presence of radicals produced photochemically, *e.g.* from hydroperoxide photolysis, which attack allylic CH_2 groups (equation 49).

Equation 49: allylic hydrogen abstraction by R• (49)

Further photooxidative reactions include the formation of hydroperoxide groups (OOH) and α- and β-unsaturated carbonyl and aldehyde groups followed by chain scission reactions,[131-135] analogous with previously described mechanisms.

Chain scissions are followed by crosslinking reactions in which all types of free macroradicals formed may participate.

1,2-Vinyl units are photooxidized by the following mechanisms shown in Scheme 8.[131,136,137]

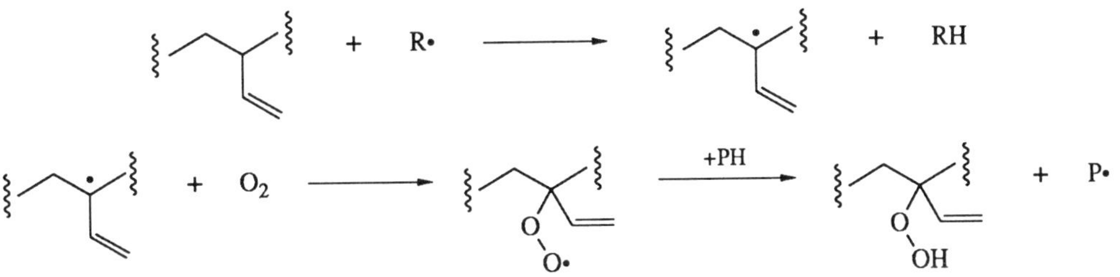

Scheme 8: photooxidation mechanism of 1,2-vinyl units

Scheme 8

Butadiene poly blends such as ABS (acrylonitrile–butadiene–styrene rubber) and copolymers such as SBR (styrene–butadiene rubber) are photodegraded by the same mechanisms as the pure polymers described earlier.[138,139]

12.3.6 Polymers Containing Carbonyl Groups

Polymers which contain carbonyl groups in the main chain, *e.g.* poly(ethylene-*co*-carbon monoxide) are photodegraded according to the Norrish type I and/or Norrish type II reactions which are competing (Scheme 9).[70,140-142]

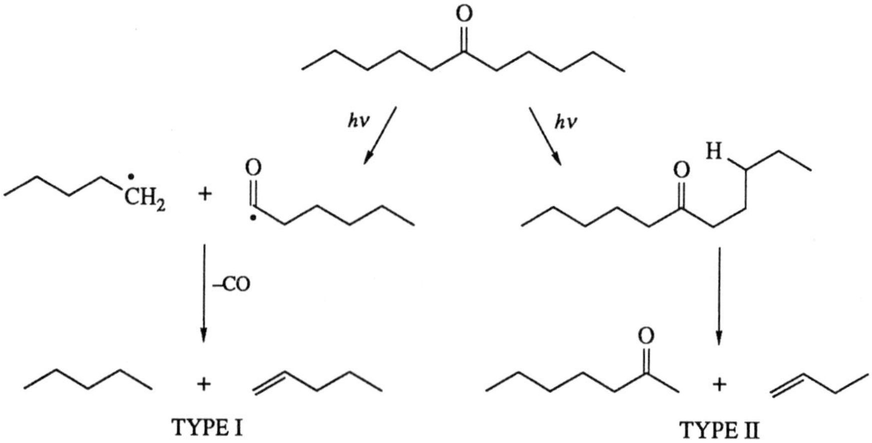

Scheme 9

Type II scission is strongly dependent on conformation and is sensitive to restrictions in motion of the polymer chain. Tension on the molecular chains may either inhibit or promote reaction rates. For example, the type II reaction rate is about 8 times faster in undrawn film of poly(ethylene-*co*-carbon monoxide) but the type I rate is at least 3 times faster in film cold drawn 400%.[70] Constraints on intercrystalline tie chains suppress formation of the cyclic intermediate in the nonradical type II process and reduce the probability of in-cage recombination of radical chain ends formed in the type I reaction. Similar effects on the reactions may be caused by thermal stress. An unsaturated side group like in poly(1,2-butadiene) may be split off by bond scission. (Scheme 10).

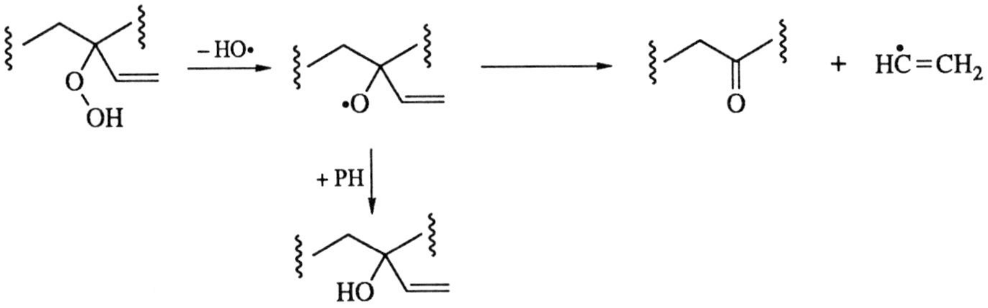

Scheme 10

Polymers which contain carbonyls in side groups, *e.g.* poly(vinylacetophenone), are photodegraded according to the Norrish type I reaction (Scheme 11).[143–147]

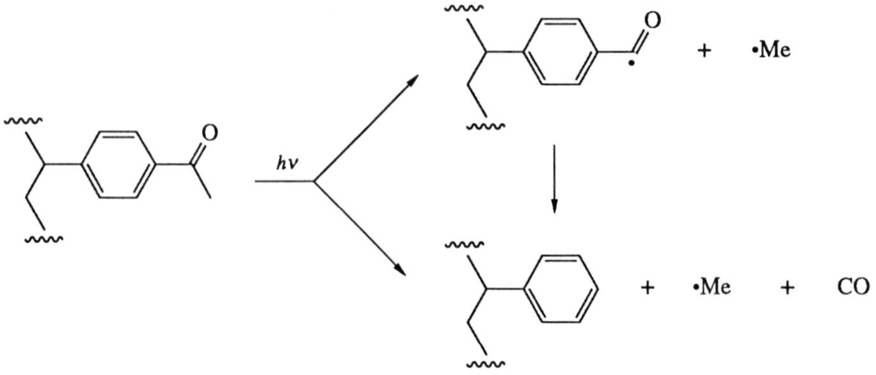

Scheme 11

In a similar way poly(*p*-propionylstyrene) is photodegraded.[148–150]

The photodegradation of polymers based on aromatic ketones, *e.g.* poly(phenyl vinyl ketone), occurs primarily *via* the Norrish type II reaction (equation 50).[151–154]

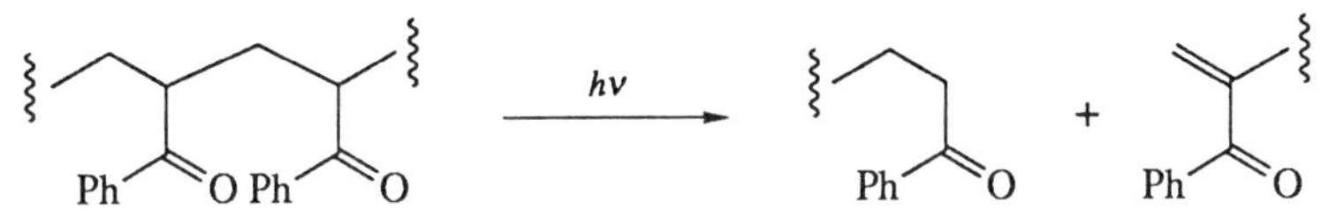

12.3.7 Polyacrylates

Photodegradation of polyacrylates, polymethacrylates and their copolymers occurs by the same mechanism for the two types of polymer:[58,61,63,155-159] (i) direct main chain scission with unzipping (monomer formation) of the chains (Scheme 12); and (ii) main chain scission following side chain scission ('deep' photolysis) (Scheme 13).

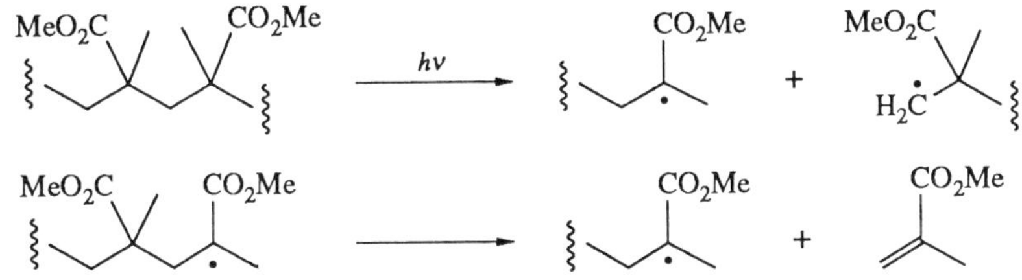

Scheme 12

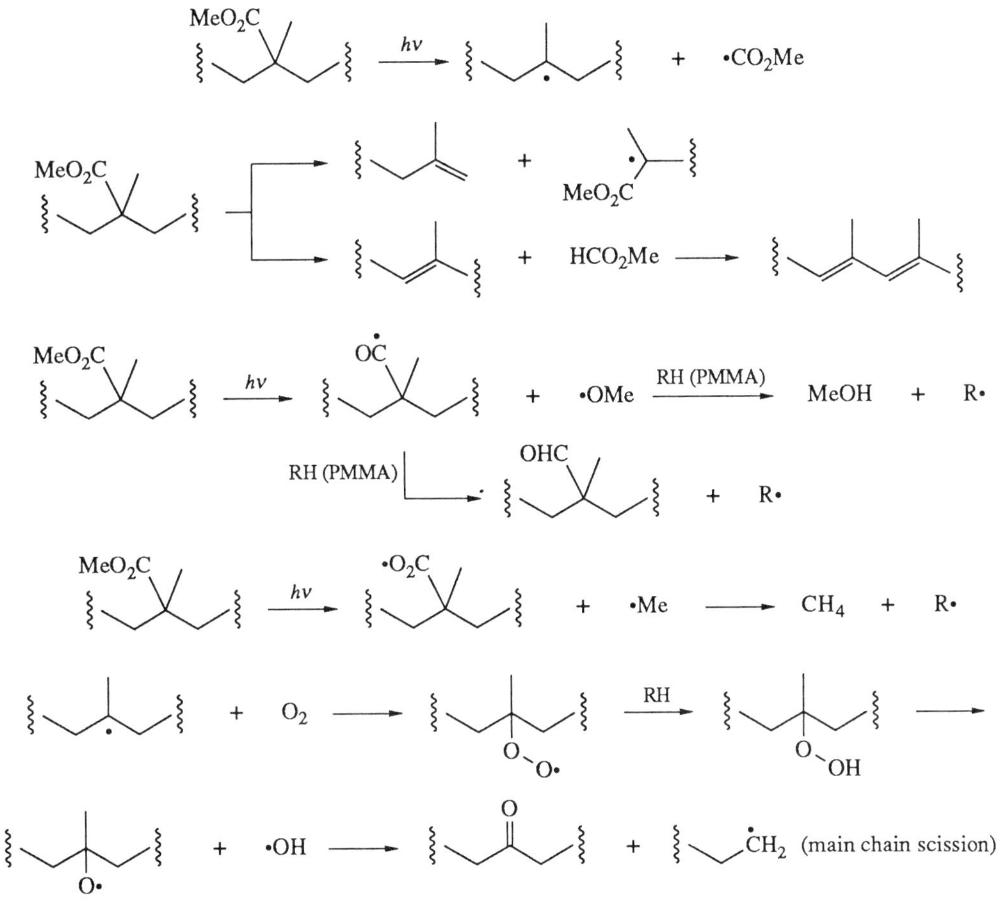

Scheme 13

The main products of deep photolysis of polymethacrylates are low molecular weight compounds such as methane, alcohols, aldehydes and acids.[58]

12.3.8 Polystyrene

The most generally favored polystyrene photooxidation mechanism is that the initial stages involve photolysis of in-chain peroxidic groups and hydroperoxides which have been incorporated into the polymer by inadvertent copolymerization of oxygen.[75,160-163] Photolysis of these species leads to the formation of polymer alkoxyl radicals (PO·) which then abstract benzylic hydrogen from the polymer (PH) to form macroradicals. The benzylic radicals react with oxygen and form acetophenone-type groups which are relatively strong UV-absorbing chromophores and thus capable of propagating the polystyrene photooxidation (Scheme 14).[64,164-168]

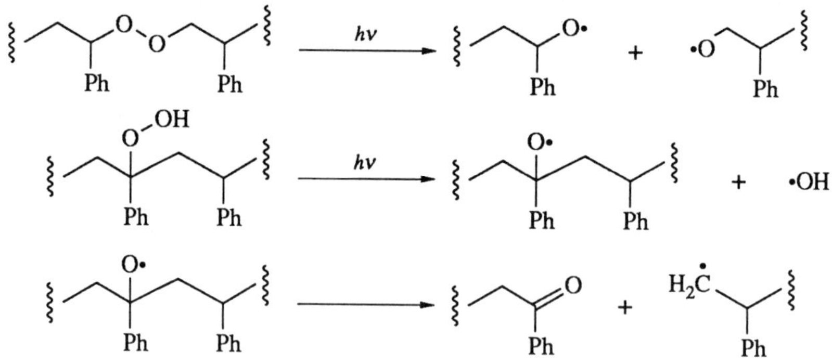

Scheme 14

The strong yellowing of polystyrene during UV irradiation can also be due to ring opening reactions with formation of muconaldehyde-type structures (equation 51).[30]

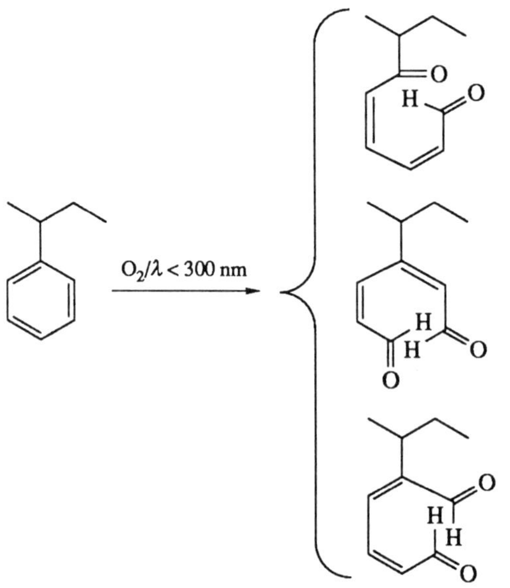

(51)

These proposed reactions are experimentally proved by IR–FTIR, NMR and ESCA spectroscopy.[169-171]

A number of studies have been made of photooxidative degradation of polystyrene analogs, e.g. poly(p-hydroxystyrene),[172,173] poly(β-nitroxystyrene),[174] poly(formylaminostyrene)[175] and styrene copolymers, e.g. styrene-co-maleic anhydride[176] and styrene–butadiene–acrylonitrile.[138,139,177-179]

12.3.9 Polyamides, Polyesters and Polycarbonates

Photooxidative degradation of polyamides proceeds with a chain scission, crosslinking and yellowing.[71,180-195] The dominant chain scission occurs at the amide linkage (Scheme 15).

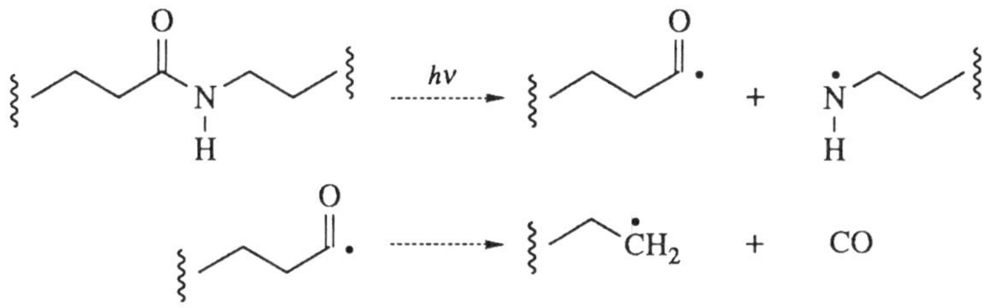

Scheme 15

The polymer alkyl end radical (P·) and polymer amino end radical (NH·) may abstract hydrogen from the same or an adjacent polymer macromolecule to form new polymer alkyl radicals (Scheme 16).

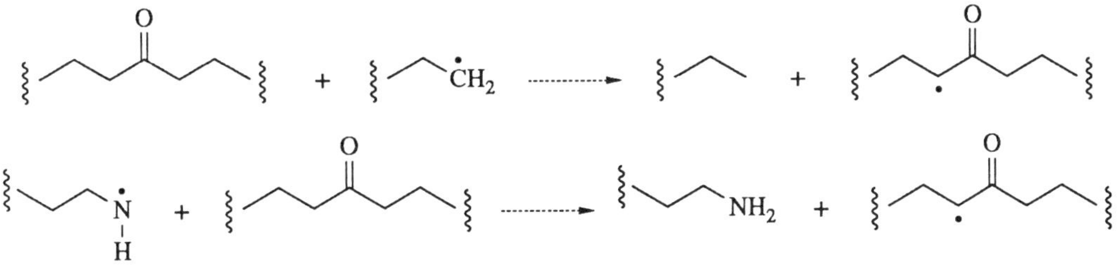

Scheme 16

These radicals add oxygen and form polymer alkylperoxy (POO·) radicals and consequently polymer hydroxyperoxide groups (POOH), which are thermally and/or photochemically decomposed to polymer alkoxyl radicals (PO·) which can be disproportionated by β-chain scission.

A special kind of reaction in some aromatic polymers is the photo-Fries rearrangement which involves a shift of an acyl from para or meta to ortho positions. Free ortho positions are therefore the requirement for the photo-Fries rearrangement which is a common reaction in aromatic polycarbonates and aromatic polyesters. For example, the photochemical yellowing of poly(4,4'-diphenol propane isophthalate) is attributed to photo-Fries rearrangement resulting in o-hydroxybenzophenone as part of the chain structure. The rate of photolysis falls off at higher exposure times because of the self-screening effect due to UV light absorption of the rearranged products.[22] A thin coating of phenyl polyesters can protect the substrate ordinarily sensitive to UV radiation. Such a polymer 'skin' containing o-hydroxybenzophenone formed in situ during the irradiation protects both the original polyester coating and the coated substrate from degradation by UV radiation.

Photodegradation of polyamides (aromatic polyamides), i.e. Kevlar fabric and fibers, in the absense of oxygen occurs mainly by the photo-Fries rearrangement, during which 1-amino- and 4-amino-benzophenone groups are formed (equation 52).[186]

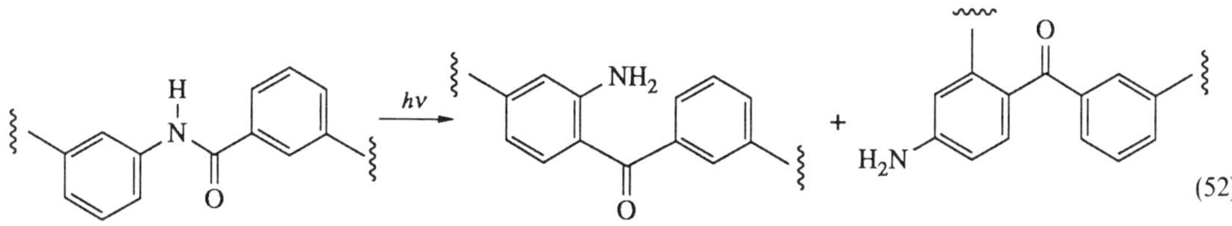

(52)

The photo-Fries reaction is decreased by the presence of oxygen, and instead of aminobenzophenone products, polymer phenylnitroso groups and polymer phenylhydroperoxy groups are formed (equation 53).

These reactions cause fast losses of mechanical properties of polyamides due to chain scission and crosslinking reactions.

Bisphenol A polycarbonate (PC; equation 54) exposed to UV irradiation becomes yellow and shows evidence of crosslinking, pitting and cracking. It loses desirable properties such as transparency, tensile strength, impact resistance and rigidity. These processes are caused by the following mechanisms:[187-196] (i) photo-Fries reactions (which occur only by irradiation below 280 nm) leading to phenylsalicylate and dihydroxybenzophenone; (ii) chain scission reactions (Scheme 17); and (iii) photooxidative reactions which are not wavelength dependent (Scheme 18).

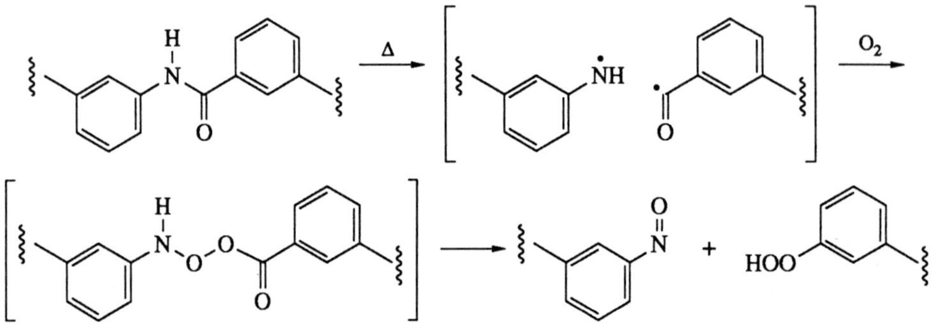

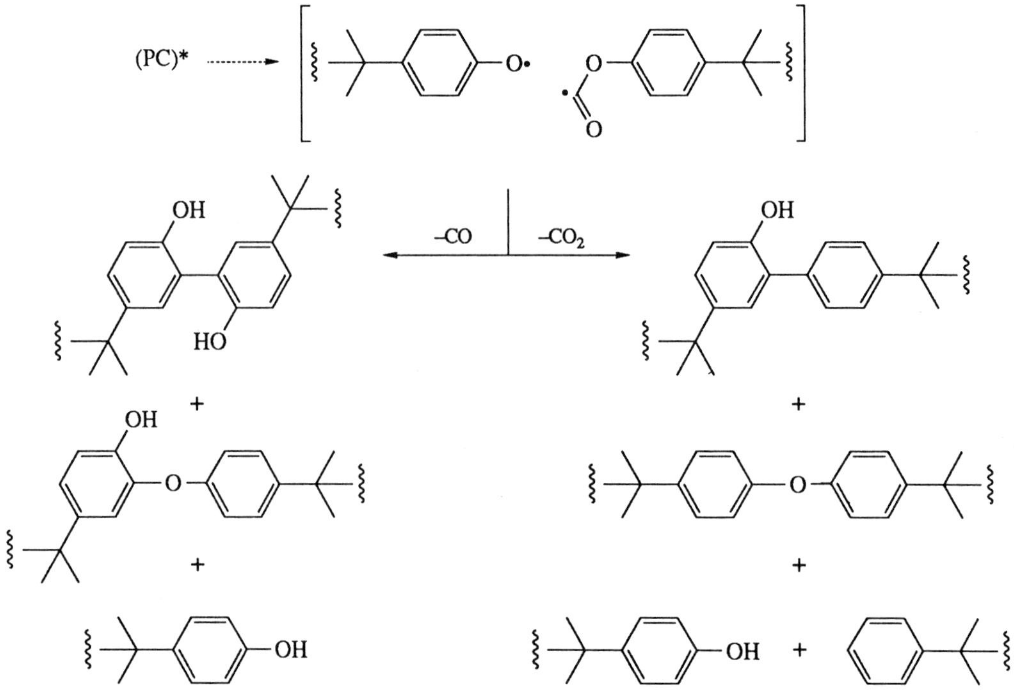

Scheme 17

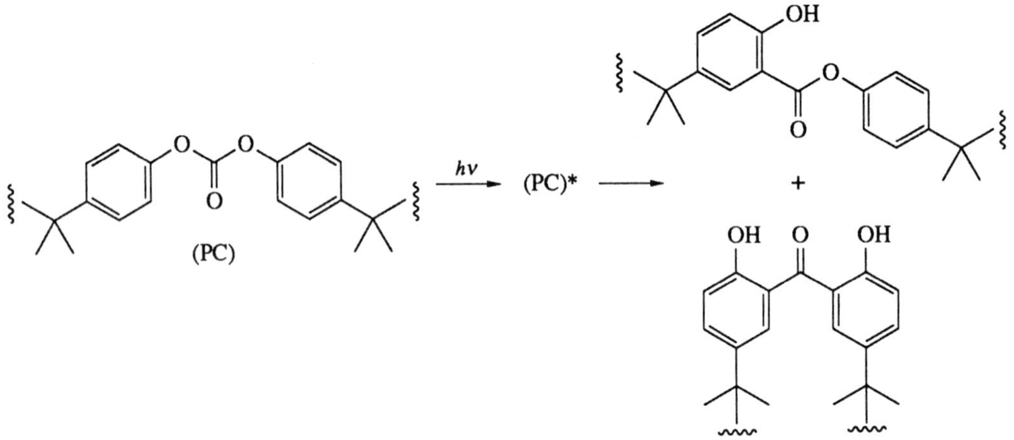

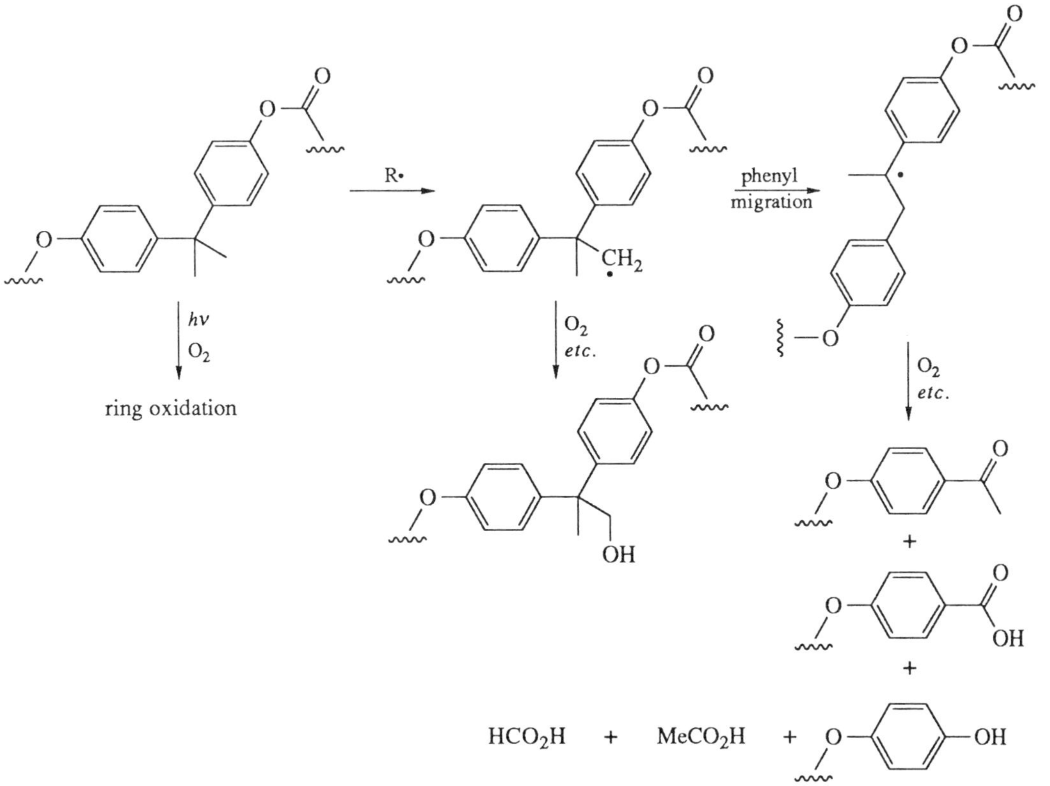

Scheme 18

The major parameters influencing outdoor photodegradation of polycarbonates, includes the wavelength of incident UV radiation, atmospheric oxygen and moisture, and monomer content.

Substitution of the four *ortho*-positions in a polycarbonate by methyl groups gives a new type of polycarbonate, *i.e.* tetramethylbisphenol A polycarbonate which is not affected by photo-Fries reaction (6).[197]

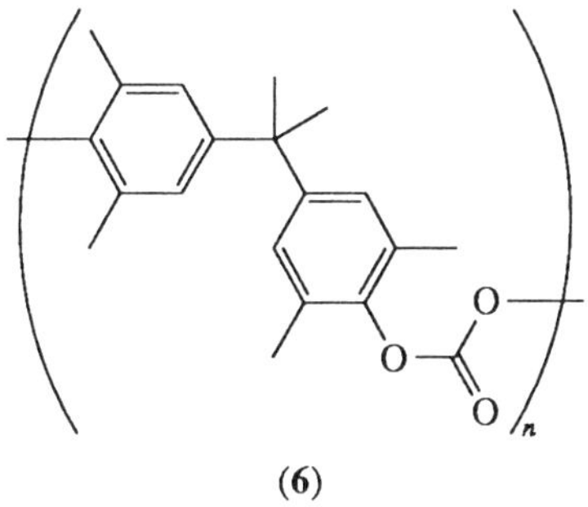

(6)

This polymer is, however, photooxidized faster than other polycarbonates due to chain scission, and the reaction is not wavelength dependent in the UV region.

12.3.10 Polyurethanes, Polysulfones, Polysilanes and Polysiloxanes

Photodecomposition of aromatic diisocyanate-based polyurethanes is, in spite of a number of investigations, still not well resolved.[198-204]

Polyurethanes based on toluene diisocyanate or 2,6-toluene diisocyanate (TDI; equation 55) and methylene 4,4′-diphenyldiisocyanate (MDI; equation 56) under UV irradiation are partially rearranged by photo-Fries reactions, which are followed by photolysis with formation of strongly yellow-colored quinone imide chromophores.[205]

[Scheme showing photodegradation reactions of TDI and MDI polyurethanes, equations (55) and (56)]

The poly(aryl sulfones) are characterized by high thermal stability but low photostability.[206-210] Bisphenol A polysulfone (PSF) and polyether sulfone (PESF) photodegrade according to the mechanism in Scheme 19.[209]

The addition to benzene rings and H-abstraction by phenyl radicals formed from C—S cleavage in diphenyl sulfone moieties are considered to give crosslinking and chain scission during the photodegradation of polysulfones at room temperature.

Silicon polymers containing disilanyl units are easily crosslinked by UV irradiation.[211,212] The main reaction observed is a photocleavage of Si—Si bonds with formation of silanyl radicals (Si·), which may further abstract hydrogen from the macromolecules (PH). The new polymer radicals formed are responsible for the crosslinking reactions (equation 57).

Polysiloxanes are photodecomposed by side chain photochemical reactions.[213-216] The direct photocleavage of the Si—O bond is improbable due to the very high dissociation energy of this bond. The main reactions observed are photodissociaton of side chains followed by crosslinking (Scheme 20).[213]

Several gaseous products, such as hydrogen, methane and ethane are evolved.

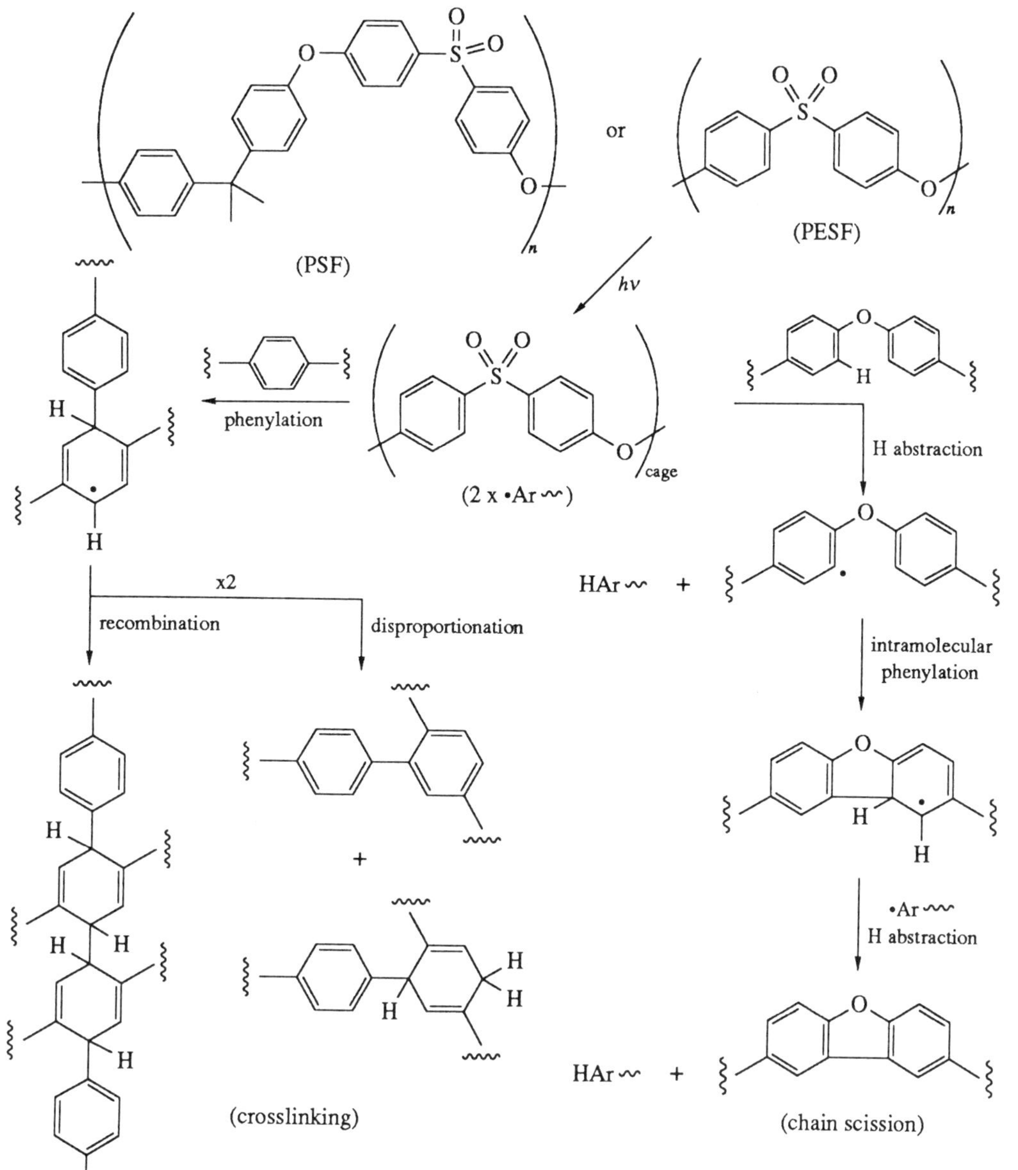

(PSF)
or
(PESF)
hν
phenylation
(2 x •Ar)
cage
H abstraction
HAr +
x2
recombination
disproportionation
intramolecular phenylation
•Ar
H abstraction
(crosslinking)
(chain scission)

Scheme 19

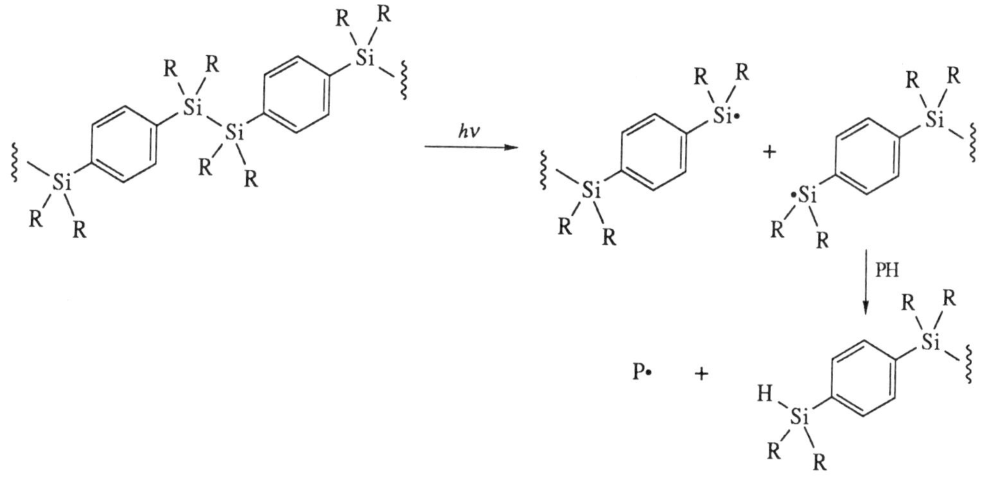

hν
PH
P•

(57)

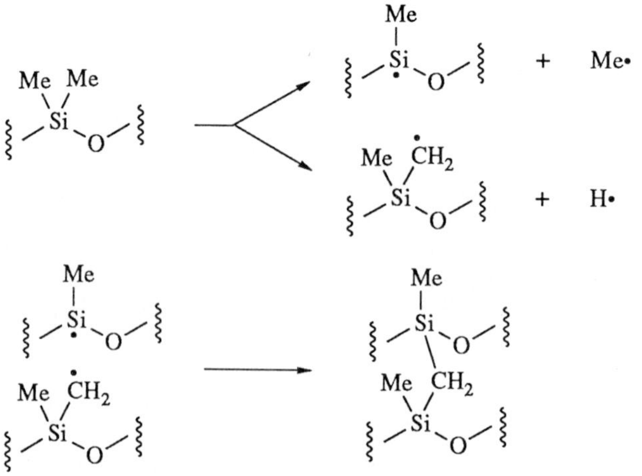

Scheme 20

12.3.11 Polyphosphazenes

Poly(organophosphazenes) (inorganic–organic polymers) with structure (**7**) are a new group of organic/inorganic polymers with inherent flame resistance and excellent fiber formation properties. They are rather resistant to photooxidative degradation processes including chain scission and crosslinking (Scheme 21).

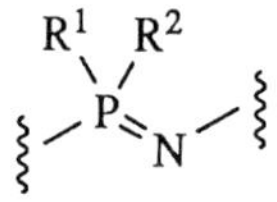

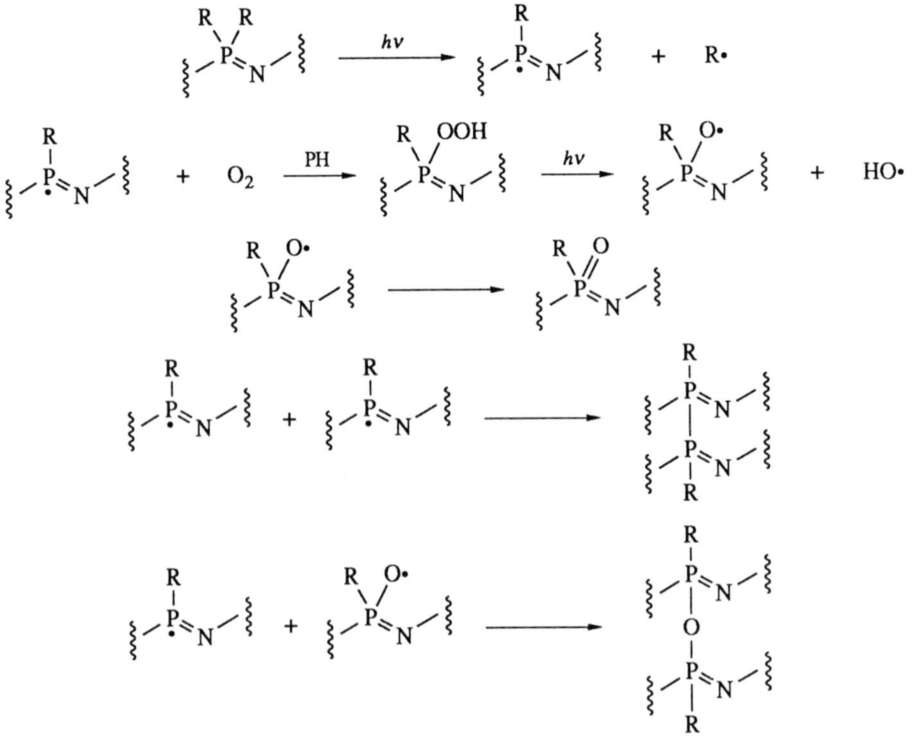

Scheme 21

Depending on the substituent group (R), spectral sensitivity and the mechanism of photolysis may differ significantly.[217-224] Photoreactivity of poly(organophosphazenes) can be strongly accelerated by irradiating these polymers in the presence of suitable external photosensitizers, *e.g.* benzophenone.[73,87]

12.3.12 Cellulose

Cellulose has a broad UV absorption peak at about 265 nm with a shoulder into the near UV of 300 to 400 nm, assigned to the glycosidic bond of the chain and an overlapping absorption of aliphatic ketones formed during isolation and purification.[225] By UV irradiation at wavelengths 250–400 nm, cellulose is photolyzed by chain scission at the glycoside bond (equation 58). The two glycopyranose radicals have been identified by ESR spectroscopy.[226] In addition, formyl radicals (HĊ=O) are formed by splitting off CH_2OH groups (the sixth carbon atom). Trapped hydrogen radicals (H·) are also formed, due to abstraction of hydrogen from OH, CH and CH_2 groups. During photolysis in air, molecular oxygen is probably added to the radical on C-4 in the glucose end group but it has not been recorded in the ESR spectra, probably due to rapid rearrangement of the expected peroxide.[227] A series of papers on the photochemical reactions of cellulose and model compounds has been published by Hon[228] who also reviewed the literature in this field that is of great practical importance for textile and paper fibers during bleaching and exposure to sunlight.

(58)

12.3.13 Degradation by Laser Radiation

During the last decade a special kind of photooxidative degradation of polymers has been developed using high energy laser radiation. Ultraviolet (150–200 nm) laser radiation of high

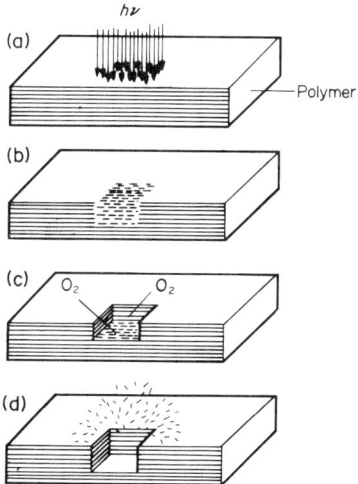

Figure 6 Ablative photodecomposition of a polymer by high energy laser beam radiation and expulsion of molecular fragments at supersonic velocities (reproduced from ref. 230 by permission of the American Association for the Advancement of Science)

intensity, emitted as pulses from an excimer laser, causes an 'ablative photodecomposition' of the surface of a polymer material. The result is a degradation of the structure of the polymeric solid by the photons and the expulsion of molecular fragments at supersonic velocities (Figure 6). An etch pattern in the solid is formed with a geometry that is defined by the laser beam. The principal advantages in using ultraviolet laser radiation is the precision (± 200 nm) with which the width and the depth of the cut can be controlled and the lack of thermal damage to the substrate at a microscopic level (μm dimensions).[229-232]

The laser ablative photodecomposition has found applications for microlithography[233] and for cutting and drilling of polymeric materials.[234]

12.4 REFERENCES

1. A. Girard, *Ann. Soc. Chim. Phys.*, 1881 (V), **24**, 337 and 382. G. Witz, *Bull Soc. Ind. Rouen*, 1883, **11**, 188.
2. G. Barr and I. H. Hadfield, *J. Text. Inst.*, 1927, **18**, T490.
3. C. Moureu and C. Dufraisse, *Bull. Soc. Chim. Fr.*, 1922, **31**, 1152.
4. H. L. Fischer, *Chem. Rev.*, 1930, **7**, 130.
5. *Cf.* W. L. Hawkins, *Encycl. Polym. Sci. Technol.*, 1989, **15**, 539, and B. Rånby and J. F. Rabek, 'Photodegradation, Photooxidation and Photostabilization of Polymers', Wiley, London, 1975, for early refs.
6. J. L. Bolland and G. Gee, *Trans. Faraday Soc.*, 1946, **42**, 236 and 244; J. L. Bolland, *Q. Rev. Chem. Soc.*, 1949, **1**, 3.
7. J. F. Rabek, 'Photostabilization of Polymers. Principles and Applications', Elsevier, London, 1990, chap. 1.
8. J. R. MacCallum, in 'Comprehensive Polymer Science', ed. G. Allen and J. C. Bevington, Pergamon Press, Oxford, 1989, vol. 6, chap. 18.
9. N. D. Searle, *Encycl. Polym. Sci. Technol.*, 1989, **17**, 796.
10. N. D. Searle, in 'Proceedings of the International Conference on Advances in the Stability and Controlled Degradation of Polymers', ed. A. V. Patsis, Technomic, Lancaster, PA, 1989, p. 62.
11. Values in Table 1 are adopted from V. J. G. Calvert and J. N. Pitts, 'Photochemistry', Wiley, New York, 1966, Table A-5.
12. B. Rånby, *J. Anal. Appl. Pyrolysis*, 1989, **15**, 237.
13. *Cf.* B. Rånby and J. F. Rabek, 'Singlet Oxygen. Reactions with Organic Compounds and Polymers', Wiley, Chichester and New York, 1978, for refs.
14. R. L. Clough, M. P. Dillon, K. K. Ju and P. R. Ogilby, *Macromolecules*, 1989, **22**, 3620.
15. B. Rånby and J. F. Rabek, 'Photodegradation. Photo-Oxidation and Photostabilization of Polymers', Wiley, Chichester, 1975.
16. B. Rånby and J. F. Rabek, 'ESR Spectroscopy in Polymer Research', Springer-Verlag, Berlin, 1977.
17. J. F. McKellar and N. S. Allen, 'Photochemistry of Man-made Polymers', Applied Science, London, 1979.
18. W. Schnabel, 'Polymer Degradation: Principles and Practical Applications', Hanser, München, 1981.
19. J. E. Guillet, 'Polymer Photophysics and Photochemistry', Cambridge University Press, Cambridge, 1985.
20. N. Grassie and G. Scott, 'Polymer Degradation and Stabilization', Cambridge University Press, Cambridge, 1985.
21. J. F. Rabek, 'Mechanisms of Photophysical Processes and Photochemical Reactions in Polymers', Wiley, Chichester, 1987.
22. J. F. Rabek, 'Photostabilization of Polymers', Elsevier, London, 1990.
23. K. Tsuji and T. Seiki, *J. Polym. Sci., Polym. Chem. Ed.*, 1971, **9**, 3063.
24. N. S. Allen, *Chem. Soc. Rev.*, 1986, **15**, 373.
25. F. Gugumus, *Makromol. Chem., Macromol. Symp.*, 1989, **27**, 25.
26. N. Grassie and N. A. Weir, *J. Appl. Polym. Sci.*, 1965, **9**, 963, 975.
27. M. Nowakowska, J. Kowal and B. Waligora, *Polymer*, 1978, **19**, 1317.
28. S. W. Bigger and O. Delatycki, *Polymer*, 1988, **29**, 1277.
29. P. R. Ogilby, M. Kristiansen and R. L. Clough, *Macromolecules*, 1990, **23**, 2698.
30. J. F. Rabek and B. Rånby, *J. Polym. Sci., Polym. Chem. Ed.*, 1974, **12**, 273.
31. J. F. Rabek, in 'Singlet Oxygen', ed. A. A. Frimer, CRC Press, Boca Raton, FL, 1985, vol. IV, p. 1.
32. J. F. Rabek and B. Rånby, *J. Polym. Sci., Polym. Chem. Ed.*, 1976, **14**, 1463.
33. H. C. Ng and J. E. Guillet, *Macromolecules*, 1978, **11**, 929.
34. H. C. Ng and J. E. Guillet, *Photochem. Photobiol.*, 1978, **28**, 571.
35. C. Tanielian and J. Chaineaux, *J. Polym. Sci., Polym. Chem. Ed.*, 1979, **17**, 715.
36. C. Tanielian and J. Chaineaux, *Eur. Polym. J.*, 1980, **16**, 619.
37. J. F. Rabek and B. Rånby, *Photochem. Photobiol.*, 1978, **28**, 557.
38. J. F. Rabek and B. Rånby, *J. Appl. Polym. Sci.*, 1979, **23**, 2481.
39. J. F. Rabek and B. Rånby, *Photochem. Photobiol.*, 1979, **30**, 133.
40. J. F. Rabek, J. Lucki and B. Rånby, *Eur. Polym. J.*, 1979, **15**, 1089.
41. J. Lucki, B. Rånby and J. F. Rabek, *Eur. Polym. J.*, 1979, **15**, 1101.
42. F. M. Peng, *J. Appl. Polym. Sci.*, 1986, **31**, 1827.
43. I. Schopov, N. Kassabova and G. Kossmehl, *Polym. Degradation Stab.*, 1989, **25**, 31.
44. A. M. Trozzolo and F. H. Winslow, *Macromolecules*, 1968, **1**, 98.
45. F. Gugumus, *Angew. Makromol. Chem.*, 1988, **158/159**, 151.
46. F. Gugumus, *Makromol. Chem., Macromol. Symp.*, 1989, **25**, 1.
47. F. Gugumus, *Angew. Makromol. Chem.*, 1990, **176/177**, 27.
48. G. Geuskens and M. Kabamba, *Polym. Degradation Stab.*, 1982, **4**, 69.
49. G. Geuskens and M. Kabamba, *Polym. Degradation Stab.*, 1983, **5**, 399.

50. G. Geuskens, F. Debie, M. Kabamba and G. Nedelkos, *Polym. Photochem.*, 1984, **5**, 313.
51. R. Geetha, A. Torika, S. Nagaya and K. Fukei, *Polym. Degradation Stab.*, 1987, **19**, 279.
52. A. Torikai, H. Shirakawa, S. Nagaya and K. Fueki, *J. Appl. Polym. Sci.*, 1990, **40**, 1637.
53. G. Scott, *ACS Symp. Ser.*, 1976, **25**, 340.
54. G. Scott, *Polym. Plast. Technol. Eng.*, 1978, **11**, 1.
55. H. Kubota, M. Kimura and Y. Ogiwara, *J. Polym. Sci., Polym. Lett.*, 1985, **23**, 21.
56. J. Lucki, J. F. Rabek, B. Rånby, and Y. C. Jiang, *Polymer*, 1986, **27**, 1193.
57. I. C. McNeill and S. Zulfiqar, *Polym. Degradation Stab.*, 1985, **10**, 147.
58. A. Torikai, M. Ohno and K. Fueki, *J. Appl. Polym. Sci.*, 1990, **41**, 1023.
59. J. F. Rabek, J. Sanetra and B. Rånby, *Macromolecules*, 1986, **19**, 1674.
60. J. F. Rabek and J. Sanetra, *Macromolecules*, 1986, **19**, 1679.
61. J. W. Martin, *J. Appl. Polym. Sci.*, 1984, **29**, 777.
62. J. F. Rabek, B. Rånby and T. A. Skowronski, *Macromolecules*, 1985, **18**, 1810.
63. H. Ito and M. Ueda, *Macromolecules*, 1988, **21**, 1475.
64. I. Mita, T. Hisano, K. Horie and A. Okamoto, *Macromolecules*, 1988, **21**, 3003.
65. B. Rånby and J. F. Rabek, *J. Appl. Polym. Sci., Appl. Polym. Symp.*, 1979, **35**, 243.
66. B. Rånby and J. F. Rabek, *ACS Symp. Ser.*, 1983, **229**, 291.
67. C. Decker and M. Balandier, *Polym. Photochem.*, 1984, **5**, 267.
68. A. Torikai, S. Asada and K. Fueki, *Polym. Photochem.*, 1986, **7**, 1.
69. A. Torikai, A. Takeuchi, S. Nagaya and K. Fueki, *Polym. Photochem.*, 1986, **7**, 279.
70. R. Gooden, D. D. Davis, M. Y. Helleman, A. J. Lovinger and F. H. Winslow, *Macromolecules*, 1988, **21**, 1212.
71. G. A. George and M. S. O'Shea, *Polym. Degradation Stab.*, 1990, **28**, 289.
72. H. Kubota, K. Takahashi and Y. Ogiwara, *Polym. Degradation Stab.*, 1990, **29**, 207.
73. M. Gleria, F. Minto, P. Bortolus, W. Porzio and A. Bolognezi, *Eur. Polym. J.*, 1989, **25**, 1039.
74. J. A. Bousquet, J. B. Donnet, J. Faure, J. P. Fouassier, B. Haidir and A. Vidal, *J. Polym. Sci., Polym. Chem. Ed.*, 1980, **18**, 765.
75. T. Wismontski-Knittel and T. Kilp, *J. Polym. Sci., Polym. Chem. Ed.*, 1983, **21**, 3209.
76. J. A. Bousquet, B. Haidar, J. P. Fouassier and A. Vidal, *Eur. Polym. J.*, 1983, **19**, 135.
77. I. Mita, T. Takagi, K. Horie and Y. Shindo, *Macromolecules*, 1984, **17**, 2256.
78. T. A. Skowronski, J. F. Rabek and B. Rånby, *Polym. Photochem.*, 1984, **5**, 77.
79. V. B. Ivanov and M. A. Zhuralev, *Polym. Photochem.*, 1986, **7**, 55.
80. J. A. Bousquet and J. P. Fouassier, *Angew. Makromol. Chem.*, 1987, **149**, 19, 45.
81. G. Geuskens and M. S. Kabamba, *Polym. Degradation Stab.*, 1987, **19**, 315.
82. S. K. Wu and J. F. Rabek, *Polym. Degradation Stab.*, 1988, **21**, 365.
83. M. A. DePaoli and G. Geuskens, *Polym. Degradation Stab.*, 1988, **21**, 277.
84. G. Geuskens and M. S. Kabamba, *Polym. Degradation Stab.*, 1987, **19**, 315.
85. H. Kubota, K. Takahashi and Y. Ogiwara, *Polym. Degradation Stab.*, 1989, **24**, 201.
86. L. Dulog, R. Kern and W. Kern, *Makromol. Chem.*, 1968, **120**, 123.
87. M. Gleria, F. Minto, P. Bortolus, W. Porzio and S. V. Melle, *Eur. Polym. J.*, 1990, **26**, 315.
88. J. F. Rabek and B. Rånby, *J. Polym. Sci., Polym. Chem. Ed.*, 1974, **12**, 295.
89. N. S. Allen, J. P. McKellar and S. A. Protopapas, *J. Appl. Polym. Sci.*, 1978, **22**, 1451.
90. P. Z. Zamotaev and S. V. Luzgarev, *Angew. Makromol. Chem.*, 1989, **173**, 47.
91. J. A. Bousquet and J. P. Fouassier, *Angew. Makromol. Chem.*, 1987, **149**, 1.
92. Y. Ogiwara, Y. Kimura, Z. Osawa and H. Kubota, *J. Polym. Sci., Polym. Chem. Ed.*, 1977, **15**, 1667.
93. A. Negishi and Y. Ogiwara, *J. Appl. Polym. Sci.*, 1980, **25**, 1095.
94. E. D. Owen and S. R. Brooks, *Polym. Photochem.*, 1985, **6**, 21.
95. D. J. Carlsson and D. M. Wiles, *Macromolecules*, 1969, **2**, 587, 597.
96. D. J. Carlsson and D. M. Wiles, *J. Macromol. Sci., Rev. Macromol. Chem.*, 1976, **C14**, 65.
97. R. Geetha, A. Torikai, S. Nagaya and K. Fueki, *Polym. Degradation Stab.*, 1987, **19**, 279.
98. F. Gugumus, *Angew. Makromol. Chem.*, 1990, **182**, 85, 111.
99. F. Gugumus, *Polym. Degradation Stab.*, 1990, **27**, 19.
100. J. A. Bousquet and J. P. Fouassier, *Polym. Degradation Stab.*, 1987, **18**, 163.
101. C. Decker, *Eur. Polym. J.*, 1984, **20**, 149.
102. J. A. Bousquet and J. P. Fouassier, *Polym. Degradation Stab.*, 1983, **5**, 113.
103. J. A. Bousquet and J. P. Fouassier, *J. Polym. Sci., Polym. Chem. Ed.*, 1984, **22**, 3865.
104. E. R. Duek, V. F. Juliano, M. Guzzo, C. Lascheres and M. A. DePaoli, *Polym. Degradation Stab.*, 1990, **28**, 235.
105. R. Lemaire and R. Arnaoud, *Polym. Photochem.*, 1984, **5**, 234.
106. N. L. Maecker and D. B. Priddy, *J. Appl. Polym. Sci.*, 1991, **42**, 21.
107. H. Schencko and J. Walker, *Eur. Polym. J.*, 1971, **7**, 1047.
108. B. Rånby, J. F. Rabek and G. Canbäck, *J. Macromol. Sci., Chem.*, 1978, **A12**, 587.
109. M. Balandier and C. Decker, *Eur. Polym. J.*, 1978, **14**, 995.
110. C. Decker and M. Balandier, *Polym. Photochem.*, 1981, **1**, 221.
111. C. Decker and M. Balandier, *Eur. Polym. J.*, 1982, **18**, 1085.
112. C. Decker and M. Balandier, *Polym. Photochem.*, 1984, **5**, 267.
113. R. Sastre, G. Martinez, F. Castillo and J. L. Millan, *Makromol. Chem., Rapid. Commun.*, 1984, **5**, 541.
114. F. Castillo, G. Martinez, R. Satre, J. Millan, V. Bellenger and J. Verdu, *Polym. Degradation Stab.*, 1985, **13**, 211.
115. N. S. Allen, J. Wooler and K. O. Fatinikun, *Polym. Degradation Stab.*, 1985, **13**, 277.
116. J. L. Gardette, S. Gaumet and J. Lemaire, *Macromolecules*, 1989, **22**, 2576.
117. A. L. Andrady, A. Torikai and K. Fueki, *J. Appl. Polym. Sci.*, 1989, **37**, 935, 2789.
118. J. L. Gardette and J. Lemaire, *Polym. Degradation Stab.*, 1989, **25**, 293.
119. A. L. Andrady, K. Fueki and A. Torikai, *J. Appl. Polym. Sci.*, 1990, **39**, 763.
120. F. Castillo, G. Martinez, R. Sastre, J. Millan, V. Bellenger, B. D. Gupta and J. Verdu, *Polym. Degradation Stab.*, 1990, **27**, 1.

121. V. Bellenger, J. Verdu, C. Martinez and J. Millan, *Polym. Degradation Stab.*, 1990, **28**, 53.
122. L. Jirackova-Audouin, V. Bellenger and J. Verdu, *Polym. Photochem.*, 1984, **5**, 283.
123. J. F. Rabek, G. Canbäck and B. Rånby, *J. Appl. Polym. Sci.*, 1977, **21**, 2211.
124. G. E. Williams and D. L. Gerrard, *J. Polym. Sci., Polym. Chem. Ed.*, 1983, **21**, 1491.
125. T. A. Skowronski, J. F. Rabek and B. Rånby, *Polym. Degradation Stab.*, 1985, **12**, 229.
126. T. A. Skowronski, J. F. Rabek and B. Rånby, *Polym. Degradation Stab.*, 1984, **8**, 37.
127. C. Decker, *Makromol. Chem., Macromol. Symp.*, 1989, **24**, 253.
128. C. Decker, *J. Polym. Sci., Polym. Lett.*, 1987, **25**, 5.
129. S. Gaumet, J. L. Gardette and J. Lemaire, *Polym. Degradation Stab.*, 1987, **18**, 135.
130. T. Skowronski, J. F. Rabek and B. Rånby, *Polym. Eng. Sci.*, 1984, **24**, 278.
131. S. W. Beavan and D. Philips, *Eur. Polym. J.*, 1974, **10**, 593.
132. J. F. Rabek, J. Lucki and B. Rånby, *Eur. Polym. J.*, 1979, **15**, 1089.
133. S. Yano, *Rubber Chem. Technol.*, 1981, **54**, 1.
134. M. A. De Paoli, *Eur. Polym. J.*, 1983, **19**, 761.
135. C. Adam, J. Lacoste and J. Lemaire, *Polym. Degradation Stab.*, 1989, **24**, 185.
136. J. Lucki, B. Rånby and J. F. Rabek, *Eur. Polym. J.*, 1979, **15**, 1101.
137. C. Adam, J. Lacoste and J. Lemaire, *Polym. Degradation Stab.*, 1990, **29**, 305.
138. T. A. Skowronski, J. F. Rabek and B. Rånby, *Polym. Degradation Stab.*, 1983, **5**, 173.
139. J. B. Adeniyi, *Eur. Polym. J.*, 1984, **20**, 291.
140. S. K. L. Li and J. E. Guillet, *J. Polym. Sci., Polym. Chem. Ed.*, 1980, **18**, 2221.
141. R. Gooden, M. Y. Hellman, R. S. Hutton and F. H. Winslow, *Macromolecules*, 1984, **17**, 2830.
142. F. A. Bovey, R. Gooden, F. C. Schilling and F. H. Winslow, *Macromolecules*, 1988, **21**, 938.
143. N. A. Weir and J. Arct, *Eur. Polym. J.*, 1987, **23**, 33.
144. N. A. Weir and J. Arct, *J. Polym. Sci., Polym. Chem. Ed.*, 1987, **25**, 191.
145. N. A. Weir, K. Whiting and J. Arct, *Polym. Commun.*, 1987, **28**, 284.
146. N. A. Weir, D. A. Jones and P. Kutok, *Polym. Commun.*, 1989, **30**, 201.
147. N. A. Weir and K. Whiting, *Eur. Polym. J.*, 1990, **26**, 991.
148. N. A. Weir and J. Arct, *J. Photochem. Photobiol., Chem. Ser.*, 1990, **53**, 1990.
149. N. A. Weir, J. Arct and K. Whiting, *Eur. Polym. J.*, 1990, **26**, 341.
150. N. A. Weir and L. Donaldson, *Polym. Commun.*, 1990, **31**, 302.
151. J. P. Bays, M. V. Encinas and J. C. Scaiano, *Macromolecules*, 1980, **13**, 815.
152. P. Hrdlovic and I. Lukac, *Dev. Polym. Degradation*, 1982, **4**, 101.
153. J. C. Scaiano and L. C. Stewart, *Polymer*, 1982, **23**, 913.
154. J. E. Guillet, *Pure Appl. Chem.*, 1980, **52**, 285.
155. A. Gupta, R. Liang, F. D. Tsay and J. Moacanin, *Macromolecules*, 1980, **13**, 1696.
156. W. J. Leigh, J. C. Scaiano, C. I. Paraskevopoulos, G. M. Charette and S. E. Sugamori, *Macromolecules*, 1985, **18**, 2148.
157. J. C. Lenain, J. C. Brosse and A. Sabet, *Makromol. Chem., Rapid. Commun.*, 1983, **4**, 1.
158. Z. Khalil, S. Michaille and J. Lamaire, *Makromol. Chem.*, 1987, **188**, 1743.
159. H. Boudevska, Ch. Bratschkov, O. Todorowa, E. Bezadea and D. Braun, *Angew. Makromol. Chem.*, 1987, **148**, 27.
160. G. A. George, *J. Appl. Polym. Sci.*, 1974, **18**, 419.
161. G. A. George and D. K. C. Hodgeman, *Eur. Polym. J.*, 1977, **13**, 63.
162. N. A. Weir and K. Whiting, *Eur. Polym. J.*, 1989, **25**, 291.
163. P. C. Lukas and R. G. Porter, *Polym. Degradation Stab.*, 1989, **26**, 203.
164. G. Geuskens, D. Baeyens-Volant, G. Delaunois, Q. Lu-Vinh, W. Piret and C. David, *Eur. Polym. J.*, 1978, **14**, 291, 299.
165. N. A. Weir, J. K. Lee and J. Arct, *Polym. Degradation Stab.*, 1986, **14**, 249.
166. N. A. Weir and T. H. Milkie, *Polym. Photochem.*, 1986, **7**, 129.
167. N. A. Weir, K. Whiting, J. Arct and G. McCulloch, *Polym. Degradation Stab.*, 1987, **18**, 293.
168. N. A. Weir, P. Kutok and K. Whiting, *Polym. Degradation Stab.*, 1989, **24**, 247.
169. J. Peeling and D. T. Clark, *Polym. Degradation Stab.*, 1981, **3**, 97.
170. P. C. Lukas and R. G. Porter, *Polym. Degradation Stab.*, 1985, **13**, 287.
171. P. C. Lukas and R. G. Porter, *Polym. Degradation Stab.*, 1988, **22**, 175.
172. N. A. Weir, J. Arct and M. Farahani, *Polym. Degradation Stab.*, 1985, **13**, 361.
173. N. A. Weir and M. Farahani, *Makromol. Chem.*, 1986, **187**, 889.
174. K. Naruchi, S. Tanaka, T. Takemori, F. Akutsu, M. Yamamoto and K. Yamada, *Makromol. Chem., Rapid. Commun.*, 1986, **7**, 607.
175. M. Shirai, M. Suzuki, T. Sato, M. Tsunooka and M. Tanaka, *Eur. Polym. J.*, 1989, **25**, 1079.
176. K. A. Holland, H. J. Griesse, D. G. Hawthorne and J. H. Hodkin, *Polym. Degradation Stab.*, 1991, **31**, 269.
177. C. Adam, J. Lacoste and J. Lemaire, *Polym. Degradation Stab.*, 1989, **26**, 269.
178. T. A. Skowronski, J. F. Rabek and B. Rånby, *Polym. Degradation Stab.*, 1983, **5**, 173.
179. C. Adam, J. Lacoste and J. Lemaire, *Polym. Degradation Stab.*, 1990, **27**, 85.
180. B. S. Stowe, R. E. Fornes and R. D. Gilbert, *Polym. Plast. Technol. Eng.*, 1974, **3**, 159.
181. S. Yano and M. Murayama, *Polym. Photochem.*, 1981, **1**, 177.
182. C. L. Renschler and F. B. Burns, *J. Appl. Polym. Sci.*, 1984, **29**, 1125.
183. A. Roger, D. Sallet and J. Lemaire, *Macromolecules*, 1985, **18**, 1771.
184. A. Roger, D. Sallet and J. Lemaire, *Macromolecules*, 1986, **19**, 579.
185. C. H. Do, E. M. Pearce, B. J. Bulkin and H. K. Reimschuessel, *J. Polym. Sci., Polym. Chem. Ed.*, 1987, **25**, 2301.
186. D. J. Carlson, L. H. Gan and D. M. Wiles, *J. Polym. Sci., Polym. Chem. Ed.*, 1978, **16**, 2353.
187. J. S. Humphrey, Jr., A. R. Schultz and D. B. G. Jacquiss, *Macromolecules*, 1973, **6**, 305.
188. A. Gupta, A. Rembaum and J. Moacanin, *Macromolecules*, 1978, **11**, 1285.
189. A. Factor and M. L. Chu, *Polym. Degradation Stab.*, 1980, **2**, 203.
190. D. T. Clark and H. S. Munro, *Polym. Degradation Stab.*, 1982, **4**, 441.
191. A. Rivaton, D. Sallet and J. Lemaire, *Polym. Photochem.*, 1983, **3**, 463.
192. D. T. Clark and H. S. Munro, *Polym. Degradation Stab.*, 1984, **8**, 195.

193. H. S. Munro and R. S. Allaker, *Polym. Degradation Stab.*, 1985, **11**, 349.
194. A. Rivaton, D. Sallet and J. Lemaire, *Polym. Degradation Stab.*, 1986, **14**, 1.
195. J. D. Webb and A. W. Czanderna, *Macromolecules*, 1986, **19**, 2810.
196. A. Factor, W. V. Ligon and R. J. May, *Macromolecules*, 1987, **20**, 2461.
197. A. Rivaton and J. Lemaire, *Polym. Degradation Stab.*, 1989, **23**, 51.
198. N. S. Allen and J. F. McKellar, *J. Appl. Polym. Sci.*, 1976, **20**, 1441.
199. J. L. Gardette and J. Lemaire, *Polym. Degradation Stab.*, 1984, **6**, 135.
200. V. Rek, H. J. Mencer and M. Bravar, *Polym. Photochem.*, 1986, **7**, 273.
201. C. E. Hoyle and K. J. Kim, *J. Polym. Sci., Polym. Chem. Ed.*, 1987, **25**, 2631.
202. C. E. Hoyle, K. J. Kim, Y. G. No and G. L. Nelson, *J. Appl. Polym. Sci.*, 1987, **34**, 763.
203. C. E. Hoyle, Y. G. No, K. G. Malone, S. F. Thames and D. Creed, *Macromolecules*, 1988, **21**, 2727.
204. C. E. Hoyle, C. P. Chawla and K. J. Kim, *J. Polym. Sci., Polym. Chem. Ed.*, 1988, **26**, 1295.
205. C. E. Hoyle and K. J. Kim, *J. Polym. Sci., Polym. Chem. Ed.*, 1986, **24**, 1879.
206. N. S. Allen and J. F. McKellar, *J. Appl. Polym. Sci.*, 1977, **21**, 1129.
207. C. Giori and T. Yamauchi, *J. Appl. Polym. Sci.*, 1984, **29**, 237.
208. H. S. Munro and D. J. Clark, *Polym. Degradation Stab.*, 1985, **11**, 211.
209. S. I. Kuroda, A. Nagura, K. Horie and I. Mita, *Eur. Polym. J.*, 1989, **25**, 621.
210. H. S. Munro and D. T. Clark, *Polym. Degradation Stab.*, 1987, **17**, 319.
211. M. Ishikawa, N. Imamura, N. Miyoshi and M. Kumada, *J. Polym. Sci., Polym. Lett.*, 1983, **21**, 657.
212. M. Ishikawa, N. Hongzhi, K. Matsusaki, K. Nate, T. Inoue and H. Yokono, *J. Polym. Sci., Polym. Lett.*, 1984, **22**, 669.
213. A. D. Delman, M. Landy and B. B. Simms, *J. Polym. Sci., Polym. Chem. Ed.*, 1969, **7**, 3375.
214. K. Nate, M. Ishikawa, N. Imamura and Y. Murakami, *J. Polym. Sci., Polym. Chem. Ed.*, 1986, **24**, 1551.
215. H. Inoue and S. Kohama, *J. Appl. Polym. Sci.*, 1986, **32**, 3853.
216. J. Devaux, J. Sledz, F. Schue, L. Giral and H. Naarmann, *Makromol. Chem.*, 1990, **191**, 139.
217. P. Bortolus, F. Minto, G. Beggiato and S. Lora, *J. Appl. Polym. Sci.*, 1979, **24**, 285.
218. J. P. O'Brien, W. T. Ferrar and H. R. Allcock, *Macromolecules*, 1979, **12**, 108.
219. M. Gleria, F. Minto, S. Lora and P. Bortolus, *Eur. Polym. J.*, 1979, **15**, 671.
220. H. Hiraoka, W. Y. Lee, L. W. Welsh, Jr. and R. W. Allen, *Macromolecules*, 1979, **12**, 753.
221. M. Gleria, F. Minto, S. Lora, L. Busulini and P. Bortolus, *Macromolecules*, 1986, **19**, 574.
222. M. Gleria, F. Minto, L. Flamigni and P. Bortolus, *Macromolecules*, 1987, **20**, 1766.
223. M. Gleria, F. Minto, S. Lora, P. Bortolus and R. Ballardini, *Macromolecules*, 1981, **14**, 687.
224. M. Gleria, F. Minto, L. Flamigni and P. Bortolus, *Polym. Degradation Stab.*, 1988, **22**, 125.
225. B. Rånby, in 'Wood Processing and Utilization', ed. J. F. Kennedy, G. O. Phillips and P. A. Williams, Horwood, London, 1989, chap. 46, p. 353.
226. H. Kubota, Y. Ogiwara and K. Matsuzaki, *J. Appl. Polym. Sci.*, 1975, **19**, 1291.
227. K. Matsuzaki, S. Nakamura and S. Shindo, *J. Appl. Polym. Sci.*, 1972, **16**, 1339.
228. N.-S. Hon, *J. Polym. Sci., Polym. Chem. Ed.*, 1975, **13**, 955, 1933, 2363, 2641 and 2653; 1976, **14**, 2497 and 2513.
229. R. Srinivasan, *J. Radiat. Curing*, 1983, **10**, 12.
230. R. Srinivasan, *Science (Washington, D.C.)*, 1986, **234**, 559.
231. R. Srinivasan and S. Lazare, *Polymer*, 1985, **26**, 1297.
232. T. Bahners and E. Schollmeyer, *Angew. Makromol. Chem.*, 1987, **151**, 1.
233. J. P. Fouassier and J. F. Rabek, 'Lasers in Polymer Science and Technology', CRC Press, Boca Raton, FL, 1990, vols. I–IV.
234. S. E. Nielsen, *Polym. Test.*, 1983, **3**, 303.

13
Biodegradable Polymers

ANN-CHRISTINE ALBERTSSON and SIGBRITT KARLSSON
The Royal Institute of Technology, Stockholm, Sweden

13.1	INTRODUCTION	285
13.2	BIODEGRADATION TESTS	287
13.3	SYNTHETIC BIODEGRADABLE POLYMERS	288
13.4	NATIVE BIODEGRADABLE POLYMERS	291
13.5	BIOMEDICAL BIODEGRADABLE POLYMERS	294
13.6	PLASTIC WASTE	294
13.7	ARE BIODEGRADABLE POLYMERS THE FUTURE?	294
13.8	REFERENCES	295

13.1 INTRODUCTION

Biodegradable polymers cover a vast area of high molecular weight compounds. Usually it is valuable to distinguish between biodegradable polymers of native and of synthetic origin. Native biodegradable polymers are the result of a synthesis developed during millions of years of evolution, leading to tailor-made materials for different applications in nature. These biopolymers include proteins, polysaccharides, nucleic acids or lipids which show completely different characteristics depending on the situation in which they are used. Synthetic polymers, on the other hand, are the result of a mere century of research and development. Synthetic polymers susceptible to biodegradation can be of different types, *e.g.* polymers containing hydrolyzable backbone polyesters. Recent research activity on biodegradable synthetic polymers has often been focused on the simulation of different biopolymers or polymers with degradable backbones, *e.g.* polyanhydrides, polycarbonates, polylactones, *etc*. Other concepts in the search for new biodegradable materials include the use of microorganisms which can produce polymers, *e.g.* poly(β-hydroxybutyrate) (PHB; **1**) and copolymers of PHB. Difficult synthesis and purification steps leading to low yields are avoided by using microorganisms. In packaging applications, a biodegradable additive is often included as a way to promote environmental degradation *e.g.* starch in polyethylene (PE).

(**1**) Poly(β-hydroxybutyrate) PHB

The term biodegradation has no clear definition. Several different definitions are being discussed and the question of a general definition is still under debate. The term 'biodegradation' has been used to encompass events taking place both in the natural environment and in the living human body, although environmental degradation by microorganisms like bacteria and fungi is probably not comparable with the degradation taking place in mammals.

Some researchers use the term 'biodegradation' to imply a clear material–organism relationship where the specific biological vector and the precise material always produce a given set of changes which is coupled to a set of specific enzymes recognizing the polymer in question.[1] In the field of biomaterials (e.g. sutures, bone reconstruction and drug delivery) the definition may simply be hydrolysis. Environmentally speaking, however, biodegradation may mean fragmentation (e.g. fungi penetrating the materials) or loss of mechanical properties or degradation through the action of living organisms (i.e. endo- and/or exo-enzymes degrade the materials).

There are at least four reasons why the discussion of the biodegradation of biodegradable polymers will continue: (i) definitions and nomenclature in this field are in some cases still vague and not generally accepted; (ii) the field covers rather wide interdisciplinary aspects ranging from mechanics and physical chemistry to enzymology and microbial taxonomy, which will seldom be covered by one researcher or even by a group of workers; (iii) all experimentation is difficult and consequently often imperfect because such work must be related to complex and extremely long-term phenomena in a wide range of natural environments which are intrinsically variable; and (iv) the immediate acceptance of interpretations of apparently opposite character will be more difficult if it interferes with patents, production or marketing interests. A general definition that excludes other degradation modes than biodegradation may be: 'transformation and deterioration of polymers solely by living organisms (including microorganisms and/or enzymes excreted by them)'.[2]

The tests for biodegradation are different and demand special knowledge and competence. Generally, biodegradable polymers are those with a relatively low molecular weight and a straight chain with a saturated structure. High molecular weight synthetic polymers containing cyclic and/or unsaturated structures are less susceptible to degradation.

The accessibility of a polymer to degradative attack by living organisms is not dependent on its origin but on its molecular composition and architecture. Complex macromolecules such as lignin and asphalt show great inertness despite being biopolymers. On the other hand, synthetic polymers with intermittent ester linkages (e.g. polyesters, polyurethanes) are readily accessible to the biodegradative action of esterases despite their usual enzymatic specificity. In addition, solitary examples of an extremely uniform chain, characteristic of a synthetic polymer molecule (for instance polyethylene with 100 to 1000 or more carbon atoms), certainly occur sporadically in nature as an artefact, intermixed throughout with countless types of other aliphatic and aromatic macromolecules of higher petroleum derivatives. How should the biodegradation of polymers of any origin be defined? Preferentially in three ways, all feasible theoretically, but unlikely to occur alone in nature as a single phenomenon:[3,4] (i) a biophysical effect, e.g. mechanical damage to a material by the swelling and bursting effect of growing cells; (ii) a secondary biochemical effect resulting from the excretion of substances other than enzymes which may act directly upon the polymer or by changing either the pH or redox conditions of the surroundings; and (iii) direct enzymatic action leading to splitting or oxidative breakdown of the material. The macroscopic or mechanically observable consequences of these three effects may be expressed in many ways, as corrosion, abrasion, deterioration, cracking, decrease in tensile strength, etc.

Fungi and bacteria can also use plasticizers and fillers as a source of nutrient and this may accelerate the ageing of the plastics.[5] The growth of a pure culture of a single, defined microbial strain on a specified accessory substance in a plastic material is unlikely to induce the ribosomal production of an entirely different enzyme in the same strain, directed towards the main molecular species in this particular plastic product.

Such a mechanism is not in agreement with existing knowledge on the mechanism of enzyme induction, which is merely a triggering of an existing gene to produce a number of molecules of the corresponding enzyme.

It is even more futile to hope that the induction of biosynthesis of yet unknown enzymes might eventually occur in the case of polyethylenes. There is instead a synergism between biodegradation and environmental degradation. The mechanism for the degradation of polyethylene is comparable with that of paraffin. In the degradation of polyethylene, abiotic steps also contribute to the total degradation.[6] There is a synergistic effect between photooxidative degradation and biodegradation.[3,6–8]

The need for biodegradable polymers has expanded and nowadays the materials are asked for in widely different situations such as packaging and as biomaterials. There are many reasons for this; an urgent question is what to do with all plastic wastes and, when using polymers in humans and animals, toxic effects must be avoided at the same time as some biomaterials must possess properties leading to their eventual removal from the body.

This article deals with the distinction between synthetic and native biodegradable polymers and discusses the biodegradation tests, biomedical biodegradable polymers and plastic waste.

13.2 BIODEGRADATION TESTS

Different approaches are taken when screening for the biodegradability of polymers. According to ASTM standard, a petri-dish test is an appropriate initial routine test. Different microorganisms can be inoculated onto the polymer depending on the type of material. The ASTM standard specifies certain standard microorganisms (usually fungi),[9,10] but it is important that the type of polymer is known so that the correct choice of microorganism can be made. After an appropriate period of time (four weeks) the polymer should be inspected for visual growth. It is also customary to determine changes in polymer mass after the incubation.

Traditional methods for testing the biodegradability of a polymer include the following techniques: (i) visual inspection of mycelium growth on the polymer surface; (ii) quantitative determination of microbial growth; (iii) quantitative determination of the weight loss of the polymer; (iv) measurement of changes in polymer properties, e.g. tensile strength; and (v) measurement of the metabolic activity of the microorganisms by oxygen uptake or CO_2 evolution. The possibility of evaluating metabolic activities is particularly important. The technique of respirometry is also a standard method in microbiology. The decomposition of the polymer can be related to the metabolically evolved gases. Respirometry is one of the most sensitive methods for calculating the degradation of polymers. The term respiration implies a biological oxidation in which molecular oxygen is the ultimate hydrogen acceptor. The product, CO_2, is the most highly oxidized form of carbon and in the absence of other carbon sources in a test environment, this CO_2 will be a degradation product of the polymer. The major drawback of respirometry is the demand for labelled polymers (^{14}C) but if ^{14}C-labelled monomers are at hand when a new polymer is synthesized the method is reliable and quite simple.[4]

Swift[11] discusses the need for standardization and the requirement for the biodegradability of polymers. He proposes a carbon balance as a convenient way of measuring biodegradability, comparing it with what is done with water soluble polymers in the detergent industry. The biodegradability should be compared with that of standards such as starch, cellulose, polycaprolactones, etc.

The biological testing of polymers demands good contact between polymer and microorganism. The hydrophobicity of most polymers often creates a major obstacle to contact between polymer and microorganism, but the addition of surfactants can promote good contact and thereby a better biodegradation.[12,13] Choosing a surfactant needs some consideration; it should neither be toxic to the microorganisms nor able to function as a second substrate so that the surfactant is degraded and not the polymer. A much larger surface area is obtained and shorter fragments of the polymer chain will be accessible to biodegradation when polymer powder is tested rather than a film.[14]

The properties of the sample are dependent in every part on the sample preparation and history. Small details like residues or solvents, catalysts or antioxidants may prevent biodegradation, and additives like lubricants, plasticizers or the low molecular weight part of the polymer may increase the rate of biodegradation.[15–17] Biodegradation is dependent on molecular weight, crystallinity, oxidation, mechanical stress and additives.[8,12,16]

Static systems, as opposed to dynamic ones, are of interest when biodegradation tests are discussed. A dynamic test more closely resembles the environment, e.g. a better contact between polymer and microorganism is obtained by stirring sets of samples. The ambition when designing a biodegradation test must be to simulate nature as closely as possible. The proper microorganisms must be chosen and the cultivation (degradation) environment must be given special attention.[12]

Microorganisms can only be activated with a favourable environment for their growth. The same thing applies to the biodegradation tests; if the microorganism is to be able to degrade a polymer the necessary nutrients must be available in a suitable form for use as building materials. A proper pH, a suitable temperature and a correct oxygen level must also be maintained.

In some cases it is better to use pure enzymes than the microorganisms. Many enzymes capable of degrading macromolecules are endoenzymes. i.e. they can only degrade materials which are actually digested by the microorganisms. It is therefore more convenient to use the pure enzyme and to follow the degradation in a solution containing the polymer and the enzyme. On the other hand it is often more difficult to handle enzymes than intact microorganisms. Enzymes are very susceptible to changes in conditions such as temperature, pH and substrate composition. A minor change in pH, for example, is sufficient to render the enzyme inactive.

In natural habitats, most microorganisms are associated with other kinds of microorganisms. A mixed population is therefore an optimal choice for biodegradation tests. In general, either bacteria, fungi or both are used for biodegradation tests.

13.3 SYNTHETIC BIODEGRADABLE POLYMERS

Almost the only high molecular weight compounds shown to be biodegradable are the polyesters. The reason for this is the extremely hydrolyzable backbone of the polyesters.

The first of the biodegradable synthetic polymers that became commercially available was poly(glycolic acid) (PGA; **2**),[18] chosen on the basis of the results of a screening of potential materials subjected to degradation *in vivo* and in physiological solution over a period of 90 d.[19] The polymer was obtained by ring-opening polymerization of glycolide, and the resultant polymer was extruded into a stiff fibre, with a high melting point of 225 °C, which degraded *in vitro* in about 25 d.[20]

(**2**) Poly(glycolic acid) PGA

The glycolide has also been copolymerized with lactide, with the result that a stiff polymer, poly(glycolic-*co*-lactic acid) (PGA/LA; **3**), was obtained.[21] Sutures have been produced from a 90:10 glycolide:lactide monomer composition[22] and this suture retains its tensile strength somewhat longer than the poly(glycolic acid) homopolymer[23] and has a lower melting point. This line of application also includes artificial tendons made of carbon fibre composite with poly(L-lactic acid) as the matrix[24-26] and bone plates of poly(L-lactic acid).[27-29] Stereoblock copolymers of L- and D-lactides were synthesized by living ring-opening polymerization in order to prepare biodegradable materials.[30] Poly(ether ester) block copolymers based on poly(ethylene oxide) and poly(lactic acid)s have been synthesized with the aim of obtaining bioabsorbable polymers.[31]

GA LA

(**3**) Poly(glycolic acid-*co*-lactic acid) PGA/LA

The degradation of poly(glycolic acid) is also used in the medicinal field to obtain slow release systems for pharmaceutical drugs. By synthesizing block copolymers, an increased elasticity is obtained. Aliphatic block copolymers containing poly(glycolic acid) and poly(ethylene glycol) have been synthesized.[32]

Degradable block copolyesters of the aromatic polyester poly(ethylene terephthalate) with poly(oxyalkenes), poly(ethylene glycol)[33,34] or poly(tetramethylene glycol)[35] are used as biodegradable materials in surgical applications. The use of poly(tetramethylene glycol) instead of poly(ethylene glycol) increases elasticity and hydrophobicity, which in turn leads to slower hydrolytic degradation.

Aliphatic homopolyesters, *e.g.* poly(tetramethylene adipate) (PTMA), and block copolymers such as poly(ethylene succinate)-*b*-poly(ethylene glycol) (PES/PEG) and poly(ethylene succinate)-*b*-poly(tetramethylene glycol) (PES/PTMG; **4**) were synthesized and the subsequent degradation studied in a pseudoextracellular fluid (PSE) buffered at pH 7.3 and maintained at 37 °C.[36] The materials obtained showed thermoplastic elastomer behaviour, the degradation rate depending on the polyether composition. Poly(β-propiolactone) has been degraded in buffered salt solution (pH 7.2) at 37 °C.[37,38] Oriented fibres and nonoriented fibres show different degradation properties, especially with regard to the changes in mechanical properties. The changes in tensile strength are slower for the oriented material than for the nonoriented.[37,38]

Polymerization of 1,5-dioxepan-2-one gave a poly(ether ester) with amorphous properties, implying its usefulness as an amorphous block in copolymers possessing elastic properties.[39]

Polycaprolactones (PCL; **5**), aliphatic polyethers without branches, have been thoroughly studied, especially in biodegradation tests. They are generally prepared by ring-opening polymerization of ε-caprolactones. Tokiwa et al. have discussed the hydrolysis of PCL and biodegradation of PCL by fungus, and shown that PCL can be degraded enzymatically.[40-42] The fungus assimilates various polyesters but in general assimilation of aliphatic polyesters is better the greater the number of carbon atoms between the ester bonds.

(4) Poly(ethylene succinate)-b-poly(tetramethylene glycol) PES/PTMG

(5) Polycaprolactone PCL

Another approach has been made using enzymes capable of degrading PCL and blends of PCL and polymers such as polyethylene (PE), nylon 6, polystyrene (PS) and poly(β-hydroxybutyrate) (PHB).[43] PCL has also been blended with low density polyethylene (LDPE) and the biodegradability of this blend was monitored,[44] showing that the degradability could be controlled by combining the melt viscosity of PCL with that of LDPE. Blends of poly(ε-caprolactones) (PCL) and poly(L-lactic acid) (PLA) with poly(glycolic acid-co-L-lactic acid) gave a material with better permeability and form stability, but at the same time a higher degradation rate, suitable for drug delivery systems.[45] Poly(ethylene oxide)-co-poly(ethylene terephthalate) systems were prepared with the aim of obtaining materials suitable for surgical applications and a composition of 60/40 appeared to be optimum for use in composite structures with current medical elastomers.[46]

The biodegradability of polyurethanes was shown to be dependent on whether the prepolymer is a polyester or a polyether.[47] The polyether–polyurethanes (6) are resistant to biodegradation, whereas the polyester–polyurethanes (7) are readily attacked. The influence of the diisocyanates is also important for the biodegradability of the polyurethanes where linear diisocyanates are less resistant than cyclic diisocyanates. Potts et al.[48] have monitored the biodegradability as a function of structure for polyesters and have shown that a linear structure has a higher biodegradability than the corresponding branched structure. Huang has also monitored the biodegradability of polymers with different structures.[49] Important factors determining the biodegradability are the presence of hydrolyzable and/or oxidizable linkages. The presence of hydrophobic and hydrophilic segments in synthetic polymers improves the degradation.[49] Poly(ester urethane) networks have been prepared[50] by ring-opening copolymerization of L-lactide or glycolide and ε-caprolactone initiated by myoinositol and the degradation products are all nontoxic, which is essential for their use as degradable biomedical materials.

(6) Polyether–polyurethane

(7) Polyester–polyurethane

Poly(α-amino acid)s (8) are often susceptible to hydrolysis and the amino acids and their copolymers are potentially useful as biodegradable biomaterials. Copolymers of L-aspartic acid and L-glutamic acid have been prepared and their degradation in vitro by papain in pseudoextracellular fluids followed.[51] The copolymers degrade by random chain scission and the nature of the side chains is important for the rate of degradation. The synthetic poly(α-amino acid)s resembling native proteins can be degraded by proteolytic enzymes and the tissue reaction of these materials is similar to those associated with native proteins.

(8) Poly(α-amino acid)

Polyamides–nylons are generally reported to be quite resistant to microorganisms.[52] Bailey, however, proposed that nylon 66 can be modified by N-acetyl substitution to give materials that are biodegradable.[53] Copolymers of ethylene containing 10% ester units have been shown to be highly biodegradable, while copolymers containing only 2% are only slowly biodegradable.[54] Huang and Kitchen[55] report on the biodegradable polymers poly(amide enamine)s (9). These materials are similar to proteins in that they produce both acidic and basic products during degradation. Poly(amide enamine)s are promising materials for controlled release applications.

(9) Poly(amide enamine)

Polyanhydrides are a group of polymers with two sites in the repeating unit susceptible to hydrolysis. These are interesting materials due to their good biocompatibility.[56] Langer et al. synthesized aliphatic–aromatic polyanhydrides for slow release formulations.[57] The bioerodible polymers, especially polyanhydrides, are useful materials for drug delivery. The degradation rates can be altered by changes in the polymer backbone. Aliphatic polyanhydrides degrade within a few days while aromatic polyanhydrides can degrade slowly over a period of several years.[58] Recently, a new synthetic route for producing linear poly(adipic anhydride) by use of ketene gas has been presented.[59] This synthetic route has the advantage of avoiding formation of acetic acid, which could lead the reaction backwards. Polyanhydrides are useful in biomedical applications due to their fibre-forming properties. Increase of the aliphatic chain length between the acid groups not only increases the molecular weight but also notably improves the hydrolytic stability.[60,61]

Aliphatic polycarbonates are polymers derived from the polymerization of six-membered ring molecules. The linear macromolecule degrades to the monomers at high temperature. McNeill and Rincon[62] report on the type of degradation products from the degradation of poly(trimethylene carbonate) (10). In this case the main product is trimethylene carbonate, whereas other workers report several different degradation products besides the six-membered molecule (the monomer).[63,64] Poly(trimethylene carbonate) was synthesized using cationic and anionic initiators, giving high molecular weight compounds with rubbery character at room temperature.[65] Since the material obtained is an aliphatic polycarbonate, it could be useful as a biodegradable polymer for medical applications.

(10) Poly(trimethylene carbonate)

Several synthetic polymers which by themselves are not biodegradable can be modified so that a biodegradability is obtained. Guillet[66] has investigated copolymers of vinyl ketones and styrene and ethylene (Ecolyte). The materials obtained were shown to biodegrade and the degradation was followed using a respirometric technique.[67,68] Bacteria belonging to species such as *Pseudomonas*, *Alcaligenes*, *Achromobacter*, *Arthrobacter* and *Aerococcus* amongst others, were isolated from samples containing Ecolyte as the sole carbon source.[69] (*Pseudomonas* is especially noted for its nutritional versatility. Species can be found that are able to degrade starch, cellulose, agar, chitin, phenols, naphthalene, hydrocarbons and resins.)

Scott[70] discusses how common packaging can be made biodegradable by photooxidation. A degradable plastic is obtained using additives which are initially antioxidants but after exposure to sunlight function as photoinitiators. Hatakeyama et al.[71] describe the biodegradation of poly(3-methoxy-4-hydroxystyrene) (11) as a simplified model of softwood lignin, having pendant guaiacyl groups. This polymer deteriorates in soil and the primary degradation product is vanillic acid.[71] The biodegradation of lignin-related polystyrenes is also reported for polymers such as poly(4-hydroxystyrene) and poly(3,5-dimethoxy-4-hydroxystyrene).[72]

Another approach to the task of making inert polymers biodegradable is to use different additives (preferably biodegradable ones) in order to promote a deterioration. Polyethylene with starch is probably the most interesting example at the moment. Starch is a cheap material and most micro-

(11) Poly(3-methoxy-4-hydroxystyrene)

organisms attack it readily. Griffin introduced the idea of mixing granular starch in its natural form with polyethylene.[73-80] The product has been improved by the addition of transition metal salts, unsaturated polymer and a thermal stabilizer in a master batch since the original development was made.[80] The biodegradation step is the initiation of the total degradation by the consumption of starch by microorganisms, leaving a brittle material more susceptible to photooxidation, *etc*. Inclusion of starch alone in LDPE does not improve the total degradation, as observed in photooxidation and thermal experiments.[81] LDPE with starch has been tested in several different environments (*e.g.* soil, refuse burial, anaerobic waste treatment, *etc.*), in which it was shown that starch is removed to varying extents in all exposure environments.[82-86]

The Scott–Gilead photobiodegradation process uses metal dithiocarbamates in LDPE, HDPE and PP with applications in packaging and in mulch films.[87] Effective antioxidant protection and heat stabilizers are provided during processing, storage and use and, after an induction period, a fast photooxidation occurs. The materials referred to above have yielded a more easily oxidized polymer, with increased hydrophilicity and an increased surface area. This could give a material more susceptible to biodegradation.

Poly(vinyl alcohol) (PVA) is a water soluble polymer with applications as thickening agent for emulsions and suspensions and as a packaging film where water solubility is desired. By chemical treatment, the final form of the polymer can be insoluble. Polyethylene glycol (PEG) and PVA are readily degraded by microorganisms. It has been shown that isolated cultures cannot grow on liquid cultures containing PVA as the sole carbon source but that mixed cultures (*Pseudomonas* sp., *Alcaligenes* sp. and *Pseudomonas putida*) can grow.[88]

Polycarboxylates containing biodegradable blocks have been prepared where the blocks consist of polysaccharides and PVA and the block polymers obtained have been shown to have improved biodegradability.[89,90] Water soluble poly(sodium carboxylate) has been subjected to biodegradation tests in activated sludge. The polymer containing more than 60% glycopyranose groups shows enhanced biodegradability.[91]

A review of water soluble polymers presented by Swift[92] contains information on the biodegradability of PVA, poly(carboxylic acid)s, poly(hydroxycarboxylic acid) (**12**) and poly(acrylic acid), among others. PVA is the only carbon–carbon backbone polymer that is biodegradable. For water soluble polymers, branching on the polymer backbone is detrimental to biodegradation, whereas heteroatomic backbone polymers (C—O—C) have enhanced biodegradability.[92]

(12) Poly(hydroxycarboxylic acid)

13.4 NATIVE BIODEGRADABLE POLYMERS

Native polymers form the basis for life and intelligence on earth. Besides being sources of nutrients, many native polymers have been used in the fabrication of clothes, for example, for thousands of years.

The polypeptides (polyamides of α-amino acids) make up fibres and occur in proteins extracted from nonfibrous natural products: important examples are wool and silk. Several of these natural, biodegradable materials have now been replaced by synthetic ones, although a material such as silk still remains a product with very competitive characteristics. Marine cuticle collagen has been investigated for use in adhesives, especially bioadhesives in aqueous environments.[93] Bioadhesives have been studied in order to prepare synthetic proteins of the same type.[94] Several research activities try to imitate natural proteins and obtain materials with novel properties.[95] Caseins are biodegradable proteins whose main applications are in the food industry, but caseins are also used as glues in plywood, for example. An undesirable side effect of the biodegradability of caseins was observed when casein was used in self-levelling concretes. The biodegradation of the materials caused malodorous low molecular weight compounds and this was one phenomenon observed in connection with the 'sick building syndrome'.[96-100]

Polysaccharides are a class of widely used compounds, *e.g.* dextrans, cellulose or chitins. Starch, another important polysaccharide, is not only used in the food industry but also in the manufacture of paper, adhesives, and so on. Recently, starch, of which amylose (13) and amylopectin (14) are the basic components, has become important as a cheap biodegradable additive in several inert synthetic polymers. The most important properties of the polysaccharides are their ability to alter the flow characteristics of fluids and their ability to act as absorbants, gel formers, ion exchange agents, *etc*. All of these polysaccharides individually perform biological functions.

(13) Amylose

(14) Amylopectin

Polysaccharides have been synthesized *via* the ring-opening polymerization of anhydro sugar derivatives, which has given materials with novel properties.[101] The polysaccharides can be degraded by dextranase and are thus biodegradable. Polysaccharides are used in encapsulation, gelling or thickening and suspending agents. Xanthan is the one major bacterial polysaccharide commercially available.[102] The bacterial polysaccharide gellan, a gelling agent, can be obtained by aerobic fermentation of *Pseudomonas elodea*.[103] Several other types of bacteria have been investigated for the production of gelation polymers.[104] *Lactic acid* bacteria and *Acetobacter* are examples of bacteria capable of synthesizing polymers with gelling characteristics.

Many polymer researchers are of the opinion that polymer chemistry had its beginning with the characterization of cellulose. Cellulose was isolated for the first time some 150 years ago. Cellulose differs in some respects from other polysaccharides produced by plants, the molecular chain being very long and consisting of only one repeating unit (the cellobiose). Naturally it occurs in a crystalline state. From the cell walls, cellulose is isolated in microfibrils by chemical extraction. Nonglucose sugars range from 15–50%.[105] When cellulose is treated with sulfuric acid and titrated back to pH 3, the microfibrils fall into short rods still giving the X-ray diagram of cellulose.[106] The biodegradation of cellulose is complicated, partly because cellulose exists together with lignin, for example, in wood cell walls. White rot fungi (attacking lignin preferentially) and brown rot fungi (attacking cellulose) are well-known microorganisms responsible for the deterioration of wood.

The degradation rate depends on both structure and chemical composition. The biodegradation of modified cellulose depends both on the type of substituents and on the degree of substitution. Cellulose triacetate is resistant to microorganisms, whereas cellulose with a low degree of acetylation readily degrades. Narayan *et al.* report different synthetic routes to well-defined, tailored cellulose/starch synthetic graft copolymers. These graft copolymers can function as compatibilizing/interfacial agents for alloying cellulose and starch with synthetic polymers to give a biodegradable/biocompatible material.[107,108]

Combining cellulose with chitosan results in various kinds of strong, gas barrier and water resistant composite films.[109] Films, nonwoven textiles and various mouldings yielding biodegradable materials are planned.[110] Chitin (15) is a macromolecule formed in the shells of crabs, lobsters, shrimps, and insects, amongst others. Chemical treatment of chitin produces chitosan. The materials are biocompatible, and have antimicrobial activity as well as the ability to absorb heavy metal ions. Because of their water-retaining and moisturizing ability they also find applications in the cosmetic industry. Using chitin and chitosan as carriers, a water-soluble prodrug has been synthesized.[111]

(15) Chitin

Modified chitosans have been prepared with various chemical and biological properties.[112] N-Carboxymethylchitosan and N-carboxybutylchitosan have been prepared for use in cosmetics and in wound treatment.[113]

Chitin derivatives can also be used as drug carriers,[114] and a report of the use of chitin in absorbable sutures shows that chitins have the lowest elongation among suture materials consisting of chitin, PGA [poly(glycolic acid)], plain catgut and chromic catgut.[115] The tissue reaction of chitin is similar to that of PGA.

Alginate, the structural biopolymer of brown kelp, is used because of its ability to thicken aqueous solutions and to form gels under mild conditions.[116] Pullulan, a bacterial polysaccharide produced by *Aeurobasidium pullulans*, is a water soluble polymer with a viscosity comparable to that of arabic gum [the repeating unit is maltobiose (glucose is the monomer)]. It has applications in materials such as moulding resins and packaging films, since it forms excellent films.

Microbial polyesters have become the subject of considerable research interest over several years. Different bacteria produce polyesters as an energy and carbon reserve. Poly(β-hydroxybutyrate) (PHB) was first isolated in 1925 by Lemoigne in France. PHB belongs to the family of poly-(β-hydroxyalkanoates) (PHA). PHB is produced by several bacteria and fungi during starvation, *i.e.* nutrient limitation, lack of essential nutrients such as nitrogen, oxygen, phosphorus, *etc*.

PHB is an interesting material due to its biodegradability which, together with its biocompatibility, makes it useful in applications such as surgical pins, sutures, wound dressings, bone replacement, stimulation of bone growth and healing by piezoelectric properties.[117,118] PHB also shows such interesting properties that it is used as a packaging, in spite of its relatively high price.

Considerable interest arose recently when Holmes *et al.* (ICI)[119] developed a large-scale, controlled fermentation process for the production of copolymers of PHB. Feeding the bacteria with a variety of carbon sources led to the production of different copolymers and a material was obtained with better mechanical properties than PHB.[120-129]

The biodegradation of PHB and copolymers has been studied using environments such as soil, activated sludge and sea water.[124] Films (0.07 mm thick) of PHB (homopolymer), a copolymer of 91% 3HB and 9% 4HB and a copolymer of 50% 3HB and 50% 3HV (16) were subject to biodegradation in soil. The fastest biodegradation rate was obtained for P(3HB-*co*-9% 4HB). In activated sludge the P(3HB-*co*-9% 4HB) was completely decomposed after two weeks.[124]

(16) Poly(3-hydroxybutyrate-*co*-3-hydroxyvalerate) PHB/HV

The native polyesters are also hydrolyzed in water at a very slow rate. *In vivo* this is the main mechanism, involving chain scission of the polymer. The hydrolytic degradation of hydroxybutyrate–hydroxyvalerate copolymers *in vitro* begins with a surface modification, accompanied by water diffusion into the matrix.[130] A progressive increase in porosity facilitates the diffusion by removal of degradation products.[130] Doi *et al.*[131] report that the hydrolytic degradation of microbial polyesters occurs by homogeneous erosion over two stages: random hydrolytic chain scission of the ester group, leading to a decrease in molecular weight, is followed by a second step ($M_n \sim 13\,000$) in which weight loss occurs.

Bacterial polyesters have also been blended with PE and PS. The goal was to expand the physical properties while retaining biodegradability.[132,133] The biodegradation of PHB and copolymers of PHB with γ-hydroxy valerate (PHV) was monitored using an accelerated test based on a chemostat-like technique.[134] Latex films of the polyesters were compared with paper coated on one side with latex and the materials were immersed in a broth containing microorganisms isolated from activated sludge. The latex films were readily degraded and the coated papers lost about 60% of their initial weight after a week of degradation.[134]

13.5 BIOMEDICAL BIODEGRADABLE POLYMERS

During the last 30 years a number of new polymers have been synthesized with characteristics making them suitable for medical applications. Recently the term biomaterial was defined as 'a nonviable material used in medical device application that is intended to interact with a biological system'.[135] In this context it is also important to define biocompatibility, which deals with how the tissue reacts to foreign materials. Williams proposed a definition of biocompatibility: 'the ability of a material to perform with an appropriate host response in a specific application.[135] An extensive survey of implant devices in use today is given in a review by Hench.[136] The review deals with polymeric materials, but also gives examples of composites and metals in use for medical applications. Surgical biomaterials can be sutures, macroscopic implants, porous composites (artificial skin, *etc.*) and drug delivery systems and depending on the use biodegradability is a necessary characteristic of these materials.[137] The degradation of surgical polymers in physiological environments is highly influenced by enzymes, lipids and bacteria.[138] Implant fixation by adhesives offers a number of advantages over conventional metal osteosynthesis in fracture treatments.[139] Examples of adhesives are collagen, epoxy resins and polyurethane forms. An understanding of the complex interactions of biomaterial surfaces is important for a successful use of a polymer in medical applications.[140] Modifications of a surface involves alteration of its chemical and/or physical nature.

13.6 PLASTIC WASTE

Using (bio)degradable polymers cannot totally eliminate the problems associated with plastic waste. Hitherto the polymers have been synthesized to be inert and bioresistant and ways have been sought to prevent all types of degradation. Now the environment is overloaded with waste and it is necessary to try to solve this situation. In the future several options will be asked for when dealing with polymers. Sanitary landfills have been used routinely but, due to lack of oxygen, moisture and microorganisms, they can give very slow degradation rates. Composting uses the natural cycle to degrade paper and food waste as well as biodegradable plastics.[141] Another approach to deal with plastic waste is incineration. The energy is utilized in this case, although the emission of air pollutants generally raises the cost of incineration considerably.[142] Recycling is also an attractive choice in this context. The poly(ethylene terephthalate) (PET) bottle is an example of a polymeric material where recycling is used. Here the wash time for the PET bottles is three times longer than is needed to clean the bottles alone due to difficulties in removing the paper/plastic label and adhesives.[143] This reflects the difficulties encountered in recycling polymers; complex plastic wastes create problems for the recycling units as well as difficulties when designing new materials from recycled polymers.

In the future, the use of biodegradable polymers will involve several different plastic waste management techniques depending on the situation. Already at the synthesis level, a new material will demand a careful consideration of possible waste management: landfill, composting, incineration or recycling, and possibly this will also decide the synthesis route, possible additives and the whole lifetime of the biodegradable polymer.

13.7 ARE BIODEGRADABLE POLYMERS THE FUTURE?

New ASTM guidelines are under preparation for testing biodegradable polymers, as opposed to photodegradable and oxidatively and hydrolytically degradable ones. Several countries have taken action against nondegradable polymers, in particular Japan, where the Ministry of International Trade and Industry conducted a feasibility study on the development of biodegradable plastics in 1989. Biopolymers will be developed by microorganisms, chemical synthesis of biopolymers and the commercial use of natural macromolecules. A biodegradable plastic research group was also organized by about 50 companies in 1989 in Japan.

Today, large efforts are being made to make existing nondegradable polymers biodegradable, photodegradable or hydrolyzable by chemical modification or by the inclusion of additives (*e.g.* sensitizers and biopolymers), and by developing methods to make it possible to use native biodegradable polymers [*e.g.* poly(β-hydroxybutyrate)]. Renewed interest in several hydrolyzable polymers (*e.g.* polyesters, polyanhydrides and polycarbonates) has resulted in new materials suitable for medical devices. Functional polymers are a concept which has become important in biomedical and environmental polymers. By homo- or co-polymerization polymers are prepared whose main property is related to the type of functional group in the main chain. Several of these polymers are of interest as biocompatible polymers and polymeric drugs.[144] Many of the polymers are very expensive solutions for packaging and mulch films, but in biomedical connections the cost is of little consequence. In the primary step, a new material can be expensive, but, for large series, even these complex biodegradable polymers can be used for applications such as packaging.

13.8 REFERENCES

1. G. S. Kumar, V. Kalpagam and U. S. Nandi, *JMS-Rev. Macromol. Chem. Phys.*, 1982, **22** (2), 225.
2. A.-C. Albertsson and S. Karlsson, in 'Degradable Materials, Perspectives, Issues and Opportunities: the First International Scientific Consensus Workshop Proceedings, ed. S. A. Barenberg *et al.*, CRC Press, Boca Raton, FL, 1990, p. 263.
3. A.-C. Albertsson, in 'Advances in Stabilization and Degradation of Polymers, ed. A. Patsis, Technomic, Lancaster, CT, 1989, vol. 1, p. 115.
4. A.-C. Albertsson, Ph.D. Thesis, The Royal Institute of Technology, Stockholm, 1977.
5. B. Dolezel, *Br. Plast.*, 1967, **40**, 105.
6. A.-C. Albertsson, S.-O. Andersson and S. Karlsson, *Polym. Degradation Stab.*, 1987, **18**, 73.
7. A.-C. Albertsson, *Eur. Polym. J.*, 1980, **16**, 623.
8. A.-C. Albertsson and S. Karlsson, *Prog. Polym. Sci.*, 1990, **15**, 177.
9. 'Recommended Practice for Determining Resistance of Synthetic Polymeric Materials to Fungi', Annual Book of ASTM Standards, ASTM, Philadelphia, PA, 1985, vol. 8.03, ASTM-G-21.
10. 'Recommended Practice for Determining Resistance of Plastics to Bacteria', Annual Book of ASTM Standards, ASTM, Philadelphia, PA, 1985, vol. 8.03, ASTM-G-22.
11. G. Swift, in 'International Symposium on Biodegradable Polymers, Tokyo, 1990', Biodegradable Plastic Society, Tokyo, 1990, p. 61.
12. A.-C. Albertsson, *J. Appl. Polym. Sci.*, 1978, **22**, 3419.
13. S. Karlsson, O. Ljungquist and A.-C. Albertsson, *Polym. Degradation Stab.*, 1988, **21**, 237.
14. A.-C. Albertsson, Z. G. Banhidi and L. L. Beyer-Ericsson, *J. Appl. Polym. Sci.*, 1978, **22**, 3435.
15. G. J. L. Griffin, *Pure Appl. Chem.*, 1980, **52**, 399.
16. A.-C. Albertsson and Z. G. Banhidi, *J. Appl. Polym. Sci.*, 1980, **25**, 1655.
17. A.-C. Albertsson and B. Rånby, *J. Appl. Polym. Sci., Appl. Polym. Symp.*, 1979, **35**, 423.
18. E. E. Schmitt and R. A. Polistina, *US Pat.* 3 297 033 (1967) (*Chem. Abstr.*, 1967, **66**, 38 656).
19. E. J. Frazza and E. E. Schmitt, *Biomed. Mater. Symp.*, 1971, **I**, 43.
20. C. C. Chu, *Ann. Surg.*, 1982, **195**, 55.
21. A. K. Schneider, *US Pat.* 3 636 956 (1972).
22. D. Wasserman and A. Levi, *Can. Pat.* 950 308 (1974).
23. R. L. Kronenthal, in 'Polymers in Medicine and Surgery', ed. R. L. Kronenthal, Z. Oser and E. Martin, Plenum Press, New York, 1975, p. 10.
24. J. Aragona, J. R. Parsons, H. Alexander and A. B. Weiss, *Clin. Orthop. Relat. Res.*, 1981, **160**, 268.
25. H. Alexander, A. B. Weiss and J. R. Parsons, *Aktuel. Probl. Chir. Orthop.*, 1983, **26**, 78.
26. J. R. Parsons, H. Alexander and A. B. Weiss, *Surg. Sci. Ser.*, 1984, **2**, 417.
27. M. Vert, F. Chabot, J. Leray and P. Christel, *Makromol. Chem., Suppl.*, 1981, **5**, 30.
28. P. Christel, F. Chabot, J. Leray, C. Morin and M. Vert, *Adv. Biomater.*, 1982, **3**, 271.
29. F. Chabot, M. Vert, S. Chapelle and P. Granger, *Polymer*, 1983, **24**, 53.
30. N. Yui, P. J. Dijkstra and J. Feigen, *Makromol. Chem.*, 1990, **191**, 481.
31. D. Cohn and H. Younes, *J. Biomed. Mater. Res.*, 1988, **22**, 993.
32. D. J. Casey and M. S. Robys, *Eur. Pat.* 108 933 (1984).
33. D. K. Gilding and A. M. Reed, *Polymer*, 1979, **20**, 1454.
34. A. M. Reed and D. K. Gilding, *Polymer*, 1981, **22**, 499.
35. V. V. Shevchenko, A. S. Chegolya and E. M. Aisenshtein, *Vysokomol. Soedin.*, 1985, **A27**, 2333 (*Chem. Abstr.*, 1985, **104**, 149 525).
36. A.-C. Albertsson and O. Ljungquist *Acta Polym.*, 1988, **39**, 95.
37. T. Mathisen, M. Lewis and A.-C. Albertsson, *J. Appl. Polym. Sci.*, 1991, **42**, 2365.
38. T. Mathisen and A.-C. Albertsson, *J. Appl. Polym. Sci.*, 1990, **38**, 591.
39. T. Mathisen, K. Masus and A.-C. Albertsson, *Macromolecules*, 1989, **22**, 3842.
40. Y. Tokiwa and T. Suzuki, *Nature*, 1977, **270**, 76.
41. Y. Tokiwa, T. Ando and T. Suzuki, *J. Ferment. Technol.*, 1976, **54**, 603.
42. Y. Tokiwa, T. Suzuki and T. Ando, *J. Appl. Polym. Sci.*, 1979, **24**, 1701.
43. Y. Tokiwa, in '33rd IUPAC, International Symposium on Macromolecules, Book of Abstracts, Montreal, Canada, 1990', IUPAC, Montreal, 1990, p. 2.6.6.
44. Y. Tokiwa, T. Ando, K. Takeda, A. Iwamoto and M. Koyama, in 'International Symposium on Biodegradable Polymers, Tokyo, 1990', Biodegradable Plastic Society, Tokyo, 1990, p. 69.
45. Y. Cha and C. G. Pitt, *Biomaterials (Guildford, Engl.)*, 1990, **11**, 108.

46. D. K. Gilding and A. M. Reed, *Polymer*, 1979, **20**, 1454.
47. R. T. Carby and A. M. Kaplan, *Appl. Microbiol.*, 1968, **16**, 900.
48. J. E. Potts, R. A. Clendinning and W. B. Ackart, *Polym. Prepr., Am. Chem. Soc., Div. Polym. Chem.*, 1972, **13**, 629.
49. S. J. Huang, *Polym. Prepr., Am. Chem. Soc., Div. Polym. Chem.*, 1990, **31**, (2), 54.
50. P. Bruin, G. J. Veenstra, A. J. Nijenhuis, A. J. Pennings, *Makromol. Chem., Rapid Commun.*, 1988, **9**, 589.
51. T. Hayashi and M. Iwatsuki, *Biopolymers*, 1990, **29**, 549.
52. J. E. Potts, in 'Kirk-Othmer Encyclopedia of Chemical Technology', ed. M. Grayson, Wiley-Interscience, New York, 1984, suppl. vol., p. 626.
53. D. M. Ennis, A Kramer, C. W. Jamesson, P. H. Mazzocchi and W. J. Bailey, *Appl. Environ. Microbiol.*, 1978, **35**, 51.
54. W. J. Bailey and B. Gapaud, ACS *Symp. Ser.*, 1985, **280**, 423.
55. S. J. Huang and O. Kitchen, *Polym. Prepr., Am. Chem. Soc., Div. Polym. Chem.*, 1990, **31** (2), 207.
56. K. W. Leong, B. C. Brott and R. Langer, *J. Biomed. Mater. Res.*, 1985, **19**, 941.
57. K. W. Leong, P. D'Amore, M. Marletta and R. Langer, *J. Biomed. Mater. Res.*, 1986, **20**, 51.
58. E. Ron, E. Mathiowitz, G. Mathiowitz and R. Langer, *Polym. Prepr., Am. Chem. Soc., Div. Polym. Chem.*, 1989, **30**, (1), 462.
59. A.-C. Albertsson and S. Lundmark, *J. Macromol. Sci., Chem.*, 1988, **A25**, 247.
60. A.-C. Albertsson and S. Lundmark, *J. Macromol. Sci., Chem.*, 1990, **A27**, 397.
61. A.-C. Albertsson and S. Lundmark, *Br. Polym. J.*, 1990, **23**, 205.
62. I. C. McNeill and A. Rincon, *Polym. Degradation Stab.*, 1989, **24**, 59.
63. I. C. McNeill and H. A. Leiper, *Polym. Degradation Stab.*, 1985, **11**, 267.
64. I. C. McNeill and H. A. Leiper, *Polym. Degradation Stab.*, 1985, **12**, 373.
65. A.-C. Albertsson and M. Sjöling, *J. Macromol. Sci., Chem.*, 1992, **A29**, 43.
66. J. E. Guillet, *Proc. ACS, Div. Polym. Mater. Sci. Eng.*, 1990, **63**, 946.
67. P. H. Jones, D. Pradas, H. Heskins, M. H. Morgan and J. E. Guillet, *Environ. Sci. Technol.*, 1974, **8**, 919.
68. J. E. Guillet, T. W. Regulski and T. B. McAneney, *Environ. Sci. Technol.*, 1974, **8**, 923.
69. L. R. Spencer, M. Heskins and J. E. Guillet, *Proc. Int. Biodegradation. Symp.*, 3rd, *1975*, 1976, 753.
70. G. Scott, *Arabian J. Sci. Eng.*, 1988, **13** (4), 605.
71. H. Hatakeyama, E. Hayashi and T. Haraguchi, *Polymer*, 1977, **18**, 759.
72. T. Haraguchi and H. Hatakeyama, in 'Lignin Biodegradation: Microbiology, Chemistry and Potential Applications', ed. T. K. Kirk *et al.*, CRC Press, Boca Raton, FL, 1980, vol II, p. 147.
73. G. J. L. Griffin, *Ger. Pat.* 2 455 732 (1975).
74. G. J. L. Griffin, in 'Conference on Degradability of Polymers and Plastics, Institute of Electrical Engineers, 1973', London, 1973, p. 15/1.
75. G. J. L. Griffin, *ACS, Adv. Chem. Ser.*, 1974, **134**, 159.
76. G. J. L. Griffin, in 'Preprints of the 15th Microsymposium on the Degradation and Stabilization of Polyolefins, Prague, 1975', Czechoslovak Academy of Sciences, Prague, 1975, p. 66.
77. G. Griffin and H. Mivetchi, *Proc. Int. Biodegradation Symp.*, 3rd, *1975*, 1976, 807.
78. G. J. L. Griffin, *J. Polym. Sci., Polym. Symp.*, 1976, **57**, 281.
79. G. J. L. Griffin, *Pure Appl. Chem.*, 1980, **52**, 399.
80. G. J. L. Griffin, *Int. Pat.* PCT/GB88/00386 (1988).
81. A.-C. Albertsson, C. Barenstedt and S. Karlsson, *Polym. Degradation Stab.*, 1992, in press.
82. R. P. Wool and M. A. Cole, in 'ASM Engineering Handbook', ed. J. N. Epel, J. M. Margolis, S. Newman and R. B. Seymour, ASM, Metals Park, OH, 1988, vol. 2, p. 783.
83. R. P. Wool, J. S. Peansky, J. M. Long and S. M. Goheen, in 'Degradable Materials; Perspectives, Issues and Opportunities: the First International Scientific Consensus Workshop Proceedings', ed. S. A. Barenberg *et al.*, CRC Press, Boca Raton, FL, 1990, p. 515.
84. J. S. Peansky, J. M. Long and R. P. Wool, *J. Polym. Sci., Polym. Phys. Ed.*, 1991, **29**, 565.
85. W. J. Maddever and P. D. Campbell, in 'Degradable Materials, Perspectives, Issues and Opportunites: the First International Scientific Consensus Workshop Proceedings', ed. S. A. Barenberg *et al.*, CRC Press, Boca Raton, FL, 1990, p. 237.
86. R. G. Austin, in 'Degradable Materials, Perspectives, Issues and Opportunities: the First International Scientific Consensus Workshop Proceedings', ed. S. A. Barenberg *et al.*, CRC Press, Boca Raton, FL, 1990, p. 209.
87. D. Gilead and G. Scott, *Br. Pat.* 1 344 388 (1978).
88. M. Shimao and N. Kato, in 'International Symposium on Biodegradable Polymers, Tokyo, 1990', Biodegradable Plastic Society, Tokyo, 1990, p. 80.
89. S. Matsumura, M. Nishioka, H. Shigeno and S. Yoshikawa, in International Symposium on Biodegradable Polymers, Tokyo, 1990', Biodegradable Plastic Society, Tokyo, 1990, p. 75.
90. S. Matsumura, S. Maeda and Y. Yoshikawa, *Makromol. Chem.*, 1990, **191**, 1269.
91. S. Matsumura, M. Nishioka and S. Yoshikawa, *Polym. Prepr. Jpn. (Engl. Transl.)*, 1990, **39** (2), E969.
92. G. Swift, *Proc. ACS, Div. Polym. Mater. Sci. Eng.*, 1990, **63**, 848.
93. T. Takimoto and H. Yamamoto, *Polym. Prepr. Jpn. (Engl. Transl.)*, 1990, **39** (2), E1087.
94. H. Yamamoto, A. Nishida, T. Takimoto, K. Ikeda and S. Yamaguchi, *Polym. Prepr. Jpn. (Engl. Transl.)*, 1990, **39** (2), E1088.
95. J. H. Richards, in 'Macromolecular Preprints, Second Euro–American Conference on Functional Polymers and Biopolymers, Oxford, 1989', ed. M. J. Epton, Oxford University Press, 1989, p. 11.
96. S. Karlsson, E. Banhidi, Z. G. Banhidi and A.-C. Albertsson, in 'Proceedings of the 3rd International Conference Indoor Air, Stockholm, 1984', ed. B. Berglund *et al.*, Swedish Council for Building Research, Stockholm, 1984, vol. 3, p. 247.
97. S. Karlsson, Z. G. Banhidi and A.-C. Albertsson, *Appl. Microbiol. Biotechnol.*, 1988, **28**, 305.
98. S. Karlsson, Z. G. Banhidi and A.-C. Albertsson, *J. Chromatogr.*, 1988, **442**, 267.
99. S. Karlsson, Z. G. Banhidi and A.-C. Albertsson, *Mater. Constr.* (Paris), 1989, **22**, 163.
100. S. Karlsson and A.-C. Albertsson, *Mater. Constr.* (Paris), 1990, **23**, 352.
101. K. Kobayashi, *Proc. ACS, Div. Polym. Mater. Sci. Eng.*, 1990, **62**, 477.
102. G. T. Colegrove, *Ind. Eng. Chem. Prod. Res. Dev.*, 1983, **22**, 456.

103. K. S. Kang, G. T. Veeder, P. J. Mirrasoul, T. Kanecko and W. Cottrell, *Appl. Environ. Microbiol.*, 1982, **43**, 1086.
104. V. J. Morris, *Proc. ACS, Div. Polym. Mater. Sci. Eng.*, 1990, **62**, 462.
105. J. Cronhaw, A. Myers and R. D. Preston, *Biochim. Biophys. Acta*, 1958, **27**, 89.
106. B. G. Rånby and G. Ribi, *Experientia*, 1950, **6**, 2.
107. R. Narayan, *Appl. Biochem. Biotechnol.*, 1988, **17**, 7.
108. C. J. Biermann, J. B. Chug and R. Narayan, *Macromolecules*, 1987, **20**, 954.
109. J. Hosokawa and M. Nishiyama, in 'International Symposium on Biodegradable Polymers, Tokyo, 1990', Biodegradable Plastic Society, Tokyo, 1990, p. 144.
110. J. Hosokawa, M. Nishiyama, K. Yoshihara, T. Kubo, *Ind. Eng. Chem. Res.*, 1990, **29**, 800.
111. K. Inosaka, Y. Ohya and T. Ouchi, *Polym. Prepr. Jpn. (Engl. Transl.)*, 1990, **39** (2), E975.
112. R. A. A. Muzzarelli, 'Chitin in Nature and Technology', Plenum Press, New York, 1986.
113. R. A. A. Muzzarelli, *Polym. Prepr., Am. Chem. Soc., Div. Polym. Chem.*, 1990, **31** (1), 626.
114. S. Tokura, Y. Miura, Y. Uraki, K. Watanabe, I. Saiki and I. Azuma, *Polym. Prepr., Am. Chem. Soc., Div. Polym. Chem.*, 1990, **31** (1), 627.
115. M. Tashibana, A. Yaita, H. Tamiura, K. Fukasawa, N. Nagasue and T. Nakanura, *Jpn. J. Surg.*, 1988, **18**, 533.
116. P. A. Sandford, *Polym. Prepr., Am. Chem. Soc., Div. Polym. Chem.*, 1990, **31** (1), 628.
117. R. Lenz, R. A. Gross, H. Brandl and R. C. Fuller, *Chin. J. Polym. Sci.*, 1989, **7**, 289.
118. R. H. Marchessault, 'Biopolyesters: Nature's Environmentally Compatible Plastic', manuscript, 1989.
119. P. A. Holmes, L. F. Wright and S. H. Collins (ICI), *Eur. Pat.* 0 052 459 (1982); *Eur. Pat.* 0 069 497 (1983).
120. H. Preusting, A. Nijenhuis and B. Witholt, *Macromolecules*, 1990, **23**, 4220.
121. S. Bloembergen, D. A. Holden, T. L. Bluhm, G. K. Hamer and R. H. Marchessault, *Macromolecules*, 1989, **22**, 1656.
122. S. Bloembergen, D. A. Holden, T. L. Bluhm, G. K. Hamer and R. H. Marchessault, *Macromolecules*, 1989, **22**, 1663.
123. Y. Doi and C. Abe, *Macromolecules*, 1990, **23**, 3705.
124. M. Kunioka, Y. Kawaguchi and Y. Doi, *Appl. Microbiol. Biotechnol.*, 1989, **30**, 569.
125. M. Kunioka, A. Tamaki and Y. Doi, *Macromolecules*, 1989, **22**, 694.
126. M. Knuioka, Y. Nakamura and Y. Doi, *Polym. Commun.*, 1988, **29**, 174.
127. Y. Doi, A. Tamaki, M. Knuioka and K. Soga, *J. Chem. Soc., Chem. Commun.*, 1987, 1635.
128. Y, Doi, M. Kunioka, Y. Nakamura and K. Soga, *Macromolecules*, 1987, **20**, 2988.
129. Y. Doi, M. Knuioka, Y. Nakamura and K. Soga, *Macromolecules*, 1986, **19**, 2860.
130. S. J. Holland, M. Yasin and B. J. Tighe, *Biomaterials*, 1990, **11**, 206.
131. Y. Doi, Y. Kanesawa, Y. Kawaguchi and M. Knuioka, *Makromol. Chem., Rapid Commun.*, 1989, **10**, 227.
132. S. N. Bhalakia, T. Patel, R. A. Gross and S. P. McCarthy, *Polym. Prepr., Am. Chem. Soc., Div. Polym. Chem.*, 1990, **31** (1), 441.
133. P. B. Dave, N. J. Ashar, R. A. Gross and S. P. McCarthy, *Polym. Prepr., Am. Chem. Soc., Div. Polym. Chem.*, 1990, **31** (1), 442.
134. R. H. Marchessault, C. Monasterios, B. Ramsey and I. Saracovan, in '33rd IUPAC, International Symposium on Macromolecules, Montreal, 1990', IUPAC, Montreal, 1990, p. 2.6.4.
135. D. F. Williams, *Mater. Sci. Technol.*, 1987, **3** (10), 797.
136. L. L. Hench, *Science (Washington, D.C.)*, 1980, **208**, 826.
137. S. Vainionpää, P. Rokkanen and P. Törmälä, *Prog. Polym. Sci.*, 1989, **14**, 679.
138. D. F. Williams, *J. Mater. Sci.*, 1982, **17**, 1233.
139. S. C. Weber and M. W. Chapman, *Clin. Orthop. Relat. Res.*, 1984, **191**, 249.
140. J. H. Brybrook and L. D. Hall, *Prog. Polym Sci.*, 1990, **15**, 715.
141. R. Narayan, in 'International Symposium on Biodegradable Polymers, Tokyo, 1990', Biodegradable Plastic Society, Tokyo, 1990, p. 35.
142. S. J. Huang, *Proc. ACS, Div. Polym. Mater. Sci. Eng.*, 1990, **63**, 633.
143. M. F. Bouzianis, in 'Degradable Materials, Perspectives, Issues and Opportunities: the First International Scientific Consensus Workshop Proceedings', ed. S. A. Barenberg *et al.*, CRC Press, Boca Raton, FL, 1990, p. 381.
144. O. Vogl, A.-C. Albertsson, D. A. Bansleben, E. Borsig, P. Grosso, Z. Janovic, M. Kitayama, S. J. Li, J. Muggee, Z. Nir, Y. Okahata, W. Padellok, M. D. Purgett and F. Xi, *Macromol. Chem., Suppl.*, 1984, **7**, 1.

14
Molecular Engineering of Liquid Crystalline Polymers

VIRGIL PERCEC and DIMITRIS TOMAZOS
Case Western Reserve University, Cleveland, OH, USA

14.1	INTRODUCTION AND HISTORICAL REVIEW		300
14.2	A BRIEF INTRODUCTION TO LIQUID CRYSTALS		301
	14.2.1 *Introduction to Low Molar Mass Liquid Crystals and Definitions*		301
	14.2.2 *Liquid Crystalline Polymers*		303
14.3	ISOMORPHISM OF LIQUID CRYSTALS		304
14.4	SOME CONSIDERATIONS ON THE 'RIGIDITY' OF 'ROD-LIKE' MESOGENS		305
	14.4.1 *Rigid Rod-like Groups*		305
	14.4.2 *Semirigid Rod-like Groups*		306
	14.4.3 *Flexible Rod-like Groups*		307
14.5	MANIPULATION OF PHASE TRANSITION TEMPERATURES THROUGH STRUCTURAL VARIATIONS: SOME THERMODYNAMIC CONSIDERATIONS		309
	14.5.1 *Equilibrium States*		309
	14.5.2 *Metastable States*		310
	14.5.3 *Influence of Molecular Weight on Phase Transition Temperatures*		313
		14.5.3.1 Case 1. Both monomeric structural unit and polymer display an enantiotropic mesophase	314
		14.5.3.2 Case 2. The structural unit displays a virtual or a monotropic mesophase; the polymer displays a monotropic or an enantiotropic mesophase	315
		14.5.3.3 Case 3. The structural unit displays a virtual mesophase; the polymer displays a virtual mesophase	315
		14.5.3.4 Case 4. Rigid rod-like polymers	315
14.6	MAIN CHAIN LIQUID CRYSTALLINE POLYMERS		317
	14.6.1 *Soluble and Fusible Main Chain Liquid Crystalline Polymers*		317
		14.6.1.1 Poly(p-phenylene) derivatives	318
		14.6.1.2 Metal-containing poly(yne)s	320
		14.6.1.3 Spinal columnar liquid crystalline polymers	322
		14.6.1.4 Aromatic polyamides and polyesters	322
		14.6.1.5 Rod-like soluble polyimides	334
		14.6.1.6 Thermotropic poly(1,4-arylenevinylene)s	335
		14.6.1.7 Polyurethanes	336
		14.6.1.8 Polycarbonates	337
		14.6.1.9 Poly(ester anhydride)s	337
		14.6.1.10 Poly(ester imide)s	339
		14.6.1.11 Polyhydrocarbons	341
		14.6.1.12 Polyethers	342
	14.6.2 *Persistence Lengths of Soluble Polyesters and Polyamides*		342
14.7	CHEMICAL HETEROGENEITY IN MAIN CHAIN LIQUID CRYSTALLINE COPOLYMERS		342
14.8	FLEXIBLE AND SEMIFLEXIBLE LIQUID CRYSTALLINE POLYETHERS		343
14.9	HYPERBRANCHED DENDRITIC LIQUID CRYSTALLINE POLYMERS		353
14.10	CYCLIC MAIN CHAIN LIQUID CRYSTALLINE POLYETHERS		356
14.11	SIDE CHAIN LIQUID CRYSTALLINE POLYMERS		356
	14.11.1 *General Considerations*		356
	14.11.2 *Molecular Engineering of Liquid Crystalline Polymers by Living Polymerization*		360
		14.11.2.1 General considerations	360

14.11.2.2	Influence of molecular weight on phase transition of poly[ω-{(4-cyano-4'-biphenylyl)oxy}alkyl vinyl ether]s	362
14.11.2.3	Molecular engineering of liquid crystalline phases by living cationic copolymerization	364
14.11.2.4	Side chain liquid crystalline polymers exhibiting a reentrant nematic mesophase	366
14.11.2.5	Influence of tacticity on phase transitions	368
14.11.2.6	Cyclic polysiloxanes containing mesogenic side groups	369
14.11.2.7	Side chain liquid crystalline polyalkynes	369
14.12	LIQUID CRYSTALLINE POLYMERS CONTAINING CROWN ETHERS AND POLYPODANTS	370
14.13	ELECTRON DONOR–ACCEPTOR (EDA) COMPLEXES OF LIQUID CRYSTALLINE POLYMERS CONTAINING DISCOTIC MESOGENS	371
14.14	MOLECULAR RECOGNITION DIRECTED SELF-ASSEMBLY OF SUPRAMOLECULAR LIQUID CRYSTALLINE POLYMERS	372
14.15	REFERENCES	376

14.1 INTRODUCTION AND HISTORICAL REVIEW

Liquid crystals were discovered in 1888 by the Austrian botanist Reinitzer[1] who observed that cholesteryl benzoate exhibits two melting points. The first melting at 145.5 °C transforms the crystal phase into an anisotropic, turbid liquid which 'melts' into an isotropic, clear liquid at 178.8 °C. This anisotropic liquid was first named by Lehmann in 1889 'flowing crystals', then 'crystalline liquid' and finally 'liquid crystal' in 1900.[2] The first synthetic nematic liquid crystals were prepared by Gattermann in 1890.[3] The first smectic liquid crystals were synthesized by Vorländer in 1902.[4]

The 58th discussion organized by the Faraday Society on the 24th and 25th of April, 1933 was devoted to 'Liquid Crystals and Anisotropic Melts',[5] thus recognizing the general scientific interest in the field of liquid crystals. The proceedings of this meeting are still providing a comprehensive entry to this field. In our opinion, the most valuable monographs describing the dependence between the molecular structure and the properties of thermotropic liquid crystals remain the book published by Gray in 1962 and his subsequent reviews.[6a–d] The history of the entire field of liquid crystals was described by Kelker.[7,8] A systematic description of the molecular structure–properties development during the first 100 years of liquid crystal chemistry was published by Demus.[9] Additional monographs which can introduce the reader to the field of both thermotropic and lyotropic liquid crystals[10–13] and provide basic information also on its history,[7,8,11] classification[14] and identification of various mesomorphic textures,[15,16] are available.

To our knowledge, the idea of performing chemical reactions in anisotropic liquids belongs to Svedberg who reported his first experimental data in 1916.[17–20] In his publications Svedberg showed that the rate of decomposition of picric acid, trinitroresorcinol and pyrogallol increases slightly with increasing temperature when the reaction is performed in the nematic phase of p-azoxyphenetole, jumps at the clearing point and continues steeply upwards in the isotropic liquid. In addition, the reaction rate exhibits a distinct change when the nematic phase is under the increasing influence of a magnetic field. Regardless of the accuracy of these results, they suggest that an anisotropic phase can play the role of a 'catalyst' in a certain reaction. Reactions and interactions in liquid crystalline media were recently reviewed.[21]

As early as 1922 Lehmann[22] recognized that some of the properties exhibited by liquid crystals may have analogies to those of the living state. At the Faraday Discussions from 1933[23,24] it was clearly recognized that the mobility and structural order of biological and synthetic liquid crystals furnish an ideal medium for catalytic action and that the living cell is actually a liquid crystal.

A significant citation from Bernal made at the Faraday Society meeting in 1933 is as follows:[24] 'Rinne was not the first to see that liquid crystals had a bearing on biological problems, but I think that he had glimpsed at the fundamental role that they played, and understood the essential nature of the properties that enabled them to do so.

The biologically important liquid crystals are plainly two or more component systems. At least one must be a substance tending to *para*-crystallinity and another will in general be water. This variable permeability of liquid crystals enables them to be as effective for chemical reactions as true liquids or gels as against the relative impenetrability of solid crystals. On the other hand, liquid crystals possess internal structure lacking in liquids, and directional properties not found in gels. These two properties have far-reaching consequences. In the first place, a liquid crystal in a cell through its own structure becomes a proto organ for mechanical or electrical activity, and when associated in specialized cells in heigher animals gives rise to true organs, such as muscle and nerve. Secondly, and probably more fundamentally, the oriented molecules in liquid crystals furnish an ideal medium for catalytic action, particularly of the complex type needed to account for growth and reproduction. Lastly, a liquid crystal has the possibility of its own structure, singular lines, rods and cones, etc. Such structures belong to the liquid crystal as a unit and not to its molecules which may be replaced by others without destroying them, and they persist in spite of the complete fluidity of the substance. These are just the properties to be required for a degree of organization between that of the continuous substance, liquid or crystalline solid and even the simplest living cell.'

Today, it is well accepted that many of the low molar mass and polymeric biological derivatives exhibit liquid crystalline phases.[25-37] Although the elucidation of the role of liquid crystalline phases in various biological processes is in an early stage of development, it should be quite instructive to recall the following statement: 'Liquid crystals stand between the isotropic liquid phase and the strongly organized solid state. Life stands between complete disorder, which is death, and complete rigidity, which is death again'.[25]

Onsager in 1933[38] and Flory in 1956[39] predicted that rigid rod-like macromolecules should display liquid crystallinity. However, the first main chain thermotropic liquid crystalline polymer was reported in open literature only in 1975.[40] This publication is independent from the flexible spacer concept publication of deGennes[41] in the same year. However, patents on thermotrophic main chain liquid crystalline polymers were published in 1973.[42] Publications dealing with the driving force behind the development of main chain thermotropic liquid crystalline polymers and various historical variants of these developments are available.[43-49]

Although the field of thermotropic side chain liquid crystalline polymers is older than that of thermotropic main chain liquid crystalline polymers, systematic investigations in this field started only after Ringsdorf et al. introduced the spacer concept in 1978.[50,51] Several monographs and review articles on side chain and main chain liquid crystalline polymers are available.[52-63] This brief introduction attempts to focus attention on the fact that both low molar mass and polymeric liquid crystals occur widely in nature.

Since the goal of this chapter is to discuss the present state of the art of the molecular engineering of liquid crystalline polymers, we will not review the entire field, but instead will consider those issues which can and should be used in the molecular design of liquid crystalline polymers with well-defined phase transitions. This will, however, require a brief and general introduction to low molar mass and polymeric liquid crystals. For additional information on this topic the reader should consult the literature cited in this introduction. This chapter is addressed mostly to those interested in the synthesis of liquid crystalline polymers with specific phase transitions.

14.2 A BRIEF INTRODUCTION TO LIQUID CRYSTALS

14.2.1 Introduction to Low Molar Mass Liquid Crystals and Definitions

A liquid crystalline or mesomorphic phase or mesophase refers to a state of matter in which the degree of order is between the almost perfect long-range positional and orientational order present in solid crystals and the statistical long-range disorder characteristic of isotropic liquids, amorphous solids and gases. Mesogens or mesogenic groups are compounds which under suitable conditions give rise to mesophases. They can be classified as nonamphiphilic (i.e. most frequently compounds exhibiting a rod-like or disc-like shape) and amphiphilic (i.e. compounds that contain within the same molecule lipophilic and hydrophilic groups which have the ability to dissolve in organic solvents and water, respectively). Thermotropic mesophases are induced by a change in temperature, while lyotropic mesophases are induced by a solvent. Some mesogens can generate amphotropic mesophases, i.e. the liquid crystalline phase can be induced either by a change in temperature, by a solvent, or by both.[64a] Other highly immiscible groups like perfluorinated and saturated hydrocarbons available within the structure of the same molecule also give rise to amphotropic mesophases.[64b]

Depending on their thermodynamic stability with respect to the crystalline phase, thermotropic mesophases can be virtual (unstable with respect to the crystalline phase), monotropic (metastable with respect to the crystalline phase), or enantiotropic (stable with respect to the crystalline phase).[65,66] Virtual mesophases exist only below the melting and crystallization temperature and therefore cannot be observed. Monotropic mesophases can be observed only during cooling due to the fact that the crystallization process is kinetically controlled and therefore is supercooled, while the liquid crystalline phase is thermodynamically controlled and is not supercooled. Enantiotropic mesophases can be observed both on heating and cooling. A detailed discussion on the conversion of a mesophase from virtual into monotropic and into enantiotropic will be presented later.

Traditionally, thermotropic mesophases exhibited by rod-like or calamitic mesogens can be classified in uniaxial, biaxial, and chiral nematic (cholesteric), and a number of different smectic phases which form either an untilted layer (smectic A, s_A etc.) or a tilted layer (smectic C, s_C etc.) structure (Figure 1). Characteristic for the nematic phase is a parallel orientation of the molecules with an axis that corresponds to the long axis of the mesogen. While uniaxial nematic phases (biaxial nematic[67] phases are not of interest for this discussion) exhibit a one-dimensional degree of order, smectic phases can exhibit a two-dimensional (s_A, s_C exhibiting unstructured layer structures) or a three-dimensional (s_B, s_G, s_E, s_H exhibiting structured layer structures) order.[16] Therefore, s_B, s_G, s_E and s_H liquid crystalline phases are in fact crystalline. Disc-like molecules exhibit various columnar or discotic mesophases (Figure 2).[68,69] Plastic crystals are a class of mesophases in which sphere-shaped molecules, though orientationally completely disordered, reside (with minor fluctuations) at the points of a spatial lattice, i.e. there is positional order but orientational disorder (e.g. methane, carbon tetrachloride, carbon tetrafluoride, camphor, cyclohexanol, etc.).

Upon increasing their concentration, amphiphilic mesogens generate isotropic, and then micellar solutions. At higher concentrations rod-like micelles generate hexagonal mesophases while spherical micelles generate cubic mesophases. Alternatively an isotropic solution of an amphiphilic molecule can assemble directly into a lamellar liquid crystalline phase (Table 1).[11,22,33,36,70,71] Cubic mesophases do not exhibit a texture since they are optically isotropic. A brief inspection of Figure 2 and Table 1 shows that the columnar hexagonal mesophase of discotic mesogens and the hexagonal mesophase of amphiphilic molecules are from a symmetry point of view identical.

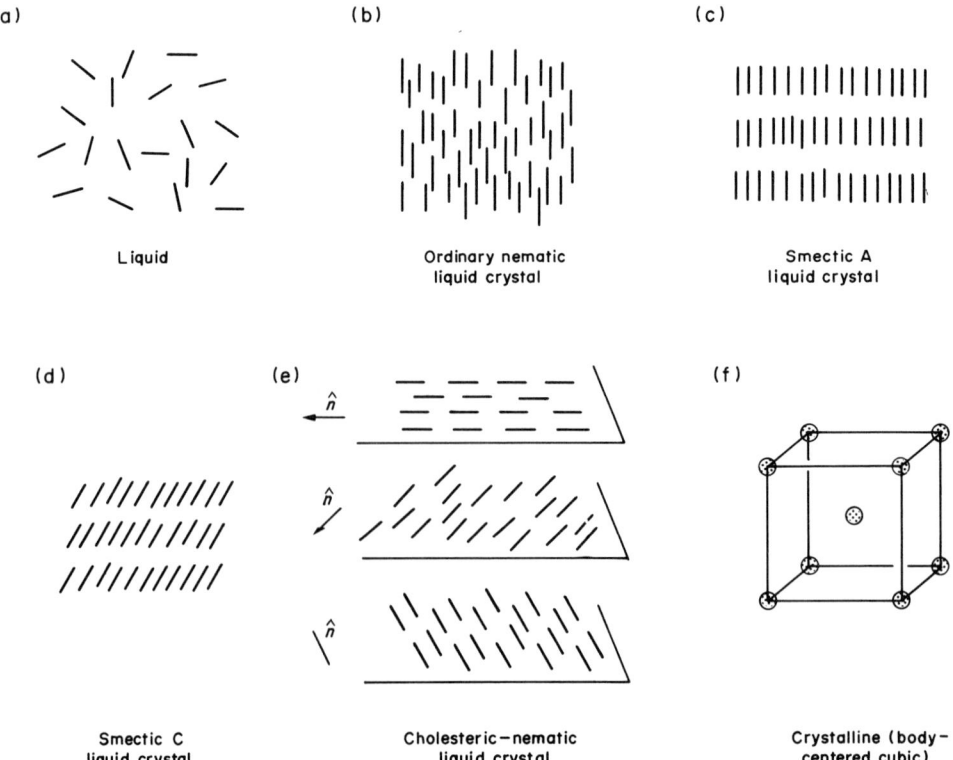

Figure 1 Schematic representation of the molecular arrangement in the: (a) isotropic phase; (b) nematic phase; (c) smectic A phase; (d) smectic C phase; (e) cholesteric or chiral nematic phase; and (f) crystalline state (dots represent molecules)

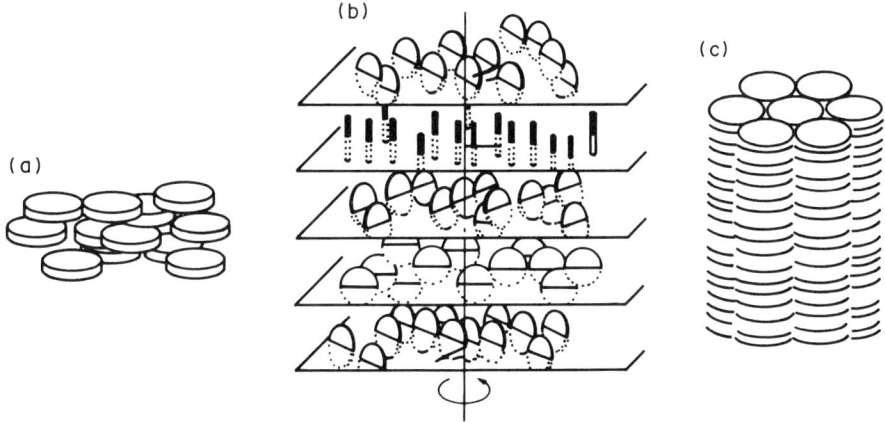

Figure 2 A schematic representation of the molecular arrangement: (a) in the nematic phase N_D; (b) in the cholesteric phase N^*_D; and (c) in the columnar phase D, of disc-like liquid crystals

Table 1 Some Properties of Lytropic Systems Composed of an Amphiphile and water

Suggested structural arrangement	% Water[a] (approximate range)	Physical state	Gross character	Freedom of movement	Microscopic properties (crossed nicols)	X-ray data	Structural order
	0	Crystalline	Opaque solid	None	Birefringent	Ring pattern 3–6 Å	3 dimensions
	5–22–50	Liquid crystalline, lamellar	Clear, fluid, moderately viscous	2 directions	Neat soap texture	Diffuse halo at about 4.5 Å	1 dimension
	23–40	Liquid crystalline, face-centered cubic	Clear, brittle, very viscous	Possibly none	Isotropic with angular bubbles	Diffuse halo at about 4.5 Å	3 dimensions
	34–80	Liquid crystalline, hexagonal compact	Clear, viscous	1 direction	Middle soap texture	Diffuse halo at about 4.5 Å	2 dimensions
	30–99.9	Micellar solution	Clear, fluid	No restrictions	Isotropic with round bubbles		None
	> 99.9	Solution	Clear, fluid	No restrictions	Isotropic		None

[a]The different percentages of water show that different amphiphiles require different amounts of water, For soaps, the lamellar structure generally occurs between 5 and 22% of water; with some lipophiles the water may be as high as 50%. The cubic structure generally occurs between 23 and 40%.

14.2.2 Liquid Crystalline Polymers

Liquid crystalline polymers can be classified into main chain, side chain and combined (Figure 3). This classification is based on the place in the polymer where the mesogen is inserted, *i.e.* within the main chain, as side groups, or both within the main chain and as side groups. More complex polymer architectures are also possible.[9,64,72] The mesogen used in the construction of the liquid crystalline polymer can be rod-like, disc-like or amphiphilic.[9,64,72–74] In addition to linear polymer

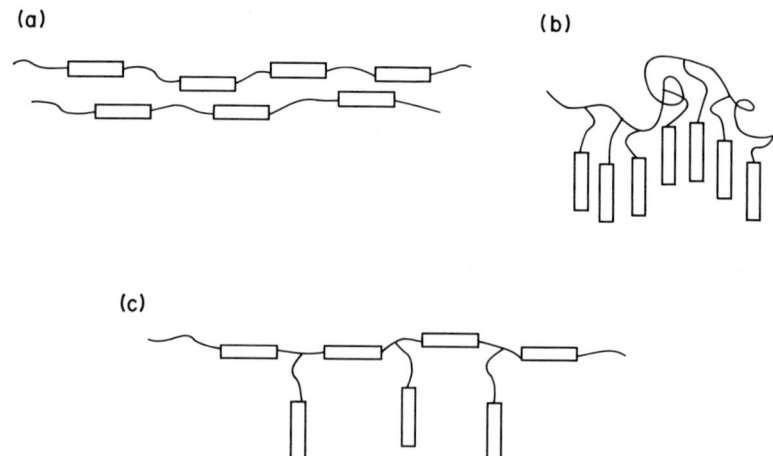

Figure 3 Schematic representation of: (a) main chain; (b) side chain; and (c) combined liquid crystal polymers

structures, cyclic,[75] branched and crosslinked[76,77] architectures have also been synthesized. Therefore, liquid crystalline polymers display thermotropic and lyotropic mesophases which are similar to those exhibited by thermotropic and lyotropic low molar mass liquid crystals. In addition, rigid rod-like main chain nonamphiphilic polymers like poly(phenyleneterephthalamide) (Kevlar®),[78] poly(p-phenylene-2,6-benzobisthiazole) and poly(p-phenylene-2,6-benzobisoxazole)[79] exhibit lyotropic nematic mesophases.

A characteristic of side chain liquid crystalline polymers is that their polymer backbone adopts a random-coil conformation only in solution and in isotropic melt. In a nematic phase the conformation of the polymer backbone is slightly distorted. Depending on the flexibility of the polymer backbone, its conformation in the smectic phase is so highly distorted that it can even be confined to the smectic layer (in the case of polymers based on very flexible backbones) or it can cross the smectic layer (in the case of rigid backbones).[80–82]

14.3 ISOMORPHISM OF LIQUID CRYSTALS

The miscibility of various liquid crystalline low molar mass substances and polymers is of great importance both for the identification of various liquid crystalline phases and for the preparation of mixtures with well-defined phase transitions. In the field of low molar mass liquid crystals, Arnold, Sackmann and Demus have developed the miscibility rules[83–86] which can be summarized as follows: if two liquid crystalline phases are miscible, they are isomorphic, and therefore belong to the same type of mesophase. However, the reverse is not true. The situation is more complicated for mixtures of liquid crystalline polymers with low molar mass liquid crystals and for mixtures of liquid crystalline polymers. Many times, similar phases of liquid crystalline polymers and low molar mass liquid crystals are not miscible.[86–89] The same is the case for similar phases of two different liquid crystalline polymers.[90] For a general discussion on this topic and a comparison of isomorphism of crystalline and liquid crystalline polymers and copolymers see ref. 90. From the preparative point of view it is valuable to know that when two compounds are isomorphic within a certain mesophase, both their thermal transition temperatures and corresponding thermodynamic parameters exhibit continuous dependences *versus* composition. The same is the case for a copolymer. When its structural units are isomorphic within a certain phase, the corresponding transition temperatures and thermodynamic parameters exhibit continuous dependences on compositon. This means that both the components of mixtures and the structural units of copolymers behave like an ideal solution. Therefore, by knowing the temperature transitions and thermodynamic parameters of the parent compounds we can apply the equations of Schroeder and Van Laar (equation 1) and predict the phase diagram of the mixture or of the copolymer[91]

$$F_1 = \left[1 - \frac{\Delta H_1^0 T_2 (T - T_1)}{\Delta H_2^0 T_1 (T - T_2)} \right]^{-1} \tag{1}$$

where F_1 is the molar fraction of component 1, T_1 and ΔH_1 are the transition temperature and the enthalpy change of pure component 1, T_2 and ΔH_2 have the same meaning for component 2, while T is the transition temperature corresponding to composition F_1. T should be lower than T_2 and higher than T_1. Thus, a phase diagram that exhibits any temperature greater or lesser than the maximum or minimum T_2 or T_1 respectively, violates the T criteria above and must therefore be discussed by the introduction of nonideality of solutions of structural units of copolymers. The negative, positive or linear dependence of T versus F_1 is determined by the parameter

$$A = \frac{\Delta H_1^0 T_2}{\Delta H_2^0 T_1} \tag{2}$$

from equation (1).

Mixtures of nematic liquid crystals with isotropic solute molecules varying in size and shape have been studied extensively over the past years both experimentally[92-96] and theoretically.[97-100] The addition of an isotropic solute to a liquid crystalline solvent decreases the nematic–isotropic transition temperature of the solvent and leads to a two-phase region. This phase behavior is very sensitive to the shape of the solute and its molecular weight. Thus a change of the shape of the solute from elongated to more isotropic and an increase of its molecular weight both lead to a sharper decrease of the nematic–isotropic transition temperature. Flexible random-coil polymers are not miscible at all with low molar mass liquid crystals or with liquid crystalline polymers. For example, a smectic low molar mass liquid crystal could theoretically accomodate a random-coil polymer only between the smectic layers.[101]

14.4 SOME CONSIDERATIONS ON THE 'RIGIDITY' OF 'ROD-LIKE' MESOGENS

The traditional pathway used to synthesize low molar mass liquid crystals[6,9] and both, main chain[42-49] and side chain[50,51,72] liquid crystalline polymers is based on the use of rigid rod-like mesogenic units. Recently, the concept of flexible rod-like mesogens or rod-like mesogens based on conformational isomerism was advanced and used in the synthesis of both side chain[72,102] and main chain[103-106] liquid crystalline polymers. Low molar mass liquid crystals based on flexible mesogens were discussed elsewhere.[72,102,103] Depending on their conformational or configurational rigidity we have suggested the classification of rod-like mesogenic groups as rigid rod-like, semirigid or semiflexible rod-like and flexible rod-like groups.[107] The flexibility of these mesogens will be discussed further.

14.4.1 Rigid Rod-like Groups

Rigid rod-like mesogenic units such as diphenylacetylene (**1**), oligo(p-phenylene) (**2**), and benzoxazole (**3**) derivatives are based on linearly substituted aromatic or heterocyclic rings. They exhibit free rotation about certain C—C bonds as in the case of (**1–3**), but this rotation does not perturb the elongated or extended shape, and the molecule retains its rigid rod-like character. Therefore, in rigid rod-like mesogenic units, rigid refers to the rigidity of the linear shape.

In oligo(p-phenylene) (**2**) the steric interaction of the *ortho* hydrogens impede this internal rotation and the conjugation effect reinforces the rigidity due to the double bond character of the C—C bond between the phenylene rings. A ^{13}C NMR study[108] of (**4**) showed an energy barrier of 5 kcal mol^{-1} for (1 kcal = 4.18 kJ) for the internal rotation of the outer rings. The PCILO (Perturbative

Configurational Interaction using Localized Orbitals, a method using semiempirical quantum mechanics) conformational calculations provided for this rotation a value of about 4.8 kcal mol^{-1} [109]

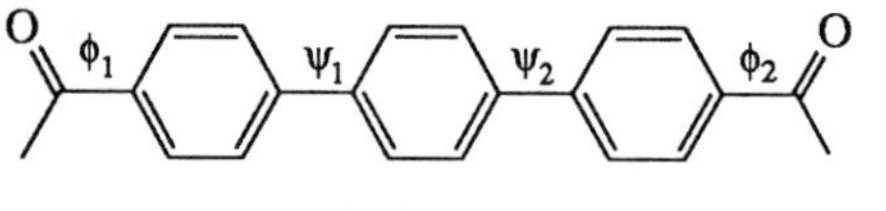

(4) 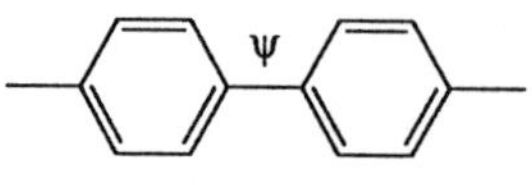   (5)

The minimum energy is at $\phi_1 = \phi_2 = 0$ or $\phi_1 = 0$, $\phi_2 = 180°$. In the case of (5), the energy barrier is of 1.9 kcal mol^{-1} and the minimum energy is at $\phi = 40°$. This value is in close agreement with the experimental value obtained in the vapor state (i.e. $\phi = 42°$).[109] If the aromatic rings are connected via a double bond as in the case of stilbene (6) the molecule exhibits cis and trans configurational isomers. The structure of the trans isomer is energetically more favorable. The trans isomer can be isomerized into the cis one, but the activation energy for this transformation is very high and, therefore, makes this molecule almost rigid. Kalinowski and Kessler[110] have reported an activation energy of 42.8 kcal mol^{-1} for the thermal, uncatalyzed isomerization of stilbene.

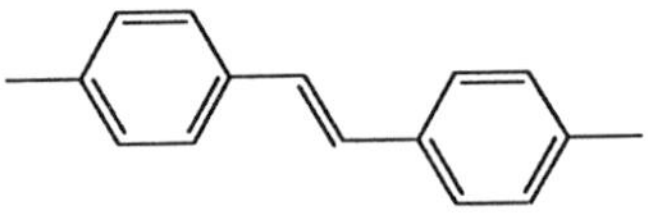

(6)

14.4.2 Semirigid Rod-like Groups

The second group of mesogenic units have both conformational and configurational character and refer to semiflexible molecules such as amides (7) and esters (8). However, due to electronic reasons, the rotation about the C—O or C—N bond of (7) and (8) is retarded and can be even prohibited. In esters like (8), there is rotation about three bonds: R, S and T.

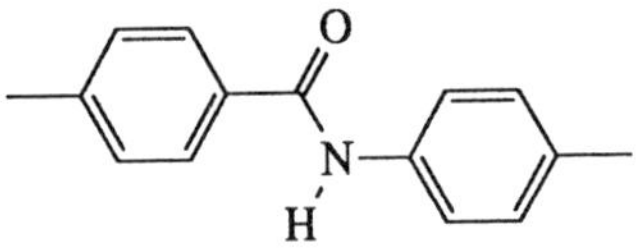

(7) 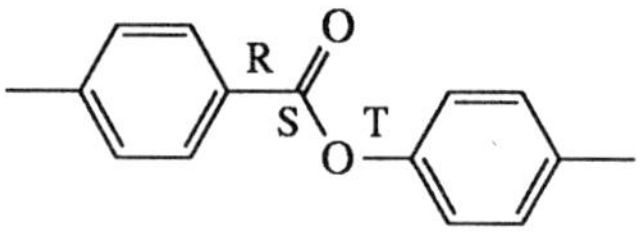   (8)

The most important rotation is about the S-bond, since it determines the kinked (cis) or extended (trans) conformation of the molecule. Theoretical calculations[111] suggest that the trans conformation represents the most stable structure. The arguments for this statement are as follows. In the trans form of esters and acids there is a better O_n—$C_\sigma*$ 'hyperconjugation' which decreases the free energy of this conformer.[111] The steric effect in the cis form makes it more unstable than the trans form. The potential energy of the cis form is higher than that of the trans form by 7.9 kcal mol^{-1}. In the energy profile there is a hump at 90° which is due to the O_p—$C_\pi*$ overlap which creates a double bond character on the O—C bond. This barrier is smaller for aromatic esters than for aliphatic esters. This is because the delocalization of the p-electrons of the oxygen on the aromatic ring attached to the oxygen gives the structure (9) which competes with the contributor (10). This conclusion is supported by the difference between the C—O bond length of aliphatic and aromatic esters. The C—O bond length is 1.33 Å in alkyl benzoate and 1.36 Å in aryl benzoate.[111] This last value is the closest to the typical ether C—O bond length of 1.42–1.43 Å. This double bond character creates a configurational character for the cis and trans conformers of esters and amides and subsequently, these compounds behave like the class of rigid mesogens. Therefore, they should be classified as semirigid or semiflexible rod-like mesogens.

(9)   (10)

The structure and conformational behavior of amides (**7**) are very close to those of esters (**8**). The different resonance structures (**11**), (**12**) and (**13**) of a typical amide molecule[112] are similar to those of esters. Due to its double bond character, the contribution of (**12**) hinders the rotation about the C—N bond.

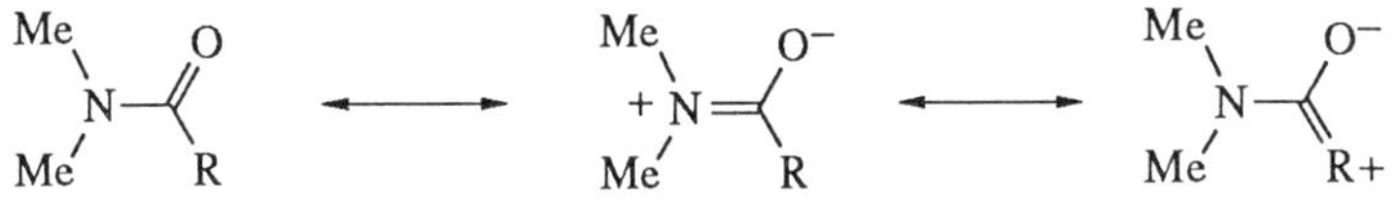

(11)　　　　(12)　　　　(13)

Pauling[112] predicted a rotational energy barrier of 21 kcal mol^{-1} which corresponds to about 40% double bond character for the C—N bond of amides. This double bond character decreases due to the participation of (**13**), especially when R = aryl. This double bond character contribution was demonstrated by NMR studies performed on Me$_2$N—C(O)—R. When R = Me[113] the rotational energy barrier is 18.9 kcal mol^{-1} and decreases to 14.4 kcal mol^{-1} for R = phenyl.[114] On the basis of an NMR study, Tadokoro et al.[115] reported a rotational energy barrier of about 20 kcal mol^{-1} for the rotation about the C—N bond of phenyl benzamide.

14.4.3 Flexible Rod-like Groups

The third class of rod-like molecule is flexible. 1,2-Disubstituted ethane (**14**) and methylenoxy or benzyl ether (**15**) derivatives belong to the class of flexible mesogenic units. Before discussing the fully flexible rod-like mesogens or the mesogens based on conformational isomerism, let us briefly recapitulate the conformational behavior of the *n*-butane molecule (**16**). *n*-Butane exhibits a number of different conformers. Out of them, the most stable are the *gauche* (**17**) and (**18**) and the *anti* (**19**). The electron diffraction study[116,117] of *n*-butane in the gas phase (at 287 K)[116] provided an energy difference of 0.6 kcal mol^{-1} between the *gauche* and the *anti* conformers.

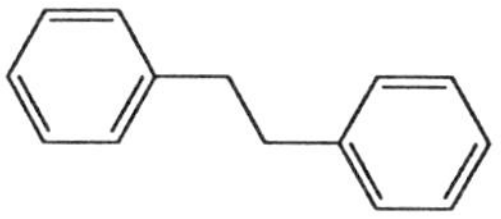

(14) 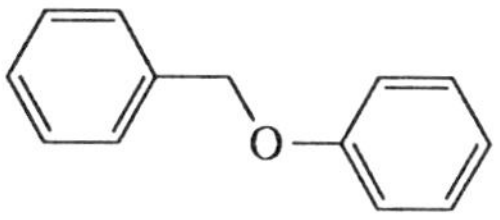　　　　(15)

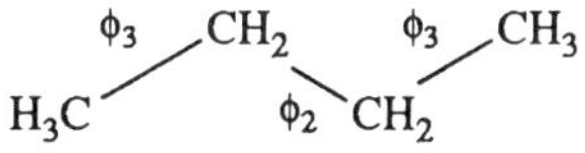

(16)

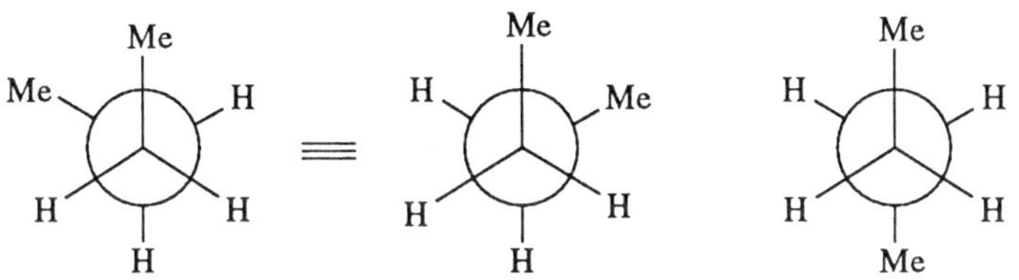

(17)　　　　(18)　　　　(19)

The same parameter was determined by Raman spectroscopy[118] (0.77 kcal mol^{-1}) and ^{1}H NMR spectroscopy[119] (0.68 kcal mol^{-1}). The most recent Raman study[120] performed on *n*-butane in the gas phase shows a potential energy difference of 0.966 ± 0.054 kcal mol^{-1} between its *gauche* and *anti* conformers. Theoretical SCF–MO (Self Consistent Field–Molecular Orbital) based calculations[121] led to a rotational energy barrier at $\phi_2 = 60°$ (when the Me group is eclipsed with a hydrogen atom) of 3.7 kcal mol^{-1} and to a potential energy difference of 1.19 kcal mol^{-1}. Flory et al.[122] have calculated a rotational energy barrier of 2.8 kcal mol^{-1} and a potential energy difference of 0.53 kcal mol^{-1}. Consequently, the theoretical and experimental results of n-butane agree.

The X-ray study[123] of diphenylethane suggests that in the solid state the phenyl groups are located exclusively in the *anti* position. IR and Raman spectroscopy studies[124] performed on 1,2-diphenylethane and 1,2-di(p-chlorophenyl)ethane (20) also demonstrated that both molecules exist in the solid state as *anti* conformers. An optical anisotropy study[125] performed by diffusion Rayleigh depolarization on solutions of 1,2-diphenylethane in CCl$_4$ and cyclohexane shows that the *anti* conformer is preferred also in solution. In solution and Nujol mulls[124] of *para*-substituted 1,2-diphenylethane derivatives, some extra bands were observed suggesting the existence of the *gauche* conformer. At 25 °C the ratio of *anti* to *gauche* is 84:16, but in case of 1,2-diphenylethane (14) the absence of any extra band suggests the possibility that the concentration of *gauche* conformer is too low to be detected by IR spectroscopy. A ^{13}C NMR spectroscopy study[126] of 1,2-di(2,6-dimethylphenyl)ethane in CCl$_4$ solution showed a dynamic equilibrium of 7:3 *anti* to *gauche* conformations.

Theoretical calculations performed on 1,2-diphenylethane gave contradictory results. Ivanov et al.[127] concluded that the *gauche* conformer of (14) is more stable than the *anti* one by 0.57 kcal mol^{-1}. Jacobus[128] also indicated that the *gauche* conformer is more stable and he obtained a difference of 1.15 kcal mol^{-1} between the *anti* and the *gauche* conformers. On the basis of an LCAO–MO (Linear Combination of Atomic Orbitals–Molecular Orbital) treatment of the benzene rings, Ivanov et al.[127] showed that there is no difference in the stretch, bend, torsion and stretch–bend interactions but there is a difference in the nonbonded interactions between the atoms of benzene rings which prefer the *gauche* conformer. A theoretical study[129] using the MM2/MMP2 (molecular mechanics based computer program) program provided data which are in agreement with the experimental results, suggesting that the *anti* form is more stable than the *gauche* form. The energy difference between these conformers is 0.95 kcal mol^{-1} and increases to 1.59 kcal mol^{-1} when point charge and electrostatic interactions are also considered. Jonsson et al.[105a] calculated the conformational energy barrier of α,α'-diphenyl-p-xylene (21) as a model compound for the polymer based on benzyl ether mesogenic units. The potential energy barrier to move one of the outer rings to the position where it becomes linear is 3.87 kcal mol^{-1}.

(20)   (21)

From this discussion it is clear that in the crystalline state only the *anti* conformer exists. However in solution both the *anti* and the *gauche* conformers exist in a dynamic equilibrium. To our knowledge, the rotational energy barrier of 1,2-diphenylethane is not available. Ivanov et al.[127] calculated the potential energy at different values of torsional angle (Ph—C—C—Ph), i.e. when they are in the *anti*, *gauche* and eclipsed forms. In the eclipsed form, the two rings are in closest geometry (the steric and other nonbonded interactions are maximum). Depending on the method used, the calculated energy of the eclipsed form is 3.61, 3.49 and 1.95 kcal mol^{-1}. This conformer has the maximum potential energy, suggesting that the conformational energy barrier should be lower than the above-mentioned values because during the *gauche* to *anti* or *anti* to *gauche* conformational change the benzene rings eclipse with the ethylenic hydrogen atoms, and this steric interaction is less than for two benzene rings.

In conclusion, based on the difference between the rotational energy barrier of different configurational isomers or conformers of rod-like groups we suggested[107] their classification into three categories: rigid, semirigid or semiflexible, and flexible. Rigid rod-like groups are rigid units whose shape is rigid as for example (1), (2) and (3), and configurational isomeric units which require a high rotational energy barrier or activation energy, like for example stilbene ($\Delta E_a = 42.8$ kcal mol^{-1}). Semirigid or semiflexible rod-like groups are conformationally flexible, but due to some electronic reasons they behave like configurational isomers of medium rotational energy barrier. Classic examples are aromatic amides and esters ($\Delta E_a = 14.4$ kcal mol^{-1} and 7.1 kcal mol^{-1} respectively). The rotational energy barrier for these molecules is less than half that for rigid molecules but is much higher than that for flexible molecules. Flexible rod-like groups are conformationally flexible groups whose rotational energy barrier is within the same range of values as those of butane ($\Delta E_a = 2.8$–3.7 kcal mol^{-1}). Classical examples include molecules based on phenyl benzyl ether ($\Delta E_a = 3.87$ kcal mol^{-1}) and 1,2-diphenylethane ($\Delta E_a = 3.61$ kcal mol^{-1}).

14.5 MANIPULATION OF PHASE TRANSITION TEMPERATURES THROUGH STRUCTURAL VARIATIONS: SOME THERMODYNAMIC CONSIDERATIONS[65,66]

14.5.1 Equilibrium States

For the considerations that follow it will be sufficient to consider the basic thermodynamic relationship

$$dG = V dP - S dT \qquad (3)$$

where G is the free energy, S the entropy, V the volume, P the pressure and T the temperature. For the scheme in question we will consider first the melting of a true crystal at constant pressure ($dP = 0$). As seen from Figure 4, and as follows from equation (3), the free energies of both crystal (G_k) and isotropic liquid (G_i) decrease with increasing temperature where the decrease in G_i is the steeper, due to $S_i > S_l > S_k$. Where G_i crosses G_k the crystal melts, which of course is at $T_{k-i} = T_m$, the melting point of the crystal. In this case, as drawn in Figure 4, the free energy of any hypothetical mesophase, G_{lc}, cannot fall below both G_k and G_i, and hence corresponds to a stable state at any temperature. While G_{lc} decreases faster with T than G_k it will only cross G_k at a point which is above G_i, i.e. where the isotropic liquid is already the stablest phase. Thus the mesophase is virtual and remains unrealizable as a stable phase.

In order to create a stable mesophase a section of the G_{lc} versus T curve will need to be brought beneath both G_k and G_i (thus to a state of greatest stability). This can be achieved either by (i) raising G_i (Figure 5), or (ii) raising G_k (Figure 6), or by a combination of both (i) and (ii). As seen in Figures 5 and 6 the mesophase will be 'uncovered' in a temperature range bounded by T_{k-lc} and $T_{lc-i} = T_i$, corresponding to temperatures of crystal melting and isotropization respectively. (The lowering of G_{lc} would have the same effect, but changes in G_{lc} are expected to be small compared to those in G_i and G_k and will be disregarded in what follows.) In general, raising of G_i (Figure 5) arises from the lowering of the melt entropy (e.g. by increasing the rigidity of the molecule *via* a permanent structural change or *via* a temporary change such as shear or any other kind of orientation, constraint or pressure), while raising of G_k (Figure 6) from the reduction in the perfection of the crystal (e.g. by copolymerization or another kind of chemical and/or physical modification) causes T_m to decrease and therefore G_k increases. Note that in the case of Figure 5 the crystal melting point T_m is raised, while in that of Figure 6 it is lowered.

We can generalize further by considering the influence of change in pressure, i.e. the $V dP$ term in equation (3). Usually the specific volume is larger for the liquid than for the crystal with the mesophase expected to lie in between; hence $(dG_k/dP)_T < (dG_{lc}/dP)_T < (dG_i/dP)_T$. In principle it could therefore happen that at some P G_{lc} falls below G_k and thus a mesophase may become 'uncovered'. It has been found that in most experimental systems[130] the effect of increased hydrostatic pressure is to promote the mesophase ('barophyllic' behavior). However, there is no fundamental reason that would make this a general rule. This is clearly illustrated by the example of the sequence of alkanes → polyethylene (as it is described in detail elsewhere[66]) where 'barophobic' behavior of short *n*-alkanes changes continuously with increasing chain length toward the 'barophyllic' behavior of polyethylene.[131]

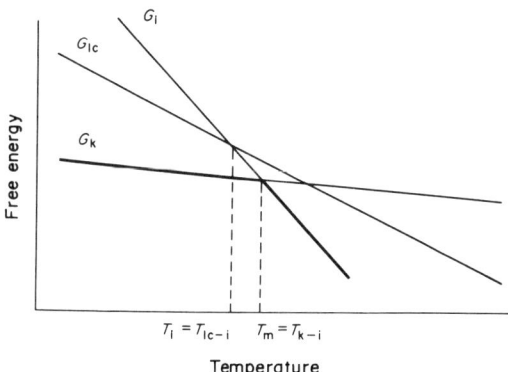

Figure 4 Schematic plot of free energies *versus* temperature for a system that does not show a mesophase. G_k, G_{lc} and G_i are, respectively, the free energies of the crystalline, mesomorphic (virtual) and isotropic liquid states. $T_{k-i} = T_m$ is the crystalline melting point. Here, as in subsequent Figures 5 and 6, the heaviest lines correspond to the stablest state at a given temperature

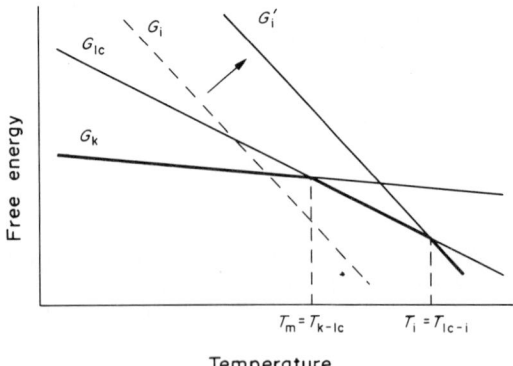

Figure 5 Schematic plot of free energies *versus* temperature for the system in Figure 4 but with G_i raised (to G_i') so as to 'uncover' the mesophase. T_{k-lc} and T_{lc-i} are the crystal–mesophase transition and the isotropization transition temperatures

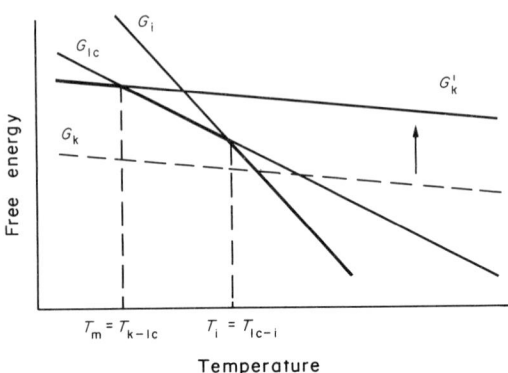

Figure 6 Schematic plot of free energies *versus* temperature for the system in Figure 4 but with G_k raised (to G_k') so as to 'uncover' the mesophase

14.5.2 Metastable States

Figures 4–6 refer to states of thermodynamic equilibrium. However, systems may not respond immediately when passing from one stable regime to another within the phase diagram, hence the metastable phases can often arise.

The most commonly encountered metastability is that arising on crystallization. As is familiar, crystallization only sets in at a certain supercooling temperature. In polymers in particular, the crystallization temperature T_{i-k} can be appreciably below the equilibrium melting point $T_{k-i} = T_m$. On the other hand, the formation of a mesophase generally requires less supercooling. Now, if the temperature for the conversion from the isotropic liquid to a normally unstable mesophase lies somewhere between T_{k-i} and T_{i-k}, such a 'virtual' mesophase may materialize on cooling. As the temperature is lowered still further, crystallization will occur. At this state two extreme situations may be envisaged, as depicted in Figures 7 and 8. On crystallization the free energy either drops from G_k' to G_k, *i.e.* the value for the perfect crystal (Figure 7, where the dotted line indicates a possible pathway), or else there is no continuous change in G, *i.e.* a highly imperfect crystal is formed with its free energy remaining at G_k' (Figure 8). The realistic path would be somewhere in between these two extremes, *i.e.* some decrease in G is expected, which may not quite reach the level of G_k.

We shall first consider the extreme situations of Figures 7 and 8. When the perfect crystal of Figure 7 is reheated, it melts directly into the isotropic liquid at T_{k-i}; thus such a system displays the mesophase only on cooling, and is called 'monotropic'. On the other hand, the imperfect crystal of Figure 8 first changes back into the mesophase at $T_{k'-lc}$ and then into the isotropic liquid at T_{lc-i} on reheating; thus the mesophase occurs both on cooling and heating and is called 'enantiotropic'. The latter case clearly illustrates that an enantiotropic mesophase does not necessarily mean stability of the mesophase, as sometimes implied, although a stable mesophase, naturally, must be

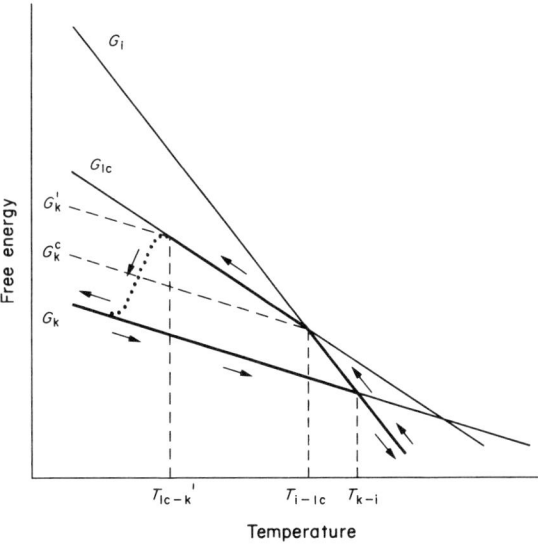

Figure 7 Schematic plot of the free energy diagram illustrating the origin of a monotropic liquid crystal. Here, as in Figure 8, the arrows represent heating and cooling pathways

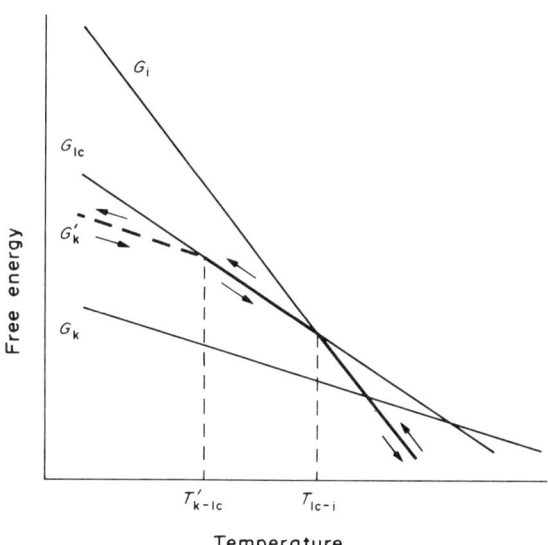

Figure 8 Schematic free energy diagram for an enantiotropic liquid crystal, where the mesophase is metastable

enantiotropic. As mentioned before, real systems are in between those described by Figures 7 and 8. Some decrease below G'_k will occur upon crystallization, the magnitude of the drop depending, among others, on crystallization kinetics. Accordingly, neglecting possible perfectioning on subsequent heating, mono- and enantio-tropic behaviors are distinguished by the magnitude of the drop in G on crystallization; if G stays above a critical value (G^c_k) the system is enantiotropic, if it falls below it is monotropic, the definition of G^c_k being apparent from Figure 7. It is easily seen how crystal perfectioning on annealing can lead to a 'conversion' of an enantiotropic into a monotropic mesophase, an effect frequently observed in both polymeric and low molar mass liquid crystals.

It is worth noting further that, under certain conditions, polymers will also display superheating effects, in which case the mesophase may only appear on heating; this can be regarded as 'monotropic' behavior in the reversed sense. An example of this in connection with polyethylene was discussed previously[66] and an example with a side chain liquid crystalline polymer was reported recently.[132]

As seen from the above, kinetics will always play a part in any phase diagram of a liquid crystalline polymer, as determinable in practice. Its influence on our present considerations will depend on its magnitude. If it is small enough, so as not to alter the sequence of appearance (or disappearance) of the different phases with changing temperature, the equilibrium situations in Figures 4–6 will be taken to apply with the kinetics as an overlay, affecting only the exact numerical values of the actual divides. If, however, the phase sequence itself is affected, and in fact new phases are created due to metastability, then the whole phase behavior will become kinetically determined and the consideration in Figures 7 and 8 will pertain.

Finally, a further kinetic factor, the glass transition (T_g), needs invoking, particularly pertinent to liquid crystalline polymers. On cooling the system becomes immobilized at T_g (more precisely, also dependent on the rate of cooling), hence phase transformations will be arrested or altogether prevented. The converse will apply when a previously immobilized system is heated above T_g when the system will again be able to follow its course towards the equilibrium state. In practice, this will lead to inaccessibility of certain portions of the phase diagram, or conversely, lead to the freezing in of the liquid or of the liquid crystalline state enabling their attainment at temperatures where, by thermodynamic criteria, they would be unobtainable otherwise (isotropic or liquid crystalline glass respectively). Even if T_g is not a thermodynamic quantity the indication of its location in the phase diagram can nevertheless serve a useful purpose.

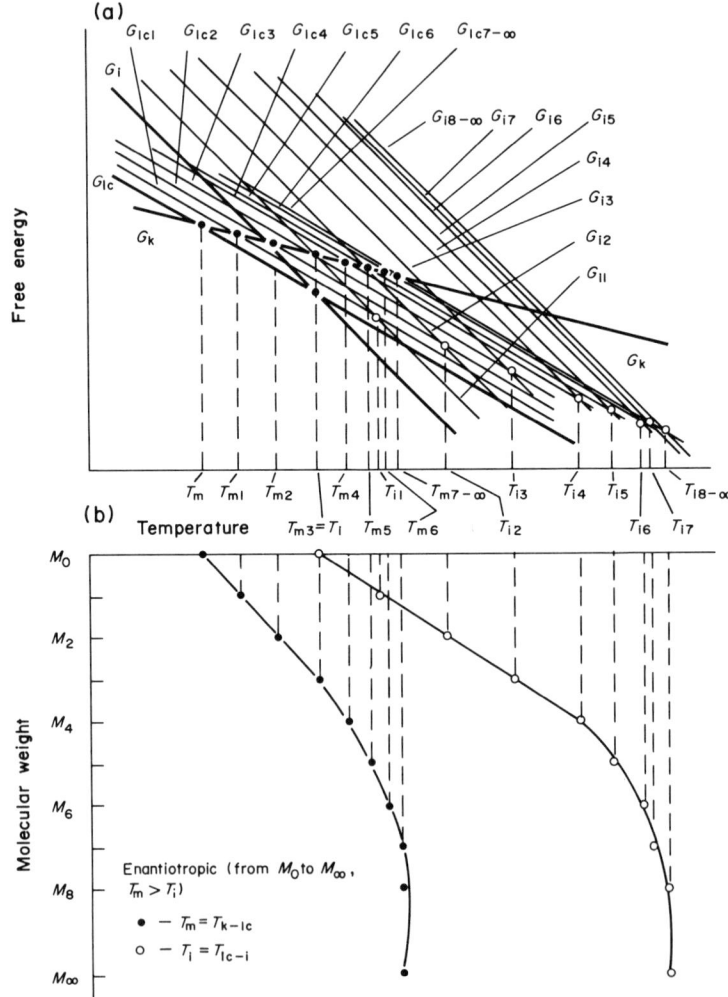

Figure 9 The broadening of the temperature range of an enantiotropic mesophase of the monomeric structural unit (M_0) by increasing the degree of polymerization. The upper part (a) describes the influence of molecular weight on the dependence between the free energies of the crystalline (G_k), liquid crystalline (G_{lc}) and isotropic (G_i) phases and transition temperatures. The translation of this dependence into the dependence of phase transition temperature–molecular weight is presented in the lower part (b)

14.5.3 Influence of Molecular Weight on Phase Transition Temperatures

We will now discuss the relationship between phase transition temperatures and polymer molecular weight for three different situations.[65,66] Upon increasing the molecular weight from monomer to polymer, the entropy of the liquid phase (S_i) decreases. The decrease of the entropies of mesomorphic and crystalline phases is lower than that of the isotropic phase. For simplicity, the decrease of the entropy of the crystalline phase will be neglected. The decrease in S_i and S_{lc} tends asymptotically to zero with increasing molecular weight (M). M_0 from Figures 9 to 11 refers to molecular weight of the polymer structural unit. M_1 to M_∞ from the same figures are arbitrary molecular weights of the corresponding polymer. It follows that G_i and G_{lc} increase with the increase of the polymer molecular weight, again asymptotically; above a certain molecular weight, we may consider both parameters as remaining constant.

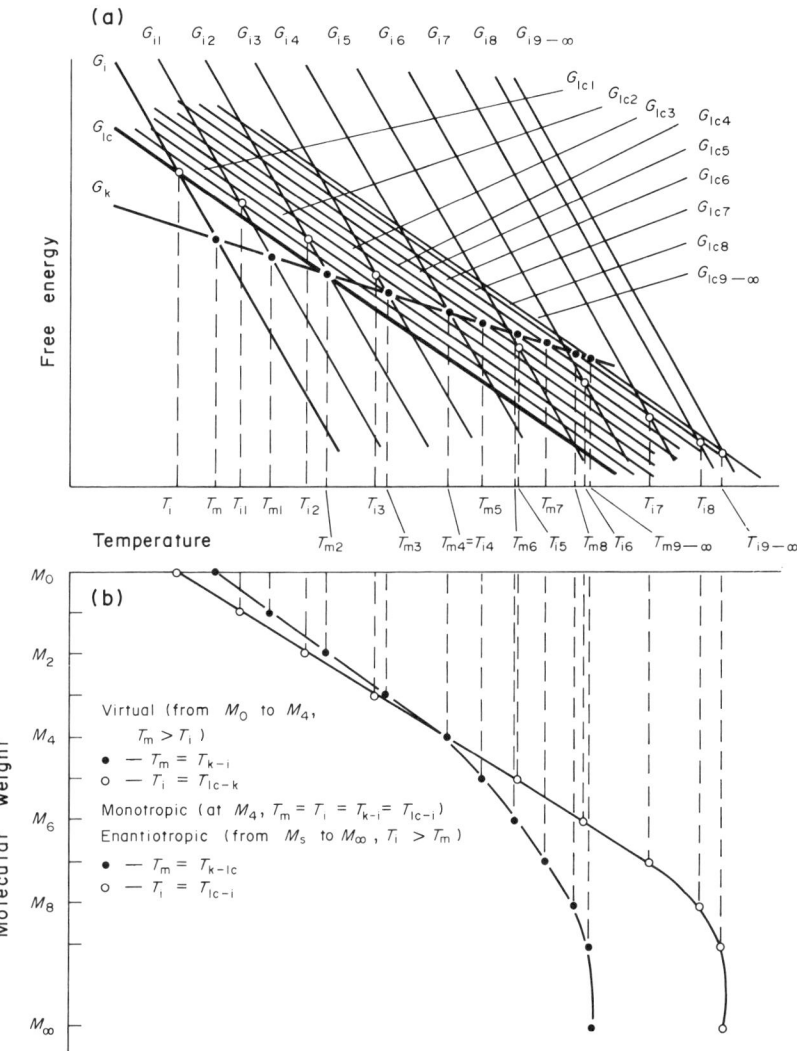

Figure 10 Transformation of a virtual or monotropic mesophase of the monomeric structural unit (M_0) into an enantiotropic mesophase by increasing the degree of polymerization. The upper part (a) describes the influence of molecular weight on the dependence between the free energies of the crystalline (G_k), liquid crystalline (G_{lc}) and isotropic (G_i) phases and transition temperatures. The translation of this dependence into the dependence of phase transition temperature–molecular weight is presented in the lower part (b)

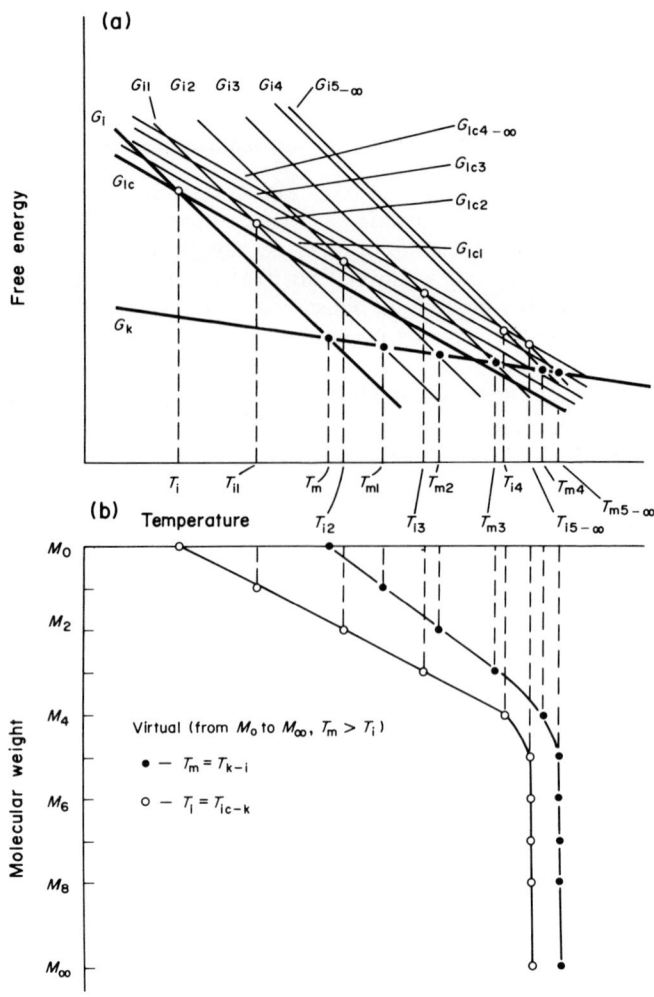

Figure 11 The narrowing of the temperature range of a virtual mesophase of the monomeric structural unit (M_0) by increasing the degree of polymerization. The upper part (a) describes the influence of molecular weight on the dependence between the free energies of the crystalline (G_k), liquid crystalline (G_{lc}) and isotropic (G_i) phases and transition temperatures. The translation of this dependence into the dependence of phase transition temperature–molecular weight is presented in the lower part (b)

14.5.3.1 Case 1. Both monomeric structural unit and polymer display an enantiotropic mesophase

The first situation we will consider refers to the case in which the monomeric structural unit displays an enantiotropic mesophase. Upon increasing its molecular weight to dimer, trimer, *etc.*, S_i decreases and, therefore, G_i increases. Beyond a certain molecular weight, G_i remains for all practical considerations constant. Figure 9 transforms the free energy (G) *versus* transition temperature (T) dependence into a transition temperature (T) *versus* molecular weight (M) dependence. The T versus M plot in Figure 9 demonstrates that both melting (T_{k-lc}) and isotropization (T_{lc-i}) temperatures increase with molecular weight up to a certain range of M values beyond which T_{k-lc} and T_{lc-i} remain approximately constant with the T_i line lying above the line for T_m for all molecular weights. However, the slope of the increase of T_{lc-i} is steeper than that of T_{k-lc}. The difference between these two slopes determines the relative thermodynamic stabilities of the mesomorphic *versus* the crystalline phases at different polymer molecular weights.

For this particular case, the higher slope of the T_{lc-i}–M *versus* that of the T_{k-lc}–M dependence leads to a widening of the temperature range between the two curves with increasing molecular weights of the polymer. This widening of the liquid crystal temperature regime with molecular weight agrees with experimental data reported for the case of both main chain[133–135] and side chain[72,136-145] liquid crystalline polymers. This effect has been repeatedly labeled the 'polymer effect', especially in the case of side chain liquid crystalline polymers.[72]

14.5.3.2 Case 2. The structural unit displays a virtual or a monotropic mesophase; the polymer displays a monotropic or an enantiotropic mesophase

The steeper slope of the $T_{lc-k(i)}$–M dependence *versus* that of the $T_{k-i(lc)}$–M dependence has even more important implications on the molecular weight–phase transition temperature dependence for the situation when the monomer structural unit displays only a monotropic or a virtual mesophase (Figure 10).

As can be seen, the two lines T_i versus M and T_m versus M intersect. This arises from the fact that $T_i < T_m$ for low molecular weights with a steeper slope for T_i. Specifically (for the illustration in Figure 10) the T_i (i.e. T_{lc-k}) values are below the corresponding T_m (i.e. T_{k-i}) values for M_0 to M_4, hence in the range M_0 to M_4 the monomer together with its low oligomers up to 4 display only a virtual mesophase. Beyond M_4 the mesophase becomes stable, hence the system is enantiotropic. In addition to the thermodynamic criterion, kinetics also influence the phase transitions. A certain amount of supercooling of the isotropic–mesomorphic and especially much more so of the mesomorphic–crystalline transitions is possible which can lead to monotropic behavior for molecular weights slightly below and at the intersection point (i.e. at M_4). In view of the fact that T_i and T_m are expected to be continuous functions of molecular weight, the intersection point can be arbitrarily closely approached (from below), hence realization of a metastable liquid crystal phase and subsequent monotropic behavior is to be expected for appropriate molecular weights in cases to which Figure 10 pertains. This effect has been observed experimentally both in the case of main chain[146-148] and side chain[149] liquid crystalline polymers and was labeled 'transformation of a monotropic mesophase into an enantiotropic mesophase by increasing the molecular weight of the polymer'. A series of quantitative experiments on this line will be described later.

14.5.3.3 Case 3. The structural unit displays a virtual mesophase; the polymer displays a virtual mesophase

The third situation is illustrated in Figure 11 and also refers to a different case in which the monomeric unit displays only a virtual mesophase. Here as before, the slope of the T_i (i.e. T_{lc-k})–M dependence is higher than that of the T_m (i.e. T_{k-i})–M dependence, the latter lies above the former throughout, hence the two curves do not cross. Therefore, the resulting polymer displays also a virtual mesophase. This thermodynamic situation was recently applied to the synthesis of virtual liquid crystal polyethers containing both flexible mesogens and flexible spacers,[103] and will be detailed in a subsequent section.

14.5.3.4 Case 4. Rigid rod-like polymers

The discussion from cases 1 to 3 refers to semiflexible or semirigid and flexible polymer systems which exhibit first-order transition temperatures (i.e. melting, isotropization) which are molecular weight dependent only up to a certain degree of polymerization. In these systems the melting temperature is determined by the length of the chain fold. Rigid rod-like polymers such as poly(p-phenylene)s, poly(p-phenylenebenzobisthiazole) (PBT) and poly(p-phenylenebenzobisoxazole) (PBO) exhibit a linear dependence of their melting transition over their entire range of molecular weights. Consequently, the dependences of their first-order transitions on molecular weight follow the pattern from Figure 12. Table 2 summarizes the dependence of the first-order transition temperatures of poly(p-phenylene)s H—(C_6H_4)—H as a function of their degree of polymerization and their axial ratio ($x = L/d$; Figure 13a).[150-152] As predicted by theory,[39,150] when the axial ratio ($x = L/d$) of rod-like molecules reaches a value of about 6.2, the compound should exhibit an enantiotropic mesophase. This is indeed the case. For n of 6 or 7 poly(p-phenylene)s exhibit an enantiotropic nematic mesophase (as predicted by Figure 12). At lower n values the nematic mesophase is virtual. However, for oligomers with n larger than 7 the nematic mesophase cannot be observed since the decomposition temperature first overlaps the melting transition and then becomes lower than the melting temperature. Other rigid rod-like polymers like PBT and

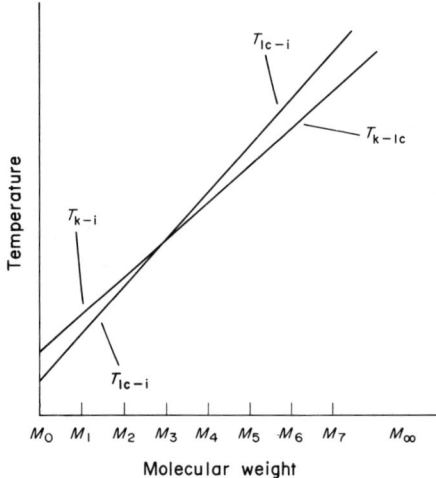

Figure 12 The dependence of various first order transitions on molecular weight for a rigid rod-like polymer

Table 2 Melting and Isotropization Transitions of Poly(p-phenylene)s

n	$x = L/d$	T_{n-i}(Theoretical) (°C)	Polymorphism (°C)
2	≈ 2	—	$T_{k-i} = 70$
3	≈ 3	—	$T_{k-i} = 210$
4	≈ 4	≈ 120	$T_{k-i} = 320$
5	≈ 5	≈ 450	$T_{n-i} = 430$
6	≈ 6	≈ 1900	$T_{n-i} = 560$
7	≈ 7	Isotropic	$T_{k-n} \approx 550$ (also decomposition)

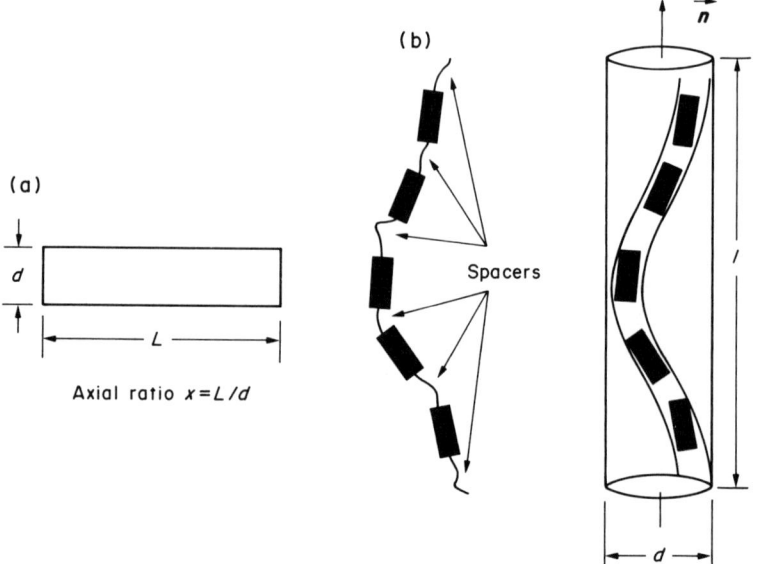

Figure 13 (a) Definition of the axial ratio in the model of rod-like mesogen; (b) model for nonrigid main chain liquid crystalline polymer; l defines the Kuhn's segment

PBO and even semirigid systems like fully aromatic and unsubstituted polyamides and polyesters decompose before melting.

While in rigid rod-like polymers the axial ratio $x = L/d$ (Figure 13a) defines the chain dimension which is responsible for the generation of an enantiotropic mesophase, in the case of semirigid and flexible polymers the ratio l/d (Figure 13b), where l is now a renormalized length (the Kuhn's segment), defines it. In this second case the higher l, the higher the stiffness of the chain.[152]

In conclusion, the steeper slope of the isotropization temperature–molecular weight dependence *versus* that of the melting temperature–molecular weight dependence should have the following effects on going from monomer to polymer. When the monomer structural unit displays an enantiotropic mesophase, the resulting polymer will display a broader enantiotropic mesophase, *i.e.* an anisotropic temperature interval. When the monomeric unit displays a monotropic mesophase, the resulting polymer will, most probably, display an enantiotropic mesophase. When the monomeric unit displays a virtual mesophase, the resulting polymer may display either an enantiotropic, monotropic or virtual mesophase. For the case of side chain[149,153–155] and main chain[133,147,148] liquid crystalline polymers containing flexible spacers and displaying a single mesophase, the nature of the mesophase displayed by the polymer is most frequently identical to that of its monomeric unit. Primarily the molecular weight–phase transition dependences are determined by the relationship between the free energies of the crystalline, liquid crystalline and isotropic phases of the monomeric structural unit yet influenced by the molecular weight of the main chain backbone through its effect on the melt entropy, as in case 1 above. Combinations of many mesomorphic phases of different thermodynamic stabilities in a monomeric structural unit will follow the same molecular weight dependence trend as those described for the monomers displaying a single mesophase.[139–145] However, there are a few examples where the nature of the mesophase displayed by the main chain and side chain liquid crystalline polymers is molecular weight dependent. Both in the case of the main chain[156,157] and side chain[139,140,142,158] liquid crystalline polymers it has been demonstrated that this change in the mesophase represents a continuous dependence of molecular weight. In the case of rigid rod-like polymers both melting and isotropization temperatures reach higher values than the decomposition temperature within the range of low molecular weights.

14.6 MAIN CHAIN LIQUID CRYSTALLINE POLYMERS

14.6.1 Soluble and Fusible Main Chain Liquid Crystalline Polymers

Amorphous polymers are soluble in conventional solvents at room temperature while crystalline polymers are soluble only at temperatures which sometimes are close to the melting point. The solubility of any polymer increases with the decrease of its ability to crystallize. A suppression of the melting temperature below the glass transition temperature (*i.e.* a kinetically prohibited crystalline phase) makes the corresponding polymer noncrystallizable. Therefore, since most rigid polymers have much higher glass transition temperatures than flexible polymers it is easier to decrease the crystallization tendency of the former polymer and, therefore, make it soluble. This is also the case for main chain liquid crystalline polymers. A decrease of the melting transition temperature can be accomplished by raising G_k, as shown in Figure 6 and discussed in Section 14.5.1. An increase in G_k can be accomplished by any synthetic technique which increases S, *i.e.* introduces defects in the polymer structure. Most conveniently, this is realized by copolymerization, addition of lateral substituents which provide a high configurational entropy (alkyl groups, t-butyl groups), insertion of flexible spacers, increasing the overall flexibility of the molecule and insertion of structural defects such as kinks, *etc.* The alternative solution is to increase both the glass transition temperature of the polymer and its entropy simultaneously. In order to maintain a high ability towards mesophase formation the increase in solubility should not disturb the ability of the backbone to achieve a linear conformation. Therefore, the insertion of kinks or any other nonlinearities is not favorable since it increases the solubility and decreases the ability to achieve the mesophase. The ability to decrease the crystallization tendency is not so much related to the size of the lateral substituent as it is to the number of different structural units which can result from the monomer which contains the side group. In fact, according to thermodynamic considerations a high entropy situation can be realized either by a chain containing a large number of different structural units with different small substituents or by a chain containing a large substituent in each structural unit. The second situation, however, reduces drastically the interchain interactions and, therefore, suppresses not only the tendency towards crystallization but also the ability to generate a liquid crystalline phase. Therefore, the most profitable synthetic technique is to use the copolymerization of a few monomers containing small substituents with different electronegativities and the ability to generate more than one constitutional isomeric structural unit from each monomer. Various research groups have used these principles, although in a nonsystematic way, and accomplished fusible and soluble polymers. In the following sections we will enumerate only selected examples of polymers solubilized by using these principles.

14.6.1.1 Poly(p-phenylene) derivatives

To our knowledge the first soluble oligo(p-phenylene)s were synthesized by Kern et al. in 1960 by adding methyl substituents on the phenyl ring.[159] The mesomorphic behavior of these oligomers

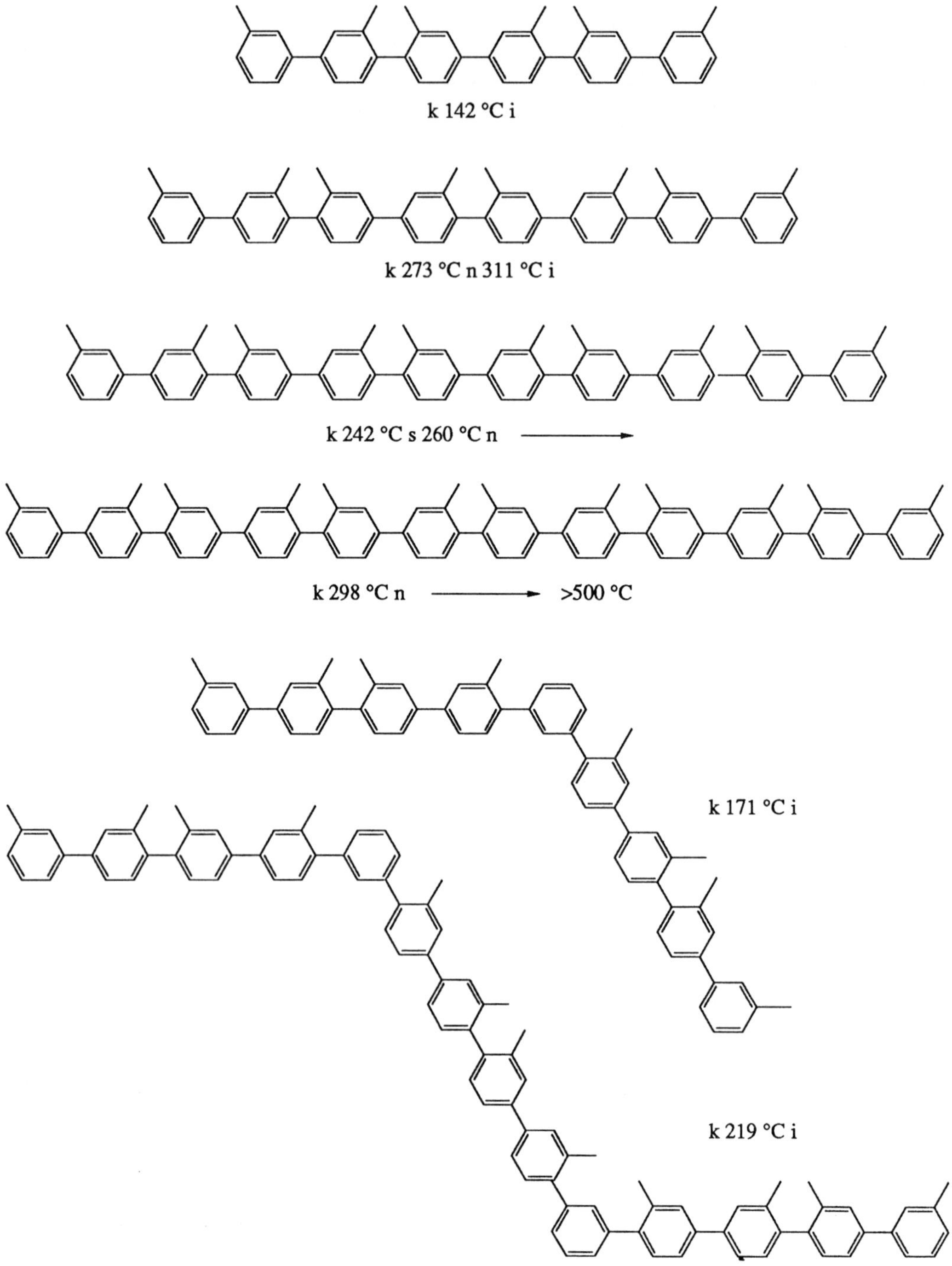

Scheme 1

was first reported by Heitz.[160-162] These oligomers and their phase transitions are outlined in Scheme 1. By substitution their solubility increases[159] and their melting points decrease. Methyl-substituted oligo(p-phenylene)s have an axial ratio (x) equal to 0.8 times the number of benzene rings. According to theoretical predictions hexamethylsexiphenyl, which has $x = 4.8$, and lower oligomers should not exhibit an enantiotropic mesophase. This is indeed the experimental case. However, octamethyloctaphenyl, with $x = 6.4$, and longer oligomers (i.e. larger than the critical value $x = 6.2$) should and indeed do show an enantiotropic nematic phase. Insertion of meta linkages suppresses their ability to form a mesophase (Scheme 1).

Phenylated poly(p-phenylene)s with and without flexible spacers were synthesized by the 1,4-cycloaddition of biscyclopentadienones (bistetracyclones) with bisethynyl compounds (Scheme 2).[163-165] Although these polymers are soluble in conventional solvents their mesomorphic behavior was not investigated.

Soluble semiladder-phenylated polyarylenes were synthesized by the Diels–Alter reaction of biscyclodienones and bispyronedienes with bisbenzynes like 3,4,3',4'-didehydrobiphenyl which was generated in situ by the aprotic diazotization of 3,3'-dicarboxybenzidine (Scheme 3).[166] No mesomorphic properties were investigated.

Soluble poly(p-2,5-di-n-alkylphenylene)s were prepared by the Yamamoto route i.e. Ni-catalyzed coupling of 1,4-dibromo-2,5-di-n-alkylbenzene[167] and via Pd^0-catalyzed coupling of 4-dibromo-2,5-di-n-alkylbenzeneboronic acid (Suzuki reaction) (Scheme 4).[168,169]

Poly(dimethylbiphenylene)s of low molecular weight were prepared by the Ullmann reaction and by the Semmelhack reaction (Scheme 5).[170] Both poly(p-di-n-alkylphenylene)s and poly(dimethylbiphenylene)s do not exhibit liquid crystallinity. This is most probably due to their low molecular weights.

Soluble phenylated poly(p-phenylene)s were also synthesized by Ni^{II}-catalyzed polymerization of 2,5-dibromobiphenyl via the Kumata reaction, and via anionic polymerization of 2-phenyl-1,3-cyclohexadiene followed by aromatization (Scheme 6).[171] Most probably, both synthetic methods lead to regioregular, phenylated poly(p-phenylene)s. The high molecular weight polymer fractions exhibit an enantiotropic nematic mesophase.

Regioirregular, substituted poly(p-phenylene)s were synthesized from substituted hydroquinones via Ni^0-catalyzed homocoupling of their bistriflates (Scheme 7).[172] All these polymers are soluble in conventional solvents; however, they have low molecular weights and do not exhibit liquid crystallinity.

Water soluble poly(p-phenylene) derivatives were synthesized in water by the Suzuki reaction, using a water soluble Pd^0 catalyst (Scheme 8).[173] The mesomorphic properties of these water soluble poly(p-phenylene) derivatives were not investigated.

This brief survey on the synthesis and properties of soluble poly(p-phenylene)s demonstrates that although the rigid rod-like character of these polymers is maintained, most of these polymers do not exhibit liquid crystalline properties. This is probably because bulky side groups attached to these polymers require a much larger x value for the generation of liquid crystalline properties and also because these bulky side groups diminish the interchain interactions which are required for the generation of a thermodynamically stable mesophase.

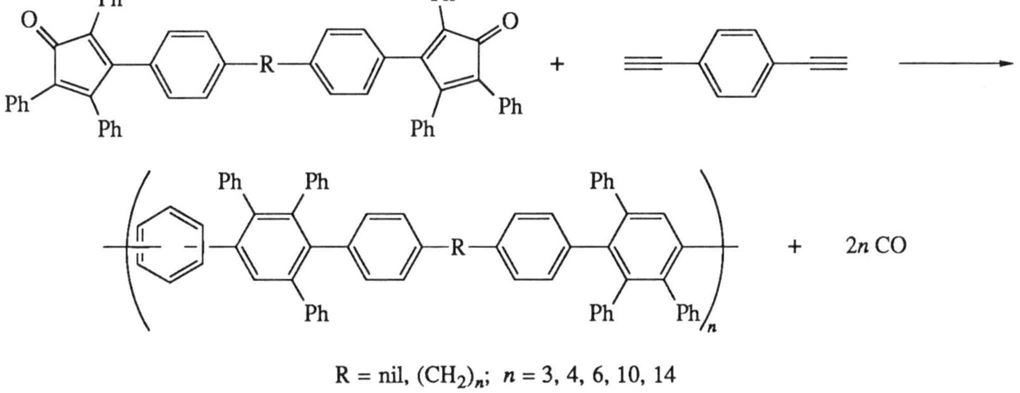

R = nil, $(CH_2)_n$; $n = 3, 4, 6, 10, 14$

Scheme 2

14.6.1.2 Metal-containing poly(yne)s

Metal-containing poly(yne)s are a relatively new class of rigid rod-like polymers. They can be prepared by a copper chloride–triethylamine coupling of the appropriate metal halide with a dialkyne (Scheme 9).[174,175] A review on their synthesis is available.[176] These polymers exhibit lyotropic liquid crystalline phases.[177] Recently, two novel methods for the synthesis of this class of polymers were reported. The first one consists of the reaction of bis(trimethylstannyl)alkynides with trans-[M(PBu$_3$)$_2$Cl$_2$],[178] while the second one involves the reaction of [Rh(PMe$_3$)$_4$Me] with

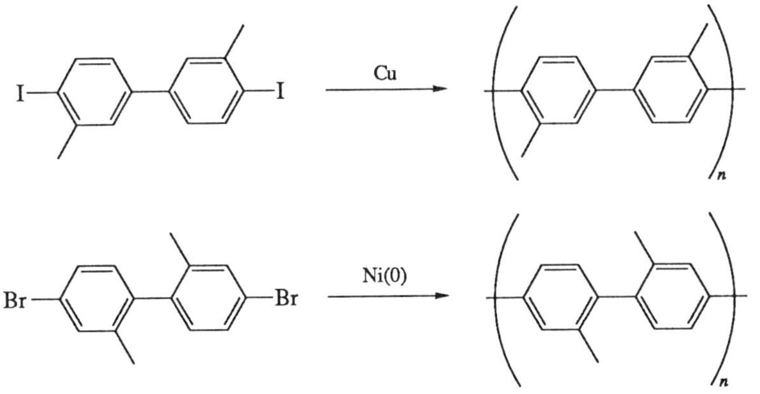

Scheme 5

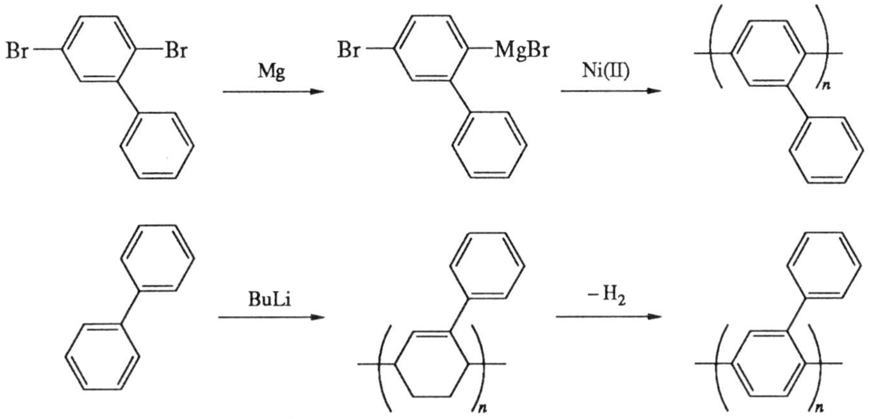

Scheme 6

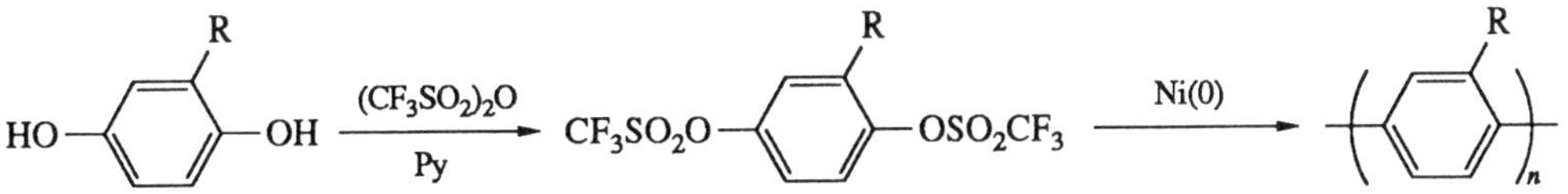

Scheme 7

R = Ph, Bu^t, CO$_2$Et, *etc.*

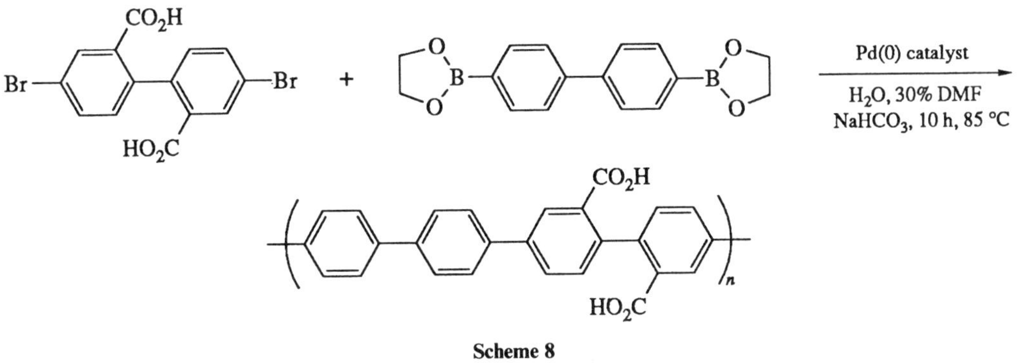

Scheme 8

dialkynes (Scheme 10).[179,180] These metal-containing poly(yne)s can be prepared with weight average molecular weights of up to 100 000[178] and are therefore of interest both for their rigid rod-like character and for their nonlinear optical properties.[180] Additional classes of metal-containing liquid crystalline polymers were discussed in a recent review.[181]

14.6.1.3 Spinal columnar liquid crystalline polymers

Dihydroxysilicon(IV) phthalocyanine can be condensed in solid state to yield insoluble phthalocyaninatopolysiloxanes.[182,183] However, octaalkyl-substituted derivatives of dihydroxytin(IV) phthalocyanine,[184] dihydroxysilicon(IV) octakis(dodecyloxymethyl)phthalocyanine[185,186] and octaalkoxy-substituted silicondihydroxyphthalocyanines[187,188] undergo polycondensation in liquid crystalline phase and yield soluble spinal columnar liquid crystalline polymers (Scheme 11). These polymers exhibit columnar hexagonal liquid crystalline phases which can range from $-7\,°C$ up to $300\,°C$.[184–187] The diameter of the column is determined by the length of the alkyl side groups.[187]

14.6.1.4 Aromatic polyamides and polyesters

Synthetic procedures for the preparation of soluble aromatic polyamides and polyesters were developed mostly by Gaudiana et al.[189–193] Their research was recently reviewed.[194] Depending on their structure, these polymers can be either amorphous or liquid crystalline.[194] A discussion of the influence of various molecular factors such as the position, polarizability and size of the substituents on the solubility of aromatic polyamides is available.[192,194] Table 3[192] summarizes some representative polymer structures. Their solubility is described in Table 4. An inspection of the

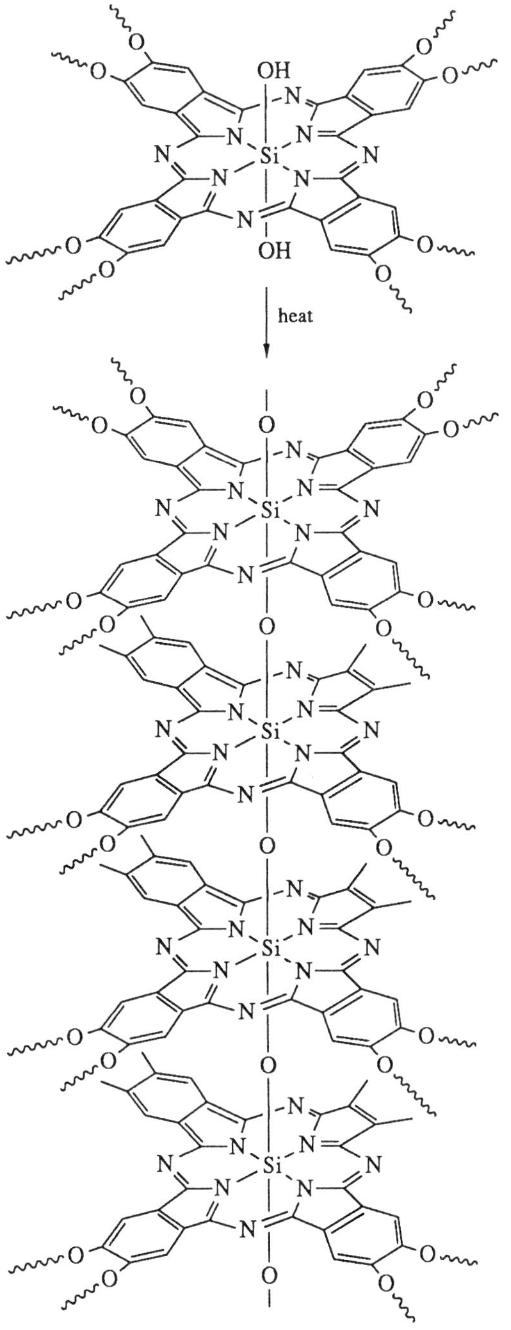

Scheme 11

structures from Table 3 and of the solubilities of these polymers presented in Table 4 suggests that the noncoplanar conformation of the biphenyl moiety (introduced by the presence of substituents in their 2,2'-positions), the random distribution of the biphenyl enantiomers, and the presence of structurally dissimilar diacids and diamines, all contribute to an enhanced solubility. The combination of these molecular factors diminishes or even eliminates interchain correlations necessary for close packing in crystals and may diminish Van der Waals forces of attraction. These orientation dependent intermolecular attractive forces are important in the formation of all classes of liquid crystals.[195] Therefore, when these forces of attraction are significantly weakened by various combinations of steric and dipolar repulsive forces the parallel allignment of polymeric chains is prevented.

Table 3 Molecular Structures of Aromatic Polyamides

Head to head phenyl and biphenyl polyamides

Table 3 (*continued*)

Head to tail biphenyl polyamides

[Structure XII: biphenyl polyamide with two CF₃ groups]   XII

[Structure XIII: copolymer with OMe and CF₃ biphenyl units], $m = 0.24, n = 0.76$ XIII

[Structure XIV: copolymer with OMe and CF₃ biphenyl units], $m = 0.10, n = 0.76$ XIV

Substituted terphenyl and quaterphenyl polyamides

[Structure XV: terphenyl polyamide with two CF₃ and OMe]   XV

[Structure XVI: terphenyl polyamide with two CF₃]   XVI

[Structure XVII: stilbene-linked terphenyl polyamide with two CF₃]   XVII

[Structure XVIII: stilbene-linked quaterphenyl polyamide with two CF₃]   XVIII

[Structure XIX: quaterphenyl polyamide with two CF₃]   XIX

Soluble aromatic polyamides with a much lower solubility were also realized by attaching phenyl substituents either on the *p*-phenylenediamine or on the terephthalic acid monomer.[196–198] Some representative structures together with their solubilities are presented in Table 5.[196] All these polymers exhibit lyotropic liquid crystalline phases and melt before decomposition.

A reversible solubilization of poly(*p*-phenyleneterephthalamide) was accomplished by the polycondensation of the chromium tricarbonyl complex of *p*-phenylenediamine with terephthaloyl chloride in *N*,*N*-dimethylacetamide[199] (Scheme 12) by applying vacuum to remove HCl. The polymer containing tricarbonylchromium-complexed structural units is soluble in dipolar aprotic solvents. Decomplexation can be achieved either by oxidation with I_2 or by heating above 150 °C.

Table 4 Viscosity and Solubility Properties of Biphenyl Polyamides

Polymer	η_{inh} ($\times 10^{-1}$ L g^{-1})[a]	LiCl/Amides[b]	Amides	Solubility Ethers[c]	Ketones[d]
I	3.23	S[e]	44–48[f]	Sw	Ins
II	1.84(DMAc)	S	S	Ins	Ins
III	2.34(DMAc)	S	S	Ins	Ins
IV	1.19	S	Sls	Ins	Ins
V	1.68	S	Ins	Ins	Ins
VI	1.27	S	S	S[g]	Ins
VII	1.10	S	S	Ins	S
VIII	1.75	S	S	36(THF)	S
IX	0.91	S	S	25(THF)	S
X	2.46	S	S	17(THF)	Ins
XI	1.68	S	S	14(THF)	Ins
XII	5.0	S	S	S	S
XIII	4.74	S	S	Ins	Ins
XIV	4.94	S	S	S	S
XV	1.39	S	S	Ins	Ins
XVI	—	Ins	Ins	Ins	Ins
XVII	2.30	S	Ins	Ins	Ins
XVIII	1.31	S	S	S	S
XIX	6.04	S	S	Sls	S

[a]0.5% polymer (w/v)/5% (w/v) LiCl/DMAc at 30 °C. [b]Amide solvents: DMAc, TMU, and NMP. [c]Ether solvents: THF, glymes, but not diethyl ether, dioxane. [d]Ketone solvents: acetone and cyclohexanone. [e]S, soluble; Sw, swells; Sls, slightly soluble; Ins, insoluble. [f]w/v%. [g]Methoxy ethanol but not THF.

Table 5 Phenyl-substituted Aromatic Polyamides[a]

Structure	Inherent viscosity η_{inh}[b] ($\times 10^{-1}$ L g^{-1}) H$_2$SO$_4$[c]	DMAc[d]	Solubility in DMAc or NMP[d] [g(100 g)$^{-1}$]
(structure 1)	0.61	1.10	50
(structure 2)	3.28	7.55	18
(structure 3)	1.45	2.52	20
(structure 4)	2.28	3.32	14–15

[a]All polymers prepared by the phosphorylation method. [b]0.5% concentration at 25 °C. [c]96% H$_2$SO$_4$. [d]Containing 4% LiCl.

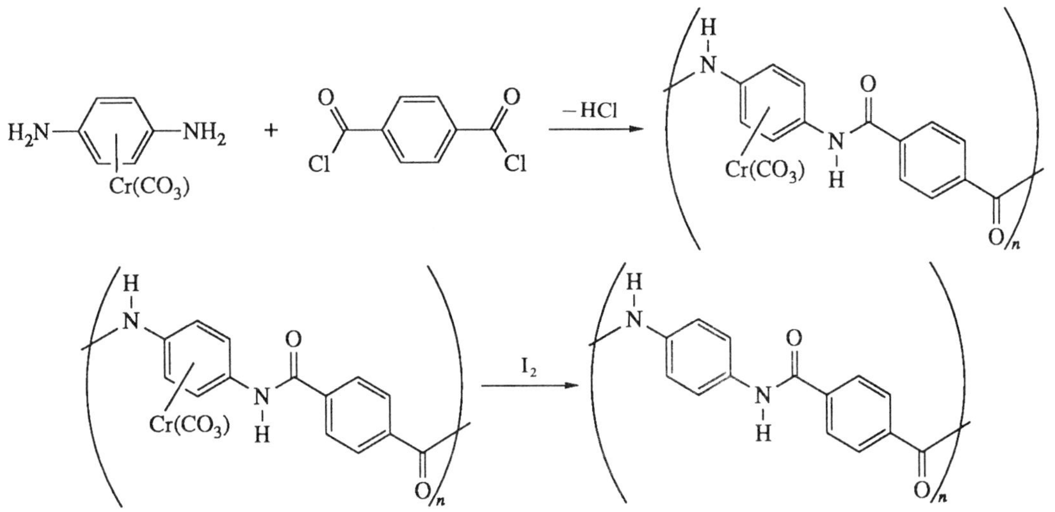

Scheme 12

Soluble poly(p-phenyleneterephthalamide)s with phenyl and biphenyl units in the terephthalic acid were also synthesized by Heitz et al.[200] The melting temperatures of these polymers are in the range of 440–490 °C and overlap their decomposition temperatures.

Aromatic polyamides containing alkyl side chains were reported by Ringsdorf et al.[201,203] and by Ballauff et al.[203] When each structural unit contains four alkoxy groups, depending on their length, these polymers exhibit a board-like, 'sanidic' mesophase (Figure 14).[201,202] However, when there are only two alkoxy groups per one of the two monomeric units these polyamides display a new layered liquid crystalline mesophase, as depicted in Figure 15.[203,204] The corresponding polyesters containing four alkoxy groups in each monomeric unit or two alkoxy groups in only one monomeric unit display similar mesophases with the polyamides.[201–204]

Soluble aromatic polyesters were synthesized by using similar synthetic procedures as those used in the synthesis of soluble aromatic polyamides. The research groups from Mainz reported on the solubilization of aromatic polyesters by the addition of alkyl side groups.[201–205] Solubilization of aromatic polyesters containing alkyl side groups is easily accomplished when the length of the alkyl groups derived from terephthalic acid monomer and hydroquinone monomer are dissimilar in length (Scheme 13).[205] For example, the polyester from the top of Scheme 13, PTA16HQ-16, is highly crystalline and melts directly into an isotropic phase. However, PTA16HQ-6 exhibits much less crystallinity, is highly soluble and melts into a liquid crystalline phase. It is also instructive to compare the phase behavior of polyesters derived from bicyclo[2.2.2]octane-1,4-dicarboxylic acid and 2,5-dialkoxyhydroquinone diacetate (PBCOHQ-n) with those obtained from terephthalic acid and 2,5-dialkoxyhydroquinones (PTAHQ-n). Representative data on these polymers are available in Tables 6 and 7. As we can observe from the data reported in Tables 6 and 7, all PTAHQ-n melt into a liquid crystalline phase while PBCOHQ-n melt directly into an isotropic phase. These results

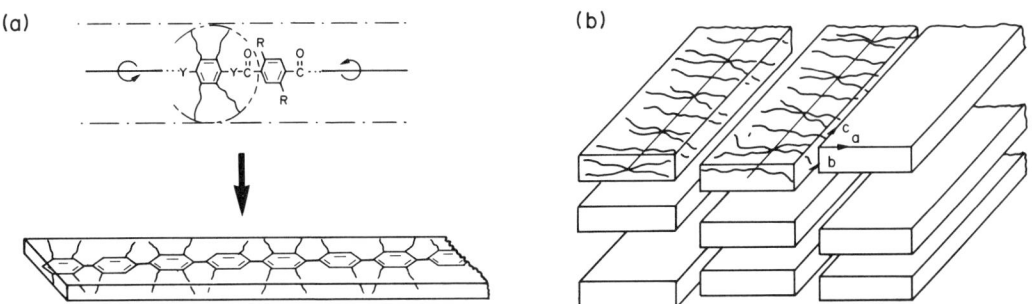

Figure 14 (a) Board formation of highly substituted rigid rod polymers, caused by hindered rotation of their mesogenic units;[201] (b) schematic representation of a sanidic (from Greek for boardlike) mesophase[202]

Polyamide 1

Polyamide 2

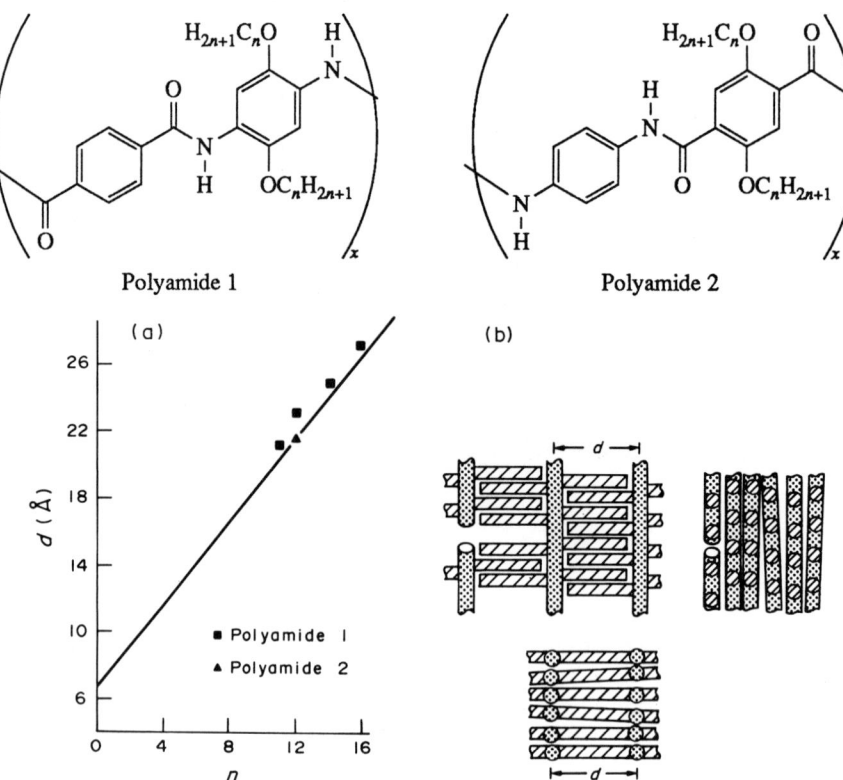

Figure 15 (a) Layer spacing *d* calculated according to Bragg's law from low angle X-ray reflections obtained from mesophases of polyamides 1 and 2 as function of *n*, the number of carbon atoms in the side chains; (b) model of molecular packing in the layered mesophase[204]

PTA16HQ-n

PTACOHQ-n

PTAHQ-n

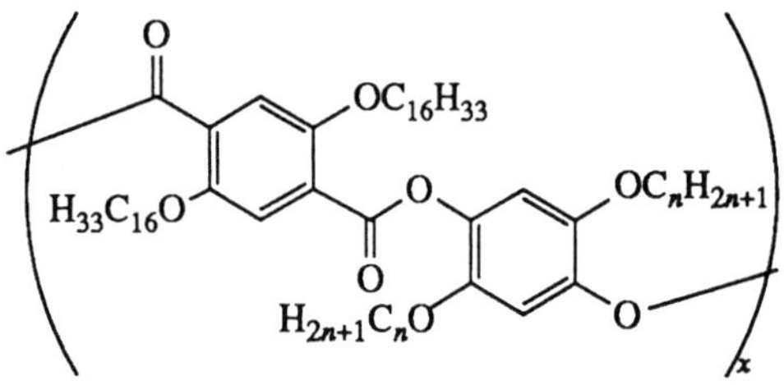

Scheme 13

suggest that interchain interactions provided by the wholly aromatic backbone are necessary for the stabilization of a mesophase in this class of polymers.[205] Soluble homologs of poly(*p*-hydroxybenzoate) were also synthesized by attaching pendant alkyl groups.[206] Thus, while poly(*p*-hydroxybenzoate) and poly(3-*n*-alkyl-4-hydroxybenzoate)s with alkyl groups shorter than six are only crystalline, the corresponding polymers with alkyl groups from six to 16 are soluble and exhibit a thermotropic mesophase.[206] It is interesting to mention that the polyesters with alkyl

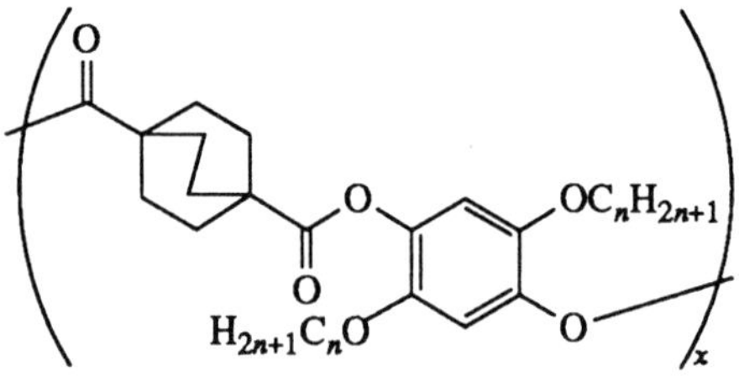

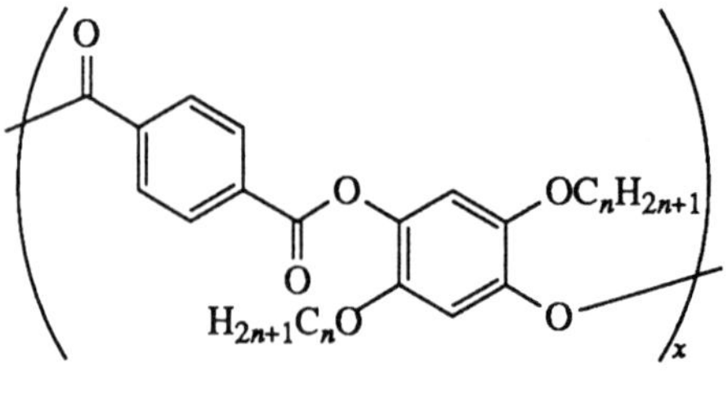

Table 6 Characterization of Polymers from Bicyclo[2.2.2]octane-1,4-dicarboxylic acid and 2,5-Dialkoxyhydroquinone Diacetates (PBCOHQ-n)

n^a	$[\eta]^b$ ($\times 10^{-1}$ L g^{-1})	Layer spacing, d (Å)	Thermal transitions^c (°C) Heating		Cooling	
			T_s	T_m	T_s	T_m
6	1.10	10.1	63^d	315	—	254
8	0.99	11.5	60^d	255	—	210
10	1.21	12.8	55^d	243	—	204
12	0.53	15.2	28	223	−2	186
14	0.87	16.5	46	206	30	166
16	0.59	17.8	60	193	49	160

^{a}Number of carbon atoms in the side chain. ^{b}Measured in CH$_2$Cl$_2$/TFA (v/v) at 20 °C. ^{c}Heating and cooling rates: 20 °C min^{-1}. ^{d}Transition observed only in the first heating scan.

Table 7 Characterization of Polymers from Terephthalic Acid and 2,5-Dialkoxyhydroquinones (PTAHQ-n)

n^a	$[\eta]^b$ ($\times 10^{-1}$ L g^{-1})	Layer spacing, d (Å)^c	(Å)^d	Thermal transitions^e (°C) Heating			Cooling		
				T_s	T_m	T_i	T_s	T_m	T_i
6	1.20	10.3	11.7	79^f	275	292	—	234	279
8	1.71	12.8	14.0	75^f	255	282	—	217	256
10	1.21	13.3	—	78^f	270	—	—	245	—
12	0.73	15.5	18.2	78	236	—	32	181	215
14	0.72^g	17.6	20.5	91	215	230	62	160	207
16	0.70^g	19.1	21.8	104	190	215	83	162	202

^{a}Number of carbon atoms in the side chain. ^{b}Measured in CH$_2$Cl$_2$/TFA 95/5 v/v at 20 °C. ^{c}Measured at room temperature. ^{d}Measured at 10 °C below the isotropization temperature. ^{e}Heating and cooling rates: 20 °C min^{-1}. ^{f}Transition observed only in the first heating scan. ^{g}Measured in CHCl$_3$/TFA (95/5 v/v) at 25 °C.

side groups display a layered mesophase which is different from the classic nematic or smectic mesophases exhibited by conventional rod-like polymers.

A large variety of soluble polyesters were prepared by Gaudiana et al.[193,194] by using the same synthetic principles as those used in the preparation of soluble polyamides. Table 8 summarizes some results on the relationship between the structure and solubility of polyesters.[194]

A large group of soluble aromatic polyesters was prepared by the group of Heitz by attaching various large substituents such as phenylalkyl,[207,208] perfluoroalkyl,[209] phenoxy and t-butyl,[210] phenyl and biphenyl,[162,211] or twisted biphenylene units such as 2,2'-dimethylbiphenylene[212,213] and 1,1'-binaphthyl-4,4'-ylene.[214] The relationship between the structure of various structural units and the solubility of the resultant polyesters was reviewed.[211,215,216a]

Table 8 The Effect of CF$_3$-substituted Biphenyl Content on the Solubility of Aromatic Polyesters in THF

Polyester	Maximum solubility in THF (% w/v)
$A = 1.0$, $B = 0$	53.0
$A = 0.5$, $B = 0.5$	49.8
$A = 0.4$, $B = 0.6$	48.5
$A = 0.2$, $B = 0.8$	11.1
$A = 0.1$, $B = 0.9$	<5.0
$A = 0$, $B = 1.0$	insoluble

Table 9 Properties of Polyesters

$$\left(O-\underset{R^1}{\bigcirc}-O-\overset{O}{\underset{}{C}}-\underset{R^2}{\bigcirc}-\overset{O}{\underset{}{C}} \right)_n$$

R^1	R^2	η_{inh} ($\times 10^{-1}$ L g^{-1})	T_g (°C)	T_m (°C)	Phase > T_m	Solubility[b] pCP	TCE	CHCl$_3$
–CH$_2$–C$_6$H$_5$	H	1.0	—	380[a]	—	+ +	+	–
–(CH$_2$)$_2$–C$_6$H$_5$	H	0.7	—	340	Anisotropic	+ +	+	–
–(CH$_2$)$_3$–C$_6$H$_5$	H	4.2	—	325	Anisotropic	+ +	+	–
–CH$_2$–C$_6$H$_5$	Br	2.0	91	(207)[c]	Nematic	+ +	+	–
–(CH$_2$)$_2$–C$_6$H$_5$	Br	2.5	79	(195)[c]	Nematic	+ +	+	–
–(CH$_2$)$_3$–C$_6$H$_5$	Br	1.0	58	(249)[c]	Nematic	+ +	+	+ +

[a] Melting in the decomposition range. [b] + +: stable solution (> 0.5 g 10^{-1} L^{-1}) at room temperature. +: soluble on heating, precipitation on cooling, –: insoluble; pCP: p-chlorophenol, TCE: 1,1,2,2-tetrachloroethane. [c] Weak endotherm on annealed samples.

Table 10 Properties of Polyesters

$$\left(O-\underset{R^1}{\bigcirc}-O-\overset{O}{\underset{}{C}}-\underset{R^2}{\bigcirc}-\overset{O}{\underset{}{C}} \right)_n$$

R^1	R^2	η_{inh} ($\times 10^{-1}$ L g^{-1})	T_g (°C)	T_m (°C)	Phase > T_m	Solubility[b] pCP	TCE	CHCl$_3$
–(CH$_2$)$_2$–C$_6$H$_5$	–(CH$_2$)$_2$–C$_6$H$_5$	0.3–3.6	60	195	Nematic	+ +	+ +	+ +
Bu^t	–(CH$_2$)$_2$–C$_6$H$_5$	0.3–5.0	100	225	Nematic	+ +	+ +	+ +
Bu^t	Br	0.2–2.6	131	—	—	+	0	–
Bu^t	CF$_3$	0.3–1.9	141	228	Nematic	+ +	+ +	+ +
–O–C$_6$H$_5$	Cl	1.8	85	241	Nematic	+ +	+ +	0
–O–C$_6$H$_5$	Br	0.5–3.0	90	220	Nematic	+ +	+ +	+ +
–O–C$_6$H$_5$	CF$_3$	0.35	88	—	Isotropic	+ +	+ +	+ +
–O–C$_6$H$_5$	C$_6$H$_5$	0.4–0.7	112	—	Isotropic	+ +	+ +	+ +

[a] Products of the highest molecular weights. [b] + +: stable solution (> 0.5 g 10^{-1} L^{-1}) at room temperature, +: completely soluble on heating, o: partially soluble at elevated temperatures, –: insoluble; pCP: p-chlorophenol. TCE: 1,1,2,2-tetrachloroethane.

Tables 9 and 10 summarize the dependence between structure, thermal properties and solubilities of polyesters based on substituted hydroquinone and substituted terephthalic acid.[216] The presence of dissimilar substituents on both monomers generates highly soluble polymers. Tables 11 and 12 show the relationship between the properties of polyesters obtained with twisted monomers and their structure.[216]

Additional examples of aromatic polyesters with lower crystallization ability and with enhanced solubility were reported by Kricheldorf *et al.* Nematic polyesters were obtained by the condensation of 2-phenylthioterephthalic acid, 2-(*p*-tolylthio)terephthalic acid and 2-(*p*-chlorophenylthio)terephthalic acid with the diacetates of hydroquinone, methylhydroquinone or 2,5-biphenyldiol.[217] 2-(2-Naphthylthio)terephthalic acid was also used in conjunction with the same dinucleophiles to obtain noncrystallizable nematic copolyesters.[218] The influence of various groups present in the structure of biphenols (e.g. O, CO, SO_2, S, CMe_2) on the phase behavior of polyesters with phenylthioterephthalic acid was also investigated.[219] Similar polyesters were synthesized from phenoxyterephthalic acid,[220] 4-fluoro-, 4-chloro-, 4-bromo-, 4-phenyl-, 4-cyclohexyl-, 4-phenoxy- and 4-cumyl-phenol[221] with hydroquinone or 4,4′-biphenyldiol. The influence of the various polyesterification reaction conditions on the final structure and properties of polymers was also investigated. The synthetic methods used most frequently are outlined in Scheme 14.[220–222]

Table 11 Polyesters with Twisted Biphenyl Units

Polymer	R	η_{inh}^a ($\times 10^{-1}$ L g^{-1})	T_g	T_m	Phase	pCP	Solubility^b TCE	CHCl$_3$
A	H	1.43			Nontractable	+	−	−
	Br	1.21	100	230	Nematic	+	−	−
	Ph	1.10	125	—	Nematic	+	+	+
	Me	0.81			Nontractable	+	+	−
B	Bu^t	1.43	150	—	Nematic	+	+	+

^{a}Measured in *p*-chlorophenol at 45 °C. ^{b}+ +: stable solution (>0.5 g 10^{-1} L^{-1}) at room temperature, +: soluble on heating, precipitation on cooling, −: insoluble; pCP: *p*-chlorophenol, TCE: 1,1,2,2-tetrachloroethane.

Table 12 Amorphous Polyesters with Binaphthyl Units

R^1	R^2	η_{inh} (10^{-1} L g^{-1})	T_g (°C)
biphenyl	binaphthyl	0.72	163
binaphthyl	biphenyl	1.90	183
phenyl (Bu^t)	binaphthyl	1.51	189

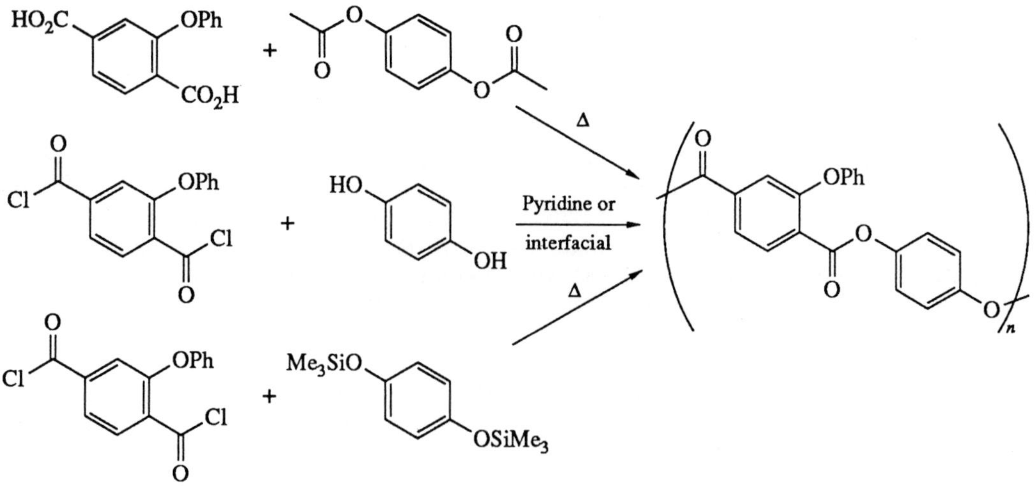

Scheme 14

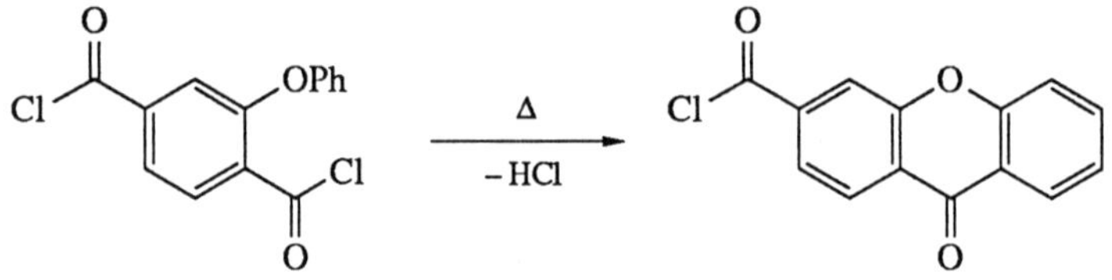

Scheme 15

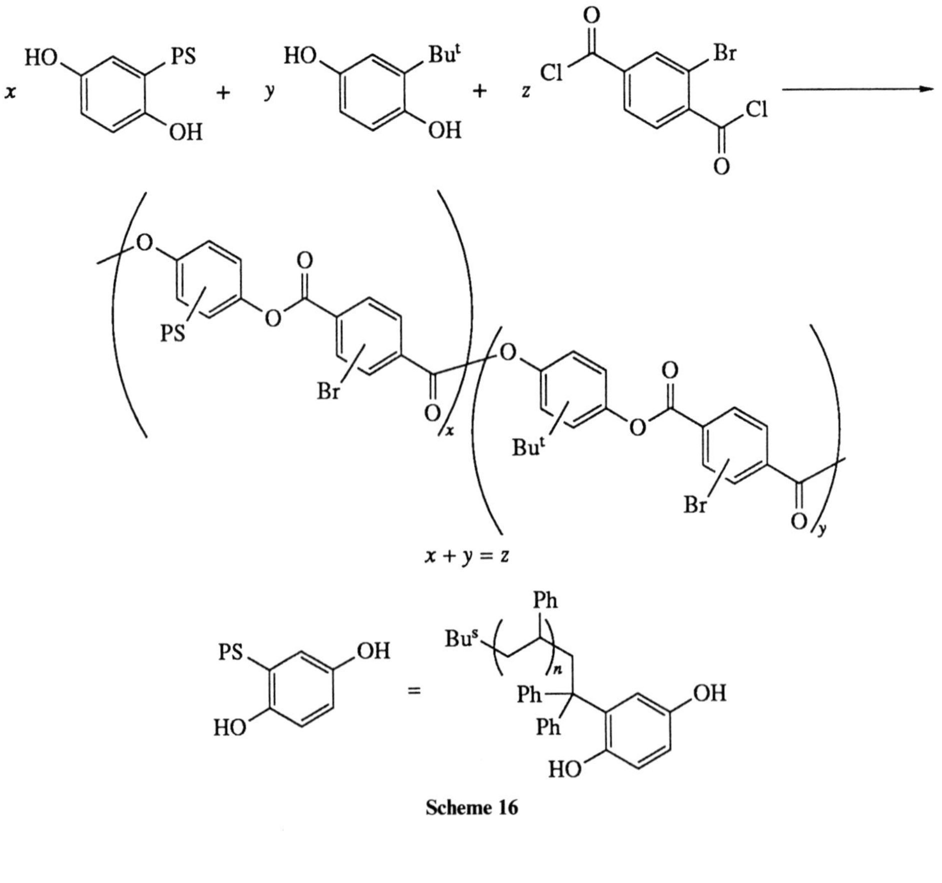

Scheme 16

For example the acetate method results in partial debromination of 4-bromophenoxyterephthalic acid units during their condensation with hydroquinone diacetate but not during the polymerizationn *via* the other two methods (Scheme 14).[221] Model reactions have shown that both phenoxyterephthalic acid and its acid chloride yield anthrone-3-carboxylic acid above 180 °C and this product terminates the polymer chain end (Scheme 15).[222] This reaction is less probable when the acid chloride is reacted with the silylated hydroquinone. The interest in aryloxy- and arylthio-substituted terephthalic acid was mainly determined by their convenient synthesis from dimethyl nitroterephthalate *via* aromatic nucleophilic displacement from the corresponding phenolate or thiophenolate.

Soluble liquid crystalline aromatic polyesters were also synthesized from terephthalic acid derivatives, *t*-butylhydroquinone and aromatic hydroxyfunctional polystyrene macromonomers.[223a] The number average molecular weight of polystyrene side chains varied from 1000 to 2000 (Scheme 16). The resulting polyesters contain polystyrene grafts and are of interest as emulsifiers for nematic aromatic polyesters with polystyrene.[223a]

An additional example of a graft copolymer containing a polyester main chain and poly(methyl methacrylate) graft units was synthesized from a hydroquinone substituted with polymethacrylate, *t*-butylhydroquinone and bromoterephthaloyl chloride.[223b] The hydroquinone containing the poly(methyl methacrylate) substituent was prepared by group transfer polymerization. Scheme 17 outlines the synthetic procedure.

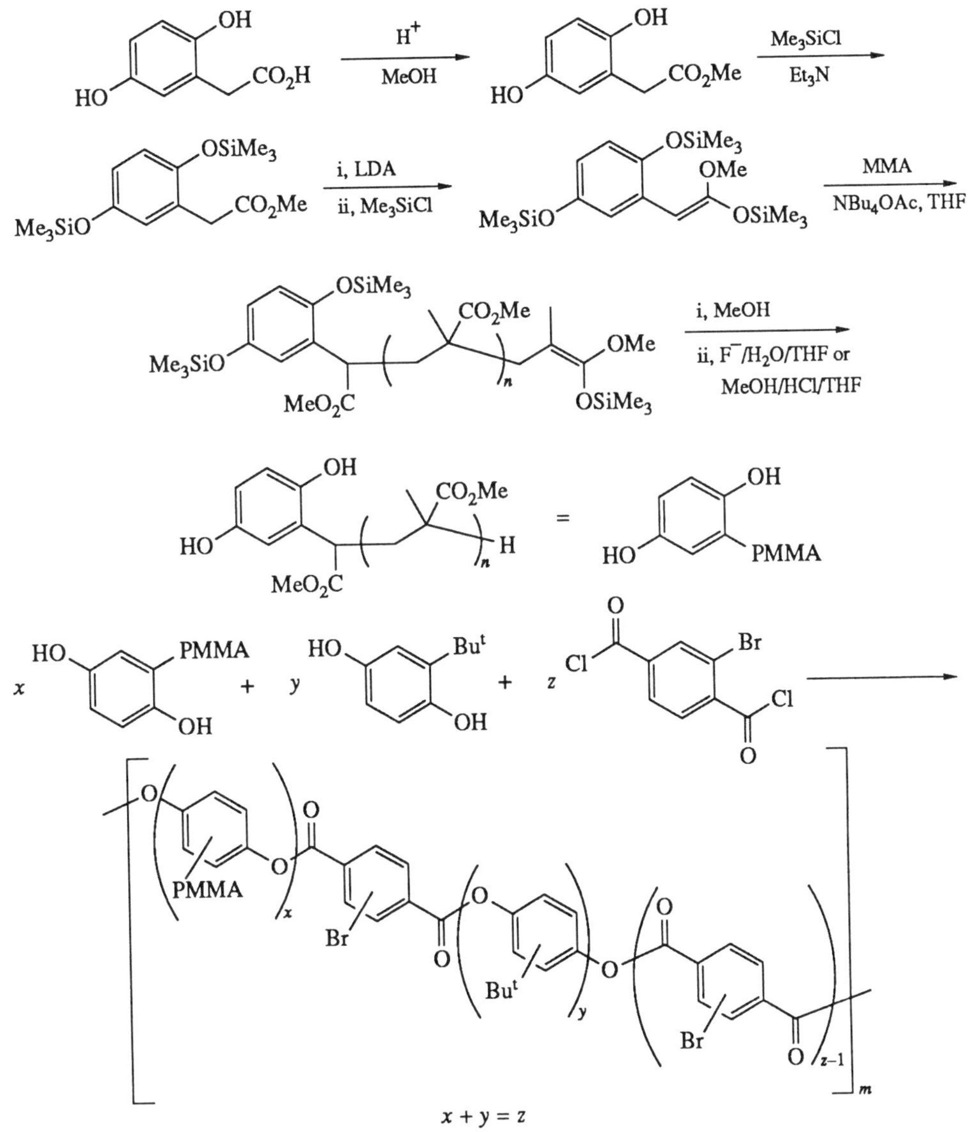

$x + y = z$

Scheme 17

14.6.1.5 Rod-like soluble polyimides

Rigid rod-like polyimides were synthesized by the sequence of reactions presented in Schemes 18 and 19.[224,225] The series of polymers from Scheme 18 contain alkyl side groups[224] while the ones from Scheme 19 contain oligo(oxyethylene) side groups.[225] The merit of these two synthetic schemes consists in the fact that the resulting polymers contain only imide rings. Soluble polyimides were also prepared by the sequence of reactions from Scheme 20 *i.e.* by conventional imidization.[226] All these rigid rod-like polyimides form layered mesophases.[224-226] It is interesting to note that polymers containing alkyl side groups of identical length are isomorphic within their crystalline phase and, therefore, form solid solutions regardless of the nature of their polymer backbone.[226] For example, the polyimide from the top part of Scheme 20 is isomorphic with the polyester which has an identical number of carbons in its side groups (bottom part of Scheme 20).[226]

Soluble polyimides were also synthesized from 3,6-diphenylpyromellitic dianhydride and various diamines.[227] A lyotropic polyimide was obtained from the twisted 2,2'-bis(trifluoromethyl)-4,4'-diaminobiphenyl and 3,6-diphenylpyromellitic dianhydride (Scheme 21).[227]

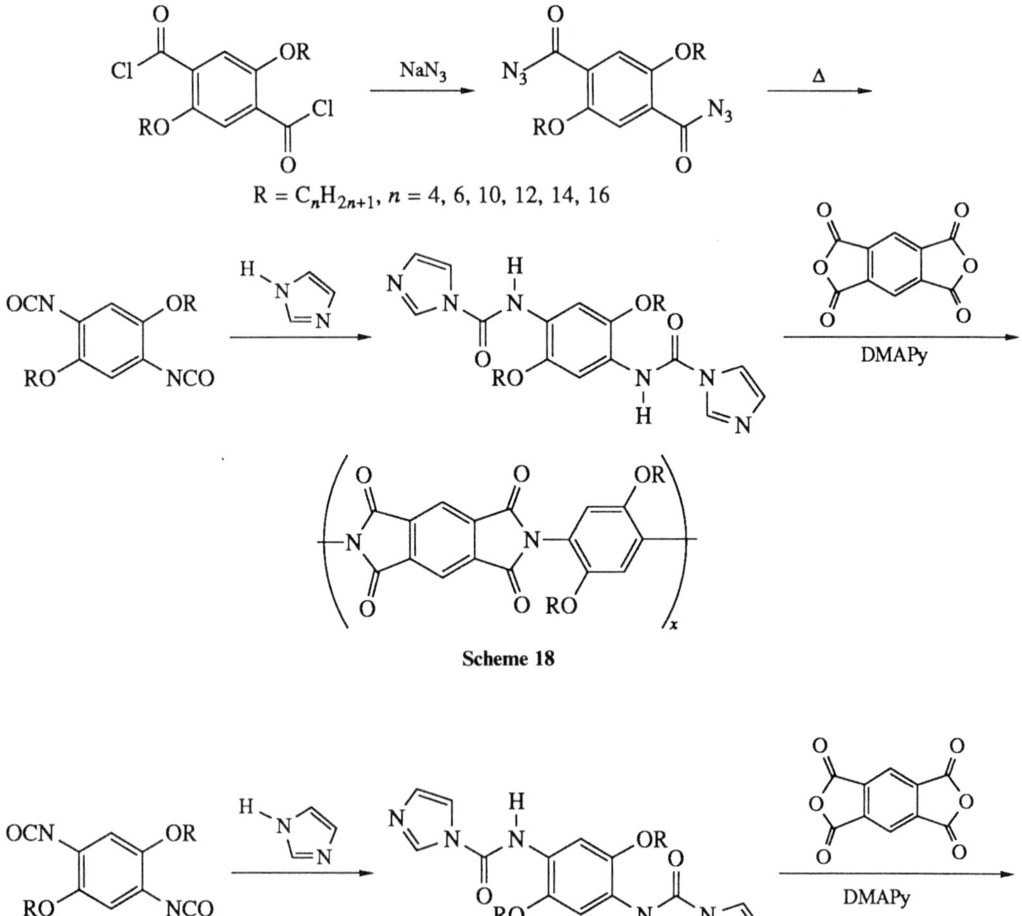

$R = C_nH_{2n+1}$, n = 4, 6, 10, 12, 14, 16

Scheme 18

$R = Me(CH_2CH_2)_y$

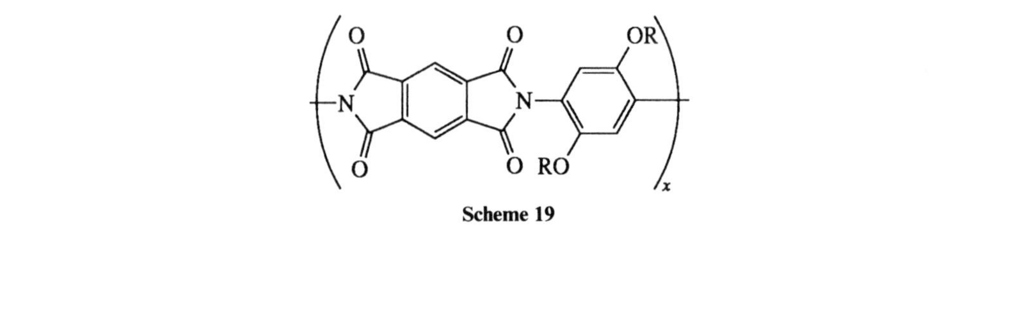

Scheme 19

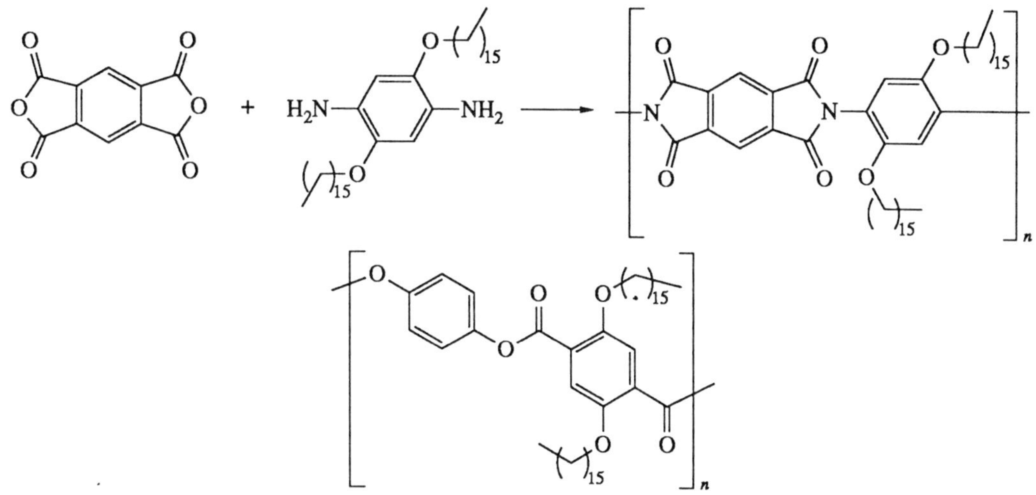

Scheme 20

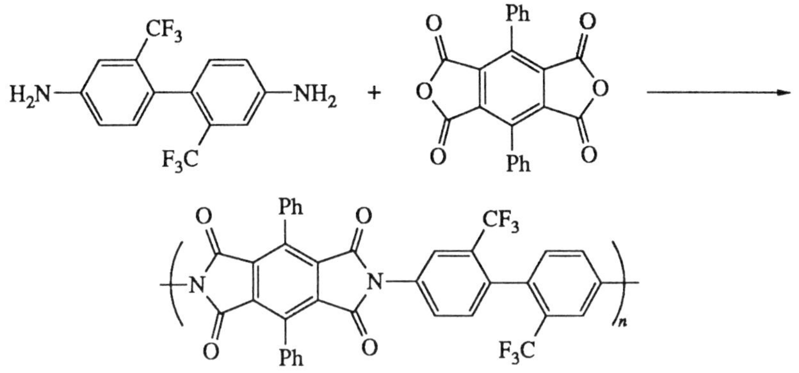

Scheme 21

14.6.1.6 Thermotropic poly(1,4-arylenevinylene)s

Poly(phenylenevinylene)s can be synthesized by a variety of reactions including the Wittig reaction,[228] Hofmann elimination,[229] the McMurry reaction,[230] the Heck reaction[231] and the Siegrist reaction.[232]

Liquid crystalline oligo(p-phenylenevinylene)s were first reported by Campbell and McDonald.[233] High molecular weight poly(p-phenylenevinylene)s are insoluble and infusible. The first thermotropic poly(phenylenevinylene)s were reported by Memeger[234] and were synthesized by the Wittig reaction. The insertion of suitable substituents such as Cl on the phenyl ring, the replacement of phenyl rings with anthracene or naphthalene and of vinylene with ethane allowed a synthesis of thermotropic poly(arylenevinylene)s.[234]

Thermotropic poly(arylenevinylene)s were also reported by Heitz et al. who developed various synthetic methods based on the Heck reaction.[211,231,235,236] Some of these procedures are outlined in Scheme 22. The highest solubility is obtained with the polymer with $R_1 = H$ and $R_2 = Ph$. The polymers with $R_1 = H$ and $R_2 = Ph$ or CF_3 exhibit thermotropic mesophases,[235] Additional examples of poly(p-arylenevinylene)s are presented in Scheme 23.[237] Both homopolymers and copolymers were synthesized by the procedure outlined in Scheme 23. As expected, copolymers with highly dissimilar structural units which can also yield constitutional isomeric structural units yield soluble, thermotropic liquid crystalline polymers which are conjugated.[237] Only qualitative information is available on their liquid crystalline behavior. Thermotropic poly(phenylenevinylene)s containing flexible spacers synthesized by the Heck reaction were reported by Saegusa et al.[238]

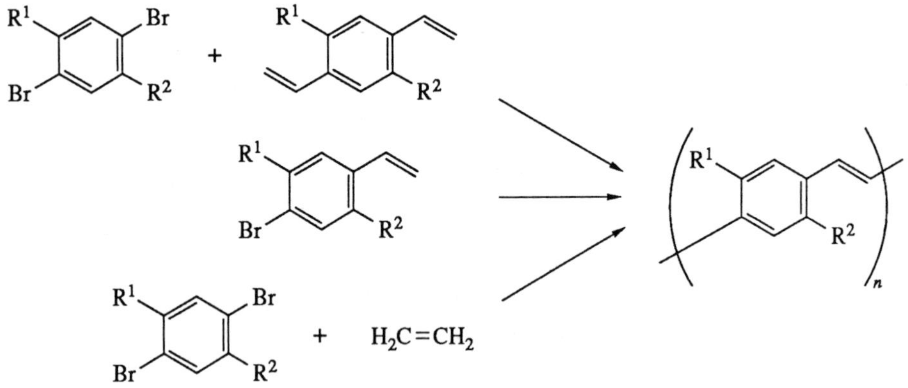

$R^1 = R^2 = H$; $R^1 = H$, $R^2 = Me$, CF_3 or Ph; $R^1 = R^2 = Me$

Scheme 22

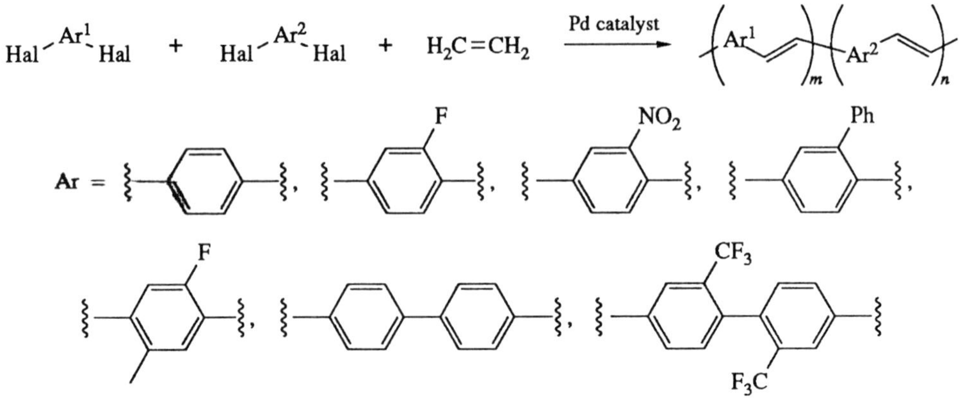

Scheme 23

14.6.1.7 Polyurethanes

The first series of thermotropic polyurethanes was synthesized in 1981.[239–243] However, they were prepared mostly from isocyanates or primary amines and were not stable above their melting temperature. The first examples of thermally stable polyurethanes were reported by Kricheldorf et al.[244–246] Smectic polyurethanes were synthesized from the bischloroformates of 4,4′-alkylenedioxydiphenols prepared from α,ω-dibromoalkanes and hydroquinone by interfacial polycondensation with piperazine, trans-2,5-dimethylpiperazine and 4,4′-bipiperidine (Scheme 24).[246] Polyurethanes without flexible spacers were synthesized from 4,4′-bipiperidine, 1,2-bis(4-piperidinyl)ethane and 1,3-bis(4-piperidinyl)propane with the bischloroformates of p-phenylene, 2-methyl-1,4-phenylene or 2,5-biphenylylene. Copolyurethanes were also synthesized.[247] The polymers derived from hydroquinone or methylhydroquinone are semicrystalline with short-term stability up to 310 °C. The polymers obtained from phenylhydroquinone are amorphous with a thermal stability up to 360 °C. Polyurethanes and copolyurethanes prepared from 4,4′-biphenylylene units form a smectic layered structure in the solid state and a nematic mesophase above the melting temperature (Scheme 25). While the research group of Kricheldorf decreases the crystallization ability by suppressing the hydrogen-bonding ability of the urethane groups, Mormann et al. have generated thermally stable polyurethanes from diisocyanate-substituted benzoates.[248–254] The resultant poly(esterurethane)s are more flexible and generate broader ranges of mesophases by decreasing the melting temperature.[253,254] Additional examples of polyurethanes were reported by MacKnight et al.[255]

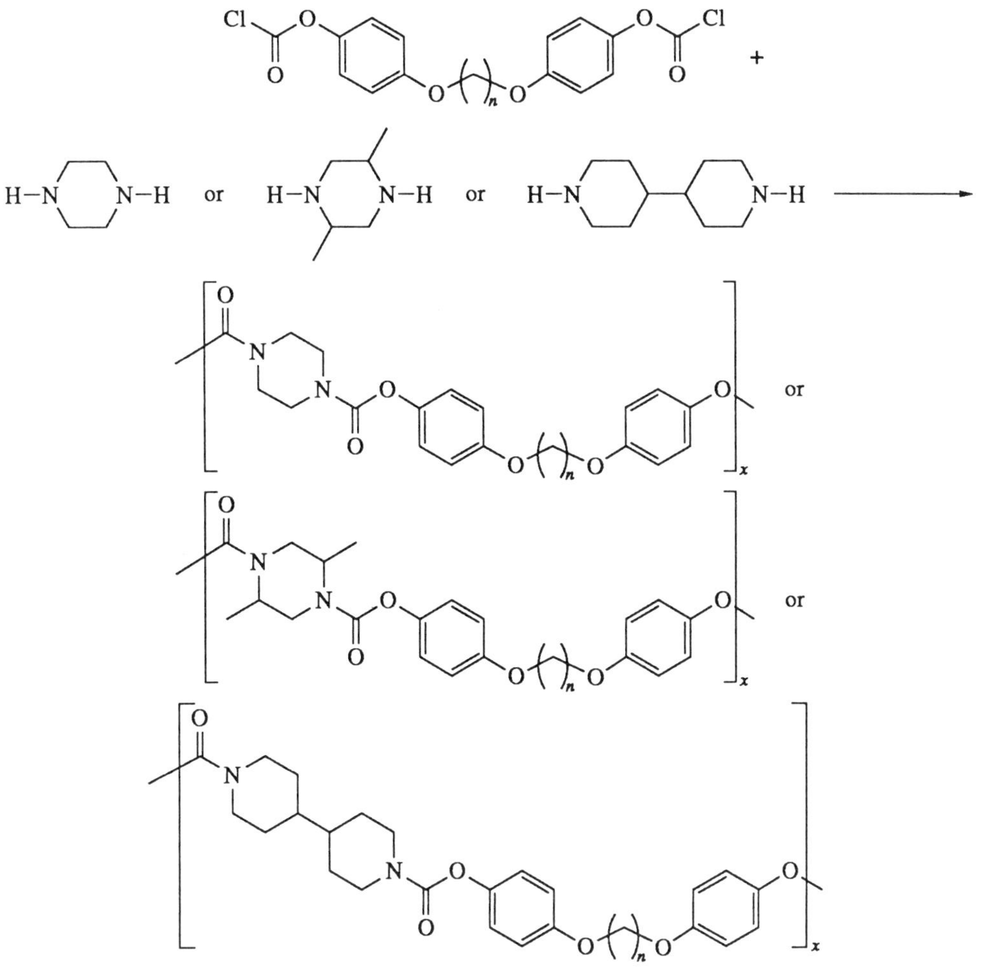

Scheme 24

14.6.1.8 Polycarbonates

The first examples of polycarbonates containing flexible spacers were reported by Roviello and Sirigu[256,257] and by Sato et al.[258,259] Poly(ester carbonate)s from t-butylhydroquinone and terephthalic acid were first reported by Prevorsek et al.[260] Soluble aromatic copolycarbonates were synthesized by Kricheldorf and Lubbers.[261,262] They are conveniently prepared by copolymerization of three or four biphenols with bis(trichloroacetyl)carbonate. For example, copolycarbonates based on 4,4'-dihydroxybiphenyl, methylhydroquinone and 4,4'-dihydroxybiphenyl ether or 4,4'-dihydroxybenzophenone are soluble in aprotic organic solvents such as dichloromethane.[262] Poly(ester carbonate)s were also synthesized by melt polycondensation of substituted hydroquinones and diphenyl terephthalate with diphenyl carbonate[263] and from p-hydroxybenzoic acid, 4,4'-dihydroxybiphenyl and diphenyl carbonate.[264]

14.6.1.9 Poly(ester anhydride)s

Thermotropic poly(ester anhydride)s were synthesized from terephthaloyl chloride with silylated hydroxy acids.[265] The polymers presented in Scheme 26 are noncrystallizable and exhibit a nematic phase over a broad range of temperatures.[265]

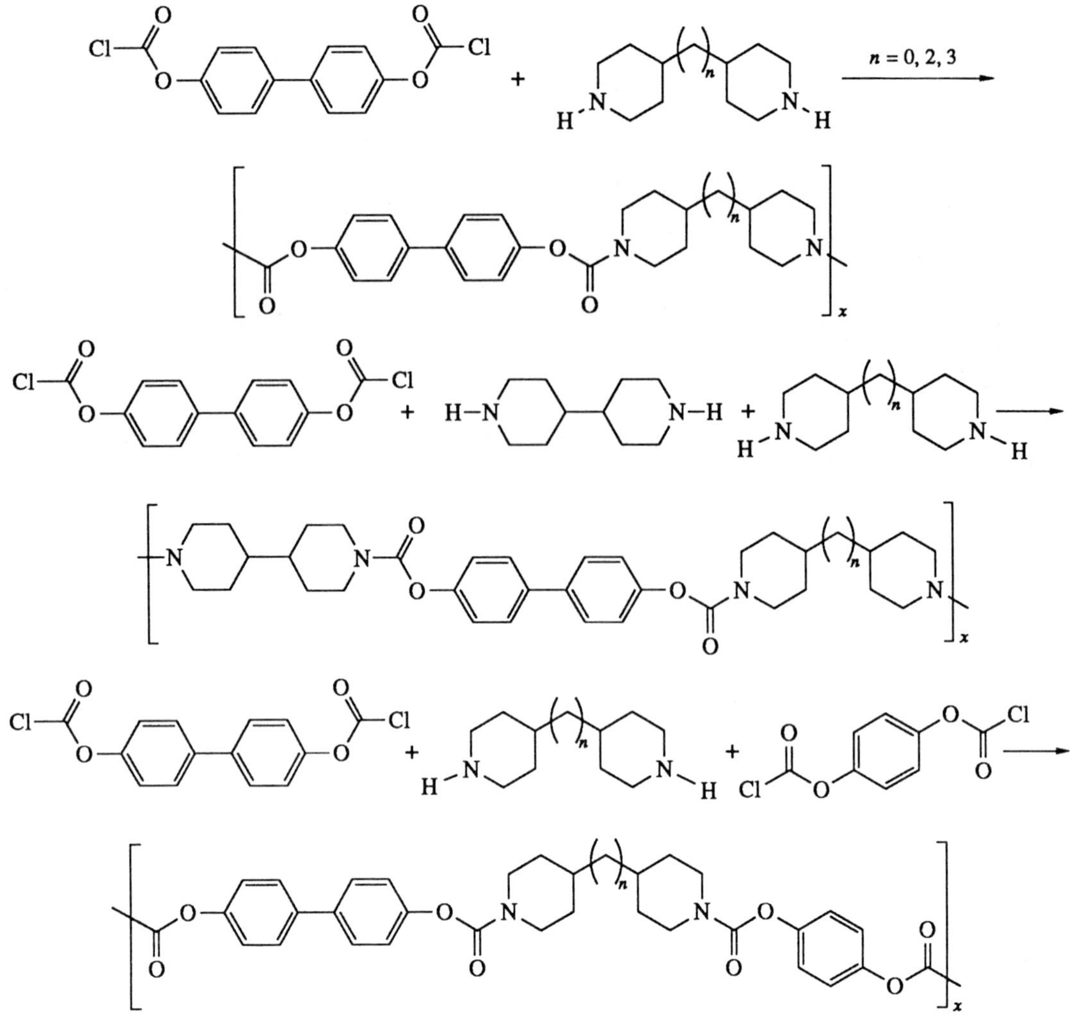

Scheme 25

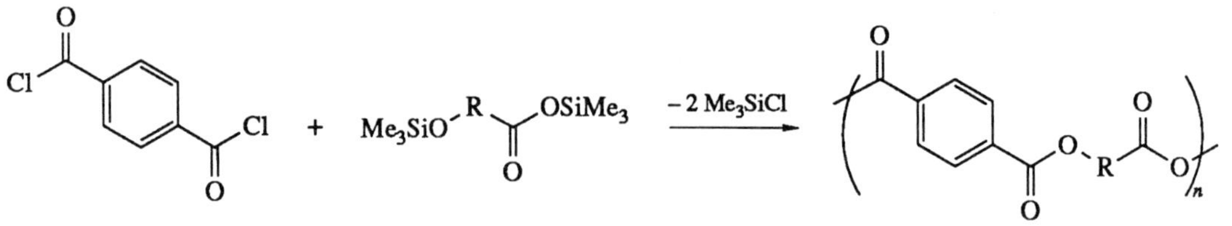

Scheme 26

R	T_g (°C)	T_i (°C)
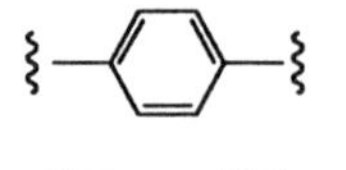 –C₆H₄–	136	360
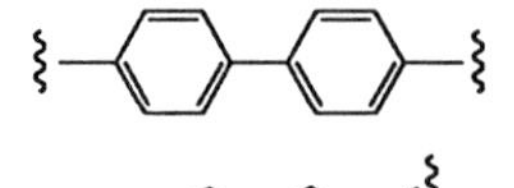 –biphenyl–	160	450
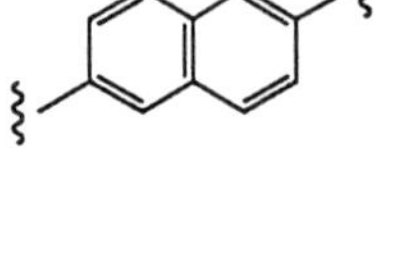 –naphthyl–	162·	440

14.6.1.10 Poly(ester imide)s

Thermotropic poly(ester imide)s were developed by Kricheldorf's group. The first series of thermotropic poly(ester imide)s were synthesized starting from trimellitic anhydride and α,ω-diaminoalkanes containing from 4 to 12 methylene units. The resulting α,ω-diaminoalkane bis(trimellitimide)s were reacted with various 4,4'-diacetoxybiphenols (Scheme 27), including those based on 4,4'-dihydroxybiphenyl,[266] hydroquinone, methyl-, chloro-, phenyl- and tetrachlorohydroquinone, and 1,5- 1,4-, 2,7- and 2,6-dihydroxynaphthalenes.[267] All these polymers crystallize in a layered morphology. The polymers with Ar = 4,4'-biphenylyl melt into a smectic mesophase.[266] The melting varies from 297 °C for the polymer with $n = 12$, to 393 °C for the polymer with $n = 4$. The liquid crystalline phase undergoes isotropization at 386 °C for the polymer with $n = 12$ and at 467 °C for the polymer with $n = 4$. The rate of crystallization of all these polymers is unusually fast. With the exception of the polymers based on Ar = 1,4-phenylene, 1,4-tetrachlorophenylene and 2,6-naphthalene, which exhibit a smectic mesophase, the other polymers melt directly into an isotropic phase.[267]

The second procedure used in the preparation of poly(ester imide)s is based on the reaction of pyromellitic dianhydride or benzophenone-3,3',4,4'-tetracarboxylic dianhydride with amino acids or lactams followed by the condensation of the resulting diacids with diacetates of hydroquinone, 2,6-dihydroxynaphthalene or 4,4'-dihydroxybiphenyl (Scheme 28).[268] All polymers presented in Scheme 28 crystallize in a layered structure. With the exception of the polymer with $n = 11$, all the other polymers based on Ar = 4,4'-biphenylyl exhibit a smectic mesophase. The polymer with $n = 10$ and Ar = 1,4-phenylene exhibits a smectic mesophase. The other polymers based on Ar = 1,4-phenylene or 2,6-naphthylene melt directly into an isotropic phase.[268]

An additional method for the synthesis of thermotropic poly(ester imide)s is based on the reaction of trimellitic dianhydride with amino acids or lactams followed by condensation with acetates of biphenols (Scheme 29). This reaction can be performed as a 'one-pot procedure'.[269] Again all homopolymers exhibit a lamellar crystalline phase which in all cases melts into an enantiotropic smectic phase. However, the corresponding copolymers exhibit a nematic mesophase. It is interesting to mention that the homopoly(ester imide)s from Scheme 29 are in fact copolymers since their structural units have two constitutional isomers. Finally, fully aromatic poly(ester imide)s were

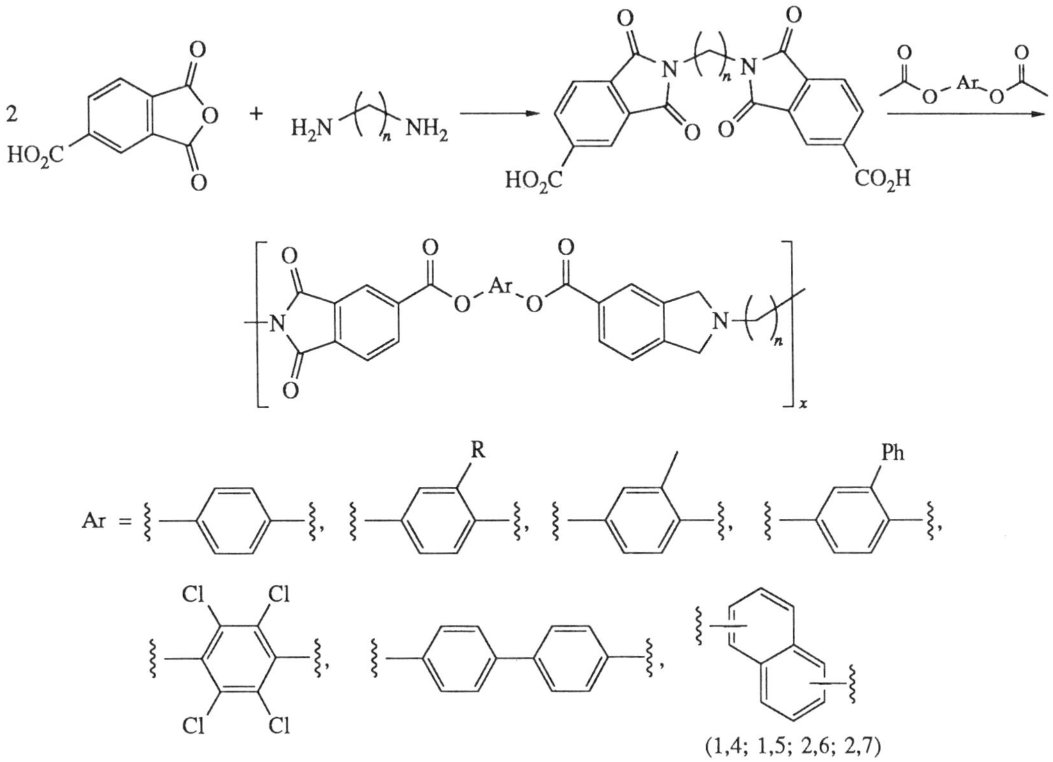

Scheme 27

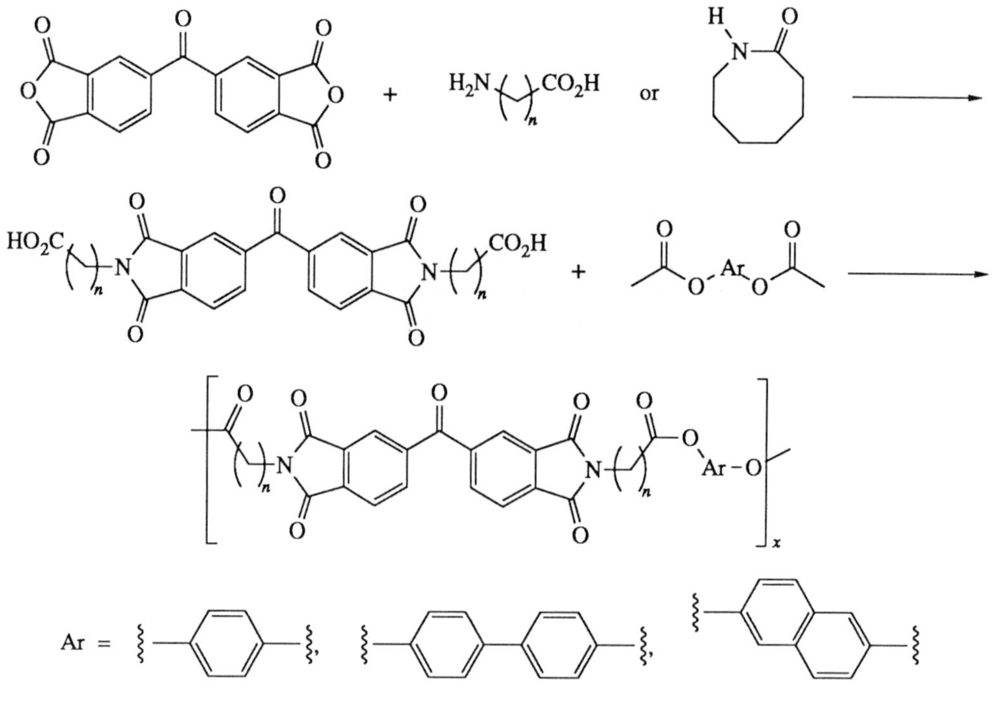

Scheme 28

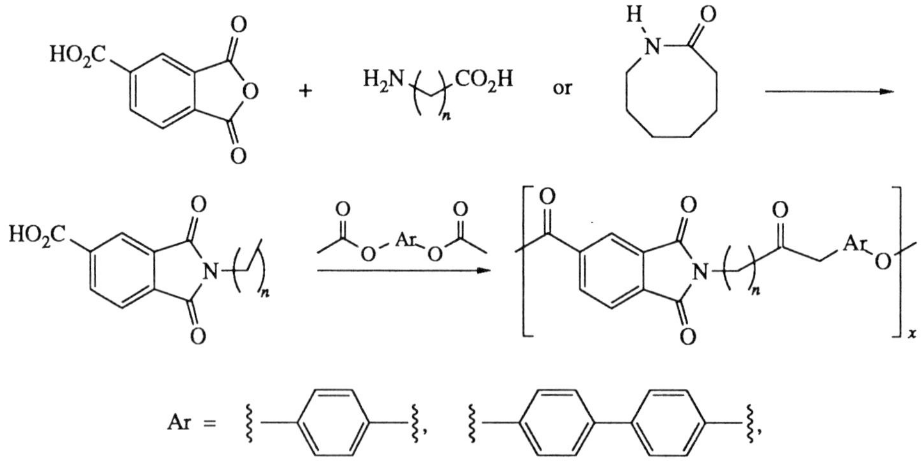

$n = 3, 4, 5, 6, 10, 11$

Scheme 29

$n = 3, 4, 5, 6, 10$

prepared by the same research group by using several different synthetic procedures which are outlined in Scheme 30.[270–273] The imide monomers from Scheme 30 were homopolymerized and copolymerized either between themselves or with other aromatic dicarboxylic acids, biphenols or hydroxy acids to give fully aromatic poly(ester imide)s with complex thermal behavior.[271,272] However, the fully aromatic poly(imide ester)s from Scheme 31 exhibit glass transition temperatures between 140 and 180 °C and nematic mesophases which undergo isotropization between 375 and 500 °C.[273]

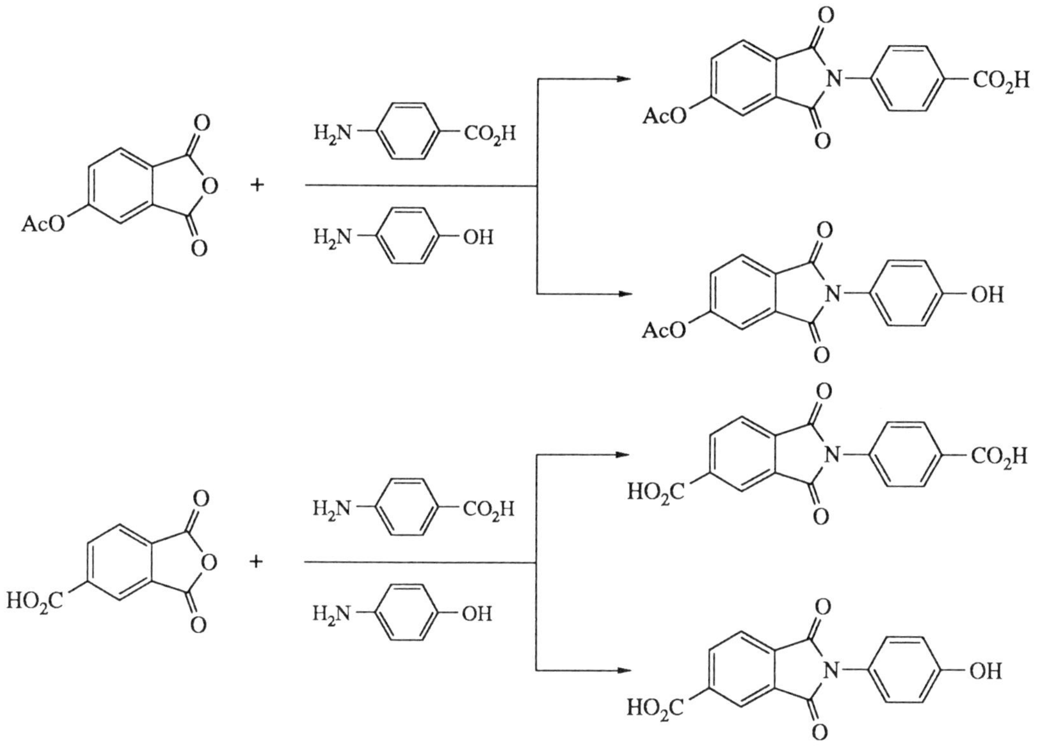

Scheme 30

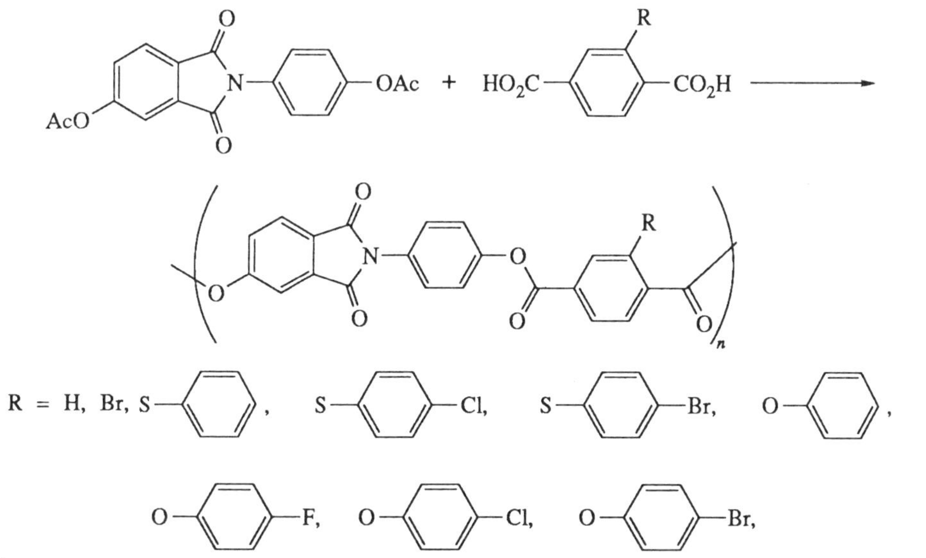

Scheme 31

14.6.1.11 Polyhydrocarbons

In addition to the class of poly(arylenevinylene)s described in Section 14.6.1.6 and poly(phenylene)s and poly(arylene)s described in Section 14.6.1.1 liquid crystalline polymers containing flexible spacers and only carbon–carbon bonds were also synthesized by homocoupling of bis(4-bromophenyl)alkanes.[274]

14.6.1.12 Polyethers

Thermotropic polyethers based on rigid rod-like mesogens and flexible spacers are of interest since they do not undergo transesterification or other rearrangement reactions during their characterization, which requires extensive annealing at high temperatures. The effect of thermally induced rearrangement reactions will be discussed in Section 14.7. Polyethers can be conveniently prepared by the phase transfer catalyzed polyetherification of mesogenic biphenols with α,ω-dibromoalkanes. The main criterion required for their synthesis is good solubility of the resulting polymer in the reaction medium.[275-278]

14.6.2 Persistence Lengths of Soluble Polyesters and Polyamides

The persistence length and the Mark–Houwink coefficient, a, were determined both for soluble polyesters and polyamides. For soluble polyesters the persistence length data obtained by various authors are 95 Å ($a = 0.95$),[216a] 120 ± 10 Å ($a = 0.85$ but estimated to be 1.0),[216b] and 110 Å ($a = 1.0–1.1$).[216c] For substituted aromatic polyamides the values of the persistence length reported are 85 ± 26 Å[198] and 200 Å ($a = 1.1–1.2$).[194] Both values are lower than those reported for poly(p-phenylenediamineterephthalic acid) (150–290 Å) and for poly(p-benzamide) (240–750 Å).[198] All these results demonstrate that both polyesters and polyamides do not resemble rigid rods but are comparable to typical semiflexible chains containing approximately 2–20 persistence lengths. In addition, the values obtained for the Mark–Houwink coefficient, a, also support the semiflexible character of these chains (a is 0.8 for flexible and 2.0 for rigid chains). While aromatic polyamides, depending on substituent, can generate lyotropic solutions, soluble polyesters do not. Films cast from solutions of aromatic polyesters are transparent (amorphous) and become turbid (anisotropic) only after annealing above their glass transition temperature. In conclusion, although due to interchain interactions the rigidity of polyesters and polyamides increases *versus* that predicted by gas phase and dilute solutions of low molar mass compounds (Section 14.4), both types of polymer belong to the class of semiflexible polymers.

14.7 CHEMICAL HETEROGENEITY IN MAIN CHAIN LIQUID CRYSTALLINE COPOLYMERS

As discussed in Section 14.6, main chain liquid crystalline polymers and copolymers are synthesized by step polymerizations that are based on reversible or irreversible reactions. A polymerization reaction is reversible depending on its mechanism and on the reaction conditions used. In contrast to chain copolymerizations, step copolymerizations performed in a homogeneous phase at high conversion and with a stoichiometric ratio between comonomers always lead to copolymer compositions that are identical with the comonomer feed. Also, the difference between the reactivities of various monomers used in step reactions is lower than that of the monomers used in chain reactions. This implies that the compositional heterogeneity of the copolymers obtained by step reactions is lower than that of copolymers synthesized by chain copolymerizations, where copolymer composition is conversion dependent. However, again in contrast to chain copolymerizations, in step copolymerizations the copolymer's sequence distribution can be either kinetically (in irreversible copolymerizations) or thermodynamically (in reversible copolymerizations) determined. In reversible step copolymerizations, the sequence distribution is thermodynamically controlled and the copolymer microstructure is determined by redistribution reactions. The copolymer sequence distribution[279,280] and the configuration of the structural units[104,281,282] are both determined by the type of phase (isotropic, liquid crystalline, or crystalline) in which copolymerization or the copolymer reorganization reaction is performed. Microheterogeneous copolymerization reactions complicate the control of the copolymer's microstructure since the concentration of the comonomers in the proximity of the growing chain is determined by the miscibility and/or the association between the growing chain and the monomers. The concept that the growing chain can control its own environment during copolymerization was explained on the basis of the 'bootstrap' model and its implications were recently reviewed.[283] Some of the most recent examples of 'bootstrap effects' were observed in the radical copolymerization of macromonomers,[284-286] in the synthesis of block copolymers from immiscible amorphous segments,[287,288] and in the synthesis of ternary copolymers from monomers that can give rise to amorphous and liquid crystalline structural units by reversible copolymerization reactions.[289]

Copolymerization reactions performed in bulk are frequently used in the preparation of liquid crystalline polyesters, poly(ester imide)s, *etc.* The starting monomers lead to isomorphic melts; nevertheless, above a certain conversion this reversible reaction is performed in the liquid crystalline phase. A heterogeneous composition is generated both by the different reactivities of the monomers and by the polydispersities of the polymers. This heterogeneous composition generates a microphase-separated reaction mixture which at a given temperature contains isotropic, liquid crystalline and crystalline phases. Each of these phases generates a polymer homologous series of copolymers which have both different compositions and, for the same compositions, different sequence distributions. The sequence distribution is determined by the phase in which the copolymer is generated *i.e.* isotropic, liquid crystalline or crystalline. This microheterogeneous copolymerization reaction enhances the chemical heterogeneity of the resultant copolymers. Both the polydispersity and the chemical heterogeneity of liquid crystalline copolymers are responsible for the biphasic or even multiphasic nature of liquid crystalline copolyesters.[45,290-296] The chemical heterogeneity of liquid crystalline copolymers is reduced or even eliminated in the case of azeotropic irreversible copolymerizations.[103d,297,298] Excellent discussions on the change in sequence distribution and molecular weight during the annealing of thermotropic polyesters are available.[45,296,299] The same processes should take place in polyamides, polycarbonates, poly(ether imide)s, polyurethanes, polyanhydrides and any other polymers containing chemical bonds which can undergo rearrangement reactions during thermal treatment.

14.8 FLEXIBLE AND SEMIFLEXIBLE LIQUID CRYSTALLINE POLYETHERS

Flexible liquid crystalline polyethers are based on flexible mesogenic or structural units which are rod-like mesogenic units based on conformational isomerism. In gas phase and dilute solution these units are as flexible as for example butane (see Section 14.4.2).[107] However, melt phase interchain interactions increase their rigidity. The simplest structural units which can be used for the synthesis of flexible liquid crystalline polymers based on conformational isomerism are diphenylethane, phenyl benzyl ether, and methyleneoxy units (Scheme 32). Flexible rod-like mesogens or rod-like mesogens based on conformational isomerism exhibit a number of different conformers which are in dynamic equilibrium. The two most stable conformers are the *anti* and *gauche*. The *anti* conformer has an extended rod-like shape and therefore is expected to display liquid crystallinity. The *gauche* conformer is similar to a 'kinked' unit which is occasionally introduced within the structure of main chain liquid crystalline copolymers based on rigid rod-like mesogens to decrease their crystallization ability. Therefore, the insertion of flexible units capable of giving rise to extended and kinked conformers within the main chain of a polymer is expected to provide a liquid crystalline polymer having a dynamic composition. To date there are no data on the dynamic equilibrium between different conformers of 'flexible' rod-like mesogens. However, it is well established that when rod-like or linear conformations are in dynamic equilibrium with random-coil conformations, the rod-like conformation is preferred in the nematic phase.[300-302] Therefore, it is not excluded that in the nematic phase the preferred conformer may be the *anti*.

Low molar mass liquid crystals based on benzyl ether and diphenylethane units are well known. Table 13 summarizes the relationship between the structure of some benzyl ether compounds and the thermodynamic stability of their mesophase. A brief inspection of Table 13 shows that all flexible compounds listed exhibit liquid crystalline mesophases. However, the thermodynamic stability of their mesophase with respect to the crystalline phase is drastically dependent on subtle changes in their structure. For example, compound 1 exhibits a virtual nematic mesophase at $-20\,°C$. Hydrogenation of one of its phenyl rings enhances its isotropization by $68.6\,°C$. Nevertheless,

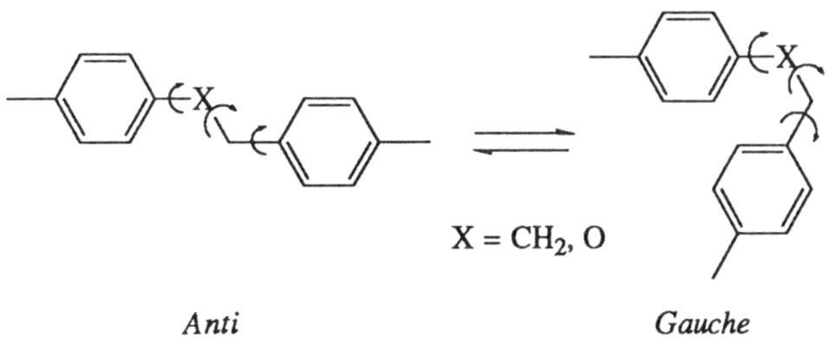

$X = CH_2, O$

Anti　　　　　　　　　　Gauche

Scheme 32

Table 13 Low Molar Mass Liquid Crystals Based on Benzyl Ether and Methyleneoxy Flexible Units[303-305]

Compound	Structure	Phase transitions (°C)
1	C_5H_{11}—⟨phenyl⟩—CH_2—O—⟨phenyl⟩—CN	k 49 n [−20][a] i
2	C_5H_{11}—⟨cyclohexyl-H⟩—CH_2—O—⟨phenyl⟩—CN	k 74.3 n (48.6)[b] i
3	C_5H_{11}—⟨bicyclooctane⟩—CH_2—O—⟨phenyl⟩—CN	k 72 n 73 i
4	C_5H_{11}—⟨cyclohexyl-H⟩—CH_2—O—⟨phenyl⟩—C_5H_{11}	k 35 s_B (31) n [21] i
5	C_5H_{11}—⟨cyclohexyl-H⟩—CH_2—O—⟨phenyl-F⟩—C_5H_{11}	k 43 n [−10] i
6	C_5H_{11}—⟨cyclohexyl-H⟩—CH_2—O—⟨phenyl⟩—OBu	k 55 s_B (47) n (53) i
7	C_5H_{11}—⟨cyclohexyl-H⟩—CH_2CH_2—⟨phenyl⟩—OBu	k 22 s_B 44.5 n 45.5 i

[a] [ ]: virtual. [b] ( ): monotropic.

the nematic phase of compound 2 is only monotropic. The replacement of the 1,4-cyclohexane ring of compound 2 with a bicyclooctane unit generates an enantiotropic mesophase. The difference between the isotropization temperatures of compounds 1 and 3 from Table 13 is 93 °C. If we replace the cyano group of compound 2 with a pentyl group the new compound 4 exhibits again a virtual mesophase. The replacement of a methylenic unit of the pentyl group of compound 4 with an oxygen generates compound 6 which displays a monotropic nematic phase. The substitution of the methyleneoxy group of compound 6 with an ethane unit generates an enantiotropic nematic mesophase (compound 7).

Table 14 provides some examples of low molar mass liquid crystals based on diphenylethane, cyclohexylphenylethane, dicyclohexylethane and dicyclooctylphenylethane.[303,306-310] The flexible mesogenic units from Table 14 follow the same kind of trend as those from Table 13. For example, the phase behaviors of compounds 1 and 2 from Table 13 are almost identical to those of compounds 8 and 9 from Table 14. Hydrogenating a phenyl ring from compound 8 increases the thermal stability of the nematic phase of compound 9 by about 75 °C *versus* that of compound 8. Replacing a phenyl ring from compound 8 with a bicyclooctane ring enhances the thermodynamic stability of the nematic phase by 137 °C.

Let us now consider that we increase the molecular weight of compound 1 or compound 8 by polymerization. According to the thermodynamic trends discussed in Section 14.5.3, the virtual mesophase of the monomeric structural unit should become at higher molecular weights enantiotropic, monotropic or remain virtual. In the last case the transition of the virtual mesophase of the polymer should, however, be shifted to higher temperatures.

The synthesis of the first flexible main chain liquid crystalline polyether without spacers is outlined in Scheme 33.[103a] As expected based on the above discussion, the resultant polyethers based on *trans*-1,4-bis[(methylsulfonyl)methyl]cyclohexane and methylhydroquinone or phenylhydroquinone display an enantiotropic nematic mesophase. The copolymers containing also some amount of *cis*-1,4-bis[(methylsulfonyl)methyl]cyclohexane also display a nematic mesophase. DSC traces of the polyethers based on methylhydroquinone and *trans*-1,4-bis[(methylsulfonyl)methyl]cyclohexane are identical to those of the corresponding polyesters based on methylhydroquinone and 1,4-cyclohexanedicarboxylic acid,[281,282,311] except that the phase transitions of the polyethers are shifted

Table 14 The Dependence Between Structure and Phase Transitions of Some Low Molar Mass Liquid Crystals Based On Disubstituted Ethane Derivatives[303,306–310]

Compound	Structure	Phase transitions (°C)
8	C_5H_{11}–Ph–CH_2CH_2–Ph–CN	k 62 n [–24]^a i
9	C_5H_{11}–Cy(H)–CH_2CH_2–Ph–CN	k 30 n 51 i
10	C_5H_{11}–bicyclo–CH_2CH_2–Ph–CN	k 76 n 113 i
11	C_5H_{11}–Cy(H)–CH_2CH_2–Ph(F)–CN	k 45 n 54 i
12	C_5H_{11}–Cy(H)–CH_2CH_2–Ph(F)–OC_5H_{11}	k 26 n (12)^b i
13	C_5H_{11}–Cy(H)–CH_2CH_2–Ph–Pr	k 4 s_B 18 i
14	C_5H_{11}–Cy(H)–CH_2CH_2–Ph–C_5H_{11}	k 46 s_B 109 i

^{a}[ ]: virtual. ^{b}( ): monotropic.

to lower temperatures.[103a] Based on the thermodynamic principles discussed previously this is an expected result since the corresponding polyesters are identical to the polyether from Scheme 33 except that the methyleneoxy units from the structure of the polyethers are replaced with ester units. This increases only the overall rigidity of the polyesters *versus* that of the polyethers.

Polyethers based on substituted hydroquinones and *p*-dibromo- or dichloro-xylene also lead to liquid crystalline polyethers.[312,313]

In 1975, Memeger[313] synthesized high molecular weight polyethers based on α,α'-dichloro-*p*-xylene and methylhydroquinone in dimethylacetamide using K_2CO_3 as base. Fibers with high modulus (233 gpd) were spun from these polymers. Unfortunately the polyetherification of substituted hydroquinones with α,α'-dichloro-*p*-xylene leads both to *C*- and *O*-alkylation reactions. Nematic polyethers with and without flexible spacers were also synthesized by cationic ring opening polymerization of *exo*-2-methyl-7-oxabicyclo[2.2.1]heptane and 7-oxabicyclo[2.2.1]heptane as well as by the copolymerization of the latter monomer with tetrahydrofuran and/or ethylene oxide (Scheme 34).[314,315]

A comprehensive series of experiments was performed with polyethers based on 1-(4-hydroxyphenyl)-2-(2-*R*-hydroxyphenyl)ethane (RBPE) and α,ω-dibromoalkanes containing from four to 20 methylenic units,[90,103b–f,316–321] particularly those based on 1-(4-hydroxyphenyl)-2-(2-methylhydroxyphenyl)ethane (MBPE). All these polyethers are in fact copolyethers since their structural unit has two constitutional isomers (Scheme 35). With the exception of MBPE-5 and MBPE-9 which exhibit two monotropic nematic mesophases, and MBPE-8 and MBPE-11 which exhibit one monotropic nematic mesophase, all the other MBPE-*X* polyethers are only crystalline.[103e] Nevertheless, again based on the thermodynamic discussion from Section 14.5 we can assume that all of them may exhibit virtual mesophases. The virtual transition temperatures and thermodynamic parameters of all MBPE-*X* polymers were determined by copolymerization experiments of MBPE with two different flexible spacers.[90,103,316–321]

The technique developed and used for the determination of their virtual transition temperatures and thermodynamic parameters is as follows. The structural units of the copolymers resulting from

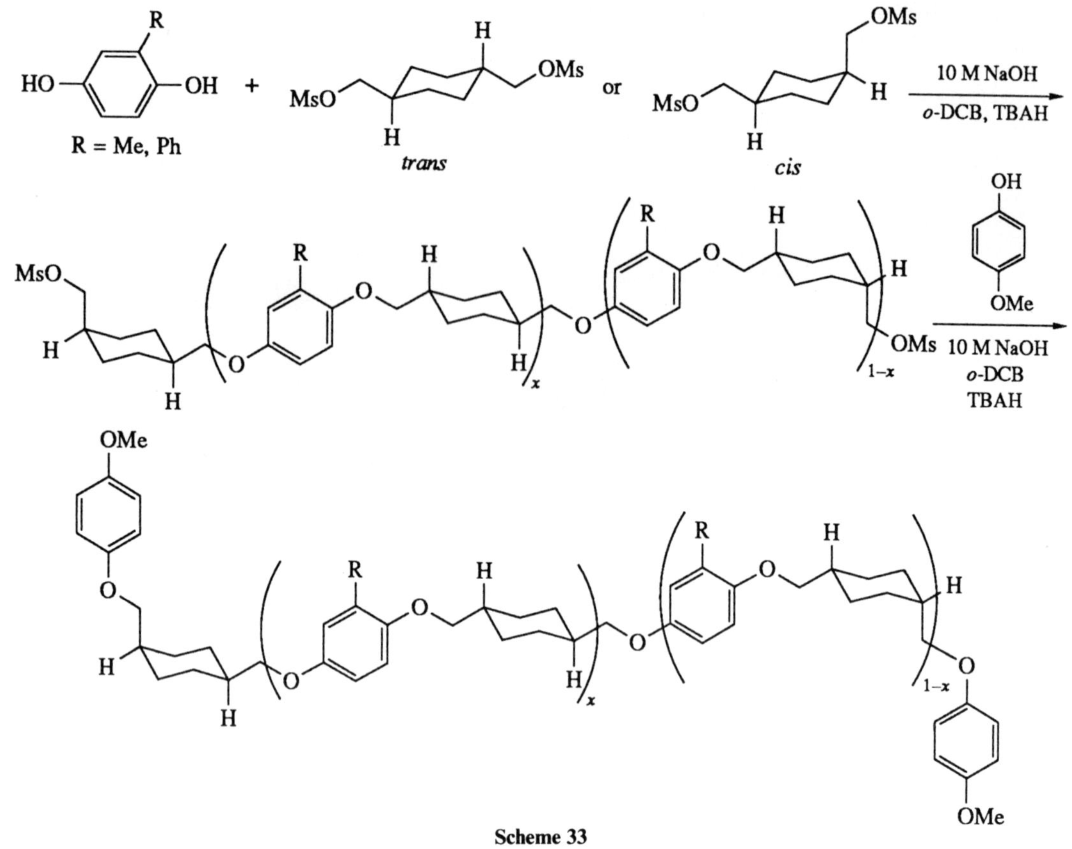

Scheme 33

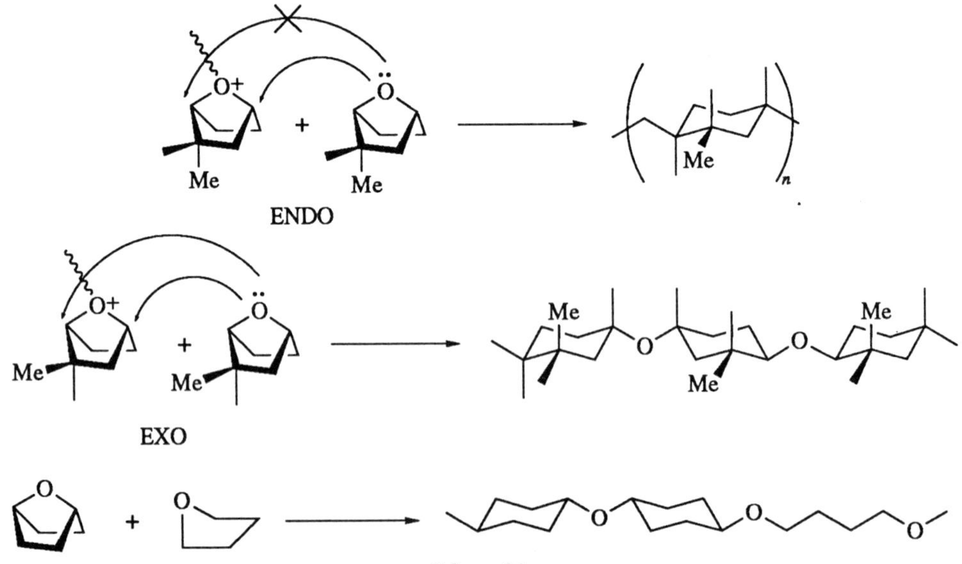

Scheme 34

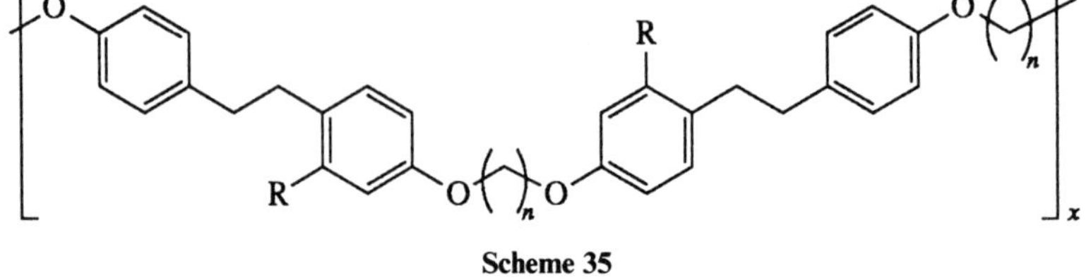

Scheme 35

two flexible spacers and one mesogen are isomorphic in their mesophases but not in their crystalline phases. As a consequence, the crystallization temperature of the copolymer is decreased and exhibits a eutectic point while both the temperature and the enthalpy associated with the mesophase display continuous dependences on composition. Therefore, in the mesophase these two structural units behave as an ideal solution and obey the Schroeder–Van Laar equations (Section 14.3).[91] Since the difference between ΔH_1 and ΔH_2 of the two homopolymers is not large and the values of ΔH_1 and ΔH_2 are proportional to their T_1 and T_2 transition temperatures, most of these dependences are linear. Therefore, upon extrapolation to the composition of the two homopolymers the 'uncovered' temperature transitions of the copolymers can be used to determine both virtual transition temperatures and their associated thermodynamic parameters of the parent homopolymers. An example of DSC traces for the copolymer MBPE-5/8 (i.e. based on five and eight methylenic units in the spacer) is presented in Figure 16. The dependences of transition temperatures and enthalpy changes of MBPE-5/8 versus composition for the results collected from Figure 16 together with their extrapolations are presented in Figure 17.[317] The virtual phase transition temperatures and thermodynamic parameters of MBPE-X determined from various pairs of flexible spacers agree very well.[103e] Figure 18 presents the plot of the dependence of the virtual isotropic–nematic transition temperatures as a function of the number of carbons (x) in the flexible spacer and its inverse ($1/x$).[103e] We can observe from Figure 18 that the virtual mesophase of these polymers does not vanish even at very long spacer length. This suggests that polyethylene should also exhibit a virtual mesophase as was indeed theoretically predicted.[322] The highest temperature nematic mesophase of MBPE-X homopolymers and copolymers was confirmed by X-ray experiments.[103f,323]

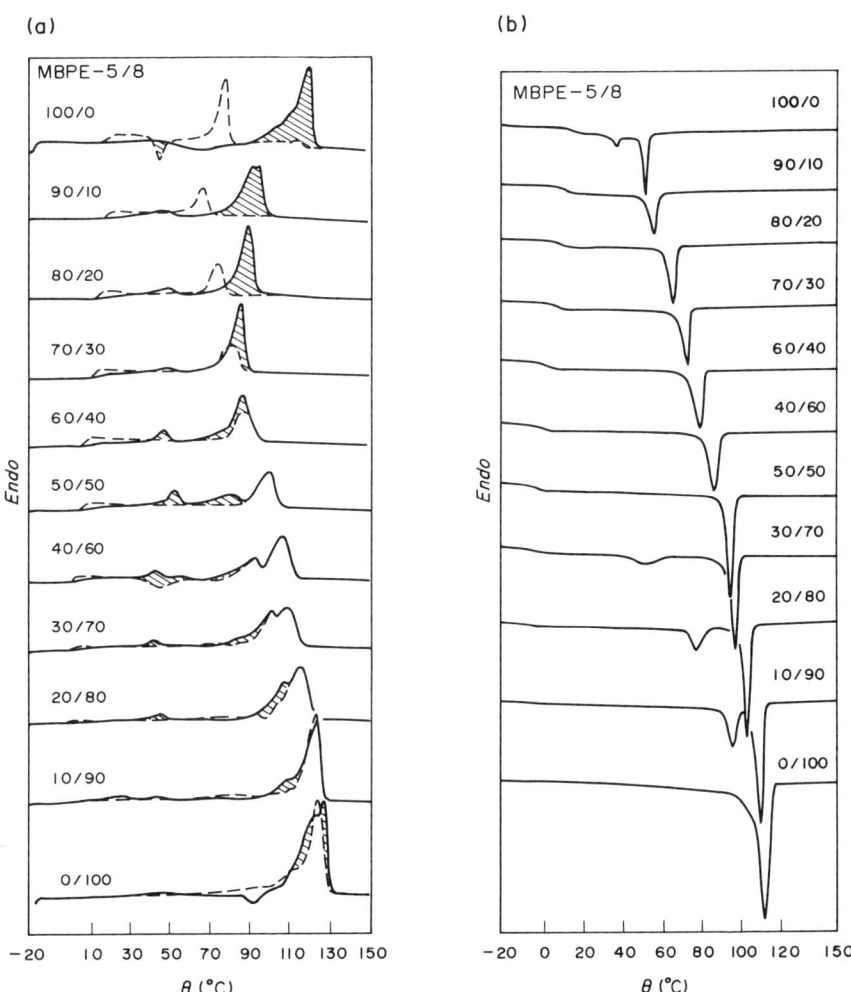

Figure 16 DSC curves of polyethers based on MBPE and 1,5-dibromopentane (MBPE-5), MBPE and 1,8-dibromooctane (MBPE-8) and corresponding copolyethers [MBPE-5/8(A/B)]. (a) First (——) and second (---) heating scans; (b) cooling scans

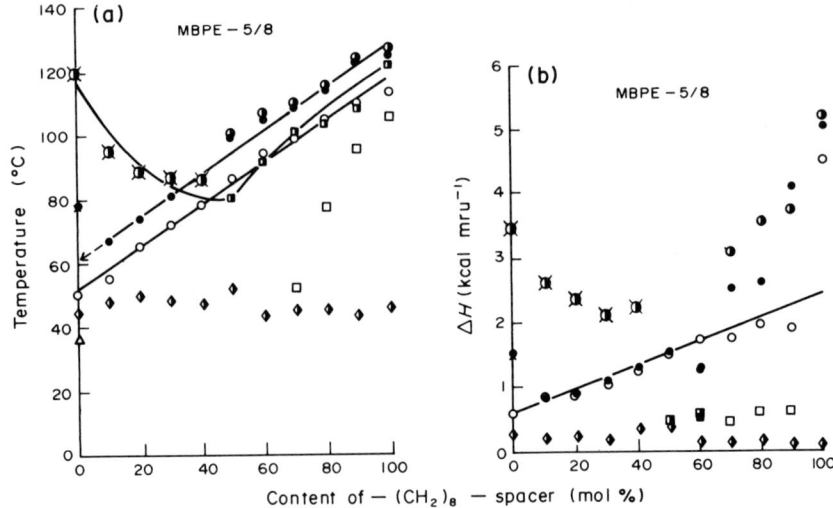

Figure 17 (a) Thermal transitions of polyethers (MBPE-5 and MBPE-8) and copolyethers [MBPE-5/8(A/B)]. First heating scan ⊗: T_{ki}, ◐: T_{ni}, ▨: T_{kn}, ◆: T_{kk}; second heating scan ⊗: T_{ki}, ●: T_{ni}, ■: T_{kn}; cooling scan ⊗: T_{in}, ◻: T_{nk}; △, T_{ns}. Arrows point to virtual transition temperatures for the homopolymers. (b) Enthalpy changes of polyethers (MBPE-5 and MBPE-8) and copolyethers [MBPE-5/8(A/B)]. First heating scan ⊗: ΔH_{ki}, ◐: ΔH_{ni}, ▨: ΔH_{kn}, ◆: ΔH_{kk}; second heating scan ⊗: ΔH_{ki}, ●: ΔH_{ni}, ■: ΔH_{kn}; cooling scan ○: ΔH_{in}, ◻: ΔH_{nk}. Arrows point to enthalpies of virtual transitions for the homopolymers

The entire series of ClBPE-X homopolymers behaves almost identically to the series of MBPE-X polymers and confirms all these data.[324]

The thermodynamic parameters associated with the isotropic–nematic phase transition temperatures of MBPE-X[103e] and ClBPE-X were plotted as a function of the number of carbons in the spacer and the orientational contribution of the mesogen and conformational contribution per CH_2 group at the isotropic–nematic transitions were determined.[103e,324] These values are summarized in Table 15 together with the corresponding data obtained for the polyethers based on 4,4'-dihydroxy-α-methylstilbene (HMS-X)[278] and of the polyesters based on 4,4'-dihydroxy-2,2'-dimethylazoxy-benzene and α,ω-alkanedioic acids (DMAB-X).[48] The polymers HMS-X and DMAB-X contain rigid rod-like mesogens, while MBPE-X and ClBPE-X contain flexible rod-like mesogens. As we can see from Table 15 both the enthalpic and entropic contributions per CH_2 are independent of the nature of the mesogen. They are higher for the polymers based on even spacers and all data are in good agreement. However, the orientational entropic contributions per mesogen are higher in the case of flexible mesogens than in the case of rigid mesogens. This result suggests that at the isotropic–nematic transition, the flexible mesogen undergoes, in addition to the orientational arrangement, a transition from its *gauche* into its *anti* conformer.

The determination of the virtual phase transition temperatures of polymers can also be done by preparing mixtures with low molar mass liquid crystals or blends with other virtual liquid crystalline polymers. However, these two techniques are less reliable than the copolymerization-based technique.[90]

It has been also shown that copolymerizations based on two and more than two flexible spacers or mesogenic units can be used to engineer phase transition temperatures and corresponding thermodynamic parameters of copolymers.[103d,321] The only requirement for these experiments is that the structural units of these copolymers should be isomorphic within the mesophase we want to tailor-make and the molecular weight of the homopolymers and copolymers should be higher than the molecular weight below which phase transitions are molecular weight dependent. When the thermal transition temperatures and the corresponding enthalpy changes of the homopolymers are not highly dissimilar both the transition temperatures of the copolymers and their associated enthalpies can be calculated from equations (4) and (5), in which X_n is the mole fraction of the structural unit n from copolymer and T and ΔH are, respectively, the transition temperature and the enthalpy change associated with the same phase transition of the parent homopolymer based on an identical structural unit.[321]

$$T = \sum_{i=1}^{n} X_n T(n) \quad (4)$$

$$\Delta H = \sum_{i=1}^{n} X_n \Delta H(n) \quad (5)$$

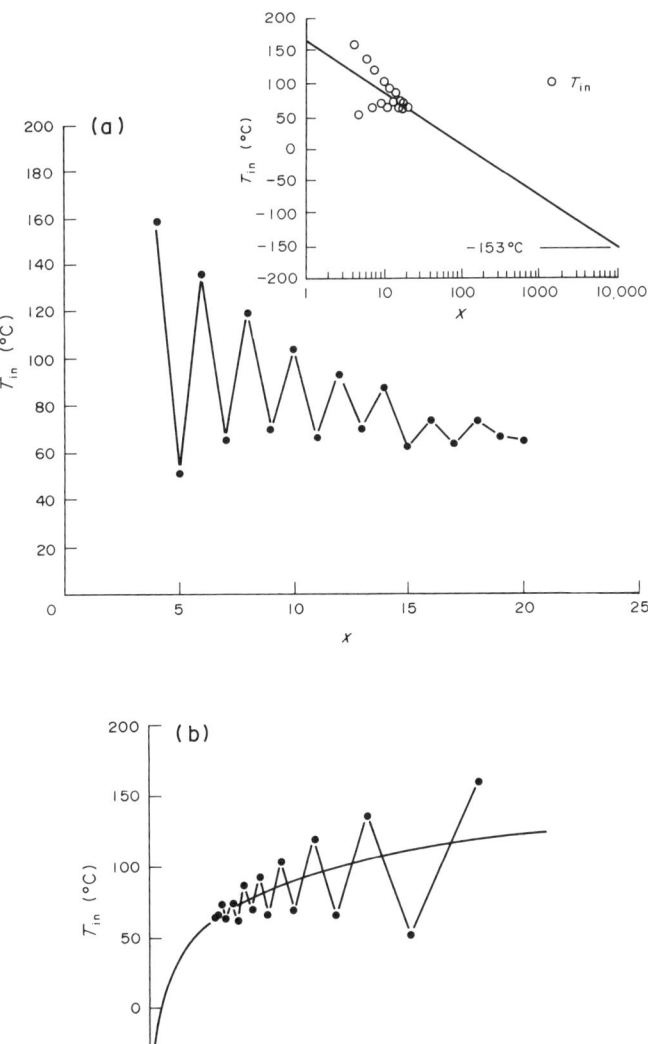

Figure 18 Dependence of the virtual isotropic–nematic (T_{in}) phase transition temperature of MBPE-X homopolymers on the number of methylene units in the flexible spacer (X). Upper right side corner presents the logarithmic plot of this dependence. (b) Dependence of the virtual isotropic–nematic (T_{in}) phase transition temperature of MBPE-X homopolymers on the inverse of the number of methylene units in the flexible spacer ($1/X$). The logarithmic dependence is presented with a continuous line

MBPE-X and ClBPE-X with $X = 5, 7, 9, 11$ and 13 exhibit a second uniaxial nematic mesophase which in certain polymers is monotropic while in others is virtual.[103e,324] This second nematic mesophase was transformed into an enantiotropic one by copolymerization experiments and was subsequently characterized by X-ray experiments.[325] This second uniaxial nematic phase was theoretically predicted.[326]

The influence of the size of various substituents on the phase transitions of RBPE-X was also investigated.[327] It has been demonstrated that phase transition temperatures of polymers depend on the nature of the lateral substituent by following the same trend as that followed by low molar mass liquid crystals. Some data are summarized in Table 16. A detailed discussion is available.[327] When the substituent of RBPE-X polymers is H. i.e. BPE, they do not exhibit a nematic mesophase anymore. For example, BPE-8/10 (i.e. the copolymer containing a 50/50 molar ratio of spacer lengths containing eight and 10 carbon atoms) displays an enantiotropic s_B mesophase.[328] At the same time, BPE-8/12 (i.e. the copolymer containing a 50/50 molar ratio of spacer lengths containing eight and 12 carbon atoms) exhibits an enantiotropic columnar hexagonal (Φ_h) mesophase.[329] By

Table 15 Thermodynamic Parameters Corresponding to the Orientational and Conformational Order Contributions of HMS-X,[278] MBPE-X,[103e] ClBPE-X,[324] and Polyesters Based on 4,4′-Dihydroxy-2,2′-Dimethylazoxybenzene and α,ω-Alkanedioic Acids (DMAB-X)[48]

Nature of Order	(ΔH_{in})even (kcal mru^{-1})				(ΔH_{in})odd (kcal mru^{-1})			
	HMS-X	MBPE-X	ClBPE-X	DMAB-X	HMS-X	MBPE-X	ClBPE-X	DMAB-X
Orientational/ mol of mesogen	1.45	2.00	1.63	1.12	0.34	0.30	0.20	0.22
Conformational/ mol of –CH$_2$–	0.03	0.01	0.03	0.38	0.04	0.03	0.04	0.045

Nature of Order	(ΔS_{in})even (cal mol^{-1} K)				(ΔS_{in})odd (cal mol^{-1} K)			
	HMS-X	MBPE-X	ClBPE-X	DMAB-X	HMS-X	MBPE-X	ClBPE-X	DMAB-X
Orientational/ mol of mesogen	2.15	4.02	3.60	1.80	0.45	0.85	0.62	0.34
Conformational/ mol of –CH$_2$–	0.15	0.17	0.12	0.21	0.13	0.11	0.07	0.14

Table 16 Thermal Transitions (°C) and Corresponding Thermodynamic Parameters, ΔH_{in} (kcal mru^{-1}) and ΔS_{in} (cal mru^{-1} K)[a] of 1-(4-Methoxyphenyl)-2-(2-R-4-methoxyphenyl)ethane (RBMPE) and 1-(4-Hydroxyphenyl)-2-(2-R-hydroxyphenyl)ethane (RBPE) and of the corresponding RBPE-8 and RBPE-5 polymers. Van der Waals' Radius of R (Å) and Electronegativity of R, Ar—R bond length (Å), and Breadth of Molecule (Å)

R	RMBPE	RBPE	Radius of R	Electroneg-ativity of R	Breadth of RBPE	Ar—R Bond Length	RBPE-8				RBPE-5		
							T_{ni}	T_{in}	ΔH_{in}	ΔS_{in}	T_{in}	ΔH_{in}	ΔS_{in}
H	125–127	200	1.2	2.19	6.69	1.08	155	146	—	—	—	—	—
F	73–74	172–173	1.35	4.0	7.03	1.30	165	152	2.82	6.8	92	0.92	2.5
Cl	62–64	126	1.8	3.0	7.82	1.70	130	118	2.27	5.83	52	0.65	2.0
Br	81–82	114–115	1.95	2.8	8.10	1.85	111	102	2.12	5.3	47	0.59	1.86
Me	52–54	148	2.0	2.3	7.76	1.53	132	119	2.33	5.9	38	0.49	1.57
CF$_3$	69–70	64–65	2.3	3.29	8.15	1.53	76	43	0.93	2.94	—	—	—

[a] Data from ref. 327

increasing the length of the mesogenic unit from BPE to 4,4′-bis[2-(4-hydroxyphenyl)ethyl]benzene (PEB; Scheme 36), the resulting homopolyethers PEB-X and the copolyethers based on PEB, BPE and various flexible spacers all exhibit columnar mesophases.[330,331] Both the phase transition temperatures and the corresponding thermodynamic parameters of the columnar hexagonal mesophases can be tailor-made by similar copolymerization experiments[330,331] as those used to tailor-make nematic mesophases.[103d,f] The columnar hexagonal mesophase of BPE and PEB copolymers

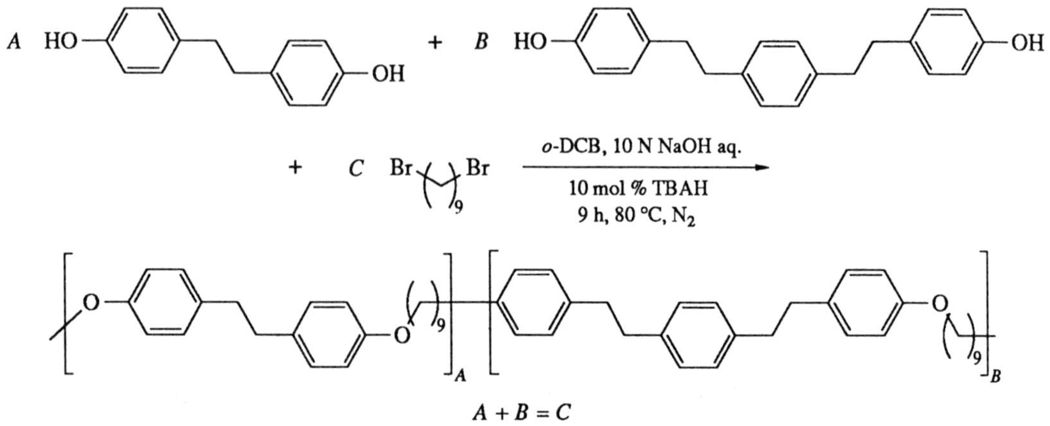

Scheme 36

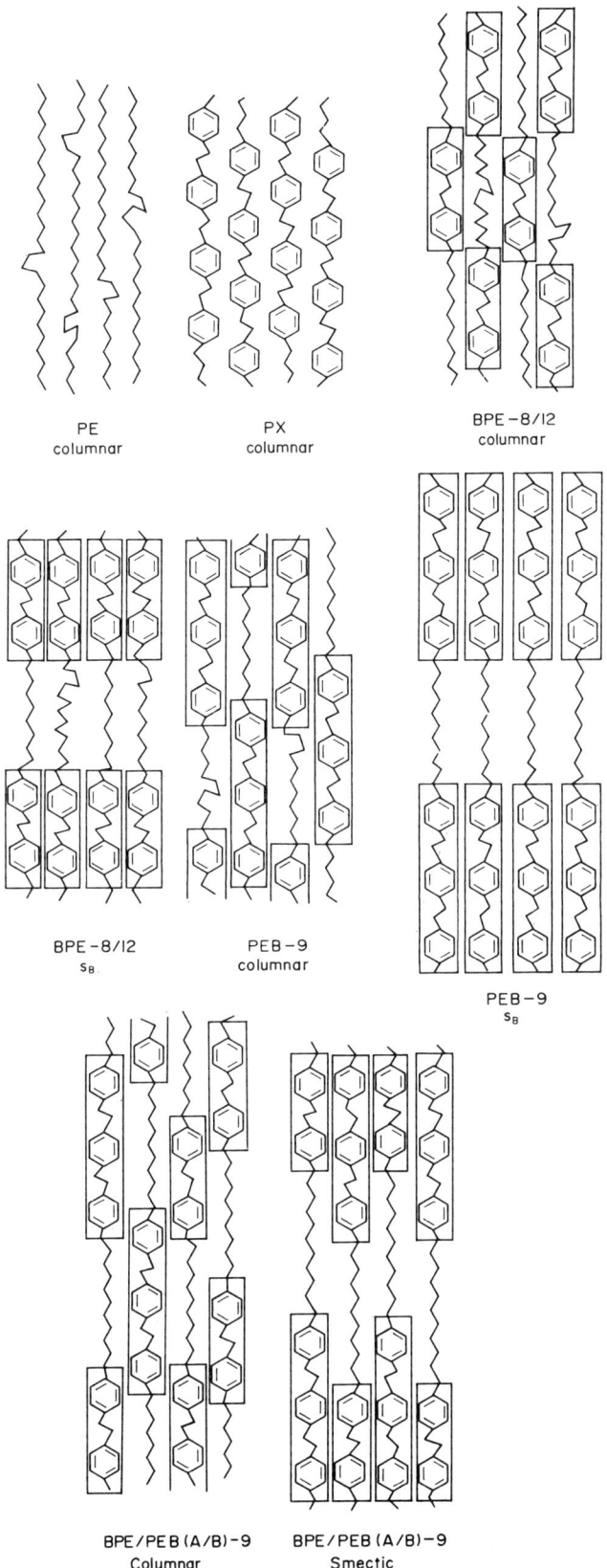

Figure 19 Schematic representation of columnar and smectic mesophases exhibited by polyethers and copolyethers based on BPE, PEB and 1,9-dibromononane

is not unexpected if we consider them as being 'copolymers' of polyethylene and poly(p-xylylene). Since both polyethylene and poly(p-xylylene) display a columnar hexagonal mesophase it is expected that RBPE-X with R = H or even F to display the same mesophases.[329] Figure 19 outlines the columnar hexagonal phases of polyethylene, poly(p-xylylene) and of some of the copolymers based on BPE and PEB.[329] As we can observe from Figure 19 the columnar hexagonal mesophase is induced by the conformational disorder of the spacer.

So far, with few exceptions, all the experiments with polyethers based on RBPE generate mesophases which are either thermodynamically unstable (virtual) with respect to the crystalline phase or metastable (monotropic). Based on the thermodynamic discussion from Section 14.5, an increase of the rigidity of the mesogen and a simultaneous increase in its disorder or entropy should decrease the ability towards crystallization and increase the ability towards the generation of a mesophase. Scheme 37 outlines the development of a series of polyethers based on this concept.[332] A 4-hydroxyphenyl group from BPE was replaced with 4-hydroxybiphenyl to increase its rigidity. The ethyl group from BPE was replaced with a 1,2-butane group. Thus a new mesogenic group 1-(4-hydroxy-4'-biphenyl)-2-(4-hydroxyphenyl)butane (TPB) results.[332] TPB has a chiral center and therefore the racemic monomer leads to TPB-X 'copolymers' based on four constitutional isomeric structural units (Scheme 37). These polymers are soluble in conventional solvents. The phase transitions of TPB-X (with X = 4 to 20) are summarized in Figure 20.[332] The dependences of glass transition, isotropization and melting transition temperatures *versus* X are different and as a consequence, they generate polymers exhibiting glassy (TPB-5), noncrystallizable nematic (TPB-4, TPB-6 to TPB-10), crystallizable nematic (TPB-11 to TPB-15), monotropic nematic (TPB-16 and TPB-17) and virtual nematic (TPB-18 to TPB-20) mesophases. This behavior is self-explanatory if we follow the plots from Figure 20. For example, some polymers are noncrystallizable since their glass transition temperature prohibits for kinetic reasons their crystallization. TPB-5 is only glassy since its glass transition temperature does not allow the formation of the mesophase, again for

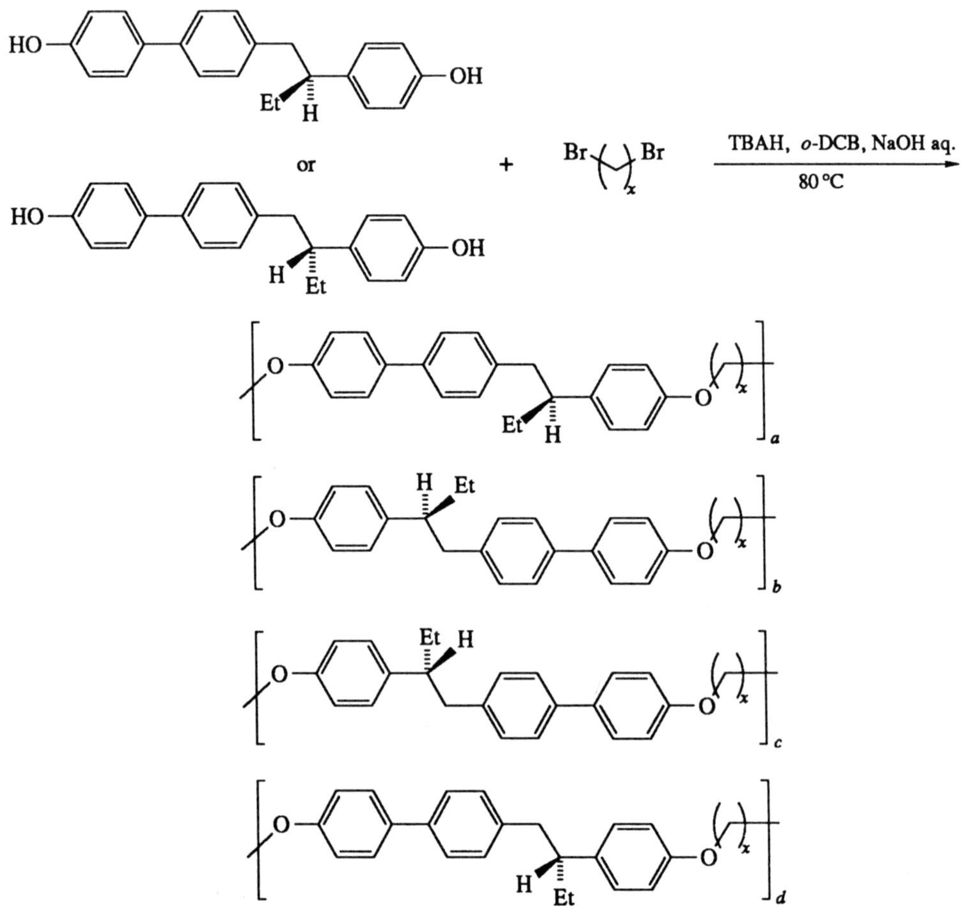

Scheme 37

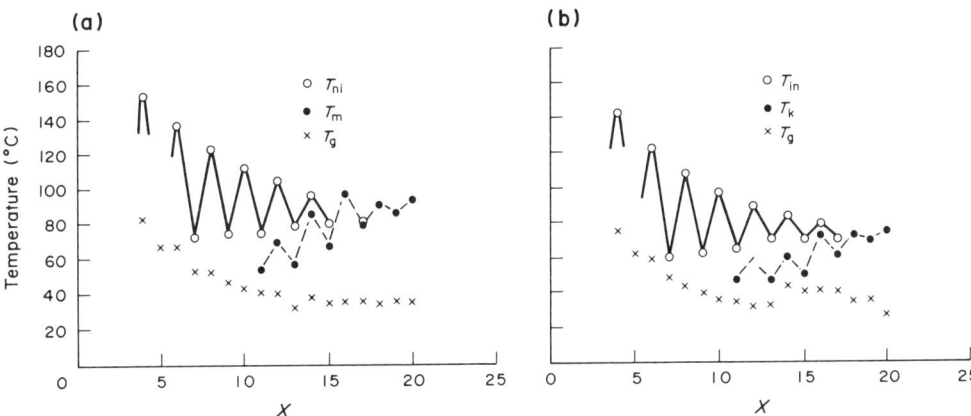

Figure 20 Dependence of phase transition temperatures of TPB-X polyethers on the number of methylenic units in the flexible spacer (X): (a) data from the second heating scan (T_g data of TPB-11 to TPB-20 are from the first heating scan); (b) data from the first cooling scan

kinetic reasons. When the melting temperature is higher than the glass transition, these polymers can crystallize, while when melting is higher than isotropization they exhibit a virtual nematic mesophase only.

TPB-10 was synthesized with a perdeuterated spacer and the conformational and orientational order in the nematic phase was determined by DNMR measurements and conformational calculations.[333] Since this polymer does not undergo redistribution reactions as the polyesters do, its behavior is as predicted by theory.[334] Beginning at the isotropic–nematic transition at 104 °C, the polymer chains in the nematic phase adopt nearly fully extended (alternate *trans*) conformations and a rather high degree of orientational order parameter of *ca.* 0.85 is attained. As the temperature is lowered to 96 °C, the isotropic peak disappears completely, indicating the absence of any isotropic phase and hairpin conformations. With increasing supercooling, the conformational order shows little change, but the orientational order parameter increases continuously to *ca.* 0.95 at 80 °C. These results are, therefore, in good agreement with the predictions of a recent theory on the nematic phase of a polymer comprising rigid and flexible sequences in alternating succession.[334] When the 1,2-butane group from TPB is replaced by 1,2-propane, the new mesogen 1-(4-hydroxy-4′-biphenyl)-2-(4-hydroxyphenyl)propane (TPP) has a more cylindrical shape. This new mesogen therefore resembles RBPE with R = H. The polyethers TPP-X with X containing an even number of methylenic units display columnar hexagonal or smectic mesophases while those with an odd number of methylenic units exhibit nematic mesophases.[335]

14.9 HYPERBRANCHED DENDRITIC LIQUID CRYSTALLINE POLYMERS

Hyperbranched dendritic macromolecules are a novel class of branched polymers which contain a branching point in each structural unit. For a review see ref. 336. Although dendritic polyphenylenes,[337,338] poly(benzyl ether)s[339,340] and aromatic polyesters[341,342] were synthesized, none of them displayed liquid crystallinity. A lamellar lyotropic liquid crystalline phase was reported for a mixture of octanoic acid and a dendrimeric poly(ethylene imine) of the third generation, $N[CH_2CH_2N\{CH_2CH_2N(CH_2CH_2N)_2\}_2]_3$.[343]

The first thermotropic dendrimer is a polyether based on a mesogenic unit which exhibits conformational isomerism (TPD-b).[344] The conformational isomerism of the mesogen allows both the synthesis of the dendrimer and the generation of a thermotropic nematic mesophase (Scheme 38).[344] The thermotropic behavior of this dendrimer is strongly dependent on the nature of its chain end R groups (Table 17).

An aromatic polyamide lyotropic dendrimer was obtained by the polycondensation of 5-aminoisophthaloyl chloride. However, the dendrimer obtained from the polymerization of 3,5-diaminobenzoyl chloride does not display a lyotropic phase.[345]

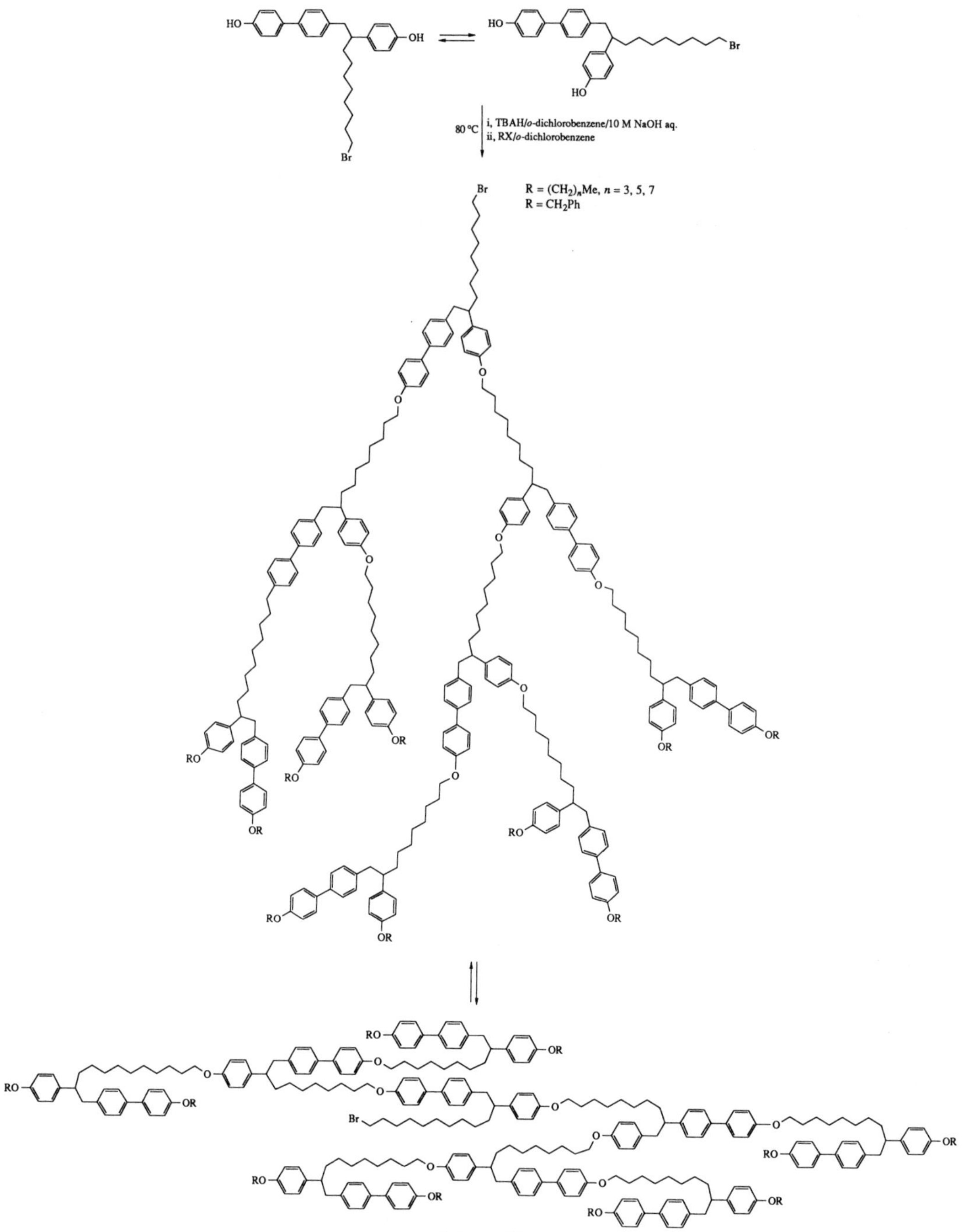

Scheme 38

Table 17 Charaterization of Hyperbranched Polyethers Based on TPD-b with Different Tail Lengths R. Data Collected from Second Heating and First Cooling DSC Scans[344]

Tail R	Yield (%)	$(M_n)_{GPC}$	$(M_w M_n)_{GPC}$	Thermal transitions (°C) and corresponding enthalpy changes (Kcal mru^{-1}) in parentheses	
				Heating	Cooling
Octyl	80.0	7900	2.6	g 20 n 40 (0.75) i	i 30 (0.70) n 15 g
Hexyl	82.2	4700	3.9	g 24 n 39 (0.15) i	i 30 (0.18) n 18 g
Butyl	70.3	4400	3.5	g 40 i	i 34 g
Benzyl	79.2	4700	3.8	g 48 n 60 (0.10) i	i 53 (0.15) n 43 g

Molecular Engineering of Liquid Crystalline Polymers 355

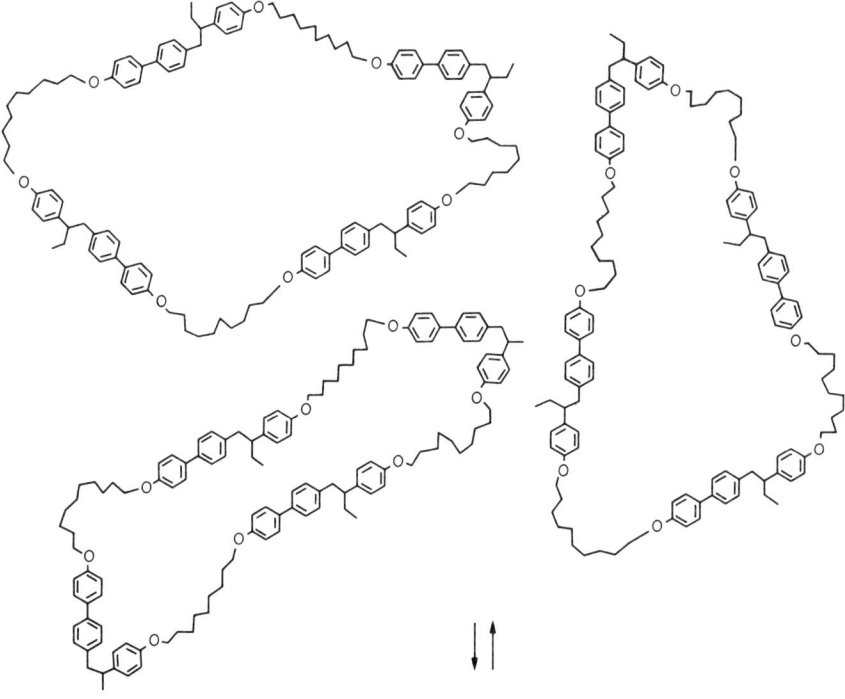

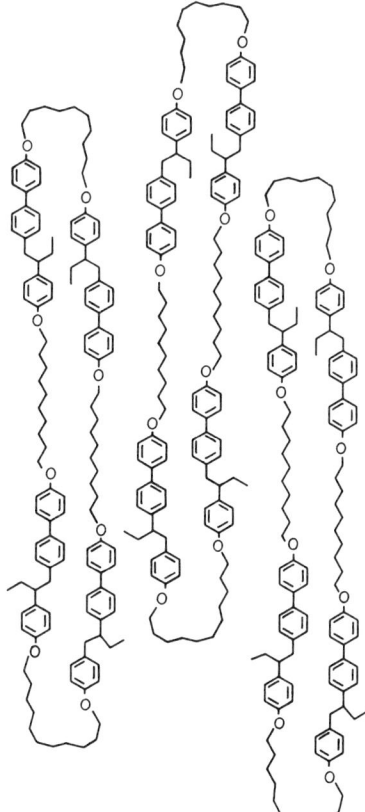

Figure 21 Structure of cyclic main chain liquid crystalline polyethers

14.10 CYCLIC MAIN CHAIN LIQUID CRYSTALLINE POLYETHERS

The cyclic monomer, dimer, trimer, tetramer, pentamer, and high molecular weight polyethers based on 1-(4-hydroxy-4'-biphenyl)-2-(4-hydroxyphenyl)butane (TPB) and 1,10-dibromodecane were synthesized and characterized and their phase behavior was compared to that of the linear TPB-10 polymer with high molecular weight.[346] Figure 21 outlines their structure. The cyclic monomer is liquid. All other cyclics, dimer and larger exhibit a thermotropic nematic phase. The cyclic tetramer and pentamer display higher isotropization temperatures than the linear polymer with high molecular weight.[346]

14.11 SIDE CHAIN LIQUID CRYSTALLINE POLYMERS

14.11.1 General Considerations

The field of side chain liquid crystalline polymers was recently reviewed.[58] Therefore, we will discuss only recent progress made on their molecular engineering. For a basic introduction to the various aspects of the field of side chain liquid crystalline polymers we suggest ref. 58 be consulted. Most of the present discussion will be made on side chain liquid crystalline polymers with mesogenic groups normally attached to the polymeric backbone.[72] Figure 22 outlines the concept of side chain liquid crystalline polymers. It has been theoretically predicted[347] that the conformation of the polymer backbone should get distorted in the liquid crystalline phase. Both small angle neutron scattering (SANS) experiments[80,81,348-350] and X-ray scattering experiments[82b,351,352] have shown

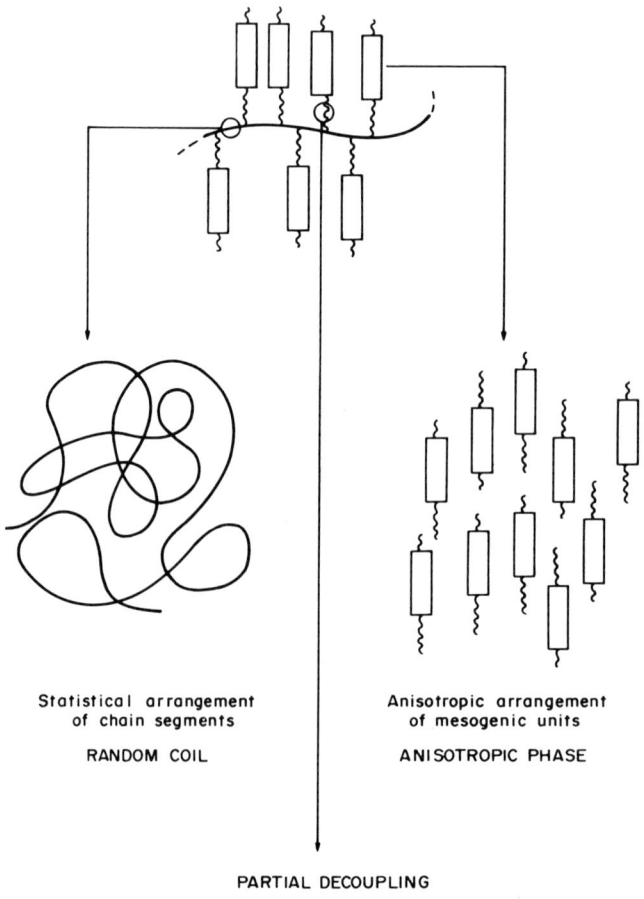

Figure 22 Schematic representation of side chain liquid crystalline polymers showing the necessity of decoupling the mesogenic groups and the polymer backbone through flexible spacers

that the statistical random-coil conformation of the polymer backbone is slightly distorted in the nematic phase and highly distorted in the smectic phase.

Let us now consider very briefly the influence of various parameters (*i.e.* nature of flexible spacer and its length, nature and flexibility of the polymer backbone and its degree of polymerization) on the phase behavior of a side chain liquid crystalline polymer. According to the thermodynamic schemes from Section 14.5, the increase of the degree of polymerization decreases the entropy of the system and therefore if the monomeric structural unit exhibits a virtual or monotropic mesophase, the resulting polymer should most probably exhibit a monotropic or enantiotropic mesophase. Alternatively, if the monomeric structural unit displays an enantiotropic mesophase, the polymer should display an enantiotropic mesophase which is broader. It is also possible that the structural unit of the polymer exhibits more than one virtual mesophase and therefore at high molecular weights the polymer will increase the number of its mesophases. All these effects were observed with various systems.[72]

The length of the flexible spacer determines the nature of the mesophase. Long spacers favor smectic phases while short spacers favor nematic phases. This effects is similar to that observed in low molar mass liquid crystals.

At constant molecular weight the rigidity of the polymer backbone determines the thermodynamic stability of the mesophase. According to the thermodynamic schemes described in Section 14.5 the isotropization temperature of the polymer with the more rigid backbone should be higher. However, experimentally this situation is reversed. The highest isotropization transition temperature is observed for polymers with more flexible backbones. This conclusion is based on systematic investigations performed with two mesogenic groups which are constitutional isomers *i.e.* 4-methoxy-4'-hydroxy-α-methylstilbene (4-MHMS)[353] and 4-hydroxy-4'-methoxy-α-methylstilbene (4'-MHMS)[354] and polymethacrylate, polyacrylate, polysiloxane and polyphosphazene backbones (Scheme 39).[353-355] This dependence can be explained by assuming that a more flexible backbone uses less energy to get distorted and therefore generates a more decoupled polymer system. In fact the more flexible backbones do not generate only higher isotropization temperatures but also higher ability towards crystallization. However, contrary to all expectations the entropy change of isotropization is higher for those polymers which are based on more rigid backbones and therefore, exhibit lower isotropiz-

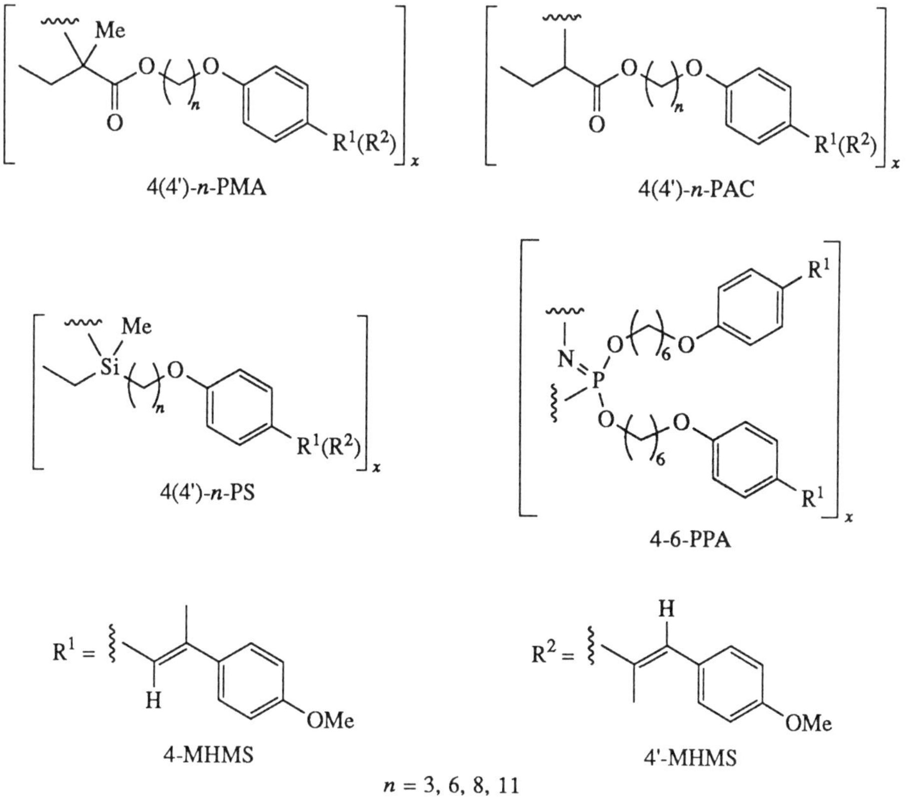

$n = 3, 6, 8, 11$

Scheme 39

ation temperatures (Figures 23a and b).[82a] This contradiction between the values of the entropy change and the isotropization temperatures can be accounted for by a different mechanism of distortion of different polymer backbones, as outlined in Figure 24; that is, while a rigid backbone gets more extended and, therefore, in the smectic phase it can cross the smectic layer, in the case of a flexible backbone it gets squeezed between the smectic layers. The higher configurational entropy of the flexible backbone *versus* that of the rigid backbone in the smectic phase can account for the difference between the entropy change of isotropization from Figure 23. At shorter spacer lengths, there is not much difference between the contribution of various backbone flexibilities since in order to generate a mesophase they should get extended. Therefore the entropy change of isotropization is less dependent on backbone flexibility (Figure 23).

Based on this discussion it is quite obvious that copolymers containing structural units with and without mesogenic groups and flexible backbones display a microphase-separated morphology of their smectic phase (Figure 25).[72,82b,352] Therefore, the highest degree of decoupling is expected for copolymers containing mesogenic and nonmesogenic structural units and highly flexible backbones, *i.e.* microphase-separated systems. In this latter case, when the monomeric structural unit of the polymer exhibits a virtual mesophase, the high molecular weight polymer might also display only a virtual or a monotropic mesophase. The transformation of a virtual and/or monotropic mesophase of the homopolymers into an enantiotropic mesophase can be most conveniently accomplished by making copolymers based on two monomers which are constitutional isomers such as monomers based on 4-MHMS and 4'-MHMS.[356,357] Since the structural units of the homopolymers based on 4-MHMS and 4'-MHMS are isomorphic within their liquid crystalline phase, but not within their crystalline phase, the crystalline-melting transition decreases while the mesophase exhibits a

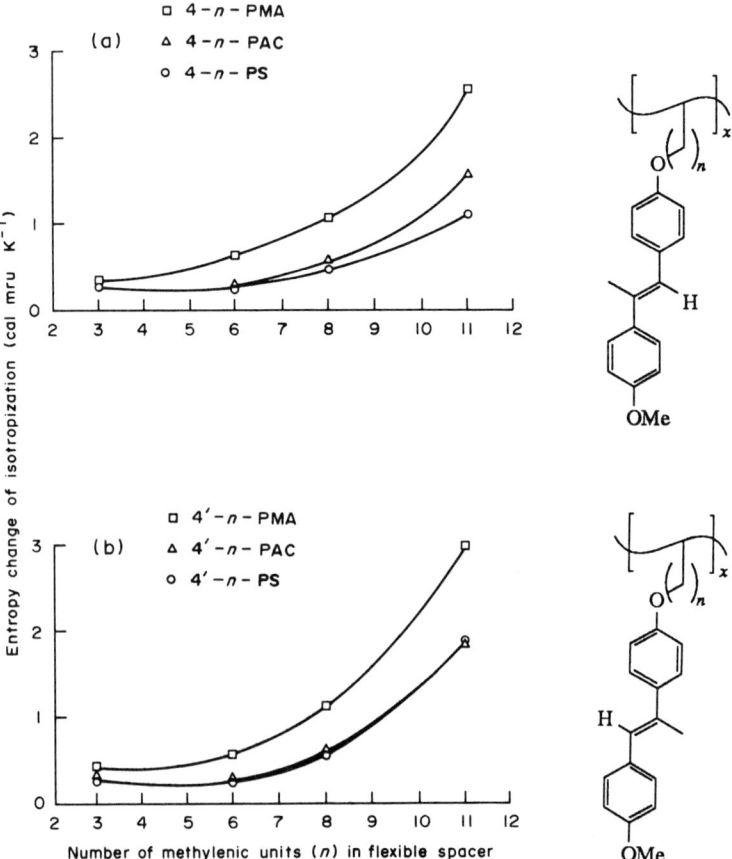

Figure 23 (a) The dependence between the entropy change of isotropization (ΔS_i) determined from the cooling DSC scans, the nature of the polymer backbone and the number of methylenic units (n) in the flexible spacer for the series of polymers based on 4-MHMS isomer. (b) The dependence between the entropy change of isotropization (ΔS_i) determined from the cooling DSC scans, the nature of the polymer backbone and the number of methylenic units (n) in the flexible spacer for the series of polymers based on 4'-MHMS isomer. Notation as for Scheme 39

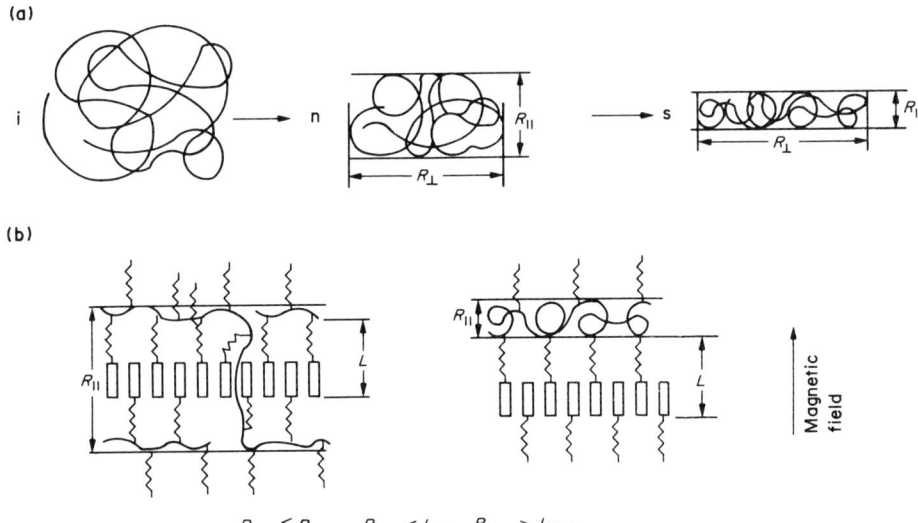

Figure 24 (a) Schematic representation of the theoretical distortion of the statistical random-coil conformation of the polymer backbone in the nematic and smectic phases (M. Warner); (b) two possible modes of distortion of the random-coil conformation of a rigid (left) and a flexible (right) polymer backbone (Saclay Group). $R_\parallel$ refers to the radius of gyration paralleled to the magnetic field. The radius of gyration perpendicular to the magnetic field is labeled as $R_\perp$

continuous, almost linear dependence on composition. As a consequence, the virtual or monotropic mesophase of the homopolymer becomes enantiotropic.[356,357] Finally, the molecular weight at which the isotropization temperature becomes independent of molecular weight should be, and indeed is, dependent on the flexibility of the polymer backbone. For example, the isotropization temperature of polysiloxanes[75,358] containing mesogenic side groups is molecular weight dependent up to much higher molecular weights than polymethacrylates containing mesogenic side groups.[72]

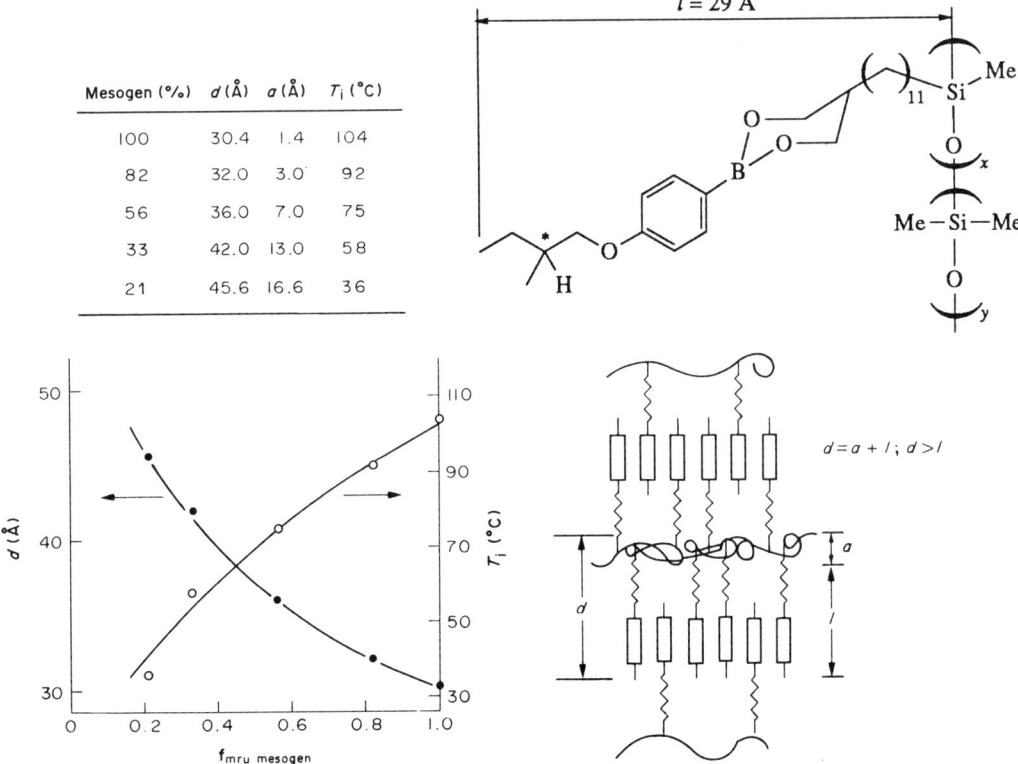

Figure 25 Microphase separated morphology of smectic copolymers

Based on this discussion and on the thermodynamic discussion from Section 14.5, we can easily consider that the 'polymer effect' can lead to the same effect *via* its molecular weight and backbone flexibility. In an oversimplified way it can be considered that it provides an overall change in the entropy of the system. Through this change, it can transform, in a reversible way, a virtual mesophase into a monotropic or enantiotropic phase. In addition, the kinetic factors provided by the glass transition and crystallization should always be considered. For example, the formation of a mesophase located in the close proximity of a glass transition temperature becomes kinetically controlled or can even be kinetically prohibited.

14.11.2 Molecular Engineering of Liquid Crystalline Polymers by Living Polymerization

14.11.2.1 General considerations

Several polymerization methods were investigated in order to develop living polymerization procedures for the preparation of side chain liquid crystalline polymers with well-defined molecular weight and narrow molecular weight distribution. They include cationic polymerization of mesogenic vinyl ethers,[359,360] cationic ring-opening polymerization of mesogenic cyclic imino ethers,[361] group transfer polymerization of mesogenic methacrylates,[153,362–364] and polymerization of methacrylates

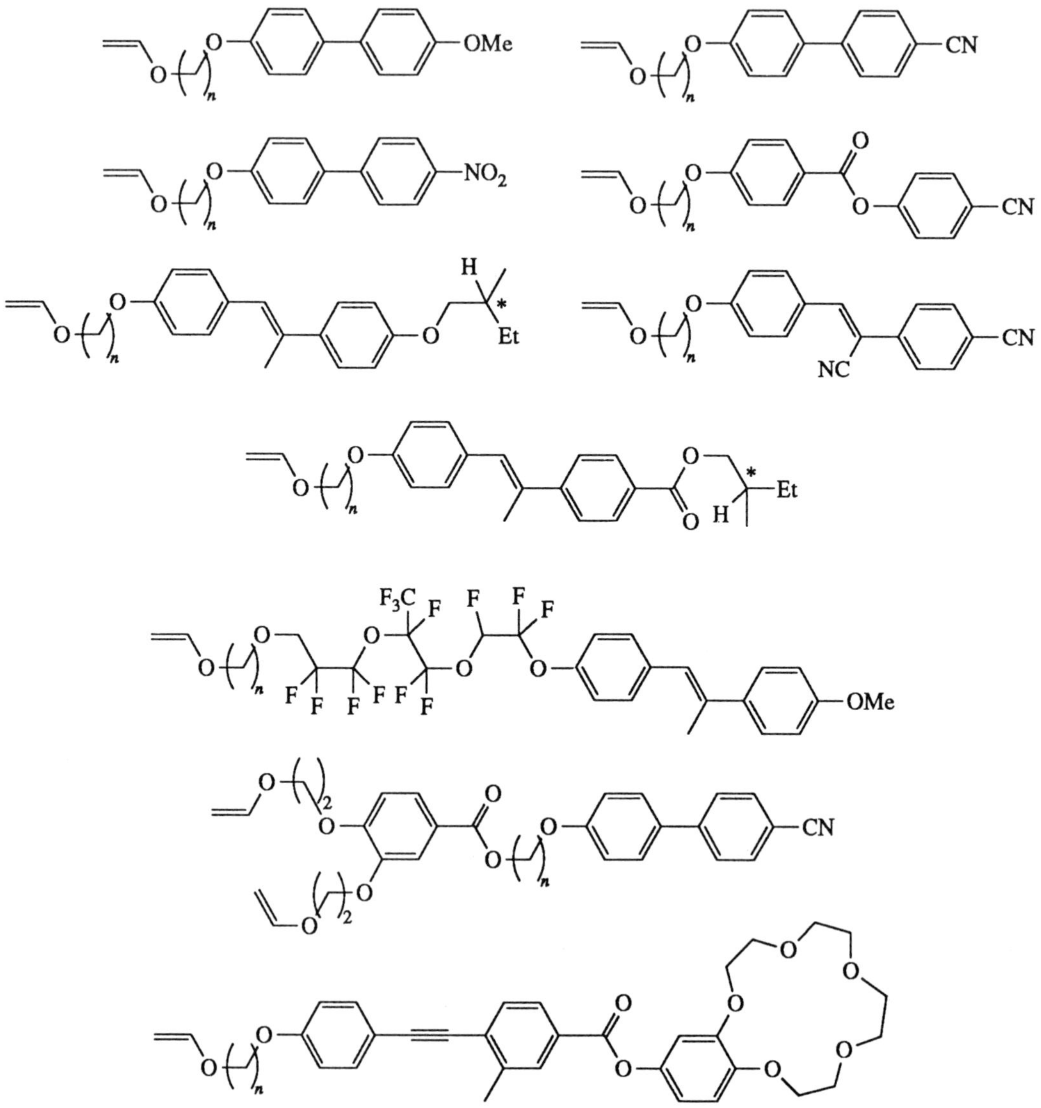

Scheme 40

with methylaluminium porphyrin catalysts.[365] Cationic polymerization has been proved to be the most successful since it can be used to polymerize under living conditions mesogenic vinyl ethers containing a large variety of functional groups.[139–145,154,158,366–378]

Scheme 40 provides some representative examples of mesogenic vinyl ethers which could be polymerized by a living mechanism by our preferred initiating system (*i.e.* CF_3SO_3H, Me_2S, CH_2Cl_2, 0 °C).[379] As we can observe from Scheme 40, vinyl ethers containing nucleophilic groups such as methoxybiphenyl,[371] electron-withdrawing groups such as cyanobiphenyl,[139–144,371,372] nitrobiphenyl and cyanophenyl benzoate,[371] double bonds like in 4-alkoxy-α-methylstilbene,[373] double bonds and cyano groups like in 4-cyano-4'-α-cyanostilbene,[143] aliphatic aromatic esters,[145] acidic protons and perfluorinated groups,[371,374] oligo(oxyethylene) and aromatic ester groups,[376] crown ethers and triple bonds,[377] can all be polymerized by a living cationic mechanism. In addition, cationic polymerization of any of these monomers can be performed in melt phase either in liquid crystalline phase or in isotropic phase by using thermal,[380] or photo cationic initiators.[381,382] When the polymerization is performed in liquid crystalline phase with aligned films of liquid crystalline monomers, perfectly aligned single crystal liquid crystalline polymer films are obtained.[381,382]

In the following two sections we will discuss two topics. The first one refers to the influence of molecular weight on the phase transitions of poly [ω-{(4-cyano-4'-biphenylyl)oxy}alkyl vinyl ether]s with alkyl groups containing from four to 11 methylene units. In the second one we will demonstrate the molecular engineering of phase transitions of side chain liquid crystalline polymers by azeotropic living copolymerization experiments.

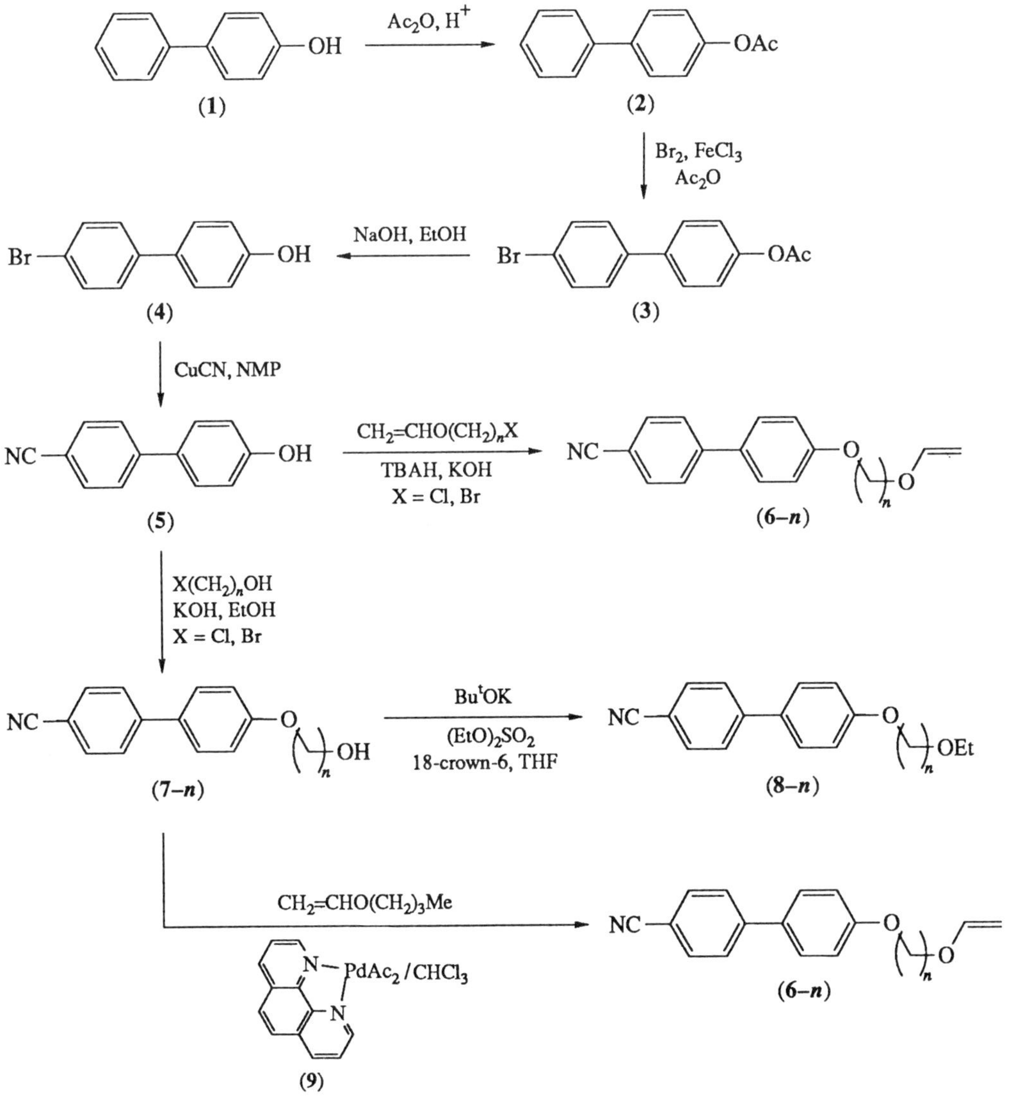

Scheme 41

14.11.2.2 Influence of molecular weight on phase transitions of poly[ω-{4-cyano-4'-biphenylyl)oxy}alkyl vinyl ether]s

Scheme 41 outlines the general method used for the synthesis of ω-[(4-cyano-4'-biphenylyl)-oxy]alkyl vinyl ethers (6-n) and of the model compound for the polymer with a degree of polymerization of one *i.e.* ω-[(4-cyano-4'-biphenylyl)oxy]alkyl ethyl ethers (8-n). We will use over the entire discussion the same short notations as in the original publications. The synthesis and characterization of poly(6-n) and poly(8-n) with n = 2, 3, 4,[140] 5, 7,[141] 6, 8,[372] 9, 10,[142] and 11[139] will be briefly discussed. Details are available in the original publications. All polymers have polydispersities of about 1.10. Scheme 42 outlines the polymerization mechanism and the structure of the resultant polymers. This structure was confirmed by 300 MHz 1D and 2D ^{1}H NMR spectroscopy.[383]

All data were classified according to their similarities. Figure 26 presents the dependence of phase transition temperatures of poly(6-n) with n = 3, 4, 7 and 9 as a function of molecular weight. These data were collected from second heating scans. The data for 8-n are not plotted. 8-3 is crystalline, 8-4 and 8-7 exhibit a monotropic nematic mesophase while 8-9 shows monotropic nematic and smectic mesophases. As we can observe from Figure 26, by increasing the molecular weight all four polymers show a broadening of the thermal stability of their mesophase. The mesophase of 8-7 and 8-9 changes from nematic to s_A by increasing the degree of polymerization from one to about three.

Figure 27 presents similar data for poly(6-2), poly(6-6) and poly(6-8). In all cases the nature of the mesophase is molecular weight dependent. Poly(6-2) has a nematic mesophase only at degrees of polymerization lower than 5. 8-2 is only crystalline. At degrees of polymerization higher than 5, poly(6-2) is only glassy. This is because its glass transition temperature becomes higher than the isotropization temperature and, therefore, the mesophase is kinetically prohibited. 8-6 exhibits an enantiotropic nematic mesophase. At low degrees of polymerization poly(6-6) and poly(6-8) exhibit nematic and s_A mesophases. Due to the difference between the slope of the dependences of the nematic phase transition temperature on molecular weight and of the s_A phase transition temperature on molecular weight, above a certain molecular weight the nematic phase disappears. Both poly(6-6)

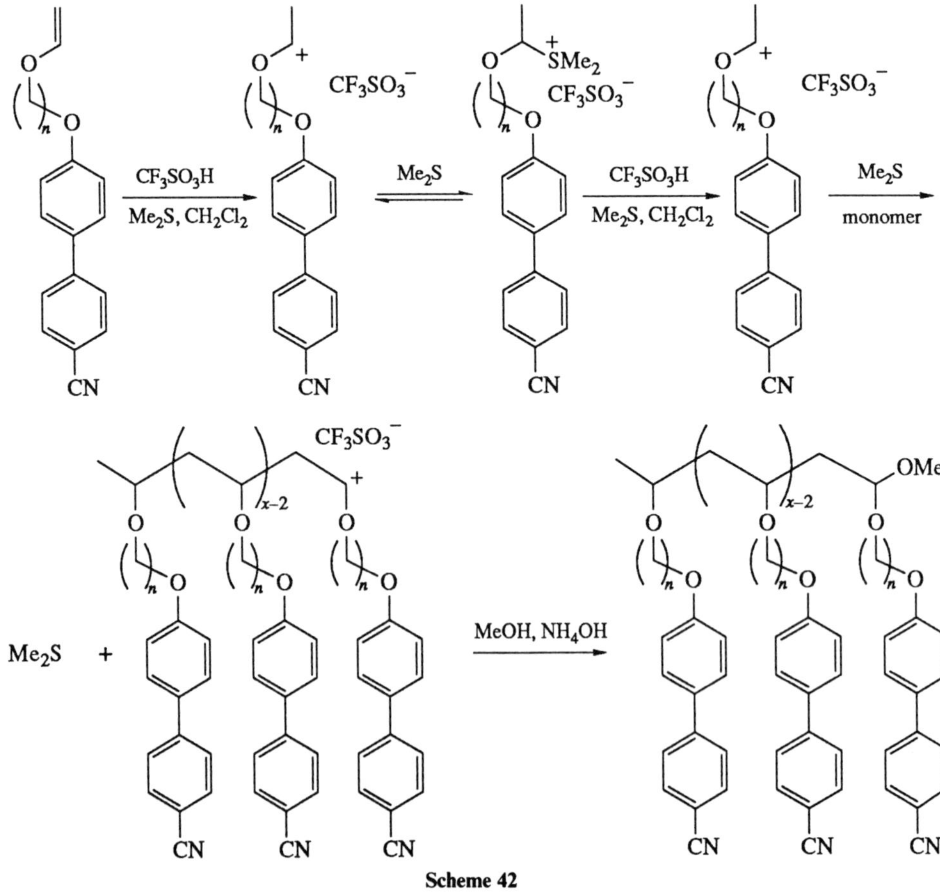

Scheme 42

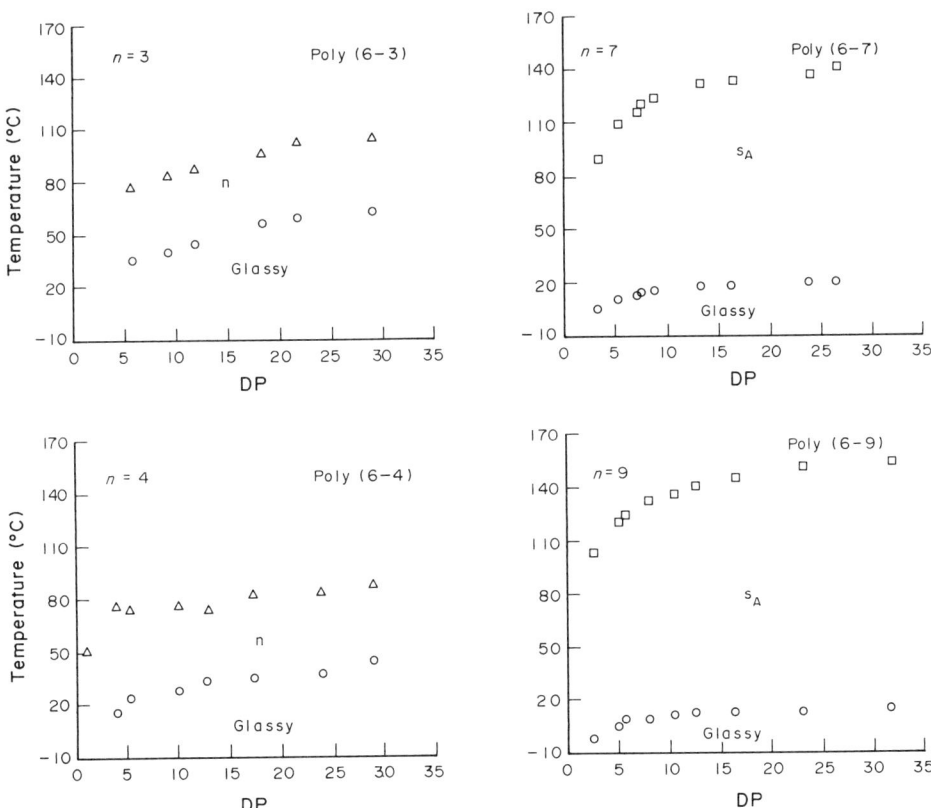

Figure 26 The influence of molecular weight on the phase behavior of poly(6-3), poly(6-4), poly(6-7) and poly(6-9) (determined from second DSC heating scans)

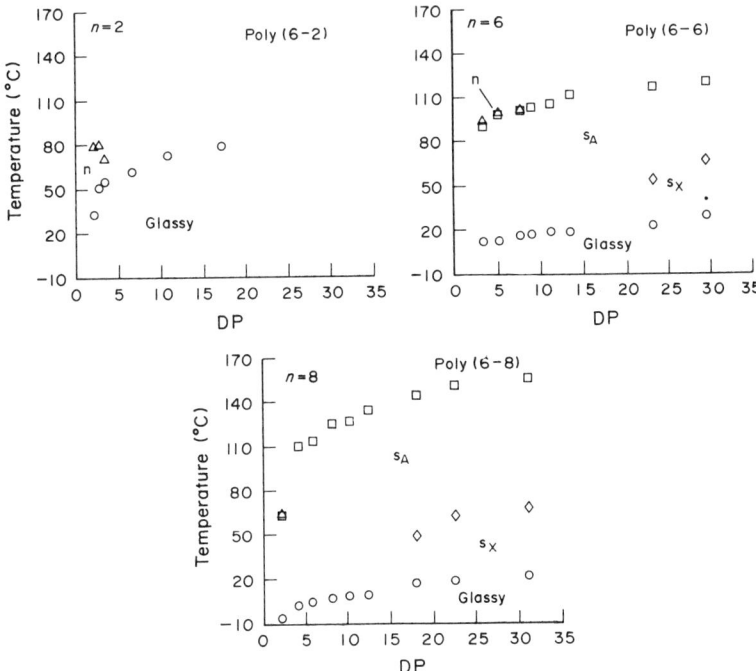

Figure 27 The influence of molecular weight on the phase behavior of poly(6-2), poly(6-6) and poly(6-8) (determined from second DSC heating scans)

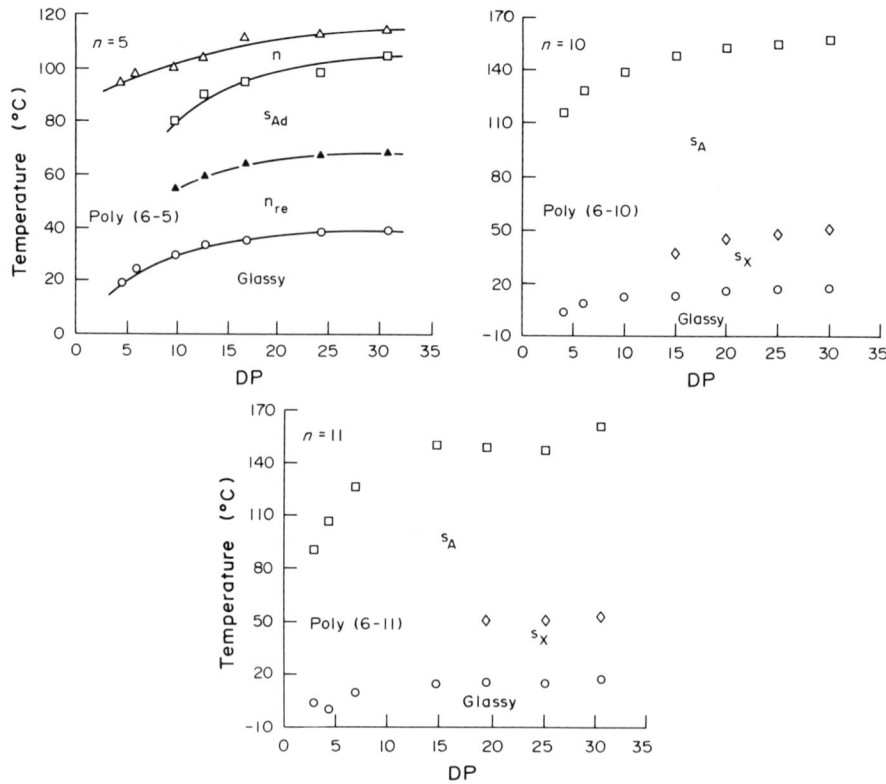

Figure 28 The influence of molecular weight on the phase behavior of poly(6-5), poly(6-10) and poly(6-11) (determined from second DSC heating scans)

and poly(6-8) show a second smectic mesophase (s_X, *i.e.* unassigned). Qualitatively, this behavior is in agreement with the influence of molecular weight on phase transitions predicted by thermodynamics (Section 14.5). Quantitative predictions of these phase diagrams require more theoretical research.

Finally, Figure 28 presents the behavior of poly(6-5), poly(6-10) and poly(6-11). 8-5 shows a monotropic nematic phase, 8-10 a monotropic s_A phase while 8-11 an enantiotropic s_A mesophase. Poly(6-5) exhibits above a degree of polymerization of 10, the unusual sequence isotropic–nematic–s_{Ad}–n_{re}–glassy.[384] This will be discussed in more detail in a subsequent section. At high molecular weights poly(6-10) and poly(6-11) exhibit s_A and s_X phases.

As a general observation we can mention that polymers with short spacers ($n = 2, 3, 4$) and medium length spacers containing an odd number of methylene units ($n = 7, 9$) do not generate polymorphism at different molecular weights. Polymers with medium length and an even number of methylene units ($n = 6, 8$), as well as longer polymers with both even and odd numbers of methylenic units ($n = 10, 11$) generate a rich polymorphism which is molecular weight dependent. The borderline polymer is poly(6-5) which is the only one displaying n and s_A mesophases over a broad range of molecular weights and, therefore, also generates the reentrant nematic mesophase.[384]

14.11.2.3 *Molecular engineering of liquid crystalline phases by living cationic copolymerization*

In order to tailor-make mesophases of side chain liquid crystalline copolymers we first need to synthesize copolymers with constant molecular weight and controllable composition. Copolymer composition is conversion dependent in all statistical copolymerizations. The only exception is provided by azeotropic copolymerizations in which the copolymer composition is identical to the monomer feed at any conversion.[385] The situation is provided by monomers with $r_1 = r_2 = 1$. Since the reactivity of the polymerizable vinyl ether groups is not spacer length dependent, all 6-*n* monomers have the same reactivity. Therefore, all 6-*n* pairs of monomers lead to azeotropic copolymerizations, and when the copolymerization is performed under living conditions they lead to copolymers with controllable molecular weight. The azeotropic copolymerization of various

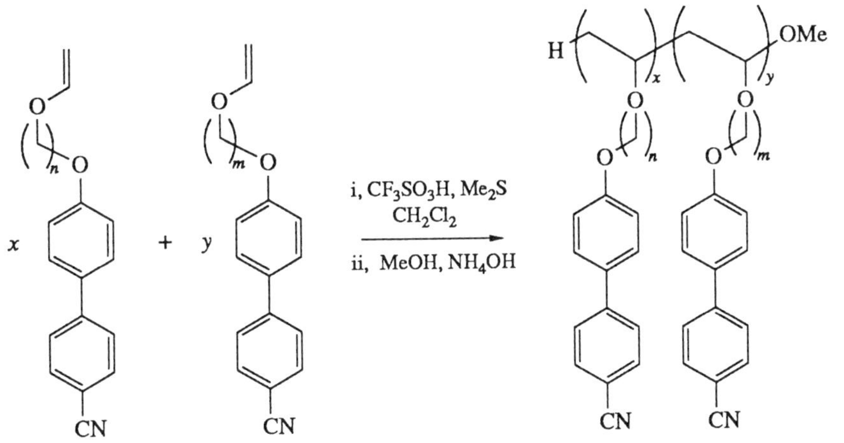

Scheme 43

pairs of 6-n monomers is outlined in Scheme 43. We will discuss selected examples of copolymers prepared from monomer pairs which give rise to homopolymers exhibiting nematic and nematic, s_A and s_A, nematic and s_A, and glassy and s_A phases as their highest temperature mesophases.

Figure 29 presents the dependence of phase transition temperatures obtained from second DSC heating scans (a, d), cooling scans (b, e) and the enthalpy changes associated with the highest temperature mesophase of copolymers poly[(6-3)-co-(6-5)]X/Y and poly[(6-6)-co-(6-11)]X/Y. The degrees of polymerization of all copolymers are equal to 20.[386] Copolymers poly[(6-3)-co-(6-5)]X/Y are based on a monomer pair which gives rise to two homopolymers displaying an enantiotropic nematic mesophase as their highest temperature mesophase. As we can observe from Figure 29 the nematic–isotropic transition temperature and its associated enthalpy change show linear dependences of composition. This means that the structural units derived from poly(6-3) and poly(6-5) are isomorphic into their nematic mesophase. However, the same two structural units are isomorphic within s_A mesophase exhibited by poly(6-5) only over a very narrow range of compositions. The linear dependence of the isotropization temperature is predictable by the Schroeder–Van Laar equations.[91] The same discussion is valid for the copolymer system poly[(6-6)-co-(6-11)]X/Y except that the isotropization temperature of these copolymers exhibit an upward curvature. This upward curvature is also predicted by the Schroeder–Van Laar equations[91] and is due to the more dissimilar enthalpy changes associated with the isotropization temperatures of the two homopolymers.

Figure 30 presents the phase diagrams of copolymers poly[(6-3)-co-(6-11)]X/Y[384,386] and poly[(6-5)-co-(6-11)]X/Y.[384] Both sets of copolymers have degrees of polymerization of 20. Both pairs of copolymers are based on monomers which give rise to homopolymers exhibiting nematic and s_A as their highest temperature mesophases. However, poly(6-5) displays a nematic and an s_A mesophase, while poly(6-3) only a nematic mesophase. Both sets of copolymers display continuous dependences of their highest temperature mesophase with a triple point at a certain composition. This triple point generates over a narrow range of compositions copolymers exhibiting the sequence isotropic–nematic–s_A–n_{re}. Again the shape of the dependences of the phase transition temperature on composition obeys the Schroeder–Van Laar equations.[91]

Figure 31 presents two sets of phase diagrams obtained from monomer pairs giving rise to homopolymers which exhibit isotropic and s_A mesophases as their highest temperature mesophases, i.e. poly[(6-2)-co-(6-8)]X/Y with degree of polymerization of 10,[384,387] and poly[(6-2)-co-(6-11)]X/Y with degree of polymerization of 15.[384,388] Both sets of copolymers display a similar phase diagram. Over a certain range of compositions the two structural units are isomorphic within the s_A phase, after which follows a triple point. After this triple point the two structural units are isomorphic within a newly generated nematic mesophase. Both copolymers generate within a certain range of compositions on the left side of the triple point the sequence isotropic–nematic–s_A–n_{re}.[384] Again the shape of the dependence of the highest temperature mesophase on composition is predictable by the Schroeder–Van Laar equations. This means that the structural units of all binary copolymers based on an identical mesogenic unit but different spacer lengths behave as an ideal solution. This behavior allows the engineering of mesomorphic phase transition temperatures and of their thermodynamic parameters in a straightforward manner by living azeotropic copolymerizations. The same behavior was demonstrated for monomer pairs which both give rise to homopolymers exhibiting a chiral smectic C mesophase.[389]

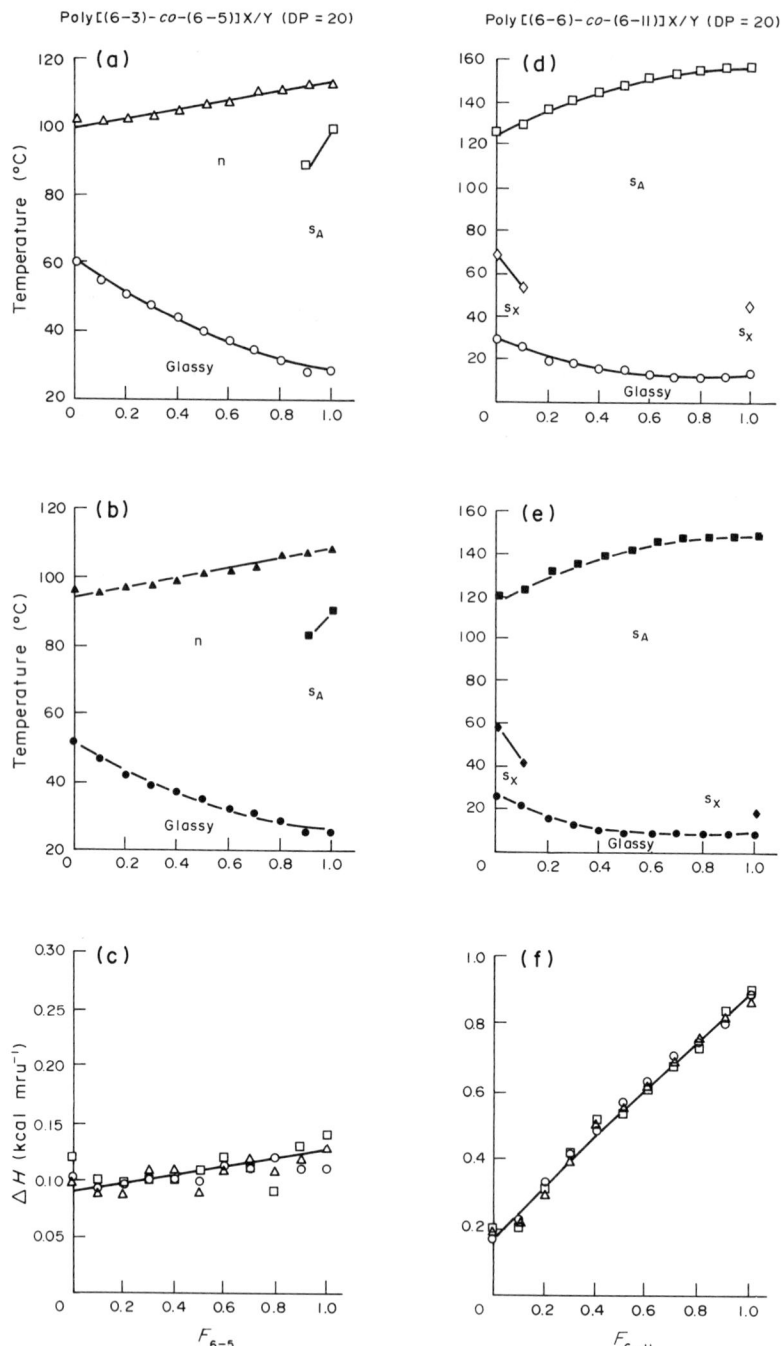

Figure 29 The dependence of phase transition temperatures obtained from second heating scan (a, d), cooling scan (b, e), and the enthalpy changes associated with their highest temperature mesophase of copolymers poly[(6-3)-co-(6-5)]X/Y and poly[(6-6)-co-(6-11)]X/Y (all with degrees of polymerization equal to 20)

14.11.2.4 Side chain liquid crystalline polymers exhibiting a reentrant nematic mesophase

The reentrant nematic phase (n_{re}) was discovered in 1975 in low molar mass liquid crystals.[390] Since then it has received substantial theoretical and experimental interest.[391-399]

The first side chain liquid crystalline polymers exhibiting an n_{re} phase were reported in 1986.[400,401] Some other examples of polymers exhibiting the sequence isotropic–nematic–s_{Ad}–n_{re} were reported in the meantime.[155,384,402-405] All these polymers are based on mesogenic units containing a cyano

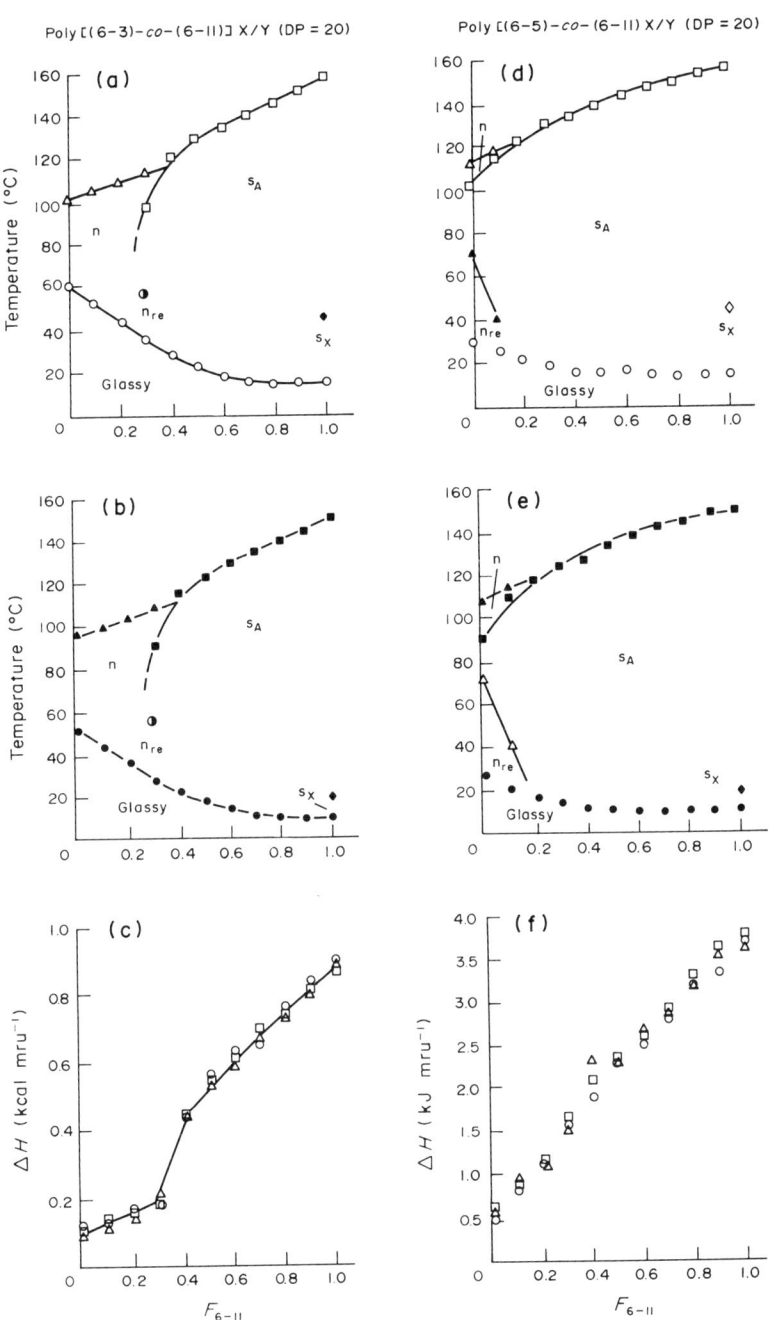

Figure 30 The dependence of phase transition temperatures obtained from second heating scan (a, d), cooling scan (b, e), and the enthalpy changes associated with their highest temperature mesophase of copolymers poly[(6-3)-co-(6-11)]X/Y and poly[(6-5)-co-(6-11)]X/Y (all with degrees of polymerization equal to 20)

group, five or six atoms in the flexible spacer and a polyacrylate or poly(vinyl ether) backbone. The replacement of these quite flexible backbones with a more rigid one like polymethacrylate does not allow the formation of the n_{re} phase. As discussed in the previous section, an n_{re} mesophase can be generated by copolymerization of two monomers which lead to homopolymers with nematic or isotropic and s_A as their highest temperature mesophases, since these copolymers exhibit a triple point on their phase diagrams.[384] According to our experimental results any polymer which exhibits the sequence isotropic–nematic–s_A should also display an n_{re} phase. The most probable mechanism for the generation of an n_{re} phase is outlined in Figure 32.[399] The most stable s_A phase of mesogens containing cyano groups is based on layers containing dimers of mesogens. On cooling, the nematic

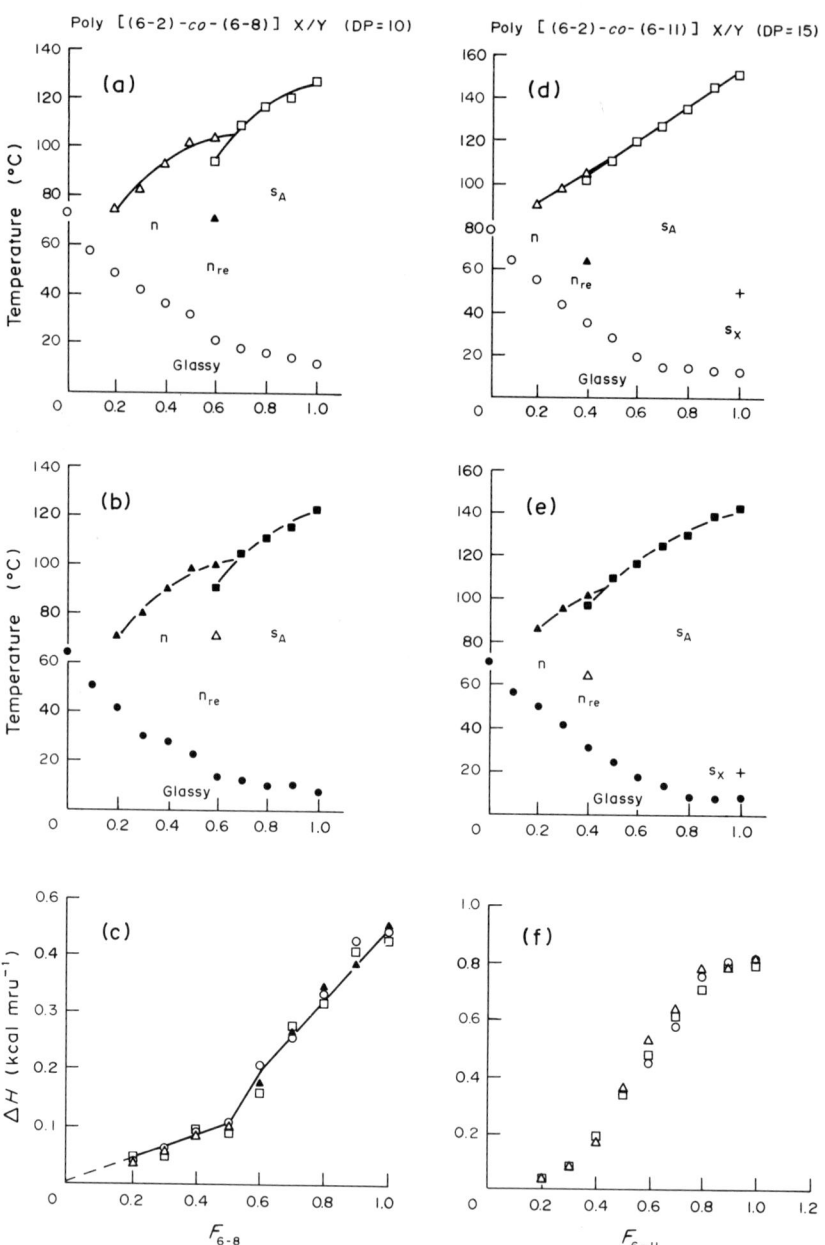

Figure 31 The dependence of phase transition temperatures obtained from second DSC heating scan (a, d), cooling scan (b, e), and the enthalpy changes associated with their highest temperature mesophase of copolymers poly[(6-2)-*co*-(6-8)]X/Y and poly[(6-2)-*co*-(6-11)]X/Y (with degrees of polymerization equal to 10 and 15, respectively)

phase formed directly from the isotropic phase contains both dimeric mesogens and monomeric mesogens and so does the first s_A phase. In order to go from the less-ordered s_A phase to the s_A phase based on dimeric mesogens, an n_{re} phase is required (Figure 32).[399]

14.11.2.5 Influence of tacticity on phase transitions

The tacticity of the main chain determines its flexibility. Therefore, tacticity should affect the phase behavior of side chain liquid crystalline polymers in the same way as different polymer backbones with different flexibilities. Liquid crystalline polymethacrylates with atactic, syndiotactic and isotactic backbones were synthesized and characterized.[406–409] Although there are differences between the phase behavior of polymers with different tacticities, there is no quantitative under-

Molecular Engineering of Liquid Crystalline Polymers

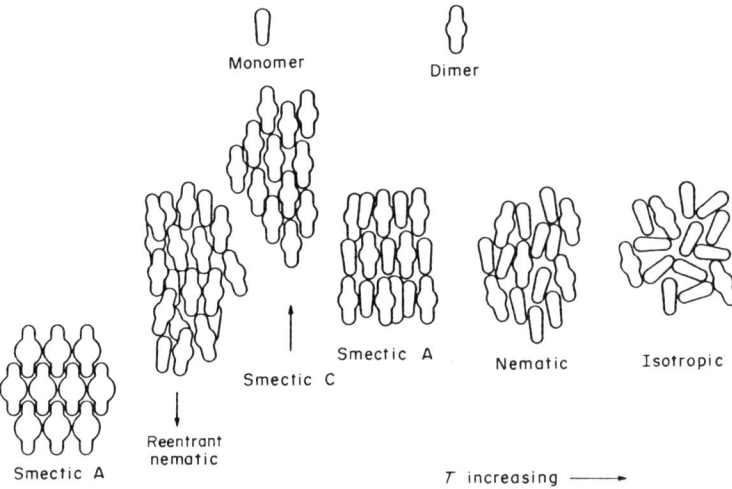

Figure 32 The mechanism of formation of the reentrant nematic mesophase[390]

satnding of this trend. This would require the investigation of the influence of both molecular weight and tacticity on phase transitions and, therefore, it still remains an open subject of research.

14.11.2.6 Cyclic polysiloxanes containing mesogenic side groups

Low molecular weight[410,411] and high molecular weight[75] cyclic polysiloxanes containing mesogenic side groups have been reported. Figure 33 presents the dependence of the isotropization temperature on the degree of polymerization for linear and cyclic polysiloxanes containing cyanophenyl benzoate groups connected to the backbone by a pentyl spacer.[75] Isotropization temperatures of cyclic polymers are higher than those of the linear polymers. However, more research is required to reach the same level of understanding between structure and properties in this class of polymers as that available presently for glassy and crystalline cyclic polymers.[412]

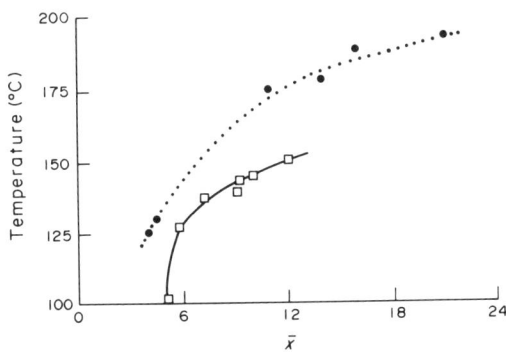

Figure 33 Isotropization transition temperatures of cyclic (●) and linear (□) polysiloxanes containing cyanophenyl benzoate mesogen connected to the backbone *via* a pentyloxy group[75]

14.11.2.7 Side chain liquid crystalline polyalkynes

Monoalkynic monomers containing mesogenic groups have been reported from several laboratories.[413-415] However, the first report on the synthesis, polymerization and characterization of a liquid crystalline polyalkyne was reported only recently.[416] 4-Methoxy-4'-(hexyloxy)biphenyl dipropargyl acetate was cyclopolymerized by various transition metal catalysts and the resulting poly(1,6-heptadiyne) derivative containing mesogenic side groups exhibits an enantiotropic smectic mesophase (Scheme 44).[416]

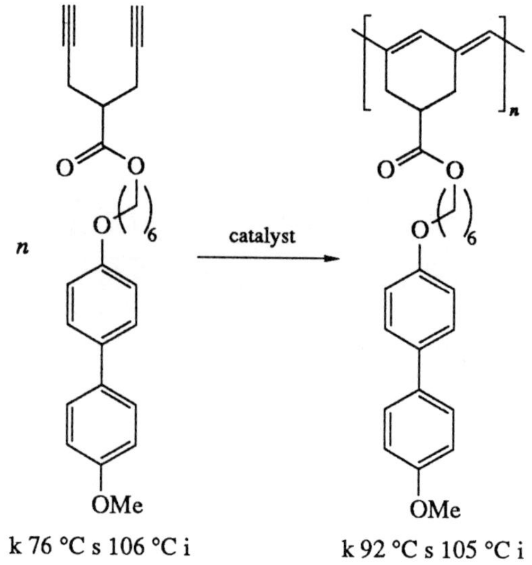

Scheme 44

k 76 °C s 106 °C i   k 92 °C s 105 °C i

14.12 LIQUID CRYSTALLINE POLYMERS CONTAINING CROWN ETHERS AND POLYPODANTS

Mesomorphic host–guest systems of low molecular weight and polymer liquid crystals containing macroheterocyclic ligands and polypodants provide a novel approach to self-assembled systems which combine selective recognition with external regulation.[417-420] Three basic architectures can be considered for liquid crystalline polymers containing crown ethers (Figure 34): main chain liquid crystalline polymers containing crown ethers in the main chain of the polymer and side chain liquid crystalline polymers containing crown ethers either in the mesogenic group or in the main chain. Alternatively, the same series of polymers with polypodants instead of crown ethers can be considered.

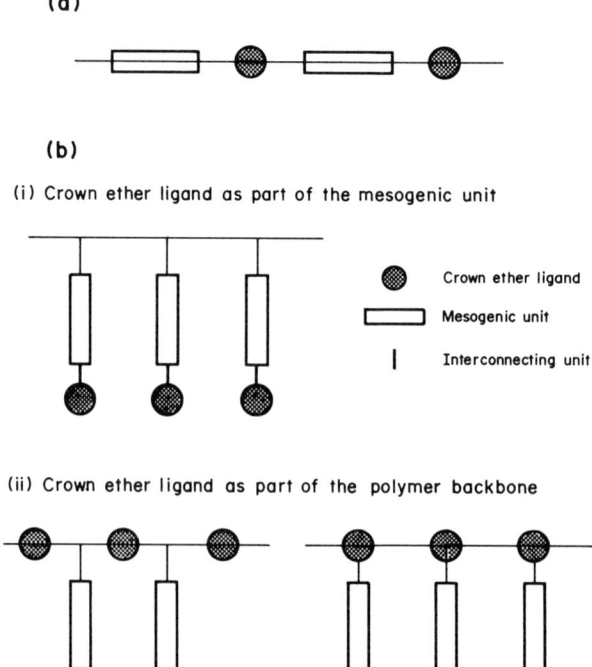

Figure 34 The molecular architecture of liquid crystalline polymers containing crown ether ligands: (a) main chain liquid crystalline polymers; (b) side chain liquid crystalline polymers

Main chain polyamides and polyethers containing crown ethers have been reported.[421,422] A variety of side chain liquid crystalline polymers containing crown ether groups at one end of the mesogenic unit were designed.[377,423–428] Side chain liquid crystalline polymers containing crown ethers in the main chain were synthesized by living cationic cyclopolymerization and cocyclopolymerization of 1,2-bis(2-ethenyloxyethoxy)benzene derivatives containing mesogenic side groups.[375,376] Polymers containing crown ethers in the side groups dissolve ion pairs and behave as copolymers containing two different mesogenic groups, *i.e.* complexed and uncomplexed. Their behavior is similar to that of copolymers derived from two different mesogenic groups. Therefore, their phase behavior is directed by molecular recognition.[429] The use of oligo(oxyethylene)-type spacers in main chain,[430] and side chain[431–433] liquid crystalline polymers leads to liquid crystalline polypodants. Both main chain[430] and side chain[433] liquid crystalline polypodants dissolve large amounts of alkali metal salts, and the resulting liquid crystalline polyelectrolytes are ionic conductors.[434]

14.13 ELECTRON DONOR–ACCEPTOR (EDA) COMPLEXES OF LIQUID CRYSTALLINE POLYMERS CONTAINING DISCOTIC MESOGENS

Ringsdorf *et al.* have studied the effect of doping with electron acceptor molecules on the phase behavior of polymers containing disc-like donor mesogens.[435–440] They found that charge transfer

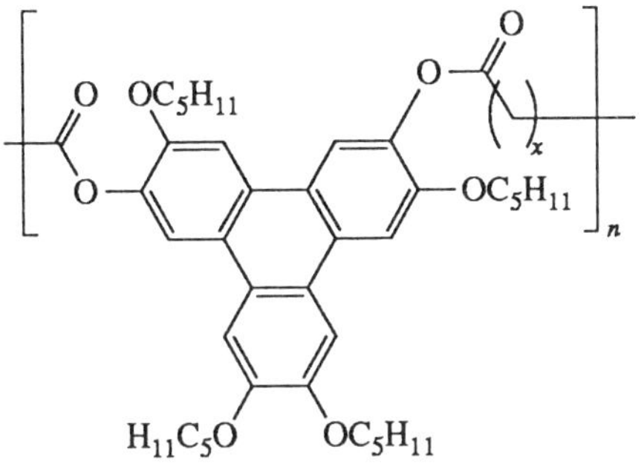

$x = 10$ g 50 °C D_{ho} 195 °C i $\xrightarrow[\text{0.25 mol}]{\text{TNF}}$ g 50 °C D_{ho} 246 °C i

$x = 20$ g 35 °C i $\xrightarrow[\text{0.30 mol}]{\text{TNF}}$ g 22 °C D_{ho} 83 °C i

Scheme 45

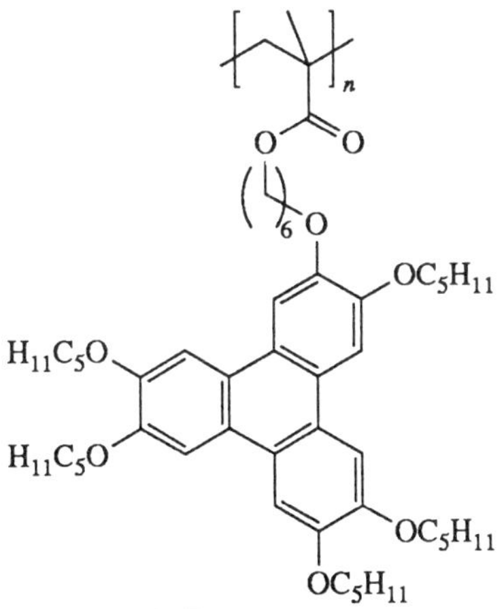

g 30 °C i $\xrightarrow[\text{0.30 mol}]{\text{TNF}}$ g 50 °C N_c 170 °C i

Scheme 46

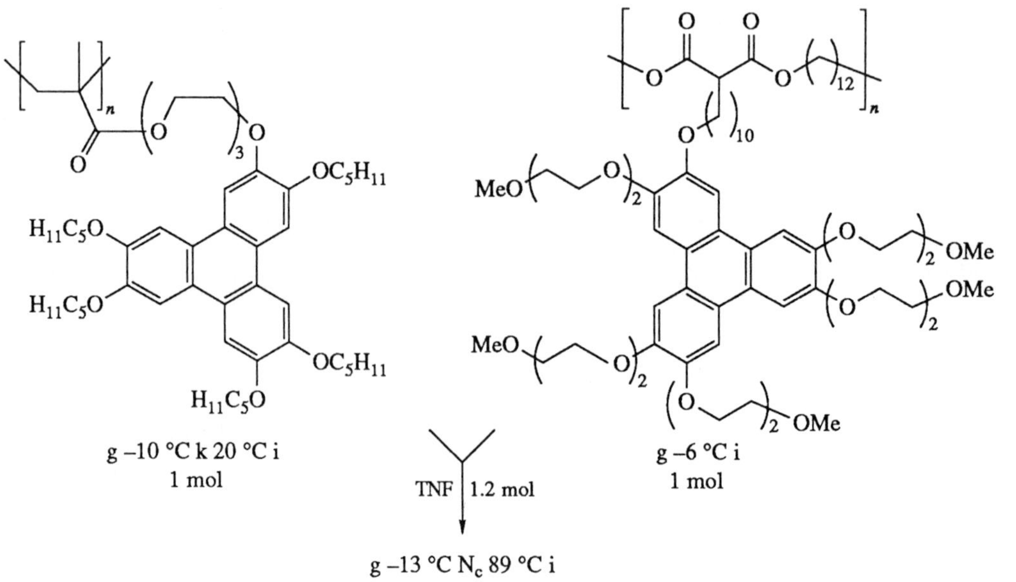

Scheme 47

interactions both increase the degree of order within the liquid crystalline phase, and also stabilize the liquid crystalline mesophase. For example, when a discotic main chain polyester was doped with 2,4,7-trinitrofluorenone (TFN), the width of the discotic columnar hexagonal mesophase was increased by ~50 °C, for molar ratios of polymer/TNF equal to 3/1 and 2/1 (Scheme 45).[436] The intermolecular spacing decreased also considerably, proportional to the amount of TNF present in the complex. The same authors doped a nonliquid crystalline main chain polyester containing disc-like mesogens and this resulted in the induction of a discotic columnar mesophase (Scheme 45). The intercolumnar spacings found were extremely large and the authors assumed that considerable amounts of TNF molecules were uncomplexed and were located in the intercolumnar spacing. Further, doping of a nonliquid crystalline polymethacrylate containing donor disc-like groups in the side chain resulted in the formation of a nematic columnar mesophase (Scheme 46). Doping also induced compatibility and liquid crystallinity in an incompatible mixture of nonmesogenic electron donor polymers, *i.e.* a polymethacrylate and a polyester. The resulting discotic polymer blend exhibited a nematic columnar phase (Scheme 47).[436]

Finally, chirality was induced in a discotic liquid crystalline polymer by doping with a chiral acceptor.[437]

14.14 MOLECULAR RECOGNITION DIRECTED SELF-ASSEMBLY OF SUPRAMOLECULAR LIQUID CRYSTALLINE POLYMERS

The molecular recognition of complementary components leads to systems able to self-assemble or self-organize *i.e.* systems capable of generating spontaneously a well-defined supramolecular architecture from their components under a well-defined set of conditions.[417,418]

Although self-assembly is a well recognized process in biological systems,[440,441] the general concept of self-assembly of synthetic molecules by molecular recognition of complementary components received a revived interest only after it was integrated by Lehn in the new field of supramolecular chemistry.[417,418,442,443]

Several examples in which molecular recognition induces the association of complementary nonmesomorphic components into a low molar mass or polymeric supramolecular liquid crystal are described below.

The principles of formation of a mesogenic supramolecule from two complementary components is outlined in Scheme 48. The particular example used by Lehn *et al.*[444] to generate a supramolecular mesogenic group which exhibits a hexagonal columnar mesophase is by formation of an array of three parallel hydrogen bonds between groups of uracil and 2,6-diaminopyridine type, as depicted in Scheme 48.

The transplant of the same concept to the generation of a supramolecular liquid crystalline polymer is outlined in Scheme 49.[445] The complementary moieties used, TU$_2$ and TP$_2$, are uracil

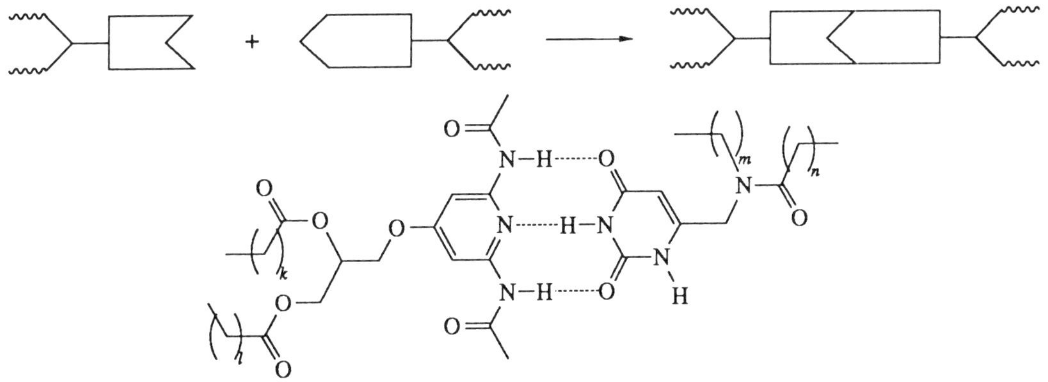

Scheme 48

(U) and 2,6-diacylaminopyridine (P) groups connected through tartaric acid esters (T). The tartaric acid (T) unit provides, in addition, the opportunity to investigate the effect of changes in chirality on the species formed. Thus the components LP_2, LU_2, DP_2, MP_2 and MU_2 are derived from L(+), D(−) and meso (M) tartaric acid respectively. Although all monomers (LP_2, LU_2, DP_2, MP_2 and MU_2) are only crystalline, the corresponding supramolecular 'polymers' obtained through hydrogen bonding ($LP_2 + LU_2$, $DP_2 + LU_2$ and MP_2 and MU_2) exhibit hexagonal columnar mesophases. These hexagonal columnar mesophases are generated from cylindrical helical suprastructures.[445]

An additional example of a supramolecular liquid crystalline polymer obtained through the hydrogen bonding of nonmesomorphic monomers was recently reported.[446]

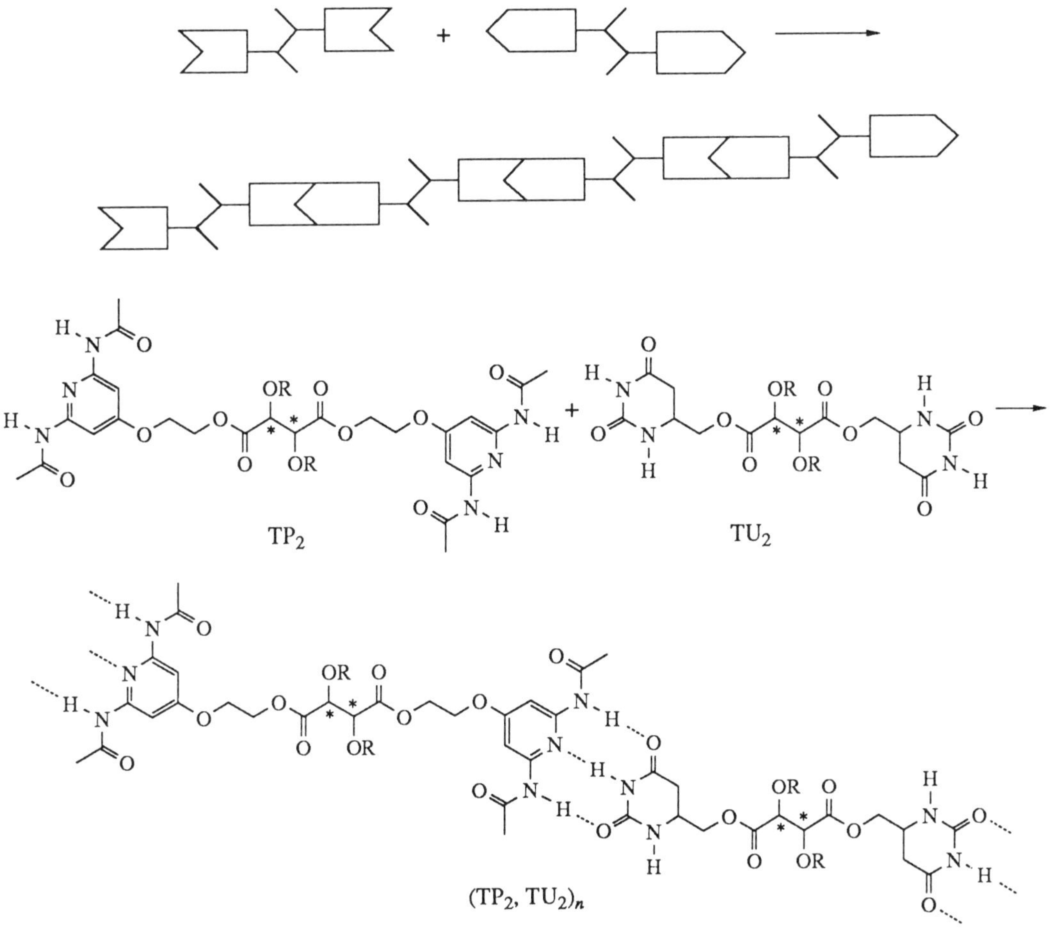

Scheme 49

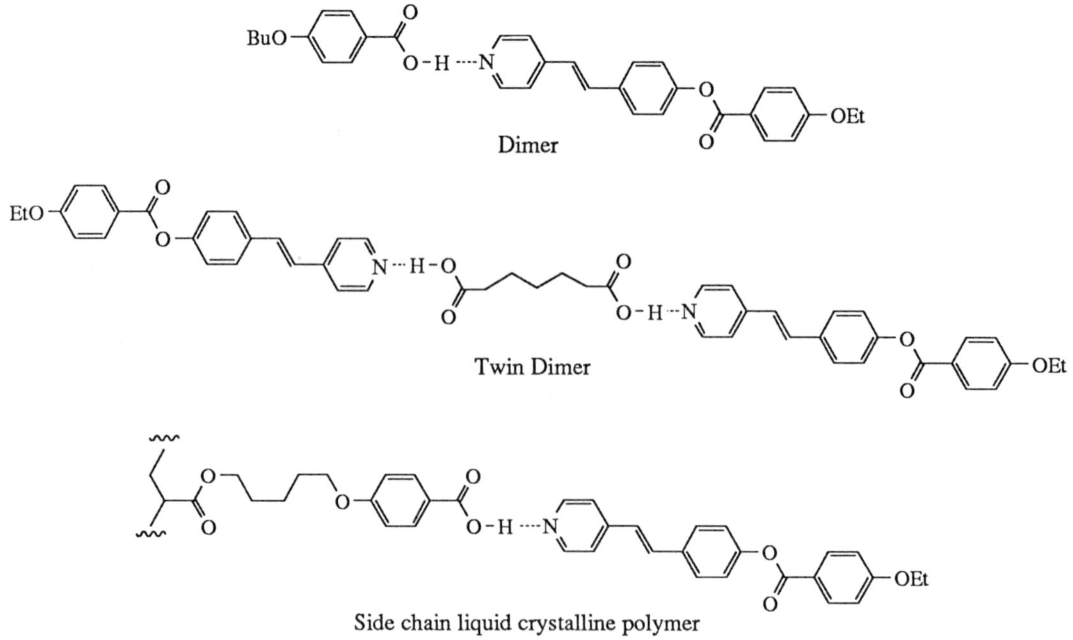

Scheme 50

Examples in which a mesophase was generated through dimerization of carboxylic acid derivatives *via* hydrogen bonding are available in the classic literature on liquid crystals and have been extensively reviewed.[6,447] New and interesting examples on the generation of nonsymmetrical liquid crystalline dimers,[448] twin dimer[449] and side chain liquid crystalline polymers[450] by specific hydrogen bonding 'reactions' continue to be reported (Scheme 50).

Recently, a new approach to molecular recognition directed self-assembly of a liquid crystalline supramolecular structure by a mechanism which resembles that of self-assembly of tobacco mosaic

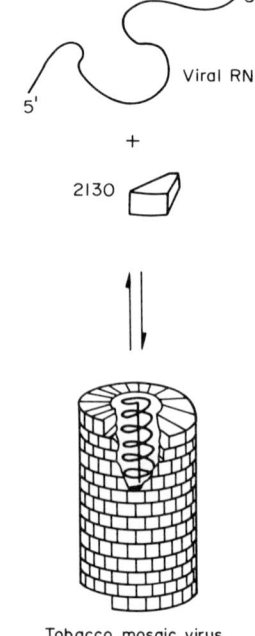

Figure 35 Self-assembly of tobacco mosaic virus (TMV). The protein subunits define the shape of the helix and the RNA defines the helix length. All information for assembly is contained within the component parts. The structure contains 16 1/3 protein subunits per turn and 2130 identical protein subunits (17 500 daltons each). The virus dimensions are 3000 Å in length, 180 Å in diameter, a helical pitch of 23 Å, a central hole of 40 Å, with 3 nucleotides bound per protein subunit[441]

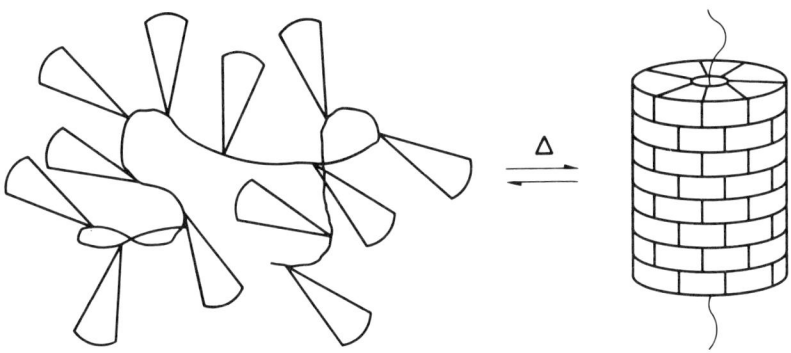

Figure 36 The self-organization of a randomly coiled flexible polymer containing tapered side groups into a rigid rod-like columnar structure

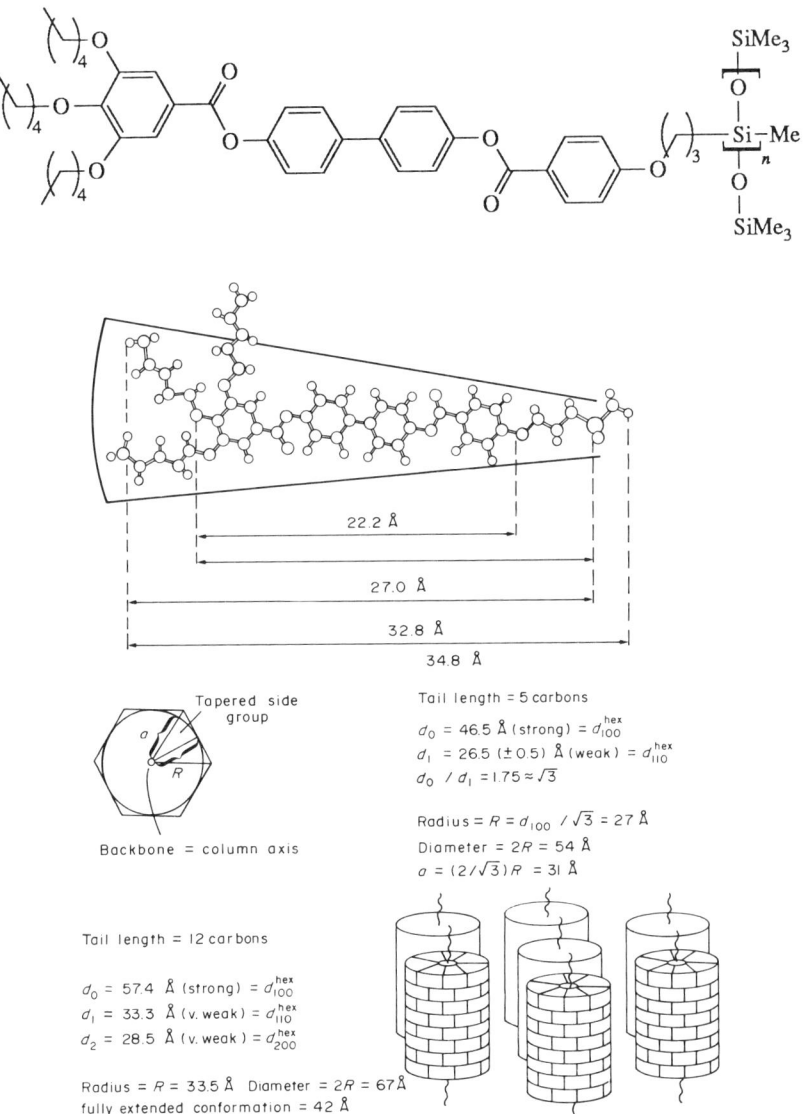

Figure 37 A representative structure of a flexible polymer containing tapered side groups and its self-assembling into a columnar structure which exhibits a hexagonal columnar mesophase[451]

virus (TMV) was reported.[451] The self-assembly mechanism of TMV is outlined in Figure 35.[441] The synthetic approach can be summarized as follows. A flexible polymer backbone containing tapered side groups self-organizes the side groups into a column which surrounds the polymer backbone (Figure 36). These polymers exhibit thermotropic columnar mesophases (Figure 37). Although the number of chains penetrating through the center of the column is not yet known and requires further research (Figure 37), it seems that this self-assembling system is complementary to those elaborated by Lehn et al.[444,445] In the model elaborated by Lehn et al.[444,445] the complementary pairs are self-orgainzed through hydrogen bonding type interactions (*endo*-recognition), while in the latter case[441,451] only the shape of tapered side groups is responsible for the generation of a polymeric column (*exo*-recognition).

ACKNOWLEDGEMENTS

Financial support by the National Science Foundation and the Office of Naval Research is gratefully acknowledged.

14.15 REFERENCES

1. F. Reinitzer, *Monatsh. Chem.*, 1888, **9**, 421. An English translation of this paper was published in *Liq. Cryst.*, 1989, **5**, 7.
2. O. Lehmann, *Vorhandl. Dtsch. Phys. Ges., Sitzungsber.* 1900, **16**, 1. Cited in H. Kelker, *Mol. Cryst. Liq. Cryst.*, 1973, **21**, 1.
3. L. Gattermann and A. Ritscheke, *Ber. Dtsch. Chem. Ges.*, 1890, **23**, 1738.
4. D. Vorländer, in 'Kristallinisch-flüssige Substanzen', Enke, Stuttgart, 1908.
5. *Trans. Faraday. Soc.*, 1933, **29(2)**.
6. (a) G. W. Gray, 'Molecular Structure and the Properties of Liquid Crystals', Academic Press, London, 1962; (b) G. W. Gray, in 'Liquid Crystals and Plastic Crystals', ed. G. W. Gray and P. A Winsor, Horwood, Chichester, 1974, p. 125; (c) G. W. Gray, in 'The Molecular Physics of Liquid Crystals', ed. G. R. Luckhurst and G. W. Gray, Academic Press, London, 1979, p. 14; (d) G. W. Gray, in 'Polymer Liquid Crystals', ed. A. Ciferri, W. R. Krigbaum and R. B. Meyer, Academic Press, New York, 1982, p. 5.
7. H. Kelker, *Mol. Cryst. Liq. Cryst.*, 1973, **21**, 1.
8. H. Kelker, *Mol. Cryst. Liq. Cryst.*, 1988, **165**, 1.
9. D. Demus, *Liq. Cryst.*, 1989, **5**, 75; D. Demus, *Mol. Cryst., Liq. Cryst.*, 1988, **165**, 45.
10. G. W. Gray and P. A. Winsor, 'Liquid Crystals and Plastic Crystals', Horwood, Chichester, 1974, vol. 1 and 2.
11. H. Kelker and R. Hatz, 'Handbook of Liquid Crystals', Verlag Chem., Weiheim, 1980.
12. 'Liquid Crystals, The Fourth State of Matter', ed. F. D. Saeva, Dekker, New York, 1979; 'Thermotropic Liquid Crystals', ed. G. W. Gray, Wiley, New York, 1987.
13. G. R. Luckhurst and G. W. Gray, 'The Molecular Physics of Liquid Crystals', Academic Press, New York, 1979.
14. D. Demus and H. Zaschke, 'Flüssige Kristalle in Tabellen', vol. II, VEB Deutscher Verlag für Grundstoffindustrie, Leipzig, 1984.
15. D. Demus and L. Richter, 'Textures of Liquid Crystals', VEB Deutscher Verlag für Grundstoffindustrie, Leipzig, 1978.
16. G. W. Gray and J. W. Goodby, 'Smectic Liquid Crystals', Leonard Hill, Glasgow, 1984.
17. T. Svedberg, *Kolloid-Z.*, 1916, **18**, 54; (*Chem. Abstr.*, 1916, **10**, 2429).
18. T. Svedberg, *Kolloid-Z.*, 1916, **18**, 101; (*Chem. Abstr.*, 1916, **10**, 2826).
19. T. Svedberg, *Kolloid-Z.*, 1917, **20**, 73; (*Chem. Abstr.*, 1917, **11**, 3145).
20. T. Svedberg, *Kolloid-Z.*, 1917, **21**, 19; (*Chem. Abstr.*, 1918, **12**, 552).
21. V. Percec, H. Jonsson and D. Tomazos, in 'Polymerization in Organized Media', ed. C. M. Paleos, Gordon and Breach, New York, in press.
22. O. Lehman, in 'Handbuch der biologischen Arbeitsmethoden', ed. E. Arberhalden, Urban and Schwarzenberg, Munich, 1922, pp. 123–352.
23. F. Rinne, *Trans. Faraday Soc.*, 1933, **29(2)**, 1016.
24. J. D. Bernal, *Trans. Faraday Soc.*, 1933, **29(2)**, 1082.
25. D. G. Dervichian, *Mol. Cryst. Liq. Cryst.*, 1977, **40**, 19.
26. G. H. Brown and J. J. Wolken (eds.), in 'Liquid Crystals and Biological Structures', Academic Press, New York, 1979.
27. D. Chapman, in 'Advances in Liquid Crystals', ed. G. H. Brown, Academic Press, New York, 1982, vol. 5, p. 1.
28. R. J. Hawkins and E. W. April, in 'Advances in Liquid Crystals', ed. G. H. Brown, Academic Press, New York, 1983, vol. 6, p. 243.
29. D. Chapman, in 'Advances in Liquid Crystals', ed. F. D. Saeva, Dekker, New York, 1979, p. 305.
30. D. Chapman, in 'Liquid Crystals and Plastic Crystals', ed. G. W. Gray and P. A. Winsor, Horwood, Chichester, 1974, vol. 1, p. 288.
31. G. T. Stewart, in 'Liquid Crystals and Plastic Crystals', ed. G. W. Gray and P. A. Winsor, Horwood, Chichester, 1974, vol. 1, p. 308.
32. Y. Bouligand, in 'Liquid Crystalline Order in Polymers', ed. A. Blumstein, Academic Press, New York, 1978, p. 261.
33. S. Friberg and R. F. Gould (eds.), 'Lyotropic Liquid Crystals and the Structure of Biomembranes'. Advances in Chemistry Series, no. 152, American Chemical Society, Washington, D.C., 1976.
34. M. J. Ostro, 'Liposomes', Dekker, New York, 1983.
35. J. H. Fendler, in 'Membrane Mimetic Chemistry', Wiley, New York, 1982.

36. (a) 'Physics of Amphiphiles: Micelles, Vesicles and Microemulsions', ed. V. DeGiorgio and M. Corti, North-Holland, Amsterdam, 1985; (b) J. Seddon and R. Templer, *New Sci.*, May 18, 1991, **130**, 45.
37. P. J. Collins, 'Liquid Crystals. Nature's Delicate Phase of Matter', Princeton Science Library, Princeton, New Jersey, 1990.
38. L. Onsager, *Ann. N. Y. Acad. Sci.*, 1949, **51**, 627.
39. P. J. Flory, *Proc. R. Soc. London, Ser. A*, 1956, **234**, 73.
40. A. Roviello and A. Sirigu, *J. Polym. Sci., Polym. Lett. Ed.*, 1975, **13**, 455.
41. P. G. deGennes, *C. R. Hebd. Seances Acad Sci., Ser. B.*, 1975, **281**, 101.
42. H. F. Kuhfuss and W. J. Jackson, Jr., *US Pat.* 3 778 410 (1973); 3 804 805 (1974); W. J. Jackson, Jr. and H. F. Kuhfuss, *J. Polym. Sci., Polym. Chem. Ed.*, 1976, **14**, 2043.
43. J. Economy, *Mol. Cryst. Liq. Cryst.*, 1989, **169**, 1.
44. W. J. Jackson, Jr., *Mol. Cryst. Liq. Cryst.*, 1989, **169**, 23.
45. J. Economy, *Angew. Chem., Int. Ed. Engl.*, 1990, **29**, 1256.
46. M. Ballauff, *Angew. Chem., Int. Ed. Engl.*, 1989, **28**, 253.
47. H. Finkelmann, *Angew. Chem., Int. Ed. Engl.*, 1987, **26**, 816.
48. R. B. Blumstein and A. Blumstein. *Mol. Cryst. Liq. Cryst.*, 1988, **165**, 361.
49. M. Gordon and N. A. Platé (eds.), *Adv. Polym. Sci.*, 1984, **59**.
50. H. Finkelmann, H. Ringsdorf and J. H. Wendorff, *Makromol. Chem.*, 1978, **179**, 273.
51. H. Finkelmann, M. Happ, M. Portugal and H. Ringsdorf, *Makromol. Chem.*, 1978, **179**, 2541.
52. M. Gordon and N. A. Platé (eds.), *Adv. Polym. Sci.*, 1984, **60/61**.
53. A. Blumstein, 'Mesomorphic Order in Polymers and Polymerization in Liquid Crystalline Media', ACS Symp. Ser., no. 74, American Chemical Society, Washington D.C., 1978.
54. A. Blumstein, 'Liquid Crystalline Order in Polymers', Academic Press, New York, 1978.
55. A. Blumstein, 'Polymeric Liquid Crystals', Plenum Press, New York, 1985.
56. L. L. Chapoy, 'Recent Advances in Liquid Crystalline Polymers', Elsevier, London, 1985.
57. A. Ciferri, W. R. Krigbaum and R. B. Meyers, 'Polymer Liquid Crystals', Academic Press, London, 1982.
58. C. B. McArdle, 'Side Chain Liquid Crystal Polymers', Blackie, Glasgow, 1989.
59. A. Ciferri (ed.), 'Liquid Crystallinity in Polymers. Principles and Fundamental Properties', VCH, New York, 1991.
60. R. A. Weiss and C. K. Ober, 'Liquid Crystalline Polymers', ACS Symp. Ser., no. 435, American Chemical Society, Washington D.C., 1990.
61. E. Chiellini and R. W. Lenz, in 'Comprehensive Polymer Science', ed. G. Allen and J. C. Bevington, Pergamon Press, Oxford, 1989, vol. 5, p. 701.
62. R. Zentel, in 'Comprehensive Polymer Science', ed. G. Allen and J. C. Bevington, Pergamon Press, Oxford, 1989, vol. 5, p. 723.
63. A. M. Donald and A. H. Windle, 'Liquid Crystalline Polymers', Cambridge University Press, Cambridge, 1991; C. Noël and P. Navard, *Prog. Polym. Sci.*, 1991, **16**, 55.
64. (a) H. Ringsdorf, B. Schlarb and J. Venzmer, *Angew. Chem., Int. Ed. Engl.*, 1988, **27**, 113; (b) W. Mahler, D. Guillon and A. Skoulios, *Mol. Cryst. Liq. Cryst. Lett.*, 1985, **2**, 111; C. Viney, T. P. Russell, L. E. Depero and R. J. Twieg, *Mol. Cryst. Liq. Cryst.*, 1989, **168**, 63; C. Viney, R. J. Twieg, T. P. Russell and L. E. Depero, *Liq. Cryst.*, 1989, **5**, 1783; J. Höpken, C. Pugh, W. Richtering and M. Möller, *Makromol. Chem.*, 1988, **189**, 911.
65. V. Percec and A. Keller, *Macromolecules*, 1990, **23**, 4347.
66. A. Keller, G. Ungar and V. Percec, in 'Advances in Liquid Crystalline Polymers', ed. R. A. Weiss and C. K. Ober, ACS Symp. Ser., no. 435, American Chemical Society, Washington D.C., 1990, p. 308.
67. Y. Galerne, *Mol. Cryst. Liq. Cryst.*, 1988, **165**, 131.
68. S. Chandrasekhar, in 'Advances in Liquid Crystals', ed. G. H. Brown, Academic Press, London, 1982, vol. 5, p. 47.
69. S. Chandrasekhar and G. S. Ranganath, *Rep. Prog. Phys.*, 1990, **53**, 57.
70. A. Skoulios and D. Guillon, *Mol. Cryst. Liq. Cryst.*, 1988, **165**, 317.
71. P. A. Winsor, *Chem. Rev.*, 1968, **68**, 1; P. S. Pershan, *J. Phys. Coll.*, 1979, **40**, C3–423.
72. V. Percec and C. Pugh, in 'Side Chain Liquid Crystal Polymers', ed. C. B. McArdle, Chapman and Hall, New York, 1989, p. 30.
73. H. Ringsdorf and R. Wustefeld, *Philos. Trans. R. Soc. London, Ser. A.*, 1990, **330**, 95.
74. H. Finkelmann, B. Luhmann, G. Rehage and H. Stevens, in 'Liquid Crystals and Ordered Fluids', ed. A. C. Griffin and J. F. Johnson, Plenum Press, New York, 1984, vol. 4, p. 715; E. Jahns and H. Finkelmann, *Colloid Polym. Sci.*, 1987, **265**, 304; B. Luhmann and H. Finkelmann, *Colloid Polym. Sci.*, 1986, **264**, 189; B. Luhmann and H. Finkelmann, *Colloid Polym. Sci.*, 1987, **265**, 506.
75. R. D. Richards, W. D. Hawthorne, J. S. Hill, M. S. White, D. Lacey, J. A. Semiyen, G. W. Gray and T. C. Kendrick, *J. Chem. Soc., Chem. Commun.*, 1990, 95.
76. W. Gleim and H. Finkelmann, in 'Side Chain Liquid Crystal Polymers', ed. C. B. McArdle, Chapman and Hall, New York, 1989, p. 287.
77. R. Zentel, *Angew. Chem., Int. Ed. Engl., Adv. Mater.*, 1989, **28**, 1407.
78. P. W. Morgan, *Macromolecules*, 1977, **10**, 1381; S. L. Kwolek, P. W. Morgan, J. R. Shaefgen and L. W. Gulrich, *Macromolecules*, 1977, **10**, 1390; T. I. Bair, P. W. Morgan and F. L Killian, *Macromolecules*, 1977, **10**, 1396; M. Panar and L. Beste, *Macromolecules*, 1977, **10**, 1401.
79. S. P. Papkov, *Adv. Polym. Sci.*, 1984, **59**, 75.
80. C. Noël, in 'Side Chain Liquid Crystal Polymers', ed. C. B. McArdle, Chapman and Hall, New York, 1989, p. 159.
81. (a) C. Noël, *Makromol. Chem., Macromol. Symp.*, 1988, **22**, 95; (b) P. Davidson, L. Noirez, J. P. Cotton and P. Keller, *Liq. Cryst.*, 1991, **10**, 111 and refs. cited therein; (c) H. Pepy, J. P. Cotton, F. Hardouin, P. Keller, M. Lambert, F. Moussa, L. Noirez, A. Lapp and C. Strazielle, *Makromol. Chem., Macromol. Symp.*, 1988. **15**, 251 and refs. cited therein.
82. (a) V. Percec and D. Tomazos, *Polymer*, 1990, **31**, 1658; (b) V. Percec, B. Hahn, M. Ebert and J. H. Wendorff, *Macromolecules*, 1990, **23**, 2092.
83. H. Arnold and H. Sackmann, *Z. Phys. Chem. (Leipzig)*, 1960, **213**, 137.
84. H. Arnold and H. Sackmann, *Z. Phys. Chem. (Leipzig)*, 1960, **213**, 145.

85. H. Sackmann and D. Demus, *Mol. Cryst. Liq. Cryst.*, 1973, **21**, 239; D. Demus, S. Diele, S. Grande and H. Sackmann, in 'Advances in Liquid Crystals', ed. G. H. Brown, Academic Press, London, 1983, vol. 6, p. 1.
86. W. R. Krigbaum, *J. Appl. Polym. Sci., Appl. Polym. Symp.*, 1985, **41**, 105.
87. C. Casagrande, M. Veyssie and H. Finkelmann, *J. Phys. Lett. (Orsay. Fr.)*, 1982, **43**, L-671.
88. H. Ringsdorf, H. W. Schmidt and A. Schneller, *Makromol. Chem., Rapid. Commun.*, 1982, **3**, 745.
89. H. Benthack-Thoms and H. Finkelmann, *Makromol. Chem.*, 1985, **186**, 1895.
90. V. Percec and Y. Tsuda, *Polymer*, 1991, **32**, 661.
91. T. Schroeder, *Z. Phys. Chem., Stoechiom. Verwandschaftsl.*, 1893, **11**, 449; J. J. Van Laar, *Z. Phys. Chem. Stoechiom. Verwandschaftsl.*, 1908, **63**, 216; G. R. Van Hecke, *J. Phys. Chem.*, 1979, **83**, 2344; M. F. Achard, M. Mauzac, M. Richard, M. Sigaud and F. Hardouin, *Eur. Polym. J.*, 1989, **25**, 593.
92. B. Kronberg, D. F. R. Gilson and D. Patterson, *J. Chem. Soc., Faraday Trans. 2*, 1976, **72**, 1673.
93. S. Ghodbane and D. E. Martire, *J. Phys. Chem.*, 1987, **91**, 6410.
94. B. Kronberg, I. Bassignana and D. Patterson, *J. Phys. Chem.*, 1978, **82**, 1714.
95. H. Orendi and M. Ballauff, *Liq. Cryst.*, 1989, **6**, 497.
96. H. Orendi and M. Ballauff, *Mol. Cryst. Liq. Cryst. Lett.*, 1991, **7**, 185.
97. B. Kronberg and D. Patterson, *J. Chem. Soc., Faraday Trans. 2*, 1976, **72**, 1686.
98. F. Brochard, J. Jouffroy and P. Levinson, *J. Phys. (Orsay. Fr.)*, 1984, **45**, 1125.
99. M. Ballauff, *Ber. Bunsenges. Phys. Chem.*, 1986, **90**, 1053.
100. D. E. Martire and S. Ghodbane, *J. Phys. Chem.*, 1987, **91**, 6403.
101. P. G. DeGennes, *Phys. Lett. A.*, 1969, **28A**, 725.
102. C. S. Hsu and V. Percec, *J. Polym. Sci., Polym. Chem. Ed.*, 1987, **25**, 2909; C. S. Hsu and V. Percec, *J. Polym. Sci., Polym. Chem. Ed.*, 1989, **27**, 453.
103. (a) V. Percec and R. Yourd, *Macromolecules*, 1988, **21**, 3379; (b) V. Percec and R. Yourd, *Macromolecules*, 1989, **22**, 524; (c) V. Percec and R. Yourd, *Macromolecules*, 1989, **22**, 3229; (d) V. Percec and Y. Tsuda, *Macromolecules*, 1990, **23**, 5; (e) V. Percec and Y. Tsuda, *Macromolecules*, 1990, **23**, 3509; (f) G. Ungar, J. L. Feijoo, A. Keller, R. Yourd and V. Percec, *Macromolecules*, 1990, **23**, 3411.
104. R. S. Irwin, S. Weeny, K. H. Gardner, C. R. Gochanour and M. Weinberg, *Macromolecules*, 1989, **22**, 1065; K. H. Gardner, C. R. Gochanour, R. S. Irwin, S. Weeny and M. Weinberg, *Mol. Cryst. Liq. Cryst.*, 1988, **155**, 239.
105. (a) H. Jonsson, P. E. Werner, U. W. Gedde and A. Hult, *Macromolecules*, 1989, **22**, 1683; (b) H. Jonsson, E. Wallgren, A. Hult and U. W. Gedde, *Macromolecules*, 1990, **23**, 1041; (c) H. Jonsson, U. W. Gedde and A. Hult, in Liquid Crystalline Polymers', ed. R. A. Weiss and C. K. Ober, ACS Symp. Ser. no. 435, American Chemical Society, Washington D.C., 1990, p. 62.
106. A. Y. Bilibin, A. V. Tenkovtsev and O. N. Piraner, *Makromol. Chem.*, 1989, **190**, 3013.
107. V. Percec and M. Zuber, *Polym. Bull.*, 1991, **25**, 695.
108. P. Tékély, F. Lauprêtre and L. Monnerie, *Macromolecules*, 1983, **16**, 415.
109. P. Meurisse, F. Lauprêtre, C. Noël, *Mol. Cryst. Liq. Cryst.*, 1984, **110**, 41.
110. H. Kalinowski and H. Kessler 'Topics in Stereochemistry', ed. E. L. Eliel and N. L. Allinger, Wiley, New York, 1973, vol. 7, p. 295.
111. P. Coulter and A. H. Windle, *Macromolecules*, 1989, **22**, 1129.
112. H. Kessler, *Angew. Chem., Int. Ed. Engl.*, 1970, **9**, 219.
113. A. Mannschreck, *Tetrahedron Lett.*, 1965, 1341.
114. A. Mannschreck, A. Mattheus and G. Rissmann, *J. Mol. Spectrosc.*, 1967, **23**, 15.
115. K. Tashiro, M. Kokayashi and H. Tadokoro, *Macromolecules*, 1977, **10**, 413.
116. L. S. Bartell and D. A. Kohl, *J. Chem. Phys.*, 1963, **39**, 3097.
117. K. Kuchitsu, *Bull. Chem. Soc. Jpn.*, 1959, **32**, 748.
118. G. J. Szasz, N. Sheppard and D. H. Rank, *J. Chem. Phys.*, 1948, **16**, 704.
119. P. B. Woller and E. W. Garbisch, Jr., *J. Am. Chem. Soc.*, 1972, **94**, 5310.
120. A. L. Verma, W. F. Murphy and H. J. Bernstein, *J. Chem. Phys.*, 1974, **60**, 1540.
121. J. A. Darsey and B. K. Rao, *Macromolecules*, 1981, **14**, 1575.
122. A. Abe, R. L. Jernigan and P. J. Flory, *J. Am. Chem. Soc.*, 1966, **88**, 631.
123. C. J. Brown, *Acta Crystallogr.*, 1954, **7**, 97.
124. K. K. Chiu, H. H. Huang and L. H. L. Chia, *J. Chem. Soc., Perkin Trans. 2*, 1972, **2**, 286.
125. A. Unanue and P. Bothorel, *Bull. Soc. Chim. Fr.*, 1965, 2827.
126. A. J. M. Reuvers, A. Sinnema, F. van Rantwijk, J. D. Ramijnse and H. van Bekkum, *Tetrahedron*, 1969, **25**, 4455.
127. P. Ivanov, I. Pojarlieff and N. Tyutyulkov, *Tetrahedron Lett.*, 1976, 775.
128. J. Jacobus, *Tetrahedron Lett.*, 1976, 2927.
129. I. Petterson and T. Liljefors, *J. Comput. Chem.*, 1987, **8**, 1139.
130. D. C. Bassett, in 'Developments in Crystalline Polymers—1', ed. D. C. Bassett, Applied Science, London, 1982, p. 115.
131. G. Ungar, *Macromolecules*, 1986, **19**, 1317.
132. V. Percec, D. Tomazos and A. E. Feiring, *Polymer*, 1991, **32**, 1897.
133. A. Blumstein, S. Vilasagar, S. Ponrathnam, S. B. Clough, R. B. Blumstein and G. Maret, *J. Polym. Sci., Polym. Phys. Ed.*, 1982, **20**, 877.
134. V. Percec, H. Nava and H. Jonsson, *J. Polym. Sci., Polym. Chem. Ed.*, 1987, **25**, 1943.
135. J. L. Feijoo, G. Ungar, A. J. Owen, A. Keller and V. Percec, *Mol. Cryst. Liq. Cryst.*, 1988, **155**, 487.
136. S. G. Kostromin, R. V. Talrose, V. P. Shibaev and N. A. Platé, *Makromol. Chem. Rapid Commun.*, 1982, **3**, 803.
137. Y. K. Godovsky, I. I. Mamaeva, N. N. Makarova, V. P. Papkov and N. N. Kuzmin, *Makromol. Chem. Rapid Commun.*, 1985, **6**, 797.
138. V. Percec and B. Hahn, *Macromolecules*, 1989, **22**, 1588.
139. V. Percec, M. Lee and H. Jonsson, *J. Polym. Sci., Polym. Chem. Ed.*, 1991, **29**, 327.
140. V. Percec and M. Lee, *J. Macromol. Sci., Chem.*, 1991, **A28**, 651.
141. V. Percec, M. Lee and C. Ackerman, *Polymer*, 1992, **33**, 703.
142. V. Percec and M. Lee, *Macromolecules*, 1991, **24**, 2780.
143. V. Percec, A. D. S. Gomes and M. Lee, *J. Polym. Sci., Polym. Chem. Ed.*, 1991, **29**, 1615.

144. V. Percec, C. S. Wang and M. Lee, *Polym. Bull.*, 1991, **26**, 15.
145. V. Percec, Q. Zheng and M. Lee, *J. Mater. Chem.*, 1991, **1**, 611.
146. J. Majnusz, J. M. Catala and R. W. Lenz, *Eur. Polym. J.*, 1983, **19**, 1043.
147. Q. F. Zhou, X. Q. Duan and Y. L. Liu, *Macromolecules*, 1986, **19**, 247.
148. V. Percec, and H. Nava, *J. Polym. Sci., Polym. Chem. Ed.*, 1987, **25**, 405.
149. H. Stevens, G. Rehage and H. Finkelmann, *Macromolecules*, 1984, **17**, 851.
150. P. J. Flory and G. Ronca, *Mol. Cryst. Liq. Cryst.*, 1979, **54**, 311.
151. P. A. Irvine, W. Dacheng and P. J. Flory, *J. Chem. Soc., Faraday Trans. 1*, 1984, **80**, 1795.
152. G. Sigaud, 'Phase Transitions and Phase Diagrams in Liquid Crystalline Polymers', to be published.
153. V. Percec, D. Tomazos and C. Pugh, *Macromolecules*, 1989, **22**, 3259.
154. T. Sagane and R. W. Lenz, *Polymer*, 1989, **30**, 2269.
155. V. Shibaev, *Mol. Cryst. Liq. Cryst.*, 1988, **155**, 189.
156. R. S. Kumar, S. B. Clough and A. Blumstein, *Mol. Cryst. Liq. Cryst.*, 1988, **157**, 387.
157. R. B. Blumstein and A. Blumstein, *Mol. Cryst. Liq. Cryst.*, 1988, **165**, 361.
158. V. Percec and M. Lee, *Macromolecules*, 1991, **24**, 1017.
159. W. Kern, W. Gruber and H. O. Wirth, *Makromol. Chem.*, 1960, **37**, 198.
160. W. Heitz, *Chem. Ztg.*, 1986, **110**, 385.
161. W. Heitz, *Makromol. Chem., Macromol. Symp.*, 1989, **26**, 1.
162. W. Heitz, *Makromol. Chem., Macromol. Symp.*, 1991, **47**, 111.
163. J. K. Stille, F. W. Harris, R. O. Rakutis and H. Mukamal, *J. Polym. Sci., Polym. Lett. Ed.*, 1966, **4**, 791.
164. H. Mukamal, F. W. Harris and J. K. Stille, *J. Polym. Sci., Part A-1*, 1967, **5**, 2721.
165. J. K. Stille, R. O. Rakutis, H. Mukamal and F. W. Harris, *Macromolecules*, 1968, **1**, 431.
166. J. M. Deneen and A. A. Volpe, in 'The 27th IUPAC International Symposium on Macromolecules, Strasbourg, 1981', Abstracts of Communications, vol. I, p. 22.
167. M. Rehahn, A.-D. Schluter, G. Wegner and W. J. Feast, *Polymer*, 1989, **30**, 1054.
168. M. Rehahn, A.-D. Schluter, G. Wegner and W. J. Feast, *Polymer*, 1989, **39**, 1060.
169. M. Rehahn, A.-D. Schluter and G. Wegner, *Makromol. Chem.*, 1990, **191**, 1991.
170. W. R. Krigbaum and K. J. Krause, *J. Polym. Sci., Polym. Chem. Ed.*, 1978, **16**, 3151.
171. A. Noll, N. Siegfried and W. Heitz, *Makromol. Chem., Rapid Commun.*, 1990, **11**, 485.
172. V. Percec, S. Okita and R. Weiss, *Macromolecules*, 1992, **25**, 1816.
173. T. I. Wallow and B. M. Novak, *J. Am. Chem. Soc.*, 1991, **113**, 7411.
174. S. Takahashi, H. Morimoto, E. Murata, S. Kataoka, K. Sonogashira and N. Higara, *J. Polym. Sci., Polym. Chem. Ed.*, 1982, **20**, 565.
175. S. Takahashi, Y. Takai, H. Morimoto and K. Sonogashira, *J. Chem. Soc., Chem. Commun.*, 1984, 3.
176. N. Hagihara, K. Sonogashira and S. Takahashi, *Adv. Polym. Sci.*, 1980, **41**, 149.
177. S. Takahashi, Y. Takai, H. Morimoto, K. Sonogashira and N. Hagihara, *Mol. Cryst. Liq. Cryst.*, 1982, **82**, 139; A. Abe, N. Kimura and S. Tabata, *Macromolecules*, 1991, **24**, 6238.
178. S. J. Davis, B. F. G. Johnson, M. S. Khan and J. Lewis, *J. Chem. Soc., Chem. Commun.*, 1991, 187.
179. H. B. Fyfe, M. Miekuz, D. Zargarian, N. J. Taylor and T. B. Marder, *J. Chem. Soc., Chem. Commun.*, 1991, 188.
180. M. H. Chisholm, *Angew. Chem., Int. Ed. Engl.*, 1991, **30**, 673.
181. A.-M. Giroud-Godquin and P. M. Maitlis, *Angew. Chem., Int. Ed. Engl.*, 1991, **30**, 375.
182. E. A. Orthmann, V. Enkelmann and G. Wegner, *Makromol. Chem., Rapid Commun.*, 1983, **4**, 687.
183. E. A. Orthmann and G. Wegner, *Makromol. Chem., Rapid Commun.*, 1986, **7**, 243.
184. C. Sirlin, L. Bosio and J. Simon, *J. Chem. Soc., Chem. Commun.*, 1987, 379.
185. C. Sirlin, L. Bosio and J. Simon, *J. Chem. Soc., Chem. Commun.*, 1988, 236.
186. C. Sirlin, L. Bosio and J. Simon, *Mol. Cryst. Liq. Cryst.*, 1988, **155**, 231.
187. T. Sauer and G. Wegner, *Mol. Cryst. Liq. Cryst.*, 1988, **162**, 97.
188. W. Caseri, T. Sauer and G. Wegner, *Makromol. Chem., Rapid Commun.*, 1988, **9**, 651.
189. H. G. Rogers, R. A. Gaudiana, W. C. Hollinsed, P. S. Kalyanaraman, J. S. Manello, C. McGowan, R. A. Minns and R. Sahatjian, *Macromolecules*, 1985, **18**, 1058.
190. H. G. Rogers and R. A. Gaudiana, *J. Polym. Sci., Polym. Chem. Ed.*, 1985, **23**, 2669.
191. H. G. Rogers, R. A. Gaudiana, R. A. Minns and D. M. Spero, *J. Macromol. Sci., Chem.*, 1986, **A23**, 905.
192. R. A. Gaudiana, R. A. Minns, H. G. Rogers, R. Sinta, L. D. Taylor, P. S. Kalyanaraman, and C. McGowan, *J. Polym. Sci., Polym. Chem. Ed.*, 1987, **25**, 1249.
193. R. Sinta, R. A. Minns, R. A. Gaudiana and H. G. Rogers, *J. Polym. Sci., Polym. Lett. Ed.*, 1987, **25**, 11.
194. R. A. Gaudiana, R. A. Minns, R. Sinta, N. Weeks and H. G. Rogers, *Prog. Polym. Sci.*, 1989, **14**, 47.
195. P. J. Flory, *Adv. Polym. Sci.*, 1984, **59**, 1.
196. J. Y. Jadhav, W. R. Krigbaum and J. Preston, *Macromolecules*, 1988, **21**, 538.
197. J. Y. Jadhav, J. J. Preston and W. R. Krigbaum, *J. Polym. Sci., Polym. Chem. Ed.*, 1989, **27**, 1175.
198. W. R. Krigbaum, T. Tanaka, G. Brelsford and A. Ciferri, *Macromolecules*, 1991, **24**, 4142.
199. W. I. Jin and R. Kim, *Polym. J.*, 1987, **19**, 977.
200. W. Hatke, H. T. Land, H. W. Schmidt and W. Heitz, *Makromol. Chem., Rapid Commun.*, 1991, **12**, 235.
201. H. Ringsdorf, P. Tschirner, O. H. Schoenherr and J. H. Wendorff, *Makromol. Chem.*, 1987, **188**, 1431.
202. O. H. Schoenherr, J. H. Wendorff, H. Ringsdorf and P. Tschirner, *Makromol. Chem., Rapid Commun.*, 1986, **7**, 791.
203. M. Ballauff, *Makromol. Chem., Rapid Commun.*, 1986, **7**, 407.
204. M. Ballauff and G. F. Schmidt, *Makromol. Chem., Rapid Commun.*, 1987, **8**, 93.
205. J. M. Rodriguez-Parada, R. Duran and G. Wegner, *Macromolecules*, 1989, **22**, 2507.
206. R. Stern, M. Ballauff and G. Wegner, *Makromol. Chem., Macromol. Symp.*, 1989, **23**, 373.
207. W. Brugging, U. Kampschulte, H.-W. Schmidt and W. Heitz, *Makromol. Chem.*, 1988, **189**, 2755.
208. W. Vogel and W. Heitz, *Makromol. Chem.*, 1990, **191**, 829.
209. L. Freund, H. Jung, N. Niessner, H. W. Schmidt and W. Heitz, *Makromol. Chem.*, 1989, **190**, 1561.
210. W. Heitz and N. Niessner, *Makromol. Chem.*, 1990, **191**, 225.
211. W. Heitz, *Makromol. Chem., Macromol. Symp.*, 1991, **48/49**, 15.

212. H. W. Schmidt and D. Guo, *Makromol. Chem.*, 1988, **189**, 2029.
213. M. S. Classen, H. W. Schmidt and J. H. Wendorff, *Polym. Adv. Technol.*, 1990, **1**, 143.
214. M. Hohlweg and H. W. Schmidt, *Makromol Chem.*, 1989, **190**, 1587.
215. H. W. Schmidt, *Makromol. Chem., Macromol. Symp.*, 1989, **26**, 47.
216. (a) W. Heitz and H. W. Schmidt, *Makromol. Chem., Macromol. Symp.*, 1990, **38**, 149; (b) H. Kromer, R. Kuhn, H. Pielartzik, W. Siebke, V. Eckhardt and M. Schmidt, *Macromolecules*, 1991, **24**, 1950; (c) B. S. Hsiao, R. S. Stein, N. Weeks and R. Gaudiana, *Macromolecules*, 1991, **24**, 1299.
217. H. R. Kricheldorf, V. Doring and V. Eckhardt, *Makromol. Chem.*, 1988, **189**, 1425.
218. H. R. Kricheldorf, V. Doring, I. Beuermann and V. Eckhardt, *Makromol. Chem.*, 1988, **189**, 1437.
219. H. R. Kricheldorf and J. Erxleben, *Polymer*, 1990, **31**, 944.
220. H. R. Kricheldorf, I. Beuermann and G. Schwarz, *Makromol. Chem., Rapid Commun.*, 1989, **10**, 211.
221. H. R. Kricheldorf and J. Engelhardt, *J. Polym. Soc., Polym. Chem. Ed.*, 1990, **28**, 2335.
222. H. R. Kricheldorf, G. Schwarz and F. Ruhser, *J. Polym. Sci., Polym. Chem. Ed.*, 1988, **26**, 1621.
223. (a) T. Heitz, P. Rohrbach and H. Hocker, *Makromol. Chem.*, 1989, **190**, 3295; (b) T. Heitz and O. W. Webster, *Makromol. Chem.*, 1991, **192**, 2463.
224. M. Wenzel, M. Ballauff and G. Wegner, *Makromol. Chem.*, 1987, **188**, 2865.
225. F. H. Metzmann, M. Ballauff, R. C. Schulz and G. Wegner, *Makromol. Chem.*, 1989, **190**, 985.
226. R. Duran, M. Ballauff, M. Wenzel and G. Wegner, *Macromolecules*, 1988, **21**, 2897.
227. F. W. Harris and S. L. C. Hsu, *High Perform. Polym.*, 1989, **1**, 3.
228. G. Kossmehl and M. Samandari, *Makromol. Chem.*, 1985, **186**, 1565; H. H. Horhold and J. Opfermann, *Makromol. Chem.*, 1970, **131**, 105; H. H. Horhold, M. Helbig, D. Raabe, J. Opfermann, U. Scherf, R. Stockmann and D. Weiss, *Z. Chem.*, 1987, **27**, 126.
229. S. Antoun, D. R. Gagnon, F. E. Karasz and R. W. Lenz, *J. Polym. Sci., Polym. Lett. Ed.*, 1986, **24**, 503.
230. W. J. Feast and I. S. Millichamp, *Polym. Commun.*, 1983, **24**, 102; M. Rehahn and A. D. Schluter, *Makromol. Chem., Rapid Commun.*, 1990, **11**, 375; A. W. Cooke and K. B. Wagener, *Macromolecules*, 1991, **24**, 1404.
231. W. Heitz, W. Brugging, L. Freund, M. Gailberger, A. Greiner, H. Jung, U. Kampschulte, N. Niesser, F. Osan, H. W. Schmidt and M. Wicker, *Makromol. Chem.*, 1988, **189**, 119.
232. A. E. Siegrist, *Helv. Chim. Acta*, 1981, **64**, 662; H. Kretzschmann and H. Meier, *Tetrahedron Lett.*, 1991, **32**, 5059.
233. T. W. Campbell and R. N. MacDonald, *J. Org. Chem.*, 1959, **24**, 1246.
234. W. Memeger, Jr., *Macromolecules*, 1989, **22**, 1577.
235. A. Greiner and W. Heitz, *Makromol. Chem., Rapid Commun.*, 1988, **9**, 581.
236. M. Brenda, A. Greiner and W. Heitz, *Makromol. Chem.*, 1990, **191**, 1083.
237. H. Martelock, A. Greiner and W. Heitz, *Makromol. Chem.*, 1991, **192**, 967.
238. M. Suzuki, J. C. Lim and T. Saegusa, *Macromolecules*, 1990, **23**, 1574.
239. K. Iimura, N. Koide, H. Tanabe and M. Takeda, *Makromol. Chem.*, 1981, **182**, 2569.
240. M. Tanaka and T. Nakaya, *Makromol. Chem.*, 1984, **185**, 1915.
241. M. Tanaka and T. Nakaya, *Makromol. Chem.*, 1986, **187**, 2345.
242. M. Tanaka and T. Nakaya, *J. Macromol. Sci., Chem.*, 1987, **A24**, 777.
243. M. Tanaka and T. Nakaya, *Makromol. Chem.*, 1988, **189**, 771.
244. H. R. Kricheldorf and J. Awe, *Makromol. Chem., Rapid Commun.*, 1988, **9**, 681.
245. H. R. Kricheldorf and J. Jenssen, *Eur. Polym. J.*, 1989, **25**, 1273.
246. H. R. Kricheldorf and J. Awe, *Makromol. Chem.*, 1989, **190**, 2579.
247. H. R. Kricheldorf and J. Awe, *Makromol. Chem.*, 1989, **190**, 2597.
248. W. Mormann and E. Hissmann, *Tetrahedron Lett.*, 1987, **28**, 3087.
249. W. Mormann and M. Brahm, *Makromol. Chem.*, 1989, **190**, 631.
250. W. Mormann and E. Hohn, *Makromol. Chem.*, 1989, **190**, 1919.
251. W. Mormann and M. Brahm, *Mol. Cryst. Liq. Cryst.*, 1990, **185**, 163.
252. W. Mormann and A. Baharifar, *Polym. Bull.*, 1990, **24**, 413.
253. W. Mormann and M. Brahm, *Macromolecules*, 1991, **24**, 1096.
254. W. Mormann and S. Benadda, *Makromol. Chem.*, 1991, **192**, 2411.
255. P. J. Stenhouse, E. M. Valles, S. W. Kantor and W. J. MacKnight, *Macromolecules*, 1989, **22**, 1467.
256. A. Roviello and A. Sirigu, *Gazz. Chim. Ital.*, 1980, **110**, 403.
257. A. Roviello and A. Sirigu, *Eur. Polym. J.*, 1979, **15**, 423.
258. M. Sato, K. Kurosawa, K. Nakatsucki and Y. Ohkatsu, *J. Polym. Sci., Polym. Chem. Ed.*, 1988, **26**, 3077.
259. M. Sato, K. Nakatsucki and Y. Ohkatsu, *Makromol Chem., Rapid Commun.*, 1986, **7**, 231.
260. C. Y. Lai, B. T. DeBona and D. C. Prevorsek, *J. Appl. Polym. Sci.*, 1988, **36**, 819.
261. H. R. Kricheldorf and D. Lubbers, *Makromol. Chem., Rapid Commun.*, 1989, **10**, 383.
262. H. R. Kricheldorf and D. Lubbers, *Macromolecules*, 1990, **23**, 2656.
263. C. Y. Lai, B. T. DeBona and D. C. Prevorsek, in 'Liquid Crystalline Polymers', ed. R. A. Weiss and C. K. Ober, ACS Symp. Ser. no. 435, American Chemical Society, Washington, D.C., 1990, p. 102.
264. M. Kawabe, I. Yamaoka and M. Kimura, in 'Liquid Crystalline Polymers', ed. R. A. Weiss and C. K. Ober, ACS Symp. Ser. no. 435, American Chemical Society, Washington, D.C., 1990, p. 115.
265. H. R. Kricheldorf and D. Lubbers, *Makromol. Chem., Rapid Commun.*, 1990, **11**, 303.
266. H. R. Kricheldorf and R. Pakull, *Makromolecules*, 1988, **21**, 551.
267. H. R. Kricheldorf and R. Pakull, *Polymer*, 1987, **28**, 1772.
268. H. R. Kricheldorf, R. Pakull and S. Buchner, *Macromolecules*, 1988, **21**, 1929.
269. H. R. Kricheldorf, R. Pakull and S. Buchner, *J. Polym. Sci., Polym. Chem. Ed.*, 1989, **27**, 431.
270. H. R. Kricheldorf and R. Pakull, *J. Polym. Sci., Polym. Lett. Ed.*, 1985, **23**, 413.
271. H. R. Kricheldorf, G. Schwarz and W. Nowatzky, *Polymer*, 1989, **30**, 936.
272. H. R. Kricheldorf and R. Pakull, *New Polym. Mater.*, 1989, **1**, 165.
273. H. R. Kricheldorf and R. Huner, *Makromol. Chem., Rapid Commun.*, 1990, **11**, 211.
274. T. C. Sung, J. J. Mallon, E. D. T. Atkins and S. W. Kantor, in 'Liquid Crystalline Polymers', ed. R. A. Weiss and C. K. Ober, ACS Symp. Ser. no. 435, American Chemical Society, Washington, D.C., 1990, p. 158.

275. V. Percec, *Mol. Cryst. Liq. Cryst.*, 1988, **155**, 1.
276. V. Percec, *Makromol. Chem., Macromol. Symp.*, 1988, **13/14**, 397.
277. V. Percec, K. Asami, D. Tomazos, J. L. Feijoo, G. Ungar and A. Keller, *Mol. Cryst. Liq. Cryst.*, 1991, **205**, 47.
278. V. Percec, K. Asami, D. Tomazos, J. L. Feijoo, G. Ungar and A. Keller, *Mol. Cryst. Liq. Cryst.*, 1991, **205**, 67.
279. R. W. Lenz, J. I. Jin and K. A. Feichtinger, *Polymer*, 1983, **24**, 327.
280. G. Chen and R. W. Lenz, *Polymer*, 1985, **26**, 1307.
281. H. R. Kricheldorf and G. Schwarz, *Makromol. Chem.*, 1987, **188**, 1281.
282. S. L. Kwolek and R. R. Louise, *Macromolecules*, 1986, **19**, 1789.
283. H. J. Harwood, *Makromol. Chem., Macromol. Symp.*, 1987, **10/11**, 331.
284. K. Muhlbach and V. Percec, *J. Polym. Sci., Polym. Chem. Ed.*, 1987, **25**, 2605.
285. V. Percec and J. H. Wang, *J. Polym. Sci., Polym. Chem. Ed.*, 1990, **28**, 1059.
286. V. Percec and J. H. Wang, *Makromol. Chem., Macromol. Symp.*, 1992, **54/55**, 561.
287. B. C. Auman, V. Percec, H. A. Schneider, W. Jishan and H. J. Cantow, *Polymer*, 1987, **28**, 119.
288. B. C. Auman, V. Percec, H. A. Schneider and H. J. Cantow, *Polymer*, 1987, **28**, 1407.
289. B. C. Auman and V. Percec, *Polymer*, 1988, **29**, 938.
290. J. F. D'Allest, P. P. Wu, A. Blumstein and R. B. Blumstein, *Mol. Cryst. Liq. Cryst. Lett.*, 1986, **3**, 103.
291. J. S. Moore and S. I. Stupp, *Macromolecules*, 1988, **21**, 1217.
292. P. G. Martin and S. I. Stupp, *Macromolecules*, 1988, **21**, 1222.
293. S. I. Stupp, J. S. Moore and P. G. Martin, *Macromolecules*, 1988, **21**, 1228.
294. C. K. Ober, S. McNamee, A. Delvin and R. H. Colby, in 'Liquid Crystalline Polymers', ed. R. A. Weiss and C. K. Ober, ACS Symp. Ser. no. 435, American Chemical Society, Washington, D.C., 1990, p. 220.
295. M. Laus, D. Caretti, A. S. Angeloni, G. Galli and E. Chiellini, *Macromolecules*, 1991, **24**, 1459.
296. J. Economy, R. D. Johnson, J. R. Lyerla and A. Muhlebach, in 'Liquid Crystalline Polymers', ed. R. A. Weiss and C. K. Ober, ACS Symp. Ser. no. 435, American Chemical Society, Washington, D.C., 1990, p. 129.
297. V. Percec and Y. Tsuda, *Polymer*, 1991, **32**, 673.
298. D. Y. Yoon, G. Sigaud, M. Sherwood, C. Wade, V. Percec and M. Kawasumi, to be published.
299. W. A. MacDonald, A. D. W. McLenaghan, G. McLean, R. W. Richards and S. M. King, *Macromolecules*, 1991, **24**, 6164.
300. P.-G. deGennes and P. Pinkus, *Polymer. Prepr., Am. Chem. Soc., Div. Polym. Chem.*, 1977, **18**, 161.
301. Y. H. Kim and P. Pinkus, *Biopolymers*, 1979, **18**, 2315.
302. P. J. Flory and R. R. Matheson, *J. Phys. Chem.*, 1984, **88**, 6606.
303. N. Carr and G. W. Gray, *Mol. Cryst. Liq. Cryst.*, 1985, **124**, 27.
304. M. A. Osman, *Mol. Cryst. Liq. Cryst.*, 1982, **82**, 295.
305. M. A. Osman, *Mol. Cryst. Liq. Cryst.*, 1982, **82**, 47.
306. N. Carr, G. W. Gray and D. G. McDonnell, *Mol. Cryst. Liq. Cryst.*, 1983, **97**, 13.
307. M. A. Osman and T. H. Ba, *Helv. Chim. Acta*, 1983, **66**, 1786.
308. S. M. Kelly and H. Schadt, *Mol. Cryst. Liq. Cryst.*, 1984, **110**, 239.
309. M. Schadt, M. Petrzilka, P. R. Gerber, A. Villiger and G. Trickes, *Mol. Cryst. Liq. Cryst.*, 1983, **94**, 139.
310. K. Praefke, D. Schmidt and G. Heppke, *Chem. Ztg.*, 1980, **104**, 269.
311. M. Kyotari and H. Manetsuna, *J. Polym. Sci., Polym. Phys. Ed.*, 1983, **21**, 379.
312. V. Percec, T. D. Shaffer and R. Yourd, unpublished data.
313. W. Memeger, Experimental Station, DuPont; personal communication, Nov. 11, 1991, copy of notebook from 1975, unpublished data.
314. D. J. Sikkema and P. Hoogland, *Polymer*, 1986, **27**, 1443.
315. J. Kops and H. Spanggaard. *Polym. Bull.*, 1986, **16**, 507.
316. V. Percec and R. Yourd, *Makromol. Chem.*, 1990, **191**, 25.
317. V. Percec and R. Yourd, *Makromol. Chem.*, 1990, **191**, 49.
318. V. Percec and Y. Tsuda, *Polym. Bull.*, 1989, **22**, 489.
319. V. Percec and Y. Tsuda, *Polym. Bull.*, 1989, **24**, 497.
320. V. Percec and Y. Tsuda, *Polym. Bull.*, 1990, **24**, 9.
321. V. Percec and Y. Tsuda, *Polymer*, 1991, **32**, 673.
322. G. Ronca and D. Y. Yoon, *J. Chem. Phys.*, 1982, **76**, 3295.
323. S. Z. D. Cheng, M. A. Yandrasits and V. Percec, *Polymer*, 1991, **32**, 1284.
324. M. Zuber, PhD Thesis, Case Western Reserve University, Cleveland, Ohio, 1991; V. Percec and M. Zuber to be published.
325. G. Ungar, V. Percec and M. Zuber, *Macromolecules*, 1992, **25**, 75.
326. S. V. Vasilenko, A. R. Khokhlov and V. P. Shibaev, *Macromolecules*, 1984, **17**, 2270.
327. V. Percec and M. Zuber, *J. Polym. Sci., Polym. Chem. Ed.*, 1992, **30**, 997.
328. G. Ungar, J. L. Feijoo, V. Percec and R. Yourd, *Macromolecules*, 1991, **24**, 1168.
329. G. Ungar, J. L. Feijoo, V. Percec and R. Yourd, *Macromolecules*, 1991, **24**, 953.
330. V. Percec, M. Zuber, S. Z. D. Cheng and A. Q. Zhang, *J. Mater. Chem.*, 1992, **2**, 407.
331. V. Percec, M. Zuber, G. Ungar and A. Alvarez-Castillo, *Macromolecules*, 1992, **30**, 439.
332. V. Percec and M. Kawasumi, *Macromolecules*, 1991, **24**, 6318.
333. D. Y. Yoon, G. Sigaud, M. Sherwood, C. Wade, V. Percec and M. Kawasumi, *Macromolecules*, to be published.
334. D. Y. Yoon and P. J. Flory, *MRS Symp. Proc.*, 1989, **134**, 11.
335. V. Percec and P. Chu, to be published.
336. D. A. Tomalia, A. M. Naylor and W. A. Goddard, III, *Angew. Chem., Int. Ed. Engl.*, 1990, **29**, 138.
337. T. M. Miller and T. X. Neenan, *Chem. Mater.*, 1990, **2**, 346.
338. Y. H. Kim and O. W. Webster, *J. Am. Chem. Soc.*, 1990, **112**, 4592.
339. C. J. Hawker and M. J. Frechet, *J. Chem. Soc., Chem. Commun.*, 1990, 1010.
340. K. L. Wooley, C. J. Hawker and J. M. J. Frechet, *J. Chem. Soc., Perkin Trans. 1*, 1991, 1059.
341. C. J. Hawker, R. Lee and J. M. J. Frechet, *J. Am. Chem. Soc.*, 1991, **113**, 4583.
342. E. W. Kwock, T. X. Neenan and T. M. Miller, *Chem. Matter.*, 1991, **3**, 775.
343. S. E. Friberg, M. Podzimek, D. A. Tomalia and D. M. Hedstrand, *Mol. Cryst. Liq. Cryst.*, 1988, **164**, 157.
344. V. Percec and M. Kawasumi, *Macromolecules*, in press.

345. Y. Kim, personal communication, Experimental Station, DuPont, Nov. 12, 1991, *J. Am. Chem. Soc.*, in press.
346. V. Percec and M. Kawasumi, *Macromolecules*, in press.
347. M. Warner, in 'Side Chain Liquid Crystal Polymers', ed. C. B. McArdle, Blackie, Glasgow, 1989, p. 1.
348. L. Noirez, J. P. Cotton, F. Hardouin, P. Keller, F. Moussa, G. Pepy and C. Strazielle, *Macromolecules*, 1988, **21**, 2889.
349. P. Davidson, L. Noirez, J. P. Cotton and P. Keller, *Liq. Cryst.*, 1991, **10**, 111.
350. F. Hardouin, S. Mery, M. F. Achard, L. Noirez and P. Keller, *J. Phys. II*, 1991, **1**, 511.
351. H. Mattoussi, R. Ober, M. Veyssie and H. Finkelmann, *Europhys. Lett.*, 1986, **2**, 233.
352. F. Kuschel, A. Madicke, S. Diele, H. Utschik, B. Hisgen and H. Ringsdorf, *Polym. Bull.*, 1990, **23**, 373.
353. V. Percec and D. Tomazos, *J. Polym. Sci., Polym. Chem. Ed.*, 1989, **27**, 999.
354. V. Percec and D. Tomazos, *Macromolecules*, 1989, **22**, 2062.
355. V. Percec, D. Tomazos and R. A. Willingham, *Polym. Bull.*, 1989, **22**, 199.
356. V. Percec and D. Tomazos, *Macromolecules*, 1989, **22**, 1512.
357. V. Percec and D. Tomazos, *Polymer*, 1989, **30**, 2124.
358. G. W. Gray, in 'Side Chain Liquid Crystal Polymers', ed. C. B. McArdle, Blackie, Glasgow, 1989, p. 106.
359. J. M. Rodriguez-Parada and V. Percec, *J. Polym. Sci., Polym. Chem. Ed.*, 1986, **24**, 1363.
360. V. Percec and D. Tomazos, *Polym. Bull.*, 1987, **18**, 239.
361. J. M. Rodriguez-Parada and V. Percec, *J. Polym. Sci., Polym. Chem. Ed.*, 1987, **25**, 2269.
362. C. Pugh and V. Percec, *Polym. Prepr., Am. Chem. Soc., Div. Polym. Chem.*, 1985, **26**, 303.
363. W. Kreuder, O. W. Webster and H. Ringsdorf, *Makromol. Chem., Rapid Commun.*, 1986, **7**, 5.
364. M. Hefft and J. Springer, *Makromol. Chem., Rapid Commun.*, 1990, **11**, 397.
365. T. Kodaira and K. Mori, *Makromol. Chem., Rapid Commun.*, 1990, **11**, 645.
366. T. Sagane and R. W. Lenz, *Polym. J.*, 1988, **20**, 923.
367. T. Sagane and R. W. Lenz, *Macromolecules*, 1989, **22**, 3763.
368. V. Heroguez, A. Deffieux and M. Fontanille, *Makromol. Chem., Macromol. Symp.*, 1990, **32**, 199.
369. V. Heroguez, M. Schappacher, E. Papon and A. Deffieux, *Polym. Bull.*, 1991, **25**, 307.
370. E. Papon, A. Deffieux, F. Hardouin and M. F. Achard, *Liq. Cryst.*, in press.
371. H. Jonsson, V. Percec and A. Hult, *Polym. Bull.*, 1991, **25**, 115.
372. V. Percec and M. Lee, *Macromolecules*, 1991, **24**, 1017.
373. V. Percec, C. S. Wang and M. Lee, *Polym. Bull.*, 1991, **26**, 15.
374. V. Percec, Q. Zheng and M. Lee, *J. Mater. Chem*, 1991, **1**, 611.
375. R. Rodenhouse, V. Percec and A. E. Feiring, *J. Polym. Sci., Polym. Lett.*, 1990, **28**, 345.
376. R. Rodenhouse and V. Percec, *Adv. Matter.*, 1991, **3**, 101.
377. R. Rodenhouse and V. Percec, *Polym. Bull.*, 1991, **25**, 47.
378. S. G. Kostromin, N. D. Cuong, E. S. Garina and V. P. Shibaev, *Mol. Cryst. Liq. Cryst.*, 1990, **193**, 177.
379. C. G. Cho, B. A. Feit and O. W. Webster, *Macromolecules*, 1990, **23**, 1918.
380. H. Jonsson, P. E. Sundell, V. Percec, U. W. Gedde and A. Hult, *Polym. Bull.*, 1991, **25**, 649.
381. H. Jonsson, H. Andersson, P. E. Sundell, U. W. Gedde and A. Hult, *Polym. Bull.*, 1991, **25**, 641.
382. H. Jonsson, V. Percec, U. W. Gedde and A. Hult, *Makromol. Chem., Macromol. Symp.*, 1992, **54/55**, 83.
383. V. Percec, M. Lee, P. L. Rinaldi and V. E. Litman, *J. Polym. Sci., Polym. Chem. Ed.*, 1992, **30**, 1213.
384. V. Percec and M. Lee, *J. Mater. Chem.*, 1991, **1**, 1007.
385. D. A. Tirrell, in 'Encyclopedia of Polymer Science and Engineering', ed. H. F. Mark, N. M. Bikales, C. G. Overberger and G. Menges, 2nd edn., Wiley, New York, 1986, vol. 4, p. 192.
386. V. Percec and M. Lee, *Macromolecules*, 1991, **24**, 4963.
387. V. Percec and M. Lee, *Polymer*, 1991, **32**, 2862.
388. V. Percec and M. Lee, *Polym. Bull.*, 1991, **25**, 131.
389. V. Percec, Q. Zheng and M. Lee, *J. Mater. Chem.*, 1991, **1**, 611, 1015.
390. P. E. Cladis, *Phys. Rev. Lett.*, 1975, **35**, 48.
391. P. E. Cladis, R. K. Bogardus, W. B. Daniels and G. N. Taylor, *Phys. Rev. Lett.*, 1977, **39**, 720.
392. D. Guillon, P. E. Cladis and J. Stamatoff, *Phys. Rev. Lett.*, 1978, **41**, 1598.
393. P. E. Cladis, R. K. Bogardus, and D. Aadsen, *Phys. Rev. A.*, 1978, **18**, 2292.
394. N. H. Tinh, *J. Chim. Phys. Phys.-Chim. Biol.*, 1983, **80**, 83.
395. J. W. Goodby, T. M. Leslie, P. E. Cladis and P. L. Finn, in 'Liquid Crystals and Ordered Fluids', ed. A. C. Griffin and J. F. Johnson, Plenum, New York, 1984, p. 203.
396. G. Sigaud, N. H. Tinh, F. Hardouin and H. Gasparoux, *Mol. Cryst. Liq. Cryst.*, 1981, **69**, 81.
397. F. Hardouin, A. M. Levelut, M. F. Achard and G. Sigaud, *J. Chim. Phys. Phys.-Chim. Biol.*, 1983, **80**, 53.
398. F. Hardouin, *Physica A. (Amsterdam)*, 1986, **140**, 359.
399. P. E. Cladis, *Mol. Cryst. Liq. Cryst.*, 1988, **165**, 85.
400. P. Le Barny, J. C. Dubois, C. Friedrich and C. Noël, *Polym. Bull.*, 1986, **15**, 341.
401. T. I. Gubina, S. G. Kostromin, R. V. Talrose, V. P. Shibaev and N. A. Platé, *Vysokomol. Soedin., Ser. B*, 1986, **28**, 394.
402. N. Spassky, N. Lacoudre, A. Le Borgne, J. P. Vairon, C. L. Jun, C. Friedrich and C. Noël, *Makromol. Chem., Macromol. Symp.*, 1989, **24**, 271.
403. T. I. Gubina, S. Kise, S. G. Kostromin, R. V. Talrose, V. P. Shibaev and N. A. Platé, *Liq. Cryst.*, 1989, **4**, 197.
404. S. G. Kostromin, V. P. Shibaev and S. Diele, *Makromol. Chem.*, 1990, **191**, 2521.
405. C. Legrand, A. Le Borgne, C. Bunel, N. Lacoudre, P. Le Barny, N. Spassky and J. P. Vairon, *Makromol. Chem.*, 1990, **191**, 2979.
406. B. Hahn, J. H. Wendorff, M. Portugal and H. Ringsdorf, *Colloid. Polym. Sci.*, 1981, **259**, 875.
407. V. Frosini, G. Levita, D. Lupinacci and P. L. Magagnini, *Mol. Cryst. Liq. Cryst.*, 1981, **66**, 21.
408. P. L. Magagnini, *Makromol. Chem. Suppl.*, 1981, **4**, 223.
409. Y. Okamoto, T. Asakura and K. Hatada, *Chem. Lett.*, 1991, 1105.
410. V. Percec and B. Hahn, *J. Polym. Sci., Polym. Chem. Ed.*, 1989, **27**, 2367.
411. B. Hahn and V. Percec, *Mol. Cryst. Liq. Cryst.*, 1988, **157**, 125.
412. V. Percec, C. Pugh, O. Nuyken and S. D. Pask, in 'Comprehensive Polymer Science', ed. G. Allen, Pergamon Press, Oxford, 1989, vol. 6, p. 281.

413. J. Le Moigne, A. Soldera, D. Guillon, and A. Skoulios, *Liq. Cryst.*, 1989, **6**, 627.
414. X. Zhang, Y. Ozcayir, C. Feng and A. Blumstein, *Polym. Prepr., Am. Chem. Soc., Div. Polym. Chem.*, 1990, **31(1)**, 597.
415. J. Le Moigne, A. Hilberer and C. Strazielle, *Polym. Prepr., Am. Chem. Soc., Div. Polym. Chem.*, 1991, **32(3)**, 96.
416. S. H. Jin, S. H. Kim, H. N. Cho and S. K. Choi, *Macromolecules*, 1991, **24**, 6050.
417. J. M. Lehn, *Angew. Chem., Int. Ed. Engl.*, 1988, **27** 89.
418. J. M. Lehn, *Angew. Chem., Int. Ed. Engl.*, 1990, **29**, 1304.
419. D. J. Cram, *Angew. Chem., Int. Ed. Engl.*, 1988, **27**, 1009.
420. C. J. Pedersen, *Angew. Chem., Int. Ed. Engl.*, 1988, **27**, 1021.
421. J. M. G. Cowie and H. H. Wu, *Br. Polym. J.*, 1988, **20**, 515.
422. V. Percec and R. Rodenhouse, *Macromolecules*, 1989, **22**, 2043.
423. V. Percec and R. Rodenhouse, *Macromolecules*, 1989, **22**, 4408.
424. G. Ungar, V. Percec and R. Rodenhouse, *Macromolecules*, 1991, **24**, 1996.
425. V. Percec and R. Rodenhouse, *J. Polym. Sci., Polym. Chem. Ed.*, 1991, **29**, 15.
426. J. S. Wen, G. H. Hsiue and C. S. Hsu, *Makromol. Chem., Rapid Commun.*, 1990, **11**, 151.
427. R. Rodenhouse and V. Percec, *Makromol. Chem.*, 1991, **192**, 1873.
428. G. H. Hsiue, J. S. Wen and C. S. Hsu, *Makromol. Chem.*, 1991, **192**, 2243.
429. V. Percec, G. Johansson and R. Rodenhouse, *Macromolecules*, in press.
430. T. D. Shaffer and V. Percec, *J. Polym. Sci., Polym. Chem. Ed.*, 1987, **25**, 2755.
431. J. M. Rodriguez-Parada and V. Percec, *J. Polym. Sci., Polym. Chem. Ed.*, 1986, **24**, 1363.
432. C. J. Hsieh, C. S. Hsu, G. H. Hsiue and V. Percec, *J. Polym. Sci., Polym. Chem. Ed.*, 1990, **28**, 425.
433. V. Percec and D. Tomazos, to be published.
434. C. J. Hsieh, G. H. Hsiue and C. S. Hsu, *Makromol. Chem.*, 1990, **191**, 2195.
435. H. Ringsdorf and R. Wustefeld, *Philos. Trans. R. Soc. London, Ser. A.*, 1990, **330**, 95.
436. H. Ringsdorf, R. Wustefeld, E. Zerta, M. Ebert and J. H. Wendorff, *Angew. Chem., Int. Ed. Engl.*, 1989, **28**, 914.
437. M. M. Green, H. Ringsdorf, J. Wagner and R. Wustefeld, *Angew. Chem., Int. Ed. Engl.*, 1990, **29**, 1478.
438. W. Paulus, H. Ringsdorf, S. Diele and G. Palze, *Liq. Cryst.*, 1991, **9**, 807.
439. H. Bengs, R. Renkel, H. Ringsdorf, C. Baehr, M. Ebert and J. H Wendorff, *Makromol. Chem., Rapid Commun.*, 1991, **12**, 439.
440. For a discussion of self-organization in biological systems see: M. Eigen and L. DeMaeyer, *Naturwissenschaften*, 1966, **53**, 50; M. Eigen, *Naturwissenschaften*, 1971, **58**, 465.
441. For a review on the self-assembly of tobacco mosaic virus (TMV) which represents the best understood self-organized biological system see: A. Klug, *Angew. Chem., Int. Ed. Engl.*, 1983, **22**, 565.
442. For general reviews on self-assembly see: J. S. Lindsey, *New J. Chem.*, 1991, **15**, 153 and D. Philp and J. F. Stoddart, *Synlett.*, 1991, 445; G. M. Whitesides, J. P. Mathias and C. T. Seto, *Science (Washington, D.C.)*, 1991, **254**, 1312; for other representative contributions in this field see: P. L. Anelli, N. Spencer and J. F. Stoddart, *J. Am. Chem. Soc.*, 1991, **113**, 5131 and refs. cited therein; C. T. Seto and G. M. Whitesides, *J. Am. Chem. Soc.*, 1991, **113**, 712; C. T. Seto and G. M. Whitesides, *J. Am. Chem. Soc.*, 1990, **112**, 6409; J. Rebek, Jr., *Angew. Chem., Int. Ed. Engl.*, 1990, **29**, 245.
443. F. Vogtle, 'Supramolekulare Chemie', Teubner, Stuttgart, 1980.
444. M. J. Brienne, J. Gabard, J. M. Lehn and I. Stibor, *J. Chem. Soc., Chem. Commun.*, 1989, 1868.
445. C. Fouquey, J. M. Lehn and A. M. Levelut, *Adv. Mater.*, 1990, **2**, 254.
446. R. Fornasier, M. Tornatore and L. L. Chapoy, *Liq. Cryst.*, 1990, **8**, 787.
447. R. Eidenschink, *Angew. Chem., Int. Ed. Engl., Adv. Mater.*, 1989, **28**, 1424.
448. T. Kato and J. M. J. Frechet, *J. Am. Chem. Soc.*, 1989, **111**, 8533.
449. T. Kato, A. Fujishima and J. M. J. Frechet, *Chem. Lett.*, 1990, 919.
450. T. Kato and J. M. J. Frechet, *Macromolecules*, 1989, **22**, 3819.
451. V. Percec, J. Heck and G. Ungar, *Macromolecules*, 1991, **24**, 4957.

15

Rheological Behaviour of Liquid Crystalline Polymers

VALERII G. KULICHIKHIN, VALERII S. VOLKOV and NICOLAI A. PLATÉ
Russian Academy of Sciences, Moscow, Russia

15.1	INTRODUCTION	385
15.2	RHEOLOGY OF LYOTROPIC LC SYSTEMS	385
	15.2.1 Viscosity Properties	386
	15.2.2 Molecular Theories	387
	15.2.3 Viscoelastic Properties	390
	15.2.4 Transient Behaviour	391
	15.2.5 Flow Curve Shapes	393
15.3	ANISOTROPIC VISCOELASTICITY OF LC POLYMERS	395
	15.3.1 Anisotropy of Viscosity	395
	15.3.2 Transversely Isotropic Liquids	396
	15.3.3 Relaxation Anisotropy	398
	15.3.4 Nematic Viscoelastic Liquids	399
	15.3.5 Linear Anisotropic Viscoelasticity	402
15.4	REFERENCES	405

15.1 INTRODUCTION

Thirty years have passed since the appearance of Robinson's first publication[1] devoted to the study of rheological properties of liquid crystal (LC) polymer systems, *i.e.* solutions of poly(γ-benzyl-L-glutamate) (PBG). During this period of time the science on the rheological behaviour of anisotropic polymer media has passed through several stages, both revolutionary and evolutionary. Therefore, it now seems appropriate to summarize some of the results and conceptions of LC polymer rheology, which is the scope of this chapter. As distinct from other numerous reviews covering this area[2-7] the first part of it will bear retrospective and to some extent chronological character, but in the second part we will describe some new theoretical approaches to the rheology of anisotropic media.

15.2 RHEOLOGY OF LYOTROPIC LC SYSTEMS

It seems difficult to even regard the first papers published on the rheology of lyotropic LC polymers as truly rheological in the real sense of the term. Rather, they are devoted to viscosity or, to be more precise, to the concentration and temperature dependences of the same effective viscosity η which is measured for isotropic polymer solutions and melts. So, it was unexpected for researchers who are accustomed to working with ordinary solutions to see that when the LC phase appears, viscosity decreases with increasing concentration. This effect is so significant that it allows us to draw the equilibrium lines of the LC state in the phase diagram for stiff chain

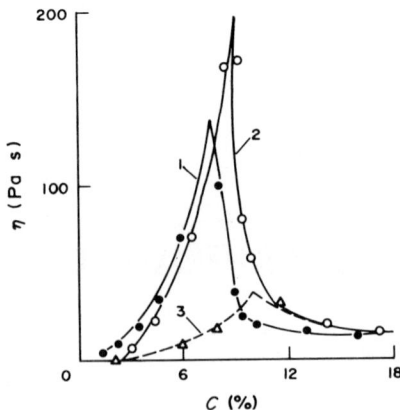

Figure 1 Viscosity *versus* concentration dependences for PBG solutions of different molecular mass,[8] which decreases from curve 1 to curve 3

polymer solutions. Such an application of the rheological data appeared to be a kind of novel approach to the description of the interrelation between properties and the phase state of polymer systems.

Here it would be appropriate to illustrate the aforementioned concentration dependence of viscosity for PBG solutions (taken from Hermans;[8] Figure 1). In this case, not only the maxima of the functions $\eta(C)$, but also their shift towards lower concentrations with increasing PBG chain length is well pronounced. It should be noted that the publications describing the first experiments made on LC solutions appeared practically at the same time as Flory's epochal theoretical work[9] predicted the principal LC phase appearance in a long, stiff rod system, uniformly distributed in athermal solution. It also affords a description of the phase diagram shape of such a system and illustrates the relationship between the critical concentration of solution transition to an anisotropic state φ^* and the rod aspect ratio $X: \varphi^* \sim 8/X$. On passing over from the rigid rods to real macromolecules it is apparent that $X \sim M$. Therefore, the dependence $\varphi^*(M)$, according to viscosity data of Hermans, who was the first to discover it, experimentally provides the prerequisite for the estimation of M using φ^* values.

15.2.1 Viscosity Properties

This prerequisite proved to be especially interesting when used with respect to a new class of lyotropic LC systems, *viz.* the solutions of aromatic polyamides with *para* structure. These systems rendered a revolutionary breakthrough in industry, presenting a new generation of super high strength and super high modulus synthetic fibres. As a reward, aromatic polyamide solutions have earned great attention from research workers from different countries, especially from the largest chemical companies. In the present chronological context it is worth mentioning the names of two outstanding women engaged in synthetic chemistry: Stephania-Loise Kwolek (du Pont, USA) and Vera Kalmykova (Research and Industrial Institute of Artificial Fibres, Russia). It was these two women who were the first to prepare anisotropic solutions of poly(*p*-benzamide) (PBA)[10,11] and subsequently other aromatic polyamides, thus placing unique solutions at the disposal of other researchers.

The concentration dependence of viscosity for PBA solutions in dimethylacetamide (DMAA) containing 3% LiCl[12] is given in Figure 2. As is seen, the aspects of 'pure viscous' data are being refined. In particular, the viscosity values of the maximum, η^*, have been determined with more accuracy, which allows us to generalize the dependence in a primitive scale version: η/η^* *versus* C/C^* (Figure 3) for PBA samples of different M. At the same time it has been shown that the exponent α of the dependence $\eta \sim M^\alpha$ for such solutions in an isotropic state can reach a value of around 8, compared with 3.4 for typical flexible macromolecules.

A further viscometric study of the lyotropic LC systems is related to the increase in the number of objects. So, side by side with PBA dissolved in an amide salt solvent appears poly(*p*-phenylene terephthalamide) (PPTA), soluble only in concentrated mineral acids. Variation of the solvent leads to a substantial change in the phase diagram, in which there now appears a crystal solvate phase

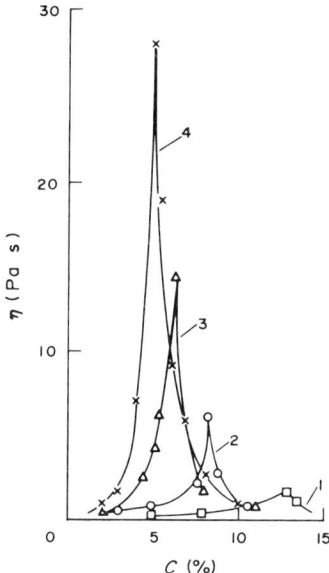

Figure 2 Influence of molecular mass on the $\eta(C)$ curves for solutions of PBA with $M = 9900$ (1), 17 500 (2), 22 200 (3) and 29 200 (4)[12]

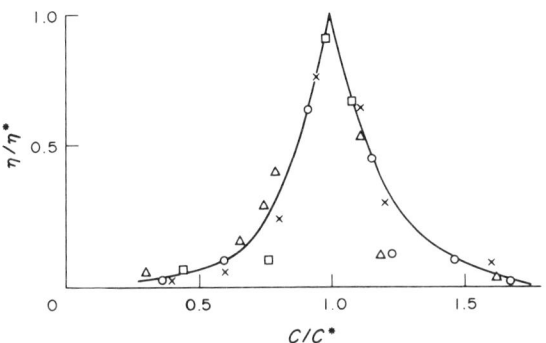

Figure 3 Reduced concentration dependence of viscosity for PBA solutions of different M (see preceding figure)

(Figure 4). This novel approach in viscometry provides a direct comparison between the concentration dependence of viscosity and the flow curve shapes at different regions of the diagram.[13] Viscosity is presented as a function of two variables, $\eta = \eta(C, T)$ (Figure 5) or $\eta = \eta(C, \sigma_{12})$ (Figure 6; σ_{12} is the shear stress), which play to some extent a practical role, as they clearly predict the region of a stable, low viscous state far from the transition temperatures.[14] The latter aspect is significant for such solutions, insomuch as commercial aromatic polyamide fibres of the Kevlar-type are prepared from PPTA solutions in H_2SO_4.

15.2.2 Molecular Theories

At present, the theoretical descriptions in Flory's style are beyond the scope of prediction of phase equilibria in solutions of extremely stiff chain polymers. Calculations are made for macromolecules involving various flexibility mechanisms, including semistiff persistent chains; this evolutionary movement towards a 'widening' of the theoretical approach allows us to get a general picture of the transition to the LC state[15,16] for a variety of model macromolecules. In addition, there are dynamic LC solution theories, thanks chiefly to the efforts of Doi and Edwards,[17] which permit the description of the rheological properties of such solutions with the help of specially developed molecular 'polymer' theory.

Let us discuss briefly the molecular approach to the dynamics of LC polymers. The phenomenological theories based on the consideration of low molecular mass liquid crystals as anisotropic

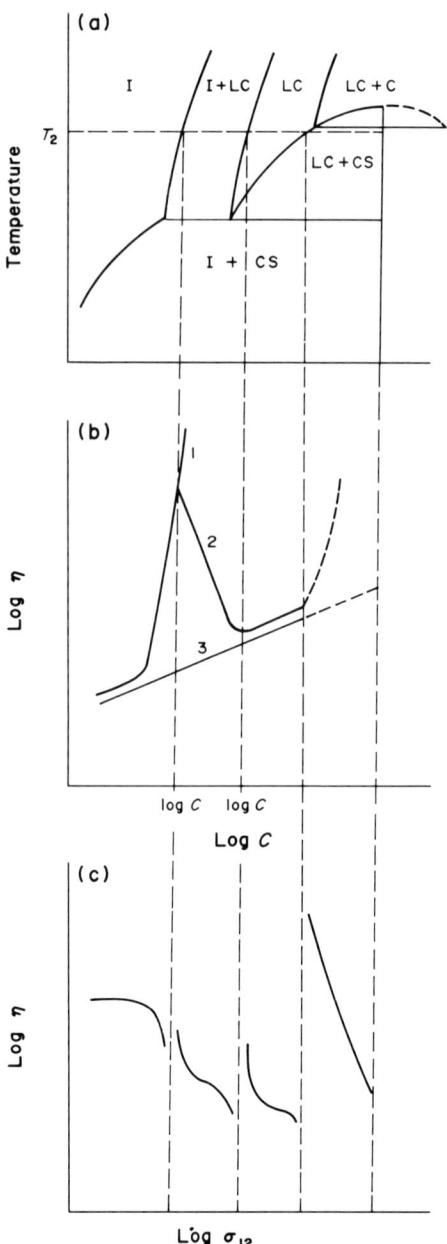

Figure 4 (a) Schematic phase diagram[13] of the aromatic polyamide–H_2SO_4 system; (b) concentration dependence of viscosity; and (c) flow curve shapes for different sections of the diagram. I—isotropic LC—liquid crystal, CS—crystal solvate and C—crystalline phases. The curve 1 relates to the viscosity at small σ_{12}; curve 2 at medium σ_{12}; and curve 3 at high σ_{12}

continuous media[18,21] failed to find the answer concerning the influence of the size, shape and interaction of LC polymer macromolecules on their symmetry and rheological properties. Doi,[19] in 1975, was the first to study the dynamic properties of LC polymers from a molecular viewpoint. Using the tube concept he later elaborated on the molecular kinetic theory of viscoelasticity for nematic solutions of ultimately stiff (rod-like) macromolecules.[20] This theory describes the properties of isotropic and anisotropic solutions. It predicts the shape of the concentration dependence of viscosity, which turns out to be coincident, in general, with the experimental results.

At small shear rates the Doi rheological equation for the anisotropic liquid reduces to a particular case of the Leslie–Ericksen phenomenological theory, in the absence of an external field[46]

$$\sigma'_{ij} = \frac{CT}{2D_r}\left[\frac{1-S}{3}\gamma_{ij} + S(\gamma_{ie}n_e n_j + \gamma_{je}n_e n_i) - 2S^2\gamma_{em}n_e n_m n_i n_j\right] \quad (1)$$

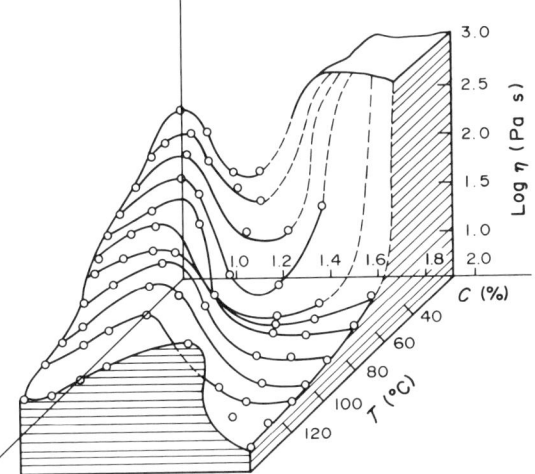

Figure 5 η–C–T relationships for PPTA solutions[14]

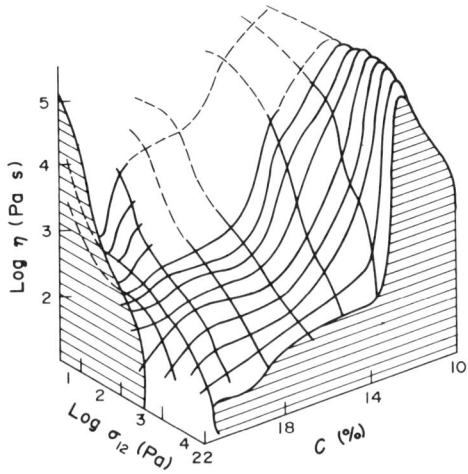

Figure 6 η–C–σ_{12} diagram for PPTA copolymer solutions at 80 °C[12]

This equation determines the nonequilibrium stress σ'_{ij} for a given velocity gradient γ_{ij}. Here C is the solution concentration, T is the temperature in energy units, D_r is the rotational diffusion constant, n_i is the director and $\gamma_{ij} = (v_{ij} + v_{ji})/2$ is the symmetric part of the velocity gradient. The equilibrium scalar order parameter S varies with the LC phase appearance as follows

$$S = \begin{bmatrix} 0 & ; \varphi < \varphi^* \\ \tfrac{1}{4} + \tfrac{3}{4}(1 - 8\varphi^*/9\varphi)^{1/2}; \varphi > \tfrac{8}{9}\varphi^* \end{bmatrix} \tag{2}$$

where φ is the solution concentration (volume fraction), φ^* is the critical concentration corresponding to the stable LC phase.

The Doi nonlinear equation for the tensor order parameter in the case of weak flows boils down to the orientation equation assuming the following form[46]

$$\dot{n}_i = \frac{1 + 2s}{3s} v_{ie} n_e + \frac{1 - s}{3s} v_{ei} n_e - \xi n_i \tag{3}$$

where s is the nonequilibrium scalar order parameter. The parameter ξ is chosen in such a way so as to satisfy the condition $\dot{n}_e n_e = 0$.

The Doi theory predicts anisotropy of viscosity and relates scalar viscosity coefficients with the order parameter. For the dependence of apparent viscosity on solution concentration the theory predicts

$$\eta/\eta^* = (\varphi/\varphi^*)^3 Z \tag{4}$$

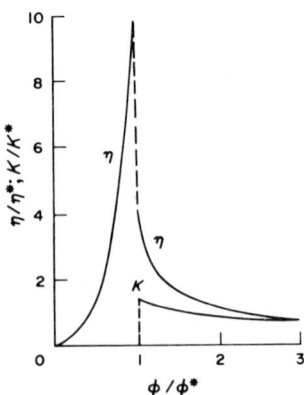

Figure 7 Reduced theoretical dependences of viscosity and coefficient of normal stresses on concentration[20]

where $Z = (1 - S)^4(1 + S)^2(1 + 2S)[(1 + 3S)/2](1 + S/2)^{-2}$. This result (Figure 7) is in accordance with the existing experimental data. Assuming that $\varphi^* \sim M^{-1}$ and $\eta^* \sim M^3/\ln M$ it is possible to write

$$\eta \sim \varphi^3 M^6 Z \tag{5}$$

Since the Z value for the isotropic solutions ($S = 0$) is equal to 1, the above-mentioned formula converts to $\eta \sim M^6 \varphi^3$, which coincides with the experimental results very well. Incidentally, Doi did not draw his attention to one experimental result in the literature,[12] namely the fact that $\eta^* \sim M^{3.2}$. It has been hypothesized that for equivalent states of PBA solutions at the φ^* point (or at an equal distance from it), the exponent value becomes just the same as the used value, 3.4. This further complicates our understanding of the rheological correspondence of equiconcentrated solutions active for flexible chain polymers.

The doubtless dignity of this theory is its prediction not only of the whole concentration dependence of viscosity but also the corresponding dependence of the first normal stress difference coefficient, K, which in the isotropic domain is proportional to the second power of the shear rate $\dot{\gamma}^2$, and in the anisotropic domain to the first power of $\dot{\gamma}$. With an increase of polymer concentration, for LC solutions the K value decreases in accordance with the equation

$$K/K^* = 3/2(\varphi/\varphi^*)^3(1 - S)^{7/2}(1 + S)^2(1 + 2S)^{1/2}[(1 + S)/2]^{-2} S \tag{6}$$

However, as distinct from the experimental data, equation (6) never corresponds to negative normal stresses.

15.2.3 Viscoelastic Properties

Other rheological characteristics of LC solutions, apart from viscosity, have attracted the attention of researchers too: normal stresses N_1, extrudate swelling, transient processes, nonstability of different kinds, *etc.* These aspects are practically all the same independent of the lyotropic LC system studied. Therefore, alongside aromatic polyamide solutions, PBG systems have also attracted the attention of scientists. In addition, novel lyotropic polymers based on cellulose derivatives have become available. Indeed, hydroxypropylcellulose, for example, is an available and inexpensive polymer, which is remarkably soluble in water to form LC solutions. This allows the use of unique precisional equipment in their study.

The most significant experimental work in the field of LC polymer melt elasticity is the investigation carried out by Kiss and Porter.[22] They have shown the presence of a shear rate region in which N_1 becomes negative (Figure 8). The occurrence of the compressive force during PBG LC solution flow in the narrow gap between the cone and the plate, according to the aforementioned authors, is due to the morphological peculiarities of the flow. Of special interest is the occurrence of banded textures, the director being oriented at a $\pm 45°$ angle toward the shear direction in the adjacent bands.

It should be pointed out that the problem concerning the change in the texture of thin layers of LC solutions under flow has long attracted the attention of researchers. The formation of banded

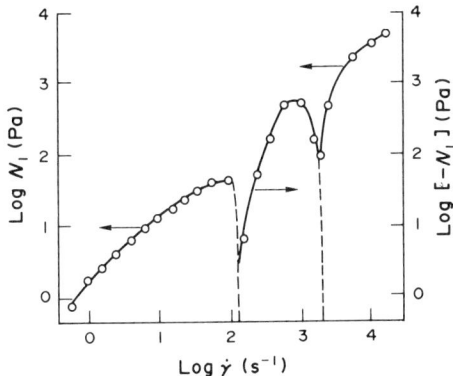

Figure 8 Dependence of the first normal stress difference on shear rate for an LC solution of PBG[22]

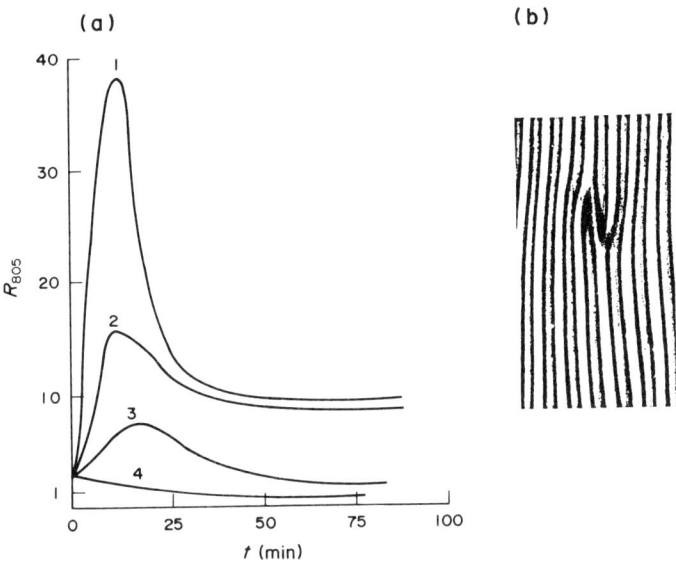

Figure 9 (a) The change in the dichroic ratio of the π-band (805 cm^{-1}) for PBA at relaxation after stopping the steady flow at $\dot{\gamma} = 1.08(1)$, 1.68(2), 16.40(3) and 96.00 s^{-1}(4). (b) The picture of the strips forming in the sample during relaxation after flow at $\dot{\gamma} < 10$ s^{-1}

texture in a weak mechanical field was first discovered in PBA solutions.[13] Of essential interest was the effect of a sharp increase in the molecular orientation angle upon relaxation preceding the stage of such macrodomain formation (Figure 9). Undoubtedly, the orientation processes are related to the LC phase elastic properties. This can be seen on studying shear at low stresses followed by their relaxation. The by-product of this phenomenon was the idea that the ideal, planar oriented texture is unstable and breaks down into a system of regular domains. This concept has since been used to explain the extrudate fibrillation effect observed for the main chain thermotropic LC polymers.[27]

More recently, research teams headed by Chapoy,[23] Ciferri,[24] Marrucci,[25] Navard[26] and others have focused their attention on the kinetic parameters and the process of banded texture formation during shear and elongation flow of LC hydroxypropylcellulose solutions. However, the real reason for the occurrence of the periodic normal orientation, besides the previously mentioned hypothesis, has not yet been found.

15.2.4 Transient Behaviour

A new approach to LC polymer rheology has been demonstrated by Moldenaers and Mewis,[28] who concentrated their efforts mainly on transient flows of anisotropic polymer solutions. By this time, lyotropic LC systems were being replaced by a new class of thermotropic LC polyesters.

Therefore rheological effects will be discussed later on for the two classes of polymeric LC systems, though some specificity remains for the solutions as two-component compatible mixtures. However, this specificity does not influence the qualitative picture of the main rheological properties. As to transient processes, it has been shown that for certain LC solutions[7] and melts[29,30] there exists a region of shear rates $\Delta\dot{\gamma}^*$ with the unusual development of time-dependent stresses (Figure 10). At $\dot{\gamma} < \Delta\dot{\gamma}^*$ we can see a typical (for polymers) curve $\sigma_{12}(t)$. At $\dot{\gamma} > \Delta\dot{\gamma}^*$, the maximum related to the polymer structure strength (yield value) disappears. When loading the LC sample it seems possible to observe a two-stage increase in the tangential stresses inside the $\Delta\dot{\gamma}^*$ region. Figure 10 represents the scheme of the stress evolution. The real picture may be more complicated, as is seen from the time-dependent torque development curves for melts of polyesters based on poly(ethylene terephthalate) and p-hydroxybenzoic acid, as obtained by Baird's group (Figure 11).[31] The process of a steady-state flow becomes exceedingly complicated with increasing shear rates, though one may find a tendency towards the appearance of a periodic variation of the function $\sigma_{12}(t)$.

Baird ascribed the first peak to the energy losses in the orientation of the ordering aggregates located in a lower ordering matrix, whilst the second peak corresponds to the straining of the matrix. As a matter of fact, an Italian group headed by Nicolais[32] subsequently arrived at the two-phase flow hypothesis even for the steady-state case.

Mewis and Moldenaers have studied in detail the above processes and have succeeded in advancing the understanding of transient phenomena in the LC systems. First of all, using PBG and hydroxypropylcellulose solutions, they duplicated the results obtained by Kiss and Porter, thus showing that there exists a region $\Delta\dot{\gamma}^*$ in the steady-state shear flow, in which the first normal stress difference is negative.[33] Further, it appeared that the development of the tangential stresses was of an oscillatory character mainly in this region (Figure 12). Moreover, close analysis of the stress relaxation curves allowed them to reveal two novel effects. The first one is that the stresses relax much faster than the dynamic storage modulus;[34] the second, that there is a two-step relaxation. The first step, proceeding fast, is responsible for the viscous energy dissipation; the second step is a cyclic one with a period close to that of stress oscillations in a regime of sample loading.

The first explanations of these effects have been based on the peculiarities of the stream morphology, in particular on the undefined role of the domain structure. Interpretation of these

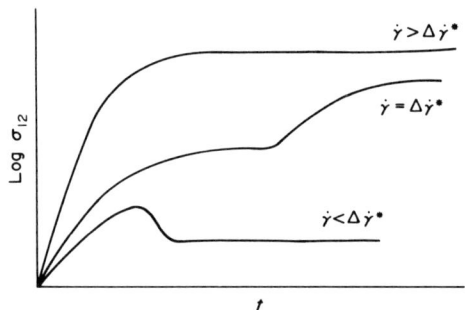

Figure 10 The shear stress development at different shear rates

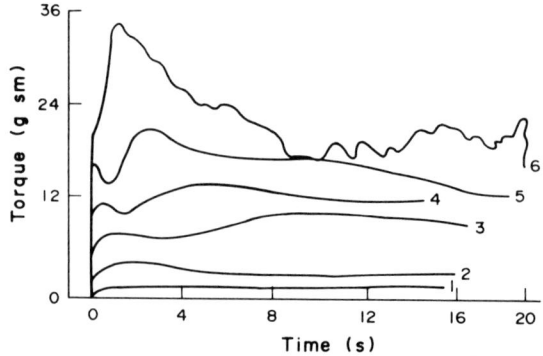

Figure 11 The change in the torque–time curve shapes for a thermotropic LC polyester at different $\dot{\gamma}$ ($\dot{\gamma}$ increases from curve 1 to curve 6)[31]

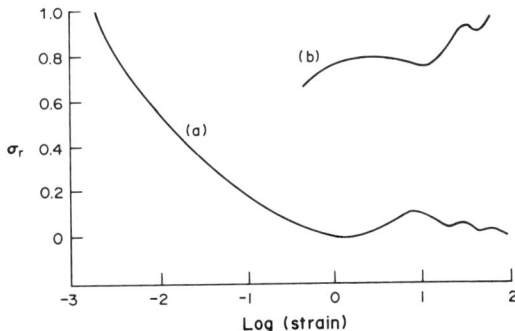

Figure 12 (a) Reduced shear stress relaxation curve at a final shear rate of 0.1 s^{-1}; and (b) shear stress development curve after a stepwise increase in shear rate[34]

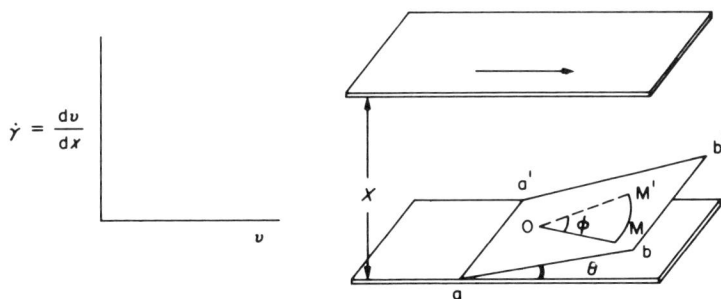

Figure 13 The location of vector OM at steady shear flow. φ is the orientation angle, θ is the incline angle of the aa'b'b plane with respect to the lower measuring plate

effects received a new impulse when the theory developed by Marrucci[35,36] arose, which predicted the existence of the $\Delta \dot{\gamma}^*$ region wherein the director loses its stability (tumbling regime). This implies that it can rotate in the stream, giving rise to an unpredicted rheological reaction particularly of negative N_1 magnitudes. In the steady-state behaviour region, the director is inclined at a small angle θ towards the shear plane (Figure 13), being equal to 2–4° for PBA solutions.[37] In a common case[38] $tg^2\theta = \alpha_3/\alpha_2$, where α_3 and α_2 are two of the Leslie anisotropic viscosity coefficients.

Earlier Semenov[39] showed that for stiff rods, ratios of α_3/α_2 can be negative and this produces indefiniteness of the θ angle with time. But in the presence of even a slightly flexible component of persistent character, in the macromolecule α_3/α_2 becomes positive; so, for such systems a steady-state flow is possible at all $\dot{\gamma}$ values.

15.2.5 Flow Curve Shapes

Let us consider the shape of the LC polymer flow curves determined by the conventional rotational or capillary techniques. As far back as 1972 evidence of a sharp viscosity decrease with increasing shear stress was observed in the region of small values of σ_{12} for anisotropic PBA solutions.[40] Further, flow curves in a wide range of $\dot{\gamma}$ were obtained and interpreted as consisting of three sections (Figure 14).[12] Of great importance for the development of the LC polymer flow mechanism in the light of the flow curve shape were the works of Onogi and Asada.[4,41]

Following the opinion of many scientists, region I was attributed to the polydomain flow of the LC polymer with different director orientations of the individual domains. Worthy of a more detailed consideration is the existence and destruction under flow of networks formed by disclinations, which play the role of topological restrictions in a uniform plastic deformation region.[7,42] For such a dynamic texture there may exist a yield stress, though we do not exclude the occurrence of a Newtonian flow at very small rates (region I*), which is identified as a creep (slow irreversible displacement in a stream of disclinations and their knots, without noticeable loss of their mutual locations). The Wissbrun liquid crystal domain model[3,43] can be applied very successfully to these flow curve sections; the domain in this case may represent not a uniformly oriented microvolume,

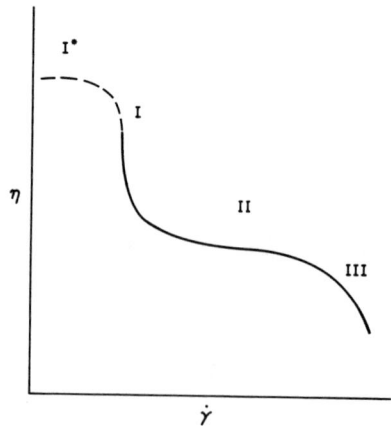

Figure 14 Schematic flow curve of LC polymers.

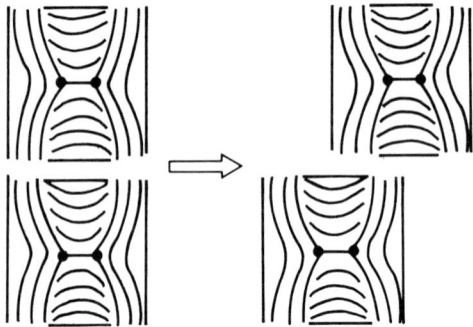

Figure 15 Domain flow of LC polymers[43]

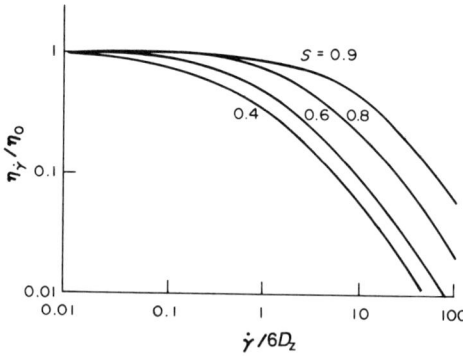

Figure 16 The theoretical predictions of the Doi theory[46] concerning the viscosity anomaly of stiff chain LC polymers

restricted by the surface disclinations, but the part of the LC sample with a system of disclinations without their advantageous orientation (Figure 15). The domain theory has also been elaborated in the works of Marrucci[44] and Larson.[45]

Section II is interpreted as a quasilinear region of monodomain flow (after completion of orientation and disappearance of discrete domains). The viscosity anomaly at high shear rates (section III) is of a relaxational character and, as will be shown below, it is due to the anisotropic nonlinear relaxation processes occurring in the LC polymers. Apropos, the Doi theory[46] describes the effect of the viscosity anomaly increasing with a decrease in the macroscopic parameter of order (Figure 16).

In general, the relation between the morphology and the flow curve shape can be represented as is indicated in Figure 17. The principal conclusion consists in the fact that in an ideally oriented

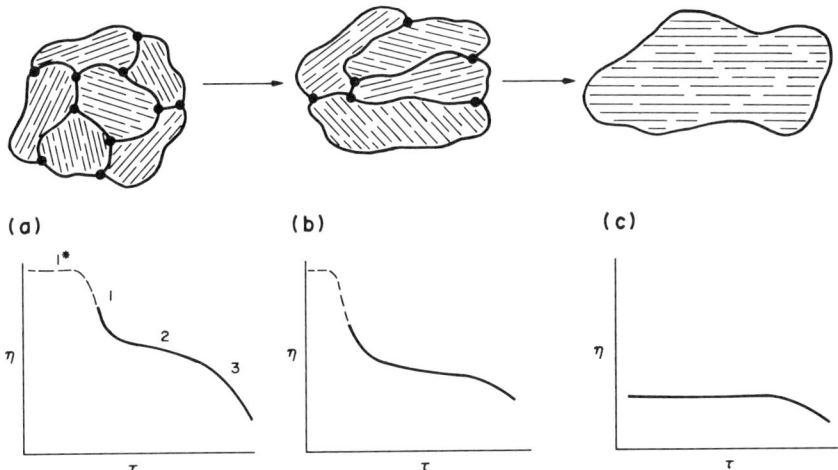

Figure 17 Orientation–flow relationships at different shear rates, increasing from (a) to (c), for LC polymers

case the sections I* and I may be absent. Thus, we approach the model of the oriented system for which recently the principles of a novel anisotropic viscoelastic theory has been developed.

15.3 ANISOTROPIC VISCOELASTICITY OF LC POLYMERS

The dual nature of LC polymers (high molar mass compounds, on one hand, and liquid crystals, on the other) leads to the fact that from the macroscopic viewpoint they represent anisotropic viscoelastic liquids which are capable of storing elastic strains and relaxing the stresses anisotropically. As discussed above, for investigations of the rheological properties of LC polymers the simple rheometric flows, such as in the case of isotropic polymers, are used. In so doing, measurements are made only of the effective rheological characteristics of the LC media from which, as will be shown later, one can obtain information about their rheological anisotropy.

15.3.1 Anisotropy of Viscosity

In general, the LC polymer viscosity should be characterized by the fourth order tensor η_{ijkl}. This tensor determines the linear connection between the stress field σ'_{ij} occurring under flow and the strain rate field γ_{kl}

$$\sigma'_{ij} = \eta_{ijk\ell} \gamma_{kl} \tag{7}$$

Equation (7) is a generalized Newton law. Clearly, the viscosity tensor can be represented in the form of the sixth order matrix

$$[\eta_{ijk\ell}] = \begin{vmatrix} \eta_{1111} & \eta_{1122} & \eta_{1133} & \eta_{1123} & \eta_{1131} & \eta_{1112} \\ \eta_{2211} & \eta_{2222} & \eta_{2233} & \eta_{2223} & \eta_{2231} & \eta_{2212} \\ \eta_{3311} & \eta_{3322} & \eta_{3333} & \eta_{3323} & \eta_{3331} & \eta_{3312} \\ \eta_{2311} & \eta_{2322} & \eta_{2333} & \eta_{2323} & \eta_{2331} & \eta_{2312} \\ \eta_{3111} & \eta_{3122} & \eta_{3133} & \eta_{3123} & \eta_{3131} & \eta_{3112} \\ \eta_{1211} & \eta_{1222} & \eta_{1233} & \eta_{1223} & \eta_{1231} & \eta_{1212} \end{vmatrix} \tag{8}$$

Because of the symmetry of the stress and strain rate tensors, the number of different viscosity coefficients does not exceed 36. Their further reduction may be possible owing to the symmetry of the liquid characterizing the anisotropy of a particular LC polymer. The situation is improved when considering the simplest rheometric flow—the Couette linear flow determined by the rate field in the form

$$v_1 = \dot{\gamma} x_2, \quad v_2 = 0, \quad v_3 = 0 \tag{9}$$

where $\dot{\gamma}$ is the shear rate. In this case the stressed state of the anisotropic liquid (equation 7) is determined, to a full measure, by the simple viscosity tensor

$$[\eta_{ijke}] = \begin{vmatrix} \eta_{1112} & \eta_{1212} & 0 \\ \eta_{2112} & \eta_{2212} & 0 \\ 0 & 0 & \eta_{3312} \end{vmatrix} \quad (10)$$

Hence it follows that, by measuring the apparent viscosity $\eta = \sigma_{12}/\dot{\gamma} = \eta_{1212}$, we actually obtain information concerning only the viscosity tensor component η_{1212}. Consequently, in LC polymer rheology the rather complicated problem of measuring the viscosity tensor arises. Such a task practically does not exist for the isotropic (flexible chain) polymers, which are characterized by the same set of properties in all directions. For incompressible isotropic liquids, the viscosity tensor has the following form

$$\eta_{ijke} = 2\eta I_{ijke} \quad (11)$$

where $I_{ijke} = (\delta_{ik}\delta_{je} + \delta_{ie}\delta_{jk})/2$ is the fourth order unit tensor. The viscosity tensor for such liquids is determined only by one of the scalar viscosity coefficients η. The type of viscosity tensor for anisotropic liquids possessing a more complicated symmetry can be determined on the basis of the general invariant form of the connection between the tensorial field characterizing their motion and physical and geometrical properties. As a result, the viscosity tensor for anisotropic liquids can be expressed through a unit tensor of the second rank δ_{ik} and internal tensor parameters characterizing the microstructure of the LC polymers.

15.3.2 Transversely Isotropic Liquids

The simplest anisotropic liquids have uniaxial anisotropy. Upon rotation to an arbitrary angle relative to a nonpolar axis of a preferred orientation and at any reflection relative to the plane containing this axis, the properties of such an anisotropic liquid will remain constant. The viscosity tensor for the simple uniaxial anisotropic liquid is determined by two scalar viscosity coefficients and is expressed by the form[52]

$$\eta_{ijkl} = \eta_\perp I_{ijkl} + (\eta_\perp - \eta_\parallel) N_{ijkl} \quad (12)$$

In this case the fourth rank tensor

$$N_{ijke} = 2(n_{ijke} - n_{(i}\delta_{j)(k}n_{e)}) \quad (13)$$

has particular properties: $N_{ijke}N_{kemn} = -N_{ijmn}$. Here we introduce the following notation[52]

$$n_{ijkl} = n_i n_j n_k n_l \quad (14)$$

$$4n_{(i}\delta_{j)(k}n_{l)} = n_i \delta_{jk} n_l + n_j \delta_{ik} n_l + n_i \delta_{jl} n_k + n_j \delta_{il} n_k \quad (15)$$

The symmetrization is made from the indices given in the parentheses.

The orientation of liquid particles with uniaxial anisotropy is characterized by one single vector, n_i (the director). It is determined from an additional equation describing the changes in its orientation induced by the flow

$$\rho \ddot{n}_i = g_i(\rho, n_i, \dot{n}_i, \gamma_{ij}) \quad (16)$$

where ρ is the density. The dot denotes the value of the material derivative. The director orientation equation reflects the unique properties of the anisotropic orientable liquids. It has no counterparts in the case of isotropic liquids. From the physical point of view the director, n_i, indicates the direction of the average molecular orientation.

The anisotropy of the uniaxial liquid is defined with respect to a single preferred direction n_i and consequently, in some sense, relative to an actual state of special physical importance. With allowance for this situation the longitudinal $\eta_\parallel$ and transversal $\eta_\perp$ (with respect to the director) viscosities are introduced. These viscosity coefficients are of basic importance. They allow us to express viscosities with respect to any other directions. The main viscosities, by their physical essence, characterize different internal resistance which the uniaxial anisotropic liquid experiences when shear is applied along and across the director (Figure 18). An important feature of the introduced viscosities is the fact that they are always positive. This follows from the positive definition of the quadratic form determining the dissipation energy. Viscosities $\eta_\parallel$ and $\eta_\perp$ are independent of the orientation of the director and the shear rate. They characterize only the properties of the liquid and are directly connected with its molecular structure. These viscosities are invariant toward the

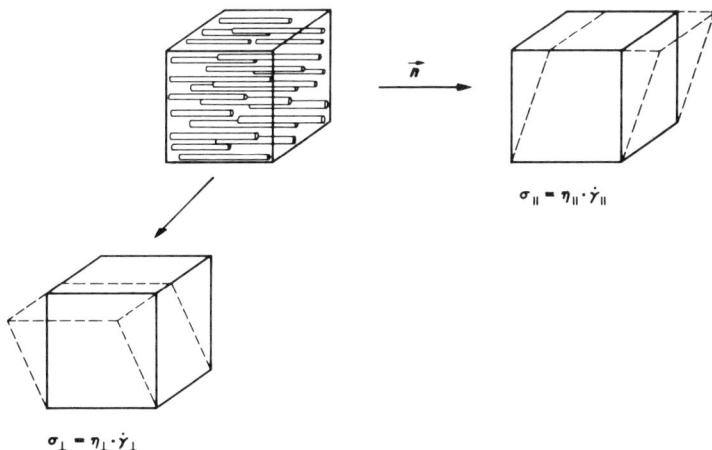

Figure 18 The principal sense of the basal anisotropic viscosity coefficients

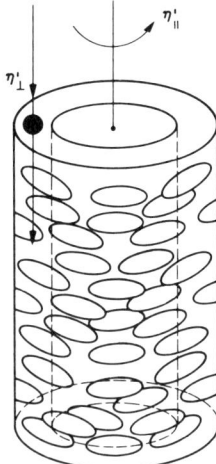

Figure 19 Sketch of apparatus used for the study of viscosity anisotropy[37]

ways and details of measurements, in accordance with the requirements for determining the true viscosity coefficients.

For anisotropic liquids with a lower degree of symmetry the viscosity tensor is more complicated than that in equation (12). The LC polymer symmetry determines the required number of the experimentally measured viscosity coefficients. These coefficient values directly characterize the viscous properties of a particular LC polymer.

A small modification in the rotational couple of an ordinary cylinder–cylinder viscometer renders it able to register the existence of viscosity anisotropy for LC solutions.[37] The modification consists of using a transparent, externally fixed cylinder, the presence of which, alongside with the measurement of the torque moment, allows the determination of the time taken for a hard, spherical ball to fall into the gap under the shear of the anisotropic solution (Figure 19).

To make a quantitative estimation of the viscosity anisotropy of nematic polymers one should use the resistance law of the ball moving in a uniaxial anisotropic liquid. For a spherical particle with radius d moving translationally at constant velocity v_i in the anisotropic viscous liquid having one preferred direction n_i, the resistance force, F_i, is determined in the following way[47]

$$F_i = -\zeta_{ik} v_k \tag{17}$$

$$\zeta_{ik} = \zeta_\perp (\delta_{ik} - n_i n_k) + \zeta_\parallel n_i n_k \tag{18}$$

where $\zeta_\perp$ and $\zeta_\parallel$ are the transverse and longitudinal (with respect to the director n_i) resistance coefficients. In an anisotropic liquid, the sphere resistance force depends on the direction of the motion of the sphere, *i.e.* we observe a mobility anisotropy of the isotropic in the body of the shape. When

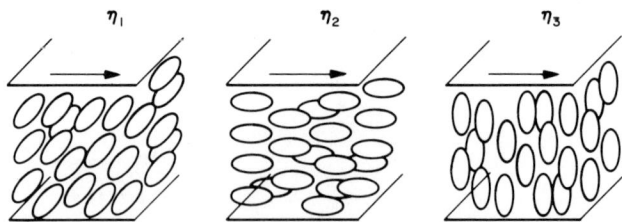

Figure 20 Representation of the Miesowicz coefficients of LC sample viscosity[48]

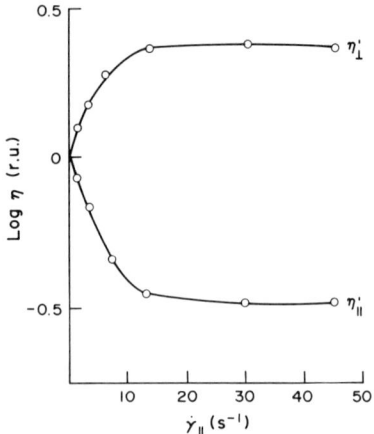

Figure 21 The difference between longitudinal $\eta'_\parallel$ and transversal $\eta'_\perp$ coefficients of viscosity, measured by the method illustrated in Figure 19, for LC PBA solutions[37]

the spherical particle moves parallel or perpendicular to n_i, the Stokes resistance law is valid

$$F_i^\parallel = -\zeta_\parallel v_i, \quad \zeta_\parallel = 6\pi d\eta_\parallel \qquad (19)$$

$$F_i^\perp = -\zeta_\perp v_i, \quad \zeta_\perp = 6\pi d\eta_\perp \qquad (20)$$

Here, $\eta_\parallel$ and $\eta_\perp$ are the longitudinal and transverse viscosity coefficients. These coefficients are related to the Ericksen–Leslie viscosity coefficients and are close to the Miesowicz viscosity coefficients η_1 and η_2 (the scheme describing the meaning of the Miesowicz viscosity coefficients is given in Figure 20).[48]

A rough estimation of the $\eta'_\perp$ value using a ball as a viscometer has been made for LC PBA solutions.[37] With increasing shear rate $\dot\gamma$ the measured value $\eta'_\perp(\eta_\parallel, \eta_\perp, n_i)$ rises achieving the constant value 1.2 Pa s at $\dot\gamma \sim 10\,\text{s}^{-1}$ (Figure 21). At this shear rate it is likely that the equilibrium value of the director orientation n_j is achieved; however, its uncertain final value, as well as the complicated stress fields created under ball motion in a viscoelastic anisotropic medium, do not prove the equivalence of $\eta'_\perp$ and $\eta_\perp$. The anisotropy viscosity effect is most pronounced when one compares $\eta'_\perp$ with the viscosity $\eta'_\parallel$ measured as usual with the rotational viscometer method, i.e. using a torque value. The dependence $\eta'_\parallel(\dot\gamma)$ is a mirror reflection of the $\eta'_\perp(\dot\gamma)$ function, but the above-mentioned considerations about the possible distinction of the angle ϕ from 0 require us to use for these terms, the special notation $\eta'_\perp$. For isotropic PBA solutions the longitudinal and transversal viscosity components are equal. The viscosity coefficient proportional to the Miesowicz viscosity coefficient η_3 has been estimated using the initial resistance to deformation of PBA solutions of homeotropic texture. This coefficient value is higher than 2.5 Pa s.[37] The difference in the Miesowicz viscosities η_1 and η_3 appears only in the presence of hard measuring surfaces with a narrow gap between them.

15.3.3 Relaxation Anisotropy

The appearance of LC polymers has aroused interest in the general problems in the theory of anisotropic viscoelastic liquids. In recent years there has been a growing necessity to study models of liquids taking into account the features of anisotropy connected with the polymer specificity of these media. The principal purpose of investigations into this area is to study the rheological behaviour regularities of a sufficiently broad class of anisotropic viscoelastic liquids in order to establish the dependence of rheological properties on the values of the groups of parameters determining the viscoelastic anisotropy. Without this it is unthinkable to use the models of anisotropic liquids for interpreting the experimental data on LC polymer rheology.

The simplest invariant theory of an anisotropic liquid with anisotropic viscosity was formulated by Ericksen in 1960. Based on Ericksen's theory of transversely isotropic liquids[21] and the Oseen–Frank LC elasticity theory, Leslie developed the theory of dynamic properties of nematic liquid crystals.[48,49] The Leslie–Ericksen continuum theory describes the main features of the flow of nematic, low molecular mass liquid crystals. The application of this theory for LC polymers is restricted to the region of low shear rates. The nonlinear effects observed in the flow of the LC polymers at high shear rates are not described by the Leslie–Ericksen theory. In a recent work Volkov and Kulichikhin formulated a simple rheological equation for monodisperse LC polymers[50,51]

$$\sigma_{ij} = \sigma_{ij}^0 + \sigma'_{ij} \tag{21}$$

$$\tau_{ijkl} \frac{\Delta}{\Delta t} \sigma'_{kl} + \sigma'_{ij} = \eta_{ijkl} \gamma_{kl} \tag{22}$$

where σ_{ij}^0 and σ'_{ij} are the equilibrium and nonequilibrium parts of the stress tensor, $\Delta/\Delta t$ is the invariant time derivative and γ_{kl} is the strain rate tensor. LC polymers in the region of slow relaxation are considered as anisotropic viscoelastic liquids with a tensor viscosity η_{ijkl} and tensor relaxation time τ_{ijkl}.

A distinct feature of the theory under consideration is the fact that the description of the anisotropy of the rheological properties of LC polymers is made in terms of their viscoelastic behaviour. The rheological equation (21) is based on the natural assumption that LC polymers as well as any real liquids are subjected not only to viscous strains but also to elastic ones. It describes the anisotropic viscoelastic liquids in which the elastic strains are small in comparison with the overall deformations. Such a case is realized in weakly elastic liquids to which, apparently, the monodisperse LC polymers should be assigned. Anisotropy of the elastic properties of LC polymers results naturally in the anisotropy of the relaxation times which is dependent on the direction of the measurements. The viscosity tensor is associated with the relaxation time tensor by the expression

$$\eta_{ijkl} = G_{ijmn} \tau_{mnkl} \tag{23}$$

where G_{ijmn} is the elastic coefficient tensor.

Thus the specific features of the LC polymer structure influence the character of the relaxation processes—they become anisotropic. In this connection, there arises a fairly difficult experimental and theoretical problem, concerning the study of the tensor relaxation processes occurring in LC polymers subjected to various mechanical actions.

In the above-mentioned work[51] consideration is restricted to the case when the Oseen–Frank LC elasticity can be neglected. It is assumed that the flow destroys the defects and, as a result of this, the initial polydomain texture is transformed into a monodomain anisotropic system. It should be noted that LC polymers are very viscous liquids if we compare them with low molecular mass liquid crystals. Therefore, as is mentioned in the literature,[52] the contribution of the Oseen–Frank elasticity of LC polymers is small in relation to the viscous stresses and it can be neglected in a flow analysis.

15.3.4 Nematic Viscoelastic Liquids

The first step in the investigation of nematic LC polymers is the study of viscoelastic liquids having one director. The existence of a single preferred direction is the principal distinguishing feature of these polymers. The distortion of the uniaxial symmetry by the flow of these media as a rule is small in the low velocity gradient region. In the case of quasilinear dynamics of nematic polymers it is natural to suppose that the influence of the flow field on the axial symmetry is also negligibly small. Taking into account the concrete kind of viscosity and relaxation time tensors, possessing uniaxial anisotropy, from equation (21) we can obtain the rheological equation for a corotational nematic liquid[51]

$$\tau_\perp \frac{D\sigma'_{ij}}{Dt} + (\tau_\perp - \tau_\parallel) \frac{D_n \sigma'_{ij}}{Dt} + \sigma'_{ij} = 2\eta_\perp \gamma_{ij} + 2(\eta_\perp - \eta_\parallel) \gamma_{ij}^n \tag{24}$$

The director $n_i (n_i n_i = 1)$ appearing in the rheological equation for a nematic liquid (equation 24) in a simple case is obtained from an additional Ericksen orientation equation[19]

$$\frac{D n_i}{Dt} = \lambda(\gamma_{il} n_l - \gamma_{lm} n_l m_i) \qquad (25)$$

where λ is the material parameter related to the final orientation of the director in a steady shear flow. Equations (24) and (25) include the following notation

$$\gamma^n_{ij} = 2 n_{ijlm} \gamma_{lm} - n_{il} \gamma_{lj} - n_{jl} \gamma_{li} \qquad (26)$$

The rheological equations (24 and 25) contain the anisotropic combination of Jaumann's derivatives D/Dt

$$\frac{D_n \sigma'_{ij}}{Dt} = 2 n_{ijlm} \frac{D \sigma'_{lm}}{Dt} - n_{il} \frac{D \sigma'_{lj}}{Dt} - n_{jl} \frac{D \sigma'_{li}}{Dt} \qquad (27)$$

The properties of a uniaxial liquid (equations 24 and 25) are different along and across the director. At any plane perpendicular to the director (plane of isotropy), the properties are independent of the direction of measurement. An anisotropic viscoelastic liquid (equations 24 and 25) is characterized by two independent relaxation times: the longitudinal $\tau_\parallel$ and the transversal $\tau_\perp$ (with respect to the director) relaxation times have a stringent physical sense, as they can be directly associated with the molecular structure. The degree of viscoelastic anisotropy of the liquid (equation 24) is defined by two dimensionless parameters

$$\beta = \eta_\perp / \eta_\parallel, \quad \alpha = \tau_\perp / \tau_\parallel \qquad (28)$$

which depend essentially on the geometric form of the molecules and the character of the intermolecular interactions.

The behaviour of any material is manifested most fully by its constitutive rheological equation, which relates the stress to the kinematic variables and, in the final run, to the rates. On the basis of the rheological equations (24 and 25) we can analyze the anisotropy of the relaxation processes taking place in the nematic polymers subjected to different mechanical actions. The most urgent problem is to elucidate the nature of the nonlinear relaxation phenomena in the LC polymers. The nonlinearity of the rheological behaviour is of general significance in physics since it is associated with the fundamental peculiarities of the structure of matter. In addition, the behaviour of the LC polymers in real situations, for instance, during operation or processing, is always nonlinear, when the deformation takes place at high stresses. Therefore the quantitative description of the nonlinear effects and the elucidation of their physical mechanism are of essential significance. Let us now consider the simple characteristics of nonlinear viscoelasticity of LC polymers—viscometric functions. A distinguishing feature of LC polymers as anisotropic media is the fact that they are capable of reaching a steady shear flow.

The equations (24) and (25) shown above lead to a nonlinear dependence of the apparent viscosity $\eta = \sigma_{12}/\dot\gamma$ and the first normal stress difference $N_1 = \sigma_{11} - \sigma_{22}$, determined in a usual rheological experiment, on shear rate in a simple shear flow

$$\eta = \frac{\eta_0 + G_p \tau_\perp \tau_\parallel \dot\gamma}{1 + \tau_\perp \tau_\parallel \dot\gamma^2}, \quad N_1 = \frac{\eta_1 + 2 G_1 \tau_\parallel \tau_\perp \dot\gamma}{1 + \tau_\parallel \tau_\perp \dot\gamma^2} \dot\gamma \qquad (29)$$

The apparent viscosity depends significantly on the angle φ between the direction of flow and the director; at low shear rates it is constant (plateau region II, Figure 14) and is determined by the relationship

$$\eta_0 = \eta_\perp \sin^2 2\varphi + \eta_\parallel \cos^2 2\varphi \qquad (30)$$

The viscosity starts to decrease at a shear rate of the order of $1/(\tau_\perp \tau_\parallel)^{1/2}$. Such a viscosity anomaly is due to the relaxation properties of the system and, in contrast to the isotropic liquid, is characterized by two relaxation times even for monodisperse media. The following notation is used in equation (29)

$$G_p = \tfrac{1}{2}(G_\parallel - G_\perp) \sin 4\varphi \qquad (31)$$

$$G_1 = G_\perp \sin^2 2\varphi + G_\parallel \cos^2 2\varphi \qquad (32)$$

where $G'_\parallel = \eta_\parallel / \tau_\parallel$, $G_\perp = \eta_\perp / \tau_\perp$.

One of the main features of anisotropic liquids is the fact that at low shear rates, the first normal stress difference distinguishes from zero and is proportional to the shear rate

$$N_1^0 = \eta_1 \dot{\gamma} \tag{33}$$

where

$$\eta_1 = (\eta_\perp - \eta_\parallel) \sin 4\varphi \tag{34}$$

This result is consistent with the experimental observation for LC systems. According to equation (31), the first normal stress difference at low shear rates is negative for an anisotropic liquid with $\beta < 1$. Such an unusual rheological feature of the anisotropic liquids is found in some LC melts and also in anisotropic polymer solutions.

Figure 22 presents the dimensionless dependences of the shear stress and the first normal stress difference on the shear rate, for different values of anisotropy parameters and a particular director orientation with $\varphi = \pi/8$. The shape and mutual position of the viscometric curves depend significantly on the degree of anisotropy of the viscous and relaxation properties of the liquids. For media with well-pronounced viscosity anisotropy ($\beta > 3$) the first normal stress difference is larger than the shear stress at all shear rates. For anisotropic liquids at $\beta < 3$, the viscometric curve intersect remains parallel in the low shear rates region. Especially important is the case with $\beta = 3$. It indicates the existence of anisotropic liquids whose first normal stress difference coincides in magnitude with the tangential stresses in the low shear rates region.

The experimental data obtained for the comb-shaped LC polymer melts show that a rise in the temperature leads to a change in the locations of the dependences of the tangential and normal stresses on shear rate (Figure 23). Typically, at low temperatures, the curve $N_1(\dot{\gamma})$ is located above

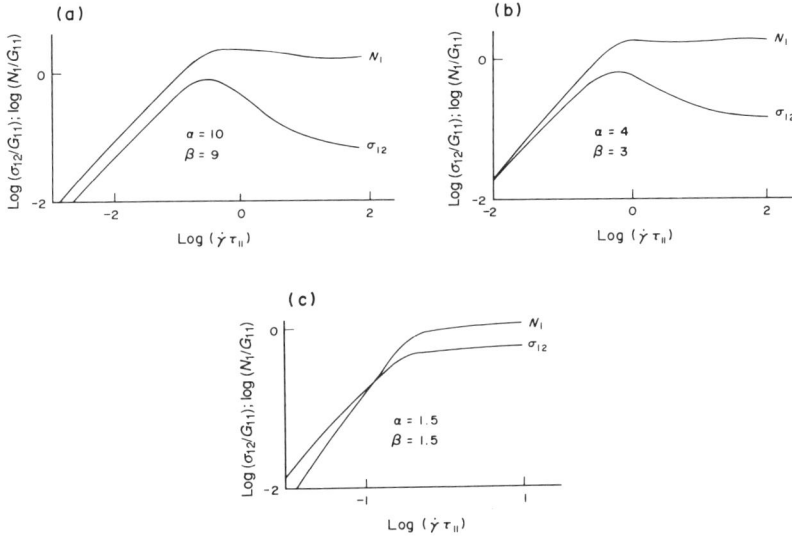

Figure 22 The theoretical predictions about the shapes and mutual locations of the $\sigma_{12}(\dot{\gamma}\tau_\parallel)$ and $N_1(\dot{\gamma}\tau_\parallel)$ curves at: (a) $\alpha = 10$, $\beta = 9$; (b) $\alpha = 4$, $\beta = 3$; and (c) $\alpha = 15$, $\beta = 1.5$[52]

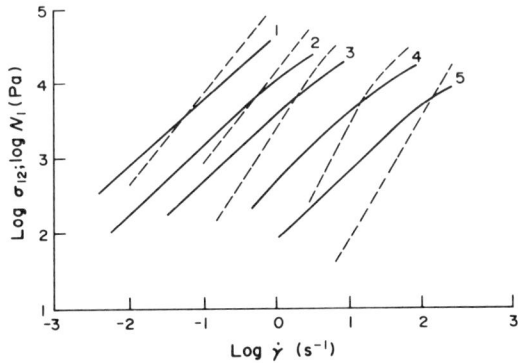

Figure 23 The experimentally determined curves for $\sigma_{12}(\dot{\gamma})$ (dotted lines) and $N_1(\dot{\gamma})$ (solid lines) for comb-shaped LC polyacrylate at temperatures of 90 °C (1), 100 °C (2), 110 °C (3), 130 °C (4) and 160 °C (5). The clearing temperature is 122 °C[57]

the curve $\sigma_{12}(\dot{\gamma})$ in a wide shear rate interval, i.e. the viscosity anisotropy for such systems is substantial. By raising the temperature, the nematic melts become less anisotropic, i.e. the viscous response approaches the elastic one and can even exceed it.

In all cases, with a rise in shear rate the effects associated with the relaxation elasticity of the anisotropic liquid significantly exceed those for the viscous response. In these conditions, when the critical elastic strains are accumulated, the liquid can be destroyed as an elastic anisotropic solid. The effect of the loss of fluidity for polymer liquids depends to a great extent on their polydispersity—it is most pronounced in quasimonodisperse polymers. The elementary anisotropic viscoelastic liquid presented here shows the instability of the shear flow. This conclusion follows from the nonlinear character of the viscometric curves.

15.3.5 Linear Anisotropic Viscoelasticity

Starting from the presentation of nonequilibrium statistical mechanics it may be shown that the linear viscoelastic behaviour of liquids with an arbitrary anisotropy can be characterized by the constitutive rheological equation, where p is the isotropic pressure

$$\sigma_{ij} + p\delta_{ij} = 2 \int_{-\infty}^{t} G_{ijke}(t-s)\gamma_{ke}(s)\,ds \tag{35}$$

This equation is the starting point in the study of the linear viscoelasticity of LC polymers. It is valid only within small strains and their rates. From the phenomenological viewpoint, equation (35) represents the most common form of linear viscoelastic relationship between the stresses and strain rates. The components $G_{ijkl}(t)$ are the relaxation functions of the liquid. They form the fourth rank tensor and depend on the relaxation properties of a particular liquid. The tensorial relaxation functions possess a number of characteristic properties, determined by symmetry conditions.

The introduction of the general dependences (equation 35) does not provide the particular form of the relaxation functions entering into them. It appears not to be a simple task to determine them for different LC polymers. This problem is the main purpose of the experimental investigations, as well as of the molecular kinetic theories. The rheological information obtained in this way enables us to get a deeper insight into the LC polymer structure. In many cases it is desirable to have analytical conceptions of the viscoelastic characteristics of $G_{ijkl}(t)$. In the case of isotropic incompressible viscoelastic liquids, the tensor of the relaxation functions assumes the following form

$$G_{ijke}(t) = G(t)I_{ijke}, \quad G(t) = \sum_{\alpha=1}^{N} \frac{\eta_\alpha}{\tau_\alpha}\exp(-t/\tau_\alpha) \tag{36}$$

Here η_α and τ_α are the relaxation viscosities and times, respectively.

Among the relaxation functions of the anisotropic liquids there exist only V independent functions which can always be written in the form[58]

$$G_{ijke}(t) = \sum_{\alpha=1}^{N} C^\alpha_{ijke} G_\alpha(t), \quad 1 \leqslant N \leqslant V \tag{37}$$

Here C^α_{ijke} are the constant tensors, whereas $G_\alpha(t)$ represents certain scalar functions which, in the case of liquids with a discrete relaxation spectrum, can be represented in the form of a finite exponent sum with relaxation times $\tau_{\alpha\gamma}$

$$G_\alpha(t) = \sum_{\gamma=1}^{N} \frac{\eta_{\alpha\gamma}}{\tau_{\alpha\gamma}}\exp(-t/\tau_{\alpha\gamma}) \tag{38}$$

The number of sum terms (equations 37 and 38) depends on the properties of the liquid.

A particular case of rheological relation (cf. equations 35 and 37) is the following general constitutive equation for linear viscoelastic nematic liquids

$$\sigma_{ij} + p\delta_{ij} = 2\int_{-\infty}^{t}\{G_1(t-s)\gamma_{ij}(s)\,ds + G_2(t-s)[\gamma_{ie}(s)n_{ej} + \gamma_{je}(s)n_{ei}] + G_3(t-s)n_{ijem}\gamma_{em}(s)\}\,ds \tag{39}$$

This equation was recently described in Larson's paper.[53] Within the steady shear flow it may be reduced to the Ericksen equation for an anisotropic liquid without yield stress

$$\sigma_{ij} + p\delta_{ij} = \eta_1\gamma_{ij} + \eta_2(\gamma_{ie}n_{ej} + \gamma_{je}n_{ei}) + \eta_3 n_{ijem}\gamma_{em} \tag{40}$$

where $\eta_1 = \int_{-\infty}^{t} G_i(t-s)\,ds$.

A viscoelastic nematic liquid (equations 24 and 25) in the framework of low strain rates obeys the equation of the linear viscoelasticity (equation 35) with a relatively simple tensor relaxation function

$$G_{ijke}(t) = G_\perp e^{-t/\tau_\perp} I_{ijke} + (G_\perp e^{-t/\tau_\perp} - G_\parallel e^{-t/\tau_\parallel}) N_{ijke} \quad (41)$$

There may be an alternative method for representing the linear anisotropic viscoelastic relationships (equation 35) involving the tensorial complex moduli. The application of the Fourier transform to it affords a compact and physical, well-grounded form of anisotropic viscoelastic relationship between the stresses and strain rates

$$\sigma'_{ij}(\omega) = G_{ijke}(\omega) \gamma_{ke}(\omega) \quad (42)$$

Here $G_{ijke}(\omega)$ is the dynamic modulus tensor

$$G_{ijke}(\omega) = G'_{ijke}(\omega) + i G''_{ijke}(\omega) \quad (43)$$

whilst $G'_{ijke}(\omega)$ and $G''_{ijke}(\omega)$ are its real and imaginary parts (storage and loss modulus tensors). Determination of the dynamic modulus tensor $G_{ijke}(\omega)$ of LC polymers is an extremely complicated experimental task. It is much easier to measure the apparent viscoelastic characteristics, *i.e.*, the effective viscoelastic moduli $G(\omega)$, $G_1(\omega)$ and $G_2(\omega)$ connected with the shear stress, the first and second normal stress differences occurring at small amplitude oscillatory shear flow. In this kind of flow, the shear rate varies according to the harmonic law $\dot{\gamma}(t) \sim \exp(-i\omega t)$. Thus, the effective complex dynamic moduli are determined by the following equations

$$\sigma_{12} = G(\omega) \gamma(\omega) \quad (44)$$

$$\sigma_{11}(\omega) - \sigma_{22}(\omega) = G_1(\omega) \gamma(\omega) \quad (45)$$

$$\sigma_{22}(\omega) - \sigma_{33}(\omega) = G_2(\omega) \gamma(\omega) \quad (46)$$

The consideration of the small amplitude oscillatory shear flow of a nematic viscoelastic liquid (equations 24 and 25) leads to the following expressions for the effective complex moduli components related to the shear stress and first normal stress difference[54]

$$G'(\omega) = G'_\perp(\omega) \sin^2 2\varphi + G'_\parallel(\omega) \cos^2 2\varphi \quad (47)$$

$$G''(\omega) = G''_\perp(\omega) \sin^2 2\varphi + G''_\parallel(\omega) \cos^2 2\varphi \quad (48)$$

$$G'_1(\omega) = [G'_\perp(\omega) - G'_\parallel(\omega)] \sin 4\varphi, \quad G''_1 = [G''_\perp(\omega) - G''_\parallel(\omega)] \sin 4\varphi \quad (49)$$

Here $G'_\parallel(\omega)$, $G'_\perp(\omega)$ and $G''_\parallel(\omega)$, $G''_\perp(\omega)$ are the longitudinal and transversal (with respect to the director) storage and loss moduli, respectively. They are expressed *via* the longitudinal $\tau_\parallel$ and transversal $\tau_\perp$ relaxation times

$$G'_\parallel = G_\parallel \frac{\omega^2 \tau_\parallel^2}{1 + \omega^2 \tau_\parallel^2}, \quad G'_\perp = G_\perp \frac{\omega^2 \tau_\perp^2}{1 + \omega^2 \tau_\perp^2} \quad (50)$$

$$G''_\parallel = G_\parallel \frac{\omega \tau_\parallel}{1 + \omega^2 \tau_\parallel^2}, \quad G''_\perp = G_\perp \frac{\omega \tau_\perp}{1 + \omega^2 \tau_\perp^2} \quad (51)$$

The director n_i (additional dynamic variable of a nematic liquid) in the oscillating medium can be represented in the form

$$n_i = n_i^0 + \delta n_i \quad (52)$$

Here n_i^0 is the constant, nondisturbed value which in the case under consideration is determined by the orientation equation (25)

$$n_1^0 = \cos \varphi, \quad n_2^0 = \sin \varphi, \quad n_3^0 = 0 \quad (53)$$

The angle φ, characterizing the steady orientation, does not depend on shear rate according to the simple orientation law equation (25). Therefore the small variable part δn_i determining the director oscillation depends on the shear rate and satisfies the conditions

$$n_i^2 = n_i^{0^2} = 1, \quad n_i^0 \delta n_i = 0 \quad (54)$$

When the anisotropic moduli (equations 47, 50 and 51) of the linear viscoelasticity are calculated, the stresses, quadratic in respect of shear rate, are discarded. Therefore the linear viscoelastic moduli (equation 47) depend on the steady director orientation.

Figures 24 and 25 present the typical dependences of the dimensionless dynamic moduli calculated in accordance with equation (47) and equations (50) and (51) at different anisotropy parameter values α, β and orientation angle $\varphi = \pi/8$. The dynamic moduli $G'(\omega)$ and $G''(\omega)$ reveal distinctly two relaxation regions related to the longitudinal and transversal relaxation times, respectively. We can visualize the splitting effect of the spectrum for the long relaxation times on the longitudinal and transversal branches. The dynamic moduli $G'_1(\omega)$ and $G''_1(\omega)$, which characterize the normal

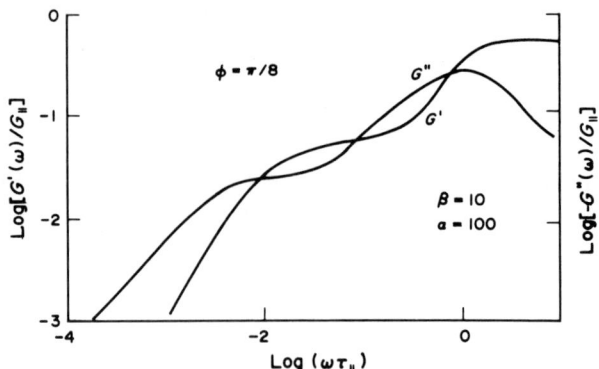

Figure 24 The dependences of the dimensionless storage and loss moduli associated with shear stress on the dimensionless frequency

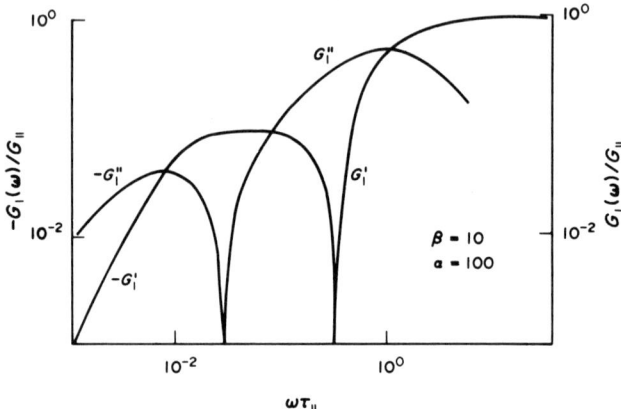

Figure 25 The dimensionless frequency dependences of components of the complex dynamic modulus associated with the first normal stress difference

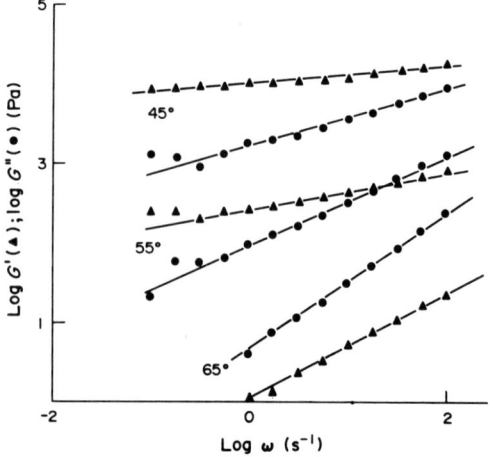

Figure 26 Frequency dependence of the complex dynamic modulus for twin LC oligomers at different temperatures[56]

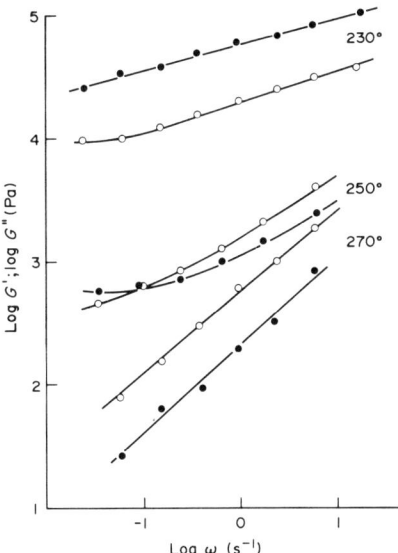

Figure 27 Frequency dependence of the complex dynamic modulus components for poly(ethylene terephthalate–*p*-hydroxybenzoic acid) (30:70) copolyester at different temperatures[57]

stresses at low frequencies, are negative. Such a peculiar rheological behaviour in the linear viscoelastic region is the main distinguishing feature of anisotropic liquids.

The experimental dependences of $G'(\omega)$ and $G''(\omega)$, particularly those obtained in Figures 26 and 27, do not refute the theoretical predictions.[55,56] They may be considered, firstly, as belonging to different frequency regions. Secondly, they are concerned with the viscoelastic behaviour of the LC systems having different degrees of anisotropy (compare with Figure 22). However, such a comparison of the theoretical predictions with the current experimental data is, of course, of subjective character, since achieving a similar orientation of an LC sample invariable at different frequencies, including its monitoring, is an experimental task of the future. The theoretical picture discussed in this chapter allows us to elaborate experiments to verify the theoretical approaches and the complex study of the rheological anisotropy of LC polymers.

15.4 REFERENCES

1. C. Robinson, *Tetrahedron*, 1961, **13**, 219.
2. S. P. Papkov and V. G. Kulichikhin, 'Liquid Crystalline State of Polymers', Khimiya, Moscow, 1977 (in Russian).
3. K. F. Wissbrun, *J. Rheol. (N.Y.)*, 1981, **25**, 619.
4. T. Asada and S. Onogi, *Polym. Eng. Rev.*, 1983, **3**, 323.
5. D. G. Baird, in 'Liquid Crystalline Order in Polymers', ed. A. Blumstein, Academic Press, New York, 1978, chap. 7.
6. V. G. Kulichikhin, A. Ya. Malkin and S. P. Papkov, *Vysokomol. Soedin., Ser. A*, 1984, **26**, 451 (in Russian).
7. V. G. Kulichikhin, in 'Liquid Crystal Polymers', ed. N. Platé, Khimiya, Moscow, 1988, chap. 8 (in Russian).
8. J. Hermans, *J. Colloid. Sci.*, 1962, **17**, 638.
9. P. J. Flory, *Proc. R. Soc. London, Ser. A*, 1956, **234**, 73.
10. S. L. Kwolek, *US Pat.* 3 600 350 (1971).
11. V. D. Kalmykova, G. J. Kudrjavtsev, S. P. Papkov, *Vysokomol. Soedin., Ser. B*, 1971, **13**, 707 (in Russian).
12. S. P. Papkov, V. G. Kulichikhin, V. D. Kalmykova and A. Ya. Malkin, *J. Polym. Sci., Polym. Phys. Ed.*, 1974, **12**, 1753.
13. V. G. Kulichikhin, *Mol. Cryst. Liq. Cryst.*, 1989, **169**, 51.
14. V. G. Kulichikhin, G. J. Kudrjavtsev and S. P. Papkov, *Int. J. Polym. Mater.*, 1982, **9**, 239.
15. A. N. Semenov and A. R. Khokhlov, *Usp. Fiz. Nauk*, 1988, **156**, 427 (in Russian).
16. A. R. Khokhlov, in 'Liquid Crystal Polymers', ed. N. Platé, Khimiya, Moscow, 1988, chap. 1 (in Russian).
17. M. Doi and S. F. Edwards, *J. Chem. Soc., Faraday Trans. 2*, 1978, **74**, 568, 918.
18. F. M. Leslie, *Arch. Ration. Mech. Anal.*, 1968, **28**, 265.
19. M. Doi, *J. Phys. (Orsay, Fr.)*, 1975, **36**, 607.
20. M. Doi, *J. Polym. Sci., Polym. Phys. Ed.*, 1981, **19**, 229.
21. J. L. Ericksen, *Kolloid-Z.* 1960, **173**, 117.
22. G. Kiss and R. S. Porter, *J. Polym. Sci., Polym. Phys. Ed.*, 1980, **18**, 361.
23. L. L. Chapoy, B. Marcher and K. H. Rasmussen, *Liq. Cryst.*, 1988, **3**, 1611.
24. E. Marsano, L. Carpaneto, A. Ciferri and Y. Wu, *Liq. Cryst.*, 1988, **3**, 1561.
25. G. Marrucci, N. Grizzuti and A. Buonario, *Mol. Cryst. Liq. Cryst.*, 1987, **153**, 263.
26. E. Peuvrel and P. Navard, in 'Proceedings of the 3rd European Rheology Conference, Edinburgh', ed. D. R. Oliver, Elsevier, London, 1990, p. 391.

27. A. M. Donald, C. Vihey and A. H. Windle, *Polymer*, 1983, **24**, 155.
28. P. Moldenaers and J. Mewis, *J. Rheol. (N.Y.)*, 1986, **30**, 567.
29. F. N. Cogswell, in 'Recent Advances in Liquid Crystal Polymers, Proceedings of the European Science Foundation 6th Polymer Workshop on Liquid Crystal Polymer Systems, Lingby, 1983', London, New York, 1984, p. 165.
30. K. F. Wissbrun, *Br. Polym. J.*, 1980, **12**, 163.
31. D. G. Baird, A. Cotsis *et al.*, 'Polymer Liquid Crystals, Proceedings of the 2nd Symposium, Division of Polymer Chemistry, Washington D.C., 1983', New York, London, 1985, p. 183.
32. A. T. Dibenedetto, L. Nicolais, E. Amendola, C. Carfagna and M. R. Nobile, *Polym. Eng. Sci.*, 1989, **29**, 153.
33. J. Mewis and P. Moldenaers, *Mol. Cryst. Liq. Cryst.*, 1987, **153**, 291.
34. P. Moldenaers and J. Mewis, in 'Proceedings of the 10th International Congress on Rheology', ed. P. H. T. Uhlerr, Australian Society of Rheology, Sydney, 1988, vol. 2, p. 134.
35. F. Cocchini, C. Aratari and G. Marrucci, *Macromolecules*, 1990, **23**, 4446.
36. G. Marrucci, *Rheol. Acta*, 1990, **29**, 523.
37. V. G. Kulichikhin, N. V. Vasil'eva, V. A. Platonov, A. Ya. Malkin and S. P. Papkov, *Vysokomol. Soedin., Ser. A*, 1979, **21**, 1407 (in Russian).
38. P. G. de Gennes, 'The Physics of Liquid Crystals', Clarendon Press, Oxford, 1974.
39. A. N. Semenov, *Zh. Eksp. Teor. Fiz.*, 1983, **85**, 549.
40. V. G. Kulichikhin, A. Ya. Malkin, L. P. Gudim and S. P. Papkov, *Vysokomol. Soedin., Ser. A*, 1972, **14**, 244 (in Russian).
41. T. Asada, T. Tanaka and S. Onogi, *J. Appl. Polym. Sci: Appl. Polym. Symp.*, 1985, **41**, 229.
42. P. G. Horn and M. Kleman, *Ann. Phys. (Paris)*, 1978, **3**, 229.
43. K. Wissbrun, *Faraday Discuss. Chem. Soc.*, 1985, **79**, 161.
44. G. Marrucci and P. L. Maffetone, *J. Rheol. (N.Y.)*, 1990, **34**, 1217.
45. R. G. Larson, in 'Proceedings of the 10th International Congress on Rheology', ed. P. H. T. Uhlerr, Australian Society of Rheology, Sydney, 1988, vol. 2, p. 64.
46. M. Doi in 'Theory and Application of Liquid Crystals', ed. J. L. Ericksen and D. Kinderlehrer, Springer-Verlag, New York, 1987, p. 143.
47. V. S. Volkov and G. V. Vinogradov, *Rheol. Acta*, 1984, **23**, 231.
48. M. Miesowicz, *Nature*, 1946, **158**, 27.
49. F. M. Leslie, *Q. J. Mech. Appl. Math.*, 1966, **19**, 357.
50. V. S. Volkov and V. G. Kulichikhin, *Vysokomol. Soedin., Ser. A*, 1990, **32**, 2358 (in Russian).
51. V. S. Volkov and V. G. Kulichikhin, *J. Rheol. (N.Y.)*, 1990, **34**, 281.
52. W. B. Vanderheyden and G. Ryskin, *J. Non-Newtonian Fluid Mech.*, 1987, **23**, 383.
53. R. G. Larson, and D. W. Mead, *J. Rheol. (N.Y.)*, 1989, **33**, 185.
54. V. S. Volkov and V. G. Kulichikhin, in 'Proceedings of the 3rd European Rheology Conference, Edinburgh', ed. D. R. Oliver, Elsevier, London, 1990, p. 492.
55. Y. G. Lin, R. Zhou, J. C. W. Chien and H. H. Winter, *Macromolecules*, 1988, **21**, 2014.
56. O. V. Vasil'eva, Yu. J. Akulin, B. H. Strelez, P. V. Khokhlow and V. G. Kulichikhin, *Vysokomol. Soedin., Ser. A*, 1990, **32**, 1461 (in Russian).
57. E. P. Plotnikova, M. A. Rogunova, R. V. Tal'roze, V. G. Kulichikhin and V. P. Shibaev, *Vysokomol. Soedin., Ser. A*, 1991, **33**, 1761 (in Russian).
58. V. S. Volkov, *Prikl. Mat. Mekh.*, 1982, **46**, 248 (in Russian).

16
Polymers for Photonic Applications

KWANG-SUP LEE, MAREK SAMOC and PARAS N. PRASAD
State University of New York at Buffalo, NY, USA

16.1	INTRODUCTION	407
	16.1.1 Optical Nonlinearity of Organics	408
	16.1.2 Molecular Basis of Second- and Third-order Nonlinearity	410
16.2	SECOND-ORDER NONLINEAR OPTICAL POLYMERS	411
	16.2.1 Blends of Glassy Polymers and Dye Molecules	413
	16.2.2 Amorphous Polymers Attached to NLO Units	414
	16.2.3 Liquid Crystalline Polymers	418
	16.2.4 SHG Polymers with NLO Units in the Main Chain	422
	16.2.5 Langmuir–Blodgett Films	425
16.3	THIRD-ORDER NONLINEAR OPTICAL POLYMERS	429
	16.3.1 Polydiacetylenes	432
	16.3.2 Polyenes and Polyacetylenes	435
	16.3.3 Conjugated Polymers Prepared by the Polyelectrolyte Precursor Technique	436
	16.3.4 Polythiophene and other Thiophene-based Conjugated Polymers	438
	16.3.5 Rigid Rod-like Polymers	439
	16.3.6 Ladder Polymers	440
	16.3.7 Organosilane Polymers	442
16.4	CURRENT STATUS AND FUTURE PROSPECTS	442
16.5	REFERENCES	443

16.1 INTRODUCTION

Photonics is a term coined to describe a set of emerging technologies which deal with various aspects of the use of light beams to store, transfer and manipulate information. In a broader sense, the term is also sometimes meant to encompass such technology as that of harvesting light energy and converting it to electricity. Often it is also used to cover many other technologies which have emerged as a result of the advent of lasers. We will, however, use a somewhat limited meaning of the term 'photonics' and make use of an analogy with the word 'electronics'. Thus, photonics is an analog of electronics in which photons rather than electrons carry the information and are made to perform various tasks, and one can think of photonic switches, gates, amplifiers, modulators and photonic integrated circuits.

The advantages of using photons instead of electrons as information carriers are obvious. First of all, the frequency of light is much higher than usable frequencies in electrical circuits. Therefore, potentially, the information-carrying capacity of a single laser beam is simply enormous. One can easily imagine squeezing millions of telephone conversations or thousands of video transmissions into a single beam of light travelling through a fiber optic cable. This capacity of a fiber optic cable becomes even more enviable on considering the fact that a cable can contain many such fibers, all of which can be used independently with no crosstalk between them and with almost complete immunity to outside interferences. Optical processing of information can also be superior to that provided by electronic circuits since interaction of light with matter can occur on a very rapid time scale, say in less than a picosecond, and therefore functions such as the opening and closing of photonic switches, *i.e.* generation of optical logic states, can be performed very quickly. This leads to the possibility of building optical computers in which the logic operations are performed by light beams.

There are additional advantages of using light rather than electric currents. Since the wavelength of light is comparable with geometrical sizes of elements in optical microcircuits, the wave nature of light can also be an asset in analog processing of signals. Phenomena such as phase conjugate reflection and image processing by four-wave mixing processes are examples of the use of light interference and the resulting transient holograms (which are related to the Fourier transform of the signals carried by the light beams) for analog information processing.

Carrying further the analogy between electronic devices and circuits and those which either have already been constructed or are suggested for the use in photonic information processing, there is a need for optimized design of passive light beam transmitting devices like waveguides and beam-steering and -shaping elements as well as those that can play an active role (*i.e.* modify the transmitted signals). Similarly, as metal wires and conducting paths of low resistance are used to interconnect active elements of electronic circuits, optical fibers, planar and channel waveguides can perform the same tasks for light beams. The obvious demand for such elements is a low transmission loss as well as low distortion of the transmitted signal. On the other hand, processes such as modulation of the carrier wave with a signal, switching, frequency mixing and signal amplification are performed in electronics with active circuit elements. Their analogs in the photonic world can be found, thanks to phenomena which are collectively characterized as nonlinear optical (NLO) processes.[1] As we shall see below, the nonlinear optical behavior of materials can be utilized to build devices that can be building blocks for photonic integrated circuits. Present day electronics relies heavily on the complicated technology of semiconductor materials and would be impossible without the tremendous effort that has been invested in the investigations of those materials. Even more so, the success of photonics will depend on the ability to devise and produce optimized NLO materials.

Polymeric materials of special design are a group of NLO materials that have been of much interest recently.[1,2] The inherent ease of modification of organic molecule structures makes it possible to synthesize tailor-made molecules and modify their properties. We will describe here the basic concepts of the nonlinear optical behavior of organics and show how these concepts can be used to understand the behavior of the optically nonlinear polymers investigated to date. We will also discuss the directions of polymer research expected to hold promise for the future.

16.1.1 Optical Nonlinearity of Organics

The nonlinear optical behavior is most often described by treating the electric polarization P induced in a material under the action of an electric field E as a power series[1,3]

$$P = \chi^{(1)} \cdot E + \chi^{(2)} : EE + \chi^{(3)} \vdots EEE + \cdots \cdots \quad (1)$$

where the first term describes the usual linear polarization due to the linear susceptibility $\chi^{(1)}$. The higher-order susceptibilities $\chi^{(n)}$ are responsible for the deviations from linear behavior. By assuming that the field E oscillates at a frequency ω one can easily verify that the presence of the second-order ($\chi^{(2)}$) term leads to the generation of a polarization component oscillating with a frequency 2ω (second harmonic generation—commonly abbreviated as SHG) as well as constant (time independent) field component (optical rectification).[1] The efficiency of second harmonic generation depends on the value of $\chi^{(2)}$ as well as on many other factors such as the refractive index of the medium and the possibility of obtaining so-called phase matching. Often in the literature instead of $\chi^{(2)}$ one quotes a coefficient d which is related to the nonlinear susceptibility by $\chi^{(2)} = 2d$.

Assuming that the field E contains both a constant component and one oscillating at ω, one can find that the presence of the constant field modifies the phase of the field oscillating at ω (electrooptic effect or linear Pockels effect) through the second-order term in the polarization expansion.[1] This is equivalent to a change in the refractive index brought about by the presence of a d.c. field (or one varying much more slowly than the field of a light beam). Traditionally, the Pockels effect is described by the electrooptic coefficient r defined by

$$\Delta\left(\frac{1}{n_{ij}^2}\right) = r_{ijk} E_k \quad (2)$$

where n_{ij} are refractive index components used to describe the optical indicatrix.

When it is assumed that the field E contains two frequencies, ω_1 and ω_2, the second-order term leads to the generation of the polarization components at $\omega_1 - \omega_2$ and at $\omega_1 + \omega_2$. These phenomena are called difference and sum frequency generation.

In a similar way, it can be shown that the third-order ($\chi^{(3)}$) term is responsible for such phenomena as third harmonic generation, quadratic electrooptic effect (Kerr effect), various frequency mixing schemes, and phenomena such as light intensity dependent refractive index.[1] The latter phenomenon is of special interest as it describes variation of the (complex) refractive index due to the action of a light beam. Therefore, the refractive index can be described as

$$n = n(0) + n_2 I \tag{3}$$

where I is the light intensity and n_2 is the so-called nonlinear refractive index. For materials that show absorption, both the refractive index n and n_2 must be taken as complex quantities. The light intensity I is proportional to the square of the electric field E. The square of the refractive index, on the other hand, is equal to the dielectric permittivity ε. In the cgs units system, this leads to the following relation between n_2 and $\chi^{(3)}$

$$n_2 = 12[\pi/n(0)]\chi^{(3)} \tag{4}$$

A change in the refractive index of an optical material may be utilized for many operations performed on a signal-carrying light beam. Modulation of a refractive index performed as a function of time brings about a phase modulation for a cw beam. If the refractive index changes are anisotropic, then one usually terms them 'induced birefringence'. The existence of such birefringence brings about a difference in the phase velocities of light beam components with different polarizations. This leads to changes in the polarization state of a beam and can be easily transformed (using appropriately positioned polarizing optics) into intensity modulation of the beam. Figure 1 shows schematically an action of an all-optical Kerr gate built using a third-order NLO polymeric material in which optically induced birefringence Δn produced by a very short, strong laser pulse allows the passage of the probe beam (optical signal through two cross polarizers).

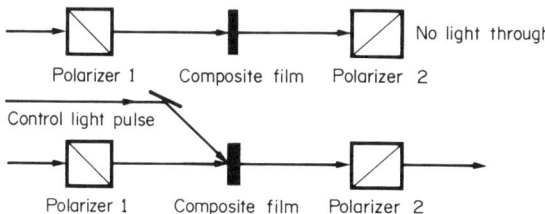

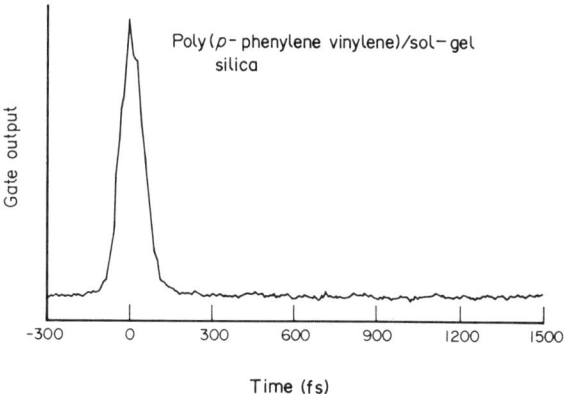

Figure 1 A scheme of the action of an optical Kerr gate and the result of an investigation of opening and closing of the gate in a composite material obtained by sol–gel processing[166]

16.1.2 Molecular Basis of Second- and Third-order Nonlinearity

The usual approach to investigating a molecular solid is an attempt to trace the properties of the solid back to those of the structural subunits of the solid: molecules or molecular groups. On the microscopic level the linear and nonlinear optical properties can be discussed using a power expansion of the induced dipole moment P in powers of the electric field:[1]

$$P = \alpha \cdot E + \beta : EE + \gamma \vdots EEE + \cdots \cdots \tag{5}$$

where α is the linear polarizability and β and γ are the second- and third-order hyperpolarizabilities, respectively. The simplest possible approach here is that of the so-called oriented gas model. In this model one assumes that molecules forming a solid preserve all their properties and their interaction can be taken account of by introducing some correcting parameters, for example, a local field factor. Then, a physical property of a solid is a result of a simple summation of the properties of the molecules. Since many physical properties of interest are not scalars but rather tensors of various ranks, the summation is not straightforward, but one must take into account the orientations of molecules as given by their directional cosines. For example, the second-order susceptibility $\chi^{(2)}$ is a third rank tensor and can be considered within an oriented gas model to be related to molecular second-order hyperpolarizabilities β, which are also tensors of the third rank, in the following manner:[1]

$$\chi^{(2)}_{IJK}(-\omega_1 = -\omega_2 - \omega_3, \omega_2, \omega_3) = \mathscr{L}(\omega_1)\mathscr{L}(\omega_2)\mathscr{L}(\omega_3) \sum_s \sum_i \sum_j \sum_k a^{(s)}_{iI} a^{(s)}_{jJ} a^{(s)}_{kK} \beta^{(s)}_{ijk} \tag{6}$$

Here, $\mathscr{L}$ is the local field factor which is often approximated with a Lorentz expression $\mathscr{L} = (n^2 + 2)/3$, n being the refractive index at a given frequency. The components of the a_{ij} matrix are direction cosines between the set of the molecular axes and the set of external axes. The summation is carried out over all molecules in a unit of volume. For a crystal, summation over s can be performed by summing over translationally identical molecules and it is only necessary to take into account different classes of molecules in a unit cell. For an amorphous polymer, one must take into account a distribution in orientations of molecules. For $\chi^{(2)}$, which is a third-rank tensor, this orientational summation in a fully amorphous or isotropic medium vanishes. Therefore, no second-order nonlinear process can be observed in a fully amorphous medium.

The above relation suggests that the properties of a solid can be predicted from the properties of the molecules building the solid if orientations of the molecules can be determined, or, better still, controlled by processing. One must be aware, however, that the correctness of the oriented gas approximation can often be put in doubt. The elements which build the solid may be macromolecules rather than simple molecules, especially in the case of polymers, and therefore it cannot be assumed that the solid consists of small noninteracting or weakly interacting elements. One can predict that some information from the field of molecular crystals may not be fully suitable for polymers. An example of such an assumption which may be totally unsuitable for polymers is that of the local field being approximated by the Lorentz expression with an average (isotropic) refractive index. In oriented films of π-conjugated polymers the strong anisotropy reflected in very large birefringence may also give rise to anomalously large anisotropy of the local field factors.

While a quantitative description of nonlinear properties of a polymer using data obtained from simple molecules may be difficult, the trends observed in hyperpolarizabilities of molecules must be reflected in the properties of polymers containing such molecules as structural units. It may therefore be useful to discuss some results obtained on simple molecules and their oligomers.

A very useful rule in the field of molecular linear polarizabilities is that of additivity. Indeed, for most organic molecules a good estimate of the molecular polarizability α may be obtained by summing up tabulated contributions due to atoms, and/or bonds and functional groups. The notion of additivity was also applied to hyperpolarizabilites.[1] However, it seems that even in the case of σ-electron molecules such as substituted methanes the additivity concept does not work well. For π-electron molecules the deviation from additivity is much stronger. In the case of second-order nonlinearities a very well-known prototype molecule is that of p-nitroaniline. The hyperpolarizability of this molecule is much bigger than that which could be deduced by combining results for nitrobenzene and aniline (see Table 1). This enhancement of the second-order nonlinearity is generally interpreted in terms of internal charge transfer in a molecule containing both electron donor and electron acceptor groups.[1] In general, the most commonly employed scheme of a second-order nonlinear molecule or molecular fragment (chromophore) is as follows.

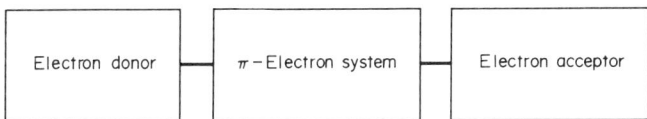

A good example of deviation from additivity in the case of third-order optical nonlinearities is provided by results obtained on oligomers of thiophene and benzene.[4,5] The experimentally determined hyperpolarizabilities of oligomers are significantly larger than those given by simple additivity of contributions due to structure subunits.

The strong enhancement of nonlinearity with the elongation of a π-electron conjugated system can be interpreted in various ways. First of all, one asymptotic behavior which may be considered is that of a free electron gas in a one-dimensional potential well. Theoretical calculations performed by Rustagi et al.[6] leads to the conclusion that such a system should exhibit a hyperpolarizability which increases as the length of the system in sixth power. Unfortunately, at the same time, the energy gap of such a system will decrease, leading to the loss of transparency.

Of course, π-electrons in molecules can hardly be considered free electrons and the free electron model is not supposed to be a good approximation of the actual molecules. It is well known from the studies of one-dimensional conductors that, even for an infinite chain, one does not necessarily obtain the energy gap equal to zero, but, due to Peierls localization, the energy levels are split to form a nonzero gap. The hyperpolarizability (per structural unit) of an oligomer chain will also not increase without a limit.

We have provided a simple explanation of the hyperpolarizability of oligomers by a coupled anharmonic oscillator model. This model is based on the concept that the NLO response of a system is derived from the anharmonic motion of the bound electrons. In other words an optically nonlinear unit is an anharmonic oscillator. According to the coupled anharmonic oscillator model, a chain molecule may be treated as an assembly of anharmonic oscillators which are coupled. Depending on the value of the coupling parameter one can obtain saturation of the hyperpolarizability for chains of different lengths.

More detailed understanding of the structure–property relationship for NLO organic molecules and polymers can be gained by performing quantum chemical analyses of these systems.[1] Quantum chemical calculations have proved useful in explaining both second- and third-order nonlinear optical properties of molecules. However, their predictive capabilities are still not satisfactory, principally due to difficulties in *ab initio* level calculations for large molecules.

16.2 SECOND-ORDER NONLINEAR OPTICAL POLYMERS

As already mentioned, the presence of second-order susceptibility $\chi^{(2)}$ in a material results in the possibility of observing frequency mixing, like second harmonic generation (SHG) and the electrooptic effect. Since both $\chi^{(2)}$ and the underlying molecular property, the second-order hyperpolarizability β, are tensors of the third rank, there are certain symmetry conditions necessary for second-order effects. A third rank tensor must be identically equal to zero (*i.e.* all its elements must be zero) in symmetry groups containing a center of symmetry. This means that only molecules without a center of symmetry can have nonzero β values and that, in order to obtain a nonzero $\chi^{(2)}$, one has to arrange the molecules (chromophores) in a noncentrosymmetric arrangement. Obviously, a random ordering of molecules as is found in solution or in a truly amorphous polymer, results in $\chi^{(2)} = 0$ and, for example, in no SHG signal. However, the statistical center of symmetry can be removed by application of an external electric field E_p. Molecules in solutions will then partially reorient, tending to align their dipole moments with the direction of the field. In effect, a nonzero $\chi^{(2)}$ is induced. When a beam of laser light is passed through such a partly ordered solution, a second harmonic of light will be produced. The effect is known as electric field induced second harmonic generation (EFISHG).[1] It can be shown that the induced second-order susceptibility is

$$\chi^{(2)} \propto N\beta_z\mu E_p$$

where N is the density of NLO active molecules, μ is their dipole moment, and E_p is the poling field. $\boldsymbol{\beta}_z$ is here a so-called vector part of the hyperpolarizability β. EFISHG is a standard method to assess molecular hyperpolarizabilities (or rather their vector parts) by measurements on solutions. Similarly, if an electric field is applied to a polymer near its glass transition, orientation of certain elements of the structure (sections of the backbone, pendant groups, low molecular weight dopant

Table 1 Second-order NLO Susceptibilities of Some Organics

Structure	λ (μm)	Solvent	β (10^{-30} esu)	Ref.
H$_2$N–C$_6$H$_5$	1.06	Neat	1.1	7
C$_6$H$_5$–NO$_2$	1.06	Neat	2.2	7
H$_2$N–C$_6$H$_4$–NO$_2$	1.06	Methanol	34.5	7
H$_2$N–C$_6$H$_4$–CH=CH–C$_6$H$_4$–NO$_2$	1.06	Acetone	260	8
Me$_2$N–C$_6$H$_4$–CH=CH–C$_6$H$_4$–NO$_2$	1.06	Acetone	450	8
Me$_2$N–C$_6$H$_4$–CH=CH–NO$_2$	1.06	Chloroform	220	8
Me$_2$N–C$_6$H$_4$–CH=CH–CH=CH–NO$_2$	1.06	Acetone	650	8
O$_2$N–(3-Me-pyridine)–N$^+$–O$^-$	1.06	Crystal	8.5	9
Me$_2$N–pyridine–N$^+$–O$^-$	1:06	Chloroform	5	10
Me$_2$N–C$_6$H$_4$–CH=CH–pyridine–N$^+$–O$^-$	1.06	Chloroform	50	10
MeO–C$_6$H$_4$–NO$_2$	1.06	1,4-Dioxane	4	11
MeO–C$_6$H$_4$–CH=CH–C$_6$H$_4$–NO$_2$	1.06	1,4-Dioxane	6	11
MeO–C$_6$H$_4$–(CH=CH)$_4$–C$_6$H$_4$–NO$_2$	1.06	1,4-Dioxane	89	11
Et$_2$N–C$_6$H$_4$–N=N–C$_6$H$_4$–C(CN)=CH–CN	1.58	DMSO	390	12
Me$_2$N–C$_6$H$_4$–NO$_2$	1.35	DMSO	21	12
Me$_2$N–C$_6$H$_4$–CH=C(CN)$_2$	1.35	DMSO	31	12
Me$_2$N–C$_6$H$_4$–C(CN)=C(CN)$_2$	1.35	DMSO	78	12
dithiolane=CH–C$_6$H$_4$–NO$_2$	1.35	DMSO	52	12
HOCH$_2$CH$_2$(Et)N–C$_6$H$_4$–N=N–C$_6$H$_4$–NO$_2$	1.35	DMSO	125	12

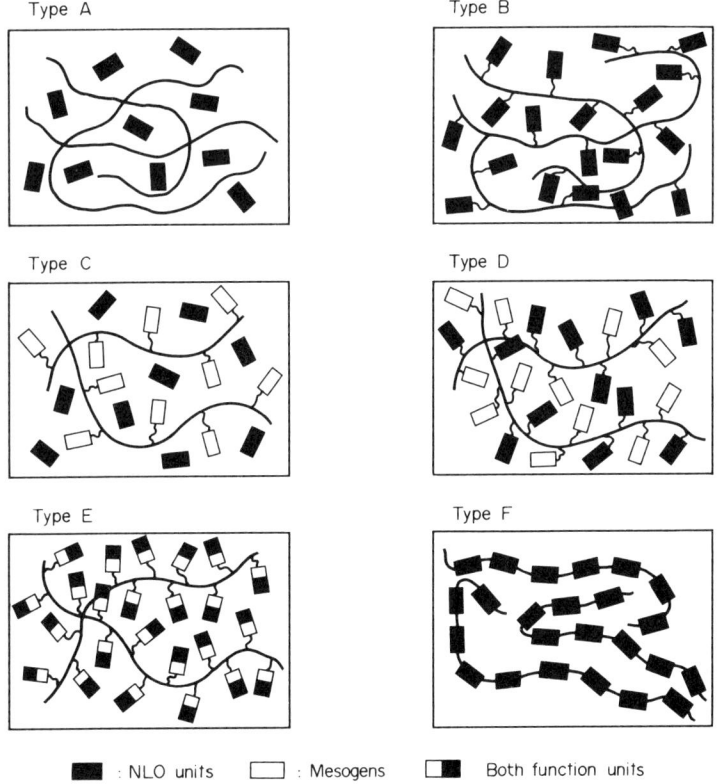

Figure 2 Schematic drawings for second-order NLO polymers. Type A: Blends of amorphous polymers and NLO units. Type B: Amorphous polymers attached to NLO units. Type C: Blends of side chain liquid crystalline units and NLO units. Type D: Copolymers of functionalized liquid crystalline units and NLO units. Type E: Liquid crystalline polymers in which dye units play role as mesogens. Type F: Polymers which have NLO units in their main chains

molecules in doped polymers) will also take place. If the temperature is then lowered well below the glass transition, with the poling field still on, the resulting polarization can be frozen-in resulting in a poled polymer that can exhibit NLO activity.

Therefore, the design of a second-order NLO polymeric material should take into account the following factors: (i) selection of proper chromophores or dye molecules with high values of the $\mu\beta_z$ factor; (ii) possibility of applying high poling fields to maximize $\chi^{(2)}$; (iii) stability of the poled polymer under usual environmental conditions without the loss of the frozen-in polarization; and (iv) good transparency and low optical losses as well as good mechanical properties.

The choice of the chromophore is relatively simple. Table 1 shows examples of organic molecules for which high values of β_z have been reported in the literature. The results have been obtained by the EFISHG technique. Several possible methods of incorporating the NLO active molecules into polymeric structures are outlined in Figure 2.

The process of poling is schematically shown in Figure 3. The polymer system is heated to a temperature near the glass transition (T_g), or to the liquid crystalline phase region, so as to increase the mobility of the polymer chains. At this point the electric field is applied which results in partial alignment of the dipolar molecules and/or molecular fragments. After waiting for an appropriate period of time to allow for alignment, the material is cooled down and the field is finally shut off at a much lower temperature. A big problem here is the possibility of randomization of the orientations which results in the loss of NLO activity. This can be avoided if poling at a higher temperature is carried out simultaneously with crosslinking, therefore stabilizing the newly formed polar structure.

16.2.1 Blends of Glassy Polymers and Dye Molecules

Since glassy polymers are in an amorphous state, they often show high optical quality and good physical properties. Therefore, these polymers are promising as matrices for NLO dyes (type A in Figure 2).

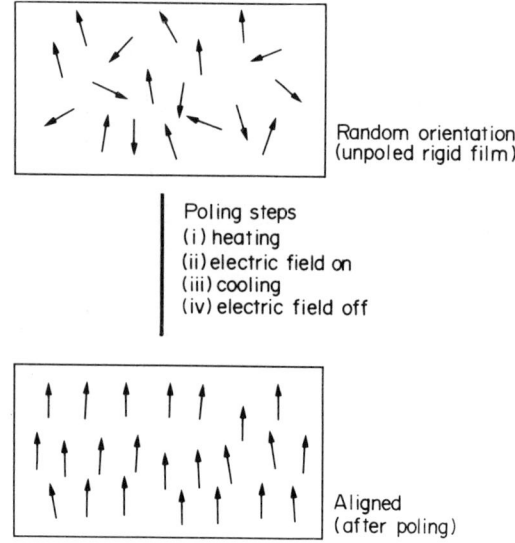

Figure 3 Schematics of electric field poling

Singer et al.[13] found SHG in a polymer system composed of poly(methyl methacrylate) (PMMA) as the matrix and an azo dye, 4-[N-ethyl-N-(2-hydroxyethyl)]amino-4'-nitroazobenzene (Disperse Red 1) as the NLO dopant. The observed second harmonic coefficient was $d_{33} = (6.0 \pm 1.3) \times 10^{-9}$ esu at 1.58 μm. Also, Hampsch et al.[14,15] examined NLO properties in blends of glassy polymers such as polystyrene (PS) or PMMA, doped with 4-(dimethylamino)-4'-nitrostilbene (DANS), 2-methyl-4-nitroaniline (MNA) or Disperse Red 1. In the case of a PS or a PMMA film doped with 4% DANS, the second-order NLO coefficient of this DANS/PMMA system was reported to be about $d_{33} = 2 \times 10^{-9}$ esu. However, the SHG intensity of this blend system is reduced by 80% in a relatively short time period (12 h) due to the mobility of the dyes and/or the polymer matrix. They also observed that the SHG intensity was directly related to the dye size and the type of polymer matrix system.

Watanabe and Miyata[16] have reported a novel approach in which the SHG-active p-nitroaniline (PNA) crystals were formed in a crystalline polymer such as poly(oxyethylene) (POE). By applying an electric field during the crystallization process, they observed the SHG efficiency of the PNA/POE mixture to be 98 times larger than that of urea. Molecular doping of poly(ε-caprolactone) (PCL) by PNA (18 mol %) has shown a much stronger SHG intensity, in fact as much as 120 times that of urea, even without any application of an electric field.[17] The high SHG intensity of the PNA/POE system decreased gradually after the electric field had been removed but, in the PNA/PCL system, the SHG intensity remained constant for over 1000 h, indicating that the noncentrosymmetric crystal structure was very stable in this system. The structures of dyes and matrix materials used in doped NLO polymer systems are shown in Scheme 1.

The aforementioned blending method of glassy polymers and dye molecules facilitates convenient preparation of NLO materials. Unfortunately, there is only a limited amount of dye which can be dissolved in the polymer matrix before segregation begins to occur. Generally, the acceptable concentration of dye to be mixed in the host polymer is set at about 10% by weight.

16.2.2 Amorphous Polymers Attached to NLO Units

Mark et al.[18] have studied a type B polymer (Figure 2) with covalent bonds connecting the NLO units to the polymer backbone. They introduced Disperse Red 1 or 4-(4-N, N-dimethylamino-styryl)pyridinium iodide (DASP) to the polymer chain by thallium-mediated esterification or by quaternization, after the conversion of iodomethyl derivatives for increased reactivity of chloromethylated PS.

The values of d_{33} for (**1**) and (**2**) have been determined to be 2.7×10^{-9} esu and 1.2×10^{-10} esu, respectively, at a poling field of 0.3 MV cm^{-1}. For the films of (**1**), the magnitude of d_{33} increases linearly with the strength of the poling field up to 0.5 MV cm^{-1}. For polymer (**2**), however, the value of d_{33} was found to increase in a linear way only up to approximately 0.3 MV cm^{-1}. At a

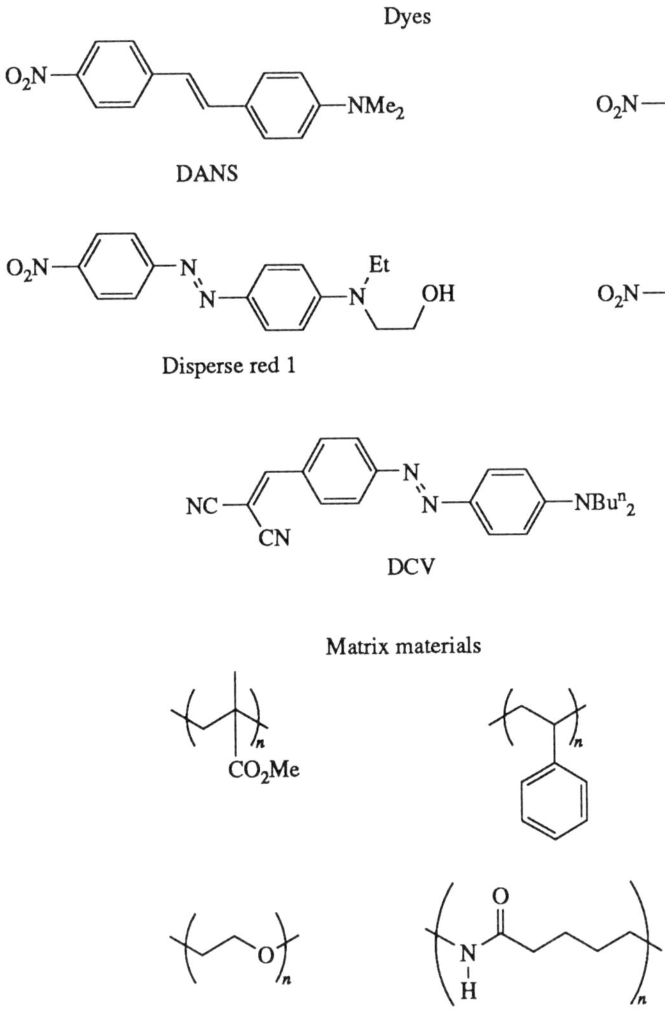

Dyes

DANS

MNA

Disperse red 1

PNA

DCV

Matrix materials

Scheme 1

higher field the second harmonic generation saturates, possibly due to the mobility of ions in this polymer. The SHG efficiency of (**1**) continued even after storage for several days at room temperature.

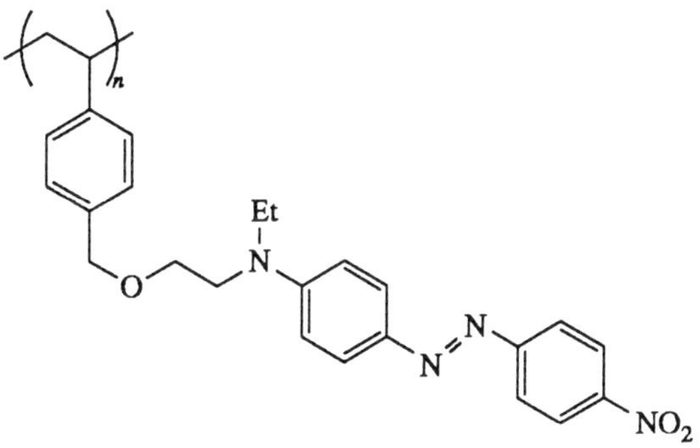

(**1**)

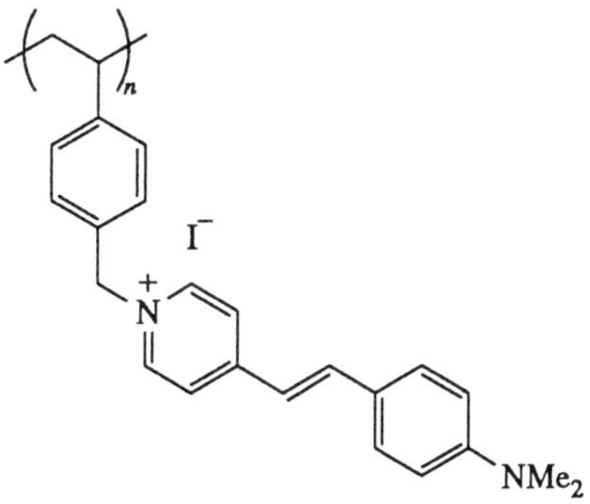

(**2**)

Strong SHG was also observed in PS functionalized with *N*-(4-nitrophenyl)-L-prolinol (NPP) units.[19] Covalent bonding of dye units to the PS backbone offers several advantages over simple doped systems. Polymers (**3**) and (**4**) can support much higher dye densities of NLO units without segregation and opacity and are able to withstand much higher poling fields. Also, dye-functionalized PS generally exhibits much longer-lived SHG performance, as shown in Table 2.

(3) (4)

Table 2 SHG Coefficients[a] for Dye-functionalized Polystyrenes[19]

Material	Functionalization level (% phenyl rings)	Poling field (MV cm^{-1})	Decay time (days)	d_{33} (10^{-9} esu)[b]
(1)	12.5	0.3		2.7
(2)	4.5	0.3		0.12
(1)	36.0	0.7		3.8(8.9)
(3)	15.0	0.7	313	5.1(3.7)
(3)	25.0	0.3	195	3.0(2.6)
(3)	48.0	0.6		11.6(10.1)
(3)	48.0	1.6	42	18.0

[a] Measured within 0.5 h of poling: $\lambda = 1.06\,\mu$m. [b] Theoretical values are given in parentheses.

Singer et al.[20,21] reported SHG in thin films of a copolymer (5) of methyl methacrylate (MMA) and 4-dicyanovinyl-4'-(dialkylamino)azobenzene-substituted methacrylate. This polymer can be synthesized by the azo coupling reaction. The sample (5) was corona poled above the glass transition temperature ($T_g = 127\,°$C). The value of d_{33} was 51×10^{-9} esu at the wavelength of 1.58 μm. For comparison, the DCV/PMMA doped system gave $d_{33} = 74 \times 10^{-9}$ esu. On storage, the SHG coefficient of the MMA–DCV copolymer (5) decayed to about 90% of its initial value while d_{33} of the doped system decayed to about 25% of the initial value (Figure 4). A similar type of copolymer (6) was investigated by Esselin et al.,[22] who instead of DCV used Disperse Red 1 in the polymer backbone. They found that the value of d_{33} depended on the dye content. For example, for the density of the chromophore equal to 0.60×10^{20} molecules cm^{-3}, the d_{33} value was 2.7 pm V^{-1} (6.4×10^{-9} esu) and at 7.42×10^{20} molecules cm^{-3} it was 57.9 pm V^{-1} (1.4×10^{-7} esu). Apparently, the investigated system offered many advantages over doped systems, especially that of a higher possible NLO chromophore content and of decreased molecular mobility of the aligned chromophores which resulted in less-pronounced relaxation of the SHG properties.

(5) X = CH=C(CN)$_2$
(6) X = NO$_2$

Eich et al.[23] reported the results of in situ measurements of corona poling induced SHG on an amorphous polymer (7), which contained PNA molecules as side groups connected to a polypropylene backbone. They performed several types of studies such as X-ray, dielectric relaxation, studies of thermal properties and waveguide and absorption spectroscopy, in order to understand the

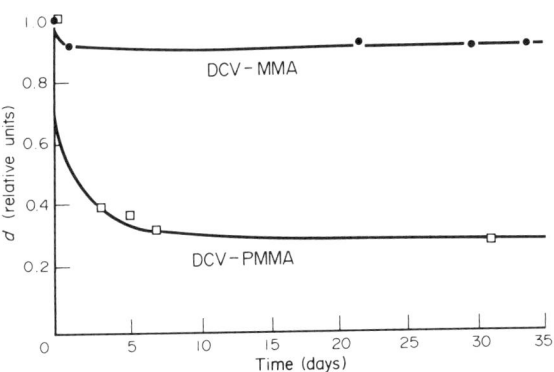

Figure 4 Decay of second harmonic coefficient of DCV–MMA (**5**) and DCV/PMMA (doped system) corona-poled films[20]

material characteristics. A nonlinear coefficient of this compound was $d_{33} = 31$ pm V^{-1} (7.4 × 10^{-8} esu). This measurement was taken directly after poling, but after five days it showed significant decay to $d_{33} = 19$ pm V^{-1} (4.5 × 10^{-8} esu).

(**7**)

When the polymer analogy reaction is used, large amounts of NLO units cannot be introduced into the amorphous polymer backbone. To overcome this problem, one needs to prepare a monomer which contains NLO units with polymerizable functional groups. Yoshida et al.[24] reported the synthesis of one of this type of monomer, (S)-2-methacryloyloxymethyl-1-(4-nitrophenyl) pyrrolidine (**8**), which was obtained by the reaction of NPP and methacryloyl chloride. This monomer could easily be polymerized by the standard procedure for radical polymerization (equation 7). The obtained polymer (**9**) shows relatively low yields (9–52%) and viscosities, the two values which are dependent on reaction times and solvents. One interpretation of these results may be that the nitro groups of NPP act as an inhibitor for radical polymerization. The SHG values of this polymer have not been measured yet.

(7)

Miyata et al.[25] prepared simple vinyl monomers like phenyl acrylamide and phenyl maleimide derivatives, both having electron-donating and electron-withdrawing substituents (**10–14**). These monomers showed sizable SHG powder efficiencies 9–14 times greater than that of urea. Poly(phenyl acrylamide) can be prepared in polar solvents such as DMSO and has a molecular weight of several thousands. However, it could not be polymerized in THF using the radical initiator. The SHG efficiency of the polymer has not been reported.

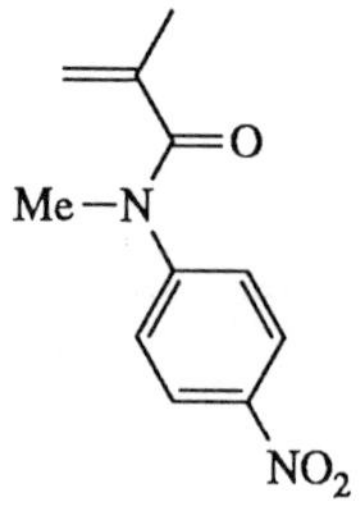

(10)

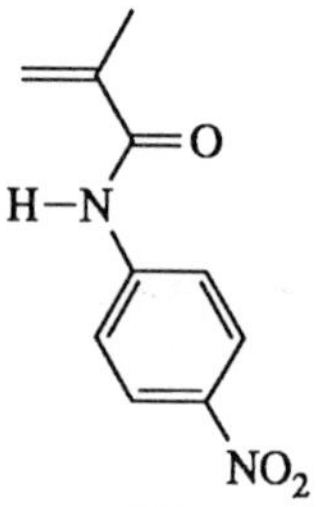

(11)

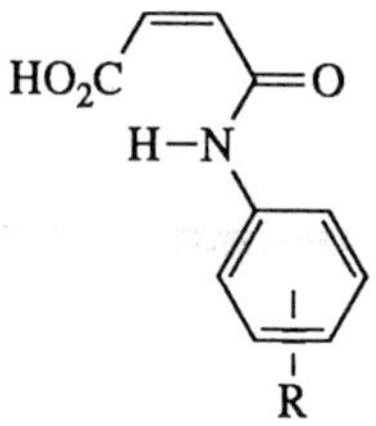

R = 2-F (12)
4-F (13)
2,4,6-tri-F (14)

16.2.3 Liquid Crystalline Polymers

A good approach to obtaining polymeric systems suitable for second-order nonlinear optical applications is that of incorporating NLO chromophores in liquid crystalline polymers (type C in Figure 2). Because liquid crystalline phases exhibit strong axial ordering of molecules, this approach can yield high second-order nonlinear optical properties. Meredith et al.[26] were the first to investigate the possibility of enhanced dipolar alignment by using host–guest interactions of a thermotropic nematic side chain liquid crystalline copolymer (15) with DANS (doped 2% by weight).

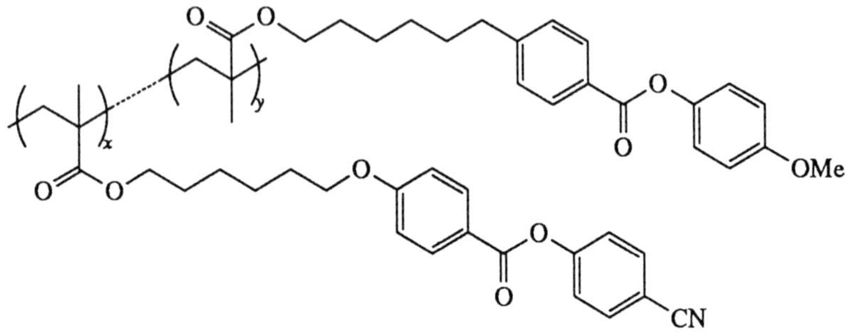

(15) $x = y = 0.5$

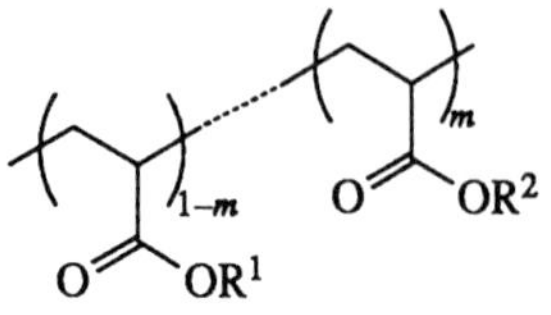

(16)

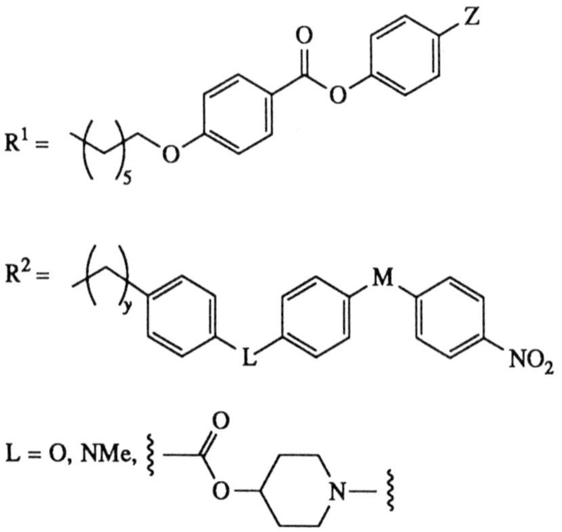

$L = O, NMe,$

$M = N=N, HC=CH$
$Z = CN, CF_3$
$y = 5,6$

The maximum value of $\chi^{(2)}_{111}$ of the film formed from this mixture was 3×10^{-9} esu when the film was poled with an applied field of $1.3\,\text{V}\,\mu\text{m}^{-1}$ at $20\,^\circ\text{C}$ for $6\,\text{h}$. This SHG intensity was 100 times larger than that which could have been obtained with a 2% DANS doped PMMA. This result indicates that the host–guest interactions in liquid crystalline polymer systems boost the SHG intensity very significantly. However, in many cases the miscibility of the dye molecules in the liquid crystalline polymers was basically poor. This problem can be solved by making a copolymer of a monomer containing the dye molecules and a monomer containing the mesogenic unit (type D in Figure 2). Noël et al.[27] have synthesized this type of copolymer, (16), which has the potential to demonstrate NLO behavior.

R^1 groups are known to yield nematic liquid crystalline polymers, and R^2 groups are dyes associated with high hyperpolarizability. However, the dyes which have strong electron acceptors, like the nitro group, generally act as inhibitors or retarders of polymerization, as already mentioned. As a result, the molecular weight (degree of polymerization) is reduced. This leads to poor mechanical properties and unsatisfactory nonlinear optical properties of the polymer system. Krongauz et al.[28] reported on a new polymer structure which has a high concentration of chromophores with high second-order hyperpolarizability β. The composition of this polymer consists of conventional mesogenic units and spiropyran units known as thermochromic and photochromic dyes. Even though the spiropyran has a nitro group, it does not directly inhibit polymerization. Here, several polymers (17), (18) and (19) have considerable SHG intensity themselves. In order to improve the $\chi^{(2)}$ values by alignment, these polymers can be mixed with quasi-liquid crystal (QLC) or high β dye molecules, like DANS and MBANS. For example, by doping 16.2% DANS in QLC1–(17)(1:4) blend, the $\chi^{(2)}_{zzz}$ value can be greatly enhanced to 60×10^{-9} esu when poled at $50\,\text{kV}\,\text{cm}^{-1}$ (6.4×10^{-9} esu at $10\,\text{kV}\,\text{cm}^{-1}$). In contrast $\chi^{(2)}_{zzz}$ of (17) alone when poled at $10\,\text{kV}\,\text{cm}^{-1}$, is considerably lower (0.6×10^{-9} esu).

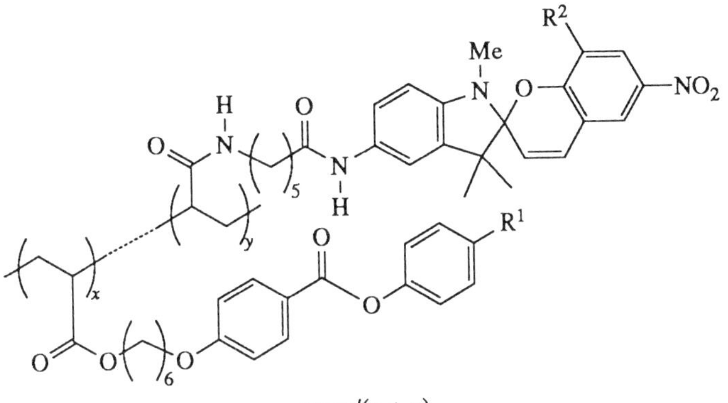

$z = y/(x + y)$

$R^1 = \text{CN}, R^2 = \text{H} \quad z = 0.15 \quad$ (17)
$R^1 = \text{OMe}, R^2 = \text{H} \quad z = 0.15 \quad$ (18)
$R^1 = \text{CN}, R^2 = \text{NO}_2 \quad z = 0.05 \quad$ (19)

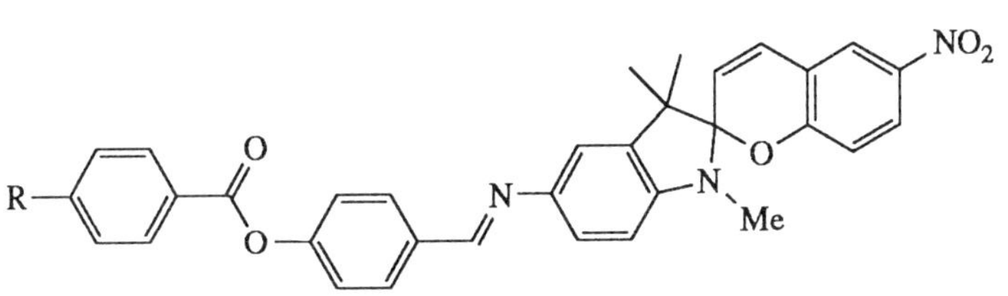

$R = \text{OC}_6\text{H}_{13}$ (QLC 1)
$R = \text{OMe}$ (QLC 2)

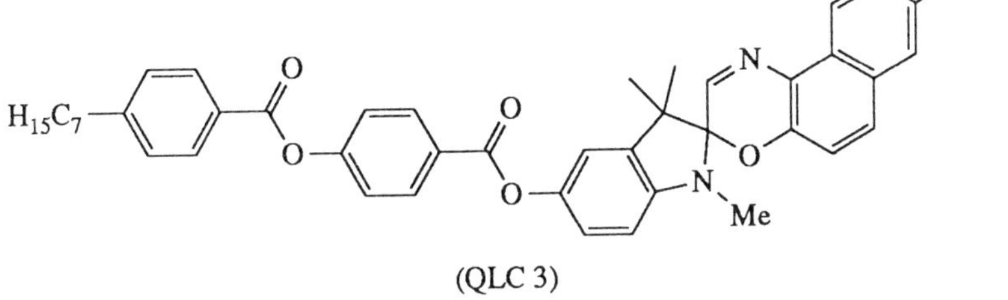

(QLC 3)

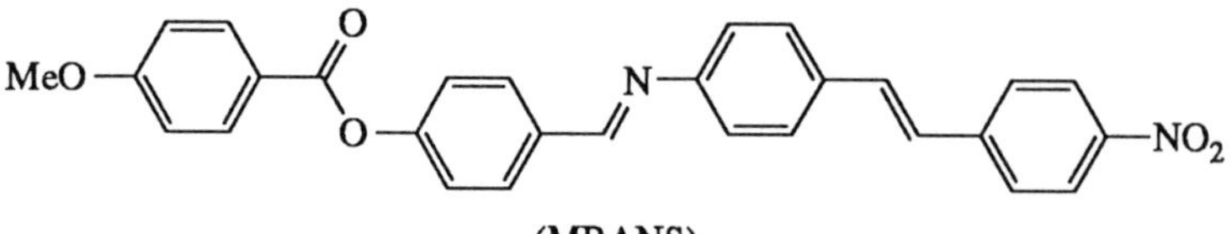

(MBANS)

An even more advantageous polymer structure can be formed if the NLO units to be attached to the polymer backbone can be made to act simultaneously as mesogens of the liquid crystal (type E in Figure 2). This type of polymer might be more stable than host–guest systems, and it can also achieve a higher concentration loading of dye molecules in the polymer backbone. Moreover, it can easily be aligned in the liquid crystalline phase. Such polymers have been prepared as homopolymers or copolymers by several research groups. The structure of one of these types of polymer is given by (20), which contains covalently bonded 4'-amino-4-nitrostilbenes and 4'-amino-4-nitroazobenzenes.[29] Homopolymers (20) and (21) displayed liquid crystalline textures when observed under a polarized microscope. The DSC curve of the stilbene dye homopolymer (20) exhibited two endothermic transitions, one at 161 °C ($T_s - T_{lc}$) and the other at 186 °C ($T_{lc} - T_i$).

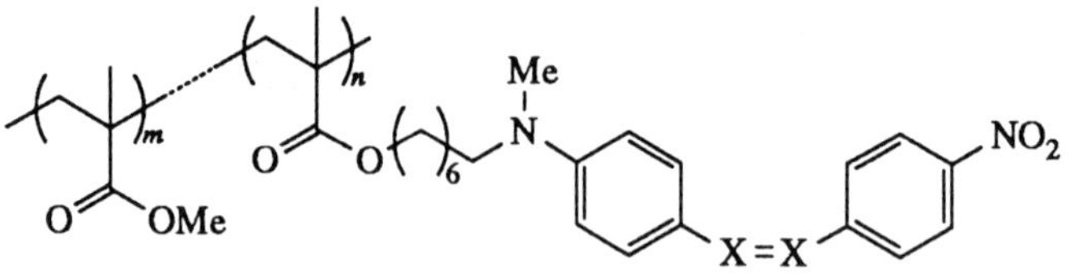

R = H, Me
X = CH, N
$0 \leq n/m \leq 40$

R = Me, X = CH, m = 0   (20)
R = Me, X = N, m = 0   (21)

Griffin et al.[30–32] also synthesized side chain liquid crystalline polymers which have nitroaromatic groups (22 and 23), pyridine N-oxide groups (24 and 25), or chiral nitroaromatic groups (35–40) as potential NLO materials. In order to overcome the retardation of free radical polymerization,

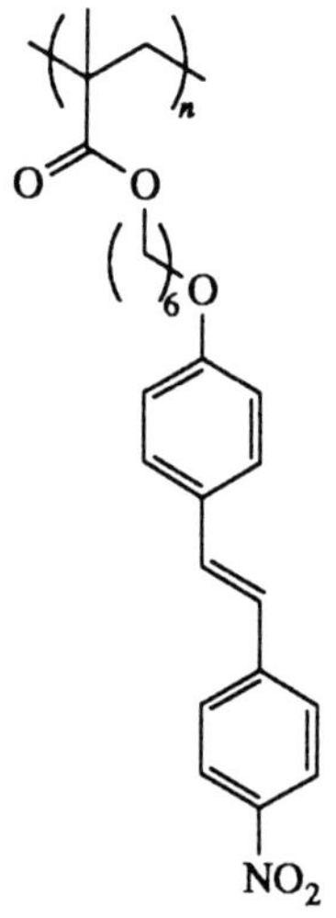

(22)

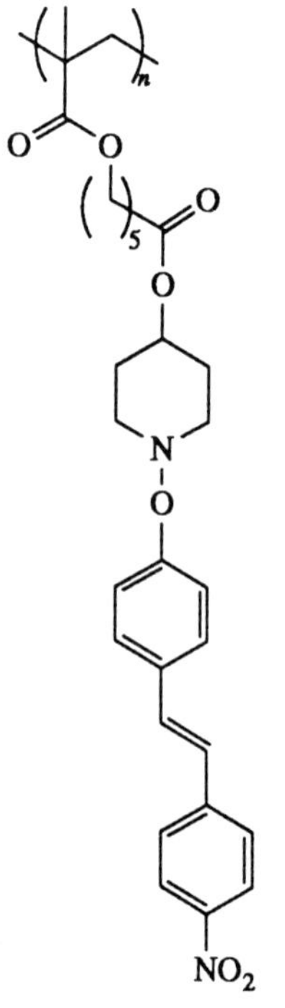

(23)

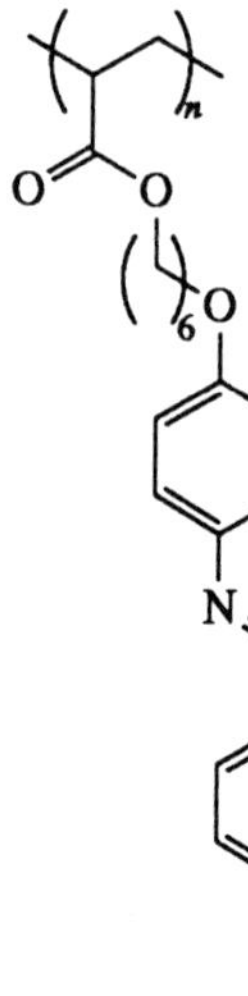

(24)

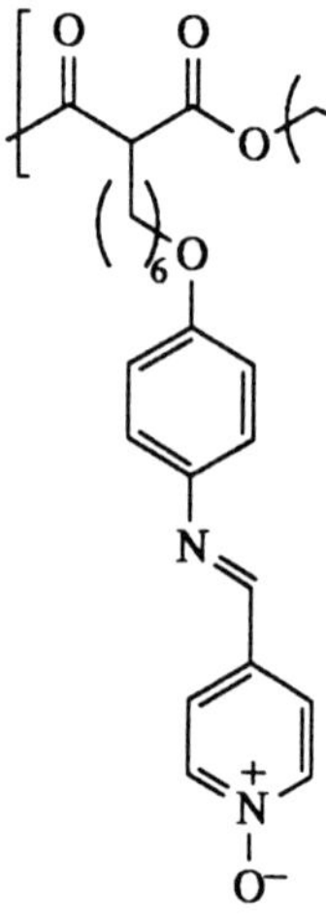

(25)

some of these liquid crystal polymers (**26–34**) were obtained by condensation polymerization.[32,33] From the group of these polymers, (**36**) was used to fabricate LB films.[34] According to the surface plasmon results, these LB films were homogeneous and their optical qualities were good when compared with orientational properties of LB films. Wijekoon et al.[35] have studied SHG in an electrically poled, thin film sample of (**36**; Figure 5). The $\chi^{(2)}_{zzz}$ value was 7.3×10^{-8} esu (monomer: 8.41×10^{-8} esu). The relaxation dynamics of this polymer shows that initial rate of decay of the SHG signal is larger in the case of the polymer than that of the monomer.

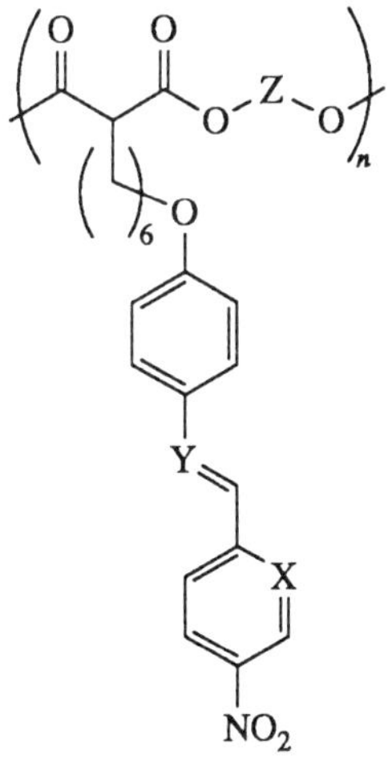

	X	Y	Z
(26)	CH	CH	$-CH_2CH_2CH_2CH_2-$
(27)	CH	CH	$-CH_2CH_2CH(Me)-$
(28)	CH	CH	$-CH_2CH_2CH(Me)CH_2CH_2CH_2-$
(29)	CH	N	$-CH_2CH_2CH_2CH_2-$
(30)	CH	N	$-CH_2CH_2CH(Me)-$
(31)	CH	N	$-CH_2CH_2CH(Me)CH_2CH_2CH_2-$
(32)	N	N	$-CH_2CH_2CH_2CH_2-$
(33)	N	N	$-CH_2CH_2CH(Me)-$
(34)	N	N	$-CH_2CH_2CH(Me)CH_2CH_2CH_2-$

Copolymer

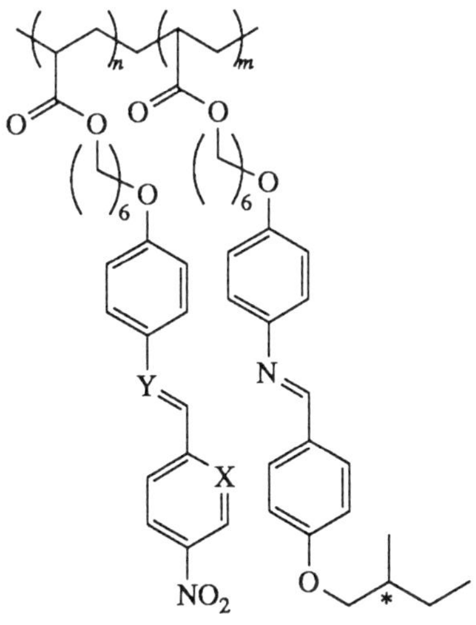

X = Y = CH (**35**)
X = CH, Y = N (**36**)
X = Y = N (**37**)

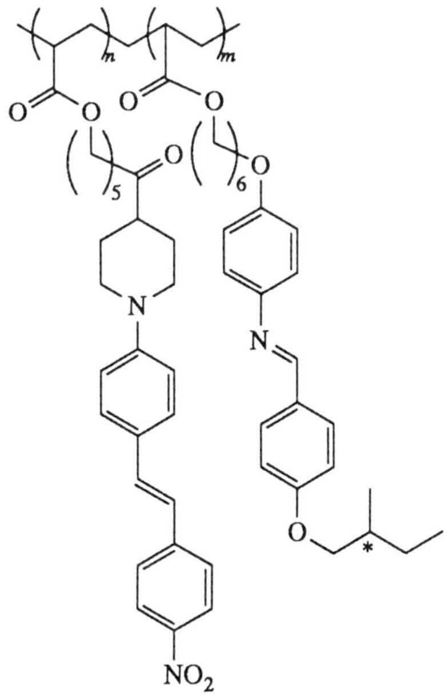

(**38**)

RP = radical polymerization
CP = condensation polymerization

The side chain liquid crystalline vinyl polymers (**41**), which contain piperazinyl nitrostilbene, have also been prepared by DeMartino et al.[36] using the radical polymerization method. They reported that one polymer from this series exhibited SHG $\chi^{(2)}$ of at least about 1×10^{-6} esu when measured at the fundamental wavelength of 1.91 μm.

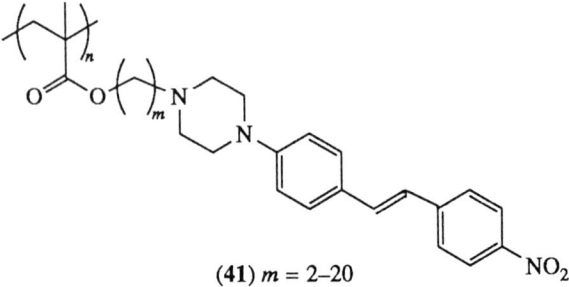

(39) RP

(40) CP

RP = radical polymerization
CP = condensation polymerization

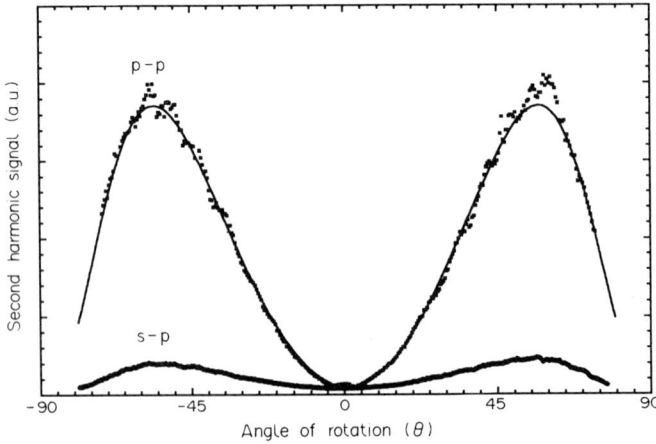

(41) $m = 2$–20

Figure 5 The SHG signal as a function of the angle of incidence for (**36**) for different input polarizations. The solid lines are the theoretical fits[35]

16.2.4 SHG Polymers with NLO Units in the Main Chain

In addition to side chain NLO polymers, both homopolymers and copolymers with the NLO units in the main chain have also been prepared (type F in Figure 2). These types of polymers may be able to provide a high dye concentration and have a potentially more stable alignment than the side chain polymers.

As a first example of main chain dye polymers, the high dipolar polymers incorporating a donor–acceptor quinodimethane have been synthesized by polyesterification using some quinodimethane monomers and methyl 12-hydroxydodecanoate (12-HDE).[37] A 5:1 copolymer of DAQ (**42a**) with 12-HDE obtained by melt polymerization was insoluble in common organic solvents and had a high T_g of 208 °C. A 1:1 copolymer was only soluble in hexafluoro-2-propanol (HFIP) and its T_g (27 °C) and T_m (62 °C) were low. Therefore, these polymers are not attractive as NLO materials. Copolyester obtained by reaction of DEQ (**42b**) and 12-HDE, using a two-step process in benzophenone with dibutyltin diacetate catalyst (equation 8), has increased solubility. This polymer (**43**) is soluble in dipolar aprotic solvents (DMF, DMSO and NMP) at elevated temperatures and can be melted. In this way it can be cast into films for fabrication of optical devices. However, due to the low T_g of this polymer, the dipolar units in the polymer backbone are easily relaxed and do not preserve their alignment after electric field poling.

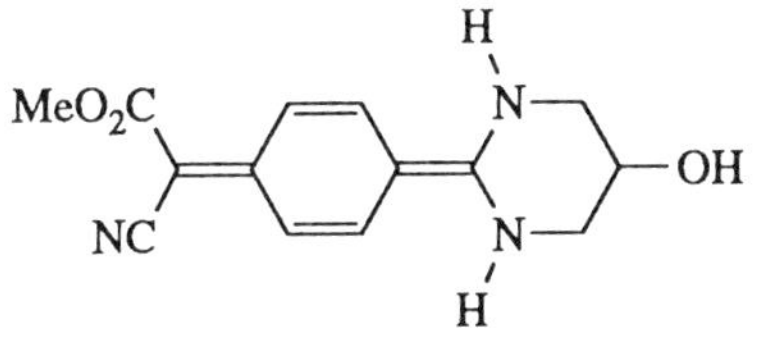

(**42a**) DAQ

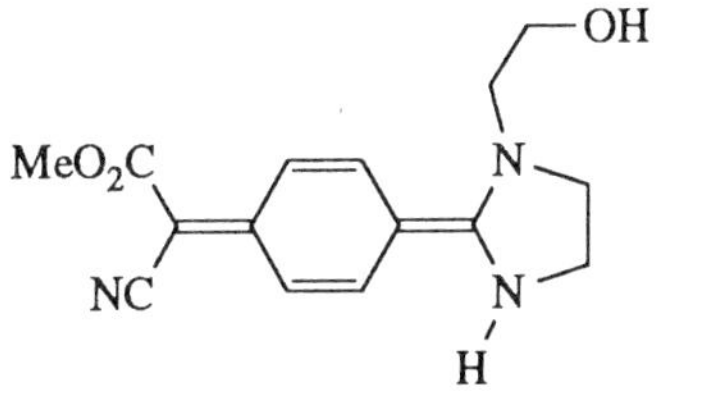

(**42b**) DEQ

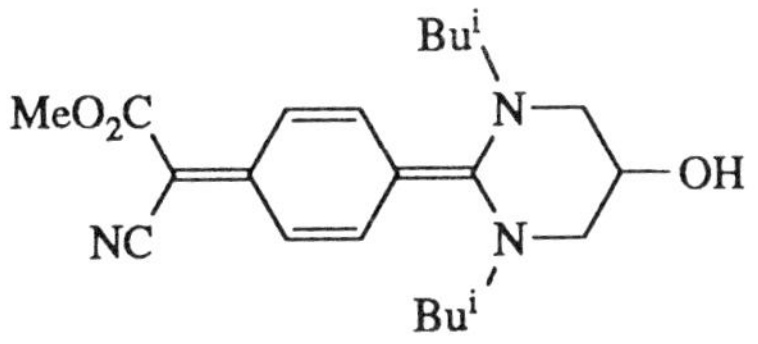

(**42c**) DIQ

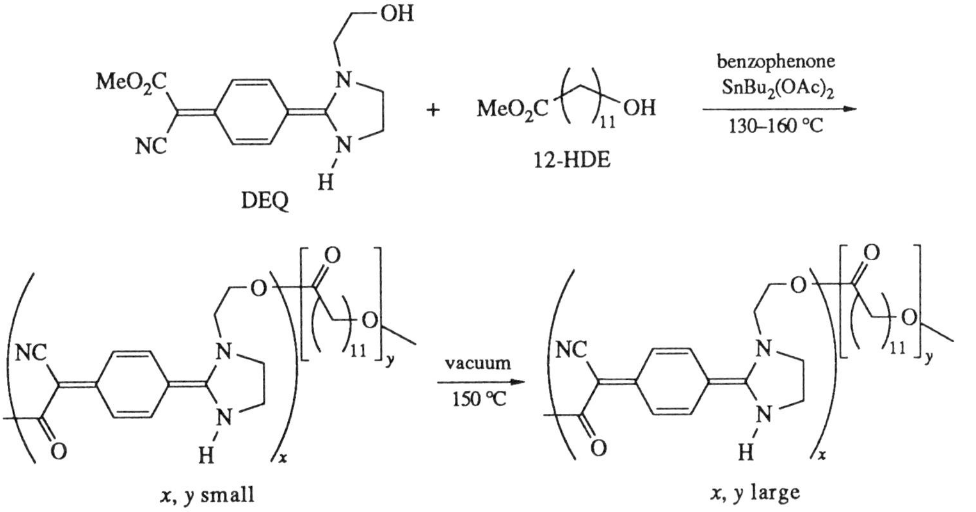

(8)

(**43**) random copolymer

Other materials of potential interest for second-order nonlinear optical properties are polymers which contain p-oxy-α-cyanocinnamate (POC) units. Such polymers were prepared by condensation polymerization of ω-hydroxy-α-cyano ester monomers with methyl 12-hydroxydodecanoate, by two-step high temperature polyesterification (equation 9).[38] The molecular weight and the density of dipolar units in the polymer backbone can be controlled by varying the second step reaction and the monomer feed ratio. These copolyesters (**44a**; equation 9) are easily soluble in a common solvent like dichloromethane. Solution-casted films are yellowish, transparent and flexible. By heating these films, they can be stretched to over 300% of their original length. When the polymer is oriented, it becomes cloudy in appearance due to formation of crystalline regions. The crystallization can be a positive factor since the mechanical properties of the polymer improve. On the other

Table 3 Average Monomer Susceptibilities for Various Molecular Weight Polymers[39,40]

Molecular weight, M_n[a]	Degree of polymerization, DP	$M_z B_z/n$ (10^{-48} esu)[b]
POC dye monomer	1	57
17 000	37	850
70 000	152	1140

[a] Determined by GPC relative to polystyrene. [b] Average monomer susceptibility.

hand, the crystalline phase formation decreases the optical quality which is a negative factor for fabrication of optical devices. One of the polymers (**44a**), which has good solubility in chloroform was chosen for the EFISHG measurement.[39,40] Data obtained for the polymers of two different molecular weights are compared to that recorded for the monomers in Table 3. The polymers show an enhanced susceptibility which is 20 times larger than that of the POC-dye monomer. The fact that the polymers are not random in solution but do have some ordering could be the cause of such a large degree of enhancement.

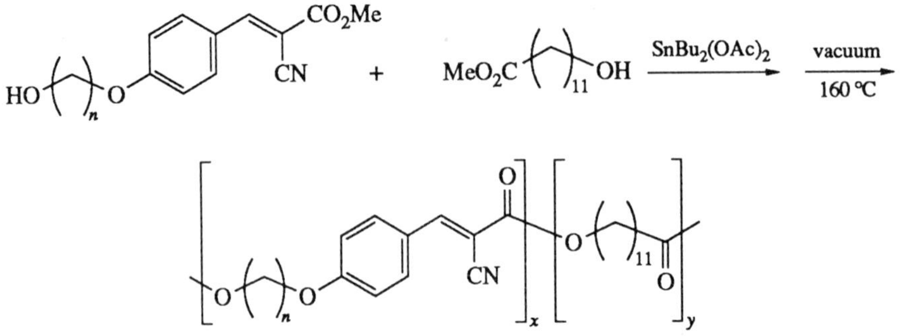

(9)

(**44a**) $n = 3$
(**44b**) $n = 6$

Stenger-Smith et al.[41] have synthesized a homopolyester (**47**) derived from 4-[N-ethyl-N-(2-hydroxy)amino]-α-cyanocinnamic acid (**45**) or ethyl-4-[N-ethyl-N-(2-hydroxy)amino]-α-cyanocinnamate (**46**). The T_g of the polymer made by solution polymerization of acid (**46**) varied between 60–90 °C, depending on polymerization conditions. The T_g of polymers made by melt polymerization of the monomer (**46**) were between 100–110 °C, due to their higher molecular weight. All polymers were soluble in chloroform, m-cresol and methylene chloride. The solution casted films from chloroform were poled near the T_g of polymers for 15–120 min. A summary of the preliminary results of the SHG studies indicates that d_{33} is about 7 ± 1.4 pm V^{-1} ($16.7 \pm 3.3 \times 10^{-9}$ esu).

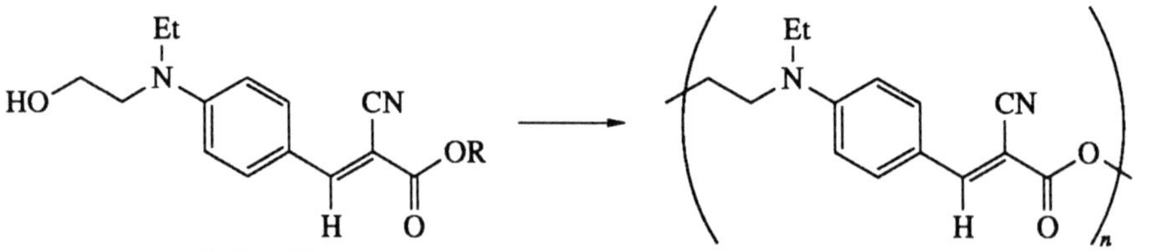

(**45**) R = H
(**46**) R = Et
(**47**)

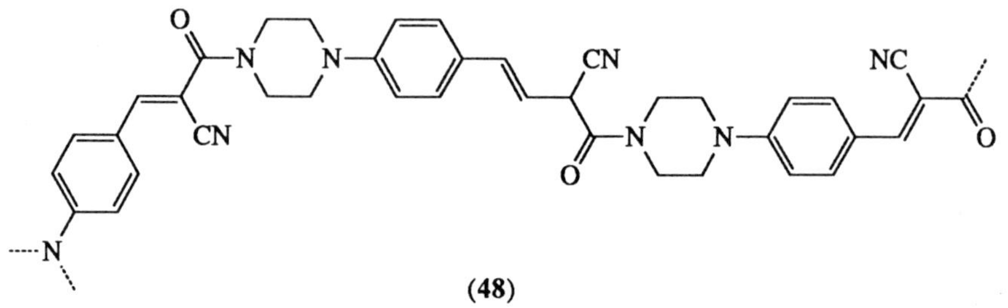

(**48**)

Finally another main chain dye oligomer which has piperazine-linked dyes has been reported by Katz and Schilling (**48**).[42] However, due to its limited solubility optical quality films of this dye have not been made.

16.2.5 Langmuir–Blodgett Films

The Langmuir–Blodgett (LB) technique of fabrication of thin oriented films has received much attention in the field of nonlinear optics since this technique enables one to build highly ordered organic thin films with control on a molecular level. Molecules that can be deposited by this technique must contain hydrophilic head groups and hydrophobic tail groups. When a solution of such molecules is spread on a surface of water and the solvent evaporates, the molecules are oriented with their heads down and tails up. The partly oriented film is then compressed with a moving barrier until it becomes a closely packed monolayer, which also improves the ordering. The film can be transferred onto a surface of a suitable substrate by dipping the substrate. Repeated dipping may serve as a way of obtaining multilayer films (Figure 6). In most cases, the monolayer deposited on the down stroke and that deposited on the up stroke will have opposite orientations which results in a head-to-head and tail-to-tail arrangement of the multilayer LB film. Such multilayer films are called Y-type LB films. There are also X-type films with all heads facing away from the substrate and Z-type films with all heads pointing towards the substrate. The Y-type films in which the polar group create a vertical dipolar arrangement (as shown in Figure 6e) possess a center of symmetry, and therefore cannot exhibit any second-order nonlinear optical properties. The X-type and Z-type films do not possess this symmetry and would therefore be suitable for NLO applications. However, X- or Z-type deposition is not very common and therefore special techniques are often used to obtain noncentrosymmetric (polar) multilayer LB films. One example is the technique of alternate deposition of two different molecules.

The concept for the synthesis of LB compounds is basically simple; the compound can usually be prepared by the reaction of highly polar dye molecules with hydrophobic chains having reactive groups. In other words, the dye is derivatized with straight alkyl chains. Generally the linkage between dye molecule and the hydrophobic chain are formed by a covalent or ionic bond. The chromophore groups can be incorporated into the general design of LB active molecules. Examples of chromophores can be those listed in Table 1. Several types of LB-forming compounds reported

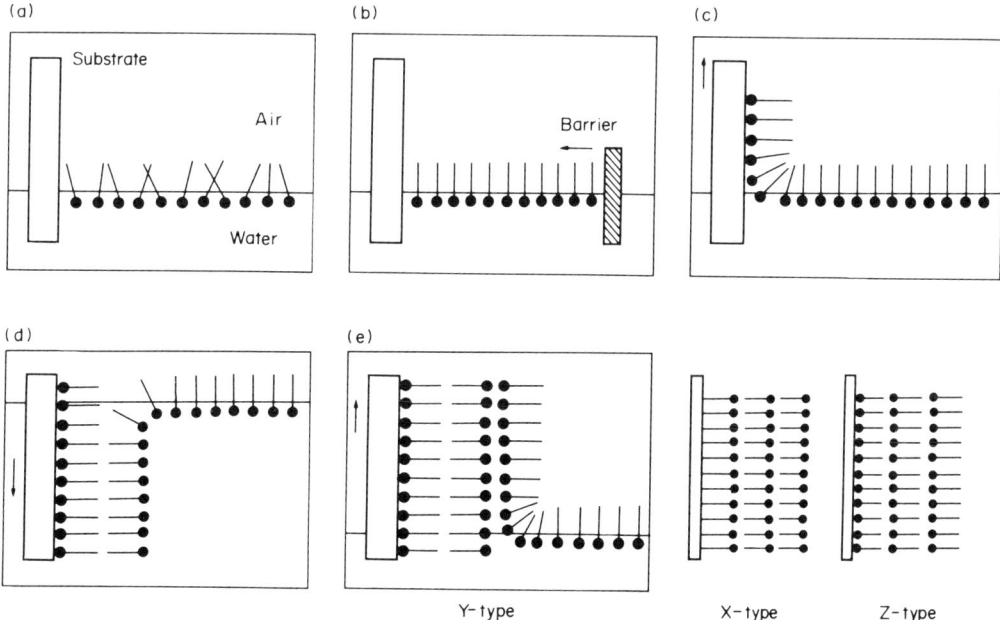

Figure 6 Deposition stages for the making of Langmuir–Blodgett films: (a) LB molecules spread on the water surface; (b) monomolecular layer compressed by the moveable barrier; (c) deposit of the monolayer to a suitable support; (d, e) building a multilayer film through repeated dipping (Y-type: head to head or tail to tail stacking, X-type: all heads up, Z-type: all heads down)

in the literature are listed in Figure 7. Among these types, the second group has been most widely utilized in the synthesis of NLO active LB film forming materials.

SHG studies of monolayers and interfaces have been described by Shen and his coworkers[43-45] and also by Aktsipetrov et al.[46-49] Since these studies were reported, many research groups have intensely studied the NLO properties of LB films fabricated from several kinds of compounds, such as merocyanine,[50,51] hemicyanine,[52-56] diazostilbene derivatives,[57-60] polyene,[61,62] nitropyridine and nitroaniline derivatives,[63-65] retinal schiff base[66] and bacteriorhodopsins.[67]

LB multilayer films obtained by the deposition of low molecular weight molecules are relatively unstable. Molecules are held together only by relatively weak Van der Waals forces and exhibit a relatively high freedom of motion. Therefore, the films are vulnerable to extreme conditions such as the thermal, chemical and mechanical stresses that can be encountered on device fabrication as well as during prolonged storage. This is especially true for polar LB multilayer films that can

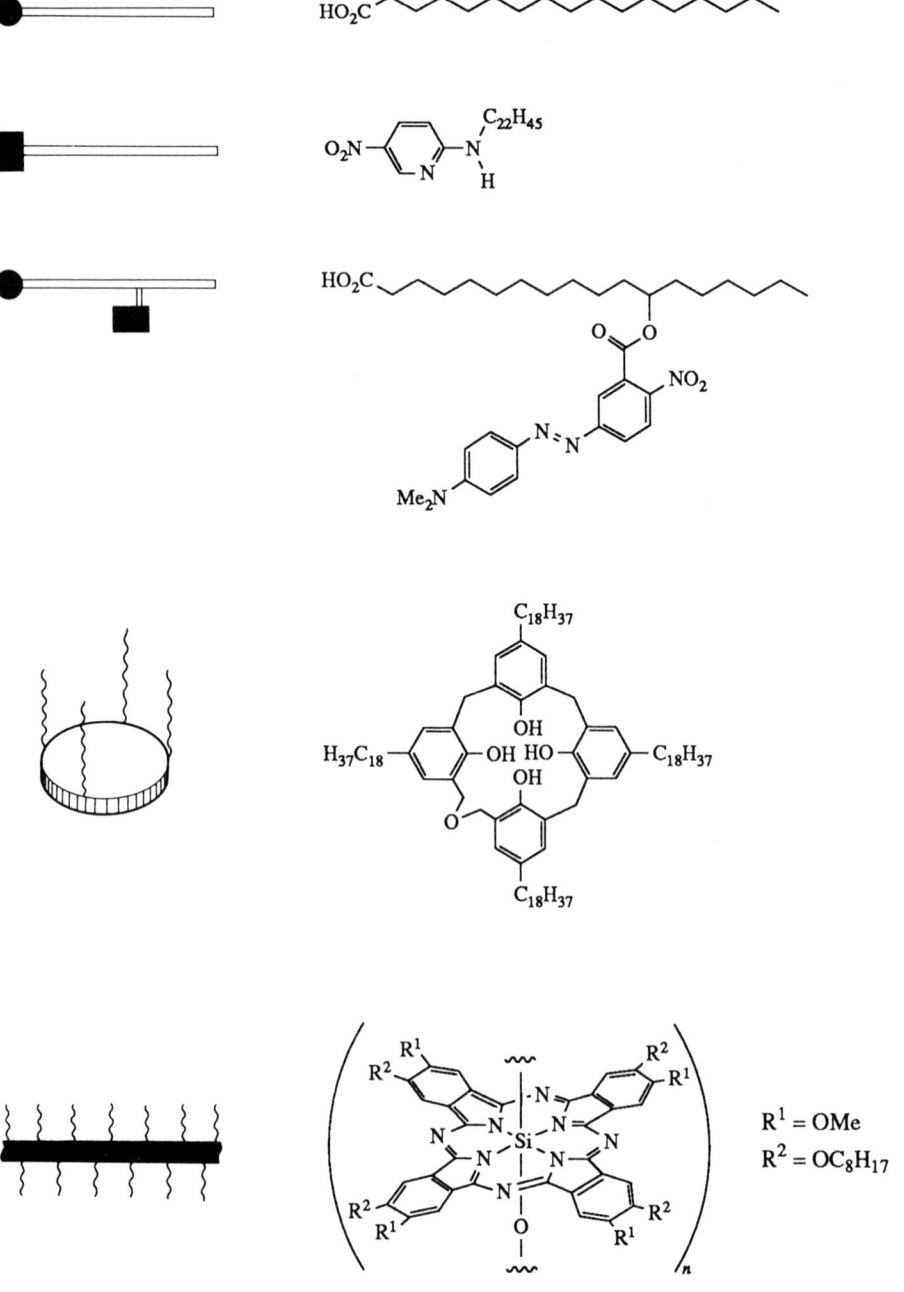

Figure 7 Various types of LB molecule

spontaneously relax to nonpolar arrangements. It is therefore more preferable to use polymeric structures to fabricate LB structures of higher environmental resistance and good mechanical strength. Following this line of reasoning, Stroeve et al.[68] carried out deposition studies of mixed monolayers of a merocyanine dye (**49**) and PMMA, which exhibited high optical quality in the solid state. From this mixture, a Z-type film with high PMMA content and a Y-type film which had high dye content were fabricated. Considerable SHG was observed from the Z-type LB film. However, the NLO activity observed in this film was drastically diminished by its exposure to air and water, as reported by Girling et al.[50] The activity of the film was recovered in an ammonia atmosphere.[69] The SHG intensities from the protonated sample (**50**) which were not kept in an ammonia vapor were 10 to 50 times weaker than those of the deprotonated sample (**49**).

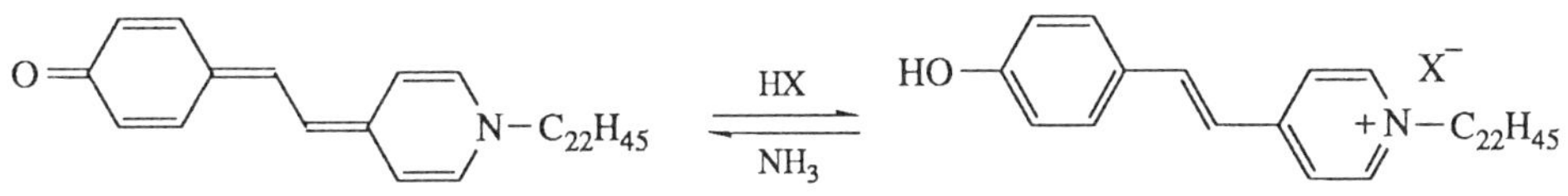

(**49**)

(**50**) X = Cl, HCO_3

The LB material was stable in air when the polymer was mixed with hemicyanine instead of merocyanine. The hemicyanine dye (**51**) can be synthesized by condensation of N,N-dimethyl-p-aminobenzaldehyde with docosyl picolinium bromide obtained by quarternization of γ-picolinol and docosyl bromide (equation 10).[52] Hayden et al.[70,71] have prepared monolayer and multilayer films from mixtures of this hemicyanine and either PMMA or poly(octadecyl methacrylate) (PODMA). The mixed system of hemicyanine and PMMA formed an almost perfect Z-type film, but hemicyanine in PODMA formed an incomplete Y-type film. Both of these films have good environmental stability and show high SHG intensity, however, the SHG intensity does not follow the expected quadratic dependence on the film thickness.

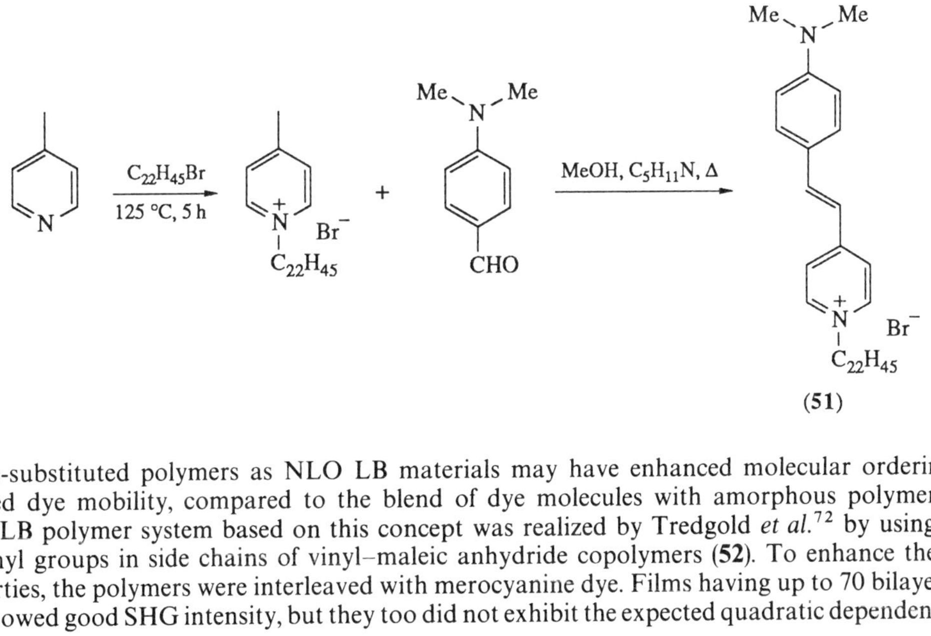

(10)

(**51**)

Dye-substituted polymers as NLO LB materials may have enhanced molecular ordering and reduced dye mobility, compared to the blend of dye molecules with amorphous polymers. The NLO LB polymer system based on this concept was realized by Tredgold et al.[72] by using alkyl-biphenyl groups in side chains of vinyl–maleic anhydride copolymers (**52**). To enhance the NLO properties, the polymers were interleaved with merocyanine dye. Films having up to 70 bilayers (300 nm) showed good SHG intensity, but they too did not exhibit the expected quadratic dependence.

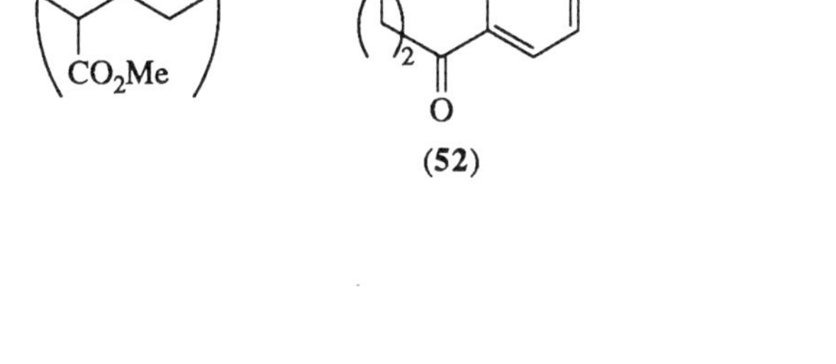

(**52**)

A similar type of polymer LB system was studied for its second harmonic activity by Anderson et al.[73] This system was created with dialkylaminostilbazolium poly(epichlorohydrin), a dye substituted polymer containing a hemicyanine-type dye which bears an aliphatic tail covalently bonded to its polyether backbone. This polymer can be synthesized by the reaction of picolinium polyepichlorohydrin with 4-ethyloctadecylaminobenzaldehyde (equation 11). This polymer dye (**53**) formed a stable monolayer that could readily be interleaved with behenic acid to form noncentrosymmetric structures. Unlike the hemicyanine dye, the acid did not show any aggregation in LB films. The SHG intensity of this polymer was comparable to that of the hemicyanine dye. Multilayers of the polymer mixed with the behenic acid showed an enhancement of SHG, even though the quadratic dependence expected was still not obtained.

An improved structure for polymeric LB films was designed by Anderson et al.[74] They constructed a type of polymer which consists of a hemicyanine dye substituted into the side chain of the polyether backbone. Polymer (**54**) has 43 repeating units per molecule with a dye attached to 47% of the units, and polymer (**55**) has 11 repeating units per molecule with a dye attached to 33% of the units. No crystalline or LC transition was detected in the thermal analysis of this material, and it was stable to about 230 °C in N_2 atmosphere. The mixture from these polymers formed noncentrosymmetric Y-type LB films, and the SHG intensity enhancement observed in the first four bilayers exhibited quadratic dependence, although successive bilayers showed a linear dependence. Then, for the prevention of hydrophilic–hydrophilic interaction between polymer–dye bilayers, behemic acid was interleaved, thus allowing the mixture to show quadratic increase up to ten layers.

Also, significant NLO effects had been observed in preformed polymer LB films which had 50% substitution by the azo dye molecules on the poly(dimethylsiloxane) polymer side chains (**56**).[75] The approximate value of second order polarizability for the monolayer of this polymer was $\beta = 8.5 \times 10^{-29}$ esu at 1.06 μm. No deposition of multilayer films has so far been reported.

(56) $n = 9 \pm 2, m = 8 \pm 2$

16.3 THIRD-ORDER NONLINEAR OPTICAL POLYMERS

In general, the complexity of third-order nonlinear optical phenomena is much higher in comparison to the second-order effect. There are numerous microscopic mechanisms which give rise to material behavior that can be formally categorized as a third-order nonlinear optical process. This complexity also contributes to common misconceptions in the assessment of material properties and comparisons between various materials. As a result the materials development for third-order nonlinear processes is not as advanced as that for the second-order processes.

As already mentioned, the third-order nonlinearity is derived from the presence of a cubic term in the power series expansion of the material polarization *versus* the electric field acting on the material. This treatment leads to various frequency mixing schemes, among which third harmonic generation is perhaps the simplest process but of rather low technological importance. Processes of greater importance are those which involve modification of material properties whereby an intense (control) beam of light can modify the propagation of other beams of light. In other words, one can control light by light. Examples of such processes are: gating and limiting action, real time holography, self phase modulation, *etc.*[1]

Within the power expansion presented in equation (1), all these processes are treated, however, as instantaneous interactions which are local in time, *i.e.* depend only on the magnitude of the electric field at a given moment of time. This is an obvious oversimplification and its rigorous fulfilment would mean, for example, the absence of any frequency dispersion of the third-order nonlinearity. In a real situation, any interaction of the electric field with matter is not rigorously instantaneous and the time scale of the phase lag of polarization with respect to the electric field depends on the microscopic processes involved. Just as electronic, atomic and orientational contributions to first-order polarization of a dielectric manifest on different time scales, the third-order polarization can also involve various contributions which are characterized by different temporal responses. In the case of purely electronic contribution to the third-order nonlinearity, the response should be practically instantaneous when the frequencies of the interacting fields are far from material resonances. One usually calls this a nonresonant nonlinearity.

In many cases, however, the interaction of electric fields yields Fourier components of low frequency that are able to induce relatively slow polarization components such as reorientation of molecules or molecular fragments. The presence or importance of such components in a third-order nonlinear response will then depend on the character of the laser beams used, *i.e.* continuous wave or pulsed and, for pulsed beams, on the pulse duration.

The picture is even more complicated if any frequency of the interacting beams lies within the range of material absorption. The resonant nonlinear response observed under such conditions will depend on the properties of excited species which may be formed upon absorption of light. A well-known example of nonlinear behavior under resonant conditions is saturable absorption. The properties of a material are simply modified because of the depletion of molecules (chromophores) in ground state and subsequent generation of excited species. If the excited species do not absorb at the frequency of ground state resonance, then one observes the phenomenon of saturable absorption. The presence of excited species may also modify the refractive index of the material, thereby contributing to the so-called nonlinear refractive index n_2. The contribution of such resonant behavior will also depend on the time scale of the experiment. If the duration of the laser pulses is longer than the lifetime of excited species then the contribution is integrated over the pulse duration and may be very large, especially for long-lived excitations.

Apart from the complications mentioned above, many authors characterize the third-order nonlinear materials simply with an effective value of $\chi^{(3)}$, which may be used to estimate the magnitude of the nonlinear response under specified conditions of the laser pulse duration and power. One should, however, use caution in comparing the values quoted by different authors, especially if relatively long laser pulses are used and there is a suspicion that resonant contributions may exist.

Among various materials which have been suggested for use in third-order NLO devices, organic materials have to compete with inorganic semiconductors and semiconductor structures such as quantum wells.[76] While these inorganic semiconductors offer very promising high nonlinearities, they rely largely on resonant effects. Therefore, there is a trade-off between the magnitude of the nonlinear response and the characteristic response time. Resonant effects also impose limits on the maximum power level of the incident laser beam and on the interaction length, due to absorption losses.

Purely electronic nonresonant third-order nonlinear properties of organics can therefore be considered attractive, even if the obtainable magnitude of $\chi^{(3)}$ is lower than the effective values reachable under resonant conditions. Also, processability of organics and the ease of modification of their structures make them interesting for development of third-order nonlinear materials.

One possible approach to building polymeric materials of optimized third-order nonlinear properties is by molecular engineering in which one uses the knowledge acquired by systematic studies of low molecular weight materials. Structure–property relationships can be identified and employed. Figure 8 shows a result of one such study[77] in which the influence of the increase of the π-electron system and the effects of substitution are investigated in a family of compounds derived from benzothiazole. Based on such studies, synthetic strategies can be developed resulting in optimized polymeric structures.

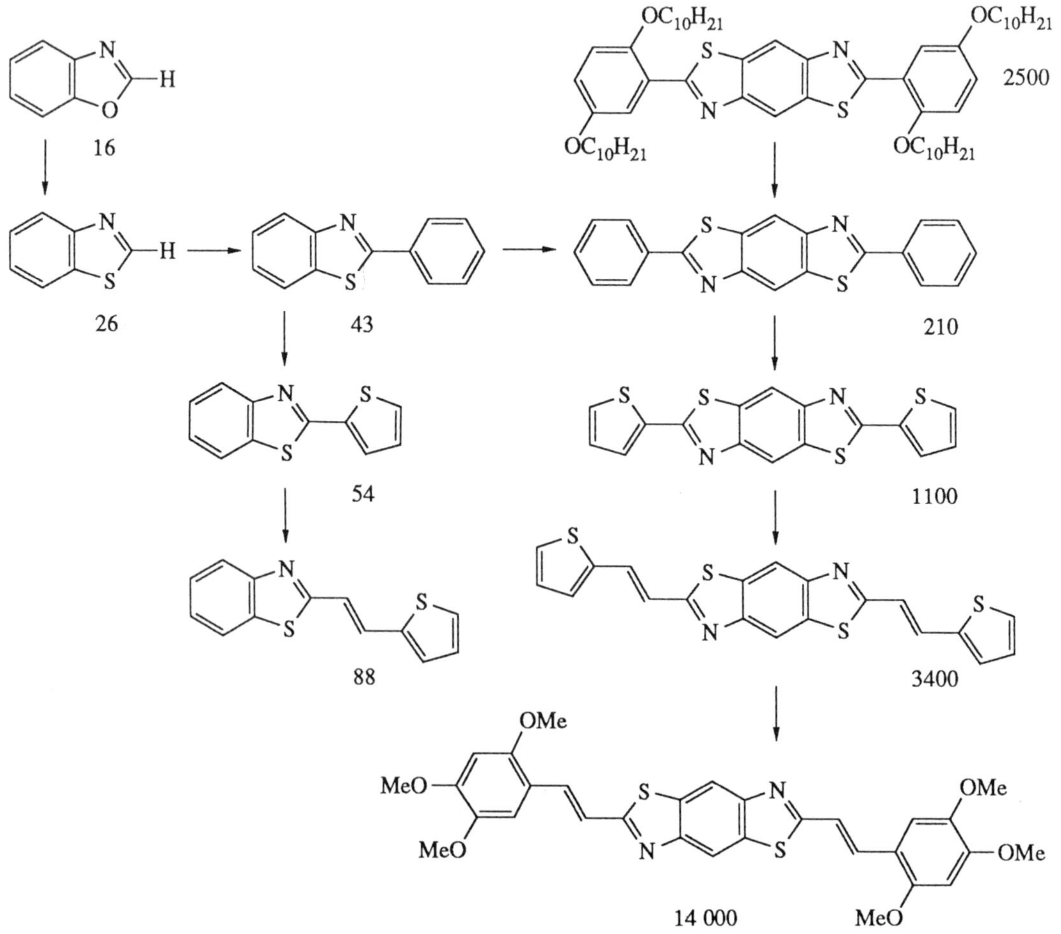

Figure 8 Third-order nonlinear susceptibility (γ) depending on structural changes in benzothiazole model compounds. Corresponding γ values are listed below the structure; all γ values are multiplied by 10^{36} esu[77]

Table 4 The $\chi^{(3)}$ Values of Some Inorganics and Polymers

Systems	Structure	λ (μm)	$\chi^{(3)}$ (10^{-12} esu)	Ref.
GaAs		a	12	81
InSb		a	50	81
Polyimide, LARC-TPI		0.602	~2	82
Poly(p-phenylene bisbenzothiazole)		0.602	~10	82
Poly(phenylacetylene)		0.602	50	82
Poly(p-phenylenevinylene)		0.602	~400	82
Polydiacetylene–PTS	R = R' = –CH$_2$OSO$_2$–C$_6$H$_4$–CH$_3$	2.620 1.890	160 850	83 83
Polythiophene (electrochemically deposited)		0.602	~10^3	84
Ladder polymer		0.532	1280	85

[a] Far from the band gap.

Treatment of properties of polymers as extrapolation of properties of the corresponding monomers and oligomers has its drawbacks. Conjugated polymers which exhibit novel properties that are not apparent from the knowledge of the behavior of the corresponding low molecular weight molecules are an important class of third-order NLO materials. The quasi-one-dimensional character of π-conjugated polymers allows for creation of novel excitations such as solitons (in systems with degenerate ground state), polarons and bipolarons. While the presence of such excitations is of utmost importance for the magnitude and dynamics of the resonant third-order NLO response of polymers, their influence can also be important off resonance.[1]

Two microscopic models have often been used to describe resonant third-order nonlinearity in π-electron conjugated polymers. One of them is a phenomenon called phase space filling.[78] Roughly speaking, the nonlinearity is considered here to arise from the reduction of the oscillator strength of the absorption transition due to the filling of the phase space by excitons. One can show that under such assumptions the nonlinearity will depend on the geometric size of an exciton, *i.e.* its length and the cross section. The other mechanism considers the effects due to the substantial shift in the oscillator strength occurring upon change of geometry of one-dimensional chains which accompanies formation of solitonic or polaronic species. Both processes can contribute and their relative importance may depend on the time scale of the interaction of light with the material.[79,80]

Table 4 shows some representative values of the third-order nonlinear susceptibility $\chi^{(3)}$ determined by various authors for several representative NLO polymers. Unfortunately, a direct comparison of the $\chi^{(3)}$ values is risky, for the reasons outlined above.

Below, we provide some detailed surveys of several classes of NLO polymers investigated to date, together with some comments on the possible microscopic mechanisms involved.

16.3.1 Polydiacetylenes

Polydiacetylenes (PDAs; **57**) were among the first conjugated polymers investigated for third-order NLO properties. One of the reasons why PDA structures have been of extreme interest to both chemists and physicists is that some PDAs can be obtained by topochemical polymerization of suitably substituted diacetylenes (equation 12).[86-88] In a favorable case the polymerization can be carried out on a single crystal of the diacetylene monomer to obtain a nearly perfect single crystal of the polymer. Therefore, investigations of perfectly ordered crystals of π-conjugated polymers are possible as are comparisons between the properties of monomers and polymers, polymerization kinetics and so on. Polymerization results in the formation of a π-electron rich chain which in most cases exhibits an electronic structure well approximated by the double–single–triple–single bond sequence as presented in equation (12). This highly delocalized electronic structure results in very high optical nonlinearities compared to the corresponding monomers or oligomers with a limited conjugation length. For example, for the poly(bisphenylurethane) of 5,7-dodecadiyne-1,12-diol (TCDU), the $\chi^{(3)}$ value of the polymer has been estimated to be 580 times larger than that of the monomer.

The $\chi^{(3)}$ values of PDAs have been measured using various techniques, such as third harmonic generation (THG), electric field induced second harmonic generation (EFISHG), degenerate four wave mixing (DFWM), surface plasmon waveguide coupling (SPWC) and the Kerr gate (KG)

Table 5 The $\chi^{(3)}$ Values of Some Dialkynes and Several PDAs

Structure	Abbreviated name	Material form	Measurement technique[a]	λ (μm)[b]	$\chi^{(3)}$ (10^{-12} esu)	Ref.
Monomers						
Et₃Si—≡—≡—SiEt₃		Solution	THG	1.064	0.11	89, 90
			THG	1.097	0.12	89, 90
Ph—≡—≡—Ph		Solution	THG	1.064	1.8	89, 90
			THG	1.907	0.92	89, 90
(C₅H₄)₂Fe—≡—≡—Fe(C₅H₄)₂		Solution	THG	1.907	0.225	89, 90
(dialkyne diurethane structure)	TCDU	Crystal	THG	1.89	0.12	83
Polymers						
R = R' = (CH₂)₄OC(O)NHPh	Poly-TCDU	Crystal	THG	2.62(∥)	37 ± 14	83
				2.62(⊥)	< 0.4	83
				1.89(∥)	70 ± 50	83
				1.89(⊥)	< 0.7	83
R = R' = CH₂OSO₂-C₆H₄-CH₃	Poly-PTS	Crystal	THG	2.62(∥)	160 ± 100	83
				2.62(⊥)	< 2	83
				1.89(∥)	850 ± 500	83
				1.89(⊥)	< 10	83
		Crystal	THG	1.94	11	91
		Crystal	DFWM	0.651	9000	92
				0.701	500	92
R = R' = (CH₂)₄OSO₂-C₆H₄-CH₃	Poly-PTS-12	Solution	EFISHG		500	93
R = R' = (CH₂)₄OC(O)NHCH₂CO₂Bu	Poly-4-BCMU	Cast film (red)	DFWM	0.602	400	94
		Cast film (yellow)	DFWM	0.602	25	94
		Cast film (doped with PMMA)	KG	$\hbar\omega_1$ = broad band	≤ 300	95, 96
			KG	$\hbar\omega_2$ = 1.99 eV	30	95, 96
		LB film (red)	THG	1.06	30	97
R = (CH₂)₁₅CH₃; R' = (CH₂)₈CO₂H	Poly-15/8-acid	LB film 500 Å	SPWC	0.65	400	98, 99
			SPWC	0.75	40	98, 99
R = R' = CH₂OC(O)NHBu	DAUH	Vacuum-deposited film (blue)	THG	1.90	14	100
				2.05	6.5	100

Table 5 (continued)

Structure	Abbreviated name	Material form	Measurement technique[a]	λ (μm)[b]	$\chi^{(3)}$ (10^{-12} esu)	Ref.
R = R' = –C$_6$H$_4$–N(H)–C(O)–C$_{17}$H$_{35}$		Cast film	THG	1.9	14	101
R = R' = –(CH$_2$)$_3$–O–C(O)–NMe$_2$		Cast film	THG	1.9	5.2	101
R = R' = –C$_6$H$_3$(CF$_3$)(CF$_3$)–	DFMP	Crystal	THG	1.83	24	91
		Crystal	THG	1.94	26	91
R = R' = –C$_6$F$_4$–Bu	BTFP	Crystal	THG	1.83	280	91
		Crystal	THG	1.94	130	91
R = R' = –CH$_2$–O–C(O)–N(H)–C$_6$H$_4$–Et					62 (calculated value)	102
R = R' = –(CH$_2$)$_3$–O–C(O)–N(H)–C$_6$H$_4$–Et					21 (calculated value)	102
(Fused silica glass)[c]					0.028	103
(Fused quartz)[c]					0.029	104

[a]Measurement techniques: THG, third harmonic generation; EFISHG, electric field-induced second harmonic generation; DFWM, degenerate four-wave mixing; SPWC, surface plasmon waveguide coupling; KG, Kerr gate. [b]Parallel ($\|$) to the polymer chain axis, or perpendicular ($\perp$) to the polymer chain axis. [c]Standard reference material for calculation of THG.

method.[1] The samples can be prepared in several forms, like crystals, solutions, solution-cast films and LB films. The reported $\chi^{(3)}$ values of several PDAs are listed in Table 5.

The first THG measurement of the copolymer of PDA and the poly[bis(p-toluene sulfonate)] of 2,4-hexadiyne-1,6-diol (PTS) was made by Sauteret et al.[83] They reported a $\chi^{(3)}$ value of $(8.5 \pm 5) \times 10^{-10}$ esu along the polymer chain direction at a wavelength of 1.89 μm. The corresponding $\chi^{(3)}$ value along the transverse direction was found to be less than 10^{-11} esu. This result is in agreement with the theoretical prediction that under nonresonant condition the largest component of the $\chi^{(3)}$ tensor for a one-dimensional conjugated polymer is along the π-electron delocalized polymer backbone.[83,102] Hermann and Smith[105] also reported $\chi^{(3)}$ measurement for PDA–PTS, but their value was 3×10^{-9} esu at 1.9 μm. For a good optical quality PTS polymer crystal obtained by the shear-growth technique, a $\chi^{(3)}$ value of 9×10^{-9} esu was reported at 0.65 μm (the absorption edge).[106] This value is the largest among the reported $\chi^{(3)}$ values for PTS.

Another interesting polymer series, PDA–BCMU, has also been widely investigated due to its solubility in common organic solvents, a characteristic which enables it to be cast into films. Moreover, the BCMU polymers can be deposited in the form of LB films. The various BCMU polymers, for example 3-BCMU, 4-BCMU and 9-BCMU differ by the number of methylene groups in the side chain. Of these polymers, 4-BCMU is the most extensively investigated. The BCMU polymers exhibit interesting chromism, for example solvatochromism. Due to the existence of a rod-to-coil transition and agglomeration of the chains, these polymers can exist in several forms, usually called the 'blue' form, the 'red' form and the 'yellow' form, which exhibit different absorption spectra and different nonlinearities. The $\chi^{(3)}$ values reported for 4-BCMU are in the range 4×10^{-10}–1.1×10^{-9} esu.[94,97,107–109] Numerous studies on BCMU PDAs have shown that the third-order nonlinearity is strongly influenced by one-photon and two-photon resonances behavior.[110,111] Even in the case of the yellow form of PDAs, the molecular hyperpolarizability and the third-order susceptibility is often found to be complex indicating a contribution from two-photon absorption. In addition to the TCDU, PTS and BCMU polymers, several PDAs with different side substituents R and R' have been prepared. Some of these compounds are available in single crystalline form, but their reported THG intensity does not exceed that of the earlier polymers. However, the potential for improvement of the NLO properties of these PDA still exists due to the possibility of structural modification through the attachment of effective side groups.

16.3.2 Polyenes and Polyacetylenes

The simplest structure of a π-conjugated polymer is that of polyacetylene. An understanding of the properties of polyacetylene and its derivatives may be facilitated by studies on model compounds with alternating single and double bonds. A considerable amount of theoretical[112-118] and experimental[119-121] work has been done on various polyenes with different conjugation lengths and *cis* or *trans* configurations.

An interesting polyene derivative is β-carotene (**58**). THG studies performed on β-carotene by Hermann et al.[122] and Meredith et al.[123] yielded $\gamma = 4.8 \times 10^{-33}$ esu at 1.89 μm and 1.1×10^{-32} esu at 1.908 μm, respectively.

(58)

In particular, cyclopolyene compounds can be polymerized by the ring-opening metathesis reaction. Perry et al.[90,124] reported that mixtures of cyclooctatetraene (**59**) and 1,5-cyclooctadiene (**60**) lead to soluble conjugated polymers (**61**) when polymerized using a tungsten carbene complex (equation 13). The composition of (**61**) can be controlled by varying the ratios of cyclopolyenes. In these copolymers, it is possible to have segments of five, nine and 12 double bonds derived from one, two or three consecutive cyclooctatetraene monomers. A small amount of segments (less than 2% by weight) which have more than 13 double bonds is also found to be present. The reported value of $\chi^{(3)} = \sim 2.4 \times 10^{-12}$ esu for the copolymer derived from 32% of (**59**) molecules may be due to a presence of the relatively long polyene segments. This value is comparable to β-carotene, which has 11 conjugated double bonds with $\chi^{(3)} = 9.1 \times 10^{-11}$ esu.

(59)   (60)   tungsten carbene complex catalyst

(13)

(61)

NLO properties of polyacetylene (PA) have been studied by many workers.[125-132] PA can be produced in the form of either a *cis* (**62a**) or a *trans* (**62b**) isomer.

(62a) *cis*-PA   (62b) *trans*-PA

The *trans* form is favored at high temperatures. Thus the conversion of *cis*-PA to the *trans* form can be easily accomplished by heating the former to about 150 °C for several hours. On the other hand, nearly pure *cis*-PA can be obtained by washing out the *trans* form at a low temperature (below −78 °C) after polymerization. The *cis-trans* content of certain samples can be checked by infrared spectroscopy. Kajzar et al.[128] studied the NLO property of *trans*-PA by THG; the nonresonant $\chi^{(3)}$ value was found to be $\sim 10^{-10}$ esu over the range of 1.17–1.50 eV and $\sim 10^{-9}$ esu at ~2.0 eV. Sinclair et al.[129,130] have also investigated the NLO behavior of *trans*-PA and *cis*-PA at 1.06 μm. The $\chi^{(3)}$ value of the *trans* form was about 4×10^{-10} esu, but the *cis*-PA value was 15 to 20 times smaller.

A derivatized PA structure, poly(phenylacetylene), was studied by Prasad et al.[133] The value of $\chi^{(3)}$ measured using the DFWM technique was $\sim 10^{-11}$ esu, lower than that of pure PA. Perhaps the poly(phenylacetylene) structure has a reduced π-conjugation character resulting from conformational distortions by the phenyl groups. In addition, Prasad et al.[134] have reported an interesting method to increase the stability of PA in air and increase its processibility. They used a graft copolymer PA–PMMA. This type of polymer can be casted in solution and shows enhanced stability in air. The $\chi^{(3)}$ value obtained by DFWM was $\sim 10^{-10}$ esu at 0.602 μm.

16.3.3 Conjugated Polymers Prepared by the Polyelectrolyte Precursor Technique

Many conjugated polymeric structures have the disadvantage of being almost totally nonprocessible. The processibility can be improved by attaching aliphatic and/or polar side groups to the polymer chain. Another approach has been developed for poly(*p*-phenylenevinylene) (PPV)[135–139] and its analogs,[140–145] poly(2,5-thienylenevinylene) (PTV),[139,146,147] poly(1,4-naphthalenevinylene) (PNV)[148–150] and their copolymers.[151–153] This approach involves a soluble precursor polymer route. The precursor polymer having good solubility and high molecular weight is cast as a film and then converted to the final conjugated polymer by chemical or thermal treatment. An appropriate precursor polymer is prepared by polymerization of the corresponding monomer, bis(sulfonium) salts, in aqueous solution. The cast precursor polymer film can be uniaxially drawn during the thermal elimination reaction. Due to stretching, the conjugated chain is expected to align along the draw direction, thereby enhancing $\chi^{(3)}$ along this direction.

Among the aforementioned polymers, PPV has been most extensively studied as a result of its easy synthesis and good stability of the corresponding precursor. A typical synthetic route to produce PPV (**64**) through the precursor polymers (**63a**) and (**63b**) obtained from the salts is shown in equation (14).

$$X = Cl, Br \qquad (63a) \; M = Me_2 \qquad (64)$$

$$(63b) \; M = \text{cyclohexyl}$$

(14)

This polymer can readily be modified by the substitution of different types of polar and/or nonpolar groups on the phenyl ring of the monomer and thereby influencing the π-electron distribution of the polymer chain. The NLO properties of the polymer can thus be varied by such substitutions. The use of electron donors as substituents on the phenyl ring leads to a reduction of band gap energy and a consequent increase of the $\chi^{(3)}$ value (Table 6).

The first $\chi^{(3)}$ measurement of an unoriented PPV obtained from dimethyl sulfonium chloride precursor polymer (**63a**) was performed by Kaino et al.[154] using the THG method. They observed a value of $\chi^{(3)} = 7.8 \times 10^{-12}$ esu at 1.85 μm. The $\chi^{(3)}$ value decreased at longer wavelength. Bradley and Mori[155] investigated the $\chi^{(3)}$ behavior at various stages in the thermal elimination conversion from the precursor (**63a**) to conjugated PPV. The $\chi^{(3)}$ value increases as the conversion proceeds. This result again confirms strong dependence of $\chi^{(3)}$ on the π-conjugation. The value of $\chi^{(3)}$ $(-3\omega:\omega,\omega,\omega)$ determined by THG for a fully converted sample was reported to be 7.5×10^{-11} esu at 1.064 μm. Bubeck et al.[156] also reported a THG study of unoriented PPV made from the tetramethylene sulfonium chloride precursor polymer (**63b**). They observed $\chi^{(3)}$ $(-3\omega:\omega,\omega,\omega) = (1.5 \pm 0.6) \times 10^{-10}$ esu at 1.064 μm. The discrepancy between the three reported values may be a result of the use of different incident wavelengths and the degree of contamination in the polymer lattice by the leaving groups.

Subpicosecond DFWM experiments at 0.58 and 0.602 μm on a free-standing film of PPV (10:1 uniaxially stretched) were performed by Singh et al.[158] The $\chi^{(3)}$ value along the draw direction was reported to be $\sim 4 \times 10^{-10}$ esu at both wavelengths. These authors also studied the anisotropy of the $\chi^{(3)}$ value in the film plane. The $\chi^{(3)}$ value was observed to be a minimum when the incident

Table 6 Band Gap Energy and $\chi^{(3)}$ Values of PPV, PPV Derivatives and Analogs

Material	Abbreviated name	Band gap energy (eV)	Measurement technique	λ (μm)	$\chi^{(3)}$ (esu)	Remarks	Ref.
$R^1 = R^2 = H$	PPV	2.5–2.7	THG	1.85	7.8×10^{-12}	Unoriented	154
			THG	1.06	7.5×10^{-11}	Unoriented	155
			THG	1.064	$(1.5 \pm 0.6) \times 10^{-10}$	Unoriented	156
			THG	1.064	$(2 \pm 1.5) \times 10^{-11}$	Stretch-oriented	157
			DFWM	0.602	4×10^{-10}	10:1 uniaxial film	158
$R^1 =$ OMe $R^2 =$ H	M-PPV	2.30	—	—	—	—	140
$R^1 =$ Br $R^2 =$ OMe	BrM-PPV	2.26	DFWM	0.602	9×10^{-10}	Unoriented	159, 160
$R^1 =$ O(CH$_2$)$_3$Me $R^2 =$ OMe	BuM-PPV	2.19	DFWM	0.602	$\sim 10^{-9}$	Unoriented	160
$R^1 = R^2 =$ OMe	DM-PPV	2.12	DFWM	0.602	4×10^{-9}	6:1 uniaxial film	161, 162
	PNA	2.05	—	—	—	—	150
	PTV	1.75	DFWM	0.602	$\sim 10^{-9}$	Unoriented	163

electric field vectors were perpendicular to the draw direction. The ratio of the parallel to the perpendicular values $[(\chi^{(3)})_\parallel / (\chi^{(3)})_\perp]$ was 37, indicating a high degree of orientational anisotropy. X-Ray scattering of such stretch-oriented PPV film also reveals a highly aligned structure of the polymer chains along the draw direction. Another measurement of stretch-oriented PPV film using THG method was performed by McBranch et al.[157] They reported a $\chi^{(3)}$ value of $(2 \pm 1.5) \times 10^{-11}$ esu at 1.06 μm. They also observed a large anisotropy in $\chi^{(3)}$. The observed anisotropy again is consistent with the theoretical prediction that the largest value of $\chi^{(3)}$ is along the π-electron conjugated chain direction. As discussed above, a similar anisotropic behavior is also exhibited by polydiacetylene single crystals, such as PDA–PTS, which possess a perfectly oriented structure.[83]

Incorporation of electron donating groups into the PPV backbone leads to a reduction of the band gap energy. Recently, polymers with this type of backbone, poly(2-bromo-5-methoxyphenylenevinylene) (BrM-PPV) and poly(2-butoxy-5-methoxyphenylenevinylene) (BuM-PPV), were synthesized by Jin et al.[141,159] Both polymers have narrower band gaps (2.20–2.30 eV) when compared to PPV (2.70 eV). Lee et al.[160] investigated the $\chi^{(3)}$ behavior of these samples employing the DFWM technique. The $\chi^{(3)}$ value reached 9×10^{-10} esu for BrM-PPV and 10^{-9} esu for BuM-PPV at 0.602 μm even though the polymer films used were unoriented. The 2,5-dimethoxy substituted PPV (DM-PPV) polymer has a narrower band gap (2.12 eV) than the above derivatized PPVs. The $\chi^{(3)}$ value along the draw direction of a 6:1 uniaxially stretch-oriented DM-PPV film was reported to be 4×10^{-9} esu at 0.602 μm using DFWM.[161,162]

To this date, the lowest band gap material among polymers obtained by the polymeric precursor route is poly(2,5-thienylenevinylene) (PTV). This polymer was prepared by Jen et al.[146] using a water soluble precursor polymer, and by Yamada et al.[147] using an organic solvent soluble precursor. The optical properties of this polymer were studied by Kaino et al.[164] using THG measurements; the $\chi^{(3)}$ value was evaluated to be 3.2×10^{-11} esu at 1.85 μm wavelength. This value is four times higher than that found for PPV (7.8 × 10^{-12} esu), when measured by the same technique. From the DFWM measurement on unoriented thin films using DFWM at 0.602 μm, Lee et al.[163] have obtained a $\chi^{(3)}$ value of ~10^{-9} esu for PTV.

A novel and elegant approach which improves the optical quality of PPV for wave guiding applications was used by Wung et al.[165] They prepared a composite of the PPV polymer with the sol gel processed silica glass with a composition up to 50% by weight. In order to create the new composite, the polymeric precursor is homogenously mixed with the sol gel precursor in a common solvent. The film prepared by spin coating or doctor blading is converted into the final conjugated polymer. This film is of better optical quality than that of pure PPV. The third-order NLO properties of the films containing about 50% by weight of PPV were investigated by Pang et al.[166] using ultrashort laser pulses (below 100 fs). It has been found that this new photonic material exhibits a very fast electronic response which can be observed in experiments like optical Kerr gate and degenerate four-wave mixing. At high laser powers, however, the nonlinear optical response can be observed in a transient absorption experiment (Figure 1) and it is equivalent to appearance of an imaginary component of the third-order susceptibility $\chi^{(3)}$. The imaginary part of $\chi^{(3)}$ has been estimated as 6.6 × 10^{-12} esu and the real part as 4.5 × 10^{-11} esu. Other investigations have been conducted by Lee et al.[160] on the $\chi^{(3)}$ values of composite films made using the same technique but using PPV derivatives such as BrM-PPV and BuM-PPV. The observed $\chi^{(3)}$ values suggest that tight trapping of polymer chains in the silica matrix may change the configuration of the polymer in such a way that the effective conjugation length achieved in the composites may be shorter than that found in the pure polymers.

16.3.4 Polythiophene and other Thiophene-based Conjugated Polymers

Polythiophene (PT) is a polymer which has drawn considerable attention in the field of third-order NLO properties. The band gap of approximately 2 eV allows for investigations of NLO properties both under resonant conditions and off-resonance. The polymer has a very good stability in air and is easily doped. In addition to these advantages, PT can be made processable by attaching long flexible chains in the 3-position of thiophene. Thus, 3-substituted PTs are possible to study in several forms such as cast film, LB film or fiber. PT and its derivatives can be synthesized chemically or electrochemically. Some known 3-substituted PTs (65) are poly(3-hexylthiophene) (P3HT), poly(3-octylthiophene) (P3OT), poly(3-dodecylthiophene) (P3DDT), poly(3-octadecylthiophene) (P3ODT) and poly(3-eicosylthiophene) (P3ET).

(65) R = , n = 5, 7, 11, 17, 19

Prasad et al.[167] investigated the third-order NLO behavior of electrochemically polymerized PT using the DFWM method with 350 fs pulses at 0.602 nm. At this wavelength the DFWM response was found to be governed by resonant nonlinearity. The effective value of $\chi^{(3)}$ was ~4 × 10^{-10} esu. A very interesting feature of the results described by Prasad et al. was the effect of doping on the third-order nonlinearity. Oxidizing (p-doping) of the PT film resulted in a drastic reduction of the effective $\chi^{(3)}$ value. Singh et al.[80] investigated chemically and electrochemically polymerized films of P3DDT using the same technique. They found that the rise time of the nonlinear response was instantaneous and the decay involved a time constant of about 200 fs. The measured effective $\chi^{(3)}$ values for the chemically and electrochemically prepared polymers were 3 × 10^{-10} esu and 5 × 10^{-10} esu respectively at 602 nm and 4.5 × 10^{-10} and 7 × 10^{-10} esu at 590 nm. Yang et al.[168] have reported $\chi^{(3)} = $ ~10^{-9} esu for unsubstituted PT around 600 nm. The observed value of $\chi^{(3)}$ at 705 nm was 4.0 × 10^{-11} esu for both samples.

Kobayashi and coworkers performed femtosecond transient absorption studies on PT and found decay times in the picosecond range.[110,111] Similar results have been obtained by Vardeny et al.[169] who observed a complicated decay of transient absorption extending over time scales from femtoseconds to nanoseconds. It follows that the resonant nonlinear behavior of PT involves many photophysical processes. Singh et al.,[80] and, more recently, Pang and Prasad[79] concluded, however, that the major part of the NLO response observed with 350 fs pulses at 602 nm[80] or with 60 fs pulses at 620 nm derives from excitonic phase space filling, and a smaller part is due to polarons.

Logsdon et al.[170] have reported that the electrochemically polymerized P3DDT forms a stable monolayer and can be deposited as a LB film. The $\chi^{(3)}$ value of this LB film, measured at 0.602 μm by DFWM, was found to be $\sim 1 \times 10^{-9}$ esu with subpicosecond response. The large $\chi^{(3)}$ value, which is a result of the resonance enhancement, is not surprising when one considers that the optical band gap taken from the onset of the absorption band is ~ 1.9 eV.

An analog of PT, poly(isothianaphthene) (PITN) (**66a**) obtained by benzannelation exhibits the narrowest band gap (~ 1 eV) among the known conjugated polymers.[171] The NLO behavior of this compound has not been investigated at this point due to its low optical quality. Other modified PTs, polythieno(3,2-*b*)thiophene (PTT) (**66b**), polydithieno(3,2-*b*, 2′, 3′-*b*)-thiophene (PDTT) (**66c**) and poly[1,4-di-(2-thienyl)benzene] (PDTB) (**66d**), have been synthesized by electrochemical polymerization.[172,173]

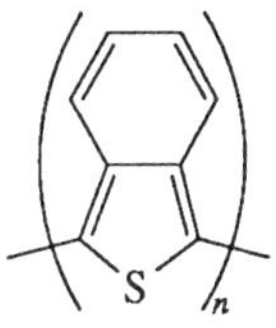

(**66a**) PITN 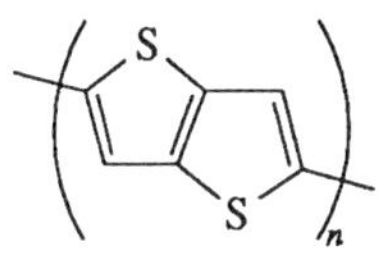   (**66b**) PTT 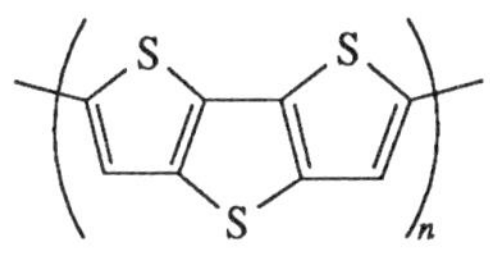   (**66c**) PDTT

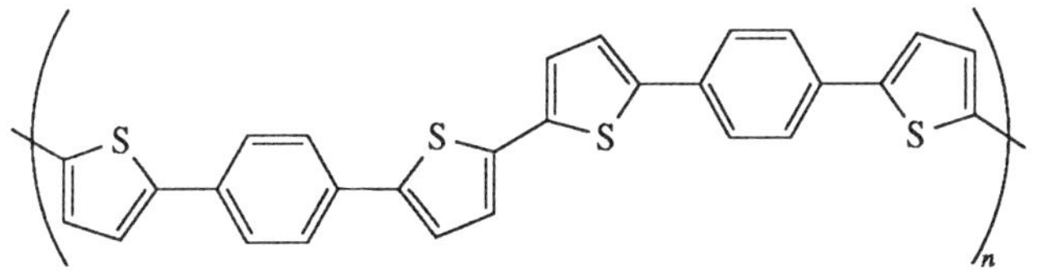

(**66d**) PDTB

The natural films for PTT and PDTT show good optical quality and their band gaps are both smaller than that of PT. The $\chi^{(3)}$ value measured for PTT and PDTT over the wavelength range 0.53–0.63 μm was determined to be of the order of 10^{-9} esu using DFWM.[168] The large value is expected due to the strong electronic resonance enhancement process. The PDTB on the other hand has a band gap of only 2.3 eV and the $\chi^{(3)}$ value obtained by THG for the fundamental wavelength in the range 1.064–1.907 μm is $1.1 \pm 0.1 \times 10^{-11}$ esu.[173]

16.3.5 Rigid Rod-like Polymers

Aromatic heterocyclic polymers possessing π-conjugation characters, such as poly-*p*-phenylene benzobisthiazole (PBT) (**67a**) and poly(*p*-phenylenebenzobisoxazole) (PBO) (**67b**), can be classified as another kind of NLO material. This type of polymer shows ultra high mechanical strength and good environmental stability after processing through the lyotropic liquid crystalline phase. Furthermore, this polymer has a very high laser damage threshold. Rao et al.[174] studied the $\chi^{(3)}$ behavior in a biaxial film of PBT using the DFWM technique. They reported $\chi^{(3)} = \sim 10^{-11}$ esu at both 0.585 and 0.605 μm, with subpicosecond response.

They also found that the DFWM signal was dependent on the orientation of the PBT film, indicating that the polymer was anisotropic. The $\chi^{(3)}$ value of PBO is of a similar magnitude. A sizeable $\chi^{(3)}$ value was also observed for LARC–TPI (**68**), a common nonconjugated polyimide.[175] This polymer can be processed into good optical quality films by using the polyamic acid precursor. However, the $\chi^{(3)}$ value for this polymer was found to be an order of magnitude less than that for PBT. This result again substantiates the importance of π-electron conjugation for NLO effects.

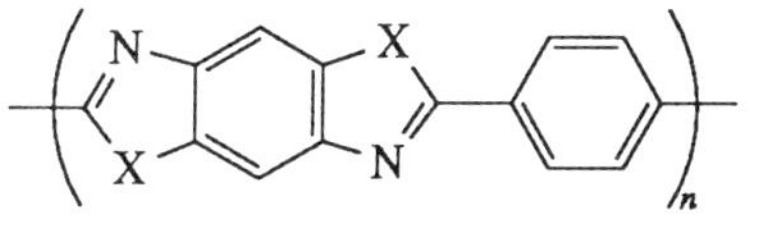

(**67a**) X = S
(**67b**) X = O

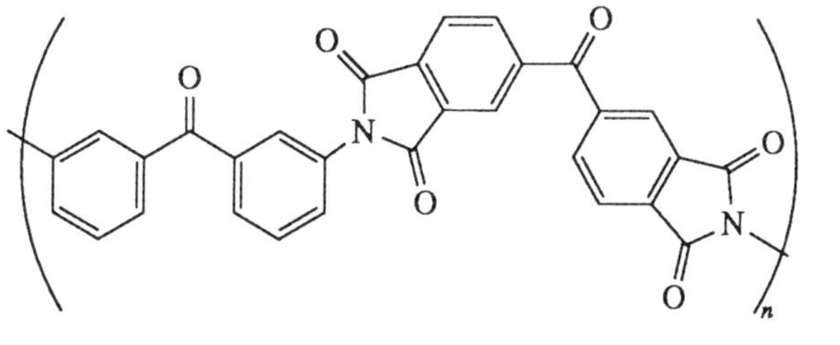

(**68**) LARC-TPI

Wholly aromatic polyazomethines are another type of rigid polymer which have been considered as high performance materials, but processing limitations resulting from their poor solubility and infusibility have hampered their practical application. To overcome this problem, Lee et al.[176] have synthesized aromatic polyazomethines with flexible side chains (**69**).

(**69**)

$R = C_nH_{2n+1}$, $n = 12, 14, 16$

These polymers can be cast from conventional solvents like chloroform, THF and toluene. It is also possible to make melt films. DFWM measurements of these samples were performed using 400 fs pulses at 0.602 μm. The $\chi^{(3)}$ values were found to be approximately $\sim 10^{-12}$ esu.[177] The relatively small $\chi^{(3)}$ value may be a result of noncoplanarity of the aromatic groups connected by the azomethine linkage.

Phthalocyaninatopolysiloxane has been known to be another type of rigid polymer. The explanation for the extreme stiffness of the chain in this polymer can be found when one considers that the Si—O—Si bond length (3.33 Å) is smaller than the intermolecular van der Waals' distance of two adjacent carbon atoms (~ 3.6 Å). By using new synthetic strategy to improve the solubility, Wegner et al.[178] and Sauer[179] have synthesized soluble phthalocyanine polymers which are symmetrically or unsymmetrically substituted with long alkoxy chains (**70**).

$R^1 = OMe$
$R^2 = OC_8H_{17}$

(**70**)

The polymer (**70**) can be transferred as an LB film to form a Y-type deposition.[180] In the LB film, the rigid rod molecules are oriented parallel to the dipping direction. Bubeck et al.[181] have performed THG studies on this film. The $\chi^{(3)}$ value was estimated to be of the order of 10^{-12} esu, assuming the refractive indices $n(\omega) = 1.7 \pm 0.3$ and $n(3\omega) = 2.3 \pm 0.7$ typical for organic materials near resonance. This $\chi^{(3)}$ value is about one to three orders of magnitude smaller than those of the monomeric phthalocyanine polycrystalline films or the LB films reported by Ho et al.[182] and Prasad and coworkers.[183-186] However, a direct comparison is not possible since the latter studies refer to effective nonlinearity involving resonant effects and are dependent on experimental conditions such as laser pulse width, peak power, etc.

16.3.6 Ladder Polymers

Ladder polymers which have a double strand structure with the following general formula (**71**; ⏀ are substituted phenyl groups or fused aromatic rings and ⏀ is a quinoid structure of substituted phenyl groups, or substituted fused aromatic groups) are generally known to be materials which exhibit exceptional mechanical strength and have good thermal stability due to the stiffness of their polymer chains. One of the major difficulties encountered with these polymers is that their solubilities are very limited, lower than those of the single strand rigid polymers discussed in the previous section. Dalton's group[85,187-190] has solved this problem by introducing side substituents into the polymer backbone to interrupt the strong van der Waals' interaction between the π-electron clouds of the neighboring polymer backbones. Moreover, a precursor technique was used during the ladder polymer preparations. A soluble flexible chain precursor polymer was used to cast film which was

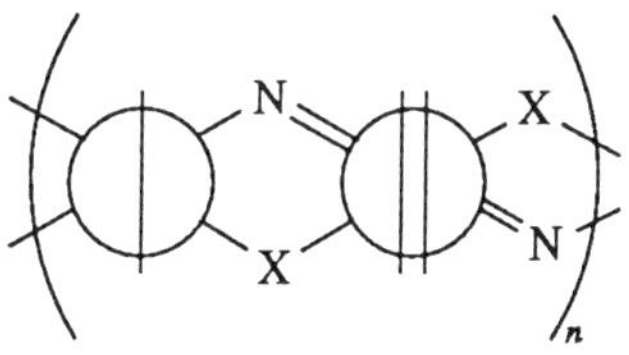

(71) X = S, O, NH

converted by thermal treatment into a fully fused-ring ladder polymer film. For example, equation (15) shows a ladder polymer (75) obtained from 3,6-disubstituted 2,5-dichloroquinone (72) and tetraaminobenzene (73).[191]

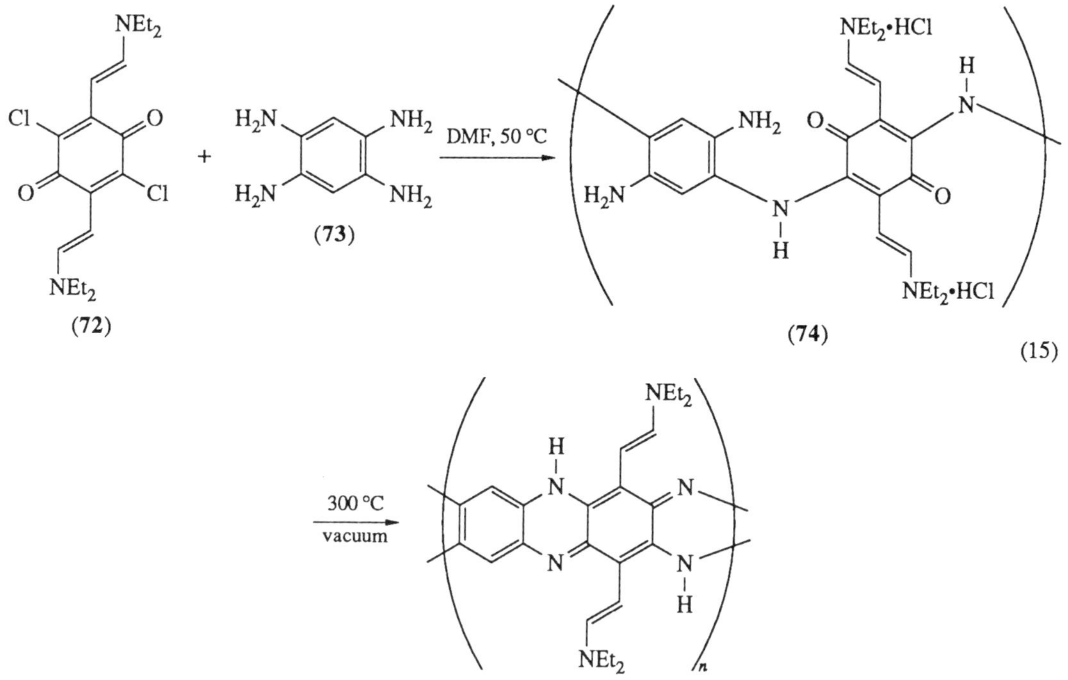

(15)

The synthesis of the polymer involved two steps. First, the substituted dichloroquinone monomer (72) and amine (73) were reacted in DMF at 50 °C to make a soluble open chain precursor polymer (74). Then the resulting precursor was heated to form a ladder polymer by a ring closure reaction. The $\chi^{(3)}$ value of this polymer obtained using DFWM was reported to be 1.6×10^{-9} esu at 0.532 μm wavelength with 25 ps pulses.

Large optical nonlinearities were also reported for copolymers which consist of ladder type segments having some finite π-conjugation and flexible chain spacers. The NLO properties of this type of polymers can be controlled by systematically varying the conjugated segment while the solubility can be controlled by the flexible chain length. One example of this type of copolymer is structure (76).[192] A preliminary report of the NLO measurement on this polymer quotes a $\chi^{(3)}$ value of 4.5×10^{-9} esu at 0.585 μm.

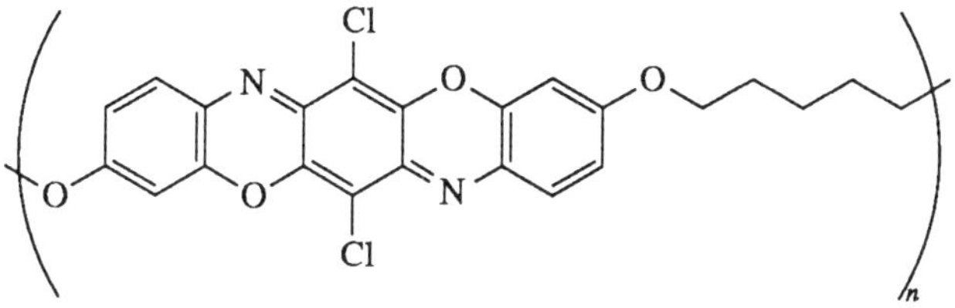

(76)

16.3.7 Organosilane Polymers

Organosilane polymers (polysilanes) are not typical NLO polymers because they do not have delocalized π-electrons in the polymer main chain. However, the σ-electrons forming the silicon–silicon single bonds do exhibit some degree of delocalization along the backbone and this σ-conjugation is the source of a rich variety of photophysical properties including the NLO behavior. Moreover, they show good optical quality in visible regions. These polymers have the general formula (77), in which R^1 and R^2 are phenyl groups, substituted phenyl groups or aliphatic chains. By attaching appropriate substituents to the polymer backbone, the T_g and solubility of these polymers can easily be controlled. However, the use of large size substituents can cause poor solubility due to their excessive hindrance of rotational freedom of the backbone silicon–silicon bond.

$$\left(\begin{array}{c} R^1 \\ | \\ Si \\ | \\ R^2 \end{array}\right)_n$$

(77)

Kajzar et al.[193] studied the NLO properties of a spin-coated polysilane $[PhMeSi]_n$ film using the THG technique. They reported a $\chi^{(3)}$ value of $(1.5 \pm 0.1) \times 10^{-12}$ esu at 1.064 μm. This sizable value of $\chi^{(3)}$ was attributed to a three-photon resonance at 1.064 μm. This polymer is transparent in the visible region. In a Kerr gate experiment on the above compound, Yang et al.[194] found that the NLO response of the polysilane film was faster than 3 ps. By using pump and probe wavelengths of 1.06 and 0.53 μm respectively in the Kerr gate experiment, they obtained a $\chi^{(3)}$ value of $(2.0 \pm 0.6) \times 10^{-12}$ esu. This value is similar to that obtained by DFWM at 0.53 μm. McGraw et al.[195] studied the nuclear and electronic contributions to the $\chi^{(3)}$ value by conducting DFWM studies on poly(n-octylmethylsilane) $[(n\text{-}C_8H_{17})MeSi]_n$. The results indicate that the $\chi^{(3)}$ value for the electronic contribution is $(1.8 \pm 0.5) \times 10^{-12}$ esu and that for the nuclear contribution is $(1.1 \pm 0.4) \times 10^{-12}$ esu at 0.532 μm.

Recently, Embs et al.[196] reported that polysilanes with flexible side groups, such as poly(bis-p-butoxyphenylsilane) (78) and poly(bis-m-butoxy-phenylsilane) (79), form monolayers, which can be deposited as LB films.

(78)   (79)

The microscopic molecular morphology of these LB films exhibits high orientation with the silicon backbone along the dipping direction and the side chains perpendicular to the axis of the main chain. The $\chi^{(3)}$ value of a film 400 layers thick was reported to be 2.8×10^{-12} esu along the polymer backbone. By annealing at 120 °C for 30 min, the $\chi^{(3)}$ value of this compound decreased to 4.3×10^{-12} esu. The $\chi^{(3)}$ value perpendicular to the polymer chain was reported to be 2–5 times smaller than that along the chain direction. This result may indicate that the σ-conjugation of polysilane contributes to the $\chi^{(3)}$ values.

16.4 CURRENT STATUS AND FUTURE PROSPECTS

For electrooptic modulation (spatial light modulator, beam deflector, wave guide electrooptic modulator), electrically poled polymers have emerged as a promising class of materials because of the ease with which a polymeric material can be fabricated into device structures such as fibers and channel wave guides. Methods utilizing selective spatial poling have been suggested for forming device structures such as Mach–Zehnder interferometers.[197] Also, polymeric structures are thermally and environmentally more stable and they do not have high vapor pressure problems associated with small organic molecules. Another advantage offered by polymeric structures is the low dielectric constant and consequently high band width and speed for electrooptic modulation. Although fabrication of prototype electrooptic modulators utilizing poled side chain polymers have been reported, no such devices have yet appeared on the market. In order for the poled polymer based electrooptic devices to successfully compete with the existing devices, future challenges to be met are the simultaneous achievement of the following features: (i) large concentration of the NLO units to maximize $\chi^{(2)}$; (ii) minimum dipolar relaxation of the poled alignment; (iii) low optical losses ($<1\,dB\,cm^{-1}$), particularly for wave guiding operations; and (iv) a broad optical transparency range.

All optical devices utilizing third-order processes, such as optical switching, optical bistability, optical gates and power limiting action, are even more distant in the future. The major hurdle is the currently achievable nonresonant $\chi^{(3)}$ ($<10^{-9}$ esu) value. This value is not sufficient to operate an all optical device at a technologically useful low power level. This value of $\chi^{(3)}$ has to be increased by at least two orders of magnitude. Another problem is, again, the optical quality of conjugated polymeric systems, the most promising present $\chi^{(3)}$ organic materials. They are optically highly lossy. Polymer processing to improve optical quality of film and fiber forms will play a very important role in the development of this area.

16.5 REFERENCES

1. P. N. Prasad and D. J. Williams, 'Introduction to Nonlinear Optical Effects in Organic Molecules and Polymers', Wiley, New York, 1991.
2. P. N. Prasad and D. R. Ulrich (eds.), 'Nonlinear Optical and Electroactive Polymers', Plenum Press, New York, 1988.
3. Y. R. Shen, 'Principles of Nonlinear Optics', Wiley, New York, 1984.
4. M. T. Zhao, B. P. Singh and P. N. Prasad, *J. Chem. Phys.*, 1988, **89**, 5535.
5. M. T. Zhao, M. Samoc, B. P. Singh and P. N. Prasad, *J. Phys. Chem.*, 1989, **93**, 7916.
6. K. C. Rustagi and J. Ducuing, *Opt. Commun.*, 1974, **10**, 258.
7. J. L. Oudar and D. S. Chemla, *J. Chem. Phys.*, 1977, **66**, 2664.
8. J. L. Oudar, *J. Chem. Phys.*, 1977, **67**, 446.
9. J. Zyss and J. L. Oudar, *Phys. Rev. A*, 1982, **26**, 2028.
10. J. Zyss, D. S. Chemla and J. F. Nicoud, *J. Chem. Phys.*, 1981, **74**, 4800.
11. R. A. Huijts and G. L. T. Hesselink, in 'Proceedings of the NATO Advanced Research Workshop on Nonlinear Optical Effects in Organics Polymers, Nice, 1988', ed. J. Messier, F. Kajzar, P. Prasad and D. R. Ulrich, Kluwer, Dordrecht, 1989, p. 101.
12. H. E. Katz, K. D. Singer, J. E. Sohn, C. W. Dirk, L. A. King and H. M. Gordon, *J. Am. Chem. Soc.*, 1987, **109**, 6561.
13. K. D. Singer, J. E. Sohn and S. J. Lalama, *Appl. Phys. Lett.*, 1986, **49**, 248.
14. H. L. Hampsch, J. Yang, G. K. Wong and J. M. Torkelson, *Macromolecules*, 1988, **21**, 528.
15. H. L. Hampsch, J. Yang, G. K. Wong and J. M. Torkelson, *Macromolecules*, 1990, **23**, 3640.
16. T. Watanabe and S. Miyata, in 'Proceedings of IUPAC International Symposium on Molecular Design of Functional Polymers, Seoul, 1989', ed. The Polymer Society of Korea, Seoul, 1989, p. 395.
17. S. Miyata, T. Watanabe and T. Miyazaki, in 'Proceedings of IUPAC International Symposium on Macromolecules, Kyoto, 1988', ed. The Society of Polymer Science (Japan), Kyoto, 1988, p. 582.
18. C. Ye, T. J. Marks, J. Yang and G. K. Wong, *Macromolecules*, 1987, **20**, 2322.
19. C. Ye, N. Minami, T. J. Marks, J. Yang and G. K. Wong, in 'Proceedings of the NATO Advanced Research Workshop on Nonlinear Optical Effects in Organic Polymers, Nice, 1988', ed. J. Messier, F. Kajar, P. Prasad and D. R. Ulrich, Kluwer, Dordrecht, 1989, p. 179.
20. K. D. Singer, M. G. Kuzyk, W. R. Holland, J. E. Sohn, S. L. Lalama, R. B. Comizzoli, H. E. Katz and M. L. Schilling, *Appl. Phys. Lett.*, 1988, **53**, 1800.
21. J. E. Sohn, K. D. Singer, M. G. Kuzyk, W. R. Holland, H. E. Katz, C. W. Dirk, M. L. Schilling and R. B. Comizzoli, in 'Proceedings of the NATO Advanced Research Workshop on Nonlinear Optical Effects in Organic Polymers, Nice, 1988', ed. J. Messier, F. Kajar, P. Prasad and D. R. Ulrich, Kluwer, Dordrecht, 1989, p. 291.
22. S. Esselin, P. LeBarney, P. Robin, D. Broussoux, J. C. Dubois, J. Raffy and J. P. Pocholle, in 'Proceedings of SPIE—The International Society for Optical Engineering on Nonlinear Optical Properties of Organic Materials, San Diego, 1988', ed. G. Khanarian, SPIE, Bellingham, 1988, p. 120.
23. M. Eich, A. Sen, H. Loser, D. Y. Yoon, G. C. Bjorklund, R. Twieg and J. D. Swalen, in 'Proceedings of SPIE—The International Society for Optical Engineering on Nonlinear Optical Properties of Organic Materials, San Diego, 1988', ed. G. Khanarian, SPIE, Bellingham, 1988, p. 128.
24. M. Yoshida, M. Asano, M. Tamada and M. Kumakura, *Makromol. Chem., Rapid Commun.*, 1989, **10**, 517.
25. A. Hayashi, Y. Goto, M. Nakayama, H. Sata and S. Miyata, in 'Proceedings of IUPAC International Symposium on Molecular Design of Functional Polymers, Seoul, 1989', ed. The Polymer Society of Korea, Seoul, 1989, p. 393.
26. G. R. Meredith, J. G. VanDusen and D. J. Williams, *Macromolecules*, 1982, **15**, 1385.

27. C. Noël, C. Friedrich, V. Leonard, P. LeBarny, G. Ravaux and J. C. Dubois, *Makromol. Chem., Macromol. Symp.*, 1989, **24**, 283.
28. S. Yitzchaik, G. Berkovic and V. Krongauz, *Macromolecules*, 1990, **23**, 3539.
29. D. R. Robello, *J. Polym. Sci., Polym. Chem. Ed.*, 1990, **28**, 1.
30. A. C. Griffin, A. M. Bhitta and R. S. L. Hung, in 'Proceedings of SPIE—The International Society for Optical Engineering on Molecular and Polymeric Optoelectronic Materials, San Diego, 1986', ed. G. Khanarian, SPIE, Bellingham, 1987, p. 65.
31. A. C. Griffin, A. M. Bhatti and G. A. Howell, in 'Proceedings of Materials Research Society Symposium on Nonlinear Optical Properties of Polymers, Boston, 1987', ed. A. J. Heeger, J. Orenstein and D. R. Ulrich, Materials Research Society, Pittsburgh, 1988, p. 115.
32. A. C. Griffin, A. M. Bhatti and R. S. L. Hung, in 'Proceedings of an American Chemical Society Symposium on Nonlinear Optical and Electroactive Polymers, Denver, 1987', ed. P. N. Prasad and D. R. Ulrich, Plenum Press, New York, 1988, p. 375.
33. A. C. Griffin, A. M. Bhatti and R. S. L. Hung, *Mol. Cryst. Liq. Cryst.*, 1988, **155**, 129.
34. M. M. Carpenter, P. N. Prasad and A. C. Griffin, *Thin Solid Films*, 1988, **161**, 315.
35. W. M. K. P. Wijekoon, Y. Zhang, S. P. Karna, P. N. Prasad, A. C. Griffin and A. M. Bhatti, *J. Opt. Soc. Am. B, Opt. Phys.*, in press.
36. R. N. DeMartino (Hoechst Celanese Corp.), *Eur. Pat.* 0 294 706 (1988) (*Chem. Abstr.*, 1989, **16**, 110).
37. G. D. Green, H. K. Hall, Jr., J. E. Mulvaney, J. Noonan and D. J. Williams, *Macromolecules*, 1987, **20**, 716.
38. G. D. Green, J. I. Weinschenk, III, J. E. Mulvaney and H. K. Hall, Jr., *Macromolecules*, 1987, **20**, 722.
39. C. S. Willand and D. J. Williams, *Ber. Bunsenges. Phys. Chem.*, 1987, **91**, 1304.
40. C. S. Willand, S. E. Feth, M. Scozzafava, D. J. Williams, G. D. Green, J. I. Weinschenk, III, H. K. Hall, Jr. and J. E. Mulvaney, in 'Proceedings of an American Chemical Society Symposium on Nonlinear Optical and Electroactive Polymers, Denver, 1987', ed. P. N. Prasad and D. R. Ulrich, Plenum Press, New York, 1988, p. 107.
41. J. D. Stenger-Smith, J. W. Fisher, R. A. Henry, J. M. Hoover, G. A. Lindsay and L. M. Hayden, *Makromol. Chem., Rapid Commun.*, 1990, **11**, 141.
42. H. E. Katz and M. L. Schilling, *J. Am. Chem. Soc.*, 1989, **111**, 7554.
43. T. F. Heinz, H. W. K. Tom and Y. R. Shen, *Phys. Rev. A*, 1983, **28**, 1883.
44. T. F. Heinz, C. K. Chen, D. Ricard and Y. R. Shen, *Phys. Rev. Lett.*, 1982, **48**, 478.
45. T. Rasing, Y. R. Shen, M. W. Kim, P. Valint and J. Bock, *Phys. Rev. A*, 1985, **31**, 537.
46. O. A. Aktsipetrov, N. N. Akhmediev, E. D. Mishina and V. R. Novak, *Pis'ma Zh. Eksp. Teor. Fiz.*, 1983, **37**, 175.
47. O. A. Aktsipetrov, N. N. Akhmediev, I. M. Baranova, E. D. Mishina and V. R. Novak, *Pis'ma Zh. Tekh. Fiz.*, 1985, **11**, 599.
48. O. A. Aktsipetrov, N. N. Akhmediev, I. M. Baranova, E. D. Mishina and V. R. Novak, *Zh. Eksp. Teor. Fiz.*, 1985, **89**, 911.
49. O. A. Aktsipetrov, N. N. Akhmediev, I. M. Baranova, E. D. Mishina and V. R. Novak, *Sov. Tech. Phys. Lett. (Engl. Transl.)*, 1985, **11**, 249.
50. I. R. Girling, N. A. Cade, P. V. Kolinsky and C. M. Montgomery, *Electron. Lett.*, 1985, **21**, 169.
51. I. R. Girling, P. V. Kolinsky, N. A. Cade, J. D. Earls and I. R. Peterson, *Opt. Commun.*, 1985, **55**, 289.
52. I. R. Girling, N. A. Cade, P. V. Kolinsky, J. D. Earls, G. H. Cross and I. R. Peterson, *Thin Solid Films*, 1985, **132**, 101.
53. J. S. Schildkraut, T. L. Penner, C. S. Willand and A. Ulman, *Opt. Lett.*, 1988, **13**, 134.
54. K. Kazikawa, K. Shirota, H. Takezoe and A. Fukuda, *Jpn. J. Appl. Phys., Part 1*, 1990, **29**, 913.
55. P. Winant, A. Scheelen and A. Persoons, in 'Proceedings of SPIE— The International Society for Optical Engineering on Nonlinear Optical Properties of Organic Materials, 2, San Diego, 1989', ed. G. Khanarian, SPIE, Bellingham, 1989, vol. 1127, p. 132.
56. P. Winant, A. Scheelen and A. Persoons, in 'Proceedings of the NATO Advanced Research Workshop on Nonlinear Optical Effects in Organic Polymers, Nice, 1988', ed. J. Messier, F. Kajzar, P. Prasad and D. R. Ulrich, Kluwer, Dordrecht, 1989, p. 219.
57. L. Ledoux, D. Josse, P. Vidakovic, J. Zyss, R. A. Hann, P. F. Gordon, B. D. Bothwell, S. K. Gupta, S. Allen, P. Robin, E. Chastaing and J. C. Dubois, *Europhys. Lett.*, 1987, **3**, 803.
58. J. W. Barton, M. Buhaenko, B. Moyle and N. M. Ratcliffe, *J. Chem. Soc., Chem. Commun.*, 1988, 488.
59. D. W. Kalina and S. G. Grubb, *Thin Solid Films*, 1988, **160**, 363.
60. I. Ledoux, D. Josse, P. Fremaux, J.-P. Piel, G. Post, J. Zyss, T. McLean, R. A. Hann, P. F. Gordon and S. Allen, *Thin Solid Films*, 1988, **160**, 217.
61. F. Kajzar and I. Ledoux, *Thin Solid Films*, 1989, **179**, 359.
62. I. Ledoux, D. Josse, J. Zyss, T. McLean, P. F. Gordon, R. A. Hann and S. Allen, *J. Chim. Phys. Phys. Chim. Biol.*, 1988, **85**, 1085.
63. G. Decher, B. Tieke, C. Bosshard and P. Guenter, *Ferroelectrics*, 1989, **91**, 193.
64. C. Bosshard, B. Tieke, M. Seifert and P. Guenter, in 'Proceedings of the Institute of Physics Conference Series on Nonlinear Electrooptic Materials', 1989, vol. 103, p. 181.
65. G. Decher, B. Tieke, C. Bosshard and P. Guenter, *J. Chem. Commun., Chem. Commun.*, 1988, 933.
66. J. Huang, A. Lewis and Th. Rasing, *J. Phys. Chem.*, 1988, **92**, 1756.
67. A. S. Alekseev, S. I. Valyanskii and V. V. Savranski, *Kratk. Soobshch. Fiz.*, 1989, **2**, 35.
68. P. Stroeve, M. P. Srinivasan, B. G. Higgins and S. T. Kowel, *Thin Solid Films*, 1987, **146**, 209.
69. M. F. Danil and G. W. Smith, *Mol. Cryst. Liq. Cryst.*, 1984, **102**, 193.
70. L. M. Hayden, S. T. Kowel and M. P. Srinivasan, *Opt. Commun.*, 1987, **61**, 351.
71. L. M. Hayden, B. L. Anderson, J. Y. S. Lam, B. G. Higgins, P. Stroeve and S. T. Kowel, *Thin Solid Films*, 1988, **160**, 379.
72. R. H. Tredgold, M. C. J. Young, R. Jones, P. Hodge, P. Kolinsky and R. J. Jones, *Electron. Lett.*, 1988, **24**, 308.
73. A. L. Anderson, J. M. Hoover, G. A. Lindsay, B. G. Higgins, P. Stroeve and S. T. Kowel, *Thin Solid Films*, 1989, **179**, 413.
74. B. L. Anderson, R. C. Hall, B. G. Higgins, G. A. Lindsay, P. Stroeve and S. T. Kowell, *Synth. Met.*, 1989, **28**, D683.
75. N. Carr, M. J. Goodwin, A. M. McRoberts, G. W. Gray, R. Marsden and R. M. Scrowston, *Makromol. Chem., Rapid Commun.*, 1987, **8**, 487.
76. H. Haug, 'Optical Nonlinearities and Instabilities in Semiconductors', Academic Press, New York, 1960.
77. P. N. Prasad and B. A. Reinhardt, *Chem. Mater.*, 1990, **2**, 660.

78. B. I. Greene, J. Orenstein, R. R. Millard and L. R. Williams, *Phys. Rev. Lett.*, 1987, **58**, 2750.
79. Y. Pang and P. N. Prasad, *J. Chem. Phys.*, 1990, **93**, 2201.
80. B. P. Singh, M. Samoc, H. S. Nalwa and P. N. Prasad, *J. Chem. Phys.*, 1990, **92**, 2756.
81. S. K. Kurtz, in 'Treatise of Quantum Electrics', ed. H. Rabin and C. L. Tang, Academic Press, 1975, vol. 1, p. 210.
82. P. N. Prasad, in 'Proceedings of the Applied Solid State Chemistry Group of the Dalton Division of the Royal Society of Chemistry on Organic Materials for Nonlinear Optics, Oxford, 1988', ed. R. A. Hann and D. Bloor, The Royal Society of Chemistry, London, 1989, p. 264.
83. C. Sauteret, J.-P. Hermann, R. Frey, F. Pradère, J. Ducuing, R. H. Baughman and R. R. Chance, *Phys. Rev. Lett.*, 1976, **36**, 956.
84. P. N. Prasad, in 'Proceedings of the NATO Advanced Research Workshop on Nonlinear Optical Effects in Organic Polymers, Nice, 1988', ed. J. Messer, F. Kajzar, P. Prasad and D. R. Ulrich, Kluwer, Dordrecht, 1989, p. 351.
85. L.Yu and L. R. Dalton, *Synth. Meth.*, 1989, **29**, E463.
86. G. Wegner, *Z. Naturforsch., Teil B*, 1969, **24**, 824.
87. V. Enkelmann, *Adv. Polym. Sci.*, 1984, **63**, 91.
88. M. Schott and G. Wegner, in 'Nonlinear Optical Properties of Organic Molecules and Crystals', ed. D. S. Chemla and J. Zyss, Academic Press, 1987, vol. 2, p. 3.
89. J. W. Perry, A. E. Stiegman, S. R. Marder and D. R. Coulter, in 'Proceedings of the Applied Solid State Chemistry Group of the Dalton Division of the Royal Society of Chemistry on Organic Materials for Nonlinear Optics, Oxford, 1988', ed. R. A. Hann and D. Bloor, The Royal Society of Chemistry, London, 1989, p. 189.
90. J. W. Perry, A. E. Stiegman, S. R. Marder, D. R. Coulter, D. N. Beratan, D. E. Brinza, F. L. Klavetter and R. H. Grubbs, in 'Proceedings of SPIE—The International Society for Optical Engineering on Nonlinear Optical Properties of Organic Materials, San Diego, 1988', ed. G. Khanarian, SPIE, Bellingham, 1988, p. 17.
91. H. Matsuda, S. Okada, H. Nakanishi, S. Takaragi and M. Kato, in 'Proceedings of IUPAC International Symposium on Macromolecules, Kyoto, 1988', ed. The Society of Polymer Science (Japan), Kyoto, 1988, p. 579.
92. G. M. Carter, M. K. Thakur, Y. J. Chen and J. V. Hryniewicz., *Appl. Phys. Lett.*, 1985, **47**, 457.
93. F. Kajar and J. Messier, *J. Opt. Soc. Am. B*, 1987, **4**, 1040.
94. D. N. Rao, P. Chopra, S. K. Ghosal, J. Swiatkiewicz and P. N. Prasad, *J. Chem. Phys.*, 1986, **84**, 7049.
95. P. P. Ho, R. Dorsinville, N. L. Yang, G. Odian, G. Eichmann, T. Jimbo, Q. Z. Wang, G. C. Tang, N. D. Chen, W. K. Zou, Y. Li and R. R. Alfano, 'Proceedings of SPIE, San Diego, 1986', ed. G. Khanarian, SPIE, Bellingham, 1986, vol. 683, p. 36.
96. P. P. Ho, N. L. Yang, T. Jimbo, Q. Z. Wang and R. R. Alfano, *J. Opt. Soc. Am. B, Opt. Phys.*, 1987, **4**, 1025.
97. G. Berkovic, Y. R. Shen and P. N. Prasad, *J. Chem. Phys.*, 1987, **87**, 1897.
98. G. M. Carter, Y. J. Chen and S. K. Tripathy, *Appl. Phys. Lett.*, 1983, **43**, 891.
99. G. M. Carter, Y. J. Chen, M. F. Rubner, D. J. Sandman, M. K. Thakur and S. K. Tripathy, in 'Nonlinear Optical Properties of Organic Molecules and Crystals', ed. D. S. Chemla and J. Zyss, Academic Press, 1987, vol. 2, p. 85.
100. S. Tomaru, K. Kubodera, S. Zembutsu, K. Takeda and M. Hasegawa, *Electron. Lett.*, 1987, **23**, 595.
101. T. Kurihara, K. Kubodera, S. Matsumoto and T. Kaino, *Polym. Prepr. (Japan)*, 1987, **36**, 1157.
102. G. P. Agrawal, C. Cojan and C. Flytzanis, *Phys. Rev. B*, 1978, **17**, 776.
103. G. R. Meredith, B. Buchalter and C. Hanzlik, *J. Chem. Phys.*, 1983, **78**, 1533.
104. J. P. Hermann, *Opt. Commun.*, 1973, **9**, 74.
105. J. P. Hermann and P. W. Smith, *Dig. Tech. Pap.-Int. Quantum Electron. Conf., 11th*, 1980, 656.
106. R. J. Seymour, G. M. Carter, Y. T. Chen, B. S. Elman, M. E. Rubner, M. K. Thakur and S. K. Tripathy, in 'Proceedings of SPIE—The International Society for Optical Engineering on Advanced Materials for Active Optics, ed. S. Musikant, SPIE, 1986, vol. 567, p. 56.
107. J. Biegajski, R. Burzynski, D. A. Cadenhead and P. N. Prasad, *Macromolecules*, 1986, **19**, 2457.
108. P. P. Ho, K. Dorsinville, N. L. Yang, G. Odian, G. Eichmann, T. Jimbo, Q. Z. Wang, G. C. Tang, N. D. Chen, W. K. Zou, Y. Li and R. R. Alfano, in 'Proceedings of SPIE—The International Society for Optical Engineering on Molecular and Polymeric Optoelectronic Materials, San Diego, 1986', ed. G. Khanarian, SPIE, Washington, 1987, p. 36.
109. J. S. Obhi, M. Samoc and P. N. Prasad, unpublished results.
110. U. Stamm, M. Taiji, M. Yoshizawa, K. Yoshino and T. Kobayashi, *Mol. Cryst. Liq. Cryst.*, 1990, **182A**, 147.
111. T. Kobayashi, M. Yoshizawa, U. Stamm, M. Taiji and M. Hasegawa, *J. Opt. Soc. Am. B, Opt. Phys.*, 1990, **7**, 1558.
112. K. Schulten, I. Ohmine and M. Karplus, *J. Chem. Phys.*, 1976, **64**, 4422.
113. O. Zamani-Khamiri and H. F. Hameka, *J. Chem. Phys.*, 1980, **73**, 5693.
114. M. G. Papadopoulos, J. Waite and C. A. Nicolaides, *J. Chem. Phys.*, 1982, **77**, 2527.
115. D. N. Beratan, J. N. Onuchic and J. W. Perry, *J. Phys. Chem.*, 1987, **91**, 2696.
116. J.-M. Amdré, C. Barbier, V. Bodart and J. Delhalle, in 'Nonlinear Optical Properties of Organic Molecules and Crystals', ed. D. S. Chemla and J. Zyss, Academic Press, 1987, vol. 2, p. 137.
117. B. M. Pierce, in 'Proceedings of the Materials Research Society Symposium on Nonlinear Optical Properties of Polymers, Boston, 1987', ed. A. J. Heeger, J. Orenstein and D. R. Ulrich, The Materials Research Society, Pittsburgh, 1988, p. 109.
118. A. F. Garito, J. R. Heflin, K. Y. Wong and O. Zamani-Khamiri, in 'Proceedings of the Materials Research Society Symposium on Nonlinear Optical Properties of Polymers, Boston, 1987', ed. A. J. Heeger, J. Orenstein, and D. R. Ulrich, The Materials Research Society, Pittsburgh, 1988, p. 91.
119. J. P. Hermann and J. Ducuing, *J. Appl. Phys.*, 1974, **45**, 5100.
120. J. F. Ward and D. S. Elliot, *J. Chem. Phys.*, 1978, **69**, 5438.
121. S. H. Stevenson, D. S. Donald and G. R. Meredith, in 'Proceedings of the Materials Research Society Symposium on Nonlinear Optical Properties of Polymers, Boston, 1987', ed. A. J. Heeger, J. Orenstein and D. R. Ulrich, The Materials Research Society, Pittsburgh, 1988, p. 103.
122. J. P. Hermann, *Opt. Commun.*, 1973, **9**, 74.
123. G. R. Meredith and S. H. Stevenson, in 'Proceedings of the NATO Advanced Research Workshop on Nonlinear Optical Effects in Organic Polymers, Nice, 1988', ed. J. Messer, F. Kajzar, P. Prasad and D. R. Ulrich, Kluwer, Dordrecht, 1989, p. 365.
124. S. R. Mardar, J. W. Perry, F. L. Klavetter and R. H. Grubbs, in 'Proceedings of the Applied Solid State Chemistry

Group of the Dalton Division of the Royal Society of Chemistry on Organic Materials for Nonlinear Optics, Oxford, 1988', ed. R. H. Hann and D. Bloor, The Royal Society of Chemistry, London, 1989, p. 288.
125. C. V. Shank, R. Yen, R. L. Fork, J. Orenstein and G. L. Baker, *Phys. Rev. Lett.*, 1982, **49**, 1660.
126. Z. Vardeny, J. Strait, D. Moses, T.-C. Chung and A. J. Heeger, *Phys. Rev. Lett.*, 1982, **49**, 1657.
127. C. V. Shank, *Science*, 1984, **219**, 1027.
128. F. Kajzar, S. Etemad, G. L. Baker and J. Messier, *Synth. Meth.*, 1987, **17**, 563.
129. M. Sinclair, D. Moses, K. Akagi and A. J. Heeger, in 'Proceedings of the Materials Research Society Symposium on Nonlinear Optical Properties of Polymers, Boston, 1987', ed. A. J. Heeger, J. Orenstein and D. R. Ulrich, The Materials Research Society, Pittsburgh, 1988, p. 205.
130. M. Sinclair, D. Moses, K. Akagi and A. J. Heeger, in 'Proceedings of the NATO Advanced Research Workshop on Nonlinear Optical Effects in Organic Polymers, Nice, 1988', ed. J. Messier, F. Kajzar, P. Prasad and D. R. Ulrich, Kluwer, Dordrecht, 1989, p. 29.
131. M. Sinclair, D. McBranch, D. Moses and A. J. Heeger, *Synth. Meth.*, 1989, **28**, D645.
132. M. Sinclair, D. Moses, D. McBranch, A. J. Heeger, J. Yu and W. P. Su, *Synth. Meth.*, 1989, **28**, D655.
133. P. N. Prasad, *Thin Solid Films*, 1987, **152**, 275.
134. P. N. Prasad, J. Swiatkiewicz and J. Pfleger, *Mol. Cryst. Liq. Cryst.*, 1987, **160**, 53.
135. F. E. Karasz, J. D. Capistran, D. R. Gagnon and R. W. Lenz, *Mol. Cryst. Liq. Cryst.*, 1985, **118**, 327.
136. D. R. Gagnon, J. D. Capistran, F. E. Karasz, R. W. Renz and S. Antoun, *Polymer*, 1987, **28**, 567.
137. I. Murase, T. Ohnishi, T. Noguchi and M. Hirooka, *Polym. Commun.*, 1984, **25**, 327.
138. I. Murase, T. Ohnishi, T. Noguchi, M. Hirooka and S. Murakami, *Mol. Cryst. Liq. Cryst.*, 1985, **118**, 333.
139. S. Tokito, T. Momii, H. Murata, T. Tsutsui and S. Saito, *Polymer*, 1990, **31**, 1137.
140. W. B. Liang, R. W. Lenz and F. E. Karasz, *J. Polym. Soc., Polym. Chem. Ed.*, 1990, 28, 2867.
141. J.-I. Jin, C. K. Park, H.-K. Shim and Y.-W. Park, *J. Chem. Soc., Chem. Commun.*, 1989, 1205.
142. T. Kaino, H. Kobayashi, K. Kubodera, T. Kurihara, S. Saito, T. Tsutsui and S. Tokito, *Appl. Phys. Lett.*, 1989, **54**, 1619.
143. W. J. Swatos and B. Gordon, III, *Polym. Prepr., Am. Chem. Soc., Div. Polym. Chem.*, 1990, **31**, 505.
144. S. H. Askari, S. D. Rughooputh, F. Wudl and A. J. Heeger, *Polym. Prepr., Am. Chem. Soc., Div. Polym. Chem.*, 1989, **30**, 157.
145. I. Murase, T. Ohnishi, T. Noguchi and M. Hirooka, *Polym. Commun.*, 1985, **26**, 362.
146. K.-Y. Jen, M. Maxfield, L. W. Shacklette, and R. L. Elsenbaumer, *J. Chem. Soc., Chem. Commun.*, 1987, 309.
147. S. Yamada, S. Tokito, T. Tsutsui and S. Saito, *J. Chem. Soc., Chem. Commun.*, 1987, 1448.
148. S. Antoun, D. R. Gagnon, F. E. Karasz and R. W. Lenz, *J. Polym. Sci., Polym. Lett. Ed.*, 1986, **24**, 503.
149. S. Antoun, D. R. Gagnon, F. E. Karasz and R. W. Lenz, *Polym. Bull.*, 1986, **15**, 181.
150. J. D. Stenger-Smith, T. Sauer and G. Wegner and R. W. Lenz, *Polymer*, 1990, **31**, 1632.
151. H.-K. Shim, C. K. Park. J.-I. Jin and R. W. Lenz, *Polym. Bull.*, 1989, **21**, 409.
152. H.-K. Shim, R. W. Lenz and J.-I. Jin, *Makromol. Chem.*, 1989, **190**, 389.
153. J.-I. Jin, H.-K. Shim and R. W. Lenz, *Synth. Meth.*, 1989, **29**, E53.
154. T. Kaino, K.-I. Kubodera, S. Tomaru, T. Kurihara, S. Saito, T. Tsutsui and S. Tokito, *Electron. Lett.*, 1987, **23**, 1095.
155. D. D. C. Bradley and Y. Mori, *Jpn. J. Appl. Phys.*, 1989, **28**, 174.
156. C. Bubeck, A. Kaltbeitzel, R. W. Lenz, D. Neher, J. D. Stenger-Smith and G. Wegner, in 'Proceedings of the NATO Advanced Research Workshop on Nonlinear Optical Effects in Organic Polymers, Nice, 1988', ed. J. Messier, F. Kajzar, P. Prasad and D. R. Ulrich, Kluwer, Dordrecht, 1989, p. 143.
157. D. McBranch, M. Sinclair, A. J. Heeger, A. O. Patil, S. Shi, S. Askari and F. Wudl, *Synth. Meth.*, 1989, **29**, E85.
158. B. P. Singh, P. N. Prasad and F. E. Karasz, *Polymer*, 1988, **29**, 1940.
159. J.-I. Jin, C. K. Park, and H.-K. Shim, *Synth. Meth.*, in press, 1991.
160. K.-S. Lee, C. J. Wung, P. N. Prasad, C. K. Park, H.-K. Shim and J.-I. Jin, unpublished results, 1990.
161. J. Swiatkiewicz, P. N. Prasad, F. E. Karasz, M. A. Druy and P. Glatkowski, *Appl. Phys. Lett.*, 1990, **56**, 892.
162. J. Swiatkiewicz, P. N. Prasad, and C. Lee, unpublished results, 1990.
163. K.-S. Lee, H. M. Kim, C. J. Wung and P. N. Prasad, unpublished results, 1990.
164. T. Kaino, K. Kubodera, H. Kobayashi, T. Kurihara, S. Saito, T. Tsutsui, S. Tokito and H. Murata, *Appl. Phys. Lett.*, 1988, **53**, 2002.
165. C. J. Wung, Y. Pang, P. N. Prasad and F. E. Karasz, *Polymer*, 1991, **32**, 605.
166. Y. Pang, M. Samoc and P. N. Prasad, *J. Chem. Phys.*, in press, 1991.
167. P. N. Prasad, J. Swiatkiewicz and J. Pfleger, *Mol. Cryst. Liq. Cryst.*, 1987, **160**, 53.
168. L. Yang, R. Dorsinville, Q. Z. Wang, W. K. Zou, P. P. Ho, N. L. Yang, R. R. Alfano, R. Zamboni, R. Danieli, G. Ruani and C. Taliani, *J. Opt. Soc. Am. B*, 1989, **6**, 753.
169. Z. Vardeny, H. T. Grahn, A. J. Heeger and F. Wudl, *Synth. Meth.*, 1989, **28**, C299.
170. P. B. Logsdon, J. Pfleger and P. N. Prasad, *Synth. Meth.*, 1988, **26**, 369.
171. A. O. Patil, A. J. Heeger and F. Wudl, *Chem. Rev.*, 1988, **88**, 183.
172. P. Di Marco, M. Mastragostino and C. Taliani, *Mol. Cryst. Liq. Cryst.*, 1985, **18**, 177.
173. F. Kajzar, G. Ruani, C. Taliani and R. Zamboni, *Synth. Meth.*, 1990, **37**, 223.
174. D. N. Rao, J. Swiatkiewicz, P. Chopra, S. K. Ghoshal and P. N. Prasad, *Appl. Phys. Lett.*, 1986, **48**, 1187.
175. P. N. Prasad, in 'Proceedings of an American Chemical Society Symposium on Electroactive Polymers, Denver, 1987', ed. P. N. Prasad and D. R. Ulrich, Plenum Press, New York, 1988, p. 41.
176. K.-S. Lee, J. C. Wong and J. C. Jung, *Makromol. Chem.*, 1989, **190**, 1547.
177. K.-S. Lee and M. Samoc, *Polym. Commun.*, in press, 1991.
178. E. Orthmann and G. Wegner, *Makromol. Chem., Rapid Commun.*, 1986, **7**, 243.
179. T. Sauer, Ph.D. Thesis, University of Mainz, 1989.
180. E. Orthmann and G. Wegner, *Angew. Chem., Int. Ed. Engl.*, 1986, **25**, 1105.
181. C. Bubeck, D. Neher, A. Kaltbeitzel, G. Duda, T. Arndt, T. Sauer and G. Wegner, in 'Proceedings of the NATO Advanced Research Workshop on Nonlinear Optical Effects in Organic Polymers, Nice, 1988', ed. J. Messier, F. Kajzar, P. Prasad and D. R. Ulrich, Kluwer, Dordrecht, 1989, p. 185.
182. Z. Z. Ho, C. Y. Ju and W. M. Hetherington, *J. Appl. Phys.*, 1987, **62**, 716.
183. P. N. Prasad, *Thin Solid Films*, 1987, **152**, 275.

184. P. N. Prasad, M. K. Casstevens, J. Pfleger and P. Logsdon, *SPIE Proc.*, 1988, **878**, 106.
185. P. N. Prasad, M. K. Casstevens and M. Samos, *SPIE Proc.*, 1989, **1056**, 117.
186. M. K. Casstevens, M. Samoc, J. Pfleger and P. N. Prasad, *J. Chem. Phys.*, 1990, **92**, 2019.
187. L. R. Dalton, in 'Proceedings of an American Chemical Society Symposium on Electroactive Polymers, Denver, 1987', ed. P. N. Prasad and D. R. Ulrich, Plenum Press, New York, 1988, p. 243.
188. L. R. Dalton, J. Thomson and H. S. Nalwa, *Polymer*, 1987, **28**, 543.
189. L. R. Dalton, in 'Proceedings of the Materials Research Society Symposium on Nonlinear Optical Properties of Polymers, Boston, 1987', ed. A. J. Heeger, J. Orenstein and D. R. Ulrich, The Materials Research Society, Pittsburgh, 1988, p. 301.
190. L. R. Dalton, in 'Proceedings of the NATO Advanced Research Workshop on Nonlinear Optical Effects in Organic Polymers, Nice, 1988', ed. J. Messier, F. Kajzar, P. Prasad and D. R. Ulrich, Kluwer, Dordrecht, 1989, p. 123.
191. L. Yu and L. R. Dalton, *Macromolecules*, 1990, **23**, 3439.
192. L. Yu and L. R. Dalton, *J. Am. Chem. Soc.*, 1989, **111**, 8699.
193. F. Kajzar, J. Messier and C. Rosilio, *J. Appl. Phys.*, 1986, **60**, 3040.
194. L. Yang, Q. Z. Wang, P. P. Ho, R. Dorsinville, R. R. Alfano, W. K. Zou and N. L. Yang, *Appl. Phys. Lett.*, 1988, **53**, 1245.
195. D. J. McGraw, A. E. Siegman, G. W. Wallraff and R. D. Miller, *Appl. Phys. Lett.*, 1988, **54**, 1713.
196. F. W. Embs, D. Neher, G. Wegner, R. D. Miller, R. Sooriyakumaran and C. G. Willson, *Polym. Prepr., Am. Chem. Soc., Div. Polym. Chem.*, 1990, **31**, 298.
197. A. Buckley, J. P. Riggs, J. B. Stamatoff and H. N. Yoon, in 'Proceedings of the NATO Advanced Research Workshop on Organic Molecules for Nonlinear Optics and Photonics, La Rochelle, 1990', ed. J. Messier, F. Kajzar and P. Prasad, Kluwer, Dordrecht, 1991, p. 447.

17
Structure, Properties and Applications of Polymeric Langmuir–Blodgett Films

KLEMENS MATHAUER, FRANK EMBS and GERHARD WEGNER
Max-Planck-Institut für Polymerforschung, Mainz, Germany

17.1	INTRODUCTION	449
17.2	POLYMERIZATION IN FILMS OF THE MONOMERS	452
	17.2.1 Polymerizations of Vinyl Monomers	452
	17.2.2 Topochemical Polymerizations	454
	17.2.3 Polycondensation in Monolayers	455
17.3	PREFORMED POLYMERS	456
	17.3.1 Conventional Linear Polymers	456
	17.3.2 Amphiphilic Polymers with Strong Hydrophilic Parts	457
	17.3.3 Polymers with Side Chains other than Alkyl Chains	458
	17.3.4 Rod-like Polymers	460
17.4	THE STRUCTURE OF POLYMERIC LB FILMS	462
17.5	PERSPECTIVES OF TECHNICAL APPLICATION	465
	17.5.1 Optoelectronic and Nonlinear Optical Effects	465
	17.5.2 Pyroelectric and Piezoelectric LB Films	465
	17.5.3 Sensors	466
	17.5.4 Separation Layers	466
	17.5.5 Surface Modification	466
	17.5.5.1 Orientation layers	466
	17.5.5.2 Lubricant layers	467
	17.5.5.3 Wettability	467
	17.5.5.4 Biocompatibility	467
	17.5.6 Microlithography	467
17.6	REFERENCES	468

17.1 INTRODUCTION

Two methods exist for the preparation of ultrathin polymer films; the self-assembly technique[1,2] and the Langmuir–Blodgett (LB) technique. For the self-assembly technique special molecular systems (α,ω-functionalized long alkyl chains) are required to build up multilayer systems, whereas for the Langmuir–Blodgett technique a large range of different molecular structures are suitable. The classical substances for the LB technique are amphiphilic molecules consisting of a long alkyl chain with a hydrophilic head group. They form monomolecular layers at the air/water interface that can be transferred onto solid substrates.

In 1917 Langmuir introduced the film balance for the study of molecular films on the water surface.[3] He had already demonstrated that fatty acids of different chain lengths form insoluble monolayers at the air/water interface and that they occupy the same cross-sectional area, *i.e.* they are oriented perpendicular to the water surface. The method for the deposition of these monomolecular films onto solid substrates to build up multilayers, today called the Langmuir–Blodgett technique, was first published by Blodgett in 1935.[4]

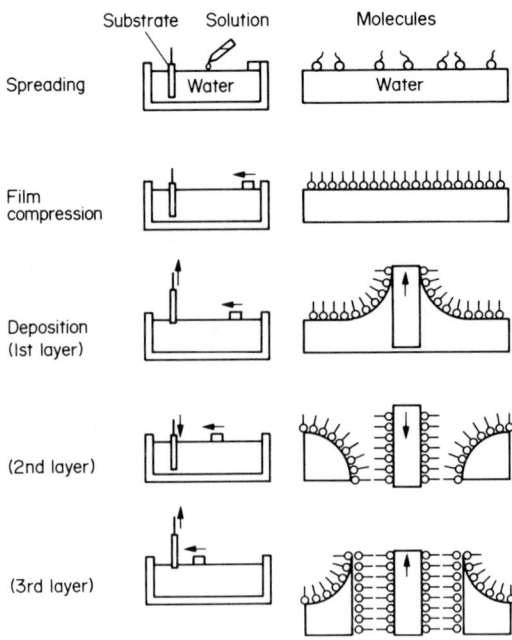

Figure 1 Schematic representation of monolayer and Langmuir–Blodgett film formation of low molecular weight amphiphilic compounds

Two different methods are used to measure the surface pressure: the Wilhelmy and the Langmuir techniques. In the Wilhelmy technique,[5] the surface tension is measured directly by weighing a hydrophilic plate which is dipped through the monolayer into the aqueous subphase. In the Langmuir method, the difference in surface pressure is measured between the pure water/air surface and the surface covered with the film.

The area for the spread film can be changed by a moving barrier, and surface pressure–area isotherms are recorded by measuring the surface tension while compressing the monolayers (Figure 1). The transfer of monofilms by dipping a solid substrate into and out of the water subphase has to be done at a constant surface pressure. Thus, during a transfer the barrier is moved continuously towards the substrate. Depending on the substance and the substrate, transfer occurs either during both the down- and up-strokes or only during one of them. So three arrangements of amphiphiles in the multilayers, the X, Y and Z films, are possible.

The type of molecules that are able to form monofilms at the air/water interface can be divided into two classes:[6,7] (i) low molecular weight, water insoluble substances which have a high attraction to the water subphase. Most of these substances consist of a long alkyl side chain and one hydrophilic head group; and (ii) polymers which are adsorbed at the liquid/gas interface.

To assess the suitability of a certain substance to form a monolayer four factors must be taken into account: (i) solubility; (ii) volatility; (iii) interaction between the molecules; and (iv) interaction with the subphase.

In the case of the classical amphiphile, a long chain compound with a carboxylic headgroup, for example, all requirements are fulfilled. Carboxylic acids with sufficiently long alkyl chains are insoluble in water and are nonvolatile. The balance of the interaction between the molecules and the subphase is very important for the spreading procedure. If the attraction to the subphase is too small in comparison with the bulk adhesion, crystallization will occur instead of spreading. Polar headgroups for alkyl chains which form monomolecular films at the air/water interphase are compared in Table 1.

The shape of the surface pressure–area isotherms of these simple molecules are all very similar (Figure 2). The monolayer exists in different states of matter, with well-defined changes between them. The three main states are named as gaseous, expanded and condensed, in analogy to the gaseous, liquid and solid states in three dimensions. A very dilute monolayer may be described by a two-dimensional equation for an ideal gas

$$\Pi/C = RT/M$$

where C is the two-dimensional concentration (mass/area). With further compression of the

Table 1 Attraction to Water of Polar Head Groups of Amphiphiles[6]

Very weak (no film)	Weak (unstable film)	Strong (stable film with C_{16} chain)	Very Strong (C_{16} chain compounds dissolve)
—Me	—CH$_2$OMe	—CH$_2$OH	—SO$_3^-$
—CH$_2$Br	—C$_6$H$_4$OMe	—CO$_2$H	—OSO$_3^-$
—CH$_2$Cl	—CO$_2$Me	—CN	—C$_6$H$_4$SO$_3^-$
—NO$_2$		—CONH$_2$	—NR$_3^+$
		—CH=NOH	
		—C$_6$H$_4$OH	
		—CH$_2$COMe	
		—NHCONH$_2$	
		—NHCOMe	

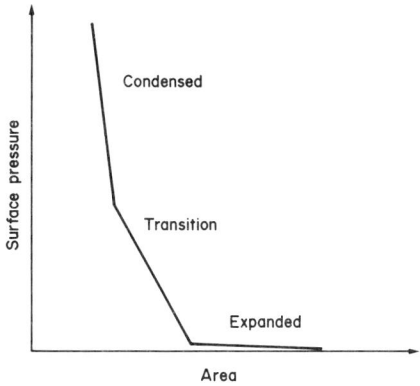

Figure 2 Generalized Π-A relation for long chain alcohols and acids

monolayer, transitions from the gaseous to the expanded and to the condensed state will occur. These transitions can either be achieved by compression, lowering the temperature or increasing the length of the hydrocarbon chain in a homologous series.

A clearer interpretation of the behaviour of these films considers the surface pressure as a two-dimensional osmotic pressure, which leads to the same equation. The interface may be regarded as the solvent, and in the Langmuir film balance the float is indeed a semipermeable 'membrane' which is impermeable for the monolayer molecules, while the solvent molecules are free to distribute between interface and bulk. For more concentrated monolayers the virial expansion of the above equation can be applied

$$\Pi/C = RT(1/M + A_2 C + \cdots)$$

where A_2 is the second virial coefficient.

In the condensed state the monolayer may exist in different phases or modifications, characterized by different modes of packing of the constituent molecules in two dimensions as functions of surface pressure and temperature. Two-dimensional phases analogous to liquid crystals or conventional solids have been described.[8]

In order to produce well-defined polymeric mono- and multi-layers, the idea of polymerizing amphiphilic materials was suggested. Polymerizable groups which are not too big so as to disturb the close packing of the long alkyl chains have to be introduced into the amphiphiles which form the layers, in order to carry this out. About 40 years after the first report on how to achieve the transfer of monolayers to solid substrates, a kind of renaissance of Langmuir–Blodgett films occurred, which was stimulated by the work of Kuhn et al. They showed that radiationless energy transfer could be conveniently studied in molecular assemblies consisting of dye-containing multilayers.[9,10]

Many detailed investigations have been undertaken since that time on the polymerization of 'classical' Langmuir–Blodgett films by the use of unsaturated amphiphilic substances as constituents for mono- and multi-layers. Two major aims seemed to be worth striving for: (i) the improvement of the stability of the films; and (ii) the possibility of using these materials for resist applications.

Two possibilities exist for carrying out polymerization: (i) polymerization of the monolayer floating at the air/water interface, followed by transfer of the then polymerized monolayer to a substrate; and (ii) building up multilayers from polymerizable amphiphiles *via* the LB technique, followed by polymerization in the form of LB assemblies. The latter is a special case of topochemical polymerization.

A further possibility for preparing multilayers of polymers is to start from preformed polymers which spread at the air/water interface to form monomolecular films. The problems related to the design and performance of such polymers are the subject of Section 17.3.

17.2 POLYMERIZATION IN FILMS OF THE MONOMERS

17.2.1 Polymerizations of Vinyl Monomers

The first experiments for carrying out the polymerization of monolayers at the air/water interface made use of long chain esters or amides of acrylic and methacrylic acid.[11,12] To monitor the reaction in the monolayer during irradiation or exposure to reactive gases, one can measure either the surface pressure (at constant area) or the area (at constant pressure) with time. Additionally, the whole pressure/area isotherm can be recorded once more after the experiment.

By analogy to the polymerization of these materials in three-dimensional solution, the reaction in the monolayers was initiated by adding radical-forming substances which could be excited either directly by UV light or *via* energy transfer from additional dissolved sensitizers. It was shown that oxygen could also participate in the reaction within the monolayers under UV irradiation to form polymeric peroxides. The analysis of the reaction products is quite difficult, because of the small amount of material (0.05–0.1 mg) obtained when conventional Langmuir troughs (typical surface area 1000–2000 cm^2) are used. Mass spectroscopic investigations served to characterize the different products formed when the polymerization of *N*-octadecylacrylamide (**1**) (Table 2) with oxygen or UV light was attempted.[11]

The first investigation of polymerization in a multilayer concerned vinyl stearate as the monomer.[13] Polymerization was initiated with ^{60}Co γ radiation and was monitored by multiple internal reflection infrared spectroscopy. X-Ray diffraction showed that the polymer film retained the basic layered structure after complete conversion.

In mixed films of ethyl stearate and vinyl stearate (**2**) (Table 2) the extent of polymerization was not lowered, even at a concentration of 30% of unreactive ethyl stearate.[14] Nor did it change much when alternating layers of the two compounds were investigated, which showed that the polymerization mainly took place within individual layers.

The mixing of photoreactive species with a normal amphiphile facilitates the polymerization in

Table 2 Some Polymerizable Amphiphiles

Structure		Ref.
(1) Me(CH$_2$)$_{17}$NHCOCH=CH$_2$		10, 11, 21, 22
(2) Me(CH$_2$)$_{16}$COCH=CH$_2$		13, 14, 26
(3) CH$_2$=CH–C$_6$H$_4$–(CH$_2$)$_{15}$CO$_2$H		15
(4) R–C(O)–C(H)=C(H)–CO$_2$H (5) R–C(O)–C(H)=C(H)–CO$_2$H	(a) R = O(CH$_2$)$_{17}$Me (b) R = NH(CH$_2$)$_{17}$Me	16
(6) Me(CH$_2$)$_{16}$CH=CHCO$_2$H		19
(7) CH$_2$=CH(CH$_2$)$_{20}$CO$_2$H		33, 34
(8) C$_n$H$_{2n+2}$C≡C–C≡CR	R = CH$_2$OH, CO$_2$H, (CH$_2$)$_m$CO$_2$H	39, 40
(9) Me(CH$_2$)$_{12}$CH=CHCH=CHR	R = CO$_2$H, CHO, CH$_2$OH	44
(10) CH$_2$=CH(CH$_2$)$_{20}$OCOCH=CHCH=CH$_2$		45
(11) epoxide–(CH$_2$)$_{18}$CO$_2$H		49

cases where the polymerizable molecule does not form stable monolayers by itself. Thus, the styrene derivative (3) (Table 2) can be polymerized in a 'solution' of stearic acid.[15]

The reactivity of a given type of molecule may depend strongly on the state of aggregation. A particularly good example was observed in the polymerization of derivatives of maleic and fumaric acid.[16] When irradiated by UV light in solution only *cis–trans* isomerization takes place; in the solid state only (5b) shows a reasonable polymerization rate, whereas all substances can be polymerized in the form of multilayers. Whether the polymerization in multilayers is controlled by topochemical, *i.e.* packing effects, or by kinetical effects, *i.e.* suppression of termination reactions, remains unclear.

As well as free carboxylic acids, their salts with mono-, bi- or tri-valent counterions can be polymerized also. These salts are usually formed *in situ* by spreading the pure acids on an aqueous subphase containing the desired counterion at the proper pH. Thus, Cd octadecyl fumarate can be polymerized in multilayers. If alternating layers with unreactive cadmium stearate were prepared, no difference in reactivity to the pure fumarate film could be detected. This means that the reaction must occur in separate layers, independent of the neighbouring layers.[17] Upon irradiation at the water surface, Cd octadecyl fumarate shows an expansion (at constant pressure), whereas Cd octadecyl maleate contracts.[18] A mixture of fumarate and maleate in a ratio of 1:3 forms a homogeneous monofilm that preserves its area upon irradiation.

A detailed investigation of the photopolymerization of 2-icosenoic acid (6) and octadecyl fumarate (5a) shows that the multilayers of both amphiphiles undergo substantial structural changes during polymerization.[19] The melting of the layer structure, that corresponds to a melting of the hydrocarbon chains, occurs in the polymer film at lower temperatures. For both monomer and polymer films a partial reorganization during cooling of the melted layers was observed.

N-Octadecyl acrylamide can be polymerized in multilayers by UV irradiation to give uniform films with a high stability against solvents.[20,21] The films are suitable for electron-beam lithography. Structural features of less then 10 nm have been reported.[22]

Of special interest are polymerizations of divinyl compounds which proceed with crosslinking. A 'gel-point' can be expected for the formation of an infinite network in two dimensions that should correlate to Flory's fundamental theories of network formation by polymerization.[23] A critical increase in surface viscosity during polymerization has been observed in monolayers of diacrylic esters,[24] although this could not be related unequivocally to the point of the formation of an infinite network. Similarly, a critical effect should be expected for the mechanical modulus of the monofilms. This has been observed for the crosslinking polymerization of monofilms of diacrylic esters adsorbed at the oil/water interphase.[25] The treatment of the experimental data in terms of a two-dimensional system is ambiguous in this case, as the system is not really two-dimensional. One end of the growing polymer chain is always free to leave the interface and can be readsorbed at another position. So the bridging of one polymer chain by another is possible.

Multilayers of amphiphilic compounds with long aliphatic chains possess a highly crystalline lamellar structure. Consequently, for polymerization reactions, each layer has to be regarded as a two-dimensional reaction space in which the reaction path follows random walk statistics. The length of the polymer repeat unit is about 2.5 Å, which is significantly smaller than the packing distance of paraffin chains which is about 4.5–5 Å. Because of this mismatch, phase separation of polymer and monomer can occur and large chain lengths cannot be expected. For the polymerization of vinyl stearate in LB films, a maximum degree of polymerization of about 6 has been reported.[26]

In the case of 'fluid' mono- and multi-layers, which were mainly investigated in the form of two-dimensional arrays of modified lipids (monolayer, bilayer, liposome and black-lipid membrane),[27,87] diffusion in two dimensions is possible.[28,29] The term 'fluid' means that the system is investigated above the phase transition temperature. Below the transition temperature the aliphatic side chains form a long range ordered lattice which shows melting phenomena at the transition temperature. However, the layered arrangement of the constituent molecules is retained above this transition temperature. The polymerization is not lattice controlled and polymers with a large degree of polymerization have been achieved.[30,31] The diffusion of polymer chains in the monolayer is rather slow, so that normal termination reactions in radical-initiated polymerizations such as disproportionation and combination are supressed. A primary termination reaction by initiator radicals has been suggested recently.[32] Although this kind of monolayer polymerization avoids difficulties arising in topochemical reactions from the packing requirements, another problem arises from the fact that the systems will usually tend to separate by phase during the polymerization.

All the substances mentioned so far have their polymerizable groups positioned close to the hydrophilic headgroup. The mismatch between the rather large packing distance of the hydrocarbon chains and the comparitively small length of the repeat unit arising from vinyl polymerization in the polymers leads to structural defects during the polymerization. There is no control with regard to tacticity either, which contributes to the low perfection of the layers. Therefore, it is advantageous to conceive amphiphilic monomers in which the polymerizable function is as far removed from the hydrophilic head group as possible. In such molecules the hydrophobic aliphatic chain has the function of a spacer, decoupling the chemically reactive moieties from the packing requirements of the hydrophilic head groups. When polymerizable groups are in the alkyl chains, and are separated from the hydrophilic head groups, the degree of freedom for forming of a polymer chain should be greater.

A good example is ω-tricosenoic acid (7), which has the polymerizable group at the end of the hydrocarbon chain. (7) can be polymerized in multilayers with an electron beam without any appreciable disorganization of the layers,[33,34] despite the fact that the intermolecular distances in the monolayer are too big for forming C—C bonds without stress.

17.2.2 Topochemical Polymerizations

A large number of dialkynes show a lattice-controlled polymerization,[35] that leads in many cases to polymer single crystals.[36-38] Polymerized LB multilayers with total retention of the layer structure and without destruction of the packing in the individual layers can be prepared from amphiphilic monomers (8), bearing the dialkynic group either near the polar headgroup[39] or in the middle of the alkyl chain (Figure 3).[40]

While dialkynic amphiphiles are highly reactive in the solid-condensed state, no reaction occurs in the liquid-expanded state.[41] The corresponding multilayers have a crystalline domain structure. Large monolayer crystallites can be produced by saturating the atmosphere above the monolayer with the spreading solvent.[42]

A topochemical photoinduced dimerization in LB multilayers has been observed to occur in multilayers of cinnamic acid derivatives. The same products are formed as in the irradiation of the crystals in bulk.[43]

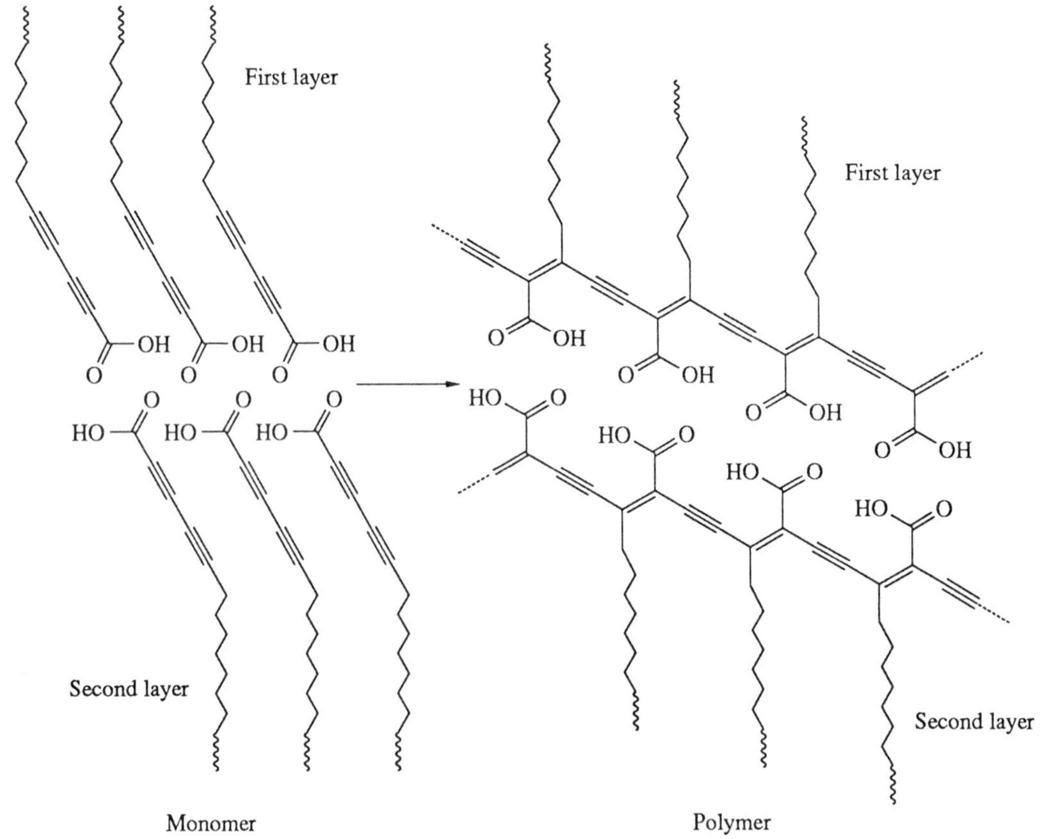

Figure 3 Solid state photopolymerization of dialkynes in multilayers[39,40]

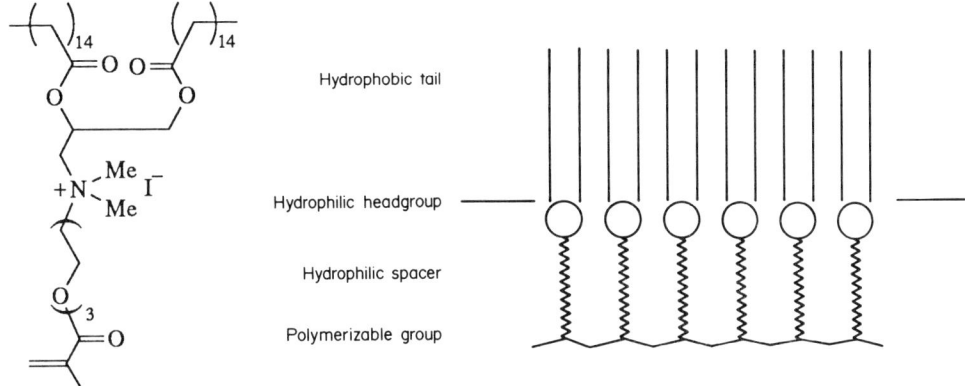

Figure 4 The 'spacer concept'; decoupling of the polymer main chain motion and the self-organization of the amphiphilic side chains[48]

For substituted butadienes (9), polymerization only occurs in the monolayer state where the molecules are obviously packed in suitable orientation relative to each other. This is in contrast to the crystalline solid state, which is inactive with regard to photopolymerization.[44]

Monolayers of ω-tricosenyl-2,4-pentadienoate (10) can undergo three different types of reactions depending on the irradiation conditions.[45] With 247 nm UV light a 3,4-photodimerization in the butadiene part takes place, yielding an acrylate function that can be polymerized at 215 nm. Finally, reaction at the ω double bond leads to crosslinks, if the sample is exposed to an electron beam.

LB mono- and multi-layers exhibit a number of features similar to lipid membranes, liposomes and vesicles. Thus, frequently the same amphiphilic molecules can be used to either prepare LB layers or vesicles.[87] In order to stabilize model membranes (monolayers, black lipid membranes, liposomes) the polymerization of phospholipid analogues bearing dialkyne and butadiene moieties has been investigated in great detail.[46,47] Ringsdorf and coworkers have reviewed this subject recently.[87]

In contrast to the butadiene derivatives, the polymers formed by dialkyne phospholipids are extremely rigid and no longer exhibit phase transitions. The best way to avoid the loss of the desired fluidity seems to be the total decoupling of the motions of the polymer chain and the amphiphilic structure by hydrophilic spacer groups (Figure 4).[48] In this case the polymers obtained by monolayer or bulk polymerization exhibit nearly the same pressure–area isotherms.

Another polymerization reaction with a small change of volume during the polymerization is that of an oxirane (11).[49] Although this offers the promise of defect-free polymeric films, the electron beam sensitivity appears to be too low for resist application.

17.2.3 Polycondensation in Monolayers

Polycondensation reactions in mono- and multi-layers are interesting mainly with regard to the influence of the spatial alignment of the molecules on the reaction rate. Long chain esters of amino acids undergo a condensation reaction in the monolayer to form the corresponding polypeptide (under elimination of the long chain alcohol) without the need of any catalyst. This reaction does not occur in the bulk.[50-52] The reactivity of N-carboxy anhydrides of amino acids (NCAs) has been recently used to obtain the polymer of γ-dodecyl L-glutamate at the air/water interface.[53] Condensation in liposomes of amino acids can be achieved by adding a water soluble carbodiimide to the liposomal solution.[54] Crosslinking of monolayers of amphiphiles with sugar head groups has been reported with bifunctional reagents such as divinyl sulfone or epichlorohydrin.[55] A monofunctional monolayer of octadecanal can be crosslinked by grafting the monolayer onto poly(L-lysine) which is initially dissolved in the aqueous subphase.[56]

A special condensation reaction occurs between the surface-active dihexylterephthalaldiimine (12) and 3,3'-diaminobenzidine (13) (Figure 5) dissolved as the tetrahydrochloride in the subphase. The long chains of the amphiphile are eliminated during the reaction and dissolve in the subphase. The resulting poly(o-aminoanil) monolayer can be deposited onto solid substrates and further annealing at high temperature gives rise to the poly(benzimidazol) (14).[57]

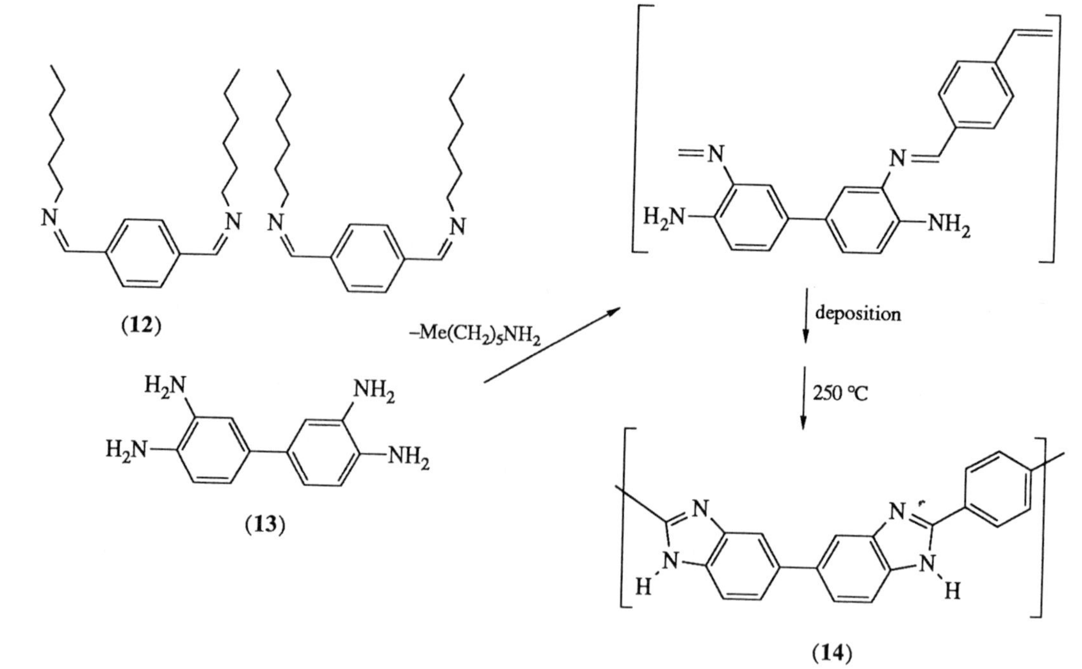

Figure 5 Reaction scheme for the production of poly[1,4-phenylene-5,5'(6,6')-bibenzimidazole diyl] Langmuir–Blodgett films[56]

17.3 PREFORMED POLYMERS

17.3.1 Conventional Linear Polymers

Although the monomer units of conventional linear polymers cannot be considered to be amphiphilic, many of them form monolayers at the air/water interface. Even water soluble polymers such as poly(ethylene oxide) tend to adsorb to the water surface.[58] As mentioned above, the surface pressure corresponds to a two-dimensional osmotic pressure and a monolayer on the water surface can be considered as a two-dimensional solution. This relationship between three- and two-dimensional solutions provoked a series of investigations concerned with the concentration and temperature dependence of surface pressure in comparison with theoretical considerations. Huggins's theory of polymeric monolayers on liquid surfaces,[59] which is an extension of his polymer solution theory,[60] can be applied to various polymers, and yields information about orientation, interaction of polymer chains and compatibility of different structures in two dimensions.

In the picture of a two-dimensional solution, the interface can be regarded as a solvent. In the case of poly(vinyl acetate)[61] the interface shows different solvent behaviour, depending on the spreading solvent. If the spreading solvent is 'poor', the interface behaves as a 'good' solvent.

Geometric factors play an important role for the compatibility of different compounds in monofilms.[62] Poly(vinyl acetate), poly(methyl methacrylate) and poly(propyl methacrylate) have, at the air/water interface, a horizontal disposition, while poly(octadecyl methacrylate) and poly(vinyl stearate) show a perpendicular orientation of their alkyl side chains. Compatibility only occurs between polymers having the same orientation. The same effect has been found for mixing of simple long chain alcohols and acids with poly(vinyl stearate).[63] The compatible systems all show a negative deviation from ideality, indicating negative values of excess free energy of mixing. This principle has been applied to form Langmuir–Blodgett films of poly(3-alkyl thiophenes) (15) which themselves do not form stable monolayers.[64]

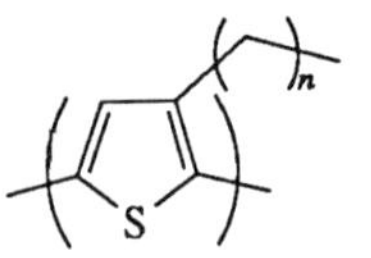

Mixtures of (**15**) with suitable proportions of stearic acid form stable, transferable monolayers. The multilayers can be rendered electrically conductive by exposure to suitable oxidizing agents.

Polymer monofilms are often divided into two classes according to their pressure/area isotherms. Those with a steeper Π/A curve are called condensed films and those with a gradual rise of pressure during compression are called expanded films, in analogy to the different phases appearing in monolayers of small amphiphiles. Rather small deviations in the structure of the polymer can cause totally different behaviour at the air/water interface. Examples of pairs of similar polymers that belong to different classes are poly(ethyl methacrylate) (condensed) and poly(ethyl acrylate) (expanded),[65] and poly(vinyl fluoride) (condensed) and poly(vinylidene fluoride) (expanded).[66]

In three-dimensional polymer solutions of intermediate concentration, when the chains begin to overlap, the virial expansion of the equation for the osmotic pressure is no longer valid. The osmotic pressure can now be expressed using scaling laws.[67,68] A discussion of the experimental results in these terms has been tried in the two-dimensional case of monolayers for poly(vinyl acetate) and poly(methyl methacrylate).[69-71]

Poly(vinyl methyl ether) monolayers were studied in order to investigate the influence of dissolved salts in the water subphase.[72] A 'salting out' effect was observed that means that there is a decreasing surface pressure and area per residue with increasing ion strength in the water.

The surface activity of poly(dialkylsiloxanes) has been known for many years.[73-75] Poly(siloxanes) are still a subject of interest,[76] because of the ease of their modification *via* hydrosilylation of poly(alkylhydrogendialkylsiloxanes). Side groups can be attached to optimize the transfer properties and to introduce chromophores for nonlinear optics[77] or mesogenic units to obtain liquid crystalline polymers, for example.[78]

The use of the Langmuir–Blodgett film deposition technique requires uniform, tightly packed but mobile monolayers of a definite thickness in order to get good quality multilayers. In the case of linear polymers such as poly(vinyl acetate) the surface pressure at the air/water interface is due to an equilibrium between segments attached to the interface (trains) and segments emerged in the adjacent phase (loops and tails). At low pressure these monolayers show high compressibilities (mainly an entropic effect), while at medium pressures the loops and tails become close packed until they are forced out of the interface at still higher pressures. There is no state during the compression where such a monofilm could be regarded as uniform.

17.3.2 Amphiphilic Polymers with Strong Hydrophilic Parts

Amphiphilic polymers are frequently designed in terms of a backbone structure, which then carries the hydrophobic moieties in the form of alkyl side chains and the hydrophilic, polar groups regularly or randomly attached, also as side groups. In some cases, sections of the main chain of polar character may serve as the hydrophilic part of the polymer.

The synthesis of such amphiphilic polymers follows the usual concepts. They are obtained by either copolymerization or copolycondensation by suitably substituted monomers or by polymer analogous reactions. The latter case has caused much interest in the recent literature. For example, poly(vinyl alcohol) can be reacted with long chain aldehydes to give cyclic acetals, where, for statistical reasons, a minimum of 13.5% of the hydroxy groups remain unreacted (Figure 6). These polymers give stable films with close packed polymer chains at the air/water interface and the side chains are directed approximately normal to the interface.[79,80]

Perfluorinated alkylisocyanates have been reacted with poly(allyl amine),[81] yielding polymers for LB deposition even at high degrees of conversion of the amino groups to the corresponding urethane function.

Maleic anhydride can be copolymerized with a variety of terminal long chain alkenes to give alternating copolymers (Figure 7).[82] The anhydride function can be converted to give different types of hydrophilic groups (acid/acid, acid/amide, acid/ester, ester/ester). The opening of the cyclic

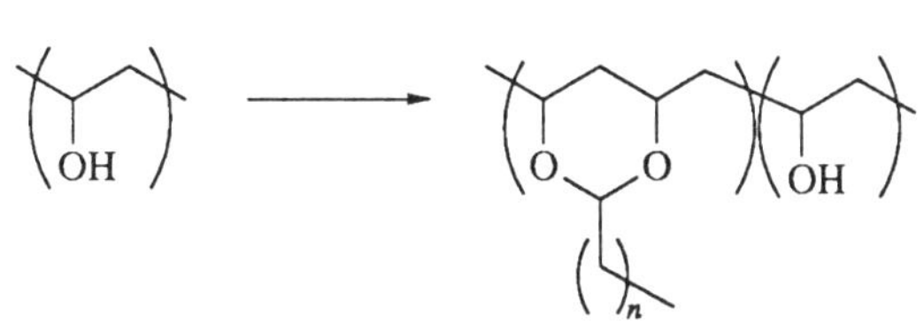

Figure 6 Synthesis of poly(vinyl alkylals)

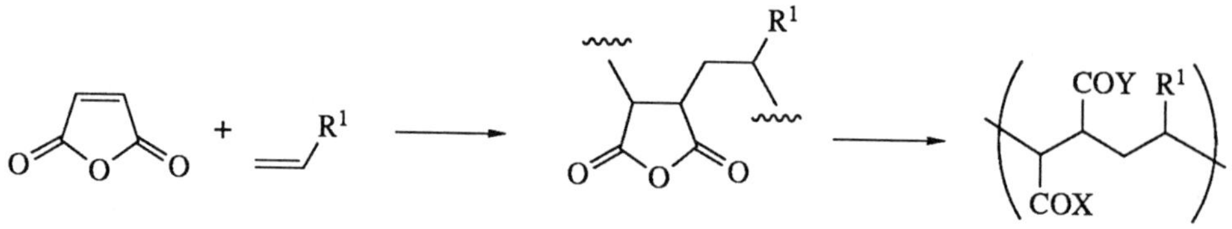

Figure 7 Synthesis of vinyl/maleic anhydride copolymers and derivatives[82]

anhydride by an alcohol provides a way to attach further functionalities to the polymer backbone.[83] The high quality of the multilayers is demonstrated by X-ray studies[84] and optical waveguide experiments.[85] However, the layer thickness derived from X-ray analysis is too small to be related to extended alkyl chains, indicating that there is no close packing of alkyl side chains in terms of a long range order within the layers.

In all of these examples, the polymer backbone is situated at the interface and most probably forms a two-dimensional coil during compression. Depending on the total chain length, the polymer may not be able to unfold from its three-dimensional coiled structure when the solvent is evaporated and, therefore, may be present as a collapsed coil. Therefore, a densely packed monolayer cannot be expected. To overcome this problem the polymer chain may be separated from the interface to provide space for the self-organization of amphiphilic structures attached *via* a spacer to the backbone.

This so-called 'spacer concept' (see Figure 4) leads to an efficient decoupling of the motions of the polymeric chain and the amphiphilic groups.[48,86] The flexible, water soluble spacer consists in most cases of an oligo(ethylene glycol) segment and the amphiphilic group is a lipid-like derivative of glycerol with two long alkanoic side chains. The self-organization of these materials in aqueous systems to form liposomes under ultrasonication works with both the monomers and the preformed polymers.[87]

The oligo(ethylene glycol) spacers can be regarded as anchors to the water subphase. The hydrophobic side chains are not disturbed by this effect, they tend to self-organize. Even polymers with side chains of different lengths that give more fluid-like monolayers can be deposited to give homogeneous, stable films. However, it should be considered that a lot of water will be transferred with the monofilm when multilayers are formed.

The hydrophilic spacer, which is just a covalent link from the organized monofilm to the polymer base, can be replaced by an electrostatic interaction between a charged amphiphile and a high molecular polyion.[88–91] This method can also be used to prepare ultrathin films of insoluble polymers by transferring a polyionic precursor polymer as a complex with an ionic amphiphile by the LB method, followed by chemical treatment of the LB film. In particular polyimides have been prepared by this method (Figure 8).[92,93] Films of the poly(amide acid) salt shown in Figure 8 have been deposited as Z type multilayers. Removal of the alkyl ammonium moieties by thermal treatment with a simultaneous ring closure produces a polyimide multilayer structure with an interlayer distance of 4 Å. Studies by scanning tunnelling microscopy on these films revealed for the first time individual polymer chain segments at a resolution which allows the identification of individual repeat units.[94]

A similar preparation method has been used to obtain poly(*p*-phenylene vinylene) multilayers.[95] A solution of the poly(sulfonium salt) (19) in a water/ethanol/trichloroethane mixture can be spread on water to form monolayers that can be transferred. Thermal treatment of the resulting multilayers leads to poly(*p*-phenylene vinylene) films which can be oxidized by SO_3 in order to obtain electrically conducting films (Figure 9).

17.3.3 Polymers with Side Chains other than Alkyl Chains

Long hydrocarbon chains are not the only structural component which allows for self-organization during compression in the monolayer. Fluorocarbon chains[96,97] or mesogenic aromatic moieties[98,99] have also been investigated as the hydrophobic parts of amphiphilic structures. The use of 'amphotropic' systems,[87] molecules which in the form of LB mono- and multi-layers also show a phase transition to a thermotropic liquid crystalline phase, could be a way to overcome the problem of having grain boundaries in LB films. However, the growth of monodomain liquid crystalline LB films has not yet been demonstrated.

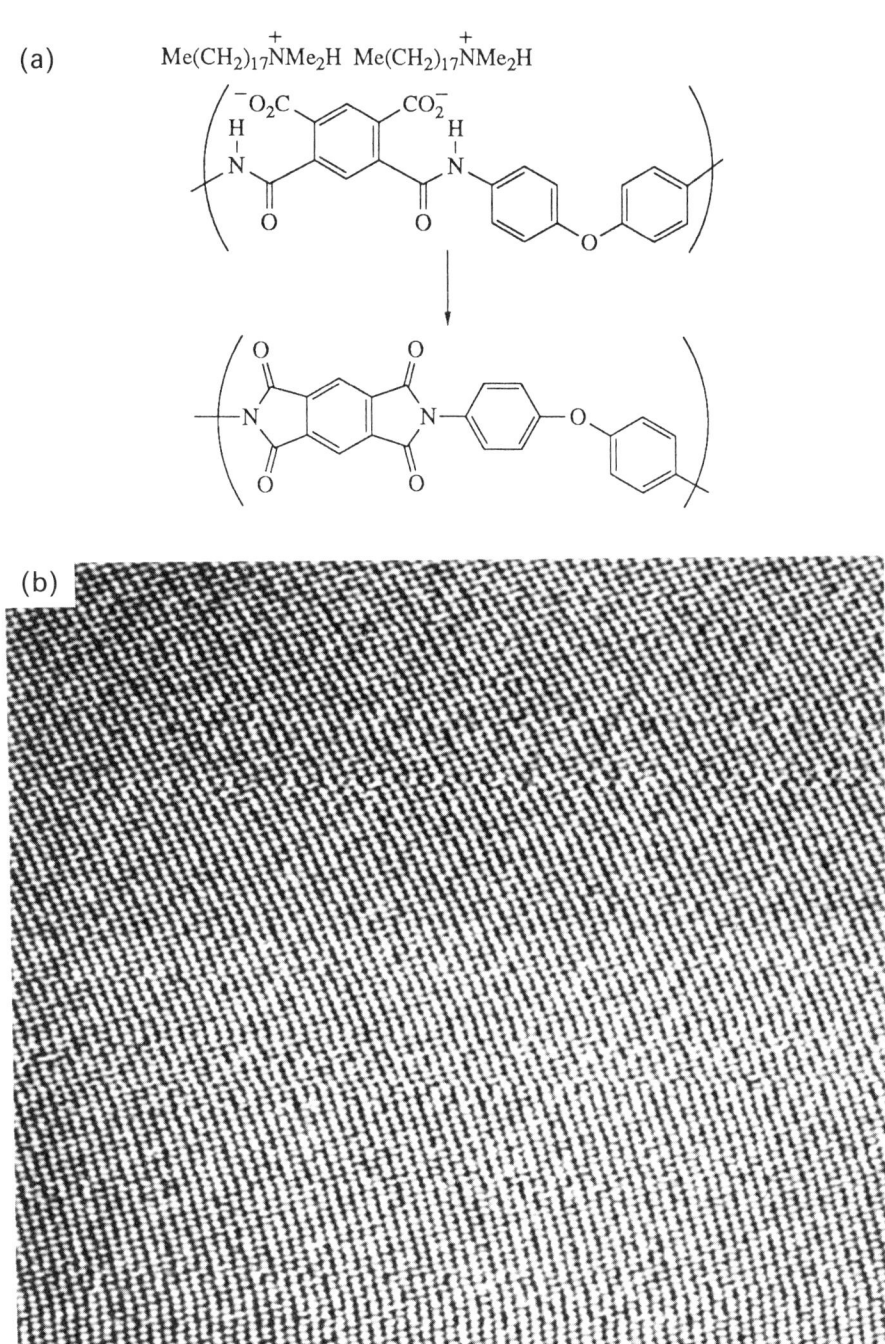

Figure 8 (a) Formation of ultrathin polyimide films; the polymeric salt is deposited as LB film and then converted to the polyimide.[92] (b) Raw top-view STM image of a monolayer of polyimide on highly oriented pyrolytic graphite in air in the constant current mode[94]

By the same token, amphiphilic liquid crystalline polymeric LB films have been prepared.[87,99,100] LB multilayers of reasonable quality were obtained, as demonstrated by optical wave guide experiments.[101] Furthermore, several mesogenic amphiphiles containing structural elements which constitute a chromophore were investigated with regard to nonlinear optical effects.[102] It is also worth mentioning that molecules forming discotic liquid crystals have been discussed for LB layer formation, which should offer new ways to novel multilayer architectures.[103]

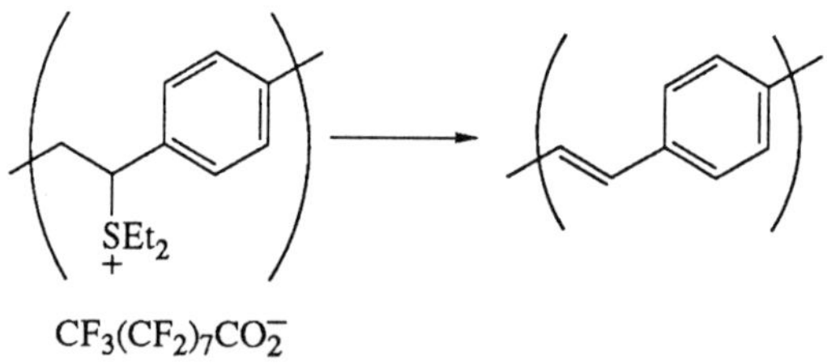

Figure 9 Formation of poly(p-phenylene vinylene) multilayers[95]

17.3.4 Rod-like Polymers

Rod-like polymers are a new class of monolayer-forming molecules which can be organized in the form of LB films for reasons related to the stiffness of their backbone. Typical examples of such polymers are phthalocyaninatopolysiloxanes,[104] several polyglutamates in the α-helical conformation,[105–114] special cellulose alkyl ethers[106,115,116] and specifically substituted polysilanes (Figure 10).[117]

These polymers cannot be classified according to the rules developed for the behaviour of amphiphilic compounds at the air/water interface. In contrast to classical amphiphilic molecules with a hydrophobic tail and a hydrophilic head which are separated topologically, the monolayer-

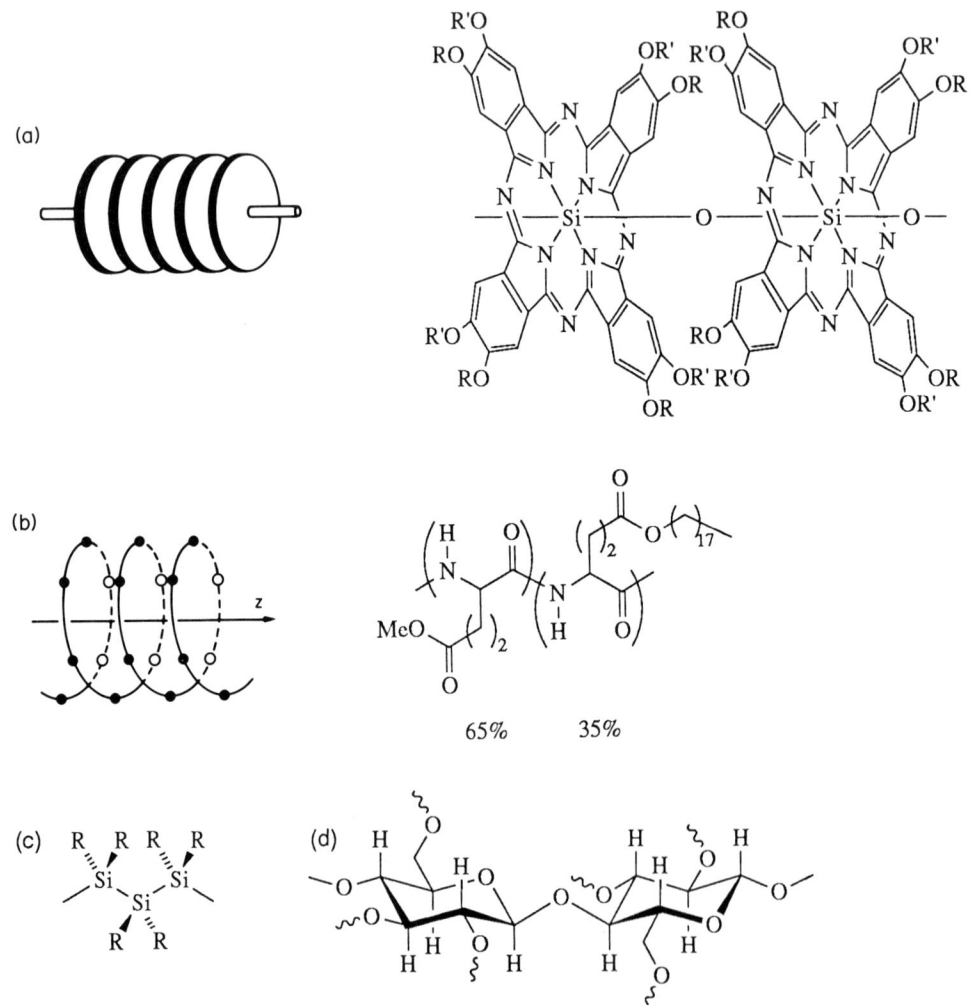

Figure 10 Chemical structures of 'hairy rod' type polymers: (a) phthalocyaninatopolysiloxanes, (b) α-helical polyglutamates, (c) polysilanes with bulky side chains and (d) cellulose alkyl ethers

forming rod-like polymers have a cylindrical symmetry with segments of little polarity distributed equally over the whole molecule. Therefore these polymers are not oriented with their molecular axis perpendicular to the air/water interface, as is known for amphiphilic molecules, but they are lying flat on the surface so that a maximum number of polar groups are in contact with the water surface. Ionogenic groups are not necessary and, therefore, specific interactions between the layer-forming molecules and the water or components dissolved in the water are not observed; specifically, ionic impurities in the water subphase do not interfere with the formation of the monolayers.[118]

Surface pressure–area diagrams of rod-like polymers are known for cellulose esters, ethyl cellulose[105,115] and polypeptides in the α-helical conformation, such as poly(benzyl aspartate) and poly(methyl glutamate).[106–108] The behaviour of polypeptides as monolayers at the air/water interface has been studied extensively in the past. However, applying the LB method to the transfer of only one such layer to quartz substrates with the α-helical conformation of the backbone persisting in the transferred layers has been reported.[109] In the case of multilayers achieved with the Schäfer technique, a hexagonal packing of the α-helical chains has been proposed based on SAXS measurements.[110] Preferential alignment of the polypeptide backbones parallel to the dipping direction was found first by circular dichroism[111] and polarized IR spectroscopy.[112]

Generally, the deposition behaviour of the polypeptides and the cellulose derivatives was reported to be very poor, and therefore these materials were not considered to be suitable for the formation of multilayers with a reasonable degree of perfection.

As a new concept to overcome this problem, rod-like polymers surrounded by conformationally mobile side chains have been developed. Side chain crystallization was prevented by mixing short and long side chains along the backbone. This provides fluidity to the monolayer at the air/water interface and screens the direct backbone interactions.[104,113,114] Thus, rod-like polymers are decorated by alkyl chains of different length wrapping the individual backbones with a skin of liquid-like segments which act as chemically attached solvent molecules. The structure of these polymers is, therefore, quite different to that of conventional amphiphilic molecules forming LB films (Figure 11).

This type of polymer is represented best by an unsymmetrical substituted phthlalocyaninatopolysiloxane.[104] More than 200 layers of these molecules have been transferred easily onto hydrophobic substrates. Stable LB films of high quality are formed, in which the backbones are aligned preferentially parallel to the dipping direction. The spreading, compressing, transfer and packing in layers of these molecules is schematically depicted in Figure 12.

In the first step (a) the rod-like molecules are spread out of a polymer solution onto the water surface, followed by the evaporation of the solvent. In the second step (b) the area of the surface is reduced by the moving barrier so that the flat-lying rod-like polymers are compressed to form a stable monolayer in terms of a two-dimensional nematic structure. In the final step (c) the

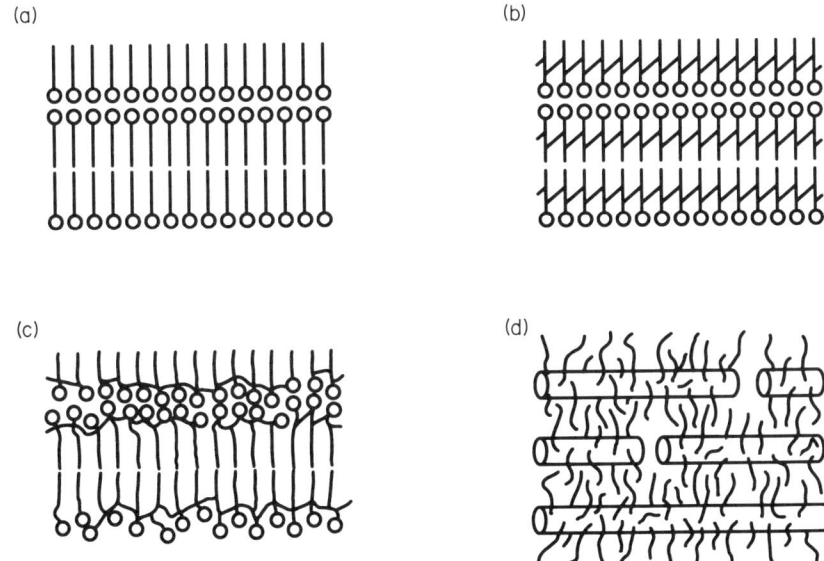

Figure 11 Comparison of LB films of different compounds: (a) classical amphiphile, *e.g.* fatty acids, (b) solid state polymerized multilayer, (c) preformed amphiphilic polymer and (d) rigid rod-like polymer with flexible side chains

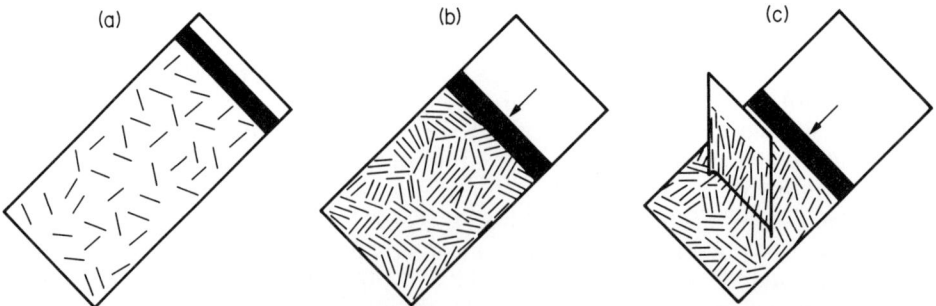

Figure 12 Schematic of Langmuir–Blodgett transfer of monolayers form rod-like polymers: (a) expanded, (b) compressed film and (c) transfer with orientation of the main axis of the molecules in dipping direction

monolayer is transferred to a solid substrate by the Langmuir–Blodgett technique, with simultaneous alignment of the stiff polymer backbones parallel to the dipping direction. Recent experiments have shown that the orientation of the backbones is induced by the flow of the monolayer on the water surface during the transfer process. So the two-dimensional orientation parameter is influenced by changes of the flow conditions on the surface.[119]

As already mentioned, high quality multilayers of α-helical poly(alkyl glutamates)[113,114,120] can only be built up when the side chains are disordered. This is achieved by selecting copolymers with an appropriate ratio of long and short alkyl chains. A detailed investigation was carried out for poly(γ-methyl-co-octadecyl L-glutamate) where a ratio of methyl/octadecyl of 1 to 2 was found to be the optimum.

17.4 THE STRUCTURE OF POLYMERIC LB FILMS

As can be seen in the preceding sections, the chemical reactivity of multilayers from polymerizable monomers is closely related to the microstructure of the multilayers. Several studies have been concerned with the detailed structure analysis of these films.[121] X-ray and electron diffraction studies, as well as electron microscopy, show that LB films of monomers based on the close packing of long alkyl chains possess a highly crystalline lamellar structure and consist of numerous domains

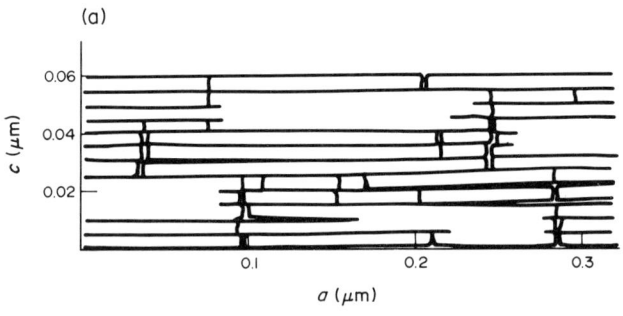

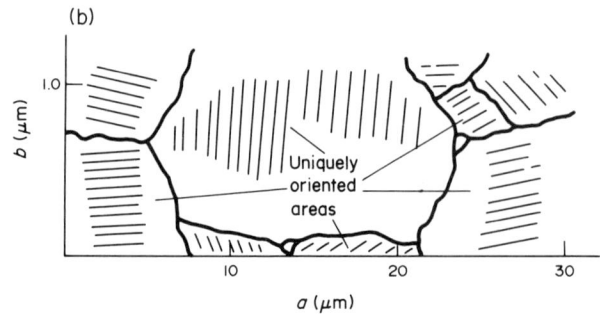

Figure 13 Model of a multilayer of amphiphilic compounds: (a) parallel and (b) perpendicular to the substrate

with diameters between 0.1 and 10 μm. Adjacent layers are often not packed in crystallographic registry, as has been shown by electron diffraction. Multilayers of vinyl stearate[122] or some dialkyne monocarboxylic acids[123] show the so-called 'hexagonal twinning' that leads to a superposition of three identical diffraction patterns each rotated with respect to the preceding one by 120°. A model of such multilayers is illustrated in Figure 13. The orientation of the alkyl chains, which are either tilted or perpendicular to the layer plane, can be determined by transmission and reflection infrared spectroscopy.[124]

The structure of the multilayers of amphiphilic polymers is also determined by the interaction of the alkyl side chains. One example is isotactic poly(octadecyl methacrylate) (i-PODMA).[113,125] The orientation of the alkyl chains perpendicular to the layer plane is derived from FTIR spectra

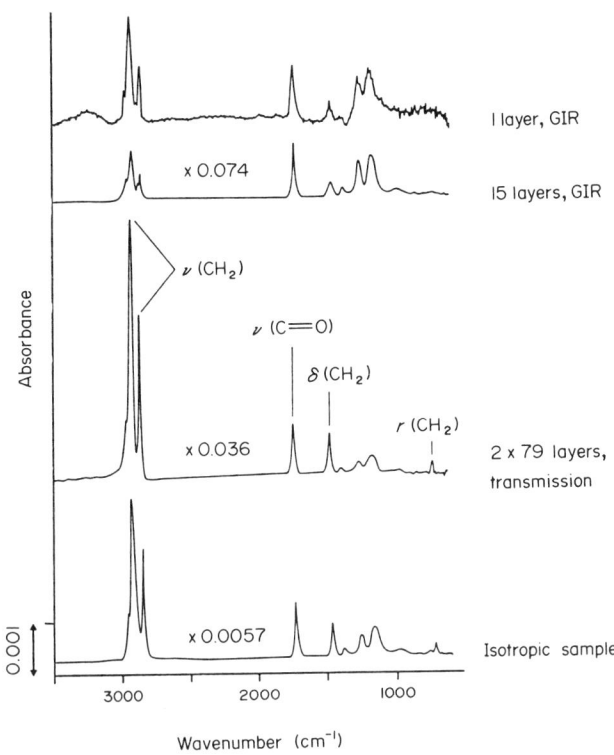

Figure 14 FTIR spectra of i-PODMA LB layers and of an isotropic film cast from solution[125]

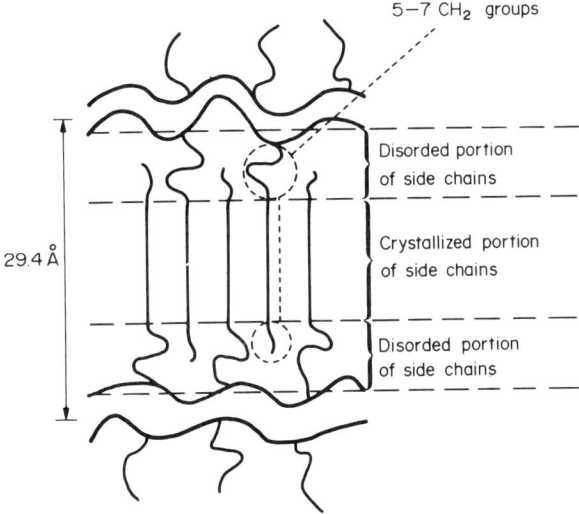

Figure 15 Proposed model of i-PODMA LB layers[113,125]

which show considerably higher relative intensity of the ν (CH_2) modes in transmission than in grazing incident reflection (GIR) (Figure 14).

Small angle X-ray scattering of i-PODMA multilayers gave a value of 30 Å for the layer periodicity; together with the results of electron diffraction studies which indicate the existence of hexagonal packed alkyl chains in these multilayers, the model illustrated in Figure 15 was proposed. Side chains of polymers situated in adjacent layers interdigitate and crystallize; however, only a fraction of all side chain CH_2 units participate in crystal formation, the others staying amorphous.

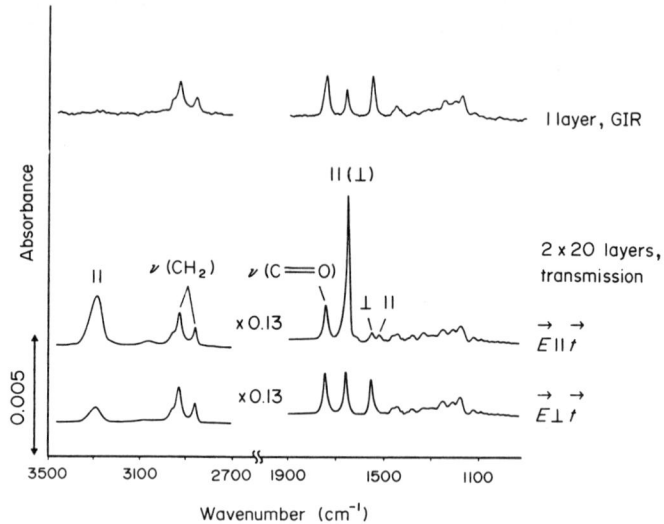

Figure 16 FTIR spectra of PMOLG LB layers[113]

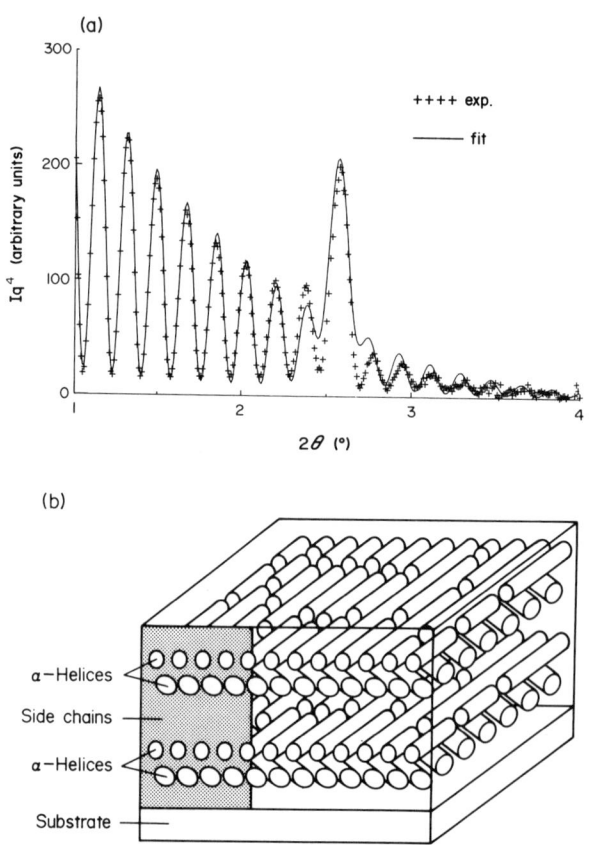

Figure 17 (a) X-ray reflection curve from 24 layers of PMOLG.[126] (b) Proposed model of PMOLG LB layers

The structure of LB films based on the rod-like polymers has been investigated by FTIR spectroscopy, X-ray reflection and electron diffraction. Figure 16 shows the transmission and GIR FTIR spectra of multilayers of α-helical poly(L-glutamate) with 35% octadecyl and 65% methyl side chains (PMOLG).

The amide A band at 3300 cm^{-1} is absent in the GIR spectrum, indicating that the helices lie flat on the substrate. In transmission, the various amide bands display dichroism, while the bands of the side chains [$v(CH_2)$ and $v(C=O)$] do not. By analyzing the intensities of the different IR bands for the different polarization directions, the average angle between the helical axis and the transfer direction turned out to be 26°, while the orientation of the side chains turned out to be random. This could be confirmed by diffraction studies, which did not give any evidence for side chain crystallization. X-Ray reflection measurements gave a periodicity of 35 Å, which is due to bilayer formation. The highly resolved so-called Kiessig fringes (arising from interference of reflected beams at the two interfaces, substrate/polymer and polymer/air) indicate a well-defined layer structure with a low surface roughness (Figure 17a).[126] The interference pattern can be used to calculate the total thickness of the film. The model illustrated in Figure 17(b) was proposed for this kind of multilayer.

17.5 PERSPECTIVES OF TECHNICAL APPLICATION

To form LB films, the molecular architecture must follow the principles of either amphiphilic molecules or rod-like polymers. The desired physical properties, such as high molecular polarizability, can be achieved by appropriate molecular design. Additionally, most technical applications require mechanical, thermal and chemical resistance, as well as long-term stability of the LB films.

Therefore, polymeric systems have a decisive advantage over monomer assemblies, as they form much more stable LB films due to their covalent bonding in the film plane. As an example, a polyamide LB film is thermally stable up to 250 °C,[127] whereas a monomeric film from 22-tricosenoic acid detoriorates at about 70 °C.

Further improvement of the stability of polymeric LB films, especially against chemical treatment, should be achieved by attaching photo-crosslinkable groups to the flexible side chains of rod-like polymer systems.

17.5.1 Optoelectronic and Nonlinear Optical Effects

Optical birefringence is frequently observed in LB multilayers, due to their anisotropic molecular structure.[104,128] Thus, optical filters for linear optical applications can be prepared. In the case of mesogenic side groups or rod-like polymers, the optical anisotropy may be induced or enhanced after film deposition, by applying an external electrical field, or by thermal treatment.

One of the most promising technological prospects for thin ordered organic films are applications involving nonlinear optical and optoelectronic effects. The physical basis for this is provided by the highly polarizable conjugated π-electron system in certain organic compounds, especially in molecules containing aromatic groups with donor/acceptor moieties, preferably in *ortho/para* position of benzene rings.

Large second-order nonlinear optical effects can be observed in noncentrosymmetric LB multilayers of chromophores which exhibit an intrinsic polarization. The second-order susceptibility $\chi^{(2)}$ of donor/acceptor systems such as hemicyanines or phenylhydrazones can be two or three orders of magnitude higher than those of more conventional inorganic compounds such as $LiNbO_3$. These films can be used for frequency doubling or to construct electrooptical switching devices, for example in waveguide configurations.

An even more fascinating application can be expected in optical data processing, such as for real time pattern recognition. In this case, third-order effects such as four-wave mixing might be used to create optical switches with femtosecond response times.

Third-order optical effects in LB films may also be applied to prepare phase-conjugated mirrors which can be used, for example, for correction of signal dispersion effects in optical transmission lines.

17.5.2 Pyroelectric and Piezoelectric LB Films

The surface potential of Langmuir monolayers of amphiphilic compounds can be varied by more than 1 V, modifying the structure of the hydrophobic end groups,[129] and thus changing the dipole moment of the molecules. So, adjustable macroscopic polarization can be achieved building up LB multilayers with a noncentrosymmetric structure out of selected monolayers.

In such LB films, pyroelectricity and piezoelectricity can be observed.[130,131] Due to the small thickness of the multilayers, the pyroelectric response is very fast. Therefore pyroelectric detectors for thermal radiation have been proposed which may be used, for example, in fast infrared TV cameras.[132-135]

17.5.3 Sensors

There is an increasing demand for improved analytical techniques using small sensor devices in pharmaceutical and biomedical industries for process control. Highly specific and selective devices which could be made of thin organic films are of particular interest. The operation of a sensor device can be divided into three steps: (i) specific recognition of the analyte; (ii) change of a physical parameter caused by recognition; and (iii) conversion of the change into an observable signal. Therefore chemical processes have to be interfaced with conventional electronic or optical devices as, for example, membrane-coated electrodes,[133] field-effect transistors (FETs) or chemically sensitive field-effect transistors (ChemFETs),[134] pH- or ion-sensitive gate electrodes (ISFETs) etc.[134,135] Optical methods which can be applied for detection are methods using energy transfer mechanisms, such as fluorescence quenching or excitation.[136] Plasmon resonance[137] or ellipsometry[138] detect changes in the refractive index or the thickness of an LB film in contact with the analyte. Further, detection of a mass change is achieved by a quartz microbalance[139-141] or surface acoustic waves.[142]

While these different readout systems can be used with isotropic coatings in the micrometer thickness range, superior chemical sensors can be made with LB films. Therefore, specially designed organic films are needed to obtain a selective response. Further, the surface composition of the films must be controlled to avoid unspecific adsorption. To achieve a maximum number of reactive species per unit surface area, the orientation of the molecules within the monolayer configuration must be optimized. Stable films of small overall thickness that are of a minimum number of layers are required to guarantee short response times and fast equilibration. Additionally local defects in the LB films have to be avoided, if the sensor element must be used under conditions of high electric field strength or in contact with electrolytes.[143]

The rigid rod polymers mentioned in this chapter fulfil these requirements very well. They exhibit excellent film-forming properties and can be used as components of FET sensors, as was shown with multilayers of phthalocyaninatopolysiloxanes.[144]

17.5.4 Separation Layers

The molecular structure of LB multilayers, which bear similarities to biological membrane systems, also suggests possible applications for highly selective separation membranes.[145] Investigations for reversed osmosis[146] and gas separation membranes show that LB films on porous supports exhibit structural defects, which lead to poorly reproducible separation properties caused by leakage. This problem can be overcome using a special composite technique, in which a very thin nonporous but highly permeable layer is installed between the porous support and the LB multilayer. On this support the number of defects in the LB film is dramatically reduced. Although this demonstrates that molecular architecture can control the film properties, the transport phenomena through these anisotropic films are not yet fully understood.

17.5.5 Surface Modification

Conventional isotropic organic surface coatings with thickness in the micrometer range play an enormous economic role as protective coatings, lacquers, lubricants, adhesive layers and interfacial layers for inducing desired structural order at phase boundaries. In some of these applications LB films may show improvements, as discussed below.

17.5.5.1 Orientation layers

In the field of liquid crystal displays, orientation layers for the alignment of liquid crystals parallel to the substrate surface are needed. Usually, isotropic surfaces are made of polyimide films which have to be rubbed mechanically to form an aligned layer. During this rubbing process the elements of the display may be destroyed. In contrast, aligned layers prepared with the LB technique obviate the need for the rubbing procedure. Experiments indicate that orientation can be induced even by a single monolayer.[147-149]

17.5.5.2 Lubricant layers

Lubrication and encapsulation are very important areas of research in the magnetic recording industry. A good permeability barrier protects the magnetic media, a good lubricant avoids wear and tear during transits of the tape across the magnetic head. For use in high-density magnetic storage media very thin layers of lubricants (no more than 10 nm) are required because of the small spatial gap between the read/write head and the magnetic layer. These can be achieved by using films prepared *via* the LB method[150-152] or *via* the self-assembly technique. A drastic reduction of the friction coefficient of pure substrate surfaces can be obtained by depositing a few monolayers or even a single monolayer.[153] Also, an increase of the friction coefficient after a few hundred transits of the tape over the head, observed on pure metal surfaces, could be prevented if the surface was coated with some monolayers of barium stearate.[154]

17.5.5.3 Wettability

In contrast to the established modification technologies, such as plasma and corona discharge, flame or chemical treatment and polymer grafting, ultrathin films offer an interesting alternative to tailor the surface properties. Phenomena of surface modification such as hydrophobization of metal or glass surfaces are well known. The effects of predefined chemical structures and controlled surface order on wettability has been studied on monolayers of long alkyl chain thiols on gold surfaces prepared *via* the self-assembly technique.

17.5.5.4 Biocompatibility

Artificial implants in contact with body fluids have to be biocompatible. Biocompatibility in itself is a complex phenomenon involving numerous biochemical interactions and reactions. The initial step in all cases of reduced biocompatibility is an interaction of proteins with a solid surface. This surface reaction of proteins is caused by dispersion, polar, electrostatic and hydrogen-bonding forces, as well as mechanical forces.

In principle, all the parameters determining these forces can be adjusted by the LB technique. Therefore, LB films represent good model systems to study the complex processes which take place on biological surfaces.

17.5.6 Microlithography

Many applications of ultrathin films, for example as sensors, optical devices or resists, require methods to write patterns of defined structure.

The finest structures would be achieved by electron beam microlithography, due to the extremely small wavelength of high energy electrons. The main disadvantage of electron beam resists is their electron-scattering characteristic. Even though it is possible to focus the electron beam to a diameter less than 10 nm, a halo of much larger dimension is formed around the point of impact caused by the scattering in the resin material on the substrate. The resolution attainable is improved only by using very thin layers of the resists. Therefore the LB technique provides a method of depositing materials onto substrates to aid in microstructure fabrication. The topochemical polymerization of dialkyne amphiphiles[35] or ω-tricosenoic acid[33,155,156] in LB films are examples for a negative resist with a resolution better than 50 nm.

In order to transfer these ultrafine structures to solid substrates, however, the etch resistance of the LB films has to be improved. Appropriate films for this purpose are based on polysiloxanes and polysilanes.[157]

17.6 REFERENCES

1. L. Netzer, R. Iscovici and J. Sagiv, *Thin Solid Films*, 1983, **99**, 235.
2. C. D. Bain and G. M. Whitesides, *Adv. Mater.*, 1989, **1**, 110; *Angew. Chem., Int. Ed. Engl., Adv. Mater.*, 1989, **28**, 506; *Angew. Chem., Adv. Mater.*, 1989, **101**, 522.
3. I. Langmuir, *J. Am. Chem. Soc.*, 1917, **39**, 1848.
4. K. B. Blodgett, *J. Am. Chem. Soc.*, 1935, **57**, 1007.
5. L. Wilhelmy, *Ann. Phys.*, 1863, **119**, 117.
6. G. L. Gaines, Jr., *Insoluble Monolayers at Liquid–Gas Interfaces*, Wiley-Interscience, New York, 1966.
7. G. L. Gaines, Jr., *J. Colloid Interface Sci.*, 1977, **62**, 191.
8. A. M. Bibo and I. R. Bibo, *Adv. Mater.*, 1990, **2**, 309.
9. H. Kuhn, S. Möbius and H. Bücher, in 'Physical Methods of Chemistry', ed. A. Weissberger and B. Rossiter, Wiley, New York, 1972, vol. 1, p. 577.
10. H. Kuhn, *Thin Solid Films*, 1983, **99**, 1.
11. R. Ackermann, O. Inacker and H. Ringsdorf, *Kolloid Z. Z. Polym.*, 1971, **249**, 1118.
12. N. Beredjick and W. J. Burlant, *J. Polym. Sci., Polym. Chem. Ed.*, 1970, **8**, 2807.
13. A. Cemel, T. Fort Jr. and J. B. Lando *J. Polym. Sci. Polym. Chem. Ed.*, 1972, **10**, 2061.
14. M. Puterman, T. Fort, Jr. and J. B. Lando, *J. Colloid Interface Sci.*, 1974, **47**, 705.
15. D. G. Whitten and P. R. Worsham, *Org. Coat. Plast. Chem.*, 1978, **38**, 572.
16. R. Ackermann, D. Naegele and H. Ringsdorf, *Makromol. Chem.*, 1974, **175**, 699.
17. D. Naegele, J. B. Lando and H. Ringsdorf, *Macromolecules*, 1977, **10**, 1339.
18. J. P. Rabe, J. F. Rabolt, C. A. Brown and J. D. Swalen, *J. Chem. Phys.*, 1986, **84**, 4096.
19. A. Laschewsky, H. Ringsdorf and G. Schmidt, *Polymer*, 1988, **29**, 448.
20. A. Banerjie and J. B. Lando, *Thin Solid Films*, 1980, **68**, 67.
21. T. Miyashita, H. Yoshida, T. Murakata and M. Matsuda, *Polymer*, 1987, **28**, 311.
22. A. N. Broers and M. Pomerantz, *Thin Solid Films*, 1983, **99**, 323.
23. P. J. Flory, 'Principles of Polymer Chemistry', Cornell University Press, Ithaca, 1975.
24. A. Dubault, C. Casagrande and M. Veyssié, *J. Phys. Chem.*, 1975, **79**, 2254.
25. H. Rehage and M. Veyssié, *Angew. Chem.*, 1990, **102**, 497.
26. K. C. O'Brien, J. Long and J. B. Lando *Langmuir*, 1985, **1**, 514.
27. D. F. O'Brien and R. Ramaswami, in 'Encyclopedia of Polymer Science and Engineering', 2nd edn., Wiley, New York, 1989, vol. 17, p. 108.
28. E. Evans and E. Sackmann, *J. Fluid Mech.*, 1988, **194**, 553.
29. R. Merkel, E. Sackmann and E. Evans, *J. Phys. France*, 1989, **50**, 1535.
30. K. Dorn, E. V. Patton, R. T. Klingbiel, D. F. O'Brien and H. Ringsdorf, *Makromol. Chem., Rapid Commun.*, 1983, **4**, 513.
31. D. Bolikal and S. L. Regen, *Macromolecules*, 1984, **17**, 1287.
32. T. D. Sells and D. F. O'Brien, *Macromolecules*, 1991, **24**, 336.
33. A. Barraud, C. Rosilio and A. Ruaudel-Teixier, *J. Colloid Interface Sci.*, 1977, **62**, 509.
34. A. Barraud, C. Rosilio and A. Ruaudel-Teixier, *Solid State Technol.*, 1979, **22**, 120.
35. B. Tieke, *Adv. Polym. Sci.*, 1985, **71**, 79.
36. G. Wegner, *Makromol. Chem.*, 1972, **154**, 35.
37. R. H. Baughman, *J. Polym. Sci., Polym. Phys. Ed.*, 1974, **12**, 1511.
38. G. Wegner and W. Schermann, *Colloid Polym. Sci.*, 1974, **252**, 655.
39. B. Tieke, H. J. Graf, G. Wegner, B. Naegele, H. Ringsdorf, A. Banerjie, D. R. Day and J. B. Lando, *Colloid Polym. Sci.*, 1977, **255**, 521.
40. B. Tieke, G. Lieser and G. Wegner, *J. Polym. Sci., Polym. Chem. Ed.*, 1979, **17**, 1631.
41. D. Day, H. H. Hub and H. Ringsdorf, *Isr. J. Chem.*, 1980, **18**, 325.
42. D. Day and J. B. Lando, *Macromolecules*, 1980, **13**, 1478, 1483.
43. V. Enkelmann, B. Tieke, H. Kapp, G. Lieser and G. Wegner, *Ber. Bunsenges. Phys. Chem.*, 1978, **82**, 875.
44. H. Ringsdorf and H. Schupp, *J. Macromol. Sci., Chem*, 1981, **A15**, 1015.
45. A. Barraud, C. Rosilio and A. Ruaudel-Teixier, *Polym. Prepr., Am. Chem. Soc. Div. Polym. Chem.*, 1978, **19**, 179.
46. D. R. Day and H. Ringsdorf, *Makromol. Chem.*, 1979, **180**, 1059.
47. B. Hupfer, H. Ringsdorf and H. Schupp, *Makromol. Chem.*, 1981, **182**, 247.
48. R. Elbert, A. Laschewsky and H. Ringsdorf, *J. Am. Chem. Soc.*, 1985, **107**, 4134.
49. B. Boothroyd, P. A. Delaney, R. A. Hann, R. A. W. Johnstone and A. Ledwith, *Br. Polym. J.*, 1985, **17**, 360.
50. A. Baniel, M. Frankel, I. Friedrich and A. Katchalsky, *J. Org. Chem.*, 1948, **13**, 791.
51. K. Fukuda, Y. Shibasaki and H. Nakahara, *J. Macromol. Sci., Chem.*, 1981, **A15**, 999.
52. K. Fukuda, Y. Shibasaki and H. Nakahara, *Thin Solid Films*, 1983, **99**, 87.
53. A. Shibata, A. Oasa, Y. Hashimura, S. Yamashita, S. Ueno and T. Yamashita, *Langmuir*, 1990, **6**, 217.
54. R. Neumann and H. Ringsdorf, *J. Am. Chem. Soc.*, 1986, **108**, 487.
55. W. N. Emmerling and B. Pfannemüller, *Colloid Polym. Sci.*, 1983, **261**, 677.
56. S. J. Valenty, *Macromolecules*, 1978, **11**, 1221.
57. A. K. Engel, T. Yoden, K. Sanui and N. Ogata, *Polym. Mater. Sci. Eng.*, 1986, **54**, 119.
58. D. J. Crisp, *J. Colloid Sci.*, 1946, **1**, 49, 161.
59. M. L. Huggins, *Kolloid Z. Z. Polym.*, 1972, **251**, 449.
60. M. L. Huggins, *J. Phys. Chem.*, 1970, **74**, 371; 1971, **75**, 1255; *Polymer*, 1971, **12**, 389; 1972, **13**, 554.
61. G. Gabrielli, P. Baglione and E. Ferroni, *Colloid Polym. Sci.*, 1979, **257**, 121.
62. G. Gabrielli, M. Pugelli and P. Baglioni, *J. Colloid Interface Sci.*, 1982, **86**, 485.
63. K. Fukuda, T. Kato, S. Machida and Y. Shimizu, *J. Colloid Interface Sci.*, 1979, **68**, 82.
64. I. Watanabe, K. Hong and M. F. Rubner, *Langmuir*, 1990, **6**, 1164.
65. D. J. Crisp, in 'Surface Phenomena in Chemistry and Biology', ed. J. F. Danielli, K. G. A. Pankhurst and A. C. Riddiford, Pergamon Press, Oxford, 1958, p. 23.
66. E. Ferroni, G. Gabrielli and M. Puggelli, *Chim. Ind. (Milan)*, 1967, **49**, 147.

67. P.-G. de Gennes, 'Scaling Concepts in Polymer Physics', Cornell University Press, Ithaca, 1979.
68. M. Daoud, J. P. Cotton, B. Farnoux, G. Jannik, G. Sarma, H. Benoit, R. Duplessix, C. Picot and P.-G. de Gennes, *Macromolecules*, 1975, **8**, 804.
69. R. Vilanove and F. Rondelez, *Phys. Rev. Lett.*, 1980, **45**, 1502.
70. A. Takahashi, A. Yoshida and M. Kawaguchi, *Macromolecules*, 1982, **15**, 1196.
71. M. Kawaguchi, A. Yoshida and A. Takahashi, *Macromolecules*, 1983, **16**, 956.
72. D. D. Eley, M. J. Hey and J. Speight, *J. Chem. Soc., Faraday Trans. 1*, 1983, **79**, 755.
73. H. W. Fox, P. W. Taylor and W. A. Zisman, *Ind. Eng. Chem.*, 1947, **59**, 1401.
74. W. Noll, H. Steinbach and C. Sucker, *Kolloid Z. Z. Polym.*, 1965, **204**, 94.
75. W. Noll, H. Steinbach and C. Sucker, *J. Polym. Sci.*, 1971, **C34**, 123.
76. S. Granick, S. J. Clarson, T. R. Formoy and J. A. Semlyen, *Polymer*, 1985, **26**, 925.
77. N. Carr and M. J. Goodwin, *Makromol. Chem., Rapid Commun.*, 1987, **8**, 487.
78. G. W. Gray, W. D. Hawthorne, J. S. Hill, D. Lacey, M. S. K. Lee, G. Nestor and M. S. White, *Polymer*, 1989, **30**, 964.
79. M. Koyama, R. Tomioka, M. Ueno and K. Meguro, *Colloid Polym. Sci.*, 1974, **252**, 372.
80. M. Watanabe, Y. Kosaka, K. Oguchi, K. Sanui and N. Ogata, *Macromolecules*, 1988, **21**, 2997.
81. M. Tamura, H. Ishida and A. Sekiya, *Chem. Lett.*, 1988, 1277.
82. P. Hodge, E. Khoshdel, R. H. Tredgold, A. J. Vickers and C. S. Winter, *Br. Polym. J.*, 1985, **17**, 370.
83. R. Jones, C. S. Winter, R. H. Tredgold, P. Hodge and A. Hoorfar, *Polymer*, 1987, **28**, 1619.
84. R. H. Tredgold, A. J. Vickers, A. Hoorfar, P. Hodge and E. Koshdel, *J. Phys. D*, 1985, **18**, 1139.
85. R. H. Tredgold, M. C. J. Young, P. Hodge and E. Khoshdel, *Thin Solid Films*, 1987, **151**, 441.
86. A. Laschewsky, H. Ringsdorf, G. Schmidt and J. Schneider, *J. Am. Chem. Soc.*, 1987, **109**, 788.
87. H. Ringsdorf, B. Schlarb and J. Venzmer, *Angew. Chem., Int. Ed. Engl.*, 1988, **27**, 113; *Angew. Chem.*, 1988, **100**, 117.
88. W. McNaughtan, K. A. Snook, E. Caspi and N. P. Franks, *Biochim. Biophys. Acta*, 1985, **818**, 132.
89. M. Shimomura and T. Kunitake, *Thin Solid Films*, 1985, **132**, 243.
90. N. Higashi and T. Kunitake, *Chem. Lett.*, 1986, 105.
91. A. Takahara, N. Morotomi, S. Hiraoka, N. Higashi, T. Kunitake and T. Kajiyama, *Macromolecules*, 1989, **22**, 617.
92. M. Suzuki, M. Kakimoto, T. Konishi, Y. Imai, M. Iwamoto and T. Hino, *Chem. Lett.*, 1986, 395.
93. Z. Nishikata, T. Konishi, A. Morikawa, M. Kakimoto and Y. Imai, *Polym. J.*, 1988, **20**, 269.
94. H. Sotobayashi, T. Schilling and B. Tesche, *Langmuir*, 1990, **6**, 1246.
95. Y. Nishikata, M. Kakimoto and Y. Imai, *J. Chem. Soc., Chem. Commun.*, 1988, 1040.
96. A. Takahara, N. Morotomi, S. Hiraoka, N. Higashi, T. Kunitake and T. Kajiyama, *Macromolecules*, 1989, **22**, 617.
97. J. Schneider, C. Erdelen, H. Ringsdorf and J. F. Rabolt, *Macromolecules*, 1989, **22**, 3475.
98. O. Albrecht, W. Cumming, W. Kreuder, A. Laschewsky and H. Ringsdorf, *Colloid Polym. Sci.*, 1986, **264**, 659.
99. R. H. Tredgold, R. A. Allen and P. Hodge, *Thin Solid Films*, 1987, **155**, 343.
100. K. A. Suresh, A. Blumstein and F. Fondolez, *J. Phys. (Orsay, Fr.)*, 1985, **46**, 453.
101. A. J. Vickers, R. H. Tredgold, E. Khoshdel, P. Hodge and I. Girling, *Thin Solid Films*, 1985, **134**, 43.
102. M. Carpenter, P. N. Prasad and A. C. Griffin, *Thin Solid Films*, 1988, **161**, 315.
103. A. Laschewsky, *Angew. Chem., Int. Ed. Engl., Adv. Mater.*, 1989, **28**, 1574.
104. E. Orthmann and G. Wegner, *Angew. Chem., Int. Ed. Engl.*, 1986, **25**, 1105; *Angew. Chem.*, 1986, **98**, 1114.
105. M. Hittmeier, L. S. Sandell and P. Luner, *J. Polym. Sci. C*, 1971, **36**, 267.
106. B. R. Malcolm, *Nature (London)*, 1962, **195**, 901.
107. B. R. Malcolm, *Proc. R. Soc. London, Ser. A*, 1968, **305**, 363.
108. D. W. Gaupil and F. C. Goodrich, *J. Colloid Interface Sci.*, 1976, **62**, 142.
109. G. I. Loeb and R. E. Baier, *J. Colloid Interface Sci.*, 1968, **27**, 38.
110. J. P. Green, M. C. Phillips and G. G. Shipley, *Biochem. Biophys. Acta*, 1973, **330**, 243.
111. D. G. Cornell, *J. Colloid Interface Sci.*, 1979, **70**, 167.
112. F. Takeda, M. Matsumoto, T. Takenaka, Y. Fujiyoshi and N. Uyeda, *J. Colloid Interface Sci.*, 1983, **91**, 267.
113. G. Duda, A. J. Shouten, T. Arndt, G. Lieser, G. F. Schmidt, C. Bubeck and G. Wegner, *Thin Solid Films*, 1988, **159**, 221.
114. G. Duda and G. Wegner, *Makromol. Chem. Rapid Commun.*, 1988, **9**, 496.
115. J. R. Katz and P. J. Samwell, *Nature (London)*, 1928, **16**, 592.
116. T. Kawaguchi, H. Nakahara and K. Fukuda, *Thin Solid Films*, 1985, **133**, 29.
117. F. W. Embs, G. Wegner, D. Neher, P. Albouy, R. D. Miller, C. G. Wilson and W. Schrepp, *Macromolecules*, 1991. **24**, 5068.
118. A. A. Kalachev, T. Sauer, V. Vogel, N. A. Plate and G. Wegner, *Thin Solid Films*, 1990, **188**, 341.
119. T. Vahlenkamp, S. Schwiegk, Xu Yuan-Ze and G. Wegner, *Macromolecules*, 1992, **25**, in press.
120. W. Hickel, G. Duda, G. Wegner and W. Knoll, *Makromol. Chem. Rapid Commun.*, 1989, **10**, 353.
121. For an overview see (a) Proc. 2nd Int. Conf. Langmuir–Blodgett Films, *Thin Solid Films*, 1985, **132–134** and (b) Proc. 3nd Int. Conf. Langmuir–Blodgett Films, *Thin Solid Films*, 1988, **159–160**.
122. D. Day and J. B. Lando, *J. Polym. Sci., Polym. Chem. Ed.*, 1978, **16**, 1431.
123. G. Lieser, B. Tieke and G. Weger, *Thin Solid Films*, 1980, **68**, 77.
124. C. Bubeck and D. Holtkamp, *Adv. Mater.*, 1991, **3**, 32.
125. T. Arndt and G. Wegner, in 'Optical Techniques to Characterize Polymer Systems', ed. H. Bässler, Elsevier, Amsterdam, 1989, p. 41.
126. M. Schaub, Diploma Thesis, University of Mainz, 1990.
127. P. Tippmann-Krayer, H. Riegler, M. Paudler, H. Möhwald, H.-U. Siegmund, J. Eickmanns, U. Scheunemann, U. Licht and W. Schrepp, *Adv. Mater.*, 1991, **3**, 46.
128. K. H. Drexhage, in 'Progress in Optics', ed. E. Wolf, North Holland, Amsterdam, 1974, vol. 12, p. 163.
129. V. Vogel and D. Möbius, *Thin Solid Films*, 1988, **159**, 73.
130. C. A. Jones, M. C. Petty and G. G. Roberts, *Thin Solid Films*, 1988, **160**, 117.
131. M. B. Bibble and S. E. Rickert, *Ferroelectrics*, 1987, **76**, 133.
132. G. G. Roberts and B. Holcroft, *Thin Solid Films*, 1989, **180**, 211.
133. H. T. Tien, *Adv. Mater.*, 1990, **2**, 316.
134. D. N. Reinhoudt and E. J. R. Sudhölter, *Adv. Mater.*, 1990, **2**, 23.

135. P. Clechet and N. Jaffrezic-Renault, *Adv. Mater.*, 1990, **2**, 293.
136. O. S. Wolfbeis and M. J. P. Leiner, *Proc. Soc. Photo-Opt. Instrum. Eng.*, 1988, **906**, 42.
137. M. F. Finlan, *Eur. Pat.* 032 629 1A1 (1988).
138. P. A. Cuypers, W. T. Hermens and H. C. Hemker, *Ann. N.-Y. Acad. Sci.*, 1977, **283**, 77.
139. Y. Okahata, X. Ye, A. Shimizu and H. Ebato, *Thin Solid Films*, 1989, **180**, 51.
140. Y. Okahata, T. Tsuruta, K. Ijkiro and K. Ariga, *Thin Solid Films*, 1989, **180**, 65.
141. Y. Okahata, H. Ebato and X. Ye, *J. Chem. Soc., Chem. Commun.*, 1988, 1037.
142. B. Holcroft and G. G. Roberts, *Thin Solid Films*, 1988, **160**, 445.
143. I. R. Peterson, *J. Mol. Electron.*, 1987, **3**, 103.
144. T. Sauer, W. Caseri, G. Wegner, A. Vogel and B. Hoffmann, *J. Phys. D, Appl. Phys.*, 1990, **23**, 79.
145. J. H. Fendler, *J. Membr. Sci.*, 1987, **30**, 323.
146. H. Cackovic, H. P. Schwengers, J. Springer, A. Laschewsky and H. Ringsdorf, *J. Membr. Sci.*, 1986, **26**, 63.
147. F. C. Saunders, J. Staromlynska, G. W. Smith and M. F. Daniel, *Mol. Cryst. Liq Cryst.*, 1985, **122**, 297.
148. H. Ikeno, A. Oh-saki, M. Nitta, N. Ozaki, Y. Yokoyama, K. Nakaya and S. Kobayashi, *Jpn. J. Appl. Phys.*, 1988, **27**, L475.
149. T. Seki, T. Tamaki, Y. Suzuki, Y. Kawanishi and K. Ichimura, *Macromolecules*, 1989, **22**, 3505.
150. V. Novotny, J. D. Swalen and J. P. Rabe, *Langmuir*, 1989, **5**, 485.
151. T. Ginnai, A. Harrington, V. Rodov, A. Matsuno and S. Saito, *Thin Solid Films*, 1989, **180**, 277.
152. E. Ando, J. Goto, K. Morimoto, K. Ariga and Y. Okahata, *Thin Solid Films*, 1989, **180**, 287.
153. T. Arndt, H. Schupp and W. Schrepp, *Thin Solid Films*, 1989, **178**, 319.
154. J. Seto, T. Nagai, C. Ishimoto and H. Watanabe, *Thin Solid Films*, 1985, **134**, 101.
155. A. Barraud, C. Rosilio and A. Ruaudel-Teixier, *Thin Solid Films*, 1980, **68**, 91.
156. A. Barraud, *Thin Solid Films*, 1983, **99**, 317.
157. F. Embs, D. Funhoff, A. Laschewsky, U. Licht, H. Ohst, W. Prass, H. Ringsdorf, G. Wegner and R. Wehrmann, *Adv. Mater.*, 1991, **3**, 25.

18

Science and Technology of Polymer Composites

JOSE M. KENNY and LUIGI NICOLAIS
University of Naples, Italy

18.1	INTRODUCTION	472
18.2	MECHANICAL PROPERTIES OF COMPOSITE MATERIALS	472
	18.2.1 *Mechanical Properties of Polymeric Matrices*	472
	18.2.2 *Properties of Reinforcements*	477
	18.2.3 *Mechanics of Composite Laminates*	480
	18.2.3.1 General aspects	480
	18.2.3.2 Laminate properties: strength and stiffness	481
	18.2.4 *Short Fiber Composites*	485
	18.2.4.1 Short fiber problems	486
	18.2.4.2 Strength of isotropic short fiber composites	488
	18.2.4.3 Dimensional stability	489
	18.2.5 *Particulate Composites*	491
	18.2.6 *Fracture Mechanics in Composite Materials*	495
	18.2.6.1 General aspects	495
	18.2.6.2 Theoretical considerations	495
	18.2.6.3 Linear elastic fracture mechanics approach	496
	18.2.6.4 Strength analysis approach	497
	18.2.6.5 Comparison between theoretical and experimental results	498
18.3	PROCESSING OF HIGH PERFORMANCE COMPOSITE MATERIALS	501
	18.3.1 *Introduction*	501
	18.3.2 *General Aspects of Processing Technologies*	501
	18.3.2.1 Autoclave lamination	501
	18.3.2.2 Pultrusion	502
	18.3.2.3 Resin transfer molding (RTM)	503
	18.3.2.4 Filament winding	504
	18.3.2.5 Physicochemical behavior	504
	18.3.3 *Modeling of Composites Processing*	505
	18.3.4 *Thermokinetic Model*	506
	18.3.4.1 Chemical aspects	506
	18.3.4.2 Reaction kinetics	506
	18.3.5 *Chemorheology*	511
	18.3.5.1 General aspects	511
	18.3.5.2 Determination of the gelation limits	512
	18.3.5.3 Viscosity modeling of thermosetting reactions	512
	18.3.6 *Heat Transfer*	514
	18.3.6.1 General aspects	514
	18.3.6.2 Dimensionless model	515
	18.3.6.3 Application of processing models	517
	18.3.7 *Resin Flow*	520
	18.3.8 *Concluding Remarks*	523
18.4	REFERENCES	523

18.1 INTRODUCTION

Composite materials with polymeric matrices have emerged as strong candidates for load-bearing structural applications in the commercial airplane and automotive industries. Furthermore, in view of the energy shortage, the composite's low weight coupled with energy efficient operations during their processing makes them one of the most desirable materials for use in the future. However, unlike the metals which the composites are replacing in structural applications, their processing characteristics and their final properties can be greatly affected by chemical composition, the load conditions and the various environments under which they perform. Consequently, the problems of understanding and predicting the influence of these effects on the performance of the composite material have attracted attention from both academic and industrial researchers, due to their complexity and relatively prominent role in providing short- and long-term solutions to many needs of the transportation industry.

The state of development of composite materials is quite unique in the scientific world with simultaneous advances being made both in their usage and basic understanding. The complexity and high technology required in manufacturing structural parts with these materials, as well as the need for fundamental description of their processing and property characteristics, necessitates a close collaboration between different scientific areas. The fact that the transportation industry, with its current international character, has a vital interest in composite materials for weight-saving applications has provided a strong incentive for extending the fundamental research in the field of mechanical properties and processing operations.

The polymeric matrix and its interaction with a reinforcing phase, in the form of continuous or discontinuous high strength and stiffness fibers, is one of the major controlling factors in the processing and property characteristics of composites. Although traditionally the matrix has been thought to play a passive role in composite performance and development, with the major emphasis being placed on the fiber properties alone, the recent demanding uses of composites require that the polymeric matrix plays an increasingly important role in composite performance. The wide variety of polymeric materials available, of both a thermoplastic and a thermosetting nature, that can be tailor-made to meet specific performance characteristics, necessitates a clear identification of the property and processing requirements of the material to be used as a matrix in the composite. Accordingly, the contents of this chapter address explicitly the effects of processing and properties of the polymeric matrix on the behavior of high performance composite materials.

The first part (Section 18.2) is dedicated to a description of the mechanical properties of polymer composites, starting with a brief description of the behavior of the components – the matrix and fibers. The properties of particulate, long fiber and laminate composites are then described through the different models generated in the literature.

The second part (Section 18.3) is focused on the processing of polymer composite materials, highlighting the fundamental operations associated with the matrix behavior. This section, which concentrates on high performance thermoset matrix composites, includes aspects related to reaction kinetics, chemorheology, fluid flow and heat transfer, proposing a general approach which is able to describe and predict their processing behavior under different processing conditions.

18.2 MECHANICAL PROPERTIES OF COMPOSITE MATERIALS

18.2.1 Mechanical Properties of Polymeric Matrices

A relevant contribution to the composite properties depends on the matrix characteristics. The major role of the matrix in a fiber-reinforced composite is to transfer stresses between the fibers, to provide a barrier to adverse environments and to protect the surface of the fibers from mechanical and chemical aggression. While the contribution of the fibers mainly affects the tensile properties of the composite, the matrix is responsible for off-axis properties and interlaminar shear as well as in-plane shear properties. Furthermore, the processability and the eventual defects of a composite depend strongly on the physical characteristics of the matrix, such as the viscosity, glass transition temperature and processing conditions.[1-4] Composite matrices traditionally are crosslinked polymers, but recently a growing interest has been devoted towards thermoplastic polymers, taking advantage of their easier processability and higher fracture toughness characteristics.

Polymeric materials are structurally much more complex than ceramics or metals. Depending on their chemical structure, polymeric matrices can be divided into two broad categories: thermoplastics and thermosets. Thermoplastic polymers are constituted of linear molecules and are held together by weak secondary bonds. On heating they can be softened, melted (if crystalline) and reshaped as many times as desired. In a thermoset polymer the molecules are chemically joined together by crosslinking, forming a rigid structure. Once these crosslinks are formed during polymerization the polymer can not be melted or reshaped anymore.

The peculiar molecular structure of polymers (thermosets or thermoplastics) is responsible for the different types of behavior under different level of stresses obtained with constant or variable loads. In particular, their viscoelastic behavior differentiates polymers from other traditional materials.

An elastic solid is defined as a body with a definite shape that can be deformed into a new equilibrium shape reversibly. On the other hand, a viscous liquid has no definite shape and is able to flow irreversibly under the action of external forces. One of the most interesting features of polymers is that they can exhibit an intermediate behavior between the two above mentioned, depending on the experimental time scale and on the test temperature.[5,6]

Newton's law defines the viscous behavior as

$$\sigma_V = \mu \, \delta\gamma/\delta t \tag{1}$$

where $\delta\gamma/\delta t$ is the strain rate, σ_V the applied stress and μ the viscosity.

Hooke's law defines the elastic behavior as

$$\sigma_E = G\gamma \tag{2}$$

where G represents the shear modulus.

One possible formulation of the linear viscoelastic behavior obtained by combining these equations is

$$\sigma = \sigma_V + \sigma_E = \mu \, \delta\gamma/\delta t + G\gamma \tag{3}$$

This equation represents the total stress in a linear viscoelastic element where a viscous and an elastic element are connected in parallel (Voigt model).[5]

This is a very simple model, but is very useful in order to picture the nature of viscoelastic materials, in particular the creep behavior of a polymer, shown in Figure 1. The change in deformation with time under constant stress (a creep experiment) is given by the sum of the immediate elastic deformation γ_1, the delayed elastic deformation γ_2 and the viscous flow γ_3. Therefore the creep modulus $G(t)$ is

$$G(t) = \sigma/\gamma(t) \tag{4}$$

The creep modulus is given by the combination of these three deformations. Usually below the glass transition temperature (T_g) the viscous deformation is negligible. Polymeric matrices are usually characterized by time dependent properties such as the creep modulus.[5]

The effect of the time scale of the experiment is one of the main consequeces of viscoelasticity and must be taken into account when mechanical properties of a viscoelastic material are measured. Figure 2 shows the creep modulus as a function of time, here at very short time the polymer behaves as glassy solid characterized by a time independent modulus. At intermediate times the polymer is in the viscoelastic region and the modulus becomes time dependent. At long times a time

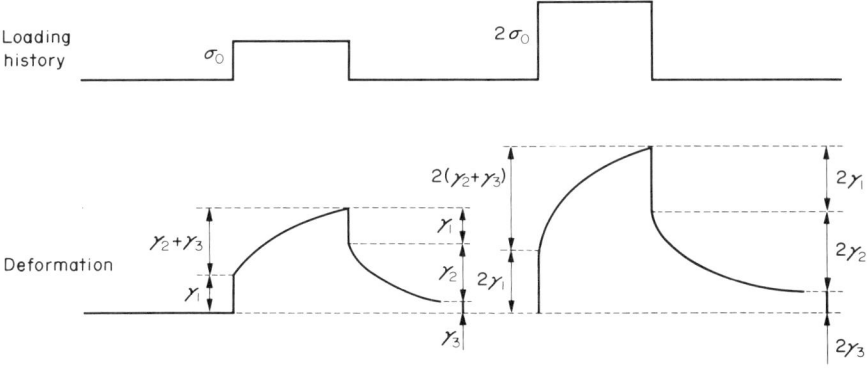

Figure 1 Creep behavior of a polymer

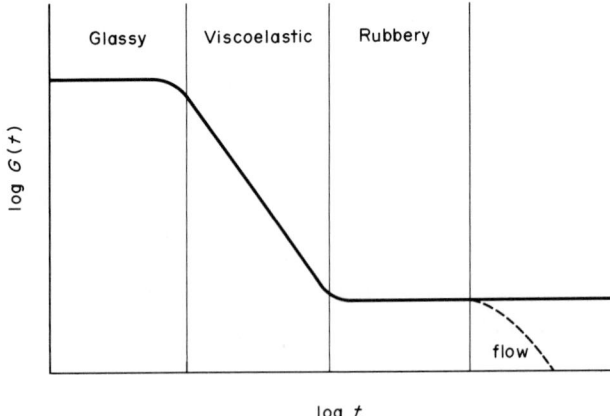

Figure 2 Creep modulus of a polymer as a function of time

independent behavior is again shown in the rubbery region. Finally at very long times thermoplastics can flow, while thermosets show a rubbery behavior.

The time dependent characteristics of viscoelastic materials are typically studied using dynamic mechanical experiments in which a sinusoidally varying strain is applied to the specimen.[7] The response of the system is a stress, also varying sinusoidally but out of phase with the strain

$$\gamma = \gamma_0 \sin \Omega t \tag{5}$$

$$\sigma = \sigma_0 \sin(\Omega t + \delta) \tag{6}$$

where δ is the phase lag and Ω is the angular frequency.

The stress–strain relation can be defined using a quantity G_1, named storage modulus, in phase with the strain input, and a quantity G_2, named loss modulus, 90° out of phase

$$G_1 = \sigma_0/\gamma_0 \cos \delta \quad G_2 = \sigma_0/\gamma_0 \sin \delta \tag{7}$$

The storage modulus is proportional to the elastic energy stored in the specimen, while the loss modulus is proportional to the viscous dissipation of energy.

For dynamic mechanical experiments the frequency acts in the same manner as the time for the creep experiments above discussed. G_1 and G_2 are reported as a function of the frequency in Figure 3. G_1 follows a pattern similar to that shown in Figure 2 (obviously frequency represents a time^{-1}), showing the typical behavior of the rubbery, viscoelastic and glassy regions.

At low and high frequencies G_2 is zero, the stress and strain being in phase for the rubbery and glassy state. The viscoelastic region is characterized by a maximum of G_2 corresponding to a maximum of viscous dissipation of energy. The dissipation occurs when the input frequency and

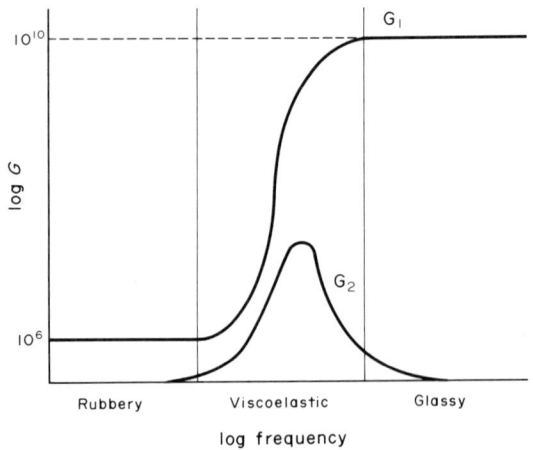

Figure 3 Loss and storage modulus as function of frequency for a polymer in a dynamic mechanical test

the relaxation times of the specimen are of the same order of magnitude. Correspondingly, the storage modulus also rapidly rises up to the values of a glassy material. An analogous behavior is shown by working at a fixed frequency and varying the relaxation times of the material by increasing the temperature. As shown in Figure 4 for thermoplastic polymers, the storage modulus changes with temperature following a pattern similar to that reported in Figures 2 and 3. At a temperature corresponding to T_g the structure changes from glass to rubber and the modulus dramatically decreases by 3 to 5 orders of magnitude for thermoplastics. At this point a further increase in temperature causes viscous flow.

The effect of crystallinity and crosslinking is also shown in Figure 4. The modulus decay at the glass transition temperature is function of the amount of crystallinity; the bigger the amount of crystallinity, the smaller is the response to the T_g.[8-10] Further increases of temperature do not cause a steep drop in the value of the modulus until the melting temperature is reached, here the modulus drops down sharply.

The modulus measured in a dynamic mechanical experiment as a function of temperature for an epoxy resin is reported in Figure 5.[5] These curves were obtained on systems cured with different amounts of amine, causing different crosslinking densities. An increase of the stoichiometric excess of amine corresponds to a shift in the glass transition temperature. Moreover, high crosslinking produces a higher value of the storage modulus above T_g than low crosslinking, while there is little effect at room temperature. Crosslinked systems show a smaller decrease of the modulus at T_g than amorphous polymers, but a larger decrease than semicrystalline ones. A further increase of temperature leads to degradation.

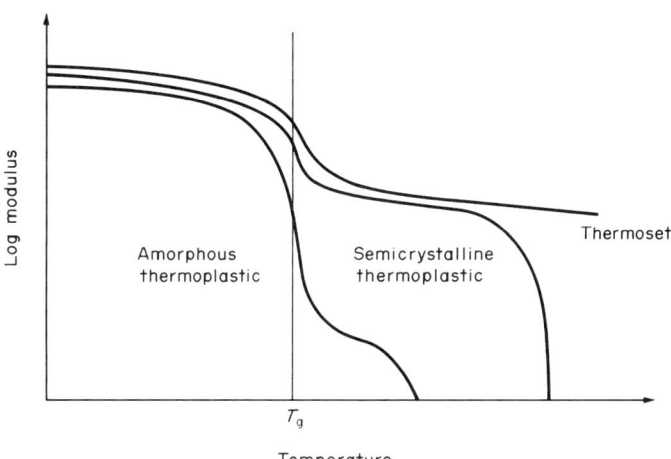

Figure 4 Modulus as a function of temperature for an amorphous and a semicrystalline thermoplastic polymer and a thermoset polymer

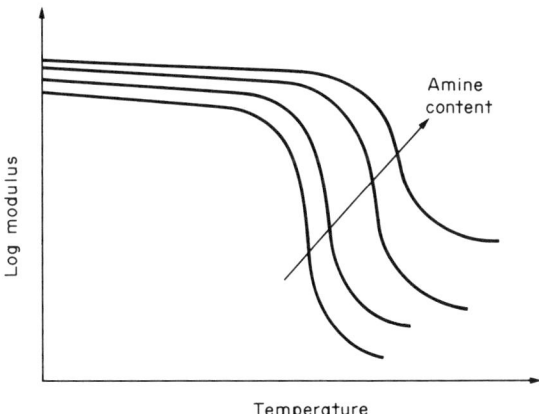

Figure 5 Modulus of a thermoset polymer as a function of temperature at different degrees of crosslinking

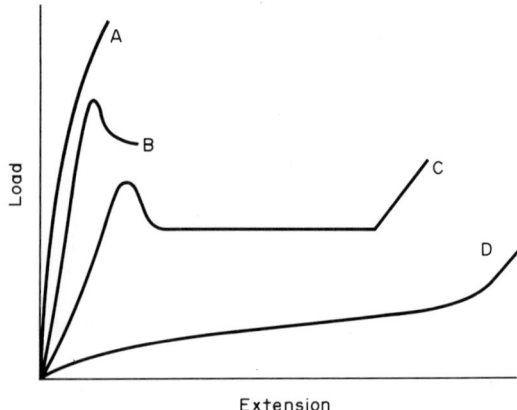

Figure 6 Different stress/strain behaviors of polymers

As shown above, the mechanical properties of polymers are greatly affected by the time scale of the experiment and by the temperature. The large variation in the stress–strain behavior of polymers as measured at a constant rate of strain is shown in Figure 6. Curve A is for a hard brittle material, curve B is typical of a hard ductile polymer showing uniform extension, curve C is also for a hard ductile polymer, but is typical of a material which cold-draws with necking and curve D is typical of elastomeric materials. All these types of curves could be obtained from a single polymer by just changing the test temperature over a sufficiently wide range. At low temperature the extension prior to breaking is low and there is no yield point (curve A). At higher temperatures the yield point appears (curve B), then necking and cold drawing are possible (curve C), and finally, above T_g, the typical rubber behavior is shown (curve D). Therefore, with increasing temperature a polymer shows a transition from brittle to ductile behavior.

The above-mentioned cold-drawing (Figure 7) manifests itself as a necking of the polymer under stretching.[6] Necking starts at a localized point in the specimen, the cross section becomes much less than that of the remaining portion of the specimen and the stress remains constant during stretching. Cold-drawing after the yield point means that there must be a strain-hardening process, otherwise the material would break without drawing at the reduced cross section where the necking started. The strain hardening is generally caused by molecular orientation or strain-induced crystallization. The necked section increases in length as stretching continues, until all the specimen becomes cold-drawn. On further stretching the stress rapidly increases and failure occurs. Cold-drawing of polymers is widely used for the fabrication of fibers which are also used as precursor fibers for high modulus reinforcement of composites.

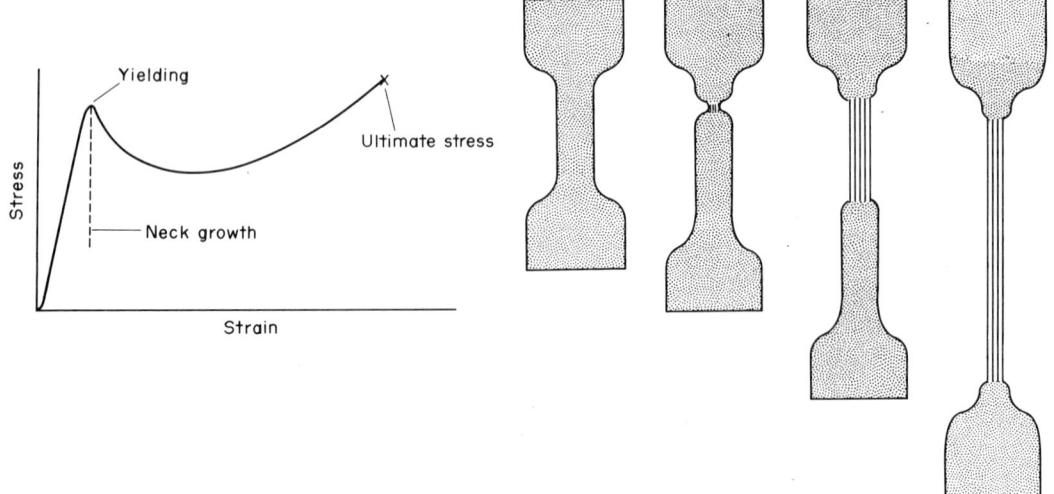

Figure 7 Cold drawing of a thermoplastic polymer

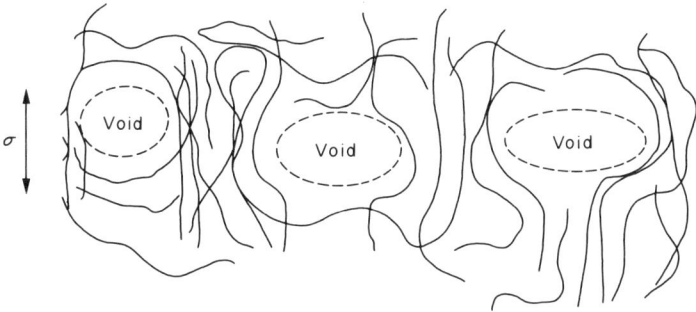

Figure 8 Craze formation in a stretched polymer

Table 1 Properties of the Most Common Resins for High Performance Composites

	Tensile strength (MPa)	Flexural modulus (MPa)	Density (g cm^{-3})	Max. service temperature (°C)	Coefficient of thermal expansion (10^{-5} °C^{-1})	Water absorption (24 h %)
Epoxy	35–85	15–35	1.38	25–85	8–12	0.1
Polyimide	120	35	1.46	380	9	0.3
PEEK	92	40	1.30	140	6–9	0.1
Polyamide/imide	95	50	1.38	200	6.3	0.3
Polyether/imide	105	35	1.27	200	5.6	0.25
Polyphenylene/sulfide	70	40	1.32	75	9.9	0.2
Phenolics	50–55	10–24	1.30	50–175	4.5–11	0.1–0.2

When a brittle polymer is stretched over the elastic limits, crazing formation occurs. A craze crack consists of two regions (Figure 8): an oriented and molecularly cold-drawn material which is not easily fractured because it contains strong bonds between polymer molecules (such as physical entanglements and oriented chain segments), and voids originated in weak regions under the stretching action. The oriented regions between voids tend to prevent coalescence of the voids which would result in a true crack and catastrophic failure.[6]

Somewhat similar phenomena are involved in the toughened epoxy resins that, due to their better impact resistance and ductile behavior, are replacing the classical epoxies as matrices. Rubber particles may act as stress concentrators, so that a large number of craze/cracks first start on the boundary of these particles. The craze/cracks grow until they run into another rubber particle. At this point the crack stops since the radius of the particle is greater than the radius of curvature of the crack tip, so the intensity of the stress concentration is reduced. Because of the large number of rubber particles, an enormous number of craze/cracks are initiated, leading to the dissipation of a large quantity of energy. The quick growth of only a few cracks could cause catastrophic failure, if not hindered by the stress field of other cracks.

Table 1 summarizes some properties of the most widely used matrices for composites, including thermosets (epoxy and phenolics), amorphous thermoplastics (polyether/imide and polyamide/imide) and semicrystalline thermoplastics (polyether/ether/ketone, polyphenylene/sulfide and polyimide).

18.2.2 Properties of Reinforcements

Reinforcements are usually in the form of long fibers, flakes, whiskers, particles, discontinuous fibers or fabric. However, in high performance composites the choice is oriented toward long fibers and fabrics.

The use of fibers for high-performance engineering materials is due to their unique characteristics.[11,12] Firstly, a diameter that is small with respect to the grain size allows a higher fraction of the strength to be attained than is possible in a bulk form. This is the so-called 'size effect'; the smaller the size, the lower the probability that there is an imperfection in the material. Figure 9 shows the decrease in strength when the diameter of the fiber increases. Therefore, the exceptional mechanical properties of fibers are a consequence of this size effect.

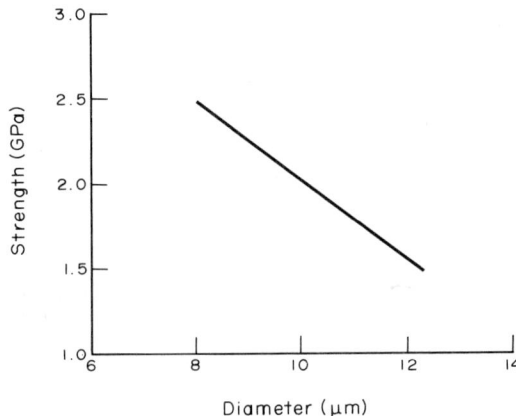

Figure 9 Strength of a fiber as a function of the diameter

Secondly, the high aspect ratio (the ratio l/d, length/diameter), enhances the capability to transfer the load to the other fibers through the matrix. The interfacial bond strength is one of the main characteristics that has to be taken into account when designing a composite with the desired mechanical properties. In fact, noncatastrophic failure requires a balance between the values of the tensile strength of the fibers and the interfacial bond strength. Too high an interfacial bond strength, for a given fiber tensile strength, will produce a more brittle composite which, of course, will be flaw sensitive and will have a lower tensile strength. Too low an interfacial bond strength will produce lower shear and tensile strength in the composite, but lower flaw sensitivity.

In order to improve the adhesion between fiber and matrix, the surface of the fibers is treated by different processes, among these the most important for carbon fibers is the oxidative process. Furthermore, fibers for high performance applications require a low density and a very high degree of flexibility, that allows the composite to be processed with the widest number of techniques so that many different shapes can be obtained.

The thermal expansion coefficient is also an important property for reinforcements. The thermal expansion coefficient of the fibers is usually lower than that of the matrix, causing residual stresses after processing. However, this difference results in lower thermal expansion of the composites and consequently a better dimensional stability after processing operations. Finally the wide range of fibers available gives the possibility of designing composites with different electrical conductivities.

Reinforcing fibers can be divided into four main groups.

Graphite fibers. Carbon fibers owe their success in high performance composites to their extremely high tensile modulus/weight and tensile strength/weight ratios, high fatigue strength and low coefficient of thermal expansion, coupled with a low ratio of cost to performance. Carbon fibers are commercially available with a variety of moduli ranging from 270 GPa to 600 GPa. They are produced following two different processes, depending on the type of precursors; namely textile precursor and pitch precursor. A flow diagram of carbon fiber processing is given in Figure 10.

Glass fibers. Glass fibers are the most common of all reinforcing fibers for polymeric matrix composites. Their main advantages are low cost, high tensile strength, high chemical resistance and good insulating properties. On the other hand they display a low tensile modulus, a relatively high density compared to the other fibers, a high sensitivity to wearing and a low fatigue resistance. Depending on the chemical composition of the glass they are commercially available in three different grades: E, S and C. Among these the E fibers are cheapest and present lower mechanical properties. The C fibers are mainly used when corrosion is the main concern.

Boron fibers. These fibers are characterized by an extremely high tensile modulus coupled with a large diameter, offering an excellent resistance to buckling that contributes to a high compressive strength of the composites. Their high cost is due to the processing operations. For this reason boron fibers find application only in aerospace and military contexts.

Aramid fibers. The most common aramid fiber available is the Kevlar 49. These fibers are composed of a highly oriented crystalline polymer and present the highest tensile strength/weight ratio. On the other hand the disadvantages they present are the low compressive strength, difficulty of manufacturing and a sensitivity to ultraviolet light and to water. However Kevlar fibers find applications in sporting goods and the aircraft industry.

In Table 2 the main properties of the commercially available reinforcing fibers are listed. The stress/strain behavior of different fibers, shown in Figure 11, is that of an elastic brittle material.[13]

Science and Technology of Polymer Composites

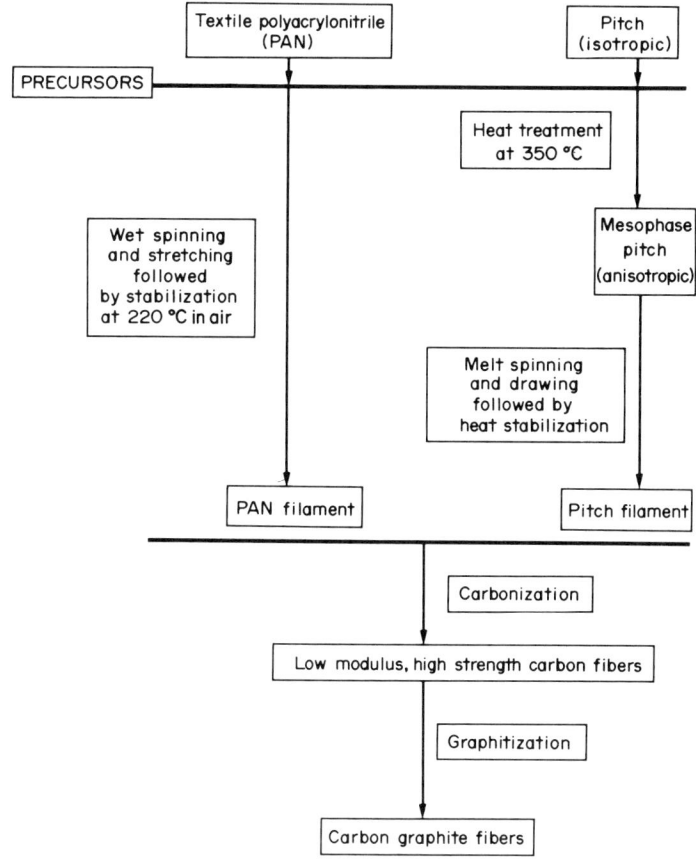

Figure 10 Flow diagram of the processing of carbon fibers

Table 2 Main Properties of Some of the Commercially Available Fibers

Fiber	Tensile modulus (GPa)	Tensile strength (GPa)	Strain to failure (%)	Density	Coefficient of thermal expansion (10^{-6} mm $°C^{-1}$)
Glass					
E glass	73	3.5	3.5	2.54	5
S glass	87	4.3	5.0	2.49	2.9
PAN-carbon					
T 300	228	3.2	1.4	1.76	−0.1 to −0.5
AS	220	3.1	1.2	1.77	0.5 to −1.2
HMS	345	2.34	0.58	1.85	−0.1 to −0.5
Pitch-carbon					
P55	380	1.9	0.5	2.0	−0.9
P100	690	2.2	0.31	2.15	−1.6
Kevlar 49	131	3.62	2.8	1.45	−2
Boron	393	3.1	0.79	2.7	5
SiC	400	3.44	0.84	3.08	1.5

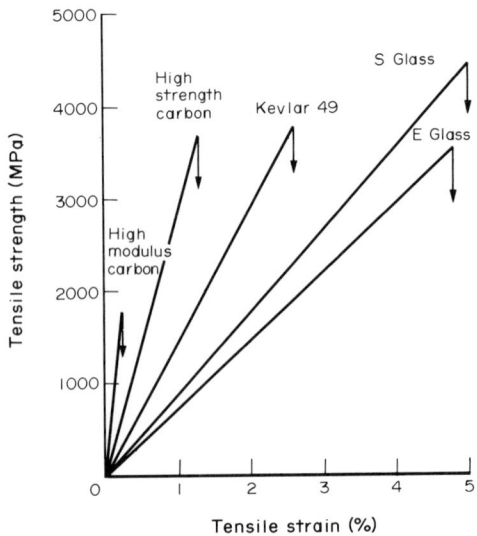

Figure 11 Stress/strain behavior of the most common fibers

18.2.3 Mechanics of Composite Laminates

18.2.3.1 General aspects

The 1960s opened with an uninhibited enthusiasm for the extraordinarily high strength and stiffness properties available primarily in single-crystal whiskers, but also in continuous filament forms. While the realization of the potential of defect-free whiskers in engineering composites has seeped away in the murk of technological reality, striking progress has been achieved in advanced filament reinforcement. The 1960s witnessed the development of boron and graphitized polymeric fibers, as well as several other potential reinforcements. In fact, the technology has developed to the point that 'proof-of-concept' is now being demonstrated in operational engineering structural items.[14] At this point, it is interesting to notice that our understanding of composite solids has altered in a subtle, but nonetheless radical manner. The initial emphasis upon extravagantly large fiber moduli and strength has been tempered by the realization that the engineer cannot, or will not, execute design based solely upon longitudinal filament properties. Technical composites are multiphased, anisotropic bodies. Efficient and reliable use of such materials necessitates optimization of a laminated composite with respect to all of its directional properties (particularly transverse tensile, shear and longitudinal compression, as well as the earlier emphasis upon longitudinal properties).

By employing current design engineering characterization and design procedures, we are now in a position to evaluate the performance capability of composites for a variety of filaments, ranging from E glass through to the very high modulus, high strength graphite systems. These observations may be summarized as follows.[15] The longitudinal strength of unidirectional composites are relatively independent of the specific reinforcement. Engineering design levels are the product of laminate stiffness and the smallest ply strain (usually the strain transverse to the fiber direction) required to damage the composite. Matrix properties and filament anisotropy exert a strong influence upon design capability. Filaments which do not exhibit a longitudinal composite strain capability at failure of at least +0.005–0.007 or more cannot be used as a general engineering material. Anisotropic graphite and organic polymeric filaments possess small but negative expansion coefficients transverse to the fibers. As a consequence, special attention must be paid to residual thermal stress in fabrication, but this penalty is offset by the potential of developing absolute dimensional stability in the plane of a composite plate. All filaments except glass are fatigue insensitive. High modulus matrix systems must produce a transverse ply strain capability of +0.003 to be of serious engineering use.

One of the more interesting conclusions of this evaluation is the suggestion that for general structural applications, economic and technical considerations will dictate only one continuous filament prepreg system, other than glass, with fibers possessing properties of roughly 19–26 GPa modulus and 2–2.5 GPa in strength. Such a system will be close to the properties exhibited by type II or HTS graphite fiber materials. Of course, other fiber systems will be available, but predominantly for special applications and at a relatively high price, reflecting low volume demand.

18.2.3.2 Laminate properties: strength and stiffness

Current attitudes regarding composite materials[16-18] emphasize the relationship of structural performance to the properties of a ply. A 'ply' is a thin sheet of material consisting of an oriented array of fibers embedded in a continuous matrix material (Figure 12). These plies are stacked one upon other, in a defined sequence and orientation, and bonded together yielding a laminate with tailored properties. The properties of the laminate are related to the properties of the ply by the specification of the ply thickness, stacking sequence, and the orientation of each ply. The properties of the ply are, in turn, specified by the properties of the fibers and the matrix, their volumetric concentration, and geometric packing in the ply. Generally, the ply material is preformed and can be purchased in a continuous compliant tape or sheet form which is in a chemically semicured condition. Fabrication of structural items involves using this 'prepreg' material, either winding it on to a mandrel or cutting and stacking it on to a mold, after which heat and pressure or tension is applied to complete the chemical hardening process.

The basis for engineering design of such a material is then the properties of a cured ply or lamina as it exists in a laminate. This ply is treated as a thin two-dimensional item and is mechanically characterized by its stress–strain response to: (i) loading in the direction of the filaments, which exhibits a nearly linear response up to a large fracture stress; (ii) loading in the direction transverse to the filament orientation, which exhibits a significantly decreased moduli and strength; and (iii) the response of the material to an in-plane shear load.

By the contrast with isotropic metallic materials, an oriented ply, in the form of a thin sheet, is anisotropic and requires four elastic (plane stress) constants[16-18] to specify its stiffness properties in its natural orientation

$$\sigma_1 = Q_{11}\epsilon_1 + Q_{12}\epsilon_2$$
$$\sigma_2 = Q_{12}\epsilon_1 + Q_{22}\epsilon_2 \qquad (8)$$
$$\sigma_6 = Q_{66}\epsilon_6$$

which $\sigma_6 = \tau_{12}$ and $\epsilon_6 = \tau_{12}$, or in matrix form

$$\begin{vmatrix} \sigma_1 \\ \sigma_2 \\ \sigma_6 \end{vmatrix} = \begin{vmatrix} Q_{11} & Q_{12} & 0 \\ Q_{12} & Q_{22} & 0 \\ 0 & 0 & Q_{66} \end{vmatrix} \begin{vmatrix} \epsilon_1 \\ \epsilon_2 \\ \epsilon_6 \end{vmatrix} \qquad (9)$$

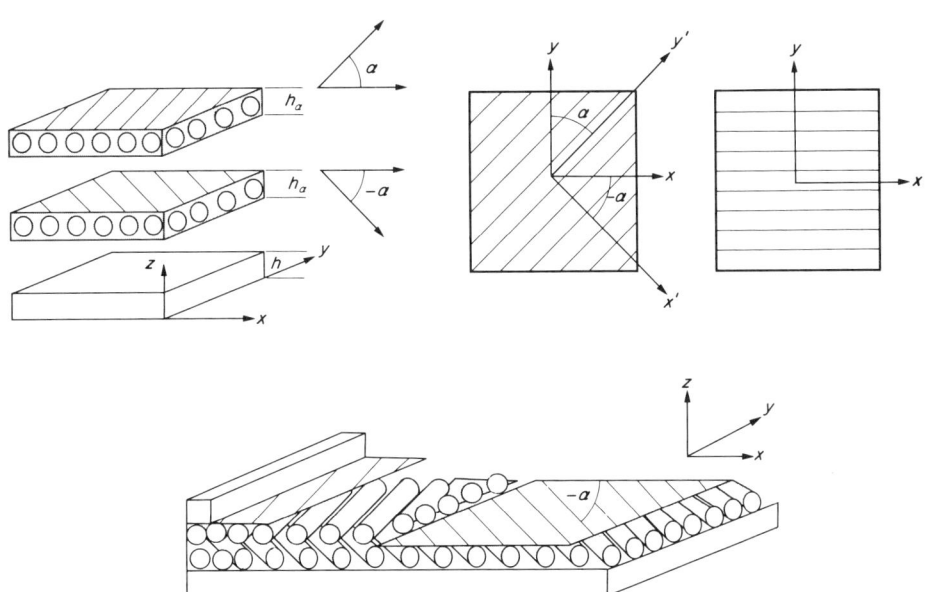

Figure 12 Illustration of a continuous fiber composite laminate

where the plane stress stiffness moduli are

$$Q_{11} = E_{11}/(1 - v_{12}v_{21})$$
$$Q_{22} = E_{22}/(1 - v_{12}v_{21})$$
$$Q_{12} = v_{21}E_{11}/(1 - v_{12}v_{21})$$
$$Q_6 = G_{12}$$
(10)

where v_{ij} is the Poisson ratio, defined as $-\epsilon_i/\epsilon_j$.

If, however, the ply is rotated with respect to the applied stress or strain direction, two additional moduli appear, which results in the direction-indicated shear coupling rotation in simple extension

$$\begin{vmatrix} \sigma_1 \\ \sigma_2 \\ \sigma_6 \end{vmatrix} = \begin{vmatrix} Q_{11}^* & Q_{12}^* & Q_{16}^* \\ Q_{12}^* & Q_{22}^* & Q_{26}^* \\ Q_{16}^* & Q_{26}^* & Q_{66}^* \end{vmatrix} \begin{vmatrix} \epsilon_1 \\ \epsilon_2 \\ \epsilon_6 \end{vmatrix}$$
(11)

where

$$Q_{11}^* = U_1 + U_2 \cos(2\theta) + U_3 \cos(4\theta)$$
$$Q_{22}^* = U_1 - U_2 \cos(2\theta) + U_3 \cos(4\theta)$$
$$Q_{12}^* = U_4 - U_3 \cos(4\theta)$$
$$Q_{66}^* = U_5 - U_3 \cos(4\theta)$$
$$Q_{16}^* = -\tfrac{1}{2}U_2 \sin(2\theta) - U_3 \sin(4\theta)$$
$$Q_{26}^* = -\tfrac{1}{2}U_2 \sin(2\theta) + U_3 \sin(4\theta)$$
(12)

the invariants U_i to the rotation are

$$U_1 = \tfrac{1}{8}(3Q_{11} + 3Q_{22} + 2Q_{12} + 4Q_{66})$$
$$U_2 = \tfrac{1}{2}(Q_{11} - Q_{22})$$
$$U_3 = \tfrac{1}{8}(Q_{11} + Q_{22} - 2Q_{12} - 4Q_{66})$$
$$U_4 = \tfrac{1}{8}(Q_{11} + Q_{22} + 6Q_{12} - 4Q_{66})$$
$$U_5 = \tfrac{1}{8}(Q_{11} + Q_{22} - 2Q_{12} + 4Q_{66})$$
(13)

In addition, lamination can result in up to 18 elastic coefficients and increased deformational complexities, but the additional coefficients can all be derived from the four primary coefficients using the concept of rotation and ply-stacking sequence.[16,17] These complications are the result of geometric variables. If the laminate is properly constructed, the in-plane stretching or stiffness properties can still be specified by four elastic coefficients. We shall consider laminates of this nature.

The flow diagram of a typical calculation is shown in Figure 13. Note that both short and continuous fibers are handled in the same manner. These calculations, while tedious, are analytically simple. The 'plane stress', the Q_{ij} terms, are employed because lamination neglects the mechanical properties through the ply thickness. These stiffnesses are sometimes regrouped into new constants called 'invariants', the U_i terms, for analytical simplicity. To compute the properties of the laminate one then sums the ply properties through the thickness of the laminate, weighted by the thickness (h_k) of each oriented ply

$$A_{ij} = \sum_{k=1}^{N} (Q_{ij})_k h_k$$

For a balanced (same number of $\pm\theta$) and symmetrical system ($+\theta$ or $-\theta$ at same distance above and below the midplane) the laminate solution is

$$A_{11} = U_1 + U_2 \cos(2\theta) + U_3 \cos(4\theta)$$
$$A_{22} = U_1 - U_2 \cos(2\theta) + U_3 \cos(4\theta)$$
$$A_{12} = U_4 - U_3 \cos(4\theta)$$
$$A_{66} = U_5 - U_3 \cos(4\theta)$$
(14)

Note the inverted terms A_{ij} yield the required elastic properties of the laminate in terms of the individual ply properties E_{11}, E_{12} and G_{12}.

$$E_{11} = (A_{11}A_{22} - A_{12}^2)/A_{22}$$
$$E_{22} = (A_{11}A_{22} - A_{12}^2)/A_{11}$$
$$v_{12}/E_{11} = A_{12}/(A_{11}A_{22} - A_{12}^2) \qquad G_{12} = A_{66}$$
(15)

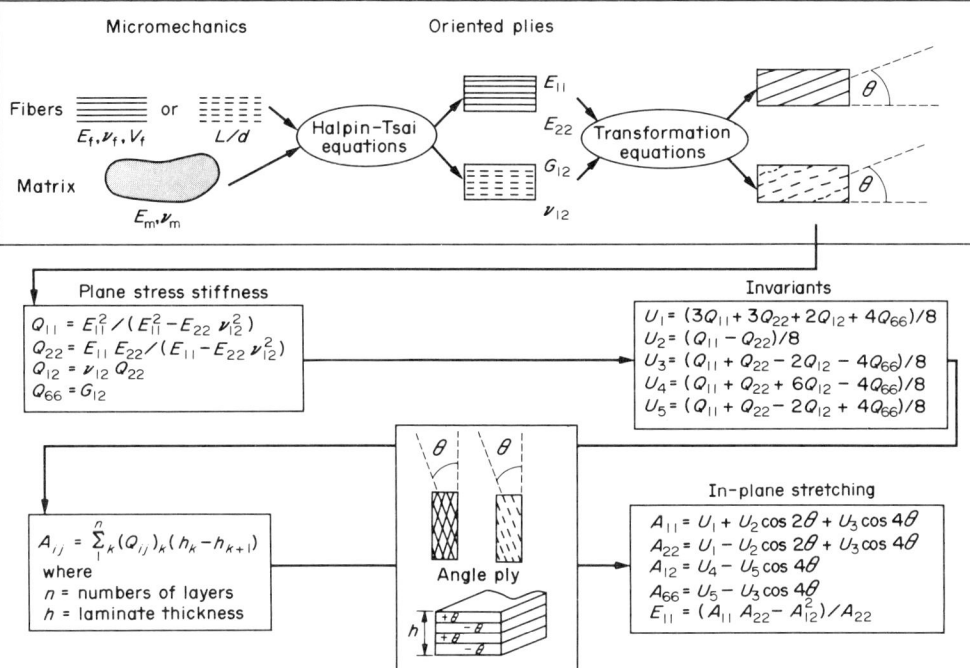

Figure 13 Laminate calculations

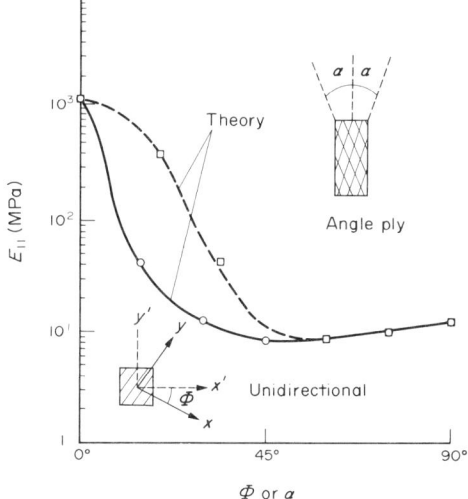

Figure 14 Young's modulus *versus* fiber orientation for nylon fiber reinforced rubber. $E_f = 2.0$ GPa, $E_m = 2.1$ MPa, $v_f = 0.2$, $v_m = 0.4999$

These calculations have been thoroughly tested and agree closely with experiment (Figure 14). In this figure the circles and squares are the experimental points and the lines are the theoretical predictions for a nylon fiber reinforced rubber. The angle ply laminate is predicted from the ply properties. The ply properties are in turn correlated with the transformation equations and the micromechanics. The micromechanics employed in this demonstration are based upon the 'self-consistent method' developed by Hill.[19] Hill rigorously modeled the composite as a single fiber, encased in a cylinder of matrix, with both embedded in an unbounded homogeneous medium which is macroscopically indistinguishable from the composite. Hermann[20] employed this model to obtain a solution in terms of Hill's 'reduced moduli'. Halpin and Tsai[21] reduced Hermann's solution to a simpler analytical form and extended its use for a variety of filament geometries

$$E_{11} = E_f V_f + E_m V_m$$
$$v_{12} = v_f V_f + v_m V_m \quad (16)$$
$$p/p_m = (1 + \eta \xi V_f)/(1 - \eta V_f)$$

where

$$\eta = (t_f/t_m - 1)/(p_f/p_m + \zeta)$$
$$\zeta_E = 2(e/d); \quad \zeta_{G12} = 1; \quad \zeta_{G23} = 1/(3 - 4v_m)$$
$$p = E_{22}, G_{12}, G_{23}; \quad p_f = E_f, G_f; \quad p_m = E_m, G_m$$

These equations are suitable for single calculation and were employed previously for the single ply and angle ply properties. The short fiber composite properties are also given by the Halpin–Tsai equations, where the moduli in the fiber orientation direction is a sensitive function of aspect ratio (l/d) at small aspect ratios, and has the same properties of a continuous fiber composite at large but finite aspect ratios.

If the ply illustrated in Figure 15 is used in the construction of a balanced and symmetrical 0/90 laminate and is mechanically tested, a bilinear stress/strain curve is obtained, and the stiffness is the sum, through the thickness, of the plane stress stiffness of each layer. As the laminate is deformed, each ply possesses the same in-plane strain, ϵ, and when the strain on the 90° layers in the laminate prevents the 90° layer from carrying its share of the load, $Q_{ij}(90°) = 0$. This load is transferred to the unbroken layers, the 0° layers for our illustration, and results in a loss of laminate stiffness or modulus. Continual loading will ultimately produce a catastrophic failure of the laminate when the strain capability of the unbroken, 0°, layers is exceeded. For a 0/90 construction, employing a glass/epoxy material, the ratio of the ultimate failure stress to the crazing stress (the knee in Figure 15) is 6.1. Experimental data and a theoretical stress/strain curve are shown in Figure 16 for a $\pi/4$ (0°/±45°/90°) glass/epoxy laminate. Note a change in stiffness as the 90° and then the 45° layers fail, and the correspondence of the theoretical ultimate strength of 356 MPa with the experimental results of 346 MPa. While the strain for transverse ply failure is constant from laminate to laminate, the stress required to craze the system as well as cause final failure is a function of laminate geometry (Figure 17), because the construction of the laminate specifies the stiffness properties (crazing stress = stiffness × allowable transverse ply strain). It must be noticed that the area under the stress/strain curve is proportional to the impact energy. Therefore, lamination permits the engineer to tailor a fixed prepreg system to meet the conflicting stress/strain demands at different points in a structure. A further point, the crazing stress threshold is generally at or below the creep fracture or fatigue limit for all classes of composites (for glass/epoxy the fatigue limit lies between 0.25 and 0.30 of static ultimate strength). Boron and graphite are fatigue insensitive filaments, thus no fatigue damage is realized below first ply failure.

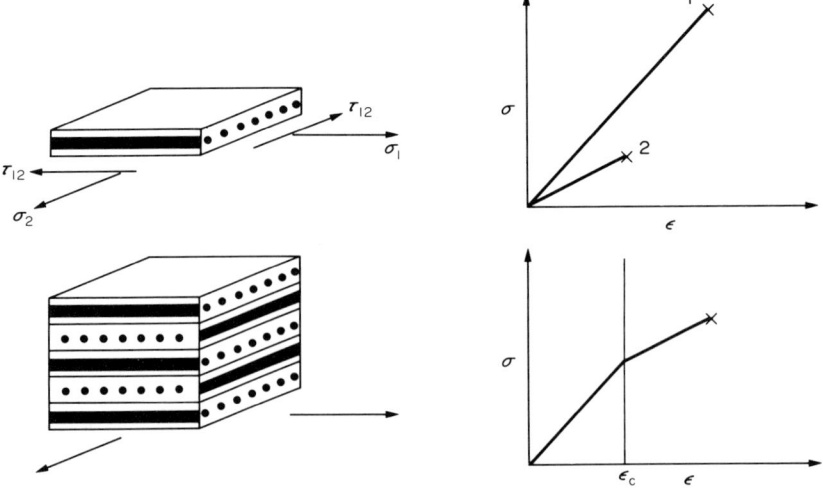

Figure 15 Determination of laminate properties

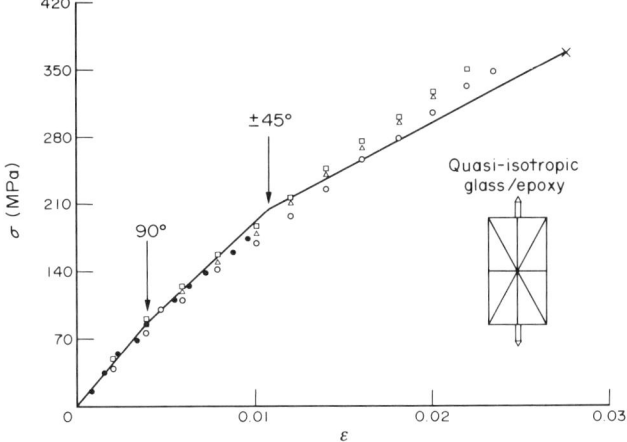

Figure 16 Stress/strain curve for a $\pi/40°$ ($0°/\pm45°/90°$) glass/epoxy laminate. The solid line is a theoretical prediction

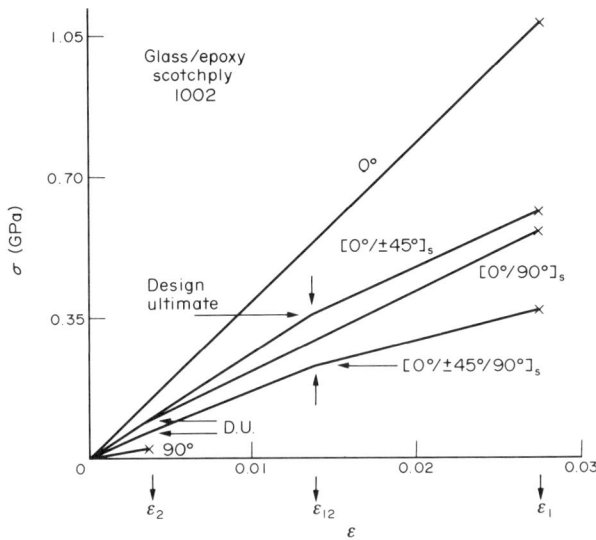

Figure 17 Effect of laminate geometry on the stress/strain curve of glass/epoxy composites

Thus, the material properties of a laminate are specified in terms of the ply engineering moduli, E_{11}, E_{22}, v_{12} and G_{12}; the engineering strains to failure, ϵ_1, ϵ_2 and ϵ_6; and the thermal expansion coefficients; e_1 and e_2.

18.2.4 Short Fiber Composites

We shall now consider the problem of short fiber composites. Random or nearly random distributions of fibers, finite in length and arranged in matrix, constitute many naturally occurring and synthetic materials. In the majority of cases, the spatial orientation of the discontinuous fibers is intermediate between a truly random array in three dimensions and a two-dimensional random array of fibers. Generally these materials possess internal orientations which are independent of the thickness. In some systems of technological interest, the distribution of fibers may vary through the thickness. Halpin and Pagano[16,21] proposed that such materials can be modeled mathematically as laminated systems. The laminate model consists of layers of unidirectional composites, in our case short fiber composites, with the fiber volume fractions in a layer oriented at an angle being governed by the percentage of fibers at angle θ in the actual material. If the material to be modeled exists in sheet form where the planar thickness is considerably less than the average lengths of

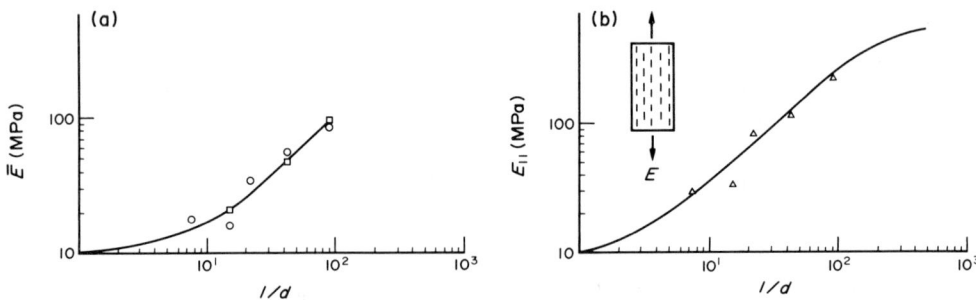

Figure 18 Dependence of longitudinal modulus on aspect ratio for random (○), quasi isotropic (□), and oriented (△) short nylon fiber/rubber composites. $E_f/E_m = 973$, $V_f = 0.35$. The lines are theoretical predictions

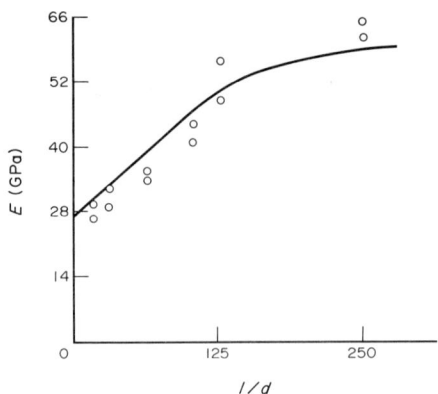

Figure 19 Stiffness of random short fiber/epoxy composites. $V_f = 0.40$

fibers, then the reinforcement may be considered as a two-dimensional random array of fibers. This situation is described in laminated plate theory as 'quasi-isotropic'.

The success of the lamination approximation is strongly dependent upon the assumption of physical volume averaging in real material systems, combined with an ability to estimate the stiffness for an oriented short fiber sheet (Figure 18b). Employing the Halpin–Tsai equation (equation 16), we have been able to predict the strong dependence of E_{11} on the aspect ratio.[22] Note that as the aspect ratio becomes large, uniaxial stiffness E_{11} becomes identical with oriented continuous filaments. The other moduli, E_{22}, G_{12} and possibly v_{12}, are not dependent on the aspect ratio and may be approximated by the continuous fiber result. This approximation is necessary as no three-dimensional elastic micromechanic solution exists for this problem. The strength properties do not approach the continuous filament composite properties, as was shown by Riley.[23]

The properties of random short fiber composites are shown in Figure 18a as circles. To test the laminate model, 'quasi-isotropic' laminates of two different aspect ratios were made and tested. The stiffness of these two laminates were comparable to the random composites. Confirmation of the model was also achieved by noting the agreement between the data and the theoretical calculation.

In Figure 19, data and predictions for random short fibers of boron in an epoxy matrix are shown. Note that a plateau is reached as the aspect ratio increases. The critical value of l/d at which the plateau occurs is strongly dependent upon the ratio E_f/E_m.

18.2.4.1 Short fiber problems

In the previous analysis it was assumed that no bias in fiber orientation was present and that all fibers have the same aspect ratio. In the fabrication of short fiber composites it is often desired to optimize properties by attempting to orient the short fibers. Usually only partial orientation is achieved. In addition, most fabrication procedures result in considerable fiber breakage,[24] producing a distribution of fiber lengths. The corresponding theory employing the lamination analogy requires that the fiber orientation be handled as a complex laminate of weighted groups of angle plies $\pm \theta$,

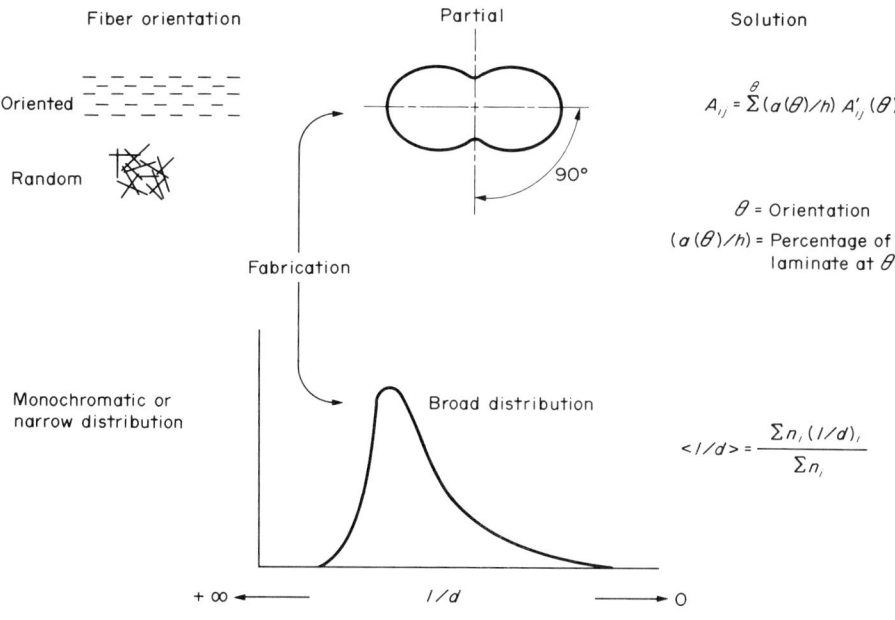

Figure 20 Short fiber fabrication

assuming a symmetrical angular distribution $f(\theta)$ to match the required fiber distribution (Figure 20). Employing the laminate analogy, it is assumed that the molded material behaves mathematically as a biased laminate fabricated from layers of oriented short fiber composite material of volume fraction, V_f.

The percentage, $a(\theta_k)/h$ of the material oriented at the angle $\pm \theta$ is obtained from the experimental angular distribution function

$$\int_0^\pi f(\theta)\,d\theta = 1.0$$

In this manner different orientations contribute to the overall response in proportion to their fractional thickness in the laminate or

$$A_{ij} = \sum_{k=1}^{N} \frac{a(\theta_k)}{h} A_{ij'}(\theta_k) \qquad (17)$$

where A_{ij} is the laminate stiffness moduli and $A_{ij'}(\theta_k)$ is the stiffness of the laminate plies oriented at the angles θ_k, and N is the number of θ orientations. The stiffness moduli are specified in terms of the moduli, E_{11}, E_{22}, v_{12} and G_{12}, for a ply, and the angle of orientation θ. In turn, the ply properties are related to the fiber and matrix properties. E_f, V_f, E_m, V_m, v_f, v_m and the aspect ratio l/d are as specified in equation (16). The 'engineering stiffness properties' of the laminate are simply obtained by the inversion of the A_{ij} matrix (equation 15).

The results of a calculation of this type, employing the data of Tock and McMackins,[25] are shown in Figure 21 for an isotropic and a nonrandom distribution of fibers in a polycarbonate polymeric matrix. In the nonrandom distribution case, the major portion of the fibers were within $\pm 20°$ to the flow direction in the extrusion operation resulting in the substantial difference between the longitudinal and transverse directions. The bumps in the curves are due to the fact that the angular distribution functions were not identical for all volume fractions.

If the laminate analogy is extended to the variable aspect ratio problem, it would be necessary to divide each laminate layer for each specific θ direction into another series of laminates having different aspect ratios in each layer.

This procedure, while rigorous, is obviously too cumbersome for practical calculations. A simpler solution is to use an average aspect ratio. The 'first moment' of the distribution is called the 'number average aspect ratio' $\langle l/d \rangle$, which is inserted into the micromechanic calculations, the Halpin–Tsai equations, as if it were a monochromatic aspect ratio. The influence of variable aspect ratio is shown in Figure 22 to be adequately treated, for random orientations at least, by the statistical

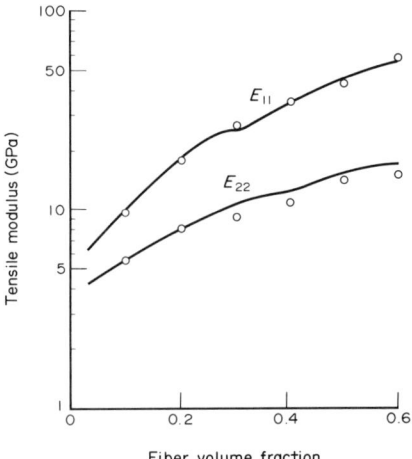

Figure 21 Stiffness of glass fiber/polycarbonate composites. $l/d = 360$

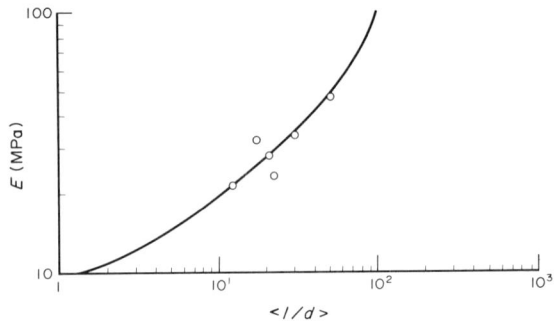

Figure 22 Stiffness of random short nylon fiber/rubber composites *versus* the number average aspect ratio. $V_f = 0.35$, $E_f/E_m = 973$

number average aspect ratio. The theory is indicated by the solid curve, which is the curve computed earlier for the monochromatic aspect ratios employed in the initial test of the laminate analogy.

18.2.4.2 Strength of isotropic short fiber composites

Maximum strain theory may be modified to predict the strength of randomly oriented short fiber composites.[26] The Halpin–Tsai equations (equation 16) have established relations for the stiffness of an oriented short fiber ply from the matrix and fiber properties. These equations show that the longitudinal stiffness of an oriented short fiber composite is sensitive to the aspect ratio.

The short fiber stiffness asymptotically approaches the continuous filament stiffness at large aspect ratios. Strength, like stiffness, is a function of aspect ratio and approaches an asymptotic limit as the aspect ratio becomes large. However, the strength limit for short fibers does not approach the continuous filament strength. Thus, the oriented short fiber material fails at an ultimate longitudinal strain which is less than the ultimate longitudinal strain of the continuous fiber material. Previous workers have shown that the asymptotic short fiber strength limit is less than the continuous filament composite strength.[27] A finite element analysis of the discontinuous fiber by Barker and McLaughlin[28] shows a stress concentration factor due to the fiber ends. This stress concentration factor becomes constant at sufficiently large aspect ratios. Chen[27] and Riley[29] also report that a plateau strength is reached once the fibers become sufficiently long.

The strength of high aspect ratio short fiber randomly oriented composites with maximum strain can be predicted by reducing the longitudinal strain allowable, so as to reflect the reduced strength in the fiber direction due to discontinuous reinforcements. Chen's data and analysis[27] show that at sufficiently large aspect ratios, the strength of short fiber glass/epoxy composites will be 60% of the strength of the continuous filament composite. This fact is incorporated into maximum strain theory by reducing the continuous filament longitudinal strain allowable by 60%. Figure 23 shows

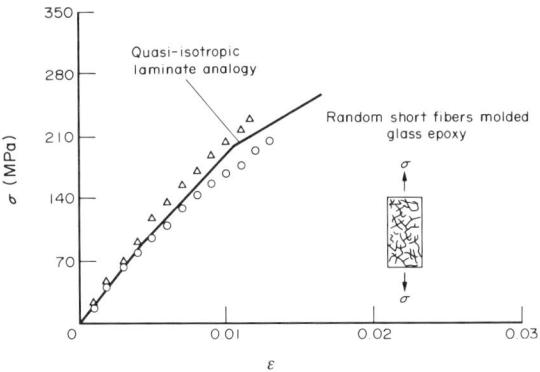

Figure 23 Stress/strain curve of randomly oriented short glass fiber/epoxy composites with the prediction from the maximum strain theory

the experimentally measured stress/strain curve for a randomly oriented short fiber composite and the prediction from maximum strain theory and the laminate analogy. The effect of fiber volume fraction of the quasi-isotropic strength can be included by using the Halpin–Tsai equations to calculate the orthotropic short fiber ply moduli.

18.2.4.3 *Dimensional stability*

Composite materials also find application where high specific properties, *e.g.* chemical inertness, low maintenance, dielectric capability, and dimensional stability, are required. Problems related to dimensional stability are particularly well suited for composite application, as currently available filaments possess a wide range of technically useful properties. Combining fibers within a matrix can generate a stiff, strong, laminated solid with exceedingly small, if not zero, in-plane expansion. This result is achieved when the fibers are of sufficient stiffness to dominate the axial response of the laminate. It can be noticed that the expansion coefficients or expansional strains are tensors[16,17] like the mechanical stress or strain and are, therefore, a function of the angular rotation of a ply. For the lamina these coefficients are given by

$$e_1 = \frac{E_m e_m V_m + E_f e_f V_f}{E_m V_m + E_f V_f} \tag{18}$$

$$e_2 = (1 + v_m) e_m V_m + (1 + v_f) e_f V_f - e_1 (v_f V_f + v_m V_m)$$

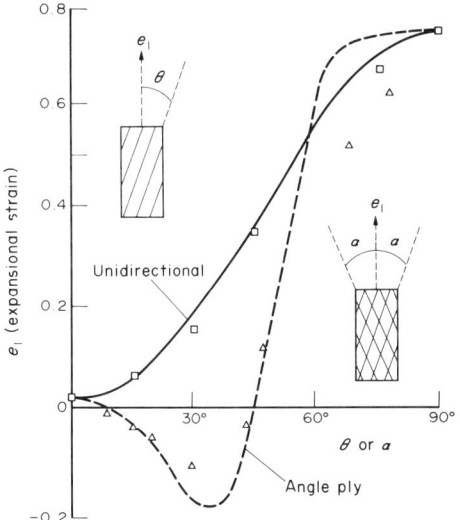

Figure 24 Angular dependence of expansion strain for unidirectional and angle ply laminates

Once the expansion coefficients of a single layer of fiber-reinforced material have been computed from constitutive material properties, the response of a laminate can be determined though the use of lamination theory[16,21]

$$e_1 = \frac{A_{22}R_1 - A_{12}R_2}{A_{11}A_{22} - A_{12}^2}$$
$$e_2 = \frac{A_{11}R_2 - A_{12}R_1}{A_{11}A_{22} - A_{12}^2}$$
(19)

where

$$R_1 = J_1 h + J_2 H_1$$
$$R_2 = J_1 h - J_2 H_1$$
$$A_{11} = U_1 h + U_2 H_1 + U_3 H_2$$
$$A_{22} = U_1 h - U_2 H_1 + U_3 H_2$$
$$A_{12} = U_4 h - U_3 H_2$$
$$J_1 = (U_1 + U_4)W_1 + 2U_2 W_2$$
$$J_1 = U_2 W_1 + 2W_2(U_1 + 2U_3 - U_4)$$
$$W_1 = (e_1^L + e_2^L)/2$$
$$W_2 = (e_1^L - e_2^L)/2$$
$$H_1 = \sum_{u=1}^{N} h_u \cos 2\theta_u$$
$$H_2 = \sum_{u=1}^{N} h_u \cos 4\theta_u$$
(20)

and U_i is as defined previously (equation 13).

The laminate calculation is somewhat surprising, in that it suggests that at certain angles the laminate will contract in the longitudinal direction when heated. This prediction was confirmed as indicated by the points in Figure 24, where the lines are the theoretical predictions.

For example, if it is desired to fabricate a plate possessing a zero expansion coefficient in any direction within the plane of the plate, one solves the equivalent problems of the 'quasi-isotropic laminate'. Halpin and Pagano[16,21] have obtained a solution for the 'quasi-isotropic expansion strain' in terms of the ply properties. This solution asserts the following general principle: for any uncoupled laminate with equal stiffness in the two in-plane directions ($A_{11} = A_{22}$), the expansional strain field is isotropic in this plane and the normal strain in all directions is simply given by

$$e = W_1 + \frac{2(E_{11} - E_{22})W_2}{E_{11} + (1 + 2v_{12})E_{22}}$$
(21)

We may now extend these results to the short fiber problem by modifying Schapery's results[22,30]

$$e_1 = \bar{e} + (\overline{Ee}/\bar{E} - \bar{e})\frac{(1/E_L - 1/E_{11})}{(1/E_L - 1/E_u)}$$
(22)

where $1/E_L = V_f/E_f + V_m/E_m$; $E_u = V_f E_f + V_m E_m$, with the bars representing volume averages: i.e.

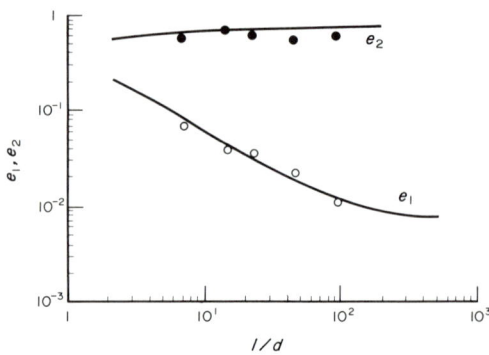

Figure 25 Dependence of expansion strains on aspect ratio for an oriented short-fiber nylon/rubber composite. The solid lines are the Schapery's equation; $e_m = 0.78$, $e_f = 0.007$, $V_f = 0.35$

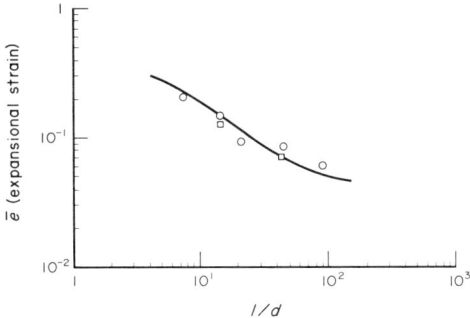

Figure 26 Expansion strain for a random (○) and a quasi-isotropic (□) short fiber nylon/rubber composite. $e_m = 0.78$, $e_f = 0.007$, $V_f = 0.35$. The solid line is a plot of equation (16)

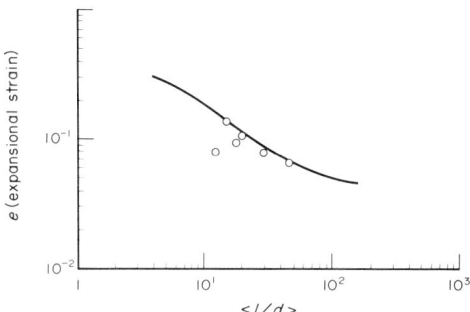

Figure 27 Expansion strain for a random short fiber nylon/rubber composite as a function of the number average aspect ratio. The solid line is a plot of equation (16)

$\bar{e} = e_f V_f + e_m V_m$. A comparison of the micromechanics with experimental results for a nylon–rubber composite is shown in Figure 25. Note the strong dependence of the longitudinal expansion on the aspect ratio. As in the case of the moduli, the longitudinal response levels out, at finite aspect ratios, with an expansion strain equal to that of continuous fiber reinforcement.

The assumptions of the laminate model are again tested in Figure 26 for randomly oriented short fibers, through the comparison of random samples (the circles), with quasi-isotropic short fiber laminate (the squares) and the theory (the solid line). The laminate calculation is that of Halpin and Pagano for the 'quasi-isotropic expansional strain'. The effect of variable aspect ratio (Figure 27) is found to correlate with the number average aspect ratio in the same manner as previously discussed for the moduli. Again the theoretical curve is the same as the previous figure for the monochromatic aspect ratio specimens.

18.2.5 Particulate Composites

The introduction of another phase in polymer, such as a particle or a simple filler, generally modifies the morphological order of the polymer. This phenomenon can be exploited in order to obtain a material with the desired properties. Generally, the main reasons for using particles in a composite are:[15] (i) to increase the rigidity of a system; (ii) to reduce the viscous flow; (iii) to manipulate the thermal expansion coefficients; (iv) to vary the permeability of the composites to gases or liquids; (v) to increase the resistance to abrasion; (vi) to modify the electrical properties of the material; (vii) to modify the rheological characteristics of the material; and (viii) to lower the cost of the material. The use of fillers modifies the material properties as a consequence of changes to the microstructure of the material.

Depending on the adhesion between the filler and the polymer, the particles can act as reinforcing agents or simply as stress concentrators. Usually the bond between the filler and the matrix is created by the pressure increase subsequent to the shrinkage during polymerization for thermosets matrices, or by the molding process for thermoplastics.

When spherical particles are included in the polymeric matrix, the behavior of such material is isotropic and the elastic properties can be easily predicted by using the Kerner equation[17] shown below or the Halpin–Tsai equations for G_{12} and E_{22} discussed earlier

$$E_c = E_m \frac{1 + ACV_f}{1 - CV_f} \tag{23}$$

where

$$A = \frac{7 - 5v_m}{8 - 10v_m} \qquad C = \frac{E_f/E_m - 1}{E_f/E_m - A}$$

v_m is the Poisson's ratio of the matrix, V_f is the volume fraction of filler and E_f and E_m are the Young's moduli of the filler and the matrix, respectively. In Figure 28 experimental data relative to styrene acrylonitrile (SAN), acrylonitrile butadiene/styrene (ABS), polystyrene (PS), epoxy and polyester/glass bead composites are reported, together with the plot of the Kerner equation. In addition to the elastic moduli, the other tensile properties of this class of materials are sensitive to the properties of the matrix and the adhesion between filler and matrix. In fact when a matrix is able to craze, it creates an inhomogeneous deformational mechanism, which leads to an increase in the elongation and energy required to break the resulting composites, due to the formation and propagation of crazes through the polymer. Accordingly to Kambour,[31] the crazes normally

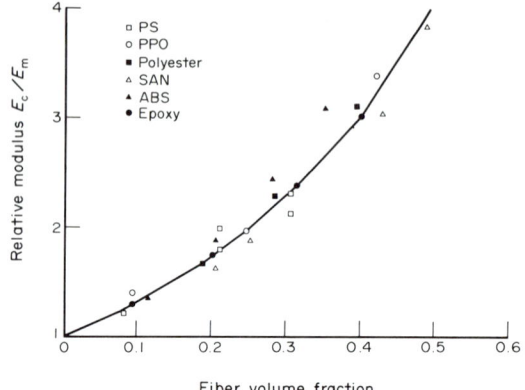

Figure 28 Relative moduli. E_c/E_m versus volume fraction of glass beads for various composites.[36] The solid line is a plot of the Kerner equation

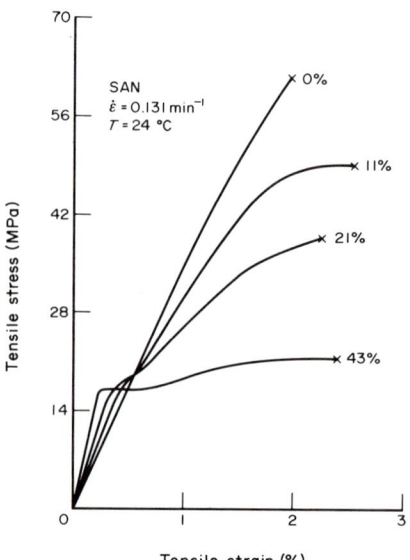

Figure 29 Stress/strain curve for SAN/glass bead composites at various filler contents

observed in thermoplastic materials are not cracks, but rather are localized regions of highly oriented polymer. The high elongation observed[32] in the glass bead filled SAN can be explained by assuming that the growth of crazes can be terminated by the glass beads or the vacuoles around them. In fact, if the propagating craze encounters a glass sphere to which the matrix is not strongly adherent, interfacial debonding can effectively blunt the tip of the craze and prevent, or at least slow down, further craze propagation. The elongation at the break is enhanced by the fact that the crazes, which in the unfilled material may be nucleated only at singular defects, in the glass bead composites nucleate throughout the whole specimen. The filler particles act as stress risers, allowing multiple volume elements to reach the critical stress for craze formation. This situation is reflected in the stress/strain response by the appearance of a knee in the deformation curves at a well-defined value of stress which is practically independent of filler content. In Figure 29 a typical stress/strain curve for such composites is reported, showing a sudden change in the slope at a fixed value of stress which is practically independent of filler content. The strength of these composites decreases as the volume content of beads increases. This phenomenon is also observed in limiting the transverse tensile strain and in-plane shear strain capabilities of the unidirectional prepreg materials discussed earlier. It is because of this phenomenon that variations in reinforcement volumetric packing is not a material design variable. This strength decrease is simply a reflection of the decreased cross sectional area of the polymer bearing the load and can be expressed as a function of concentration by the following equation[33]

$$\sigma_c = \sigma_m(1 - 1.21\, V_f^{2/3}) \qquad (24)$$

where σ_c and σ_m are the strengths of the composite and the polymer respectively. In Figure 30 the experimental values of the strength are compared with those calculated for various composites.

Figure 31 shows the ultimate elongation of the ABS/glass bead composites as a function of filler content. It can be seen that, in contrast to theoretical prediction,[34,35] the ultimate elongation of these

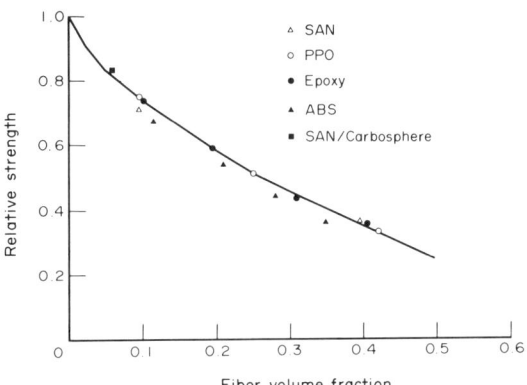

Figure 30 Relative strength σ_c/σ_m *versus* volume fraction of glass beads for various composites.[36] The solid line is a plot of equation (24)

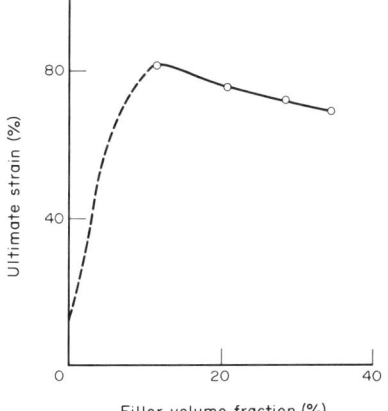

Figure 31 Ultimate elongation of ABS/glass bead composites *versus* the filler content

composites increases sharply when the filler is added, up to about 10% filler volume, and then decreases slowly. However, it should be recalled that the theories are only applicable to simple two-phase systems, not to those which are dilatable (crazing) and can form a third mechanical phase.

The result of this combined dewetting and crazing leads to greatly improved toughness by the simple fact that a large volume of the specimen is involved in the deformation and hence this increases the work-to-break, *i.e.* the area under the stress/strain curve. These results are very important practically in the fact that the addition of these rigid inclusions to the ABS leads to a new material with better mechanical properties. The interesting point is that only a small quantity of glass beads increases the work-to-break very sharply, as is shown in Figure 32.

In the case of good adhesion the deformational mechanism is different, as also reported by Broutman,[36,37] in the sense that for a very brittle matrix the strength can be predicted using a finite element method. In Figure 33, the stress/strain behavior of SAN/glass bead composites at very low temperature is reported. The final strength of the SAN/glass bead composite seems independent of filler content, but is lower than that of the unfilled polymer. The ultimate elongation, in this case, decreases with filler content, as shown in Figure 34.

In general, particulate composites, even if they do not have structural applications, can be used in many applications, due to the fact that the addition of rigid particulate fillers to a polymeric matrix leads to a material with a higher modulus, lower creep, more resistance to abrasion, different rheological and electrical properties, a different deformational mechanism and lower cost.

For example, permeable membranes can be obtained as films of thermoplastic glassy polymers/glass bead composites, in which permeability is achieved by prestressing the membrane before

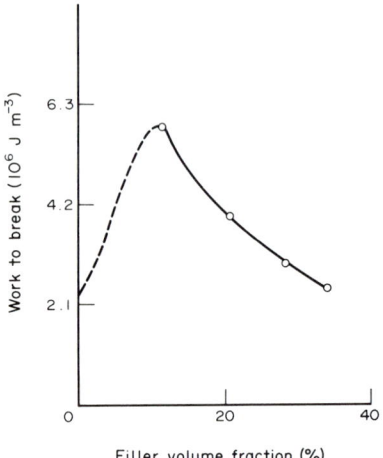

Figure 32 Work to break of ABS/glass bead composites *versus* filler content

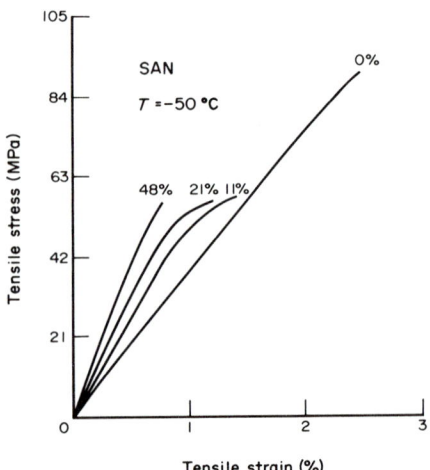

Figure 33 Stress/strain curves of SAN/glass bead composites at $T = -50\,°C$, $\epsilon = 0.131\,\text{min}^{-1}$

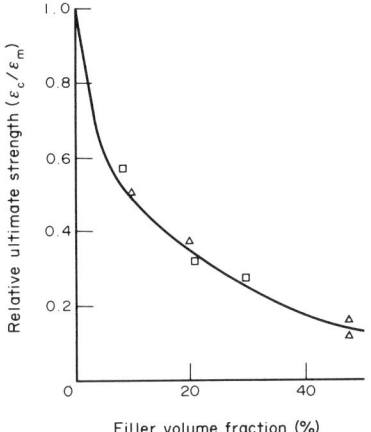

Figure 34 Relative ultimate elongation ϵ_c/ϵ_m *versus* filler content of SAN and PS/glass bead composites at $T = -50\,°C$

permeability testing, thereby developing crazes in the polymeric matrix.[38] These films can be used to support dynamic membranes or generally as a support for different kinds of membranes (thin fiber membranes or plasma-formed membranes).

18.2.6 Fracture Mechanics in Composite Materials

18.2.6.1 *General aspects*

Fiber-reinforced composites have emerged as viable structural materials due to their advantageous stiffness, thermal expansion, strength and density properties. These properties are derived in a composite through the dominance of high stiffness/high strength fibers over the relatively soft, low strength matrix. Good fibrous reinforcements are generally brittle in character; they deform linearly to failure without yielding. This attribute creates a situation in which, in the presence of a notch or hole under static tension or compression test conditions, the fiber-reinforced composite behaves more like a brittle material than metal. This issue has been a source of concern in materials development and selection activities, as well as in engineering design.[39,40]

Theoretical investigations into the phenomenon of the notch sensitivity of composites have tended to rely on classical fracture concepts. These efforts have taken two forms: micro- and macromechanical representations. In the micromechanics format, local fracture processes (debonding, matrix cracking, fiber breaking, *etc.*) are studied in the hope that a mechanical treatment can be developed for a nonhomogeneous multiangular laminate or moulded part. This approach faces serious difficulties and has not yet matured. The macromechanical approaches use a simplified model of the composite and classical fracture mechanics for homogeneous isotropic materials. The simplified composite model is the plane stress laminate theory,[16,41,42] which converts the nonhomogeneous laminated anisotropic solid into an anisotropic homogeneous solid. Within the macroscopic approaches, two lines of activity exist: a fracture mechanics argument,[43-45] and a blending of classical fracture concepts and notch theory.[46,47] These two approaches will be outlined and tested in the subsequent sections. The concepts will be evaluated against data for a series of graphite/epoxy laminates and compared with other literature data to illustrate the necessary points.

18.2.6.2 *Theoretical considerations*

Strength concepts generally imply that rupture occurs because the spatial average stress or strain exceeded some critical value (generally an empirical criterion) which characterizes the mechanical stability of a solid. Such an attitude is useful if the microstructural perturbations to the local stress/strain fields inside a solid are of a small mean dimension with a low dispersion around the mean size. On the other hand, there are conditions in which discrete flaws, substantially larger than the uniform size distribution normally present, can exist in a material. Because they are discrete, usually relatively sharp, and larger than the surrounding disturbances, they induce additional stress concentrations and provide the site of cohesive fracture initiation. If these inherent flaws are crack-like, ordinary elastic stress concentration factors cannot be used because theoretical results predict an infinite concentration factor multiplying the average stress in the crack vicinity. Thus, the local stress value will exceed the finite allowable stress (or strain) experimentally measured for the base material containing only a reasonable uniform distribution of inherent flaws. The strength degradation in such a situation is illustrated by region II of Figure 35(c). Fracture is then defined as the cleavage of a solid by crack extension process.

18.2.6.3 *Linear elastic fracture mechanics approach*

Based upon the above rationale, as in Figure 35, it is apparent that when a failure surface for a lamina (and through lamination theory for a laminate) is measured and defined, it is, on a spatial average basis, the critical stress (or strain) space of the naturally occurring flaws that is actually measured. Since a body containing randomly distributed microflaws is observed to sustain a set of finite loads without observable fracture, it appears that a characteristic zone of cubical dimension, r_c, must encapsulate a flaw, thus insulating the body from the infinite, and hence fracture-producing, stresses which would otherwise exist at the microflaw tip. Fracture (that is, the rupture of a laminate in the presence of a flaw) results when a volume element encapsulating the flaw tip experiences a local load equal to or greater than that required to rupture the unnotched material. This critical element, characterized by the dimension r_c, serves as the link between fracture and strength analysis in solids (including composites). The hypothesis is that it is only necessary to calculate stresses outside a region r_c from a geometric discontinuity, and then view these stresses as applied to the characteristic critical zone in the same manner as occurs in the case of unflawed laminates.

This hypothesis requires that if a latent microscopic flaw is an inherent material characteristic, then the above finite volume which contains the microscopic flaw must also possess a characteristic dimension. Within this context, a measure of fracture susceptibility (or resistance) would be defined in terms of a characteristic critical zone r_c for a lamina/laminate material system.

For example Wu[48] and Sih[49] have shown that the stress $\sigma_{yy}(x, 0)$ near the tip of a straight line crack of length $2c$ in an isotropic plate is approximated by the expression

$$\sigma_{xy}/K_1 = 1/[2\pi(x - c)]^{0.5} \tag{25}$$

where K_1 is the Mode 1 stress intensity factor defined as

$$K_1 = \sigma(\pi c)^{0.5} \tag{26}$$

The term σ is a uniform tensile stress applied parallel to the y axis at infinity. At fracture, the stress σ_{yy} is equal to or less than the unnotched material strength σ^* at a distance $(x^* - c) > r_c$ ahead of the crack tip (see Figure 35b). This dimension represents the distance over which the material must be critically stressed in order to find a flaw sufficiently large to initiate failure.

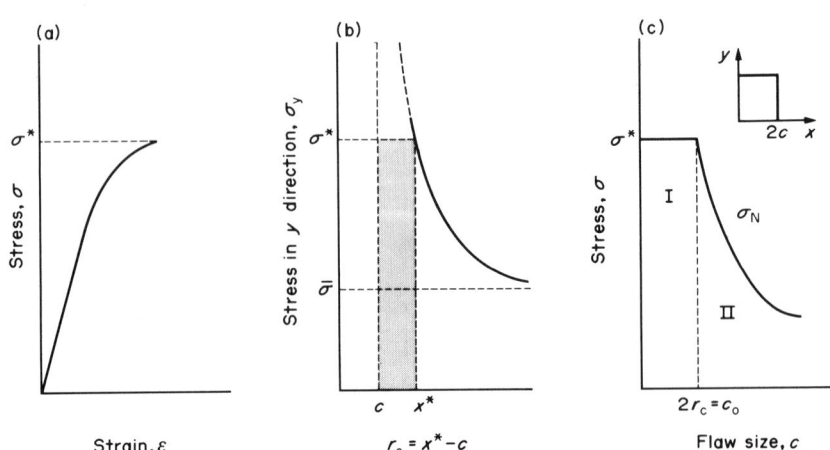

Figure 35 Illustration of fracture criteria: (a) stress/strain curve for a laminate showing progressive failure of the three ply orientations; (b) conditions for fracture of a material volume r_c^3 in the neighborhood of a crack tip under uniform tensile stress in the y direction; (c) dependence of strength on flaw size

The stress distribution ahead of a real crack of half length c may be regarded as identical to the elastic stress (or strain) field ahead of a 'notional' crack of half length $(c + r_c)$

$$\sigma_{yy}/K_1 \to \sigma^*/K_1 \text{ at } 1/(\pi r_c)^{0.5} = 1/(\pi c_0)^{0.5} \tag{27}$$

and

$$K_{1c} \to \sigma_n[\pi(c + r_c)]^{0.5} = \sigma_n[\pi(c + c_0)]^{0.5} \tag{28}$$

where $c_0 = r_c$. As the intentionally implanted flaw of half length c goes to zero, the apparent notched strength, σ_n, goes to the unnotched strength, σ^*, defined as σ_0, thus

$$\sigma_n(c \to 0) \to \sigma_0(c < c_0) = K_{1c}/(\pi c_0)^{0.5} \tag{29}$$

The apparent stress concentration K_t, will then be given as

$$\sigma_0/\sigma_n = K_t = [(c_0 + c)/c_0]^{0.5} \tag{30}$$

and the critical zone dimension will be

$$c_0 = r_c = c/[(\sigma^*/\sigma_n)^2 - 1] \tag{31}$$

The assumption that the critical dimension r_c is a material constant is supported by the nature of equation 25, which states that the geometry of the stress field ahead of a line crack in an isotropic body is independent of the degree of anisotropy of the body or of the absolute dimension of the crack.

Similar arguments may be advanced for the estimation of the critical stress for a laminate pierced by a circular hole. The above analysis attributes the contribution of a damage zone of size c_0 to be analogous to Bowie's solution[50] of symmetrical cracks of dimension C emanating from circular hole of radius r in an isotropic, homogeneous material

$$K = \sigma_n^\infty (c_0 \pi)^{0.5} f(c_0/r) \tag{32}$$

or, at fracture the critical stress is

$$\sigma_n^\infty K/[(c_0 \pi)^{0.5} f(c_0/r)] \tag{33}$$

For a specimen with no hole

$$\sigma_0 = \sigma_n^\infty |c/r \to \infty = K_c/(c_0 \pi)^{0.5}(1.00)$$

thus

$$\sigma_0/\sigma_n^\infty = f(c_0/r) \tag{34}$$

From the ratio of the unflawed strength to the flawed specimen strength, an estimate of $f(c_0/r)$ is made. Employing tables of $f(c/r)$ a value of c/r and then c_0 is obtained. In the above expression, c_0 for the slit and hole should be of comparable dimension for a given laminate and K_c must be the same numerical quantity if the 'fracture toughness' is a material parameter. In addition, the above statements require that the apparent stress concentration $K_T = \sigma_0/\sigma_n^\infty$ is dependent upon absolute hole size, varying between

$$1.0 > \sigma_0/\sigma_n^\infty |r \to 0 \text{ to } \sigma_0/\sigma_n^\infty \to 1.13 |r \to \infty$$

in a continuous sigmoidal fashion. Note that in the limit of large hole size, effective static stress concentration factors, as estimated by anisotropic theory, are observed in composites and that a hole and slit give comparable strength reduction at small to moderate size dimensions.

18.2.6.4 Strength analysis approach

A stress criteria for fracture of laminates containing notches or holes has been developed,[46,47] based upon considerations of the exact stress gradients adjacent to either the hole or the notch. It is assumed that failure occurs when the stress over some distance away from the discontinuity is equal to or greater than the strength of the unnotched material, σ_0. It is further assumed that this characteristic distance, d_0, is a material property independent of hole or notch dimension. This dimension represents the distance over which the material must be critically stressed in order to find a sufficient flaw size to initiate failure. This dimension is analogous to the approach used for predicting the plastic zone found in metals[51] and is comparable to the arguments advanced above. Using this criterion and the approximate stress gradient for a hole in an isotropic plate, the ratio of the notched to the unnotched strength is

$$\sigma_n/\sigma_0 = 2/[2 + \xi_1^2 + 3\xi_1^4 - (K_T - 3)(5\xi_1^6 - 7\xi_1^8)] \tag{35}$$

where

$$\xi_1 = r/(r + d_0)$$

and K_T, the stress concentration factor, is given by

$$K_T = 1 + 2\{2[(E_{11}^*/E_{22}^*)^{1/2} - v_{12}^*] + E_{11}^*/G_{12}^*\}^{1/2} \qquad (36)$$

The asterisk over the moduli denotes effective elastic moduli of the laminate and the subscript one denotes that the axis is parallel to the applied load. For very large holes, $\epsilon \to 1$ and the classical stress concentration result $\sigma_n/\sigma_0^\infty = 1/K_T$, is recovered. For vanishingly small holes the ratio $\sigma_n/\sigma_0^\infty \to 1.0$, as would be expected. As the anisotropic laminate goes to quasi-isotropic conditions

$$E^* = E_{11}^* = E_{22}^*, \quad E^* = 2(1 + v_{11}^*)G_{12}^*$$
$$\sigma_n/\sigma_0^\infty = 2/(2 + \xi_1^2 + 3\xi_1^4)$$

Similar arguments for a center-cracked geometry yield a ratio of the notched to the unnotched strength of

$$\sigma_n^\infty/\sigma_0 = (1 - U_3^2) \qquad (37)$$

where

$$U_3 = c/(c + d_0) \qquad (38)$$

The predicted crack size effect on the measured value of the fracture toughness, K_q, can be developed by noting that

$$K_q = \sigma_n(\pi c)^{1/2}. \qquad (39)$$

Using equation (37)

$$K_q = \sigma_0[(\pi c(1 - U_3^2)]^{1/2} \qquad (40)$$

In this expression, the expected limit of $K_q = 0$ for vanishingly small crack lengths is reached, while for a large crack length, K_q asymptotically approaches a constant value. This asymptotic value is

$$K_q = \sigma_0(2\pi d_0)^{1/2} \qquad (41)$$

which compares with the previous fracture mechanics asymptote

$$K_q = \sigma_0(\pi c_0)^{1/2} \qquad (42)$$

It is apparent that

$$c_0 = 2d_0 \qquad (43)$$

18.2.6.5 Comparison between theoretical and experimental results

Experimental results reported on commercial graphite/epoxy laminates are reported in Table 3,[40] where the angular constructions of the laminates are also indicated. Lamination sequence is denoted by the subscript which indicates the number of times the angular pattern was repeated, and 's' indicates that the stacking sequence was symmetrical around the laminate midplane. From the failure load of each unnotched coupon, the failure stress, σ_0, was computed. Similarly, the gross failure stress, σ_n, of the notched specimens was obtained from the failure load of these specimens. The values of the gross failure stresses, σ_n, were adjusted for all laminates by multiplying σ_n by the isotropic finite width correction factors for holes[52,53]

$$\sigma_n^\infty = \sigma_n[2 + (1 - 2R/w)(w/\pi c)^3]/[3(1 - 2R/w)] \qquad (44)$$

and for cracks

$$\sigma_n^\infty = \sigma_n[(w/\pi c)\tan(\pi c/w)]^{1/2} \qquad (45)$$

(where w = width of the test coupon) to obtain the notched, infinite width failure stress, σ_n^∞. Equation (44) is very accurate for $2R/w < 0.5$, while equation (45) is highly accurate and notched data were normalized by using the average of three specimens and then computing the ratio σ_n^∞/σ_0.

The ratio σ_n^∞/σ_0 was used to estimate the apparent fracture toughness K_q and the characteristic dimension c_0. These estimates, using equations (31) and (34), are outlined in Table 3.[40] Table 4 summarizes data reported by Whitney and coworkers[46,47] for a comparable graphite/epoxy system, and for a glass/epoxy system. These data were analyzed in accordance with the strength concept outlined above. Table 5 is a reproduction of early data[44] for graphite, boron, and glass fiber-reinforced epoxy composites. These data were analyzed using the fracture mechanics concept outlined earlier. It is interesting to note that a variety of materials evaluated by several workers at different locations and time indicate comparable values or trends of the fracture toughness parameter K_q and the characteristic dimension c_0 or d_0. The average characteristic dimensions from Tables 3, 4 and 5 are $c_0 = 2.1$ mm and $d_0 = 1.05$ mm.

Table 3 Experimental Values for Strength, Toughness and Critical Dimension for the Graphite/Epoxy Material System AS/3501

Laminate construction angles	Tensile strength, σ_0 (kg mm^{-2})	Fracture toughness, K_q (kg mm$^{-3/2}$)	Critical dimension, c_0 (mm)
$[0°]_{8s}$	121.1	292.7	1.86
$[0°/90°]_{4s}$	51.0	76.3	0.77
$[0°/\pm 60°]_{4s}$	46.1	90.4	1.22
$[0°/\pm 45°/90°]_{2s}$	48.9	105.8	1.49
$[0°/\pm 36°/\pm 72°]_{2s}$	42.5	116.7	2.40
$[0°/\pm 30°/\pm 60°/90°]_{2s}$	47.5	87.9	1.09

Table 4 Comparison of Experimental Estimate of the Tensile Strength and Fracture Toughness Parameters

Laminate construction angles	Tensile strength, σ_0 (kg mm^{-2})	Fracture toughness, K_q (kg mm$^{-3/2}$)	Damage dimension (mm)
T300/5208 graphite/epoxy			
$[0/90°]_{4s}$	60.0	135.7	2.03
$[0/\pm 45°]_{4s}$	55.3	128.5	2.03
$[0/\pm 45°/90°]_{2s}$	47.6–50.4	107.1	2.03
E-glass/epoxy (Scotch ply 1002)			
$[0/90°]_{4s}$	43.2	99.1	2.03
$[0/\pm 45°/90°]_{2s}$	32.7	78.5	2.03

Table 5 Comparison of Experimental Estimates of the Fracture Toughness Parameters

Angular content, θ	Tensile strength, σ_0 (kg mm^{-2})	Fracture toughness, K_q (kg mm$^{-3/2}$)	Critical dimension, c_0 (mm)
Graphite/epoxy			
0	106.4	117.8	—
$[0°/90°]_{2s}$	53.9	135.7	1.78
$[0°/\pm 45°]_{2s}$	47.1	107.1	1.02
$[0°/\pm 45°/90^2]_{2s}$	34.9	78.5	—
With round hole	27.9	97.1	2.79
With slit	28.0	91.7	2.29
90°	3.7	5.4	—
Boron/epoxy			
0	135.2	—	—
$[0°_2/\pm 45°]_{2s}$	71.1	199.9–239.1	2.54–3.56
$[0°_3/\pm 45°/90°]_{2s}$	70.4	192.8	2.54
$[0°/\pm 45°]_{2s}$	62.0	160.6	2.54
$[0°/\pm 45°/90°]_{2s}$	42.9	124.9	2.79
E-glass/epoxy			
0	105.6	—	—
$[0°/90°]_{2s}$	54.9	—	—
$[0°_2/\pm 45°]_{2s}$	62.1	—	—
$[0°/\pm 45°/90°]_{2s}$	35.5	—	—
With hole	19.7	121.4	2.54

Table 6 Comparison of Predicted Strength and Fracture Toughness for Different Materials and Laminates

Laminate construction angles	π/n	Tensile strength (kg mm^{-2})	Fracture toughness, K_q (kg mm$^{-3/2}$)
AS/350/and T300/5208 graphite/epoxy			
$0°$	$\pi/1$	133.8	339.2
$0°/90°$	$\pi/2$	65.9	167.1
$0°/\pm 45°$	—	53.3	135.3
$0°/\pm 60°$	$\pi/3$	48.2	122.5
$0°/\pm 45°/90°$	$\pi/4$	48.2	122.5
$0°/\pm 36°/\pm 72°$	$\pi/5$	48.2	122.5
$0°\pm 30°/\pm 60°/90°$	$\pi/6$	48.2	122.5
HTS graphite/epoxy			
$0°$	$\pi/1$	106.4	269.9
$0°/90°$	$\pi/2$	53.9	136.7
$0°/\pm 45°$	—	45.0	115.0
$0°/\pm 45°/90°$	$\pi/4$	38.0	96.4
Fiberglass (Scotch ply 1002)			
$0°$	$\pi/1$	105.6	267.8
$0°_2/\pm 45°$	—	63.1	159.9
$0°_2/90°$	$\pi/2$	55.6	141.0
$0°/\pm 45°/90°$	$\pi/4$	38.0	96.4
Boron/epoxy			
$0°$	$\pi/1$	135.2	342.7
$0°_2/\pm 45°$	—	73.9	187.7
$0°_2/\pm 45°/90°$	$\pi/4$	49.3	125.0

In Table 6, the strength predictions (plane stress laminated plate theory) for each material system and laminate geometry are presented with an estimate of the expected fracture toughness K_q. The fracture toughness was estimated using the following equation[27]

$$K_q/\sigma_0 = (\pi c_0)^{1/2} \text{ or } K_q = 0.5\sigma_0 \qquad (46)$$

for $c_0 = 2.0$ mm. These theoretical projections correlate with the trends reported in Tables 3, 4 and 5.

The following comments are appropriate from the data.

(i) The current macroscopic approach correctly predicts the declining strength with either increasing notch dimensions, or hole radius.

(ii) The characteristic dimensions c_0 or d_0 appear to be insensitive to variations in laminate construction and possibly to different fibrous reinforcements. This observation is limited to what are called 'fiber or filament-dominated laminates'.

(iii) Both approaches for treating data outlined above appear to yield comparable results. In fact, the approximate results of equation (35) can be used to estimate the $f(c_0/r)$ function in the Bowie expression (equation 34).

(iv) Use of plane stress laminated plate theory for performing strength analysis and stress analysis outside the critical zones c_0 or d_0 appears to be more than adequate for both materials development and engineering design. The known problems associated with the free-edge effect,[54] matrix crazing[55] and material nonlinearity do not compromise this capability.

Table 7 Quasi-isotropic Laminate Properties for Different Materials

Material	Extension stiffness (kg mm^{-2} × 10^2)	Strength (kg mm^{-2})	Fracture toughness (kg mm$^{-3/2}$)	Density (g cm^{-3})
Aluminum 2024	74	43.6	114.2	2.77
Graphite/epoxy	56	47.9	107.1	1.49
Boron/epoxy	80	42.2	116.4	1.99
E-glass/epoxy	19	38.0	100.0	1.77
Epoxy	1.8	8.4	2.68	—

(v) As the laminates of any given material family approach the plane stress quasi-isotropic construction π/n when n > 3, they reach a constant value for strength and fracture toughness independent of detailed angular content. Furthermore, the absolute properties are comparable to metallic systems. A comparison of quasi-isotropic laminate properties for different material systems with a typical aluminum alloy is given in Table 7.

(vi) Fiber-reinforced systems, although quasi-brittle in terms of fracture response, are unique in that the fracture toughness increases with increasing stiffness and strength. In most metals the toughness decreases with increasing yield strength. In reinforced plastics, the general rule is that if a particulate reinforcement raises the ultimate strength, it will lower the average strain at fracture, and the toughness.[15,56-58]

18.3 PROCESSING OF HIGH PERFORMANCE COMPOSITE MATERIALS

18.3.1 Introduction

The widespread use of polymer matrix composite materials has been facilitated by the development of several specific technologies for the processing of parts with different geometries using different raw materials. However, the understanding of the fundamental engineering principles associated with each technology still represents a challenge for research activities in the field of the processing of high performance composite materials characterized by improved material quality and reliability requirements. Today the choice of processing parameters in the industrial environment is often based on extensive and costly testing. In addition, specific testing results may not be applicable if raw materials or part geometries are changed. In the last decade several studies focused, on the analysis of the fundamental aspects of the processing of high performance composites, have been reported,[2,3,59-63] allowing a more rational choice of processing conditions and reducing the experimental work needed to determine the proper fabrication cycle. From a materials point of view, resin chemistry, storage and processing conditions should be selected to give the most uniform and reliable fabrication process, with low defects and a composite part obtained in the shortest possible time. In practice, factory operations, material quality and cost must be considered in a compromise to design an optimum processing cycle.

The principal findings reported in the literature on the understanding of the basic phenomena governing the behavior of high performance thermoset matrix composites are presented in this section. The physicochemical changes of the matrices are related to the processing behavior of the composites through the integration of the material-dependent models into more general models. These general models consider the specific technological aspects of each fabrication process and the effect of the different processing variables, in order to analyze their influence on the development of the polymeric structure.

18.3.2 General Aspects of Processing Technologies

Thermoset-based composites are commonly fabricated using epoxy, polyester or other thermosetting resins reinforced with carbon, glass or aramidic fibers. The main advanced processing technologies include lamination, resin transfer molding, pultrusion and filament winding. A brief description of these technologies, based on the available literature[64] may be useful to understand the role of the polymeric matrix on the processing behavior of high performance composites.

18.3.2.1 Autoclave lamination

Thermoset based composite laminates are generally produced by the autoclave/vacuum degassing lamination process.[62,63] The characteristics of this process are shown in Figure 36. In this process, prepreg plies of desired shape are laid up in a prescribed orientation to form a laminate. The laminate is covered with successive layers of an absorbent material (glass bleeder fabric), a fluorinated film to prevent sticking, and, finally, with a vacuum bag. The entire system is placed on a smooth metal tool surface in an autoclave, vacuum is applied to the bag and the temperature is increased at a constant rate in order to promote the resin flow and polymerization. The autoclave process will be used in this section as a case study to describe the influence on the matrix characteristics on the processing behavior of high performance composites.

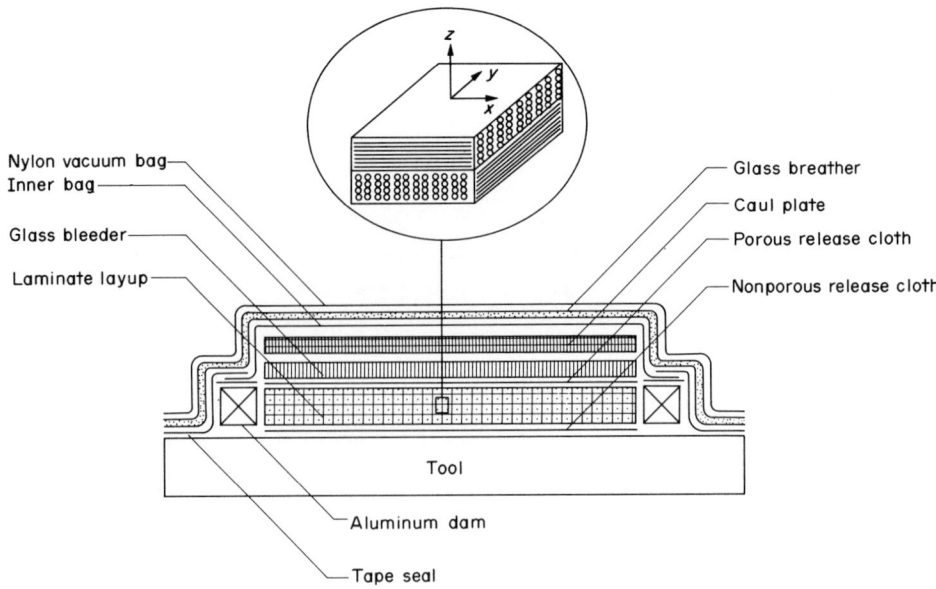

Figure 36 Schematic diagram of a laminate lay-up. Insert shows a microscopic view of the fiber orientation within the laminate (after Dave' et al.[62])

18.3.2.2 *Pultrusion*

In the last few years a growing interest has been devoted to the fabrication of high performance composites applying the pultrusion technology. This offers the advantages, among others, of continuous production and the integration of fiber impregnation and composite consolidation in the same process. The simplicity, the efficiency and the flexibility of this process makes it one of the most interesting methods for the fabrication of continuous fiber composite products with constant cross section. As shown in Figure 37, during pultrusion, fibers in tape, woven and/or mat form are driven through a resin bath where a good impregnation can be achieved using a resin of the correct viscosity. After the resin excess is removed in preforming guides, the fiber/resin system acquires the desired shape and passes through the cure process in a heated die, which acts as a continuous reactor. Usually, different heating zones are provided along the die depending on several factors, such as the type of resin, the pulling speed and the length of the die. The application of the dragging force to the pultruded parts is performed using a pulling device able to impose the desired processing speed. Finally, the pultruder is equipped with a sawing system to cut off the continuous composite produced. The viscosity changes of the matrix determines the pulling force and the fiber wettability, while the polymer reactivity determines the pulling speed in order to obtain a completely cured composite at the end of the process.

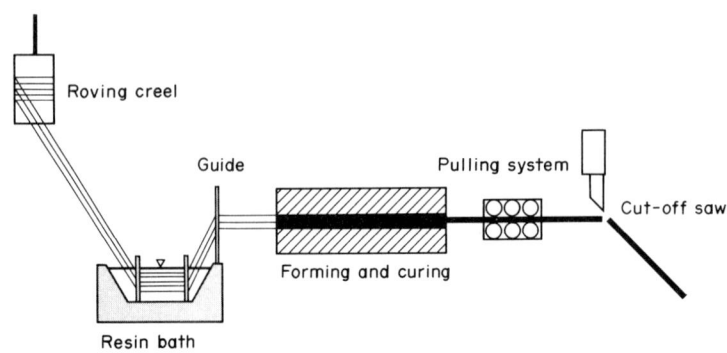

Figure 37 Schematic diagram of the pultrusion process

18.3.2.3 *Resin transfer molding (RTM)*

This technology is characterized by preplacement of dry reinforcement in the mold before the mold is closed and resin is injected (Figure 38). Normally a low viscosity resin is used, and either low pressure or low vacuum employed to assist resin flow and wetout of reinforcement. Complex preform shapes are generated by the shaping of mats or fabrics, pretreated with thermoplastic binder, with heat or pressure. RTM, presenting the ability to combine precise control in the placement of a high concentration of fibers with rapid processability, provides many of the fundamental requirements to meet the economic needs of a mass production industry. The main characteristics of RTM as an alternative to autoclave processing of high performance composites are given by the capability to produce large integrated parts with complex geometries, including box sections using foam cores, and other sandwich structures. In addition, lower investment costs and no storage problems of unstable B staged prepregs are involved.

Current research in RTM technology is addressed to further development of preform technology to optimize cycle times, fiber wetout, mechanical properties and surface appearance. A wide range

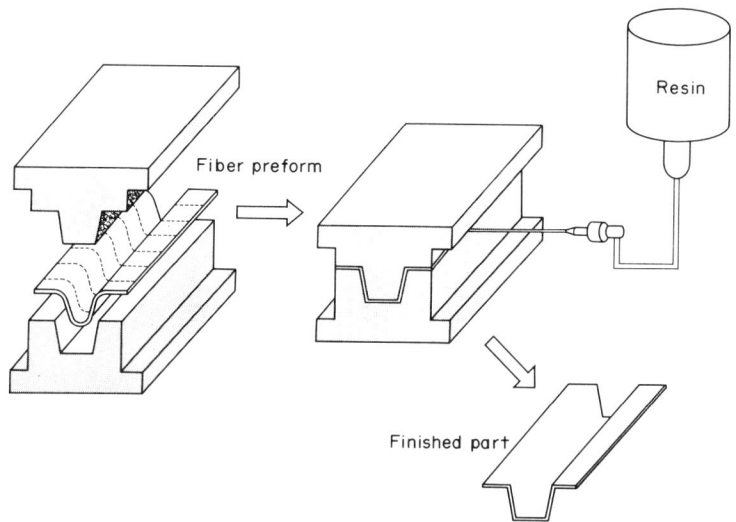

Figure 38 Schematic diagram of the RTM process

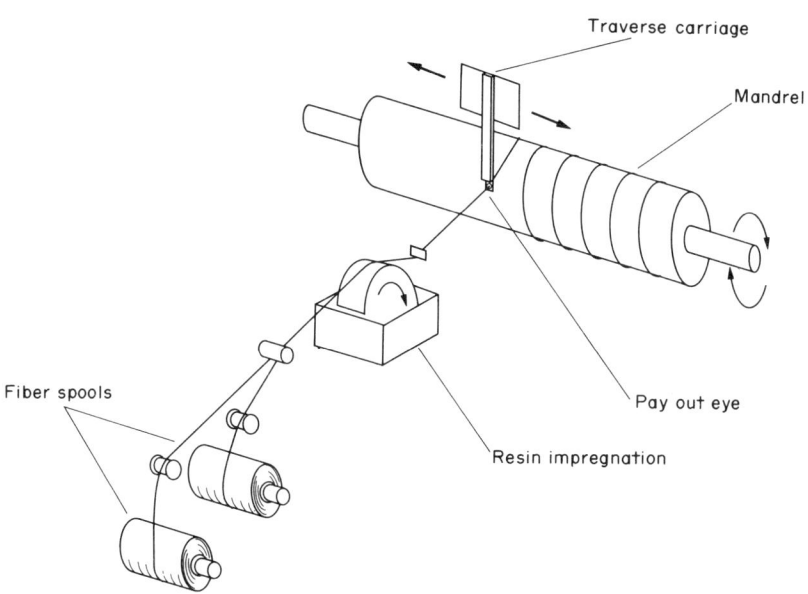

Figure 39 Schematic diagram of the filament winding process

of reinforcements and resins has been developed recently, fulfilling the specific requirements of RTM: high reactivity and low viscosity systems for fast fill, complete penetration of the preform and good wetout of the fibers, low void formation, lack of movement of the reinforcement fibers and avoidance of resin-rich areas.

18.3.2.4 Filament winding

In filament winding (Figure 39), a filamentous yarn or tow is first wetted by a resin and then uniformly and regularly wound around a rotating mandrel along a predescribed path. Also preimpregnated tapes can be used as an alternative. After the wound stage, the composite is cured by heating at a given temperature in an oven or autoclave or by exposure to IR radiation, and the mandrel is removed. Typical products range from a simple pipe to an aircraft fuselage, while typical materials include glass, carbon or aramid fibers coupled to polyester, vinyl ester or epoxy resin. The principal advantage of filament winding over other methods for composite fabrication is the possibility of adopting automation and robotic procedures. The greatest disadvantage is the geometric limitation of available tools, including the inability to wind on negatively curved (concave) surfaces. Also in filament winding the processing behavior is strongly dependent on resin characteristics. The final fiber content is a function of the radial motion of the fiber with respect to the resin during winding. This motion is a consequence of the forces acting on the fibers, *i.e.* the imposed tension and the friction between the fiber and the resin (which is a function of matrix viscosity).

18.3.2.5 Physicochemical behavior

Although the described technologies have very different characteristics, the behavior of the thermoset matrix can be described by applying the same fundamental principles that are discussed in this section. During the processing of thermoset based composites, shaping operations are accompanied by polymerization reactions (curing) and rheological changes of the matrix that strongly influence the final properties and the quality of the composite part. Moreover, the cure process is not only associated with significant variations in the material viscosity, but is also coupled with a strong heat generation due to the exothermic nature of the thermosetting reactions. The relative rates of heat generation and transfer determine the values of the temperature, and therefore, the values of the advancement of the reaction and the viscosity through the thickness of the composite.

An uncontrolled polymerization reaction may cause undesired and excessive thermal and rheological variations that could induce microscopic defects in the network structure of the matrix phase, such as different crosslinking densities and macroscopic defects such as voids, bubbles and debonded and broken fibers.[59,65] Processing of polymeric composites based on thermoset matrices needs, therefore, optimization of the cure cycle parameters, as well as adequate formulation of the reacting system as a function of the geometry of the part. The analysis presented in this review is based on the production of carbon fiber/epoxy matrix laminates for aeronautic applications. However, the approach used is completely general, and can be easily applied to other processing technologies for thermoset composites, such as resin transfer-like molding, pultrusion and filament winding.

As discussed before, the role of the matrix on the processing behavior of high performance composites fabricated in the autoclave process will be analyzed in detail, as a case study, in this section.

The principal physical events occurring during a typical cure cycle are illustrated in Figure 40.[61,63,64] The viscosity initially decreases with time as the temperature is increased, but the reactions are still not activated. During this first stage, sorbed volatiles start to diffuse out of the resin. When the viscosity reaches its minimum trapped bubbles must be allowed to leave the composite. Pressure, which drives the resin flow, should be correctly applied over a limited viscosity range in order to properly remove the excess resin and trapped bubbles, and to consolidate the plies. Figure 40 qualitatively shows the viscosity limits (A and B) that must be matched by the resin. The lower limit (B) is imposed by the flow characteristics of the system to avoid an excessive loss of resin from the composite and to ensure the flow forces necessary to mobilize the bubbles. A too viscous liquid (upper limit A), on the other hand, does not allow a sufficient flow of resin and consolidation of the system. Other important physical events, such as void nucleation and diffusion and desorption of volatiles, must be considered.

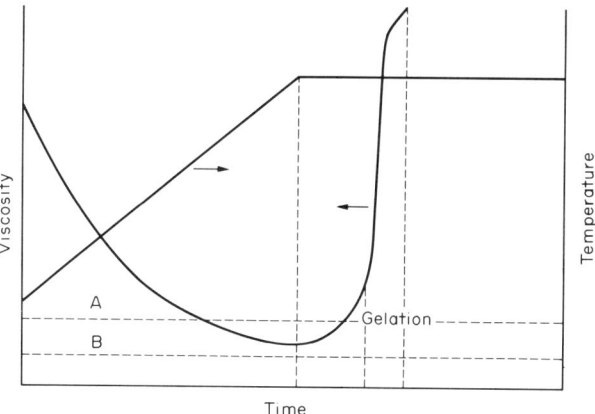

Figure 40 Temperature and pressure as a function of time for a typical TGDDM–DDS prepreg cure cycle and the corresponding change in viscosity. Lines A and B correspond to the maximum and minimum viscosity that are allowed at the low viscosity part of the cure cycle (after Kenny et al.[2])

18.3.3 Modeling of Composites Processing

As mentioned before, in the early 1980s, research activities on process modeling started in the field of processing of thermosets and thermoset matrix composites.[59–66] The final objective of these activities has been the construction of a general processing model that could be adapted to different specific processes. In order to develop such general model several submodels are needed, as shown in Figure 41. The first submodel should describe the kinetics of the matrix chemical transformations, responsible for the final structure of the composite. The thermokinetic model predicts the exothermal heat of reaction and the degree of cure as a function of process time and temperature.

Also needed is a rheological model to describe the viscosity evolution as function of time and temperature. Viscosity will also depend on the degree of cure. Therefore, the rheological model has to be combined with the thermokinetic model. Together they form the chemorheological model, allowing the prediction of the viscosity of the reactive matrix as a function of the degree of cure and of the temperature during the polymerization process. The chemorheological model includes also the prediction of time required for the resin to reach the gelation point once the processing conditions have been fixed.

A flow model is the third submodel needed. This model should be able to predict resin content distribution and final composite thickness.

Since the final objective of this approach is concerned with the cured composite characteristics, void formation should also be included in the modeling effort. The void model should ideally be able to predict the conditions needed to avoid the formation of voids. In addition, it is also of

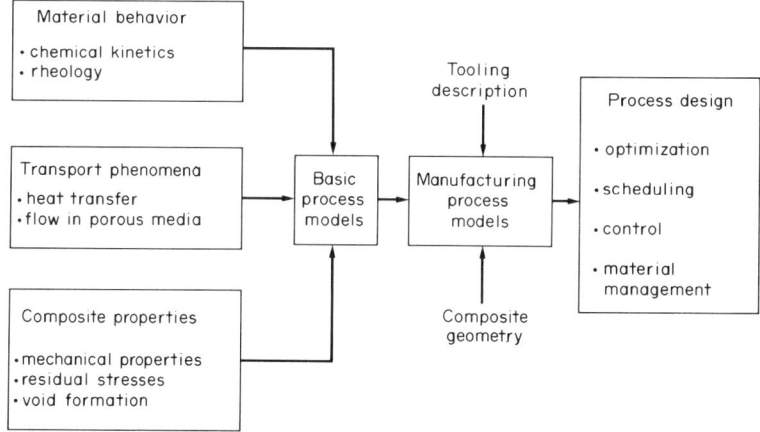

Figure 41 Schematic diagram of the constituent submodels in a general composite processing model

interest for the prediction of the volume fraction and of the size distribution of voids in the cured composite, for specified processing conditions.

For nonisothermal conditions a heat transfer model is also needed. If the heat transfer model is combined with the chemorheological and cure kinetics models, the degree of cure, temperature and viscosity, as functions of time and position in the composite, can be predicted.

A more detailed examination of the modeling of composite processing will be given in the following sections, where each of the submodels mentioned in this section, integrated into a general master model, will be discussed.

18.3.4 Thermokinetic Model

18.3.4.1 Chemical aspects

In the cure of thermosetting polymers, a series of independent reactions occur, involving monomers and molecules composed of a sequence of two (dimers), three (trimers) and more structural units. The interlinking capacity of a monomeric molecule is described by the concept of functionality. In the case of highly branched polymers, the growth of macromolecules is not only restricted in two directions. The increasing degree of branching leads, at the gel point, to an essentially infinite network and to a dramatic increase in the viscosity. The synthesis of a thermoset needs, therefore, the presence of a certain number of units of functionality greater than two, in order to produce the conditions for the formation of an infinitely large branched structure. An outline of the mechanisms of reaction leading to the crosslinking of thermosetting polymers may be useful to enable appropriate modeling of network formation.

Epoxy resins for high performance composites are commonly based on mixtures of tetraglycidyldiaminodiphenylmethane (TGDDM) epoxy and diaminodiphenyl sulfone (DDS). Polymerization of epoxy systems follows a typical reaction between functional groups. Reported investigations[67,68] have indicated that three reactions dominate the cure behavior of the TGDDM–DDS systems; namely, epoxide–primary amine addition, epoxide–secondary amine addition and epoxide–hydroxyl etherification. The kinetic constant of the former reaction is at least one order of magnitude higher than the other two constants. For this reason the DDS molecule is considered as a bifunctional curing agent, while the TGDDM is tetrafunctional.

18.3.4.2 Reaction kinetics

The thermokinetic model describes the degree of cure, α, as a function of temperature and time, and is the first step in the construction of the master model being a prerequisite for all the other submodels. Considerable research has been devoted to the kinetic characterization of thermosetting polymers using differential scanning calorimetry (DSC).[69–71] DSC has been employed for determination of the progress of curing, by assuming that the heat evolved during the polymerization reaction is proportional to the overall extent of reaction given by the fraction of reactive groups consumed. Following this approach, the degree of reaction, α, has been defined as[69]

$$\alpha = H(t)/H_T \qquad (47)$$

where $H(t)$ is the heat developed between the starting point and a given time t, and H_T is the total heat developed, calculated by integrating the total area under the DSC curve. The reaction rate is thus given by the following expression

$$d\alpha/dt = 1/H_T \, dH/dt \qquad (48)$$

where dH/dt is the rate of generation of heat as measured directly from the DSC thermogram. This information can be processed to construct a kinetic model for α as a function of time and temperature. Although processing of thermosetting matrices involves very complex reactions, several simple equations have been proposed to describe their general behavior as an overall kinetic process, of the form

$$d\alpha/dt = K f(\alpha) \qquad (49)$$

where K is the temperature dependent rate constant and $f(\alpha)$ is a function to be determined by best fitting of the experimental results. The temperature dependence is normally considered through the rate constant K given by an Arrhenius type equation

$$K = K_0 \exp(E_a/RT) \tag{50}$$

where K_0 is the preexponential factor (frequency factor), R is the gas constant and T the absolute temperature. Physically, K is a measure of the velocity of the reaction, and E_a is the activation energy whose magnitude depends on the operating chemical reaction mechanism and on catalyst chemistry.

On the other hand, the form of $f(\alpha)$ depends on the particular reaction mechanism. The research work on reaction kinetics of thermosets has been extensively dedicated to the study of high performance epoxy systems. Horie et al.[72] used isothermal DSC to investigate the curing kinetics of epoxies with amines and concluded that the curing reaction of diglycidyl ether of bisphenol A (DGEBA) with aliphatic tetrafunctional diamines proceeds through a third order mechanism followed by a diffusion-controlled mechanism. Prime[73] employed an empirical single nth order equation to fit the isothermal DSC data for DGEBA resin cured with amines.

$$d\alpha/dt = K(1-\alpha)^n \tag{51}$$

Souror and Kamal[74] assumed that both primary amine hydrogen atoms have equal reactivity so that the total rate of consumption of epoxies could be expressed by modifying the model of Horie et al.[72]

$$d\alpha/dt = (K_1 + K_2\alpha)(1-\alpha)(B-\alpha) \tag{52}$$

The model fitted the isothermal experimental data, except in the later stages where the reaction mechanism is diffusion controlled. Ryan and Dutta[75] have proposed a method for estimating the parameter values of the following kinetic expression applied to the reaction kinetics of epoxy systems

$$d\alpha/dt = (K_1 + K_2\alpha^m)(1-\alpha)^n \tag{53}$$

The TGDDM–DDS system has been the subject of specific research.[76-80] Stark et al.[76] used Prime's model (equation 51) to analyze the reaction kinetics of high performance commercial TGDDM–DDS matrices. They found that the kinetic parameter values determined by dynamic DSC depend on the heating rate of the scan, though their direct determination was prevented by the complexity of the reaction mechanism. Mijovic et al.[77] used equation (53) to describe the autocatalytic behavior of the polymerization reaction of commercial TGDDM–DDS formulations. Isothermal DSC data were employed to calculate the model parameters and check the kinetic model. The influence of the reactive kinetics of TGDDM–DDS matrices (equation 51) on the chemorheological behavior of commerical prepregs in the autoclave process has been described by Kenny et al.[2]

Moreover, Kenny and Trivisano have recently proposed[71] a new model for the kinetic behavior of TGDDM–DDS systems that takes into account the later diffusion-controlled effects, and they have analyzed the correlation between isothermal and dynamic DSC results by testing the kinetic model under the complex thermal conditions characteristic of the processing of epoxy-based composites. Normally, isothermal tests are performed at different temperatures to obtain a complete DSC characterization. It has been reported that the total heat of reaction values calculated by integration of the isothermal thermograms (H_i), are significantly lower than those obtained in dynamic tests (H_T), since during low temperature isothermal tests polymerization is not completed and the final α value is an increasing function of the test temperature. This behavior is illustrated in Figure 42 for a commercial high performance epoxy system.

As mentioned earlier, incomplete reaction during isothermal processes is attributed to diffusion control, owing to the loss of mobility of the reacting molecules within the developed network. Structural changes produced by polymerization are associated with an increase of the glass transition temperature, T_g, of the reactive polymer. When the T_g approaches the isothermal cure temperature, mobility is strongly reduced. When the system reaches vitrification, the reaction becomes diffusion controlled and eventually ceases, or perhaps decreases to a minimum value.[71,78,79] For practical purposes, however, considering the time scale associated with the normal processing of thermoset based materials, it can be assumed to stop. This assumption means that equation (51) cannot be used to interpret isothermal results, since it predicts that the reaction rate becomes zero when $\alpha = 1$, i.e. when the system is fully cured. The model can be modified mathematically in two ways in order to overcome this limitation. The first way is to modify K when diffusion control becomes dominant. Here the overall reaction rate is a function of temperature, concentration of reactants, reaction mechanism, and the local viscosity (which in turn is a function of the molecular weight and temperature). Then, equation (49) can be rewritten as

$$d\alpha/dt = K_0 f(T) f(\alpha) f(\mu_L) \tag{54}$$

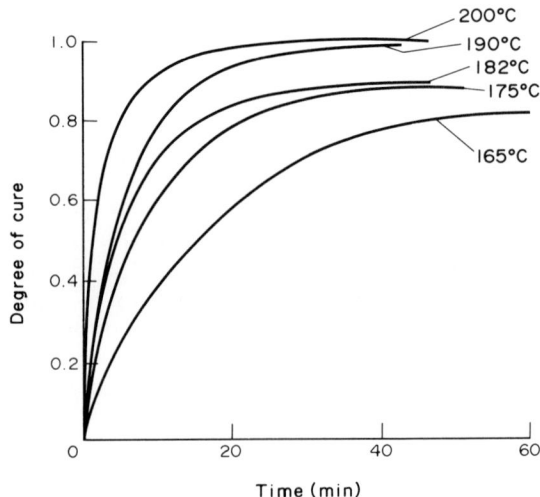

Figure 42 Degree of cure as a function of time for a toughened TGDDM–DDS epoxy system cured isothermally at different temperatures (after Kenny et al.[71])

where K_0 is the preexponential factor of the rate constant, $f(T)$ is a function of temperature, generally given by an Arrhenius type equation, $f(\alpha)$ is unchanged, and $f(\mu_L)$ is a function of the local viscosity. Diffusion control, manifested through dependence of the reaction rate on the local viscosity, is sometimes considered by including a conversion dependent term in the rate constant.[80,81] On the other hand, equation (51) can also be modified empirically to provoke zero prediction of the reaction rate at the vitrification point. Equation (51) can be rewritten to allow the behavior represented by equation (54)

$$d\alpha/dt = K(\alpha_m - \alpha)^n \qquad (55)$$

This clearly predicts the expected behavior: the reaction rate during an isothermal process will be zero when the degree of reaction equals α_m. The average value of the total heat H_T developed in dynamic tests was used as a reference to determine the final degree of reaction during the isothermal tests ($\alpha_m = H_i/H_T$). For modeling purposes, it is convenient to determine the behavior of α_m as a function of the isothermal test temperature. The linear dependence of α_m on T illustrated in Figure 43 recalls the dependence between T_g and α for a reactive polymer. As discussed above, the T_g value reached by the polymeric matrix can be assumed to be of the same order as the isothermal test temperature. It has been found that a simple linear dependence expresses the empirical dependence of α_m on T

$$\alpha_m = pT + q \qquad (56)$$

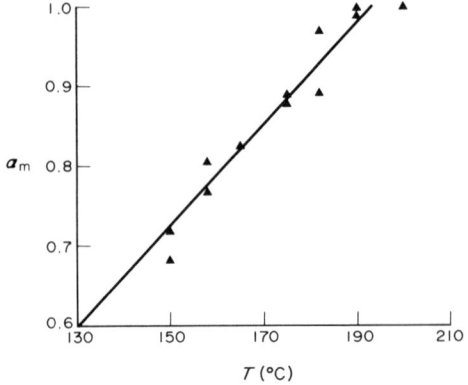

Figure 43 Maximum degree of cure obtained in isothermal experiments (reported in Figure 42) *versus* isothermal test temperature (after Kenny et al.[71])

Table 8 Kinetic Parameters of the Model Represented by Equations (55) and (56), Obtained from the Thermal Characterization of a Typical TGDDM–DDS Matrix Prepreg[71]

H_r (kJ g^{-1})	E_a (kJ mol^{-1})	$\ln K_0$ (s^{-1})	n	q	p (K^{-1})
456	62.4	10.4	1.07	−1.96	0.00635

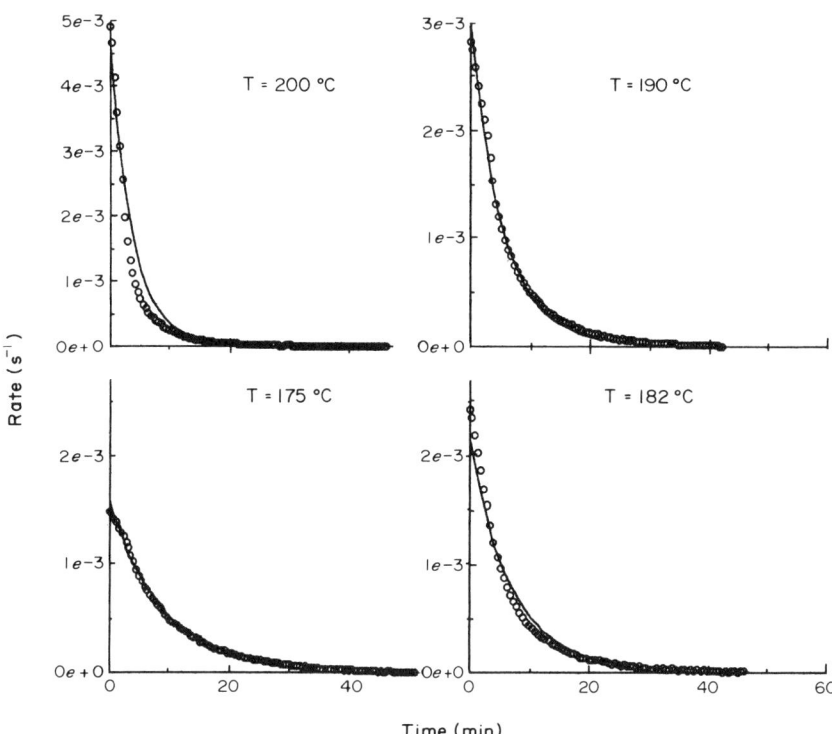

Figure 44 Reaction rate *versus* time for different temperature isothermal tests: comparison between experimental DSC data (points) and model predictions (full lines) (after Kenny et al.[71])

The values of the kinetic parameters of the general model given by equations (55) and (56) are listed in Table 8. The ability of the model to represent the kinetic behavior has been tested by comparison with experimental results.[71] Reaction rate data plotted as a function of time, obtained from isothermal DSC experiments and from equation (56) predictions, are shown in Figure 44. Theoretical curves (full lines) were computed using the average parameter values listed in Table 8. The good fit corroborates the soundness of the procedure used to formulate the model.

The shape and parameters of the equations (55) and (56) model were obtained from isothermal tests. The complete model, however, should also describe dependence of the reaction rate on temperature during a dynamic test. Therefore, the complete model was tested by comparison with experimental dynamic thermograms. Close fits can be seen in Figure 45, where the reaction rate is shown as a function of time for different heating rates. Dotted lines correspond to DSC experimental results, full lines correspond to model predictions from the values in Table 8.

Practical employment of the model was investigated through a mixed isothermal–dynamic DSC experiment on the prepregs with the same processing conditions as used in autoclave. The results expressed as the reaction rate as a function of time are shown in Figure 46 (dotted line). The left part of the figure corresponds to the dynamic process performed at a specified heating rate, the right to the isothermal process. These processes were also mathematically simulated by applying the developed model (Figure 46, full line). The excellent agreement obtained between the experimental and model results indicates that the model can reproduce and predict the behavior of epoxy matrices during simple isothermal and dynamic tests, and also under more complex conditions, such as those characteristic of autoclave lamination. However, the application of the

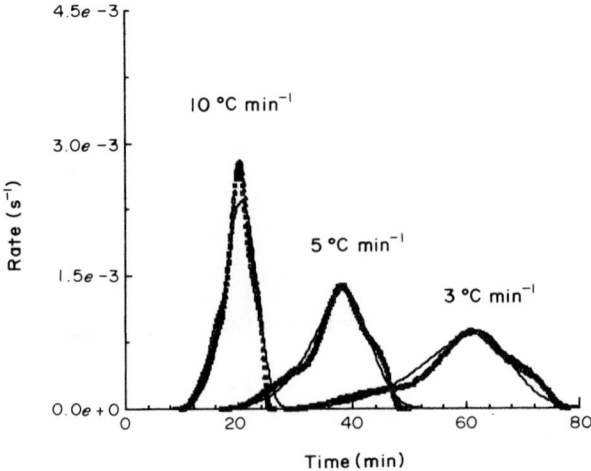

Figure 45 Reaction rate *versus* time for different heating rate dynamic tests: comparison between experimental DSC data (points) and model predictions (full lines) (after Kenny et al.[71])

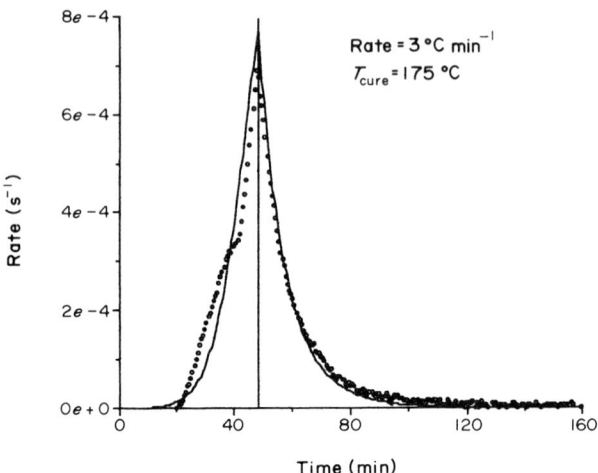

Figure 46 Reaction rate *versus* time for cure simulation: comparison between experimental DSC data (points) and model predictions (full lines) (after Kenny et al.[71])

model to industrial processes should also consider the complex thermal processes associated with the polymerization reaction as a consequence of the geometry of the laminate, and of the heat transfer from the air and tool to the prepregs. In this case, a simple scaling of DSC results is not sufficient, and the kinetic model must be included in a general model considering the characteristics of the lamination process.[2]

A practical application of the thermokinetic model is the prediction of the cure time at a given temperature to reach a specified degree of cure, for example full cure. If the degree of cure at the gel point has been determined, for instance by viscosity measurements coupled with DSC, the gel point can be predicted since it is a constant conversion event (it occurs at the same degree of cure, independent of temperature). This, of course, is only true as long as the reaction mechanism is unchanged. The thermokinetic model can be also used, associated with the chemorheological model, to describe the evolution of the polymeric structure and its viscosity. Finally, the thermokinetic model also predicts the amount of heat given off during the reaction. Then, if it is coupled with a heat transfer model (as described later), the temperature and degree of cure distribution inside a composite can be described for nonisothermal conditions (thick composites).

18.3.5 Chemorheology

18.3.5.1 General aspects

The processing and final properties of thermosets depend on their composition as well as on the network structure generated before the gel point. The viscosity reflects the molecular distribution and can be considered to be one of the most important properties in polymer processing. An accurate prediction of the material properties, such as the polymerization kinetics and related changes in viscosities, implies the knowledge of the basic phenomena occurring during the overall process.

Generally, the cure of a low molecular weight prepolymer involves the transformation of a fluid resin into a rubber and then into a solid glass. This is a result of the chemical reactions of the reactive groups present in the system, which develop a progressively denser polymeric network. The growth and branching of the polymeric chains are due to intramolecular reactions that initially occur in the liquid state until a critical degree of branching is reached and an infinite network and an insoluble material is formed. After gelation, successive crosslinking reactions increase the crosslink density of the network and the stiffness of the polymer is steadily increased, leading, at the end of the process, to the glassy structure of a fully cured thermoset. In thick composites, different physical states (namely liquid, rubber and gelled or ungelled glasses) can be reached at the same time in different sections as a function of temperature and degree of cure.

From the physical point of view, two major events must be analyzed in relation to their technological influence: gelation and vitrification. The gelation event is of great technological importance since after that no flow is possible and void diffusion and further consolidation of the composite can no longer occur. Gelation is a constant conversion event, meaning that irrespective of temperature, gelation always occurs at the same degree of cure. This can be determined theoretically from Flory's branching theory,[82] provided stepwise polymerization is the only mechanism and the functionalities of the reactive molecules are known.

The second physical event of great importance is vitrification. As the curing reaction progresses, the mobility of the participating molecules is steadily decreased, raising the T_g of the system. If the T_g approaches the curing temperature the reaction stops before completion. The principal physical phenomena occurring during the cure of thermoset matrices can be schematically observed in Figure 47, where the time–temperature–transformation diagram (TTT), first presented by Gillham,[83] is shown. This diagram, similar to the TTT diagrams used for metals, has been constructed from torsional braid analysis measurements and illustrates transformations in thermosets and how they occur as a function of time and temperature. The TTT diagram provides an intellectual framework for the interpretation of thermoset cure behavior. T_{g0} is the glass transition temperature of the unreacted resin mixture. T_{ggel} is the temperature at which gelation and vitrification occur simultaneously. $T_{g\infty}$ is the glass transition temperature for a fully cured thermoset. In the diagram, lines for gelation and vitrification are given (at very low temperatures, vitrification may actually occur before gelation). At high temperatures, degradation may occur and 'char' is obtained. The solid lines are isoviscous lines computed using the WLF equation.[84] As can be observed, curing

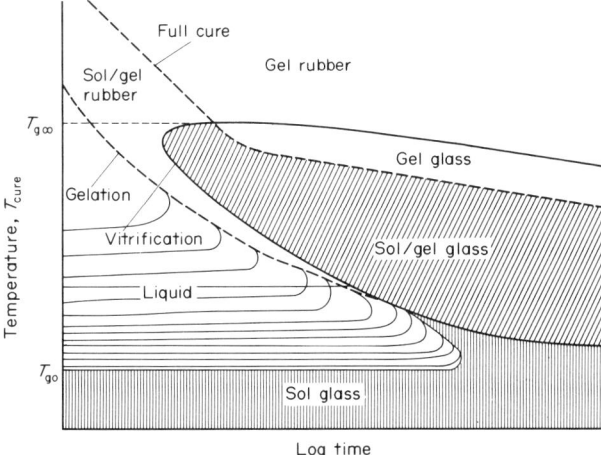

Figure 47 Time–temperature–transformation (TTT) diagram of thermoset cure (after Gillham et al.[83])

at temperatures significantly lower than $T_{g\infty}$ gives rise to a glassy thermoset which, although it appears fully cured, still has unreacted molecules (sol/glass). A safe way to achieve full cure is to cure above $T_{g\infty}$. However, for some thermosets this is not possible, since thermal degradation occurs at temperatures below $T_{g\infty}$.

The viscosity model should be able to predict the viscosity at any combination of temperature and time. Since the viscosity depends on the degree of cure, the viscosity model has to be combined with the thermokinetic model to really become useful. The combination of the thermokinetic and the viscosity model describes the chemorheological behavior of the thermoset.

18.3.5.2 Determination of the gelation limits

Gelation, associated with a dramatic increase in viscosity, occurs at a degree of reaction that can be derived by the theory first proposed by Flory.[82] In order to determine the conditions for the formation of an infinitely large branched structure, Flory introduced a convenient quantity, the branching coefficient (β), defined as the probability that a given reactive group of a branched unit of functionality greater than two is connected, *via* a chain of bifunctional units, to another branched molecule. Generally, if P_e and P_a are the fraction of the resin and hardener active groups which have reacted, respectively, then the branching coefficients is given by

$$\beta = P_e P_a = r P_a^2 = P_e^2/r \tag{57}$$

where r is the ratio of the hardener (m_a) to resin (m_e) groups initially present in the reactive mixture

$$r = m_a/m_e = f_a M_a / f_e M_e \tag{58}$$

where f_e and f_a are the functionalities and M_e and M_a are the moles of the epoxy resin and hardener molecules. A convenient critical condition describing the incipient formation of an infinite network is derived by the theoretical dependence of the molecular weight on the branching coefficient. In fact, when the branching coefficient gets its critical value (β_c), the network leads to an infinite molecular weight

$$\beta_c = 1/[(f_e - 1)(f_a - 1)] \tag{59}$$

or, referring to a function of the reagent degree of conversion when

$$P_{ec}^2 = r/[(f_e - 1)(f_a - 1)] = M_a f_a / [M_e f_e (f_e - 1)(f_a - 1)] \tag{60}$$

The value of P_{ec} is the critical conversion of the epoxy groups at the gel point.

18.3.5.3 Viscosity modeling of thermosetting reactions

The processing and final properties of thermosets depend on their composition as well as on the network structure generated before the gel point. The rheological behavior of a reacting system is governed by two effects: the first related to the molecular structural changes induced by the cure reactions, and the second associated with the variation of the segment mobility determined by temperature variations. Several authors have considered theoretically the rheological behavior of a curing thermoset through proposed models of different levels of complexity. Reviews on the chemorheology of thermosets are available in the literature.[85,86]

One possible approach is to model the chemorheological behavior of thermoset matrices by relating the viscosity changes to the fundamentals of the molecular structure and development and to the thermal history of the system. At a fixed temperature, any variation in viscosity is determined by the changes in the molecular distribution. For a linear polymer, the viscosity is related to the molecular weight[87] by the equation

$$\mu = K M_w a \tag{61}$$

where $1 < a < 2.5$ for M_w lower than a critical value, and $a = 3.4$ for M_w higher than the critical value. A similar relationship has been proposed by Valles and Macosko[88] for a nonlinear copolymerization such as that considered here

$$\mu = K (g M_w)^{3.4} \tag{62}$$

where K is a constant, M_w is the weight-average molecular weight of the polymer and g is the ratio of the radii of gyration of a branched chain to a linear chain of the same molecular weight. The relationship derived by Stockmayer[89] for three-dimensional condensation polymers is applied here for the computation of the M_w value to use in equation (61). In the case of the simplest condensation reaction, M_w becomes

$$M_w = \frac{(M_e M_{we}^2 + M_a M_{wa}^2 + Z)}{(M_e M_{we} + M_a M_{wa})} \quad (63)$$

where M_{we} and M_{wa} are the initial molecular weight of the reagents and Z is given by

$$Z = \frac{P_e f_e M_e [P_e(f_e - 1)M_{me}^2 + P_a(f_a - 1)M_{wa}^2 + 2M_{we}M_{wa}]}{[1 - P_e P_a (f_e - 1)(f_a - 1)]} \quad (64)$$

The gel point will occur when the molecular weights becomes infinite, i.e. when the denominator of the last equation becomes zero

$$P_{ec} = [r/(f_e - 1)(f_a - 1)]^{1/2} \quad (65)$$

This result coincides with the condition expressed by Flory in equation (60).

The value of g in equation (62) is related to the reagent functionality and branching coefficient.[90] For $f_e = 4$, g becomes

$$g = [(1 - 3\beta)/2\beta] \ln[(1 - \beta)/(1 - 3\beta)] \quad (66)$$

On the other hand, the influence of the temperature on the viscosity for systems of constant glass transition and fixed branching coefficient (i.e. for a nonreacting system such as a TGDDM–DDS epoxy system at $T < 120\,°C$) is described by the William, Landel and Ferry (WLF) equation[91]

$$\ln[\mu(T)/\mu(T_r)] = \frac{-C_1(T - T_g)}{(C_2 + T - T_g)} + \frac{C_1(T_r - T_g)}{(C_2 + T_r - T_g)} \quad (67)$$

where T_g is the glass transition temperature of the system of fixed branching coefficient and T_r a reference temperature. Then, the viscosity changes in a reacting system continuously heated up to the final cure temperature can be derived by considering both the temperature dependence of the viscosity, according to the WLF equation where the glass transition temperature of the system is a function of the branching coefficient, and the molecular weight dependence of the viscosity. One of the main assumptions of this approach is represented by the equivalence between the extent of reaction of the epoxy groups, P_e, and the degree of reaction measured from calorimetric results, α, considering that all the epoxy groups initially present in the system react with the amine groups in excess. Then the final chemorheological model can be written in the following way

$$\mu(T, \alpha)/\mu(T_0) = \frac{gM_w(\alpha)^{3.4}}{M_{w0}} \frac{\exp\{C_1(T_r - T_{g0})/(C_2 + T_r - T_{g0})\}}{\exp\{C_1(T_r - T_g(\alpha))/(C_2 + T_r - T_g(\alpha))\}} \quad (68)$$

The rate of variation of the glass transition temperature with the degree of cure can be experimentally determined by differential scanning calorimetry, and is illustrated in Figure 48.

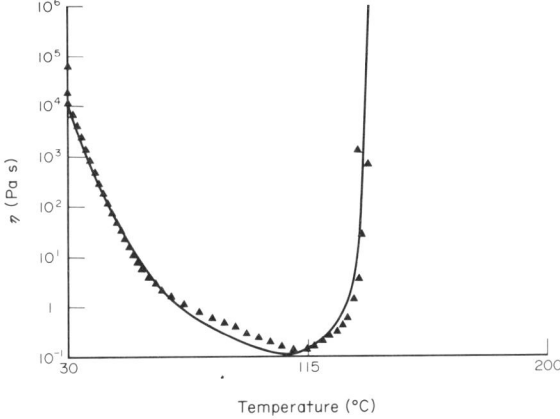

Figure 48 Relative viscosity versus time for a typical TGDDM–DDS system during a dynamic test: comparison between experimental data (points) and model predictions (full lines) (after Kenny et al.[2])

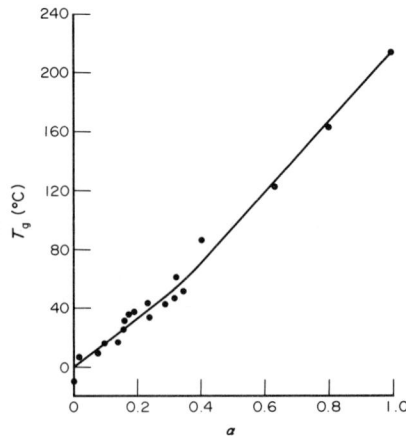

Figure 49 Glass transition temperature, T_g, *versus* degree of cure, α, for a typical TGDDM–DDS system (after Kenny et al.[2])

Theoretical models accounting for the influence of the degree of cure on the polymer glass transition can be also used[92]

$$(T_g(\alpha) - T_{g0})/T_{g0} = K\alpha/(1-\alpha) \tag{69}$$

where K is a constant reported to be between 1.0 and 1.2.[92]

Viscosity measurements are preferably performed in a dynamic mechanical spectrometer for fluids provided with disposable parallel plates. The location of the gelpoint can be determined from infinite viscosities obtained in constant shear or in dynamic tests.[86] Figure 49 shows model and experimental values of the viscosity as a function of time for a typical commercial epoxy system. Excellent agreement was found in the first part of the curve (unreacted resin) where the viscosity depends only on the temperature and is described by the unmodified WLF equation. Also, the minimum of the viscosity and the gelation limit are well described by the proposed model, which gives an important tool to predict and/or control the rheological behavior during processing of epoxy laminates. The difference between the numerical and experimental results can be mainly attributed to the approximations in the simple kinetic equation utilized.

For thermosets with complicated reaction mechanisms or where the composition and functionalities of the molecules in the resin mixture are unknown, an empirical approach is necessary. In order to describe the viscosity of polyester matrices Kenny et al.[3] adopted a model similar to the one originally used by Castro and Macosko for polyurethane viscosity[66]

$$\mu(\alpha, T) = A_\mu \exp[E_\mu/RT][\alpha_g/(\alpha_g - \alpha)]^{(A+B\alpha)} \tag{70}$$

where α_g is the extent of reaction at the gel point and A_μ, E_μ, A and B are constants that have to be determined by regression analysis of experimental data.

18.3.6 Heat Transfer

18.3.6.1 General aspects

For isothermal processing conditions, the models presented so far will give all the information needed to predict the behavior of the material at a given processing temperature. However, as discussed earlier, the cure of a thermoset is always associated with a significant development of heat. The temperature distribution inside the composite will depend on the competition between heat generation and heat diffusion through the thickness. While for thin laminates it can be assumed that the predominant heat diffusion will favor isothermal conditions, for thicker laminates the conditions will clearly be nonisothermal. The reacting system must be viewed as a nonisothermal bulk reactor with volumetric heat generation and transfer for the initial heating and for the dissipation of the heat of reaction. The temperature and degree of cure profile inside the composite can be computed taking into account the system geometry, the thermal diffusivity of the composite, and the resin reaction rate. Then, a heat transfer model for the prediction of the temperature distribution in the composite during curing must be provided. Material properties, boundary and

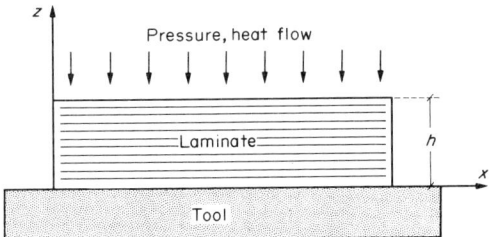

Figure 50 Schematic diagram of laminate geometry and reference axes

initial conditions (*e.g.* temperatures) and the thermokinetic behavior are the input data of the heat transfer model. Indeed, the kinetic and heat transfer models are coupled and a numerical solution method must be used.

During cure, heat is generated in the material due to the exothermic nature of the curing reaction. For thick laminates, not all the heat can be dissipated fast enough to maintain isothermal conditions. The processing conditions and the material characteristics will determine whether the highest temperature is in the core, at the skin or at some intermediate position. Several authors have described how the temperature inside the composite can be calculated.[2,60,93] This can be done by solving the energy balance, together with an appropriate expression for the cure kinetics. The following assumptions are normally introduced.

(i) The simple case of absence of resin flow is considered and then only heat transfer by conduction is assumed.

(ii) The laminate thickness is small compared to the other two dimensions. Then, only conduction of heat in the transverse direction is considered. The geometry of the system is shown in Figure 50.

(iii) The density ρ and the specific heat C_p are computed as proper averages of the single resin and fiber property values

$$\rho = (1 - V_f)\rho_r + V_f \rho_f \tag{71}$$

$$C_p = (\rho_r/\rho)(1 - V_f)C_{pr} + (\rho_f/\rho)V_f C_{pf} \tag{72}$$

where V_f is the volume fraction of fibers and the subscripts r and f refer to the resin and the fiber, respectively.

(iv) The composite thermal conductivity, k, in the direction perpendicular to the plane of the laminate can be computed using the mechanical analogy reported by Halpin[16]

$$k = k_r(1 - B_1 B_2 V_f)/(1 - B_1 V_f) \tag{73}$$

$$B_1 = (k_f/k_r - 1)/(k_f/k_r + B_2) \tag{74}$$

$$B_2 = 1/[4 - 3(1 - V_f)] \tag{75}$$

(v) Experimental evidence,[2,3] moreover, suggests that no variation of these properties with the temperature and/or degree of cure should be considered. The structure of the model, the resolution and the results are not significantly affected by this assumption.

With these assumptions the law of conservation of energy takes the form

$$\rho C_p \delta T/\delta t = k \delta^2 T/\delta z^2 + \rho \, dH/dt \tag{76}$$

where dH/dt is the rate of the heat generated by the chemical reactions and is defined, from equation 48, as follows

$$dH/dt = d\alpha/dt \, H_T \tag{77}$$

where $d\alpha/dt$ can be obtained from the thermokinetic model, using, for example, equations (51), (52), (53) or (56). The solution of the heat transfer equation can be obtained once the initial and boundary conditions have been determined.

18.3.6.2 *Dimensionless model*

Kenny *et al.*[2] introduced dimensionless numbers to facilitate the numerical solution and to generalize the model to other processing technologies. A characteristic time of the material behavior is defined: the isothermal gel time, t_g, represents the time interval in which the material is changing from the liquid to the rubbery state. Taking the imposed processing temperature T_e as the reference temperature, t_g is obtained by integrating the kinetic model. If equation (51) is considered, the following equation for t_g is obtained

$$t_g = [(1 - \alpha_g)^{(1-n)} - 1]/[(n - 1) A \exp(- E_a/RT_e)] \tag{78}$$

The dimensionless variables of the model are defined as

$$\theta = (T - T_0)/(T_e - T_0) \quad t^* = t/t_g \quad Z = \frac{z}{(h/2)} \tag{79}$$

The final equations then become

$$\delta\theta/\delta t = \text{De}\, \delta^2\theta/\delta Z^2 + \text{St}\, d\alpha/dt^* \tag{80}$$

$$d\alpha/dt^* = A \exp[E(\theta - 1)(T_e - T_0)/(\theta + T_0)](1 - \alpha^n) \tag{81}$$

In equation (80) De is a modified dimensionless diffusion Deborah number. The relevance of a diffusion Deborah number in problems where morphological changes are induced in a polymer by either heat or mass transfer phenomena has been recognized for a long time. The concept was first introduced by Duda and Vrentas.[94] Astarita and Kenny[95] have reviewed the use of the diffusion Deborah number and have studied its influence in polymer crystallization problems. The concept is related to the relaxation phenomena associated with morphological changes in solid polymers. In this case a relaxation time can be defined as a yardstick of the natural time scale for the morphological change. The ratio of such a time scale to the characteristic time for diffusion is the Deborah number. In the case of thermoset processing the morphological change in the polymeric matrix is associated with the transition originated by the crosslinking reactions, and the time scale is characterized by the isothermal gel time t_g. The heat diffusion time scale is the square of the characteristic dimension of the laminate, $h/2$, divided by the heat diffusivity $k/\rho C_p$. Hence the Deborah number is expressed as

$$\text{De} = K\, t_p/\rho C_p (h/2)^2 \tag{82}$$

The Deborah number also represents the relationship between the heat transferred by conduction and the accumulation of heat in the material. Owing to the high conductivity of the carbon fibers and the small thickness of the laminates, typical De numbers for high performance epoxy laminates are of the order of 10.[2]

In equation (80) the Stefan number, St, is also introduced

$$\text{St} = H_T/[(T_e - T_0)C_p] \tag{83}$$

and may be considered to be the relationship between the latent heat associated with the chemical reaction and the accumulation of heat in the material. Typical Stefan numbers for this process are on the order of 1.[2] The influence of the Deborah and Stefan numbers on thermoplastics processing has been reported in the literature.[95] Finally, the dimensionless kinetic constant, A, and activation energy, E, were used in equation (81)

$$A = [(1 - \alpha_g)^{(1-n)} - 1]/(n - 1) \tag{84}$$

$$E = E_a/RT_e \tag{85}$$

The advantage of dimensionless numbers can be summarized as follows. A simpler numerical method can be used when the concept of dimensionless numbers is applied. Since first order parameters are used to a larger extent with this approach, the problems associated with convergence and stability become small. But perhaps the most attractive advantage is that by changing the dimensionless numbers one order of magnitude at a time, say from 1 to 10 to 100, effects of four parameters at a time can be observed. It is, for instance, easy to observe effects of conductivity (carbon fiber compared with glass fiber reinforcement), material thickness (thick *versus* thin laminates) and characteristic time (comparison between epoxy and polyester or different epoxies). Also the same model can easily be extended to model pultrusion, RTM and filament winding.[3,96]

According to the material and geometrical characteristics of the laminates, there are some cases in which the heat transfer and the kinetic behavior can be described by more simple models. Broyer and Macosko[97] have shown that, if the thickness of a thermosetting material is increased above a critical value, the imbalance between the heat generation and the thermal diffusivity of the system during processing leads to a virtually adiabatic situation. Following their approach a laminate presents adiabatic-like behavior when the half thickness is greater than

$$h_{ad} = [k/\rho C_p K_p \exp(-E)]^{1/2} \qquad (86)$$

and the maximum temperature rise in this case is given by

$$\Delta T = H_T/C_p \qquad (87)$$

On the other hand, isothermal-like conditions can be derived from the energy balance assuming a quasi-stationary state and a maximum temperature difference between the core and the skin of 1 °C. The laminate half thickness which fulfils these requirements is

$$h_{ts} = [4k/\rho H_T K_0 \exp(-E))]^{1/2} \qquad (88)$$

For carbon–epoxy composite laminates, typical values of the critical half thickness for both conditions are

$$h_{ad} > 1.0 \text{ cm} \qquad (89)$$

$$h_{ts} < 0.1 \text{ cm} \qquad (90)$$

The third situation ($h_{ad} < h < h_{is}$) is characterized by a nonuniform distribution of the temperature inside the laminate and must be described by solving the described heat transfer model.

18.3.6.3 Application of processing models

The usefulness of processing models may be better appreciated by the presentation of a few applications. As described in Section 18.3.5, the chemorheological model by Apicella et al.[2,86] was able to model accurately the viscosity as a function of time for an actual autoclave cure cycle. Since the heat transfer model gives temperature as a function of position and time, the viscosity gradient through the thickness can also be described for thick laminates.

Numerical results reported[2] on a typical TGDDM–DDS matrix laminate, assuming that the prepregs are suddenly exposed to the cure temperature, are shown in Figure 51 in the variation of

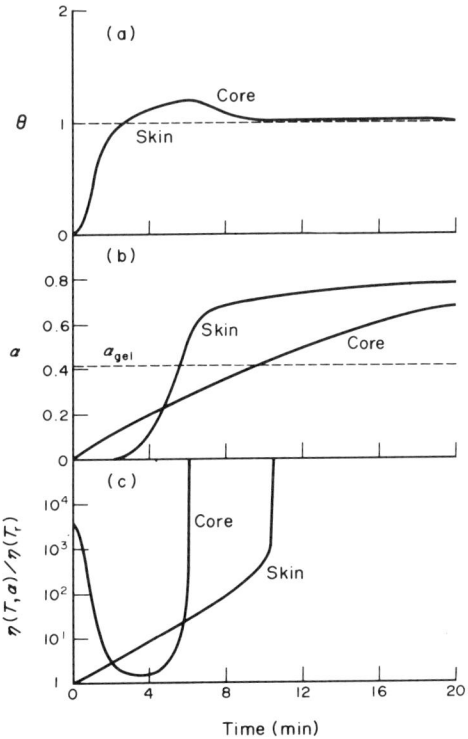

Figure 51 Dimensionless temperature (a), degree of reaction (b), and viscosity (c), *versus* processing time, computed at the skin and at the center of a typical epoxy matrix/carbon fiber laminate of 7 mm half thickness, cured isothermally at 177 °C (after Kenny et al.[2])

Table 9 Parameters of the Thermochemorheological Model of the Autoclave Lamination Process[2]

Heat of reaction (J g^{-1})	H_T	473
Activation energy (kJ mol^{-1})	E_a	138
Kinetic constant (s^{-1})	ln K_0	30.0
Reaction order	n	1.7
1st constant of equation (68)	C_1	40.5
2nd constant of equation (68) (K)	C_2	52
Epoxy molecular weight	M_{we}	422
Amine molecular weight	M_{wa}	218
Epoxy functionality	f_e	2
Amine functionality	f_a	4
Epoxy content	M_e	1
Amine content	M_a	1
T_g for the unreacted polymer (K)	T_{g0}	273
Composite density (kg m^{-3})		1500
Thermal conductivity (W m^{-1} K^{-1})	k_x	0.4
Specific heat (J g^{-1} °C^{-1})	C_p	1.67

the temperature, degree of reaction and viscosity as a function of the processing time, both on the skin and on the core of the laminate. Input data of the full model are given in Table 9.[2] Due to the contribution of the thermal conductivity of the fibers the temperature at the center of the laminate rapidly reaches the external imposed temperature and increases as a consequence of the imbalance between the rate of heat generation and the thermal diffusivity of the composite (Figure 51a). When these two quantities are comparable, the temperature profile reaches a maximum. Numerical results also indicate that the temperature increases up to a maximum near the center of the laminate. This behavior is different from that described by Williams et al.[98] for unfilled thermoset resins in isothermal wall molds. The lower thermal conductivity of that system, in fact, reduced the amount of heat transferred to the center of the specimen and consequently the reaction was gradually activated starting from the skin and moving to the core. They suggested, however, that increasing the thermal diffusivity of the system, a temperature behavior similar to the one described by Kenny et al.[2] and reported here, should be observed. Then, the value and the position of the maximum temperature reached in the laminate during the isothermal cure does not only depend on the heat of reaction but is also strongly influenced by the value of the kinetic parameters and the thermal diffusivity of the system.

The values of temperature (Figure 51a) and degree of cure (Figure 51b) have been used to compute the viscosity as a function of cure time by means of equation (68) (Figure 51c). The heat accumulated at the center leads to higher reaction rates locally and, consequently, to a faster increase of the viscosity. Therefore, in most of the cases, it is observed that the gelation moves from the core to the skin. The high temperatures developed and the coexistence of gelled and ungelled regions could induce, on cooling, undesired stresses and, therefore, the adiabatic-like condition should be avoided.

In normal fabrication processes, the laminate is placed in the autoclave at room temperature and the system is heated at a controlled rate to the final processing temperature. Figure 52(a) shows for the same commercial system the variation of temperature at the center of the laminate as a function of time in the case of a typical heating rate of 3 °C min^{-1}. Once again, due to the contribution of the thermal conductivity of the fibers, the temperature in the center follows the external imposed

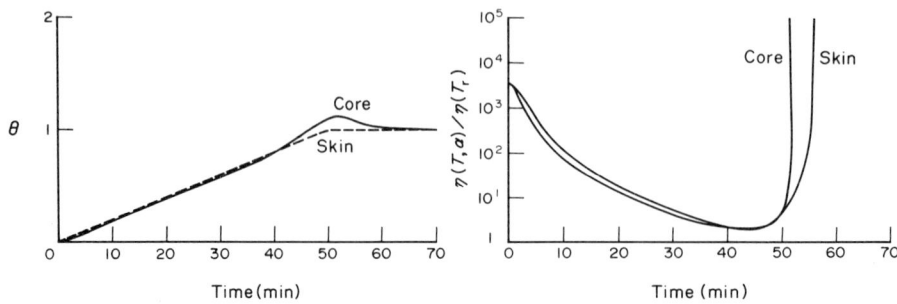

Figure 52 Modeling results for the system reported in Figure 51 cured in a nonisothermal process with a heating rate of 3 °C min^{-1} and with a final temperature of 177 °C (after Kenny et al.[2])

temperature until the reaction starts. Then the laminate shows an intense temperature maximum at the center as a consequence of the sudden acceleration of the reaction.

When the imposed autoclave temperature is increased gradually, more regular viscosity profiles are obtained (Figure 52b). Following the initial increase of the temperature, the viscosity of the unreacted resin decreases uniformly throughout the laminate, from a relatively high value at room temperature, to a minimum before the activation of the cure reactions, enabling the bubbles to leave the prepreg. Once the reaction has started, the viscosity begins to grow and, due to the higher reaction rate, the viscosity becomes higher at the center than at the external surface. The material reaches the gel point first at the center and, as in the isothermal case, the gelation moves from the center to the skin.

The effect of aging on prepreg has also been analyzed through modeling applications, as illustrated in Figure 53 using the input data reported in Table 10.[99] 22 days of aging at room temperature raised the activation energy of the resin without changing the overall reactivity of the system. Reactions are now activated at higher temperatures so the molecular weight increases more slowly initially. This causes a lower minimum viscosity and potential starvation problems. Gelation will also occur later for the aged resin when the delayed reaction is finally developed. Considerable differences in chemorheological behavior exist between different resins, and such differences can also be induced by aging.

Using the modeling approach, unexpected cure cycles can also be modeled. If, through some unexpected processing disturbance, the process is disrupted, the result can be easily modeled. This can provide valuable information as a basis for decisions on whether valuable composite parts can be used or not.

The processing of polyester matrix composites has also been modeled.[3,96,100] In ref. 3 the resin transfer moulding process (RTM) was modeled and the predictions were experimentally verified. In a manner different from epoxy-based materials, in this case the polyester resin gelled from the skin to the core, as shown in Figure 54.

Other processing technologies have been modelled using a similar approach.[96,101-103] In the case of pultrusion,[96] for most conditions under which carbon fiber/epoxy composites are pultruded, gelation moves from the center and outwards, whereas in pultrusion of glass fiber/polyester,

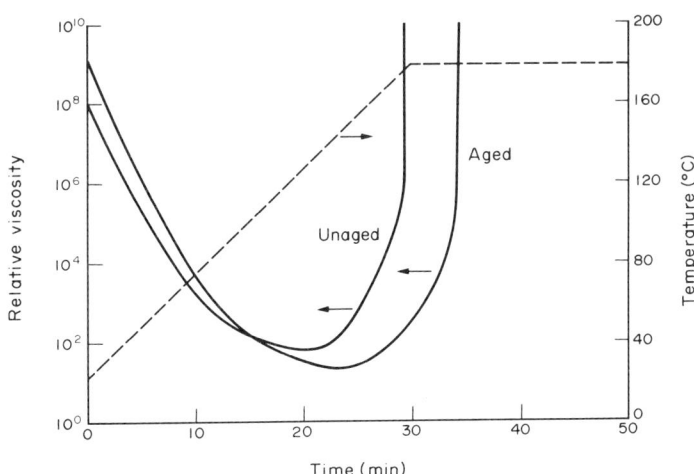

Figure 53 Viscosity *versus* time for aged and unaged TGDDM–DDS epoxy matrices (after Kenny et al.[99])

Table 10 Parameters of the Thermochemorheological Model Used in the Analysis of the Influence of the Aging Time on the Processing of Epoxy Prepregs Shown in Figure 53[99]

Storage time (days)	0	22
Heat of reaction (J g^{-1})	420	420
Activation energy (kJ mol^{-1})	67	107
Kinetic constant (s^{-1})	12.6	23.3
Reaction order	1.93	3.10
T_g for the unreacted polymer (K)	273	288

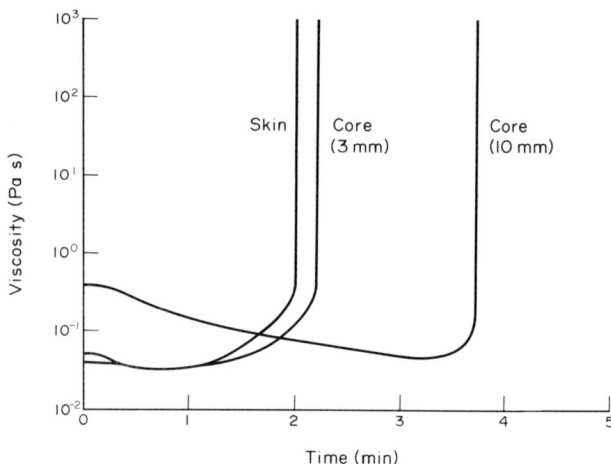

Figure 54 Viscosity *versus* time at the skin and at the core for two glass fiber/polyester composites with different thickness cured in a mold at 60 °C (after Kenny *et al.*[3])

processing conditions can be chosen in order to invert the process, provoking gelation to move from the skin to the core. The principal reasons for this behavior are given by the lower thermal conductivity of glass fibers as compared with carbon fibers, and by the induction characteristic of polyester resins.

18.3.7 Resin Flow

The thermal and chemorheological models can be considered as inputs for the description of the consolidation of the prepregs, during the fabrication of laminate composites. The consolidation process is mainly governed by air elimination during the vacuum bag stage[104] and by resin excess flow.[105] The flow of liquids on porous beds has been traditionally applied to fluidynamic phenomena associated with filtration, water draining through soils and oil extraction. The relation between the applied pressure (ΔP) and the flow rate (Q) in a unidirectional flow is generally described by the empirical 'Darcy's law'.[106,107]

$$Q = -A(K/\mu)(\Delta P/L) \tag{91}$$

where A is the area traversed by the flow, L is the length of the flow channel, μ is the viscosity and K is the permeability characteristic of the porous material, depending on its porosity and geometric characteristics. Normally, the permeability of fiber beds is described through the 'Carman–Kozeny' equation[106-108]

$$K = (R_f^2/4k)(1 - V_f)^3/V_f^2 \tag{92}$$

Here R_f is the fiber radius, V_f is the volume fraction and k is the so-called 'Kozeny constant'.

The term consolidation deserves some extra explanation. Figure 55[62] shows a spring immersed in an incompressible liquid. The stopcock holding the piston in (a) is opened in (b) and the liquid begins to flow through the escape channel. As time passes from (b) to (c) to (d), more and more liquid escapes due to the mass on top of the piston, and the piston compresses the spring to an increasing extent. After a long time full consolidation is obtained (d) and the length of the spring is similar to the case when no liquid is present. When pressure is applied to a composite bleeder system in an autoclave, resin flows from the composite in the directions normal and parallel to the laminate surface. The resin flow rate depends on: (i) the magnitude of the portion of the applied pressure which is transmitted to the resin; (ii) the gas pressure in the bag; (iii) the viscosity of the resin; (iv) the specific permeability of the fiber network in each direction; (v) the porosity of the fiber network; and (vi) the dimensions of the composite laminate. The flow is not a steady state flow, but is transient since there is a continuous depletion of resin from the composite with time.

Springer[105] presented a model for resin flow during prepreg consolidation in autoclave processing. The geometrical description of the fluid flow problem during autoclave lamination is shown in Figure 56. A porous material (referred to as 'bleeder') is placed on one or both sides of the prepregs. Restraints are mounted around the composite to prevent lateral motion and to minimize resin flow

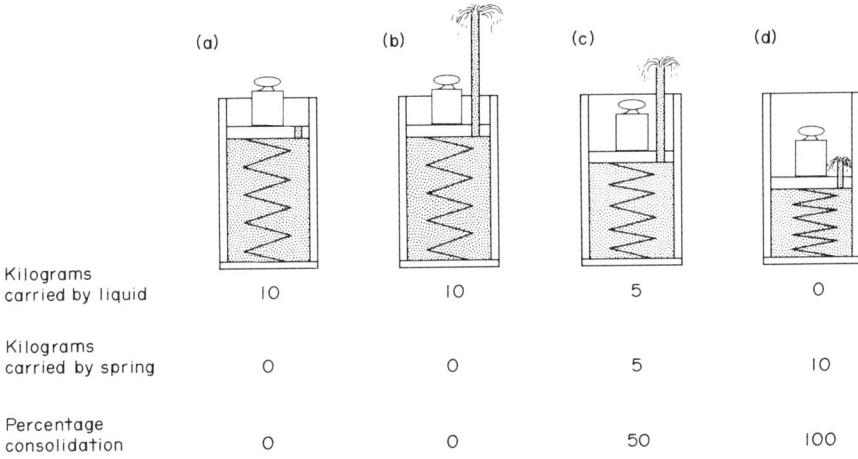

Figure 55 Diagram explaining the term consolidation (after Dave et al.[62])

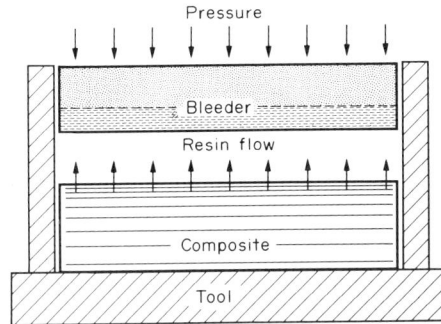

Figure 56 Illustration of the composite bleeder system for the autoclave lamination system[105]

through the edges. When pressure is applied, resin may flow in normal (z) and parallel (x) directions. Springer[105] assumed that the flow in the x direction could be neglected and that the pressure acts on the liquid matrix only. Then the consolidation process moves as a wave from the upper surface to the bottom (Figure 56).

The consolidation process was also described by Gutowski et al.[109,110] who considered that the pressure is also supported by the deformable fiber bed. With this assumption and applying Darcy's law, the following general relationship between the resin pressure (P_r) and the local fiber volume fraction (V_f) was obtained

$$\frac{K_{xx}}{V_f}\frac{\delta^2 P_r}{\delta x^2} + \frac{K_{zz}}{V_f}\frac{\delta^2 P_r}{\delta y^2} + \frac{1}{V_0^2}\frac{\delta}{\delta z}\left(V_f K_{zz}\frac{\delta P_r}{\delta z}\right) = \mu\frac{\delta}{\delta t}\left(\frac{1-V_f}{V_f}\right) \quad (93)$$

In equation (93) V_f varies only as a function of z, μ is the resin viscosity and K_{xx} and K_{zz} are the axial and transverse permeabilities. Moreover, Gutowski et al.[109] considered the 'blocking effect' of fiber rearrangement on the resin flow. They adopted a modified Carman–Kozeny equation which allows $K_{zz} \to 0$ for $V_f \to V_a$, where V_a is the volume fraction for which all flow is essentially shut off

$$K_{zz} = (R_f^2/4k')[(V_a/V_f)^{0.5} - 1]^3/[(V_a/V_f)^{0.5} + 1] \quad (94)$$

Using equation (94) a better description of experimental results were obtained, as is shown in Figure 57.

Kardos and coworkers[62,111] presented a three-dimensional consolidation and resin flow model which predicts: (i) resin pressure and velocity profiles inside the composite as a function of position and time; (ii) the consolidation profile of the laminate as a function of position and time; and (iii) the resin content profile as a function of position and time. They used analogies from the consolidation and flow theory used in soil mechanics, considering that the porous bed of fibers behaves like a spring, bearing part of the pressure. The 'viscoelastic' resin flow model presented predicts the direction of resin flow during cure and the pressure gradients. Nonlinear pressure

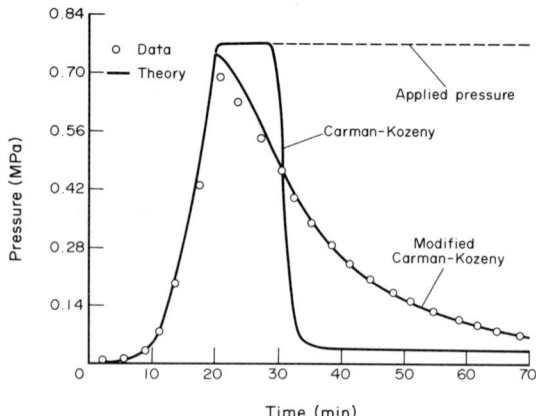

Figure 57 Comparison of measured and predicted values of the resin pressure for simulated bleeder ply molding. Values used in the model are $V_0 = 0.47$, $k = 150$ for the Carman–Kozeny equation (equation 92); and $k' = 0.2$ and $V'_a = 0.76$ for the modified Carman–Kozeny equation (equation 94). The fluid viscosity was 9.62 Pa s, the composite was 10 plies thick and AS-4 fibers were used (after Gutowski et al.[110])

gradients were considered in both vertical and horizontal directions, promoting resin flow and consolidation processes. The governing equations of this model are

$$\delta e/\delta t = -(1+e)[\delta V_z/\delta z + \delta V_x/\delta x] \qquad (95)$$

$$\delta P_r/\delta x = -\delta(t)[V_x/K_{xx} - \delta^2 V_x/\delta z^2] \qquad (96)$$

$$\delta P_r/\delta z = -\delta(t)[V_z/K_{zz}] \qquad (97)$$

where e is the resin/fiber volume ratio; and V_x and V_z are the local volume average resin flow velocities in the x and z directions.

The model is able to predict the distribution of the resin in the prepreg as a function of the processing conditions. In order to solve equations (91)–(93), K_{zz}, K_{xx} and P_r have to be expressed as a function of V_f. An expression for P_r can be determined experimentally using an approach valid for a large number of consolidated soils.

Most of the work performed in the area of resin flow through fiber beds has been dedicated to the theoretical and experimental determination of the permeability.[109–114] Kardos et al.[111] determined experimentally the permeabilities K_{zz} and K_{xx}, correlating the results with the Kozeny–Carman relation for different geometries and porosities of the fiber beds. They found, as expected, a highly anisotropic behavior with as much as a factor of 19 for K_{xx}/K_{zz} for a unidirectional laminate. A typical relation between permeability and volume fraction of fibers for different fiber systems is shown in Figure 58, where the 'blocking behavior' at high volume fraction of fibers, reported previously by Gutowski et al.[109] is clearly shown.

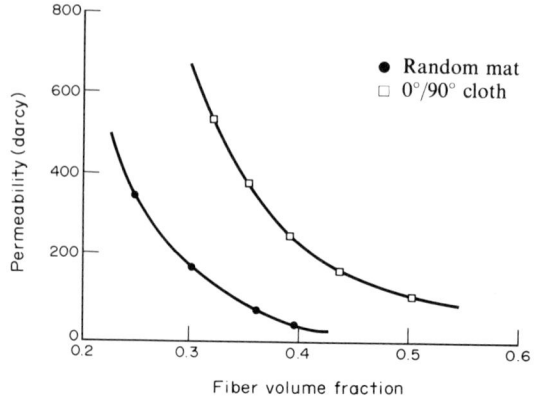

Figure 58 Effect of fiber volume and orientation on permeability (after Kim et al.[112])

The current research work in this field is mainly devoted to the analysis of the resin transfer molding (RTM) process, where the fluid flow through fiber preforms placed in the mold is the most important operation.[115-117]

18.3.8 Concluding Remarks

This chapter has described the general features of models for composites processing. A detailed treatment has been given, presenting the individual submodels that are integrated into a general processing model. The strong influence of processing conditions on composite laminate structure and properties, as well as the economy of the composites product is not apparent. Recent work has demonstrated how the use of a scientific approach can be a valuable tool for the solution of practical processing problems, such as the choice of matrix characteristics and processing conditions.

Recently developed processing models are based on an understanding of the physics and chemistry ruling the behavior of the polymeric matrix. In order to solve practical processing problems, an approach is suggested which combines resin characterization and modeling with experiments in the actual processing environment.

18.4 REFERENCES

1. J. C. Seferis and L. Nicolais, 'The Role of the Polymeric Matrix in the Processing and Structural Properties of Composite Materials', Plenum Press, New York, 1983.
2. J. M. Kenny, A. Apicella and L. Nicolais, *Polym. Eng. Sci.*, 1989, **29**, 973.
3. J. M. Kenny, A. Maffezzoli and L. Nicolais, *Compos. Sci. Tech.*, 1990, **38**, 339.
4. J. M. Kenny and A. Maffezzoli, *Polym. Eng. Sci.*, 1991, **31**, 607.
5. I. M. Ward, 'Mechanical Properties of Solid Polymers', Wiley, London, 1971.
6. L. E. Nielsen, 'Mechanical Properties of Polymers and Composites', Dekker, New York, 1974.
7. T. Murayama, 'Dynamic Mechanical Analysis of Polymeric Material', Elsevier, New York, 1978.
8. A. Maffezzoli, J. M. Kenny, L. Torre and L. Nicolais, in 'Proceedings of the 11th SAMPE European Chapter Conference, Basilea, Switzerland, 1990', p. 307.
9. A. Maffezzoli, J. M. Kenny and L. Nicolais, in 'Proceedings of the 49th SPE–ANTEC, Montreal, 1991', p. 2079.
10. J. M. Kenny and L. Torre, in 'Proceedings of the 49th SPE–ANTEC, Montreal, 1991', p. 2108.
11. R. J. Diefendorf, in 'Carbon Fibers and their Composites', ed. E. Fitzer, Springer-Verlag, New York, 1985.
12. K. K. Chawla, 'Composite Science and Engineering', Springer-Verlag, New York, 1987.
13. P. K. Mallick, 'Fiber-reinforced composites', Dekker, New York, 1988.
14. J. L. Kardos, in 'High Performance Polymers', ed. E. Baer and A. Moet, Hanser, Munich, 1991, p. 199.
15. L. Nicolais, *Polym. Eng. Sci.*, 1975; **15**, 137.
16. J. E. Ashton, J. C. Halpin and P. H. Petit, 'Primer on Composite Materials: Analysis', Technomic, Stamford, CT, 1984.
17. S. W. Tsai, J. C. Halpin and N. J. Pagano, 'Composite Materials Workshop', Technomic, Stamford, CT, 1968.
18. J. C. Halpin and L. Nicolais, *Quad. Ing. Chim. Ital.*, 1971, **7**, 173.
19. R. Hill, *J. Mech. Phys. Solids*, 1964, **12**, 199.
20. J. J. Hermans, 'Proc. Konigl. Nederl. Akad. Weteschappen Amsterdam', 1970, **B70**, 1.
21. J. C. Halpin and N. J. Pagano, *J. Compos. Mater.*, 1969, **3**, 720.
22. J. C. Halpin, *J. Compos. Mater.*, 1969, **3**, 732.
23. V. R. Riley, *J. Compos. Mater.*, 1968, **2**, 4366.
24. R. G. Schierding and O. des Deex, *J. Compos. Mater.*, 1969, **3**, 618.
25. R. W. Tock and D. E. McMackins, Monsanto-Washington U. ONR/ARPA Association Report HPC69-97, 1969.
26. K. L. Jerina, J. C. Halpin and L. Nicolais, *Quad. Ing. Chim. Ital.*, 1973, **9**, 94.
27. P. E. Chem, *Polym. Eng. Sci.*, 1971, **11**, 51.
28. R. M. Barker and T. F. MacLaughlin, *J. Compos. Mater.*, 1971, **5**, 492.
29. R. M. Jones, 'Mechanics of Composite Materials', McGraw–Hill, New York, 1975.
30. R. A. Schapery, *J. Compos. Mater.*, 1968, **2**, 380.
31. R. P. Kambour, *J. Polym. Sci., Polym. Phys. Ed.*, 1965, **3**, 1713.
32. R. E. Lavengood, L. Nicolais and M. Narkis, *J. Appl. Polym. Sci.*, 1973, **17**, 1173.
33. L. Nicolais and M. Narkis, *Polym. Eng. Sci.*, 1971, **11**, 3.
34. L. F. Nielsen, *J. Appl. Polym. Sci.*, 1966, **10**, 97.
35. A. S. Kenyon and H. J. Duffey, *Polym. Eng. Sci.*, 1967, **7**, 1.
36. S. Sahu and L. J. Broutman, *Polym. Eng. Sci.*, 1972, **12**, 91.
37. L. Nicolais and L. Nicodemo, *Polym. Eng. Sci.*, 1973, **13**, 469.
38. E. Drioli, L. Nicolais and A. Ciferri, *J. Polym. Sci., Polym. Chem. Ed.*, 1973, **11**, 3327.
39. G. Caprino, J. C. Halpin and L. Nicolais, *Composites*, 1979, **4**, 223.
40. J. C. Halpin, 'Primer on Composite Materials: Analysis', Technomic, Lancaster, 1984.
41. P. H. Petit and M. E. Waddoups, *J. Compos. Mater.*, 1969, **3**, 3.
42. J. C. Halpin and J. L. Kardos, *Polym. Eng. Sci.*, 1978, **18**, 496.
43. M. E. Waddoups, J. R. Eisenmann and B. E. Kaminski, *J. Compos. Mater.*, 1971, **5**, 446.

44. J. C. Halpin, K. L. Jerina and T. A. Johnson, 'Analysis of the Test Methods for High Modulus Fibers and Composites', *ASTM Spec. Tech. Publ.*, 1973, **521**, 5.
45. M. E. Waddoups and J. C. Halpin, *Comput. Struct.*, 1974, **4**, 1.
46. J. M. Whitney and R. J. Nuismer, *J. Compos. Mater.*, 1974, **8**, 253.
47. R. J. Nuismer and J. M. Whitney, 'France Mechanics of Composites', *ASTM Spec. Tech. Publ.*, 1975, **593**, 117.
48. E. M. Wu, *J. Appl. Mech.*, 1967, **34**, 967.
49. G. C. Sih, *Int. J. Fract. Mech.*, 1974, **10**, 279.
50. O. L. Bowie, *J. Math. Phys. (Cambridge, Mass.)*, 1956, **35**, 60.
51. G. R. Irwin, 'Fracture Dynamics, Fracturing of Metals', ASTM, Cleveland, 1948.
52. R. E. Peterson, in 'Stress Concentration Factors', Wiley, New York, 1974, p. 110.
53. P. C. Paris and G. C. Sih, 'Fracture Toughness Testing and its Application', *ASTM Spec. Tech. Publ.*, 1965, **381**, 84.
54. N. J. Pagano and R. B. Pipes, *J. Compos. Mater.*, 1971, **5**, 50.
55. S. S. Wang, J. F. Mandel and F. J. McGarry, 'Fracture Mechanics of Composites', *ASTM Spec. Tech. Publ.*, 1975, **593**, 1.
56. A. D. Wambach, K. L. Trachte and A. T. Di Benedetto, *J. Compos. Mater.*, 1968, **2**, 226.
57. A. T. Di Benedetto and A. D. Wambach, *Int. J. Polym. Sci.*, 1972, **1**, 159.
58. K. L. Trachte and A. T. Di Benedetto, *Int. J. Polym. Mater.*, 1971, **1**, 75.
59. J. C. Halpin, G. I. Kardos and M. P. Dudukovic, *Pure Appl. Chem.*, 1983, **55**, 893.
60. A. C. Loos and G. S. Springer, *J. Compos. Mater.*, 1983, **17**, 135.
61. J. C. Halpin, A. Apicella and L. Nicolais, in 'Polymer Processing and Properties', ed. G. Astarita and L. Nicolais, Plenum Press, New York, 1984, p. 143.
62. R. Dave, J. C. Kardos and M. P. Dudukovic, *Polym. Compos.*, 1987, **8**, 29.
63. J. M. Kenny, A. Trivisano and L. A. Berglund, *SAMPE J.*, 1991, **27**, 39.
64. P. K. Mallick and S. Newman (eds.), 'Composite Materials Technology, Processing and Properties', Hanser, Munich, 1990.
65. J. L. Kardos, M. P. Dudukovic and R. Dave', in 'Advances in Polymer Science, 80: Epoxy Resins and Composites IV', ed. K. Dusek, Springer-Verlag, Berlin, 1986, p. 101.
66. J. M. Castro and C. Macosko, *AIChE J.*, 1982, **28**, 251.
67. A. Apicella, L. Nicolais, M. Iannone and P. Passerini, *J. Appl. Polym. Sci.*, 1984, **29**, 2083.
68. R. J. Morgan and E. T. Mones, *J. Appl. Polym. Sci.*, 1987, **33**, 999.
69. R. B. Prime, in 'Thermal Characterization of Polymeric Materials', E. A. Turi, Academic Press, New York, 1981, chap. 5.
70. J. M. Barton, *Makromol. Chem.*, 1973, **171**, 247.
71. J. M. Kenny and A. Trivisano, *Polym. Eng. Sci.*, 1991, **31**, 1426.
72. K. Horie, H. Hiura, M. Sawada, I. Mita and H. Kambe, *J. Polym. Sci., Polym. Chem. Ed.*, 1970, **8**, 1357.
73. R. B. Prime, *Polym. Eng. Sci.*, 1973, **13**, 365.
74. S. Sourour and M. Kamal, *Thermochim. Acta*, 1976, **14**, 41.
75. M. E. Ryan and A. Dutta, *Polymer*, 1979, **20**, 203.
76. E. B. Stark, J. Seferis, A. Apicella and L. Nicolais, *Thermochim. Acta*, 1984, **77**, 19.
77. J. Mijovic, J. Kim and J. Slaby, *J. Appl. Polym. Sci.*, 1984, **29**, 1449.
78. J. B. Enns and J. K. Gillham, *J. Appl. Polym. Sci.*, 1983, **28**, 2567.
79. K. P. Pang and J. K. Gillham, *J. Appl. Polym. Sci.*, 1990, **39**, 909.
80. C. S. Chern and G. W. Poehlein, *Polym. Eng. Sci.*, 1987, **27**, 788.
81. I. Havlicek and K. Dusek, in 'Crosslinked Epoxies', ed. B. Sedlacek and J. Kahovec, de Gruyter, New York, 1987, p. 417.
82. P. J. Flory, 'Principles of Polymer Chemistry', Cornell University Press, Ithaca, NY, 1953.
83. M. T. Aronhime and J. K. Gillham, in 'Advances in Polymer Science, 78: Epoxy Resins and Composites III', ed. K. Dusek, Springer-Verlag, Berlin, 1986, p. 83.
84. X. Peng and J. K. Gillham, *J. Appl. Polym. Sci.*, 1985, **30**, 4685.
85. M. B. Roller, *Polym. Eng. Sci.*, 1986, **26**, 432.
86. A. Apicella, in 'Developments in Reinforced Plastics–5', ed. G. Pritchard, Elsevier, New York, 1986, p. 151.
87. M. Bodanecky and J. Kovar, 'Viscosity of Polymer Solutions', Elsevier, Amsterdam, 1982.
88. E. M. Valles and C. W. Macosko, *Macromolecules*, 1979, **12**, 521.
89. W. H. Stockmayer, *J. Chem. Phys.*, 1944, **12**, 125.
90. B. H. Zimm and W. H. Stockmayer, *J. Chem. Phys.*, 1949, **17**, 1301.
91. J. D. Ferry, 'Viscoelastic Properties of Polymers', 2nd edn., Wiley, New York, 1970.
92. A. T. Di Benedetto, *J. Appl. Polym. Sci.*, 1987, **B25**, 1949.
93. J. Mijovic and H. T. Wang, *SAMPE J.*, 1988, **24**, 42.
94. J. S. Vrentas, C. M. Jarzbeski and J. L. Duda, *AIChE J.*, 1974, **21**, 894.
95. G. Astarita and J. M. Kenny, *Chem. Eng. Commun.*, 1987, **53**, 69.
96. A. Trivisano, A. Mafezzoli, J. M. Kenny and L. Nicolais, *Adv. Polym. Technol.*, 1990, **10**, 251.
97. E. Broyer and C. W. Macosko, *AIChE J.*, 1976, **22**, 268.
98. R. J. Williams, A. Rojas, J. H. Marciano, M. Ruzzo and H. G. Hack, *Polym. Plast. Technol. Eng.*, 1985, **24**, 243.
99. J. M. Kenny, A. Apicella, L. Nicolais and M. Iannone, in 'Composite Structures-4, Volume 2: Damage Assessment and Material Evaluation', ed. I. H. Marshall, Elsevier, London, 1988, p. 2230.
100. H. Han and K. W. Lem, *J. Appl. Polym. Sci.*, 1983, **28**, 3155.
101. C. D. Han, D. S. Lee and H. B. Chin, *Polym. Eng. Sci.*, 1986, **26**, 393.
102. S. Y. Lee and G. S. Springer, *J. Compos. Mater.*, 1990, **24**, 1270.
103. G. S. Springer, *J. Compos. Mater.*, 1990, **24**, 1299.
104. K. J. Ahn, J. C. Seferis and J. C. Berg, *Polym. Compos.*, 1991, **12**, 146.
105. G. S. Springer, *J. Compos. Mater.*, 1982, **16**, 400.
106. J. Bear, 'Dynamics of Fluids in Porous Media', Elsevier, New York, 1972.
107. R. E. Collins, 'Flow of Fluids through Porous Media', Van Nostrand-Reinhold, Princeton, NJ, 1961.
108. P. C. Carman, *Trans. Inst. Chem. Eng.*, 1937, **15**, 150.
109. T. G. Gutowski, T. Morigaki and Z. Cai, *J. Compos. Mater.*, 1987, **21**, 172.
110. T. G. Gutowski, Z. Cai, S. Bauer, D. Boucher, S. Kingery and S. Wineman, *J. Compos. Mater.*, 1987, **21**, 650.

111. R. C. Lam and J. L. Kardos, *Polym. Eng. Sci.*, 1991, **31**, 1064.
112. Y. R. Kim, S. P. McCarthy, J. P. Fanucci, S. C. Nolet and C. Koppernaes, *SAMPE Q.*, 1991, **20**, 16.
113. K. L. Adams and L. Rebenfeld, *Polym. Compos.*, 1991, **12**, 179.
114. K. L. Adams, and L. Rebenfeld, *Polym. Compos.*, 1991, **12**, 186.
115. J. A. Molnar, L. Trevino and L. J. Lee, *Polym. Compos.*, 1989, **10**, 414.
116. J. P. Coulter and S. I. Guceri, *Compos. Sci. Tech.*, 1989, **35**, 317.
117. R. S. Parnas and F. R. Phelan, *SAMPE Q.*, 1991, **20**, 53.

19
Polymers From Renewable Resources

ALESSANDRO GANDINI
Institut National Polytechnique de Grenoble, France

19.1	INTRODUCTION		528
19.2	GENERAL ASPECTS		529
	19.2.1	*Polymers*	529
	19.2.2	*Oligomers*	531
	19.2.3	*Resins*	532
	19.2.4	*Monomers*	533
	19.2.5	*Composites (Wood)*	534
19.3	THE CHEMICAL MODIFICATION OF NATURAL POLYMERS		535
	19.3.1	*Cellulose*	535
		19.3.1.1 *Inorganic esters*	535
		19.3.1.2 *Organic esters*	536
		19.3.1.3 *Ethers*	536
		19.3.1.4 *Xanthates*	536
		19.3.1.5 *Other derivatives*	536
		19.3.1.6 *Cellulosic liquid crystals*	537
		19.3.1.7 *Cellulosic membranes*	537
		19.3.1.8 *Crosslinking*	537
		19.3.1.9 *Grafting and blocking*	537
	19.3.2	*Starch*	540
	19.3.3	*Chitin*	540
	19.3.4	*Carbon*	541
19.4	THE POLYMERIZATION OF OLIGOMERS ARISING FROM RENEWABLE RESOURCES		541
	19.4.1	*Drying Oils*	541
	19.4.2	*Rosins*[12]	543
	19.4.3	*Lignins*	543
		19.4.3.1 *Fillers or additives for polymers*	544
		19.4.3.2 *Macromonomers*	545
	19.4.4	*Tannins*	549
19.5	MONOMERS FROM THE BIOMASS AND THEIR POLYMERIZATION		550
	19.5.1	*Lignin Degradation Products*	550
	19.5.2	*Terpenes*	551
	19.5.3	*Glycerol, Ethylene Glycol and Propylene Glycol*	552
	19.5.4	*Monosaccharides and Disaccharides*	553
		19.5.4.1 *Condensation reactions*	553
		19.5.4.2 *Ring-opening polymerization*	553
	19.5.5	*Biodegradable Polymers from Renewable Resources*	555
		19.5.5.1 *Bacterial polymerization*	555
		19.5.5.2 *Chemical polymerization*	556
	19.5.6	*Pentoses and Hexoses as Sources of Furanic Monomers*	556
		19.5.6.1 *Furfural and derivatives*	556
		19.5.6.2 *Hydroxymethylfurfural and derivatives*	560
	19.5.7	*The Polymerization of Furanic Monomers*	561
		19.5.7.1 *Chain polymerization*	561
		19.5.7.2 *Stepwise polymerization*	564
	19.5.8	*The Chemical Modification of Furanic Polymers*	568
19.6	THE CHEMISTRY OF UNREFINED LIGNOCELLULOSICS		569
19.7	CONCLUSION		569
19.8	REFERENCES		570

19.1 INTRODUCTION

Nature provides an incessant output of small molecules, oligomers and polymers, through the biological syntheses performed by the animal and vegetal realms. Most structures are organic, but organometallic compounds and complexes also play a key role.

Of course the very survival of humanity is based on the sustenance some of these products offer. Others have been exploited to improve the quality of human life throughout the ages in terms of both improved comfort (skins, fibres, wood *etc*.) and pleasure or culture (paper, oils, dyes *etc*).

The advent of industrial chemistry greatly improved the technologies associated with the isolation, purification, chemical modification and specific processing of these substances which today find applications in such diverse materials and products as foodstuffs, textiles, pharmaceuticals, tyres, paper and cardboard, resins and inks, leather, the vast array of articles from wood manufacturing, plastics and membranes *etc*.

Many of these applications involve polymers or oligomers made by nature, or made by man but using monomers and macromonomers or monomer precursors arising from the biomass.[1] Many others can be envisaged both in terms of the chemical modification of naturally available polymers or of the synthesis of new structures.[1,2]

What interest is there in the development of this type of new polymeric material in a technological context dominated by decades of brilliant advances in the synthesis, characterization, modification, and blending of macromolecules derived from petrochemistry?

The major justification for such a strategy is the ubiquitous and renewable character of most of these natural products whose output is guaranteed by solar energy. Every society, irrespective of its climate, geographic context, economic situation or mineral resources, can count upon some vegetation in the form of agriculture and forestry, or simply wild growths. In the same way, animal resources are universal, even if variable in quantity and kind. A rational use of this biomass towards the elaboration of polymeric materials other than those already exploited would in many instances reduce the dependence on products derived from petroleum or coal, thus solving some of the typical problems associated with their availability (economic burden for those who have to import them, price fluctuations, periodic regional or world crises) and preparing for a not too distant future in which these nonrenewable commodities will become globally scarce.

These general features are complemented by the existence of some less common or even exotic natural products from renewable sources in specific areas, which can be used in an original fashion to generate added value materials.

It seems, therefore, important to stimulate a growing interest into these matters in conjunction with the parallel search for new forms of energy harvesting as, for example, the Brazilian policy of alcohol production from biomass. Of course the latter aspect falls outside the scope of this work, but in the same way that petroleum and coal are mostly a source of energy and only marginally used, *viz*. about 7%, for making chemicals, the biomass should be considered as playing the same double role.

The production of terrestrial biomass through solar energy amounts to about 3×10^{11} t per year, with a relatively even global distribution, certainly much more even than mineral resources in general and petrol in particular. It is by all standards an impressive figure in terms of the tremendous potential it offers for rational and profitable utilizations which could be applied and exploited practically everywhere.

The aim of this chapter is to describe the state of the art and recent research on topics related to the chemical exploitation of the biomass for the production of new polymers. However, this review cannot possibly touch upon all aspects of polymer science related to natural renewable resources.

The choices made as to restrictions are, as always, partly objective and partly subjective. Since this chapter is devoted to polymers from natural renewable resources and not to polymers actually ready-made by nature, the objective exculsions comprise all technologies dealing with polymers which are used as such, *i.e.* without any major chemical modification, although important chemical operations aimed at their separation or purification, might intervene. Examples of major deliberate omissions include, therefore, cotton, wool and silk manufacture and properties, papermaking as regards the isolation and conditioning of cellulosic fibres, natural rubber (here vulcanization and other chemical modifications do not differ in essence from those carried out on synthetic elastomers which have been discussed elsewhere in this treatise), the elaboration of wood-based materials, the manufacture of leather articles, the isolation and purification of other natural polymers like starch and chitin *etc*.

Subjective omissions have to do with minor aspects of the very large spectrum covered by the polymer chemistry of natural products. It was in fact felt that too many brief citations of episodic or sporadic investigations might risk spoiling the main thread of this review.

Obviously, the present context excludes all topics dealing with the biochemistry and biology of natural polymers.

Finally, the chemical modification of cellulose will only receive short comments aimed at bringing up to date the chapter already devoted to this topic in Volume 6 of *Comprehensive Polymer Science*.[3]

What is then left for inclusion in this survey? Three major items, namely: (i) the principles and aims of the chemical modification of natural high polymers, mostly through the grafting of small or large chains, from specific reactive sites, originally present on the substrate backbone, or specifically introduced thereon; (ii) the oligomerization, polymerization, or crosslinking of monomers and macromonomers which occur naturally, or arise from minor chemical operations performed upon natural products; and (iii) the synthesis and polymerization of monomers derived from renewable resources, through chemical operations which are the counterpart of petrochemistry, *i.e.* substantial structural modifications. The interest of these operations in terms of properties of the ensuing materials will also be emphasized.

19.2 GENERAL ASPECTS

Before proceeding to a coverage of the three topics enumerated above, it seems advisable to provide a more general outlook on the major structures synthesized by living organisms and related to the contents of this chapter, *i.e.* polymeric materials and their possible sources.

19.2.1 Polymers

Biological activities produce a large variety of high polymers: they can be linear, branched or crosslinked, very regular and crystalline or highly irregular and amorphous.[4]

Polysaccharides are by far the most abundant among natural polymers.[5-7] As implied in their generic name, they are made up of monosaccharide residues joined by glycosidic bonds. They are present in plants, animals and microbes where they can play various fundamental roles, such as structural materials (*e.g.* in wood) and carbon- or energy-storage products. In terms of natural abundance and thus applications towards useful materials, three high polymeric renewable products dominate: two arising mostly from plants, cellulose (**1**) (linear structure) and starch (linear amylose plus highly branched amylopectin) and one from an animal source, chitin (**2**) (and chitosan).

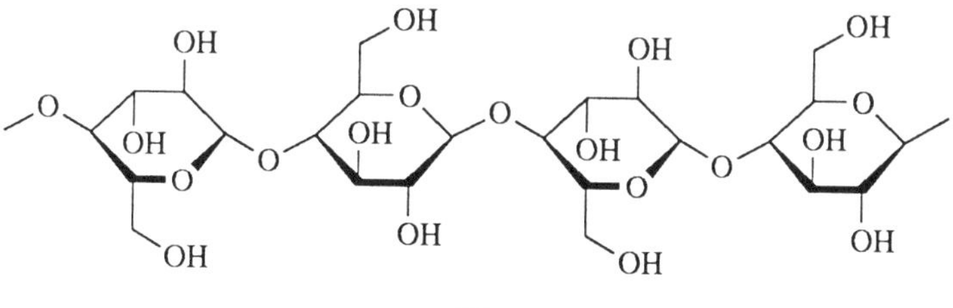

(1)

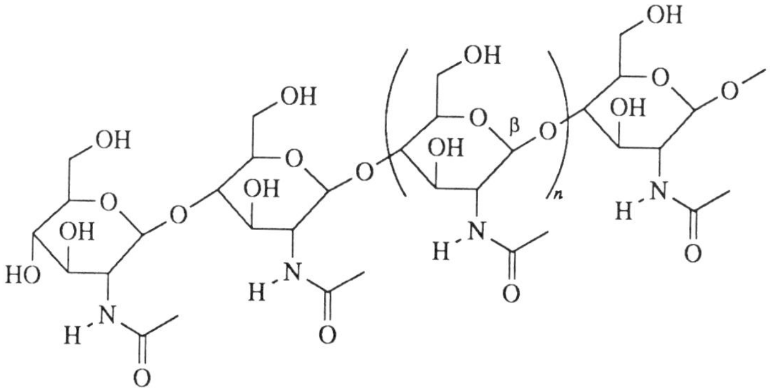

(2)

The high structural regularity of cellulose, the most abundant component of the biomass, and the easy establishment of a vast array of hydrogen bonds among its chains, promote ready crystallization and some remarkable physical properties which translate into the familiar features of cotton, paper and wood *etc.* Starch is the main food reserve polysaccharide in plants, but it is being increasingly used for elaborating new materials. Many other polysaccharides are found in vegetal matter, but they are less relevant to the present topic. Polysaccharides contained in algae include alginic acid, a polyuronide extensively used in the food industry and other structures like agar and carrageenan.

The basic animal polysaccharides are glycogen and chitin. The former is present in muscle and in liver. The latter, which is much more abundant, can be extracted in large amounts from crustaceans and is a major source of polymeric applications with cellulose and starch.

Proteins follow:[8] in terms of 'materials' rather than biochemical activity and specificity, collagen, keratin and fibroin are the most representative structures. Conformations involving multiple helices or pleated sheets are associated with these polypeptides, as indicated by the structure of keratin (**3**) in wool.

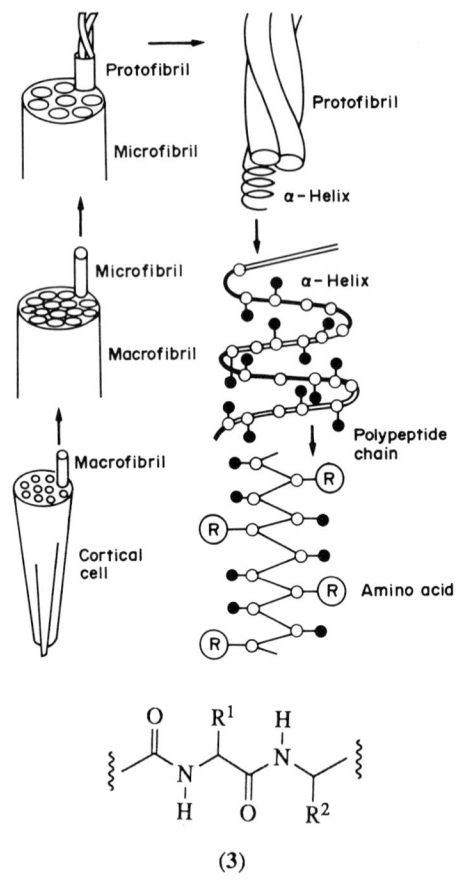

(**3**)

Keratin is also a major constituent of hair, furs, feathers, and horns *etc.*

Collagen is another triple helix macromolecule which occurs in hides, skins, bones and teeth. The tanning of some of these products, a treatment involving crosslinking of the protein polymer molecules, yields leather. Fibroin is the bulk substance of silk which owes its strength to the high order and strong interchain binding energy through hydrogen bonds as in polyamides. The silk fibres are coated with sericin, another protein.

Natural rubber[9] is another abundant and very important renewable material made up of macromolecular chains of poly(1,4-isoprene), a linear unsaturated structure synthesized by a few tropical plants in Malaysia, Indonesia, Brazil and elsewhere.

Different plants produce different forms of the polymer which is extracted as an aqueous emulsion (latex) or as a sap-like dispersion from which it is coagulated. The *cis* form (**4**) of the polymer tends to be amorphous. It has a glass transition temperature of about $-70\,°C$ and is therefore suitable for elastomeric applications after vulcanization. Its world production is still very large, in spite of the rapid development of synthetic rubbers which became more important than natural ones in the early 1960s and is estimated at about 5×10^6 t per year. The *trans* form (**5**), called gutta percha or balata, crystallizes readily forming a rigid material melting at 70 °C.

cis-1,4 (**4**)   *trans*-1,4 (**5**)

The most abundant renewable crosslinked polymers, omnipresent in the phytomass, are lignins.[10] Structure (**6**) is an attempt at representing the complexity of these macromolecules. Whereas it is legitimate to have a specific name for well-defined polymers such as cellulose and natural rubber, it is not so for other natural polymers like lignin because its structure is not unique, but depends instead on a number of factors including the plant species, its age and the season during which it was produced. It seems therefore correct to use the generic term of 'lignins' for these polymers, rather than the singular form which suggests a unique chemical structure.

(**6**)

Another crosslinked natural polymer is elastin, a complex protein present in ligaments.

19.2.2 Oligomers

Living organisms produce a large variety of oligomers and potential macromonomers.

Glycerol esters,[11,12] *viz.* the majority of fats and oils, and the corresponding long chain carboxylic acids are among the best examples of these structures. Saturated chain homologues are nonreactive with respect to polymerization. Multiple unsaturations as in linoleic (**7**) and linolenic (**8**) acid are a source of polymerizability through the action of oxygen. The drying oils used as vehicles of paints and printing inks are mixtures of such unsaturated and saturated structures, with a predominance of the latter ones (*e.g.* linseed oil) and owe their name to their susceptibility to crosslink upon air exposure.

Of course fats and oils, mostly the saturated or the weakly unsaturated structures, are also the major source of soaps, surfactants, fatty acids and alcohols and thereby of glycerol. Among these products, only the latter will be discussed below as a monomer in polyester or polyurethane formulations.

(7)    (8)

Hemicelluloses[5-7] represent a vast array of oligomeric cellulose-like structures produced by the biomass. The major differences with respect to the high molecular weight polysaccharides mentioned above are their relatively low DPs (less than 300) and their structural irregularities and chain branching. Examples of typical hemicellulose molecules and of their monosaccharide precursors are given in Scheme 1.

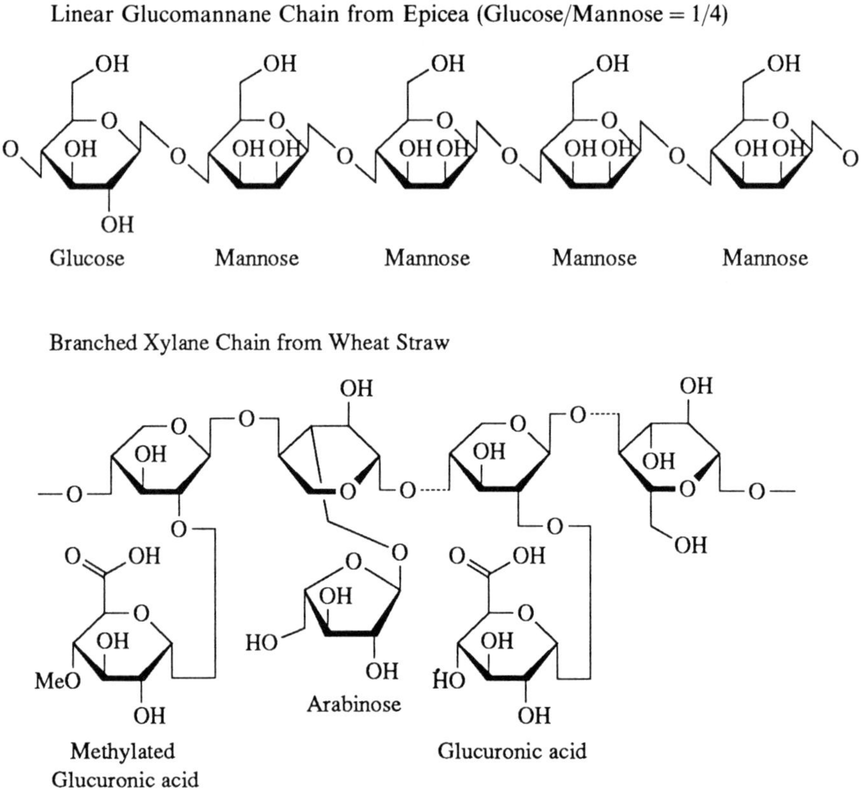

Linear Glucomannane Chain from Epicea (Glucose/Mannose = 1/4)

Glucose   Mannose   Mannose   Mannose   Mannose

Branched Xylane Chain from Wheat Straw

Methylated
Glucuronic acid   Arabinose   Glucuronic acid

Scheme 1

19.2.3 Resins

Natural resins[12,13] are produced by plants with shellac as the only animal counterpart, and consist of a more or less complex mixture of condensation products of various precursors such as terpenoids or flavonoids. The molecular weights of these materials are relatively low (a few hundred to a few thousand), hence a low melt viscosity, but their glass transition temperatures can be as high 100 °C in certain instances. Chemical modification can of course increase their molecular weight and change specific properties.

Rosins are the most important representatives of this family, with a yearly production of about 10^6 t. They are extracted from pine trees and have been used for centuries in various traditional sealing and lubricating applications. They are made up of a mixture of unsaturated polycyclic carboxylic acids of which abietic acid (9) is the major component. Various chemical modifications of rosins have been perfected and are discussed below.

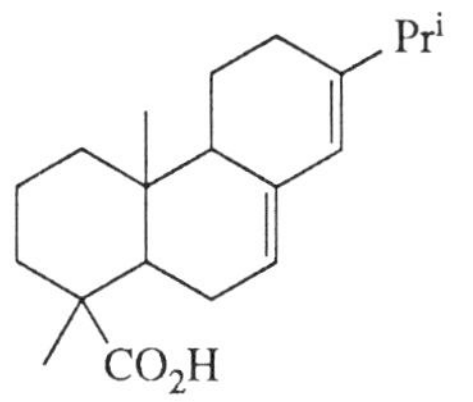

(9)

Other resins obtained from vegetal renewable resources include traditional materials used throughout the ages such as varnishes, coatings, paint vehicles and balsams *etc.* Although they usually require little processing before their utilization, chemical modification procedures have been applied to them, following the same principles as with rosins (see below). Examples of commercial resins are the dammars from Indonesia and Malaysia, used in protective coatings and lacquers; mastic from Chios, a relatively flexible film-forming material again used for coatings, but also for printing inks and as an adhesive; and gum elemi from the Philippines, with a glass transition below room temperature which favours applications as a plasticizer and in adhesive formulations. The chemical compositions of these natural products have not been investigated in detail, but they are all complex mixtures containing polyterpenes, aromatic acids, phenolics and essential oils. These traditional resins have lost some of their importance and are often replaced today by synthetic counterparts. They remain, however, in application for specific uses, sometimes as high added value products for aesthetic and artistic purposes.

Shellac is the only animal resin widely used by man. It is secreted by an insect and produced in India and Southeast Asia. It is a mixture of polyhydroxy acids, lactones and esters, with some unsaturations. High contents (about 45%) of 9,10,16-trihydroxypalmitic acid are accompanied by lesser amounts (about 25%) of shelloic acid (10), a cyclic sesquiterpenic carboxylic structure.

Shellac is still used for protective coatings and sealing in housing, food, pharmaceutical and dielectric applications.

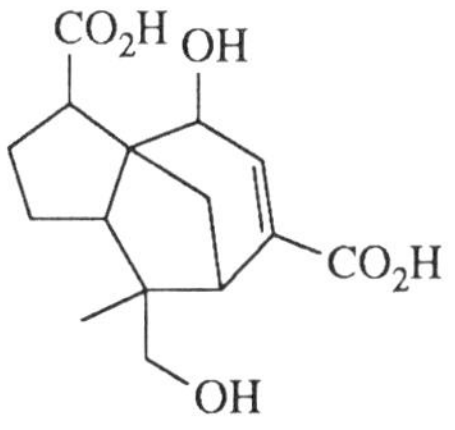

(10)

19.2.4 Monomers

Finally, natural biochemical activities produce a large number of monomers, potential monomers or monomer precursors.

Monosaccharides can be polymerized to linear polysaccharides by various techniques.[7] These monomers are mostly modified (*e.g.* ethers and orthoesters) sugars, but of course free mono- and di-saccharides are also present in plants and animals.

Essential oils often contain terpenes,[14] *i.e.* unsaturated hydrocarbons of general formula $(C_5H_8)_n$. Indeed, polyisoprene is a polymeric terpene. Many monomeric terpenes are produced by trees and plants. Their polymerization has been studied and exploited. α-Pinene (11), β-pinene (12), limonene (13) and myrcene (14) are typical representatives of such monomers.

Tannins[15] are polymeric flavonoids present in the bark of many trees. As their name suggests, they have been extensively used for the tanning of leather, *i.e.* as crosslinking agents for collagen-type proteins.

Other monomeric structures can be obtained from more substantial chemical transformations of natural products. Thus lignin degradation under controlled conditions can lead all the way down

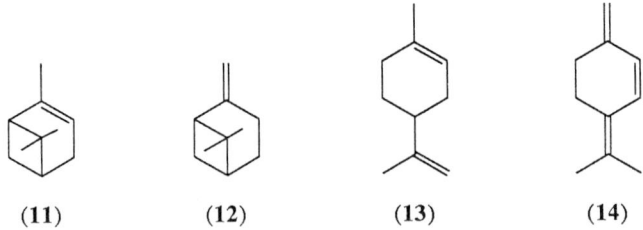

(11) (12) (13) (14)

to phenolic derivatives. The acid-catalyzed dehydration of certain monosaccharides is a source of furan derivatives which are or can be converted into interesting monomers.

19.2.5 Composites (Wood)

To complete this survey, some superstructural assemblies should be mentioned. Wood[16] is a particularly pertinent one because it is an excellent example of a composite material. The cellulosic fibres, which are the component providing the mechanical strength through the crystalline character of the macromolecules gathered into filaments, are surrounded by lignin which is the amorphous matrix that cements these fibres together and holds the overall shape. The compatibilization of the interface between these two basic elements is provided by hemicelluloses which are the 'wetting agent' operating in part through actual covalent chemical bonds linking some lignin chain segments to the cellulose backbone. The cellulose fibres are oriented in a specific direction, usually that of the trunk of the tree or the stem of the plant, and this, as with man-made composites, establishes large differences in modulus as a function of the direction of the stress applied. As a consequence, tree branches sway when submitted to side forces (wind), but are stiff with respect to longitudinal traction or compression. Figure 1 is a schematic representation of a typical wood cell with the composition distribution of its three basic components.

Other phytomass products, like straw, sugar-cane bagasse, corn cobs and many other agricultural wastes, have similar morphologies and contain basically the same components, although their relative abundance varies considerably from one material to another.

Delignification in pulping processes consists, in its early stages, in the breaking of the connecting bonds thus allowing the separation of lignin from cellulose. Later, the lignin macromolecules are partly degraded into soluble fragments and the cellulose is recovered for papermaking.

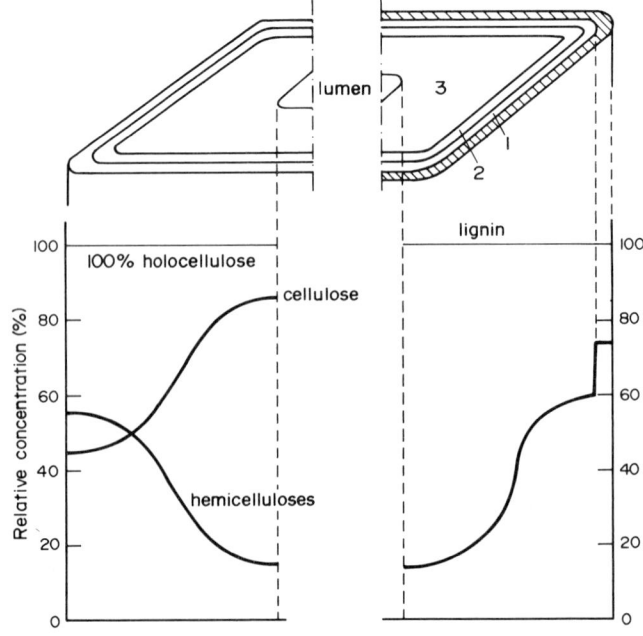

Figure 1 A pine cell wall. 1. Primary wall. 2. Compound middle lamella. 3. Secondary wall. The lower part of the figure shows the relative concentration profiles of the three major components

Recent developments in the chemical exploitation of the biomass, such as the steam explosion process and the organosolv techniques, aim at a rational and economical separation of the three basic macromolecules of most bulk structures (cellulose, hemicelluloses and lignins) and at finding optimal utilizations for each one of them. These are perhaps the most important challenges within the present context and they will be dealt with accordingly.

As stated in the introduction, the natural polymeric materials used as such (wood,[16] cotton,[3,17] wool,[18] silk,[18] and some resins *etc*.) or following minor chemical or technological modifications, as with leather,[19] natural rubber[9] and papermaking[20] *etc*., fall outside the scope of this survey. The relevant topics are treated below following an organization based on the three categories given above, namely: (i) polymer modification; (ii) oligomers and macromonomers; and (iii) monomers. Both naturally occurring monomers and those arising from major chemical operations will be considered.

19.3 THE CHEMICAL MODIFICATION OF NATURAL POLYMERS

19.3.1 Cellulose

Cellulose[5,6,21–23] is a remarkable polymer under all its native forms, but also under the various applications man has found for it. Cotton textiles[17] and paper-based products[20] provide a constant proof of its qualities: good mechanical properties due to its crystallinity, good thermal stability for most common applications, high water affinity (which can of course be viewed also as a drawback) and ease of dyeing *etc.*, but also (slow) biodegradability. It is moreover the most abundant natural polymer and its ready availability is synonymous of low cost.

These positive features and the fact that textile and papermaking technologies had provided an excellent opportunity for getting to understand this macromolecule well (even if its macromolecular character was in fact unknown until this century) were the reasons for attempting to achieve its chemical modification in search of new materials at a time when synthetic polymers were still unthought of. Indeed, such classical operations as mercerization, bleaching and dyeing introduced the chemists into the realm of cellulose reactivity during the last two centuries.

Moreover, another driving force towards trying to modify cellulose was to minimize some of its drawbacks, particularly its poor thermoplasticity, resistance to abrasion, dimensional stability and solubility in most organic solvents. Cellulose esters and ethers (including the xanthates) were thus born over a century ago and are still today relevant materials which have fulfilled a key role particularly up to the second world war, *i.e.* before the petrochemical boom and the corresponding explosion of synthetic polymer chemistry.

Present interest in modifying cellulose is more in the direction of grafting, the purpose of these investigations being the achievement of special properties which could give cellulose novel applications in areas where synthetic polymers dominate but are costly or, better still, where they have proved inadequate.

It is clear that any reaction aimed at altering the structure of these polysaccharides without affecting their chain length can only call upon the hydroxy groups borne by each monomer unit. In the case of cellulose, total substitution implies the reaction of all three OH groups of each unit, but, of course, chemically modified celluloses also include derivatives with lower degrees of substitution. After a brief survey on esters, ethers and crosslinking,[3] a little more consideration will be devoted to recent trends in grafting. Degradation reactions[23] will not be considered here except in the context of the photolytic and radiolytic generation of free radicals for grafting purposes.

19.3.1.1 Inorganic esters

Among these derivatives, nitrocelluloses[24] are the oldest materials prepared synthetically from cellulose. They are still an industrial commodity today, but their importance has decreased very considerably and their uses relegated mostly to explosives and propellants and to a minor degree to components of inks, lacquers and adhesives. This decline is of course due to the flammability and explosive properties of the dry material which has been superseded by safer and more performant polymers, *e.g.* for cinematographic films.

The cellulose sulfates and phosphates,[7,21,22] the latter usually in the form of sodium or ammonium salts, are even less widespread than the nitrate and find applications in medical and pharmaceutical products, as flame retardants and as food additives.

Phosphorylated celluloses,[7,21,22] e.g. in the form of ammonium salts, are fire-resistant and have good nonsoiling properties.

Cellulose methane- and trifluoromethane-sulfonates have also been prepared and characterized.

19.3.1.2 Organic esters

Cellulose carboxylates[7,21,22] are another family of traditional polymers dating back from the beginning of the 'plastics age'. Their properties depend on: (i) the degree and type of substitution (including mixed carboxylates); and (ii) the average molecular weight of the starting material. Contrary to the nitrate, their industrial production has not faltered and is today some ten times higher than that of cellulose nitrate and growing. The three major materials are the triacetate, the acetate propionate and the acetate butyrate which possess a number of typical plastics applications. The triacetate is also used to manufacture fibres thanks to its ease of crystallization. The diacetate is receiving renewed attention and its production has been increasing in the last few years, one of its applications being for cigarette filters.

Cellulose trifluoroacetates and tosylates are interesting esters of strong organic acids.[7,21,22]

19.3.1.3 Ethers

Substitution of the hydroxy group(s) of cellulose by alkylation leads to ether groups. Some of these products[7,21,22,25,26] have been industrial commodities for many decades and are still economically important. Sodium carboxymethylcelluloses (known as CMC) are synthesized from cellulose and sodium chloroacetate in an alkaline medium. Depending on the degree of substitution and purity, these water-soluble anionic ethers find applications as thickeners, films and food additives etc. Hydroxyethylcellulose and hydroxypropylcellulose are manufactured using ethylene oxide and propylene oxide, respectively, as insertion reagents. They are also water soluble and film forming, but nonionic. Their numerous uses include adhesives, paper and food additives, stabilizers and encapsulation materials. Methyl and ethyl derivatives are other well-known cellulose ethers possessing similar properties and finding similar applications. Products bearing ether and hydroxyalkyl groups, such as ethylhydroxyethylcellulose are also available commercially. Finally, cationic cellulose ethers are manufactured: the positive charge is typically held by an ammonium moiety.

19.3.1.4 Xanthates

The reaction of cellulose with carbon disulfide in the presence of sodium hydroxide yields ionic xanthogenates (viscose) which have been known since the end of last century.[21,22] Cellophane is the most typical material derived from viscose.

19.3.1.5 Other derivatives

Deoxycelluloses[7,21,22] (15) are modified celluloses in which part or all the OH groups have been replaced by other functionalities, e.g. a hydrogen atom, a halogen atom, a nitrogen- or sulfur-containing group, etc. Some recent work[27] on these derivatives seems quite promising in terms of novel applications like biomedical materials, flame resistance and textile complexation agents.

Anhydrocelluloses[7,21,22,27] (16) are modified celluloses in which intramolecular condensation has left a 2,3-oxirane or a 3,6-oxepane ring.

The synthesis of cellulose carbamates (urethanes) has been optimized[28] in order to attain a maximum and reproducible degree of substitution which allows precise molecular weight distribution determinations.

Many other recent reports of studies on cellulose derivatives have appeared in books and conference proceedings.[29-32]

19.3.1.6 Cellulosic liquid crystals

Cellulose and many of its derivatives form cholesteric liquid crystals. After the first report concerning a lyotropic mesophase formation from aqueous solutions of hydroxypropylcellulose 15 years ago, considerable attention has been devoted to these phenomena extended to other derivatives (ethers, esters) and to thermotropic behaviour. These studies[33-35] concern fundamental aspects about the origin of the liquid crystal properties, structural investigations and the preparation and characterization of fibres and films.

Microfibrils of cellulose also display nematic features in appropriate aqueous conditions. All these materials certainly show very promising applications[36] in the fields of optics, high modulus fibres and other advanced technologies.

19.3.1.7 Cellulosic membranes

Membrane-based separation processes play a paramount role in a number of industrial, analytical and medical activities. Cellulose (*e.g.* regenerated) and some of its derivatives (mostly acetates or graft copolymers) are choice materials, either as such or blended with other polymers, for many of these applications. This explains the vigorous research activities related to this topic and the notable progress achieved, as reported in recent events and monographs.[37-40]

19.3.1.8 Crosslinking

The reaction of hydroxy groups of cellulose with di- or poly-functional compounds induces the formation of a network structure. This treatment is applied to cellulosic fibres in textiles, paper and cardboard to improve their crease and shrink resistance. Apart from formaldehyde and urea–formaldehyde linear and cyclic oligomers, several other reagents are used for this purpose.[3,7,21,22]

19.3.1.9 Grafting and blocking

Graft copolymerization of cellulose has stimulated much research in past decades, starting in the early 1950s. The vast majority of these studies call upon grafting from procedures by radical polyaddition and operate heterogeneously.[41-45] The most important problems which need to be overcome to optimize these reactions are: (i) the minimization of parallel homopolymerization of the grafting monomer(s); (ii) the maximization of the number of cellulose molecules grafted; (iii) a good control over the chain length of the inserted polymer; (iv) good reproducibility of the grafting reaction and of the properties of the resulting materials; and (v) improvement of the economy and the simplicity of the processes.

Indeed, despite the deep involvement of many laboratories, very few products arising from cellulose grafting, whether for paper or textile applications, have reached the stage of industrial production. Undoubtedly, economic considerations have been the most important factor behind this (temporary) failure. Cotton-based textiles and paper or cardboard materials are for the most part cheap commodities which cannot withstand important additional production costs such as those introduced by major chemical modifications (here grafting processes) unless of course these modifications bring about a substantial improvement of certain properties thus providing high added value. The trend is therefore to find the simplest possible routes to good quality grafting or to search for very special products for advanced technologies such as electronics, optics, energy storage, superabsorbance and medicine.

Many chemical and physical methods of activation of the cellulose substrate have been tested in order to produce a good yield of free radical sites usually in aqueous media. Persulfates probably act through equation (1), whereas with the redox system Fe^{II}/H_2O_2 the HO· radicals are responsible for the activation of cellulose, as shown in equation (2).

$$\text{Cell–OH} + SO_4^{-\cdot} \longrightarrow \text{Cell–O}\cdot + HSO_4^- \qquad (1)$$

$$\text{Cell–OH} + OH\cdot \longrightarrow \text{Cell–O}\cdot + H_2O \qquad (2)$$

With both these systems free radicals are generated before the interaction with cellulose and can also attack the monomer to be grafted thus giving homopolymer. The grafting efficiency will therefore depend, among other things, upon the relative rate of the two alternative processes.

The so-called xanthate method of grafting involves the preparation of the iron(II) salt of cellulose xanthate which is then made to react with hydrogen peroxide in the presence of the grafting monomer, as shown in Scheme 2.

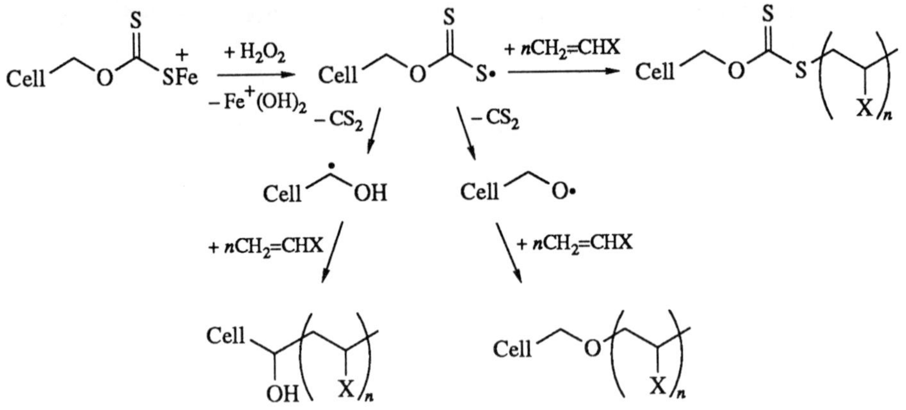

Scheme 2

However, the xanthate anion is also present in the system and HO· radicals are formed in conjunction with the cellulosic radicals (see Scheme 2) and unless conditions are optimized the problem of primary noncellulosic radicals reappears.

The use of cerium(IV) ions as initiators eliminates this duality because free radicals are generated directly on the cellulose structure through a redox mechanism which seems to operate according to Scheme 3.

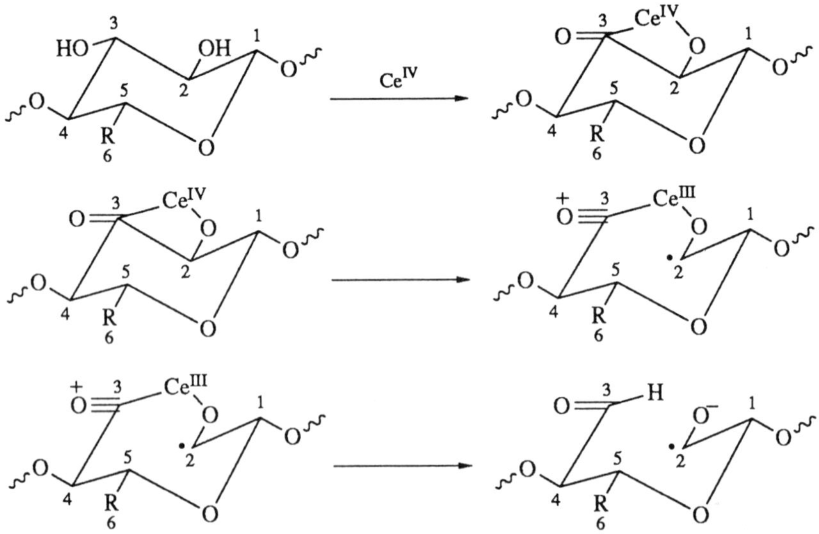

Scheme 3

Very high grafting efficiencies have been obtained with this chemistry, on both cellulose and some of its derivatives.

Manganese (III and IV) ions seem to operate through a similar direct redox process.

Other methods of chemical initiation have been tested with varying degrees of success.[41–45]

Photochemical activation to form cellulose macroradicals has also attracted the attention of polymer chemists.[41,42] One approach consists in photosensitizing the reaction in a classical fashion.

Benzophenone and phenyl acetophenone derivatives were shown to operate through hydrogen abstraction from the cellulosic OH groups to give the corresponding ketyl radicals which initiate the grafting of the added monomer. Alternatively, cellulose can be irradiated directly with an ultraviolet source. Radicals are formed even when the incident light has a wavelength higher than 280 nm and the photochemistry involves chain scission, dehydrogenation and dehydroxymethylation. These cellulose macroradicals can initiate the polymerization of added monomers and grafting, but also blocking from the split macromolecules occurs. Similar irradiation procedures on cellulose ethers and esters tend to induce cleavage reactions at the substituents.

High energy radiation, such as gamma rays, has been extensively used to generate free radicals onto the cellulose substrate.[41,42] These have been studied by ESR and the typical structures postulated include primary and tertiary carbon radicals as well as $CH_2O\cdot$ radicals. All these species can initiate the radical polymerization of vinyl monomers giving grafted/blocked configurations. Although the added monomer can also suffer radiolysis, it has been shown that the initiation from the cellulose macroradicals dominates because these are formed in higher yields and are longer lived than the radicals arising from the fragmentation of the monomer molecules. Also, one can circumvent monomer radiolysis by postirradiation grafting, *i.e.* by submitting cellulose to gamma rays and adding the monomer to be grafted only after this treatment. The macroradicals formed in the irradiation period have a long lifetime because they are trapped and immobilized within the rigid cellulosic structure. They can 'survive' for several days if the irradiated cellulose is kept in an inert atmosphere at low temperature.

Free radical grafting of cellulose in a homogeneous organic medium, rather than in aqueous suspensions, is feasible, as illustrated by experiments in the solvent system DMSO–paraformaldehyde with acrylonitrile as grafting monomer.[46]

Grafting by anionic polymerization has received some attention.[41-45] For example, acrylonitrile can be made to polymerize from sodium cellulosate in an aprotic medium. The hydroxy functions of cellulose can initiate the polymerization of ethylene oxide, but only short grafts are obtained. Ethylene imine reacts likewise. More recently, another route based on termination onto cellulose has been developed.[45,47] It consists in preparing a monofunctional living carbanionic polymer which is made to react with, say, carbon dioxide to introduce a carboxylate moiety and then adding this to cellulose esters with good leaving properties, *e.g.* mesylate, to attain grafting.

Cationic polymerization has also been used recently to graft cellulose.[48,49] These studies exploited the very pronounced reactivity of certain nucleophilic monomers towards ester-type initiators. Cellulose tosylate and other esters of strong acids were shown to induce the pseudocationic polymerization of 2-oxazolines to yield polyamide-grafted products.

Block copolymers of trimethylcellulose with polytetrahydrofuran have also been prepared by cationic initiation.[50] Partial degradation of fully substituted cellulose under controlled anhydrous conditions yielded monofunctional chains bearing a terminal C(1)—Cl bond. This was the source of activation by $AgSbF_6$ in the presence of tetrahydrofuran, as shown in Scheme 4.

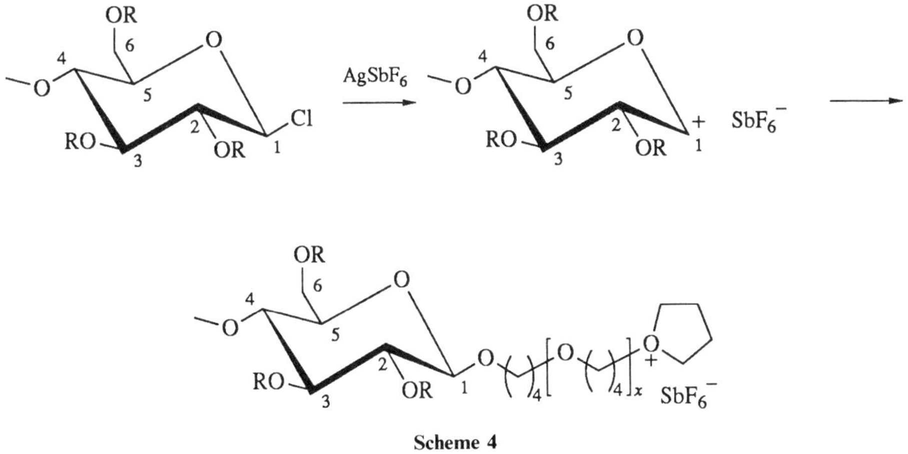

Scheme 4

Even Ziegler–Natta polymerization has been attempted on cellulose in the form of a sheet of paper. A catalytic mixture of triethyl aluminium and titanium tetrachloride was deposited on the fibres and thereafter the sheet was exposed to gaseous ethylene. The monomer readily polymerized thus forming a layer of polyethylene on the sheet.[51]

Grafting by polycondensation has also been envisaged and carried out. Lactones and lactams can be ring opened by OH or COCl groups on cellulose to give, respectively, polyester and polyamide grafts.[41] Alternatively, cellulose can be used as a comacromonomer in conjunction with a diacid chloride and a diamine to yield polyamide grafts.[41] Finally, linear preformed polyesters, polyamides and polyepoxides have been grafted onto cellulose through the condensation of end groups with the hydroxy moieties of the substrate.[41]

The grafting procedures described for cellulose have also been applied to many of its derivatives with similar results but with quantitative features which often depend on the nature of the groups appended to the substrate.

Cellulose triesters bearing a single hydroxy group have been coupled with bis(4-isocyanatophenyl)-disulfide to give a cellulosic macroinitiator for the free radical synthesis of block copolymers.[52]

A large number of monomers have been used to graft cellulosic polymers. In fact, virtually all common vinylic and acrylic monomers known to polymerize *via* a free radical mechanism have been tested. The properties of the grafted products, such as their mechanical performances, thermal resistance, dimensional stability, affinity for water, solubility in organic solvents, resistance to light, fire and aggressive chemicals, ion-exchange capacity, dye fixation and wet strength, have been extensively tested in view of novel applications of these fibres in textile, paper and cardboard products.[41-44]

The most promising areas of development of grafted cellulosics are those connected with extremely high water absorbancy and polyelectrolyte activity. Superabsorbency[53] has become a sought-after property for sanitary uses, soil conditioning, medical applications and the drying of organic liquids. Some cellulose graft copolymers bearing polyacrylate anions give very good performances and can replace the totally synthetic crosslinked polyacrylates which are nonbiodegradable.

The above discussion on the grafting of cellulose and cellulose derivatives applies today more and more to lignocellulosic materials such as lignin-containing pulps and even wood itself. The main purpose is to achieve physical compatibilization between the grafted materials and synthetic polymers in order to develop new blends and composites.[42,54]

19.3.2 Starch

There is nothing specific to starch with respect to chemical modification in the sense that the same concepts, reaction mechanisms and strategies have been applied to this polysaccharide as to cellulose. The only difference is from the point of view of the amount of work carried out. Cellulose, with its major applications, cotton and paper, is a much more important polymeric material than starch and has spurred therefore a much wider interest, both in fundamental studies and in the search for practical realizations. This has not stopped researchers from looking into the potentials of starch derivatives and graft copolymers,[55,56] often as a complementary study to an original investigation on cellulose.

Thus, carboxymethylstarch, starch xanthates and starch graft copolymers, *e.g.* with polyelectrolytic structures for superabsorbing materials, have all found their way into the market.

Moreover, in recent years, the natural susceptibility of starch to biodegradation has sparked a considerable amount of research in view of developing polymeric materials capable of undergoing such ecological destruction. The first generation of these products[57] was in fact based on a deceiving concept, namely the mere mixing of a nonbiodegradable polymer, such as polyethylene in a fibre form, with starch (typically 10–20% of the composition). The claim of biodegradability was shakily based upon the mechanical disintegration of the macrostructure due to the destruction of the starch: of course the main constituent remains chemically unaffected and is therefore not recycled by natural events. More recently, a polymer blend of starch (up to 80%) with another undisclosed biodegradable polymer has been manufactured and tested.[58] It seems a promising material in that it is claimed to combine biodegradability with good mechanical properties.

The importance of research aimed at biodegradable polymers is too obvious to need further emphasis: suffice it to quote an estimated present market for such material of about 400 kt per year (see also polyhydroxybutyrate below).

19.3.3 Chitin

This polysaccharide is the most abundant organic component of invertebrates. The native material possesses a variable extent of *N*-acetyl groups and can be totally deacetylated to give chitosan. This polymer finds several applications as in wound healing and metal complexation.

Chitin derivatives other than chitosan are usually prepared from native or incompletely deacetylated precursors and give rise therefore to terpolymers consisting of (typically) some 40% chitin (N-acetylglucosamine) units, some 10% chitosan units and the remainder of substituted moieties. These include N-acyl and N-carboxyalkyl groups arising from typical reactions of amines applied to the chitosan units, O-carboxyalkyl groups, formed in reactions of the hydroxy groups as with cellulose, and metal-ion chelates.

Their applications span from textile adjuvants to molecular sieves, to blood-dialysis membranes, to slow release pharmaceuticals. The strong chelating power of some of these derivatives make them also particularly well suited as ion collectors (recovery of trace metals from solutions) and chromatographic substrates. A recent symposium monograph[59] gives a clear picture of the growing importance of these materials. Indeed, chitin and chitosan are today a major renewable source of polymers with special properties. Several monographs[59-61] testify to this effect and give a wide array of specific examples.[62]

19.3.4 Carbon

It should be remembered that certain natural polymers are the source of carbon black and activated carbon. The chemical modification is here quite drastic. Thus, combustion in an oxygen-poor atmosphere of bones and other animal residues produces these carbonaceous materials, which can be viewed as polymers (oligomers) in so far as their molecular structure is that of a polycyclic aromatic hydrocarbon including tens of fused rings and some residual functional groups at the periphery of these large molecules. The fundamental role of such products is typified by their use as additives in tyres, pigments in printing inks and specific scavengers in purification or separation procedures. Of course mineral resources such as paraffins are also employed for making the same products and therefore renewable resources are not structure specific for this purpose.

An interesting recent development is the possibility of grafting macromolecular chains by step polymerization from these carbon particles or by condensation onto them.[63,64]

19.4 THE POLYMERIZATION OF OLIGOMERS ARISING FROM RENEWABLE RESOURCES

The term oligomers is used here to define two types of structures which are susceptible to undergo polymerization and have been studied and used in this manner: the unsaturated oils derived mostly from vegetable matter and the lignin fragments as obtained from the various delignification processes. Hemicelluloses are oligosaccharides, but have not interested polymer chemists as potential macromonomers to any relevant extent.

19.4.1 Drying Oils

These natural substances[11,12] are never in the form of a pure compound, but rather a mixture of several triglycerides which usually include saturated, monounsaturated and polyunsaturated aliphatic chains. Linseed oil, one of the major representatives of this family has the following approximate composition expressed in percentages of fatty acid residues: palmitic, 4–8; stearic, about 3; oleic, 15–30; linoleic (7), 15–25; and linolenic (8), 45–60.

The capacity to 'dry', i.e. polymerize, in air to a crosslinked solid film, arises from the reactivity of the C—H bonds situated next to a C=C unsaturation. The mechanism of this oxido-polymerization[65] is, not surprisingly, essentially the same as that of the oxidation of elastomers containing a large number of unsaturations, such as polyisoprene (natural rubber) and polybutadiene. Free radicals are formed by hydrogen atom abstraction from the reactive sites mentioned above, followed by oxygen addition to give peroxy radicals. The sequence of reactions that follows is given in Scheme 5 with linoleic acid.

Obviously, the higher the double bond content of the oil the higher its susceptibility to 'dry'. This is usually expressed as the iodine value, or number, which reflects the analytical method used to determine the degree of unsaturation of the oil and represents the grams of iodine consumed in a titration with I_2Cl_3 per 100 g of oil. Iodine values below about 150 are indicative of modest drying capacities and correspond to less than two unsaturations per ester chain. Above 150 the oil dries well and is generally apt to be used in formulations. Thus, whereas oleates are inadequate

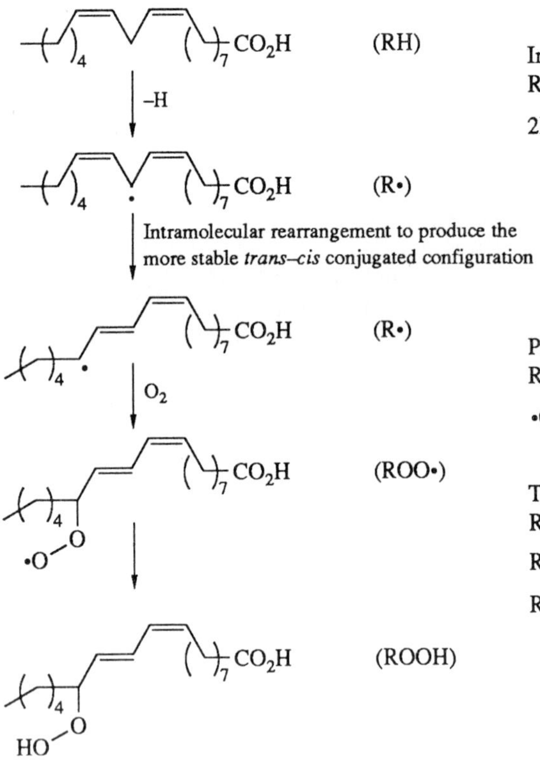

Initiation

ROOH ⟶ RO• + •OH

2ROOH ⟶ RO• + ROO• + H_2O

Propagation

RO• + RH ⟶ ROH + R•

•OH + RH ⟶ R• + H_2O

Termination

R• + R• ⟶ R_2

R• + RO• ⟶ ROR

RO• + RO• ⟶ ROOR

Scheme 5

drying oils, with only one double bond per chain, linoleates, and, better still, linolenates are appropriate structures. The other parameter relevant to the drying rate is the specific state of the unsaturations, *viz.* whether they are conjugated or not, the former structures being more reactive. Thus, linolenates dry less rapidly than olaeostearates under the same conditions.

In fact, this reactivity is not very pronounced if the oil is in bulk, for the obvious reasons that little oxygen is in actual contact with the sensitive structures. Only when the material is spread as a thin film on a substrate, *e.g.* paper or canvas, does it expose itself to a more radical intervention of the atmosphere because now the proportion of molecules at or near the surface is high with respect to those still inside the bulk of the liquid.

The basic applications of these oils are therefore in printing ink vehicles, paints and surface treatment (wood, floors, *etc.*). In some instances a catalyst is added to accelerate the polymerization: typically, organic transition metal salts or complexes are used, such as cobalt(II) naphthenate. The role of these cations (the organic part only improves the solubility in the oil) is to enhance the production of free radicals through a propogation loop, as shown by equations (3) and (4).

$$Co^{2+} + ROOH \longrightarrow Co^{3+} + RO\bullet + OH^- \quad (3)$$

$$Co^{3+} + RH \longrightarrow Co^{2+} + R\bullet + H^+ \quad (4)$$

The desired viscosity with these oils is obtained by a controlled prepolymerization in bulk at relatively high temperature, a process which has kept the old name of standolization from the time when the oils were left for long periods to thicken in contact with air.

Castor oil belongs to this family but is not, as such, a drying oil since its major component is the glycerine triester of a monounsaturated monohydroxylated acid, ricinoleic acid (**17**). Only a specific thermal dehydration treatment can convert this structure into a doubly unsaturated ester chain (as shown in equation 5) and this new substance (**18**) can be used in typical drying oil applications.

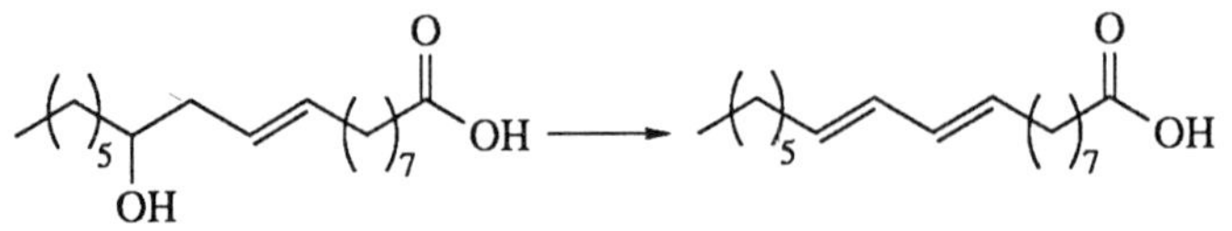

(5)

(17) **(18)**

Castor oil itself on the other hand is also a macromonomer thanks to the three OH groups of the glycerol ricinoleate. It is used for the synthesis of elastomeric polyurethanes in conjunction with diisocyanates and other polyester or polyether diols.

Drying rates can be enhanced by using esters of these unsaturated naturally occurring acids, obtained from the hydrolysis of the triglycerides, with polyols with functionalities higher than three, *e.g.* pentaerythritol. Other fast-drying compositions involve the same acids with alkyl resins, *viz.* aromatic oligoesters, bisphenol-A epoxy resins and styrene–allyl alcohol co-oligomers.

Both triglycerides and their acids can be modified by the Diels–Alder reaction of their double bonds with maleic anhydride here too to enhance the drying rate and introduce water mixibility after hydrolysis of the anhydride function.

Recently, interest in euric acid(**19**), a monounsaturated C-22 carboxylic acid present in rapeseed and *crambe abyssinica* oils, has been revived by studies of its transformation into the saturated difunctional brassylic acid (**20**) which is a potential comacromonomer in the synthesis of polyesters by partial substitution of adipic acid.[66] This leads to better flexibility of the resulting polymers used as coatings.

(19) (20)

19.4.2 Rosins[12]

The two conjugated double bonds in abietic acid (**9**) and some of the other components of rosins are sensitive to oxidative polymerization, following a mechanism entirely similar to that sketched above for drying oils. Thus, these rosins applied as coatings tend to evolve with time towards cross-linked structures. This property can be useful in certain applications such as with printing inks, but in others long term chemical stability is preferred. Therefore, as mentioned above, natural rosins are frequently subjected to chemical modifications.

Catalytic hydrogenation of at least one double bond is an example of such procedures: not only does one decrease or eliminate the susceptibility of the rosin to oxidation, but also the concentration of chromophores and potential chromophores by reduction and with it the possibility of colouration with ageing.

Thermal treatment of rosins with appropriate catalysts is an alternative stabilization process: abietic acid is disproportionated to dehydro (aromatic) and dihydro (monounsaturated) homologues which are obviously more resistant to oxidation.

A third approach involves the acid-catalyzed dimerization or oligomerization of the abietic-type acids in rosins to reduce the degree of unsaturation.

Finally, treatment of rosins with dienophiles such as maleic anhydride gives useful mono-unsaturated polycarboxylic structures through cycloaddition (Diels–Alder reaction). These have been used as such or after hydrolysis, and also in the synthesis of polyimides. The Diels–Alder adduct of rosin with acrylic acid has been prepared and converted into a diacid chloride in order to generate polyesters and polyamides.

Other chemical modifications, involving this time the carboxylic groups, are common practice with rosins, whether natural or modified as above. Thus, esters with alcohols, glycols, glycerol and pentaerythritol, and metal salts like sodium and potassium soaps and calcium 'resinate', are readily prepared to impart specific properties to the rosins, but alcohol and amine functions are also appended in certain instances.

The applications of these materials range from paper sizing, printing ink vehicles and varnishes, to adhesives, emulsifiers and food additives.

19.4.3 Lignins

Native lignins in vegetal matter represent an extremely abundant renewable resource, second only to cellulose. Very large quantities of lignins are produced by the papermaking industry which usually burns them *in situ* to recover energy and pulping chemicals. The recent development of novel pulping technologies, such as the organosolv processes, and of original biomass refineries, such as steam explosion, are further sources of lignins. It is clear that all sectors concerned, whether traditional or emerging, would profit if lignins were to find interesting uses and valorizations.

Among the various potential uses of lignins, other than fuel, polymers are undoubtedly a promising alternative and the last decade has witnessed a considerable research effort in that general direction.

Lignins vary very considerably in structure, molecular shape, average molecular size and molecular weight distribution according to the plant species, its age, the location within the cell wall and the delignification technique used to isolate them. Structure (6) is a typical series of enchainments which purports to show all the characteristic moieties in lignins. Two major sites determine, by their relative abundance, the nature of lignins: those from softwood possess predominantly guaiacyl units, *i.e.* trisubstituted rings (21), whereas those from hardwood contain guaiacyl and syringyl (tetrasubstituted rings) units (22). A third unit is found in grass lignin, *i.e.* a 1,4-disubstituted phenolic structure like (21) and (22), but without methoxy groups.

The aliphatic groups appended to these typical moieties are mostly made up of three carbon atoms bearing alcohol, ether and, less frequently, carbonyl functions, as shown in structure (6). It is therefore a standard practice to express the relative content of specific groups as their number per C-9 (aromatic ring + 3 aliphatic carbon atoms) unit, as in the generic 'monomer' unit (23). The structural characterization of lignins, both *in situ* and preferably after extraction from the lignocellulosic substrate, has received considerable attention and a powerful set of procedures is available today.[67] These range from classical chemical analyses to specific degradation processes, aimed at identifying the typical substructures, to detailed UV, IR and NMR probings of unmodified and chemically modified, *e.g.* acetylated, samples. To the search and quantitative assessment of each moiety one must add the physicochemical determinations related to average molecular weights and their distribution and to molecular shape including the degree of branching. Finally, the residual content of saccharides and the ionic functions left from the original structure in the plant, or introduced by the extraction technique, must be evaluated.

Following the delignification procedure adopted, lignins can appear in various forms: (i) the sulfonates, from the sulfite pulping processes, with various cations are water-soluble materials commercially available in large quantities; (ii) kraft or soda pulping processes yield materials which are soluble in alkali and in some organic solvents; and (iii) lignins from organosolv, steam explosion and acid hydrolysis processes vary in structure and molecular size according to the type of operation and the choice of parameters and chemicals used.

Among the uses, actual or potential, of lignins, several concern polymer science and technology, namely the degradation to monomers and the use as fillers or macromonomers. Whereas the former is discussed in the appropriate section below, the others are dealt with here, with particular emphasis on the chemical reactivity. The use of lignins, as obtained or chemically modified, as surfactants, viscosity control agents, flocculants, papermaking additives, *etc.*, *i.e.* their roles outside the realm of polymeric materials, will not be discussed here. The details of the chemical modifications of lignins to be employed as fillers, additives or macromonomers will also be avoided in most instances. A recent monograph constitutes an excellent survey of these general topics.[68]

19.4.3.1 *Fillers or additives for polymers*

Numerous studies have been conducted on the possible use of lignins as fillers. The addition of lignins to formaldehyde-based resins, particularly phenol–formaldehyde adhesives, has been a topic of considerable interest. The results are not particularly encouraging if the lignins are not pretreated to render them chemically reactive towards the resin. In fact, a mere physical participation of the lignins into the final network tends to reduce the performances of the adhesive and to slow down curing.[69,70] As explained below, these drawbacks are due to the very limited number of sites potentially available for cocondensation, particularly in a basic reaction medium. Poor solubility in the resin is of course another important negative factor: conventional pulping techniques give lignins which are insoluble in phenol–formaldehyde prepolymers, but organosolv lignins, with lower molecular weights and narrower DP distributions show a better affinity and hence increase the possibility of chemical participation at the curing stage. Compatibility with styrene–butadiene rubber is not particularly good if the lignins are added by dry milling. On the other hand, coprecipitation from a latex containing the lignin in solution gives good physical properties, but poor performances in the processing stage.[68] A more recent investigation has shown that with chlorobutadiene–butadiene latexes and a lignin which had been treated with epichloridrin to convert phenolic OH groups into aliphatic ones, coprecipitation leads to less viscous and therefore more readily processable compounds.[71] Moreover, the mechanical properties of the vulcanized product were as good as those of a test compounding recipe in which the standard carbon black additive was used instead of the molified lignin.

Blends of modified lignins with various synthetic polymers have also been studied.[72] The chemical modification of kraft lignin involved the oligomerization of methyl oxirane (propylene oxide) from both aliphatic and phenolic OH groups, a procedure which had been used originally to 'condition' lignin in view of its reaction with diisocyanates (see below). The hydroxypropyl lignin thus obtained was mixed with polyethylene, poly(methyl methacrylate), poly(vinyl alcohol) and with ethylene–vinyl acetate copolymers of different compositions. Whereas the blends with polyethylene displayed phase imcompatibility, increasing proportions of vinyl acetate in ethylene–vinyl acetate copolymers, while maintaining immiscibility, improved the tensile strength of the polyblends. This suggests that as the copolymer gains polarity it also becomes more compatible with the modified lignin. Blends with poly(methyl methacrylate) were also biphasic under all tested conditions, with some degree of compatibility illustrated by the observation of a single glass transition temperature. Their mechanical properties depended on the composition and the molecular weight of the hydroxypropyl lignin. When poly(vinyl alcohol)s with OH contents higher than 75% were used as synthetic polymers, the blends were homogeneous in all the compositions studied. Clearly, the similarity in the polar character of the two components determined this compatibility.

In another set of studies,[73] blends of organosolv lignins, as obtained or chemically modified, with hydroxypropyl cellulose were prepared and characterized. An optimum compatibility was obtained with 20 to 40% substitution of the lignin's OH groups by ethylation or acetylation.

In another vein, which calls upon a strategy involving a much more reduced utilization, the possible role of lignins as stabilizers against various forms of polymer degradation has been investigated. Claims to the stabilizing effect of lignins and lignosulfonates against the thermal, oxidative and photolytic degradation of polyalkenes have been published sparsely during the last 30 years.[74–76] It seems logical to postulate that these additives would play a role similar to that of substituted (hindered) phenols, *i.e.* radical traps. Organosolv lignins, characterized by low molecular weights, *viz.* 1000 to 2000, might prove interesting for this application.

19.4.3.2 Macromonomers

The use of lignins from various origins as macromonomers in the synthesis of polymeric materials has received renewed and sustained attention in the last decade or so. Two approaches have been followed: (i) either the lignin oligomers are used as obtained from the delignification process; or (ii) some chemical modification is carried out upon them before the actual polymerization is conducted. The reactive sites which can be exploited for both situations are, in order of importance and with reference to sturcture (**6**), the hydroxy groups, usually both aliphatic and phenolic, the mobile hydrogen atoms attached to aromatic carbons and those on aliphatic carbons. The occasional alkenyl, carbonyl and carboxylic functions play a very minor role in this context, except perhaps the CO_2H groups in specific lignins arising from strongly oxidizing delignification processes which are still in their infancy.

One of the possible uses of lignins as macromonomers is their transformation into phenolic-type networks. This has been achieved by oxidative coupling promoted by hydrogen peroxide or by acid catalysis, applied to both kraft lignins and lignosulfonates. The mechanisms proposed[70] for each of these condensation processes are given in Schemes 6 and 7.

Another important mode of lignins exploitation as macromonomers is their incorporation as coreactants in phenol–formaldehyde resins. It is well known that the addition reactions of formaldehyde with phenol in both basic and acidic media lead to the formation of methylol groups

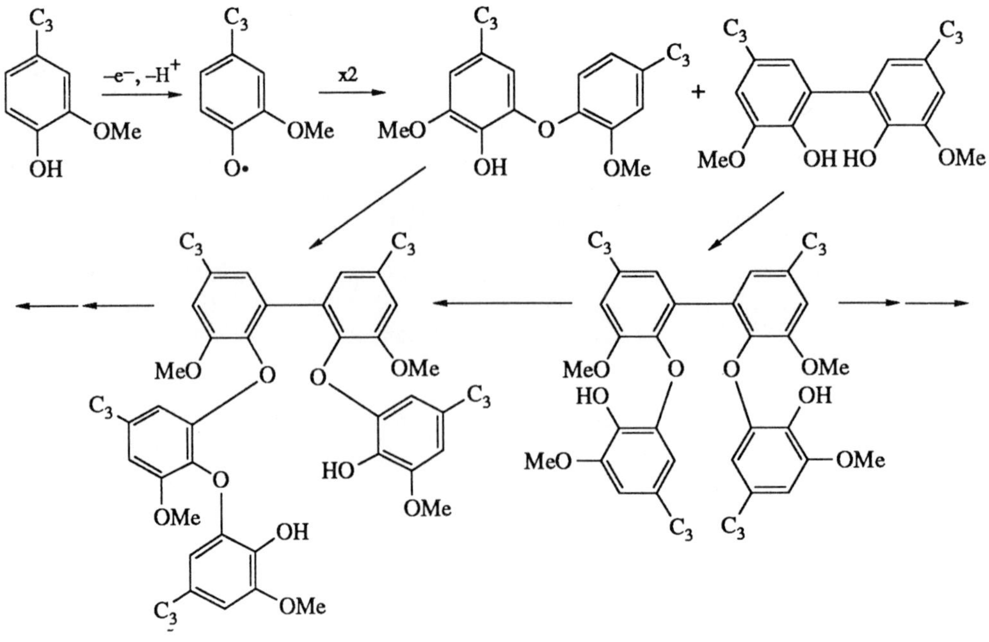

Scheme 6

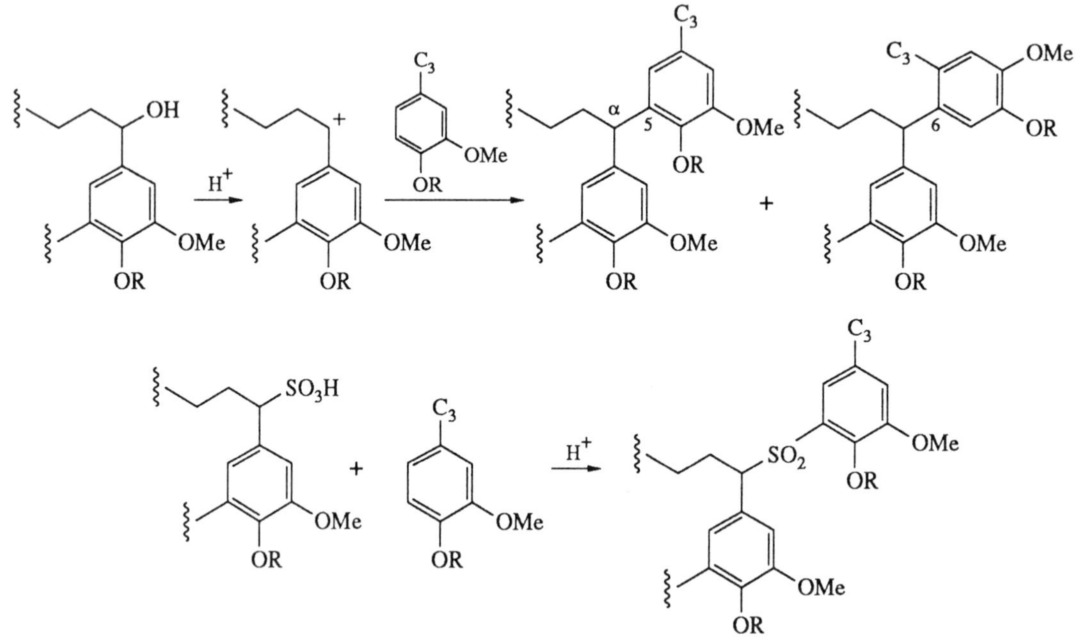

Scheme 7

at the two *ortho*, or 2 and 6, and/or at the *para*, or 4, positions with respect to the phenolic OH. Successive condensation reactions of these methylol functions between themselves or again with free *ortho* and *para* positions, produce chain growth with the formation of CH_2OCH_2 and CH_2 bridges between the phenolic rings. Given the potential trifunctionality of phenol, this polycondensation system generates branched structures (resins or prepolymers) and ultimately crosslinking at the curing stage. An inspection of structures (**21**) and (**22**) suggests that lignins from softwood should have few reactive sites to offer for the addition of formaldehyde or the condensation with methylol groups from phenol–formaldehyde or lignin–formaldehyde resins and that lignins from hardwood ought to be even less prone to such reactions. If one considers moreover the existence of steric hindrance displayed by lignin macromolecules towards these condensation reactions, the reactivity appears reduced to a minimum expression. This is indeed the case of typical systems with basic catalysis. As pointed out above, addition of traditional lignins to phenol–formaldehyde prepolymers and subsequent cure, shows little or no chemical participation of lignins, i.e. their role is reduced to that of a filler. However, organosolv lignins seem to give good results when used in typical phenol–formaldehyde compositions in partial replacement of phenol.[77,78] Their performance as wood adhesives was better than that of similar systems involving kraft lignins and even better than that of standard control resins. The reasons for these positive results are probably to be attributed to the higher solubility of organosolv lignins in the reaction medium comprising the resin and sodium hydroxide.

It has been shown by very detailed studies involving model compounds[79] that both guaiacyl and syringyl units are susceptible to electrophilic substitution reactions at the *meta*, or 3 and 5, positions with respect to the phenolic OH groups, i.e. the permanently free sites in structures (21)–(23). In particular, formaldehyde readily condenses with the hydrogen atoms on these sites in acidic media, whereas with basic catalysis only the few available *ortho* and *para* positions are activated. The application of these findings to lignins, i.e. their activation by acid-catalyzed methylolation, opens the way to the development of lignin-containing resins of the classical phenol–formaldehyde type in which the lignins are incorporated chemically into the final network.

A similar approach calls upon the acid-catalyzed phenolation of lignins. This has been achieved either after reaction of lignins with formaldehyde,[80] by condensation of the resulting methylol groups with the *ortho* and *para* positions of phenol, or by direct condensation[81] of the same sites with lignin functional groups (OH, OR *etc.*) on the carbon atom in the α-position with respect to the aromatic ring. The aim of phenolation is to introduce reactive sites in the modified lignins so that they can participate in polycondensation reactions as with classical phenol–formaldehyde systems, except that the phenol is now replaced by the phenolated lignin. Good adhesive properties have indeed been attained,[80,81] particularly with steam explosion lignins.[81]

In conclusion, the possibilities of utilizing lignins as chemical components in the formulation of phenol–formaldehyde type resins and thus achieving good results in terms of adhesive properties, seem to be enhanced if an appropriate but quite simple functionalization procedure is applied and/or if more reactive and more soluble lignins from novel separation processes are used.

Grafting reactions from lignin core molecules *via* free radical polymerizations were first studied in the 1960s.[82] Monomers like styrene, methyl methacrylate and acrylonitrile were used and initiation was either chemical or through gamma rays. On the whole, chemical initiation provides a better means of grafting, but both approaches have to face the inevitable retarding or even inhibiting role of the phenolic moieties towards free radicals.

It seems established that the branching sites are generated on the lignin molecules, i.e. that grafting occurs from the substrate, and not onto it, by coupling of a homomacroradical of the monomer with a lignin radical. Lignin radicals can be generated by hydrogen abstraction from various lignin sites or by first appending radical sources onto the lignin structure and then decomposing them appropriately. Thus, for example, one can work with a classical initiator such as benzoyl peroxide or pretreat the lignin with ozone or hydrogen peroxide to introduce labile O—O bonds.

The possibility of initiation from a free radical generated on the lignin substrate depends on: (i) its relative stabilization, which implies that the more delocalized the unpaired electron, the less likely the initiation with the added monomer. Thus, reactions carried out in the presence of oxygen yield lower grafting efficiencies because the aromatic lignin radicals readily form highly stabilized quinoid radicals. In an inert atmosphere, some of the lignin radicals, namely the benzylic ones, can be rather effective for giving rise to grafts; (ii) the nature of the monomer, i.e. its reactivity towards free radicals and the degree of stabilization of its own ensuing radical: and (iii) the extent of accessibility of the radicals in terms of steric factors, with homogeneous, but much more with heterogeneous systems in which the degree of swelling and chain expansion are obviously very important. On the whole, these studies tended to show rather poor performances and gave materials of little interest.

More recent work has focussed on the grafting of acrylamide onto various types of lignins.[83–85] Typical initiating systems comprise an oxidizing agent and a cerium(IV) salt in an acidic medium in the presence of chloride ions.[83,84] Initiation with this complex mixture was shown to need the chloride ions and not the cerium ions, but no exhaustive mechanism could be proposed. A further investigation[85] has provided means to control the product properties through the choice of lignin and of the experimental parameters. It is interesting to note that no crosslinked materials are obtained in these syntheses, i.e. that coupling of grafted macroradicals between themselves or with lignin radicals are not dominant termination reactions.

In a similar vein, acrylamide has been grafted onto lignosulfonates using the classical

Fe^{2+}/hydrogen peroxide initiating system in water.[86] However, the simple mechanism proposed for the initiation has been questioned.[85]

The polyacrylamide-grafted lignins are water-soluble materials with promising applications, both as such or as the corresponding acrylates, as thickening and dispersing agents or as superabsorbers.

An alternative way of promoting free radical grafting from lignins consists in condensing acrylate groups onto them through the reaction of acryloyl chloride or acrylic anhydride with the hydroxy groups.[88] This chemical modification is followed by addition of an acrylic monomer and a typical radical initiator: homopolymerization is accompanied by some copolymerization with the prepared macromonomers and some grafting thus occurs. The functionalization of lignin with polymerizable moieties will be discussed further below in the context of the polymerization of macromonomers.

Grafting of lignins with chain reactions other than radical polymerization has not been reported with vinyl-type monomers, but anionic grafting with oxiranes has become an important research topic.[89,90] Indeed, considerable efforts have been devoted to the base-catalyzed reaction of propylene oxide with lignin model compounds and various kinds of lignins.[91] Most of this work is aimed at preparing precursors for lignin-based crosslinked polyurethanes. A considerable knowledge has been acquired on the reactivity of the different OH groups present in lignins, and star-like structures with known average branch numbers, as in (24), have been prepared by reducing the available hydroxy groups through preliminary partial ethylation with diethyl sulfate.[90]

(24)

Oxypropylated lignins have been used extensively as macromonomers for the preparation of polyurethanes.[92–95] The chain extension was intended to bring the OH groups away from the core of the lignin molecules and thus enhance their reactivity by eliminating steric hindrance, but also to introduce flexible segments of variable length with respect to the stiff lignin structure. Aliphatic and aromatic diisocyanates were employed for these nonlinear polycondensations.

The properties (glass transition temperature, mechanical behaviour) of these materials were investigated as a function of structural parameters related to the comonomers used. This included the reactions of hydroxypropyl or unmodified lignins with diisocyanates in the presence of diols, typically polyethylene oxide glycol.[96] The structure/properties relationships established that T_g, elasticity modulus and ultimate tensile strength increased with increasing lignin content in the networks. Conversely, ultimate tensile strain decreased with increasing lignin content. The role of the type of diisocyanate employed to build the network and the choice between chain extension and the use of a polyether comonomer were also inspected closely together with the results obtained from unmodified but fractionated lignins.

Polyurethane foams based on lignin, polyethylene oxide glycol and a diisocyanate have also been prepared and evaluated.[96]

Other polyurethanes based on lignin as comonomer have been described recently.[97–99]

Epoxy lignins can be readily prepared by appropriate reactions of kraft, hydroalkyl, or phenolated lignins.[100] Reactions of these modified macromonomers with diamines or dianhydrides yield tough crosslinked materials which can be softened by the use of comacromonomers like NH_2-terminated elastomeric structures.[100] The aim of this study was to arrive at rubber-modified tough epoxy networks based on lignin with mechanical properties and glass transitions which could be modulated as a function of the composition of hard and soft regions.

Other epoxy resins compositions loaded with as much as 40% of lignins (alkali, kraft or ozonized lignin) have been studied in view of developing new adhesives.[101,102] It seems clear that lignins are not simple physical additives (fillers) in these homogeneous mixtures, but the specific reactions involving their chemical participation in the curing processes are not well established.

A different approach to lignin macromonomers is based on the introduction of polymerizable C=C unsaturations, such as vinylic[87] or acrylic[88] moieties, by specific reactions involving the OH groups. One of these reactions has been mentioned earlier, viz. the esterification with acrylic chloride.

An alternative method calls upon more specific reagents such as isocyanates (**25**) and (**26**) bearing the appropriate unsaturations at the other end of the molecule.[103]

(**25**)   (**26**)

The lignin derivatives thus synthesized[87,88,103] inevitably bear, on average, more than one polymerizable function per lignin molecule and display therefore strong branching capacities in both homopolymerizations and copolymerizations with standard monomers or other macromonomers. Only radical polyadditions have been applied to these systems which always lead ultimately to network formation. No applications seem envisageable for the time being for these products.

Polyesters derived from lignins have been investigated in the last few years.[104,105] When kraft lignin is dissolved in a polar solvent like dimethylacetamide and mixed with a diacid chloride, such as sebacoyl or terephthaloyl chloride, in the presence of a proton acceptor like pyridine or triethylamine, polycondensation reactions between the aliphatic and phenolic OH groups of lignin and the COCl functions readily take place and lead to gellation.[105] Under optimized conditions favouring a maximum involvement of the OH groups, *i.e.* with an excess of diacid chloride, most of the product was crosslinked. Clearly, however, no special measure, such as chain extension, was found necessary to ensure the participation of the OH groups of lignin in the esterification reactions. Both aliphatic and phenolic hydroxy functions intervened in the polycondensation as shown by the presence of the characteristic IR peaks for the corresponding ester carbonyl groups.[105]

These networks do not show a clear-cut glass transition by DSC analysis because of the high polydispersity of the starting lignin. The thermal degradation of these crosslinked polyesters occurs above 200 °C.

In a further study on lignin-based polyesters,[106] the same diacid chlorides were made to react with the same unmodified lignin, but without solvent and base. Instead, polyethylene oxide glycols (PEO) were added in order to have a comacromonomer and also to achieve homogeneity since kraft lignins are soluble in these oligomers. It was possible to vary the relative proportion of lignin with respect to the total hydroxylated macromonomers from zero to 35% and to study the changes in properties of the materials. Reactions proceeded smoothly up to gel formation and maximum yields of crosslinked products were obtained with the correct COCl/OH ratio of unity.

This time the glass transition became manifest on the DSC tracings, but it covered a temperature range which was the wider the higher the lignin content in the original reaction mixture. The T_g values also increased with the lignin content for a given diacid moiety and were obviously higher with terephthalates than with sebacates. Another parameter studied was the molecular weight of the PEO.[106]

These elastomeric polyesters constitute a promising new family of lignin-based polymeric materials which are now under study in terms of mechanical properties.

Lignosulfonates and 'sulfur' lignins doped with certain Lewis acids have been reported to exhibit semiconductor properties with conductivities of 0.006 S cm^{-1}.[107] This is an interesting result which should encourage further studies.

Finally, it is noteworthy to quote an investigation in which hydroxypropyl lignin was condensed with cellulose triacetate oligomers bearing NCO groups.[108] The resulting networks had crystalline and amorphous block-like substructures with rather poor mechanical properties.

Although other applications have been suggested for lignins in view of obtaining novel macromolecular materials,[109] the above survey is already eloquent proof of the wide scope of the research activities in this field. As with cellulose grafting, the problem of the economic incidence of the chemical or physicochemical transformations discussed above has to be considered very carefully against the gain in added value achieved, *i.e.* the improvement of some specific properties of these polymers. The trend is in the right direction, but much more research involvement is needed to cover the ground ahead.

19.4.4 Tannins

Tannins are phenolic-based natural products. They are found mostly in the bark of pine, the wattle of mimosa and hemlock and in the wood of certain trees such as quebracho and sumach. The extraction of these substances leads to a mixture of oligo- and poly-flavonoids which are known as condensed tannins, with number average molecular weights ranging from 1000 to 4000 depending on the species which generated them. Two typical structures of flavonoid units are those from mimosa (4–6 flavonoid) **(27)** and pine (4–8 flavonoid) **(28)** tannins, respectively.

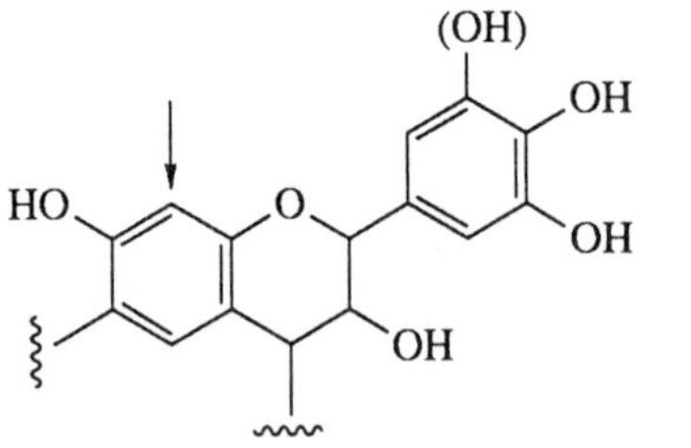

(27) 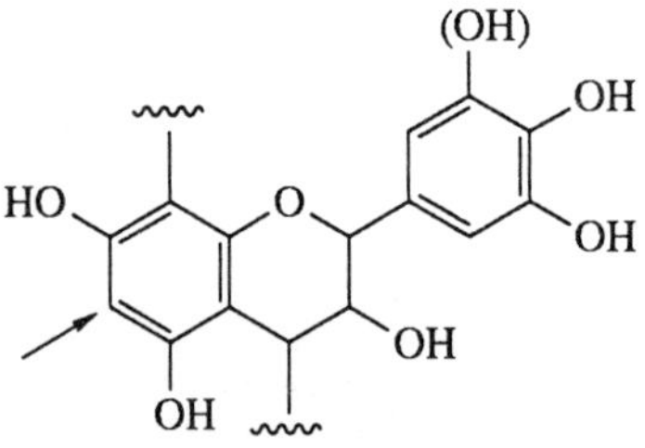   (28)

Tannins are the major reagents for the tanning of leather and are manufactured at a yearly tonnage of several hundred thousand. Given the phenolic-type structures borne by these oligomers, it is quite logical to have envisaged their use as macromonomers in formulations involving the characteristic phenol–formaldehyde condensation reactions.[15] The nucleophilic sites of possible condensation are the 8-positions in the 4–6 units of mimosa tannin and the 6-positions in the 4–8 units of pine tannin, as indicated by the arrows on the respective structures. Other possible condensation sites are in fact much less reactive and do not participate extensively in the chain extension and crosslinking mechanisms, except if the pH is increased above about 10.

Much experience has been gained on the making and properties of tannin-based resins.[15,110] These include a number of combinations, such as phenol–formaldehyde, resorcinol–formaldehyde, urea–formaldehyde prepolymers and also mixtures therefrom to which tannins or tannin–formaldehyde resols are added: in a basic medium these mixtures cure at room temperature to networks which possess good adhesive properties, particularly for plywood. Although cold curing ensures most of the network formation, it was shown that further condensation occurs when one heats these materials at about 130 °C.

An interesting novel way[111] to employ tannins in these materials consists in hydrolyzing the condensed structures with acids in the presence of resorcinol in order to generate protonated monomeric flavonoids which add onto resorcinol. The resulting tris(biphenolic) structures are then crosslinked by polycondensation with formaldehyde (Scheme 8).

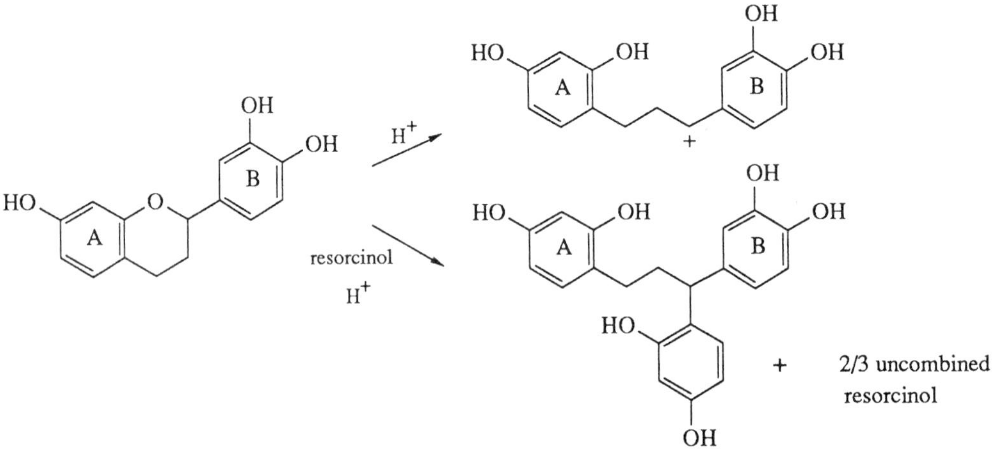

Scheme 8

The percentage of tannins in the various adhesive compositions described above can reach very high proportions, up to 65%.

19.5 MONOMERS FROM THE BIOMASS AND THEIR POLYMERIZATION

19.5.1 Lignin Degradation Products

Depolymerization of lignins all the way down to phenolics can be carried out in a variety of ways: catalytic oxidation or hydrogenation and pyrolysis.[68] Little has been done to exploit these processes in the direction of monomers, the products being instead considered in the context of fuels and commodity or fine chemicals. It does not seem unlikely, however, that specific degradation routes followed by functionalization should become possible in order to synthesize aromatic monomers possessing original structures, not readily available from petrochemistry, *e.g.* vinyl phenols, *p*-methoxystyrene for polyaddition reactions and difunctional aromatic compounds for polycondensations leading to polyesters *etc.* Some of these potential applications of monomers from lignin have been discussed in detail in two recent reviews.[109,112]

19.5.2 Terpenes

Unsaturated hydrocarbons from essential oils constitute an important family of monomers which have been used for a long time for the manufacture of resinous oligomers. Hydrocarbon resins are loosely defined as low molecular weight (typically below 2000) thermoplastic polymers which can range in appearance from viscous liquids to amorphous solids.

α-Pinene (**11**) and β-pinene (**12**) are among the most important terpenes used as monomers. They are susceptible to cationic activation and Lewis acids ($AlCl_3$, $BF_3 \cdot Et_2O$, $SnCl_4$ *etc.*) are the more frequently employed initiators. The mechanism of the polymerization of β-pinene is better understood[113,114] than that of its α isomer. Protonation of the monomer is the preliminary initiation step. This takes place following a classical cocatalytic route in which the Lewis acid–water complex is the protonating species. The tertiary carbenium ion (**29**) thus formed rearranges to the more stabilized structure (**30**) before propagation (Scheme 9).

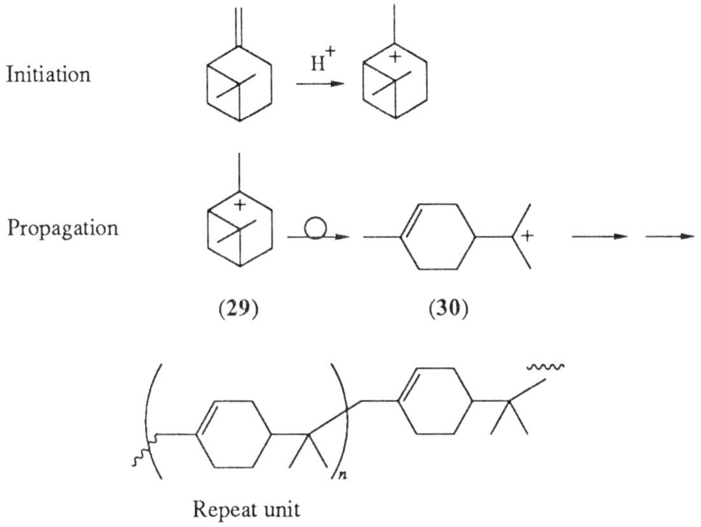

Scheme 9

The resulting macromolecules are thus in the first approximation a sequence of alternating 'isobutene' and 'cyclohexene' units. Chain transfer, very probably to the incoming monomer, occurs, as suggested by the low molecular weights of the products. Also, solvent transfer by alkylation involving the incorporation of the aromatic moiety is likely to take place.

The actual synthesis of these resins is usually carried out in xylene at 40–50 °C. The monomer is in fact a mixture containing about 75% β-pinene, the rest being α-pinene and limonene.

The structure of polymers from α-pinene is less well defined and certainly not the same as that arising from β-pinene according to Scheme 9. In fact, after initiation, which gives the same carbenium ion as with β-pinene, propagation is sluggish on account of the steric hindrance associated with the internal double bond of the monomer. The oxidative degradation of the oligomers suggests that two units characterize their structure with about two-thirds of unsaturated moieties.[114] It was shown that the number average molecular weight is somewhat smaller with resins from α-pinene, but that the size distribution of these products is narrower than that of the β-pinene oligomers. Therefore, the resins from the two isomeric pinenes have different properties, such as solubility, adhesion and tack performance.[115]

Limonene (**13**) is also used as a source of resins through Lewis acid initiated cationic polymerization.[113–115] The oligomers bear two major structural units, one unsaturated, the other bicyclic.

Other less common terpene monomers include myrcene (**14**), 3-carene (**31**) and camphene (**32**). They yield resinous materials by cationic polymerization through mechanisms similar to those illustrated above.

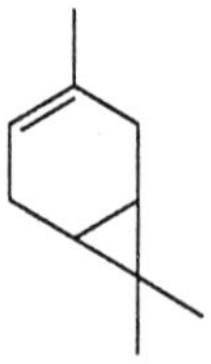

(**31**) 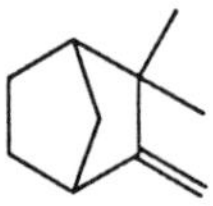   (**32**)

The cationic copolymerization of terpenes among themselves or with petroleum-based monomers such as styrene is frequently practised to obtain resinous products with specific properties.[115] Also, partly oxidized limonene can be treated with maleic anhydride and heated to give a solid red resin bearing hydroxy functions which impart to it solubility in alcohols.[115] Copolymers of phenol with terpenes are also well-known resinous materials.[115]

The applications of all the above materials[113] cover mostly the realms of: (i) adhesives, and more specifically tackifying additives to elastomers (pressure sensitive and hot melt compositions), mastics and sealants; (ii) printing ink vehicles, mainly in the letterpress, gravure and lithographic compositions; and (iii) paints and varnishes for protective coatings.

Recently, myrcene (**14**) was polymerized by free radical initiation induced by hydrogen peroxide to give unsaturated oligomers with molecular weights of a few thousand.[116] They contained the two major structural units (**33**) and (**34**).

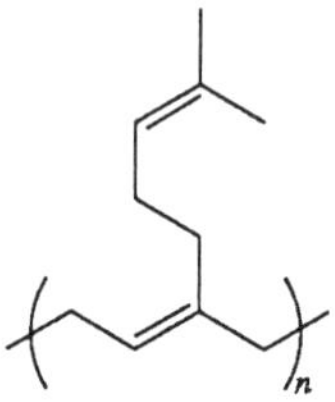

1,4-addition (**33**) 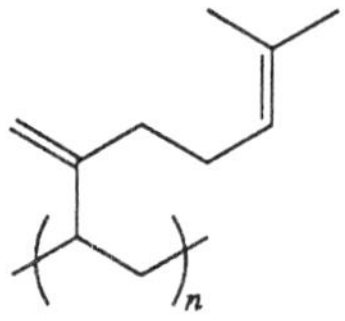   3,4-addition (**34**)

Hydroxy groups were present in these oligomers as end-functions arising from the initiator. The average OH functionality ranged between 1.3 and 2.3 depending on the operating conditions and the polydispersity was about 1.3. These polyols were used to synthesize different types of polyurethanes.[117,118] The first group of such materials[117] consisted of elastomers prepared with the polyols from myrcene, 1,4-butanediol and 4,4'-methylene diphenylene diisocyanate. A second set of materials[118] were glassy commercial polyurethane resins modified by the myrcene polyols in order to reduce brittleness.

19.5.3 Glycerol, Ethylene Glycol and Propylene Glycol

Glycerol, obtained from the various industrial processes involving natural fats and oils (triglycerides) is an important commodity provided by renewable resources. Its use in the synthesis of polyesters and polyurethanes is presently the most important application in the context of polymeric materials.

The present market situation for glycerol is one of overproduction and new utilizations are being sought in both direct intervention as a reagent or a source of other chemicals which might find higher added value. Thus for example, partial dehydroxylation by a catalyzed reaction in a reducing atmosphere of H_2 or $H_2 + CO$ leads to *n*-propanol with high yields and selectivity.[119]

The conversion of sugars into diols, namely ethylene and propylene glycols is another domain of research involving renewable resources. It has been shown that these processes, based on catalytic hydrogenation, have a potential competitiveness with traditional petrochemical technologies and constitute therefore a promising way of exploiting such raw materials as starch and sugars.[120]

The importance of these diols in various nonpolymer applications such as antifreeze agents and their role as monomers and comonomers in polycondensation reactions to give polyesters and polyurethanes, is too well known to be tackled here.

19.5.4 Monosaccharides and Disaccharides

Synthetic polysaccharides possessing well-defined structures are not readily prepared: nature still works much better than man in this domain.[7]

19.5.4.1 Condensation reactions

Acid-catalyzed polycondensation is the oldest method of saccharide polymerization.[121] Here, the difficulties reside with the variety of modes of chain growth associated with the condensation reactions of simple sugars. It is known that these condensations always involve the hydroxy group at C-1 (hemiacetal function) and a hydrogen atom of any other of the OH groups of another sugar molecule. The specific monosaccharide used and the conditions of polycondensation determine the relative reactivity of the C-2 to C-5 or C-6 OH moieties. If one considers moreover that all the four isomeric forms of aldoses coexisting in equilibrium (*e.g.* the α- and β-furanose and α- and β-pyranose forms of D-galactose in Scheme 10) can participate in the condensation reactions (although again with different reactivity as a function of the polymerization conditions), it becomes immediately obvious that the ensuing polysaccharide, even if prepared from a single sugar, will tend to be extremely heterogeneous in its sequence of monomer units and highly branched.

It follows that if linear polymers are to be obtained from sugars, only the C-1 position and another hydroxy group must remain available for condensation on the monomer molecule. This entails the blocking or masking of the other three OH, *e.g.* by esterification or etherification. Urethanes have also been used in this context.

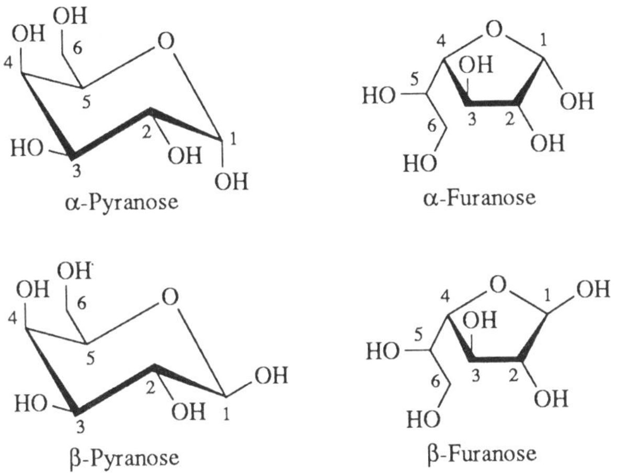

Scheme 10

Some steric control can be achieved, particularly by an appropriate substitution at C-2. However, the polycondensation seldom attains degrees of polymerization higher than about 50. Low DPs are also obtained when sugar orthoesters are used as monomers because the alcohol generated in the condensation reaction is a terminating agent. Cyclization reactions are an additional source of problems.

Di- and even tri-saccharides can be used as monomers, because polycondensations are not disturbed by the fact that the starting materials are already of an oligomeric nature.

19.5.4.2 Ring-opening polymerization

This mode of polymerization implies a chain propagation mechanism and is therefore more likely to produce high degrees of polymerization, particularly if transfer reactions are not too important. The ring-opening polymerization of appropriately modified saccharides is a cationic process. Two types of monomeric structures are used to promote such reactions: (i) internal orthoesters;[121] and (ii) anhydrosugars.[122]

The former class requires the presence of hydroxy groups and an acid catalyst to assist the propagation step. It seems likely that the monomer molecule, rather than the polymer chain, is activated and the mechanism in Scheme 11 would then apply to chain growth, as shown here with the orthoester of a pentose polymerizing *via* 1,5-additions.

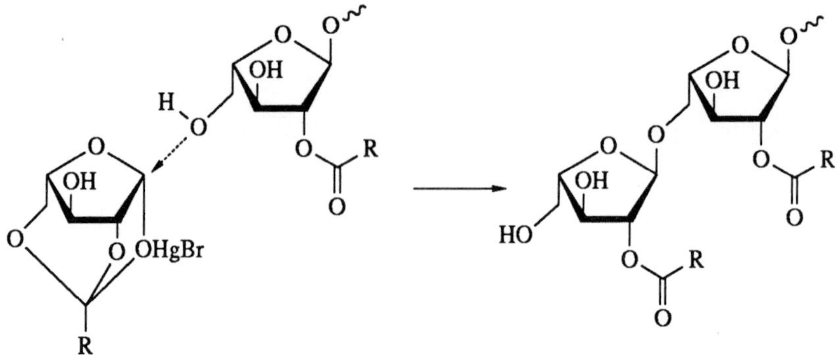

Scheme 11

This type of cationic polymerization can also occur without free OH groups, but in the presence of a trityl salt acting as catalyst.

The second, and more efficient way of conducting a cationic ring-opening polymerization leading to polysaccharides uses anhydrosugars as monomers. The most interesting materials are those formed in the polymerization of fully substituted or masked anhydrosugars. The mechanism of these polymerizations resembles that of the cationic polymerization of dioxolane and involves an activated polymer chain adding onto a 'neutral' monomer molecule, but whereas 5-substituted dioxolanes do not polymerize because of unfavourable thermodynamic parameters (ΔH too low), the anhydrosugars do because of the release of the ring strain associated with its opening, except in the case of 1,6-anhydrofuranoses which possess an unstrained conformation. The polymerizable (and available) structures are the 1,2-, 1,3-, 1,4- and 1,6-anhydropyranoses and it is with the latter monomers that most of the fundamental studies have been carried out. The propagation mechanism shown in Scheme 12 illustrates the mode of chain growth envisaged for these polymerizations.

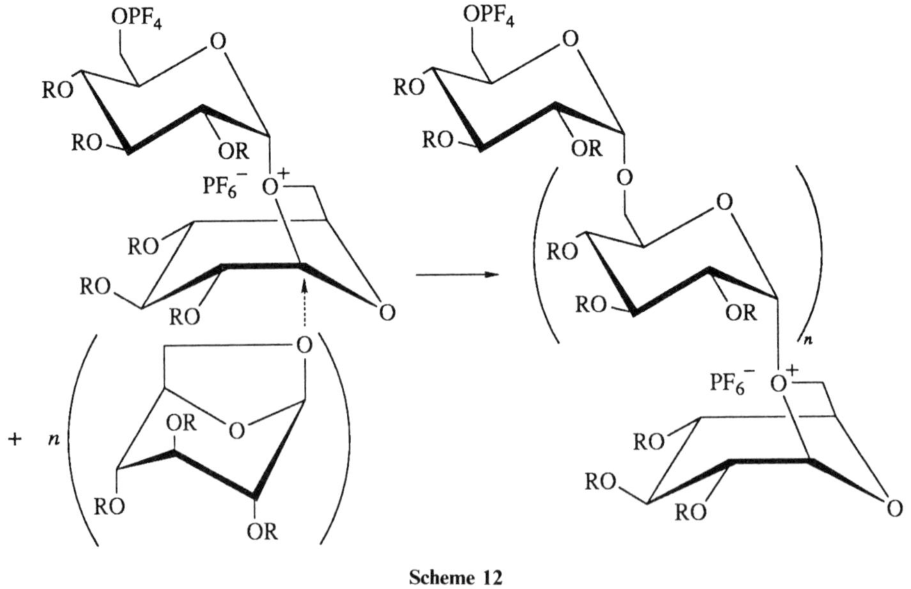

Scheme 12

Most of the time propagation is accompanied by configuration inversion at C-1, as shown. The conditions favouring stereospecificity and DPs up to 5000 are not the same for all monomers, but the general guidelines call upon Lewis acids like PF_5 and boron fluoride etherate in chlorinated hydrocarbons at low temperature.

When the substituents are removed to regenerate the polysaccharide backbone, the DPs tend to decrease somewhat due to some chain scission accompanying deprotection.

Various copolymerization systems involving different anhydrosugars have been studied in terms of monomer reactivities and polymer structure. Substituted anhydrodisaccharides can also be polymerized, although more slowly than their single sugar analogues, to give comb-like macromolecules.

It seems reasonable to expect that these polymers and copolymers will find further applications in areas related to medical and pharmaceutical products and for specific biological functions, *i.e.* the type of uses they have already found,[123] quite apart from their importance as reference structures for a better understanding of natural polysaccharides.

Sugars like sorbitol (**35**) have been used as polyols for the preparation of polyurethanes (foams). Usually, the OH groups are chain extended with ethylene or propylene oxide to give a viscous noncrystalline monomer. Glucose has also been used as a source of monomers for polyurethanes by making it react first with polytetrahydrofuran glycols to obtain liquid oligomeric polyols which are then mixed with diisocyanates.[124] The resulting materials have promising thermal and mechanical properties.

Another way of exploiting mono- and di-saccharides to generate polymeric structures is based on their preliminary conversion into diols and their subsequent use in polycondensation reactions. Thus, sorbitol (**35**) is readily dehydrated by acid catalysis to give isosorbide (**36**; equation 6).

(6)

(**35**)   (**36**)

A recent report[125] describes the use of this diol with nonequivalent OH groups to prepare linear polyurethanes with various diisocyanates. Similar experiments were also conducted with isomannide from mannitol and with the corresponding lactone arising from gluconic acid. Cellobiose[125] has also been tested for the synthesis of polyurethanes: it seems that the two primary hydroxy groups were much more reactive than the six secondary groups and this allowed a linear polymerization with diisocyanates without the need for preliminary masking. The interest of this approach is of course that the secondary OH groups can be made to react at a second stage, *e.g.* to crosslink a thermoplastic material after moulding.

Diepoxy derivatives of hexoses have also been prepared and tested in typical epoxy resin compositions and appear to give good networks.[125]

In another context, sugar moieties have been appended to typical vinylic and acrylic structures and then polymerized or copolymerized. A whole variety of such monomers have been studied.[126,127] Their polymers are generally water soluble.

Similar polymeric structures can be prepared by condensing the saccharide moiety onto a reactive group of a starting polymer, *e.g.* poly(vinyl phenol)[128] (etherification) or poly(methyl methacrylate)[129] (transesterification). Again, these materials are interesting in the context of biochemical and medical (*e.g.* drug delivery) applications.

19.5.5 Biodegradable Polymers from Renewable Resources

19.5.5.1 *Bacterial polymerization*

This brief incursion into the realm of bacterial polymers is justified because the products in question are becoming a novel class of commercial materials. The main representative of this family of polyesters is poly(3-hydroxybutyrate) (**37**), but copolymers (**38**) of 3-hydroxybutyrate with 3-hydroxyvalerate have also been prepared.

(**37**)   (**38**)

Numerous bacteria can fabricate (**37**): these microbes store it as a source of energy and a supply of carbon. Copolymers such as (**38**) are much less widespread and have only been identified recently. The fact that these sturctures are sensitive to environmental hydrolytic degradation, and give first water-soluble oligomers by the action of some extracellular enzymes of microorganisms, then carbon dioxide and water by the action of intracellular enzymes, has spurred much interest in academic and industrial laboratories.[130,131] This research has resulted in the development of a controlled fermentation process in which the polymerization activity is strongly enhanced, compared to natural biosynthesis, to yield as much as 80% polymer with respect to the weight of the dry cell. The feedstocks used for these syntheses can arise from both petrochemicals and renewable resources. 'The first thermoplastic from biotechnology'[130] (indeed today both the homopolymer (**37**) and copolymers (**38**) with up to about 30% of 3-hydroxyvalerate units) is available as a commercial commodity.

Recent investigations on these polymers include on the one hand basic studies related to detailed structural determinations[132] and on the other hand more technological aspects related to applications[133] of bacterial products, but also of high molecular weight materials prepared by nonbacterial catalytic processes.

The utilization of these materials in medical and pharmaceutical applications is of much interest. Their hydrolytic degradation and biodegradation[134] have been studied and shown to vary quite substantially with the methods employed to synthesize them.

A recent monograph deals in detail with microbial polyesters from the points of view of both synthesis and properties.[135]

Another type of polymer synthesized by bacteria grown in a sugar medium is dextran,[136] a water-soluble branched polysaccharide based on D-glucopyranose units. It is produced industrially and used as an additive in foods, cosmetics, pharmaceuticals, *etc.* Other microbial polysaccharides have been described.[135]

Bacterial polymerization of a lignin model compound to macromolecular pigments suggests a new possible exploitation of lignin.[137]

19.5.5.2 Chemical polymerization

Lactic acid (**39**), obtained from the fermentation of hexoses and their polymers, is a hydroxy acid which polycondenses to the biodegradable polyester structure (**40**).[138]

(39)   (40)

Copolymers with ε-caprolactone have also been studied. Glycolic acid, $HOCH_2CO_2H$, is also a promising monomer in this context.

These materials are used in surgery, temporary protheses, disposable plastics and for the controlled release of medicines.

19.5.6 Pentoses and Hexoses as Sources of Furanic Monomers

Hemicelluloses (the poor relatives of celluloses), starch, and also, directly, certain mono- and di-saccharides are sources of furan derivatives through simple chemical operations.

19.5.6.1 Furfural and derivatives

The acid-catalyzed hydrolysis of polymeric pentoses all the way down to their corresponding monosaccharides (aldopentoses) followed by acid-catalyzed dehydration[139] leads to the formation of 2-furancarbaldehyde (furfural) (**41**). D-Xylose gives the highest yields: Scheme 13 illustrates the accepted mechanism of this triple dehydration.

Rhamnose, formed from the corresponding methylated units in hemicelluloses, yields concurrently small amounts of 5-methyl-2-carbaldehyde (methylfurfural) (**42**), which is therefore a typical

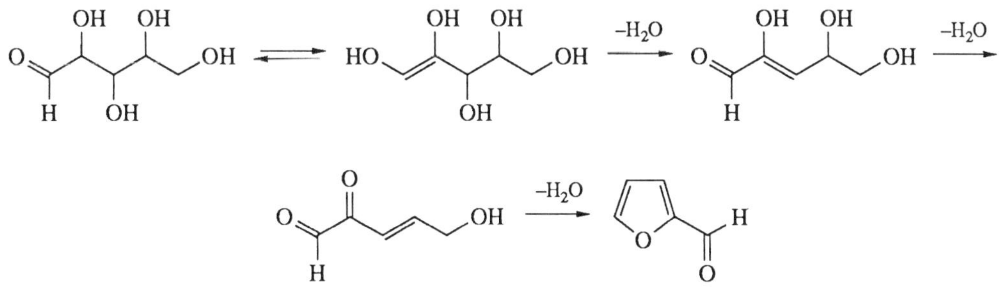

Scheme 13

'impurity' in the manufacture of furfural. If needed, the separation of these two homologues is readily achieved by fractional distillation.

This process was industrialized more than 50 years ago and is today applied to a large variety of pentosans-rich raw materials, mostly agricultural wastes such as corn cobs, rice hulls, sugar-cane bagasse, olive residues and cotton seeds, but also wood and by-products of paper mills.[140] It can be envisaged that the newer technologies of biomass refining, steam explosion and organosolv processes for example, will select this chemical post-treatment as one of the major alternatives for the rational use of the hemicelluloses arising from them. The yearly production of furfural amounts to about 200 000 tons. Most of it is catalytically reduced to furfuryl alcohol (**43a**) which constitutes therefore the most important industrial chemical of the furan series presently available on the market.[141] This is due to the usefulness of furfuryl alcohol as monomer in the preparation of various resinous materials.[141] However, many other monomers can be prepared from furfural and thus be the source of new polymers.[2,141,142,144]

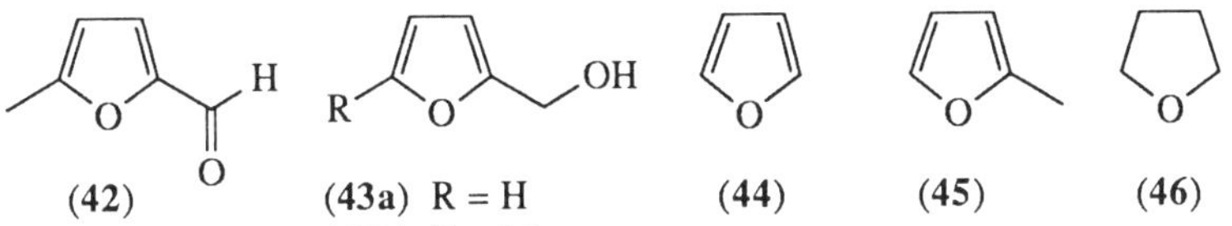

(**42**)   (**43a**) R = H   (**44**)   (**45**)   (**46**)
            (**43b**) R = Me

The catalytic decarbonylation of (**41**) and (**42**) is the standard procedure for the synthesis of furan (**44**) and 2-methylfuran (**45**), both available industrially. Furan is one of the two industrial sources of tetrahydrofuran (**46**), the other being 1,4-butanediol. Both furan and tetrahydrofuran are monomers, but for very different materials, as discussed below.

Certain hexoses used as such, or formed in the hydrolysis of polysaccharides such as starch or

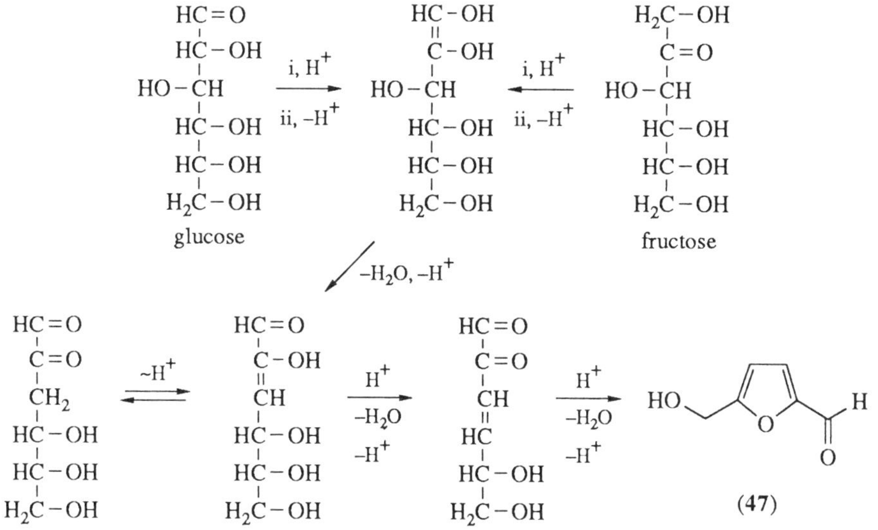

Scheme 14

inulin, can be exploited in a manner similar to that described for pentoses and their acid-catalyzed dehydration[139,145-147] affords then 5-hydroxymethyl-2-furancarbaldehyde (hydroxymethylfurfural) (**47**) or 5-chloromethyl-2-furancarbaldehyde, the latter being the specific product arising when HCl is used as catalyst. The mechanism of formation of (**47**) from fructose[145] is shown in Scheme 14.

Compound (**47**) is not an industrial commodity as yet, but a pilot plant based on an original aqueous process using a chromatographic separation procedure has been operating for a few years.[148] Other processes are being actively studied[147] with the aim of reaching a viable industrial realization. The use of (**47**) as precursor to furanic monomers[143,144] is the second object of this section.

Glucose can be conveniently converted into furanic tetrols by condensation with dicarbonylic compounds in the presence of mild Lewis acids such as calcium chloride,[149] as shown by equation (7).

$$\text{glucose} + \text{OMe OMe dicarbonyl} \longrightarrow \text{furanic tetrol} + 2H_2O \quad (7)$$

The interest of this process is that starting from sucrose and proceeding to its hydrolysis into fructose and glucose, one can envisage the use of the latter for the preparation of furanic tetraols and take advantage of the selectivity of reaction (7)[149] to recuperate the untouched fructose for other utilizations, *e.g.* the preparation of (**47**). Reaction (7) also applies to certain pentoses and then gives furanic triols. All these polyols constitute interesting monomers for the synthesis of polyurethanes.

The monomers arising from (**41**) and (**42**) can be first- or second-generation derivatives, as sketched in the general pattern given in Scheme 15.

All the above monomers generate in principle macromolecular structures in which the furan ring is pendant to the chain and not part of it. This is so only if no side reactions occur which implicate the heterocycle. It will be shown that such anomalies do take place in some systems.

2-Alkenylfurans (**48**) are among the oldest furanic monomers studied. Their syntheses follow three alternative routes: (i) the classical condensation of the furanic aldehyde with malonic acid to give the 2-furanacrylic acids, which are then decarboxylated at relatively high temperature;[150] (ii) the Wittig reaction between the same aldehydes and phosphonium salts which has been optimized with an original application of phase-transfer catalysis;[151] (iii) the Grignard reaction applied to both furanic aldehydes and ketones to produce the respective alcohols, followed by their dehydration;[152] and (iv) the lithiation of furan (**44**) and 2-methylfuran (**45**) followed by the addition of acetaldehyde or acetone to give the secondary or tertiary alcohols[153] which are dehydrated as in (iii). Whereas methods (i) and (ii) provide ways to prepare the vinyl derivatives (**48a**) and (**48b**) only, methods (iii) and (iv) apply to all four alkenylfurans (**48**). However, method (ii) is by far the best in terms of ease of application, economy and yields.[154]

Another well-known family of monomers is that of 2-furfurylidene ketones (**49**). Their preparation involves a simple aldol-type condensation of furanic aldehydes with ketones or aldehydes in a basic (aqueous) medium. The specific reaction of (**41**) with acetone to give the parent compound (**49a**) has been thoroughly investigated and optimized.[155]

The 2-furyloxiranes (**50**) on the other hand, are a more recent acquisition. They are prepared by the reaction of furanic aldehydes with triphenylmethylphosphonium bromide, with a basic solid providing phase-transfer catalysis. A thorough investigation of these syntheses[156] has defined the parameters allowing a near-quantitative yield under mild conditions and short reaction times.

2-Furyl ethenyl ketones are best prepared from furanic ketones.[157,158]

Furfuryl alcohol (**43a**) and its 5-methyl homologue (**43b**), prepared by the reduction of (**42**), are precursors to other furanic monomers. Their esterification with acrylic and methacrylic acid chlorides or anhydrides give the corresponding furfuryl derivatives (**51**) in a straightforward manner. Another synthesis has also been reported for these monomers. The vinyl exchange reaction of (**43**) with, *e.g.*, isobutyl vinyl ether provides a route to furfuryl vinyl ethers (**52**).[159]

The 2-furoic acids (**53**), prepared by oxidation of (**41**) and (**42**), are also intermediates for the synthesis of furanic monomers. Their chlorides are readily transformed into azides which in turn yield the corresponding isocyanates (**54**) by the Curtius reaction.[160] Alternatively, vinyl exchange reactions of these acids with, *e.g.*, vinyl acetate give the corresponding vinyl furoates (**55**).[161]

The 2-furfurylamines (**56**), of which (**56a**) is a commercial derivative of furfural, provide another set of monomers, *viz.* 2-furfurylisocyanates (**57**),[160] through their reaction with phosgene, or, better still, with bis(trichloromethyl) carbonate, known as triphosgene.

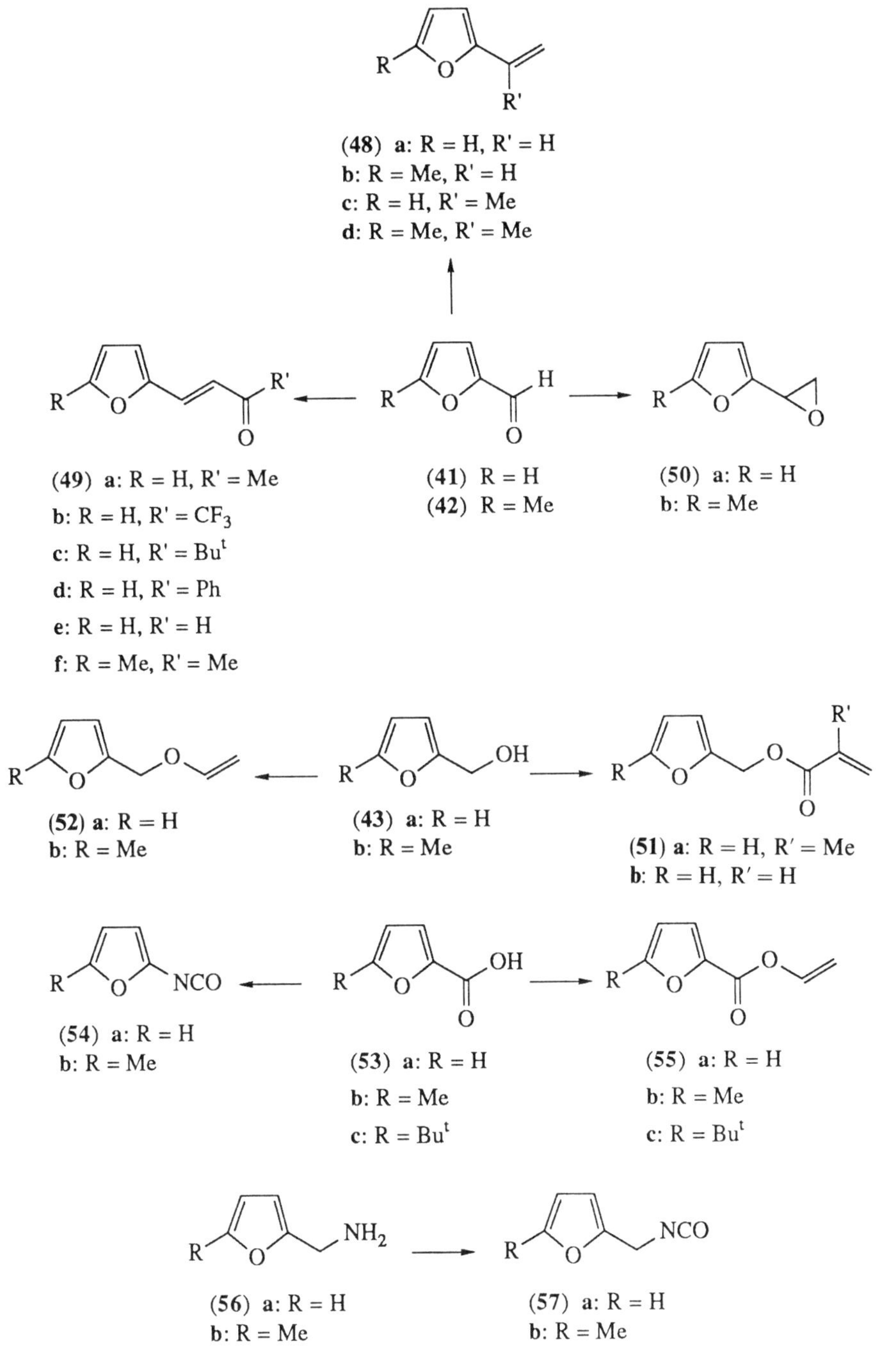

(**48**) a: R = H, R' = H
b: R = Me, R' = H
c: R = H, R' = Me
d: R = Me, R' = Me

(**49**) a: R = H, R' = Me
b: R = H, R' = CF$_3$
c: R = H, R' = Bu^t
d: R = H, R' = Ph
e: R = H, R' = H
f: R = Me, R' = Me

(**41**) R = H
(**42**) R = Me

(**50**) a: R = H
b: R = Me

(**52**) a: R = H
b: R = Me

(**43**) a: R = H
b: R = Me

(**51**) a: R = H, R' = Me
b: R = H, R' = H

(**54**) a: R = H
b: R = Me

(**53**) a: R = H
b: R = Me
c: R = Bu^t

(**55**) a: R = H
b: R = Me
c: R = Bu^t

(**56**) a: R = H
b: R = Me

(**57**) a: R = H
b: R = Me

Scheme 15

Whereas all the above reactions provide monomers which are suited for chain polymerization reactions, esters of (**53a**) and (**56a**) can also be used as precursors to difuranic difunctional monomers for polycondensation reactions. The well-known acid-catalyzed condensation of furan derivatives with aldehydes or ketones shown in equation (8) is applied successfully to this specific context.[124,161]

The diesters can be used as such in polytransesterification reactions, or transformed into diacids, dicacid chlorides, their nitriles and eventually their furanic diisocyanates. The diamine salts

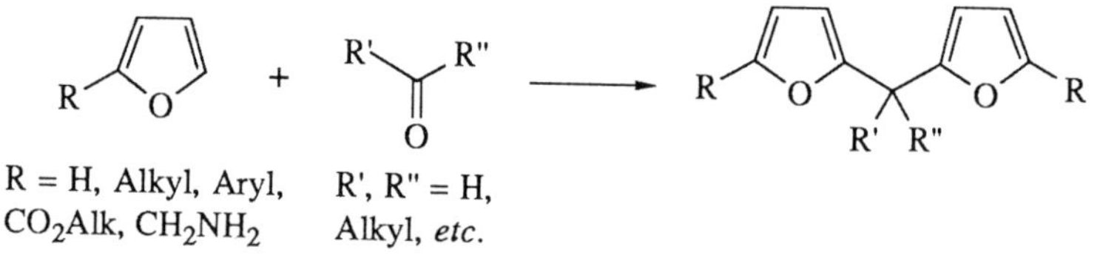

(8)

are also a source of diisocyanates by phosgenation, but this time one is dealing with furfuryl diisocyanates.

19.5.6.2 Hydroxymethylfurfural and derivatives

The major source of difunctional monomers for stepwise polymerizations is, however, **(47)**, which already bears two reactive sites, although of a different nature. These transformations are sketched in Scheme 16.

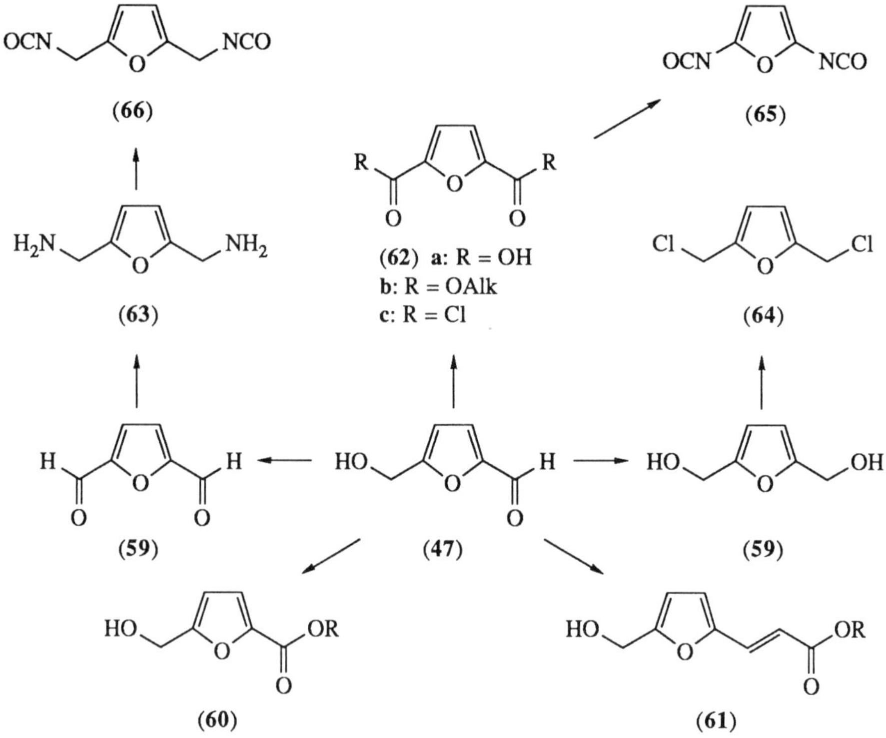

Scheme 16

The first group of syntheses involves those reactions which transform only one of the sites on **(47)**. Thus, the dialdehyde **(58)**, the diol **(59)**, the hydroxy acid and its esters **(60)** represent classical monomers obtained by selective oxidations and reductions. These processes have been and are being keenly studied.

The transformation of only the aldehyde function of (**47**) into an acrylate one (**61**) is a further interesting process which has been optimized[162] and which provides a structure adapted to polytransesterification.

The second type of strategy is to modify both reactive groups of (**47**). It has been applied to the synthesis of: (i) the diacid, its dichloride and its esters (**62**) by straightforward oxidation[163] followed by the appropriate functionalization; (ii) the diamine (**63**) *via* the dialdehyde and the dioxime;[164] (iii) the dichloride (**64**) from the diol and SO_2Cl_2;[165] and (iv) the two diisocyanates (**65**) and (**66**), respectively by the Curtius reaction of the diazide[166] of (**62a**) and the treatment of (**63**) with triphosgene.[167]

It is well known that (**47**) can be readily transformed into a condensed dialdehyde through the reaction of its hydroxy groups. This dialdehyde (**67**),[168] but also the corresponding diamine (**68**),[169] have been tested as monomers in polycondensation reactions.

(67)   (68)

All these difunctional monomers are naturally suited for stepwise polymerization and generate macromolecular structures which contain the furan ring within the polymer backbone. This is a second major difference with respect to most monomers formed from (41), which on the one hand are suited for chain polymerization and, on the other, give macromolecules bearing the furan heterocycles pendant to them.

It is obvious that quite apart from the different mechanisms required to polymerize these two families of monomers, the structure–properties relationships of the resulting materials will differ considerably on account of the relative emplacement of the ring in the polymer structure.

The above list of furanic monomers is by no means exhaustive. It purports to show, however, the wide variety of structures already explored, the often very simple synthetic routes perfected and the scope of things done but also of things to come.

The chemistry of furan derivatives was very thoroughly reviewed in a monumental work published a few decades ago.[170] Since then the topic has been updated[171,172] and reassessed following a more modern approach.[173] Some of the peculiarities in the behaviour of furanic monomers in polyaddition and polycondensation reactions stem from the specific chemical features imposed by the heterocycle. The sections below will emphasize these aspects whenever they introduce a major deviation with respect to the classical patterns encountered with the corresponding aliphatic or aromatic monomers.

19.5.7 The Polymerization of Furanic Monomers

The present review of furan polymers will be restricted to a reminder of the results obtained from systems studied up to a few years ago[2,141,142,144] and to a more-detailed account of very recent work.

19.5.7.1 Chain polymerization

Furan itself (44) and 2-methylfuran (45) have been polymerized by acids or copolymerized by free radical initiators with maleic anhydride. Whereas the cationic process yields highly conjugated oligomers with a complex structure,[142,174] a clear-cut sequence of units characterizes the copolymers obtained radically.[2,175] Other copolymers with acrylic monomers give less clear-cut structures.

Recently, interest in polyfuran has been revived by the more general search for electronically conducting polymers. The publications dealing with these novel electrochemical polymerizations[176–178] of (44) are, however, disappointing because of the lack of thoroughness concerning the structural determinations and the hasty conclusions reached on the basis of very meagre experimental data. More work is therefore needed to establish clear criteria about the interest of polyfurans compared with the much better characterized polypyrroles and polythiophenes.

Furan (44) has recently been used in a different macromolecular synthetic approach. Its Diels–Alder adduct with maleic anhydride was either hydrolyzed to the dicarboxylic acid (69a) or reduced to the dimethyl ether (69b) bicyclic derivatives. Both compounds were then polymerized[179,180] by ring-opening metathesis polymerization to give the corresponding structures (70a) and (70b). A similar study[181] on the unmodified Diels–Alder adducts of furan with maleimides indicates that high molecular weight, heat resistant polymers are obtained.

The polymerization of 2-alkenylfurans has been reviewed.[2,142,144] In particular, the peculiarities of radical, cationic, Ziegler–Natta and anionic systems have been underlined. An interesting consequence of the behaviour of 2-vinylfuran (48a) in cationic polymerization, viz. the importance of side reactions such as the electrophilic substitution at C-5 and the formation of polyunsaturated sequences on the polymer chains by multiple hydride ion plus proton losses, was the preparation of highly conjugated crosslinked materials prepared from that monomer in bulk and with strong acids added in fairly high concentrations.[182] The resins showed a very strong Lewis base character, particularly in organic media and were utilized as powerful acid scavengers. Their activity was much larger than that of commercial ion-exchange products used to fix acids.

Another useful fallout of previous detailed studies of the cationic polymerization of 2-alkenylfurans (48) and of the preparation of model compounds to elucidate those mechanisms came from the extreme ease with which the C-5 position of certain 2-substituted furans are alkylated. Addition

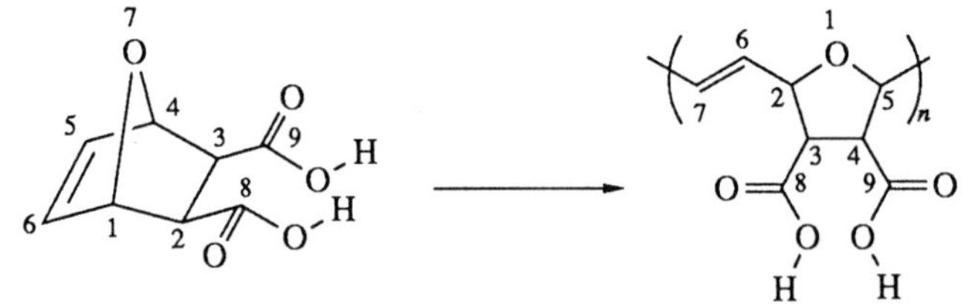

(69a) → (70a) II

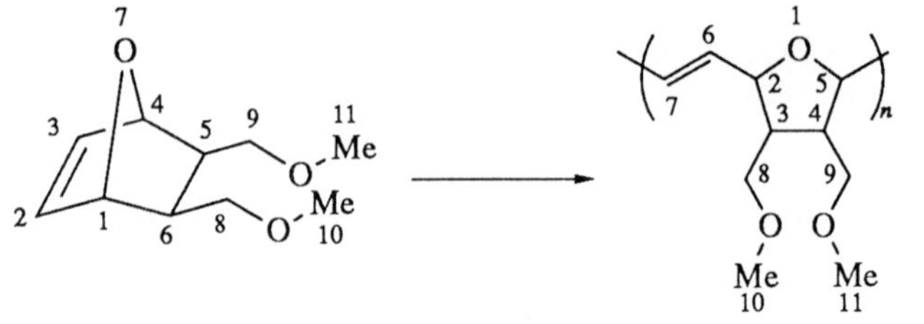

(69b) → (70b)

of 2-alkylfurans to typical cationic polymerization systems[183] with monomers such as isobutene, styrene, the vinyl ethers, *etc.* produces an important transfer reaction by electrophilic substitution on the furan compound. By adjusting the conditions, this reaction can be made to dominate over other chain-breaking events and therefore virtually all polymer (oligomer) chains will bear a furan ring at one end. The interest of these materials resides in the specific reactivity of the furan ring and in the terminal structures one can thus introduce (see also below). For example, the Diels–Alder reaction with maleic anhydride can give rise to two carboxylic groups at the end of a nonpolar hydrocarbon-type chain like oligoisobutene, as shown by structure (71).

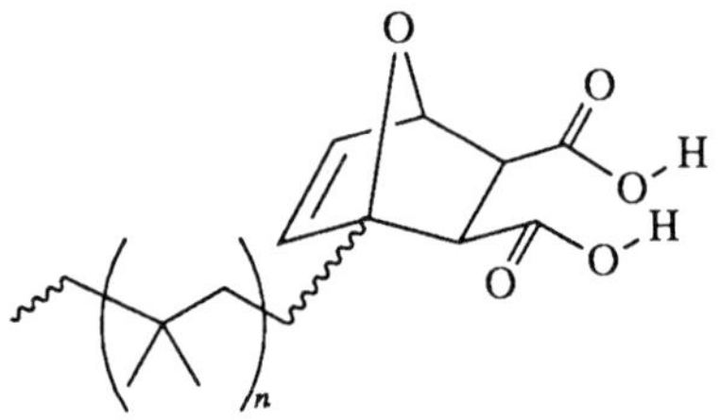

(71)

A stimulating extension of these concepts consists in switching to a difuranic derivative as transfer agent. Then, after a careful optimization of the reaction conditions, one can synthesize block copolymer by a simple one-pot cationic operation. A first monomer A is introduced in the medium containing the difuran compound and polymerized: the product will be mostly composed of poly-A terminated by the monoalkylated difuran moiety. A second monomer B is now introduced: its cationic polymerization will also be submitted to chain transfer by electrophilic substitution onto the second C-5 position remaining at the extremity of the poly-A chains formed in the first phase of the process, thus forming block copolymers like (72).[184]

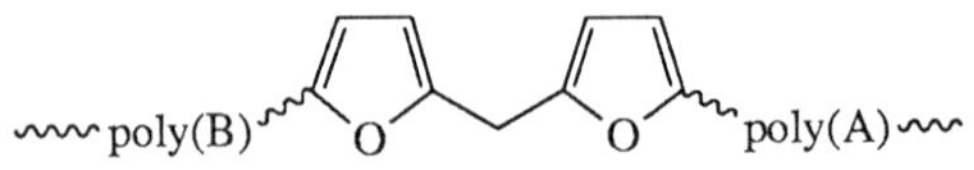

∼∼∼poly(B)∼∼∼O∼∼∼O∼∼∼poly(A)∼∼∼

(72)

Vinyl ethers of the furan series such as (52) are extremely prone to cationic initiation, not surprisingly if one considers the combination of two factors enhancing the nucleophilic character of the C=C unsaturation: the oxygen atom of the ether bond and the furan ring. Like other very basic monomers,[185] these were shown[186] to copolymerize with furfural (41) and 5-methylfurfural (42), these aldehydes only entering as single ether units in the copolymer, by the opening of the carbonyl group, because of the thermodynamic restrictions to their homopolymerization.[185] Under appropriate experimental conditions, alternating copolymers like (73) were obtained.

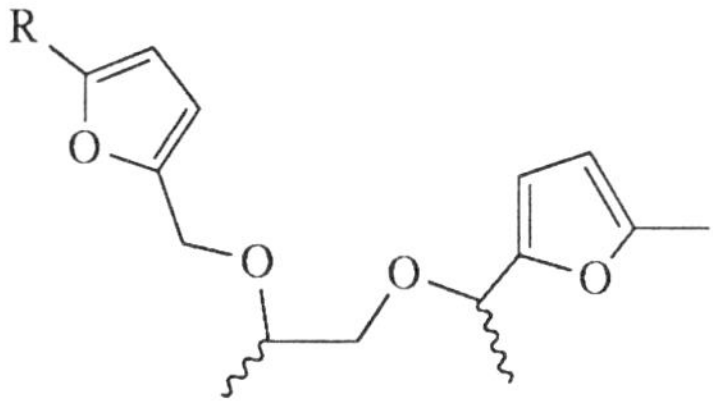

(73)

The cationic polymerization of tetrahydrofuran to give poly(tetramethylene oxide)glycols is too well documented to be rediscussed here.

Turning to anionic polymerization, three monomers are worth some comments. 2-Furfurylidene acetone (**49a**), a compound much studied in acidic media where it gives rise to black crosslinked resins, was submitted instead to the action of nucleophilic initiators.[187] The resulting polymerization was shown to occur with a proton shift from the δ-C to the β-C position at each propagation event and this produced the 'phantom' polymer (**74**) with the carbonyl group within the backbone.

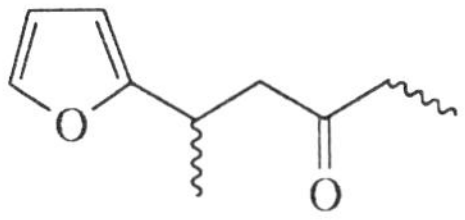

(74)

It was possible to prove this mechanism both from a careful analysis of the polymer structure and by using homologues of the monomer: those which did not possess a mobile hydrogen atom at δ-C, such as the trifluoro (**49b**), the phenyl (**49d**) and the t-butyl (**49c**) derivatives, simply did not polymerize.

2-Furyloxirane (**50**) is best polymerized anionically.[188] Under certain conditions, it is possible to synthesize telechelic oligomers with DP up to about 100 bearing OH terminal groups. These products have been tested in polyurethane compositions and produced interesting properties.[189] In this study,[188] it was found that alcohols and also traces of water can initiate the polymerization even in the absence of any stronger nucleophile. With water, the mechanism involves a first (slow) step of oxirane ring opening to give the corresponding furanic diol which propagates (more rapidly) essentially by ring opening at the β-carbon, as shown in Scheme 17.

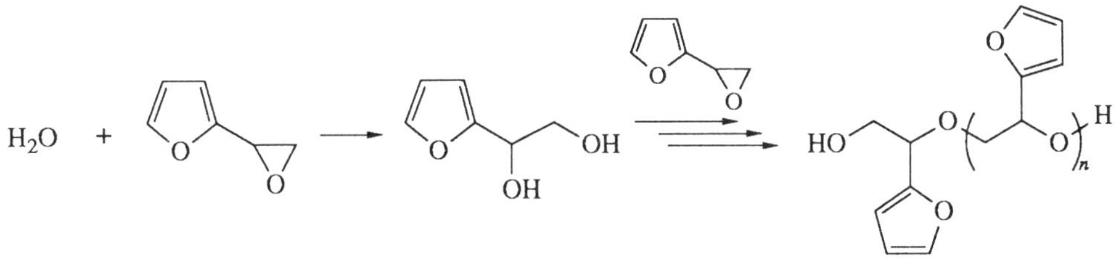

Scheme 17

Diols lead to chain growth in both directions and triols and tetraols to star-shaped poly(2-furyloxirane)s.

2-Furyl isocyanate (**54a**) and 2-furfuryl isocyanate (**57a**) were only recently tested as monomers in typical anionic systems[190] adapted to the activation of the C=N bond. The cyanide anion proved inactive, probably because of a specific complexation with the furan ring that distracted it from its well-known role of initiator with aliphatic and aromatic isocyanates. Sodium naphthalene and Bu^{t}OK manifested the expected activity, but gave a mixture of cyclic trimer and polymer, both highly crystalline. The conditions favouring polymer formation involved the use of a solvent of high polarity, low temperature and initiation by sodium naphthalene. Thus (**57a**) gives predominantly polymer (**75**) at $-78\,°\text{C}$ in DMF with that catalyst.

Free radical initiation is often followed by anomalous interactions which can perturb or even spoil the desired propagation reaction. The causes of these side events are to be found in the propensity of the furan ring to react with free radicals through addition or substitution modes.[142-144,191,192]

(75)

The radical polymerization of vinyl 2-furoate (**55a**) is a typical example of a limiting situation leading to autoinhibition due to the fact that the primary radicals formed by the initiator, and the rare oligomeric radicals which manage to develop, are much more likely to add onto the C-5 position of the furan ring rather than to the vinylic unsaturation because the furyl radicals ensuing from the former reaction are much better stabilized than those generated by the latter. This situation is only marginally improved by bulky substituents at C-5, as with 5-t-butylvinyl-2-furoate. In other words, vinyl 2-furoates (**55**) are not real monomers, at least within typical conditions for free radical polymerization, since they cannot generate their corresponding polymers. This also applies to copolymers because these compounds do not enter in such composition; they just inhibit the polymerization of the comonomer or slow it down by scavenging all or some of the free radicals present.

2-Furfuryl acrylate (**51a**) and methacrylate (**51b**) are much more prone to polymerize by free radical means since the radical formed at the acrylic site is now sufficiently stabilized to compete effectively with the furyl radical resulting from addition at the C-5 position of the heterocycle.

Various simple furanic compounds have been shown to retard or inhibit the radical polymerization of common monomers.[191]

Furfuryl alcohol has been copolymerized with acrylonitrile by free radical initiation[193] and the resulting products modified by heat and acid treatments.

19.5.7.2 *Stepwise polymerization*

As mentioned earlier, furfuryl alcohol (**43a**) is presently still the most important furanic monomer,[2,141,142] and indeed derivative, in industrial terms. Its resins have found a wide range of applications, mostly for foundry cores and moulds, corrosion-resistant materials, precursors to graphitic compositions and adhesives. Important investigations in the last 30 years have dealt with the mechanisms of this acid-catalyzed polycondensation in order to unravel its main features, but also the origin of the black colour and the reasons for branching and crosslinking. Indeed, if the major condensation reaction is that between a hydroxy group and a hydrogen atom at C-5 to give structure (**76**) and if this is only accompanied by sporadic OH–OH condensation reactions to give occasional CH_2OCH_2 linkages, then one would expect polymeric chains which are both linear and lacking any chromphore since difurylmethane, 2,5-dimethylfuran, furfuryl alcohol and furfuryl ether do not absorb beyond 250 nm.

(76)

Attempts to explain these anomalies have on the whole failed in the face of later evidence and only very recent work[194] seems to reach an acceptable picture of the overall process and rationalize the various features and phenomena observed. This required the use of model systems with pseudomonomers and pseudopolymeric structures.

The fact that 2-furfuryl acetate gives essentially the same phenomenology and the same products as (**43a**) when submitted to acid-catalyzed resinification[195] leads to a first important conclusion: the products arising from the acid-catalyzed hydrolysis of furan rings (*e.g.* carboxylic acids, ketones, *etc.* resulting from ring-opening reactions) cannot be responsible for any of the major peculiarities of these systems, contrary to previous suggestions, because with 2-furfuryl acetate no water, but acetic acid instead, is formed during the polycondensation (which was carried out in dry organic media) and hence no hydrolytic transformations are possible.

The fact that 2-(2-furyl)propanol (**77**) gives under conditions typically applied to (**43a**), a linear, regular yellowish oligomer (DP up to 12) instead of a black crosslinked mass, points to two further basic conclusions: (i) the C-3 and C-4 positions of the furan rings are not responsible for any of the 'side reactions' (*e.g.* electrophilic substitution causing branching), contrary to previous suggestions, because otherwise they would have occurred also with the dimethylated homologue (**77**); (ii) conversely, the hydrogen atoms on the methylene group separating the furan rings in (**76**) are, directly or indirectly, the culprit of both chain branching leading eventually to network formation and the generation of chromophores responsible for the dark colour of these polymers.

In fact, one mobile hydrogen atom is sufficient to spur these degeneracies, since 2-(2-furyl)ethanol (**78**) resinifies under acidic conditions following a pattern similar to that exhibited by (**43a**).

The next important task was to elucidate the detailed mechanisms leading to these structural anomalies. Again, part of the answer came from the use of model compounds. The 2,5'-dimethylated 'dimers' (**79**), (**80**) and (**81**) of the three furanic alcohols mentioned above in the context of polycondensation experiments were prepared together with the corresponding 'trimer' (**82**).

(**77**)   (**78**)

(**79**)   (**80**)   (**81**)

(**82**)

Treatment of these molecules with a hydride-ion abstractor, such as a dioxolenium salt, reveals that (**79**) and (**80**) give rise to the same UV–visible spectrum, characterized by a strong peak at 495 nm, whereas (**81**) does not react. It seems reasonable to postulate that the loss of H⁻ creates the corresponding secondary and tertiary carbenium ions from (**79**) and (**80**), which are very similar in structure and which are responsible for the visible absorption associated with the high delocalization of the positive charge over the entire difuranic structure. With (**81**), the lack of mobile hydrogen atoms makes the system unreactive towards the hydride ion abstractor.

The 'trimer' (**82**) behaves in this context according to a more complex pattern: first, a peak at 495 nm appears, then, later, a second less intense absorption slowly rises at about 650 nm and the solution turns from brown to green.

These phenomena mimic the colour evolution during the polycondensation of (**34a**) which indeed goes from brown to green and then to almost black. The mechanism giving rise to conjugated sequences is typified by the set of equilibria shown in Scheme 18.

During the polycondensation, both carbenium ion and neutral species coexist. Neutralization of the reaction medium leads to partial decolouration and only the polyconjugated uncharged structures remain. Reacidification reestablishes the equilibria as indicated by the bathochromic shifts in the UV–visible spectra which become again those of the polymerizing solutions. Of course, the relative position of these equilibria will depend upon the amount of acid reintroduced.

Having established the causes and the reaction pattern leading to chromophores, and thus deeply coloured resins, it remains to ascertain how the ramifications arise in these systems. The most likely sites for these branching points are the C=C groups formed in the chromogenic sequences discussed above, *i.e.* both the unsaturated carbon atoms between the heterocycles and the C-3 and C-4 positions of the dihydrofuranic structures, but not the unmodified C-3 and C-4 furanic moieties. These reactive sites constitute alternative condensation points for a CH_2OH group, whether from the monomer or the oligomers, and therefore the events producing conjugated sequences are also at the origin of branching, but the former mechanisms precede the latter.

That this is indeed the branching mode was suggested by the fact that model (**79**), which simulates a polymer chain before hydride-ion abstraction and which cannot lead to conjugated patterns for

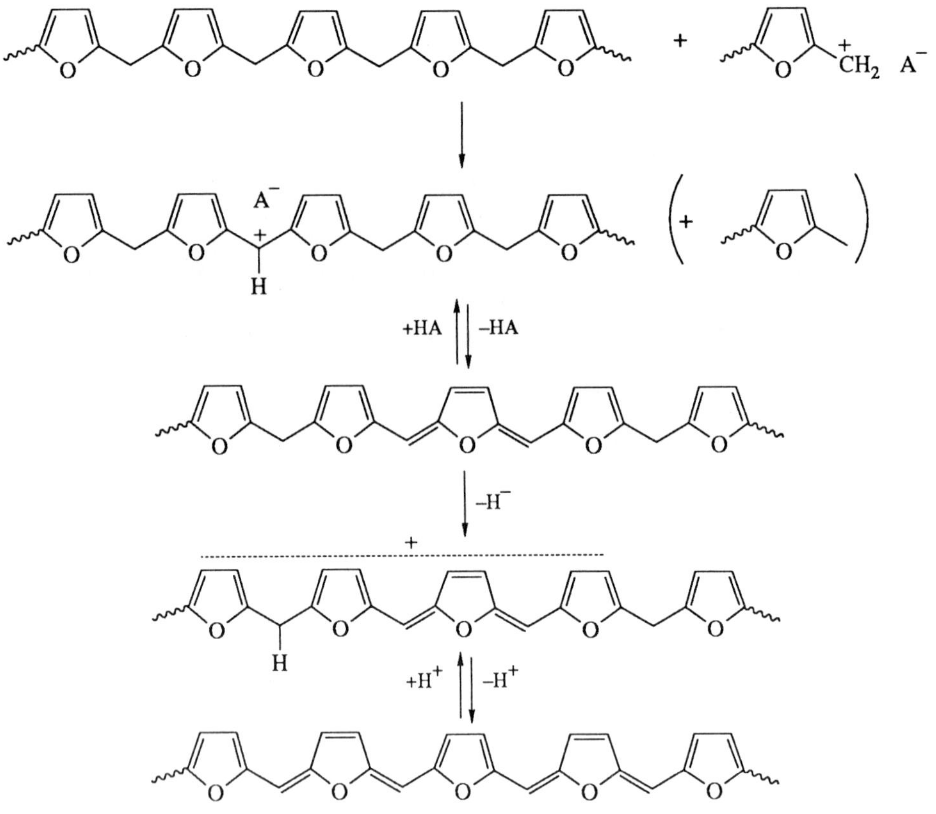

Scheme 18

lack of sufficient precursor length, does not react with 5-methylfuryl alcohol (**43b**), which simulates a monomer or polymer end-group. Contrary to what had recently been suggested,[196] it is therefore not an acid-catalyzed condensation between CH_2OH and CH_2 groups that engenders the branches, but instead the equivalent reactions involving CH_2OH and $-CH=$ groups (except the H-3 and H-4 of furanic moieties). The extent of branching is very important in the typical systems involving furfuryl alcohol, as testified by both the structure of the oligomers isolated in the early stages of the reaction and the fact that gel formation is a sudden event occurring at relatively low conversions with respect to the OH and H-5 consumed. In other words, the high functionality of the highly branched oligomers induces premature crosslinking as in all polycondensations.

Polyesters of the furan series were thoroughly investigated a decade ago.[197] This study led to rather disappointing conclusions as to the properties of the many products obtained which exhibited poor thermal performances and were never reported to crystallize, suggesting rather irregular structures. These drawbacks were probably a consequence of too agressive experimental conditions for the polymerizations. The use of milder catalysts and relatively low temperatures affords products with the expected regular structure and interesting features. An example of these syntheses is the bulk polymerization of monomer (**61**) by transesterification catalyzed by potassium carbonate.[198] The resulting polymer is a crystalline material, melting at 180 °C, with an NMR spectrum reflecting a highly regular structure. This polyester possesses an interesting feature in that it can be photo-crosslinked by near-UV irradiation thanks to the furanacrylic chromophore it bears on each monomer unit.

Rigid-chain polyesters from 2,5-furandicarboxylic acid chloride (**62c**) and hydroquinone are readily formed by classical syntheses similar to those used for all-aromatic polyesters, but they are intractable because they are neither meltable nor soluble.[199] The furanic diacid chloride (**62c**) also gives copolyesters with mixtures of hydroquinone and aliphatic comonomers added to reduce chain stiffness.

Model structures with three units, alternatively one furan ring between two aromatic ones, or one phenyl moiety between two furan heterocycles, have been constructed to test their possible mesogenic properties.[200] None were observed, even when specific substituents were appended on the furan C-5 positions with the latter structures.

Other furanic polyesters and polyethers based on monomers (**59**) and (**62**) and aromatic or aliphatic comonomers have been prepared by phase-transfer catalysis.[201] Their molecular weights are not high, but there does not seem to be any major anomaly in their structures.

A series of polyamides built from 2,5-furandicarboxylic acid (62a) and either aromatic or furanic diamines have been recently prepared and characterized.[169] The direct polycondensation technique was optimized with the furanic–aromatic systems and high molecular weights obtained. Structural studies were quite satisfactory in all instances. These polymers crystallized readily and had very good thermal properties with decomposition thresholds well above most thermoplastic furan polymers prepared previously. One of these polyamides, viz. that resulting from the condensation of (62a) with p-phenylenediamine to give structure (83), displayed a lyotropic liquid crystal behaviour above a certain molecular weight. The furan ring is not conducive to mesogenic structures in small size molecules. Therefore, the appearance of a mesophase with polyamide (83) must be related to its polymeric character coupled with the chain stiffening due to hydrogen bonding.

(83)

The all-furanic polyamides, such as (84) formed from (62c) and the furanic diamine (63), had lower DPs than the furanic–aromatic counterparts and showed a somewhat higher thermal fragility. This is due to the CH_2 groups attached to the furanic moiety coming from the diamine. In fact, one cannot prepare amines in which the NH_2 group is attached directly to the furan ring because such structures tautomerize to imines. This is a drawback from the points of view of both the chemical reactivity (easy loss of hydrogen atoms) of the methylene groups which are interposed between the ring and the amine function, and the increased flexibility of the polymer chains introduced by these junctions.

(84)

Polyurethanes bearing furan rings have received scanty attention in the past[124] and the few published studies provide little or no detailed information concerning structures, molecular weights or physical properties. The subject has now been tackled in a systematic way in order to ascertain first of all the feasability and stability of various furanic urethane structures before investigating the polymerizations themselves and the characterization of the ensuing products. This involved a preliminary study on the synthesis of furanic monoisocyanates[160] (54) and (57) and diisocyanates[166,167] (65) and (66) little known or unknown previously as already mentioned above.

Furanic urethanes and diurethanes are stable compounds which are readily prepared following common procedures.[202,203] The reactivity of furanic isocyanates and alcohols obeys straightforward considerations of electronic and steric factors: the furan ring, with its dienic character, exerts a strong influence on the reactivity of the NCO group, stronger than that of the phenyl ring, even when a methylene group separates them. Indeed, 2-furyl isocyanate (54) is probably the most reactive isocyanate reported to date in the context of urethane formation. Steric effects manifest themselves strongly with the furanic alcohols, just like with aliphatic and arylalkyl alcohols. Thus, for example, the relative reactivity of (43a), (78) and (77) towards a given isocyanate (or diisocyanate) decreases because of the steric hindrance brought in by the methyl group(s), despite the fact that the basicity of these alcohols increases with progressive methyl substitutions.

The polyurethanes prepared after this preliminary investigation include three families of products, namely: (i) those prepared from furanic diisocyanates and aliphatic or benzylic diols; (ii) those from aliphatic or aromatic disocyanates and furanic diols; and finally (iii) those derived entirely from furanic monomers.[167,204] Moreover, according to the structure of the furanic diols used, the furan rings appeared either in the main polymer backbone or as pendant groups, or in both situations.

High DPs could be obtained with all compositions and the regular structures of the polymers suggested that no important side reactions occurred during the syntheses. All these thermoplastic materials are chemically stable up to about 200 °C. Their glass transition temperatures vary within a large interval as a function of the specific structures of the two monomers. The furan ring, as observed previously in other macromolecules, plays a 'stiffening' role both when it is present within the chain or as a side group, but of course the importance of this effect is more notable in the former type of sturctures. Compared with the phenyl group, the furan heterocycle is somewhat less voluminous (both are planar) and thus contributes correspondingly less to an enhancement of T_g. As an example, the all-furanic polyurethane (**85**) displayed a T_g of 56 °C compared with a value of 84 °C for its homologue prepared with diphenylmethane diisocyanate and of 24 °C for the polymer prepared with hexamethylene diisocyanate.

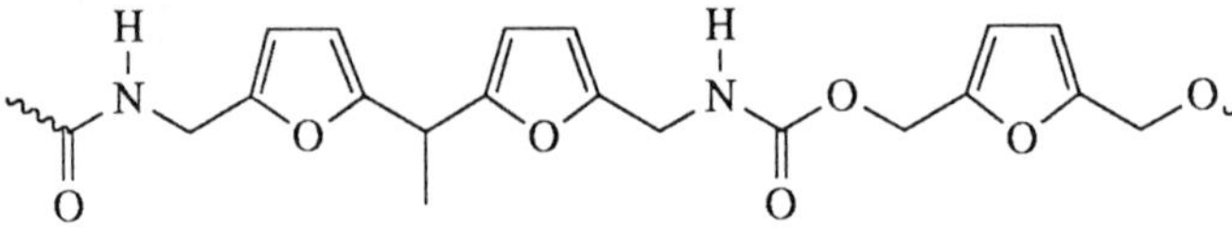

(**85**)

Crosslinked furanic polyurethanes, both glassy and elastomeric at room temperature, derived from monomers possessing higher functionalities, such as the polyol formed in reaction (7), have also been prepared and characterized.[205]

2,5-Bis(chloromethyl)furan (**64**) has been used as precursor to poly(2,5-vinylenefurylene) (**86**) in a sequence of steps simulating those used to prepare the phenylene homologue.[206,207] The latter has spurred much interest as an electronically conducting polymer. The properties of the furanic polyconjugated structure are similar: conductivities as high as 36 S cm^{-1} were claimed after doping the polymer with iodine.

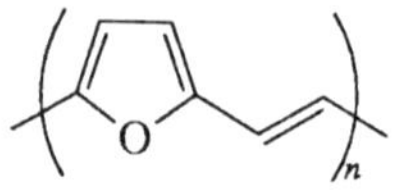

(**86**)

2,5-Furandiacrylic acid (**87**), readily prepared from the dialdehyde (**58**) and its esters can be photo-dimerized or photopolymerized by UV irradiation.[208] The strong chromophore in the monomers allows an easy transition to the cyclobutane dimer using even the near-UV component of a simple tungsten lamp, as shown in equation (9). Photopolymerization from the dimers requires shorter wavelengths and proceeds less readily, as indicated by preliminary studies.

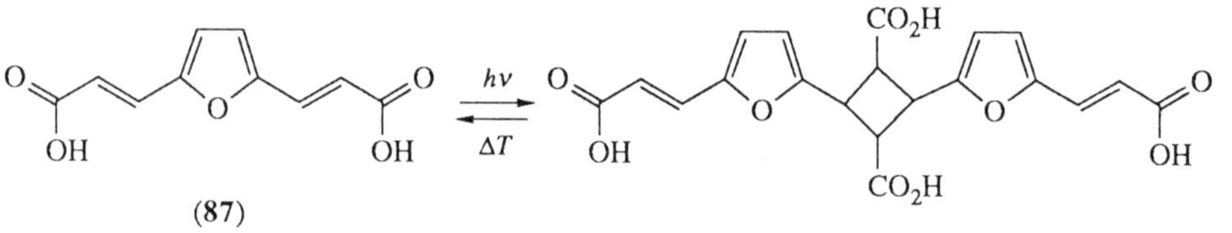

(9)

(**87**)

Polybenzimidazoles prepared from 2,5-furandicarboxylic acid chloride (**62c**) and 1,5-dinaphthyl-amino-2,4-diaminobenzene with M_n of about 10 000 and stable up to 400 °C have been prepared and their properties compared with those of the 1,4-phenylic homologues.[209]

Other recent studies on furanic polymers prepared by step reactions include the synthesis of polysilylfurylenes,[210] polyimides prepared from furfuryl alcohol oligomers and bis(maleimides)[211] and condensation resins from furfural[212] (**41**) and hydroxymethylfurfural (**47**).[213]

Ribbon-shaped polymers derived from difuranic compounds and bisdienophiles have been reported.[214]

19.5.8 The Chemical Modification of Furanic Polymers

The furan heterocycle displays a peculiar chemical behaviour based on mixed aromatic–dienic properties.[173] Compared with the sulfur (thiophene) and nitrogen (pyrrole) homologues, furan is the least aromatic in character and thus the most dienic member of the series. This property is clearly demonstrated by the readiness with which furan and some of its derivatives undergo the Diels–Alder cycloaddition reaction[215] with such dienophiles as maleic anhydride and maleimides.

The substituents on the furan cycle which preserve this reactivity are mostly alkyl groups, provided steric problems do not intervene. Carbonyl moieties attached to the ring tend to deactivate it with respect to this cycloaddition. Of course the Diels–Alder reaction is reversible and the adducts regenerate their constituents upon heating.

Furan polymers can therefore be modified by the Diels–Alder reaction, particularly if the ring is pendant to the chain, *i.e.* more accessible to dienophiles. These modifications can be made permanent if the adducts obtained are transformed irreversibly by such reactions as hydrolysis. In this way various polar or reactive groups can easily be introduced on the furan polymer or copolymer. As an example a copolymer of isobutene and 5-methyl-2-vinylfuran prepared by cationic polymerization could readily be converted into an isobutene copolymer bearing CO_2H groups along the chain by reacting the furan moieties with maleic anhydride and then hydrolyzing the cyclic adducts.[216] Of course, these considerations also apply to furan-terminated structures, as those described above in the context of cationic polymerization with dominant transfer by alkylation at C-5.

It is precisely this electrophilic substitution reaction at the C-2 or C-5 positions of the furan ring that must be considered another facile reaction which provides a means of chemical modification of polymers and copolymers bearing furan moieties, as already mentioned on several occasions. Again, polymers or copolymers possessing furan rings as side groups, but this time with the C-5 position unsubstituted, can be grafted by monomers which can polymerize cationically, *i.e.* the furan ring can be used as a source of grafting or of appending specific side groups.

These examples are by no means exhaustive, but were quoted to emphasize the potentials brought about by the furan heterocycle, introduced into a polymer even in small doses, towards original structural modifications.

19.6 THE CHEMISTRY OF UNREFINED LIGNOCELLULOSICS

Considerable interest has been revived during the last few years in using wood and other biomass products as a source of materials without proceeding to separate their components, *viz.* lignin, cellulose and hemicelluloses.

One important series of investigations within this context deals with the thermoplasticization of wood[217] by specific chemical modifications capable of destroying most of the crystallinity and hydrogen bonds in the cellulose fibres and at the same time partly degrading the lignin networks. If, moreover, this is done with reagents which have plasticizing structrual properties, such as long aliphatic chains, then wood can be transformed into a plastic mass from which films can be cast, objects moulded and foams blown.

The use of amine oxides as solvent for lignocellulosics[218] is another approach to the search of novel materials based on polymers from the biomass.

Composites between naturally occurring and synthetic polymers are also becoming an important research topic[219-221] both in terms of physical blending and chemical reactivity at the interface.

Adhesive compositions suitable for preparing wood-based composites can be prepared from renewable resources such as lignin, tannins and carbohydrates.[222] The resulting materials are thus made up entirely of components arising from the biomass.

Further development in this research field needs a better understanding of the interaction mechanisms taking place in such complex mixtures, *i.e.* more fundamental studies and interactions with laboratories working on topics dealing with the corresponding situations involving the individual components.

19.7 CONCLUSION

As pointed out in the opening remarks, it is impossible to cover the field of polymers from renewable resources in a comprehensive manner within such a short space. It was deemed necessary to convey at least a primary message which would reflect the vast array of topics being pursued actively at present and their interest both in scientific and technological terms. Also, and not necessarily in contradiction with that message, it was hoped to show that very much remains to be done and that a body of results and materials constituting a real alternative with respect to polymers from petrochemical sources can only be offered to industry if the research effort is multiplied in the next few years. Comparatively few laboratories are involved in these fields for them to be truly competitive with the massive past and present investigation efforts within the context of petroleum-based polymers, yet the raw materials are definitely there and will not cease to be available for mankind for a long time to come.

19.8 REFERENCES

1. L. H. Sperling and C. E. Carraher, in 'Encyclopedia of Polymer Science and Engineering', ed. H. F. Mark, N. M. Bikales, C. G. Overberger and G. Menges, Wiley, New York, 1988, vol 12, p. 658.
2. A. Gandini, in 'Encyclopedia of Polymer Science and Engineering', ed. H. F. Mark, N. M. Bikales, C. G. Overberger and G. Menges, Wiley, New York, 1986, vol. 7, p. 454.
3. J. C. Arthur, Jr., in 'Comprehensive Polymer Science', ed. G. Allen and J. C. Bevington, Pergamon Press, Oxford, 1989, vol. 6, p. 49.
4. E. A. MacGregor and C. T. Greenwood, 'Polymers in Nature', Wiley, New York, 1980.
5. J. F. Kennedy and C. A. White in 'Comprehensive Organic Chemistry', ed. D. Barton and W. D. Ollis, Pergamon Press, Oxford, 1979, vol. 5, p. 755.
6. G. O. Aspinall (ed.), 'The Polysaccharides', vols. 1–3, Academic Press, New York, 1982, 1983 and 1985.
7. M. Yalpani, 'Polysaccharides', Elsevier, New York, 1988.
8. P. S. Katsayannis, 'The Chemistry of Polypeptides', Plenum Press, New York, 1977.
9. P. W. Allen, 'Natural Rubber and Synthetics', Halsted, New York, 1972.
10. K. V. Sarkanen and C. H. Ludvig (eds.), 'Lignins: Occurrence, Formation, Structure and Reactions', Wiley, New York, 1971.
11. F. D. Gunstone, 'Chemistry and Biochemistry of Fatty Acids and Their Glycerides', Chapman and Hall, London, 1967.
12. D. H. Solomon, 'The Chemistry of Organic Film Formers', Krieger, Huntington, NY, 1977.
13. R. L. Whistler and J. N. BeMiller, 'Industrial Gums', Benjamin, New York, 1966.
14. A. R. Pinder, 'The Chemistry of the Terpenes', Wiley, New York, 1961.
15. A. Pizzi, in 'Wood Adhesives', ed. A. Pizzi, Dekker, New York, 1983, chap. 4.
16. D. Fengel and G. Wegener, 'Wood: Chemistry, Ultrastructure, Reactions', de Gruyter, Berlin, 1984.
17. H. B. Brown and J. O. Ware, 'Cotton', McGraw-Hill, New York, 1958.
18. R. S. Asquith (ed.), 'Chemistry of Natural Protein Fibers', Plenum Press, New York, 1977.
19. T. C. Thorstensen, 'Practical Leather Technology' Krieger, Huntington, NY, 1975, D. G. Bailey, in 'Encyclopedia of Polymer Science and Engineering', Wiley, New York, 1989, suppl. vol., p. 362.
20. J. P. Casey, 'Pulp and Paper Chemistry and Chemical Technology', Wiley, New York, 1980, vol. 1.
21. T. P. Nevell and S. H. Zeronian (eds.), 'Cellulose Chemistry and its Applications', Horwood, Chichester, UK, 1985.
22. N. M. Bikales and L. Segal (eds.), 'Cellulose and Cellulose Derivatives', Wiley, New York, 1971.
23. R. M. Brown (ed.), 'Cellulose and Other Natural Polymer Systems: Biogenesis, Structure and Degradation', Plenum, New York, 1982.
24. J. Qunichon and T. Tranchant, 'Nitrocelluloses', Horwood, Chichester, UK, 1989.
25. R. L. Davidson (ed.), 'Handbook of Water-Soluble Polymers', McGraw-Hill, New York, 1980.
26. R. Dönges, Br. Polym. J., 1990, 23, 315.
27. T. L. Vigo, in 'Encyclopedia of Polymer Science and Engineering', ed. H. F. Mark, N. M. Bikales, C. G. Overberger and G. Menges, Wiley, New York, 1985, vol. 3, p. 124.
28. J.-M. Lauriol, P. Froment, F. Pla and A. Robert, Holzforschung, 1987, 41, 109.
29. R. M. Rowell and R. A. Yound (eds.), 'Modified Cellulosics', Academic Press, New York, 1978.
30. C. Schuerch (ed.), 'Cellulose and Wood: Chemistry and Technology', Wiley, New York, 1989.
31. J. F. Kennedy, G. O. Phillips and P. A. Williams (eds.), 'Cellulose: Structural and Functional Aspects', Horwood, Chichester, UK, 1989.
32. J. P. Kennedy, G. O. Phillips and P. A. Williams (eds.), 'Lignocellulosics–Science, Technology, Development and Uses', Horwood, Chichester, 1992.
33. Various authors, ref. 31, pp. 345–409.
34. D. G. Gray, in 'Cellulosics Utilization', ed. H. Inagaki and G. O. Phillips, Elsevier, London, 1989, p. 123.
35. R. D. Gilbert, ACS Symp. Ser., 1990, 433, 259.
36. R. H. Marchessault in ref. 30, p. 1.
37. Various authors, ref. 31, pp. 415–431.
38. I. Cabasso, ref. 30, p. 1361.
39. P. Konstantin, ref. 30, p. 1333.
40. K. Kamide, ref. 34, p. 38.
41. A. Hebeish and J. T Guthrie, 'The Chemistry and Technology of Cellulosic Copolymers', Springer-Verlag, Berlin, 1981.
42. ACS Symp. Ser., 187, 1982.
43. R. K. Samal, P. K. Sahoo and H. S. Samantaray, J. Macromol. Sci., Rev. Macromol. Chem., 1986, C26, 81.
44. V. T. Stannett, ref. 31, p. 19.
45. R. Narayan, in 'Cellulosics Utilization', ed. H. Inagaki and G. O. Phillips, Elsevier, London, 1989, p. 110.
46. D. I. Yoo, Y. J. Lee and W. S. Ha, ref. 31, p. 425, and refs. therein.
47. R. Narayan, ref. 30, p. 945.
48. H. Cheradame, A. U. Tadjang and A. Gandini, Makromol. Chem., Macromol. Symp., 1986, 6, 261.
49. H. Cheradame, A. U. Tadjang and A. Gandini, Makromol. Chem., Rapid Commun., 1988, 9, 255.
50. C. Feger and H. J. Cantow, Polym. Bull. (Berlin), 1980, 3, 407.
51. R. H. Marchesault, B. Fisa and J. F. Revol, ACS Symp. Ser. 1975, 10, 147.
52. T. Mezger and H. J. Kantow, Angew. Makromol. Chem., 1983, 116, 13.
53. P. K. Chatterjee (ed.), 'Absorbency' Elsevier, Amsterdam, 1985.
54. B. V. Kokta, F. Dembélé and C. Daneault, in 'Renewable-Resource Materials', ed. C. E. Carraher, Jr. and L. H. Sperling, Plenum Press, New York, 1986, p. 85.
55. G. F. Fanta and W. M. Doane, ACS Symp. Ser., 1990, 433, 288.

56. G. F. Fanta and E. B. Bagley, in 'Encyclopedia of Polymer Science and Technology', ed. H. F. Mark, N. M. Bikales, C. G. Overberger and G. Menges, Wiley-Interscience, New York, 1977, Suppl. vol. 2.
57. J. M. Gould, S. H. Gordon, L. B. Dexter and C. L. Swanson, *ACS Symp. Ser.*, 1990, **433**, 65; M. A. Cole, *ACS Symp. Ser.*, 1990, **433**, 76.
58. MATER-BI, a polymer developed by Ferruzzi Ricerca e Tecnologia S.r.l., Novara, Italy.
59. G. Skjak-Braek, T. Anthonsen and P. Sandord (eds.), 'Chitin and Chitosan', Elsevier, London, 1988.
60. R. A. A. Muzzarelli, 'Chitin', Pergamon Press, Oxford, 1977.
61. R. A. A. Muzzarelli, in 'Encyclopedia of Polymer Science and Engineering', ed. H. F. Mark, N. M. Bikales, C. G. Overberger and G. Menges, Wiley, New York, 1986, vol. 3, p. 430.
62. S. Tokura, in 'Cellulosics Utilization', ed. H. Inagaki and G. O. Phillips, Elsevier, London, 1989, p. 63.
63. N. Tasubokawa, A. Kuroda and Y. Sone, *J. Polym. Sci., Polym. Chem. Ed.*, 1989, **27**, 1701.
64. N. Tsubokawa, S. Ohshima, Y. Sone and T. Endo, *J. Polym. Sci., Polym. Chem. Ed.*, 1989, **27**, 4413.
65. N. A. Porter, L. S. Lehman, B. A. Weber and K. J. Smith, *J. Am. Chem. Soc.*, 1981, **103**, 6447.
66. D. E. Chubin, J. P. Kaczmarski, Z. Ma, D. Wang and F. N. Jones, *ACS Symp. Ser.*, 1990, **433**, 220.
67. Various authors, *ACS Symp. Ser.*, 1989, **397**, 2–177; B. Monties, in 'Methods in Plant Biochemistry', vol. 1 'Plant Phenolics', ed. J. B. Harborne, Academic Press, London, 1989, p. 113.
68. W. G. Glasser and S. S. Kelley, in 'Encyclopedia of Polymer Science and Engineering', ed. H. F. Mark, N. M. Bikales, C. G. Overberger and G. Menges, Wiley, New York, 1987, vol. 8, p. 795.
69. S. Y. Lin, *Prog. Biomass Convers.*, 1983, **4**, 31.
70. H. H. Nimz, ref. 15, chap. 5.
71. S. Yamashita and S. Kohjiya, in 'Wood Processing and Utilization', ed. J. F. Kennedy, G. O. Phillips and P. A. Williams, Horwood, Chichester, 1989, p. 187.
72. S. L. Ciemniecki and W. G. Glasser, *ACS Symp. Ser.*, 1989, **397**, 452.
73. T. G. Rials and W. G. Glasser, *ACS Symp. Ser.*, 1989, **397**, 464.
74. M. A. De Paoli and L. T. Furlan, *Polym. Degradation Stab.*, 1985, **13**, 129.
75. I. Chodàk, R. Brezny and L. Rychlà, *Chem. Zvesti*, 1986, **40**, 461.
76. K. Levon, J. Huhtala, B. Malm and J. J. Lindberg, *Polymer*, 1987, **28**, 745.
77. J. H. Lora, C. F. Wu, E. K. Pye and J. J. Balatinecz, *ACS Symp. Ser.*, 1989, **397**, 312.
78. P. M. Kook and T. Sellers, Jr., *ACS Symp. Ser.*, 1989, **397**, 324.
79. G. H. Van der Klashorst, *ACS Symp. Ser.*, 1989, **397**, 346.
80. P. C. Muller and W. G. Glasser, *J. Adhes.*, 1984, **17**, 157 and 185.
81. H. K. Ono and K. Sudo, *ACS Symp. Ser.*, 1989, **397**, 334.
82. J. J. Meister, ref. 54, p. 305.
83. J. J. Meister, D. R. Patil, L. R. Field and J. C. Nicholson, *J. Polym. Sci., Polym. Chem. Ed.*, 1984, **22**, 1963.
84. J. J. Meister and D. R. Patil, *Macromolecules*, 1985, **18**, 1559.
85. J. J. Meister, D. R. Patil, C. Augustin and J. Z. Lai, *ACS Symp. Ser.*, 1989, **397**, 294. J. J. Meister, A. Lathia and F. Chang, *J. Polym. Sci., Polym. Chem. Ed.*, 1991, **29**, 1465.
86. R. L. Chen, B. V. Kokta, C. Daneault and J. L. Valade, *J. Appl. Polym. Sci.*, 1986, **32**, 4815.
87. P. Dournel, E. Randrianalimanana, A. Deffieux and M. Fontanille, *Eur. Polym. J.*, 1988, **24**, 843.
88. H. P. Naveau, *Cell. Chem. Technol.*, 1975, **9**, 71.
89. L. C. F. Wu and W. G. Glasser, *J. Appl. Polym. Sci.*, 1984, **29**, 1111.
90. W. de Oliveira and W. G. Glasser, *ACS Symp. Ser.*, 1989, **397**, 414.
91. J. A. Hyatt, *ACS Symp. Ser.*, 1989, **397**, 425.
92. T. G. Rials and W. G. Glasser, *Holzforschung*, 1984, **38**, 263.
93. W. H. Newman and W. G. Glasser, *Holzforschung*, 1985, **39**, 345.
94. T. G. Rials and W. G. Glasser, *Holzforschung*, 1986, **40**, 353.
95. S. S. Kelley, W. G. Glasser and T. C. Ward, *ACS Symp. Ser.*, 1989, **397**, 402.
96. K. Nakamura, R. Mörck, A. Reimann, K. Kringstad and H. Hatakeyama, ref. 71, p. 187.
97. M. Detoisien, F. Pla, A. Gandini, H. Cheradame and P. Monzie, *Br. Polym. J.*, 1985, **17**, 260.
98. S. Hirose, S. Yano, T. Hatakeyama and H. Hatakeyama, *ACS Symp. Ser.*, 1989, **397**, 382.
99. R. Mörck, A. Reimann and K. P. Kringstad, *ACS Symp. Ser.*, 1989, **397**, 390.
100. W. G. Glasser, ref. 71, p. 163.
101. D. Feldman and D. Banu, *J. Polym. Sci., Polym. Chem. Ed.*, 1988, **26**, 973.
102. D. Feldman and D. Banu, A. Natansohn and J. Wang, *J. Appl. Polym. Sci.*, 1991, **42**, 1537.
103. W. G. Glasser and H. X. Wang, *ACS Symp. Ser.*, 1989, **397**, 515.
104. H. Struszczyk, *ACS Symp. Ser.*, 1989, **397**, 245.
105. Z. X. Guo, A. Gandini and F. Pla, *Polym. Int.*, 1992, **27**, 17.
106. Z. X. Guo and A. Gandini, Paper presented at the 6th European Conference on Biomass, Athens, April 1991, *Eur. Polym. J.*, 1991, **27**, 1177.
107. T. A. Kuusela, J. J. Lindberg, K. Levon and J. E. Österholm, *ACS Symp. Ser.*, 1989, **397**, 219.
108. V. Demaret and W. G. Glasser, *Polymer*, 1989, **30**, 570.
109. J. J. Lindberg, T. A. Kuusela and K. Levon, *ACS Symp. Ser.*, 1989, **397**, 190; S. S. Kelley, T. C. Ward and W. G. Glasser, *Polymer*, 1989, **30**, 2265; *J. Appl. Polym. Sci.*, 1990, **41**, 2813.
110. G. Vazquez, G. M. Antorrena, J. C. Parajo and J. L. Francisco, *Holz Roh Werkst.*, 1989, **47**, 491.
111. A. Pizzi, ref. 54, p. 323.
112. H. Hatakeyama, S. Hirose and T. Hatakeyama, *ACS Symp. Ser.*, 1989, **397**, 205.
113. B. Keszler and J. P. Kennedy, *Adv. Polym. Sci.*, 1991, **100**, 1.
114. E. M. Ruckel and H. G. Arlt, Jr., in 'TAPPI Proceedings on Paper Synthetics Conference, 1983', TAPPI, Raleigh, NC, 1983, p. 173; E. R. Ruckel, in 'Carbocationic Polymerization', ed. J. P. Kennedy and E. Maréchal, Wiley, New York, 1982, p. 491.
115. W. Vredenburgh, K. F. Foley and A. N. Scarlatti, in 'Encyclopedia of Polymer Science and Engineering', ed. H. F. Mark, N. M. Bikales, C. G. Overberger and G. Menges, Wiley, New York, 1988, vol. 7, p. 758.
116. J. L. Cawse, J. L. Stanford and R. H. Still, *J. Appl. Polym. Sci.*, 1987, **33**, 2217.

117. J. L. Cawse, J. L. Stanford and R. H. Still, *J. Appl. Polym. Sci.*, 1987, **33**, 2231.
118. J. L. Cawse, J. L. Stanford and R. H. Still, *Polymer*, 1987, **28**, 368.
119. G. Braca, G. Sbrana, A. M. Raspolli-Galletti and G. Gagliardi, in 'Biomass for Energy, Industry and Environment', ed. G. Grassi, A. Collina and H. Zibetta, Elsevier, London, 1992, p. 1235.
120. L. Marini and B. Casale, in ref. 119, p. 1269.
121. A. F. Bochkov and G. E. Zaikov, 'Chemistry of the O-Glycosidic Bond', Pergamon Press, Oxford, 1979.
122. C. Schuerch, *Adv. Carbohydr. Chem. Biochem.*, 1981, **39**, 157.
123. C. Schuerch, in 'Encyclopedia of Polymer Science and Engineering', ed. H. F. Mark, N. M. Bikales, C. G. Overberger and G. Menges, Wiley, New York, 1988, vol. 13, p. 147.
124. J. L. Cawse, J. L. Stanford and R. H. Still, *Makromol. Chem.*, 1984, **185**, 697 and 709.
125. S. K. Dirlikov, *ACS Symp. Ser.*, 1990, **433**, 176.
126. K. Kobayashi, H. Sumimoto and Y. Ina, *Polym. J.*, 1983, **15**, 667.
127. J. Klein, D. Herzog and A. Hajibegli, *Makromol. Chem., Rapid Commun.*, 1985, **6**, 675.
128. Y. Koyoma, A. Yoshida and K. Kurita, *Polym. J.*, 1986, **18**, 479.
129. M. A. Pastoriza and H. E. Bertorello, *Polym. Bull.*, 1986, **15**, 241.
130. E. R. Howell, *Chem. Ind. (London)*, **1982**, 508.
131. N. D. Miller and D. F. Williams, *Biomaterials (Guildford, Engl.)*, 1987, **8**, 129.
132. A. Ballistreri, G. Montaudo, D. Garozzo, M. Giuffrida and M. S. Montaudo, *Macromolecules*, 1991, **24**, 1231.
133. S. J. Holland, A. M. Jolly, M. Yesin and B. J. Tighe, *Biomaterials (Guildford, Engl.)*, 1987, **8**, 289.
134. N. D. Miller, in 'Biocompatibility of Degradable Polymers', ed. D. F. Williams, CRC Press, Boca Raton, FL, 1989.
135. E. A. Dawes, (ed.), 'Novel Biodegradable Microbial Polymers', Kluwer, Dordrecht, 1990.
136. Y. Naoshima, C. E. Carraher, Jr. and K. Matsumoto, ref. 54, p. 63.
137. T. Atarhouch, A. Daro and C. David, *Eur. Polym. J.*, 1991, **27**, 527.
138. E. S. Lipinsky and R. G. Sinclair, *Chem. Eng. Prog.*, 1986, **82**(8), 26.
139. O. Theander and D. A. Nelson, *Adv. Carbohydr. Chem. Biochem.*, 1988, **46**, 273.
140. 'Making and Marketing of Furfural: Added Value for Agro-Industrial Wastes', UNCTAD/GATT, Geneva, 1979.
141. W. J. McKillip, *ACS Symp. Ser.*, 1989, **385**, 408.
142. A. Gandini, *Adv. Polym. Sci.*, 1977, **25**, 47.
143. A. Gaset, J. P. Gorrichon and E. Truchot, *Inf. Chim.*, 1981, **212**, 179.
144. A. Gandini, *ACS Symp. Ser.*, 1990, **433**, 195.
145. H. E. van Dam, A. P. G. Kieboom and H. van Bekkum, *Starch*, 1986, **38**, 95.
146. A. Fuchs, *Starch*, 1987, **39**, 335.
147. B. F. M. Kuster, *Starch*, 1990, **42**, 314.
148. K. Rapp (Südzücker), Ger. Pat. 3 601 281 (1987).
149. L. F. Jo, Doctorate Thesis, National Polytechnic Institute, Toulouse, France, 1987.
150. R. Paul and S. Tchelithceff, *Bull. Soc. Chim. Fr.*, 1947, 453.
151. Y. LeBigot, M. Delmas and A. Gaset, *Synth. Commun.*, 1982, **11**, 107.
152. G. B. Bachman and L. V. Heisey, *J. Am. Chem. Soc.*, 1949, **71**, 1985.
153. R. Alvarez, A. Gandini and R. Martinez, *Makromol. Chem.*, 1982, **183**, 2399.
154. J. Arekion, M. Delmas and A. Gaset, *Biomass*, 1983, **3**, 59.
155. D. A. Isacescu and F. Avramescu, *Rev. Roum. Chim.*, 1978, **23**, 873.
156. M. E. Borredon, M. Delmas and A. Gaset, *Tetrahedron*, 1987, **43**, 3945.
157. J. Sam and J. R. Mozingo, *J. Pharm. Sci.*, 1969, **58**, 1030.
158. A. Labidi, M. C. Salon,, A. Gandini and H. Cheradame, *Polym. Bull.*, 1985, **14**, 271.
159. W. H. Watanabe and L. E. Conlon, *J. Am. Chem. Soc.*, 1956, **79**, 2828; M. H. Brunache, Doctorate Thesis, National Polytechnic Institute, Grenoble, France, 1984.
160. J. Quillerou, M. N. Belgacem, A. Gandini, J. Rivero and G. Roux, *Polym. Bull.*, 1989, **21**, 555.
161. S. Pennanen and G. Nyman, *Acta Chem. Scand.*, 1972, **26**, 1018.
162. Z. Mouloungui, M. Delmas and A. Gaset, *Synth. Commun.*, 1984, **14**, 701.
163. P. Vinke, H. E. van Dam and H. van Bekkum, in 'New Developments in Selective Oxidation', ed. G. Centi and F. Trifiro, Elsevier, Amsterdam, 1990, p. 147.
164. T. El Hajj, A. Masroua, J. C. Martin and G. Descotes, *Bull Soc. Chim. Fr.*, 1987, 856.
165. O. Moldenhauer, G. Trautmann, R. Pfluger and H. Döser, *Justus Liebigs Ann. Chem.*, 1953, **580**, 180.
166. S. Nielek and T. Lesiak, *J. Prakt. Chem.*, 1988, **330**, 825.
167. J. Quillerou, Doctorate Thesis, National Polytechnic Institute, Gernoble, France, 1991.
168. D. Chudury and H. H. Szmant, *Ind. Eng. Chem. Prod. Res. Dev.*, 1981, **241**, 203; *Polym. Prepr.*, 1983, **24**(2), 50.
169. A. Mitiakoudis and A. Gandini, *Macromolecules*, 1991, **24**, 830.
170. A. P. Dunlop and F. N. Peters, 'The Furans', Reinhold, New York, 1953.
171. P. Bosshard and C. H. Eugster, *Adv. Heterocycl. Chem.*, 1966, **7**, 377.
172. F. M. Dean, *Adv. Heterocycl. Chem.*, 1982, **30**, 167 and **31**, 237.
173. C. W. Bird and G. W. H. Cheeseman, in 'Comprehensive Heterocyclic Chemistry', ed. A. R. Katritzky and C. W. Rees, Pergamon Press, Oxford, 1984, vol. 4, chaps. 1, 2 and 3; F. M. Dean and M. V. Sargent, in 'Comprehensive Heterocyclic Chemistry' ed. A. R. Katritzky and C. W. Rees, Pergamon Press, Oxford, 1984, vol. 4, chaps. 10 and 11; D. M. X. Donnelly and M. J. Meagan, in 'Comprehensive Heterocyclic Chemistry', ed. A. R. Katritzky and C. W. Rees, Pergamon Press, Oxford, 1984, vol. 4. chap. 12.
174. B. S. Lamb and P. Kovacic, *J. Polym. Sci., Polym. Chem. Ed.*, 1980, **18**, 2423.
175. N. G. Gaylord, N. Martan and A. B. Deshpande, *J. Polym. Sci., Polym. Chem. Ed.*, 1978, **16**, 1527.
176. G. Tourillon and F. Garnier, *J. Electroanal. Chem. Interfacial Electrochem.*, 1982, **135**, 173.
177. F. Tediar, *Eur. Polym. J.*, 1985, **21**, 317.
178. G. Zotti, G. Schiavon, N. Comisso, A. Berlin and G. Pagani, *Synth. Metals*, 1990, **36**, 337.
179. W. J. Feast and D. B. Harrison, *Polym. Bull.*, 1991, **25**, 343.
180. W. J. Feast and D. B. Harrison, *Polymer*, 1991, **32**, 558.
181. M. A. Hillmyer, C. Lepetit, D. V. McGrath and R. H. Grubbs, *Polym. Prepr., Am. Chem. Soc., Div. Polym. Chem.*, 1991, **32**(1), 162.

182. M. C. Salon, Doctorate Thesis, National Polytechnic Institute, Grenoble, France, 1985.
183. H. Razzouk, K. Bouridah, A. Gandini and H. Cheradame, in 'Cationic Polymerization and Related Processes', ed. E. J. Goethals, Academic Press, New York, 1984, p. 355.
184. M. C. Salon and A. Gandini, unpublished results.
185. A. Gandini and J. Reiumont, *Br. Polym. J.*, 1977, **9**, 28.
186. M. H. Brunache, Doctorate Thesis, National Polytechnic Institute, Grenoble, France, 1984.
187. M. C. Salon, A. Gandini and H. Cheradame, *Polym. Bull.*, 1984, **12**, 441.
188. M. C. Salon, H. Amri and A. Gandini, *Polym. Commun.*, 1990, **31**, 210.
189. H. Amri, Doctorate Thesis, National Polytechnic Institute, Grenoble, France, 1990.
190. Z. Hui and A. Gandini, *Polym. Bull.*, in press.
191. A. Gandini and J. Rieumont, *Tetrahedron Lett.*, 1976, 2101. A. Gandini, in 'Preprints of the IUPAC Symposium Radical Polymerization, Santa Margherita Ligure, 1987', ed. S. Russo, AIM, Genova, 1987, p. 93.
192. J. Rieumont, R. Vega, N. Davidenko and J. A. Paz, *Eur. Polym. J.*, 1988, **24**, 909.
193. D. C. Apperley, D. M. Curran and A. H. Fawcett, *Polymer*, 1988, **29**, 909.
194. M. Choura, Doctorate Thesis, National Polytechnic Institute, Grenoble, France, 1991.
195. S. L. Buchwalter, *J. Polym. Sci., Polym. Chem. Ed.*, 1985, **23**, 2897.
196. I. Chuang, G. E. Maciel and G. E. Myers, *Macromolecules*, 1984, **17**, 1087.
197. J. A. Moore and J. E. Kelly, *Macromolecules*, 1978, **11**, 568; *J. Polym. Sci., Polym. Chem. Ed.*, 1978, **16**, 2407; 1984, **22**, 863; *Polymer*, 1979, **20**, 627.
198. J. Roudet, Doctorate Thesis, National Polytechnic Institute, Grenoble, France, 1987; A. Gandini and J. Roudet, submitted.
199. D. Maccio', G. Costa, B. Valenti and A. Gandini, submitted.
200. G. Costa, S. Morinelli, B. Valenti and A. Gandini, Paper Presented to the 2nd Italy–USSR Polymer Meeting, Leningrad, 1991.
201. S. Boileau, R. Dorigo, M. Majdoub, F. Menchin and A. Gandini, submitted.
202. M. Belgacem, J. Quillerou, A. Gandini, J. Rivero and G. Roux, *Eur. Polym. J.*, 1989, **25**, 1125.
203. N. Belgacem, J. Quillerou and A. Gandini, *Eur. Polym. J.*, in press.
204. N. Belgacem, Doctorate Thesis, National Polytechnic Institute, Grenoble, France, 1991.
205. C. Signoret, Doctorate Thesis, National Polytechnic Institute, Grenoble, France, 1989.
206. Z. Hui and A. Gandini, submitted.
207. K. Y. Jen, T. R. Jow and R. L. Elsenbaumer, *J. Chem. Soc., Chem. Commun.*, 1987, 1113.
208. A. Viallet and A. Gandini, *J. Photochem.*, 1990, **A54**, 129.
209. A. H. Kehayoglou and G. P. Karayannidis, *J. Macromol. Sci., Chem.*, 1982, **A18**, 237.
210. H. H. Hong and W. P. Weber, *Polym. Bull.*, 1989, **22**, 363.
211. H. S. Patel and B. D. Lad, *Makromol. Chem.*, 1989, **190**, 2055.
212. P. S. Patel and S. R. Patel, *Eur. Polym. J.*, 1986, **22**, 525.
213. H. Koch and J. Pein, *Polym. Bull.*, 1985, **13**, 525.
214. K. Blatter and A. D. Schlüter, *Macromolecules*, 1989, **22**, 3508.
215. F. Fringuelli and A. Taticchi, 'Dienes in the Diels–Alder Reaction', Wiley, New York, 1990.
216. D. André and A. Gandini, unpublished results.
217. N. Shiraishi, in 'Cellulosics Utilization', ed. H. Inagaki and G. O. Phillips, Elsevier, London, 1989, p. 97.
218. A. Péguy, in Cellulosics Utilization', ed. H. Inagaki and G. O. Phillips, Elsevier, London, 1989, p. 19.
219. Various authors, ref. 71, pp. 219–293.
220. F. Mora, F. Pla and A. Gandini, *Angew. Makromol. Chem.* 1989, **173**, 137.
221. R. M. Rowell, *ACS Symp. Ser.*, 1990, **433**, 242.
222. *ACS Symp. Ser.*, 1989, **385**.

20

Modeling and Computer Simulation of Reactive Processing

L. JAMES LEE
Ohio State University, Columbus, OH, USA

20.1	INTRODUCTION	575
20.2	MATERIAL MODELING	576
	20.2.1 Reaction Kinetics	576
	20.2.1.1 Isocyanate-based resins [i.e. polyurethanes, poly(urethane urea)s, polyureas and poly(urethane isocyanurate)s]	577
	20.2.1.2 Epoxies	580
	20.2.1.3 Styrenic/polyester resins	582
	20.2.1.4 Nylon and nylon block copolymers	584
	20.2.1.5 Bismaleimide resins	587
	20.2.1.6 Hybrid resins	588
	20.2.2 Rheological Changes	589
	20.2.3 Structure and Property Changes	596
	20.2.3.1 Crystallization during polymerization	596
	20.2.3.2 Phase separation during polymerization	596
	20.2.3.3 Thermal expansion and polymerization shrinkage during polymerization	598
	20.2.3.4 Void formation and foaming during polymerization	600
	20.2.3.5 Thermophysical properties during polymerization	600
20.3	PROCESS SIMULATION	600
	20.3.1 Reactive Injection Molding Processes	600
	20.3.1.1 RIM	602
	20.3.1.2 RTM/SRIM	603
	20.3.1.3 Transfer molding	604
	20.3.2 Reactive Compression Molding	605
	20.3.3 Reactive Processing of Continuous Fiber Reinforced Thermoset Composites	609
20.4	NOTATION	611
20.5	REFERENCES	613

20.1 INTRODUCTION

A large number of polymer products involve polymerization in fabricating the final shape. Examples include compression, transfer and injection molding of rubbers; compression and injection molding of unsaturated polyester resins; reaction injection molding of polyurethanes and nylons; autoclave curing of epoxy/graphite prepregs; and encapsulation of electronic parts using epoxy or polyimide resins. A common feature in processing these materials is that polymerization (or a portion of polymerization) and processing take place at the same time. Accordingly, the term 'reactive processing' has been used by many researchers to describe these processes.

Due to the reactive nature of these systems they are often more complicated than the traditional thermoplastic processing operations and conventional polymerization methods. For instance, combination of reaction into the processing step usually means reaction must occur in the bulk state. In order to compete with thermoplastic processing operations, the reaction rate must be high to thus reduce the cycle time (i.e. a cure time of minutes or seconds instead of the times of hours

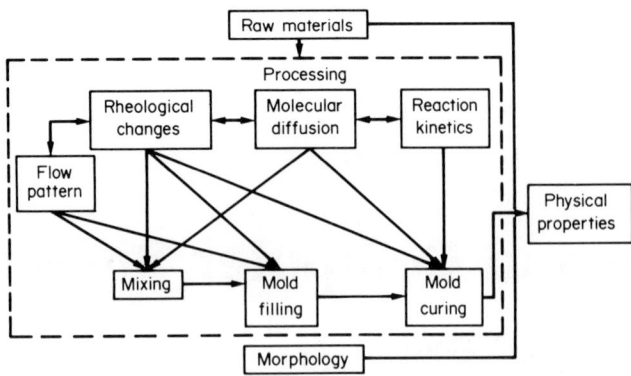

Figure 1 A schematic diagram of the reactive processing of polymers

common in a typical polymerization reactor). Quite often, the reactions are highly exothermic, making temperature control a difficult task. Most reactions are further complicated by the occurrence of multiple physical changes during molding. These changes include phase (or domain) formation in which materials change from a single phase to multiple phases because of the thermodynamic incompatibility of polymer segments or blocks; gelation in which materials change from a viscous fluid to a network gel with chemical or physical cross-linking; and glass transition or crystallization in which the reacting mixture changes from a rubbery material to a glassy polymer or from an amorphous material to a semicrystalline polymer with increasing degree of polymerization and packing of polymer segments.

These physical changes occur during the course of reaction and may certainly interact with the ongoing chemical reaction. For example, the Trommsdorff, or gel effect, which is well known for free radical polymerizations, may result in a rapid reaction with an excessive temperature rise. When coupled with the low thermal diffusivity of the polymer, these conditions may lead to the thermal runaway problem. The glass transition effect, where the material changes from the rubbery state to the glassy state, may cause premature cessation and thus an incomplete reaction. Quite often, physical properties of the finished products depend not only on the raw materials used, but also on the processing method and processing conditions.

In each process, raw materials go through a set of elementary processing steps such as mixing, mold filling and curing. These steps are governed by the transport phenomena of the process and the behavior of the reactive materials such as resin flow, molecular diffusion, rheological changes and reaction kinetics, as shown in Figure 1. For instance, the change of resin viscosity can affect the flow pattern during mold filling and the molecular diffusion during curing. The flow pattern and the molecular diffusion, in turn, may determine the reaction rate and the final conversion. Much experimental and modeling work has been carried out by researchers in order to understand the elementary processing steps and governing phenomena in various reactive processing operations.

This work includes an introduction to the major principles and a literature review of the relevant research work in the last 10–15 years in computer modeling and simulation of reactive polymer processing. To narrow the scope, only processes which involve significant polymerization during fabrication are considered. Reactive compounding in mixers or extruders (*i.e.* reactive extrusion), which deals mainly with polymer modification through chemical grafting, is not covered. Rubber cross-linking and processing are also not discussed. Section 20.2 presents basic material models for reaction kinetics, rheological changes, structure development and thermophysical properties. Section 20.3 discusses process simulation for major reactive processing operations. Two fundamental processing steps, mold filling and curing, are discussed. Because the mixing method varies from process to process and has not been well modeled except for the reactive injection molding process, it will not be covered in this work.

20.2 MATERIAL MODELING

20.2.1 Reaction Kinetics

Reactive processing necessitates a thorough understanding of the kinetics governing the cure reaction. A knowledge of the degree of cure and the thermal history of the material is required in order to interpret the structural and physical property changes. A knowledge of the heat of reaction and the rate of heat generation as a function of time and temperature is necessary for quantifying the heat transfer in the molding process.

The reactions usually involve either a step growth polymerization, a chain growth polymerization, or a combination of both. An ideal kinetic model should be simple enough to be combined with a process simulation model for predicting the mold filling and curing behavior. However, the model should be able to handle both the kinetic changes of the reacting material and the physical phenomena such as vitrification and the Trommsdorff effect. Furthermore, if the model is to be used for material design, the effect of major chemical ingredients on the reaction kinetics, and consequently, the physical properties or the defects of the molded product, need to be addressed. Obviously, it would be difficult to develop a single model which could fulfill all of the above-mentioned requirements. An alternative approach is to have several levels of models. The simpler ones have a limited application range, but may be easier to use. The more sophisticated ones may solve more complicated problems, but may be more difficult to implement and may have more model parameters which need to be determined experimentally. In the following section, several simple, general purpose kinetic models are mentioned first. More complicated kinetic models dealing with specific resins are discussed next.

The simplest kinetic model for describing the isothermal cure of reactive resins is given by the familiar nth order rate expression

$$\frac{d\alpha}{dt} = k(1-\alpha)^n \tag{1}$$

where α is conversion, $d\alpha/dt$ is rate of reaction, k is the temperature dependent rate constant and n is the kinetic exponent. This model assumes a maximum initial reaction rate, and consequently, is not capable of realistically describing the curing reaction when a rate peak in the isothermal cure is observed (i.e. a bell-shaped rate profile). In many studies of reactive processing,[1-7] the following kinetic expression proposed by Kamal and Sourour[1] has been found to correlate well with isothermal kinetic data

$$\frac{d\alpha}{dt} = (k_1 + k_2\alpha^m)(1-\alpha)^n \tag{2}$$

The kinetic exponents, m and n, are usually assumed to be independent of temperature and the rate constants, k_1 and k_2, are assumed to have an Arrhenius temperature dependence.

The parameters of these two models can be easily determined from rate profiles measured by conventional analytical tools, such as a differential scanning calorimeter (DSC). For applications where the primary objective of modeling is to estimate the overall curing rate, reaction exotherm and temperature profiles for a given resin, these models are often adequate because they are simple, yet provide a reasonable prediction. The limitation of these general purpose models is that they are not based on the resin chemistry, and therefore their parameters may not reflect the functions of the resins and curing agents in the reaction. Consequently, the predictive capability is limited.

The second level of kinetic models is based on the reaction mechanism of a given resin. The resin chemistry and the major elementary reactions are considered in the modeling. Several step growth type polymerizations (e.g. isocyanate-based resins, epoxies), chain growth type polymerizations (e.g. styrenic/unsaturated polyester resins, nylon), and the combination of both (e.g. bismaleimides, hybrid resins) are discussed.

20.2.1.1 Isocyanate-based resins [i.e. polyurethanes, poly(urethane urea)s, polyureas and poly(urethane isocyanurate)s]

The basic reaction which forms the urethane linkage is shown in Scheme 1.

To date, the kinetics of polyurethane reactions in the absence of solvents are still not completely understood. In particular, whether phase separation and/or gelation affect kinetics and where

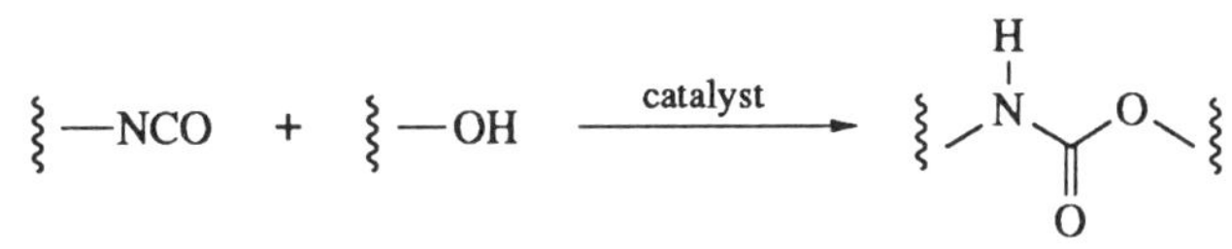

Scheme 1

cross-link density has any effect are still questions which remain unanswered.[8,9] Macosko[10] recently summarized the modeling effort in this area. His work is briefly stated here.

Robins et al.[11] proposed a kinetic mechanism for metal ion catalyzed urethane formation, as shown in Scheme 2.

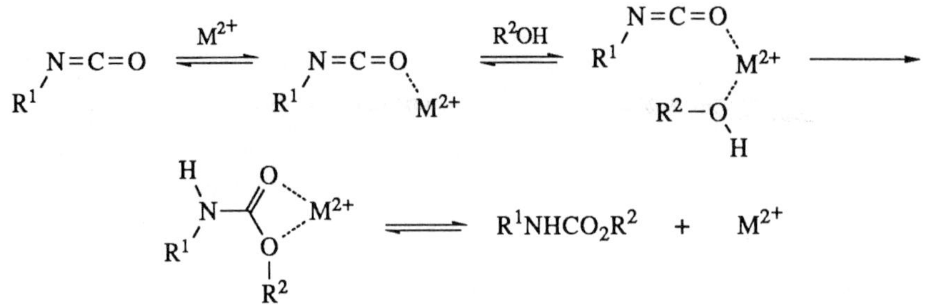

Scheme 2

This leads to a rate expression with an overall order which varies from 1 to 2 and a first order expression in the catalyst. Richter and Macosko[12] proposed a Michealis–Menten-type kinetic scheme for the RIM urethane polymerization. The key reaction steps are shown in Scheme 3.

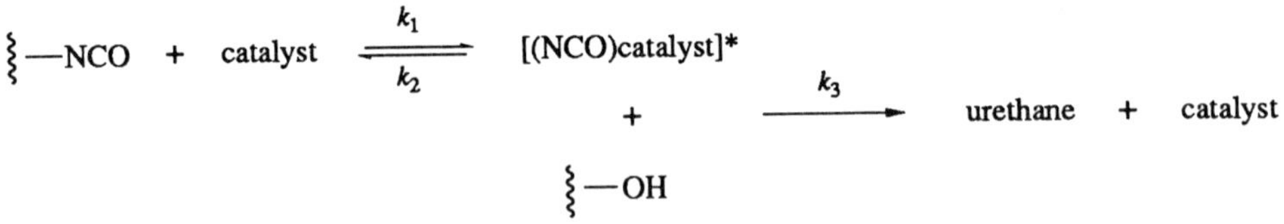

Scheme 3

This scheme results in a hyperbolic kinetic model[13] as follows

$$-\frac{dC_{NCO}}{dt} = \frac{k_1 C_{NCO} C_{OH} C_{cat}}{1 + KC_{OH}} \tag{3}$$

where C is concentration. Experimental results found the overall order of reaction and the effect of catalyst concentration changed with varying temperature.

Urethanes in reactive processing have often been modeled by a simplified nonmechanistic approach[14]

$$-\frac{dC_{NCO}}{dt} = kC_{cat}^{n_1} C_{NCO}^{n_2} C_{OH}^{n_3} \tag{4}$$

with

$$k = A \exp(-E_a/RT) \tag{5}$$

where $n_1 (= \frac{1}{2}-1)$ is the order with respect to catalyst, $n_2 + n_3 (= 1-2)$ is the overall order of the reaction. The expression can be further simplified by lumping the catalyst concentration into the rate constant. Frequently the isocyanate concentration has been considered equal to the hydroxy concentration at stoichiometry. The kinetic expression is further simplified to equation (1) with the reaction order varying from 1.5 to 3 depending on the catalyst level and resin conversion.[5,15–19] Sometimes a second rate expression is added for the slow reaction that occurs without a catalyst.[18–22] This leads to an expression with two rate constants as

$$-\frac{dC_{NCO}}{dt} = k_1 C_{cat}^{n_1} C_{NCO}^{n_2} C_{OH}^{n_3} - k_2 C_{OH}^{n_4} C_{NCO}^{n_5} \tag{6}$$

In order to increase the reaction rate and to provide better physical properties, many urethane products, especially those in the automotive body panel applications, have shifted from segmented polyurethanes to segmented poly(urethane urea)s. The main difference is that the latter system uses a low molecular weight diamine as a chain extender for hard segments instead of a short chain diol.[23] Since diamines are more reactive than diols, the formation rates of hard (urea) and soft (urethane) segments can be quite different. Since the urea linkage is thermally more stable than the urethane linkage, many researchers have also been developing total polyurea resin systems; i.e. amine-terminated polyether resins with amine chain extenders.[24-28] Among various amines, aliphatic types are more reactive than aromatic types, which often creates different reaction rates in the segmented polyurea system. This is because the hard segment is usually based on an aromatic amine, while the soft segment is mostly an aliphatic diamine or triamine. A general kinetic expression for poly(urethane urea)s or segmented polyureas can be written as[29]

$$-\frac{dC_1}{dt} = k_1 C_1^{n_1} C_{NCO}^m = A_1 \exp^{-E_1/RT} C_1^{n_1} C_{NCO}^m \qquad (7)$$

and

$$-\frac{dC_2}{dt} = k_2 C_2^{n_2} C_{NCO}^m = A_2 \exp^{-E_2/RT} C_2^{n_2} C_{NCO}^m \qquad (8)$$

where C_1, C_2 and C_{NCO} are concentrations of aliphatic amine, aromatic amine and isocyanate functional groups; k_1 and k_2 are rate constants with Arrhenius temperature dependence; and n_1, n_2 and m are reaction orders of aliphatic amine, aromatic amine and isocyanate, respectively. The activation energy E_1 is less than $2\,\text{kcal mol}^{-1}$, while E_2 is between 2–$4\,\text{kcal mol}^{-1}$.[29-31] Both are smaller than the activation energy of urethane formation, i.e. 6–$15\,\text{kcal mol}^{-1}$.[5,10,32] The constants n_1 and n_2 were found to be near 2, while m was chosen as 1.[29] Because of the extremely high reaction rate of aliphatic amine with isocyanate, some researchers[31] considered this reaction instantaneous and did not treat it in the kinetic model.

Urethane isocyanurate resins have been used in rigid thermal insulation foams for a number of years.[33] With a slight modification, this chemistry has also been used to produce fiber mat reinforced composites in the structural reaction injection molding (SRIM) process.[28] When used at high ratios of equivalents of isocyanate to hydroxy functional groups, a rigid, high-modulus material results with outstanding thermomechanical properties.[34] The basic reaction is shown in Scheme 4, where possible trimerization catalysts include potassium salt[28] and sodium cyanide.[35]

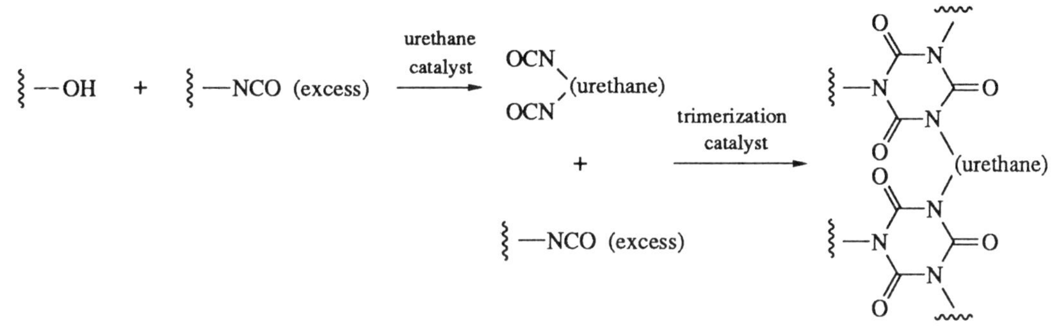

Scheme 4

In general, urethane formation occurs before the trimerization and the presence of urethane is catalytic in isocyanurate formation.[35,36] Isocyanurate kinetics have not been studied in detail. Kresta and Hsieh[36] proposed a reaction mechanism shown in Scheme 5.

Vespoli and Alberino[28] used this mechanism to derive a simple rate equation as

$$-\frac{dC_{NCO}}{dt} = \frac{3k_1 C_{NCO}^2 C_{cat}}{1 + k_2 C_{NCO}} \qquad (9)$$

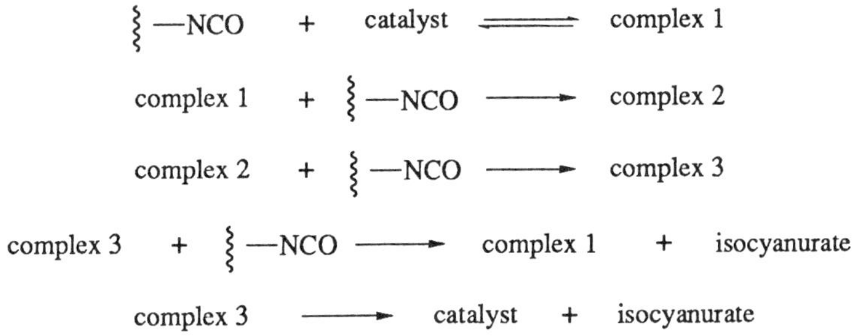

—NCO + catalyst ⇌ complex 1

complex 1 + —NCO ⟶ complex 2

complex 2 + —NCO ⟶ complex 3

complex 3 + —NCO ⟶ complex 1 + isocyanurate

complex 3 ⟶ catalyst + isocyanurate

Scheme 5

Combined with the urethane reaction (i.e. equation 4), they were able to predict the temperature profile during the curing step in the urethane isocyanurate RIM process.

20.2.1.2 Epoxies

There are many epoxy resins, hardeners (or curing agents) and catalysts.[37] Common epoxy RIM systems consist of a diglycidyl ether of bisphenol A (DGEBA) epoxy with amine curing agents.[10] The most widely used epoxy system in epoxy/graphite prepreg consists of an N,N,N',N'-tetra-glycidyl-4,4'-diaminodiphenylmethane (TGDDM) epoxy with 4,4'-diaminodiphenyl sulfone (DDS) as the curing agent.[38] Anhydride curing agents have much lower reactivity and are mainly used as accelerators in reactive processing operations. Catalysts such as Lewis acids are often added to increase the reaction rate.

The sequence of reactions shown in Scheme 6 have been suggested[39-44] for the epoxy resin cured with a primary amine, namely: (i) reaction of the primary amine with an epoxy group to form a secondary amine; (ii) reaction with another epoxy group to form a tertiary amine; and (iii) reaction of the hydroxy so formed with an epoxy group (etherification).

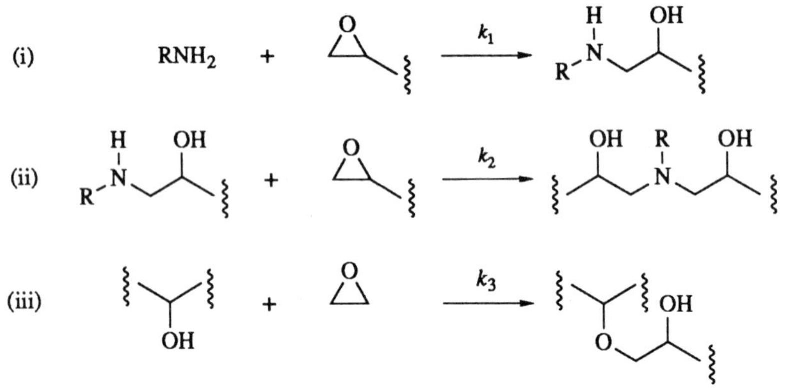

Scheme 6

In general, the reaction rate constant k_3 is much smaller than k_1 and k_2,[41-43] and the reactivities of primary amine and secondary amine are similar for many systems.[43,44] Since hydroxy groups and other proton donor species in the system (i.e. impurity additives) act as catalytic sites in the cure reaction, an overall kinetic equation may be written in terms of epoxide conversion[41,45] as

$$\frac{d\alpha}{dt} = (k_1 + k_2\alpha)(1 - \alpha)(B - \alpha) \tag{10}$$

where k_1 and k_2 are rate constants, and B is the initial ratio of diamine equivalents to epoxide equivalents.

In many epoxy systems, especially the ones catalyzed by Lewis acids, such as boron trifluoride complexes, an overlapping multiple-peaked exotherm was observed.[46,47] Lee et al.[46] modeled a two-peak rate profile in their study of the Hercules 3501-6 epoxy resin system as

$$\frac{d\alpha}{dt} = (k_1 + k_2\alpha)(1 - \alpha)(B - \alpha) \quad \text{for } \alpha \leqslant 0.3 \tag{11}$$

$$\frac{d\alpha}{dt} = k_3(1 - \alpha) \quad \text{for } \alpha > 0.3 \tag{12}$$

where k_1, k_2, k_3 and B were determined by segmental curve fitting. A similar model for a three-peak rate profile was used by Perry et al.[48]

Equations (1) and (2) have also been used to fit the epoxy kinetic data.[1,49-52] Osinski,[51] however, found that activation energy changed with temperature for epoxy reactions. Some researchers[52] explained this result as the autocatalytic reaction is only important at low temperature, while the noncatalytic one is alive throughout the entire reaction.

A common feature of epoxy resin cure is the retardation of reaction rate. This was particularly observed above the gel point and as the resin glass transition temperature, T_g, approached the cure temperature. It is likely that the reaction becomes controlled by the limited diffusion rates at high viscosity and high levels of chain entanglement and cross-linking. Barton[53] assumed that the apparent rate constants decrease linearly with increasing conversion, giving a modified equation

$$\frac{d\alpha}{dt} = (k_1^* + k_2^*\alpha)(1-\alpha)^2 \tag{13}$$

where

$$k_i^* = k_i - \left(\frac{dk_i}{d\alpha}\right)\alpha \tag{13a}$$

This give a quadratic dependence of reduced rate on conversion, i.e.

$$\frac{d\alpha}{dt} = (B_0 + B_1\alpha + B_2\alpha^2)(1-\alpha)^2 \tag{14}$$

where

$$B_0 = k_1, \quad B_1 = k_2 + \frac{dk_1^*}{dt} \quad \text{and} \quad B_2 = \frac{dk_2^*}{dt}$$

A more rigorous treatment of diffusion-affected epoxy reaction was carried out by several researchers.[54-60] Klein and coworkers[57,58] showed that diffusional effects can be incorporated into polymerization reaction rate expressions *via* the Rabinowitch equation, which leads to modified rate constants

$$k_i = \frac{A_i \exp(-E_a/RT)}{1 + (a/D)\exp(-E_a/RT)} \tag{15}$$

where D is the diffusivity, and a includes the vibrational frequency and a geometric factor. The diffusivity may be described by free volume theory[61,62] as

$$D \propto \exp(-b/V_f) \tag{16}$$

where b is a constant and V_f is the fractional free volume, which depends linearly on the glass transition temperature of the reacting mixture,[61] i.e.

$$V_f = V_{fg} + \beta(T - T_g) \tag{17}$$

where V_{fg} is the fractional free volume at T_g and β is the thermal expansion coefficient of free volume. T_g is assumed to be a function of the number average molecular weight,[63] M_n, as

$$T_g = T_g(\infty) - \left[\frac{c'}{1 + (c''/M_n)}\right]\frac{1}{M_n} \tag{18}$$

with $T_g(\infty)$, c' and c'' as constants for the given epoxy system. Since M_n is a function of conversion, this type of kinetic model has the general form of

$$\frac{d\alpha}{dt} = f[\alpha, T_g(\alpha), T] \tag{19}$$

Similar diffusion-controlled kinetic models for epoxy curing have also been proposed by others.[59,60]

In the high conversion range where the reactants are nearly depleted and after vitrification where the reaction is limited by diffusion of chain segments, T_g may be more sensitive to the changes in chemical conversion and may be measured more accurately.[64-67] Gillham and coworkers[55,56] developed a diffusion-controlled kinetic model which had the same general form as equation (19). The rate constant is expressed as[68]

$$\frac{1}{k(\alpha, T)} = \frac{1}{k_0(T)} + \frac{1}{k_D(\alpha, T)} \tag{20}$$

where k is the overall rate constant, k_0 is the true rate constant and k_D is the diffusion rate constant. The constant k_D is expected to be inversely proportional to the relaxation time of the polymer segments. This suggests that k_D can be described by the well-known WLF equation in the following form

$$\log \frac{k_D(T)}{k_D(T_g)} = \frac{c_1(T - T_g)}{c_2 + |T - T_g|} \qquad (21)$$

where $k_D(T_g)$ is assumed to be constant and can be determined using c_1 and c_2 from either experimental T_g data or a theoretical model.[56] The model assumes that there is one rate constant for bond formation. Prior to gelation, finite molecules are present in the system, and the T_g of the material varies with the increasing M_n (i.e. equation 18). After gelation, however, the system consists of both sol and gel fractions, both contributing to the increase in the T_g of the material. The T_g of the sol is a function of M_n in the sol fraction only, whereas the T_g of the gel is a function of cross-link concentration, depending more strongly on higher functional cross-linking units than on the lower ones. Gillham's model relies on the assumption that there is a unique one-to-one relationship between T_g and conversion.

The diffusion-controlled kinetic models gave a much more accurate prediction of conversion and T_g after vitrification for a thermoset epoxy resin.[56]

They were also found to be more accurate in predicting molecular weights of a noncross-linking epoxy system, especially at high conversions.[58]

20.2.1.3 Styrenic/polyester resins

The chemistry of styrenic/polyester resins is mainly a chain growth copolymerization between unsaturated polyester prepolymers and styrene or other liquid vinyl monomers. The polyester prepolymers have molecular weights ranging from a few hundred to several thousands, and carbon–carbon double bonds ranging from 4 to 10 per molecule. Styrene serves as a chain-linking agent for adjacent polyester molecules. Peroxide compounds are introduced to initiate the reaction. For high temperature processes such as compression molding of sheet molding compounds (SMC) and bulk molding compounds (BMC), injection molding of BMC, t-butyl perbenzoate and t-butyl peroctoate are often used. They may initiate polymerization only at elevated temperatures. For low temperature or fast processes such as resin transfer molding (RTM) and reaction injection molding (RIM), metallic soaps and tertiary amines are customarily added to accelerate the polymerization, and the peroxides used are low temperature initiators like methyl ethyl ketone peroxide or benzoyl peroxide.

Literature data on copolymerization of styrene and diethyl fumarate[69,70] suggest that homopolymerization of styrene monomer is significant relative to the copolymerization between styrene and polyester vinylenes. This is why the styrene monomer in unsaturated polyester resins is always present in stoichiometric excess to the polyester vinylenes. Homopolymerization of polyester vinylenes is more difficult than other reactions because of the relative immobility of the long polyester chains.

Mathematical modeling of the curing of unsaturated polyester resins has been studied in recent years. A good discussion of various kinetic models was given by Stevenson.[71] In general, there are three levels of kinetic models which have been used for simulating the curve of unsaturated polyester resins. Equation (2) has been used to fit the reaction rate profile obtained from differential scanning calorimetry (DSC). The second level of kinetic models is based on the generally accepted free radical polymerization mechanism. The major elementary reactions considered in the modeling are

$$\text{initiation: } I \xrightarrow{k_d} 2R\cdot \qquad (22)$$
$$R\cdot + M \xrightarrow{k_i} M\cdot \quad \} \; k_i \gg k_d \qquad (23)$$

$$\text{inhibition: } M_n^\cdot + Z \xrightarrow{k_z} P_n - Z \qquad (24)$$

$$\text{propagation: } M_n^\cdot + M \xrightarrow{k_p} M_{n+1}^\cdot \qquad (25)$$

$$\text{termination: } M_n^\cdot + M_m^\cdot \xrightarrow{k_t} P_{n+m} \qquad (26)$$

where I is initiator, M monomer, $R\cdot$ free radical, $M\cdot$ primary radical, $M_i^\cdot$ polymeric radical, Z inhibitor, and P_i terminated polymer. The constants k_d, k_i, k_z, k_p and k_t are rate constants of initiator decomposition, primary radical formation, inhibition, propagation and termination.

Several simplified models based on this mechanism have been proposed.[71] A typical one which can be expressed by a single equation is shown as follows[72] where f is initiator efficiency, $\overline{C}_{I_0}$ is the effective initiator concentration and t_z is inhibition time.

$$\frac{d\alpha}{dt} = k_p(1-\alpha)\left\{2f\overline{C}\left[1-\exp\left(-\int_{t_z}^{t} k_d\, dt\right)\right]\right\} \tag{27}$$

Several molding simulations based on this model have been published in the literature.[72–74] Unlike equation (2), this model not only fits the reaction profiles but can also elucidate the functions of initiators, inhibitors and monomers in the reaction. However, the effect of diffusion on the reaction was not considered in the model. This limitation often results in an overprediction of reaction rate and conversion at the later stage of cure.

This has led to the development of several kinetic diffusion models in recent years. Huang and Lee[75] used film theory in which the rate constants reflect both reaction and diffusion resistances. The diffusion resistance is determined by the free volume concept and the glass transition, i.e.

$$\frac{1}{k_p} = \frac{1}{k_{p0}} + C_M \cdot A_M \exp\left(\frac{B_M}{V_f}\right) \tag{28}$$

$$\frac{1}{k_t} = \frac{1}{k_{t0}} + C_M \cdot A_p \exp\left(\frac{B_p}{V_f}\right) \tag{29}$$

where k_{p0} and k_{t0} are true propagation and termination rate constants, C_M is radical concentration, A_M and B_M are parameters for monomer diffusion, A_p and B_p are parameters for polymer diffusion, and V_f is the free volume fraction. Their model can be expressed by a single equation as follows

$$\frac{d\alpha}{dt} = \frac{1-\alpha}{\{[\pi_1 + (\pi_2\pi_3)^2]^{1/2} + \pi_2\pi_3\}^{-1} + \pi_4} \tag{30}$$

where π_1, π_2, π_3 and π_4 are parameters accounting for kinetic and diffusion effects on propagation and termination. This model has successfully predicted the gel effect and the vitrification effect in the nonisothermal polymerizations of styrene and styrene–unsaturated polyester resins.[75–78] Similar models based on molecular entanglement,[79] free volume concept,[62,80–82] and reptation theory[83] have also been proposed for bulk vinyl polymerizations.

In the above-mentioned kinetic diffusion models, several material properties, such as glass transition temperatures and thermal expansion coefficients of monomer and polymer,[62,75,80–82] need to be known in order to determine rate constants from measured reaction rate or conversion profiles. For systems like styrene or methyl methacrylate homopolymerization, such information can be found in the literature or can be determined experimentally. For more complicated reaction systems, this may present some practical difficulties. Several researchers[84,85] have proposed simplified, empirical expressions to relate the diffusion effect to the reaction conversion. An example is shown as follows[71,84]

$$k_p = k_{p0}\left[\frac{C_M - C_{M\infty}}{C_{M0} - C_{M\infty}}\right]^m \tag{31}$$

and

$$k_t = k_{t0}\left[\frac{C_M - C_{M\infty}}{C_{M0} - C_{M\infty}}\right]^n \tag{32}$$

in which C_{M0} is the initial concentration of monomer and $C_{M\infty}$ is the monomer concentration after the cure. Although equations (31) and (32) are simpler than equations (28) and (29), they sometimes may not fit the kinetic data well because of their empirical nature.[86]

The above-mentioned models are useful for the prediction of overall reaction rate. However, the models neglect the individual reaction of styrene monomer and polyester C=C bonds. For unsaturated polyester resins, the mechanical strength of the molded compound depends on the extent of reaction of the polyester C=C bonds since this conversion affects the cross-link density of reacted polymer. The conversion of styrene monomer may also affect the properties of a molded compound because residual styrene tends to reduce the mechanical strength of the product and is the main reason for surface porosity and blisters often found in undercured products. A mechanistic copolymerization model would be very valuable as an analytical tool for the better design of molding compound and the optimization of molding conditions. Recently, Huang et al.[87] extended equation (30) to a copolymerization model with the following forms.

Styrene vinyls

$$\frac{d\alpha_1}{dt} = \frac{1-\alpha_1}{\{[\pi_1+(\pi_2\pi_3)^2]^{1/2}+\pi_2\pi_3\}^{-1}+\pi_4} \qquad (33)$$

and polyester vinylenes

$$\frac{d\alpha_2}{dt} = \frac{1-\alpha_2}{\{[\pi'_1+(\pi'_2\pi'_3)^2]^{1/2}+\pi'_2\pi'_3\}^{-1}+\pi_4} + \frac{1-\alpha_2}{(k_{P0_1}/k_{P0_2})(1/2\pi_2\pi_3)+\pi'_4} \qquad (34)$$

where

$$\pi_1 = \frac{2f\overline{C_{I0}}k_d k_{p01}^2}{k_{t0}} \exp\left(-\int_{t_z}^{t} k_d dt\right) \qquad (35)$$

(kinetic effect on propagation and termination of styrene), and

$$\pi'_1 = 0 \qquad (36)$$

(kinetic effect on propagation and termination of polyester). Because of the low mobility of polyester chains, polyester vinylenes do not exhibit any true kinetic effect.[75]

$$\pi_2 = \pi'_2 = \frac{A_t}{F}\exp(B_t/V_f) \qquad (37)$$

(diffusion effect on termination)

$$\pi_3 = f\overline{C_{I0}}k_{P0_1} F k_d \exp\left(-\int_{t_z}^{t} k_d dt\right) \qquad (38)$$

(kinetic effect on propagation of styrene)

$$\pi'_3 = f\overline{C_{I0}}k_{P0_2} F k_d \exp\left(-\int_{t_z}^{t} k_d dt\right) \qquad (39)$$

(kinetic effect on propagation of polyester)

$$\pi_4 = A_{M_1}\exp(B_{M_1}/V_f) \qquad (40)$$

(diffusion effect on styrene propagation of polyester)

$$\pi'_4 = A_{M_2}\exp(B_{M_2}/V_f) \qquad (41)$$

(diffusion effect on polyester vinylene propagation), and F is a parameter defined as

$$F = \frac{k_d A_t \exp(B_t/V_f)}{[(1-\alpha)\exp k_d(t-t_z)-1]} \qquad (42)$$

This model can predict not only the overall reaction rate and conversion profiles but also the consumption of individual reacting species (see Figure 2). The parameters can be estimated from several isothermal DSC and FTIR experiments[87] as shown in Figure 3. The above-mentioned models provide a useful package for analyzing reactive processing of unsaturated polyester resins and other radical chain growth polymerizations. Users may select any of these kinetic models based on their own needs. The more sophisticated model may provide more information and more accurate prediction. It does, however, require more detailed material characterization and more complicated parameter determination.

20.2.1.4 Nylon and nylon block copolymers

ε-caprolactam-based nylon and nylon block copolymers have been used in reactive processing, especially in RIM applications. The reaction is a base-catalyzed polymerization requiring anhydrous conditions. When a cocatalyst or initiator is present, high reaction rates are possible in the range of 80–150 °C.[88–92] Various imides such as acyllactam and bisimide are used as initiators. Alkali and alkaline earth metals and some of their hydrides, oxides or organic derivatives are used as catalysts. The major reactions responsible for linear polymer chain growth are given in Scheme 7.

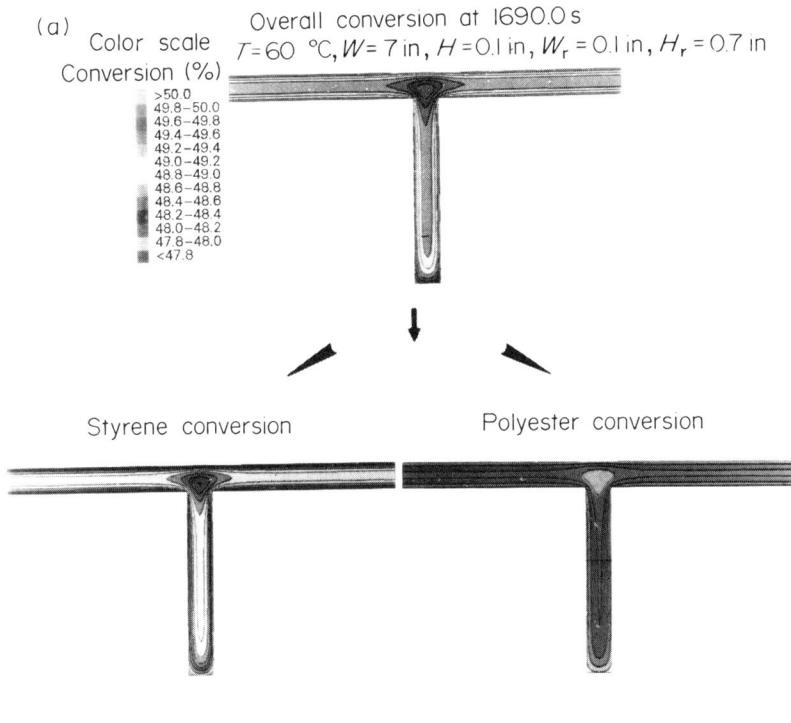

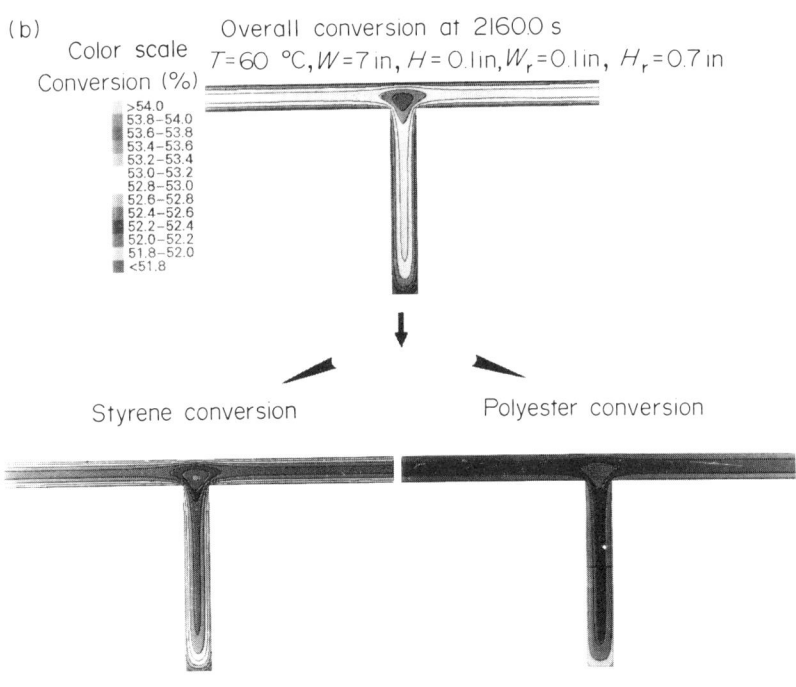

Figure 2 Simulated conversion distributions for a typical unsaturated polyester/styrene resin near the intersection area at (a) 1690 s and (b) 2160 s. $T_{\text{mold}} = 60\,°C$[87]

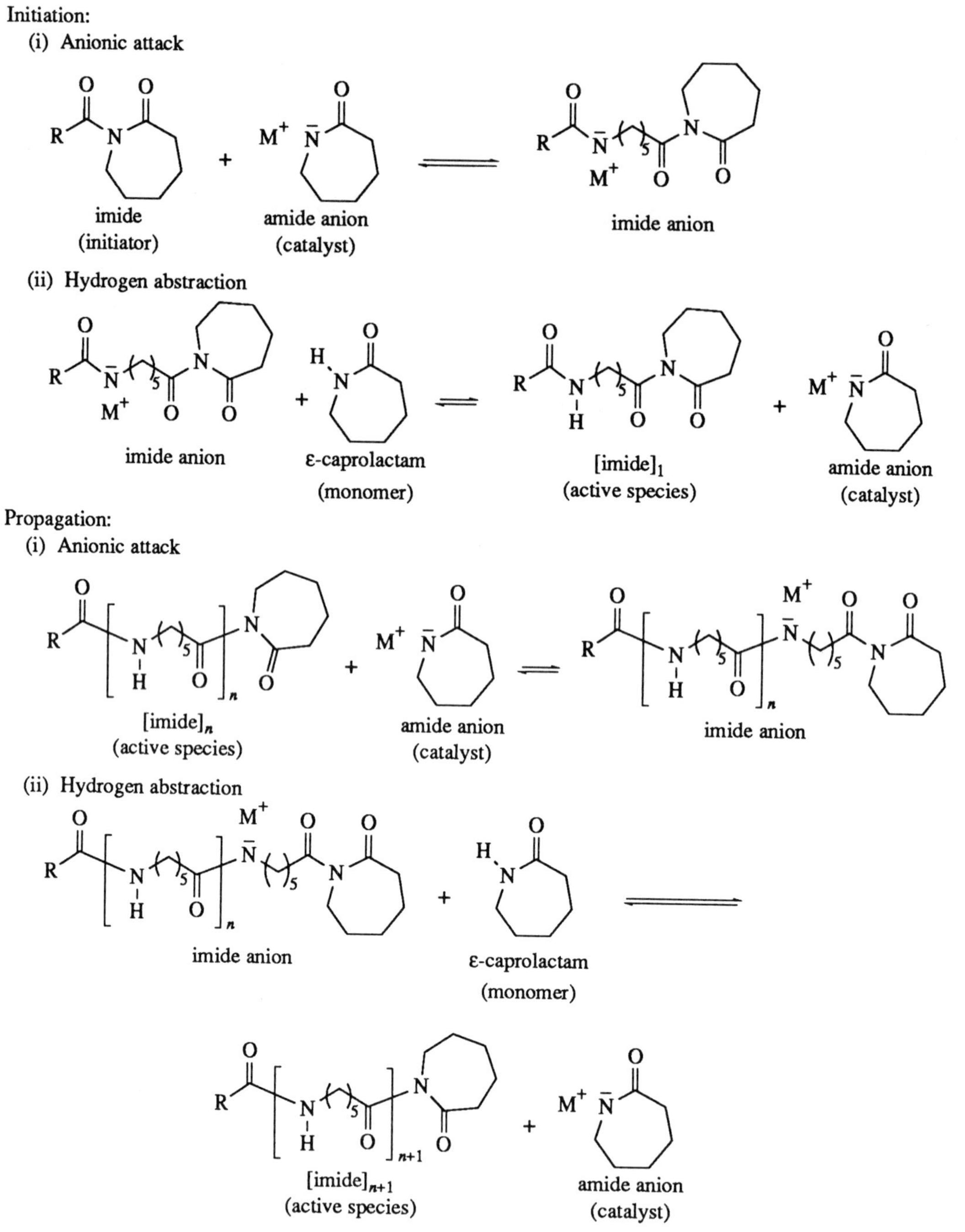

Initiation:
(i) Anionic attack

imide (initiator) + amide anion (catalyst) ⇌ imide anion

(ii) Hydrogen abstraction

imide anion + ε-caprolactam (monomer) ⇌ [imide]$_1$ (active species) + amide anion (catalyst)

Propagation:
(i) Anionic attack

[imide]$_n$ (active species) + amide anion (catalyst) ⇌ imide anion

(ii) Hydrogen abstraction

imide anion + ε-caprolactam (monomer) ⇌ [imide]$_{n+1}$ (active species) + amide anion (catalyst)

Scheme 7

where M^+ is a metal ion such as sodium or magnesium. The imide group is the main active species for the two-step initiation and propagation mechanism. In addition to the linear chain growth reaction, branching reaction and transfer reactions may also occur during polymerization.[89]

Kinetic models for the nylon-RIM polymerization have been studied by several researchers.[92–95] It was found that an autocatalytic-type model originally proposed by Malkin and coworkers[96] fits the data well and can be expressed in a relatively simple form as

$$\frac{d\alpha}{dt} = A_0 \exp\left(-\frac{E_a}{RT}\right)(1-\alpha)^n(1+\bar{B}\alpha) \qquad (43)$$

where $\bar{B}$ is called the intensity of autocatalysis. The reaction order, n, is found to be very near unity.

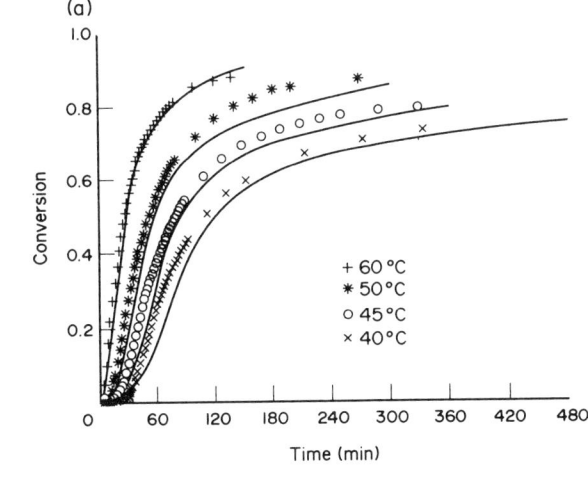

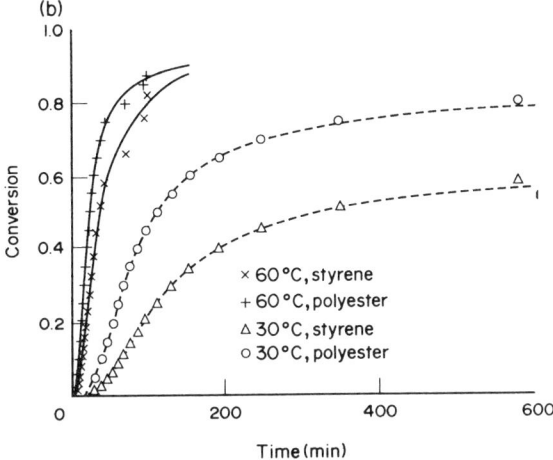

Figure 3 Conversion profiles from (a) isothermal DSC measurements at various temperatures as denoted; and (b) isothermal FTIR measurements. Symbols represent experimental data; lines are simulation results[87]

20.2.1.5 Bismaleimide resins

Bismaleimide (BMI) resins possess excellent thermal, mechanical and chemical properties, and have a number of applications in the aerospace and electronics industries. They can be processed at relatively low temperatures (i.e. < 175 °C) and then postcured at high temperatures to yield highly cross-linked networks with high glass transition temperatures. Since the bismaleimide resins are cured by addition polymerization, the curing process does not result in the evolution of volatile by-products. A major drawback of bismaleimide resins is their brittleness, which is attributed to the high cross-link density of the cured resin. In order to increase the flexibility of the cured resin, bismaleimides are reacted with aromatic diamines which reduce the cross-link density. The addition of diamines, however, results in a slight decrease in the thermal stability of bismaleimide resins. A widely used commercial BMI resin is Kerimid 601 developed by the Rhone-Poulenc Company, which consists of 2.5 mol of 4,4'-bis(maleimidodiphenylmethane) and 1 mol of 4,4'diaminodiphenylmethane. The curing reaction may follow one of two different pathways. As shown in Scheme 8, maleimide double bonds can react either by free radical homopolymerization of bismaleimide C=C bonds or by step growth polymerization of the diamine and bismaleimide C=C bonds.[97-100] The first reaction occurs around 200–250 °C, while the second reaction occurs at around 150–180 °C. At lower temperatures, the two reactions are in a sequential order, but at high temperatures, both of them occur simultaneously.

Cure kinetics of bismaleimide resins have been modeled by several researchers.[100-102] Amine addition to the maleimide double bond occurs by a second order reaction mechanism and results in an extension of the polymer chains. The secondary amine hydrogens resulting from the reaction

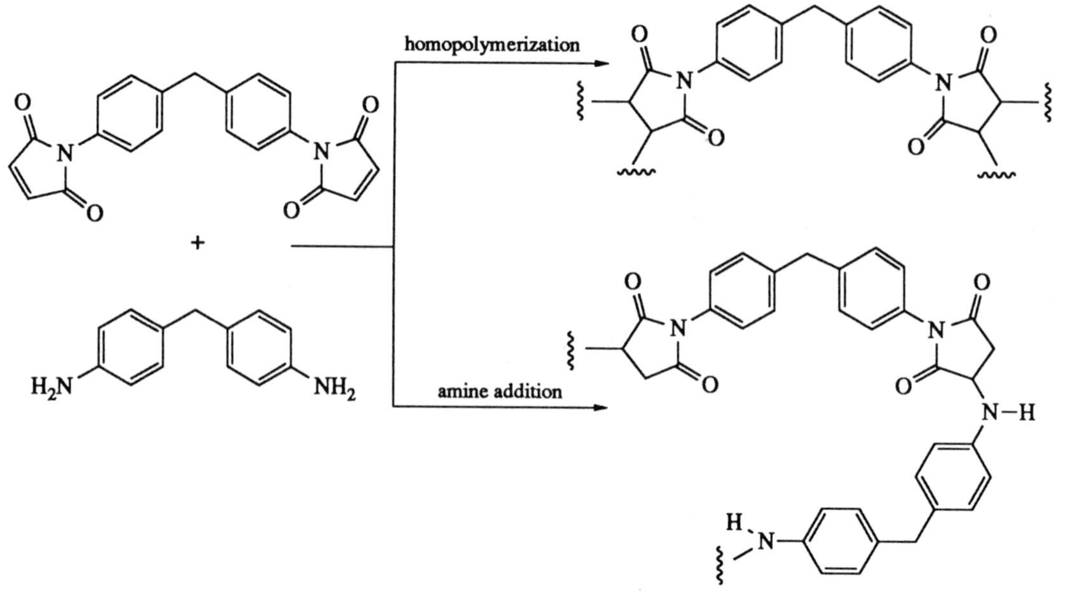

Scheme 8

of the primary amines are much less reactive with the maleimide double bonds.[101] Homopolymerization of BMI resin can be modeled by a thermally initiated, multi-step radical polymerization leading to cross-linking and network formation. A kinetic model developed by Tungare[100] has the following form

$$\frac{dC_A}{dt} = -k_1 C_M C_A \qquad (44)$$

$$\frac{dC_M}{dt} = k_{2t} C_{M\cdot} - k_{2i} C_M - k_{2p} C_M C_{M\cdot} - k_1 C_M C_A \qquad (45)$$

$$\frac{dC_{M\cdot}}{dt} = k_{2i} C_M - k_{2t} C_{M\cdot} - k_{2p} C_M C_{M\cdot} \qquad (46)$$

$$\frac{dC_P}{dt} = k_{2p} C_M C_{M\cdot} \qquad (47)$$

where C_A, C_M, $C_{M\cdot}$ and C_P are the respective concentrations of the amine groups, maleimide groups, maleimide radicals and polymer formed; and k_1, k_{2i}, k_{2t} and k_{2p} are rate constants of the second order reaction, and of initiation, termination and propagation of homopolymerization, respectively. Interactions between the two reactions and the diffusion effect on reaction rate are not considered in this model.

Tungare[100] found that the chain propagation reaction is an order of magnitude slower than the initiation and termination reactions, and is the rate controlling step. The amine addition reaction occurs more rapidly than the chain propagation reaction.

20.2.1.6 Hybrid resins

In recent years, hybrid systems consisting of more than one reactive resin have found increasing applications in reactive processing. These systems often provide improved physical properties and sometimes better processibility than the existing materials. Some researchers named such systems as interpenetrating polymer networks (IPN).[103-105] The two reaction systems are often one step-growth type and one chain-growth type. For instance, epoxy resins (step growth) can be added to acrylic solutions (chain growth),[106] while polyesters, acrylates, styrenic monomers and other vinyl systems (chain growth) can be blended into urethane resins (step growth).[107-112] Two step-growth polymerizations may also be combined to form an IPN such as polyurethane and epoxy.[113,114]

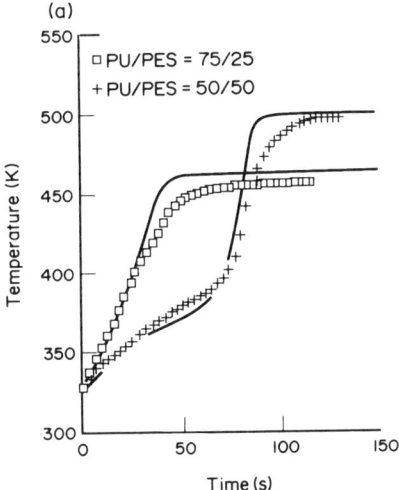 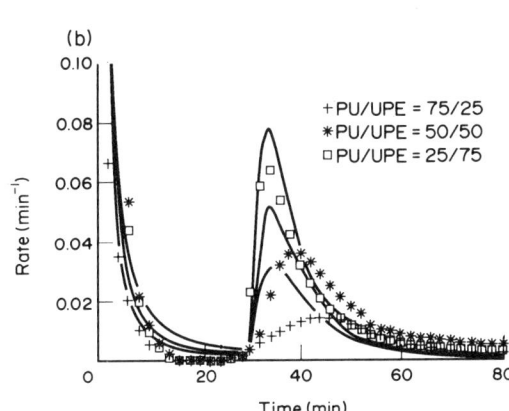

Figure 4a Adiabatic temperature rises in RIM for PU/PES hybrid resins with different compositions. Symbols represent experimental data, lines are simulation results[115]

Figure 4b Comparison of experimental data and model prediction of reaction rate *versus* time for PU/UPE hybrid resins at three compositions (80 °C). Symbols represent experimental data, lines are simulation results[115]

The cure kinetics of these systems are often modeled by adding the individual reactions together, assuming no interactions among them. Hsu and Lee[115] studied the kinetics and heat transfer during the curing of a polyurethane–unsaturated polyester hybrid resin. They used equation (1) for the urethane reaction and equation (27) for the reaction of polyester C=C bonds. Compared with the adiabatic temperature rise measured during a RIM experiment, the model prediction was close to the experimental data, as shown in Figure 4(a). Deviations of model prediction from experimental results, however, were found in the comparison of reaction rate profiles measured by differential scanning calorimetry (see Figure 4b). This suggests that reaction interactions may exist in the polymerization system. Recently, Chou and Lee[116] found that the formation of the urethane linkage between isocyanates and the hydroxy or carboxy groups at the ends of unsaturated polyester chains may increase the reactivity of C=C bonds.

20.2.2 Rheological Changes

In reactive processing, the reaction system generally starts at a low viscosity. The viscosity increases with conversion and reaches an infinite level when the material solidifies either by chemical cross-linking or physical changes, such as phase separation or crystallization. Further reaction occurs in the solid state, which may increase the polymer modulus to a desired level before demolding. During the process, mold filling must be finished before the system viscosity reaches too high a value. This process is then followed by a curing stage in which the material is reacted to a sufficient mechanical strength before being ejected. For some materials, such as polyurethanes, a postcure stage is needed so the product physical properties can reach an optimal level. A schematic diagram, as given in Figure 5, shows the relationship between rheological changes (viscosity and modulus) and kinetic changes (polymer formation) during the RIM process. A detailed understanding of rheological changes before gelation is essential in determining the mold filling characteristics. Additionally, a similar study after gelation is important in determining the demold time, polymer structure, and mechanical properties of the final product.

For epoxies, bismaleimides and styrenic/polyester resins, the liquid–solid transition is caused primarily by chemical cross-linking. The reaction temperature has to be above the glass transition temperature, T_g, of the network polymer for complete conversion to be achieved. Therefore, the material in the mold will be in a rubbery state at the end of the reaction. Generally, a high level of filler reinforcement is needed to provide enough green strength to the product at demolding. Otherwise, the part will not have sufficient modulus and strength to be removed from the mold quickly so the cycle time must be increased to allow the part to cool below the T_g.

Although there is some cross-linking in urethane and urea systems, structure develops mainly through domain formation. During polymerization, the urethane or urea hard segments associate

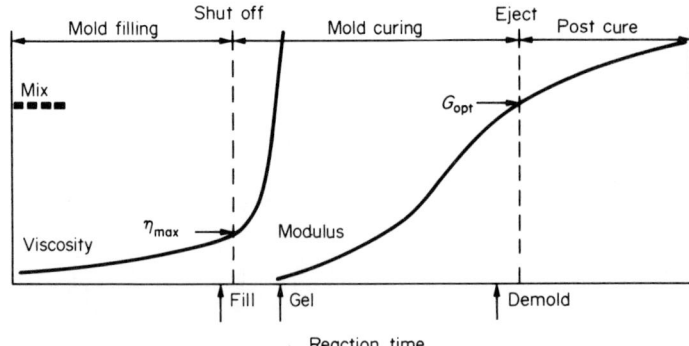

Figure 5 Changes in viscosity and modulus during reactive processing

into domains that sometimes crystallize and form spherulites. The domains act like rigid particles which rapidly raise the viscosity of the system to a gelation very similar to that which occurs in cross-linking polymerizations. In the nylon systems, it appears that viscosity increases initially by molecular weight build-up until a minimum molecular weight is reached for crystallization to occur. After that, the growth of crystallization is the main reason for the onset of physical gelation and modulus development.[90-92]

For a polymer melt, the zero shear rate viscosity, η_0, is usually expressed as a function of a friction factor and a structure factor. If the polymer is linear then

$$\eta_0 = EM_w^a \tag{48}$$

where E is a temperature-dependent friction factor and the weight-average molecular weight, M_w, is the structure factor.

A linear dependence of viscosity on the weight-average molecular weight was found for polydispersed systems at low molecular weights.[117] At high molecular weights, chain entanglement is encountered and the power of molecular weight dependence increases from 1.0 to 3.5. The value of a equals 3.5 if M_w is greater than a critical molecular weight, M_c, beyond which the effect of chain entanglement dominates the system viscosity. The value of a lies between 1.0 and 2.5 if M_w is less than M_c.

For polymer solutions, the structure parameter, M_w, is modified by multiplication with the polymer concentration, i.e. CM_w.[113-121] The solvent can effectively decrease the extent of polymer–polymer interactions and delay the onset of chain entanglement to a higher molecular weight. The critical molecular weight for chain entanglement in the polymer solution is generally expressed as M_c/ϕ, where ϕ is the volume fraction of polymer in the solution with $\phi = C/\rho$. The friction factor is also a function of polymer concentration. Thus, for polymer solutions the zero shear rate viscosity can be expressed as

$$\eta_0 = E(C)(CM_w)^a \tag{49}$$

Using molecular theories as a basis, Bueche et al.[122,123] considered the chain entanglement effect in polymer solutions and theoretically derived the viscosity expression as follows

$$\eta_0 = EC^4 M_w^{3.5} = EC^{0.5}(CM_w)^{3.5} \quad \text{for } M_w > M_c \tag{50}$$

Experimentally, Graessley et al.[120] found that for polyisoprene in tetradecane with polymer concentrations ranging from 0.02 to 0.33 g mL^{-1}, the zero shear rate viscosity could be expressed as

$$\eta_0 = EC^5 M_w^{3.5} = EC^{1.5}(CM_w)^{3.5} \tag{51}$$

The size of polymer molecules has a dominant effect on the viscosity change in a polymer solution. Solvent dilution and solvent thermodynamic effects on the viscosity come from the change in size of polymer coils under different environments.[124,125]

Different viscosity dependencies on molecular weight were observed for polymer systems with long branched chains, such as regular stars, regular combs and random trees.[120,126,127] At low and intermediate concentrations, the viscosity of a branched polymer is less than that of a linear polymer with the same molecular weight. This is due to the fact that the radius of gyration of the branched system is smaller than that of the linear system. The ratio of the gyration radii can be expressed as $g = S_B^2/S_L^2$. Thus, $g \leq 1$ and it decreases with increased branching. For branched polymers, the structure parameter is modified as gM_w. Physically, the quantity gM_w is the weight-average molecular weight of a linear chain with the same radius of gyration as the branched chain. This approach has been used to describe the viscosity for branched polymers, such as polyisoprenes with star-branches,[120,126] and polybutadiene.[128]

In summary, the zero shear rate viscosity of a branched polymer system can generally be expressed as

$$\eta_0 = EC^a(CgM_w) \quad \text{if } M_w < M_c/\phi \tag{52}$$

and

$$\eta_0 = EC^a(CgM_w)^b \quad \text{if } M_c/\phi < M_w \tag{53}$$

where the constant a ranges from 0.5 to 1.5, and b equals 3.5. The effect of solvent concentration on the polymer size is not always clear. Often, an empirical value of a is required for a given system.

Equations (52) and (53) have been derived to describe the viscosity of nonreactive polymer systems at an equilibrium state. For a reactive system, the overall viscosity increases with conversion. It is important to know whether or not the relationship derived for nonreactive systems at steady state can be employed to describe the dynamic reactive systems.

Non-Newtonian effects appear to be small for polymers built up from branched monomers, except near the gel point.[129,130] Most viscosity models for reactive resins neglect the non-Newtonian effects because the temperature and polymer (or network) formation effects tend to dominate the rheological changes. However, when cross-linking of long chains or particulate fillers are present, shear-thinning and time-dependent effects can be pronounced even at the beginning of the reaction.[131]

Roller[132] has reviewed several models that characterize the rheological changes of thermosetting resins as a function of time and temperature. In the case of resins that exhibit overall first-order curing kinetics, a phenomenological relation between the viscosity and the reaction time is given by[133]

$$\eta(t) = \eta_0 \exp(kt) \tag{54}$$

Here, $\eta(t)$ is the time-dependent viscosity, η_0 is the zero-time viscosity, k is the first-order reaction rate constant, and t is the reaction time. Roller[134] incorporated the temperature history of the resin into the model by assuming Arrhenius-type relations to describe the temperature dependence of the zero-time viscosity and the kinetic term in equation (54), i.e.

$$\eta_0 = \eta_\infty \exp(-E_\eta/RT) \tag{55}$$

and

$$k = A \exp(-E_a/RT) \tag{56}$$

where η_∞ and A are the Arrhenius preexponential factors, while E_η and E_a are the activation energies for the flow and the curing reaction. In the case of isothermal cure, equations (54)–(56) can be combined into the dual Arrhenius viscosity model given by

$$\ln \eta(t, T) = \ln \eta_\infty + \frac{E_\eta}{RT} + A \exp\left(\frac{-E_a}{RT}\right) t \tag{57}$$

For nonisothermal cure, where the resin temperature history is given by $T = f(t)$, the dual Arrhenius viscosity model is written as

$$\ln \eta(t, T) = \ln \eta_\infty + \frac{E_\eta}{Rf(t)} + A \int_0^t \exp\left(\frac{-E_a}{Rf(t)}\right) dt \tag{58}$$

Although equation (58) was derived to model the chemorheological behavior of resins that exhibited overall first-order curing kinetics, the equation was also found to be valid for some commercial epoxy resins that did not exhibit overall first-order curing behavior.[135]

While useful for selecting optimum curing conditions, the dual Arrhenius viscosity model does not provide any insight into the network structure and properties of the resin. Since the viscosity model is based on the overall curing kinetics, the model parameters are batch specific; i.e. once the resin composition is changed, the parameters have to be reevaluated.

The next level of modeling converts viscosity *versus* reaction time to viscosity *versus* conversion using independently measured kinetic data. Such a correlation corrects for the effect of reaction rate due to various catalyst concentrations and reaction temperature (see Figure 6). A typical expression

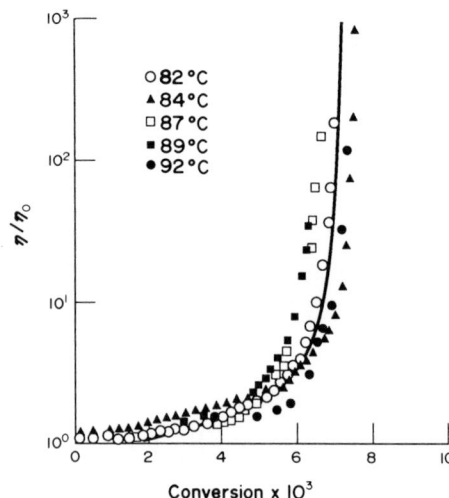

Figure 6 Reduced viscosity *versus* conversion for a free radical copolymerization of dimethacrylate and styrene at different temperatures (reproduced by permission of *J. Rheology*, 1985, **29**, 259)

is[16,136,137]

$$\eta = \eta_\infty \exp(E_\eta/RT)\left(\frac{\alpha_{gel}}{\alpha_{gel} - \alpha}\right)^{f(\alpha)} \quad (59)$$

where α_{gel} is the gel conversion.

In an alternative approach, Tajima and Crozier[138-140] used the WLF equation to describe the temperature dependence of the resin viscosity, and then related the parameters of the WLF equation to the kinetics of the curing reaction. The shift factor of the mechanical or dielectric relaxation time at temperature T to that at temperature T_s may be approximated as

$$a_T \sim \frac{\eta(T)}{\eta(T_s)} \quad (60)$$

where $\eta(T)$ is the viscosity at temperature T, and $\eta(T_s)$ is the viscosity at the reference temperature, T_s. Equation (60) can then be written as

$$\log \eta(T) = \log \eta(T_s) + \frac{c_1(T - T_s)}{c_2 + |T - T_s|} \quad (61)$$

Tajima and Crozier introduced the polymerization kinetics of amine-cured epoxies into equation (61) by writing the reference temperature as a function of the concentration of the amine hardener[138] and as a function of the conversion of epoxy groups.[139] Since the glass transition temperature changes with cure, it is sometimes used as the reference temperature.

Lee and Han[141,142] used this approach to characterize the chemorheological behavior of unsaturated polyester resins. They related the glass transition temperature to the resin conversion through several empirical relations.

For some commercial epoxy resins, it was observed[143] that the viscosity at the glass transition temperature can be modeled by

$$\log \eta(T_g) = c_3 + c_4 T_g \quad (62)$$

Equations (61) and (62) can be combined to give

$$\log \eta(T) = c_3 + c_4 T_g + \frac{c_1(T - T_g)}{c_2 + |T - T_g|} \quad (63)$$

where c_1, c_2, c_3 and c_4 are constants. Equation (63) relates the viscosity to the glass transition temperature and can be used to model the viscosity variation with time and temperature for bismaleimide resins.[100] The time dependence in equation (63) is implicitly contained in the glass transition temperature, which increases as the resin curves.

From conversion data, the corresponding molecular weight can be calculated using appropriate theories. Thus the viscosity changes during polymerization can be expressed as a function of molecular weight and temperature. Some researchers used weight-average molecular weight and Arrhenius temperature dependence[129-131,144,145]

$$\eta = \eta_\infty \exp(E_\eta/RT)\left(\frac{M_w}{M_{w0}}\right)^{f(T,M_w)} \quad (64)$$

This model brings the structure parameter, M_w, into the viscosity correlation. The function f, however, needs to be fitted empirically. For different systems, the maximum $f(T, M_w)$ values differed from 2.1–2.6 for an HDI-based polyurethane,[129] 2.6 for poly(dimethylsiloxanes),[130] and 3.4–6.7 for an MDI-based polyurethane.[144]

For cross-linkers with differing functionalities, the weight-average molecular weight does not correlate well with the viscosity data. Valles et al.[130] studied the viscosity changes during bulk reactions of poly(dimethylsiloxanes) with random branches. When plotting the viscosity versus weight-average molecular weight, the curves for branched systems do not follow the curve for the linear system. Instead, if the gyration effect is considered, the plots of η_0 versus gM_w fall onto a single curve for data taken from various branched systems. This result is similar to the star-branched polyisoprene system studied by Graessley.[126] Valles and Macosko[130] also showed that the molecular weight of the longest linear chain in the system, $M_{L,w}$, can correlate well with the viscosity data. For polyurethane reactions in solution, Lee and Lee[145] found that intramolecular reactions may affect the viscosity changes. They developed the following model for system viscosity

$$\eta = E_V(C)^a \left(\frac{M_{we}}{M_{w0}}\right)^{b+c\alpha_e} \quad (65)$$

where M_{w0} is the initial weight-average molecular weight and M_{we} is the effective molecular weight calculated from the polymer conversion when intramolecular reaction is taken into consideration.

For high temperature resins, such as epoxies and BMI, most researchers used the free volume approach[59,100,146] or WLF-type equations[66,147,148] for temperature-dependent viscosity. The free volume models have been among the most successful in describing properties near the glass transition point. The Cohen and Turnbull equation[149] describes the free volume dependence of diffusivity as shown in equation (16). Since diffusivity is expected to be inversely related to viscosity, equation (16) is equivalent to the Doolittle viscosity equation[150]

$$\eta = A \exp(b/V_f) \quad (66)$$

where A is a constant. The fractional free volume, V_f, may be expressed as a linear function of the difference between the resin temperature and the glass transition temperature as shown in equation (17).

Equations (66) and (17) are equivalent to the WLF equation, i.e.

$$\log \frac{\eta(T)}{\eta(T_g)} = \frac{(-b/2.303\,V)(T-T_g)}{(V_{fg}/\beta) + |T-T_g|} \approx \log a_T = \frac{c_1(T-T_g)}{c_2 + |T-T_g|} \quad (67)$$

with $c_1 = -b/2.303\,V_{fg}$ and $c_2 = V_{fg}/\beta$.

Combining the molecular weight and free volume dependencies described above, the viscosity can be expressed as

$$\eta = \eta_0 \exp\left(\frac{b}{V_{fg} + \beta|T-T_g|}\right) M_w^a \quad (68)$$

Macedo and Litovitz[151] have argued that an activation energy is also needed to describe η_0, i.e.

$$\eta_0 \sim \exp(E_\eta/RT) \quad (69)$$

Highly branched polymers would be expected to exhibit behavior similar to that of linear polymers above their entanglement threshold in linear systems. Unlike entanglements, however, branch points are fixed, i.e. no slippage occurs. Bueche[122,152] developed a model for the viscosity of long chain linear polymers in which he determined that above the entanglement threshold, the viscosity should be related to the molecular weight by

$$\eta \sim M_w^{2.5} \quad (70)$$

if the points of entanglement are fixed, and

$$\eta \sim M_w^{3.5} \quad (71)$$

if the entangled polymer chains can slip at their junctions.

For thermosetting systems, branch points are fixed, but some entanglements will occur as well, so the 'power law' dependence of viscosity on molecular weight should be bounded between 2.5 and 3.5.

Molecular weight based viscosity models have a better theoretical foundation than the time (*i.e.* equation 57), conversion (*i.e.* equation 59) or glass transition temperature (*i.e.* equation 63) based models because they take into consideration the structure of the resin. However, molecular weights of thermosets are difficult to measure accurately. Most thermosets become insoluble very early in the cure and hence, molecular weight data can be obtained only in a narrow range of conversion. Network formation theories, such as the statistical relations of Flory[153] and Stockmayer[154,155] or the recursive approach of Miller and Macosko[156-158] may be used to predict the molecular weight.[145,146] However, for resins with multiple reactives or more than one curing reaction, the molecular weight models become very complicated. An example is given by Wang *et al.*[159] for a polyurea reaction.

Glass transition temperatures of the reacting resins need to be measured or calculated in WLF-type or free volume type viscosity and kinetic models. They can be determined more easily than molecular weights using various analytical methods such as differential scanning calorimetry and dynamic mechanical analysis. Modeling of T_g, however, requires knowledge of the number-average molecular weight, M_n or cross-link density, X_c. Fox and Flory[160] related T_g to M_n for noncross-linking polymers as

$$T_g = T_{g\infty} - \frac{A'}{M_n} \tag{72}$$

Fox and Loshaek[63] extended the relation to include the effects of cross-linking, *i.e.*

$$T_g = T_{g\infty} - \frac{A'}{M_n} + KX_c \tag{73}$$

DiBenedetto[161] used the principle of corresponding states to derive relations between T_g and X_c. His equation has the following form

$$\frac{T_g - T_g^0}{T_g^0} = \frac{[(E_x/E_m) - (f_x/f_m)]X_c}{1 - [1 - (f_x/f_m)]X_c} \tag{74}$$

where T_g^0 is T_g of the noncross-linked polymer, E_x/E_m is the ratio of the lattice energy of the cross-linked and the noncross-linked polymer, and f_x/f_m is the corresponding ratio of the segmental mobilities.

Because M_n and X_c are difficult to measure, Tungare[100] proposed an empirical model to relate them to resin conversion. For linear step growth polymerizations, M_n is inversely proportional to $(1 - \alpha)$, while for cross-linking systems, he assumed that X_c can be substituted by α and the mobility of the cross-linked units is approximately zero (*i.e.* $f_x/f_m = 0$). For the BMI resins, his T_g model has the following form

$$T_g = [T_{g\infty} - a_1(1 - \alpha_a)]\left(1 + \frac{a_2\alpha_h}{1 - \alpha_h}\right) \tag{75}$$

where a_1 and a_2 are constants, α_a is the conversion of amine addition, and α_h is the conversion of BMI homopolymerization.

The conversion and molecular weight based viscosity models have been applied extensively in step growth type polymerizations, but not in chain growth type polymerizations. For unsaturated polyester resins, gelation often occurs at very low conversions (less than 1% in many cases[162]). This makes it very difficult to relate the viscosity changes to resin conversion. The high functionality of these resins and the nature of free radical polymerization (*i.e.* the reaction is limited to the neighborhood of highly reactive radicals) result in strong intramolecular reactions and local inhomogeneity.[163-165] These conditions cannot be easily simulated by the existing molecular weight models.[166,167] In reactive processing of these resins, several simplified models have been proposed to determine the system gel point. Gonzales-Romero and Macosko[137] assumed that gel time, t_{gel}, is equal to the inhibition time, which can be expressed as

$$t_{gel} = \frac{qC_{z0}}{2fC_{I0}k_d} \tag{76}$$

where q and f are efficiencies of the inhibitor and initiator, C_{z0} and C_{I0} are initial concentrations of inhibitor and initiator, and k_d is the initiator decomposition rate constant. This model was found

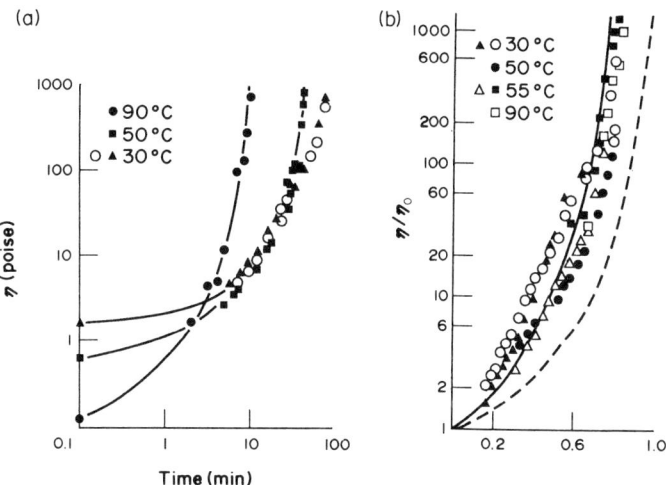

Figure 7 (a) Viscosity *versus* time at several temperatures. MDI-BDO P(PO-EO) 47% hard segment, uncatalyzed. (b) Same data plotted as reduced viscosity *versus* conversion, α. Solid curve (———) based on equation (59), dashed curve (-----) based on equation (64) with f(T, M_w) = 3.4 (reproduced by permission of *Polym. Commun.*, 1984, **25**, 82)

to overpredict the actual gel time by a large margin.[86] Muzumdar and Lee[86] developed a semiempirical model by assuming that gelation occurs at a critical radical concentration which depends on the kinetics of initiation and inhibition (*i.e.* equations 22–24). Their model agreed with experimentally measured gel times at different temperatures and curing agent concentrations. However, the critical radical concentration must be determined experimentally, and it varies from resin to resin.

Today, most theoretical modeling regarding the viscosity changes of reactive polymeric systems is based on the polymer formation. The effect of physical changes, such as domain formation or crystallization, on rheology has not been studied in detail. Castro *et al.*[168,169] studied a segmented linear polyurethane system and found that equation (59) can be used to fit the viscosity data (see Figure 7), where α_{gel}, the conversion of physical gelation, is due to the formation of a network of hard segment domains. Viscosity rises much faster than molecular weight even to a 3.5 power, which indicates that chain entanglements and branching are not the main cause of viscosity increase. Equation (59) indicates that α_{gel} is independent of temperature. This can only be true over a limited range in physical gelation. At high temperatures, α_{gel} increases[170,171] because of the decrease of domain formation rate. Higher catalyst concentration also tends to increase α_{gel}[170] because a high catalyst concentration is able to accelerate reaction kinetics more than phase separation.

Beyond the gel point, the reacting mixture becomes a soft solid. Figure 5 indicates that the modulus of this solid must build up high enough to permit demolding. If the reaction is carried out at a temperature above T_g, rubber elasticity theory can be used to predict how the modulus evolves with cross-linking.[172–176] If there are no interactions or entanglements between the network chains, the modulus of an ideal elastomer, G, is[172]

$$G = (v - \mu)RT \tag{77}$$

where v is the concentration of chains between cross-links and μ is the concentration of cross-links. For a completely reacted network[10] with tetrafunctional junctions, $v = 2\mu$, while with trifunctional junctions, $v = (3/2)\mu$. Entanglements can suppress the mobility of the cross-links (*i.e.* the μ term in equation 77) and can also add an additional contribution to the modulus due to interactions between the network chains.[176] This results in

$$G = (v - H\mu)RT + G_n^0 P_e \tag{78}$$

where H expresses the effect of entanglements on the cross-links, G_n^0 is the plateau modulus and P_e is the probability that a pairwise interaction between two network chains is trapped. For a $A_3 + B_2$ reaction system

$$\mu = C_{A_3}\left(\frac{2r\alpha^2 - 1}{r\alpha^2}\right)^3; \quad P_e = \left(\frac{2r\alpha^2 - 1}{r^2\alpha^3}\right)^4 \tag{79}$$

where r is C_A/C_B, the stoichiometric ratio.

With a value of α from reaction kinetics and the estimates of H and G_n^0, the evolution of modulus with time and temperature during molding can be predicted.

20.2.3 Structure and Property Changes

Unlike reaction kinetics and rheological changes, structure and property development during reactive processing have not been modeled in detail. While the former is essential for predicting the molding process, the latter is important for determining the physical and mechanical strength of molded products. Limited work reported in the literature regarding reaction-induced crystallization, phase separation, shrinkage, foaming, void formation and thermophysical properties is summarized in this section.

20.2.3.1 Crystallization during polymerization

The kinetics of crystallization during anionic polymerization of ε-caprolactam was studied by Lee and Kim.[177] Instead of using the Avrami equation,[178,179] which is written in the integral form, they developed a differential form for the rate of crystallization, $d\psi/dt$

$$\frac{d\psi}{dt} = A \exp\left(\frac{-E_D}{RT}\right) \exp\left(\frac{BT_m^0}{T(T_m^0 - T)}\right) \psi^{2/3}(\psi_{eq} - \psi) \tag{80}$$

where ψ is the degree of crystallinity varying from zero to the equilibrium value ψ_{eq}, E_D is an activation energy for diffusion, B is a negative constant and T_m^0 is an equilibrium melting temperature. Both E_D and T_m^0 are functions of conversion. The first exponential term indicates the diffusion of the polymer chain, the second exponential term describes the supercooling, and the $(\psi_{eq} - \psi)$ term accounts for the gradual restriction of spherulitic growth by the impingement of the neighboring spherulites during crystal growth. Figure 8 shows the calculated temperature rise at different levels of catalyst. At low catalyst levels, the rate of polymerization is slow, thus delaying the crystallization. At high catalyst levels, the reaction rate is higher than the crystallization rate and the temperature rise occurs in two stages. The first rise is attributed to the reaction exotherm and the second rise is due to the crystallization of the polymer formed.

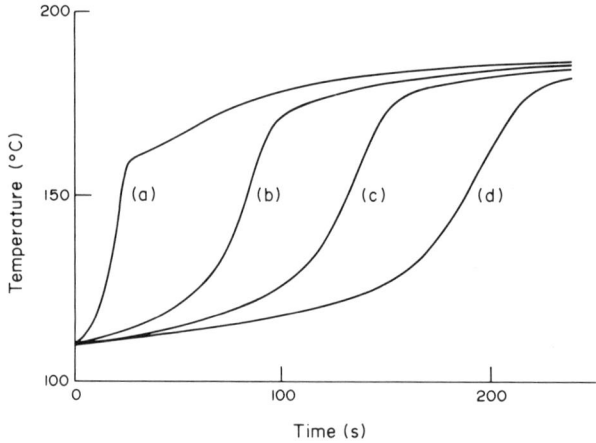

Figure 8 Adiabatic temperature rise calculated from equation (80) at different value of k: (a) 1.0×10^7; (b) 2.5×10^6; (c) 1.5×10^6; (d) 1.0×10^6 (initial temperature is $110\,°C$) (reproduced by permission of *Polym. Eng. Sci.*, 1988, **28**, 13)

20.2.3.2 Phase separation during polymerization

Phase separation or domain formation is an important feature in many reactive systems such as segmented polyurethanes and ureas, rubber-modified epoxies and low-shrink unsaturated polyester resins. It not only controls property development in reactive processing, but is also highly coupled with the reaction kinetics and rheological changes. For thermoplastic blends and solutions,

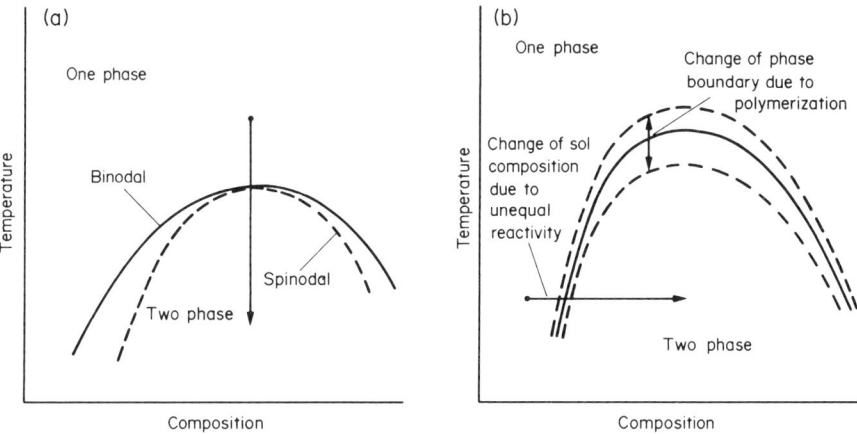

Figure 9 Phase separation due to (a) temperature jump and (b) polymerization

temperature jump is a common way to initiate and to control the phase separation (see Figure 9a). For reactive systems, phase separation may be affected by not only the temperature changes but also the change of phase boundary due to polymerization and the change of reactant compositions due to unequal resin reactivity (see Figure 9b).

Because of its complexity, polymerization-induced phase separation has not been well understood. The onset of phase separation can be predicted using the Flory–Huggins relation.[153] For a mixture of two polymers A and B

$$\chi_c = \frac{1}{2}\left(\frac{1}{\sqrt{N_A}} + \frac{1}{\sqrt{N_B}}\right)^2 \tag{81}$$

where N_A and N_B are the number of repeat units on the A and B chains, χ is a measure of the interaction between chains A and B and can be estimated from their solubility parameters, δ

$$\chi = \frac{\bar{V}}{RT}(\delta_A - \delta_B)^2 \tag{82}$$

Here $\bar{V}$ is a reference volume based on chain A. χ_c is the critical value of χ for the onset of phase separation. Macosko[10] has examined these relationships with several segmented polyurethanes and found that they agreed with experimental observations. Ryan[180] has used these relationships to elucidate phase separation in segmented polyureas and poly(urethane urea)s.

Williams and coworkers[181–183] developed a model to predict the fraction and composition of the dispersed phase segregated during the cure of rubber-modified epoxies. They assumed that phase separation begins at a critical resin conversion (i.e. the cloud point) and ends at the gel conversion. The final particle size distribution is attained at the gel point when phase separation is prevented by diffusional restrictions. The initial system is regarded as a solution of rubber (component 2) in an epoxy solvent (component 1) with a free energy per unit volume, ΔG, described by the Flory–Huggins equation

$$\Delta G = \frac{RT}{\bar{V}_1}[(1-\phi_2)\ln(1-\phi_2) + (\phi_2/Q)\ln\phi_2 + \chi\phi_2(1-\phi_2)] \tag{83}$$

where ϕ_i is the volume fraction of component i, and $Q = V_2/V_1$. The parameters $\bar{V}_1$, Q and χ vary during polymerization as

$$\begin{aligned} \bar{V}_1 &= \bar{V}_{10}(\bar{M}_n/\bar{M}_{n0}) = \bar{V}_{10}(1-2\alpha) \\ Q &= \bar{V}_2/\bar{V}_1 = Q_0(1-2\alpha) \\ \chi &= \chi_0(\bar{V}_1/\bar{V}_{10}) = \chi_0/(1-2\alpha) \end{aligned} \tag{84}$$

By plotting $(\bar{V}_{10}\Delta G)/RT$ versus ϕ_2 at different levels of resin conversion, the binodal and spinodal curves are established[181] as shown in Figure 10. They found that for typical rubber-modified epoxies, phase separation proceeds through a classic nucleation–growth mechanism rather than through spinodal decomposition. The latter would require the use of high rubber content, low curing temperature and high reaction rates. Phase separation during the cure of rubber-modified epoxies has also been modeled using the Avrami equation.[184]

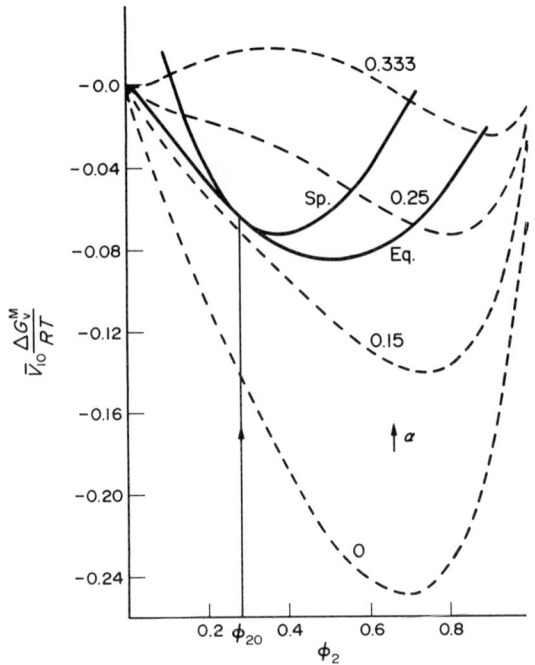

Figure 10 Dimensionless free energy of solution per unit volume as a fraction of rubber fraction volume, for different thermoset conversions. Eq. = equilibrium or binodal and Sp. = spinodal. $Q = 10$, $T = 300\,\text{K}$ and $\chi = 0.35 + (90/T)$ (reproduced by permission of *Am. Chem. Soc., Adv. Chem. Ser.*, 1984, **208**, 195)

20.2.3.3 *Thermal expansion and polymerization shrinkage during polymerization*

Volume expansion and shrinkage of resins during reactive processing may affect dimensional accuracy, surface quality and physical properties of molded parts. Although some experimental results are available for epoxies,[185,186] unsaturated polyester resins[187–192] and vinyl polymerizations[193–195], modeling of polymerization shrinkage for thermoset resins has not been well studied. Kau[190] expressed the overall volume change of a sheet-molding compound as a combination of the stages observed during compression molding

$$dV = \underbrace{\left(\frac{\partial V}{\partial P}\right)dP + \left(\frac{\partial V}{\partial T}\right)dT}_{\text{pre-cure}} + \underbrace{\left(\frac{\partial V}{\partial \alpha}\right)d\alpha}_{\text{cure}} + \underbrace{\left(\frac{\partial V}{\partial T}\right)dT}_{\text{post-cure}} \tag{85}$$

The first term on the right-hand side represents the compound compressibility. The second and fourth terms are thermal expansion coefficients before and after cure. The third term accounts for the polymerization shrinkage. No attempt was made to actually implement this expansion to solve for the material volume changes.

Recently, Lee and coworkers[191,192,196,197] found that for unsaturated polyester resins, most polymerization shrinkage occurs during the liquid–solid transition stage where the resin conversion is still very low (see Figure 11). This shrinkage seems to follow the change of radical concentration. The rest of the polymerization shrinkage is linearly proportional to the resin and is independent of temperature, as shown in Figure 12. A shrinkage model was proposed based on this observation[197] as

$$s = 1 - \{[\alpha - s_0][1 + \beta_p(T - T_0)] + [1 + \beta_M(T - T_0)](1 - \alpha)\} \tag{86}$$

and

$$s_0 = s_g(1 - e^{-a\bar{R}}) + (s_f - s_g)(\alpha/\alpha_f) \tag{87}$$

where s_0 is the isothermal polymerization shrinkage at temperature T_0; β_p and β_M are the coefficients of volume expansion for the polymer and monomer, respectively; s_g is the polymerization shrinkage due to liquid–solid transition; s_f is the final polymerization shrinkage; and $\bar{R} = C_R/(C_{Rf} - C_R)$ where C_R and C_{Rf} are the radical concentration and final radical concentration.

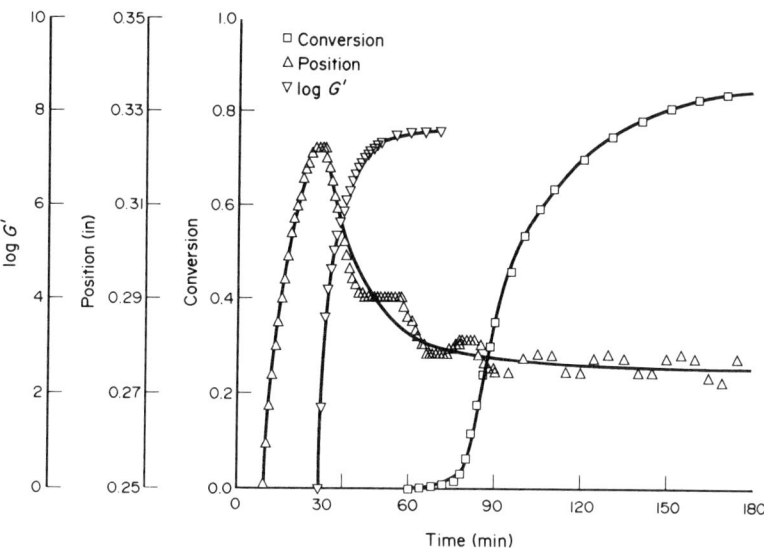

Figure 11 Interactions of gelation, conversion and volume change of an unsaturated polyester resin. $T = 80\,°C$[191]

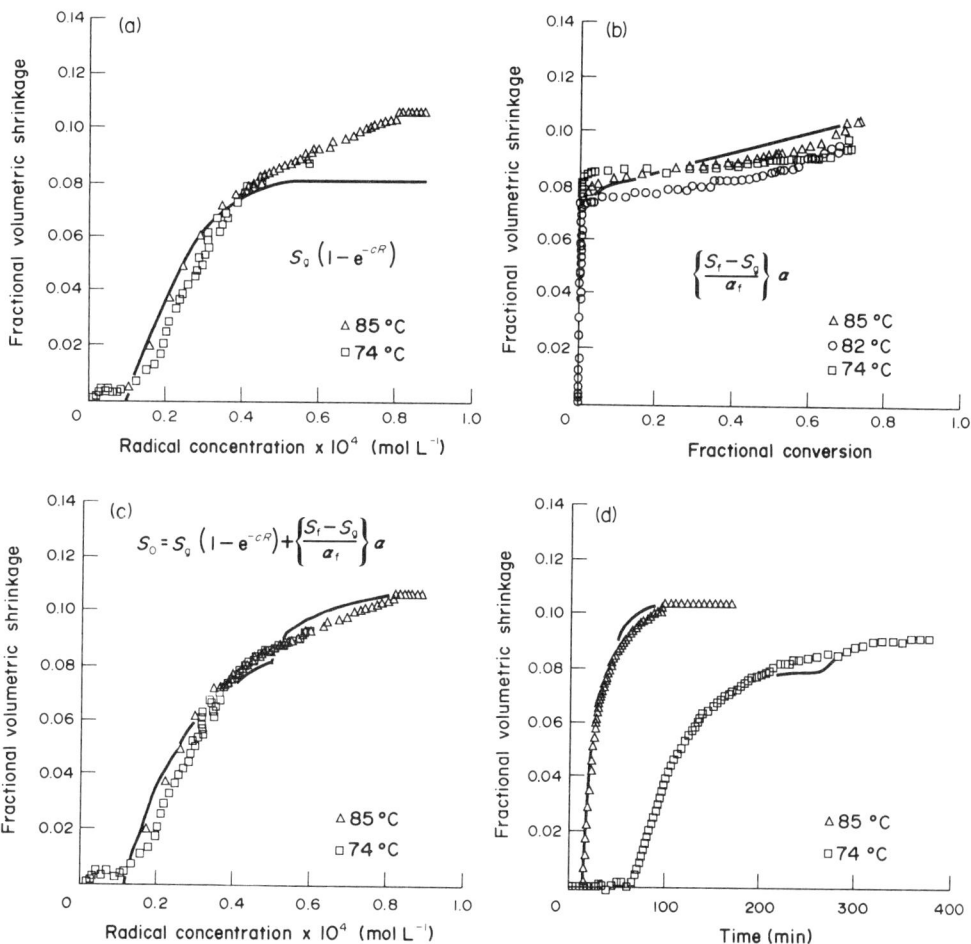

Figure 12 Polymerization shrinkage data and model prediction: (a) shrinkage *versus* radical concentration; (b) shrinkage *versus* resin conversion; (c) shrinkage *versus* radical concentration and resin conversion; (d) shrinkage *versus* time. Symbols represent experimental data and curves represent the model prediction

20.2.3.4 Void formation and foaming during polymerization

The occurrence of voids in cured parts is a major concern in many reactive processing operations. For high-performance structural composites such as epoxy/graphite laminates, void formation is detrimental to composite properties. On the other hand, integral-skin polyurethane foams are extensively used in automotive and furniture industries. Kardos and coworkers[198–200] have developed a model describing void formation, growth, and transport in thermoset composites as a function of molding pressure and temperature. Several researchers[201–203] have tried to incorporate the foaming phenomenon in the modeling of the polyurethane foaming process.

20.2.3.5 Thermophysical properties during polymerization

Changes in thermophysical properties during polymerization have not been studied in detail. For linear vinyl polymerizations, the resin density is treated as a linear function of conversion.[62,81] For thermoset resins, the density change during reaction is often neglected in modeling. As discussed in Section 20.2.3.3 a significant change of resin density may occur due to thermal expansion and polymerization shrinkage. Specific heat and thermal conductivity of an epoxy resin were measured by Sourour.[204] He found that thermal conductivity increased linearly as a function of conversion until the glass transition stage where a more pronounced increase might occur. Specific heat of the resin remained constant at low and high conversions, but increased sharply at intermediate conversions. Kim and Kim,[205] however, found that heat capacity varies linearly with conversion for an epoxy resin. Because of the limited data base, thermophysical properties are often assumed to be unchanged in most modeling of reactive processing.

20.3 PROCESS SIMULATION

Reactive resins can be processed by many different methods. These include constant volume, closed-mold processes such as reaction injection molding (RIM), resin transfer molding (RTM), and transfer molding; constant pressure, moving-mold processes such as compression molding; and mold-free processes such as autoclave curing and filament winding processes. In addition to batch-type operations, continuous processes such as reactive extrusion, continuous casting and pultrusion are also being used in industry. To narrow the scope of this work, process simulation of reactive extrusion and continuous casting are not covered. The basic techniques of process modeling and numerical analysis have been explained in detail in a recent book for thermoplastic polymer processing.[206] These include common assumptions and simplifications, and various numerical methods. Most of them can be directly applied to reactive processing and will not be repeated here. The main objective of this section is to discuss the effect of reaction on process simulation and to review the recent literature in this area.

20.3.1 Reactive Injection Molding Processes

Simulation of the thermoplastic injection molding process has been undertaken by many investigators.[207–214] Several commercial simulation programs, *e.g.* 'Mold Flow' and 'C-Flow' are available. They are increasingly being used to help resolve practical molding and mold design problems. Much less can be said for reactive injection molding, even though a commercial program, 'C-Set', has been recently developed.[215] The fundamental difference in modeling reactive processing is that the reactive resin goes through an irreversible chemical change during the forming process. Therefore, the process simulation must trace each resin element and account for its 'history' during the forming process. There are three aspects that distinguish the simulation of reactive processing from that of thermoplastic processing:[216] (i) the equation of reaction kinetics including heat of reaction; (ii) the dependence of the rheological properties, especially the shear viscosity, on the degree of cure; and (iii) the numerical technique to deal with the Lagrangian nature of the cure development.

Neglecting the inertial and gravity terms, the momentum equation for an incompressible resin in the quasi-steady state is

$$0 = -\nabla P + \nabla \cdot \eta [\nabla v + (\nabla v)^*] \tag{88}$$

where the superscript '*' denotes the conjugate of the dyad.

The energy equation for the time-dependent temperature field is

$$\rho C_p \left(\frac{\partial T}{\partial t} + v \cdot \nabla T \right) = \nabla \cdot k_T \nabla T + \Phi + \Delta H_R \left(\frac{\partial \alpha}{\partial t} + v \cdot \nabla \alpha \right) \quad (89)$$

where Φ is the viscous dissipation term, and ΔH_R is heat of reaction. The kinetic equation, $\partial \alpha / \partial t$, and the viscosity expression, η, have been discussed in previous sections. Depending on the type of resin and the objective of modeling, an appropriate kinetic model and viscosity model can be chosen for equations (88) and (89). Models for structure development and physical properties discussed in Sections 20.2.2 and 20.2.3 may also be added in the simulation.

Flow into an empty cavity is characterized by a complex flow field, known as fountain flow,[217] near the advancing flow front. This flow has a strong influence on the conversion and temperature fields, and cannot be neglected in reactive injection molding. Coyle et al.[218] studied the kinematics of this flow for an isothermal Newtonian liquid. Castro and Macosko[16] modeled this flow for a simple end-gated rectangular mold following a procedure proposed by Bhattacharji and Savic.[219] In order to handle complicated mold geometry, several approximations have been proposed to simulate the fountain flow effects.[220] Method A sets the temperature and conversion profiles at the free surface to be uniform and equal to the temperature and conversion values at the centerline of a location immediately upstream from the front. This approach was used in thermoplastic injection molding[221] and RIM.[222] Method B uses an average of the profile at the location immediately upstream.[220] Method C relocates the particles that overtake the front according to a mass balance.[223-225] A schematic representation of these three fountain flow approximations is shown in Figure 13. Using Method C, Shen and Gonzalez[226] carried out a numerical analysis of mold filling for an epoxy resin in a thin rectangular mold. They found that in the core region, both temperature and conversion patterns are similar, and are dominated by convection effects. Near the wall, the thermal boundary layer is similar to that in the thermoplastic case, but the conversion contours distinctly show no boundary layer (see Figure 14). Garcia et al.[220] recently showed that Method A captures the essence of the kinematics of the fountain flow because the material approaching the front is stretched and distributed almost uniformly across the cavity gap by the extensional flow field. This method produced the best results in the simulation of RIM mold filling.

In the following sections, a brief review of the recent literature on process simulation of RIM, RTM and transfer molding is presented.

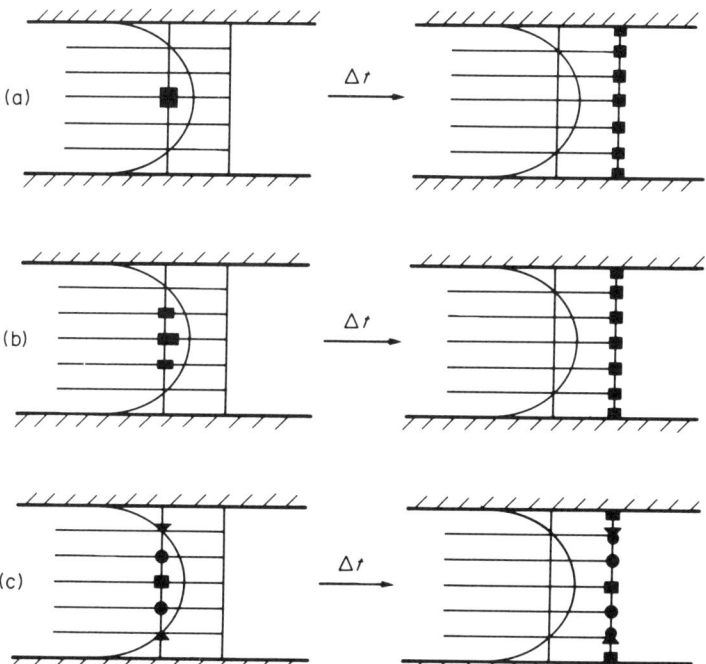

Figure 13 Schematic representation of the fountain flow approximations: (a) simplified front (Method A); (b) averaged front (Method B); (c) mass balanced front (Method C) (reproduced by permission of *Int. Polym. Proc.*, 1991, **6**, 73)

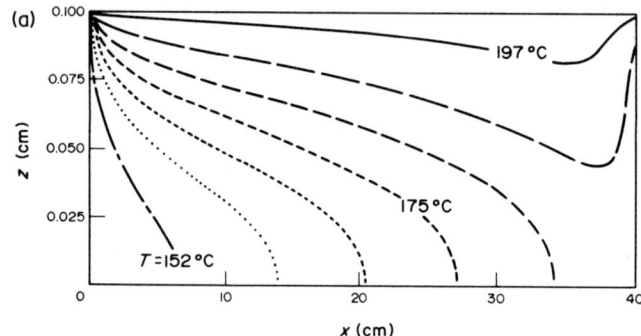

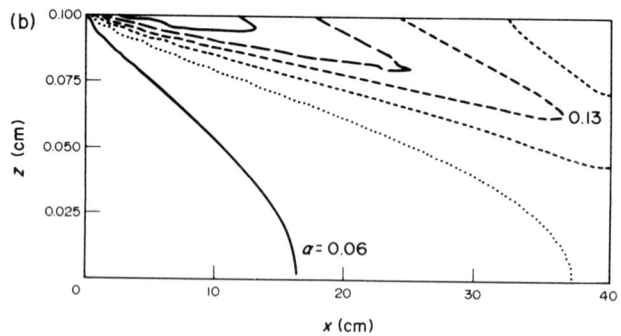

Figure 14 Calculated (a) temperature and (b) conversion contours for epoxy filling in a rectangular cavity. $T_{wall} = 200\,°C$, $T_{inlet} = 150\,°C$, $\alpha_{inlet} = 0.05$ and $v = 5.0\,cm\,s^{-1}$ (reproduced by permission of *NUMIFORM '89*, 1989, 55)

20.3.1.1 RIM

Kamal and Ryan[3,227] studied the injection molding process for an epoxy resin. They considered the cases with[227] and without[3] the flow front region. Domine and Gogos[223] carried out numerical simulation of RIM mold filling using the finite difference method. The fountain flow region was treated using Method C. They did not compare their prediction with experimental results. Manzione[222] presented a process model based on the marker and cell approach. The flow front was treated using Method A. Castro and Macosko[16,17] developed a more detailed model for the RIM process, which included both filling and curing stages. They compared the theoretical prediction with their own experimental results[16] and carried out a series of moldability analyses.[228,229] Kim and Kim[205] followed the same approach as Castro and Macosko[16] for a molding analysis of epoxy resin in a disc-type mold. Lack and Silebi[230] also carried out a numerical simulation of the RIM process in a radial flow geometry. Their model was solved by the method of lines and the fountain flow was treated using Method C.

The above-mentioned works dealt with simple flow geometries, *e.g.* rectangular or disk shape thin cavities. For complicated mold geometry, Hilger *et al.*[231] used the control volume method to analyze RIM mold filling in three-dimensional thin cavities for a Newtonian fluid. Shen[216] used the Hele-Shaw approximation to derive a Darcy's law flow equation to simulate RIM mold filling in thin cavities. Similar modeling based on numerical mapping and the finite difference method was recently developed by Garcia *et al.*[220] In the Hele-Shaw approximation, the inertia terms and the flow in the gapwise direction are neglected. For an incompressible fluid, the steady state momentum equation can be integrated in the gapwise (*i.e.* z) direction

$$\bar{v}_x = \frac{1}{h}\int_0^h v_x\,dz; \quad \bar{v}_y = \frac{1}{h}\int_0^h v_y\,dz \tag{90}$$

where *h* is the gap width. When combined with the continuity expression, this gives

$$0 = \frac{\partial}{\partial x}\left(S\frac{\partial P}{\partial x}\right) + \frac{\partial}{\partial y}\left(S\frac{\partial P}{\partial y}\right) \tag{91}$$

where the fluid conductance S is defined as

$$S = \int_0^h \frac{z^2}{\eta} dz \tag{92}$$

and

$$\bar{v}_x = -\frac{S}{h}\frac{\partial P}{\partial x}; \quad \bar{v}_y = -\frac{S}{h}\frac{\partial P}{\partial y} \tag{93}$$

Equation (91) allows for a direct calculation of the pressure field. Average velocities in the planar directions can then be determined from equation (93). The Hele–Shaw approximation greatly simplifies the momentum equation and is very useful for computer simulation of mold filling in thin cavities with complicated geometry. For thick cavities, where the flow in the gapwise direction cannot be neglected, the Hele–Shaw approximation is no longer valid. The mold filling simulation becomes very complicated because of the coupling of the pressure field and the velocity field. Although inertia terms have always been neglected in the RIM simulation, some researchers found that these effects may affect the flow pattern, especially near the area where inserts are located.[232] A modeling effort is being carried out in this area.[233]

The first curing analysis of the urethane RIM process was performed by Broyer and Macosko.[234] Their model predicted the temperature profile near the mold center reasonably well, but too low a temperature was predicted near the mold wall because of the failure of an isothermal wall assumption. Domine and Gogos[223] also developed a computer model to simulate the filling and curing steps of RIM. However, there was no experimental verification. Lee and Macosko[5] developed a more realistic heat transfer model of RIM curing, which included the heat generation of polymerization, heat build-up in the mold and heat transfer to the temperature control fluid.

The curing step in the RIM process was also analyzed by Castro and coworkers[228,229] and Osinski et al.[235] In addition to the experimental measurement and numerical analysis, they mentioned several dimensionless parameters which appeared explicitly in the balance equations and tended to govern the curing behavior. They are the Damkohler number, the dimensionless initial and wall temperatures, and the dimensionless activation energy.

Most work on curing in the RIM process is based on polyurethane systems. There are also a few studies associated with nonurethane systems such as polyureas,[159,236] isocyanurates,[28] nylons,[93,177] styrenic resins,[237] epoxy resins,[51,205,238] silicone rubbers[239] and polyurethane–polyester hybrids.[115,240] In general, the polymerizations can be categorized as either thermally activated (i.e. reaction is initiated by heat transfer from heated mold walls) or mixing activated (i.e. reaction is started by rapid mixing of two or more components which are highly reactive at the mix temperature). The latter reveals a high reaction rate at the beginning of polymerization. The former requires an induction period during which heat transferred from the mold wall raises the resin temperature to a level at which significant polymerization may occur. For thermally activated chain growth polymerizations, such as polyester and styrenic resins, there is a sharp temperature rise following the induction period, while for the thermally activated step growth polymerizations such as epoxy resins and isocyanurates, the temperature rise is much slower. Silicone rubbers, on the other hand, have very low reaction exotherms and the maximum temperature in the mold seldom exceeds the wall temperature. Cure always starts from the wall and may continue after demolding. Thick parts may not be totally cured upon demolding, even though the material near the wall has reached a desirable mechanical strength.[239]

For nylons and polyurethane–polyester hybrids, the thermogram during curing has two exotherms as shown in Figures 4(a) and 8. For nylon RIM, the combination of heat transfer from the mold plus the reaction exotherm causes the first temperature rise, while the second exotherm occurs due to crystallization and possibly additional polymerization. For polyurethane–polyester hybrids, the first temperature rise is due to the urethane formation, while the second temperature rise results from the polyester reaction exotherm.

Property and structure development during RIM molding has not been modeled in detail. There is some effort on simulating modulus growth[239] and foaming phenomenon.[201–203]

20.3.1.2 RTM/SRIM

Resin transfer molding (RTM) and structural reaction injection molding (SRIM) are processes in which two liquid monomers or prepolymers are mixed and injected into a mold cavity where fiber mats are preplaced. Due to the long fiber reinforcement and the low viscosity fluid used in processing, RTM/SRIM processes have a great potential to become major mass production techniques for producing low weight and high strength composite parts with complicated geometry. During processing, the reinforcement fibrous networks, initially in an unimpregnated form, must be completely impregnated by the resin. The flow pattern in the mold can be very complicated because of the anisotropic nature of the reinforcement fibers.

Most RTM/SRIM modeling dealt with isothermal mold filling.[241-246] These models considered two-dimensional isothermal mold filling in RTM by applying Darcy's law and either the finite element method,[241,242] the finite difference method with numerical mapping[243] or the boundary element method.[244] Young et al.[245,246] used the control volume approach to simulate the 3-D mold filling for complicated geometry, taking Darcy's equation as a base. Their model can also handle different fiber orientation, permeability changes and multiple gates. The 3-D momentum equation takes the following form

$$\begin{bmatrix} v_x \\ v_y \\ v_z \end{bmatrix} = -\frac{1}{\eta} \begin{bmatrix} K_{xx} & K_{xy} & K_{xz} \\ K_{yx} & K_{yy} & K_{yz} \\ K_{zx} & K_{zy} & K_{zz} \end{bmatrix} \begin{bmatrix} \partial P/\partial x \\ \partial P/\partial y \\ \partial P/\partial z \end{bmatrix} \quad (94)$$

where K_{ij} are the components of the pemeability tensor. Since Darcy's law can be applied to all three flow directions, 3-D mold filling simulation can be more easily achieved in RTM than in RIM or thermoplastic injection molding.

There was very little literature available for the nonisothermal mold filling in the RTM process. Gonzalez[247] considered a simple one-dimensional thin cavity problem. He used nonlocal thermal equilibrium (i.e. two energy equations) to get two dimensionless temperature expressions in the integral form. He then used experimental temperature profiles to fit the heat transfer coefficient between the resin and the fiber mat. For complicated mold geometry, Lin et al.[248] developed a numerical model for nonisothermal mold filling and curing simulation in thin cavities with preplaced fiber mats based on the control volume method. Both lumped temperature systems (i.e. local thermal equilibrium between the resin and the fiber) and unlumped temperature systems (i.e. nonlocal thermal equilibrium) were considered. Heat transfer in the gapwise direction was solved by the Chebyshev collocation spectral method.[249] A Lagrangian coordinate system was used in the flow front region to improve the energy transfer calculation. The governing equations for the unlumped temperature system are given as

$$\frac{\partial \bar{v}_x}{\partial x} + \frac{\partial \bar{v}_y}{\partial y} = 0 \quad \text{(continuity)} \quad (95)$$

$$\begin{bmatrix} \bar{v}_x \\ \bar{v}_y \end{bmatrix} = - \begin{bmatrix} S_{xx} & S_{xy} \\ S_{yx} & S_{yy} \end{bmatrix} \begin{bmatrix} \partial P/\partial x \\ \partial P/\partial y \end{bmatrix} \quad \text{(momentum)} \quad (96)$$

where

$$\begin{bmatrix} S_{xx} & S_{xy} \\ S_{yx} & S_{yy} \end{bmatrix} = \frac{1}{h} \int_{-h/2}^{h/2} \frac{1}{\eta} \begin{bmatrix} K_{xx} & K_{xy} \\ K_{yx} & K_{yy} \end{bmatrix}$$

$$\phi \rho_r c_{pr} \frac{\partial T_r}{\partial t} + \rho_r c_{pr} \left(\bar{v}_x \frac{\partial T_r}{\partial x} + \bar{v}_y \frac{\partial T_r}{\partial y} \right) = \phi k_{T_r} \left(\frac{\partial^2 T_r}{\partial x^2} + \frac{\partial^2 T_r}{\partial y^2} + \frac{\partial^2 T_r}{\partial z^2} \right) + \phi h_v (T_f - T_r) + \phi \Delta H \dot{m} \quad \text{(heat)} \quad (97)$$

$$(1-\phi)\rho_f c_{pf} \frac{\partial T_f}{\partial t} = (1-\phi) k_{T_r} \left(\frac{\partial^2 T_f}{\partial x^2} + \frac{\partial^2 T_f}{\partial y^2} + \frac{\partial^2 T_f}{\partial z^2} \right) + \phi h_v (T_r - T_f) \quad (98)$$

$$\frac{\partial \alpha}{\partial t} + \bar{v}_x \frac{\partial \alpha}{\partial x} + \bar{v}_y \frac{\partial \alpha}{\partial y} = \dot{m} \quad \text{(mass)} \quad (99)$$

where ϕ is porosity, h_v is the volume heat transfer coefficient from the resin to the fiber bed, r and f represent resin phase and fiber phase, and $\dot{m}$ is the mass generation rate of polymer (i.e. a kinetic expression). Figure 15 shows the simulated temperature profiles of the resin and the fiber during the filling stage. They also measured h_v experimentally and found results similar to those reported by Gonzalez.[247]

20.3.1.3 Transfer molding

Multicavity transfer molding of thermoset compounds was studied by Manzione and coworkers.[250,251] They used the finite difference method to simulate the temperature profile within

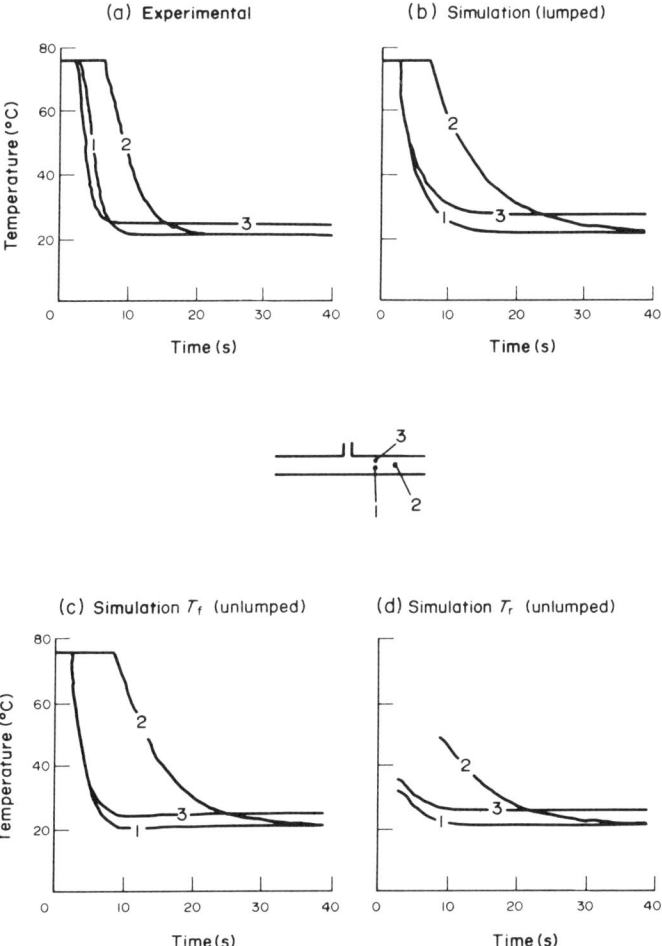

Figure 15 Temperature profiles of the resin and the fiber during the filling stage in a circular cavity with preplaced random fiber mats. $T_{r0} = 75.5\,°C$, $T_{f0} = 19.5\,°C$, $\phi = 0.81$[248]

the transfer pot when the flow pattern was assumed to be a stagnation point flow.[250] A semiempirical flow algorithm was developed for flow balancing in the case of multicavity transfer molding.[251]

20.3.2 Reactive Compression Molding

The need to mass-produce composite parts with large surface areas and high mechanical strength has made reactive compression molding an important process in the polymer industry. A widely used material in automotive applications is sheet molding compound (SMC). SMC consists of roughly one third unsaturated polyester resin, one third filler (mainly $CaCO_3$) and one third chopped glass strand (1.27–5.08 cm length). Small amounts of other ingredients are added to control the moldability of the compound and to adjust the physical properties of molded parts.[258] In this section the basic principles and major literature regarding computer simulation of SMC compression molding will be presented. Most of the results can also be applied to other systems such as bulk molding compound (BMC) compression molding and in-mold coating (IMC).

The process of compression molding of SMC involves four steps: material preparation, mold filling, curing, and ejection and cooling. Before molding, a room temperature charge is placed on an open preheated mold with the top mold-half rapidly closing down. The mold is preheated to a high temperature of around 150 °C. Mold filling starts as the top mold-half touches the surface of the charge. Because of the difference in contact time, the amount of heat conduction between the bottom mold-half and the charge may differ from that between the top mold-half and the charge. This may lead to a nonuniform and asymmetric flow with respect to the midplane of the charge during mold filling. The mechanics of SMC compression molding and the interaction between the squeezing flow and heat transfer have been the subjects of many studies in recent years. Consequently, extensive reviews are available in several articles.[252-254]

The flow kinematics across the charge thickness and the flow boundary condition at the mold–charge interface have drawn a great deal of discussion over the past 10 years. Experimentally, several earlier works[255-257] reported significant preferential flow near the mold surface when SMC was under slow compression rates. This preferential flow was attributed to the heat transfer through the mold surface and has a great effect on the flow pattern during mold filling. More recent works,[258,259] however, showed that preferential flow occurred only in thick charges molded at very slow closing speeds. Under production conditions (i.e. closing speeds of 4–8 mm s^{-1}), the transverse velocity profile resembles a plug flow with material slipping at the mold surface. In a visualization study using carbon black as a tracer, Kau[260] suggested a partial slip boundary condition at the mold surface based on the observed length of the carbon black streaks and the location of their end points.

Most current computer models for flow prediction assume that the charge and the molded part are thin in comparison to their lateral dimensions. This assumption transforms the 3-D flow problem into a 2-D one and makes the governing equation very similar to the ones in reactive injection molding. A computer code for the latter case can be easily converted to a code for the former case with a few adjustments. The earliest models for the flow analysis of thin SMC parts were based on an isothermal lubrication theory and were motivated by the successful application of the generalized Hele–Shaw equations to thermoplastic injection molding.[261-264] In those papers, SMC was modeled as an incompressible, isotropic, Newtonian or power law fluid. Normal stresses and inertia terms were neglected and the no-slip boundary condition was applied at the mold surface. The first step in modeling a thin part is to conceptually unfold it so that it lies in the x–y plane with z being the thickness direction.[265] This transforms a curvature shape into an equivalent flat part, with local closing speed adjusted according to the curvature angle as shown in Figure 16. Based on the Hele–Shaw approximation, the steady state momentum equation can be expressed as

$$\frac{\partial}{\partial x}\left(S'\frac{\partial P}{\partial x}\right) + \frac{\partial}{\partial z}\left(S'\frac{\partial P}{\partial y}\right) = -\dot{h} \tag{100}$$

where $\dot{h}$ is closing speed and S' is given by

$$S' = \int_0^h \frac{z - \lambda}{\eta} dz \tag{101}$$

Here, λ is the value of z at which the shear stresses vanish. For cases in which the thermal boundary conditions are the same at the upper and lower mold surfaces, λ is at the midpoint of the thickness. For the case when λ is not $h/2$, it can be found from[266]

$$\lambda = \left(\int_0^h \frac{z\,dz}{\eta}\right) \Big/ \left(\int_0^h \frac{dz}{\eta}\right) \tag{102}$$

For an isothermal Newtonian fluid and a part of uniform thickness, equation (102) reduces to Poisson's equation

$$\frac{\partial^2 P}{\partial x^2} + \frac{\partial^2 P}{\partial y^2} = -\frac{12\eta \dot{h}}{h^3} \tag{103}$$

with the gapwise average velocities given by

$$\bar{v}_x = -\frac{h^2}{12\eta}\frac{\partial P}{\partial x}; \quad \bar{v}_y = -\frac{h^2}{12\eta}\frac{\partial P}{\partial y} \tag{104}$$

These models yielded good agreement between calculated flow front progression and experimental results for thin charges. For thicker charges, experimental results showed a substantially different flow front progression than the calculated profile, which was supposed to be independent of cavity thickness. In addition to this discrepancy, these models may not be appropriate for pressure prediction because of the isothermal assumption and the lack of consideration for the friction force at the mold surface. Recently, Lee and Tucker[253] included normal stresses and heat transfer in their simulation. In their analysis, the effect of heat transfer on preferential flow was discussed. However, the effect of friction at the mold surface was still not considered. Several researchers also imposed a transverse viscosity gradient to simulate the formation of preferential flow.[267,268]

Barone and Caulk proposed another modeling approach.[254,269] They represented SMC as a two-dimensional sheet which stretches uniformly through the cavity thickness and slips relative to the mold surface. The momentum balance equation was expressed by three terms in a two-dimensional

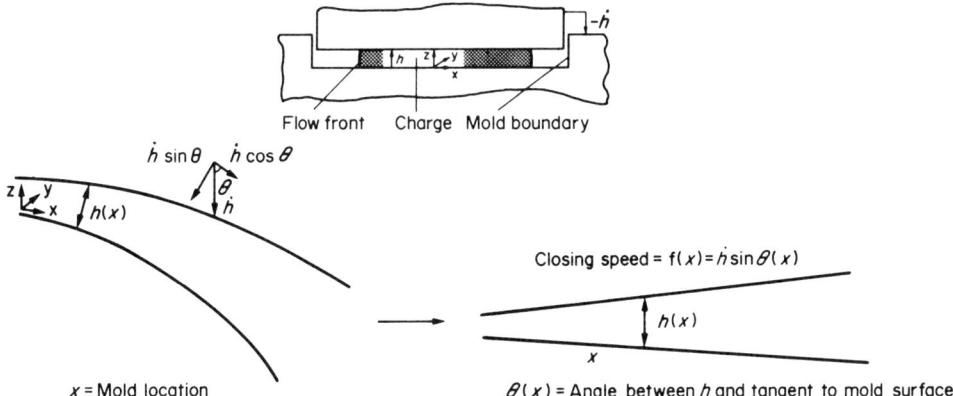

Figure 16 Schematic representation of compression molding and Hele–Shaw model[252]

frame: a hydrodynamic pressure resultant through the cavity thickness, a planar stress resultant embodying the material response to extensional deformation and the friction force between the SMC and the mold. This approach is more consistent with the observed kinematics by representing SMC as planar isotropic sheets instead of a three-dimensional isotropic fluid. A plug-like flow front profile also agrees better with experimental observation than a profile which results from the no-slip boundary condition. In a thin charge, the surface friction 'κ' dominates the momentum balance, and the material resistance may be neglected. For an isothermal Newtonian fluid and a charge of uniform thickness, the momentum and continuity equations may be combined to give

$$\frac{\partial^2 P}{\partial x^2} + \frac{\partial^2 P}{\partial y^2} = -\frac{2\kappa \dot{h}}{h^2} \tag{105}$$

with the average velocities for the thin charge given by

$$\bar{v}_x = -\frac{h}{2\kappa}\frac{\partial P}{\partial x}; \quad \bar{v}_y = -\frac{h}{2\kappa}\frac{\partial P}{\partial y} \tag{106}$$

The thin charge limit of Barone and Caulk's model is identical in form to the Hele–Shaw model. The only difference in the models lies in the interpretation of the rheological parameters and the way pressure depends on h. Recent experimental data[270–272] showed that Barone and Caulk's model is more appropriate for SMC, while the generalized Hele–Shaw model would be appropriate for the IMC process.

Thick charges of SMC do not obey the thin charge formulations. Equation (107) illustrates a charge that is thin enough for the thin charge models to apply

$$\frac{h\eta}{L^2\kappa} \ll 1 \tag{107}$$

where L is the charge diameter. Barone and Caulk's full model can be used to deal with thick charges.[271]

The transient heat conduction and reaction exotherm through SMC during mold filling and curing can be easily incorporated into Barone and Caulk's model since the assumption of plug flow can reduce the energy balance equation into a pseudo-conduction problem with a warped time. The energy balance equation is

$$\frac{\partial T}{\partial t} + v_z \frac{\partial T}{\partial z} = \frac{k_T}{\rho C_p}\frac{\partial^2 T}{\partial z^2} + \frac{\Delta H}{C_p}\frac{d\alpha}{dt} \tag{108}$$

and

$$v_z = \left(\frac{\partial \ln h}{\partial t}\right) z \tag{108a}$$

Defining dimensionless space ξ and warped time τ as

$$\xi = \frac{z}{h(t)}; \quad \tau = \frac{k_\tau}{\rho C_p}\int_0^t \frac{1}{h(t')^2}\,dt' \tag{109}$$

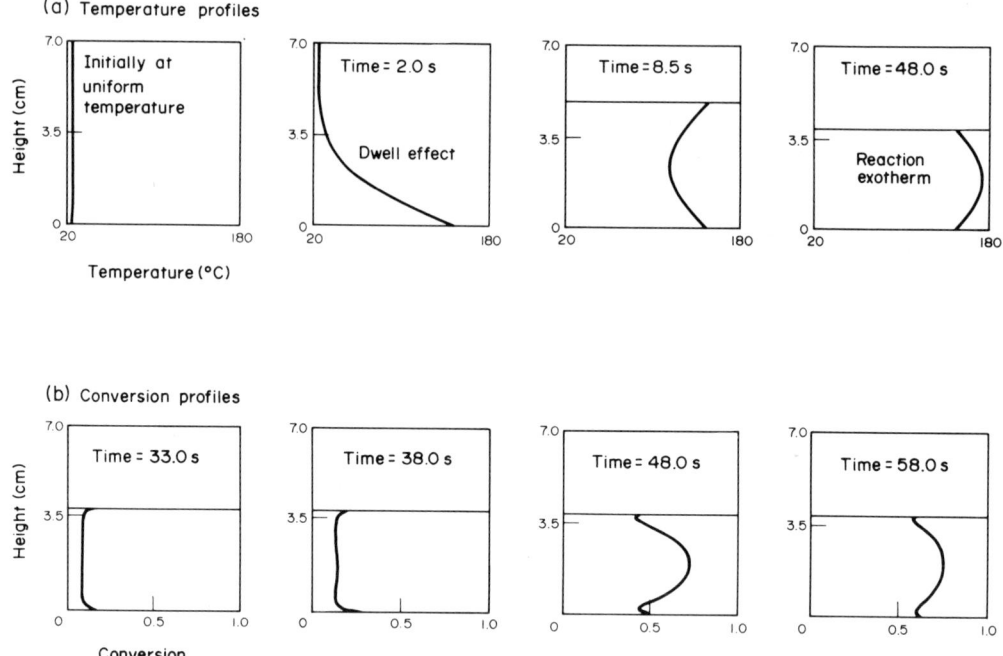

Figure 17 Calculated (a) temperature and (b) conversion profiles in the gapwise direction during compression molding of SMC[197]

Equation (108) becomes

$$\frac{\partial T}{\partial \tau} = \frac{\partial^2 T}{\partial \xi^2} + \left(\frac{h^2 \rho \Delta H}{k_T}\right)\frac{d\alpha}{dt} \tag{110}$$

Some applications based on this approach have been reported in the last several years.[4,197,252,254,257] Figure 17 shows simulated temperature and conversion profiles during molding.[197]

There are, however, several limitations associated with Barone and Caulk's model. Firstly, a total slip boundary condition at mold surface may not be accurate, as mentioned in a previous report.[260] Secondly, although the macroscopic view of the flow front profile resembles a plug flow, the small amount of prefential flow near the mold surface may play an important role in determining the surface quality or defects. Finally, one would prefer that the velocity profile arises as a consequence of the interaction between heat transfer, flow and temperature dependent flow properties, rather than as an assumption in the model. Recently, Fan et al.[273] used a 2-D finite element code to simulate the flow front shape and temperature profile in the thickness direction for a nonisothermal SMC compression molding. They were able to predict the heat transfer induced preferential flow near the mold surface at various molding conditions. The pressure change obtained by numerical simulation also compared well with experimental results. Their model, however, can be applied only to plane strain cases.

Most modeling studies treated SMC as a homogeneous isotropic continuum. For thin charges deforming between flat plates, these isotropic models are acceptable. However, when there are substructures, the mold-filling modeling becomes much more difficult as the flow behavior of SMC varies greatly from planar directions to the thickness direction. This is due to the presence of glass fibers. Some modeling effort for anisotropic flow in a plate–rib-type geometry has been carried out.[274-276] The flow pattern of short fiber (6.35 mm length) reinforced composites could be predicted reasonably well.[276] Longer fibers tended to interact with each other and caused a 'bridging effect' in the intersection area of the plate and rib.[277] This phenomenon has not been modeled yet.

Nonisothermal mold filling and curing of SMC in molds with substructures has also been studied by Fan et al.[74] They used a finite element code to simulate the material and temperature distributions during filling by assuming that the compound was isotropic and would not react during filling. The concentrations of initiator and inhibitor, however, did change during filling. Temperature, initiator and inhibitor distributions at the end of filling were used as the initial conditions for the curing stage, which was modeled using a 2-D finite difference method. Figure 18 shows a comparison

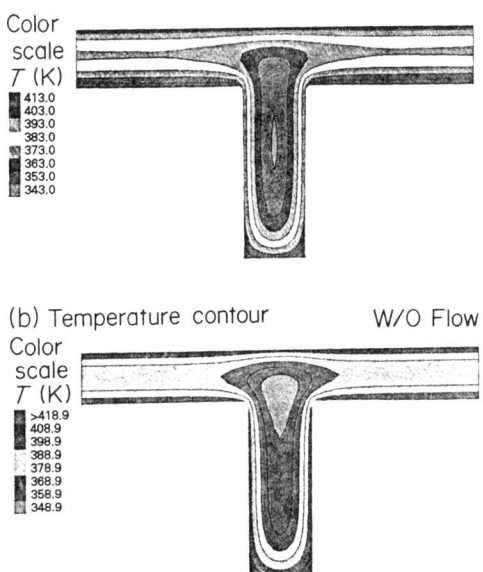

Figure 18 Comparison of simulated temperature distribution during compression molding of SMC (a) considering, and (b) without considering mold filling[174]

of temperature distributions before any reaction occurred, both with and without considering mold filling. Because the material touched the bottom mold surface first, the dwelling effect caused an asymmetric temperature distribution during filling.

Many in depth cure analyses have been completed by assuming that the mold was filled with SMC before any heat transfer and reaction. By neglecting the mold-filling stage, these works emphasized the reaction kinetics,[72,87,278] material design[73] and thermal design of molds.[4,279,280]

20.3.3 Reactive Processing of Continuous Fiber Reinforced Thermoset Composites

Many high performance composites, especially those used in the aerospace and defense industries, are made of thermoset resins reinforced with long oriented fibers. Except for the resin transfer molding process (which has been discussed in Section 20.3.1.2), most manufacturing processes consist of two steps. In the first step, fiber filaments are impregnated with uncured resin. These resin-impregnated fibers can then be pulled through a heated die, *i.e.* the pultrusion process, or wound on a suitably shaped mandrel, *i.e.* the filament-winding process. In the former case, the resin is cured during pultrusion, while in the latter case, the mandrel assembly is placed in the oven after the winding is complete and the resin is cured. Another approach to treating the resin-impregnated fibers is to align them into laminates and to partially cure the resin to form prepregs. The prepregs are then shaped and cured in an autoclave[281] or by bleeder ply molding.

A common feature of these processes is that the purpose of the applied pressure during the final stage of shaping is to squeeze out excess resin from the fiber plies (*i.e.* consolidation) instead of forcing the resin into a mold as in the case of reactive injection molding and reactive compression molding. The mechanism and modeling of composite consolidation for these processes are similar and have been discussed by several researchers.[282,283] Resin chemistry and rheological changes are also the same for these processes. Therefore, only the modeling of the autoclave process will be presented. Filament-winding[284] and pultrusion[285-287] modeling follow similar approaches.

The most common way to manufacture parts made of high performance thermoset matrix composites is by autoclave curing. In the autoclave process, the part is formed into the desired shape (see Figure 19) and is then placed in an autoclave. Inside the autoclave, heat and pressure are applied to the part. Heat accelerates the chemical reactions occurring during the cure, while the applied pressure squeezes out excess resin, compacts the plies and reduces the void content. Loos and Springer[46,281] developed a complete model applicable to autoclave curing of thermoset composites. Their model consists of five submodels as shown in Figure 20. Both thermochemical and void submodels have already been discussed in an earlier section. The flow submodel deals with the resin flow during consolidation. The consolidation theory considers the prepreg stack to be a skeleton of a solid network enclosing voids which may be filled with gas, liquid, or both. The stack is considered to behave like a nonlinear elastic spring.[288] When the fiber–resin composite is placed under stress causing its volume to decrease, there exist two possible routes by which this decrease may occur, *i.e.* by compression of the stack or by escape of the resin across the boundaries of the sample. By assuming that the spring-like fiber–fiber interaction is negligible and the flow in different directions can be decoupled, Loos and Springer[281] used the simple Darcy's law to describe the resin flow, *i.e.*

$$v_i = \frac{K_i}{\eta} \frac{\partial P}{\partial i} \tag{111}$$

where i represents the horizontal and vertical directions.

Subsequent modification of this simple flow model has been proposed by Dave *et al.*[283,289–291] Their model takes the following form

$$\frac{\partial P}{\partial t} - \frac{\partial \sigma}{\partial t} = \frac{1}{\eta m_v} \left[\frac{\partial}{\partial x}\left(K_x \frac{\partial P}{\partial x}\right) + \frac{\partial}{\partial y}\left(K_y \frac{\partial P}{\partial y}\right) + \frac{\partial}{\partial z}\left(K_z \frac{\partial P}{\partial z}\right) \right] \tag{112}$$

where P is the hydraulic pressure of the resin in the composite, σ is the total external applied stress and m_v is the coefficient of volume change, which depends on the stress–strain behavior of the

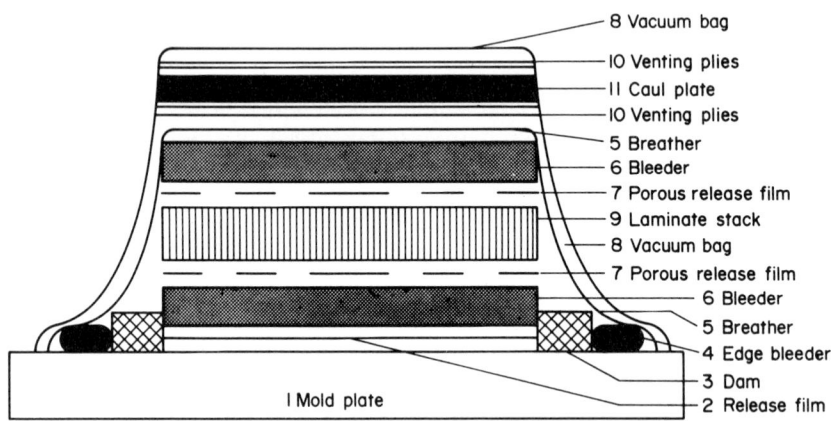

Figure 19 Schematic of the prepreg lay-up used in autoclave cure

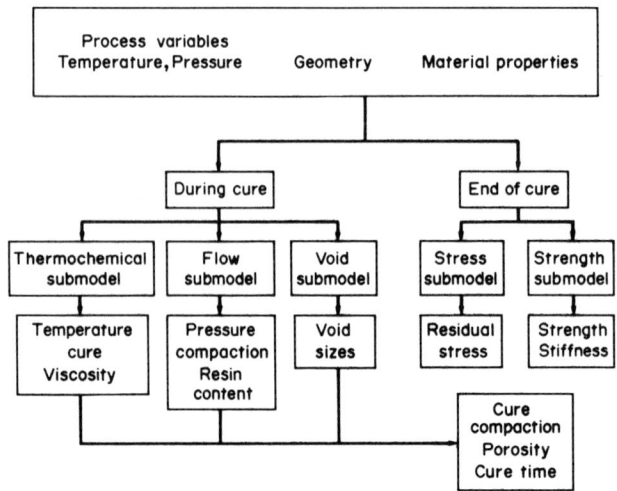

Figure 20 An autoclave cure model developed by Loos and Springer[281]

prepreg stack in confined compression,[290] i.e.

$$m_v = -\left(\frac{de}{dP}\right)(1+e)^{-1} \tag{113}$$

where e is the void ratio defined as the volume of pore space per unit volume of solids in the stack.

Gutowski and coworkers[288] proposed a consolidation model which accounts for the motion of the moving boundary layer in the thickness direction of the stack. Their model takes the following form

$$\frac{\partial}{\partial t}\left(\frac{\phi}{1-\phi}\right) = \frac{1}{\eta}\left[\frac{K_x}{(1-\phi)}\frac{\partial}{\partial x}\left(\frac{\partial P}{\partial x}\right) + \frac{K_y}{(1-\phi)}\frac{\partial}{\partial y}\left(\frac{\partial P}{\partial y}\right) + \frac{1}{(1-\phi_0)^2}\frac{\partial}{\partial z}\left(K_z(1-\phi)\frac{\partial P}{\partial Z}\right)\right] \tag{114}$$

where ϕ_0 is the initial porosity of the fiber stack.

Numerical simulation of autoclave curve of flat plate composites has been carried out by several researchers using a 1-D and 2-D finite difference method.[281,290-292] The results include temperature distribution, resin conversion, viscosity changes, resin pressure, void sizes, velocity profiles and the amount of resin flow out of the composite.

20.4 NOTATION

a, a_i	constants
a_T	shift factor in WLF equation
A, A_i	frequency coefficients in kinetic rate constant
A_M, A_P	parameters for monomer and polymer diffusion
A'	constant
b, B	constants
B_i	rate constant in equation (14)
B_M, B_P	parameters for monomer and polymer diffusion, respectively
$\bar{B}$	intensity of autocatalysis
c_i	constants in WLF equation
c'	constant in equation (18)
c_p	heat capacity
$\bar{C}_i$	concentration of reactive species i at time t
$\bar{C}_{I0}$	initiator concentration after all inhibitors have been consumed
D	mass diffusivity
e	void ratio
E	friction factor for viscosity
E_a	reaction activation energy
E_η	viscosity activation energy
E_m, E_x	lattice energies of noncross-linked and cross-linked polymer, respectively
f	initiator efficiency
f()	function
f_m, f_x	segmental mobilities of noncross-linked and cross-linked polymer
F	parameter defined in equation (42)
g	ratio of gyration radii
G	modulus
G_n^0	plateau modulus
ΔG	free energy
h	mold cavity thickness
h_v	volume heat transfer coefficient
$\dot{h}$	mold closing speed
H	effect of entanglements on the cross-links
ΔH_R	heat of reaction
I	initiator
k_i^*	kinetic rate constant
k_0	true rate constant
k_D	diffusion rate constant
k_T	thermal conductivity
k_i	apparent rate constants in equation (13)

K_{ii}	components of the permeability tensor
L	sample diameter
m	reaction order
$\dot{m}$	mass generation rate of polymer
m_v	coefficient of volume change defined in equation (112)
M	monomer
M_c	critical molecular weight for chain entanglement
$M\cdot$	primary radical
$M_i\cdot$	polymeric radical
M_n	number average molecular weight
M_w	weight average molecular weight
M_{we}	effective weight average molecular weight considering intramolecular reaction
n, n_i	reaction orders
N_i	number of repeating units on the i chain
P	pressure
P_e	probability that a pairwise interaction between two network chains is trapped
P_i	terminated polymer
q	inhibitor efficiency
Q	$Q = \bar{V}_2 / \bar{V}_1$
r	stoichiometric ratio
$R\cdot$	free radical
$\bar{R}$	reduced radical concentration
s	shrinkage
s_0	isothermal polymerization shrinkage at T_0
s_f	final shrinkage
s_g	shrinkage due to liquid–solid transition
S, S'	fluid conductances
S_B	radius of gyration of branched polymer
S_L	radius of gyration of linear polymer
t	time
t_{gel}	gel time
t_z	induction time
T	temperature
T_f	fiber temperature
T_g	glass transition temperature
T_m^0	equilibrium melting temperature
T_r	resin temperature
T_s	reference temperature
v	velocity
V	volume
V_f	fractional free volume
$\bar{V}$	specific volume
X_c	cross-link density
z	thickness direction
Z	inhibitor
α	conversion
α_e	effective conversion considering intramolecular reaction
α_f	final conversion
α_{gel}	gel conversion
β	thermal expansion coefficient
δ	solubility parameter
η	viscosity
κ	surface friction
π_i, π_i'	lumped parameter in equations (30) and (34)
ρ	density
σ	total externally applied stress
τ	warped time
ϕ	volume fraction or porosity
ϕ_0	initial porosity
Φ	viscous dissipation term

χ	interaction coefficient between chains A and B
χ_c	critical χ for the onset of phase separation
ψ	degree of crystallinity
λ	value defined in equation (102)
μ	concentration of cross-links
ν	concentration of chains between cross-links
ξ	dimensionless space

20.5 REFERENCES

1. M. R. Kamal and S. Sourour, *Polym. Eng. Sci.*, 1973, **13**, 59.
2. S. Y. Pusatcioglu, A. L. Fricke and J. C. Hassler, *J. Appl. Polym. Sci.*, 1979, **24**, 937.
3. M. E. Ryan, Ph.D. Thesis, McGill University, Montreal, 1978.
4. M. R. Barone and D. A. Caulk, *Int. J. Heat Mass Transfer*, 1979, **22**, 1021.
5. L. J. Lee and C. W. Macosko, *Int. J. Heat Mass Transfer*, 1980, **23**, 1479.
6. C. D. Han and K. W. Lem, *J. Appl. Polym. Sci.*, 1983, **28**, 743, 763, 3144, 3207.
7. K. W. Lem and C. D. Han, *Polym. Eng. Sci.*, 1984, **24**, 175.
8. C. Feger, S. E. Molis, S. L. Hsu and W. J. Macknight, *Macromolecules*, 1984, **17**, 1830.
9. C. Feger and W. J. Macknight, *Macromolecules*, 1985, **18**, 280.
10. C. W. Macosko, 'Fundamentals of Reaction Injection Molding', Hanser, Munich, 1989.
11. J. Robins, B. H. Edwards and S. K. Tokach, *Adv. Urethane Sci. Technol.*, 1984, **9**, 65.
12. E. B. Richter and C. W. Macosko, *Polym. Eng. Sci.*, 1978, **18**, 1012.
13. J. P. Casey, B. Milligan and M. J. Fasolka, *Proc. S.P.I. Ann. Tech. Mark. Conf. 28th*, 1984, 218.
14. R. E. Camargo, Ph.D. Thesis, University of Minnesota, 1984.
15. E. Broyer and C. W. Macosko, *AIChE J.*, 1976, **22**, 268.
16. J. M. Castro and C. W. Macosko, *AIChE J.*, 1982, **28**, 250.
17. J. M. Castro, S. D. Lipshitz and C. W. Macosko, *AIChE J.*, 1982, **28**, 973.
18. J. W. Baker and J. B. Holdsworth, *J. Am. Chem. Soc.*, 1947, **71**, 713.
19. J. W. Baker and J. Gaunt, *J. Am. Chem. Soc.*, 1949, **73**, 9.
20. J. W. Baker and D. N. Bailey, *J. Am. Chem. Soc.*, 1957, **81**, 4649.
21. G. Borkent, *Adv. Urethane Sci. Tech.*, 1974, **3**, 1.
22. E. C. Steinle, F. E. Critchfield, J. M. Castro and C. W. Macosko, *J. Appl. Polym. Sci.*, 1980, **25**, 2317.
23. D. Nissen and R. A. Markovs, *J. Elastomers Plast.*, 1983, **15**, 96.
24. R. J. G. Dominquez, *J. Cell. Plast.*, Nov./Dec. 1984, **20**, 433.
25. J. P. Casey, B. Milligan and M. J. Fasolka, *J. Elastomers Plast.*, 1985, **17**, 218.
26. J. H. Ewen, *J. Elastomers Plast.*, 1985, **17**, 281.
27. R. A. Grigsby and R. J. G. Dominguez, *Proc. S.P.I. Ann. Tech. Mark. Conf. 29th*, 1985, 248.
28. N. P. Vespoli and L. M. Alberino, *Polym. Proc. Eng.*, 1985, **3**, 127.
29. T. J. Hsu and L. J. Lee, *Polym. Eng. Sci.*, 1988, **28**, 955.
30. M. C. Pannone and C. W. Macosko, *J. Appl. Polym. Sci.*, 1987, **34**, 2409.
31. M. C. Pannone and C. W. Macosko, *Polym. Eng. Sci.*, 1988, **28**, 660.
32. C. W. Macosko, *Plast. Eng.*, Apr. 1983, **39**, 21.
33. H. E. Reymore, P. S. Carleton, R. A. Kolakowski and A. A. R. Sayight, *J. Cell. Plast.*, 1975, **11**, 328.
34. P. S. Carleton, D. J. Breidenbach and L. M. Alberino, *Soc. Plast. Eng.*, *Natl. Tech. Conf.*, 1979, 52.
35. K. C. Frisch, *Rubber Chem. Tech.*, 1980, **53**, 126.
36. J. E. Kresta and K. H. Hsieh, *Makromol. Chem.*, 1978, **179**, 2779.
37. H. Lee and K. Neville, 'Hand Book of Epoxy Resins', McGraw-Hill, New York, 1972.
38. R. J. Morgan, *Adv. Polym. Sci.*, 1985, **72**, 1.
39. L. Schechter and J. Wynstra, *Ind. Eng. Chem.*, 1956, **48**, 86.
40. S. A. Bidstrup and C. W. Macosko, *SPE ANTEC Tech. Pap.*, 1984, 278.
41. J. M. Barton, *Adv. Polym. Sci.*, 1985, **72**, 111.
42. A. Gupta, M. Cizmecioglu, D. Coulter, R. H. Liang, A. Yavrouian, F. D. Tsay and J. Moacanin, *J. Appl. Polym. Sci*, 1983, **28**, 1011.
43. K. Horie, H. Hiura, M. Sawada, I. Mita and H. Kambe, *J. Polym. Sci.*, Part A-1, 1970, **8**, 1357.
44. J. F. Harrod, *J. Appl. Polym. Sci.*, 1962, **6**, S63.
45. S. Sourour and M. R. Kamal, *Thermochimica Acta*, 1976, **14**, 41.
46. W. I. Lee, A. C. Loos and G. S. Springer, *J. Comp. Mat.*, 1982, **16**, 510.
47. C. M. Walkup, R. J. Morgan and T. H. Hoheisel, *ACS Polym. Prepr.*, 1984, **25** (1), 187.
48. M. J. Perry, L. J. Lee and C. W. Lee, *J. Comp. Mat.*, 1992, **26** (2), 274.
49. M. E. Ryan and A. Dutta, *Polymer*, 1979, **20**, 203.
50. J. Mijovic, J. Kim and J. Slaby, *J. Appl. Polym. Sci.*, 1984, **29**, 1449.
51. J. S. Osinksi, *Polym. Eng. Sci.*, 1983, **23**, 756.
52. C. C. Riccardi, H. E. Adabbo and R. J. J. Williams, *J. Appl. Polym. Sci.*, 1984, **29**, 2481.
53. J. M. Barton, *Polymer*, 1980, **21**, 603.
54. D. H. Kim and S. C. Kim, *Polym. Bull.*, 1987, **18**, 533.
55. M. T. Aronhime and J. K. Gillham, *J. Coatings Tech.*, 1984, **56** (718), 35.
56. G. Wisanrakkit and J. K. Gillham, *J. Coatings Tech.*, 1990, **62** (783), 35.
57. F. G. A. E. Huguenin and M. T. Klein, *IEC Prod. Res. Dev.*, 1985, **24**, 166.
58. D. F. Rohr and M. T. Klein, *Ind. Eng. Chem. Res.*, 1988, **27**, 1361.

59. W. M. Sanford and R. L. McCullough, *J. Polym. Sci., Polym. Phys. Ed.*, 1990, **28**, 973.
60. C. S. Chern and G. W. Poehlein, *Polym. Eng. Sci.*, 1987, **27**, 788.
61. F. Bueche, 'Physical Properties of Polymers', Interscience, New York, 1962.
62. F. L. Marten and A. E. Hamielec, *ACS Symp. Ser.*, 1979, **104**, 43.
63. T. G. Fox and S. Loshaek, *J. Polym. Sci.*, 1955, **15**, 371.
64. J. K. Gillham, in 'Developments in Polymer Characterization – 3' ed. J. V. Dawkins, Applied Science, London, 1982, chap. 5.
65. J. K. Gillham, *Polym. Eng. Sci.*, 1979, **19**, 670.
66. J. B. Enns and J. K. Gillham, *J. Appl. Polym. Sci.*, 1983, **28**, 2567.
67. J. K. Gillham, *Polym. Eng. Sci.*, 1986, **26**, 1429.
68. I. Havlicek and K. Dusek, in 'Crosslinked Epoxies', ed. B. Sedlacek and J. Kahovec, de Gruyter, New York, 1987, p. 417–424.
69. F. M. Lewis, C. Walling, W. Cummings, E. R. Briggs and F. R. Mayo, *J. Am. Chem. Soc.*, 1948, **70**, 1519.
70. K. Horie, I. Mita and H. Kambe, *J. Polym. Sci., Part A–1*, 1969, **7**, 2561.
71. J. F. Stevenson, *Polym. Eng. Sci.*, 1986, **26**, 746.
72. L. J. Lee, *Polym. Eng. Sci.*, 1981, **21**, 483.
73. J. D. Fan and L. J. Lee, *Polym. Composites*, 1986, **7**, 250.
74. J. D. Fan, L. J. Lee, J. Kim and Y.-T. Im, *Polym. Eng. Sci.*, 1989, **29**, 740.
75. Y. J. Huang and L. J. Lee, *AIChE J.*, 1985, **31**, 1585.
76. Y. J. Huang, J. D. Fan and L. J. Lee, *J. Appl. Polym. Sci.*, 1987, **33**, 1315.
77. Y. J. Huang and L. J. Lee, *Chem. Eng. Sci.*, 1989, **44**, 363.
78. Y. J. Huang and L. J. Lee, *J. Appl. Polym. Sci.*, 1990, **39**, 2353.
79. J. N. Cardenas and K. F. O'Driscoll, *J. Polym. Sci., Polym. Chem. Ed.*, 1977, **15**, 1883, 2097.
80. S. K. Soh and D. C. Sundberg, *J. Polym. Sci., Polym. Chem. Ed.*, 1982, **20**, 1299, 1315, 1331, 1345.
81. W. Y. Chiu, G. M. Garratt and D. S. Soong, *Macromolecules*, 1983, **16**, 348.
82. C. S. Chern and D. C. Sundberg, *Am. Chem. Soc., Div. Polym. Chem. Polym. Prepr.*, 1985, **26** (1), 296.
83. T. J. Tulig and M. Tirrell, *Macromolecules*, 1982, **15**, 459.
84. C. D. Han and D. S. Lee, *J. Appl. Polym. Sci.*, 1987, **33**, 2859.
85. G. L. Batch and C. W. Macosko, *SPE ANTEC Tech. Papers*, 1987, 33.
86. S. V. Muzumdar and L. J. Lee, *Polym. Eng. Sci.*, 1991, **31**, 1647.
87. Y. J. Huang, J. D. Fan and L. J. Lee, *Polym. Eng. Sci.*, 1990, **30**, 692.
88. H. K. Reimschuessel, in 'Ring Opening Polymerization', ed. K. C. Frisch and S. L. Reegen, Dekker, New York, 1969, chap. 7.
89. P. D. Coates and A. F. Johnson, *Plast. Rubber Proc. Appl.*, 1981, **1**, 223.
90. R. M. Hedrick and D. Gabbert, *AIChE Symp. Ser.*, 1981.
91. D. Gabbert and R. M. Hedrick, *AIChE Symp. Ser.*, 1981.
92. P. O. Sibal, R. E. Camargo and C. W. Macosko, *Polym. Proc. Eng.*, 1983, **1**, 147.
93. D. J. Lin, J. M. Ottino and E. L. Thomas, *Polym. Eng. Sci.*, 1985, **25**, 1155.
94. K. H. Lee and S. C. Kim, *Polym. Eng. Sci.*, 1988, **28**, 477.
95. T. Limtasiri, S. J. Grossman and J. C. Huang, *Polym. Eng. Sci.*, 1988, **28**, 1145.
96. A. Y. Malkin, V. G. Frolov, A. N. Ivanova and Z. S. Andrianova, *Polym. Sci., USSR*, 1979, **21**, 691.
97. L. T. Pappalardo, *J. Appl. Polym. Sci.*, 1977, **21**, 809.
98. C. Di Giulio, M. Gautier and B. Jasse, *J. Appl. Polym. Sci.*, 1984, **29**, 1771.
99. R. Fullerton, D. Roylance and R. Allred, *Polym. Eng. Sci.*, 1988, **28**, 372.
100. A. V. Tungare, Ph.D. Dissertation, Syracuse University, 1990.
101. J. V. Crivello, *J. Polym. Sci., Polym. Chem. Ed.*, 1973, **11**, 1185.
102. C. M. Tung, *Polym. Prepr., Am. Chem. Soc., Div. Polym. Chem.*, 1987, **28** (1), 7.
103. G. Allen, M. J. Bowden, D. J. Blundell, F. G. Hutchinson, G. M. Jeffs, J. Vyvoda and T. White, *Polymer*, 1973, **14**, 597, 604.
104. K. C. Frisch, D. Klempner and S. K. Mukhorijee, *J. Appl. Polym. Sci.*, 1974, **18**, 689.
105. L. H. Sperling, *Polym. Eng. Sci.*, 1985, **25**, 517.
106. R. R. Touhsaent, D. A. Thomas, and L. H. Sperling, *J. Polym. Sci., Polym. Symp.*, 1974, **46**, 175.
107. K. C. Frisch, D. Klempner, S. Migdal, H. L. Frisch and Ghiradella, 'Recent Advances in Polymer Blends, Grafts and Blocks', ed. L. H. Sperling, Plenum Press, New York, 1974.
108. K. C. Frisch, D. Klempner, S. Migdal, H. L. Frisch and Ghiradella, *Polym. Eng. Sci.*, 1974, **14**, 76.
109. K. Kircher, W. Mrotzek and G. Menges, *Polym. Eng. Sci.*, 1984, **24**, 974.
110. T. C. Wilkinson, D. Borgnaes, S. F. Chappel and W. L. Kelly, *Proc. Am. Chem. Soc. Polym. Mat. Sci. Eng. Div.*, 1983, **49**, 469.
111. W. L. Kelly, *Plast. Eng.*, Feb. 1986, 9.
112. H. R. Edwards, *SPE ANTEC Tech. Pap.*, 1986, 1326.
113. K. C. Frisch, D. Klempner and H. L. Frisch, *SAE Int. Congr. Expo.*, Feb. 1982, paper 820 422.
114. R. Pernice, K. C. Frisch and R. Navane, *J. Cell. Plast.*, Mar./Apr. 1982, **18**, 121.
115. T. J. Hsu and L. J. Lee, *J. Appl. Polym. Sci.*, 1988, **36**, 1157.
116. Y. C. Chou and L. J. Lee, *Proc. Am. Chem. Soc., Polym. Mat. Sci. Eng. Div.*, 1991, in press.
117. J. D. Ferry, M. L. Williams and D. M. Stern, *J. Chem. Phys.*, 1954, **58**, 987.
118. V. R. Allen and T. G. Fox, *J. Chem. Phys.*, 1964, **41**, 337.
119. W. W. Graessley, R. L. Hazleton and L. R. Lindeman, *Trans. Soc. Rheol.*, 1967, **11**, 267.
120. W. W. Graessley, T. Masuda, J. E. L. Roovers and N. Hadjichristidis, *Macromolecules*, 1976, **9** (1), 127.
121. D. Gupta and W. C. Forsman, *Macromolecules*, 1969, **2**, 304.
122. F. Bueche, *J. Chem. Phys.*, 1956, **25**, 599.
123. F. Bueche, *J. Chem. Phys.*, 1964, **40**, 484.
124. R. Simha and J. L. Zakin, *J. Colloid Sci.*, 1962, **17**, 270.
125. M. Fixman and J. M. Peterson, *J. Am. Chem. Soc.*, 1964, **86**, 3524.

126. W. W. Graessley, *Acc. Chem. Res.*, 1977, **10**, 332.
127. D. S. Pearson and E. Helfand, *Macromolecules*, 1984, **17** (4), 888.
128. J. M. Carella, J. T. Gotro and W. W. Graessley, *Macromolecules*, 1986, **19** (3), 659.
129. S. D. Lipshitz and C. W. Macosko, *Polym. Eng. Sci.*, 1976, **16**, 803.
130. E. M. Valles and C. W. Macosko, *Macromolecules*, 1979, **12** (3), 521.
131. A. Hale, M. Garcia, C. W. Macosko and L. T. Manzione, UMSI Report 89/69, University of Minnesota, Apr. 1989.
132. M. B. Roller, *Polym. Eng. Sci.*, 1986, **26**, 432.
133. F. G. Mussatti and C. W. Macosko, *Polym. Eng. Sci.*, 1973, **13**, 236.
134. M. B. Roller, *Polym. Eng. Sci.*, 1975, **15**, 406.
135. A. V. Tungare, M.S. Thesis, Syracuse University, 1986.
136. E. B. Richter and C. W. Macosko, *Polym. Eng. Sci.*, 1980, **20**, 921.
137. V. M. Gonzalez-Romero and C. W. Macosko, *J. Rheol.* (N.Y.), 1985, **29** (3), 259.
138. Y. A. Tajima and D. Crozier, *Polym. Eng. Sci.*, 1983, **23**, 186.
139. Y. A. Tajima and D. Crozier, *Polym. Eng. Sci.*, 1986, **26**, 427.
140. Y. A. Tajima and D. Crozier, *Polym. Eng. Sci.*, 1988, **28**, 491.
141. D. S. Lee and C. D. Han, *Polym. Eng. Sci.*, 1987, **27**, 955.
142. D. S. Lee and C. D. Han, *J. Appl. Polym. Sci.*, 1987, **34**, 1235.
143. N. F. Sheppard, Ph.D. Thesis, Massachusetts Institute of Technology, 1986.
144. C. W. Macosko, *Br. Polym. J.*, 1985, **17** (2), 239.
145. Y. M. Lee and L. J. Lee, *Polymer*, 1987, **28**, 2302.
146. A. Hale and C. W. Macosko, *Proc. Polym. Mat. Sci. Eng.*, 1988, **59**, 1196.
147. S. A. Bidstrup, Ph.D. Thesis, University of Minnesota, 1985.
148. S. A. Bidstrup and C. W. Macosko, 'Crosslinked Epoxies', ed. B. Sedlacek and J. Kahovec, de Gruyter, Berlin, 1987, p. 253.
149. D. Turnbull and M. H. Cohen, *J. Chem. Phys.*, 1969, **42**, 245.
150. A. K. Doolittle, *J. Appl. Phys.*, 1951, **22**, 1471.
151. P. B. Macedo and T. A. Litovitz, *J. Chem. Phys.*, 1965, **42**, 245.
152. F. Bueche, *J. Chem. Phys.*, 1952, **20**, 1959.
153. P. J. Flory, 'Principles of Polymer Chemistry', Cornell University Press, Ithaca, NY, 1953.
154. W. H. Stockmayer, *J. Chem. Phys.*, 1943, **11**, 45.
155. W. H. Stockmayer, *J. Chem. Phys.*, 1944, **12**, 125.
156. D. R. Miller and C. W. Macosko, *Macromolecules*, 1976, **9**, 206.
157. D. R. Miller and C. W. Macosko, *Macromolecules*, 1978, **11**, 656.
158. D. R. Miller and C. W. Macosko, *Macromolecules*, 1980, **13**, 1063.
159. K. J. Wang, Y. J. Huang and L. J. Lee, *Polym. Eng. Sci.*, 1990, **30**, 654.
160. T. G. Fox and P. J. Flory, *J. Polym. Sci.*, 1954, **14**, 315.
161. A. T. DiBenedetto, *J. Polym. Sci., Polym. Phys.*, 1987, **25**, 1949.
162. C. P. Hsu and L. J. Lee, *Polymer*, 1991, **32**, 2263.
163. Y. S. Yang and L. J. Lee, *Polymer*, 1988, **29**, 1793.
164. K. Dusek, 'Developments in Polymerization-3', ed. R. N. Haward, Applied Science, London, 1982, chap. 4.
165. K. Dusek, *Br. Polym. J.*, 1985, **17**, 185.
166. H. M. J. Boots and R. B. Pandey, *Polym. Bull.*, 1984, **11**, 415.
167. H. M. J. Boots, J. G. Kloosterboer, G. M. M. van de Hei and R. B. Pandey, *Br. Polym. J.*, 1985, **17**, 219.
168. J. M. Castro, C. W. Macosko and S. J. Perry, *Polymer Commun*, 1984, **25**, 82.
169. J. M. Castro and C. W. Macosko, *Polymer Commun*, 1985, **26**, 158.
170. K. J. Wang, T. J. Hsu and L. J. Lee, *J. Appl. Polym. Sci.*, 1990, **41**, 1055.
171. J. W. Blake, R. D. Anderson, W. P. Yang and C. W. Macosko, *J. Rheol.* (N.Y.), 1987, **31**, 1236.
172. W. W. Graessley, *Adv. Polym. Sci.*, 1982, **47**, 67.
173. E. M. Valles and C. W. Macosko, *Macromolecules*, 1979, **12**, 673.
174. M. Gottlieb, C. W. Macosko and T. C. Lepsch, *J. Polym. Sci., Polym. Phys. Ed.*, 1981, **19**, 1603.
175. M. Gottlieb, C. W. Macosko, G. S. Benjamin, K. O. Meyers and E. W. Merrill, *Macromolecules*, 1981, **14**, 1039.
176. M. Gottlieb and C. W. Macosko, *Macromolecules*, 1982, **15**, 535.
177. K. H. Lee and S. C. Kim, *Polym. Eng. Sci.*, 1988, **28**, 13.
178. L. Mandelkern, 'Crystallization of Polymers', McGraw-Hill, New York, 1964.
179. B. Wunderlich, 'Macromolecular Physics', Academic Press, New York, 1976, vol. 2.
180. A. J. Ryan, *Polymer*, 1990, **31**, 707.
181. R. J. J Williams, J. Borrajo, H. E. Adabbo and A. J. Rojas, *ACS Adv. Chem. Ser.*, 1984, **208**.
182. A. Vazquez, A. J. Rojas, H. E. Adabbo, J. Borrajo and R. J. J. Williams, *Polymer*, 1987, **28**, 1156.
183. D. Verchere, H. Sautereau, J. P. Pascault, S. M. Moschiar, C. C. Riccardi and R. J. J. Williams, *Polymer*, 1989, **30**, 107.
184. M. Pollard and J. L. Kardos, *Polym. Eng. Sci.*, 1987, **27**, 830.
185. G. A. Hunter, *Proc. S.P.I. Tech. Conf.*, Composites Institute, 1988, session 6–C.
186. J. Kulawik, Z. Szeglowski, T. Czapla and J. P. Kulawik, *Colloid Polym. Sci.*, 1989, **267**, 970.
187. E. J. Bartkus and C. H. Kroekel, *Appl. Polym. Symp.*, 1970, **15**, 113.
188. T. Mitani, H. Shiraishi, K. Honda and G. E. Owen, *Proc. S.P.I. Tech. Conf.*, Composites Institute, 1989, session 12–F.
189. J. M. Castro and E. J. Straus, *Polym. Eng. Sci.*, 1989, **29**, 308.
190. H. T. Kau, *Polym. Eng. Sci.*, 1989, **29**, 1286.
191. M. Kinkelaar, C. P. Hsu and L. J. Lee, *SAMPE Quartly*, 1990, **21** (4), 40.
192. M. Kinkelaar and L. J. Lee, *J. Appl. Polym. Sci.*, in press.
193. L. C. Rubens and R. E. Skochdopole, *J. Appl. Polym. Sci.*, 1965, **9**, 1487.
194. J. Niezette and V. Desreux, *J. Appl. Polym. Sci.*, 1971, **15**, 1981.
195. C. F. Liu and C. D. Armeniades, *Annu. Tech. Conf.–Soc. Plast. Eng. 47th*, 1989, 834.
196. F. Tollens, M. S. Thesis, The Ohio State University, 1991.
197. R. R. Hill, Ph.D. Dissertation, The Ohio State University, 1992.

198. J. L. Kardos, M. P. Dudukovic, E. L. McKague and M. W. Lehman, *ASTM Spec. Tech. Publ.*, 1983, **797**, 96.
199. J. L. Kardos, M. P. Dudukovic and R. Dave, *Adv. Polym. Sci.*, 1986, **80**, 101.
200. J. L. Kardos, R. Dave and M. P. Dudukovic, in 'Proceedings of Manufacturing Int. '88: The Manufacturing Science of Composites', ed. T. G. Gutowski, ASME, New York, 1988, vol. 6, p. 41.
201. H. Yokono, S. Tsuzuku, Y. Hira, M. Gotoh and Y. Miyano, *Polym. Eng. Sci.*, 1985, **25**, 959.
202. J. H. Marciano, M. M. Reboredo, A. J. Rojas and R. J. J. Williams, *Polym. Eng. Sci.*, 1986, **26**, 717.
203. S. C. Tighe and L. T. Manzione, *Polym. Eng. Sci.*, 1988, **28**, 949.
204. S. Sourour, Ph.D. Thesis, McGill University, Montreal, 1978.
205. D. H. Kim and S. C. Kim, *Polym. Eng. Sci.*, 1989, **29**, 456.
206. C. L. Tucker (ed.), 'Fundamentals of Computer Modeling for Polymer Processing', Hanser, New York, 1989.
207. J. R. A. Pearson, 'Mechanical Principles of Polymer Melt Processing', Pergamon Press, New York, 1966.
208. H. A. Lord and G. Williams, *Polym. Eng. Sci.*, 1975, **15**, 569.
209. E. Broyer, C. Cutfinger and Z. Tadmor, *Trans. Soc. Rheol.*, 1975, **19**, 423.
210. Y. Kuo and M. R. Kamal, *Am. Inst. Chem. Eng. J.*, 1976, **22**, 661.
211. C. A. Hieber and S. F. Shen, *J. Non-Newtonian Fluid Mech.*, 1980, **7**, 1.
212. S. F. Shen, *Int. J. Numer. Methods Fluids*, 1984, **4**, 171.
213. V. W. Wang, C. A. Hieber and K. K. Wang, *J. Polym. Eng.*, 1986, **7**, 21.
214. D. L. Trafford, M. S. Thesis, University of Delaware, 1987.
215. AC Technology, Newsletter, 1990.
216. S. F. Shen, *Proc. NUMIFORM '89*, 1989.
217. W. Rose, *Nature*, 1961, **191**, 242.
218. D. J. Coyle, J. W. Blake and C. W. Macosko, *AIChE J.*, 1987, **33**, 1168.
219. S. Bhattacharji and P. Savic, *Proc. Heat Transfer Fluid Mech. Inst.*, 1965, 248.
220. M. A. Garcia, C. W. Macosko, S. Subbiah and S. I. Guceri, *Int. Polym. Proc.*, 1991, **6** (1), 73.
221. V. W. Wang, C. A. Hieber and K. K. Wang, in 'Applications of Computer Aided Engineering in Injection Molding', ed. L. T. Manzione, Hanser, New York, 1987.
222. L. T. Manzione, *Polym. Eng. Sci.*, 1981, **21**, 1234.
223. J. D. Domine and C. Gogos, *Polym. Eng. Sci.*, 1980, **20**, 847.
224. I. Manas-Zloczower, J. W. Blake and C. W. Macosko, *Polym. Eng. Sci.*, 1987, **27**, 1229.
225. F. Dupret and L. Vanderschuren, *AIChE J.*, 1988, **34**, 1959.
226. S. F. Shen and U. F. Gonzalez, paper presented at the 60th Annual Meeting of the Society of Rheology, Gainesville, FL, Feb. 1989.
227. M. S. Kamal and M. E. Ryan, *Polym. Eng. Sci.*, 1980, **20**, 859.
228. J. M. Castro, 'Development in Plastics Technology', ed., A. Whelan and J. L. Craft, Applied Science, London, 1985, vol. 2, chap. 2.
229. S. P. Estevez and J. M. Castro, *Polym. Eng. Sci.*, 1984, **24**, 428.
230. C. D. Lack and C. A. Silebi, *Polym. Eng. Sci.*, 1988, **28**, 434.
231. H. Hilger and W. Michaeli, paper presented in 32nd Annual Polyurethane Technical/Marketing Conference, Oct. 1989, 275.
232. P. Häring, Dimplom-Ingenieur at the University of Stuttgart, Germany, 1990.
233. T. A. Osswald and Y. S. Jim, paper presented at the 7th Annual Meeting of the Polymer Processing Society, Hamilton, Canada, Apr. 1991.
234. E. Broyer, C. W. Macosko, F. E. Critchfield and L. F. Lawler, *Polym. Eng. Sci.*, 1978, **18**, 382.
235. J. S. Osinski, L. T. Manzione and C. Chan, *Polym. Proc. Eng.*, 1985, **3**, 97.
236. N. P. Vespoli and L. M. Alberino, *J. Elastomer Plast.*, 1985, **17**, 173.
237. V. M. Gonzalez-Romero, J. M. Castro and C. W. Macosko, in *Proc. World Congr. Chem. Eng.*, 2nd, 1981, **VI**, 519.
238. L. T. Manzione and J. S. Osinski, *Polym. Proc. Eng.*, 1983, **1**, 171.
239. C. W. Macosko and L. J. Lee, *Rubber Chem. Tech.*, 1985, **58**, 436.
240. J. H. Kim and S. C. Kim, *Polym. Eng. Sci.*, 1987, **27**, 1243.
241. W. Michaeli, V. Hammes, L. Kirberg, R. Kotte, T. A. Osswald and O. Specker, 'Process Simulation in the RTM Technique', Hanser, Munich, 1989.
242. M. V. Bruscheke and S. G. Advani, *Annu. Tech. Conf.-Soc. Plast. Eng. 47th*, 1989, 1769.
243. J. P. Coulter and S. I. Guceri, *J. Reinf. Plast. Compos.*, 1988, **7**, 200.
244. M. K. Um and W. I. Lee, *Proc. Int. SAMPE Symp., 35th*, 1990, 1905.
245. W. B. Young, K. Rupel, K. Han, L. J. Lee and M. J. Liou, *Polym. Compos.*, 1991, **12**, 30.
246. W. B. Young, K. Han, L. H. Fang, L. J. Lee and M. J. Liou, *Polym. Compos.*, 1991, **12**, 391.
247. V. M. Gonzalez, Ph.D. Dissertation, University of Minnesota, 1983.
248. R. J. Lin, L. J. Lee and M. J. Liou, *J. Int. Polym. Proc.*, 1991, **4**, 356.
249. R. G. Voigt, D. Gottlieb and M. Y. Hussaini, 'Spectral Methods for Partial Differential Equations', SIAM, Philadelphia, PA, 1984.
250. L. T. Manzione, G. W. Poelzing and R. C. Progelhof, *Polym. Eng. Sci.*, 1988, **28**, 1056.
251. L. T. Manzione, J. S. Osinski, G. W. Poelzing, D. L. Crouthamel and W. G. Thierfelder, *Polym. Eng. Sci.*, 1989, **29**, 749.
252. E. G. Melby and J. M. Castro, in 'Comprehensive Polymer Science', ed. S. L. Aggarwal, Pergamon Press, 1989, vol. 7, chap. 3, p. 51.
253. C. C. Lee and C. L. Tucker, *J. Non-Newtonian Fluid Mech.*, 1987, **24**, 245.
254. M. R. Barone and D. A. Caulk, in 'Proceedings of Manufacturing Int. '88: The Manufacturing Science of Composites', ed. T. Gutowski, ASME, New York, 1988, vol. 6, p. 63.
255. O. Walter, *Kunstoffe*, 1969, **59**, 827.
256. L. F. Marker and B. Ford, *Mod. Plast.*, 1977, **54**, 64.
257. L. J. Lee, L. F. Marker and R. M. Griffith, *Polym. Compos.*, 1981, **2**, 209.
258. D. L. Denton, *Proc. Annu. Conf.-Reinf. Plast./Compos. Inst., Soc. Plast. Ind., 36th*, 1981, 16-A.
259. M. R. Barone and D. A. Caulk, *Polym. Compos.*, 1985, **6**, 105.
260. H. T. Kau, *Annu. Tech. Conf.-Soc. Plast. Eng. 45th*, 1987, 27.

261. R. J. Silva-Neito, B. C. Fisher and A. W. Birley, *Polym. Compos.*, 1980, **1**, 14.
262. G. Menges and H. Derek, *Proc. Annu. Conf.–Reinf. Plast./Compos. Inst., Soc. Plast. Ind.*, 36th, 1981, 23-C.
263. C. L. Tucker and F. Folgar, *Polym. Eng. Sci.*, 1983, **23**, 69.
264. C. C. Lee, F. Folgar and C. L. Tucker, *ASME Trans. J. Eng. Ind*, 1984, **106**, 114.
265. T. A. Osswald and C. L. Tucker, *J. Int. Polym. Proc.*, 1990, **5**, 79.
266. C. C. Lee, Ph.D. Thesis, University of Illinois, 1984.
267. S. J. Lee, M. M. Denn, M. J. Crochet and A. B. Metzner, *J. Non-Newtonian Fluid Mech.*, 1982, **10**, 3.
268. C. C. Lee and C. L. Tucker, *Annu. Tech. Conf.–Soc. Plast. Eng. 41st, 1987*, 1983, 740.
269. M. R. Barone and D. A. Caulk, *ASME Trans. J. Appl. Mech.*, 1986, **53**, 361.
270. S. M. Davis and C. L. Tucker, *Annu. Tech. Conf.–Soc. Plast. Eng. 46th, 1988*, 1988, 524.
271. J. M. Castro and R. M. Griffith, *Proc. Annu. Conf.–Reinf. Plast./Compos. Inst., Soc. Plast. Ind.*, 43rd, 1988, sect. 17-E.
272. M. R. Barone and T. A. Osswald, *Polym. Compos.*, 1988, **9**, 158.
273. J. D. Fan, J. Kim, Y.-T. Im and L. J. Lee, *J. Int. Polym. Proc.*, 1991, **6**, 61.
274. T. Hirai, T. Katayama and H. Hamada, *J. JSTP*, 1984, **25**, 1113.
275. T. Hirai and T. Katayama, *Proc. Int. Conf. Compos. Mater.*, 1978, 1283.
276. J. Kim, Y. C. Shiau, L. J. Lee and Y.-T. Im, *Polym. Compos.*, in press.
277. J. Kim, J. Y. Xu, L. J. Lee and Y.-T. Im, *Polym. Compos.*, in press.
278. J. F. Stevenson, *Polym. Proc. Eng.*, 1983, **1**, 203.
279. M. R. Barone and D. A. Caulk, *Polym. Eng. Sci.*, 1981, **21**, 2239.
280. J. M. Castro and C. C. Lee, *Polym. Eng. Sci.*, 1987, **27**, 218.
281. A. C. Loos and G. S. Springer, *J. Compos. Mater.*, 1983, **17**, 135.
282. T. G. Gutowski and Z. Cai, in 'Proceedings of Manufacturing Int. '88: The Manufacturing Science of Composites', ed. T. Gutowski, ASME, New York, 1988, vol. 6, p. 13.
283. R. Dave, *J. Compos. Mater.*, 1990, **24**, 22.
284. E. P. Calius and G. S. Springer, in 'Proceedings of Manufacturing Int. '88: The Manufacturing Science of Composites', ed. T. Gutowski, ASME, New York, 1988, vol. 6, p. 49.
285. C. D. Han, D. S. Lee and H. B. Chin, *Polym. Eng. Sci.*, 1986, **26**, 393.
286. C. D. Han and H. B. Chin, *Polym. Eng. Sci.*, 1988, **28**, 321.
287. G. L. Batch and C. W. Macosko, in 'Proceedings of Manufacturing Int. '88: The Manufacturing Science of Composites', ed. T. Gutowski, ASME, New York, 1988, vol. 6, p. 57.
288. T. G. Gutowski, Z. Cai, S. Bauer, D. Boucher, J. Kingery and S. Wineman, *J. Compos. Mater.*, 1987, **21**, 650.
289. R. Dave, J. L. Kardos and M. P. Dudukovic, *Polym. Compos.*, 1987, **8**, 29.
290. R. Dave, J. L. Kardos and M. P. Dudukovic, *Polym. Compos.*, 1987, **8**, 123.
291. R. Dave, A. Mallow, J. L. Kardos and M. P. Dudukovic, *SAMPE J.*, 1990, **26** (3), 31.
292. G. S. Springer, in 'Proceedings of Manufacturing Int. '88: The Manufacturing Science of Composites', ed. T. Gutowski, ASME, New York, 1988, vol. 6, p. 1.

21

Reactive Processing of Thermoplastic Polymers

MORAND LAMBLA
Université Louis Pasteur, Strasbourg, France

21.1	INTRODUCTION	619
	21.1.1 Processing Equipment Characteristics	620
	21.1.2 Reaction Types	621
21.2	FREE RADICAL REACTIVITY	622
	21.2.1 Polymerization in Bulk	622
	21.2.2 Chain Degradation Continuous Processes	623
	21.2.3 Grafting Reactions	624
	21.2.3.1 Vinylsilane grafting onto polyalkenes	624
	21.2.3.2 Carboxylic monomer grafting reactions	626
	21.2.3.3 Other monomers and reactive species	627
	21.2.3.4 Recent developments in free radical grafting	628
	21.2.4 Crosslinking Reactions	628
21.3	OTHER TYPES OF BULK POLYMERIZATION	629
	21.3.1 Addition Polymerization	629
	21.3.2 Polycondensation Reactions	630
21.4	FUNCTIONAL MODIFICATION OF POLYMERS	630
	21.4.1 Halogenation of Butyl Rubbers and Other Polymers	630
	21.4.2 Chemical Modification of Poly(vinyl chloride)	631
	21.4.3 Condensation Reactions	632
	21.4.4 Exchange Reactions	632
	21.4.4.1 Alcoholysis of EVA copolymers	633
	21.4.4.2 Alcoholysis of copolyacrylates	633
	21.4.4.3 Aminolysis of copolyacrylates	634
21.5	REACTIVE BLENDING OF POLYMERS	634
	21.5.1 Grafting by Interpolymeric Reactivity	635
	21.5.2 Functional End Group Reactivity	636
	21.5.3 Interpolymeric Ion Bonding	636
21.6	CONCLUSION	638
21.7	LIST OF SYMBOLS	638
21.8	REFERENCES	639

21.1 INTRODUCTION

Scientific and industrial interest in the chemical modification of polymers has obviously increased since the middle of the last century. Some important discoveries include the vulcanization of natural rubbers (by Goodyear in 1839) and cellulose nitration (by Braconnot in 1833). This latter modified natural polymer was used as a substitute for ivory, by Hyatt in 1870, after plasticizing by camphor and lead to the first easily processable material: celluloid. Besides these examples based on chemically modified natural macromolecules, it is worthwhile noticing how many possibilities exist for chemically modifying synthetic polymers, as indicated in the exhaustive review published by Fettes.[1] Generally, the use of solvents or dispersed media facilitates the control and adjustment of reactivity between polymers and other components of the system. The rate of conversion as well as the final structure of the modified polymer depend to a large extent on the reaction scheme and on the process parameters, including the nature of the continuous phase (solvent or dispersing agent) and such processing parameters as temperature and pressure. It is important to point out that the low concentration of polymer (around 10%) and the related separation and purification processes, which have a great influence on the final costs of modified polymers, are among the main disadvantages of reactions conducted in solvent media. Another approach is to perform these modifications of thermoplastic polymers, leading mainly to grafting and crosslinking reactions, by reactive processing in a discontinuous or continuous mixing equipment, which will be the chemical reactor, involving problems of high temperature, viscosity, risks of corrosion, *etc.* The main medium is the molten polymer, with an associated polarity related to its chemical composition, and the corresponding reaction and processing parameters are very different from those in solution. Adjustment of the reactivity requires specific basic research on the kinetic behaviour under these conditions, even if the main reaction is well known in classical organic or polymer chemistry.[2]

The chapter title, 'Reactive Processing of Thermoplastic Polymers', summarizes the many possibilities of producing new materials by chemical reactions carried out in the bulk, including thermoset and elastomeric products, reaction injection moulding (RIM), and reactive extrusion. This presentation will deal mainly with this last speciality, which has not been adequately represented in the literature but is of greatest industrial interest, as shown by the hundreds of patents published since 1980. Reactive extrusion is now being viewed as an efficient means of continuously polymerizing monomers as well as chemically modifying existing polymers, combining two traditionally separate operations: the chemical reactions for the formation or the modification of macromolecules and the processing of the polymer for the purpose of structuring into shaped plastic products.[3] Furthermore, new developments of polymer blends and alloys of thermoplastic species have been sharply increasing, and for technical and economic reasons, reactive processing may provide viable mechanisms for the creation and preservation of desired blends, with controlled structure and morphology. In addition, the resulting microstructure, which results in the right balance of mechanical properties, has to be maintained throughout the subsequent stages of manufacturing.

Before giving more details on the chemical systems, it is important to say a few words on the continuous reactors (mixers, co-kneaders or extruders) which are mainly used.

21.1.1 Processing Equipment Characteristics

Discontinuous equipment is widely used for formulation processes, for instance, in elastomer processing or poly(vinyl chloride) stabilization and plasticizing, and it seems useful to consider the mixing chamber as a potential reactor for highly viscous systems. Adjustable residence time, power consumption control, mixing efficiency related to blade geometry and rotation speed, are basic advantages of the equipment which exists on a laboratory and industrial scale. However, a few inconveniences should be mentioned: temperature and pressure control of the viscous mixture, continuous feeding of liquid or volatile products, material extraction, *etc.*

Continuous equipment, like single or twin screw extruders has to be adapted to simultaneously process reactions in the melt, using sealed vent chambers, liquid injection ports, various kinds of pressure and temperature sensors, secondary solid or melt-fed ports, and must deal with the continuously changing nature of the reactive melt. Single screw extruders are best suited to the simpler jobs like melting, plasticizing and discharging melt for the continuous production of pipes, films, profiles, *etc.* Although both single and twin screw extruder configurations are used in reactive extrusion processes, twin screw extruders are increasingly being favoured over the single screw counterparts, as they can tackle the more complex tasks involved in reactive systems. Frund and Brown have listed the following process advantages for twin screw extruders:[4–6] (i) increased surface/volume ratio by the continuous creation of new, thin surface layers which enhance mixing, reaction, and heat-transfer; (ii) improved thermal control due to surface renewal and short residence time; (iii) solventless polymerization and other chemical modifications of polymers are possible, since the extruder can handle the high viscosity of polymers, leading to a high concentration in reactive species,[7] and to savings in raw materials and solvent recovery systems; (iv) sequential reaction initiation by multiple injection ports, which is of greatest interest with highly reactive species like free radical initiators. This is related to the possibility of segregation of the different barrel sections, through dynamic sealing devices, which prevent back-mixing of a product with its reactants; (v) adjustment of the screw design, especially in modular equipment, which allows different residence times (between 1 and 15 min maximum) and degrees of mixing; (vi) reactions can be conducted using a large range of pressures (0 to 500 atm) and temperatures (70 to 500 °C); (vii) continuous production of ready-to-use materials as a result of bulk processes.

On the other hand, there are a few inconveniences in using an extruder as a chemical reactor. Most extruders are custom made, as they have to be designed specifically for a defined chemistry, taking into account the resistance to corrosive reagents, and the screws' optimized design is insuffficiently understood. Engineering considerations, based on the type of extruder (single or twin screw, self-wiping or not, length over diameter ratio (L/D), number of reactive zones, *etc.*) and process parameters also have to be taken into account. Limited residence time, characteristic of these continuous devices, is one of the main disadvantages regarding kinetic considerations. There is a need for 'fast chemistry' in apolar systems, where beneficial solvent effects are difficult to achieve.

The applications of twin screw extruders are manifold, but in the area of reactive processing most of the equipment is modular, taking into account the required flexibility for the reactive continuous process. A comprehensive history of the development of classical extruders was given recently by White[8] and the main applications in reactive processes were summarized by Herrmann.[9] However, it is still difficult to have a complete overview of this domain, as most of the research is published in the patent literature. But it seems obvious that the next steps for the improvement of reactive extrusion will be related to close cooperation among chemists and chemical and mechanical engineers, in order to develop machinery for specific and various chemistries, carried out in molten polymer mixtures and containing products with widely differing viscosities. This last point is critical in continuous polymerization (introduction of low viscous monomer) and somewhat in chemical modification, when we feed a low molecular weight species to molten polymers.

21.1.2 Reaction Types

The hydrocarbon nature of most polymers leads to relatively poor chemical reactivity and dictates the low polarity of the medium, which is not favourable for the chemical activation of the system, for instance, polyalkenes like polyethylene and polypropylene, are relatively inert *versus* most of the chemical reactants, like acids and bases, but sensitive to oxidative and/or photochemical degradation. It appears that the degradation proceeds through free radical mechanisms and is not surprising that the same chemistry has been used in various chemical modifications of polyalkenes for grafting, crosslinking and controlled chain degradation.

The alkenic copolymers containing polar comonomers, like acrylic or acetic esters, show a complementary reactivity, which is related to the breakdown of the ester bond, through hydrolytic or exchange reactions. The corresponding homopolymers, based on the same or other functionalized vinylic monomers, and their copolymeric derivatives with different backbones (styrenic or vinylic), can also be modified by these reactions.

Substitution reactions are possible as well and some of them, like halogenation, conducted in solution or in gas phase processes, improve the great resistance of, for instance, poly(vinyl chloride), or increase the adhesion performance of elastomers like polyisobutylene.

This signifies that most of the commercial thermoplastic polymers prepared by addition polymerization can be used as raw materials for further chemical modifications, which have to be carried out in the melt, enabling flexible and economic processes and a scale-down of reactor costs.

There is another family of thermoplastic polymers which could be of interest in terms of chemical reactivity: the polycondensates. Polyamides and polyesters, for example, are widely used as engineering materials and they show good mechanical properties at higher temperatures, in comparison to commodity plastics. Due to the presence of many ester or amide bonds along the polymer chain, the possibility of transreaction has been studied in many laboratories, leading to new routes for preparation of novel copolymers with various degrees of randomness and composition.[10-14] But in the case of polycondensates, it is also important to take into account the possibility of modifying the chain ends by direct functionalization. The newly created functionality is reactive, allowing further coupling reactions with other reactive polymeric species, which have a terminal functionality on the chain. Similar condensation reactions can also be made on copolymers containing, for instance, carboxylic pendant side groups. The reaction between cyclic anhydride and primary amines is used industrially for the preparation of high impact polyamide blends, but it was shown that the reactivity of the same anhydride with hydroxylic species, low molecular weight alcohols, diols, or oligmers, is a reversible one.[15]

The aim of the present review is to describe the most important systems and to emphasize the factors influencing the chemical reactivity in melt processes. Work performed by various research groups has shown that the following types of reactions can be carried out in extruders: (i) anionic,

Table 1 Examples of Reactive Extrusion Processes

Reaction	Reactive species	Final product
Polycondensation	Polyamide precondensate	Polyamide (PAm-6,6)
Polycondensation	Polyester precondensate	Polyester (PET + PBT)
Polyaddition	Polyols + polyisocyanates	Polyurethane elastomer
Cationic polymerization	Trioxane + —	Polyoxymethylene
Anionic polymerization	Styrene (+ butadiene)	Polystyrene or SBS
Anionic polymerization	Caprolactam	PAm-6
Free radical polymerization	Methyl methacrylate	Polymethacrylate
Free radical grafting	Maleic anhydride + P	Grafted copolymers
Hydrolysis	Polyurethane scraps	Polyol, amine
Hydrolysis	Sawdust, straw	Glucose
Alcoholysis	EVA + alcohol	EVAL
Alcoholysis	EAR + alcohol	New copolyester
Aminolysis	EAR + amine	New copolyamide
Imidification	EAM + NH_3	Imidified copolymer

cationic, free radical polymerizations, coordination or metathesis growing reactions, polycondensations, all leading to high molecular weight polymers, possibly crosslinked; (ii) controlled degradation of polymers (usually polyalkenes) by means of free radical initiating species for the purpose of producing polymeric materials with adjusted molecular weight distribution; (iii) functionalization of commodity polymers by free radical grafting; (iv) chemical modification of polymeric backbones by substitution or exchange reactions; (v) interpolymeric reactivity by grafting or coupling reactions; and (vi) intra- and inter-polymeric ion bonding.

Typical reactive extrusion applications are summarized in Table 1. But in order to simplify the detailed presentation of the main chemical systems, these different areas were assembled, taking into account the main chemical mechanisms. The presentation of industrial achievements and laboratory results is not exhaustive, but the chosen examples illustrate the main trends and important key issues for future developments.

21.2 FREE RADICAL REACTIVITY

Preliminary studies conducted prior to those described in this chapter were based on polymer modifications induced by mechanochemistry, which leads to direct conversion of mechanical energy, on a molecular scale, to homolytic chain breakdown, resulting in free radical creation.[16–18] But it appeared rapidly that mechanochemistry alone is not powerful enough for efficient free radical initiation of grafting reactions.[19] Therefore, most of the free radical processes for polymerization, chain degradation, grafting and crosslinking reactions are based on the addition of free radical generators, decomposed by thermal activation.

21.2.1 Polymerization in Bulk

High molecular weight polymers can be synthesized from monomer(s) and/or low molecular weight prepolymers. As the residence time of the material in the extruder is very short, the bulk polymerization temperature must be relatively high in order to provide a significant reactivity. A maximum rate of reaction is favoured by the use of free radical initiators such as peroxides, especially by an association of peroxides with markedly different rates of decomposition. The free radical polymerization is highly exothermic and needs to be thermally well controlled within the extruder. Heat transfer is possible through the barrel wall, through the screw itself *via* the circulation of a fluid, or by the mixture of an inert material which will consume a considerable amount of energy by volatilization.

Economic applications of free radical initiated bulk polymerization in extruders are actually restricted to acrylates, homo- and co-polymers and, for some applications, to styrene.

Acrylates are polymerized in extruders[20–33] for the preparation of adhesives[20,21,29] or water-soluble copolymers through amine,[33] amide,[27,32] hydroxyalkyl[27] or acid[28] functionalities. Thiols are added to control the molecular weight of the polymers.[20,23,25] Part of the monomer(s) may be injected downstream in the reacting mixture.[20] The continuous polymerization of methyl methacrylate was studied, for instance, by Stuber and Tirrell in a Leistritz twin screw extruder (34 mm), a fully intermeshing counterrotating piece of equipment. The relationship between processing conditions and molecular weight distribution was examined and compared to the predictions based on a series of continuous stirred tank reactor (CSTR) models.[26]

Copolymers involving acrylates and containing acrylonitrile are also polymerizable in extruders following the procedures described in the previous paragraph.[20,34,35] ABS needs a temperature of 200 °C for a reactive processing polymerization.[35,36] An ABS functionalized with maleic anhydride has been produced in an extruder.[31]

Since styrene has a relatively low reactivity, higher temperatures of polymerization or longer residence times are recommended. Three twin screw extruders have been interconnected for the synthesis of styrene/acrylate copolymers to provide a much longer time of reaction.[37] Styrene is used in copolymers with acrylates or acrylonitrile.[37-39] Addition of butanone results in the formation of a charge-transfer complex with maleic anhydride, increasing drastically the reactivity of this latter monomer for the synthesis of styrene/maleic anhydride (molar ratio 8/1) copolymer in extruder.[40]

21.2.2 Chain Degradation Continuous Processes

The equilibrium between polymerization and depolymerization is governed by thermodynamics, which signifies that if a polymer can be obtained by reactive processing in an extruder, reducing the molecular weight of a polymer is also possible in the same device. This will be related to the ceiling temperature of the polymer. On the other hand, degradation initiated by free radicals in the melt could lead to specific products, with improved processability for instance.

Staudinger showed in 1929 that cool mastication of natural rubber leads to chain degradation of these high molecular weight polyisoprene molecules.[16] The polystyrene degradation in the extruder gives a decrease of the molecular weight, but an increase of the molecular weight distribution is simultaneously obtained.[41]

Moreover, studies and applications of controlled chain degradation by free radical reactive extrusion are restricted to polyalkenes, especially polypropylene. Macromolecular cleavage may be obtained using mechanochemistry involving a synergy between heat, shear stress and oxygen. However, for higher reactivity and continuous production of modified products, peroxides give better results.[42]

In the case of polypropylene (PP), the free radicals, generated by peroxide decomposition (and partly by mechanochemistry, in the presence of oxygen), abstract the labile proton on the tertiary carbon sites, leading to macromolecular cleavage by the well-known β-scission. Pure mechanochemical degradation of polypropylene in an extruder is also known, even under nitrogen atmosphere, for a residence time of only 3 to 4 s at 260 °C.[43] Exxon prefers to activate the degradation of the PP in a single screw extruder in the presence of oxygen.[44-46] Air is injected into the feed hopper and is thoroughly mixed with the PP powder before extrusion at 290 °C. For a controlled process and for economical applications the polymer needs to be free of any kind of stabilizers. Most of the studies and patents concerning the controlled degradation of PP turn to the use of small amounts of peroxide (0.1 wt% or less) for quick and reproducible reactive extrusion.[42,47-51] On-line rheological equipment is of greatest help in controlling the degradation process. Even the presence of antioxidant stabilizers no longer remains an obstacle. The controlled chain degradation of a graft copolymer, PP-*graft*-MAH, has succeeded in providing a functionalized polymer, whose low viscosity allows it to be used in RIM processes, through crosslinking of the cyclic anhydride functions with 1,6-hexanediol.[52]

Polyethylene (PE) does not contain so many tertiary carbon atoms and the resulting PE macroradicals generated by hydrogen abstraction are less reactive for chain degradation than those of PP. Under similar conditions PE tends to crosslink, while PP macromolecular chains tend to cleave. The controlled chain modification of PE has not found the same successful development as the one developed for PP. Some patents can nevertheless be mentioned, for example mechanochemical degradation of PE was obtained by reactive extrusion in the presence of 0.5 wt% of a chain-splitting agent such as tris(laurylthio)phosphite or tris(laurylthio)phosphate.[53] At a temperature of 270 °C and for a residence time of 10 min, the molecular weight of HDPE decreased from 70×10^3 g mol^{-1} to 40×10^3 g mol^{-1} and the polydispersity from 5 to 3.5.

Controlled degradation of polyisobutylene was also studied.[54-57] Decreases of molecular weight and polydispersity by reactive extrusion have essentially been provided by mechanochemistry. Polysar's patent emphasizes the influence of oxygen on degradation and claims that the resulting product contains no oxygen-containing (*e.g.* carbonyl) groups.[57] At a temperature of 220 °C, with air injection into the feed polymer and a residence time of 0.9 min, the molecular weight of a butyl rubber is reduced to a third, from 185×10^3 g mol^{-1} to 62×10^3 g mol^{-1}.

21.2.3 Grafting Reactions

Grafting reactions performed in the presence of one or more monomers lead to chemical and physical modifications of the original polymeric backbone. Previously, free radical grafting was carried out in solution, especially for low tonnage modified products, but there is an increasing interest for melt-grafting reactions performed in continuous extruding. When long branches are grafted onto the main skeleton, a new material is obtained with physical properties different from those of the original homo- or co-polymer. With short length grafts (up to five monomer units), the substrate polymer has relatively unchanged mechanical properties, but presents markedly differing chemical properties. An example, which is widely used, is the functionalization of the relatively chemically inert polyalkenes (PA) for improvement of the polarity and/or reactivity of the macromolecular backbone. However, free radical grafting in extruders is essentially restricted to the reaction of three kinds of monomers (acrylic acid and esters, maleic and other cyclic anhydrides, vinylsilanes) onto PA macromolecules.

21.2.3.1 *Vinylsilane grafting onto polyalkenes*

Crosslinking of polyalkenes (PA) provides improved mechanical properties, for instance better impact strength, higher heat and flame resistance, improved creep behaviour and better environmental stress cracking. Also, crosslinked products can incorporate high amounts of fillers without a remarkable deterioration of the material's properties. The manufacture of pipes and especially the insulation of electric cables and wires have gained a lot in performance by the crosslinking of PA. These crosslinking reactions may be obtained by direct recombination of macromolecular free radicals, induced by a chemical attack or by irradiation processes. This is currently done with polyethylene, but even in this case, the indirect crosslinking systems based on two-step reactions (grafting of a reactive monomer, which then leads to crosslinking by condensation in the presence of water and catalyst), are also very common way for the continuous production of crosslinked materials (Scheme 1).

In principle, all silane compounds which incorporate one vinyl group for the grafting and one hydrolyzable alkoxy, acyloxy, amine or chlorine segment can be used. Currently vinyltrimethoxy-silane is the most commonly used compound. A mixture of PA, peroxide and unsaturated silanes is fed into an extruder. Within this extruder the decomposition of the peroxide generates free radicals that create reactive sites by hydrogen abstraction on the PA macromolecules. This allows the silane molecules to be grafted due to being unsaturated. The product may be extruded to yield a pipe and passed through a conformator, or the polymer may be extruded onto a wire to provide it with an insulating layer. Under steam pressure the condensation reaction based on the silane groups is activated by tin catalysts leading to the crosslinking of the PA. The simple grafting of silanes onto PP and EPR has been used to produce an adhesive material,[58] but these modified products can also be used in reactive blending with other polyalkenes.

Two techniques are commonly used for the vinylsilane grafting. The Sioplas™ process has been developed by Midland Silicones and is accomplished in two steps.[59,60] In a first extruder silane is grafted onto PA by using a mixture of unsaturated silane and peroxide. In a second extruder the graft polymer is thoroughly mixed with a masterbatch of PA and a catalyst for improved hydrolysis of the silane grafts, such as dibutyltin dilaurate (DBTDL), before the wiring and the crosslinking. The Monosil™ process was later introduced by BICC Cables Limited and Maillefer.[61] In one-step grafting, incorporation of DBTDL and wiring are accomplished on the same extruder. Another one-step technique has been recently proposed, where the grafting reaction is carrried out in a Cavity Transfer Mixer™.[62-64] The PA is melted and homogenized in a single screw extruder while silane, peroxide and DBTDL are mixed together and injected into the PA at the entrance of the Cavity Transfer Mixer™.

Scores of patents have been disclosed on this topic, but they are essentially simple variations of the two basic Sioplas™ and Monosil™ processes. Grafting improvement could be obtained by the premixing of the components, especially for a better impregnation of the polymer. In the feed zone of the extruder, peroxide and silane can react before the PA has entirely melted, preventing the

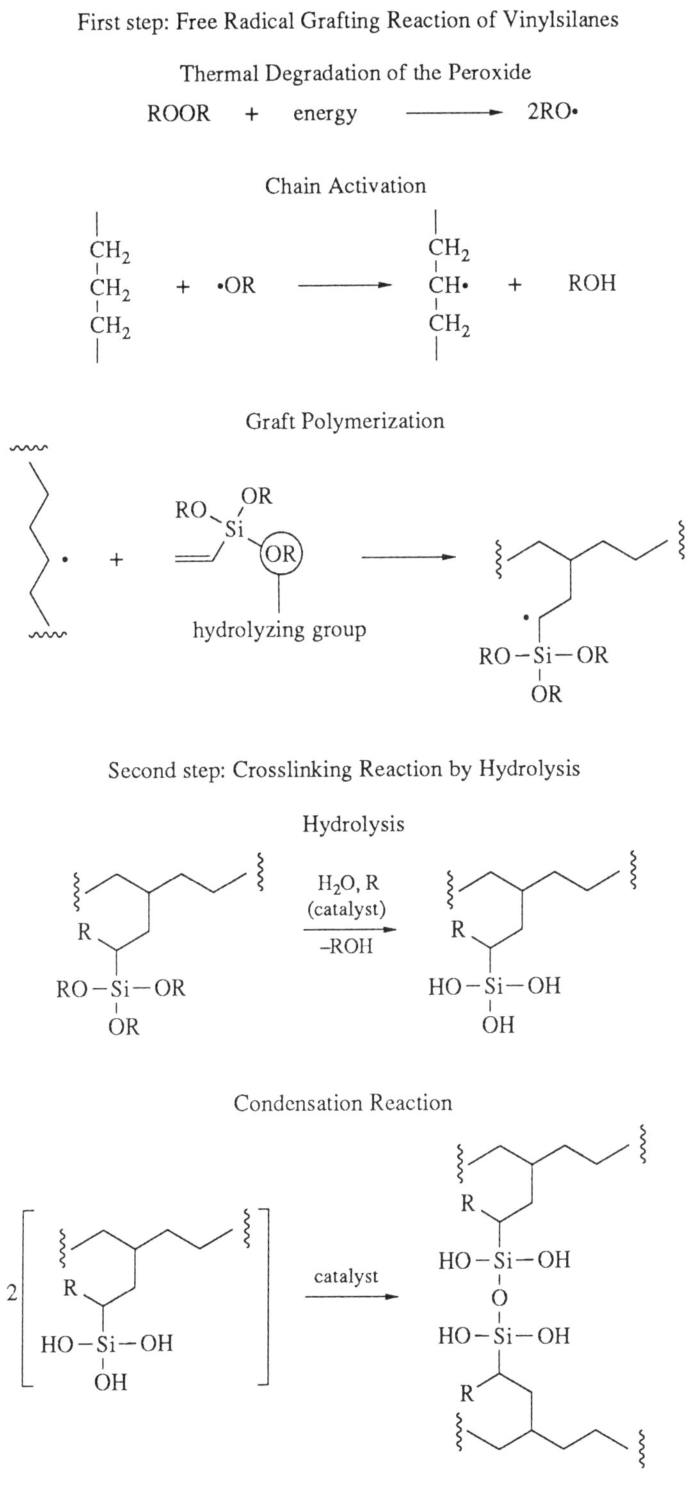

First step: Free Radical Grafting Reaction of Vinylsilanes

Thermal Degradation of the Peroxide

ROOR + energy ⟶ 2RO•

Chain Activation

—CH$_2$—CH$_2$—CH$_2$— + •OR ⟶ —CH$_2$—CH•—CH$_2$— + ROH

Graft Polymerization

[structure: polymer radical + CH$_2$=CH—Si(OR)$_3$ (hydrolyzing group) ⟶ grafted polymer with pendant —Si(OR)$_3$]

Second step: Crosslinking Reaction by Hydrolysis

Hydrolysis

R—Si(OR)$_3$ —[H$_2$O, R (catalyst), –ROH]→ R—Si(OH)$_3$

Condensation Reaction

2 [R—Si(OH)$_3$] —catalyst→ R—Si(OH)$_2$—O—Si(OH)$_2$—R

Scheme 1

formation of a homogeneous reactive medium. As a result of this observation one may consider a delayed introduction of the reagents to the melted PA after the feed zone of the extruder[65] or the introduction of only the peroxide and the DBTDL.[66] Recently, Fritz and Ultsch have published some results on continuous control by FTIR (see Figure 1) of the silane's grafting efficiency.[65,67]

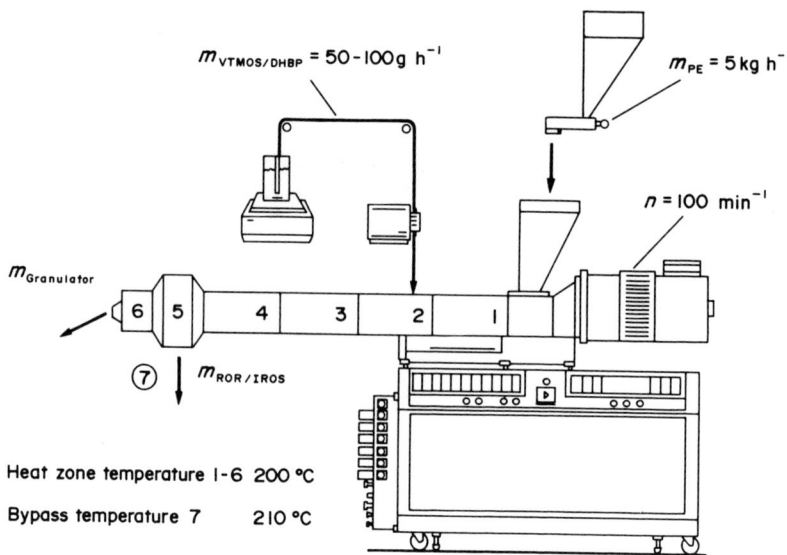

Figure 1 Machine settings for the grafting reaction of vinyltrimethoxysilane (VTMOS) in a twin screw self-wiping extruder with on-line analysis (reproduced from S. Ultsch et al. Plast. Rubber, 1990, **13** (2), 87)

21.2.3.2 Carboxylic monomer grafting reactions

The grafting of maleic anhydride (MAH) onto polyalkenes and rubbers by reactive extrusion provides macromolecules with mainly short grafts, due to the low reactivity of the unsaturated cyclic anhydride towards the homopolymerization. Consequently, the physical and mechanical properties of the polymer substrate do not change very much, but the chemical properties change drastically. However, the reactivity of the MAH and the yield of the grafting reaction remain low and there is a need for improvement in this field, despite the numerous studies by Gaylord (see Section 21.2.3.4). Mechanochemistry has been used to generate reactive sites on the macromolecules,[68] but peroxides are much more efficient. Delayed incorporation of the reagents into the molten polymer enables more homogeneous grafting when the injection port is followed by an efficient mixing zone.

Most of the patents concerning reactive extrusion address the grafting of PA. The low reactivity of the MAH demands a nonnegligible addition of peroxide. Consequently, the grafting of the MAH has to compete with parasitic reactions such as crosslinking in the case of PE, or chain degradation with PP. Improvements of the MAH grafting have been looked for and a technique like the charge-transfer complex activated system, where the anhydride is associated with another monomer or an activator, provides a suitable increase of the grafting efficiency, as will be shown later on. The manufacture of PA-graft-MAH by reactive extrusion shows a great economical development, especially for the production of PP-graft-MAH. The grafted polymer keeps the mechanical and physical properties of the PP, its hydrophobicity, a great part of its chemical resistance and the ability to be additionally functionalized. PP-graft-MAH is increasingly used as a reactive product leading to in situ creation of a compatibilizing agent for immiscible polymer blends, such as blends of PP and polyamide (PAm). The grafted PP remains compatible with PP, while the anhydride functions can react with the primary amino end groups of the PAm component, in order to give a graft copolymer PP-graft-PAm. This helps in reducing the interfacial tension and improving the dispersion of the secondary phase in the main matrix.[69]

MAH grafting is also used with other types of polymers, essentially to provide them with a reactive functionality, like the EPDM rubbers for the synthesis of 'tough nylons'. MAH grafting has been obtained with ABS,[70,71] with SBR[72,73] and hydrogenated SBS.[74] This kind of functionalization yields a marked improvement of impact properties when the thermoplastic resin can react and be linked with the elastomeric nodules. The methyl groups of poly(phenylene ether) (PPE) can be grafted with the anhydride by reactive extrusion[75,76] as well as a mixture of PPE/SBS,[77] leading to the compatibilization of blends with PAm. Grafting of maleate monoesters would provide the PPE with a higher resistance towards the discolouration and thermal degradation;[78] N-phenylmaleimide grafts will give adhesive properties to PPE.[74]

The hindered amine light stabilizers (HALS) have shown great promise in preventing polymer degradation by UV radiation. Nevertheless, they are difficult to use with films and fibres from which they migrate easily. By grafting onto molten polymers, masterbatches of compounds with active grafts, like a PA-*graft*-bis(hindered piperidinyl) maleate, can be produced and incorporated as stabilizers. A good example is the radical grafting of bis(2,2,6,6-tetramethyl-4-piperidinyl) maleate onto molten PP, which produces a photoantioxidant masterbatch with 12% graft maleate.[79,80] With the same objective, PP has been grafted with diphenylamine maleimide to give a thermal antioxidant.[81] The grafting offers a good means for the synthesis of nonmigrating polymeric stabilizers that can keep their stabilizing properties throughout the life of the manufactured items, and that will be miscible with the polymer as a masterbatch.

The development of PA grafting with polar monomers has been rapid with the objective of reducing some problems which are met in their use and that result from their lack of polarity. Paints, adhesion with fibres and fillers, and especially the compatibilization of polymer blends are affected. Acrylic monomers provide an almost unlimited choice of polarities, functionalities and sizes. They are much more reactive than the maleic monomers, providing higher degrees of grafting. On the other hand, this reactivity leads to competition between grafting and the unwanted and substantial homopolymerization. The grafting of acrylic monomers aims preferentially at the binding of reactive functions and, according to this orientation, one essentially grafts (meth)acrylic acid and, to a lesser extent, glycidyl methacrylate. Peroxides are commonly used to increase the yield of the reaction, even if grafting by mechanochemistry can be done.[82] Delayed introduction of the mixture of monomer and peroxide is frequent with its injection into a molten stream of polymer. The screw geometry and the reaction zone of the extruder must be designed to provide a high surface area of polymer for grafting. The unreacted monomer can be removed through a devolatilizing vent.

The grafting of a PA has become routine for many plastic producers. It improves the adhesive properties of the polymers and their compatibilizing abilities. The linkage of (meth)acrylic acid allows the transformation of a PA into an ionomer resin which, blended with zinc salts, provides thermoreversible crosslinking.[83] Many studies are currently focusing on the grafting onto PP since this polymer offers superior mechanical and thermal properties. Nevertheless, the reaction has to compete with the unwanted chain degradation. Elastomers are also grafted with reactive acrylic monomers, and melt blended (often in extruders) with polymers to produce 'impact resistant plastics'.[84,85] The binding of elastomer inclusions to the thermoplastic matrix, through the acrylic grafts, enhances the impact resistance of the materials.

PPE can be functionalized with acrylic monomers through reactive extrusion with methacryloyl-ε-caprolactam, methacrylamide[86] or glycidyl methacrylate.[76,77] As a result of this grafting the PPE can be compatibilized by reactive melt blending with PAm.

Melt grafting of PP with a hindered piperidinyl acrylate allows the synthesis of a photoantioxidant masterbatch.[81,87] The tendency of the acrylic monomer to homopolymerize produces an oligopolyacrylate more sensitive to migration.

21.2.3.3 Other monomers and reactive species

Styrenic monomers can be grafted onto a polymer by reactive extrusion. Large amounts of these monomers are often introduced into the extruder in order to compensate for their low grafting efficiency, which is the consequence of noticeable homopolymerization. Polyalkenes are grafted to improve their abrasion resistance and to increase the autolubricating properties.[88-92] PPE has been grafted with styrene for the improvement of the electrical properties of the polymer and to make its processing easier.[93] The American Cyanamid Co. proposes a difunctional monomer, *m*-isopropenyl-α,α-dimethylbenzyl isocyanate (TMI), to be grafted onto PA to provide the polymer with isocyanate reactive functions.[94] PAs are also grafted with vinyl monomers like vinyl stereate[95] or vinyl acetate, leading to high levels of grafting (10.8%) in an extruder.[96,97] Ethylene–propylene copolymers have been grafted with nadic anhydride, once more for the compatibilization of PA/PAm-6 blends.[98]

Combining the mechanochemistry in extrusion and a Kharasch-type addition, unsaturations of elastomers[99] and thermoplastics[100] can react with molecules possessing thiol groups. Polymers are consequently grafted with stabilizing functions.

The same reaction is known with saturated PAs, such as PE and PP.[101] Mechanochemistry generates in molten polymers alkyl macroradicals which are able to react with the thiol groups of the stabilizers through the labile hydrogen atoms. Oxygen can participate through the generation of hydroperoxide functions on the polymer backbone.

21.2.3.4 Recent developments in free radical grafting

Maleic anhydride (MAH) is widely used for polyalkene functionalization due to its low tendency to homopolymerize and the formation of short grafts which allow it to keep most of the physical and mechanical properties of the support polymer. MAH is, however, bothered by too low a reactivity, as the anhydride group impoverishes the double bond of its electrons. To compensate for this effect, higher amounts of peroxides are introduced for the melt grafting in extruders. According to this higher level of initiator, the grafting has to compete with nonnegligible parasitic reactions like crosslinking (mainly with the PE) and chain degradation (PP). Many research groups are trying to overcome these problems by increasing the grafting efficiency, which allows a lower initiator concentration and reduces the risk of parasitic events.

In order to increase the reactivity of the MAH, Gaylord has introduced in melt grafting the concept of a 'charge-transfer complex' (CTC), associating the unsaturated anhydride with an electron-rich molecule or activator. A donor/acceptor couple is generated, the MAH double bond becomes richer in electrons and consequently more reactive. Many different families of additives have been proposed: oxygenous, nitrogenous, sulfurous or phosphorous compounds.[102] The use of some amines or amides has also been patented,[103] as well as ketones[104] like acetone[105] or butanone,[36,106] for the grafting of MAH onto polyalkenes.

This orientation has been reinforced with the use of a 'mixed monomer system' CTC: MAH is associated with an electron donor comonomer which increases its reactivity and is involved in the grafting. A crowd of comonomers have been tested for this combined grafting reaction. Vinyl comonomers which copolymerize well with MAH and have reduced steric hindrance give better results, like alkenes, acrylates, styrene or α-methylstyrene.[107] The proposed classification is isobutylene > propylene > styrene = vinyl acetate,[108] but styrene[27,104,105,109,110] and, to a lesser extent, vinyl acetate and ethylene, are the most used.

Combination of the two techniques of mixed system and activator CTC is also in progress; for example, the association MAH/styrene/DMF does not look bad. The trinome MAH/styrene/ketones has been patented.[104,105] Improvement in the grafting ratio may be obtained with polymers containing polar groups ($C{=}N$ or CO_2R) when the latter are complexed before the addition of comonomers and peroxide. For example, the grafting of a styrene/acrylonitrile mixture onto poly(butyl acrylate) is increased when the polymer is complexed by $ZnCl_2$. This reaction has been done in solution but extrapolation to reactive processing is currently looked for.

Research of the optimal conditions for the best grafting of MAH onto polyalkenes during melt processing is a very tricky study with a lot of factors involved in the process. Even the polarity of the peroxide appears to affect the reaction.[110,111] A more polar peroxide would concentrate in the MAH phase, resulting in the generation of fewer reactive sites on the polymer and a higher amount of poly(maleic anhydride) homopolymer. A low polarity peroxide would stay in the PA phase, favouring polymer crosslinking (PE) and degradation (PP).

21.2.4 Crosslinking Reactions

Crosslinking reactions involve the covalent binding between macromolecules of a single polymer, while coupling is generated in blends of two or more polymers. Polymer free radical crosslinking by reactive extrusion is not the most commonly used process, as it implies great difficulties in the control of the reaction, since one must carefully avoid premature gelling within the extruders. Some industrial processes are based on mixing of comonomers (and eventually peroxides) in the extruder, followed by further crosslinking, after the die, in a heated salt bath (at around 200 °C) or through high energy irradiation techniques. On the other hand, the widely used crosslinking of polyalkenes with vinylsilanes is mainly a two-step process, where only the grafting reaction is carried out in the extruder, whereas the extruded profile is later crosslinked by a condensation reaction in the presence of water. LDPE used in wire coating is often crosslinked in the presence of one peroxide,[112,113] or better, with a mixture of peroxides of different reactivities.[114] HDPE on-line crosslinking is also performed in single or twin screw extruders for the continuous production of crosslinked pipes, which are used in central heating. The control of the extrusion process requires formulations combining inhibitors and peroxides.[115]

The molecular weight of polypropylene is easily lowered by extrusion in the presence of free radicals. Therefore, its crosslinking is only possible when polyfunctional stabilizers are added to the peroxides. Thiourea, hydroquinone and polyallylic molecules are preferred[116,117] and these stabilizers react with the alkyl macroradicals, reducing the chain degradation by β-scission. EVA copolymers are also crosslinkable, but as this reaction involves preferentially the methyl groups of the acetic ester, the obtained polymer network may be destroyed by hydrolysis.[118]

Classical synthesis of poly(phenylene sulfide) (PPS) produces low molecular weight products, unfit for industrial applications.[119] Higher molecular weight branched and crosslinked species are obtained by oxidative curing. This could be done in a twin screw extruder at temperatures largely above 300 °C.[120] Injection of oxygen stimulates the curing and, after a residence time of about 30 s, the melt-flow index (MFI: 316 °C and 5 kg) falls from 2500 to 200.

PEK and PEEK have been crosslinked with a few percent of elemental sulfur.[121,122] The mechanism of this reaction is not well understood, as it could be related to the formation of a three-dimensional network of sulfur or to free radical reactivity of S*.

21.3 OTHER TYPES OF BULK POLYMERIZATION

In a previous Section (21.2.1), it was indicated that extruder reactors can be used for free radical initiated bulk polymerization of various vinyl and acrylic monomers. Modular extruders are designed to handle the extreme difference in viscosity between starting materials and converted products and to control within narrow limits the temperature gradient in the reacting mixture arising from the exothermic heat of polymerization. Sometimes a volatile inert material is added, which helps in cooling the reacting mixture by volatilization through vacuum venting at an appropriate barrel segment. Twin screw extruders with intermeshing, self-wiping screws provide more efficient heat transfer, improved mixing and higher reaction rates, compatible with shorter residence times. Besides free radical initiated systems, there is an increasing interest in fast addition bulk polymerizations, based on ionic or coordination complex, active-initiating species.

On the other hand, polycondensation reactions arise through a repeated condensation process of two distinct coreactive functional species to give high molecular weight polymers and a low molecular weight by-product such as water or low-boiling alcohols. In a recent review, Ignatov et al.[123] have indicated that violation of the principle of unchanged activity of the functional groups with an increase of the chain length, advanced previously by Flory,[124,125] may be related to the action of physical (excluded volume effect, diffusion control etc.) and chemical (far order effect) factors. These points have to be taken into account in the search for further developments in continuous bulk polymerizations in extruders.

21.3.1 Addition Polymerization

Among a number of works carried out in the area of anionic addition bulk polymerization of vinyl monomers, it is of interest to mention the pioneering research conducted by Muller,[126] early in the 1970s. Using a two-step process, it was possible to polymerize styrene in the presence of an organometallic initiator — butyllithium. The living polymerization begins in a thin film rotating device and the final product is obtained after extrusion. During this second step, it is also possible to add a second monomer for block copolymerization, or various types of deactivating agents, like carbon dioxide, ethylene oxide, etc. in order to create functional end groups.

The continuous anionic polymerization of caprolactam in an extruder was published first by Illing,[127] who proposed specific screw configurations for a building block, twin screw extruder, including also an adjustable valve to control variable shear forces. The initiating species are based on alkaline products, like sodium lactamate (0.2 to 0.3 wt %), possibly combined with an accelerator, for instance acyllactam. This chemistry is well known and currently used in monomer casting and nylon reaction injection moulding techniques (NYRIM). The anionic polymerization of ε-caprolactam produces high molecular weight polyamide-6 (PAm-6) in just a few minutes, in contrast to the classical polycondensation process which requires reaction times of up to 10 h. The work carried out more recently by Menges, Bartilla and Berghaus[128-130] has confirmed the versatility of the process, whereby reactive extrusion is coupled with integrated processing to produce reinforced and nonreinforced nylon-6 rods.

As the preceding reaction does not produce a volatile by-product, this ring-opening polymerization is classified by IUPAC as a polyaddition. The cationic polymerization of trioxane is another example of the direct production of homo- and co-polymers (polyacetals) by reactive extrusion. These reactions, as well as the polymerization of cyclic siloxanes, are carried out industrially in twin screw extruders or in co-kneaders.[131,132]

Both polyurethanes and polyureas have been prepared in extruder reactors.[133,134] Because these reactions proceed by step-growth polymerization, stoichiometric control of the reactants is important for production of high molecular weight polymers. In a representative process,[135] a corotating, self-wiping twin screw extruder is used and the key to the success of this process is the screw design including two or three kneading zones, preceded by short metering segments. The kneading screw elements provide sufficient mixing and shearing of the reaction mixture to prevent the formation of gel inhomogeneities in the extruded product.

21.3.2 Polycondensation Reactions

Extruders with specific screw design are used in many polycondensation processes[136] but polyesters are also made in extruder reactors from low molecular weight prepolymers of, for instance, bisphenol A and a mixture (75/25) of isophthalate and terephthalate diphenyl esters. The prepolymer is prepared in a continuous stirred tank reactor (CSTR) and the polycondensation is completed in a five-stage, twin screw extruder equipped with an efficient vacuum venting system to remove the phenol by-product.[137]

An interesting example of polycondensation for the continuous production of thermostable poly(ether imides) in corotating modular twin screw equipment was published by Schmidt and coworkers (Scheme 2).[138]

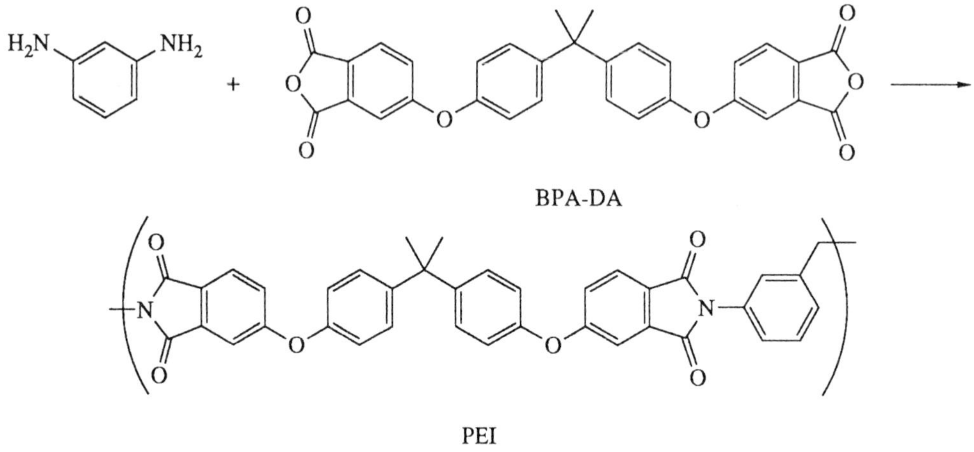

Scheme 2 Polycondensation reaction for the continuous production of poly(ether imides)

The average residence time in the extruder was around 4.5 min. The main problem was related to the controlled stoichiometric feeding of the reactive species, bisphenol A dianhydride and *m*-phenylenediamine, in the extruder. For this reason, the two products are fed to the extruder as separate melt streams *via* a concentric tube feed inlet, which also promotes better product homogeneity.

21.4 FUNCTIONAL MODIFICATION OF POLYMERS

21.4.1 Halogenation of Butyl Rubbers and Other Polymers

Butyl rubber, a copolymer of isobutylene (98%) and isoprene (2%), is almost exclusively used as the inner liner for tubeless tyres and in tyre tubes because of its very low air permeability compared to natural or other synthetic rubbers. For use in inner liners, the butyl is modified through halogenation with chlorine and bromine to improve its compatibility and ensure covulcanization with the other elastomeric materials. The classical industrial halobutyl process is based on direct halogenation of the butyl in a dilute hexane solution. The recovery of the water slurry, resulting from further neutralization with dilute aqueous caustic solution and precipitation by steam, needs a series of skimming, squeezing, and vaporization extrusion steps. The whole process is both complex and highly energy intensive.

Several years ago, Exxon conducted pioneering work in order to study whether halogenation could be carried out directly on a molten butyl rubber in an extrusion process. They succeeded in solving difficult chemical and processing problems.[139,140] The competing reactions in the halogenation of isoprene and isobutylene units are as shown in Scheme 3.

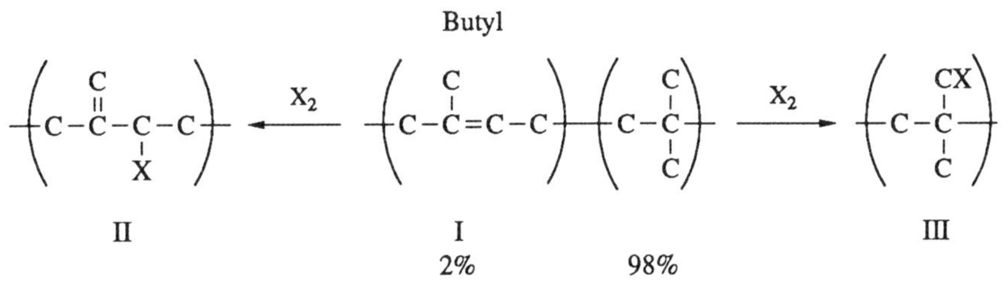

$$
\left(\begin{array}{c} \\[-2pt] C\\[-2pt] \| \\[-2pt] C-C-C-C\\[-2pt] | \\[-2pt] X \end{array}\right) \xleftarrow{X_2} \left(\begin{array}{c} C\\[-2pt] | \\[-2pt] C-C=C-C \end{array}\right) \left(\begin{array}{c} C\\[-2pt] | \\[-2pt] C-C\\[-2pt] | \\[-2pt] C \end{array}\right) \xrightarrow{X_2} \left(\begin{array}{c} CX\\[-2pt] | \\[-2pt] C-C\\[-2pt] | \\[-2pt] C \end{array}\right)
$$

II   I   III
           2%   98%

Scheme 3 Halogenation reaction of butyl rubbers

The halogenation is done in a twin screw, nonintermeshing extruder built with a special anticorrosive alloy such as Hostelloy™. The reaction zone of the extruder is almost isolated by restrictive melt seals formed from reverse-flighted screw elements (see Figure 2). In the reaction zone the screw is designed to partially fill the channels, so that only 20 to 35% are filled with polymer. This provides an adequate mixing of the polymer with the halogenating reagent: a mixture of chlorine and nitrogen (30% Cl_2) or a solution of HSO_3Cl/CCl_4. At a temperature <170 °C the intensive mixing splits up the polymer into microparticles of less than 10 μm, generating an ever-renewed, huge surface in contact with the reactant — this favours the halogenation. The halogenated rubber is passed into the following zones where it is scrubbed and sometimes neutralized with an alkaline aqueous solution. The polymer is also mixed with nitrogen and vacuum vented for the removal of unreacted chlorine, hydrogen chloride and other volatile by-products.

The chlorination process has also been used with other polyalkenes: LDPE, HDPE, EVA, EPR, EPDM, PIB and BR. The use of a 50 mm counterrotating nonintermeshing twin screw extruder allows a production of around 10 kg h^{-1} with a chlorination level between 0.38% (HDPE) and 4.74% (EVA).

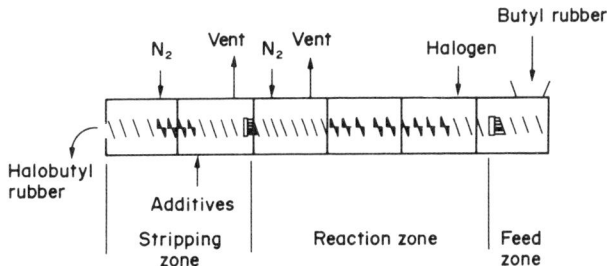

Figure 2 Continuous process of butyl rubber halogenation in a twin screw counterrotating extruder (reproduced from R. Kowalski in *Chem. Eng. Prog.*, 1989, **85**, 71)

21.4.2 Chemical Modification of Poly(vinyl chloride)

PVC is a widely used commodity plastic with some drawbacks limiting its applications. Among these, rigidity related with low impact and weak thermal behaviour have to be mentioned. By introduction of additives, like latex particles, it is possible to improve impact behaviour. On the other hand, plasticizing of PVC is easy by the introduction of polar molecules like alkyl benzoates. However, these micromolecular species migrate too easily and there is an interest for permanent plasticizing by direct grafting in the melt. This opportunity is reinforced by the fact that C—Cl bonds are relatively weak, which has the drawback that secondary reactions can occur.

Michel, Mijangos and coworkers have studied internal plasticizing by grafting aromatic and aliphatic thiolates onto PVC.[141] For example, labile chlorine of PVC reacts with sodium thiosalicylate, in the molten state, producing NaCl and creating C—S—(aromatic ester) hanging groups. The modified PVC shows decreased T_g values in correlation with the grafting efficiency. This kind of reactivity was used previously for PVC crosslinking by dithiol derivatives of triazine.[142,143] The hydrochloric acid produced was neutralized by magnesium oxide. Analogous crosslinking can be provided by aromatic diamines for a PVC/CPE (chlorinated polyethylene) blend with reactions between the C—Cl groups and the amine functions.[144]

Improvement of impact resistance can be obtained by reactive melt blending and compatibilization with PA thermoplastic or elastomeric copolymers. PVC/LDPE interchain copolymerization, initiated by peroxides at 160 °C, leads to compounds with intermediate impact properties, lower than those obtained with PVC/elastomer blends.[145] The addition of a polyfunctional monomer like triallyl isocyanurate increases noticeably the PVC/LDPE blend reactivity, producing a reciprocal crosslinked copolymer with improved impact resistance.[146]

The thermal resistance of PVC is improved by reciprocal blending with styrenic copolymers containing maleic anhydride, like Dylark™ from Arco or Cadon™ from Monsanto. In this case, the final cohesion is mostly related to polar association rather than to effective covalent bonding.

21.4.3 Condensation Reactions

Amines have been grafted on ionomer resins to increase their stiffness. However, when this modification is done in a reactor, the modified polymer loses its grafts during extrusion processing. Reactive extrusion has circumvented this drawback. With the help of Welding Engineers, the US company Advanced Glass Systems has developed an extrusion line to produce grafted ionomer sheets.[147-149] Few details have been presented other than the use of a longer (1/3 more than normal to increase the residence time) twin screw extruder and an amine injection port at mid barrel to impart the heat resistance of the grafting. The transparent modified polymer keeps the adhesion properties of the ionomer resin and becomes tougher and more impact resistant and can be used as bullet-proof glass. This is an example of a reaction which is impossible outside an extruder.

Imidification is more important than amidification, as it increases the water resistance of the polymers. Imidification has a low reactivity, but the high temperature and pressure of reactive extrusion processes are adequate for providing a suitable rate. A simple single screw extruder has been sufficient to imidify 50 to 80% of the anhydride functions of ethylene/maleic anhydride copolymers.[150] In a two-step process, the copolymer is mixed with ammonia under pressure in a reactor at 80 °C for 6 h, resulting in amidification of anhydride groups. The amidified material is then imidified in an extruder at 180 °C and the yield may be as high as 80%. In a one-step process the copolymer is directly melt blended in the single screw extruder with a compound like ammonium carbamate. The thermal decomposition of the carbamate generates ammonia. The imidification is relatively efficient since the carbamate has already been well mixed with the polymer. The yield is lowered to 55%, since the effective time of polymer/ammonia contact is much shorter. Methacrylic polymers, like poly(methyl methacrylate) (PMMA), have also been imidified in extruders with ammonia or a primary amine.[151,153]

Imidification of random styrene/maleic anhydride copolymers has been obtained in a one-step reactive extrusion process with a primary amine or ammonia. Yields as high as 80% have been claimed when anhydride was reacted with aniline in a molar ratio of 1 to 2,[154] and even 100% claimed in a single screw extruder with octadecylamine.[155] Nevertheless, it seems difficult to reach total imidification; a too high increase of the barrel temperature may lead to polymer degradation. The imidification process may be a convenient way to fix some stabilizers, like the sterically hindered amines, onto a polymer backbone.

Copolymers possessing ester and/or acid (meth)acrylic units have gained a higher thermal stability by the formation of anhydride rings.[156] Extrusion is done at 300 °C, with 0.1% of an alkaline catalyst and with high vacuum. It follows the same idea as the imidifiaction but the anhydride functions are more sensitive to moisture.

21.4.4 Exchange Reactions

Pendant ester groups on ethylene–vinyl acetate copolymers (EVA) have been saponified under controlled conditions in extruder reactors.[157,158] Injection of sodium methoxide in methanol into a molten stream of EVA in a twin screw extruder results in different degrees of saponification, depending on temperature, composition and vacuum venting efficiency of the by-product. But it seems important to reduce the amount of alkaline products, which are needed for high levels of saponification, because other types of ester breakdown reactions are introduced.

21.4.4.1 Alcoholysis of EVA copolymers

The conversion of vinyl acetate polymers (PVA) or ethylene–vinyl acetate copolymers (EVA) to the corresponding poly(vinyl alcohol)s by hydrolytic or exchange reactions is an industrial process carried out mostly in solution, emulsion or suspension, using a classical stirred tank reactor (CSTR). For copolymers containing high levels of vinyl acetate, these chemical modifications of the ester side groups (VA) can follow the polymerization process, in the same CSTR. On the contrary, copolymers of ethylene and vinyl acetate (EVA) and other terpolymers, based on lower amounts of acetate monomer, generally between 5 and maximum 40% by weight, are prepared by the high pressure free radical polymerization process, which leads to direct extruding and pelletizing. Chemical modification in solution would require numerous and costly operations of dissolution, reaction, polymer recovery and purification. Continuous modification by reactive extrusion should be an economically competitive process, if the need for a simple chemistry is fulfilled. Various patents report on this processes based on direct hydrolysis, in the presence of basic catalysts.[159–161]

For a high conversion level, it seems preferable to modify the copolymers by alcoholysis, in the presence of alkyl alcohols and basic or organometallic catalysts.[162,163] The equilibrated reaction is carried out mostly in a homogeneous medium, for instance when octanol is used as a reactive alcohol and dibutyltin dilaurate (DBTDL) as the organometallic catalyst, which are both soluble in the molten EVA. Preliminary studies were carried out in laboratory scale discontinuous equipment, like Haake Rheocord, but the best results were obtained in a corotating modular twin screw extruder (ZSK 30 from Werner and Pleiderer). Dependence of the conversion on temperature, reaction time and composition of the reacting mixture has been thoroughly investigated. With an amount of 1% sodium methoxide as catalyst, at 170 °C and an average residence time of 80 s, the resulting conversion to the secondary alcohol reaches approximately 60%. Kinetic studies were performed in the presence of dibutyltin dilaurate, an efficient catalyst, without enhancing any side reactions, and the corresponding results led to a fair evaluation of the rate constants. Beside this, it seems interesting to compare kinetic calculations and experimental data for various alkyl alcohols and for EVA copolymers containing between 5 and 28% vinyl acetate. All these experiments were carried out using the discontinuous mixing device and DBTDL as a catalyst at 170 °C. It was easy to confirm that chemical reactivity depends on the nature of the alcohol as, for instance, octan-2-ol leads to poor efficiency in conversion (32% at equilibrium), in comparison to its primary homologue octan-1-ol (60%) and surprisingly hexane-1,6-diol (62%). With different amounts of vinyl acetate in the copolymer, it appears that the rate constants remain unchanged for the random copolymer (5 to 14% by weight VA) but, for higher VA contents, the final conversion seems to be enhanced by proximity catalytic effects due to the presence of diads and triads.

21.4.4.2 Alcoholysis of copolyacrylates

In order to develop further modification of copolyesters, it was interesting to investigate the potential reactivity of various copolymers, mainly alkenic and styrenic, containing different pendant acrylic ester groups.[164,165] At the same time, the authors verified whether classical kinetics, as they occur in solution, were applicable in melt by comparing results obtained in concentrated solutions (30 wt%) and homogeneous bulk processes (in melt) at the same temperature. A series of three alkenic copolymers containing various amounts of pendant esters, based respectively on methyl, ethyl and n-butyl alcohols, were tested in exchange reactions with a well-defined secondary alcohol, phenylisopropyl alcohol. This latter product was chosen for its specific UV and IR absorption characteristics, which permitted a precise following of the kinetics. Several organometallic compounds were tested for catalytic activation. It was found that the efficiency did not only depend on the nature of the metal atom, but also on that of the organic ligands. Tin derivatives were shown to be very efficient, but further basic studies confirmed that the activated species resulted from a preliminary reaction between the tin compound and the alcohol. Results from reactions carried out in solution confirmed that alcoholysis reactions are reversible, characterized by equilibrium constants, whereas melt processes indicated irreversibility, at temperatures between 170 °C and 190 °C, regardless of the nature of the alcohol formed. This notably occurred, due to the instant volatilization of the low molecular weight alcohol. Had this not occurred, it could have brought about the reverse reaction, and may have been the reason why increasing the mixing intensity made little contribution to the final conversion.[166] The rate constants of alcoholysis of acrylates are high and depend on the nature of the pendant ester, decreasing in the order methyl > ethyl > n-butyl. Since this exchange reaction is a nucleophilic substitution, electron donation and steric hindrance of alkyl substituents

Table 2 Comparison of the Reactivity of Various Copolyacrylates in the Melt

Copolyacrylate	Apparent constants of direct reactions K_1 at temperature shown (L mol^{-1} min^{-1})			E (kcal mol^{-1})
	170 °C	180 °C	190 °C	
EMA	1.27×10^{-2}	1.97×10^{-2}	2.71×10^{-2}	15.4 ± 1.5
EEA	1.19×10^{-2}	1.90×10^{-2}	2.60×10^{-2}	19.5 ± 2.0
EBA	0.46×10^{-2}	0.93×10^{-2}	1.44×10^{-2}	23.1 ± 2.0

mainly affect this order, which was confirmed by the observed values of activation energies: 15.4, 19.4 and 23.1 kcal mol^{-1}, respectively for the three pendant esters indicated above (Table 2). With respect to these kinetic data, it was possible to reach 60–70% conversion in a corotating self-wiping twin screw extruder, with a residence time of less than three minutes.

21.4.4.3 Aminolysis of copolyacrylates

The aminolysis of poly(styrene-co-methyl acrylate) by octadecylamine was studied in solution and in the melt at temperatures close to 200 °C. It rapidly appeared that this reaction is rather slow. Several classes of catalysts were tested to accelerate it and, after systematic studies, it has been observed that the activation of aminolysis requires efficient proton transfer, which leads to the choice of tautomeric compounds, like 2-pyridone. The mechanism of the catalytic reaction was carefully elucidated and confirmed by the kinetic data determined both in solution and in the melt. Reaction mechanisms and absolute rate constants are comparable in both media. This signifies that the higher conversion levels observed in bulk processes are only related to higher concentrations of the functional species.

Comparable results were obtained with alkenic copolymers (ethylene and methyl acrylate), showing that aminolysis of EMA (alkenic) copolymers is slightly faster than for SMA (styrenic) compounds, mainly because of the steric hindrance due to phenyl groups in SMA (Table 3). In order to characterize the influence of mixing efficiency on the global reactivity, various experiments were carried out in two mixing devices: a Haake Rheocord discontinuous mixer and a Werner and Pfleiderer ZSK twin screw extruder. Increasing the rotational speed of the blades in the Haake did not influence the rate of conversion. This is typical of a homogeneous reacting system and very close to the behaviour observed in solution. Furthermore, the conversion values obtained by continuous extrusion on the twin screw satisfactorily agreed with those calculated from the kinetic data from the experiments conducted in the Haake.[166]

Table 3 Comparison of the Reactivity of Different Copolyacrylates in the Melt: S 17 = Poly(styrene-co-methyl acrylate) and EMA = poly(ethylene-co-methyl acrylate)

Temperature (°C)	Absolute rate constants (L^2 mol^{-2} h^{-1})	
	S 17	EMA
187	0.34	1.50
197	0.54	1.90
207	0.80	2.40
Activation energy (kcal mol)	18.4	11.7

21.5 REACTIVE BLENDING OF POLYMERS

Current trends are not turned to the research of new monomers and related polymers resulting from polymerization, as much as they are converging on polymer blends and alloys. The most interesting combinations are based on the association of commodity and engineering plastics, with the objective of providing the right level of properties for a defined application. The final cohesion is dependent on the composition and adjustment of interfacial tension by the introduction of an amphiphilic copolymer. In reactive blending there is a possibility of preparing *in situ* the amphiphilic compound by covalent or ionic bonding. The following presentation will deal with these different possibilities.

21.5.1 Grafting by Interpolymeric Reactivity

Interpolymeric grafting reactions combine the reactive groups of two (or more) polymers in order to form copolymeric species by melt processing in the extruder. High levels of covalent bonding must be provided by high concentrations of reactive functions and fast reactions, as it seems difficult to reach residence times over a few minutes. Catalytic activation could also be helpful, as it was shown previously for interpolymeric crosslinking.[167] Even if no monomers are involved, incorporation of a third reagent, acting like a crosslinker, may be necessary in some cases in order to enhance the reactivity. However, the constraints of a continuous process do not permit a great extent of crosslinking in the extruder and the resulting material needs to be processable by injection moulding. Depending on the system and especially on the reactants' composition, the process requires viscosity control along the screw. In addition, the resulting microstructure is mainly affected by the reactivity at the interface between the polymer phase and the related decrease of interfacial tension. This signifies that high performance reactive melt blending could be obtained by adjusting the right level of reactivity only at the interface, taking into account the possible introduction of low concentrations of reactive polymeric species, respectively compatible with the two immiscible homopolymers forming the major part of the blend. Reactive extrusion is a process for the economic production of amphiphilic copolymers. An example is the direct condensation of hydroxy- or amino-terminated polymers onto pendant carboxylic groups, on a second polymer.

Most polymers are not chemically suited for interpolymeric grafting, as they are not equipped with the right reactive functions. At least one of the polymers of a blend has to be modified prior to reactive melt blending. Polyalkenes are mostly modified, as already discussed, by free radical grafting of polar monomers. Nevertheless, it is also important to take into account the potential reactivity, by exchange reactions, of alkenic copolyesters like EVA (ethylene-*co*-vinyl acetate) and various types of ethylene copolyacrylates, which are prepared by high pressure polymerization.

Among commercially available reactive polymers, besides copolymers based on maleic anhydride like Dylark™ (Arco Chemicals) and Cadon™ (Monsanto), and acrylic acid grafted PP (Polybond™ from BP Chemicals or CXA™ from Du Pont), the versatile reactive polystyrene RPS™ (Dow Chemicals) has also to be mentioned. This polystyrenic product contains 1% vinyloxazoline and this function is able to react with various functionalities, like thiol, hydroxy, epoxy, amine, carboxylic acid and anhydride. Different studies have confirmed the potential of this reactivity, for instance in blends of RPS and carboxylated polyalkenes.[168] A similar reactive PS has been proposed for radical interpolymeric grafting especially with PA, the functionality resulting from the use of *o*-vinylbenzaldehyde (with a labile hydrogen) as comonomer.[169] Mitsubishi has introduced another kind of reactive functionality in copolymerizing ethylene with a few percent of 5-methyl-1,4-hexadiene.[170-172] These substituted unsaturations located randomly along the chain are used for further chemical modifications. The trade name of the product is Fundalon.

There is an almost unlimited number of possible polymer blends, which can be prepared through reactive processing of functionalized derivatives. Many blends have already been studied and it is difficult to present all of them.

PAm and PA are a good example of polymer incompatibility. Nevertheless, there is a great interest in combining the good mechanical or electrical properties and the hydrophobicity of polypropylene with the thermal and solvent resistance of polyamide. Dexter Corp. in the USA, Mitsui and Shisso in Japan and Atochem in France are now selling such blends. Even if it is impossible to be introduced in the detailed process, melt reactivity is certainly needed for final cohesion and adjusted morphology. PAm is already functionalized with an amine group at one end of each macromolecule, while PP is chemically inert before grafting with carboxylic acid or anhydride groups (acrylic acid or MAH, essentially). During reactive processing, the amine end groups of PAm are condensed onto the carboxylic pendant groups of the polyalkene, which leads at high temperature to cyclic imides in the case of cyclic anhydride functions. This process is close to the one used for high impact PAm, like supertough nylon from Du Pont (Zytel™). Hundreds of patents and articles on this topic exist worldwide. More general data can be found in the reviews published by Brown.[5,6]

21.5.2 Functional End Group Reactivity

Polymers synthesized by polycondensation, and some resulting from polyaddition (e.g. PAm-6, PUR), retain functional end groups that are potentially reactive. For some applications, this reactivity is a drawback since it introduces a site of instability for a possible depolymerization. PET has been stabilized in extruders by melt reactions with monofunctional molecules, such as phenyl glycidyl ether,[173,174] carbodiimide[175] or oxazoline derivatives,[176] which provide an inert end to the macromolecule. Instead of the neutralization of the end groups, a change in the functionality may be promoted. For example PET can react in an extruder with diallylglycidyl isocyanurate[177] or bisphenol A diallylglycidyl ether.[176,177] Allylic double bonds are in this way introduced into the macromolecules and the PET becomes available for thermoset applications.

The functional end groups are essentially used in reactive extrusion for the enhancement of polymer molecular weight. As the concentration of these reactive species is very low, one has to encourage the coupling by addition of an activator or a condensing reagent. The activator enchances the reactivity of the end group but is not incorporated in the modified polymer, whereas the condensing reagent acts as a link between two reactive macromolecular end groups. The molecular weight enhancement concerns essentially PAm and saturated polyesters. Typical activators are phosphonic acid and phosphite derivatives. A greater choice is possible when using linking reagents. Every di- or poly-functional molecule which may react with carboxylic acid, amine or alcohol is usable. Stoichiometric control is important because a molar excess of coupling agent may result in capping of most of the polymer functional groups, resulting in low levels of chain extension. The most-used coupling agents are bis(2-oxazoline)s, polyisocyanates, polyepoxides, polycarbodiimides, bis(acyllactam)s and diesters like diphenyl isophthalate. In the latter case, the manufacturer has to deal with phenol by-product generation.

One may wonder: why is this functional end group reactivity not used for block copolymer formation? By melt blending in an extruder it could be possible to associate in one copolymer two distinct polymers with the same or complementary end group functions. The process has been used for the synthesis of a copolymer of PAm and carboxylic acid terminated polymers[6] with coupling agents like the diphenyl esters of dicarboxylic acids for the preparation of poly(ester amide)s.[178] Also for the synthesis of poly(ester amide)s, the condensation reactivity in the melt has been enhanced through treatment of the polyester with ethylene glycol and later with trimellitic anhydride to provide polymers with almost 100% carboxylic end groups.[179] The process for the synthesis of a polycarbonate/polyamide (PC/PAm) copolymer is analogous: the acid chloride of the trimellitic anhydride is condensed on the PC phenolic end groups, providing this polymer with terminal anhydride functions.[180] By using reactive extrusion PC/PAm block copolymers are then obtained; but can we reject possible carbonate/amide exchanges? Multiblock PAm-6,6/PAm-6 copolymers have also been obtained by Allied Chemical.[181] The cyclic oligomer process studied by General Electric is related to this technique; and many opportunities may result from the associations of different cyclic oligomers towards the postpolymerization in multiblock macromolecules.[182] Indeed the reactive extrusion of mixtures of polymers with functional end groups has brought to light the exchange reactions between ester bridges. Accordingly, multiblock copolymer synthesis by ester bond exchange has superseded the opportunity provided by reactive end groups on macromolecules.

21.5.3 Interpolymeric Ion Bonding

The first development of ion bonding in polymer reinforcement was introduced through well-known ionomers, which are thermoplastic copolymers containing less than 15 mol% of an ionic comonomer. The presence of ionic sites drastically changes polymer properties via ionic interactions and it seems possible to use such kinds of dipole association resulting in clusters for the reinforcement of polymer blends. For example, blends of ionomers with PAm-6,6 give stable materials characterized by high toughness, excellent mouldability and good resistance to cutting.[183] More recently, various research groups, following the pioneering work of Eisenberg[184] have developed new approaches in polymer blending, taking into account the association of ionic groups present at the end of polymer chains. Teyssié and Jerôme have confirmed that the improvement of mechanical properties in blends of immiscible homopolymers containing basic and acidic end groups occurs from the formation of ionic bonds by reciprocal reactivity.[185] These mutual interactions, which lead to a better degree of miscibility, are controlled by proton transfer from the acid groups polymer A to the aliphatic tertiary amines of polymer B. The ion pair formation is nothing but a block polymerization process, since the extremities of the two immiscible polymers are now held together by an ionic bond instead of a covalent bond as in the traditional block copolymers.[186] In a recent review, Eisenberg has shown how ion–ion interactions affect miscibility even at very low concentrations.[187] This is especially true for cases in which one material is polycationic, while the other is polyanionic.

In industrial blends the compatibilizers (block or graft copolymers) are generally prepared separately by direct synthesis or by grafting or coupling reactions carried out in solution or melt processes. The needed amounts of these amphiphilic species are in the range of a few percent (3 to 10) and they are introduced during the melt-mixing process. Most of the resulting polyphasic materials exhibit a classical morphology of nearly spherical particles dispersed in a continuous matrix. The size of the dispersed phase is related to the mixing efficiency. The final stability of this microstructure, throughout the subsequent stages of manufacturing, is connected with the interfacial tension due to the presence of the compatibilizer. It appears of greatest interest to develop continuous processes for immiscible blends, in which the compatibilizer is formed *in situ* during the main mixing process. As indicated above in Sections 21.4.1 and 21.4.2, this approach was claimed in recent patents. The following examples will illustrate the same idea.

Polyamides (PAm) and polyalkenes (PA) are a typical example of polymer incompatibility, which prevents the effective blending leading to enhanced properties like increased thermal and solvent resistance due to the PAm, and hydrophobicity or dielectric properties due to the cheaper commodity plastics based on PA. The well-known industrial products sold by Du Pont as impact-reinforced PAm, containing grafted EP rubbers, and more recent blends of PP and PAm developed by Atochem (Orgalloy™) are both prepared by melt mixing, but it is not easy to know if the amphiphilic copolymers were prepared *in situ* or not. Lambla and coworkers[188] have shown that it is possible to reach good mechanical properties by direct blending of PAm (48.5%) and PP (48.5%) in a ZSK 30 twin screw extruder, in the presence of a small amount (3%) of reactive PP, grafted with maleic anhydride. Comparison of the variable morphologies by scanning microscopy, and the mechanical properties of nonreactive and reactive mixtures clearly showed the capacity of *in situ* compatibilization in blend reinforcement. An interesting observation was that the final properties were not only related to the level of grafted maleic anhydride onto PP, but also to the molecular weight of the reactive PP, which determines the reciprocal solubility of this product in the blend.

Various other systems have been mentioned in the literature, where ABS and PPE seem to play an important role. For instance, epoxy-modified ABS has been blended with carboxylic-terminated PS in extruders,[189] whereas the maleic anhydride grafted ABS was used in blends with polyamides.[190,191] Other systems containing polyurethane elastomers were also based on carboxylic acid modified ABS.[192,193] Blends of PPE and PAm show excellent mechanical and thermal properties, but the optimization of these mixtures was only possible by reactive blending, on the base of complex formulations. In recent patents,[194,195] Sumitomo claimed the same achievement by one-step grafting/compatibilizing procedures. The main advantage of *in situ* compatibilization is that the amphiphilic species are created and stay at the interface. This signifies that there is no further need for a high level of compatibilizer concentration, as the interfacial area of the dispersed phase is not very large and most of the prepared amphiphilic copolymers do not dissolve in the main phases.

Future trends will certainly deal more and more with end-coupling reactions. But this orientation requires the prior preparation of the modified polymer with multipurpose reactive end groups and the right adjustment of the level of reciprocal reactivity, taking into account the possible modification of the local medium in the interfacial area (viscosity, catalyst concentration, *etc.*).

Among other new possibilities, it is important to mention crossreactivity between two immiscible polymers, taking into account free radical processes and other kinds of potential reactivity related to the introduction of multifunctional reactive species. Recent work published by Rätzsch and coworkers[196] is a first example of this approach. On the other hand, bismaleimide and its derivatives were used for limited reciprocal crosslinking in PS/PA blends.[187]

Finally, it appears that all these new trends are based on systems where the final morphology was adjusted by mechanical shearing and interfacial tension, and the resulting particle size and shape are variable in a limited range. Future developments will also deal with the possibility of preparing blended materials with a well-defined morphology based on melt blending a thermoplastic matrix with different kinds of latex particles. It also seems feasible to introduce some reactivity in these systems and, in order to do that with the greatest efficiency, it is worthwhile to develop specific catalysis to reach high levels of conversion.

The final contribution for the increasing development of reactive blending could be related to computer aided reactive processing techniques. On-line viscosity measurements are used in controlled PP molecular weight degradation, by continuous extrusion in the presence of free radicals.[41,197,198] On-line rheometers could be helpful in many reactive systems and the coupling with FTIR spectrometry should be a good opportunity for on-line evaluation.

21.6 CONCLUSION

From this review of current developments of chemical reactions carried out in melt processes, it is reasonable to foresee a new and complementary source of diversification in polymeric materials. Continuous polymerization or polycondensation techniques have been developed on twin screw extruding devices. On the other hand, there is a need for the improvement of commodity plastics, which includes the association with engineering plastics in cohesive blend systems. Such an evolution requires the contribution of chemistry in industrially flexible and economic continuous processes. Reactive extrusion could be one of the most attractive techniques, but this new challenge will require the combined efforts of various specialists like chemists and chemical and mechanical engineers in order to convert classical extruders used commonly in processing and compounding to continuous chemical reactors, well adapted to reactions in molten polymers. Future research in this area will also be devoted to modelling of reactive extrusion systems, because from a practical point of view this can assist in choosing favourable operating conditions.

21.7 LIST OF SYMBOLS

L/D	Length over diameter ratio in extruders
ABS	Acrylonitrile–butadiene–styrene copolymer
PP	Polypropylene
MAH	Maleic anhydride
PP-*graft*-MAH	Polypropylene grafted with maleic anhydride
RIM	Reaction injection moulding process
PE	Polyethylene
HDPE	High density polyethylene
PA	Polyalkene
DBTDL	Dibutyltin dilaurate
FTIR	Fourier transform infra-red spectrometry
PA-*graft*-MAH	Polyalkene grafted with maleic anhydride
PAm	Polyamide
PP-*graft*-PAm	Grafted copolymers of PP and PAm
PPE	Poly(phenylene ether)
SBS	Styrene–butadiene tri-block copolymer
UV	Ultraviolet radiation
CTC	Charge transfer complex
DMF	Dimethylformamide
LDPE	Low density polyethylene
EVA	Ethylene–vinyl acetate copolymer
PPS	Poly(phenylene sulfide)
PEEK	Poly(ether ether ketone)
NYRIM	Nylon reaction injection moulding
IUPAC	International Union of Pure and Applied Chemistry
CSTR	Continuous stirred tank reactor
EPR	Ethylene–propylene copolymer
EPDM	Ethylene–propylene–diene monomer copolymer
PIB	Polyisobutylene
BR	Butyl rubber
PVC	Poly(vinyl chloride)
CPE	Chlorinated polyethylene
PMMA	Poly(methyl methacrylate)
EMA	Ethylene–methyl acrylate copolymer
SMA	Styrene–methyl acrylate copolymer
RPS	Reactive polystyrene
PUR	Polyurethane
PET	Poly(ethylene terephthalate)
PC	Polycarbonate

21.8 REFERENCES

1. E. M. Fettes (ed.), in 'Chemical Reactions of Polymers', Interscience, New York, 1964.
2. D. C. Sherrington, in 'Encyclopedia of Polymer Sciences & Engineering', ed. J. I. Kroschwitz, Wiley, 1988, vol. 14, pp. 101–169.
3. J. A. Biesenberger and D. H. Sebastian (eds.), in 'Principles of Polymerisation Engineering', Wiley-Interscience, New York, 1983.
4. Z. N. Frund, *Plast. Compd.*, 1986, **9**(5), 26.
5. S. B. Brown and C. M. Orlando, in 'Encyclopedia of Polymer Science & Engineering', ed. J. I. Kroschwitz, Wiley, 1988, vol. 14, pp. 169–189.
6. S. B. Brown, in 'Polymer Blends and Alloys', ed. A. V. Patsis, State University of New York, 1990, pp. 23–81, 83–116.
7. C. Tzoganakis, *Adv. Polym. Technol.*, 1989, **9**, 321.
8. J. L. White, W. Szydlowski, K. Min and M. H. Kim, *Adv. Polym. Technol.*, 1987, **7**, 295.
9. H. Herrmann, in 'Polymerreaktionen und Reaktives Aufbereiten in Kontinuierlichen Maschinen', VDI-Verlag, Düsseldorf, 1988, p. 1.
10. R. Sridharan and I. M. Mathai, *J. Sci. Ind. Res.*, 1974, **33**, 178.
11. D. C. Wahrmund, D. R. Paul and J. W. Barlow, *J. Appl. Polym. Sci.*, 1978, **22**, 2155.
12. A. M. Kotliar, *Macromol. Rev.*, 1981, **16**, 367.
13. J. Devaux, P. Godard and J.-P. Mercier, *Polym. Eng. Sci.*, 1982, **22**, 217.
14. R. S. Porter, *Thermochimica*, 1988, **134**, 251.
15. M. Lambla, J. Druz and F. Mazeres, *Plast. Rubber Process. Appl.*, 1990, **13**, 75.
16. H. Staudinger, *Ber. Dtsch. Chem. Ges.* 1930, **63**, 931.
17. M. Pike and W. F. Watson, *J. Polym. Sci.*, 1952, **9**, 229.
18. W. J. Kautzmann and H. Eyring, *J. Am. Chem. Soc.*, 1940, **62**, 3113.
19. N. K. Baremboim, in 'Mechanochemistry of Polymers', 1961.
20. Nitto Electrical Industrial Co. Ltd., *Jpn. Pat.* 58/53 901, 58/53 969, 58/53 970, 58/53 973, 58/53 974 (1983) (*Chem. Abstr.*, 1983, **99**, 159 039).
21. Nitto Electrical Industrial Co. Ltd., *Jpn. Pat.* 58/168 610 (1983) (*Chem. Abstr.*, 1984, **100**, 104 060).
22. T. A. Kotnour, R. L. Barber and W. L. Krueger (Minnesota Mining and Manufacturing Co.), *Eur. Pat.* 160 394 (1985) (*Chem. Abstr.*, 1986, **104**, 89 229).
23. S. A. Plastugil, *Fr. Pat.* 1 562 038 (1969) (*Chem. Abstr.*, 1970, **72**, 32 467).
24. N. Matsuoka, H. Matsumoto, Y. Hori, Y. Miki, K. Sano and I. Ijichi (Nitto Electrical Industrial Co. Ltd.), *Ger. Pat.* 3 305 727, *US Pat.* 4 487 897 (1984).
25. Mitsubishi Rayon Co. Ltd., *Jpn. Pat.* 59/48 121 (1984) (*Chem. Abstr.*, 1984, **101**, 24 678).
26. N. P. Stuber and M. Tirrell, *Polym. Process Eng.*, 1985, **3**, 71.
27. J. M. Lucas and A. C. Perricone (Milchem Inc.), *Ger. Pat.* 3 621 429 (1985) (*Chem. Abstr.*, 1987, **106**, 138 949).
28. H. Deibig and R. K. Belz (Belland AG.), *Eur. Pat.* 143 935, *US Pat.* 4 612 355 (1985) (*Chem. Abstr.*, 1985, **103**, 142 536).
29. D. Engler and A. Maistrovich (Minnesota Mining & Manufacturing Co.), *Eur. Pat.* 160 397 (1985) (*Chem. Abstr.*, 1986, **104**, 89 759).
30. R. K. Belz and H. Deibig (Belland AG.), *Ger. Pat.* 3 435 468 (1986) (*Chem. Abstr.*, 1986, **105**, 43 559).
31. C. L. Bodolus and D. A. Woodhead (Standart Oil Co. Ohio), *US Pat.* 4 542 189 (1983) (*Chem. Abstr.*, 1984, **100**, 87 034).
32. T. M. Sopko and R. E. Lorentz (Lubrizol Corp.), *World Pat.* 87/1 377, *US Pat.* 4 812 544 (1987) (*Chem. Abstr.*, 1987, **107**, 40 597).
33. H. Diebig and A. Dinkelaker (Belland AG.) *Eur. Pat.* 320 757 (1989) (*Chem. Abstr.*, 1989, **111**, 215 015).
34. R. W. Lee and W. J. Miloscia (Standart Oil Co. Ohio), *Eur. Pat.* 28 914 (1981) (*Chem. Abstr.*, 1981, **95**, 63 164).
35. C. L. Bodolus and W. J. Miloscia (Standart Oil Co. Ohio), *Eur. Pat.* 96 555; *US Pat.* 4 463 137 (1983) (*Chem. Abstr.*, 1984, **100**, 87 034).
36. C. L. Bodolus and D. A. Woodhead (Standart Oil Co. Ohio), *US Pat.* 4 692 498 (1987) (*Chem. Abstr.*, 1988, **108**, 168 553).
37. G. Illing (Werner & Pfleiderer), *US Pat.* 3 536 680 (1970) (*Chem. Abstr.*, 1970–1971, **74**, 32 174).
38. Resines & Vernis Artificiels, *UK Pat.* 871 686 (1961) (*Chem. Abstr.*, 1960–1961, **55**, 25 373).
39. BASF AG., *Belg. Pat.* 845 834 (1977) (*Chem. Abstr.*, 1978–1979, **88**, 7877).
40. ICI, *Neth. Pat.* 66/5 797 (1966) (*Chem. Abstr.*, 1967, **66**, 86 238).
41. K. Arisawa and R. S. Porter, *J. Appl. Polym. Sci.*, 1970, **14**, 879.
42. M. Dorn, *Adv. Polym. Technol.*, 1985, **5**, 87.
43. H. Schott and W. S. Kaghan, *SPE Trans.*, 1963, **3** (2), 145.
44. J. C. Staton, J. P. Keller and R. C. Kowalski (Esso Research & Engineering Co.), *Fr. Pat.* 1 547 299; *US Pat.* 3 551 943 (1968) (*Chem. Abstr.*, 1968–1969, **71**, 51 182).
45. R. C. Kowalski, J. W. Harrison, J. C. Staton and J. P. Keller (Esso Research & Engineering Co.), *US Pat.* 3 563 972, 3 608 001 (1971) (*Chem. Abstr.*, 1972–1973, **76**, 25 988).
46. A. T. Watson, H. L. Wilder, K. W. Bartz and R. A. Steinkamp (Exxon Research and Engineering Co.), *Ger Pat.* 2 454 650 (1975) (*Chem. Abstr.*, 1974–1975, **83**, 148 090).
47. K. Babba, T. Shiota, K. Murakami and K. Ono (Sumitomo Chemical Co.), *Jpn. Pat.* 48/79 851 (1973) (*Chem. Abstr.*, 1974–1975, **80**, 109 283).
48. D. Suwanda, R. Lew and S. T. Balke, *J. Appl. Polym. Sci.*, 1988, **35**, 1019.
49. C. Tzoganakis, J. Vlachopoulos and A. E. Hamielec, *Polym. Eng. Sci.*, 1988, **28**, 170; *Chem. Eng. Prog.*, 1988, **84** (11), 47.
50. C. Tzoganakis, Y. Tang, J. Vlachopoulos and A. E. Hamielec, *Polym. Plast. Techanol. Eng.*, 1989, **28**, 319.
51. A. Pabedinskas, W. R. Cluett and S. T. Blake, *Polym. Eng. Sci.*, 1989, **29**, 993.
52. P. Nogués (Atochem), *Eur. Pat.* 177 401; *US Pat.* 4 735 992; *Fr. Pat.* 2 572 417; *US Pat.* 4 727 120 (1986) (*Chem. Abstr.*, 1986, **105**, 61 271).
53. A. Shimizu and M. Yuzawa (Mitsui Petrochemical Industries Ltd.), *Jpn. Pat.* 50/6019, 6020 (1975) (*Chem. Abstr.*, 1974–1975, **83**, 80 026).
54. R. Polster, C. Alt, H. Lautenschlager and G. Schmidt-Thomee (BASF AG.), *Ger. Pat.* 1 158 264 (1963) (*Chem. Abstr.*, 1962–1966, **61**, 10 840).

55. BASF AG., *Fr. Pat.* 2 007 125, 2 033 134 (1970) (*Chem. Abstr.*, 1970–1971, **73**, 131 693).
56. K. B. Abbas and R. S. Porter, *J. Appl. Polym. Sci.*, 1976, **20**, 1289.
57. D. C. Edwards and D. Padliya (Polysar Ltd.), *US Pat.* 4 614 772 (1986) (*Chem. Abstr.*, 1987, **106**, 19 230).
58. K. Kirimoto, H. Matsunaga and K. Wakida (UBE Industries Ltd.), *Jpn. Pat.* 50/148 471 (1975) (*Chem. Abstr.*, 1976–1977, **84**, 91 807).
59. H. G. Scott (Midland Silicones), *Fr. Pat.* 2 030 899 (1970) (*Chem. Abstr.*, 1970–1971, **75**, 37 298).
60. H. G. Scott and J. F. Humphries, *Mod. Plast.*, 1973, **50**(3), 82.
61. BICC Ltd. and Maillefer S. A., *Neth Pat.* 75/14 222; *US Pat.* 4 117 195 (1976) (*Chem. Abstr.*, 1976–1977, **86**, 30 589).
62. G. M. Gale, *Soc. Plast. Eng., [Tech. Pap.]*, 1983, **29**, 109.
63. G. M. Gale, *Soc. Plast. Eng., [Tech. Pap.]*, 1984, **30**, 2.
64. R. R. Wall and B. G. Willoughby, in 'Reactive Processing: Practice and Possibilities', RAPRA Seminar, Sept. 1989.
65. S. Ultsch and H. G. Fritz, *Plast. Rubber Process. Appl.*, 1990, **13**, 81.
66. Fujikura Cable Work Ltd., *Jpn. Pat.* 60/6045 (1985) (*Chem. Abstr.*, 1985, **102**, 222 210).
67. H. G. Fritz and S. Ultsch, *Kunststoffe*, 1989, **79**, 1051.
68. S. Al-Malaika, *ACS Symp. Ser.*, 1988, **364**, 409.
69. M. Lambla, R. X. Yu and S. Lorek, *ACS Symp. Ser.*, 1989, **395**, 67.
70. T. S. Grant and D. V. Howe (Borg-Warner Chemicals, Inc.), *US Pat.* 4 740 552 (1988).
71. C. Carrot, Ph.D. Dissertation, Université Jean Monnet, Saint-Etienne, France (1990).
72. Asahi Chemical Industry Co. Ltd., *Jpn. Pat.* 56/115 307 (1981) (*Chem. Abstr.*, 1981, **95**, 220 812).
73. Y. Akiyama and T. Shiraki (Asahi Chemical Industry Co. Ltd.), *Jpn. Pat.* 62/250 018 (1987) (*Chem. Abstr.*, 1988, **108**, 151 512).
74. C. Taubitz, E. Seiler and L. Schlemmer (BASF AG.), *Ger. Pat.* 3 540 118, 3 540 119 (1987) (*Chem. Abstr.*, 1987, **107**, 177 118 and 154 994).
75. D. J. McFay, P. D. Sybert, W. L. Gately, C. Y. Han, J. A. Tyrell, S. B. Brown and R. A. Florence (General Electric Co.), *Eur. Pat.* 248 263; *US Pat.* 4 732 937 (1987) (*Chem. Abstr.*, 1988, **108**, 205 668).
76. S. B. Brown and D. J. McFay (General Electric Co.), *World Pat.* 87/7 281 (1987) (*Chem. Abstr.*, 1988, **109**, 74 179).
77. R. van der Meer and J. B. Yates (General Electric Co.), *World Pat.* 87/540 (1987) (*Chem. Abstr.*, 1987, **106**, 197 429).
78. C. Taubitz, H. Hausepohl, E. Seiler and L. Schlemmer (BASF AG.), *Ger. Pat.* 3 621 207 (1988) (*Chem. Abstr.*, 1989, **111**, 79 241).
79. J. W. Rekers and G. Scott (Millichem Research Corp.), *US Pat.* 4 743 657 (1988) (*Chem. Abstr.*, 1988, **109**, 111 571).
80. G. Scott, *Makromol. Chem., Macromol. Symp.*, 1989, **28**, 59.
81. S. Al-Malaika, *Polym. Plast. Technol. Eng.*, 1990, **29**, 73.
82. H. Korber (Bayer AG.), *Ger. Pat.* 2758785 (1979) (*Chem. Abstr.*, 1978–1979, **91**, 92281).
83. V. Aslan, M. Ionescu, V. Vasut, P. Viorel, C. Popa and I. Tincul (Intreṕrinderea de Prelucrare Mase Plastice), *Rom. Pat.* 82 857 (1984) (*Chem. Abstr.*, 1986, **104**, 69 886).
84. R. W. Lee and W. J. Miloscia (Standart Oil Co. Ohio), *US Pat.* 4 410 659 (1983) (*Chem. Abstr.*, 1981, **95**, 63 164).
85. W. H. Gift (Plaskolite Inc.), *US Pat.* 4 384 077 (1983) (*Chem. Abstr.*, 1983, **99**, 39 293).
86. C. Taubitz, E. Seiler and L. Schlemmer (BASF AG.), *Ger. Pat.* 3 540 116, 3 540 117 (1987) (*Chem. Abstr.*, 1987, **107**, 176 760).
87. S. Al-Malaika, A. Q. Ibrahim and G. Scott, *UK Appl.* 87/6 569 (1987).
88. G. D. Jones and R. M. Nowak (Dow Chemical Co.), *US Pat.* 3 177 270 (1966) (*Chem. Abstr.*, 1965, **63**, 1953).
89. A. Nakanishi, S. Shinkai, H. Shimizu and H. Kawasaki (Asahi Dow Ltd.), *Jpn. Pat.* 47/43 186 (1972) (*Chem. Abstr.*, 1974–1975, **80**, 121 981).
90. T. Nomura, T. Kuwahara and H. Tashiro (Asahi Dow Ltd.), *Jpn. Pat.* 49/112 945 (1974) (*Chem. Abstr.*, 1974–1975, **82**, 99 493).
91. G. Zeitler, L. Hoerh and H. Mueller-Thamm (BASF AG.), *Ger. Pat.* 2 323 030 (1974) (*Chem. Abstr.*, 1974–1975, **82**, 99 305).
92. S. Shimizu and K. Kugimiya (Mitsui Petrochemical Industries Ltd.), *Jpn. Pat.* 52/84 276 (1977) (*Chem. Abstr.*, 1976–1977, **87**, 185 833).
93. E. N. Peters, D. C. Bookbinder, D. W. Fox and G. F. Smith (General Electric Co.), *Eur. Pat.* 264 623 (1988) (*Chem. Abstr.*, 1988, **109**, 150 667).
94. *Mod. Plast. Int.* 1990, **20**(5), 36.
95. Asahi Chemical Industry Co. Ltd., *Fr. Pat.* 1 533 691 (1968) (*Chem. Abstr.*, 1968–1969, **71**, 40 016).
96. V. Aslan, D. Munteanu and I. Tincul (Combinatul Petrochemia Pitesti), *Rom. Pat.* 60 243 (1976), (*Chem. Abstr.*, 1978–1979, **89**, 44 599).
97. C. S. Wong (Du Pont Canada Inc.), *Eur. Pat.* 266 994 (1988) (*Chem. Abstr.*, 1989, **110**, 39 672).
98. Unitika Ltd., *Jpn. Pat.* 57/174 229 (1982) (*Chem. Abstr.*, 1983, **98**, 161 971).
99. S. Al-Malaika, S. Honggokosomo and G. Scott, *Polym. Degradation Stab.*, 1986, **16**, 25.
100. M. Ghaemey and G. Scott, *Polym. Degradation Stab.*, 1981, **3**, 405.
101. G. Scott and E. Setoudeh, *Polym. Degradation Stab.*, 1983, **5**, 1.
102. N. G. Gaylord and R. Metha, *J. Polym. Sci., Polym. Chem. Ed.*, 1988, **26**, 1189; 1903.
103. N. G. Gaylord, *US Pat.* 4 506 056 (1985) (*Chem. Abstr.*, 1985, **103**, 6897).
104. V. Banzi and R. Fabbri (Ausimont S.p.a.), *Eur. Pat.* 187 660 (1986).
105. H. J. Vroomans (Stamicarbon BV.), *Eur. Pat.* 286 734 (1988) (*Chem. Abstr.*, 1989, **110**, 76 354).
106. C. A. Starit, G. M. Lancaster and R. L. Tabor (Dow Chemical Co.), *US Pat.* 4 762 890 (1988) (*Chem. Abstr.*, 1989, **109**, 191 143).
107. Mitsui Polychemicals Co. Ltd., *Ger. Pat.* 2 023 154 (1970).
108. B. Schulz and M. Hartmann, *Plaste Kautsch.*, 1985, **32**, 84.
109. C. S. Wong (Du Pont Canada Inc.), *Eur. Pat.* 280 454 (1988) (*Chem. Abstr.*, 1989, **110**, 39 572).
110. T. Sakai, *Soc. Plast. Eng., [Tech. Pap.]*, 1988, **34**, 1853.
111. A. Hogt, Polymer Processing Society, Autumn Meeting ROLDUC, Netherlands, 1989.
112. Licentia Patent-Verwaltungs GmbH, *Fr. Pat.* 1 541 962 (1968) (*Chem. Abstr.*, 1968–1969, **71**, 13 680).
113. C. J. Arnaud, M. Quemner and G. P. Roche (Soc. de Télécommunications), *Belg. Pat.* 837 488 (1976) (*Chem. Abstr.*, 1976–1977, **86**, 73 927).

114. S. G. Foord, W. E. Simpson and R. K. Stapleton (Standard Telephones Cables, Ltd.), *UK Pat.* 1 092 225 (1967) (*Chem. Abstr.*, 1968–1969, **68**, 22 536).
115. R. Zouikri, Thesis Dissertation D. III, Universitié Louis Pasteur, Strasbourg, France, 1982.
116. I. Chodak, K. Fabianova, E. Borsig and M. Lazar, *Angew. Makromol. Chem.*, 1978, **69**, 107.
117. M. Raetzsch, *Makromol. Chem., Macromol. Symp.*, 1989, **28**, 11.
118. M. Raetzsch and M. Schönefeld, *Plaste Kautsch.*, 1972, **19**, 565.
119. H. W. Hill and D. G. Brady, in 'Encyclopedia of Polymer & Science Engineering', ed. J. I. Kroschwitz, Wiley, 1988, vol. 11, p. 531.
120. D. C. Christensen and R. V. Voss, *US Pat.* 4 532 310 (1984) (*Chem. Abstr.*, 1985, **103**, 142 883).
121. C. M. Chan and S. Venkatrama, *J. Appl. Polym. Sci.*, 1986, **32**, 5933.
122. C. M. Chan and A. C. Tannous (Raychem Corp.), *US Pat.* 4 616 056 (1986) (*Chem. Abstr.*, 1987, **106**, 6033).
123. V. N. Ignatov, V. A. Vasnev and S. V. Vinogradova, *Polym. Sci. USSR (Engl. Transl.)*, 1987, **29**, 993.
124. P. J. Flory, *J. Am. Chem. Soc.*, 1939, **61**, 3334.
125. P. J. Flory, 'Principles of Polymer Chemistry', Cornell, Ithaca, NY, 1953, p. 672.
126. D. Muller (SNPA), *Fr. Pat.* 70 300 91 (1970).
127. G. Illing, *Mod. Plast*, 1969, **46**, 70.
128. G. Menges and T. Bartilla, *Polym. Eng. Sci.*, 1987, **27**, 1216.
129. T. Bartilla, in 'Polymerreaktionen und Reaktives Aufbereiten in Kontinuierlichen Maschinen', VDI-Verlag, Düsseldorf, 1988, p. 117.
130. U. Berghaus, in 'Polymerreaktionen und Reaktives Aufbereiten in Kontinuierlichen Maschinen', VDI-Verlag, Düsseldorf, 1988, p. 143.
131. H. D. Aeppli, in 'Polymerreaktionen und Reaktives Aufbereiten in Kontinuierlichen Maschinen', VDI-Verlag, Düsseldorf, 1988, p. 105.
132. P. Franz, in 'Polymerreaktionen und Reaktives Aufbereiten in Kontinuierlichen Maschinen', VDI-Verlag, Düsseldorf, 1988, p. 211.
133. U. Barth, in 'Polymerreaktionen und Reaktives Aufbereiten in Kontinuierlichen Maschinen', VDI-Verlag, Düsseldorf, 1988, p. 179.
134. B. Quiring, G. Niederdellmann and H. Wagner (Bayer AG.), *US Pat.* 4 245 081 (1981).
135. M. Ullrich, E. Meisert and A. Eitel, (Bayer AG.), *US Pat.* 3 963 679, (1976) (*Chem. Abstr.*, 1974–1975, **82**, 5219).
136. H. Herrmann, in 'Proceedings of Conference on Reactive Production of Polymeric Goods, Zagreb, 1990', pp. 05-1–05-10.
137. G. M. Kosanovich and G. Salee (Occidental Chemical Co.), *US Pat.* 4 415 721 (1983) (*Chem. Abstr.*, 1982, **96**, 52 927).
138. L. R. Schmidt, E. Lovgren and P. G. Meissner, in 'Polymerreaktionen und Reaktives Aufbereiten in Kontinuierlichen Maschinen', VDI-Verlag, Düsseldorf, 1988, p. 193.
139. R. C. Kowalski and N. F. Newman (Exxon Research & Engineering Co.), *Eur. Pat.* 76 173; *US Pat.* 4 384 072, 4 486 575, 4 501 859 (1983); *Eur. Pat.* 124 278; *US Pat.* 4 548 995 (1984); *Eur. Pat.* 124 279 (1984) (*Chem. Abstr.*, 1985, **102**, 47 175 and 46 471).
140. N. F. Newman and R. C. Kowalski (Exxon Research & and Engineering Co.), *Br. Pat.* 83/1 822 (1984) (*Chem. Abstr.*, 1984, **101**, 172 842).
141. C. Mijangos, A. Martinez and A. Michel, *Eur. Polym. J.*, 1986, **22**, 417.
142. K. Mori, Y. Nakamura and T. Hayakari, *Angew. Makromol. Chem.*, 1978, **66**, 169.
143. A. Michel, M. Gonnu and E. Koerper, in Proceedings of the 3rd Annual Meeting of the Polymer Processing Society, Stuttgart, 1987', abstracts 1–9.
144. C. Song, K. Li and S. Li, *Int. Polym. Process.*, 1987, **2**, 83.
145. A. Ghaffar, C. Sadrmohaghegh and G. Scott, *Polym. Degradation Stab.*, 1981, **3**, 342.
146. Y. Nakamura, A. Watanabe, K. Mori, K. Tamura and H. Miyazaki, *J. Polym. Sci., Polym. Lett. Ed.*, 1987, **25**, 127.
147. J. A. Sneller, *Mod. Plast. Int.*, 1985, **15** (8), 42.
148. *Mod. Plast. Int.*, 1987, **17** (4), 27.
149. C. S. Tucker and R. J. Nichols, *Plast. Eng.*, 1987, **43** (5), 27.
150. M. Lambla, *Polym. Process Eng.*, 1988, **5**, 297.
151. R. M. Kopchik (Rohm and Haas Co.), *Ger. Pat.* 2 562 118; *US Pat.* 4 246 374 (1977) (*Chem. Abstr.*, 1976–1977, **87**, 53 884).
152. Toray Industries Inc., *Jpn. Pat.* 58/84 855 (1983) (*Chem. Abstr.*, 1983, **99**, 106 322).
153. M. P. Hallden-Abberton, N. M. Bortnick, L. A. Cohen, W. T. Freed and H. C. Formuth (Rohm and Hass Co.), *Eur. Pat.* 216 505, *US Pat.* 4 727 117 (1987) (*Chem. Abstr.*, 1987, **107**, 134 892).
154. O. Koch and H. Waniczek (Bayer AG.), *Ger. Pat.* 3 430 802 (1986) (*Chem. Abstr.*, 1986, **104**, 207 938).
155. L. G. Bourland, M. E. London and T. A. Cooper, in 'Reactive Processing: Practice and Possibilities', RAPRA Seminar, Sept. 1989.
156. Y. Inoue, T. Koyama, S. Date and M. Moritani, in 'Proceedings of the 6th Annual Meeting of the Polymer Processing Society, Nice, 1990', abstracts 1–15.
157. M. Raetzsch, U. Hofmann, M. Gebauer, G. Hoffmann, G. Bergmann and H. Schade (VEB), *Br. Pat.* 2 116 981 (1983) (*Chem. Abstr.*, 1983, **99**, 213 095).
158. R. L. Saxton (Du Pont), *US Pat.* 4 338 405 (1982) (*Chem. Abstr.*, 1982, **97**, 110 582).
159. T. Kumimoto, M. Morikawa, K. Aoki and Y. Urata (Toray Industries Inc.), *Jpn. Pat.*. 47/30 932 (1972) (*Chem. Abstr.*, 1972–1973, **78**, 137 824).
160. M. Raetzsch, J. Geyer and J. Oswald, *Ger. (East) Pat.* 107 938 (1974) (*Chem. Abstr.*, 1974–1975, **82**, 140 828).
161. D. M. McClain, B. L. Vest (National Distillers and Chemical Corp.), *US Pat.* 3 972 865 (1976) (*Chem. Abstr.*, 1976–1977, **85**, 144 021).
162. A. Bouilloux, J. Druz and M. Lambla, *Polym. Process. Eng.*, 1986, **3**, 235.
163. M. Lambla, J. Druz and Bouilloux, *Polym. Eng. Sci.*, 1987, **27**, 1221.
164. G. H. Hu and M. Lambla, Polymer Processing Society, Autumn Meeting ROLDUC, Netherlands, 1989.
165. S. Lorek, Ph.D. Dissertation, Université Louis Pasteur, Strasbourg, France, 1989.
166. G. H. Hu, Ph.D. Dissertation, Université Louis Pasteur, Strasbourg, France, 1990.
167. M. Lambla, J. Druz and N. Satyanarayana, *Macromol. Chem.*, 1988, **189**, 2703.
168. *Mod. Plast.*, 1985, **62**, 56.

169. P. van Ballegooie and A. Rudin, *Polym. Mater. Sci. Eng.*, 1988, **58**, 835; *Polym. Eng. Sci.*, 1988, **28**, 1434; *J. Polym. Sci., Polym. Chem. Ed.*, 1988, **26**, 2449.
170. S. Kitagawa and I. Okada, *Polym. Bull.*, 1983, **10**, 109, 196.
171. S. Kitagawa and I. Okada (Mitsubishi Petrochemical Co. Ltd.), *Ger. Pat.* 3 021 273; *US Pat.* 4 366 296 (1981) (*Chem. Abstr.*, 1981, **94**, 104 354).
172. S. Gotoh, M. Fujii and S. Kitagawa, *Soc. Plast. Eng., [Tech. Pap.]*, 1986, **32**, 54.
173. R. Tizmann, H. Thaler and J. Walter (Hoechst AG.), *Ger. Pat.* 1 929 149; *US Pat.* 3 657 191 (1970) (*Chem. Abstr.*, 1970–1971, **74**, 65 437).
174. G. L. Korver (Goodyear Tire & Rubber Co.), *Ger. Pat.* 2 615 156; *US Pat.* 4 071 504 (1976) (*Chem. Abstr.*, 1976–1977, **86**, 44 674).
175. J. M. Barnewall and A. M. Scheibelhoffer (Goodyear Tire Rubber Co.), *Ger. Pat.* 2 458 701; *US Pat.* 3 975 329 (1975) (*Chem. Abstr.*, 1974–1975, **83**, 180 939).
176. S. Matsumara, H. Inata, M. Ogasawara, T. Morinaga and A. Norike (Teijin Ltd.), *Eur. Pat.* 30 350; *US Pat.* 4 351 936 (1981) (*Chem. Abstr.*, 1981, **95**, 170 399).
177. H. Inata, M. Ogasawara, T. Morinaga and A. Norike (Teijin Ltd.), *US Pat.* 4 196 066 (1980), 4 269 947 (1981) (*Chem. Abstr.*, 1978–1979, **90**, 138 629).
178. V. S. Levin, V. I. Korostelev, I. V. Romanov, I. M. Nosalevich and V. V. Spesiotseva, *Sov. Pat.* 658 142 (1979) (*Chem. Abstr.*, 1978–1979, **91**, 40 322).
179. British Industrial Plastics Ltd., *Neth. Pat.* 77/6 820 (1977) (*Chem. Abstr.*, 1978–1979, **89**, 25 319).
180. S. J. Hathaway and R. A. Pyles (General Electric Co.), *US Pat.* 4 732 934 (1988); 4 800 218 (1989) (*Chem. Abstr.*, 1988, **109**, 55 972).
181. Y. P. Khanna, E. A. Turi, S. M. Aharoni and T. Largman (Allied Chemical), *US Pat.* 4 417 032 (1983) (*Chem. Abstr.*, 1984, **100**, 52 525).
182. *Mod. Plast. Int.*, 1979, **19**(12), 5.
183. C. U. Pittman, *J. Polym. News*, 1990, **15**, 7.
184. P. Smith and A. Eisenberg, *J. Polym. Sci,, Polym. Lett. Ed.*, 1983, **21**, 223.
185. R. Fayt, R. Jerôme and P. Teyssié, *ACS Symp. Ser.*, 1989, **395**, 38.
186. M. Rutkowska and A. Eisenberg, *Macromolecules*, 1984, **17**, 821.
187. A. Natanshon, R. Murali and A. Eisenberg, *CHEMTECH.*, 1990, 418.
188. M. Lambla, R. X. Yu and S. Lorek, *ACS Symp. Ser.*, 1989, **395**, 67.
189. C. Taubitz, E. Seiler, R. Bruessau and D. Wagner (BASF AG.), *Eur. Pat.* 285 968 (*Chem. Abstr.*, 1989, **110**, 96 556).
190. M. K. Akkapeddi, B. van Buskirk and A. C. Brown (Allied Signal Co.), *World Pat.* 88/8 433 (1988) *Chem. Abstr.*, 1989, **110**, 193 994).
191. O. Nakazima and S. Izawa (Asahi Chemical Industry Co. Ltd.), *Eur. Pat.* 282 664; *US Pat.* 4 863 996 (1988) (*Chem. Abstr.*, 1989, **110**, 39 930).
192. W. Goyert, W. Grimm, A. Awater, H. Wagner and B. Krieger (Bayer AG.), *Ger. Pat.* 2 854 406 (1980) (*Chem. Abstr.*, 1980, **93**, 115 402).
193. W. Goyert, A. Awater, W. Grimm, K. H. Ott, W. Oberkirch and H. Wagner (Bayer AG.), *Ger. Pat.* 2 854 407 (1980) (*Chem. Abstr.*, 1980, **93**, 115 403).
194. H. Abe, T. Nishio, Y. Suzuki, T. Sanada, S. Hosada and T. Okada (Sumitomo Chemical Co. Ltd.), *Eur. Pat.* 295 103 (1988) (*Chem. Abstr.*, 1989, **110**, 76 830).
195. H. Abe, T. Nishio, Y. Suzuki and T. Sanada (Sumitomo Chemical Co. Ltd.), *Eur. Pat.* 308 255 (1989) (*Chem. Abstr.*, 1989, **111**, 234 322).
196. M. Raetzsch and J. Pointek, in 'Proceedings of the Conference on Reactive Production of Polymeric Goods, Zagreb, 1990'.
197. H. G. Fritz, *Kunststoffe*, 1985, **75**, 785.
198. H. G. Fritz, H. Herrmann, M. J. Nettelnbreker, H. Ockert and B. Stoehrer (Werner & Pfleiderer), *Ger. Pat.* 3 610 159 (1987) (*Chem. Abstr.*, 1988, **108**, 57 221).

Guide to the Polymer Literature

RUTH MURRAY

University of Akron, OH, USA

1	MAJOR WORKS IN POLYMER CHEMISTRY	643
2	HANDBOOKS AND ENCYCLOPEDIAS	646
3	GENERAL AND TEXTBOOKS	647
4	POLYMER PROPERTIES	648
5	POLYMER STRUCTURE	650
6	POLYMER PROCESSING AND SPECIALTY POLYMERS	652
7	POLYMERIZATION	655
8	POLYMER REACTIONS	655
9	HISTORIES	656

1 MAJOR WORKS IN POLYMER CHEMISTRY

The polymer industry is the fastest-growing segment of the world economy. It is difficult and time consuming to keep track of new developments. The literature in this field is voluminous with an abundance of review books and publications on specialized material appearing regularly.

There are, however, major works without which the polymer field would not have progressed to the high degree in which it exists today. Some of these publications will be identified below.

Alfrey, T., 'Mechanical Behavior of High Polymers', Interscience, New York, 1948. Fundamental principles of mechanical behavior and how this behavior correlates with the molecular structure (*High Polym.*, vol. 6).

Birshtein, T. M. and O. B. Ptitsyn, 'Conformations of Macromolecules', Interscience, New York, 1966. A presentation of the principal concepts and mathematical methods which form the basis of macromolecular conformations (*High Polym.*, vol. 22).

Bovey, F. A., I. M. Kolthoff, A. I. Medallia and E. J. Meehan, 'Emulsion Polymerization', Interscience, New York, 1955. A study of the kinetics and mechanism of emulsion polymerization, of tests for structure, properties and quality of the rubber formed. The standard GR-S recipe is used (*High Polym.*, vol. 9).

Carswell, T. S., 'Phenoplasts: Their Structure, Properties and Chemical Technology', Interscience, New York, 1947. History and development of the synthetic resin industry beginning with Baekelite's patent in 1909, and the chemical structure, properties and applications of resins (*High Polym.*, vol. 7).

Chapiro, A., 'Radiation Chemistry of Polymeric Systems', Interscience, New York, 1962. A survey of the polymerization processes that are initiated by radiation, the effects of radiation on polymers both in solid form and in solution, and the preparation of graft copolymers by radiation techniques (*High Polym.*, vol. 15).

Davis, C. C. and J. T. Blake (eds.), 'The Chemistry and Technology of Rubber', Reinhold, New York, 1937. The existing knowledge of the time is collected and discussed critically in the light of theory and practice by way of 25 chapters authored by eminent rubber chemists.

Fettes, E. M. (ed.), 'Chemical Reactions of Polymers', Interscience, New York, 1964. A survey of the various types of chemical reactions in which at least one of the reacting species is polymeric (*High Polym.*, vol. 19).

Flory, P. J., 'Principles of Polymer Chemistry', Cornell University Press, Ithaca, NY, 1953. Pioneer study in the theory of polymer configuration, polymer crystallization, liquid crystals and light scattering.

Frisch, K. C. (ed.), 'Cyclic Monomers', Wiley–Interscience, New York, 1972. A discussion of the starting cyclic monomers, their synthesis, reactions and properties (*High Polym.*, vol. 26).

Gaylord, N. G. (ed.), 'Polyethers', Interscience, New York, 1963. Part I, Polyalkylene Oxides and Other Polyethers; Part II, Epoxy Resins; Part III, Polyalkylene Sulfides and Other Poly(thioethers) (*High Polym.*, vol. 13).

Ham, G. E. (ed.), 'Copolymerization', Interscience, New York, 1964. The theory of copolymerization is developed, and its practical application to industrial copolymers is pointed out (*High Polym.*, vol. 18).

Kennedy, J. P. and E. G. M. Tornquist (eds.), 'Polymer Chemistry of Synthetic Elastomers', Interscience, New York, 1968 and 1969 (2 parts). A collection of treatises showing various aspects of the chemistry involved in the formation of common elastomers by different mechanisms—redox, cationic, anionic, condensation, alkali metal catalysts. A history and status of synthetic elastomers, and the structure–property relationship for elastomer materials is also presented (*High Polym.*, vol. 23).

Kline, G. M. (ed.), 'Analytical Chemistry of Polymers', Interscience, New York, 1962 (3 parts). A summary of the knowledge at that time of measurement techniques for the determination of structure and composition of macromolecules (*High Polym.*, vol. 12).

Leonard, E. C. (ed.), 'Vinyl and Diene Monomers', Wiley–Interscience, New York, 1970 (three volumes). These volumes emphasize the commercial manufacture, laboratory synthesis, analysis and purification of properties (*High Polym.*, vol. 24).

Mark, H., 'Physical Chemistry of High Polymeric Systems', Interscience, New York, 1940. A comprehensive description of substances and the relationships of their properties to structure (*High Polym.*, vol. 2).

Mark, H. and R. Raff, 'High Polymeric Reactions', Interscience, New York, 1941. (G. M. Burnett's review edition of this book was published in 1954.) The first part presents the kinetics of forming high polymers, methods of preparation, and the quantitative measurement of polyreactions; the second part collects the literature on the polymerization of organic and inorganic compounds (*High Polym.*, vol. 3).

Mark, H. and G. S. Whitby (eds.), 'Collected Papers of Wallace Hume Carothers on High Polymeric Substances', Interscience, New York, 1940. A collection of the papers of Carothers' studies of polymerization and polymers. One section is devoted to his studies on polymerization and ring formation, the other to condensation polymers (*High Polym.*, vol. 1).

Meyer, K. H., 'Natural and Synthetic High Polymers: A Textbook and Reference Book for Chemists and Biologists', 2nd edn., Interscience, New York, 1950. A theoretical review of the subject since the appearance of the first edition in 1940 (*High Polym.*, vol. 4).

Morawetz, H., 'Macromolecules in Solution', 2nd edn., Wiley, New York, 1975. A study of all solutions of macromolecules is presented in updating the first edition (*High Polym.*, vol. 21).

Odian, G., 'Principles of Polymerization', 2nd edn., Wiley, New York, 1970. Details of the three types of polymerization—step, chain and ring-opening—are presented.

Ott, E., H. M. Spurlin and M. W. Grafflin (eds.), 'Cellulose and Cellulose Derivatives', 2nd edn., parts 1, 2 and 3; Bikales N. M., and L. Segal (eds.), parts 4 and 5, Interscience, New York, 1971. The first edition published in 1943 was considered the most authoritative description of the field in the English language. The second edition covers the important advances in cellulose science and technology (*High Polym.*, vol. 5).

Raff, R. A. V. and J. B. Allison, 'Polyethylene', Interscience, New York, 1956. A summary of the then known information on the subject—preparation, structure, properties, analysis and testing, properties and applications (*High Polym.*, vol. 11).

Raff, R. A. V. and K. W. Doark (eds.), 'Crystalline Olefin Polymers', Interscience, New York, 1963 (two volumes). Presentation of the fundamental principles of polyethylenes and crystalline alkene polymers, as well as their properties, processing technology and applications (*High Polym.*, vol. 20).

Saunders, J. H. and K. C. Frisch, 'Analytical Chemistry of the Polyurethanes, Parts I & II'; D. J. David and H. B. Staley, 'Analytical Chemistry of the Polyurethanes, Part III', Wiley–Interscience, New York, 1969. The first volumes cover the chemistry, preparation of raw materials, reactions between the components to form the polymers and practical applications of the products. The third volume compiles the analytical procedures which have been applied to urethane chemistry (*High Polym.*, vol. 16).

Schildknecht, C. E., 'Allyl Compounds and their Polymers (Including Polyolefins)', Wiley–Interscience, New York, 1973. A survey of 30 allyl compounds, their preparation and properties, and their polymers and applications in plastics, fibers, synthetic rubbers and adhesives.

Schildknecht, C. E. (ed.), 'Polymer Processes: Chemical Technology of Plastics, Resins, Rubbers, Adhesives and Fibers', Interscience, New York, 1956. The basic processes used in the polymer field are outlined and theory is related to practice (*High Polym.*, vol. 10).

Schildknecht, C. E. and I. Skeist (eds.), 'Polymerization Processes', Wiley–Interscience, New York, 1970. Presents the advances of the last 20 years in the fields of graft and block copolymerizations, polymerizations with Ziegler-type catalysts, by oxidative coupling, and the use of radiation (*High Polym.*, vol. 29).

Stille, J. K. and T. W. Campbell (eds.), 'Condensation Monomers', Wiley–Interscience, New York, 1972. Outlines the kinds of polymerization reactions that the monomers will enter into, their properties, preparation and commercial applications (*High Polym.*, vol. 27).

Tobolsky, A. V., 'Properties and Structure of Polymers', Wiley, New York, 1960. An outline of the basic principles of the mechanical behavior of polymers, showing how the formulations were developed from the experimental facts.

Treloar, L. R. G., 'Physics of Rubber Elasticity', 3rd edn., Clarendon Press, Oxford, 1975. Presents the main developments in the field of the equilibrium elastic properties of rubber, and the associated theoretical background.

Turner, A., J. J. Bohrer and H. Mark, 'Copolymerization', Interscience, New York, 1953. The mechanism of copolymerization itself is explained (*High Polym.*, vol. 8).

Volkenstein, M. V., 'Configurational Statistics of Polymeric Chains', Interscience, 1963. The general concept of a macromolecule as a rotational–isomeric cooperative system is presented (*High Polym.*, vol. 17).

Wall, L. A. (ed.), 'Fluoropolymers', Wiley–Interscience, New York, 1972. A review of all known chemical and physical aspects of these materials (*High Polym.*, vol. 25).

Recognizing that the field of polymers was progressing very quickly, Mark and Immergut initiated a series of 'Polymer Reviews' to emphasize the principles which caused the new branch, and to review the actual state of knowledge while the field was still being developed. The publications of this series are listed below. They continued for almost a decade.

Andrianov, K. A., 'Metalorganic Polymers' (translation from the Russian), Interscience, New York, 1965. An evaluation of the known data with respect to the synthesis and properties that relate to composition and structure (*Polym. Rev.*, vol. 8).

Battaerd, H. A. J. and G. W. Tregear, 'Graft Copolymers', Interscience, New York, 1969. A compilation of all known data on many polymers and methods of preparation (*Polym. Rev.*, vol. 16).

Bovey, F. A., 'The Effects of Ionizing Radiation on Natural and Synthetic High Polymers', Interscience, New York, 1958. A review of the fundamental aspects of how crosslinking and other reactions of polymer chains under radiation come about. Polymers reviewed: hydrocarbon, acrylate and methacrylate, chloro and fluoro, dialkene, condensation and natural polymers (*Polym. Rev.*, vol. 1).

Cassidy, H. G. and K. A. Kun, 'Oxidation–Reduction Polymers (Redox Polymers)', Interscience, New York, 1965. An introduction to the subject (*Polym. Rev.*, vol. 11).

Frazer, A. H., 'High Temperature Resistant Polymers', Interscience, New York, 1968. Thermal stability is defined, how this is evaluated, and the utilization of many polymers in various applications that require high temperature resistance is discussed (*Polym. Rev.*, vol. 17).

Furukawa, J. and T. Saegusa, 'Polymerization of Aldehydes and Oxides', Interscience, New York, 1963. The history of research on polyethers is reviewed and recent studies of new polymerization methods, new catalysts, polymerization mechanisms and properties of the resulting polymers are discussed (*Polym. Rev.*, vol. 3).

Gaylord, N. G. and H. F. Mark, 'Linear and Stereoregular Addition Polymers: Polymerization with Controlled Propagation', Interscience, New York, 1959. Information on the then new branch of polymer chemistry, stereospecificity, following Ziegler's discovery (*Polym. Rev.*, vol. 2).

Geil, P. H., 'Polymer Single Crystals', Interscience, New York, 1963. A critical review of the current state of morphological information on crystalline polymers (*Polym. Rev.*, vol. 5).

Hummel, D. O., 'Infrared Spectra of Polymers in the Medium and Long Wavelength Regions', Interscience, New York, 1966.

Ke, B. (ed.), 'Newer Methods of Polymer Characterization', Interscience, New York, 1964. A comprehensive presentation of the newer methods, which included infrared, optical fluorescence, small angle diffraction, NMR and thermal analysis (*Polym. Rev.*, vol. 6).

Madorsky, S. L., 'Thermal Degradation of Organic Polymers', Interscience, New York, 1964. A compilation of the existing knowledge on polymers and copolymers of styrene, alkenes, halocarbons, vinyl acetate, acrylonitrile, butadiene, isoprene, poly(ethylene terephthalate), polybenzyl, polyxylene, phenol, formaldehyde resin and cellulosic polymers (*Polym. Rev.*, vol. 7).

Morgan, P. M., 'Condensation Polymers: By Interfacial and Solution Methods', Interscience, New York, 1965. A review of the field of condensation, polymerization and hydrogen transfer polymerizations carried out below 100 °C (*Polym. Rev.*, vol. 10).

Peebles, L. H., 'Molecular Weight Distributions in Polymers', Interscience, New York, 1971. The book provides a summary of the various molecular weight distributions that have been derived hereto (*Polym. Rev.*, vol. 18).

Reich, L. and A. Schindler (eds.), 'Polymerization by Organometallic Compounds', Interscience, New York, 1966. A presentation of all of the literature on linear monomers with a single, double bond or two conjugated double bonds (*Polym. Rev.*, vol. 12).

Schnell, H., 'Chemistry and Physics of Polycarbonates', Interscience, New York, 1964 (*Polym. Rev.*, vol. 9).

Tobolsky, A. V. and W. J. MacKnight, 'Polymeric Sulfur and Related Polymers', Interscience, New York, 1965. A survey of the polymers containing polysulfide linkages rather than monosulfide and disulfide linkages (*Polym. Rev.*, vol. 13).

Varga, O. H., 'Stress–Strain Behavior of Elastic Materials: Selected Problems of Large Deformations', Interscience, New York, 1966 (*Polym. Rev.*, vol. 15).

Voiutskii, S. S., 'Autohesion and Adhesion of High Polymers', (translation from the Russian), Interscience, New York, 1963. A summary of the knowledge in this field at that time written from his point of view of the diffusion theory of autohesion and adhesion (*Polym. Rev.*, vol. 4).

2 HANDBOOKS AND ENCYCLOPEDIAS

Alger, M. S. M., 'Polymer Science Dictionary', Elsevier, New York, 1989. This book is devoted to explaining the terminology of polymer science—its chemical and physical terms, polymerization, polymer structure, properties and individual polymer materials. Polymer processing terms have been excluded.

Allen, G. and J. C. Bevington (eds.), 'Comprehensive Polymer Science', Pergamon Press, Oxford, 1989. A series of volumes that show the relationship between methods of preparation, treatment, structure and properties. It is organized into seven volumes: Volume 1, Polymer Characterization; Volume 2, Polymer Properties; Volumes 3 and 4, Chain Polymerization; Volume 5, Step Polymerization; Volume 6, Polymer Reactions; Volume 7, Specialty Polymers and Polymer Processing. A cumulative subject index completes the final volume.

Ash, M. and I. Ash, 'Encyclopedia of Plastics, Polymers and Resins', Chemical Publishing, New York, 1980–1981. A desktop, three volume publication that provides useful information of many plastic, polymer and resin products. In general, the chemical description, application, form and color, general, mechanical, thermal and electrical properties of each product is given.

Brandrup, J. and E. H. Immergui (eds.), 'Polymer Handbook', 3rd edn., Wiley, New York, 1989. An eight section, 70 table, 1852 page handbook, that compiles the fundamental constants and data needed in theoretical and experimental polymer research.

Cheremisinoff, N. P. (ed.), 'Handbook of Polymer Science and Technology', Dekker, New York, 1989. A series of volumes 'aimed at providing a comprehensive guide to the field of applied polymer science ... and technologies related to its engineering applications. Its purpose is to help provide unification between the theoretical ... and practical manufacturing concepts.... The series is organized by topic into five volumes of in depth discussion: Volume 1, Synthesis and Properties; Volume 2, Performance Properties of Plastics and Elastomers; Volume 3, Applications and Processing Operations; Volume 4, Composites and Specialty Applications.

Elias, H.-G. and R. Pethrick (eds.), 'Polymer Yearbook', Harwood, New York, 1984. An annual reference book that contains a collection of useful data, such as rules of nomenclature, symbols, units and formulae, also a number of short reviews on specific aspects of polymer science, and a survey of recent publications of monographs, conference proceedings, data compilations, journals and dissertations.

Kroschwitz, J. (ed.), 'Polymers: An Encyclopedic Sourcebook of Engineering Properties', ACS, Washington, DC, 1987. One of a series of reprints from the 'Encyclopedia of Polymer Science and Engineering' which brings together in one volume articles covering all aspects of the engineering properties of polymers and composites.

Mark, H. F., N. M. Bikales, C. G. Overberger and G. Menges, (eds.), 'Encyclopedia of Polymer Science and Engineering', 2nd edn., Wiley–Interscience, 1985–1989. A comprehensive account of polymer science and technology covering chemical substances, polymer properties, methods and processes and broad uses of polymers. The new edition has 17 volumes, one supplement volume and one index volume. It is dedicated to H. F. Mark on the occasion of his 90th birthday, and volume 1 includes a bibliography of his 593 publications.

Molyneux, P., 'Water-Soluble Synthetic Polymers: Properties and Behavior', CRC Press, Boca Raton, FL, 1984 (two volumes). The general features of water-soluble polymers are discussed, and their properties and behavior are compared.

Overberger, C. G. (ed.), 'Macromolecular Syntheses', Wiley, New York, vols. 1–10, 1963–1990. A series of periodic publications on macromolecular syntheses containing detailed directions for polymerization and characterization.

Roff, W. J. and J. R. Scott, 'Fibres, Films, Plastics and Rubbers', Butterworth, London, 1971. A ready reference book containing data on 42 specific materials: their structure, general, thermal, electrical and mechanical properties, and specific properties and related information.

Saechtling, H., 'International Plastics Handbook: for the Technologist, Engineer and User'. 2nd edn., Hanser, Munich (distributed in USA by Macmillan, New York), 1987. The English version of 'Kunststoff-Taschenbuch' which first appeared in 1936. General concepts, literature references, plastics synthesis, compounding, processing and manufacturing to detailed descriptions of these individual plastics are all included. Several thousand trade names and standards compilations also appear.

Seymour, R. B., 'Engineering Polymer Sourcebook', McGraw-Hill, New York, 1990. The design, processing, fabrication and application of many high performance thermosets and high performance, thermoplastics is presented, including elastomers, styrenic, acrylic, polycarbonate, acetal polyamides, isocyanate, poly(phenylene oxide), fluorocarbons, polyimides and polyethers.

Zweig, G. and J. Sherma (eds.), 'CRC Handbook of Chromatography: Polymers', CRC Press, Boca Raton, FL, vol. 1, 1982. One part describes the techniques applicable to polymeric materials, and the second part gives data tables for gas, pyrolysis gas, liquid and thin layer chromatography.

3 GENERAL AND TEXTBOOKS

The list of books below was compiled on the basis of currency, most of which were published within the last decade, and availability. The literature ever expands, with the tendency today for collaborative and special subject treatments.

Allport, D. C. and W. H. Janes (eds.), 'Block Copolymers', Wiley, New York, 1973. A review of the processes used to obtain block copolymers, and a discussion of their properties in terms of their structures. Applications of the important classes are also given. References are included for each chapter.

Billmeyer, F. W., 'Textbook of Polymer Science', Wiley, New York, 1984. An up-to-date textbook that has chapters on polymerization, characterization, properties and polymer processing.

Elias, H.-O., 'Macromolecules', 2nd edn., Plenum Press, New York, 1984. Volume 1, Structure and Properties; Volume 2, Syntheses and Materials. A survey of the field, the first volume considers macromolecular compounds and their properties. The second emphasizes the principles of syntheses and reaction and polymer technology.

Feldman, F., 'Polymeric Building Materials', Elsevier, New York, 1989. The book is designed as an introduction to polymer applications for the building industry. It provides information on different classes of polymers, composites, polymer–concrete systems, foams and their functions, adhesives, the use of macromolecules in solar energy, conservation, roofing and flooring, and the changes in properties due to weather, aging, *etc.*

Kleintjens, L. A. and P. J. Lemstra (eds.), 'Integration of Fundamental Polymer Science and Technology', Elsevier, New York, vol. 1, 1986 and vol. 2, 1988. Proceedings of international meetings held in Roldac, The Netherlands to stimulate discussions between academic and industrial polymer scientists, and engineers.

McCrum, N. G., C. P. Buckley and C. B. Pucknall, 'Principles of Polymer Engineering', Oxford University Press, New York, 1988. A text that outlines the fundamentals of polymer chemistry and physics, various forming operations, and factors in the successful design of polymers.

Mandelkern, L., 'An Introduction to Macromolecules', 2nd edn., Springer-Verlag, New York, 1983. For the beginning student, an introduction to the structure and properties of macromolecular substances.

Mark, J. E. and B. Erman, 'Rubberlike Elasticity: A Molecular Primer', Wiley, New York, 1988. An elementary presentation of the important aspects of rubberlike elasticity.

Rodriquez, F., 'Principles of Polymer Systems', 3rd edn., Hemisphere, New York, 1989. A comprehensive text, beginning with basic structure, through properties, fabrication processes and polymer analysis.

Rosen, S. L., 'Fundamental Principles of Polymeric Materials', 2nd edn., Wiley, New York, 1982. A text that reviews the fundamental principles of synthesis and polymerization that are of practical relevance in processing current materials.

Saegusa, T., T. Higashimura and A. Abe, (eds.), 'Frontiers of Macromolecular Science', Blackwell, London, 1988. Proceedings of the 32nd International Symposium on Macromolecules, the book presents up-to-date developments on research into macromolecules.

Saunders, K. J., 'Organic Polymer Chemistry', 2nd edn., Chapman and Hall, New York, 1988. An introduction to the organic chemistry of adhesives, fibers, paints, plastics and rubbers.

Tess, R. W. and G. W. Poehlein (eds.), 'Applied Polymer Science' 2nd edn., ACS, Washington, DC, 1985. An expanded and revised edition, covering all aspects of polymers: polymerization, physical phenomena, additives, analysis, products and their uses.

Ulrich, H., 'Raw Materials for Industrial Polymers', Hanser, New York, 1988. As a companion volume to his earlier 'Introduction to Industrial Polymers', this book describes methods of synthesis of major polymer raw materials, some of the emerging technologies and the relationship of major polymers to basic feedstocks. It also gives consumption figures and prices.

4 POLYMER PROPERTIES

Aklonis, J. J. and W. J. MacKnight (eds.), 'Introduction to Polymer Viscoelasticity', 2nd eds., Wiley, New York, 1983. This edition adds the new developments, such as the theory of repetition and dielectric relaxation.

Andrade, J. D., (ed.), 'Polymer Surface Dynamics', Plenum Press, New York, 1988. Shows the dynamic aspects of polymer surfaces.

Astarita, G. and L. Nicolais (eds.), 'Polymer Processing and Properties', Plenum Press, New York, 1984. Proceedings of the European Meeting in Italy on this subject, the whole area is discussed.

Barlow, F. W., 'Rubber Compounding: Principles, Materials and Techniques', Dekker, New York, 1988. A book that deals exclusively with rubber compounding, describing the materials involved as well as why and how they are used to build up a compound.

Bely, V. A., A. I. Sviridenok, M. I. Petrokovets and V. G. Savkin, 'Friction and Wear in Polymer-Based Materials', Pergamon Press, New York, 1982. The basic principles of friction and wear, the function of structure and the application of these materials in friction assemblies is shown.

Bhowmick, A. K. and H. L. Stephens (eds.), 'Handbook of Elastomers', Dekker, New York, 1988. Covers the recent developments in the commercially important synthetic and natural elastomers in current use.

Bird, R. B., R. C. Armstrong and O. Hassager, 'Dynamics of Polymeric Liquids', 2nd edn., Wiley, New York, 1987. In two volumes: fluid mechanics and kinetic theory.

Brown, R. P., 'Physical Testing of Rubber', 2nd edn., Elsevier, New York, 1986. Up-to-date procedures used by suppliers and users to characterize and investigate the physical properties of rubber materials.

Brown, R. P. and B. E. Read (eds.), 'Measurement Techniques for Polymeric Solids', Elsevier, New York, 1984. Proceedings of the National Physical Laboratory Conference on Measurement Techniques for Polymeric Solids, to review techniques for characterizing the structure and physical properties of rubbers and plastics.

Bueche, F., 'Physical Properties of Polymers', Krieger, New York, 1979. A presentation of the fundamental concepts governing the physical behavior of high polymers.

Comyn, J. (ed.), 'Polymer Permeability', Elsevier, New York, 1985. Some of the applied aspects of permeability are described—drug delivery, coatings and encapsulants, and packaging.

Cotts, D. B. and Z. Reyes, 'Electrically Conductive Organic Polymers for Advanced Applications', Noyes, Park Ridge, NJ, 1986. The properties of about 250 electrically conducting, semiconducting and semi-insulating polymers were surveyed and their conduction mechanism, mechanical properties and suitability for space-based use was evaluated.

Dyson, R. W. (ed.), 'Engineering Polymers', Chapman and Hall, New York, 1990. The book discusses polymeric materials currently in use as engineering materials. They include thermoplastics, elastomers, polymer foams, fiber-reinforced composites, films, and polymers in telecommunications and power transmission.

Feast, W. J. and H. S. Munro (eds.), 'Polymer Surfaces and Interfaces', Wiley, New York, 1987. Current and practical studies on polymer surfaces and interfaces.

Ferry, J. D., 'Viscoelastic Properties of Polymers', 3rd edn., Wiley, New York, 1980. The book relates the viscoelasticity of polymers to molecular structure and modes of molecular motion. The emphasis is on the linear viscoelasticity of amorphous polymers as it was with the earlier editions. Applications to practical problems are also given.

Flory, P. J., 'Statistical Mechanics of Chain Molecules', Wiley, New York, 1969. Reprinted edition with corrections by Flory and additional remarks, Hanser, New York, 1989. A presentation of the spatial configuration of chain molecules and their properties by mathematical methods. The reprinted edition includes an autobiographical sketch by Flory, and his honors and awards.

Ford, W. T. (ed.), 'Polymeric Reagents and Catalysts', ACS, Washington, DC, 1986. Reviews, of special materials—soluble, polymer bound reagents, ion exchange resins, transition metal catalysts, photosensitizers and polymer bound oxidizing agents.

Hearle, W. S., 'Polymers and Their Properties', Horwood, Chichester, 1982. A three volume study of polymers beginning with the fundamentals of structure and mechanics, and then on to other physical properties.

Kesting, R. E., 'Synthetic Polymeric Membranes: A Structural Perspective', 2nd edn., Wiley–Interscience, New York, 1985. Membrane types are emphasized—polymer membranes, the polymeric films, phase inversion membranes, liquid and dynamically formed membranes, biological membranes.

Kroschwitz, J. I. (ed.), 'Electrical and Electronic Properties of Polymers: A State-of-the-Art Compendium', Wiley, New York, 1988. Selected reprints from the 'Encyclopedia of Polymer Science and Engineering' on conductive and insulating polymeric materials, which include syntheses, properties and uses.

Ku, C. C. and R. Liepins, 'Electrical Properties of Polymers', Hanser, New York, 1987. The four parameters of electrical properties in polymers are discussed: dielectric constant, tangent of dielectric loss angle, dielectric breakdown and electrical conduction.

Kurata, M., 'Thermodynamics of Polymer Solutions', Harwood, New York, 1982. Thermodynamic theories of osmotic equilibrium, phase separation, concentration fluctuations and sedimentation equilibrium in polymer solutions.

Kuzmany, H., M. Mehring and S. Roth (eds.), 'Electronic Properties of Conjugated Polymers III', Springer-Verlag, New York, 1989. Proceedings of an International Winter School, Kirchberg, Tirol, the papers concentrate on basic models and application potentials of polymers.

Lee, L.-H. (ed.), 'Polymer Wear and Its Control', ACS, Washington, DC, 1985. Mechanisms and control and measurement of polymer wear, tribological behavior of polymers, and the degradation of biomaterials, films and filaments are presented.

Linford, R. G. (ed.), 'Electrochemical Science and Technology of Polymers 1', Elsevier, New York, 1987. This volume explains the concepts of electrochemistry, reviews some of the areas of application of polymers in such systems, and describes their use in the production of chlorine and caustic soda.

McCrum, N. G., C. P. Buckley and C. B. Bucknall, 'Principles of Polymer Engineering', Oxford University Press, New York, 1988. 'Solutions Manual', 1989. Textbook and manual to accompany text with solutions to each of the problems.

Makhlis, P. A., 'Radiation Physics and Chemistry of Polymers', Halsted Press, New York, 1975. The book shows the general structural changes and changes in properties resulting from irradiation.

Margolis, J. M. (ed.), 'Conductive Polymers and Plastics', Chapman and Hall, New York, 1989. Electrically conductive and ionically conductive polymers are discussed in the first part, and the metallic plating and coating on plastics, and electrically conductive plastics themselves in the second part.

Mark, J. E., A. Eisenberg, W. W. Graessley, L. Mandelkern and J. L. Koenig, 'Physical Properties of Polymers', ACS, Washington, DC, 1984. Basic concepts, current topics and projections for future research are given for rubber-like elasticity, the glassy, viscoelastic and crystalline states in polymers, and polymer spectroscopy.

Meeten, G. H. (ed.), 'Optical Properties of Polymers', Elsevier, New York, 1986. Practical and review material is presented showing relationship of function and operation of products to its optical properties.

Middleman, S., 'The Flow of High Polymers: Continuum and Molecular Rheology', Interscience, New York, 1968. The book discusses three subjects: the definition and measurement of material properties, the prediction of these properties and correlation techniques.

Moiseev, Y. V. and G. E. Zaikov (translation by R. J. Moseley), 'Chemical Resistance of Polymers in Aggressive Media', Consultants Bureau, New York, 1987. The aggressive media are solutions of acids, bases and salts, and the influences on mechanical, chemical properties and the service life of polymer articles are discussed.

Myerson, A. S. and K. Toyokura (eds.), 'Crystallization as a Separation Process', ACS, Washington, DC, 1990. The current research covering basic studies in crystal growth, interactions, solution structure and crystallization of organic and inorganic materials is presented.

Perepechko, I., 'Low-Temperature Properties of Polymers' (translation from the Russian), Pergamon Press, New York, 1980. Shows how the chemical constitution and the supramolecular structure influence the physical properties of polymers in the low temperature region.

Richards, E. G., 'An Introduction to the Physical Properties of Large Molecules in Solution', Cambridge University Press, New York, 1980. A description of the nature of large molecules in general terms relating their behavior to principles of classical physics and chemistry.

Scott, G. (ed.), 'Developments in Polymer Stabilisation', Applied Science, London, vols. 1–7, 1980. A series of books that intend to review the progress in the development of practical stabilizing systems for many polymers.

Utracki, A. and R. A. Weiss (eds.), 'Multiphase Polymers: Blends and Ionomers', ACS, Washington, DC, 1989. Some current trends in the kinds of materials under development, their characterization, theory and processing.

Van Krevelen, D. W. and P. J. Hoftyzer, 'Properties of Polymers: Their Estimation and Correlation with Chemical Structure', Elsevier, New York, 1976. This is a book for practical use for those who need to orientate numerical information on polymer properties.

Ward, I. M., 'Mechanical Properties of Solid Polymers', 2nd edn., Wiley, New York, 1983. The book develops the links between structure and properties.

Williams, J. G., 'Stress Analysis of Polymers', 2nd edn., Wiley, New York, 1980. A course in stress analysis.

Williams, J. G., 'Fracture Mechanics of Polymers', Horwood, New York, 1984. An introduction to the phenomenon of fracture, called fracture mechanics, for polymers and how it is applied to a range of materials and problems.

Woodward, A. E., 'Atlas of Polymer Morphology', Hanser, New York, 1989. Useful book for the practising morphologist. It discusses crystallization from the melt, block copolymers, processing effects, blends, fracture and effects of chemicals and other agents.

Wypych, J., 'Weathering Handbook', Chemtec, Toronto, 1990. A guide to weathering studies of materials such as coatings, paints, textiles, geosynthetics, composites, building materials, sealants, wood, paper and polymers. Degradation and methods of testing and the effect of many factors is discussed.

Zachariades, A. E. and R. S. Porter (eds.), 'High Modulus Polymers: Approaches to Design and Development', Dekker, New York, 1988. Studies are presented on liquid crystalline polymers and semicrystalline polymers. Literature is reviewed, the techniques used and properties are shown.

5 POLYMER STRUCTURE

Allen, R. S. (ed.), 'Developments in Polymer Photochemistry', Applied Science, London, vols. 1 and 2, 1980–1982.

Bark, L. S. and N. S. Alled (eds.), 'Analysis of Polymer Systems', Applied Science, London, 1982. Each chapter deals with a particular technique for the analysis, characterization of additives, molecular weight and structure, and identification of polymer systems.

Bassett, C. (ed.), 'Developments in Crystalline Polymers', Applied Science, London, vols. 1 and 2,

1982 and 1988. Contributions to the understanding of the structure of these polymers are presented.

Benham, J. L. and J. F. Kinstle (eds.), 'Chemical Reactions on Polymers', ACS, Washington, DC, 1988. A presentation of the recent and emerging technology on polymer modification and specialty polymers by chemical reactions.

Bower, D. I. and W. F. Maddams, 'The Vibrational Spectroscopy of Polymers', Cambridge University Press, New York, 1989. An introduction to the theory of vibrational spectroscopy and to its application in the study of synthetic polymers.

Brown, P. P. and B. E. Read (eds.), 'Measurement Techniques for Polymeric Solids', Elsevier, New York, 1984. Selective publication of papers presented at the NPL conference which review techniques for characterizing the structures and physical behavior of plastics and rubbers and their suitability to industrial needs.

Brydson, J. A., 'Flow Properties of Polymer Melts', 2nd edn., Goodwin, London, 1981. The book particularly intends to relate melt rheology to industrial polymer processing for the plastics technologist.

Cheremisinoff, N. P., 'Product Design and Testing of Polymeric Materials', Dekker, New York, 1990. The book reviews test methods and gives an evaluation of the physical and processing performance properties of polymeric materials.

Ciardelli, F. and P. Giusti (eds.), 'Structural Order in Polymers', Pergamon Press, New York, 1981. Proceedings of the International Symposium on Macromolecules, Florence, Italy, September 1980 dedicated to the memory of Natta with an introduction by Mark. The symposium outlines the main scientific and industrial achievements in the stereospecific polymerization of alkenes and dialkenes, and the relationship between physical properties and molecular structure.

Cooper, A. R. (ed.), 'Determination of Molecular Weight', Wiley, New York, 1989. The status of the older methods of molecular weight determination and the theory and applications of some newer methods—phase distribution chromatography, field flow fractionation, supercritical fluid chromatography, among others—are reviewed.

Cowie, J. M. G. (ed.), 'Alternating Copolymers', Plenum Press, New York, 1985. A compilation of the information on the mechanisms and kinetics of the reactions of alternation copolymers.

Dawkins, J. V. (ed.), 'Developments in Polymer Characterisation', Applied Science, London, vols. 1–5, 1978–1986. This series presents critical reviews of the methods and techniques developed in this field in the last decade.

Doi, M. and S. F. Edwards, 'The Theory of Polymer Dynamics', Clarendon Press, Oxford, 1986. The theories and dynamics of interacting polymers in the liquid state is explained.

Dwight, D. W., T. J. Fabish and H. R. Thomas (eds.), 'Photon, Electron, and Ion Probes of Polymer Structure and Properties', ACS, Washington, DC, 1981. A summary of the fundamentals and applications of photon, electron and ion probes to polymers up to this time.

El-Nokaly, M. A. (ed.), 'Polymer Association Structures: Microemulsions and Liquid Crystals', ACS, Washington, DC, 1989. The book discusses formation and characterization of microemulsions and the liquid crystalline phase of polymer association structures.

Hall, L. H. (ed.), 'Structure of Crystalline Polymers', Elsevier, New York, 1984. The book emphasizes the experimental techniques used to study polymer structure, including use of the digital computer, wide angle X-ray diffraction, electron microscopy, neutron scattering and X-ray scattering.

Hiltner, A. (ed.), 'Structure–Property Relationships of Polymeric Solids', Plenum Press, New York, 1981. A symposium at the 55th American Chemical Society Meeting, Atlanta, March 1981 presented in honor of Baer who attempted to 'find relationships of solid state structure and hierarchy to the resultant properties from which specific functions are derived'.

Hoyle, C. E. and J. M. Torkelson (eds.), 'Photophysics of Polymers', ACS, Washington, DC, 1987. Presents the current status of this subject.

Hummel, D. O. and F. Scholl, 'Atlas of Polymer and Plastics Analysis', 2nd edn., Hanser, Deerfield Beach, FL, vols. 1 and 2, 1978; vol. 3, 1980. The subject areas covered by each volume are as follows: volume 1, polymers—structures and spectra; volume 2, plastics, fibers, rubbers and resins; and volume 3, additives and processing aids.

Kaelble, D. H., 'Computer-aided design of Polymers and Composites', Dekker, New York, 1985. This book shows how the effects of temperature, stress, strain, time and environment on composite properties can be generated by mathematical simulation from atomic and molecular properties.

Kaminsky, W. and H. Sinn (eds.), 'Transition Metals and Organometallics as Catalysts for Olefin Polymerization', Springer-Verlag, New York, 1987. The papers present an overview of, and further development into, the mechanism of catalysis to provide for new polyalkenes and copolymers with different properties.

Kleintjens, L. A. and P. J. Lemstra (eds.), 'Integration of Fundamental Polymer Science and Technology', Elsevier, New York, 1986.

Labana, S. S. and R. A. Dickie (eds.), 'Characterization of Highly Cross-linked Polymers', ACS, Washington, DC, 1984. Recent progress in areas of light scattering, fracture, molecular structure relationships, morphology and the dependence of properties of thermal history are among the topics discussed.

Ladik, J. J., 'Quantum Theory of Polymers as Solids', Plenum Press, New York, 1988. The book summarizes the different quantum mechanical methods developed in the last 20 years on the structure of polymers.

Lemstra, P. J. and L. A. Kleintjens (eds.), 'Integration of Fundamental Polymer Science and Technology—2', Elsevier, New York, 1988. Discussions by academic, industrial and engineering participants on many aspects of polymer technology from chemistry to processing.

McArdle, C. B. (ed.), 'Side Chain Liquid Crystal Polymers', Chapman and Hall, New York, 1989. A description of these materials, through polymer formation and characterization, and potential end uses.

McGrath, J. E. (ed.), 'Ring-Opening Polymerization', ACS, Washington, DC, 1985. A current view of kinetics, mechanisms and synthesis.

Mark, J. E., A. Eisenberg, W. W. Graessley, L. Mandelkern and J. L. Koenig, 'Physical Properties of Polymers', ACS, Washington, DC, 1984. Basic concepts, current topics and projections for future research are given for rubber-like elasticity, the glassy, viscoelastic and crystalline states in polymers, and polymer spectroscopy.

Messier, J., F. Kajzar, P. Prasad and D. Ulrich (eds.), 'Nonlinear Optical Effects in Organic Polymers', Kluwer, Norwell, MA, 1989. A NATO Advanced Research Workshop, the theory, properties and application of organic polymers to optical phenomena are discussed.

Nagasawa, M. (ed.), 'Molecular Conformation and Dynamics of Macromolecules in Condensed Systems', Elsevier, New York, 1987. A collection of contributions based on lectures presented at the 1st Toyota Conference, Japan.

Phillips, D. (ed.), 'Polymer Photophysics', Chapman and Hall, New York, 1985. Studies on luminescence spectroscopy to show molecular motion, order and excitation diffusion and energy transfer are presented.

Prasad, P. N. and S. R. Ulrich (eds.), 'Nonlinear Optical and Electroactive Polymers', Plenum Press, New York, 1988. The use of these polymers for electronics and optical applications.

Provder, T. and C. D. Craver (eds.), 'Polymer Characterization: Physical Property, Spectroscopic, and Chromatographic Methods, ACS, Washington, DC, 1990. Polymer fraction and particle size distribution, dynamic mechanical analysis, spectroscopy and morphology are discussed.

Sawyer, L. C. and D. T. Grubb, 'Polymer Microscopy', Chapman and Hall, New York, 1987. The book provides basic microscopy techniques and specimen preparation methods that apply to polymers.

Takemoto, K., Y. Inaki and R. M. Ottenbrite, 'Functional Monomers and Polymers', Dekker, New York, 1987. The chemistry and technology involved in functionalizing monomers, and the preparation and processing of polymers to serve specific needs.

Tong, H.-M. and L. T. Nguyen (eds.), 'New Characterization Techniques for Thin Polymer Films', Wiley–Interscience, New York, 1990. This book provides a description of the instrumentation, principle of operation, areas of application and data analysis methods of the new techniques.

Turi, E. A., 'Thermal Characterization of Polymeric Materials', Academic Press, New York, 1981. A reference source which details the scope of thermal analysis as applied to polymers.

Vilcu, R. and M. Leca, 'Polymer Thermodynamics by Gas Chromatography', Elsevier, New York, 1989. The book presents the use of direct and inverse gas chromatography as tools for determining thermodynamic properties for macromolecular substances.

Weiss, R. A. and C. K. Ober (eds.), 'Liquid-Crystalline Polymers', ACS, Washington, DC, 1990. This book studies the creation, physics and engineering aspects of LCPs.

6 POLYMER PROCESSING AND SPECIALTY POLYMERS

Adams, R. D. and W. C. Wake, 'Structural Adhesive Joints in Engineering', Elsevier, New York, 1984. The book discusses stresses in joints, standard methods for the testing and preparation of adhesives and guides to their selection.

Allen, K. W. (ed.), 'Adhesion', Elsevier, New York, vols. 1–13, 1977–1988. Proceedings of the Annual Conference on Adhesion and Adhesives held at City University, London. The papers address contemporary developments and issues in this field.

Astarita, G. and L. Nicolais (eds.), 'Polymer Processing and Properties', Plenum Press, New York, 1984. Chapters deal with the shaping and nonshaping operations, properties and the relationship between properties and processing, and structural modeling of properties.

Bassett, D. C. (ed.), 'Developments in Crystalline Polymers', Elsevier, New York, vols. 1–2, 1981 and 1988. Technical developments which contribute to their syntheses, liquid crystalline polymers and their applications are discussed.

Bhattacharya, W. K. (ed.), 'Metal-Filled Polymers', Dekker, New York, 1986. The latest technologies in the field, properties and applications.

Brydson, J. A., 'Rubber Materials and their Compounds', Elsevier, New York, 1988. The book reviews the polymers and the grades available, and the additives used in compounds.

Carraher, C. E. and C. G. Gebelein (eds.), 'Biological Activities of Polymers', ACS, Washington, DC, 1982. Research into the medical and nonmedical applications of the biological activities of macromolecules (*ACS Symp. Ser.*, no. 186).

Chasin, M. and R. Langer (eds.), 'Biodegradable Polymers as Drug Delivery Systems', Dekker, New York, 1990. The book reviews the properties, synthesis and formulations of many polymers.

Cheremisinoff, N. P., 'Polymer Mixing and Extrusion Technology', Dekker, New York, 1987. The fundamental concepts of mixing and extrusion are explained, mixers and mixing practices are outlined.

Dick, J. S., 'Compounding Materials for the Polymer Industries', Noyes, Park Ridge, NJ, 1987. An overview of the materials and principles used to compound polymers for various industries, including plastics, rubber, adhesives and coatings.

Dickie, R. A. and F. L. Floyd (eds.), 'Polymeric Materials for Corrosion Control', ACS, Washington, DC, 1986. This concerns the evaluation and selection of materials for use in corrosive environments.

Elias, H.-G. and F. Vohwinkel, 'New Commercial Polymers 2', Gordon and Breach, New York, 1986. The book contains reports on the history, synthesis, properties and applications of over 80 new polymers developed between 1976 and 1984.

El-Nokaly, M. A. (ed.), 'Polymer Association Structures: Microemulsions and Liquid Crystals', ACS, Washington, DC, 1989. The book discusses the formation and characterization of microemulsions and the liquid crystalline phase of polymer association structures.

Feldman, D., 'Polymeric Building Materials', Elsevier, New York, 1989. The book is an introduction to polymer applications in the building industry for the civil and chemical engineering fields.

Fouassier, J.-P. and J. F. Rabek (eds.), 'Lasers in Polymer Science and Technology: Applications', CRC Press, Boca Raton, FL, 1989. A four volume compilation of information on current research and developments in this field.

Franta, I. (ed.), 'Elastomers and Rubber Compounding Materials', Elsevier, New York, 1989. Fundamental information on the production, properties and application of all basic materials used for formulation rubber compounds.

Glass, J. E. (ed.), 'Polymers in Aqueous Media', ACS, Washington, DC, 1989. The structures and use of polyelectrolytes in a variety of applications. Surfactant-modified polymers are among subjects discussed.

Hilyard, N. C. (ed.), 'Mechanics of Cellular Plastics', MacMillan, New York, 1982. The mechanical behavior of rigid and flexible cellular polymers is discussed, and the relationship between mechanical properties, structure and composition is shown.

Hsieh, D. (ed.), 'Controlled Release Systems: Fabrication Technology', CRC Press, Boca Raton, FL, 1988 (two volumes). Drug delivery systems for currently marketed products and procedures still under development are discussed.

Isayev, A. I. (ed.), 'Injection and Compression Molding Fundamentals', Dekker, New York, 1987. The book primarily discusses various aspects of molding thermoplastic polymers, then the molding process accompanied by polymerization or crosslinking, and the computer-aided design, engineering and manufacturing of molds.

Ise, N. and I. Tabuchi, 'An Introduction to Speciality Polymers', Cambridge University Press, New York, 1983. The chemistry of speciality polymers is described, and some of their catalysts and transport phenomena.

Lee, P. I. and W. R. Good (eds.), 'Controlled-Release Technology', ACS, Washington, DC, 1987. Delivery systems for pharmaceutical applications are described.

Legge, N. R., G. Holden and H. Schroeder (eds.), 'Thermoplastic Elastomers, A Comprehensive Review', Oxford University Press, New York, 1988. Structure, properties and future trends are emphasized in the chapters each written by researchers who have contributed significantly to the science.

Lutz, J. T. (ed.), 'Thermoplastic Polymer Additives', Dekker, New York, 1989. The modification of polymers through the use of external additives—colorants, fillers, plasticizers, smoke suppressants, *etc.*—is brought up to date.

Macosko, C. W., 'RIM Fundamentals of Reaction Injection Molding', Hanser, New York, 1989. The book emphasizes fundamentals and discusses the operations of material delivery, mixing, mold filling, curing, for urethanes and nonurethanes, and reinforced RIM. Additional references are provided with each chapter.

Margolis, J. M., 'Instrumentation for Thermoplastics Processing', Hanser, New York, 1988. Principles and practices for materials suppliers, processors and equipment companies.

Moore, G. R. and D. E. Kline, 'Properties and Processing of Polymers for Engineers', Prentice–Hall, Englewood Cliffs, NJ, 1984. An introduction to polymers for mechanical and industrial engineering students.

Morton-Jones, D. H., 'Polymer Processing', Chapman and Hall, New York, 1989. An introduction to polymer processing.

Noble, R. D. and J. D. Way (eds.), 'Liquid Membranes', ACS, Washington, DC, 1987. State of the art papers on theory and applications.

Ottenbrite, R. M., L. A. Utracki and S. Inoue (eds.), 'Current Topics in Polymer Science', Hanser, New York, 1987. Selections by the organizers of the three sponsoring societies of the International Congress of Pacific Basin Societies in areas of prime developments in polymer science: biological application of polymers, ionic polymerization, spectroscopy, molecular dynamics, rheology and processing of polymers and multiphase systems.

Pearson, J. R. A., 'Mechanics of Polymer Processing', Elsevier, New York, 1985. Continuum mechanics is emphasized—complex flow fields, continuous processes and cyclic processes are given.

Pearson, J. R. A. and S. M. Richardson (eds.), 'Computational Analysis of Polymer Processing', Applied Science, New York, 1983. Various physical phenomena that occur in polymer treatment processes are simulated and methods for overcoming these difficulties are described.

Peppas, N. A. (ed.), 'Hydrogels in Medicine and Pharmacy', CRC Press, Boca Raton, FL, 1986 (three volumes). This work covers the following areas: Volume 1, Fundamentals; Volume 2, Polymers; Volume 3, Properties and Applications.

Powell, P. C., 'Engineering with Polymers', Chapman and Hall, New York, 1983. The book introduces the terminology and technology, describes properties and materials, and the effects of processing.

Provder, T. (ed.), 'Computer Applications in Applied Polymer Science II', ACS, Washington, DC, 1989. The focus is on the implementation of new computer technologies in research and development and problem solving.

Randell, D. R. (ed.), 'Radiation Curing of Polymers', CRC Press, Boca Raton, FL, 1987. A review of the progress in this field.

Rauwendaal, C., 'Polymer Extrusion', Hanser, New York, 1986. Explains the extrusion theory, and describes different types of extruders and their instrumentation, the character of the polymers being processed, and some practical applications.

Reichmanis, E., S. A. MacDonald and T. Iwayanagi (eds.), 'Polymers in Microlithography: Materials and Processes', ACS, Washington, DC, 1989. Advances in radiation chemistry are discussed, and their application to resist materials for optical, electron and X-ray lithography.

Seymour, R. B. and H. F. Mark (eds.), 'Organic Coatings: Their Origin and Development', Elsevier, New York, 1990. Proceedings of the International Symposium on the History of Organic Coatings, held in Miami Beach, September 1989.

Sorbie, K. S., 'Polymers in Improved Oil Recovery', CRC Press, Boca Raton, FL, 1990. Discusses polymer flooding, and the types and structures of polymers used.

Stevens, M. J., 'Extruder Principles and Operation', Elsevier, New York, 1985. The book summarizes the principles of single screw extrusion processes for plastics and rubbers, practical performance and operating procedures.

Takemoto, K., Y. Inaki and R. M. Ottenbrite (eds.), 'Functional Monomers and Polymers: Procedures, Synthesis, Applications', Dekker, New York, 1987. The book is an overview of some of the unique polymer functions achieved, and the basic procedures for the preparation of these materials.

Ward, I. M. (ed.), 'Developments in Oriented Polymers', Applied Science, New Jersey, vols. 1–2, 1982 and 1987. Areas of particular progress in oriented polymers are selected for review. Additional references accompany each chapter.

Whelan, A. and K. S. Lee (eds.), 'Developments in Rubber Technology', Applied Science, London, vols. 1–4, 1979–1985. The volumes review the position in major areas of the manufacture and use of rubber products.

Zeigler, J. M. and F. W. G. Fearon (eds.), 'Silicon-Based Polymer Science: A Comprehensive Resource', ACS, Washington, DC, 1990. A reference volume that covers silicones, polysilanes, ceramic precursor polymers, polymers with pendant silicon and silicon-based supramolecular structures.

7 POLYMERIZATION

Biesenberger, J. A. (ed.), 'Devolatilization of Polymers', Macmillan, New York, 1983. Equipment, process mechanisms, models and applications.

Biesenberger, J. A. and D. H. Sebastian, 'Principles of Polymerization Engineering', Wiley, New York, 1983. Concepts and principles that are used in the design, scaling and modification of polymerization processes are described.

Bovey, F. A. and F. H. Winslow (eds.), 'Macromolecules: An Introduction to Polymer Science', Academic Press, New York, 1979. A treatment of all phases of macromolecular chemistry and physics at the undergraduate level.

Calvert, K. O. (ed.), 'Polymer Latices and their Applications', Macmillan, New York, 1983. Written by industrial experts, it presents the latest commercial aspects of international polymer latex technology.

Chien, J. C. W., 'Coordination Polymerization', Academic Press, New York, 1975. A memorial to Ziegler who discovered the catalyst system that made it possible to polymerize ethylene at low pressure to linear, high density polyethylene.

Dragutan, V., A. T. Balaban and M. Dimonie, 'Olefin Metathesis and Ring-Opening Polymerization of Cyclo-Olefins', Wiley–Interscience, New York, 1985. The scientific and applied aspects of this topic are discussed and the literature is reviewed.

Erusalimskii, B. L., 'Mechanisms of Ionic Polymerization: Current Problems', Consultants Bureau, New York, 1986. An analysis of modern views on the mechanism of formation of macromolecules in ionic systems.

Fontanille, M. and A. Guyot (eds.), 'Recent Advances in Mechanistic and Synthetic Aspects of Polymerization', Kluwer, Norwell, MA, 1987. Proceedings of a NATO Advances Workshop to discuss the latest polymerization methods, catalysis and media.

Henderson, J. N. and T. C. Bouton (eds.), 'Polymerization Reactors and Processes', ACS, Washington, DC, 1979. Based on a symposium of the ACS and AIChE, the papers mainly discuss models which enable processes to be predicted and controlled, and the hardware for the processes.

Hoyle, C. E. and J. F. Kinstle (eds.), 'Radiation Curing of Polymeric Materials', ACS, Washington, DC, 1990. Developments in cationic polymerization, laser-initiated polymerization and high energy radiation curing are discussed.

Ivin, J. and T. Saegusa (eds.), 'Ring-Opening Polymerization', Elsevier, New York, 1984 (three volumes). The first two volumes review ring-opening polymerization and the third volume includes author and subject indices.

Morton, M., 'Anionic Polymerization: Principles and Practice', Academic Press, New York, 1983. Describes the nature of the anionic mechanism of polymerization, and applications of the process in polymer synthesis.

Piirma, I. (ed.), 'Emulsion Polymerization', Academic Press, New York, 1982. The current work on this subject is outlined.

Platzer, M. A. J., (ed.), 'Polymer Manufacturing, Technology and Health Effects', Noyes, Park Ridge, NJ, 1986.

Randall, J. C. (ed.), 'NMR and Macromolecules: Sequences, Dynamic, and Domain Structure', ACS, Washington, DC, 1984. The book honors Frank Bovey of Bell Laboratories, and presents the latest techniques in this field.

Riechert, K. H. and W. Geiseler (eds.), 'Polymer Reaction Engineering: Influence of Reaction Engineering on Polymer Properties', MacMillan, New York, 1983. Shows the effect of polymer reaction engineering on the properties of the polymers produced.

8 POLYMER REACTIONS

Benham, J. I. and J. F. Kinstle (eds.), 'Chemical Reactions on Polymers', ACS, Washington, DC, 1988. Presents the recent achievements in reactive polymers, new synthesis routes, surface modification, specialty polymers, modification for analytical characterization and for functionalization (*ACS Symp. Ser.*, no. 364)

Carraher, C. E. and J. A. Moore (eds.), 'Modification of Polymers', Plenum Press, New York, 1983. The book includes reactions and preparations of copolymers, block and graft copolymers and condensation reactions.

Cullis, C. F. and M. M. Hirschiler, 'The Combustion of Organic Polymers', Clarendon Press, Oxford, 1981. Some of the properties of natural and synthetic polymers are outlined, the hazards associated with burning polymers, the different stages of burning and ways of reducing polymer flammability are described.

Dickie, R. A., S. S. Labana and R. S. Bauer (eds.), 'Cross-Linked Polymers: Chemistry, Properties and Applications', ACS, Washington, DC, 1988. The book deals with current topics in network formation, deformation, fatigue, fracture and recent advances in crosslinking chemistry.

Eisenberg, L. A. and F. E. Bailey (eds.), 'Coulombic Interactions in Macromolecular Systems', ACS, Washington, DC, 1986. The structure and application of ion-containing polymers and their effect on the properties of macromolecules.

Goddard, E. D. and R. Vincent (eds.), 'Polymer Adsorption and Dispersion Stability', ACS, Washington, DC, 1984. Papers are presented on the general topic of polymer adsorption and stabilization/destabilization. They include aqueous and nonaqueous systems, and natural and synthetic polymers.

Grassi, N. (ed.), 'Developments in Polymer Degradation', Applied Science, London, vols. 1–6, 1979–1985. Current developments in polymer degradation by prominent and active investigators are reviewed. They include a range of topics.

Guven, O. (ed.), 'Crosslinking and Scission in Polymers', Kluwer, Boston, 1990. This main lectures given at the NATO ASI held in Kemer, Antalya, Turkey, September 1988, to survey the available techniques and alternative ways of using conventional methods of analysis.

Hoyle, C. E. and J. F. Kinstle (eds.), 'Radiation Curing of Polymeric Materials', ACS, Washington, DC, 1990. Sections cover photoinitiators, photocurable systems, properties of radiation-cured materials, photodegradation, laser-initiated polymerization and high energy radiation curing.

Jones, F. R. (ed.), 'Interfacial Phenomena in Composite Materials '89', Butterworth, London, 1989, Proceedings of an international conference, it consists of papers on interfaces in polymer matrix composites.

Klemchuk, P. P. (ed.), 'Polymer Stabilization and Degradation', ACS, Washington, DC, 1985. A review of the development of the hindered amine stabilizers.

Kramer, O. (ed.), 'Biological and Synthetic Polymer Networks', Elsevier, New York, 1988. Selected papers from the 8th Polymer Networks Group Meeting, Elsinore, Denmark, on the formation, characterization and properties of polymer networks, with special attention to biological networks and to the swelling of polymer networks.

Lazar, M., T. Bleha and J. Rychly, 'Chemical Reactions of Natural and Synthetic Polymers', Wiley, New York, 1989. An introduction to the main principles of chemical reactions of macromolecules in order to open the door for new technologies.

Reichmanis, E. and J. H. O'Donnell (eds.), 'The Effects of Radiation on High Technology Polymers', ACS, Washington, DC, 1989. Emphasizes the ongoing basic and applied research in this area.

Riew, C. K. (ed.), 'Rubber-Toughened Plastics', ACS, Washington, DC, 1989. The book discusses the development of toughness improvements for both thermosets and thermoplastics.

Scott, G. (ed.), 'Developments in Polymer Stabilisation', Applied Science, London, vols. 1–8, 1979–1987. A series of books that review the progress in the development of practical stabilizing systems for many polymers.

Winnik, M. A. (ed.), 'Photophysical and Photochemical Tools in Polymer Science', Reidel, Dordrecht, 1986. Conformation, dynamics and morphology.

9 HISTORIES

Alper, J. and G. L. Nelson, 'Polymeric Materials: Chemistry for the Future', ACS, Washington, DC, 1989. The book discusses the growth of the polymer industry over the past 50 years, what the basic polymers are and why they developed, some of its problems, and the trend toward internationalization of the industry.

Carra, S., F. Parisi, I. Pasquon and P. Pino (eds.), 'Giulio Natta: Present Significance of His Scientific Contribution', Ed Chim, Milan, 1982. The contributions of Natta, known particularly for his work in stereospecific polymerization, is detailed in the order in which his research was initiated.

Elias, H.-G. (ed.),'Trends in Macromolecular Science', Gordon and Breach, New York, 1973. On the occasion of the dedication of this symposium series, this publication contains the special lectures presented by Flory, Overberger and Sannes, Calvin, Lyman, Andrews and Alfrey (*Midl. Macromol. Monogr.*, no. 1).

McMillan, F. M., 'The Chain Straighteners', MacMillan Press, London, 1981. An account of the discoveries and developments which have resulted in the growth in the production of high polymers.

Martuscelli, E., C. Marchetta and L. Nicolais (eds.), 'Future Trends in Polymer Science and Technology—Polymers: Commodities or Specialties?', Technomic, Lancaster, 1987. An International Workshop concludes there will be a growing demand for specialty polymers and their modified compounds as structural replacements for metals.

Morawetz, H., 'Polymers: The Origins and Growth of a Science', Wiley, New York, 1985. The material is divided into three parts: the early part up to World War I that deals with the early observations of polymerization; next the classic period from 1914–1942; and the third period from 1942–1960 when the science reached full maturity.

Morris, P. J. T., 'The American Synthetic Rubber Research Program, University of Pennsylvania Press, Philadelphia, 1989. Details of the history of the government-funded synthetic rubber research program, 1942–1956.

Proskauer, E. S. (ed.), 'Benchmark Papers in Polymer Chemistry', Dowden, Hutchinson & Ross, Stroudsburg, PA, 1978. The series sets out to provide the practitioner with the original literature in the study of polymer structure, behavior and properties. Volume 1: Hermans, J. J. (ed.), 'Polymer Solution Properties'; Part I, Statistics and Thermodynamics; Part II, Hydrodynamics and Light Scattering.

Seymour, R. B. (ed.), 'Pioneers in Polymer Science', Kluwer, Boston, 1989. The proceedings of an ACS Symposium that is dedicated to the 12 Nobel Laureate polymer scientists and 11 other scientists who have been largely responsible for the development of polymer science. Chapters are written by Stahl, Mark, Sperling, Fisher, Marvel, Carraher and Pauling, and detail the contributions made by pre 20th century polymer pioneers to the modern scientists in the areas of plastics, fibers, inorganic polymers, elastomers and engineering polymers.

Seymour, R. B. and G. S. Kirshenbaum (eds.), 'High Performance Polymers: Their Origin and Development', Elsevier, New York, 1986. Proceedings of the Symposium on the History of High Performance Polymers of the American Chemical Society Meeting held in New York, April 1986. Reviews include: engineering thermoplastics—polyamides, polyesters, acetal styrenics, sulfur-containing polymers, poly(aryl ether ketones), poly(ether imides); blends and alloys; liquid crystalline polymers; fluoroplastics; thermosets; fibers, carbon fibers, polybenzimidazoles; elastomers; high barrier packaging materials.

Seymour, R. S. (ed.), 'History of Polymer Science and Technology', Dekker, New York, 1982. A series of articles first published in the 'Journal of Macromolecular Science' that details the events leading to the age of polymers. Accounts of fibers, elastomers, plastics, coatings, adhesives and foams are also presented.

Stahl, G. A. (ed.), 'Polymer Science Overview: A Tribute to Herman F. Mark', ACS, Washington, DC, 1981. Proceedings of a Symposium at the 180th American Chemical Society Meeting, Las Vegas, August 1980 to honor Herman F. Mark's 85th birthday. Special papers were presented in tribute to Professor Mark, as well as overviews of many topics of polymer science.

Staudinger, H., 'From Organic Chemistry to Macromolecules: A Scientific Autobiography Based on My Original Papers', Wiley–Interscience, New York, 1970. (A translation of Herman Staudinger's 'Arbeitserinnerungen', published by Guthig, Heidelberg, 1961.) An historical review of the beginnings of macromolecular chemistry in the collected papers of Staudinger, Nobel Prize winner, teacher and 'apostle' for the 'new branch of organic chemistry'.

Vogl, O. and E. H. Immergut (eds.), 'Polymer Science in the Next Decade: Trends, Opportunities, Promises', Wiley, New York, 1987. An international symposium honoring Herman F. Mark on his 90th birthday, it emphasizes trends in the development of polymer science in the future and forecasts new developments and new directions.

SUBJECT INDEX

Abietic acid
   oxidative polymerization, 543
   synthesis
      by living organisms, 533
Ablative photodecomposition
   laser radiation
      polymer photodegradation, 280
Acenaphthylene
   radical polymerization
      self retarding, 19
Acetone, 2-furfurylidene-
   anionic polymerization, 563
Acetophenone, phenyl-
   photoinitiator, polymer photodegradation, 263
Acrylamide, N-octadecyl-
   film polymerization, polymeric Langmuir–Blodgett film formation, 452, 453
Acrylamide, phenyl-
   derivatives, monomers, nonlinear optical polymers, 417
Acrylates
   bulk polymerization
      free radical initiated, reactive processing, 622
Acrylic acid
   grafting agent
      polyethylene, 136
   graft polymerization
      polypropylene, via trapped radical technique, 147
   molten state grafting
      polyalkenes, 149
   polypropylene grafted
      interpolymeric reactivity, 635
Acrylic acid amides
   film polymerization
      polymeric Langmuir–Blodgett film formation, 452
Acrylic monomers
   polyalkene grafting
      reactive extrusion, 627
Acrylonitrile
   batch emulsion copolymerization
      butyl acrylate, 53
   radical polymerization
      copper(II) chloride retarded, 11
      iron(III) chloride retarded, 8
Acrylonitrile–butadiene–styrene
   polystyrene blends
      via interpolymeric ion bonding, 637
Acrylonitrile–butadiene–styrene rubber
   photodegradation, 269
Activated monomer polymerization
   oxiranes
      synthesis of functional polymers, 118
      synthesis of macromolecular diols, 124
Active centers
   1-alkene polymerization
      heterogeneous versus homogeneous systems, 79
   transition metal complex catalysts
      ethylene polymerization, 69
Addition polymerization
   reactive extrusion, 629
Additives
   use in biodegradation, 290
Adhesion
   particulate composites
      mechanical properties, 494
Agar
   synthesis
      by living organisms, 530
Aging
   polymers
      consequences, 257

Ahmad–Stepto theory
   gelation, 213
Alcoholysis
   copolyacrylates
      reactive extrusion, 633
   ethylene/vinyl acetate copolymers
      reactive extrusion, 633
Aldehyde end groups
   poly(vinyl ether)s
      synthesis via cationic polymerization, 122
Alginate
   natural biodegradable material
      uses, 293
Alginic acid
   synthesis
      by living organisms, 530
Alkali metals
   catalysts
      ε-caprolactam base-catalyzed polymerization, 584
Alkaline earth metals
   catalysts
      ε-caprolactam base-catalyzed polymerization, 584
Alkanes, α-amino-ω-halo-
   amination reagents
      functionalized polymer synthesis, 94
Alkanol groups, fluoro-
   functionalized polyethylene, synthesis via radical addition, 140
1-Alkenes
   copolymerization with ethylene
      via homogeneous catalysis, 76
   heterogeneous polymerization
      titanium trichloride catalyst, 67
   polymerization
      homogeneous versus heterogeneous catalysts, 78
   polymerization via homogeneous catalysis, 70–76
      isotactic chain formation, 68
Alkenyl end groups
   polypropylene
      synthesis, 142
Alkenyl monomers
   catalytic polymerization
      living systems, 113
      obstacles to, 107
      synthesis of difunctional polymers, 121–124
      synthesis of functional polymers, 110–117
      synthesis of polyfunctional polymers, 127
Alkyl isocyanates, perfluoro-
   poly(allylamine) derivatives, amphiphilic polymers, 457
Alloys
   polymer
      functionalized polypropylene application, 153
Allyl derivatives
   self retardation
      radical polymerization, 2
Allylic groups
   polyisobutenes
      via functional termination, 117
Allyloxy groups
   poly(ethyl vinyl ether)
      via functional termination, 117
Alumina
   catalyst support
      heterogeneous polymerization, 67
Aluminoxane, methyl-
   transition metal activator, ethylene polymerization, 69
Aluminoxanes, alkyl-
   transition metal complexes, homogeneous catalysis, 68
Aluminum, triethyl-
   homogeneous cocatalyst, propylene polymerization, 71
Aluminum, trimethyl-

homogeneous cocatalyst, propylene polymerization, 71
Amination
    anionic synthesis
        functionalized polymers, 92–94
Amine groups
    end functionalized polymer
        synthesis via 1,1-diphenylethylene, 99
    functionalized polymers
        TLC characterization, 86
Amines
    curing agents
        epoxy resins, 580
    tertiary
        polymerization catalysts, low temperature reactive processing, 582
Amino acids
    N-carboxyanhydrides
        monolayer polycondensation, poly(γ-dodecyl-L-glutamate) formation, 455
    esters
        monolayer polycondensation, polypeptide formation, 455
Aminolysis
    copolyacrylates
        reactive extrusion, 634
Ammonium end groups
    synthesis
        via ring-opening cationic polymerization, 119
Amorphous polymers
    attached to dye units
        nonlinear optical polymers, 414–418
    sorption and transport phenomena, 179–184
        gases, 179–181
Amphiphiles
    polar head groups
        water attraction, 451
    polymerization, 452
Amphiphilic polymers
    multilayer, 462
        structure, 463
    strong hydrophilic parts
        Langmuir–Blodgett films, 457
Anhydrides
    curing agents
        epoxy resins, 580
Aniline, p-lithio-N,N-bis(trimethylsilyl)-
    initiator and amination reagent
        butadiene polymerization and functionalization, 93
Aniline, 2-methyl-4-nitro-
    blends with glassy polymers, nonlinear optical polymers, 414
Aniline, 4-nitro-
    attached to polymer backbone
        nonlinear optical polymer, 416
    blends with glassy polymers
        nonlinear optical polymers, 414
Anionic polymerization
    living
        synthesis of functionalized polymers, 84
        polymers with functional groups, 83–103
Anisotropic polymer solutions
    transient flow behavior
        liquid crystal polymers, 391
Anisotropic viscoelasticity
    liquid crystal polymers, 395–405
Anisotropy
    composite laminates, 480
    viscosity
        liquid crystal polymers, 395
Anticorrosive coatings
    functionalized polypropylene application, 152
Aqueous phase nucleation
    emulsion copolymerization, 46
Aramid fibers
    reinforcements
        composite materials, mechanical properties, 478
Aromatic hydrocarbons
    retarders
        radical polymerization, 2
Aromatic nitro compounds

    retarders
        radical polymerization, 3
Asphalt
    biodegradation, 286
ASTM standard
    biodegradation test, 287
Autoaccelerating mode
    oxidation
        polyethylene photodegradation, 260
Autocatalytic behavior
    epoxy resins
        polymerization reactions, 507
Autoclave lamination
    composite materials
        high performance, 501
    thermochemorheological model
        parameters, 518
Azido-functionalized polymers
    synthesis
        via inifer method, 112
Aziridine, N-t-butyl-
    activated monomer polymerization
        synthesis of difunctional polymers, 126
    ring-opening cationic polymerization
        synthesis of functional polymers, 118
Azo alcohols
    functional terminating agents
        living polymerization, vinyl ethers, 116
Azobenzene, 4'-amino-4-nitro-
    dye unit
        nonlinear optical liquid crystalline polymer, 420
Azobenzene, 4-(dicyanovinyl)-4'-(dialkylamino)-
    methacrylate substituted
        copolymer from, nonlinear optical polymer, 416
Azobenzene, 4-[N-ethyl-N-(2-hydroxyethyl)]amino-4'-nitro-
    blends with glassy polymers
        nonlinear optical polymers, 414
Azobis(isobutyronitrile)
    reaction with DPPH, 5
Azo dyes
    blends with glassy polymers
        nonlinear optical polymers, 414
Azomethanes, phenyl-
    iniferters
        radical polymerization, 34

Back biting
    intramolecular exchange reactions
        pyrolysis, condensation polymers, 240
Bacteria
    biodegrading ability, 290
    use
        biodegradation, 286
Bacterial polymerization
    uses, 555
Bacteriorhodopsins
    dyes in Langmuir–Blodgett films
        nonlinear optical polymers, 426
Barone–Caulk model
    flow prediction
        sheet molding compound, 606
Barrier polymers
    sorption
        small molecules, 182
    transport
        small molecules, 182
Base-catalyzed polymerization
    nylon reaction injection molding
        kinetic models, 584
Batch emulsion polymers
    microstructure, 52–55
Behenic acid
    Langmuir–Blodgett films
        nonlinear optical polymers, 428
Benzene, 1,4-diisopropenyl-
    cationic polymerization
        synthesis of difunctional polymers, 124

Subject Index

Benzene, 1,4-divinyl-
  cationic polymerization
    synthesis of difunctional polymers, 123
Benzene, 1-isopropenyl-4-glycidyloxy-
  selective activation
    living cationic polymerization, 128
Benzenes, tetraamino-
  monomers for ladder polymers, nonlinear optical materials, 441
Benzidine, 3,3'-diamino-
  monolayer polycondensation
    with dihexylterephthalaldiimine, 455
Benzoate, 4-oxy-
  liquid crystal copolymer with poly(ethyleneterephthalate)
    diffusion coefficient, 183
Benzoic acid, 4-hydroxy-
  liquid crystal copolyester with 6-hydroxy-2-naphthoic acid
    diffusion coefficient, 182
Benzophenones
  photoinitiators
    polymer photodegradation, 263
Benzothiazoles
  nonlinear optical polymers
    structure–property relationships, 430
Benzyl ethers
  liquid crystals based on
    flexible mesogens, 343
Bicyclo[2.2.2]octane-1,4-dicarboxylic acid
  polymers with 2,5-dialkoxyhydroquinone diacetates
    characterization, 329
Bimodality
  copolymer chemical composition distribution
    batch emulsion copolymerization, 54
Biocompatibility
  protein surface reaction
    polymeric Langmuir–Blodgett films, 467
Biodegradable polymers
  from renewable resources, 555
Biodegradation
  definition, 285
  factors influencing, 287
  polymers, 285–295
    biophysical effect, 286
    consequences, 286
    direct enzymatic action, 286
    secondary biochemical effect, 286
    susceptibility, 286
  starch
    uses, 540
  synergism
    environmental degradation, 286
  tests for, 287
Biomass
  chemical exploitation, 535
  monomers from
    polymerization, 550–569
Bisimide
  initiators
    ε-caprolactam base-catalyzed polymerization, 584
Bismaleimide resins
  curing kinetics, 587
  homopolymerization
    kinetic models, 588
  reactive processing
    kinetics, 587
  structure development
    chemical crosslinking, 589
4,4'-Bis(maleimidodiphenylmethane)
  bismaleimide resin component
    curing kinetics, 587
Bisphenol A
  diglycidyl ether
    curing reaction, 507
Bisphenol A diglycidyl ether
  epoxy resin, 580
Bisphenol A polycarbonate
  amorphous
    Theodorou–Suter model, 165

poly(ethyleneterephthalate) binary blend
  pyrolysis, intermolecular exchange reaction, 239
Bis(2,2,6,6-tetramethyl-4-piperidinyl)maleate
  hindered amine light stabilizer
    polypropylene grafting, 627
Block copolymers
  biodegradable
    applications, 288
  synthesis
    from furanic monomers, 562
    polyperoxide initiation, 32
  use
    synthesis of functionalized polymers, 84
Bond dissociation energy
  polymers
    photodegradation and, 255
Bone plates
  artificial
    biodegradable, 288
Boron/epoxy composites
  properties, 499, 500
Boron fibers
  epoxy matrix, 486
  reinforcements
    composite materials, mechanical properties, 478
Boron trifluoride
  Lewis acid catalyst
    epoxy resin curing, 580
Branching coefficient
  gelation
    thermoset composites, 512
Bulk functionalization
  polyalkenes, 137–139
    synthesis of functionalized polyalkenes, 135
Bulk molding compound
  compression molding
    computer simulation, 605
  high temperature compression molding
    peroxide initiators, 582
Bulk polymerization
  free radical initiated
    reactive processing, 622
  reactive extrusion, 629
Bulk properties
  glassy polymers
    rotational isomeric state theory, 161
Butadiene
  polymerization and amination
    via functionalized initiator, 93
  radiation grafting
    polyethylene, 146
  substituted
    film polymerization, polymeric Langmuir–Blodgett film formation, 454
1-Butene
  polymerization
    via dichloro(ethyl)bis(tetrahydroindenyl)zirconium catalytic system, 75
Butyl acrylate
  batch emulsion copolymerization
    acrylonitrile, 53
    styrene, 61
    vinyl acetate, 53, 55–57, 60
  solution grafting
    polyethylene, 150
Butylene–bisphenol A copolycarbonate
  pyrolysis
    intramolecular exchange reaction, 243
t-Butyl perbenzoate
  initiator
    high temperature reactive processing, 582
t-Butyl peroctoate
  initiator
    high temperature reactive processing, 582
Butyl rubbers
  halogenation
    reactive extrusion, 630

Camera-Roda–Sarti model
  diffusion
    penetrant in polymer, 190
ε-Caprolactam
  anionic polymerization
    crystallization kinetics, 596
    reactive extrusion, 629
  nylon monomer
    reaction injection molding, kinetics, 584
ε-Caprolactone
  activated monomer polymerization
    synthesis of difunctional polymers, 126
Carbazole, N-vinyl-
  polymerization
    via living cationic system, 113
Carbene insertion
  polyalkenes
    functionalization, 140
Carbon
  from living organisms
    uses, 541
Carbonation
  anionic synthesis
    functionalized polymers, 87–90
  polymeric organolithium compounds
    product characterization, 89
    product separation, 89
Carbon dioxide
  carbonation reagent
    anionic synthesis, functionalized polymers, 87–89
Carbonell–Sarti model
  diffusion
    penetrant in polymer, 189
Carbon fiber/epoxy composites
  properties, 499
  pultrusion
    processing model, 519
Carbon fiber/epoxy matrix laminates
  aeronautical applications
    physicochemical behavior, 504
  heat transfer
    composite materials processing, 517
Carbon fibers
  reinforcements
    composite materials, mechanical properties, 478
Carbonyl compounds
  hydroxylation reagents
    anionic synthesis, functionalized polymers, 91
Carbonyl groups
  polymers containing
    photodegradation, 269–271
Carbonyl index
  embrittlement
    polyalkene photodegradation, 261
Carboxy groups
  end-functionalized polymer
    synthesis via 1,1-diphenylethylene, 100
  functionalized polymers
    TLC characterization, 86
Carboxylated latices
  copolymer latex properties
    emulsion copolymerization, 59
Carboxylic monomers
  grafting reactions
    reactive extrusion, 626
Carman–Kozeny equation
  permeability
    fiber beds, 520
β-Carotene
  nonlinear optical material
    third order, 435
Carothers' criterion
  gelation, 204
Carrageenan
  synthesis
    by living organisms, 530

Cascade theory
  gelation
    network formation, 210, 211
  network formation
    applications, 221
Caseins
  natural biodegradable material
    use as adhesives, 292
Castor oil
  oxidopolymerization, 542
Catalytic centers
  formation
    heterogeneous versus homogeneous systems, 79
Cationic polymerization
  activated monomers
    nucleophilic character, 108
  alkenyl monomers
    living systems, 113
  initiation
    requirements, 108
  mesogenic vinyl ethers
    liquid crystalline polymer synthesis, 360
  synthesis of functional polymers, 107–129
    basic requirements, 109
    comparison with anionic polymerization, 110
Cellobiose
  from biomass
    polyurethane synthesis, 555
Cellulose
  chemical modification
    uses, 535–540
  crosslinking, 537
  ethers
    uses, 536
  grafting
    via anionic polymerization, 539
    via cationic polymerization, 539
    via polycondensation, 540
    via radical polymerization, 537–540
  inorganic esters
    uses, 535
  natural biodegradable material, 292
  organic esters
    uses, 536
  photodegradation, 279
  pyrolysis
    OH hydrogen transfer, 234
  starch copolymer
    biodegradable, 292
  synthesis
    by living organisms, 529
  xanthates
    uses, 536
Cellulose, hydroxyethyl-
  uses, 536
Cellulose, hydroxypropyl-
  viscoelastic properties, 390
Cellulose, sodium carboxymethyl-
  uses, 536
Cellulose, trimethyl-
  block copolymers
    polytetrahydrofuran, synthesis, 539
Cellulose alkyl ethers
  rod-like polymers
    monolayers, Langmuir–Blodgett films, 460
Cellulose carboxylates
  uses, 536
Cellulose esters
  rod-like polymers
    monolayers, Langmuir–Blodgett films, 461
Cellulose triacetate
  uses, 536
Celluloses, deoxy-
  uses, 536
Celluloses, nitro-
  uses, 535
Cellulosic liquid crystals

Subject Index

uses, 537
Cellulosic membranes
  uses, 537
Chain degradation
  continuous processes
    free radical initiated, 623
  controlled
    polyalkenes, reactive processing, 621
Chain end control mechanism
  propylene polymerization
    via homogeneous catalysis, 70
Chain end structure
  effect on carbonation
    polymeric organolithium compounds, 88
Chain entanglement effect
  polymer solutions
    viscosity, 590
Chain growth copolymerization
  styrenic/polyester resins
    reactive processing, 582
Chain growth polymerizations
  reactive processing
    kinetic models, 577
Chain interactions
  network materials
    properties, 200
Chain lengths
  mean kinetic
    inhibited/retarded polymerization, 12
Chain propagation
  polymer photodegradation, 258
Chain scissions
  photooxidative degradation
    polymers, 261
Chain termination
  polymer photodegradation, 259
Chain transfer
  degradative
    radical polymerization, 2
Chain transfer polymerization
  polyalkenes
    chemically initiated grafting, 148–150
Chain transfer rate
  emulsion copolymerization
    kinetic factors, 49
Chain transfer reactions
  cationic polymerization, 109
Characterization methodology
  functionalized polymers, 85–87
Charge transfer complex activated system
  maleic anhydride/polyalkene grafting
    reactive extrusion, 626
Charge transfer complexes
  melt grafting
    maleic anhydride, 628
  polymers
    photooxidative degradation, 257
Chemical functionalities
  networks
    definition, 200
Chemical ionization mass spectrometry
  condensation polymers
    product detection, 231
Chemical modification
  natural polymers, 535–541
  polyalkenes
    synthesis of functionalized polyalkenes, 134
Chemical reactions
  characterization
    functionalized polymers, 87
Chemical reactivity
  thermoplastic polymers
    melt processes, 621
Chemorheological model
  processing
    composite materials, 505
Chemorheology
  processing
    thermosets, 511–514
Chiral active sites
  propylene polymerization
    via homogeneous catalysis, 71
Chiral metallocenes
  stereorigid
    homogeneous catalysts, 71
Chitin
  chemical modification
    uses, 540
  natural biodegradable material
    uses, 292
  synthesis
    by living organisms, 529
Chitosan
  cellulose composite films
    biodegradable, 292
Chlorination
  polyethylene
    bulk functionalization, 137
Chlorine-functionalized polymers
  synthesis
    via inifer method, 112
Chlorocarboxylation
  polyethylene
    bulk functionalization, 137
Chlorosulfonation
  polyethylene
    bulk functionalization, 139
Chromium oxide
  heterogeneous polymerization catalyst, 67
Chromium(VI)
  surface functionalization
    polypropylene, 137
Chromophores
  polymers
    photodegradation and, 256
    photooxidative degradation, 257
  selection
    for nonlinear optical polymers, 413
Cinnamic acid
  derivatives
    film polymerization, polymeric Langmuir–Blodgett film formation, 454
Coagulative nucleation
  emulsion copolymerization, 46
Cold drawing
  polymeric matrices, 476
Cold plasma
  surface functionalization
    polypropylene, 136
Collagen
  synthesis
    by living organisms, 530
Collision-induced dissociation mass spectrometry
  condensation polymers
    product detection, 231
Colloidal aspects
  factors influencing
    emulsion copolymerization, 60
Column chromatography
  characterization
    functionalized polymers, 86
Combination polymerizations
  reactive processing
    kinetic models, 577
Commercial polymers
  photodegradation
    molecular mechanisms, 259
Compatibility
  polymeric products
    emulsion copolymerization, 58
Composite laminates
  mechanics, 480–485
Composite materials
  fracture mechanics, 495–501

high performance
    processing, 501–523
    mechanical properties, 472–501
    processing, 472
        modeling, 505
    reinforcement, 472
    short fiber, 485–491
        strength, 488
    thick
        chemorheology, 511
    wood
        synthesis by living organisms, 534
Composition drift
    effect on copolymerization rate
        emulsion copolymerization, 55–57
    emulsion copolymerization, 42
Computer modeling
    polymer structure
        fundamental properties, 159–165
        motivation, 159
Concentration
    external
        gelation, 214
        intramolecular reactions, network formation, 209
    influence of
        on mutual diffusion coefficient, 170–176
    internal
        intramolecular reactions, network formation, 209
Condensation polymers
    thermal degradation, 227–249
Condensation reactions
    functional modification
        reactive extrusion, 632
    silane-grafted polyalkenes
        reactive extrusion, 624
Condensed films
    polymer monolayers
        Langmuir–Blodgett films, 457
Conjugated polymers
    nonlinear optical materials
        structure–property relationships, 432
        via polyelectrolyte precursor technique, 436–438
    thiophene-based
        nonlinear optical materials, 438
Controlled composition reactor
    semicontinuous polymerization
        emulsion copolymerization, 61
Copolyacrylates
    alcoholysis
        reactive extrusion, 633
    aminolysis
        reactive extrusion, 634
Copolyamides
    pyrolysis
        intramolecular exchange reaction, 243
Copolyesters
    with main chain dye units
        nonlinear optical polymers, 423
Copolyethers
    liquid crystalline polymers
        columnar and smectic mesophases, 351
Copoly(imide amine)
    pyrolysis
        depolymerization, 237
Copolymerization
    alkenes and polar comonomers
        synthesis of functionalized polyalkenes, 133
    ethylene with 1-alkenes
        via homogeneous catalysis, 76
    radical emulsion, 41–63
    substituted 1,1-diphenylethylenes
        with various monomers, 101
Copolymerization model
    styrene/polyester resins
        reactive processing, 583
Copolymerization rate
    effect of composition drift
        emulsion copolymerization, 55–57
Copolymer latex properties
    control
        semicontinuous processes, 59
    intrinsic characteristics influencing
        emulsion copolymerization, 58
    process–structure–property relationships, 60, 61
Copolymer microstructure
    emulsion copolymerization, 51
    characterization, 42
Copolymers
    alternating
        pyrolysis, 242–244
    batch emulsion
        microstructure, 52–55
    bulk polymerization
        free radical initiated, reactive processing, 623
    liquid crystalline
        thermodynamic parameters, 348
    liquid crystalline smectic
        microphase separated morphology, 359
    main chain dye units
        nonlinear optical polymers, 422
    synthesis
        from furanic monomers, 562
Copolymer sequence distribution
    batch emulsion copolymerization, 54
Copper(II) chloride
    retarder
        radical polymerization, 11
Coupling reactions
    polycondensates
        reactive processing, 621
Crazing
    polymeric matrices, 477
Crazing stress
    composite laminates, 484
Creep modulus
    viscoelastic behavior
        polymeric matrices, 473
Critical crossover concentration
    polystyrene/toluene
        polymer molecular weight influence, 172
Crosslinking
    polyalkenes
        reactive processing, 621
    polymer chains
        gel point, 203
        network formation, 200
    polymeric matrices, 475
Crosslinking number
    photooxidative degradation
        polymers, 262
Crosslinking reactions
    reactive extrusion, 628
Crown ethers
    liquid crystalline polymers, 370
    molecular architecture, 370
Crystallinity
    polymeric matrices, 475
Crystallization
    during polymerization
        reactive processing, models, 596
Cutoff limit
    photostability
        polymers, 255
Cyclic acetals
    activated monomer polymerization
        synthesis of difunctional polymers, 125
    ring-opening cationic polymerization
        synthesis of functional polymers, 119
Cyclic amines
    ring-opening cationic polymerization
        synthesis of functional polymers, 118
Cyclic carbonates
    ring-opening cationic polymerization
        synthesis of functionalized polymers, 120

Cyclic ethers
  ring-opening cationic polymerization
    synthesis of difunctional polymers, 124
    synthesis of functional polymers, 118
Cyclic macromolecules
  use
    synthesis of functionalized polymers, 84
Cyclic oligomer process
  synthesis *via* functional end group reactivity
    via reactive extrusion, 636
Cyclic siloxanes
  activated monomer polymerization
    synthesis of difunctional polymers, 127
  ring-opening cationic polymerization
    synthesis of functionalized polymers, 121
Cycloaddition
  conjugated diene polymers
    with oxygen, 256
Cycloalkenes
  homopolymerization
    via metallocene homogeneous catalysis, 77
1,5-Cyclooctadiene
  copolymers with cyclooctatetraene
    nonlinear optical materials, 435
Cyclooctatetraene
  copolymers with 1,5-cyclooctadiene
    nonlinear optical materials, 435
Cyclooctene
  copolymerization with ethylene
    via dichloro(ethyl)bis(indenyl)zirconium homogeneous catalysis, 78
Cyclopentene
  copolymerization with ethylene
    via zirconocene homogeneous catalysis, 78
Cyclopropane, vinyl-
  polymerization
    via homogeneous catalysis, 75

Dammars
  synthesis
    by living organisms, 533
Darcy's law
  resin flow
    laminate composite fabrication, 520
Deborah number
  diffusion
    penetrant in polymer, 191
  heat transfer
    composite materials processing, 516
Decomposition reaction constants
  radical polymerization
    peroxide polyinitiators, 30
Degradation
  environmental
    packaging, 285
    synergism with biodegradation, 286
  photooxidative
    synergism with biodegradation, 286
Degradative addition
  radical polymerization
    slow reinitiation, 17
Degree of polymerization
  number average
    gelation, 203
    inhibited/retarded polymerization, 12
    polyfunctional initiators, 26
Delignification
  wood
    pulping processes, 534
Dextran
  bacterial synthesis
    uses, 556
Diacrylic esters
  film polymerization
    polymeric Langmuir–Blodgett film formation, 453
Diacyl peroxides
  initiators
    block copolymer synthesis, 32
    radical polymerization, 24
Dialkynes
  film polymerization
    polymeric Langmuir–Blodgett film formation, 454
  multilayers
    solid state photopolymerization, 454
Diamines
  functional terminating agents
    living polymerization, vinyl ethers, 116
Diastereoface selectivity
  alkene polymerization
    chiral homogeneous catalysis, 76
Diazo esters
  functionalized polyalkenes
    synthesis, 141
Diazostilbenes
  dyes in Langmuir–Blodgett films
    nonlinear optical polymers, 426
Dibenzoyl peroxide
  free radical initiator
    synthesis of functionalized polyalkenes, 134
  low temperature initiator
    reactive processing, 582
Dicarboxylic oligomers
  synthesis
    via activated monomer polymerization, 125
Dicumyl peroxide
  free radical initiator
    synthesis of functionalized polyalkenes, 134
Diels–Alder reaction
  furan-functionalized polymers
    synthesis of carboxylic acid derivatives, 111
Diene copolymers
  photooxidative degradation, 263–267
Diene elastomers
  poorly photostable, 254
Differential scanning calorimetry
  kinetic characterization
    thermosetting polymers, 506
  melting points
    polypropylenes from homogeneous catalysis, 73
Diffusion
  mutual
    importance, 167
    polymeric systems, 167–196
    unifying analytical fundamentals, 193–196
  nomenclature, 191
  polymers in solvents
    infinite dilution, 168–170
  problems
    solution, 194
  single polymer chain in solution
    models, 168
  small penetrant molecules
    polymers, 179–191
Diffusion coefficient
  infinite dilution
    penetrant size effect, 180
  penetrants
    factors influencing, 181
  relationship
    polymer molecular weight, 170
Diffusion control
  epoxy resin curing
    kinetics, 581
  radical polymerization, 18
Diffusion effects
  copolymerization model
    styrene/polyester resins, 584
Diffusion flux models
  penetrant in polymer, 188–191
Difunctional macromolecules
  synthesis
    via cationic polymerization, 121–127
α,ω-Dilithio polymers
  carbonation

complications, minimization, 88
Dimensional stability
  short fiber composites, 489
Dimensionless model
  heat transfer
    composite materials processing, 515
Diols
  macromolecular
    synthesis via activated monomer polymerization, 125
1,3-Dioxolane
  activated monomer polymerization
    synthesis of difunctional polymers, 125
  cationic polymerization
    synthesis of polyfunctional polymers, 129
  ring-opening cationic polymerization
    synthesis of functional polymers, 119
Diphenylamine maleimide
  thermal antioxidant
    polypropylene grafting, 627
Direct pyrolysis mass spectrometry
  condensation polymers
    product detection, 230
Disaccharides
  from biomass
    polymerization, 553–555
Discotic mesogens
  liquid crystalline polymers
    electron donor–acceptor complexes, 371
Disperse phase segregation
  rubber modified epoxy resins
    reactive processing, models, 597
Disperse Red 1
  attached to polymer backbone
    nonlinear optical polymer, 414, 416
  blends with glassy polymers
    nonlinear optical polymers, 414
Distribution functions
  macromolecules
    polyinitiator polymerization, 30
Disulfides
  iniferters
    radical polymerization, 34
Dithiocarbamate, S-benzyl-N,N-diethyl-
  iniferter
    styrene photopolymerization, 34, 36
Dithiocarbamates
  iniferters
    radical polymerization, 34
1-Dodecanethiol
  transfer agent
    emulsion copolymerization, 50
Doi nonlinear equation
  rheology
    lyotropic liquid crystal systems, 389
Doi rheological equation
  anisotropic liquids, 388
Domain formation
  during polymerization
    reactive processing, models, 596
Doolittle viscosity equation
  resins
    free volume approach, 593
Double bond opening
  via titanocene catalyst system
    stereochemistry, 71
DPPH
  reaction with azobis(isobutyronitrile), 5
  stable free radical
    inhibitor, 5
Drying oils
  from renewable resources
    polymerization, 541–543
Dual mode concept
  sorption
    penetrant in polymer, 185
Durning model
  diffusion

    penetrant in polymer, 189
Dye molecule monomer
  copolymer with mesogenic monomer
    nonlinear optical liquid crystalline polymer, 419
Dye molecules
  blends with glassy polymers
    nonlinear optical polymers, 413
Dynamic rotational isomeric state model
  computer modeling
    polymer structure/properties, 162–164
Dynamics
  liquid crystal polymers, 387
  long times
    dynamic rotational isomeric state model, 163

Ecolyte
  biodegradation
    with bacteria, 290
E-glass/epoxy composites
  properties, 499
Elastic behavior
  polymeric matrices
    Hooke's law, 473
Elastic force
  entropy of mixing
    polymer chain compaction, 169
Elasticity theories
  networks
    tests of, 224
Elastin
  synthesis
    by living organisms, 531
Electron impact ionization mass spectrometry
  condensation polymers
    product detection, 230
Electrooptic modulation
  poled polymers, 442
Embrittlement
  photodegradation
    polyethylene, 261
Emulsion copolymerization
  advantages, 42
  applications, 42
  modelling, 42–52
  models, 43
  radical, 41–63
Emulsion copolymers
  intermolecular microstructure, 52
  intramolecular microstructure, 53
Emulsion polymerization
  conventional radical, 41
Emulsion terpolymerization
  emulsion copolymerization, 57
Enantioselectivity
  alkene polymerization
    via chiral homogeneous catalysis, 76
End biting
  intramolecular exchange reactions
    pyrolysis, condensation polymers, 240
End capping
  functional
    living polymerization, vinyl ethers, 116
Endlinking polymerization
  network formation, 200
  network properties, 216
  stoichiometric
    polymer networks, 206
Endoperoxides
  products
    polymer photooxidation, 257
Ene reaction
  alkene polymers
    with oxygen, 256
Energetic interaction
  polymer chain swelling, 169
Enzymes
  use

biodegradation tests, 287
Epoxy resins
    acrylic solutions
        hybrid resin system, 588
    chemorheological behavior
        time dependence, models, 591
    curing
        diffusion controlled, kinetic models, 581
        kinetic models, 580
    matrix laminate
        processing models, 517
    polyurethanes
        hybrid resin system, 588
    primary amine cured
        mechanism, 580
    reaction injection molding
        common systems, 580
        curing analysis, simulation, 603
        simulation, 602
    reactive processing
        kinetic models, 580–582
    structure development
        chemical crosslinking, 589
    synthesis
        via lignin macromonomers, 548
Equivalent induction period
    ideal retarded polymerization, 10
Ericksen–Leslie viscosity coefficients
    liquid crystal polymers, 398
Ericksen orientation equation
    nematic viscoelastic liquids
        liquid crystal polymers, 400
Ericksen's transversely isotropic liquid theory
    viscosity
        liquid crystal polymers, 399
Esterases
    use
        biodegradation, 286
Ethane, diphenyl-
    liquid crystals based on
        flexible mesogens, 344
Ethanes
    disubstituted
        low molar mass liquid crystals, 345
Ethyl acrylate
    batch emulsion copolymerization
        styrene, 53
Ethylbenzene, 1-chloro-1-methyl-
    inifer
        synthesis of chlorine-functionalized polymers, 112
Ethylene
    biodegradable copolymers
        with vinyl ketones, 290
    copolymerization with 1-alkenes
        via homogeneous catalysis, 76
    copolymerization with propylene
        via different catalyst systems, 77
    heterogeneous polymerization
        titanium trichloride catalyst, 67
    1-hexene copolymer
        properties compared with polyethylene, 77
    living anionic oligomers
        electrophilic attack, 140
    5-methyl-1,4-hexadiene copolymer
        interpolymeric reactivity, 635
    polymerization
        via dichlorobis(cyclopentadienyl)zirconium/methylaluminoxane catalyst, 69
        via different zirconocenes/methylaluminoxane catalyst, 69
        via homogeneous catalysis, 68–70
    propylene copolymer
        functionalization, 143
Ethylene, 1,1-diphenyl-
    end-capping reagent
        sulfonated polymer synthesis, 96
    polymer functionalization reactions
        advantages in use, 97–102

Ethylene–bisphenol A copolycarbonate
    pyrolysis
        intramolecular exchange reaction, 243
Ethylene glycol
    from biomass
        uses, 552
Ethylene oxide
    activated monomer polymerization
        synthesis of macromolecular diols, 125
    end-capping reagent
        sulfonated polymer synthesis, 95
    hydroxylation reagent
        anionic synthesis, functionalized polymers, 90
    ring-opening cationic polymerization
        synthesis of functional polymers, 118
Ethylene–propylene copolymers
    photodegradation, 266
Ethylene–propylene–ethylidenenorbornene
    photodegradation, 267
Ethylene–propylene–1,4-phenylhexadiene
    photodegradation, 267
Ethylene–propylene rubber
    crosslinking
        mechanism, 144
    molten state grafting
        functionalization, 149
Ethylene–vinyl acetate copolymers
    alcoholysis
        reactive extrusion, 633
Euric acid
    derivatives
        polymerization, 543
Exchange reactions
    alkenic copolymers
        reactive processing, 621
    functional modification
        reactive extrusion, 632–634
    pyrolysis
        condensation polymers, 239–244
Excluded volume parameter
    polymer chains
        elastic force *versus* energetic interaction, 169
Expanded films
    polymer monolayers
        Langmuir–Blodgett films, 457
Extrudate fibrillation effect
    thermotropic liquid crystal polymers, 391
Extruders
    single screw
        reactive processing, thermoplastic polymers, 620
    twin screw
        reactive processing, thermoplastic polymers, 620

Fiberglass composites
    properties, 500
Fibers
    commercially available
        properties, 479
    reinforcements
        composite materials, mechanical properties, 477
Fibroin
    synthesis
        by living organisms, 530
Fickian transport model
    diffusion
        penetrant in polymer, 188
Fick's first law
    mutual diffusion
        binary system, 194
Filament winding
    composite materials
        high performance, 504
Film
    functionalized polypropylene application, 153
Flash pyrolysis
    condensation polymers
        product detection, 230

Flooded conditions
  copolymer latex properties
    control, 60
Flory–Huggins interaction parameter
  relationship
    excluded volume parameter, 169
Flory–Huggins model
  sorption
    penetrant in polymer, 185
Flory principle
  violation
    iniferter presence in polymerization, 36
Flory–Stockmayer gel point
  network formation, 202
Flow curve shape
  liquid crystal polymers, 393–395
Flow model
  processing
    composite materials, 505
Fluorescence labeling
  polyethylene
    synthesis via carbonyl hydrazide functionalization, 141
Fluorescent group labeling
  via 1,1-diphenylethylenes, 102
Fluorination
  polyethylene
    bulk functionalization, 138
Fluorocarbon chains
  hydrophobic
    amphiphilic polymers, 458
Foaming
  during polymerization
    reactive processing, models, 600
Fourier transform infrared spectra
  poly(L-glutamate) multilayers, 464
  poly(octadecyl methacrylate), 463
Fracture stress
  composite laminates, 481
Fracture toughness
  graphite/epoxy laminates, 498
Free radical functionalization
  polyalkenes
    conditions, 134
    synthesis of functionalized polyalkenes, 134
Free radical processes
  pyrolysis
    condensation polymers, 244–248
Free radical reactivity
  thermoplastic polymers
    melt processes, 622–629
Free radicals
  grafting
    polyalkenes, 144–150
Free volume theory
  kinetic diffusion model
    bulk vinyl polymerization, 583
  mutual diffusion
    concentrated polymer solutions, 176–179
  mutual diffusion coefficient
    polystyrene/toluene, 179
Friction coefficient
  single polymer chain in solution, 168
Friction factor
  concentrated polymer solutions
    importance of solvent contribution, 178
  mutual diffusion coefficient
    polymer concentration effect, 178
  polymer chains
    mutual diffusion relationship, 172
Fumaric acid
  derivatives
    film polymerization, polymeric Langmuir–Blodgett film formation, 453
Functional end group reactivity
  polymer reactive blending, 636
Functional groups
  copolymers containing multiple
    synthesis via 1,1-diphenylethylenes, 101
  polymers containing
    anionic synthesis, 83–103
    thermal reactivity, 228
    via polymerization of protected monomers, 102
Functional group titration
  characterization
    functionalized polymers, 85
Functionality
  network junction points
    definition, 201
Functionalization reaction
  living
    via 1,1-diphenylethylene, 98
Functionalized initiator
  cationic polymerization
    synthesis of difunctional polymers, 122
    synthesis of functionalized polymers, 115
Functionalized latices
  copolymer latex properties
    emulsion copolymerization, 58
Functionalized nucleophile
  quencher
    cationic polymerization, synthesis of difunctional polymers,
Functionalized polyalkenes
  synthesis, 133–155
Functionalized polymers
  characterization methodology, 85–87
  synthesis via anionic polymerization
    molecular architecture, 84
Functionalized reactions
  utilization
    anionic synthesis, functionalized polymers, 87–97
Functional modification
  polymers
    reactive processing, 630–634
Functional monomers
  applications
    synthesis of functionalized polymers, 103
Functional polymers
  synthesis
    via cationic polymerization, 107–129
Functional termination
  living polymerization
    vinyl ethers, 116
Fungi
  use
    biodegradation, 286
    biodegradation tests, 287
Furan
  maleic anhydride adduct
    ring-opening metathesis polymerization, 561
Furan, 2-methyl-
  transfer agent
    cationic polymerization, synthesis of functional polymers, 11
Furan, tetrahydro-
  activated monomer polymerization
    synthesis of macromolecular diols, 125
  ring-opening cationic polymerization
    synthesis of difunctional polymers, 124
    synthesis of functional polymers, 119
Furan, 2-vinyl-
  cationic polymerization
    synthesis of functional polymers, 111
Furanic monomers
  polymerization, 561–568
    chain methods, 561–564
    stepwise methods, 564–568
  renewable source
    5-hydroxymethylfurfural, 560
Furanic polymers
  chemical modification, 568
Furanic tetrols
  synthesis
    via glucose condensation, 558
Furans, 2-alkenyl-
  cationic polymerization, 561

Subject Index

Furfural, 5-hydroxymethyl-
   industrial synthesis
      via pentose hydrolysis/dehydration, 556–560
   renewable source
      difunctional monomers, 560
   synthesis
      via hexose hydrolysis/dehydration, 558
2-Furfuryl acrylate
   free radical polymerization, 564
Furfuryl alcohols
   acid-catalyzed polycondensation
      mechanism, 564
   intermediates
      monomers from renewable resources, 558
   resin monomer, 557
2-Furfurylamines
   intermediates
      monomers from renewable resources, 559
2-Furfurylidene ketones
   monomers from renewable resources
      syntheses, 558
2-Furoic acids
   intermediates
      monomers from renewable resources, 558
2-Furyl isocyanate
   anionic polymerization, 563

Gas chromatography–mass spectrometry
   condensation polymers
      product detection, 230
Gel effect
   kinetic diffusion model
      styrene/polyester resins, 583
   vinyl monomer bulk polymerization
      cause, 37
Gelation
   Carothers' criterion, 204
   composite materials
      thermosets, 511
   delayed
      intramolecular reaction, 212
   epoxy resin curing
      kinetic models, 582
   ideal
      network formation, 202
   intramolecular reactions and
      network formation, 207, 212–215
   network formation, 201
   processing
      composite materials, 519
   theories
      approximate, 213
Gellan
   natural biodegradable material, 292
Gel point
   composite materials
      thermosets, 512
   delayed
      network formation, 207
   epoxy resins
      thermokinetic model, 510
   Flory–Stockmayer
      network formation, 202–204
   network formation
      definition, 201
   resin reactive processing
      determination, models, 594
Geometric mean assumption
   retarded radical polymerization, 18
Glass fiber/polyester composites
   pultrusion
      processing model, 519
Glass fibers
   reinforcements
      composite materials, mechanical properties, 478
Glass transition temperature
   copolymer chemical composition distribution
      batch emulsion copolymerization, 54
   epoxy resins
      degree of cure, 513
      polymerization reactions, 507
   polyester resins
      conversion dependence, models, 592
   resins
      measurement, 594
Glassy polymers
   blends with dye molecules
      nonlinear optical polymers, 413
   bulk density
      computer modeling, 164
Glucose
   from biomass
      polyurethane synthesis, 555
Gluing
   functionalized polypropylene, 154
Glycerol
   from biomass
      polyester synthesis, 552
Glycerol esters
   synthesis
      by living organisms, 531
Glycidol
   comonomer and transfer agent
      synthesis of polyfunctional polymers, 129
Glycidyl acrylate
   grafting agent
      polyethylene, 136
Glycogen
   synthesis
      by living organisms, 530
Grafting
   ethylene–propylene copolymer
      via dibutyl maleate, 143
   ethylene–propylene rubber
      via 2-(dimethylamino)ethyl methacrylate, 143
   free radicals
      polyalkenes, 144–150
   polyalkenes
      reactive processing, 621
   polyethylene surfaces
      agents for, 136
   via interpolymeric reactivity, 635
Grafting reactions
   continuous processes
      free radical initiated, 624–628
Graft polymerization
   peroxide initiated
      via chain transfer, 148
   polyalkenes
      chemically initiated, 147–150
Graphite/epoxy composites
   properties, 499, 500
Graphite/epoxy laminates
   composite materials
      fracture mechanics, 498
Graphite fibers
   reinforcements
      composite materials, mechanical properties, 478
Grotthus–Draper law
   photochemistry
      polymer photodegradation, 256
Group termination coefficient procedure
   retarded radical polymerization, 18
Gum elemi
   synthesis
      by living organisms, 533
Gutta percha
   synthesis
      by living organisms, 531

Hafnium, dichloro(cyclopentadienyl-1-fluorenyl)isopropyl-
   chiral homogeneous catalyst
      alkene polymerization, 76

Hafnium metallocenes
  ethylene polymerization, 69
Halogenation
  butyl rubbers
    reactive extrusion, 630
Halpin–Tsai equations
  particulate composites
    mechanical properties, 492
  short fiber composites
    laminates, 484
Hardware
  computer modeling
    polymers, 160
Head–head enchainments
  propylene polymerization
    via chiral homogeneous catalysis, 72
Heat removal
  copolymer latex properties
    control, 59
Heat transfer model
  processing
    composite materials, 506, 514–520
Hele–Shaw model
  flow prediction
    sheet molding compound, 606
Hemicelluloses
  synthesis
    by living organisms, 532
Hemicyanines
  dyes in Langmuir–Blodgett films
    nonlinear optical polymers, 426
Henry's law
  solubility
    penetrant in polymer, 184
Heparin
  polyethylene surface grafted, 136
Heterocyclic monomers
  cationic polymerization
    synthesis of difunctional polymers, 124–127
    synthesis of functional polymers, 117–121
    synthesis of polyfunctional polymers, 129
Heterogeneous catalysis
  comparison
    homogeneous catalysis, 78–80
    monoalkene polymerization, 67–80
  transition metal complexes
    mechanism, 67
1,5-Hexadiene
  terpolymerization with ethylene/propylene
    via dichloro(ethyl)bis(indenyl)zirconium homogeneous catalysis, 77
1-Hexene
  copolymerization with ethylene
    via bis(cyclopentadienyl)dimethylzirconium homogeneous catalysis, 76
Hexoses
  biomass furanic monomers
    polymerization, 556–561
High performance liquid chromatography
  copolymer analysis
    emulsion copolymerization, 53
Hindered amine light stabilizers
  polyalkene grafting
    reactive extrusion, 627
Homogeneous catalysis
  comparison
    heterogeneous catalysis, 78–80
    monoalkene polymerization, 67–80
  transition metal complexes, 68
Homogeneous nucleation
  emulsion copolymerization, 46
Homopolyesters
  aliphatic
    biodegradation, 288
    with main chain dye units
      nonlinear optical polymers, 424
Homopolymers

main chain dye units
  nonlinear optical polymers, 422–425
Huggin's theory
  polymeric monolayers
    liquid surfaces, 456
Hybrid resins
  cure kinetics
    models, 589
  reactive processing
    kinetics, 588
Hydrodynamic radius
  relationship
    mutual diffusion coefficient, 169
Hydrogen transfer
  α-CH
    condensation polymers, pyrolysis, 237–239
  β-CH
    condensation polymers, pyrolysis, 234–237
  NH
    condensation polymers, pyrolysis, 232
  OH
    condensation polymers, pyrolysis, 234
Hydrogen transfer polymerization
  propylene
    via homogeneous catalysis, 74
Hydrolytic reactions
  alkenic copolymers
    reactive processing, 621
Hydrooligomerization
  1-pentene
    via chiral homogeneous catalysis, 76
Hydroperoxidation
  polyalkenes
    chemically initiated grafting, 147
Hydroperoxides
  products
    polymer photooxidation, 257
Hydrophilic spacers
  electrostatic interaction
    charged amphiphile/high molecular polyion, 458
Hydrophobicity
  polymers
    and biodegradation tests, 287
Hydroquinones
  retarders
    radical polymerization, 2
Hydroxy end groups
  polypropylene
    synthesis, 142
Hydroxy groups
  functionalized polymers
    TLC characterization, 86
Hydroxylation
  anionic synthesis
    functionalized polymers, 90–92
Hydroxy-terminated polymers
  synthesis
    via hydroxy-protected organolithium initiators, 90
  synthesis from polymeric organolithium compounds
    via ethylene oxide hydroxyethylation, 90

2-Icosenoic acid
  film polymerization
    polymeric Langmuir–Blodgett film formation, 453
Imidazole, 2-methyl-1-vinyl-
  radical polymerization
    degradative addition, 18
Imidazole, 1-vinyl-
  radical polymerization
    degradative addition, 18
Imides
  initiators
    ε-caprolactam base-catalyzed polymerization, 584
Imidification
  ethylene/maleic anhydride copolymers
    reactive extrusion, 632
Imines

protected
    amination reagents, functionalized polymer synthesis, 92
Iminolization equilibrium
    pyrolysis
        condensation polymers, 232
Impurities
    polymers
        photooxidative degradation, 257
Induced unequal reactivities
    gel point
        polymerization, 203
Inelastic chain formation
    networks, 201
Inelastic loops
    via intramolecular reactions
        network formation, 208
Infrared spectroscopy
    Fourier transform
        characterization, functionalized polymers, 86
Inhibition
    ideal
        kinetics, 3–6
    radical polymerization, 1–20
        chemistry, 2
Inhibition period
    radical polymerization, 4
Inhibitors
    radical polymerization, 1, 2
Inifer technique
    cationic polymerization
        synthesis of difunctional polymers, 122
        synthesis of functionalized polymers, 112
        synthesis of polyfunctional polymers, 127
Iniferters
    kinetics of polymerization, 34–37
    mechanism of polymerization, 34–37
    radical polymerization, 33–39
Initiation
    polymer radicals
        photooxidative degradation, 257
Initiation efficiencies
    radical polymerization
        peroxide polyinitiators, 30
Initiators
    functionalized and protected
        amination reagents, functionalized polymer synthesis, 93
    nontraditional
        radical polymerization, 23–39
    polyfunctional
        polymerization mechanism, 24
        radical polymerization, 24–33
    protected
        hydroxylation reagents, polybutadienes, 90
In-mold coating
    computer simulation, 605
Intermolecular dynamics
    individual chains
        rotational isomeric state theory, 161
Intermolecular exchange reactions
    pyrolysis
        condensation polymers, 239
Interpenetrating polymer networks
    reactive processing
        kinetics, 588
Interpolymeric ion bonding
    polymer reinforcement
        via reactive blending, 636–638
Intrachain dynamics
    long timescales, 162–164
Intramolecular decomposition
    polyethylene photooxidation, 264
Intramolecular exchange reactions
    pyrolysis
        condensation polymers, 240
Intramolecular reactions
    chain stiffness, 215
    characteristics

network formation, 207–210
gelation
    increase in extent of reaction, 214
    network formation, 207, 212–215
inelastic chains
    network properties, 216
network formation, 201
    modulus, 216
post-gel
    network formation, 207
ring size, 215
Intrinsic characteristics
    copolymer latex properties, 58
Intrinsic unequal reactivities
    gel point
        polymerization, 203
Ionic cleavage
    condensation polymers
        thermal degradation, 228
Ionic processes
    pyrolysis
        condensation polymers, 231
Ionomer resins
    amine grafting
        reactive extrusion, 632
Iron(III) chloride
    photoinitiator
        polymer photodegradation, 263
    retarder
        radical polymerization, 8
Isobutene
    cationic polymerization
        synthesis of difunctional polymers, 122
    living cationic polymerization, 113
        synthesis of polyfunctional polymers, 127
    living polymerization
        via functionalized initiator systems, 116
    polymerization *via* inifer method
        synthesis of chlorine-functionalized polymers, 112
    weak nucleophile
        cationic polymerization, 110
Isobutyl vinyl ether
    polymerization
        via living cationic system, 113
Isocyanate-based resins
    reactive processing
        kinetic models, 577–580
Isocyanurates
    reactive injection molding process
        curing analysis, simulation, 603
Isomorphism
    liquid crystals, 304
Isotactic chains
    1-alkene polymerization
        via homogeneous catalysis, 68
Isotacticity index
    polypropylenes
        from homogeneous catalysis, 74

Jacobson–Stockmayer treatment
    intramolecular reactions
        network formation, 208
Jaumann's derivatives
    nematic viscoelastic liquids
        liquid crystal polymers, 400
Joint reinforcement
    functionalized polypropylene application, 151
Junction points
    networks
        definition, 199

Keratin
    synthesis
        by living organisms, 530
Kerner equation
    particulate composites

mechanical properties, 492
Kilb's theory
  gelation, 213
Kinetic diffusion models
  styrenic/polyester resins
    reactive processing, 583
Kinetic effects
  copolymerization model
    styrene/polyester resins, 584
Kinetic parameter
  inhibited/retarded polymerization, 13
Kinetic theories
  gelation
    network formation, 210
    nonlinear polymerization, 212
Koelsch radical
  inhibitor
    radical polymerization, 5

Lactam, acyl-
  initiators
    ε-caprolactam base-catalyzed polymerization, 584
Lactic acid
  polycondensation
    biodegradable polyester synthesis, 556
Lactones
  activated monomer polymerization
    synthesis of difunctional polymers, 126
  ring-opening cationic polymerization
    synthesis of functional polymers, 119
Ladder polymers
  nonlinear optical materials, 440
Laminate model
  short-fiber composites, 485
Lamination process
  epoxy resins
    thermokinetic model, 510
Lamination theory
  polymer composites, 481
  short-fiber composites, 490
Langmuir–Blodgett films
  comparison, 461
  fabrication
    deposition stages, 425
  formation
    low molecular weight amphiphilic compounds, 450
  noncentrosymmetric multilayer
    alternate deposition technique, 425
  nonlinear optical polymers, 425–429
  polymeric, 449–467
    amphiphilic liquid crystalline, 459
    applications, 465–467
    structure, 462–465
  rod-like polymers
    monolayer transfer, 462
    structure determination, 465
  synthesis, 425
Langmuir technique
  surface tension measurement
    Langmuir–Blodgett films, 450
Laser radiation
  polymer photodegradation, 279
Leslie anisotropic viscosity coefficients
  transient flow behavior
    liquid crystal polymers, 393
Leslie–Ericksen continuum theory
  viscosity
    liquid crystal polymers, 399
Leslie–Ericksen phenomenological theory
  rheology
    lyotropic liquid crystal systems, 388
Levoglucosane
  primary product
    cellulose pyrolysis, 234
Lewis acids
  catalysts
    epoxy resin curing, 580

Lewis bases
  effect on carbonation
    polymeric organolithium compounds, 87, 88
Lignins
  anionic grafting
    oxiranes, 548
  biodegradation, 286
  chemically modified
    additives for polymers, 544
    fillers for polymers, 544
  co-reactants
    phenol–formaldehyde resins, 545
  degradation
    synthesis of phenolic derivatives, 533
  degradation products
    monomers for polymerization, 550
  from renewable resources
    polymerization, 543–549
  macromonomers, 545–549
  oxidative coupling, 545
  polyacrylamide grafted
    via free radical polymerization, 547
  polystyrene grafted
    via free radical polymerization, 547
  synthesis
    by living organisms, 531
Lignocellulosics
  unrefined
    chemical modification, 569
Limonene
  from biomass
    cationic polymerization, 551
  synthesis
    by living organisms, 533
Linear elastic fracture
  composite materials, 496
Linear polymers
  preformed
    Langmuir–Blodgett films, 456
Linoleic acid
  oxidopolymerization, 541
  synthesis
    by living organisms, 531
Linolenic acid
  synthesis
    by living organisms, 531
Lipscomb model
  sorption
    penetrant in polymer, 186
Liquid crystalline copolymers
  main chain
    chemical heterogeneity, 342
Liquid crystalline phases
  molecular engineering
    living cationic copolymerization, 364–366
Liquid crystalline polymers
  applications, 182
  flexible
    self assembly, 375
  host–guest interaction with dyes
    nonlinear optical polymers, 418
  hyperbranched dendritic, 353
  main chain
    soluble and fusible, 317–342
  main chain/side chain, 304
  molecular engineering, 299–376
    living polymerization, 360
  permeation
    gases, 182
  properties, 182
  rheological behavior, 385–405
  side chain, 356–370
    mesogenic group decoupling, 356
    phase behavior, parameters, 357
    reentrant nematic mesophase, 366–368
  sorption
    gases, 182

small molecules, 182
spinal columnar
　synthesis, 322
thermodynamic properties, 309, 310
transport
　gases, 182
　small molecules, 182
with chromophores
　nonlinear optical polymers, 418–422
Liquid crystals
　cellulosic
　　uses, 537
　disc-like
　　molecular arrangement, 303
　enantiotropic
　　thermodynamic properties, 311–314
　isomorphism, 304
　low molar mass
　　definitions, 301–303
　monotropic
　　thermodynamic properties, 311, 313
　phases
　　molecular arrangement, 302
Living polymerization
　isobutene
　　synthesis of difunctional polymers, 123
　liquid crystalline phases
　　molecular engineering, 364–366
　liquid crystalline polymers
　　molecular engineering, 360
Living systems
　cationic polymerization
　　fundamental issues, 114
　　synthesis of functionalized polymers, 113–117
Loop formation
　networks, 201
Loss modulus
　viscoelastic behavior
　　polymeric matrices, 474
Lubricant layers
　magnetic media
　　polymeric Langmuir–Blodgett films, 467
Lyotropic liquid crystal systems
　rheology, 385–395
Lyotropic systems
　amphiphile–water
　　properties, 303

Macrocyclization
　pyrolysis
　　condensation polymers, 242
Macromolecular chain
　growth
　　heterogeneous *versus* homogeneous systems, 79
　termination
　　heterogeneous *versus* homogeneous systems, 79
Macromonomers
　functional
　　synthesis *via* cationic polymerization, 117
Magnesium chloride
　catalyst support
　　heterogeneous polymerization, 67
Maleation
　polyethylene
　　solution grafting, 150
Maleic acid
　derivatives
　　film polymerization, polymeric Langmuir–Blodgett film
　　　formation, 453
Maleic anhydride
　alkene derivatives
　　amphiphilic polymers, 457
　functionalized polyethylene
　　synthesis, 141
　grafting reactions
　　reactive extrusion, 628
　molten state grafting

polyalkenes, 149
polyalkene grafting
　reactive extrusion, 626
reactive copolymers
　interpolymeric grafting, 635
solution grafting
　ethylene–propylene rubber, 150
vinyl copolymers
　Langmuir–Blodgett films, synthesis, 458
Maleimide, phenyl-
　derivatives
　　monomers, nonlinear optical polymers, 417
Malonate end groups
　poly(isobutyl vinyl ether)s
　　synthesis *via* cationic polymerization, 122
Marine cuticle collagen
　natural biodegradable material
　　use as bioadhesive, 291
Mark–Houwink coefficient
　soluble liquid crystalline polymers
　　polyamides, 342
　　polyesters, 342
Markov chain
　homopolymerization
　　polyperoxide initiation, 30
Mass fraction
　sol
　　perfect network formation, 205
Mastic
　synthesis
　　by living organisms, 533
Material modeling
　reactive processing, 576–600
Mechanical properties
　polymeric matrices
　　effects of time and temperature, 475
Membrane osmometry
　molecular weight determination
　　functionalized polymers, 85
Membranes
　cellulosic
　　uses, 537
Merocyanines
　dyes in Langmuir–Blodgett films
　　nonlinear optical polymers, 426
Mesogenic aromatic groups
　hydrophobic
　　amphiphilic polymers, 458
Mesogenic monomer
　copolymer with dye monomer
　　nonlinear optical liquid crystalline polymer, 419
Mesogens
　rod-like
　　rigidity, liquid crystalline polymers, 305–308
Metallic soaps
　polymerization catalysts
　　low temperature reactive processing, 582
Metallocene soluble catalysts
　homogeneous alkene polymerization, 68
Metastable states
　liquid crystalline polymers
　　thermodynamics, 310
Methacrylic acid amides
　film polymerization
　　polymeric Langmuir–Blodgett film formation, 452
Methacrylic polymers
　imidification
　　reactive extrusion, 632
Methacrylonitrile
　radical polymerization
　　copper(II) chloride retarded, 11
　　iron(III) chloride retarded, 19
Methacryloyl chloride
　premonomer
　　nonlinear optical polymer, 417
Methane, 4,4′-diaminodiphenyl-
　bismaleimide resin component

curing kinetics, 587
Methane, tetraglycidyl-4,4'-diaminodiphenyl-epoxy
    high performance composites, 506
    reactive processing, 580
Methoxyamine
    amination reagent
        functionalized polymer synthesis, 93
Methyl acrylate
    batch emulsion copolymerization
        styrene, 53, 55–57
    emulsion copolymerization
        styrene, 62
    radical polymerization
        slow reinitiation, 14
    styrene emulsion copolymers
        chemical composition distributions, 62
Methyl methacrylate
    bulk polymerization
        free radical initiated, reactive processing, 622
        presence of iniferter, 37
    copolymer with azobenzene-substituted methacrylate
        nonlinear optical polymer, 416
    emulsion copolymerization models
        styrene, 47
    monomer partitioning
        emulsion copolymerization, 44
    polymerization
        DPPH inhibition, 5
    radiation grafting
        polypropylene, 146
    radical polymerization
        carbon monoxide inhibited, 13
        copper(II) bromide retarded, 11
        copper(II) chloride retarded, 11
        iron(III) bromide retarded, 11
        slow reinitiation, 14
Methylene 4,4'-diphenyldiisocyanate
    polyurethanes from
        photodegradation, 275
Methyleneoxy groups
    flexible
        low molar mass liquid crystals, 344
Micellar nucleation
    emulsion copolymerization, 46
Microlithography
    polymeric Langmuir–Blodgett films, 467
    polymeric materials
        via laser ablative photodecomposition, 280
Microorganisms
    use
        biodegradable polymer production, 285
        biodegradation tests, 287
Miesowicz viscosity coefficients
    liquid crystal polymers, 398
Model networks
    network formation, 202
Mode–mode coupling theories
    critical phenomena, 169
Molar mass chemical composition distribution
    batch emulsion copolymers
        microstructure, 53
    emulsion copolymerization
        models, 51
    three-dimensional
        emulsion copolymerization, 42
Molecular dynamics
    static model development, 160–162
Molecular engineering
    liquid crystalline phases
        living cationic copolymerization, 364–366
    liquid crystalline polymers
        living polymerization, 360
Molecular entanglement model
    kinetic diffusion model
        bulk vinyl polymerization, 583
Molecular recognition
    supramolecular liquid crystalline polymers
        self assembly, 372
Molecular theories
    rheology
        lyotropic liquid crystal systems, 387–390
Molecular weight
    functionalized polymers
        determination, 85
    influence of
        on mutual diffusion coefficient, 169
    number average
        polyperoxide initiation, 32
Molecular weight dispersity
    polyethylene
        from homogeneous catalysis, 69
Molecular weight distribution
    factors influencing
        emulsion copolymerization, 60
    iniferter mechanism of polymerization
        characteristics, 37–39
    radical polymerization
        polyperoxide initiation, 26
Molten polymers
    reactive processing, 620
Molten state grafting
    peroxide initiated
        polyalkenes, 149
Monoalkene end groups
    polypropylene
        synthesis, 143
Monoalkenes
    polymerization
        heterogeneous *versus* homogeneous catalysis, 67–80
        via homogeneous catalysts, 68–78
Monofunctional macromolecules
    synthesis
        via cationic polymerization, 110–121
Monolayer formation
    Langmuir–Blodgett films
        factors influencing, 450
Monolayer polycondensation
    film polymerization
        polymeric Langmuir–Blodgett film formation, 455
Monolayers
    formation
        low molecular weight amphiphilic compounds, 450
Monomer droplet nucleation
    emulsion copolymerization, 46
Monomer film polymerization
    polymeric Langmuir–Blodgett film formation, 452–455
Monomer partitioning
    theoretical aspects
        emulsion copolymerization, 43–46
Monomer radial distribution function
    relationship
        mutual diffusion coefficient, 169
Monomers
    activated
        cationic polymerization, 108
    synthesis
        by living organisms, 533
Monosaccharides
    from biomass
        polymerization, 553–555
    synthesis
        by living organisms, 533
Monte Carlo simulation
    extent of reaction
        gelation, 213
    gelation
        network formation, 210, 211
    gel points
        interpretation, 212
    network formation
        applications, 222
    nonlinear polymerization, 211
Morphology

polymer particles
  homogeneous catalysis, 76
Morton model
  monomer partitioning
    emulsion copolymerization, 43
Multilayer sheets
  functionalized polypropylene application, 153
Multiple minimum problem
  ordered polymers, 160
Mutual diffusion
  polymeric systems, 167–196
Mutual diffusion coefficient
  concentrated solutions
    free volume theory, 176–179
  dilute solutions
    factors influencing, 170–176
  polystyrene/cyclohexane
    concentration effect, 174
    temperature effect, 174
  polystyrene/toluene
    polymer concentration influence, 171, 173
    pressure effect, 176
    renormalization group theory comparison, 172
  polystyrene/trans-decalin
    concentration dependence, 173
    concentration effect, 175
    temperature effect, 175
  relationship
    monomer radical distribution function, 169
  single polymer chain in solution, 169
Myrcene
  from biomass
    cationic polymerization, 552
    free radical polymerization, 552
  synthesis
    by living organisms, 533
Myrcene, dihydro-
  retarder
    radical polymerization, 14

2-Naphthoic acid, 6-hydroxy-
  liquid crystal copolyester with 4-hydroxybenzoic acid
    diffusion coefficient, 182
Naphthyl end group labeling
  fluorescent
    (polystyryl)lithium, via 1,1-diphenylethylenes, 102
Nematic viscoelastic liquids
  liquid crystal polymers, 399–402
Neogi model
  diffusion
    penetrant in polymer, 190
Networks
  covalent
    formation, 199
  formation
    fundamentals, 199–225
    gelation, 201
    ideal gelation, 202
    modulus, 216
    perfect, 205–207
    polymerization statistics, 202
    properties, 216
    structure, 216
  imperfections, 217, 218
    compensating effects, 220
    universal occurrence, 219
  materials, 199–201
  physical
    formation, 199
  polymerizations forming
    theories, applications of, 220
  properties
    comparison with linear polymers, 200
    fundamentals, 199–225
  stress–strain behavior, 223–225
  structure
    fundamentals, 199–225

modulus, 216
shear modulus, 216
Nitroaromatic groups
  dye unit
    nonlinear optical liquid crystalline polymers, 420
Nitrobenzene
  retarder
    radical polymerization, 2
NMR analysis
  characterization
    functionalized polymers, 86
  polypropylene
    via homogeneous catalysis, 70
Nonlinear optical effects
  polymeric Langmuir–Blodgett films, 465
Nonlinear optical polymers
  second-order, 411–429
    design, 413
  third-order, 429–442
    nonlinear susceptibilities, 431, 432
    structure–property relationships, 430
Nonlinear optical susceptibilities
  second-order
    organic compounds, 412
Nonlinearity
  organic compounds
    molecular basis, 410
Norrish type I reaction
  polymer photodegradation, 258
Norrish type II reaction
  polymer photodegradation, 258
Notch sensitivity
  composite materials, 495
Nylon
  reaction injection molding polymerization
    autocatalytic model, 586
  reactive processing
    kinetics, 584–587
  structure development
    viscosity changes, 590
Nylon 3
  pyrolysis
    β-CH hydrogen transfer, 235
Nylon 66
  acetyl substituted
    biodegradability, 290
Nylon 6,6
  moderately photostable, 254
  pyrolysis
    α-CH hydrogen transfer, 239
    ionic versus radical mechanism, 229
Nylon block copolymers
  reactive processing
    kinetics, 584–587
Nylons
  reactive injection molding process
    curing analysis, simulation, 603

Octadecanal
  monolayer
    cross-linked, via grafting on to poly(L-lysine), 455
Octadecyl fumarate
  film polymerization
    polymeric Langmuir–Blodgett film formation, 453
Oils
  from renewable resources
    polymerization, 541–543
Oligo(ethylene glycol)s
  hydrophilic spacers
    amphiphilic polymers, 458
Oligomers
  from renewable resources
    polymerization, 541–550
  living
    ethylene, electrophilic attack, 140
  synthesis
    by living organisms, 531

Oligo(vinyl ether) macromonomers
  synthesis
    via functional end capping, living polymerization, 117
    via functional initiators, living polymerization, 117
Optical nonlinearity
  organic compounds, 408
Optimal addition profile
  copolymer latex properties
    control, 61–63
Optoelectronic effects
  polymeric Langmuir–Blodgett films, 465
Ordered polymers
  multiple minimum problem, 160
Organolithium compounds
  polymeric
    carbon dioxide carbonation, 87
    hydroxylation *via* carbonyl compounds, 91
Organosilane polymers
  nonlinear optical materials, 442
Orientation layers
  liquid crystal displays
    polymeric Langmuir–Blodgett films, 466
Oseen–Frank elasticity theory
  viscosity
    liquid crystal polymers, 399
Oxazoline
  protecting group initiator
    synthesis, carboxy functionalized polymers, 90
2-Oxazolines
  activated monomer polymerization
    synthesis of difunctional polymers, 126
  cationic polymerization
    synthesis of polyfunctional polymers, 129
  ring-opening cationic polymerization
    synthesis of functionalized polymers, 120
Oxidation
  anionic synthesis
    functionalized polymers, 97
Oxirane
  derivative
    film polymerization, polymeric Langmuir–Blodgett film formation, 455
Oxirane, 2-chloromethyl-
  activated monomer polymerization
    synthesis of macromolecular diols, 125
Oxirane, 2-furyl-
  anionic polymerization, 563
  monomers from renewable resources, syntheses, 558
Oxirane, methyl-
  ring-opening cationic polymerization
    synthesis of functional polymers, 118
Oxiranes
  activated monomer polymerization
    synthesis of macromolecular diols, 124
p-Oxy-α-cyanocinnamates
  main chain dye units
    nonlinear optical polymers, 423
Oxygen
  molecular
    reaction with polymers, 256
Oxygen/ozone mixture
  hydroperoxidation
    polyalkenes, 148
Ozone
  surface functionalization
    polypropylene, 137
Ozonolysis
  hydroperoxidation
    polyalkenes, 148

Painting
  functionalized polypropylene, 154
Palmitic acid, 9,10,16-trihydroxy-
  synthesis
    by living organisms, 533
Particle formation
  emulsion copolymerization, 46
Particle morphology
  emulsion copolymers
    properties, 58
Particle structure
  copolymer latex properties
    control, 60
Particulate composites
  mechanical properties, 491–495
Penetrant diffusion coefficient
  rubbery polymers
    factors influencing, 180
1-Pentene
  hydrooligomerization
    via chiral homogeneous catalysis, 76
1-Pentene, 4-methyl-
  hydrooligomerization
    via chiral homogeneous catalysis, 76
Pentoses
  biomass furanic monomers
    polymerization, 556–561
Penultimate model
  emulsion copolymerization, 47
Percolation
  models
    gelation, 215
Peroxidation
  polyalkenes
    radiation induced, 144
Peroxides
  initiators
    reactive processing, 582
Persistence lengths
  soluble liquid crystalline polymers
    polyamides, 342
    polyesters, 342
Phase separation
  during polymerization
    reactive processing, models, 596
Phase transitions
  liquid crystalline polymers
    tacticity, 368
Phase transition temperature
  liquid crystalline polymers
    molecular weight and, 313–317
    thermodynamics, 309–317
Phenol, 2,6-di-*t*-butyl-
  chain transfer agent
    isobutene cationic polymerization, 111
Phenol end group
  oligo(isobutenes)
    via cationic polymerization, 111
Phenol groups
  end functionalized polymer
    synthesis *via* 1,1-diphenylethylene, 99
  functionalized polymers
    TLC characterization, 86
Phenyl end groups
  polypropylene
    synthesis, 142
Phosphine, diphenyl-
  functionalized polyethylene
    synthesis *via* anionic oligomerization, 140
Phospholipids
  butadiene derivatives
    film polymerization, polymeric Langmuir–Blodgett film formation, 455
  dialkyne derivatives
    film polymerization, polymeric Langmuir–Blodgett film formation, 455
Phosphonium end groups
  synthesis
    via ring-opening cationic polymerization, 119
Photochlorination
  polypropylene, 139
Photodegradation
  polymer materials, 253–280
    general mechanisms, 256–259

molecular mechanisms, 259–263
sensitized
  polymers, 263
Photo-Fries rearrangement
  photooxidative degradation
    aromatic polycarbonates, 273
    aromatic polyesters, 273
Photoinitiators
  photodegradation
    polymers, 263
Photonic applications
  polymers, 407–443
Photooxidation
  use in biodegradation, 290
Photostability
  polymers
    definition, 254
    differences, 254
Photostabilization
  polymer materials, 254
Phthalic anhydride
  carbonation reagent
    anionic synthesis, functionalized polymers, 89
Phthalocyaninatopolysiloxanes
  field-effect transistor sensors, 466
  rigid polymers
    nonlinear optical materials, 440
  rod-like polymers
    monolayers, Langmuir–Blodgett films, 460, 461
Physicochemical behavior
  composite materials
    high performance, 504
Piezoelectric effects
  polymeric Langmuir–Blodgett films, 465
α-Pinene
  from biomass
    cationic polymerization, 551
  synthesis
    by living organisms, 533
β-Pinene
  from biomass
    cationic polymerization, 551
Plane stress
  composite laminates, 481
Plane stress laminate theory
  composite materials, 495
Plasma
  fluorocarbon
    surface functionalization, polypropylene, 137
Plastic waste
  biodegradation, 294
Poisson ratio
  lamination theory
    polymer composites, 482
Polar end groups
  polypropylene
    synthesis, 142
Polar properties
  transition states
    ideal retarded polymerization, 10
Poled polymers
  electrooptic modulation, 442
Poling fields
  high
    for nonlinear optical polymers, 413
Polyacetylenes
  nonlinear optical materials
    third order, 435
  oxygen permeability, 184
  substituted
    permeation, gases, 183
    sorption, gases, 183
    transport, gases, 183
Polyacrylates
  photodegradation
    mechanism, 271
Poly(acrylonitrile)
  glassy polymer
    diffusion coefficient, 182
Poly(N-acylaziridine)s
  synthesis
    via ring-opening cationic polymerization, 121
Polyalkenes
  backbone attachment
    low molecular weight compounds, 140–144
  functionalization, 135–144
  functionalized
    synthesis, 133–155
  grafting reactions
    reactive extrusion, 627
  photodegradation
    initiation, 261
  photooxidative degradation, 263–267
  poorly photostable, 254
  vinylsilane grafting
    reactive extrusion, 624–626
Poly(3-alkylthiophene)s
  Langmuir–Blodgett films
    preformed monolayers, 456
Polyalkynes
  liquid crystalline
    side chain, 369
Poly(allylamine)
  perfluoroalkyl isocyanate derivatives
    amphiphilic polymers, 457
Poly(amide enamine)s
  biodegradability, 290
Polyamides
  acrylonitrile–butadiene–styrene blends
    via interpolymeric ion bonding, 637
  aliphatic
    moderately photostable, 254
  aromatic
    liquid crystalline polymers, synthesis, 322–333
    molecular structures, 324
    phenyl substituted, 326
    poorly photostable, 254
    pyrolysis, free radical process, 244
    viscosity of solutions, 386
  aromatic–aliphatic
    pyrolysis, β-CH hydrogen transfer, 235
  biodegradability, 290
  biphenyl
    viscosity and solubility, 326
  o-chloro-
    pyrolysis, iminolization–cyclization process, 233
  from furanic monomers, 567
  molecular weight enhancement
    via reactive extrusion, 636
  photooxidative degradation
    mechanism, 272
  poly(phenylene ether) blends
    via interpolymeric ion bonding, 637
  polypropylene blend
    via interpolymeric ion bonding, 637
    via interpolymeric reactive grafting, 635
  pyrolysis
    α-CH hydrogen transfer, 239
  soluble liquid crystalline polymers
    persistence lengths, 342
  succinic acid containing
    pyrolysis, α-CH hydrogen transfer, 239
  thermal degradation
    ionic cleavage, 228
Poly(α-amino acid)s
  biodegradability, 289
  pyrolysis
    β-CH hydrogen transfer, 236
Poly(o-aminoanil)
  synthesis
    via monolayer polycondensation, 455
Polyanhydrides
  biodegradability, 290
  biomedical applications, 290

Polyaramides
  photodegradation
    via photo-Fries rearrangement, 273
Poly(1,4-arylenevinylene)
  thermotropic, liquid crystalline polymers
    synthesis, 335
Poly(aryl sulfone)s
  photodegradation
    mechanism, 276
Poly(azomethine)s
  pyrolysis
    free radical process, 245
  rigid conjugated polymers
    nonlinear optical materials, 440
Poly(p-benzamide)
  solutions in dimethylacetamide
    viscosity, 386
Polybenzimidazole
  synthesis
    via monolayer polycondensation, 455
Poly(benzyl aspartate)
  rod-like polymers
    monolayers, Langmuir–Blodgett films, 461
Poly-γ-benzyl-L-glutamate
  viscosity of solutions
    concentration dependence, 386
Poly[(bis-(p-butoxyphenylsilane)]
  nonlinear optical material
    Langmuir–Blodgett film, 442
Poly(bisphenylurethane)
  polydiacetylene derivative
    nonlinear optical material, 432
Poly[bis(p-toluenesulfonate)]
  polydiacetylene derivative
    nonlinear optical material, 434
Poly(2-bromo-5-methoxyphenylenevinylene)
  nonlinear optical materials
    third-order susceptibilities, 437
Polybutadiene
  branched
    viscosity, 591
Poly(1,3-trans-butadiene)
  chain in crystalline perhydrotriphenylene
    molecular dynamics application, 163
Polybutadienes, α,ω-dihydroxy-
  synthesis
    via protected initiator hydroxylation, 91
Polybutadienes, α-dimethylamino-
  synthesis
    via functionalized initiator, 93
(Polybutadienyl)lithium
  amine functionalization
    via α-amino-ω-haloalkanes, 94
  carbonation
    via carbon dioxide, 88
  hydroxylation
    via 4,4'-bis(diethylamino)benzophenone, 92
Poly(1-butene)
  via homogeneous catalysis
    titanocene complexes, 75
Poly(2-butoxy-5-methoxyphenylenevinylene)
  nonlinear optical materials
    third-order susceptibilities, 437
Poly(butylene terephthalate)
  bisphenol A polycarbonate binary blend
    pyrolysis, intermolecular exchange reaction, 239
Poly(ε-caprolactone)
  blend with p-nitroaniline
    nonlinear optical polymer, 414
Polycaprolactones
  biodegradation, 288
  in polymer blends
    biodegradation, 289
Polycarbonate matrix
  short-fiber composites, 487
Polycarbonate/polyamide copolymer
  synthesis via functional end group reactivity

via reactive extrusion, 636
Polycarbonates
  aliphatic
    pyrolysis, molecular rearrangement, 248
  aromatic
    pyrolysis, molecular rearrangement, 248
  biodegradation, 290
  liquid crystalline polymers
    synthesis, 337
  photooxidative degradation
    via photo-Fries rearrangement, 272
  pyrolysis
    end biting, intramolecular exchange reactions, 241
  thermal degradation
    ionic cleavage, 228
Polycarboxylates
  biodegradation, 291
Poly(carboxypiperazine)
  pyrolysis
    α-CH hydrogen transfer, 238
Polycondensation reactions
  reactive extrusion, 630
Poly[ω-(4-cyano-4'-biphenylyl)oxyalkyl vinyl ether]s
  liquid crystalline polymers
    synthesis, properties, 362–364
Polycycloalkenes
  via metallocene homogeneous catalysis, 77
Polycyclopentene
  via dichloro(ethyl)bis(indenyl)zirconium homogeneous catalysis,
Polydiacetylenes
  nonlinear optical materials
    third order, 432–435
  nonlinear susceptibilities
    third order, 433
Poly(dialkylsiloxane)s
  monolayers
    Langmuir–Blodgett films, 457
Polydienes
  photooxidative degradation, 269
Poly[2-(dimethylamino)ethyl methacrylate] side chains
  ethylene–propylene rubber
    synthesis, 143
Poly(2,6-dimethylphenylene oxide)
  pyrolysis
    molecular rearrangement, 249
Poly(dimethylsiloxane)s
  amine functionalized
    synthesis via functionalized initiator, 94
  difunctional
    synthesis via cationic polymerization, 127
  internal dynamics of chains
    dynamic rotational isomeric state model, 163
  Langmuir–Blodgett films
    nonlinear optical polymers, 428
  oxygen permeability, 184
  realistic statistical weights
    rotational isomeric state theory, 162
Polydispersity coefficient
  molecular weight distribution
    iniferter mechanism of polymerization, 38
  radical polymerization
    polyperoxide initiation, 27
Polydisulfides
  aromatic
    pyrolysis, molecular rearrangement, 248
Poly(dithieno[3,2-b:2',3'-b]thiophene)
  conjugated polymer
    nonlinear optical material, 439
Poly[1,4-di(2-thienyl)benzene]
  conjugated polymer
    nonlinear optical material, 439
Poly(3-dodecylthiophene)
  conjugated polymer
    nonlinear optical material, 438
Polyelectrolyte precursor technique
  synthesis
    conjugated polymers, 436–438

Polyenes
　nonlinear optical materials
　　third order, 435
Poly(ester amide)s
　synthesis via functional end group reactivity
　　via reactive extrusion, 636
Poly(ester anhydride)s
　liquid crystalline polymers
　　synthesis, 337
Poly(ester imide)s
　liquid crystalline polymers
　　synthesis, 339
Polyester resins
　unsaturated
　　curing, kinetic models, 582
Polyesters
　amorphous
　　with binaphthyl units, 331
　aromatic–aliphatic
　　pyrolysis, β-CH hydrogen transfer, 234
　aromatic
　　liquid crystalline polymers, synthesis, 322–333
　　solubility, 329
　biodegradable, 288
　biodegradation, 286
　from furanic monomers, 566
　liquid crystalline polymers
　　thermodynamic parameters, 350
　matrices
　　viscosity model, 514
　matrix composites
　　processing models, 519
　microbial
　　biodegradability, 293
　molecular weight enhancement
　　via reactive extrusion, 636
　native
　　biodegradation, 293
　photooxidative degradation
　　via photo-Fries rearrangement, 272
　properties, 330
　pyrolysis
　　macrocyclization, 242
　soluble liquid crystalline polymers
　　persistence lengths, 342
　synthesis
　　via lignin macromonomers, 549
　synthesis via polycondensation reactions
　　reactive extrusion, 630
　thermal degradation
　　ionic cleavage, 228
　with twisted biphenyl units, 331
Poly(α-esters)
　pyrolysis
　　β-CH hydrogen transfer, 237
Poly(ether ester)
　block copolymers
　　bioabsorbable, 288
Poly(ether imide)s
　synthesis via polycondensation reactions
　　reactive extrusion, 630
Polyethers
　aromatic
　　pyrolysis, free radical process, 245
　hyperbranched liquid crystalline
　　characterization, 354
　liquid crystalline polymers
　　columnar and smectic mesophases, 351
　　cyclic main chain structure, 355, 356
　　DSC curves, 347
　　enthalpy changes, 348
　　flexible and semiflexible, 343–353
　　high molecular weight, synthesis, 345
　　phase transition temperatures, 349, 353
　　synthesis, 342
　　thermal transitions, 348
　　thermodynamic parameters, 350

Poly(ether sulfone)s
　photodegradation
　　mechanism, 276
Poly(ethyl acrylate)
　expanded film
　　polymer monolayer, 457
Polyethylene
　backbone attachment
　　low molecular weight compounds, 140
　bulk functionalization, 137–139
　carboxy functionalized
　　synthesis via anionic oligomerization, 140
　carboxylic acid functionalized
　　chemical modifications, 141
　chain degradation
　　free radical initiated, 623
　degradation
　　compared with paraffin degradation, 286
　diphenylphosphine functionalized
　　synthesis via anionic oligomerization, 140
　end functionalized
　　synthesis via anionic polymerization, 102
　fluoroalkanol functionalized
　　synthesis via radical addition, 140
　functionalization
　　via 2-(dimethylamino)ethyl methacrylate, 140
　　with diethyl maleate/dicumyl peroxide, 135
　internal dynamics of chains
　　dynamic rotational isomeric state model, 163
　low density
　　photodegradation, mechanism, 260
　maleic anhydride functionalized
　　synthesis, 141
　molten state grafting
　　functionalization, 149
　photooxidation, 264
　radiation grafting
　　styrene, 146
　radical addition
　　hexafluoroacetone, 140
　semicrystalline
　　transport of gases, 181
　surface functionalization, 135
　surface grafted
　　reactions, 136
　variable characteristics
　　via homogeneous catalysis, 69
　via homogeneous catalysis
　　different catalyst systems, 70
Poly(ethylene-co-carbon monoxide)
　photodegradation
　　via Norrish reactions, 269
Poly(ethylene glycol)
　biodegradable block copolymers
　　with poly(ethylene terephthalate), 288
Poly(ethylene oxide)
　biodegradable copolymers
　　with poly(ethylene terephthalate), 289
　pyrolysis
　　free radical process, 245
Poly(ethylene succinate)
　biodegradable block copolymers
　　with poly(ethylene glycol), 288
Poly(ethylene terephthalate)
　biodegradable block copolymers
　　with poly(oxyalkenes), 288
　bisphenol A polycarbonate binary blend
　　pyrolysis, intermolecular exchange reaction, 239
　glassy polymer
　　diffusion coefficient, 182
　liquid crystal copolymer with 4-oxybenzoate
　　diffusion coefficient, 183
　moderately photostable, 254
　stabilized
　　via reactive extrusion, 636
Poly(ethyl methacrylate)
　condensed film

polymer monolayer, 457
Polyfunctional macromolecules
    synthesis
        via cationic polymerization, 127–129
Polyfuran
    synthesis
        via electrochemical polymerization, 561
Poly(L-glutamate)
    rod-like polymers
        multilayers, structure, 465
Polyglutamates
    rod-like polymers
        monolayers, Langmuir–Blodgett films, 460
Poly(glycolic acid)
    biodegradable
        synthesis from glycolide, 288
Poly(glycolic-*co*-lactic acid)
    biodegradable
        synthesis from glycolide/lactide, 288
Poly(3-hexylthiophene)
    conjugated polymer
        nonlinear optical material, 438
Polyhydrazides
    pyrolysis
        α-CH hydrogen transfer, 239
        NH hydrogen transfer, 233
Polyhydrazides, *N*-methyl-
    pyrolysis
        α-CH hydrogen transfer, 238
Polyhydrocarbons
    liquid crystalline polymers
        synthesis, 341
Poly(*p*-hydroxybenzoate)s
    soluble
        liquid crystalline polymers, synthesis, 328
Poly(3-hydroxybutyrate)
    bacterial synthesis
        uses, 555
    microbial polyester
        use in biodegradation, 293
Poly(3-icosylthiophene)
    conjugated polymer
        nonlinear optical material, 438
Polyimide films
    ultrathin
        formation, 459
Polyimides
    amphiphilic polymers
        electrostatic spacers, 458
    glassy
        penetrant diffusion coefficients, 181
    soluble rod-like
        liquid crystalline polymers, synthesis, 334
Polyinitiators
    one type of labile group
        radical polymerization, 28
    several types of labile group
        radical polymerization, 30–32
    thermostability, 25
Polyisobutenes
    diallylic
        synthesis *via* living cationic polymerization, 123
    difunctional
        synthesis *via* inifer technique, 122
    functionalized
        synthesis *via* cationic polymerization, 111
Polyisobutylene
    chain degradation
        free radical initiated, 623
Poly(isobutyl vinyl ether)s
    dimalonate functionalized
        via cationic polymerization, 122
Poly(1,4-isoprene)
    synthesis
        by living organisms, 530
Polyisoprene, α,ω-bis(dimethylamino)-
    synthesis

via α-amino-ω-haloalkanes, 94
Polyisoprene, α,ω-dilithio-
    dihydroxyethylation
        via ethylene oxide, 90
Polyisoprene, α,ω-dipotassio-
    diamine functionalization
        via α-halo-ω-aminoalkanes, 94
Polyisoprenes, α,ω-diamino-
    star-branched
        viscosity, 591
    synthesis
        via functionalized initiator, 93
    tetradecane solution
        viscosity, 590
(Polyisoprenyl)lithium
    amination
        via N-benzylidenetrimethylsilylamine, 92
    amine functionalization
        via α-halo-ω-aminoalkanes, 94
    carbonation
        via carbon dioxide, 87
Poly(isothianaphthene)
    conjugated polymer
        nonlinear optical material, 439
Polylactams
    pyrolysis
        intramolecular exchange process, 236
Poly(L-lactic acid)
    biodegradable
        applications, 288
Polylactones
    pyrolysis
        intramolecular exchange process, 236
Poly(L-lysine)
    monolayer polycondensation
        with octadecanal, 455
Polymer backbone
    liquid crystals
        rigid *versus* flexible, entropy differences, 358, 359
Polymer blends
    thermoplastic species
        reactive processing, 620
Polymer concentration
    effect of
        on mutual diffusion coefficient, 170–174
Polymer degradation temperature
    thermal stability
        evaluation, 230
Polymer materials
    important
        photodegradation, 259–280
        photodegradation, 253–280
Polymer networks — *see* Networks
Polymer structure
    fundamental properties
        computer modeling, 159–165
Polymeric composites
    science and technology, 471–523
Polymeric matrices
    mechanical properties, 472–477
Polymeric stabilizers
    nonmigrating
        synthesis *via* grafting, 627
Polymeric systems
    mutual diffusion, 167–196
Polymerization
    network formation
        models, 211
    solventless
        reactive processing, thermoplastic polymers, 620
Polymerization kinetics
    epoxy resins
        chemorheological behavior, 592
Polymerization shrinkage
    during polymerization
        reactive processing, models, 598
Polymers

bioabsorbable
   synthesis, 288
biodegradable, 285–295
   biomedical, 294
   from renewable resources, 555
   future research, 294
   native, 285, 291–294
   synthetic, 285, 288–291
chemical modification, 619
from renewable resources, 527–570
functional biodegradable
   biomedical uses, 295
functional group containing
   anionic synthesis, 83–103
natural
   chemical modification, 535–541
nondegradable
   modification to make biodegradable, 294
synthesis
   by living organisms, 529–531
Polymethacrylates
   photodegradation
      mechanism, 271
Poly(3-methoxy-4-hydroxystyrene)
   biodegradation
      model for softwood lignin, 290
Poly(methylene oxide)
   pyrolysis
      free radical process, 244
Poly(methylene sulfide)
   pyrolysis
      free radical process, 246
Poly(methyl glutamate)
   rod-like polymers
      monolayers, Langmuir–Blodgett films, 461
Poly(γ-methylglutamate)
   pyrolysis
      NH hydrogen transfer, 232
Poly(methyl methacrylate)
   blend with Disperse Red 1
      nonlinear optical polymer, 414
   highly photostable, 254
   Langmuir–Blodgett films
      nonlinear optical polymers, 427
Poly(γ-methyl-co-octadecyl L-glutamate)
   rod-like polymer
      multilayers, Langmuir–Blodgett films, 462
Poly(4-methyl-1-pentene)
   semicrystalline
      sorption/transport of small molecules, 181
   via homogeneous catalysis
      titanocene complexes, 75
Poly(α-methylstyryl)lithium
   sulfonation
      via 1, 3-propanesultone, 95
Poly(1,4-naphthalenevinylene)
   nonlinear optical materials
      synthesis via soluble precursor polymers, 436
Poly(neopentylene carbonate)
   EI/CI/DCI mass spectrometry
      pyrolysis product detection, 231
Poly(octadecyl methacrylate)
   amphiphilic polymer
      multilayers, structure, 463
   Langmuir–Blodgett films
      nonlinear optical polymers, 427
   multilayers, 463
Poly(3-octadecylthiophene)
   conjugated polymer
      nonlinear optical material, 438
Poly(n-octylmethylsilane)
   nonlinear optical material, 442
Poly(3-octylthiophene)
   conjugated polymer
      nonlinear optical material, 438
Polyoxamides
   aromatic–aliphatic

pyrolysis, β-CH hydrogen transfer, 235
   pyrolysis
      NH hydrogen transfer, 233
Poly(oxyethylene)
   blend with p-nitroaniline
      nonlinear optical polymer, 414
   internal dynamics of chains
      dynamic rotational isomeric state model, 163
Poly(oxymethylene)
   internal dynamics of chains
      dynamic rotational isomeric state model, 163
Polypeptides
   natural biodegradable materials, 291
   rod-like polymers
      monolayers, Langmuir–Blodgett films, 461
Polyperoxides
   different peroxide thermostabilities
      radical polymerization, 30
   initiators
      radical polymerization, 24
   thermolysis
      kinetic parameters, 24
Poly(phenylacetylene)
   graft copolymer with poly(methyl methacrylate)
      nonlinear optical material, 436
   nonlinear optical material
      third order, 436
Poly(p-phenylenebenzobisoxazole)
   rigid conjugated polymer
      nonlinear optical material, 439
Poly(p-phenylenebenzobisthiazole)
   rigid conjugated polymer
      nonlinear optical material, 439
Poly[1,4-phenylene-5,5'(6,6')-bibenzimidazolediyl]
   Langmuir–Blodgett film
      production, 456
Poly(phenylene ether)s
   acrylic monomer grafting
      reactive extrusion, 627
   polyamide blends
      via interpolymeric ion bonding, 637
Poly(phenylene oxyxylyl ether)
   pyrolysis
      free radical process, 245
Poly(p-phenylene)s
   isotropization transitions, 316
   liquid crystalline polymers
      synthesis, 318–320
Poly(phenylene sulfide)s
   crosslinking
      reactive extrusion, 629
   pyrolysis
      macrocyclization, 242
Poly(p-phenyleneterephthalamide)s
   soluble
      liquid crystalline polymers, synthesis, 325
   solutions in concentrated acids
      viscosity, 386
Poly(p-phenylenevinylene)s
   amphiphilic multilayers
      electrostatic spacers, 458
   multilayers
      formation, 460
   nonlinear optical materials
      synthesis via soluble precursor polymers, 436
      third-order susceptibilities, 437
   sol gel silica glass composite
      improved nonlinear optical material, 438
Poly(phenylmethylsilane)
   nonlinear optical material, 442
Poly(phenyl vinyl ketone)
   photodegradation
      via Norrish reaction, 270
Polyphosphazenes
   photodegradation
      mechanism, 278
Polypodants

liquid crystalline polymers, 370
Poly(β-propiolactone)
    biodegradation, 288
    pyrolysis
        β-CH hydrogen transfer, 235
Polypropylene
    acrylic acid grafted
        interpolymeric reactivity, 635
    acrylic monomer grafting
        reactive extrusion, 627
    atactic
        functionalization, 142
        Theodorou–Suter model, 164
        via homogeneous catalysis, 70
    backbone attachment
        low molecular weight compounds, 141–143
    bulk functionalization, 139
    chain degradation
        free radical initiated, 623
    chlorinated
        grafting, 139
    crosslinking
        reactive extrusion, 628
    from homogeneous catalytic polymerization
        microstructure, 72
    functionalization, 150–155
    functionalized
        applications, 151
    glassy
        Theodorou–Suter model, 164
    isotactic
        functionalization, 142
        via chiral homogeneous catalysts, 70
    molten state grafting
        functionalization, 149
    monofunctional
        synthesis via monoalkenes, 143
    photodegradation, 266
    plastics
        via chiral hafnocene catalysts, 73
    polyamide blend
        via interpolymeric ion bonding, 637
        via interpolymeric reactive grafting, 635
    radiation grafting
        styrene, 146
    surface functionalization, 136
    syndiotactic
        via vanadium-based catalysis, 68
    terminally functionalized
        synthesis, 141
    via zirconocene catalytic system
        characterization of fractions, 74
    waxes
        via chiral zirconocene catalysts, 73
    with attached dye unit
        nonlinear optical polymer, 416
Polysaccharides
    natural biodegradable material
        uses, 292
    synthesis
        by living organisms, 529
Polysilanes
    nonlinear optical materials, 442
    photodegradation
        mechanism, 275–278
    rod-like polymers
        monolayers, Langmuir–Blodgett films, 460
Polysiloxanes
    cyclic liquid crystalline
        with mesogenic side groups, 369
    liquid crystalline polymers
        isotropization transition temperatures, 369
    photodegradation
        mechanism, 275–278
    thermal degradation
        ionic cleavage, 228
Polystyrene
    acrylonitrile–butadiene–styrene blends
        via interpolymeric ion bonding, 637
    amide end functionalized
        synthesis via 1,1-diphenylethylene, 101
    attached to dye units
        nonlinear optical polymer, 414, 415
    α,ω-bis(dimethylamino) terminated
        synthesis via 1,1-diphenylethylene, 99
    blends with dye molecules
        nonlinear optical polymers, 414
    carboxy end functionalized
        synthesis via 1,1-diphenylethylene, 101
    chain degradation
        free radical initiated, 623
    concentrated ethylbenzene solution
        mutual diffusion coefficient, 177
    concentrated toluene solution
        mutual diffusion coefficient, 177, 178
    cyclohexane solvent
        mutual diffusion coefficient, 170
        mutual diffusion relationship, 173, 174
    trans-decalin solvent
        mutual diffusion relationship, 173, 174
    dye functionalized
        second harmonic generation coefficients, 416
    functionalized
        synthesis via cationic polymerization, 111
    photodegradation, 254
    photooxidative degradation
        mechanism, 272
    poorly photostable, 254
    reactive
        interpolymeric grafting, 635
    tetrahydrofuran/hexafluorobenzene solution
        tracer diffusion coefficient, 177
    toluene solvent
        mutual diffusion coefficient, 170, 172, 175
        mutual diffusion coefficient, concentration dependence, 171
Polystyrene, α,ω-dilithio-
    diamination
        via methoxyamine, 93
Poly(styrene-block-butadiene)
    amine functionalization
        synthesis via 1,1-diphenylethylene, 100
Polystyrenes
    sulfonated
        column chromatography characterization, 87
(Polystyryl)lithium
    amination
        via N-benzylidenetrimethylsilylamine, 92
        via methoxyamine, 93
    amine end functionalization
        synthesis via 1,1-diphenylethylene, 99
    amine functionalization
        via α-halo-ω-aminoalkanes, 94
    carbonation
        via carbon dioxide, 87
        via phthalic anhydride, 89
    fluorescent labeling
        via 1,1-diphenylethylenes, 102
    hydroxyethylation
        via ethylene oxide, 90
    hydroxylation
        via benzophenone, 92
    oxidation
        via molecular oxygen, 97
    phenol end functionalization
        synthesis via 1,1-diphenylethylene, 99
    solid state carbonation
        via carbon dioxide, 88
    sulfonation
        via 1,1-diphenylethylene/alkanesultone, 96
        via ethylene oxide/alkanesultone, 95
        via 1,3-propanesultone, 95
Polysuccinamides
    aromatic
        low thermal stability, 233

Subject Index

Polysulfides
  pyrolysis
    β-CH hydrogen transfer, 237
    end biting, intramolecular exchange reactions, 241
  thermal degradation
    ionic cleavage, 228
Polysulfones
  aliphatic
    pyrolysis, free radical process, 246
  aromatic
    pyrolysis, molecular rearrangement, 248
  photodegradation
    mechanism, 275–278
Poly(tetrafluoroethylene)
  highly photostable, 254
Poly(tetrahydrofuran)s
  three-branch star-shaped
    synthesis *via* cationic polymerization, 129
Poly(tetramethylene adipate)
  biodegradation, 288
Poly(tetramethylene glycol)
  biodegradable block copolymers
    with poly(ethylene terephthalate), 288
Poly(tetramethylene oxide) diols
  synthesis
    via ring-opening cationic polymerization, 124
Poly(thieno[3,2-b]thiophene)
  conjugated polymer
    nonlinear optical material, 439
Polythiophenes
  conjugated polymers
    nonlinear optical materials, 438
Poly(trimethylene carbonate)
  biodegradation, 290
Polyureas
  N,N-disubstituted
    pyrolysis, free radical process, 247
  pyrolysis
    NH hydrogen transfer, 232
  reactive injection molding process
    curing analysis, simulation, 603
  reactive processing
    kinetic models, 577–580
  segmented
    reactive processing, kinetics, 579
  structure development
    domain formation, 589
  synthesis *via* step-growth polymerization
    reactive extrusion, 629
  thermal degradation
    ionic cleavage, 228
Polyurethane
  unsaturated polyester hybrid resin
    cure kinetics, 589
Poly(urethane isocyanurate)s
  reactive processing
    kinetic models, 577–580
Polyurethanes
  aliphatic
    moderately photostable, 254
  aromatic
    poorly photostable, 254
  biodegradability
    polyether versus polyester, 289
  biodegradation, 286
  from furanic monomers, 567
  liquid crystalline polymers
    synthesis, 336
  matrices
    viscosity model, 514
  photodegradation
    via photo-Fries rearrangement, 275–278
  piperazine containing
    pyrolysis, α-CH hydrogen transfer, 238
  pyrolysis
    NH hydrogen transfer, 232
    tautomerization equilibrium, 232
  reactive injection molding process
    curing analysis, simulation, 603
  reactive processing
    kinetic models, 577–580
  structure development
    domain formation, 589
  synthesis
    via lignin macromonomers, 548
  synthesis *via* step-growth polymerization
    reactive extrusion, 629
  thermal degradation
    ionic cleavage, 228
Polyurethanes, N-methyl-
  pyrolysis
    α-CH hydrogen transfer, 237
    β-CH hydrogen transfer, 234
Poly(urethane urea)s
  reactive processing
    kinetics and kinetic models, 577–580
Poly(vinyl acetate)
  monolayers
    Langmuir–Blodgett films, 457
Poly(vinylacetophenone)
  photodegradation
    via Norrish reaction, 270
Poly(vinyl alcohol)
  biodegradation, 291
  cyclic acetal derivatives
    amphiphilic polymers, 457
Poly(vinylalkylals)
  Langmuir–Blodgett films
    synthesis, 457
Poly(vinyl chloride)
  aromatic/aliphatic thiolate grafting
    reactive extrusion, 631
  atactic
    Theodorou–Suter model, 165
  chemical modification
    reactive extrusion, 631
  photodegradation, 254
  photodehydrochlorination
    iron(III) chloride initiated, 263
    mechanism, 267
  photooxidative degradation, 267–269
  poorly photostable, 254
Poly(vinyl ether)s
  difunctional
    synthesis *via* cationic polymerization, 122
  functionalized
    synthesis *via* cationic polymerization, 111
  liquid crystalline polymers
    phase behavior, 363, 364
    phase transition temperatures, 366–368
  monofunctional
    synthesis *via* living polymerization, 115
Poly(vinyl fluoride)
  condensed film
    polymer monolayer, 457
Poly(vinylidene fluoride)
  expanded film
    polymer monolayer, 457
Poly(vinyl methyl ether)
  monolayers
    Langmuir–Blodgett films, 457
Poly(vinylphenol)s
  synthesis
    via living cationic polymerization, 128
Poly(yne)s
  metal containing
    liquid crystalline polymers, synthesis, 320
Potassium persulfate
  hydroperoxidation
    polyalkenes, 148
Predominant transfer
  alkenyl monomers
    cationic polymerization, 111
Preformed polymers

Langmuir–Blodgett films, 456–462
Pregel intramolecular reaction
    network formation, 202
Pressure
    influence of
        on mutual diffusion coefficient, 170–176
Process conditions
    copolymer latex properties
        emulsion copolymerization, 58
Processing equipment
    reactive processing
        thermoplastic polymers, 620
Processing models
    composite materials
        applications, 517–520
Process simulation
    reactive resin processing, 600–611
Process–structure–property relationships
    copolymer latex properties
        control, 60
L-Prolinol, N-(4-nitrophenyl)-
    attached to polymer backbone
        nonlinear optical polymer, 415, 417
Propagation
    cationic polymerization, 109
Propagation mechanisms
    pseudocationic polymerization
        living systems, 114
Propagation rate models
    emulsion copolymerization, 47–49
Propane, 2-azido-2-phenyl-
    inifer
        synthesis of azido-functionalized polymers, 112
1,3-Propanesultone
    sulfonation reagent
        functionalized polymer synthesis, 95
Property changes
    polymers and resins
        reactive processing, 596–600
Propylene
    ethylene copolymer
        functionalization, 143
    heterogeneous polymerization
        titanium trichloride catalyst, 67
    oligomers
        optical rotation of fractions, 75
    polymerization
        heterogeneous *versus* homogeneous systems, 79
        stereochemistry *via* different metallocenes, 72
        via dichloro(ethyl)bis(indenyl)zirconium catalyst, 72
        via dichloro(ethyl)bis(tetrahydroindenyl)zirconium catalyst, 72
        via different chiral stereorigid metallocene catalyst systems, 73
        via different chiral stereorigid zirconocene catalyst systems, 73
    polymerization *via* homogeneous catalysis
        cis addition mechanism, 70
    syndiotactic polymerization
        via metallocene catalyst systems, 76
Propylene–bisphenol A copolythiocarbonate
    pyrolysis
        intramolecular exchange reaction, 243
Propylene glycol
    from biomass
        uses, 552
Propylene oligomers
    via homogeneous catalysis
        optical activity, 75
Propyllithium, dimethylamino-
    initiator and amination reagent
        butadiene polymerization and functionalization, 93
        styrene polymerization and functionalization, 93
Proteins
    synthesis
        by living organisms, 530
Pseudocationic polymerization
    alkenyl monomers
        living systems, 114
Pseudo homopolymerization approach

emulsion copolymerization
    kinetic factors, 51
Pseudomonas spp.
    biodegrading ability, 290
Pullulan
    natural biodegradable material
        uses, 293
Pultrusion
    composite materials
        high performance, 502
    processing model
        carbon fiber/epoxy composites, 519
Pyrene end-labeled polymers
    fluorescent
        synthesis *via* substituted 1,1-diphenylenes, 102
Pyridine N-oxide groups
    dye unit
        nonlinear optical liquid crystalline polymers, 420
Pyridinium iodide, 4-(4-N,N-dimethylaminostyryl)-
    attached to polymer backbone
        nonlinear optical polymer, 414
Pyroelectric effects
    polymeric Langmuir–Blodgett films, 465
Pyrolysis
    condensation polymers
        free radical processes, 244–248
        ionic cleavage, 228
        ionic processes, 231
        molecular rearrangements, 248
Pyrolysis products
    condensation polymers
        detection, 230
        primary, 230
Pyrrolidine, 2-methacryloyloxymethyl-1-(4-nitrophenyl)-
    dye-containing monomer
        nonlinear optical polymer, 417
Pyrrolidone, vinyl-
    graft polymerization
        polypropylene, *via* trapped radical technique, 147

Quantum yield
    chain scission
        polymer photodegradation, 262
Quinodimethane monomers
    copolymers with methyl 12-hydroxydodecanoate
        nonlinear optical polymers, 423
Quinones
    3,6-disubstituted 2,5-dichloro-
        monomers for ladder polymers, nonlinear optical materials, 441
    photoinitiators
        polymer photodegradation, 263
    retarders
        radical polymerization, 2

Radiation grafting
    polyalkenes
        simultaneous method, 145–147
        techniques, 144–147
Radical abstraction
    polymerization
        inhibition and retardation, 1
Radical addition
    polymerization
        inhibition and retardation, 1
Radical desorption
    emulsion copolymerization
        kinetic factors, 49
Radical entry rate
    emulsion copolymerization
        kinetic factors, 50
Radical polymerization
    inhibition, 1–20
    nontraditional initiators, 23–39
    polyfunctional initiators
        mechanism, 24–26
        presence of iniferters

features, 39
presence of polyfunctional initiators
   features, 39
   retardation, 1–20
Radical production rate
   emulsion copolymerization
      kinetic factors, 49–51
Radicals
   formation
      via photooxidative polymer degradation, 257
Radius of gyration
   polymer chains
      mutual diffusion relationship, 171
Rate of initiation
   radical polymerization
      inhibition and, 4
Rate of polymerization
   ethylene
      transition metal homogeneous catalysis, 69
   polyperoxide initiators, 25
   retarded reactions, 7
Rate theory
   gelation
      network formation, 210, 211
   gel points
      interpretation, 212
   network formation
      applications, 223
Reaction injection molding
   low temperature
      amine catalysts, 582
   network formation, 201
   rheological changes *versus* kinetic changes, 589
Reaction kinetics
   processing
      composite materials, 506
   reactive processing
      material modeling, 576–589
Reaction types
   reactive processing
      thermoplastic polymers, 621
Reactive blending
   polymers, 634–638
   silane-grafted polyalkenes, 624
Reactive compression molding process
   thermoset resins
      simulation, 605–609
Reactive extrusion
   polymers
      polymerization/chemical modification, 620
Reactive injection molding process
   simulation, 600–605
      commercial programs, 600
   thermoset resins
      simulation, 602
Reactive latices
   copolymer latex properties
      emulsion copolymerization, 59
Reactive processing
   computer simulation, 575–611
   modeling, 575–611
   thermoplastic polymers, 619–638
   thermosets
      network formation, 201
Reactive processing simulation
   thermoplastic processing simulation
      differences, 600
Reactive resins
   isothermal cure
      kinetic models, 577
Reactivity ratios
   ultimate model
      emulsion copolymerization, 47
Rearrangements
   molecular
      pyrolysis, condensation polymers, 248
Recombination

geminate
   inhibition and, 5
Reduced rate
   ideal retarded polymerization, 10
Reentrant nematic mesophase
   liquid crystalline polymers
      mechanism of formation, 369
      side chain, 366–368
Regioirregularities
   propylene polymerization
      via chiral homogeneous catalysis, 72
Reinforcements
   composite materials
      properties, 477–480
Reinitiation
   slow
      radical polymerization, 13–18
Relaxation anisotropy
   viscosity
      liquid crystal polymers, 398
Renewable resources
   polymers from, 527–570
Renormalization group methods
   polymer solution thermodynamics, 169
Reptation theory model
   kinetic diffusion model
      bulk vinyl polymerization, 583
Residence time
   extruders
      reactive processing, thermoplastic polymers, 621
Resin flow
   fabrication
      laminate composites, 520–523
Resins
   for high-performance composites
      properties, 477
   synthesis
      by living organisms, 532
Resin transfer molding
   composite materials
      high performance, 503
   fluid flow
      fiber preforms, 523
   low temperature
      amine catalysts, 582
   thermoset resins
      simulation, 603
Respirometry
   biodegradation test, 287
Retardation
   ideal
      kinetics, 6–11
   radical polymerization, 1–20
      chemistry, 2
Retarders
   radical polymerization, 1, 2
Rheological behavior
   liquid crystalline polymers, 385–405
Rheological changes
   reactive processing
      material modeling, 589–596
Ring–chain equilibria
   intramolecular reactions
      network formation, 208
Ring-forming parameter
   gelation, 213
   intramolecular reactions
      network formation, 210
Ring-opening polymerizations
   cationic
      living character, 117
Rod-like groups
   flexible mesogenic units
      liquid crystalline polymers, 307
   rigid mesogenic units
      liquid crystalline polymers, 305
   semirigid mesogenic units

liquid crystalline polymers, 306
Rod-like polymers
    chemical structures, 460
    conformationally mobile side chains, 461
    monolayers
        Langmuir–Blodgett films, 460–462
    rigid
        nonlinear optical materials, 439
Rosins
    chemical modification
        uses, 543
    from renewable resources
        polymerization, 543
    synthesis
        by living organisms, 532
Rotational isomeric state theory
    computer modeling
        polymer structure/properties, 161
Rubber
    natural
        synthesis by living organisms, 530
    photodegradation, 254

Saccharides
    polymerization
        via acid-catalyzed polycondensation, 553
        via cationic ring opening, 553–555
Sanchez–Lacombe model
    sorption
        penetrant in polymer, 186–188
Screw design
    extruders
        reactive processing, thermoplastic polymers, 620
Secondary nucleation
    batch emulsion copolymerization
        composition drift, 56
Self-functionalized polymerizations
    living systems
        isobutene, 117
Self retardation
    degradative chain transfer
        radical polymerization, 2
Semicrystalline polymers
    sorption and transport phenomena, 179–184
    gases, 181
Sensors
    polymeric Langmuir–Blodgett films, 466
Separation membranes
    selective
        polymeric Langmuir–Blodgett films, 466
Sericin
    synthesis
        by living organisms, 530
Shear modulus
    elastic behavior
        polymeric matrices, 473
    network materials
        definition, 200
Sheet molding compounds
    compression molding
        computer simulation, 605
    high temperature compression molding
        peroxide initiators, 582
Shellac
    synthesis
        by living organisms, 533
Shelloic acid
    synthesis
        by living organisms, 533
Silane, vinyltrimethoxy-
    polyalkene grafting
        reactive extrusion, 624
Silarylene–siloxane copolymers
    pyrolysis
        intramolecular exchange reaction, 244
Silica
    catalyst support

heterogeneous polymerization, 67
Silicon polymers
    photodegradation
        mechanism, 276
Silicone rubbers
    reactive injection molding process
        curing analysis, simulation, 603
Silyl ketene acetal
    functional terminating agents
        living polymerization, vinyl ethers, 116
Size exclusion chromatography
    copolymer separation
        emulsion copolymerization, 53
    molecular weight determination
        functionalized polymers, 85
Slow release systems
    pharmaceutical drugs
        based on poly(glycolic acid) degradation, 288
Smith–Ewart equations
    emulsion copolymerization, 50
Sodiomalonates
    functional terminating agents
        living polymerization, vinyl ethers, 116
Software
    computer modeling
        polymers, 160
Solid state carbonation
    (polystyryl)lithium
        via carbon dioxide, 88
Solid state precursors
    chain transfer polymerization
        polyalkenes, 148
Solution grafting
    peroxide initiated
        polyalkenes, 150
Solvent fractionation
    polypropylene
        obtained *via* homogeneous catalysis, 74
Solvent viscosity
    relationship
        mutual diffusion coefficient, 169
Sorbitol
    from biomass
        polyurethane synthesis, 555
Sorption
    amorphous/semicrystalline polymers, 179–184
Sorption models
    penetrant in polymer, 184–188
Spacer concept
    amphiphilic polymers
        Langmuir–Blodgett films, 458
Spanning-tree approximation
    polymerization model, 211
Spin labeling
    polyethylene
        synthesis *via* carbonyl hydrazide functionalization, 141
Spiropyrans
    thermochromic dyes
        nonlinear optical liquid crystalline polymers, 419
Stabilizers
    rubber photodegradation, 254
Star-branched polymers
    use
        synthesis of functionalized polymers, 84
Starch
    as additive
        use in biodegradation, 290
    chemical modification
        uses, 540
    natural biodegradable material
        use as additive, 292
    synthesis
        by living organisms, 529
Starved conditions
    copolymer latex properties
        control, 60
Static models

development
    molecular dynamics, 160–162
Stationary state assumption
    radical inhibition, 6
Stefan number
    heat transfer
        composite materials processing, 516
Step growth polymerizations
    reactive processing
        kinetic models, 577
Stereochemical inversions
    propylene polymerization
        via chiral homogeneous catalysis, 73
Stereocontrol
    heterogeneous catalysis
        monoalkene polymerization, 67
Stereoselectivity
    alkene polymerization
        via chiral homogeneous catalysis, 76
Stereospecificity
    propylene polymerization
        via homogeneous catalysis, 71
    racemic zirconocene complexes
        homogeneous catalysis, 73
Stilbene, 4'-amino-4-nitro-
    dye unit
        nonlinear optical liquid crystalline polymer, 420
Stilbene, 4-(dimethylamino)-4'-nitro-
    blends with glassy polymers
        nonlinear optical polymer, 414
Storage modulus
    viscoelastic behavior
        polymeric matrices, 474
Strength analysis approach
    composite materials, 497
Stress–strain behavior
    networks, 223–225
Structural reaction injection molding process
    thermoset resins
        simulation, 603
Structure
    condensation polymers
        thermal stability and, 229
Structure changes
    polymers and resins
        reactive processing, 596–600
Structure determination
    functional groups
        polyethylenes, 140
Styrene
    anionic addition polymerization
        reactive extrusion, 629
    batch emulsion copolymerization
        butyl acrylate, 61
        ethyl acrylate, 53
        methyl acrylate, 53, 55–57
    biodegradable copolymers
        with vinyl ketones, 290
    bulk copolymerization with acrylonitrile
        presence of initiators/iniferters, 39
    bulk copolymerization with methacrylate
        presence of initiators/iniferters, 39
    bulk polymerization
        free radical initiated, reactive processing, 623
        presence of iniferter, 37
    copolymerization
        with substituted 1,1-diphenylethylenes, 101
    copolymerization with diethyl fumarate
        kinetics, 582
    derivatives
        film polymerization, polymeric Langmuir–Blodgett film formation, 453
    emulsion copolymerization
        methyl acrylate, 62
    emulsion copolymerization models
        ethyl methacrylate, 47
    methyl acrylate emulsion copolymers

chemical composition distributions, 62
molten state grafting
    polyalkenes, 149
monomer partitioning
    emulsion copolymerization, 44
photopolymerization
    presence of iniferter, 36
polymerization
    DPPH inhibition, 5
polymerization and carboxylation
    via protected oxazoline initiator, 90
radiation grafting
    polyethylene, 146
radical polymerization
    iron(III) chloride inhibited, 9
Styrene, 4-*t*-butoxy-
    living cationic polymerization
        synthesis of polyfunctional polymers, 128
Styrene, *p*-methoxy-
    polymerization
        via living cationic system, 113
Styrene–butadiene rubber
    photodegradation, 269
Styrenes
    living polymerization
        synthesis of functionalized polymers, 117
    polymer grafting
        reactive extrusion, 627
Styrenic/polyester resins
    reactive processing
        kinetics, 582–584
    structure development
        chemical crosslinking, 589
Styrenic resins
    reactive injection molding process
        curing analysis, simulation, 603
Substitution effects
    gel point
        polymerization, 203
Substitution reactions
    polyalkenes
        reactive processing, 621
Sulfonation
    anionic synthesis
        functionalized polymers, 95–97
Sulfone, 4,4'-diaminodiphenyl
    curing agent
        reactive processing, 580
Sulfone, diaminodiphenyl
    epoxy resin component
        high performance composites, 506
Sulfur
    surface functionalization
        polyethylene, 135
Sultones
    sulfonation reagents
        functionalized polymer synthesis, 95
Superposition principle
    block copolymer composition distribution
        polyperoxide initiation, 33
    methyl methacrylate polymerization, 28
    molecular weight distribution
        iniferter mechanism of polymerization, 37
        polyperoxide initiation, 26–28
Supramolecular liquid crystalline polymers
    self assembly
        molecular recognition directed, 372
Surface functionalization
    polyalkenes, 135–137
Surface modification
    polyethylene
        mechanism, 136
        methods, 136
    polymeric Langmuir–Blodgett films, 466
Sutures
    biodegradable
        synthesis, 288

Tacticity
   liquid crystalline polymers
      phase transitions, 368
Tail–tail enchainments
   propylene polymerization
      via chiral homogeneous catalysis, 72
Tannins
   from renewable resources
      polymerization, 549
   macromonomers
      phenol–formaldehyde resins, 550
   synthesis
      by living organisms, 533
Tautomerization equilibrium
   pyrolysis
      polyurethanes, 232
Temperature
   influence of
      on mutual diffusion coefficient, 170–176
Tendering
   natural textile fabrics, 253
Tendons
   artificial
      biodegradable, 288
Terephthalaldiimine, dihexyl-
   monolayer polycondensation
      with 3,3'-diaminobenzidine, 455
Terephthalic acid
   polymers with 2,5-dialkoxyhydroquinones
      characterization, 329
Terminal model
   emulsion copolymerization, 47
Termination
   cationic polymerization, 109
   size dependent
      radical polymerization, 18–20
Termination rate constant
   emulsion copolymerization
      kinetic factors, 50
Terpenes
   from biomass
      polymerization, 551
   synthesis
      by living organisms, 533
Terpolymerization
   emulsion copolymerization, 57
Terpolymers
   ethylene/propylene/ethylidenenorbornene
      via bis(cyclopentadienyl)dimethylzirconium homogeneous
         catalysis, 77
Textile fabrics
   tendering, 253
Theodorou–Suter model
   glassy polypropylene, 164
Thermal conductivity
   processing
      composite materials, 515
Thermal degradation
   condensation polymers, 227–249
Thermal expansion
   during polymerization
      reactive processing, models, 598
   short-fiber composites, 489
Thermal reactivity
   functional groups
      in polymeric structures, 228
Thermal stability
   condensation polymers
      structure and, 229
Thermal volatilization analysis
   condensation polymers
      product detection, 230
Thermodynamic factor
   polymer chains
      mutual diffusion relationship, 172
Thermodynamic interaction
   influence of
      on mutual diffusion coefficient, 169
Thermodynamic parameters
   copolymer liquid crystals
      determination, 348
   polyester liquid crystals
      determination, 348
   polyether liquid crystals
      determination, 345
Thermodynamic treatment
   monomer partitioning
      emulsion copolymerization, 44
Thermokinetic model
   processing
      composite materials, 505, 506–510
Thermophysical properties
   during polymerization
      reactive processing, models, 600
Thermoplastic injection molding process
   simulation
      commercial programs, 600
Thermoplastic polymers
   reactive processing, 619–638
Thermoplastics
   polymeric matrices, 472
Thermoset composites
   continuous fiber reinforced
      reactive processing, 609–611
   manufacture
      autoclave curing, models, 609
Thermosets
   polymeric matrices, 472
Thermosetting polymers
   kinetic characterization
      differential scanning calorimetry, 506
Thermosetting reactions
   composite materials
      viscosity modeling, 512–514
Thermosetting resins
   rheological changes
      time dependence, models, 591
Theta solvent
   polymer/solvent system, 170
Thin-layer chromatography
   characterization
      functionalized polymers, 86
Thiolanium end groups
   via ring-opening cationic polymerization
      use in synthesis of functionalized polymers, 119
Thiuram disulfide, tetraethyl-
   iniferter
      radical polymerization, 34, 37
Time–temperature–transformation diagram
   networks, 201
   thermoset matrices
      composite materials, 511
Titanium, dichloro(ethyl)bis(indenyl)-
   homogeneous catalyst
      propylene polymerization, 74
Titanium, dichloro(ethylene)bis(indenyl)-
   chiral homogeneous catalyst
      propylene polymerization, 70
Titanium metallocenes
   homogeneous catalysts
      ethylene polymerization, 69
Titanium trichloride
   heterogeneous polymerization catalyst, 67
Tobacco mosaic virus
   self assembly, 374
2,6-Toluene diisocyanate
   polyurethanes from
      photodegradation, 275
Topochemical polymerizations
   film polymerization
      polymeric Langmuir–Blodgett film formation, 454
Topological chain entanglements
   network materials

properties, 200
Tracer diffusion coefficient
  binary system, 195
  polystyrene/tetrahydrofuran
    concentration dependence, 177
  relationship
    mutual diffusion coefficient, 177
Transfer molding process
  thermoset resins
    simulation, 604
Transient behavior
  anisotropic polymer solutions
    liquid crystal rheology, 391–393
Transition temperatures
  virtual
    polyether liquid crystals, determination, 345
Transport amorphous/semicrystalline polymers, 179–184
Transport behavior
  penetrant in polymer
    theoretical approaches, 184
Transreactions
  polycondensates
    reactive processing, 621
Transversely isotropic liquids
  viscosity
    liquid crystal polymers, 396–398
Trapped radical techniques
  radiation grafting
    polyalkenes, 147
Triad fractions
  copolymer sequence distribution
    batch emulsion copolymerization, 55
ω-Tricoenyl-2,4-pentadienoate
  film polymerization
    polymeric Langmuir–Blodgett film formation, 455
ω-Tricosenoic acid
  film polymerization
    polymeric Langmuir–Blodgett film formation, 454
Triesters
  initiators
    living cationic polymerization, synthesis of polyfunctional polymers, 127
Triflate end groups
  synthesis
    via ring-opening cationic polymerization, 120
Triglycerides
  oxidopolymerization, 543
Triols
  macromolecular
    synthesis via activated monomer polymerization, 125
Trioxane
  cationic addition polymerization
    reactive extrusion, 629
Triperoxides
  initiators
    block copolymer synthesis, 33
Triphenylmethyl radical
  inhibitor
    radical polymerization, 5
Trommsdorf–Norrish gel effect
  batch emulsion copolymerization
    composition drift, 57

Ultimate model
  emulsion copolymerization, 47
Ultraviolet radiation
  free radical initiator
    synthesis of functionalized polyalkenes, 135
Ultraviolet–visible spectroscopy
  characterization
    functionalized polymers, 86
Unidirectional composites
  longitudinal strength, 480
Unit fraction
  gel
    perfect network formation, 205
  sol

perfect network formation, 205
Urethane–isocyanurate resins
  structural reaction injection molding
    kinetics, 579
Urethane resins
  polyesters
    hybrid resin system, 588
Urethanes
  formation
    metal ion catalyzed, kinetic mechanism, 578
  polymerization
    reaction injection molding, kinetic model, 578
    reactive processing, simplified model, 578
  reactive injection molding process
    curing analysis, simulation, 603

Vapor phase osmometry
  molecular weight determination
    functionalized polymers, 85
Vinyl acetate
  batch emulsion copolymerization
    butyl acrylate, 53, 55–57, 60
  polymerization
    DPPH inhibition, 5
  radical polymerization
    slow reinitiation, 14
Vinyl chloride
  radical polymerization
    iron(III) chloride retarded, 11
    slow reinitiation, 16
Vinyl ethers
  cationic polymerization
    synthesis of difunctional polymers, 122
  living cationic polymerization
    synthesis of polyfunctional polymers, 127
  polymerization
    via living cationic system, 113
  strong nucleophiles
    cationic polymerization, 110
Vinyl 2-furoate
  free radical polymerization, 564
Vinyl ketones
  biodegradable copolymers
    with styrene and ethylene, 290
Vinyl monomers
  bulk polymerization
    presence of iniferters, 37
  film polymerization
    polymeric Langmuir–Blodgett film formation, 452–454
Vinyl stearate
  film polymerization
    polymeric Langmuir–Blodgett film formation, 452
Viscoelastic behavior
  polymeric matrices, 473
Viscoelasticity
  linear anisotropic
    liquid crystal polymers, 402–405
  nematic solutions
    molecular kinetic theory, 388
  nonlinear
    liquid crystal polymers, 400
  resin flow model, 521
Viscoelastic nematic liquids
  linear
    constitutive equation, 402
Viscoelastic properties
  lyotropic liquid crystal systems
    rheology, 390
Viscosity
  epoxy resins
    glass transition temperature dependence, models, 592
  liquid crystalline polymers
    concentration dependence, 385
  long chain linear polymers
    molecular weight dependence, models, 593
  polydisperse systems
    linear dependence, weight-average molecular weight, 590

polymer systems
  dependence on molecular weight, 590
properties
  lyotropic liquid crystal systems, 386
reactive polymeric systems
  polymer formation dependence, models, 595
resins
  conversion dependence, models, 591
  temperature dependence, models, 592
  weight-average molecular weight dependence, models, 592
step growth polymerizations
  conversion dependence, models, 594
  molecular weight dependence, models, 594
thermoset composites
  molecular weight dependence, 512
thermosets
  molecular weight dependence, models, 594

Viscosity measurements
  on-line
    polypropylene degradation, 637
Viscous behavior
  polymeric matrices
    Newton's law, 473
Vitrification
  composite materials
    thermosets, 511
  epoxy resin curing
    diffusion control, 581
Vitrification effect
  kinetic diffusion model
    styrene/polyester resins, 583
Void formation
  during polymerization
    reactive processing, models, 600
Voigt model
  viscoelastic behavior
    polymeric matrices, 473

Weight fraction
  sol
    perfect network formation, 205
Wettability
  surface modification
    polymeric Langmuir–Blodgett films, 467
Wilhelmy technique
  surface tension measurement
    Langmuir–Blodgett films, 450
William–Landel–Ferry equation
  viscosity
    thermoset composites, 513
Wissbrun liquid crystal domain model
  flow curve shaws
    lyotropic polymers, 393
Wood
  synthesis
    by living organisms, 534

Xanthan
  natural biodegradable material, 292
X-ray reflection
  poly(L-glutamate) multilayers, 464
D-Xylose
  hydrolysis/dehydration
    furfural synthesis, 556

Ziegler–Natta catalysts
  use
    synthesis of functionalized polyalkenes, 134
Zirconium, bis(cyclopentadienyl)dichloro-
  homogeneous catalyst
    ethylene polymerization, 68
Zirconium, bis(cyclopentadienyl)dimethyl-
  homogeneous catalyst
    ethylene polymerization, 68
Zirconium, dichlorobis[(1-phenylethyl)cyclopentadienyl]-
  chiral homogeneous catalyst
    propylene polymerization, 71
Zirconium, dichloro(cyclopentadienyl-1-fluorenyl)isopropyl-
  chiral homogeneous catalyst
    alkene polymerization, 76
Zirconium, dichloro(ethyl)bis(indenyl)-
  homogeneous catalyst
    cycloalkene polymerization, 78
Zirconium, dichloro(ethylene)bis(4,5,6,7-tetrahydro-1-indenyl)
  chiral homogeneous catalyst
    propylene polymerization, 71
Zirconium, dichloro(ethyl)(3-methylindenyl)-
  chiral homogeneous catalyst
    propylene polymerization, 72
Zirconium, ethylbis(tetrahydroindenyl)-
  1,1'-bis-2-naphtholate
    chiral homogeneous catalyst, 76
Zirconium, ethylbis(tetrahydroindenyl)dimethyl-
  chiral homogeneous catalyst
    propylene polymerization, 75